Foundations of Mechanics

Foundations of Mechanics

Second Edition

Revised, enlarged, and reset

A mathematical exposition of classical mechanics with an introduction to the qualitative theory of dynamical systems and applications to the three-body problem

Ralph Abraham and Jerrold E. Marsden
University of California
Santa Cruz and Berkeley

with the assistance of
Tudor Ratiu and Richard Cushman

Routledge
Taylor & Francis Group
LONDON AND NEW YORK

First published 1978 by Westview Press

Published 2018 by Routledge
52 Vanderbilt Avenue, New York, NY 10017
2 Park Square, Milton Park, Abingdon, Oxon OX14 4RN

Routledge is an imprint of the Taylor & Francis Group, an informa business

Library of Congress Cataloging-in-Publication Data
Abraham, Ralph.
Foundations of mechanics.

Bibliography: p.
Includes index.
1. Mechanics, Analytic. 2. Geometry, Differential. 3. Mechanics, Celestial. I. Marsden, Jerrold E., joint author. II. Title.
QA805.A2 1977 531'.01'51 77-25858

ISBN 13: 978-0-367-00509-2 (hbk)
ISBN 13: 978-0-367-15496-7 (pbk)

IN MEMORIAM

RUFUS BOWEN
1947–1978

KAREL DE LEEUW
1930–1978

Contents

Preface to the Second Edition

Since the first edition of this book appeared in 1967, there has been a great deal of activity in the field of symplectic geometry and Hamiltonian systems. In addition to the recent textbooks of Arnold, Arnold–Avez, Godbillon, Guillemin–Sternberg, Siegel–Moser, and Souriau, there have been many research articles published. Two good collections are "Symposia Mathematica," vol. XIV, and "Géométrie Symplectique et Physique Mathématique," CNRS, Colloque Internationaux, no. 237. There are also important survey articles, such as Weinstein [1977b]. The text and bibliography contain many of the important new references we are aware of. We have continued to find the classic works, especially Whittaker [1959], invaluable.

The basic audience for the book remains the same: mathematicians, physicists, and engineers interested in geometrical methods in mechanics, assuming a background in calculus, linear algebra, some classical analysis, and point set topology. We include most of the basic results in manifold theory, as well as some key facts from point set topology and Lie group theory. Other things used without proof are clearly noted.

We have updated the material on symmetry groups and qualitative theory, added new sections on the rigid body, topology and mechanics, and quantization, and other topics, and have made numerous corrections and additions. In fact, some of the results in this edition are new.

We have made two major changes in notation: we now use f^* for pull-back (the first edition used f_*), in accordance with standard usage, and have adopted the "Bourbaki" convention for wedge product. The latter eliminates many annoying factors of 2.

A. N. Kolmogorov's address at the 1954 International Congress of Mathematicians marked an important historical point in the development of the theory, and is reproduced as an appendix. The work of Kolmogorov, Arnold, and Moser and its application to Laplace's question of stability of the solar system remains one of the goals of the exposition. For complete details of all the theorems needed in this direction, outside references will have to be consulted, such as Siegel–Moser [1971] and Moser [1973a].

We are pleased to acknowledge valuable assistance from

Paul Chernoff   Wlodek Tulczyjew
Morris Hirsh   Alan Weinstein

and our invaluable assistant authors

Richard Cushman and Tudor Ratiu

who all contributed some of their original material for incorporation into the text.

Also, we are grateful to

Ethan Akin	Kentaro Mikami
Judy Arms	Harold Naparst
Michael Buchner	Ed Nelson
Robert Cahn	Sheldon Newhouse
Emil Chorosoff	George Oster
André Deprit	Jean-Paul Penot
Bob Devaney	Joel Robbin
Hans Duistermaat	Clark Robinson
John Guckenheimer	David Rod
Martin Gutzwiller	William Satzer
Richard Hansen	Dieter Schmidt
Morris Hirsch	Mike Shub
Michael Hoffman	Steve Smale
Andrei Iacob	Rich Spencer
Robert Jantzen	Mike Spivak
Therese Langer	Dan Sunday
Ken Meyer	Floris Takens
	Randy Wohl

for contributions, remarks, and corrections which we have included in this edition.

Further, we express our gratitude to Chris Shaw, who made exceptional efforts to transform our sketches into the graphics which illustrate the text, to Peter Coha for his assistance in organizing the Museum and Bibliography, and to Ruthie Cephas, Jody Hilbun, Marnie McElhiney, Ruth (Bionic Fingers) Suzuki, and Ikuko Workman for their superb typing job.

Theoretical mechanics is an ever-expanding subject. We will appreciate comments from readers regarding new results and shortcomings in this edition.

RALPH ABRAHAM
JERROLD E. MARSDEN

Preface to the First Edition

In the Spring of 1966, I gave a series of lectures in the Princeton University Department of Physics, aimed at recent mathematical results in mechanics, especially the work of Kolmogorov, Arnold, and Moser and its application to Laplace's question of the stability of the solar system. Mr. Marsden's notes of the lectures, with some revision and expansion by both of us, became this book.

Although the lectures were attended equally by mathematicians and physicists, our goal was to make the subject available to the nonspecialists. Therefore, the mathematical background assumed was dictated by the physics graduate students in the audience. Hoping this would be typical of the people interested in this subject, I have made the same assumptions in the book.

Thus, we take for granted basic undergraduate calculus and linear algebra, and a limited amount of classical analysis, point set topology, and elementary mechanics. Then we begin with modern advanced calculus, and go on to a complete and self-contained treatment of graduate level classical mechanics. The later chapters, dealing with the recent results, require an ever-increasing adeptness in general topology, and we have collected the topological topics required in Appendix A.

To further aid the nonmathematician, the proofs are unusually detailed, and the text is replete with cross-references to earlier definitions and propositions, all of which are numbered for this purpose. The extent of these is testimony of Mr. Marsden's patience.

As our goal is to make a concise exposition, we prove propositions only if the proofs are easy, or are not to be found readily in the literature. This results in an irregular collection of proofs—in the first four chapters nearly everything is proved, being easy, and in the last three chapters there are several longer proofs included and many omitted. Some of those included are necessary because the propositions are original, and can be omitted in a first reading or an elementary course.

For the mathematical reader, the proofs we have omitted can easily be found in books or journals, and we give complete references for each (References in square brackets refer to the Bibliography.) For this reason, the book, although not self-contained, gives a complete exposition.

In this connection we are grateful to Al Kelley for the opportunity of publishing two research articles of his, as Appendixes B and C, which have not appeared elsewhere. In each of these he proves an original theorem that is important to our development of the subject. As Kolmogorov's address at the 1954 International Congress of Mathematicians (in Russian), which inspired the most important of the recent results, has not been available in English, we include a translation of it in Appendix D. The exercises at the end of each section are nearly all used in a later section, and may be read as part of the text.

I am indebted to Arthur Wightman for his enthusiasm in making arrangements for my lectures and the publication of the book, to René Thom for discussions on structural stability and a preliminary manuscript of part of his book on that subject, to Jerrold Marsden for his energetic collaboration in the writing of this book, and to many colleagues for valuable suggestions. Some of these are acknowledged in the Notes at the end of each chapter, which also give general historical and bibliographical information.

We are both happy to express our gratitude to June Clausen for editing and typing the bulk of the manuscript, and to Patricia Clark, Bonnie Kearns, Elizabeth Epstein, Elizabeth Margosches, and Jerilynn Christiansen for their valuable assistance.

RALPH ABRAHAM

Princeton, New Jersey
June 1966

Museum

Archimedes, 287 B.C.-212 B.C. *Courtesy of the Library of Congress, Washington, D.C., U.S.A.*

Nicholas Copernicus, 1473-1543.
Courtesy of the Library of Congress, Washington, D.C., U.S.A.

Galileo Galilei, 1564-1642.
D. J. Struik, A Concise History of Mathematics. Dover Publications, New York (1948).

Johannes Kepler, 1571-1630. *Kepler, Gesammelte Werke. Beck, München (1960).*

Isaac Newton, 1642-1727. *Courtesy of the Trustees of the British Museum.*

Gottfried Wilhelm Leibnitz, 1646-1716. *Courtesy of the Trustees of the British Museum.*

Pierre Louis Moreau de Maupertuis, 1698-1759. *Courtesy of the Bibliothèque Nationale, Paris, France.*

Leonhard Euler, 1707-1783. *E. T. Bell, Men of Mathematics. Simon and Schuster, New York (1937).*

Joseph-Louis Lagrange, 1736-1813.
Courtesy of the Bibliothèque Nationale, Paris, France.

Pierre-Simon de Laplace, 1749-1827.
Courtesy of the Bibliothèque Nationale, Paris, France.

Adrien Marie Legendre, 1752-1833.
A. Legendre, with an introduction by K. Pearson, F.R.S. Tables of the Complete and Incomplete Elliptic Integrals. Cambridge University Press, Cambridge, England (1934).

Simeon-Denis Poisson, 1781-1840. *Courtesy of the Bibliothèque Nationale, Paris, France.*

Carl Gustav Jakob Jacobi, 1804-1851. *C. W. Borchart, C.G.J. Jacobi's Gesammelte Werke. Verlag Von G. Reimer, Berlin (1881).*

William Rowan Hamilton, 1805-1865. *Courtesy of the Royal Irish Academy.*

Joseph Liouville, 1809-1882. *L. J. Gino, Liouville and his Work. Scripta Math.* **4**: *147-154, 257-262 (1936).*

Georg Friedrich Bernhard Riemann, 1826-1866. *Courtesy of the Deutsches Museum, Munich.*

Gaston Darboux, 1842-1917.
G. Darboux, Éloges Académiques et Discours Librairie Scientifique. A. Hermann et Fils, Paris (1912).

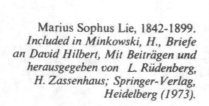

Marius Sophus Lie, 1842-1899.
Included in Minkowski, H., Briefe an David Hilbert, Mit Beiträgen und herausgegeben von L. Rüdenberg, H. Zassenhaus; Springer-Verlag, Heidelberg (1973).

Sonya Kovalevsky, 1850-1891.
H. Minkowski, Briefe an David Hilbert, Mit Beiträgen und herausgegeben von L. Rüdenberg, H. Zassenhaus; Springer-Verlag, New York (1976).

Henri Poincaré, 1854-1912. *Courtesy of the Library of Congress, Washington, D.C., U.S.A.*

Aleksandr Mikhailovich Liapounov, 1857-1918. *Akademija Nauk, SSSR (1954).*

Constantin Carathéodory,
1873-1950. *H. Tietze, Constantin Carathéodory. Archiv der Mathematik* **2***: 241-245 (1950).*

Edmond Taylor Whittaker,
1873-1956. *G. Temple, Edmond Taylor Whittaker, Biographical Memoirs of Fellows of the Royal Society 2: 299-325 (1956).*

Albert Einstein, 1879-1955. *Courtesy of the Library of Congress, Washington, D.C., U.S.A.*

George David Birkhoff, 1884-1944.
G. D. Birkhoff, Collected Mathematical Papers, American Mathematical Society, New York (1950).

Elie Cartan, 1869-1951. *Selecta, Jubilé Scientifique de M. Elie Cartan, Gauthier-Villars, Paris (1939).*

Amalie Emmy Noether, 1882-1935.
Constance Reid, Courant in Göttingen and New York. The Story of an Improbable Mathematician, Springer-Verlag, New York (1976).

Carl Ludwig Siegel, 1896- *C. L. Siegel, Gesammelte Abhandlungen, Springer-Verlag, Berlin (1966).*

Andrei Nikolaevic Kolmogorov, 1903-
Photograph by Jürgen Moser.

Jürgen Moser, 1928- . *Photograph by Caroline Abraham.*

Stephen Smale, 1930- . *Photograph by Caroline Abraham.*

Vladimir I. Arnold, 1937- . *Photograph by Jürgen Moser.*

Introduction

Mechanics begins with a long tradition of qualitative investigation culminating with KEPLER and GALILEO. Following this is the period of quantitative theory (1687–1889) characterized by concomitant developments in mechanics, mathematics, and the philosophy of science that are epitomized by the works of NEWTON, EULER, LAGRANGE, LAPLACE, HAMILTON, and JACOBI. Both of these periods are thoroughly described in DÜGAS [1955].

Throughout these periods, the distinguished special case of *celestial mechanics* had a dominant role (see MOULTON [1902] for additional historical details). Formalized in the quantitative period as the *n-body problem*, it recurs in the writings of all of the great figures of the time. The question of *stability* was one of main concerns, and was analyzed with series expansion techniques by LAPLACE (1773), LAGRANGE (1776), POISSON (1809), and DIRICHLET (1858), all of whom claimed to have proved that the solar system was stable.

As DIRICHLET died before writing down this proof, KING OSCAR of Sweden offered a prize for its discovery, which was given to POINCARÉ in 1889. The results of HARETU (1878) and POINCARÉ (1892), suggest that the series expansions of LAPLACE et al. diverge, and the discovery by BRUNS (1887) that no quantitative methods other than series expansions could resolve the *n*-body problem brought the quantitative period to an end. (See MOSER [1973a] for additional historical information.) For celestial mechanics this situation represented a great dilemma, comparable to the crises associated with relativity and quantum theory in other aspects of mechanics. The resolution we owe to the genius of Poincaré, who resurrected the qualitative point of view, accompanied by completely new mathematical methods. The inventions of Poincaré, culminating in modern differential geometry and topology, constitute a recent and lesser known example of concomitant development of mathematics and mechanics, comparable to calculus, differential equations, and variational theory.

The neoqualitative period in mechanics, that is, from Poincaré to the present, consists primarily in the amplification of the qualitative, geometric methods of Poincaré, the application of these methods to the qualitative questions of the previous period—for example, stability in the n-body problem—and the consideration of new qualitative questions that could not previously be asked.

Poincaré's methods are characterized first of all by the global geometric point of view. He visualized a dynamical system as a field of vectors on phase space, in which a solution is a smooth curve tangent at each of its points to the vector based at that point. The qualitative theory is based on geometrical properties of the *phase portrait*: the family of solution curves, which fill up the entire phase space. For questions such as stability, it is necessary to study the entire phase portrait, including the behavior of solutions for all values of the time parameter. Thus it was essential to consider the entire phase space at once as a geometric object. Doing so, Poincaré found the prevailing mathematical model for mechanics inadequate, for its underlying space was Euclidean, or a domain of several real variables, whereas for a mechanical problem with angular variables or constraints, the phase space might be a more general, nonlinear space, such as a generalized cylinder. Thus the global view in the qualitative theory led Poincaré to the notion of a *differentiable manifold* as the phase space in mechanics. In mechanical systems, this manifold always has a special geometric structure pertaining to the occurrence of phase variables in canonically conjugate pairs, called a *symplectic structure*. Thus the new mathematical model for mechanics consists of a *symplectic manifold*, together with a *Hamiltonian vector field*, or global system of first-order differential equations preserving the symplectic structure.

This model offers no natural system of coordinates. Indeed a manifold admits a coordinate system only locally, so it is most efficient to use the intrinsic calculus of Cartan rather than the conventional calculus of Newton in the analysis of this model. The complete description of this model for mechanics comes quite a bit after Poincaré, as the intrinsic calculus was not fully developed until the 1940s. One advantage of this model is that by suppressing unnecessary coordinates the full generality of the theory becomes evident.

The second characteristic of the qualitative theory is the replacement of analytical methods by differential-topological ones in the study of the phase portrait. For many questions, for example the stability of the solar system, one is interested finally in qualitative information about the phase portrait. In earlier times, the only techniques available were analytical. By obtaining a complete or approximate quantitative solution, qualitative or geometric properties could be deduced. It was POINCARÉS idea to proceed directly to qualitative information by qualitative, that is, geometric methods. Thus POINCARÉ, BIRKHOFF, KOLMOGOROV, ARNOLD, and MOSER show the existence of periodic solutions in the three-body problem by applying differential-topological theorems to the phase portraits in addition to analytical methods. No analytical description of these orbits has been given. In some cases the orbits have been plotted approximately by computers, but of course the computer cannot prove that these solutions are periodic.

A third aspect of the qualitative point of view is a new question that emerges in it—the problem of *structural stability*, the most comprehensive of many different notions of stability. This problem, first posed in 1937 by Andronov–Pontriagin, asks: If a dynamical system X has a known phase portrait P, and is then perturbed to a slightly different system X' (for example, changing the coefficients in its differential equation slightly), then is the new phase portrait P' close to P in some topological sense? This problem is of obvious importance, since in practice the qualitative information obtained for P is to be applied not to X, but to some nearby system X', because the coefficients of the equation may be determined experimentally or by an approximate model and therefore approximately.

The traditional mutuality of mechanics and philosophy has declined in recent years, perhaps because of the justifiable interest in the problems posed by relativity and quantum theory. But current problems in mechanics give new insight into the structure of physical theories.

At the turn of this century a simple description of physical theory evolved, especially among continental physicists—DUHEM, POINCARÉ, MACH, EINSTEIN, HADAMARD, HILBERT—which may still be quite close to the views of many mathematical physicists. This description—most clearly enunciated by DUHEM [1954]—consisted of an *experimental domain*, a *mathematical model*, and a *conventional interpretation*. The model, being a mathematical system, embodies the logic, or axiomatization, of the theory. The interpretation is an agreement connecting the parameters and therefore the conclusions of the model and the observables in the domain.

Traditionally, the philosopher-scientists judge the usefulness of a theory by the criterion of *adequacy*, that is, the verifiability of the predictions, or the quality of the agreement between the interpreted conclusions of the model and the data of the experimental domain. To this DUHEM adds, in a brief example [1954, pp. 138 ff.], the criterion of *stability*.

This criterion, suggested to him by the earliest results of qualitative mechanics (HADAMARD), refers to the stability or continuity of the predictions, or their adequacy, when the model is slightly perturbed. The general applicability of this type of criterion has been suggested by RENÉ THOM [1975].

This stability concerns variation of the model only, the interpretation and domain being fixed. Therefore, it concerns mainly the model, and is primarily a mathematical or logical question. It has been studied to some extent in a general logical setting by the physicologicians BOULIGAND [1935] and DESTOUCHES [1935], but probably it is safe to say that a clear enunciation of this criterion in the correct generality has not yet been made. Certainly all of the various notions of stability in qualitative mechanics and ordinary differential equations are special cases of this notion, including LAPLACE's problem of the stability of the solar system and structural stability, as well as THOM's stability of biological systems.

Also, although this criterion has not been discussed very explicitly by physicists, it has functioned as a tacit assumption, which may be called the *dogma of stability*. For example, in a model with differential equations, in which stability may mean structural stability, the model depends on parameters, namely the coefficients of the equation, each value of which corresponds to a different model. As these parameters can be determined only approximately, the theory is useful only if the equations are structurally stable, which cannot be proved at present in many important cases. Probably the physicist must rely on faith at this point, analogous to the faith of a mathematician in the consistency of set theory.

An alternative to the dogma of stability has been offered by THOM [1975]. He suggests that stability, precisely formulated in a specific theory, be added to the model as an additional hypothesis. This formalization, despite the risk of an inconsistent axiomatic system, reduces the criterion of stability to an aspect of the criterion of adequacy, and in addition may admit additional theorems or predictions in the model. As yet no implications of this axiom are known for celestial mechanics, but THOM has described some conclusions in his model for biological systems.

A careful statement of this notion of stability in the general context of physical theory and epistemology could be quite useful in technical applications of mechanics as well as in the formulation of new qualitative theories in physics, biology, and the social sciences.

Most of this book is devoted to a precise statement of mathematical models for mechanical systems and to precise definitions of various types of stability in this narrow context. These are illustrated by a number of examples, but by one example in depth, namely, the restricted three-body problem in Chapter 10.

Preview

To motivate the introduction of symplectic geometry in mechanics, we briefly consider Hamilton's equations. The starting point is Newton's second law, which states that a particle of mass $m>0$ moving in a potential $V(q)$, $q=(q^1,q^2,q^3)\in \mathbf{R}^3$, moves along a curve $q(t)$ in $\mathbf{R}^3$ in such a way that $m\ddot{q}=-\operatorname{grad} V(q)$. If we introduce the momentum $p_i=m\dot{q}^i$ and the energy $H(q,p)=(1/2m)\|p\|^2+V(q)$, then Newton's law is equivalent to Hamilton's equations:

$$\begin{cases} \dot{q}^i=\partial H/\partial p_i \\ \dot{p}_i=-\partial H/\partial q^i, \qquad i=1,2,3 \end{cases}$$

One proceeds to study this system of first-order equations for a general $H(q,p)$. To do this, we introduce the matrix $J=\begin{pmatrix} 0 & I \\ -I & 0 \end{pmatrix}$, where I is the 3×3 identity, and note that the equations become $\dot{\xi}=J\cdot \operatorname{grad} H(\xi)$, where $\xi=(q,p)$. (In complex notation, setting $z=q+ip$, they may be written as $\dot{z}=-2i\,\partial H/\partial \bar{z}$.)

Set $X_H=J\cdot \operatorname{grad} H$. Then $\xi(t)$ satisfies Hamilton's equations iff $\xi(t)$ is an integral curve of X_H, that is, $\dot{\xi}(t)=X_H(\xi(t))$. The relationship between X_H and H can be rewritten as follows: introduce the skew-symmetric bilinear form ω

on $R^3 \times R^3$ defined by

$$\omega(v_1, v_2) = v_1 \cdot J \cdot v_2$$
$$v_1, v_2 \in R^3 \times R^3$$
$$v_1 = (x_1, y_1)$$
$$v_2 = (x_2, y_2)$$

[In complex notation on $C^3 \cong R^3 \times R^3$, $\omega(v_1, v_2) = -Im\langle v_1, v_2\rangle$, where $v_1 = x_1 + iy_1$, $v_2 = x_2 + iy_2$, and $\langle\ ,\ \rangle$ is the Hermitian inner product.]

Then we have, for all $\xi \in R^3 \times R^3$ and $v \in R^3 \times R^3$,

$$\omega(X_H(\xi), v) = dH(\xi) \cdot v$$

where $dH(q,p) = (\partial H/\partial q^i, \partial H/\partial p^i)$, a row vector in $R^3 \times R^3$, as is easily checked. One calls ω the *symplectic form* on $R^3 \times R^3$, and X_H the *Hamiltonian vector field* with energy H.

Suppose we make a change of coordinates $\eta = f(\xi)$, where $f: R^3 \times R^3 \to R^3 \times R^3$ is smooth. If $\xi(t)$ satisfies Hamilton's equations, the equations satisfied by $\eta(t) = f(\xi(t))$ are $\dot{\eta} = A\dot{\xi} = AJ\, grad_\xi H(\xi) = AJA^*\, grad_\eta H(\xi(\eta))$, where $(A)^{ij} = (\partial \eta^i / \partial \xi^j)$ is the Jacobian of f, and A^* is the transpose of A. The equations for η will be Hamiltonian with energy $K(\eta) = H(\xi(\eta))$ if and only if $AJA^* = J$. A transformation satisfying this condition is called *canonical* or *symplectic*, (or a symplectomorphism). In terms of the symplectic form ω, this condition, denoted $f^*\omega = \omega$, says the transformation f leaves ω unchanged.

The space $R^3 \times R^3$ of the ξ's is called the *phase space*. For a system of N particles we would use $R^{3N} \times R^{3N}$.

For many fundamental physical systems, the phase space is a manifold rather than Euclidean space. For instance, manifolds often arise when constraints are present. For example, the phase space for the motion of the rigid body is the tangent bundle of the group $SO(3)$ of 3×3 orthogonal matrices with determinant $+1$. (See Sect. 4.4 for details.) Not only are manifolds important in these examples, but their terminology and notation lead to a clearer understanding of the basic structure of mechanics.

PART 1

PRELIMINARIES

The basic tools needed for our study of mechanics are developed here. More specialized tools are developed later as needed. Those with the requisite mathematical training can of course skip this part after familiarizing themselves with our notation. Obviously one cannot hope to master all the preliminaries if one is starting from scratch, without a massive effort. Therefore, it seems wise first to go through this part quickly and then, starting with Part 2, to come back when the occasion arises for a more serious second study.

CHAPTER 1

Differential Theory

The categories of differentiable manifolds and vector bundles provide a useful context for the mathematics needed in mechanics, especially the new topological and qualitative results. This chapter develops these categories. The tools for this development—topology and calculus in linear spaces—are studied first.

1.1 TOPOLOGY

One of the greatest difficulties this book presents to the nonmathematician is the reliance on point set topology. Although excellent references are available, the topics required cannot all be found in a single text, and the variation of notations and order in the different books presents a difficult challenge to an inexperienced reader. We assemble here for reference the topics needed, in a consistent notation used throughout this book. A number of more technical proofs that are not relevant for us are omitted and the reader referred to a standard text.

This section is not meant to replace a full course in topology and the reader should not expect to master it on first reading.

Ralph Abraham and Jerrold E. Marsden, Foundation of Mechanics, Second Edition

ISBN 0-8053-0102-X

The reader is assumed to be familiar with usual notations of set theory such as $\in$, $\cup$, $\cap$ and with the concept of a mapping. If A and B are sets and $f: A \to B$ is a mapping, we often write $a \mapsto f(a)$ for the effect of the mapping on the element $a \in A$; "iff" stands for "if and only if" (="*if*" in definitions).

1.1.1 Definitions. *A **topological space** is a set S together with a collection $\mathcal{O}$ of subsets called **open sets** such that*

(T1) *$\varnothing \in \mathcal{O}$ and $S \in \mathcal{O}$;*
(T2) *If $U_1, U_2 \in \mathcal{O}$, then $U_1 \cap U_2 \in \mathcal{O}$;*
(T3) *The union of any collection of open sets is open.*

*For such a topological space the **closed sets** are the elements of*

$$\Gamma = \{A \mid \complement A \in \mathcal{O}\}$$

where $\complement$ denotes the complement, $\complement A = S \backslash A = \{s \in S \mid s \notin A\}$. (The closed sets then obey rules dual to those for open sets.)

*An **open neighborhood of a point** u in a topological space S is an open set U such that $u \in U$. Similarly, for a subset A of S, U is an **open neighborhood of** A if U is open and $A \subset U$.*

*If A is a subset of a topological space S, the **relative topology** on A is defined by*

$$\mathcal{O}_A = \{U \cap A \mid U \in \mathcal{O}\}$$

(which is a topology on A).

*Let S be a topological space. Then a **basis** for the topology is a collection $\mathcal{B}$ of open sets such that every open set of S is a union of elements of $\mathcal{B}$. The topology is called **first countable** if for each $u \in S$, there is a countable collection $\{U_n\}$ of neighborhoods of u such that for any neighborhood U of u, there is an n so $U_n \subset U$. The topology is called **second countable** if it has a countable basis.*

*Let S and T be topological spaces and $S \times T = \{(u, v) \mid u \in S \text{ and } v \in T\}$. The **product topology** on $S \times T$ consists of all subsets that are unions of sets of the form $U \times V$, where U is open in S and V is open in T. Thus, these **open rectangles** form a basis for the topology.*

*Let S be a topological space and $\{u_n\}$ a sequence of points in S. The sequence is said to **converge** if there is a point $u \in S$ such that for every neighborhood U of u, there is an N such that $n > N$ implies $u_n \in U$. We say that $\{u_n\}$ **converges to** u, or u is a **limit point** of $\{u_n\}$.*

1.1.2 Example. On the real line $\boldsymbol{R}$, the ***standard topology*** consists of the sets that are unions of open intervals (a,b). Then $\boldsymbol{R}$ is second countable (and hence first countable) with a basis

$$\left\{\left(r_n - \frac{1}{m}, r_n + \frac{1}{m}\right) \middle| \, r_n \text{ is rational, } m \in \boldsymbol{N}, \text{the positive integers}\right\}$$

ISBN 0-8053-0102-X

The topology on the plane R^2 is the product topology $R \times R$. In R, the sequence $\{1/n\}$ converges to 0, but in the subspace $(0, 1]$, the sequence does not converge.

1.1.3 Definition. *Let S be a topological space and $A \subset S$. Then the* **closure** *of A, denoted $cl(A)$ is the intersection of all closed sets containing A. The* **interior** *of A, denoted $int(A)$ is the union of all open sets contained in A. The* **boundary** *of A, denoted $bd(A)$ is defined by*

$$bd(A) = cl(A) \cap cl(\mathcal{C}A)$$

Thus, $bd(A)$ is closed, and $bd(A) = bd(\mathcal{C}A)$. Note that A is open iff $A = int(A)$ and closed iff $A = cl(A)$.

1.1.4 Proposition. *Let S be a topological space and $A \subset S$. Then*

(i) *$u \in cl(A)$ iff for every neighborhood U of u, $U \cap A \neq \varnothing$;*
(ii) *$u \in int(A)$ iff there is a neighborhood U of u such that $U \subset A$;*
(iii) *$u \in bd(A)$ iff for every neighborhood U of u, $U \cap A \neq \varnothing$ and $U \cap (\mathcal{C}A) \neq \varnothing$.*

This proposition follows readily from the definitions.

1.1.5 Definition. *Let S be a topological space. A point $u \in S$ is called* **isolated** *iff $\{u\}$ is open. The unique topology in which every point is isolated is called the* **discrete topology** *($\mathcal{O} = 2^S$, the collection of all subsets). The topology in which $\mathcal{O} = \{\varnothing, S\}$ is called the* **trivial topology**.

A subset A of S is called **dense** *in S iff $cl(A) = S$ and is called* **nowhere dense** *iff $\mathcal{C}(cl(A))$ is dense in S.*

Thus, A is nowhere dense iff $int(cl(A)) = \varnothing$.

1.1.6 Definition. *A topological space S is called* **Hausdorff** *iff each two distinct points have disjoint neighborhoods (that is, with empty intersection). Similarly, S is called* **normal** *iff each two disjoint closed sets have disjoint neighborhoods.*

Equivalent forms of Hausdorff are the following.

1.1.7 Proposition. *(i) A space S is Hausdorff iff $\Delta_S = \{(u,u) | u \in S\}$ is closed in $S \times S$ in the product topology.*

(ii) A first countable space S is Hausdorff iff all sequences have at most one limit point.

Proof. If Δ_S is closed and u_1, u_2 are distinct, there is an open rectangle $U \times V$ containing (u_1, u_2) and $U \times V \subset \mathcal{C}\Delta_S$. Then in S, U and V are disjoint. The converse is similar, and we leave (*ii*) as an exercise. ■

ISBN 0-8053-0102-X

One of the most important ways topological spaces are constructed is by means of a distance function. We turn to this idea next.

1.1.8 Definition. *Let $\hat{\mathbf{R}}^+$ denote the nonnegative real numbers with a point $\{+\infty\}$ adjoined, and topology generated by the open intervals of the form (a,b) or $(a,+\infty]$. Let M be a set. A* ***metric*** *on M is a function $d\colon M\times M\to\hat{\mathbf{R}}^+$ such that*

(M1) $d(m_1,m_2)=0$ *iff* $m_1=m_2$;
(M2) $d(m_1,m_2)=d(m_2,m_1)$;
(M3) $d(m_1,m_3)\leqslant d(m_1,m_2)+d(m_2,m_3)$, *(triangle inequality).*

For $\varepsilon\in\hat{\mathbf{R}}^+$, $\varepsilon>0$, and $m\in M$, the ε ***disk*** *about m is defined by*

$$D_\varepsilon(m)=\{m'\in M\,|\,d(m',m)<\varepsilon\}$$

The collection of subsets of M that are unions of such disks is the ***metric topology*** *of the* ***metric space*** *(M, d). (It is easily verified that it is a topology on M.)*

Two metrics on a set are called ***equivalent*** *if they induce the same metric topology.*

Let M be a metric space with metric d and $\{u_n\}$ a sequence in M. Then $\{u_n\}$ is a ***Cauchy sequence*** *iff for all $\varepsilon>0$, $\varepsilon\in\hat{\mathbf{R}}^+$, there is an $N\in\mathbf{N}$ such that n, $m\geqslant N$ implies $d(u_n,u_m)<\varepsilon$. (It is easily seen that a convergent sequence is a Cauchy sequence.) The space M is called* ***complete*** *if every Cauchy sequence converges.*

A ***pseudometric*** *on a set M is a function $d\colon M\times M\to\hat{\mathbf{R}}^+$ that satisfies* (M2), (M3), *and*

$$\text{(PM1)}\qquad d(m,m)=0\quad\text{for all } m$$

The ***pseudometric topology*** *is defined exactly as the metric space topology.*

If M is a metric space (or pseudometric space) and $u\in M$, $A\subset M$, we define $d(u,A)=\inf\{d(u,v)\,|\,v\in A\}$ if $A\neq\varnothing$, and $d(u,\varnothing)=\infty$.

Clearly, metric spaces are first countable and Hausdorff; in fact:

1.1.9 Proposition. *Every metric space is normal.*

Proof. Let A and B be closed, disjoint subsets of M, and let

$$U=\{u\in M\,|\,d(u,A)<d(u,B)\}$$

$$V=\{v\in M\,|\,d(v,A)>d(v,B)\}$$

It is verified that U and V are open, disjoint, and $A\subset U$, $B\subset V$. ■

ISBN 0-8053-0102-X

1.1.10 Definition. *The **standard metric** on $\mathbf{R}^n$ is defined by*

$$d(x,y)=\left(\sum_{i=1}^{n}(x^i-y^i)^2\right)^{1/2}$$

where $x=(x^1,\ldots,x^n)$.

Next we study continuity of mappings.

1.1.11 Definition. *Let S and T be topological spaces and $\varphi\colon S\to T$ be a mapping. Then φ is **continuous at** $u\in S$ if for every neighborhood V of $\varphi(u)$ there is a neighborhood U of u such that $\varphi(U)\subset V$. If, for every open set V of T, $\varphi^{-1}(V)=\{u\in S\,|\,\varphi(u)\in V\}$ is open in S, φ is **continuous**. (Thus, φ is continuous iff φ is continuous at each $u\in S$.)*

*If $\varphi\colon S\to T$ is a **bijection** (that is, one-to-one and onto), φ and φ^{-1} are continuous, then φ is a **homeomorphism** and S and T are **homeomorphic**.*

It follows at once that $\varphi\colon S\to T$ is continuous iff the inverse image of every closed set is closed. The following is also useful.

1.1.12 Proposition. *Let S and T be topological spaces and $\varphi\colon S\to T$. Then φ is continuous iff for every $A\subset S$, $\varphi(cl(A))\subset cl(\varphi(A))$.*

Proof. If φ is continuous, then $\varphi^{-1}cl(\varphi(A))$ is closed. But $A\subset\varphi^{-1}cl(\varphi(A))$ and hence $cl(A)\subset\varphi^{-1}cl(\varphi(A))$, or $\varphi(cl(A)\subset cl(\varphi(A))$. Conversely, let $B\subset T$ be closed and $A=\varphi^{-1}(B)$. Then $cl(A)\subset\varphi^{-1}(B)=A$, so A is closed. ■

From 1.1.4 we obtain the following.

1.1.13 Proposition. *Let S be a first countable space and $A\subset S$. Then $u\in cl(A)$ iff there is a sequence of points of A that converge to u (in the topology on S).*

Continuity may be expressed in terms of sequences as follows:

1.1.14 Proposition. *Let S and T be topological spaces with S first countable and $\varphi\colon S\to T$. Then φ is continuous iff for every sequence $\{u_n\}$ converging to u, $\{\varphi(u_n)\}$ converges to $\varphi(u)$, for all $u\in S$.*

We leave this to the reader. In fact, the result follows from 1.1.12 and 1.1.13.

For metric spaces, note that $\varphi\colon M_1\to M_2$ is continuous at $u_1\in M_1$ iff for all $\varepsilon>0$ there is a $\delta>0$ such that $d(u_1,u_1')<\delta$ implies $d(\varphi(u_1),\varphi(u_1'))<\varepsilon$.

1.1.15 Proposition. *Let M and N be metric spaces with N complete. Then the collection $C(M,N)$ of all continuous maps $\varphi\colon M\to N$ forms a complete metric*

ISBN 0-8053-0102-X

space with the metric

$$d^0(\varphi,\psi)=\sup\{d(\varphi(u),\psi(u))|u\in M\}$$

Proof. It is readily verified that d^0 is a metric. Convergence of a sequence $f_n\in C(M,N)$ to $f\in C(M,N)$ in the metric d^0 is the same as *uniform convergence*, that is, for all $\varepsilon>0$ there is an N such that if $n\geq N$,

$$d(f_n(x),f(x))<\varepsilon$$

for all $x\in M$. Now if f_n is a Cauchy sequence in $C(M,N)$, then since

$$d(f_n(x),f_m(x))\leq d^0(f_n,f_m)$$

$f_n(x)$ is Cauchy for each $x\in M$. Thus f_n converges pointwise, defining a function $f(x)$. We must show that $f_n\to f$ uniformly and that f is continuous. First of all, given $\varepsilon>0$, choose N such that $d^0(f_n,f_m)<\varepsilon/2$ if $n,m\geq N$. Then for any $x\in M$, pick $N_x\geq N$ so that $d(f_m(x),f(x))<\varepsilon/2$ if $m\geq N_x$. Thus with $n\geq N$ and $m\geq N_x$,

$$d(f_n(x),f(x))\leq d(f_n(x),f_m(x))+d(f_m(x),f(x))$$

$$<\varepsilon/2+\varepsilon/2=\varepsilon$$

so $f_n\to f$ uniformly. The reader can similarly verify that f is continuous (look in any advanced calculus text such as Marsden [1974a] for the case in $\boldsymbol{R}^n$ if you get stuck). ■

We now study some deeper properties of topological spaces and then some topics that will be of use later in our study of manifolds.

1.1.16 Definition. *Let S be a topological space. Then S is called* ***compact*** *iff for every covering of S by open sets U_α (that is, $\cup_\alpha U_\alpha=S$) there is a finite subcovering. A subset $A\subset S$ is called* ***compact*** *iff A is compact in the relative topology. A space is called* ***locally compact*** *iff each point has a neighborhood whose closure is compact.*

It follows easily that a closed subset of a compact space is compact and that the continuous image of a compact space is compact.

The following is often convenient.

1.1.17 Theorem (Bolzano–Weierstrass). *If S is a first countable space and is compact, then every sequence has a convergent subsequence.*

(The converse is also true in a metric space.)

Proof. Suppose $\{u_n\}$ contains no convergent subsequences. Then we may assume all points are distinct. Each u_n has a neighborhood $\mathcal{O}_n$ that contains no other u_m. From 1.1.13 $\{u_n\}$ is closed, so that $\{\mathcal{O}_n\}$ together with $\mathcal{C}\{u_n\}$ forms an open covering of S, with no finite subcovering. ■

ISBN 0-8053-0102-X

In a metric space, every compact subset is closed and bounded. In $\boldsymbol{R}^n$, the converse is also true (Heine Borel theorem; see Marsden [1974a], p. 62).

1.1.18 Proposition. *Let S be a Hausdorff space. Then every compact subset of S is closed. Also, every compact Hausdorff space is normal.*

Proof. Let $u \in \complement A$ and $v \in A$, where A is compact in S. There are disjoint neighborhoods of u and v and, since A is compact, there are disjoint neighborhoods of u and A. Thus $\complement A$ is open. We leave the second part as an exercise. ■

1.1.19 Proposition. *Let S be a Hausdorff space that is locally homeomorphic to a locally compact Hausdorff space (that is, for each $u \in S$, there is a neighborhood of S homeomorphic, in the subspace topology, to an open subset of a locally compact Hausdorff space). Then S is locally compact. In particular, Hausdorff spaces locally homeomorphic to $\boldsymbol{R}^n$ are locally compact.*

Proof. Let $U \subset S$ be homeomorphic to $\varphi(U) \subset T$. There is a neighborhood V of $\varphi(u)$ so $cl(V) \subset \varphi(U)$ and $cl(V)$ is compact. (We leave this as an exercise; locally compact Hausdorff spaces are regular.) Then $\varphi^{-1}(cl(V))$ is compact, and hence closed in S. By 1.1.12, $\varphi^{-1}(cl(V)) \subset cl(\varphi^{-1}V)$. Thus $\varphi^{-1}(V)$ has compact closure $cl\varphi^{-1}(V) = \varphi^{-1}cl(V)$. ■

1.1.20 Definition. *Let S be a topological space. A covering $\{U_\alpha\}$ of S is called a **refinement** of a covering $\{V_\iota\}$ iff for every U_α there is a V_ι such that $U_\alpha \subset V_\iota$. A covering $\{U_\alpha\}$ of S is called **locally finite** iff each point $u \in S$ has a neighborhood U such that U intersects only a finite number of U_α. A space is called **paracompact** iff every open covering of S has a locally finite refinement of open sets, and S is Hausdorff.*

1.1.21 Theorem. *Second countable, locally compact Hausdorff spaces are paracompact.*

Proof. S is the countable union of open sets U_n such that $cl(U_n)$ is compact and $cl(U_n) \subset U_{n+1}$. If W_α is a covering of S by open sets, and $K_n = cl(U_n) - U_{n-1}$, then we can cover K_n by a finite number of open sets each of which is contained in some $W_\alpha \cap U_{n+1}$, and is disjoint from $cl(U_{n-2})$. The union of such collections yields the desired refinement of $\{W_\alpha\}$. ■

1.1.22 Theorem. *Every paracompact space is normal.*

Proof. We first show that if A is closed and $u \in \complement A$, there are disjoint neighborhoods of u and A (regularity). For each $v \in A$ let U_u, V_v be disjoint neighborhoods of u and v. Let W_α be a locally finite refinement of the covering V_v, $\complement A$, and $V = \cup W_\alpha$, the union over those α so $W_\alpha \cap A \neq \varnothing$. A neighborhood U_0 of u meets a finite number of W_α. Let U denote the

ISBN 0-8053-0102-X

intersection of U_0 and the corresponding U_u. Then V and U are the required neighborhoods. The case for two closed sets proceeds somewhat similarly, so we leave the details for the reader. ■

Later on, the notion of paracompactness will be important because it will guarantee the existence of partitions of unity, a tool useful to us. We will be mostly interested in the differentiable case and will discuss it at the appropriate time. For comparison, we state the continuous results and refer to J. Kelley [1975] and Choquet [1969, Sec. 6] for proofs.

1.1.23 Theorem. *If S is a Hausdorff space, the following are equivalent:*

(i) *S is normal;*

(ii) *For any two closed nonempty disjoint sets A, B there is a continuous function $f: S \to [0, 1]$ such that $f(A)=0$ and $f(B)=1$ (Urysohn's lemma)*

(iii) *For any closed set $A \subset S$ and continuous function $f: A \to [a, b]$, there is a continuous extension $\tilde{f}: S \to [a, b]$ of f (Tietze extension theorem).*

These results are important for the rich supply of continuous functions they provide.

1.1.24 Definition *The **support** of a real-valued function $f: S \to \mathbf{R}$ is*

$$supp(f) = cl\{x \in S \mid f(x) \neq 0\}$$

*A **partition of unity** on S is a family of continuous mappings $\{\varphi_i: S \to [0, 1]\}$ such that*

(i) *$\{supp(\varphi_i)\}$ is locally finite.*

(ii) *$\Sigma_i \varphi_i(x) = 1$ for each $x \in S$.*

*We say that a partition of unity $\{\varphi_i\}$ is **subordinate** to a covering $\{A_\alpha\}$ of S if $supp(\varphi_i)$ is a refinement of $\{A_\alpha\}$.*

The main result on partitions of unity then is the following consequence of 1.2.22 and 1.1.23:

1.1.25 Theorem *Let S be paracompact and $\{U_i\}$ be any open covering of S. Then there is a partition of unity $\{\varphi_i\}$ (with the same index set) subordinate to $\{U_i\}$.*

Later in the book we will use these ideas to prove a C^∞ version of this, so we will not pursue it further here. We turn now to other basic notions from topology that will be needed.

1.1.26 Definition. *A topological space S is **connected** if $\varnothing$ and S are the only subsets of S that are both open and closed. A subset of S is connected iff it is connected in the relative topology. A **component** A of S is a nonempty connected*

ISBN 0-8053-0102-X

subset of S such that the only connected subset of S containing A is A; S is called ***locally connected*** *iff each point x has an open neighborhood containing a connected neighborhood of x.*

It follows easily that the continuous image of a connected set is connected. Equivalent forms of the definition follow.

1.1.27 Proposition. *A space S is not connected iff either of the following holds.*

(*i*) *There is a nonempty proper subset of S that is both open and closed.*
(*ii*) *S is the disjoint union of two nonempty open sets.*
(*iii*) *S is the disjoint union of two nonempty closed sets.*

Also, we have the following.

1.1.28 Proposition. *Let S be a connected space and $f: S \to R$ be continuous. Then f assumes every value between any two values $f(u), f(v)$.*

Proof. Suppose $f(u) < a < f(v)$ and f does not assume the value a. Then $U = \{u_0 | f(u_0) < a\}$ is both open and closed. ■

Intervals in R may be shown to be the only connected sets in R.

1.1.29 Proposition. *Let S be a topological space and $B \subset S$ be connected. Then*

(*i*) *if $B \subset A \subset cl(B)$, then A is connected;*
(*ii*) *if B_α are connected and $B_\alpha \cap B \neq \varnothing$, then $B \cup (\cup_\alpha B_\alpha)$ is connected.*

Proof. If A is not connected, A is the disjoint union of $U_1 \cap A$ and $U_2 \cap A$ where U_1 and U_2 are open in S. Then from 1.1.4(*i*), $U_1 \cap B \neq \varnothing$, $U_2 \cap B \neq \varnothing$, so B is not connected. We leave (*ii*) as an exercise. ■

1.1.30 Corollary. *The components of a topological space are closed. Also, S is the disjoint union of its components. If S is locally connected, the components are open as well as closed.*

1.1.31 Proposition. *Let S be a first countable compact Hausdorff space and $\{A_n\}$ a sequence of closed, connected subsets of S with $A_n \subset A_{n-1}$. Then $A = \cap_1^\infty A_n$ is connected.*

Proof. As S is normal, if A is not connected, A lies in two disjoint open subsets U_1 and U_2 of S. If $A_n \cap \complement U_1 \cap \complement U_2 \neq \varnothing$ for all n, then there is a sequence $u_n \in A_n \cap \complement U_1 \cap \complement U_2$ with a subsequence converging to u. As A_n, $\complement U_1$, $\complement U_2$ are closed, $u \in A \cap \complement U_1 \cap \complement U_2$, a contradiction. Hence some A_n is not connected. ■

ISBN 0-8053-0102-X

An intuitively more appealing, but less convenient, definition of connectedness is the following.

1.1.32 Definition. *Let S be a topological space and $I=[0,1]\subset \boldsymbol{R}$. An **arc** φ in S is a continuous mapping $\varphi: I\rightarrow S$. If $\varphi(0)=u$, $\varphi(1)=v$, we say φ **joins** u and v; S is called **arcwise connected** iff every two points in S can be joined by an arc in S. A space is called **locally arcwise connected** iff each point has an arcwise connected neighborhood (in the relative topology).*

The relationship with connectedness is the following.

1.1.33 Proposition. *Every arcwise connected space is connected. If a space is connected and locally arcwise connected, it is arcwise connected. In particular, a space locally homeomorphic to $\boldsymbol{R}^n$ is connected iff it is arcwise connected.*

Proof. If S is arcwise connected and not connected, write $S=U_1\cup U_2$ where U_1 and U_2 are nonempty, disjoint, and open. Let $u_1\in U_1$ and $u_2\in U_2$ and let φ be an arc joining u_1 and u_2. Now $\varphi(I)$ is connected, and since $\varphi(I)\cap U_i\neq\varnothing$, $\varphi(I)\cap U_1\cap U_2\neq\varnothing$. Hence $U_1\cap U_2\neq\varnothing$, a contradiction. For the second part, let $u\in S$ and U denote all points that can be joined to u by an arc. An easy argument shows U and $\mathcal{C}U$ are open and so $U=S$, by 1.1.27. ∎

For example, disks in $\boldsymbol{R}^n$ are arcwise connected and hence are connected.

In a metric space we already defined the distance $d(u,A)$ from a point to a set. To measure the distance between two sets, we use the following.

1.1.34 Definition. *Let S be a metric space with metric d, and 2^S denote the set of all subsets of S. For $a\in S$ and $B\subset S$, $B\neq\varnothing$, define*

$$d(a,B)=\inf\{d(a,b)\,|\,b\in B\}$$

and for $A,B\subset S$, $A,B\neq\varnothing$

$$\vec{d}(A,B)=\sup\{d(a,B)\,|\,a\in A\}$$

As this is not symmetric, we further define

$$\bar{d}(A,B)=\sup\{\vec{d}(A,B),\vec{d}(B,A)\}$$

*If $A\neq\varnothing$ and $B=\varnothing$, we define $d(a,B)=\infty$ and $\bar{d}(A,B)=\infty$. Finally, define $\bar{d}(\varnothing,\varnothing)=0$. We call $\bar{d}$ the **Hausdorff metric**.*

1.1.35 Proposition. *If d is a metric on S, then $\bar{d}$ is a pseudometric on 2^S.*

Proof. Clearly, $\bar{d}: 2^S\times 2^S\rightarrow\hat{\boldsymbol{R}}^+=[0,\infty]$, $\bar{d}(A,A)=0$ and $\bar{d}$ is symmetric. For the triangle inequality, it is sufficient to show $\vec{d}(A,C)\leqslant\vec{d}(A,B)+\vec{d}(B,C)$ for

ISBN 0-8053-0102-X

$A,B,C\in 2^S$. But $d(a,c)\leqslant d(a,b)+d(b,c)$ implies $d(a,C)\leqslant d(a,b)+d(b,C) \leqslant d(a,b)+\vec{d}(B,C)$. Hence $d(a,C)\leqslant d(a,B)+\vec{d}(B,C)$, which yields the result. The cases A, B, or $C=\varnothing$ are easily checked. ■

As metric spaces are normal, it follows easily that on the closed subsets of S, $\bar{d}$ is a metric. For further details, see Hausdorff [1962], J. Kelley [1975, p. 131], or Michael [1951, p. 152].

Continuity of a map $f\colon S\to 2^S$ can be rephrased as follows.

1.1.36 Proposition. *Let S be a metric space and $\bar{d}$ the Hausdorff metric on 2^S. Then $f\colon S\to 2^S$ is continuous at $u_0\in S$ iff for all $\varepsilon>0$ there is a $\delta>0$ such that $d(u,u_0)<\delta$ implies:*

(i) for all $a\in f(u)$, there is a $b\in f(u_0)$ such that $d(a,b)<\varepsilon$; that is

$$f(u)\subset \bigcup_{b\in f(u_0)} D_\varepsilon(b)$$

and

(ii) for all $b\in f(u_0)$, there is an $a\in f(u)$ such that $d(b,a)<\varepsilon$; that is, $f(u_0)\subset \bigcup_{a\in f(u)} D_\varepsilon(a)$.

This proposition follows at once from the definitions of continuity and the Hausdorff metric.

In a number of places later in the book we are going to form new topological spaces by collapsing old ones. We define this process now and give some examples.

1.1.37 Definition. *Let S be a set. An* ***equivalence relation*** *$\sim$ on S is a binary relation such that for all $u,v,w\in S$,*

(i) $u\sim u$,
(ii) $u\sim v$ iff $v\sim u$, and
(iii) $u\sim v$ and $v\sim w$ implies $u\sim w$.

The ***equivalence class*** *containing u, denoted $[u]$, is defined by*

$$[u]=\{v\in S\,|\,u\sim v\}$$

The set of equivalence classes is denoted $S/\sim$, and the mapping $\pi\colon S\to S/\sim$; $u\mapsto[u]$ is called the ***canonical projection.***

It follows easily that S is the disjoint union of its equivalence classes.

1.1.38 Definition. *Let S be a topological space and $\sim$ an equivalence relation on S. Then $\{U\subset S/\sim\,|\,\pi^{-1}(U)$ is open in $S\}$ is called the* ***quotient topology*** *on $S/\sim$. Similarly, if $S/\sim$ has a topology, we can induce one on S by $\{\pi^{-1}(U)\,|\,U$ is open in $S/\sim\}$. (These are clearly topologies.)*

ISBN 0-8053-0102-X

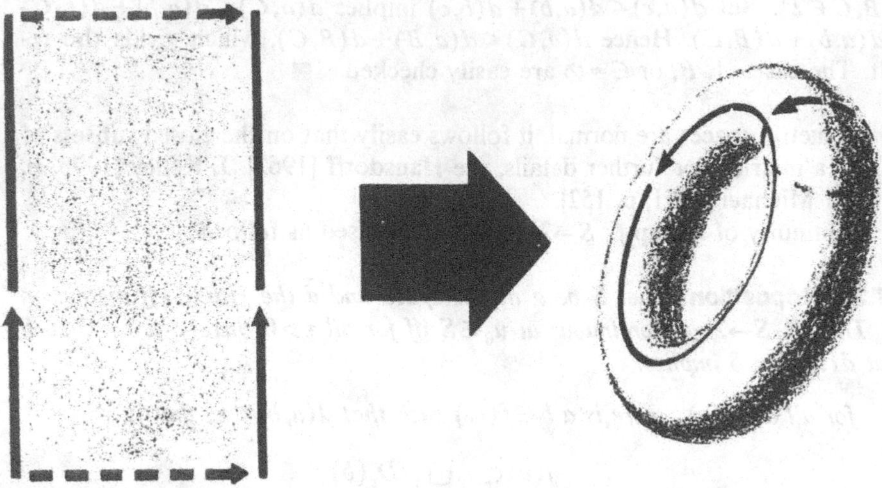

Figure 1.1-1

1.1.39 Example Consider R^2 and the relation $\sim$ defined by

$$(a_1,a_2)\sim(b_1,b_2) \quad \textit{iff} \quad a_1-b_1\in Z \text{ and } a_2-b_2\in Z.$$

(Z denotes the integers).

Then $T^2=R^2/\sim$ is called the 2-**torus**. In addition to the quotient topology, it inherits a group structure in the usual way:

$$[(a_1,a_2)]+[(b_1,b_2)]=[(a_1,a_2)+(b_1,b_2)]$$

The n-dimensional torus T^n is defined in a similar manner.

The torus T^2 may be obtained in two other ways. First, let $\square$ be the unit square in R^2 with the subspace topology. Define $\sim$ by $x\sim y$ iff any of the following hold:

(i) $x=y$;
(ii) $x_1=y_1, x_2=0, y_2=1$;
(iii) $x_1=y_1, x_2=1, y_2=0$;
(iv) $x_2=y_2, x_1=0, y_1=1$; or
(v) $x_2=y_2, x_1=1, y_1=0$;

as indicated in Fig. 1.1-1.

Then $T^2=\square/\sim$.

Second, define $T^2=S^1\times S^1$, as shown in Fig. 1.1-1.

1.1.40 Example. The Klein bottle is obtained by reversing one of the orientations on $\square$, as indicated in Fig. 1.1-2.

Then $K^2=\square/\sim$ (the equivalence relation indicated) is the Klein bottle. Although it is realizable as a subset of R^4, it is convenient to picture it in R^3

ISBN 0-8053-0102-X

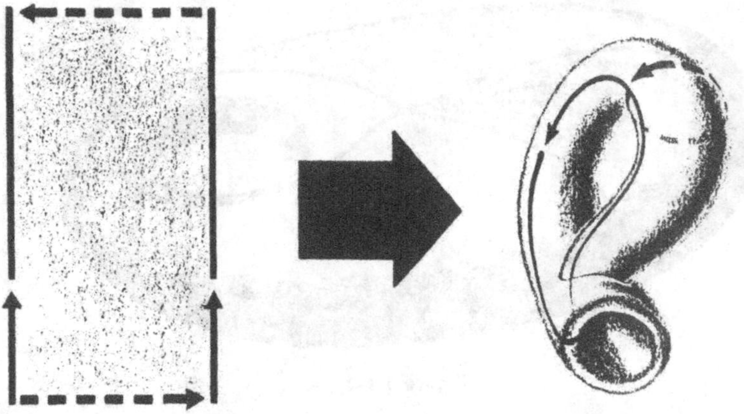

Figure 1.1-2

as shown. Notice that K^2 is not "orientable" and does not inherit a group structure from R^2, as did T^2.

1.1.41 Example. On $R^n\backslash\{0\}$ define $x \sim y$ if there is a nonzero real constant λ such that $x=\lambda y$. Then $R^n\backslash\{0\}/\sim$ is called *real projective* $(n-1)$ space and is denoted by RP^{n-1}. Alternatively, RP^{n-1} can be defined as S^{n-1} (the unit sphere in R^n) with antipodal points x and $-x$ identified. (It is easy to see that this gives a homeomorphic space.) One defines *complex projective space* CP^{n-1} in an analogous way (see Exercise 1.1H).

In mechanics almost every branch of mathematics gets used. Algebraic topology is no exception (see, e.g., Weinstein [1973b]). However, in this book only the simplest notions from this subject are needed.

1.1.42 Definition. *Let Z be a topological space and c: $[0,1] \to Z$ a continuous map such that $c(0)=c(1)=p \in Z$. We call c a* ***loop*** *in Z based at p. The loop c is called* ***contractible*** *if there is a continuous map H: $[0,1]\times[0,1]\to Z$ such that $H(t,0)=c(t)$ and $H(t,1)=p$ for all $t\in[0,1]$. (See Fig. 1.1-3.)*

Roughly speaking, a loop is contractible when it can be shrunk continuously to p. The study of loops leads naturally to homotopy theory. In fact, the loops at p can easily be made into a group called the **fundamental group**.

1.1.43 Definition. *A space Z is called* ***simply connected*** *iff every loop in Z is contractible.*

In the plane R^2 there is an alternate approach to simple connectedness, by way of the Jordan curve theorem: namely, that every simple (nonintersecting) loop in R^2 divides R^2 (that is, its complement has two components). The

ISBN 0-8053-0102-X

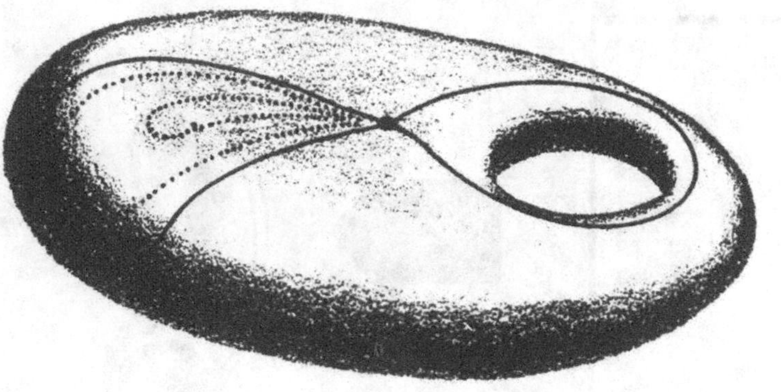

Figure 1.1-3

bounded component of the complement is called the interior, and a subset A of R^2 is simply connected iff the interior of every loop in A lies in A.

Finally, we shall need some notions that will enable us to talk about the idea of "genericity."

1.1.44 Definition. *Let X be a topological space and $A \subset X$. Then A is called* ***residual*** *iff A is the intersection of a countable family of open dense subsets of X. A space X is called a* ***Baire space*** *iff every residual set is dense.*

Recall from 1.1.5 that $B \subset X$ is nowhere dense iff $int(cl(B)) = \varnothing$, so that $X \backslash A$ is residual iff A is the union of a countable collection of nowhere dense closed sets. Clearly, a countable intersection of residual sets is residual.

1.1.45 Lemma. *Let X be a locally Baire space; that is, each point $x \in X$ has a neighborhood U such that $cl(U)$ is a Baire space. Then X is a Baire space.*

Proof. Let $A \subset X$ be residual;

$$A = \bigcap_{1}^{\infty} O_n$$

where $cl(O_n) = X$. Then

$$A \cap cl(U) = \bigcap_{1}^{\infty} (O_n \cap cl(U))$$

Now $O_n \cap cl(U)$ is dense in $cl(U)$ for if $u \in cl(U)$ and $u \in O$ then $O \cap U \neq \varnothing$ and $O \cap U \cap O_n \neq \varnothing$. Hence $cl(U) \subset cl(A)$ and so $cl(A) = X$. ■

Then the *Baire category theorem* is as follows.

1.1.45 Theorem. *Every complete pseudometric space is a Baire space.*

ISBN 0-8053-0102-X

Proof. Let $U \subset X$ be open and $A = \cap_1^\infty O_n$ be residual. We must show $U \cap A \neq \varnothing$. Now as $cl(O_n) = X$, $U \cap O_n \neq \varnothing$ and so we can choose a disk of diameter less than one, say V_1, such that $cl(V_1) \subset U \cap O_1$. Proceed inductively to obtain $cl(V_n) \subset U \cap O_n \cap V_{n-1}$, where V_n has diameter $< 1/n$. Let $x_n \in cl(V_n)$. Clearly $\{x_n\}$ is a Cauchy sequence, and by completeness has a convergent subsequence with limit point x. Then

$$x \in \bigcap_1^\infty cl(V_n) \quad \text{and so} \quad U \cap \bigcap_{n=1}^\infty O_n \neq \varnothing \qquad \blacksquare$$

For the case of a locally compact regular space, see J. Kelley [1975, p. 200], and for applications of Baire spaces, see Choquet [1969].

EXERCISES

1.1A. Let S and T be sets and $f\colon S \to T$. Show that f is a bijection iff there is a mapping $g\colon T \to S$ such that $f \circ g$ and $g \circ f$ are the identity mappings on T and S, respectively.

1.1B. Let X and Y be topological spaces with Y Hausdorff. Then show that, for any continuous maps $f, g\colon X \to Y$, $\{x \in X \mid f(x) = g(x)\}$ is closed. [*Hint:* Consider the mapping $x \mapsto (f(x), g(x))$ and use 1.1.7.]

1.1C. Prove that in a Hausdorff space, single points are closed.

1.1D. Define a *topological manifold* as a space locally homeomorphic to $\boldsymbol{R}^n$. Find a topological manifold that is not Hausdorff and not locally compact. (*Hint:* Consider $\boldsymbol{R}$ with "extra origins.")

1.1E. Show that the continuous image of a connected (resp. arcwise connected) space is connected (resp. arcwise connected).

1.1F. Let M be a topological space and $H\colon M \to \boldsymbol{R}$ be continuous and surjective. Suppose $e \in int(H(M))$. Then show $H^{-1}(e)$ divides M; that is, $M \setminus H^{-1}(e)$ has at least two components.

1.1G. (i) Show that $\bar{d}$ in 1.1.34 is not symmetric, by an example.
(ii) Prove that $\bar{d}$ is a metric on the closed subsets (including $\varnothing$).
(iii) Express the definition of uniform convergence of a sequence of (real) functions in terms of the Hausdorff metric.
(iv) If X is a compact metric space and d and d' are equivalent metrics, then show $\bar{d}$ and $\bar{d}'$ are equivalent.
(v) In case X is not compact show that (iv) can fail.

1.1H. Show that $\boldsymbol{CP}^1$ is homeomorphic to the two-sphere $S^2 \subset \boldsymbol{R}^3$.

1.2 FINITE-DIMENSIONAL BANACH SPACES

We shall be dealing largely with *finite-dimensional real* vector spaces, denoted $\boldsymbol{E}, \boldsymbol{F}, \ldots$ However, much of the following carries over to Banach spaces; see Dieudonné [1960]. In this section we review the basic properties without proofs.

ISBN 0-8053-0102-X

1.2.1 Definition. *A* ***norm*** *on a vector space* $\mathbf{E}$ *is a mapping from* $\mathbf{E}$ *into the real numbers* $\|\cdot\|: \mathbf{E} \to \mathbf{R}$, *such that*

(N1) $\|\cdot\| \geq 0$ *for all* $e \in \mathbf{E}$ *and* $\|e\| = 0$ *iff* $e = 0$;
(N2) $\|\lambda e\| = |\lambda| \|e\|$ *for all* $e \in \mathbf{E}$ *and* $\lambda \in \mathbf{R}$;
(N3) $\|e_1 + e_2\| \leq \|e_1\| + \|e_2\|$ *for all* $e_1, e_2 \in \mathbf{E}$.

If $\|\cdot\|$ is a norm on $\mathbf{E}$, $\mathbf{E}$ becomes a *metric space*. That is, the map d: $\mathbf{E} \times \mathbf{E} \to \mathbf{R}$ defined by $d(e,f) = \|e - f\|$ satisfies (M1), (M2), and (M3) of 1.1.8. A normed space whose induced metric is complete is a ***Banach space.***

1.2.2 Definition. *Two norms on a vector space* $\mathbf{E}$ *are* ***equivalent*** *iff they induce the same topology on* $\mathbf{E}$.

See Exercise 1.2C for another characterization.

1.2.3 Theorem. *Let* $\mathbf{E}$ *be a finite-dimensional real vector space. Then*

(*i*) *there is a norm on* $\mathbf{E}$,
(*ii*) *all norms on* $\mathbf{E}$ *are equivalent,*
(*iii*) *all norms on* $\mathbf{E}$ *are complete.*

For a proof, see Dieudonné [1960, §5.9]. Regarding (iii), recall that $(\mathbf{E}, \|\cdot\|)$ is complete iff every Cauchy sequence converges.

We emphasize *real and finite-dimensional*, for 1.2.3 is false in the general case. For example, the rationals are not complete relative to the absolute value norm. For the necessity of finite dimension, the space of continuous functions on $[0,1]$ has two inequivalent norms (Dieudonné [1960, p. 102]).

Since we are dealing with finite-dimensional real vector spaces, Theorem 1.2.3 tells us that a unique topology is determined by norms. Also, a mapping $f: A \subset \mathbf{E} \to \mathbf{F}$ is continuous (that is, inverse images of open sets are open) iff for all $e_0 \in A$ and any $\varepsilon > 0$ and norm $|||\cdot|||$ on $\mathbf{F}$, there is a $\delta > 0$ and a norm $\|\cdot\|$ on $\mathbf{E}$ such that $f(D_{\delta,\|\cdot\|}(e_0) \cap A) \subset D_{\varepsilon,|||\cdot|||}(f(e_0))$, where

$$D_{\delta,\|\cdot\|}(e_0) = \{ e \in \mathbf{E} \,|\, \|e - e_0\| < \delta \}$$

Recall that $f: \mathbf{E}_1 \times \mathbf{E}_2 \times \cdots \times \mathbf{E}_k \to \mathbf{F}$ is multilinear iff it is linear in each variable separately. Note that this does *not mean* f is linear on the product vector space.

1.2.4 Theorem. *For finite-dimensional real vector spaces, linear and multilinear maps are continuous.*

Again, we do not need to specify the norm because of 1.2.3. The proof is a consequence of Dieudonné [1960, p. 99].

The following is an immediate corollary of this, but is also true more generally (Dieudonné [1960, p. 89]).

ISBN 0-8053-0102-X

1.2.5 Corollary. *Addition and scalar multiplication in a (normed) vector space are continuous maps from* $E\times E\to E$ *and* $R\times E\to E$, *respectively.*

1.2.6 Definition. *Given* E,F *we let* $L(E,F)$ *denote the set of all linear maps from* E *into* F *together with the natural structure of finite-dimensional real vector space. Similarly,* $L^k(E,F)$ *denotes the space of multilinear maps from* $E\times\cdots\times E$ *(k copies) into* F, $L^k_s(E,F)$, *the subspace of symmetric elements of* $L^k(E,F)$ [*that is, if* π *is any permutation of* $\{1,2,\ldots,k\}$, *we have* $f(e_1,\ldots,e_k)=f(e_{\pi(1)},\ldots,e_{\pi(k)})$] *and* $L^k_a(E,F)$ *the subspace of skew symmetric elements of* $L^k(E,F)$ [*that is, if* π *is any permutation, we have* $f(e_1,\ldots,e_k)=(\operatorname{sign}\pi)f(e_{\pi(1)},\ldots,e_{\pi(k)})$, *where* $\operatorname{sign}\pi$ *is* ±1 *according as* π *is an even or odd permutation*].

1.2.7 Theorem. *There is a natural isomorphism*

$$L(E,L^k(E,F))\approx L^{k+1}(E,F)$$

Proof. For $\varphi\in L(E,L^k(E,F))$ we define $\tilde\varphi\in L^{k+1}(E,F)$ by

$$\tilde\varphi(e_1,\ldots,e_{k+1})=\varphi(e_{k+1})(e_1,\ldots,e_k)$$

It is easy to check that the association $\varphi\mapsto\tilde\varphi$ is an isomorphism (that is, a linear map which is bijective, or one-to-one and onto). ■

In a similar way we can identify $L(R,F)$ with F: to $\varphi\in L(R,F)$ we associate $\varphi(1)\in F$.

It is important to realize that although $L(E,R)$ and E have the same dimension, and are therefore isomorphic, any such isomorphism requires a basis for its description. Hence we regard E and $L(E,R)$ as distinct; they are not *naturally* isomorphic.

EXERCISES

1.2A. Let $f\in L(E,F)$ so that f is continuous.
- (a) Show that there is a constant K such that $\|f(e)\|\leq K\|e\|$ for all $e\in E$. Define $\|f\|$ as the greatest lower bound of such K.
- (b) Show that this is a norm on $L(E,F)$.
- (c) Prove that $\|f\circ g\|\leq\|f\|\cdot\|g\|$.

1.2B. Suppose $f\in L(E,F)$ and $\dim E=\dim F$. Then f is an isomorphism iff it is a monomorphism (one-to-one) and iff it is surjective (onto).

1.2C. Show that two norms $\|\cdot\|$ and $|||\cdot|||$ on E are equivalent iff there is a constant M such that $M^{-1}|||e|||\leq\|e\|\leq M|||e|||$ for all $e\in E$.

1.2D. Let E be the set of all C^1 functions $f\colon[0,1]\to R$ with the norm

$$\|f\|=\sup_{x\in[0,1]}|f(x)|+\sup_{x\in[0,1]}|f'(x)|$$

Prove that E is a Banach space.

ISBN 0-8053-0102-X

1.3 LOCAL DIFFERENTIAL CALCULUS

The usual approach to elementary calculus is not suitable for generalization to manifolds. Thus, in this section, we reinterpret the differentiation process in a way that will be useful for manifolds. Easy proofs which are standard in multivariable calculus will be omitted.

For a differentiable function $f: U \subset \mathbf{R} \to \mathbf{R}$, the usual interpretation of the derivative at $u_0 \in U$ is the slope of the line tangent to the graph of f at u_0.

The idea which generalizes is to interpret $Df(u_0) = f'(u_0)$ as a linear map acting on the vector $(u - u_0)$. Then we can say that $Df(u_0)$ is the unique linear map from $\mathbf{R}$ into $\mathbf{R}$ such that the mapping

$$g: U \to \mathbf{R}: u \mapsto g(u) = f(u_0) + Df(u_0) \cdot (u - u_0)$$

is tangent to f at u_0 (see Fig. 1.3-1). This motivates the following

1.3.1 Definition. *Let $\mathbf{E}$, $\mathbf{F}$ be two (finite-dimensional, real) vector spaces with maps*

$$f, g: U \subset \mathbf{E} \to \mathbf{F}$$

*where U is open in $\mathbf{E}$. We say f and g are **tangent** at $u_0 \in U$ iff*

$$\lim_{u \to u_0} \frac{\|f(u) - g(u)\|}{\|u - u_0\|} = 0$$

where $\|\cdot\|$ represents any norm on the appropriate space.

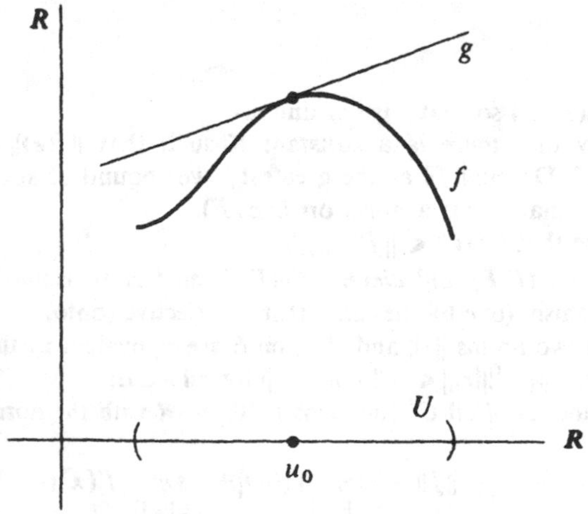

Figure 1.3-1

ISBN 0-8053-0102-X

1.3.2 Theorem. *For $f: U \subset \mathbf{E} \to \mathbf{F}$ and $u_0 \in U$ there is at most one $L \in L(\mathbf{E},\mathbf{F})$ so that the map $g_L: U \subset \mathbf{E} \to \mathbf{F}$ given by $g_L(u) = f(u_0) + L(u - u_0)$ is tangent to f at u_0.*

We leave the proof as an exercise on limits.

1.3.3 Definition. *If, in 1.3.2, there is such an $L \in L(\mathbf{E},\mathbf{F})$ we say f is* ***differentiable at*** u_0, *and define the* ***derivative*** *of f at u_0 to be $Df(u_0) = L$. If f is differentiable at each $u_0 \in U$, the map*

$$Df: U \to L(\mathbf{E},\mathbf{F});\ u \mapsto Df(u)$$

is the ***derivative*** *of f. Moreover, if Df is a continuous map we say f is* ***of class*** $\mathbf{C}^1$ *(or is* ***continuously differentiable****).*

1.3.4 Definition. *Suppose $f: U \subset \mathbf{E} \to \mathbf{F}$ is of class C^1. Define the* ***tangent*** *of f to be the map*

$$Tf: U \times \mathbf{E} \to \mathbf{F} \times \mathbf{F}$$

given by

$$Tf(u,e) = (f(u), Df(u)\cdot e)$$

where $Df(u)\cdot e$ is $Df(u)$ applied to $e \in \mathbf{E}$ as a linear map.

From a geometrical point of view, T is more natural than D. If we think of (u,e) as a vector with base point u, then $(f(u), Df(u)\cdot e)$ is the image vector *with its base point*. See Fig. 1.3-2. Another reason for this is its behavior under composition, as given in the next theorem. (This theorem expresses the fact that T is a *covariant functor*.)

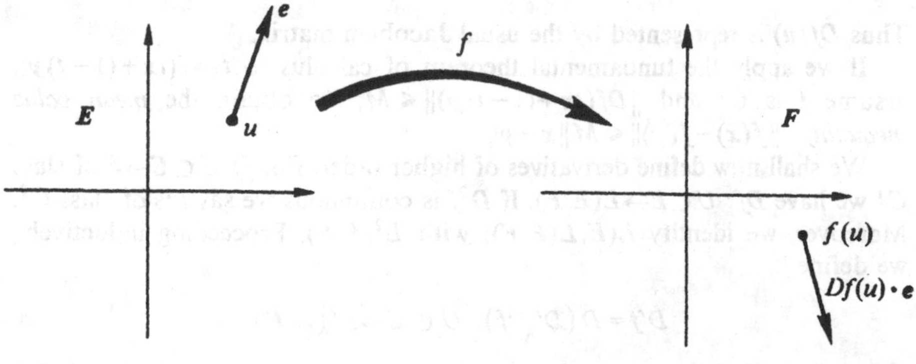

Figure 1.3-2

ISBN 0-8053-0102-X

1.3.5 Theorem (C^1 composite mapping theorem). *Suppose $f: U \subset \boldsymbol{E} \to V \subset \boldsymbol{F}$ and $g: V \subset \boldsymbol{F} \to \boldsymbol{G}$ are C^1 maps. Then the composite $g \circ f: U \subset \boldsymbol{E} \to \boldsymbol{G}$ is also C^1 and*

$$T(g \circ f) = Tg \circ Tf$$

In terms of D, this formula is equivalent to the chain rule

$$D(g \circ f)(u) \cdot e = Dg(f(u)) \cdot (Df(u) \cdot e)$$

For a proof, see Dieudonné [1960, p. 145] or Marsden [1974a, p. 168]. For the validity of this chain rule, f and g need only be differentiable.

We will now show how the derivative Df is related to the usual directional derivative. A **curve** in $\boldsymbol{E}$ is a C^1 map from I into $\boldsymbol{E}$, where I is an open interval of $\boldsymbol{R}$. Thus, for $t \in I$ we have $Dc(t) \in L(\boldsymbol{R}, \boldsymbol{E})$, by definition. We identify $L(\boldsymbol{R}, \boldsymbol{E})$ with $\boldsymbol{E}$ by associating, in this case, $Dc(t)$ with $Dc(t) \cdot 1$ ($1 \in \boldsymbol{R}$). Let

$$\frac{dc}{dt}(t) = Dc(t) \cdot 1$$

For $f: U \subset \boldsymbol{E} \to \boldsymbol{F}$ of class C^1 we consider $f \circ c$, where $c: I \to U$. It follows from 1.3.5 that

$$Df(u) \cdot e = \frac{d}{dt}\{f(u + te)\}\Big|_{t=0}$$

For let c be defined by $c(t) = u + te (u, e \in \boldsymbol{E},\ t \in \boldsymbol{R})$ for suitable $I = (-\lambda, \lambda)$, and apply the chain rule to $f \circ c$. On Euclidean space the d/dt defined this way coincides with the usual directional derivative. More specifically, suppose we have $f: U \subset \boldsymbol{R}^m \to \boldsymbol{R}^n$ of class C^1. Now $Df(u)$ is a linear map from $\boldsymbol{R}^m \to \boldsymbol{R}^n$ and so it is represented by its components relative to the standard basis $\boldsymbol{e}_1, \ldots, \boldsymbol{e}_m$ of $\boldsymbol{R}^m$. By the above formula we see

$$Df(u) \cdot e_i = \left(\frac{\partial f^1}{\partial x^i}(u), \ldots, \frac{\partial f^n}{\partial x^i}(u) \right)$$

Thus $Df(u)$ is represented by the usual Jacobian matrix.

If we apply the fundamental theorem of calculus to $t \mapsto f(t\boldsymbol{x} + (1-t)\boldsymbol{y})$, assume f is C^1 and $\|Df(t\boldsymbol{x} + (1-t)\boldsymbol{y})\| \leqslant M$, we obtain the *mean value inequality*: $\|f(\boldsymbol{x}) - f(\boldsymbol{y})\| \leqslant M\|\boldsymbol{x} - \boldsymbol{y}\|$.

We shall now define derivatives of higher order. For $f: U \subset \boldsymbol{E} \to \boldsymbol{F}$ of class C^1 we have $Df: U \subset \boldsymbol{E} \to L(\boldsymbol{E}, \boldsymbol{F})$. If $D^2 f$ is continuous we say f is of class C^2. Moreover, we identify $L(\boldsymbol{E}, L(\boldsymbol{E}, \boldsymbol{F}))$ with $L^2(\boldsymbol{E}, \boldsymbol{F})$. Proceeding inductively, we define

$$D^r f = D(D^{r-1} f): U \subset \boldsymbol{E} \to L^r(\boldsymbol{E}, \boldsymbol{F})$$

if it exists. If $D^r f$ exists and is continuous we say f is **of class C^r**. The symmetry of second partial derivatives appears here in the following form.

ISBN 0-8053-0102-X

1.3.6 Theorem. *If $f: U \subset E \to F$ is C^r, then*

(*i*) $D^r f(u) \in L_s^r(E, F)$;
(*ii*) *f is C^q, $q = 0, \ldots, r$. (C^0 = continuous.)*

For a proof of this, see Dieudonné [1960, p. 176] or Marsden [1974a, p. 178].

In a similar way we can define $T^r f$, and by induction on the C^1 composite mapping theorem we obtain

1.3.7 Theorem (C^r composite mapping theorem). *Let $f: U \subset E \to V \subset F$ and $g: V \subset F \to G$ be C^r maps. Then $g \circ f$ is C^r and*

$$T^r(g \circ f) = T^r g \circ T^r f$$

Note that a corresponding statement in terms of D is a good deal more complicated. (Exercise 1.3D.)

For computation of higher derivatives we have, by repeated application of the computational rule for $Df(u) \cdot e$,

$$D^r f(u)(e_1, \ldots, e_r) = \frac{d}{dt_r} \cdots \frac{d}{dt_1} \left\{ f\left(u + \sum_{i=1}^{r} t_i e_i \right) \right\} \Bigg|_{t_1 = \cdots = t_r = 0}$$

In particular, for $f: U \subset \mathbf{R}^m \to \mathbf{R}^n$ the components of $D^r f(u)$ in terms of the standard basis are

$$\frac{\partial^r f^\alpha(u)}{\partial x^{i_1} \cdots \partial x^{i_r}} \quad \begin{cases} 1 \leqslant \alpha \leqslant n \\ 1 \leqslant i_k \leqslant m \end{cases}$$

Thus f is of class C^r iff all its rth-order partial derivatives exist and are continuous.

Suppose $U \subset E$ is an open set. Then as $+: E \times E \to E$ is continuous, there exists an open set $\tilde{U} \subset E \times E$ with (i) $U \times \{0\} \subset \tilde{U}$, (ii) $u + \xi h \in U$ for all $(u, h) \in \tilde{U}$ and $0 \leqslant \xi \leqslant 1$, and (iii) $(u, h) \in \tilde{U}$ implies $u \in U$. For example, let

$$\tilde{U} = \{(+)^{-1}(U)\} \cap \{U \times E\}$$

Let us call such a set $\tilde{U}$, temporarily, a **thickening** of U.

1.3.8 Theorem (Taylor's theorem). *A map $f: U \subset E \to F$ is of class C^r iff there are continuous mappings*

$$\varphi_p: U \subset E \to L_s^p(E, F), \qquad p = 1, \ldots, r$$

$$R: \tilde{U} \to L_s^r(E, F)$$

ISBN 0-8053-0102-X

where $\tilde{U}$ is a thickening of U, such that, for all $(u,h)\in\tilde{U}$,

$$f(u+h)=f(u)+\frac{\varphi_1(u)}{1!}\cdot h+\frac{\varphi_2(u)}{2!}\cdot h^2+\cdots+\frac{\varphi_r(u)}{r!}\cdot h^r+R(u,h)h^r$$

where $h^r=(h,\ldots,h)$ (r times) and $R(u,0)=0$.

For the *only if* part choose $\varphi_p(u)=D^pf(u)$ and see Dieudonné [1960, p. 185] or Marsden [1974a, p. 177]. For the converse, see Abraham–Robbin [1967, §2] and Nelson [1969, p. 7–8].

If f is C^∞ (that is, is C^r for all r) then we may be able to extend the above formula into a convergent power series. If we can, we say f is of class C^ω, or **analytic**. The standard example of a C^∞ function that is not analytic is the following function from $\boldsymbol{R}$ to $\boldsymbol{R}$:

$$\theta(x)=\begin{cases} exp\{-1/(1-|x|^2)\}, & |x|<1\\ 0, & |x|\geq 1\end{cases}$$

This function is C^∞, and all derivatives are 0 at $x=\pm 1$. Hence all coefficients of the Taylor series around these points vanish. Since the function is not identically 0 in any neighborhood of ± 1, it cannot be analytic there.

Leibniz' rule for derivatives has the following general form.

1.3.9 Proposition. *For $f\colon U\subset \boldsymbol{E}\to\boldsymbol{E}'$ and $g\colon U\subset\boldsymbol{E}\to\boldsymbol{F}'$ of class C^1 define $f\times g\colon U\to\boldsymbol{E}'\times\boldsymbol{F}'$ by $(f\times g)(u)=(f(u),g(u))$. Suppose $B\colon \boldsymbol{E}'\times\boldsymbol{F}'\to\boldsymbol{G}$ is a bilinear map, and $f\cdot g=B\circ(f\times g)$. Then $f\cdot g$ is of class C^1 and $D(f\cdot g)=f\cdot Dg+Df\cdot g\colon U\to L(\boldsymbol{E},\boldsymbol{G})$ [where $(f\cdot Dg)(u)\cdot e=B(f(u),Dg(u)\cdot e)$].*

This follows easily by the composite mapping theorem, the formula $D(f\times g)=Df\times Dg$ and the fact that the derivative of B at (e_0',f_0') is $(e',f')\mapsto B(e',f_0')+B(e_0',f')$. See Dieudonné [1960, p. 144]. In the case $\boldsymbol{E}'=\boldsymbol{F}'=\boldsymbol{R}$ and B is multiplication, 1.3.9 reduces to the usual product rule for derivatives.

It will also be convenient to consider partial derivatives in this context.

1.3.10 Definition. *Let $U_1\subset\boldsymbol{E}_1$, $U_2\subset\boldsymbol{E}_2$ be open, and suppose $f\colon U_1\times U_2\to\boldsymbol{F}$ and f is differentiable. Then the **partial derivative** of f with respect to the first factor $\boldsymbol{E}_1$ denoted D_1f is defined by*

$$D_1f(u_1,u_2)\colon \boldsymbol{E}_1\to\boldsymbol{F}\colon e_1\mapsto D_1f(u_1,u_2)\cdot e_1=Df(u_1,u_2)\cdot(e_1,0)$$

We similarly define D_2f.

ISBN 0-8053-0102-X

If we identify $E_1 \times E_2$ with $E_1 \oplus E_2$, $(e_1, 0)$ with e_1, and $(0, e_2)$ with e_2, we may write Df as the sum of its partial derivatives

$$Df = D_1 f + D_2 f$$

1.3.11 Definition. *A map $f: U \subset E \to V \subset F$ (U, V open) is a C^r* ***diffeomorphism*** *iff f is of class C^r, is a bijection (that is, one-to-one and onto V), and f^{-1} is also of class C^r.*

The following is a major result that will be used many times later in the book. Since it is one of the main pillars of nonlinear analysis, we give a complete proof.

1.3.12 Theorem (Inverse mapping theorem). *Let $f: U \subset E \to F$ be of class C^r, $r \geq 1$, $x_0 \in U$, and suppose $Df(x_0)$ is a linear isomorphism. Then f is a C^r diffeomorphism of some neighborhood of x_0 onto some neighborhood of $f(x_0)$.*

It is essential to have Banach spaces in this result, rather than more general spaces such as topological vector spaces or Fréchet spaces, as simple examples show. The following example of the failure of Theorem 1.3.12 in Fréchet spaces was kindly pointed out by M. McCracken.

Let $\mathcal{H}(\Delta)$ denote the set of all analytic functions on the open unit disk with the topology of uniform convergence on compact subsets. Let $F: \mathcal{H}(\Delta) \to \mathcal{H}(\Delta)$ be defined by

$$\sum_0^\infty a_n z^n \mapsto \sum_0^\infty a_n^2 z^n$$

Then clearly F is C^∞ and by 1.3.9,

$$DF\left(\sum_0^\infty a_n z^n\right) \cdot \left(\sum_0^\infty b_n z^n\right) = \sum_0^\infty 2a_n b_n z^n$$

If $a_0 = 0$ and $a_n = 1/n$, $n \geq 1$, then $DF(\Sigma_1^\infty z^n/n)$ is a bounded, linear isomorphism. However, since

$$F\left(z + \frac{z^2}{2} + \cdots + \frac{z^{k-1}}{k-1} - \frac{z^k}{k} + \frac{z^{k+1}}{k+1} + \cdots\right) = F\left(\sum_1^\infty \frac{z^n}{n}\right)$$

F is not locally 1-1.

(Consult Schwartz [1967] for more sophisticated versions of the inverse function theorem valid in Fréchet spaces.)

ISBN 0-8053-0102-X

Although our main interest is the finite-dimensional case, for Banach spaces keep in mind the open mapping theorem: if $T: \boldsymbol{E} \rightarrow \boldsymbol{F}$ is linear, bijective and continuous, then T^{-1} is continuous. (Choquet [1969], p. 322).

Proof of Theorem 1.3.12. We begin by assembling a few standard lemmas:

1.3.13 Lemma. *Let M be a complete metric space with distance function $d: M \times M \rightarrow \mathbf{R}$. Let $F: M \rightarrow M$ and assume there is a constant λ, $0 \leqslant \lambda < 1$ such that for all $x, y \in M$,*

$$d(F(x), F(y)) \leqslant \lambda d(x, y)$$

Then F has a unique fixed point $x_0 \in M$; that is, $F(x_0) = x_0$.

This result is usually called the *contraction mapping principle* and is the basis of many important existence theorems in analysis. The other fundamental fixed point theorem in analysis is the Schauder fixed point theorem which states that a continuous map of a compact convex set (in a Banach space, say) to itself, has a fixed point—not necessarily unique however.

The proof of Lemma 1.3.13 is as follows. Pick $x_1 \in M$ and define x_n inductively by $x_{n+1} = F(x_n)$. By induction we clearly have

$$d(x_{n+1}, x_n) \leqslant \lambda^{n-1} d(F(x_1), x_1)$$

and so

$$d(x_n, x_{n+k}) \leqslant \left(\sum_{j=n-1}^{n-1+k} \lambda^j \right) d(F(x_1), x_1)$$

Thus x_n is a Cauchy sequence. Since F is obviously uniformly continuous, $x_0 \equiv \lim_{n\to\infty} x_n = \lim_{n\to\infty} x_{n+1} = \lim_{n\to\infty} F(x_n) = F(x_0)$. Since $\lambda < 1$ it follows that F has at most one fixed point. ▼

(The symbol ▼, a modification of the Halmos symbol ■ indicates that the lemma is proved but the proof of the theorem goes on).

1.3.14 Lemma *Let $GL(\boldsymbol{E},\boldsymbol{F})$ denote the set of linear isomorphisms from $\boldsymbol{E}$ onto $\boldsymbol{F}$. Then $GL(\boldsymbol{E},\boldsymbol{F}) \subset L(\boldsymbol{E},\boldsymbol{F})$ is open.*

Proof. Let

$$\|\alpha\| = \sup_{\substack{e \in E \\ \|e\|=1}} \|\alpha(e)\|$$

be the norm on $L(\boldsymbol{E},\boldsymbol{F})$ relative to given norms on $\boldsymbol{E}$ and $\boldsymbol{F}$.

ISBN 0-8053-0102-X

We can assume $E=F$. Indeed, if $\varphi_0 \in GL(E,F)$, the map $\psi \mapsto \varphi_0^{-1} \circ \psi$ from $L(E,F)$ to $L(E,E)$ is continuous and $GL(E,F)$ is the inverse image of $GL(E,E)$.

For $\varphi \in GL(E,E)$, we shall prove that any ψ sufficiently near φ is also invertible, which will give the result. More precisely, $\|\psi-\varphi\| < \|\varphi^{-1}\|^{-1}$ implies $\psi \in GL(E,E)$. The key is that $\|\cdot\|$ is an algebra norm. That is $\|\beta \circ \alpha\| \leqslant \|\beta\| \, \|\alpha\|$ for $\alpha \in L(E,E)$ and $\beta \in L(E,E)$ (see Exercise 1.2A). Since $\psi = \varphi \circ (I - \varphi^{-1} \circ (\varphi - \psi))$, φ is invertible, and our norm assumption shows that $\|\varphi^{-1} \circ (\varphi - \psi)\| < 1$, it is sufficient to show that $I - \xi$ is invertible whenever $\|\xi\| < 1$. (I is the identity operator). Consider the following sequence (called the Neumann series):

$$\xi_0 = I$$

$$\xi_1 = I + \xi$$

$$\xi_2 = I + \xi + \xi \circ \xi$$

$$\vdots$$

$$\xi_n = I + \xi + \xi \circ \xi + \cdots + (\xi \circ \xi \circ \cdots \circ \xi)$$

Using the triangle inequality and the above norm inequality, we can compare this sequence to the sequence of real numbers, 1, $1+\|\xi\|$, $1+\|\xi\|+\|\xi\|^2, \ldots$, which we know is a Cauchy sequence since $\|\xi\| < 1$. Because $L(E,E)$ is complete, ξ_n must converge. The limit, say ρ, is the inverse of $I-\xi$. Indeed $(I-\xi)\xi_n = I - (\xi \circ \xi \circ \ldots \circ \xi)$, so the result follows. ▼

1.3.15 Lemma *Let $\mathcal{I}^{-1}: GL(E,F) \to GL(F,E): \varphi \mapsto \varphi^{-1}$. Then $\mathcal{I}^{-1}$ is of class C^∞ and $D\mathcal{I}^{-1}(\varphi) \cdot \psi = -\varphi^{-1}\psi\varphi^{-1}$.*

Proof. We may assume $GL(E,F) \neq \varnothing$. If we can show that $D\mathcal{I}^{-1}(\varphi)\cdot\psi = -\varphi^{-1} \circ \psi \circ \varphi^{-1}$, then it will follow from Leibniz' rule that $\mathcal{I}^{-1}$ is of class C^∞. Since $\psi \mapsto -\varphi^{-1}\psi\varphi^{-1}$ is linear ($\psi \in L(E,F)$) we must show that

$$\lim_{\psi \to \varphi} \frac{\|\psi^{-1} - (\varphi^{-1} - \varphi^{-1}\psi\varphi^{-1} + \varphi^{-1}\varphi\varphi^{-1})\|}{\|\psi - \varphi\|} = 0$$

Note

$$\psi^{-1} - (\varphi^{-1} - \varphi^{-1}\psi\varphi^{-1} + \varphi^{-1}\varphi\varphi^{-1}) = \psi^{-1} - 2\varphi^{-1} + \varphi^{-1}\psi\varphi^{-1}$$

$$= \psi^{-1}(\psi-\varphi)\varphi^{-1}(\psi-\varphi)\varphi^{-1}$$

ISBN 0-8053-0102-X

Again, using $\|\beta \circ \alpha\| \leqslant \|\alpha\|\,\|\beta\|$ for $\alpha \in L(E,F)$, $\beta \in L(F,G)$,

$$\|\psi^{-1}(\psi-\varphi)\varphi^{-1}(\psi-\varphi)\varphi^{-1}\| \leqslant \|\psi^{-1}\|\,\|\psi-\varphi\|^2\|\varphi^{-1}\|^2$$

With this inequality, the limit above is clearly zero. ▼

To prove the theorem it is useful to note that it is enough to prove it under the simplifying assumptions that $x_0=0$, $f(x_0)=0$, $E=F$, and $Df(0)$ is the identity. (Indeed, replace f by $h(x)=Df(x_0)^{-1}\circ[f(x+x_0)-f(x_0)]$.)

Now let $g(x)=x-f(x)$ so $Dg(0)=0$. Choose $r>0$ such that $\|x\|\leqslant r$ implies $\|Dg(x)\|\leqslant \frac{1}{2}$, which is possible by continuity of Dg. Thus by the mean value inequality, $\|x\|\leqslant r$ implies $\|g(x)\|\leqslant r/2$. Let $\overline{B}_\varepsilon(0)=\{x\in E \mid \|x\|\leqslant \varepsilon\}$. For $y\in \overline{B}_{r/2}(0)$, let $g_y(x)=y+x-f(x)$. By the mean value inequality, if

$$y\in \overline{B}_{r/2}(0) \quad \text{and} \quad x_1,x_2\in \overline{B}_r(0)$$

then

(a) $\|g_y(x)\|\leqslant \|y\|+\|g(x)\|\leqslant r$ and

(b) $\|g_y(x_1)-g_y(x_2)\|\leqslant \frac{1}{2}\|x_1-x_2\|$

Thus by Lemma 1.3.13, $g_y(x)$ has a unique fixed point x in $\overline{B}_r(0)$. This point x is the unique solution of $f(x)=y$. Thus f has an inverse

$$f^{-1}: V_0=B_{r/2}(0)\to U_0=f^{-1}(B_{r/2}(0))\subset B_r(0)$$

From (b) above, $\|f^{-1}(y_1)-f^{-1}(y_2)\|\leqslant 2\|y_1-y_2\|$, so f^{-1} is continuous.

From Lemma 1.3.14 we can choose r small enough so that $Df(x)^{-1}$ will exist for $x\in B_r(0)$. Moreover, by continuity, $\|Df(x)^{-1}\|\leqslant M$ for some M and all $x\in B_r(0)$ can be assumed as well. If y_1, $y_2\in B_{r/2}(0)$, $x_1=f^{-1}(y_1)$, $x_2=f^{-1}(y_2)$, then

$$\begin{aligned}
&\left\|f^{-1}(y_1)-f^{-1}(y_2)-Df(x_2)^{-1}(y_1-y_2)\right\|\\
&=\left\|x_1-x_2-Df(x_2)^{-1}\left[f(x_1)-f(x_2)\right]\right\|\\
&=\left\|Df(x_2)^{-1}\left\{Df(x_2)\cdot(x_1-x_2)-f(x_1)-f(x_2)\right\}\right\|\\
&\leqslant M\left\|f(x_1)-f(x_2)-Df(x_2)(x_1-x_2)\right\|
\end{aligned}$$

This, together with (b) above, shows f^{-1} is differentiable with derivative $Df(x)^{-1}$ at $f(x)$. By continuity of inversion (Lemma 1.3.15) we see that f^{-1} is C^1. Also from Lemma 1.3.15 and $Df^{-1}(y)=[Df(f^{-1}(y))]^{-1}$ we see that if f is C^2, then Df^{-1} is C^1 so f^{-1} is C^2. The general case follows by induction. ■

ISBN 0-8053-0102-X

In the study of manifolds and submanifolds, the argument used in the following is of central importance.

1.3.16 Theorem (Implicit Function Theorem). *Let $U \subset E$, $V \subset F$ be open and $f: U \times V \to G$ be C^r, $r \geqslant 1$. For some $x_0 \in U$, $y_0 \in V$ assume $D_2 f(x_0, y_0)$: $F \to G$ is an isomorphism. Then there are neighborhoods U_0 of x_0 and W_0 of $f(x_0, y_0)$ and a unique C^r map $g: U_0 \times W_0 \to V$ such that for all $(x, w) \in U_0 \times W_0$,*

$$f(x, g(x, w)) = w$$

Proof. Consider the map $\Phi: U \times V \to E \times G$, $(x, y) \mapsto (x, f(x, y))$. Then $D\Phi(x_0, y_0)$ is given by

$$D\Phi(x_0, y_0)\cdot(x_1, y_1) = \begin{pmatrix} I & 0 \\ D_1 f(x_0, y_0) & D_2 f(x_0, y_0) \end{pmatrix}\begin{pmatrix} x_1 \\ y_1 \end{pmatrix}$$

which is easily seen to be an isomorphism of $E \times F$ with $E \times G$. Thus Φ has a unique C^r local inverse, say Φ^{-1}: $U_0 \times W_0 \to U \times V$, $(x, w) \mapsto (x, g(x, w))$. The g so defined is the desired map. ■

Let E be a Banach space (or more generally, a topological vector space) and $F \subset E$ a closed subspace. Then F is said to *split* or be *complemented* if there is a closed subspace $G \subset E$ such that $E = F \oplus G$ (with the topology on E coinciding with the product topology on $F \oplus G$).* In finite-dimensional spaces, which is our main concern, any subspace is closed and splits. If E is a Hilbert space any closed subspace F splits, for we can choose G to be the orthogonal complement of F.

1.3.17 Corollary. *Let $U \subset E$ be open and $f: U \to F$ be C^r, $r > 1$. Suppose $Df(x_0)$ is surjective and $\ker Df(x_0)$ is complemented. Then $f(U)$ contains a neighborhood of $f(x_0)$.*

Proof. Let $E_1 = \ker Df(x_0)$ and $E = E_1 \times E_2$. Then $D_2 f(x_0)$: $E_2 \to F$ is an isomorphism. Thus the hypotheses of Theorem 1.3.16 are satisfied and so $f(U)$ contains W_0 provided by that theorem. ■

We conclude with an example of the use of the implicit function theorem to prove an existence theorem for differential equations. For this and related examples we choose the spaces to be infinite dimensional. In fact, $E, F, G, \ldots$ will be suitable spaces of functions. The map f will often be a

*For example $c_0 \subset l_\infty$ is not complemented. If every closed subspace of a Banach space is complemented, the space must be isomorphic to a Hilbert space. See J. Lindenstrauss and L. Tzafriri [1971].

ISBN 0-8053-0102-X

nonlinear differential operator. The linear map $Df(x_0)$ is called the *linearization* of f about x_0. (Phrases like "first variation," "first-order deformation," and so forth are also used.)

1.3.18 Example. Let E = all C^1 functions f: $[0,1] \to \mathbf{R}$ with

$$\|f\|_1 = \sup_{x\in[0,1]} |f(x)| + \sup_{x\in[0,1]} \left|\frac{df(x)}{dx}\right|$$

and F = all C^0 functions with $\|f\|_0 = \sup_{x\in[0,1]}|f(x)|$. These are Banach spaces (see Exercise 1.2D). Let $F: E \to F$, $F(f) = df/dx + f^3$. It is easy to check, using 1.3.9, that F is C^∞ and $DF(\mathbf{0}) = d/dx$: $E \to F$. Clearly $DF(\mathbf{0})$ is surjective (fundamental theorem of calculus). Also, *ker* $DF(\mathbf{0})$ consists of E_1 = all constant functions. This is complemented because it is finite-dimensional; explicitly, a complement consists of functions with zero integral. Thus Corollary 1.3.17 yields:

There is an $\varepsilon > 0$ *such that if* g *is any continuous function,* g: $[0,1] \to \mathbf{R}$, $|g(x)| < \varepsilon$, *there is a* C^1 *function* f: $[0,1] \to \mathbf{R}$ *such that*

$$\frac{df}{dx} + f^3(x) = g(x)$$

EXERCISES

1.3A. Prove Theorem 1.3.2; that is, that the derivative is unique if it exists. Also prove that the derivative does not depend on the choice of equivalent norm.

1.3B. For f: $U \subset E \to F$, show that

$$T^2 f: (U\times E)\times(E\times E) \to (F\times F)\times F\times F:$$
$$(u,e_1,e_2,e_3) \mapsto \left(f(u), Df(u)\cdot e_1, Df(u)\cdot e_2, D^2f(u)\cdot(e_1,e_2) + Df(u)\cdot e_3\right)$$

1.3C. Define a map f: $U \subset E \to F$ to be of class T^1 if f is differentiable and its tangent Tf: $U \times E \to F \times F$ is continuous.

(i) For E and F finite-dimensional, show that this is equivalent to C^1.

(ii) (L. Rosen) Let E be the space of real sequences $x = (x_1, x_2, \ldots)$ such that $|n^{3/2}x_n|$ is bounded and set $\|x\| = \sup_{n\geq 1}|n^{3/2}x_n|$. Check that E is a Banach space. Define f: $E \to E$ by $f(x)_n = f_n(x_n)$, where f_n: $\mathbf{R} \to \mathbf{R}$ is a smooth convex function satisfying $f_n(y) = 0$ if $y \leqslant 1/n$ and $f_n(y) = y - 2/n$ if $y \geqslant 3/n$. Show that f is T^1, $\|Df(x)\|$ is locally bounded, but f is not C^1.

1.3D. (L. E. Fraenkel and T. Ratiu) Develop a formula for $D^r(f\circ g)$ and $D^r(fg)$ and find the error in the formula proposed in Abraham and Robbin [1967, p. 3]. (See Quart. J. Math *1* [1900] p. 359 and Math. Proc. Camb. Phil. Soc. **83** [1978] p. 159 for the solution).

ISBN 0-8053-0102-X

1.4 MANIFOLDS AND MAPPINGS

The basic idea of a manifold is to introduce a local object that will support differentiation processes and then to patch these local objects together smoothly.

Before giving the formal definitions it is good to have in mind an example. In $\boldsymbol{R}^{n+1}$ consider the n-sphere S^n; that is, all $x \in \boldsymbol{R}^{n+1}$ such that $\|x\|=1$ ($\|\ \|$ is the usual Euclidean norm). We can construct, locally, bijections from S^n to $\boldsymbol{R}^n$. One way is to project stereographically from the south pole onto a hyperplane tangent to the north pole. This is a bijection from S^n, with the south pole removed, onto $\boldsymbol{R}^n$. Similarly we can interchange the roles of the poles to obtain another bijection. (See Fig. 1.4-1.) These bijections are quite well behaved. With the usual relative topology on S^n as a subset of $\boldsymbol{R}^{n+1}$, they are homeomorphisms from their domain to $\boldsymbol{R}^n$. Each takes the overlap of the domains to an open subset of $\boldsymbol{R}^n$. If we go from $\boldsymbol{R}^n$ to the sphere by one of them, then back to $\boldsymbol{R}^n$ by the other, we get a smooth map from $\boldsymbol{R}^n$ to $\boldsymbol{R}^n$.

In this way we can assign coordinate systems to S^n. Note, however, that no single homeomorphism can be used between S^n and $\boldsymbol{R}^n$, but we can cover S^n using two of them. We demand that these be compatible; that is, in a region covered by both coordinate systems we must be able to change coordinates smoothly.

For some studies of the sphere, two coordinate systems will not suffice. We thus allow all other coordinate systems compatible with these.

1.4.1 Definition. *Let S be a set. A **local chart** on S is a bijection φ from a subset U of S to an open subset of some (finite-dimensional, real) vector space F. We sometimes denote φ by (U,φ), to indicate the domain U of φ; F also may*

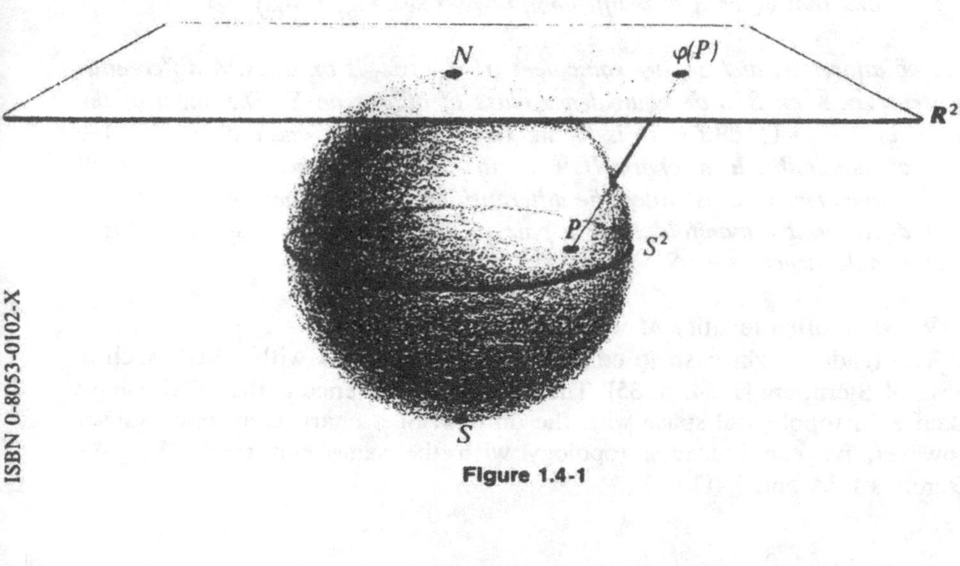

Figure 1.4-1

ISBN 0-8053-0102-X

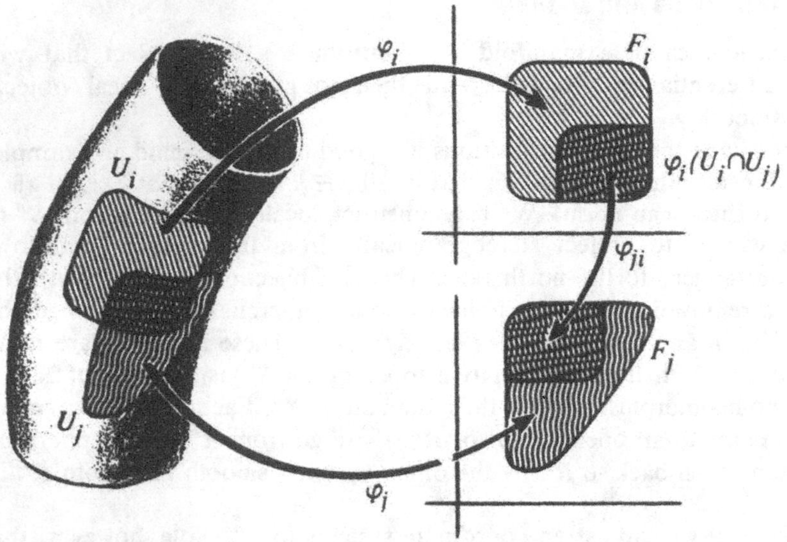

Figure 1.4-2

depend on φ. An **atlas** *on S is a family $\mathcal{A}$ of charts $\{(U_i,\varphi_i): i \in I\}$ such that*

(MA1) $S = \cup\{U_i | i \in I\}$;

(MA2) *Any two charts in $\mathcal{A}$ are compatible in the sense that the overlap maps between members of $\mathcal{A}$ are C^∞ diffeomorphisms: for two charts (U_i,φ_i) and (U_j,φ_j) with $U_i \cap U_j \neq \varnothing$ we form the* **overlap maps**: $\varphi_{ji} = \varphi_j \circ \varphi_i^{-1} | \varphi_i(U_i \cap U_j)$, *where $\varphi_i^{-1} | \varphi_i(U_i \cap U_j)$ means the restriction of φ_i^{-1} to the set $\varphi_i(U_i \cap U_j)$. We require that $\varphi_i(U_i \cap U_j)$ is open in F_i, and that φ_{ji} be a C^∞ diffeomorphism. (See Fig. 1.4-2.)*

Two atlases $\mathcal{A}_1$ and $\mathcal{A}_2$ are **equivalent** *iff $\mathcal{A}_1 \cup \mathcal{A}_2$ is an atlas. A* **differentiable structure** *$\mathcal{S}$ on S is an equivalence class of atlases on S. The union of the atlases in $\mathcal{S}$, $\mathcal{A}_{\mathcal{S}} = \cup\{\mathcal{A} | \mathcal{A} \in \mathcal{S}\}$ is the* **maximal atlas** *of $\mathcal{S}$, and a chart $(U,\varphi) \in \mathcal{A}_{\mathcal{S}}$ is an* **admissible local chart**. *If $\mathcal{A}$ is an atlas on S, then the union of all atlases equivalent to $\mathcal{A}$ is called the differentiable structure* **generated** *by $\mathcal{A}$.*

A **differentiable manifold** *M is a pair $(S, \mathcal{S})$, where S is a set and $\mathcal{S}$ is a differentiable structure on S.*

We shall often identify M with the underlying set S.

The reader might wish to compare these definitions with others, such as those of Sternberg [1964, p. 35]. The principal difference is that S is usually taken as a topological space with the domain of a chart as an open subset. However, we can induce a topology with the same end result. See also Exercises 1.4A and 1.4D.

ISBN 0-8053-0102-X

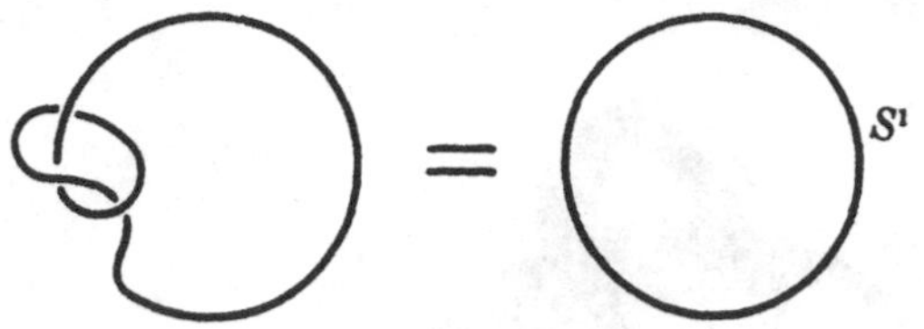

Figure 1.4-3

1.4.2 Definition. *Let M be a differentiable manifold. A subset $A \subset M$ is* ***open*** *if for each $a \in A$ there is an admissible local chart (U,φ) such that $a \in U$ and $U \subset A$, so M becomes a topological space (Exercise 1.4A).*

A differentiable manifold M is an ***n-manifold*** *iff for every point $a \in M$ there exists an admissible local chart (U,φ) with $a \in U$ and $\varphi(U) \subset \boldsymbol{R}^n$.*

A ***manifold*** *will always mean a Hausdorff, second countable, differentiable manifold.*

1.4.3 Examples

(a) S^n with a maximal atlas generated by the atlas described previously makes S^n into a n-manifold. The topology resulting is the same as that induced on S^n as a subset of $\boldsymbol{R}^{n+1}$.

(b) A set can have more than one differentiable structure. For example, $\boldsymbol{R}$ regarded as a set has the following incompatible charts

$$(U_1,\varphi_1)\colon\ U_1 = \boldsymbol{R}, \qquad \varphi_1(r) = r^3 \in \boldsymbol{R}$$

$$(U_2,\varphi_2)\colon\ U_2 = \boldsymbol{R}, \qquad \varphi_2(r) = r \in \boldsymbol{R}$$

They are not compatible since $\varphi_2 \circ \varphi_1^{-1}$ is not differentiable at the origin.

Nevertheless, these two resulting structures turn out to be diffeomorphic, but two structures can be essentially different on more complicated sets (e.g., S^7).*

(c) Essentially the only one-dimensional connected manifolds are $\boldsymbol{R}$ and S^1. This means that all others are diffeomorphic to $\boldsymbol{R}$ or S^1 (diffeomorphic will be precisely defined later). For example, the circle with a knot is diffeomorphic to S^1. (See Fig. 1.4-3.). See Milnor [1965] for proofs.

(d) A general two-dimensional connected manifold is the sphere with "handles" (see Fig. 1.4-4). This includes, for example, the torus.†

---

*That S^7 has two nondiffeomorphic differentiable structures is a famous result of J. Milnor [1956].

†The classification of two-manifolds is described in Massey [1967] and Hirsch [1976].

ISBN 0-8053-0102-X

Figure 1.4-4

1.4.4 Definition. *Let* $(S_1, \mathcal{S}_1)$ *and* $(S_2, \mathcal{S}_2)$ *be two manifolds. The* ***product manifold*** $(S_1 \times S_2, \mathcal{S}_1 \times \mathcal{S}_2)$ *consists of the set* $S_1 \times S_2$ *together with the differentiable structure* $\mathcal{S}_1 \times \mathcal{S}_2$ *generated by the atlas* $\{(U_1 \times U_2, \varphi_1 \times \varphi_2) | (U_i, \varphi_i)$ *is a chart of* $(S_i, \mathcal{S}_i)\}$.

That this is an atlas follows from the fact that if ψ_1: $U_1 \subset \mathbf{E}_1 \rightarrow V_1 \subset \mathbf{F}_1$; ψ_2: $U_2 \subset \mathbf{E}_2 \rightarrow V_2 \subset \mathbf{F}_2$, then $\psi_1 \times \psi_2$ is a diffeomorphism iff ψ_1 and ψ_2 are, and in this case $(\psi_1 \times \psi_2)^{-1} = \psi_1^{-1} \times \psi_2^{-1}$. Note that, from 1.3.9, $D(\psi_1 \times \psi_2) = D\psi_1 \times D\psi_2$. It is also clear that the topology on the product manifold is the product topology.

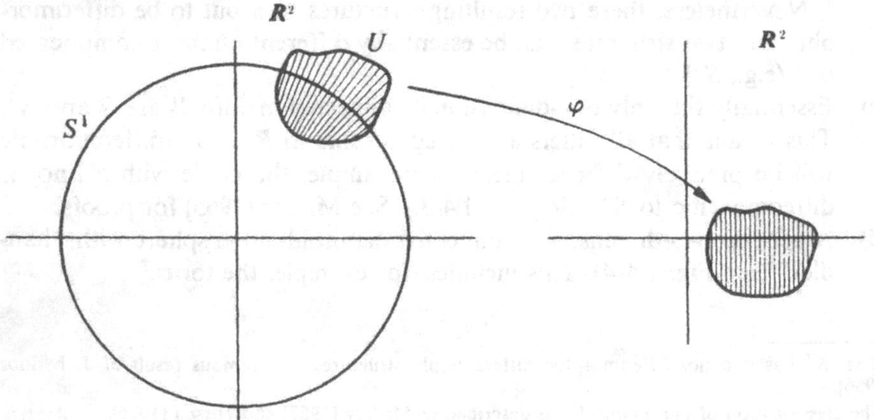

Figure 1.4-5

ISBN 0-8053-0102-X

If M is a manifold and $A \subset M$ is an open subset of M, the differentiable structure of M naturally induces one on A. We call A an **open submanifold** of M. We would also like to say that S^n is a submanifold of $\mathbf{R}^{n+1}$, although it is a closed subset. To motivate the general definition notice that there are charts in $\mathbf{R}^{n+1}$ in which S^n appears as $\mathbf{R}^n$, locally. (See Fig. 1.4.5.)

1.4.5 Definition. *A* ***submanifold*** *of a manifold M is a subset $B \subset M$ with the property that for each $b \in B$ there is an admissible chart (U, φ) in M with $b \in U$ which has the* ***submanifold property****, namely,*

(SM) $\varphi: U \to \mathbf{E} \times \mathbf{F}$, *and* $\varphi(U \cap B) = \varphi(U) \cap (\mathbf{E} \times \{\mathbf{0}\})$

An open subset of M is a submanifold in this sense. Here we merely take $\mathbf{F} = \{\mathbf{0}\}$, and use any chart.

Let B be a submanifold of a manifold M. Then B becomes a manifold with differentiable structure generated by the atlas

$$\{(U \cap B, \varphi | U \cap B) | (U, \varphi) \text{ is an admissible chart in } M \text{ having property (SM) for } B\}$$

Thus the topology on B is the relative topology.

Now $S^n \subset \mathbf{R}^{n+1}$ is, in this sense, a submanifold of $\mathbf{R}^{n+1}$. Furthermore, it is true that any n-manifold can be realized (embedded) as a closed (in the topological sense) submanifold of $\mathbf{R}^{2n+1}$. For Whitney's proof, see Hirsch [1976].

1.4.6 Definition. *Suppose we have $f: M \to N$, where M and N are manifolds (that is, f maps the underlying set of M into that of N). We say f is* ***of class C^r*** *if for each x in M and admissible chart (V, ψ) of N with $f(x) \in V$, there is a chart (U, φ) of M with $x \in U$ and $f(U) \subset V$ and the* ***local representative*** *of f, $f_{\varphi\psi} = \psi \circ f \circ \varphi^{-1}$, is of class C^r (see Fig. 1.4-6).*

For $r = 0$, this is consistent with the definition of continuity of f, regarded as a map between topological spaces (with the manifold topologies). If f is continuous, the requirement $f(U) \subset V$ can always be satisfied. The importance of property (MA2) for the differentiable structure is seen from the following.

1.4.7 Proposition. *Given $f: M \to N$ where M and N are manifolds, we have:*

(*i*) *If (U, φ) and (U, φ') are charts in M while (V, ψ) and (V, ψ') are charts in N with $f(U) \subset V$, then $f_{\varphi\psi}$ is of class C^r if and only if $f_{\varphi'\psi'}$ is of class C^r;*

(*ii*) *If (U, φ) and (V, ψ) are charts in M and N with $f(U) \subset V$ and if φ'' (and ψ'') are restrictions of φ (and ψ) to open subsets of U (and V) then $f_{\varphi\psi}$ is of class C^r implies $f_{\varphi''\psi''}$ is of class C^r;*

(*iii*) *If f is of class C^r on open subsets of M (as submanifolds) it is of class C^r on their union.*

ISBN 0-8053-0102-X

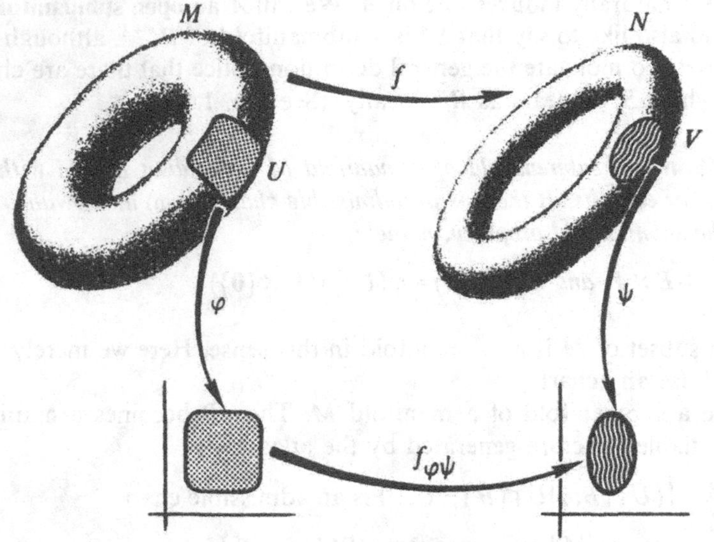

Figure 1.4-6

The proof of this is straightforward, from the smoothness of the overlap maps and the C^r composite mapping theorem. We leave the details as an exercise. Note that 1.4.7(i) implies that if $f_{\varphi\psi}$ is not C^r, then f is not C^r, while without (MA2) this might not be the case.

1.4.8 Definition. *A map $f: M \to N$, where M and N are manifolds, is called a (C^r)* ***diffeomorphism*** *if f is of class C^r, is a bijection, and $f^{-1}: N \to M$ is of class C^r.*

EXERCISES

1.4A. Show that the class of open sets on a manifold given in Definition 1.4.2 is a topology and with this topology the manifold is second countable iff it has an atlas with a countable family of local charts. (See Sect. 1.1.)

1.4B. Prove that S^1 is a submanifold of $\boldsymbol{R}^2$. Complete the details of examples 1.4.3A and 1.4.3B.

1.4C. Prove 1.4.7 and show that (i) if (U, φ) is a chart of M and $\psi: \varphi(U) \to V \subset \boldsymbol{F}$ is a diffeomorphism, then $(U, \psi \circ \varphi)$ is an admissible chart of M and (ii) admissible local charts are diffeomorphisms (in the manifold sense).

1.4D. Let $\mathcal{Q}$ be an atlas on S. Show that the differentiable structure generated by $\mathcal{Q}$ consists of all charts on S whose overlap maps with members of $\mathcal{Q}$ are C^∞.

1.4E. Let $S = \{(x, y) \in \boldsymbol{R}^2 | xy = 0\}$. Construct two "charts" by mapping each axis to the real line by $(x, 0) \mapsto x$ and $(0, y) \mapsto y$. What fails in the definition of a manifold?

1.4F. Let $S = (0,1) \times (0,1) \subset \boldsymbol{R}^2$ and for each s, $0 < s < 1$ let $\mathcal{V}_s = \{s\} \times (0,1)$ and $\varphi_s: \mathcal{V}_s \to \boldsymbol{R}$, $(s, t) \mapsto t$. Does this make S into a one-manifold?

ISBN 0-8053-0102-X

1.5 VECTOR BUNDLES

Roughly speaking, a vector bundle is a manifold with a vector space attached to each point. During the formal definitions we may keep in mind the example of the n-sphere $S^n \subset \mathbb{R}^{n+1}$. The collection of tangent planes to S^n (regarded as vector spaces) form a vector bundle. Similarly, the collection of normal lines to S^n form a vector bundle.

The definitions will follow the pattern of those for a manifold. Namely, we obtain a vector bundle by smoothly patching together local vector bundles.

1.5.1 Definition. *Let $\mathbf{E}$ and $\mathbf{F}$ be (finite-dimensional, real) vector spaces with U an open subset of $\mathbf{E}$. We call the Cartesian product $U \times \mathbf{F}$ a* ***local vector bundle****. We call U the* ***base space****, which can be identified with $U \times \{\mathbf{0}\}$, the* ***zero section****. For $u \in U$, $\{u\} \times \mathbf{F}$ is called the* ***fiber*** *over u, which we can endow with the vector space structure of $\mathbf{F}$. The map $\pi\colon U \times \mathbf{F} \to U$ given by $\pi(u, f) = u$ is called the* ***projection*** *of $U \times \mathbf{F}$. [Thus, for $u \in U$, the fiber over u is $\pi^{-1}(u)$. Also note that $U \times \mathbf{F}$ is an open subset of $\mathbf{E} \times \mathbf{F}$ and so is a local manifold.]*

Suppose we have a map $\varphi\colon U \times \mathbf{F} \to U' \times \mathbf{F}'$ where $U \times \mathbf{F}$ and $U' \times \mathbf{F}'$ are local vector bundles. We say that φ is a ***local vector bundle isomorphism*** *iff φ is a C^∞ diffeomorphism, and φ has the form $\varphi(u, f) = (\varphi_1(u), \varphi_2(u) \cdot f)$, where $\varphi_2(u)$ is a linear isomorphism for each $u \in U$. (See Fig. 1.5-1.)*

1.5.2 Definition. *Let S be a set. A* ***local bundle chart*** *of S is a pair (U, φ) where $U \subset S$ and $\varphi\colon U \subset S \to U' \times \mathbf{F}'$ is a bijection onto a local bundle $U' \times \mathbf{F}'$. (U', $\mathbf{F}'$ depend on φ.) A* ***vector bundle atlas*** *on S is a family $\mathcal{B} = \{(U_i, \varphi_i)\}$ of local bundle charts satisfying:*

(VBA1) = (MA1 *of* 1.4.1) (*it covers* S) *and,*

(VBA2) *for any two local bundle charts (U_i, φ_i) and (U_j, φ_j) in $\mathcal{B}$ with $U_i \cap U_j \neq \varnothing$, $\varphi_i(U_i \cap U_j) = U_i'' \times \mathbf{F}_i'$, and the overlap map $\psi_{ji} = \varphi_j \circ \varphi_i^{-1} | \varphi_i(U_i \cap U_j)$ is a local vector bundle isomorphism.*

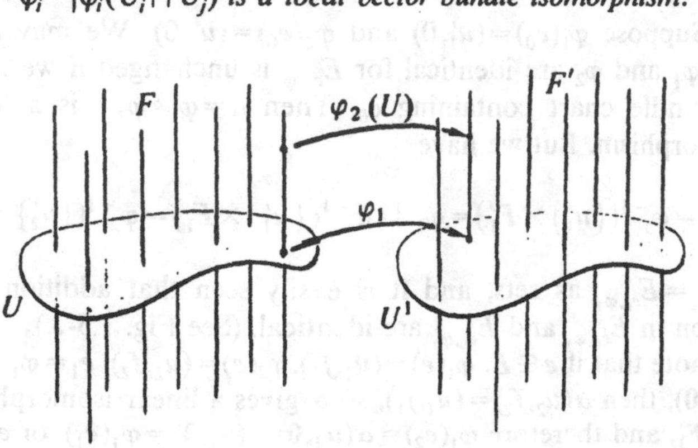

Figure 1.5-1

ISBN 0-8053-0102-X

If $\mathcal{B}_1$ and $\mathcal{B}_2$ are two vector bundle atlases on S, we say they are ***VB-equivalent*** *iff $\mathcal{B}_1 \cup \mathcal{B}_2$ is a vector bundle atlas. A* ***vector bundle structure*** *on S is an equivalence class of vector bundle atlases. A* ***vector bundle*** *E is a pair $(S, \mathcal{V})$, where S is a set and $\mathcal{V}$ is a vector bundle structure on S. A chart in an atlas of $\mathcal{V}$ is an* ***admissible vector bundle chart*** *of E.*

As in Sect. 1.4 we will often identify E with the underlying set S. Also, for a vector bundle structure $\mathcal{V}$ on S, (MA 1) and (MA2) hold, so $\mathcal{V}$ induces a differentiable structure on S. In addition, we shall assume that the differentiable structure on a vector bundle gives rise to a Hausdorff, second countable topology and that the induced manifold is of constant dimension.

For a vector bundle $E=(S, \mathcal{V})$ we define the ***zero section*** *by*

$$E_0=\{e\in E \,|\, \text{there exists } (U,\varphi)\in\mathcal{V} \text{ with } e=\varphi^{-1}(u',0)\}$$

Hence E_0 is the union of all the zero sections of the local vector bundles (identifying U with a local vector bundle via $\varphi: U\to U'\times F'$).

If $(U,\varphi)\in\mathcal{V}$ is a vector bundle chart, and $e_0\in U$ with $\varphi(e_0)=(u',0)$, let $E_{e_0,\varphi}$ denote the subset $\varphi^{-1}(\{u'\}\times F')$ of S together with the structure of a real vector space induced by the bijection φ.

The next few propositions derive basic properties of vector bundles that are sometimes included in the definition.

1.5.3 Proposition. *(i) If e_0 lies in the domains of two local bundle charts φ_1 and φ_2, then $E_{e_0,\varphi_1}=E_{e_0,\varphi_2}$, where the equality means equality as sets, and also as real vector spaces.*

(ii) For $e\in E$, there is exactly one $e_0\in E_0$ such that $e\in E_{e_0,\varphi_1}$, for some (and therefore all) (U,φ_1).

(iii) E_0 is a submanifold of E.

(iv) The map π, defined by $\pi: E\to E_0$, $\pi(e)=e_0$ [in (ii)] is surjective and C^∞.

Proof. (i) Suppose $\varphi_1(e_0)=(u_1',0)$ and $\varphi_2(e_0)=(u_2',0)$. We may assume the domains of φ_1 and φ_2 are identical for $E_{e_0,\varphi}$ is unchanged if we restrict φ to any local bundle chart containing e_0. Then $\alpha=\varphi_1\circ\varphi_2^{-1}$ is a local vector bundle isomorphism. But we have

$$E_{e_0,\varphi_1}=\varphi_1^{-1}(\{u_1'\}\times F_1')=\varphi_2^{-1}\circ\alpha^{-1}(\{u_1'\}\times F_1')=\varphi_2^{-1}(\{u_2'\}\times F_2')$$

Hence $E_{e_0,\varphi_1}=E_{e_0,\varphi_2}$ as sets, and it is easily seen that addition and scalar multiplication in E_{e_0,φ_1} and E_{e_0,φ_2} are identical. (See Fig. 1.5-2.)

For (ii) note that if $e\in E$, $\varphi_1(e)=(u_1,f_1)$, $\varphi_2(e)=(u_2,f_2)$, $e_1=\varphi_1^{-1}(u_1,\mathbf{0})$, and $e_2=\varphi_2^{-1}(u_2,\mathbf{0})$, then $\alpha(u_2,f_2)=(u_1,f_1)$, so α gives a linear isomorphism $\{u_2\}\times F_2'\to\{u_1\}\times F_1'$, and therefore $\varphi_1(e_2)=\alpha(u_2,\mathbf{0})=(u_1,\mathbf{0})=\varphi_1(e_1)$, or $e_2=e_1$.

To prove (iii) we must verify that for $e_0\in E_0$ there is an admissible chart with the submanifold property (1.4.5). For such a manifold chart we choose

ISBN 0-8053-0102-X

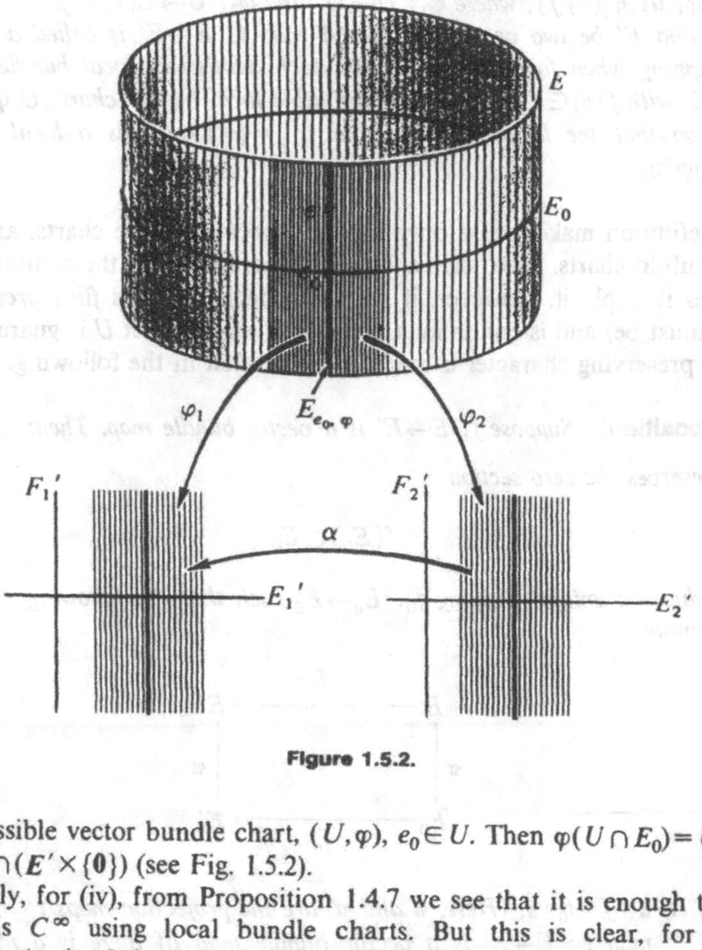

Figure 1.5.2.

an admissible vector bundle chart, (U,φ), $e_0 \in U$. Then $\varphi(U \cap E_0) = U' \times \{\mathbf{0}\} = \varphi(U) \cap (E' \times \{\mathbf{0}\})$ (see Fig. 1.5.2).

Finally, for (iv), from Proposition 1.4.7 we see that it is enough to check that π is C^∞ using local bundle charts. But this is clear, for such a representative is of the form $(u_1', f') \mapsto (u_1', 0)$. That π is onto should be clear. ∎

The following summarizes the basic properties of a vector bundle.

1.5.4 Theorem. *Let E be a vector bundle. The **zero section** E_0 of E, is a submanifold of E and there is a map $\pi: E \to E_0$ called the **projection** that is of class C^∞, and is surjective (onto). Moreover, for each $e_0 \in E_0$, $\pi^{-1}(e_0)$, called the **fiber** over e_0, has a (finite-dimensional, real) vector space structure induced by any admissible vector bundle chart, with e_0 the zero element.*

Because of these properties we sometimes write "the vector bundle $\pi: E \to E_0$" instead of "the vector bundle $(E, \mathcal{V})$." We now define vector bundle mappings analogously.

ISBN 0-8053-0102-X

1.5.5 Definition. *A map $\varphi\colon U\times F\to U'\times F'$ between local vector bundles is called a **local vector bundle mapping** when φ is of class C^∞ and has the form $\varphi(u,f)=(\varphi_1(u),\varphi_2(u)\cdot f)$, where $\varphi_1\colon U\to U'$ and $\varphi_2\colon U\to L(F,F')$.*

*Let E and E' be two vector bundles. A map $f\colon E\to E'$ is called a **vector bundle mapping** when for each $e\in E$ and each admissible local bundle chart (V,ψ) of E' with $f(e)\in V$ there is an admissible local bundle chart (U,φ) with $f(U)\subset V$ so that the local representative $f_{\varphi,\psi}=\psi\circ f\circ\varphi^{-1}$ is a local vector bundle mapping.*

This definition makes sense only for local vector bundle charts, and not for all manifold charts. Also, such a U is not guaranteed by the continuity of f, nor does it imply it. However, if we first check that f is *fiber preserving* (which it must be) and is continuous, then such an open set U is guaranteed. This fiber preserving character is made more explicit in the following.

1.5.6 Proposition. *Suppose $f\colon E\to E'$ is a vector bundle map. Then:*

(*i*) *f preserves the zero section*

$$f(E_0)\subset E_0'$$

(*ii*) *f induces a unique mapping $f_0\colon E_0\to E_0'$ such that the following diagram commutes:*

$$\begin{array}{ccc} E & \xrightarrow{\quad f\quad} & E' \\ {\scriptstyle\pi}\downarrow & & \downarrow{\scriptstyle\pi'} \\ E_0 & \xrightarrow[\quad f_0\quad]{} & E_0' \end{array}$$

that is, $\pi'\circ f=f_0\circ\pi$. (Here, π and π' are the projection maps.)

(*iii*) *A C^∞ map $g\colon E\to E'$ is a vector bundle map iff there is a map $g_0\colon E_0\to E_0'$ such that $\pi'\circ g=g_0\circ\pi$ and g restricted to each fiber is a linear map into a fiber.*

Proof. (i) Suppose $e_0\in E_0$. We must show $f(e_0)\in E_0'$. That is, for a vector bundle chart (V,ψ) and $f(e_0)\in V$ we must show $\psi f(e_0)=(v',\mathbf{0})$. But we have a chart (U,φ) so $e_0\in U$, $f(U)\subset V$. Here $\varphi(e_0)=(u',\mathbf{0})$. Hence $\psi f(e_0)=\psi\circ f\circ\varphi^{-1}(u',\mathbf{0})$. But this is of the form $(v',\mathbf{0})$ since $f_{\varphi\psi}$ is linear on each fiber.

For (ii) let $f_0=f|E_0\colon E_0\to E_0'$. With the notations above, then

$$\psi\circ\pi'\circ f(e)=\psi\circ f_0\circ\pi(e)$$

Also, if $f_{\varphi\psi}=(\alpha_1,\alpha_2)$, then $(f_0)_{\varphi\psi}=\alpha_1$, so f_0 is C^∞.

ISBN 0-8053-0102-X

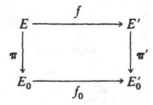

One half of (iii) is clear from (i) and (ii). For the converse we easily see that in local representation g has the form

$$g_{\varphi\psi}(u', e') = (\alpha_1(u'), \alpha_2(u')e')$$
$$= \psi \circ g \circ \varphi^{-1}(u', e')$$

where

$$u' = \varphi \circ \pi(u), \quad \alpha_1 = \psi \circ g_0 \circ \varphi^{-1} \qquad \text{and} \qquad \alpha_2(u') = (\psi \circ g \circ \varphi^{-1}) | \{u'\} \times E'$$

is obviously linear. Thus, the local representatives of g with respect to admissible local bundle charts are local bundle mappings. ■

We shall now define a second generalization of a local C^r mapping, $f: U \subset E \to F$, which globalizes not f but rather its graph mapping $\lambda_f: U \to U \times F: u \mapsto (u, f(u))$.

1.5.7 Definition. *Let $\pi: E \to B$ be a vector bundle. A C^r* ***section*** *of π is a map $\xi: B \to E$ of class C^r such that for each $b \in B$, $\pi(\xi(b)) = b$. Let $\Gamma^r(\pi)$ denote the set of all C^r sections of π, together with the obvious real (infinite-dimensional) vector space structure.*

The condition on ξ merely says that $\xi(e_0)$ lies in the fiber over e_0. The C^r sections form a linear function space suitable for global linear analysis. This differs from the more general class of global C^r maps from one manifold to another, which is a nonlinear function space. (See, for example, Eells [1958], Palais [1968], Eliasson [1967], or Ebin–Marsden [1970] for details.)

EXERCISES

1.5A. (i) Give a precise definition of the Möbius band as a vector bundle, and construct a vector bundle atlas [a cylinder with a half twist (see Fig. 2.5.1). Compare with the torus (Sect. 1.1).].
(ii) Complete the details of 1.5.3 and show that the differentiable structure on a vector bundle is larger than the vector bundle structure.

1.5B. Find an example of a fiber preserving diffeomorphism between vector bundles that is not a vector bundle isomorphism.

1.5C. Let $\pi: E \to B$ and $\pi': E' \to B'$ be two vector bundles. Define the *sum* as $\pi \times \pi': E \times E' \to B \times B'$. Show that this is a vector bundle in a natural way, and construct a vector bundle atlas.

1.5D. If $\pi: E \to B$ and $\pi': E' \to B$ are vector bundles over B, define the *Whitney sum* by $\pi \oplus \pi': E \oplus E' \to B$, where $E \oplus E'$ has fiber over $b \in B$ equal to the direct sum $E_b \oplus E'_b$. Show this is a vector bundle over B.

1.5E. (i) Let $\pi: E \to B$ be a vector bundle and $f: B' \to B$ a smooth map. Define the

ISBN 0-8053-0102-X

pull-back bundle $f^*\pi: f^*E \to B'$ by

$$f^*E = \{(e,b') \mid \pi(e) = f(b')\}, \qquad f^*\pi(e,b') = b'$$

and show that it is a vector bundle over B'.

(ii) If $\pi: E \to B$, $\pi': E' \to B$ are vector bundles and if $\Delta: B \to B \times B$ is the diagonal map $b \mapsto (b,b)$, show that $\pi \oplus \pi' = \Delta^*(\pi \times \pi')$ (see Exercises 1.5C and D).

1.6 THE TANGENT BUNDLE

Recall that for $f: U \subset E \to V \subset F$ of class C^{r+1} we define $Tf: TU \to TV$ by $TU = U \times E$, $TV = V \times F$, and $Tf(u,e) = (f(u), Df(u) \cdot e)$. Hence, Tf is a local vector bundle mapping of class C^r. Also $T(f \circ g) = Tf \circ Tg$. Moreover, for each open set U in some vector space E let $\tau_U: TU \to U$ be the projections (as in Sect. 1.5 we identify U with the zero section $U \times \{0\}$). Then the diagram

$$\begin{array}{ccc} TU & \xrightarrow{\;Tf\;} & TV \\ {\scriptstyle \tau_U}\downarrow & & \downarrow{\scriptstyle \tau_V} \\ U & \xrightarrow[\;f\;]{} & V \end{array}$$

is commutative, that is, $f \circ \tau_U = \tau_V \circ Tf$.

We will now extend the tangent functor T from this local context to the category of differentiable manifolds and mappings. During the definitions it may be helpful to keep in mind the example of the family of tangent spaces of the sphere $S^n \subset R^{n+1}$.

A major advance in differential geometry occurred when it was realized how to define the tangent space to an abstract manifold independent of any embedding in R^n.* There are several alternative ways to do this, which can be used according to taste.

(a) *Coordinate approach.* Using transformation properties of vectors under coordinate changes, one defines a tangent vector to be an equivalence class of triples (U, e, φ), where $\varphi: U' \to U \subset E$ is a chart, $u \in U$ and $e \in E$ with two triples identified if they are related by the tangent of the corresponding overlap map (see Lang [1972] for details).

(b) *Derivation approach.* This approach characterizes a vector by specifying a map that gives the derivative of a general function in the direction of that vector (see, e.g., Bishop and Goldberg [1968] and Sect. 2.2 below).

(c) *The ideal approach.* This is a variation of (b). Here T_mM is the dual

*The history is not completely clear to us, but this idea seems to be primarily due to Riemann, Weyl and Levi-Cevita and was "well known" by 1920.

ISBN 0-8053-0102-X

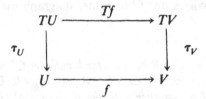

of $(\mathcal{J}_m^{(0)}/\mathcal{J}_m^{(1)})$, where $\mathcal{J}_m^{(j)}$ is the ideal of functions on M vanishing up to order j at m.

(d) *The curves approach.* This is the method followed here. We abstract the idea that a tangent vector to a surface is the velocity vector of a curve in the surface.

1.6.1 Definition. *Let M be a manifold and $m \in M$. A **curve at** m is a C^1 map $c: I \to M$ from an open interval $I \subset \mathbf{R}$ into M with $0 \in I$ and $c(0) = m$. Let c_1 and c_2 be curves at m and (U, φ) an admissible chart with $m \in U$. Then we say c_1 and c_2 are **tangent at** m **with respect to** φ if and only if $\varphi \circ c_1$ and $\varphi \circ c_2$ are tangent at 0 (in the sense of Sect. 1.3; we may restrict the domain of c_i such that $\varphi \circ c_i$ is defined; see Fig. 1.6-1).*

Thus two curves are tangent with respect to φ if they have identical tangent vectors (same direction and speed) in the chart φ.

1.6.2 Proposition. *Let c_1 and c_2 be two curves at $m \in M$. Suppose (U_β, φ_β) are admissible charts with $m \in U_\beta$, $\beta = 1, 2$. Then c_1 and c_2 are tangent at m with respect to φ_1 if and only if they are tangent at m with respect to φ_2.*

Proof. Note that c_1 and c_2 are tangent at m with respect to φ_1 iff $D(\varphi_1 \circ c_1)(0) = D(\varphi_1 \circ c_2)(0)$. By taking restrictions if necessary we may suppose $U_1 = U_2$. Hence we have $\varphi_2 \circ c_i = (\varphi_2 \circ \varphi_1^{-1}) \circ (\varphi_1 \circ c_i)$. From the C^1 composite mapping theorem it follows that $D(\varphi_2 \circ c_1)(0) = D(\varphi_2 \circ c_2)(0)$. ■

Proposition 1.6.2 guarantees that the tangency of curves at $m \in M$ is independent of the chart used. Thus we say c_1, c_2 are **tangent at** $m \in M$ if c_1, c_2 are tangent at m with respect to φ, for any local chart φ at m. It is evident

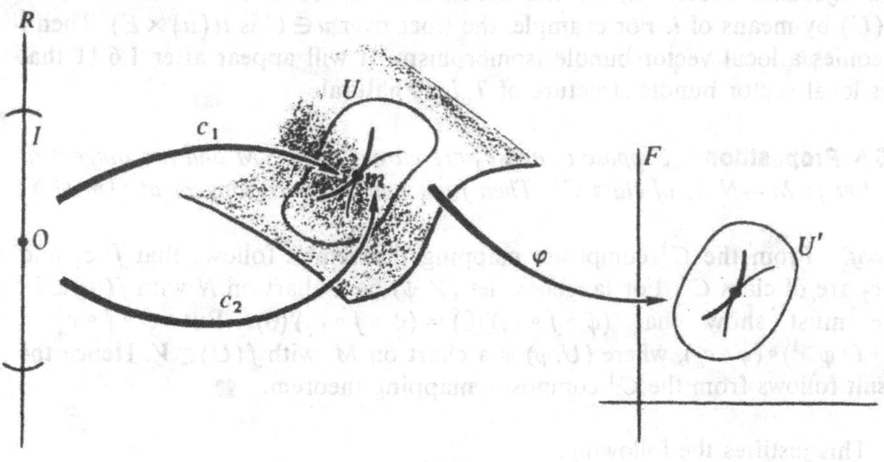

Figure 1.6-1

ISBN 0-8053-0102-X

that tangency at $m \in M$ is an equivalence relation among curves at m. An **equivalence class** of such curves is denoted $[c]_m$, where c is a representative of the class.

1.6.3 Definition. *For a manifold M and $m \in M$ the* ***tangent space of M at m*** *is the set of equivalence classes of curves at m*

$$T_m(M) = \{[c]_m \mid c \text{ is a curve at } m\}$$

For a subset $A \subset M$, let $TM|A = \cup_{m \in A} T_m(M)$. We call $TM = TM|M$ the ***tangent bundle*** *of M.*

The mapping τ_M: $TM \to M$ defined by $\tau_M([c]_m) = m$, is the ***tangent bundle projection*** *of M.*

Next we must show that the definition above corresponds to $TU = U \times \mathbf{E}$ for the case of a local manifold. We let T_3 denote T as defined in Sect. 1.3, so $T_3 U = U \times \mathbf{E}$, and let T_6 denote T as defined in this Section. That $T_3(U)$ and $T_6(U)$ can be identified is justified by the following.

1.6.4 Proposition. *Let U be an open subset of $\mathbf{E}$, and c a curve at $u \in U$. Then there is a unique $\mathbf{e} \in \mathbf{E}$ such that the curve $c_{u,e}$ defined by $c_{u,e}(t) = u + te$ [on some interval I such that $c_{u,e}(I) \subset U$] is tangent to c at u.*

Proof. From Sect. 1.3, $Dc(0)$ is the unique linear map in $L(\mathbf{R}, \mathbf{E})$ such that the curve g: $\mathbf{R} \to \mathbf{E}$ given by $g(t) = u + Dc(0) \cdot t$ is tangent to c at $t = 0$. If $e = Dc(0) \cdot 1$, then $g = c_{u,e}$. ■

Define a map i: $T_3(U) \to T_6(U)$ by $i(u,e) = [c_{u,e}]_u$. Proposition 1.6.4 says i is a bijection. Moreover, we can define a local vector bundle structure on $T_6(U)$ by means of i. For example, the fiber over $u \in U$ is $i(\{u\} \times \mathbf{E})$. Then i becomes a local vector bundle isomorphism. It will appear after 1.6.11 that this local vector bundle structure of $T_6 U$ is natural.

1.6.5 Proposition. *Suppose c_1 and c_2 are curves at $m \in M$ and are tangent at m. Let f: $M \to N$ be of class C^1. Then $f \circ c_1$ and $f \circ c_2$ are tangent at $f(m) \in N$.*

Proof. From the C^1 composite mapping theorem it follows that $f \circ c_1$ and $f \circ c_2$ are of class C^1. For tangency, let (V, ψ) be a chart on N with $f(m) \in V$. We must show that $(\psi \circ f \circ c_1)'(0) = (\psi \circ f \circ c_2)'(0)$. But $\psi \circ f \circ c_\alpha = (\psi \circ f \circ \varphi^{-1}) \circ (\varphi \circ c_\alpha)$, where (U, φ) is a chart on M, with $f(U) \subset V$. Hence the result follows from the C^1 composite mapping theorem. ■

This justifies the following.

ISBN 0-8053-0102-X

1.6.6 Definition. *If $f: M \to N$ is of class C^1, we define $Tf: TM \to TN$ by*

$$Tf([c]_m) = [f \circ c]_{f(m)}$$

*We call Tf the **tangent** of f.*

Tf is well defined, for if we choose any other representative from $[c]_m$, say c_1, then c and c_1 are tangent at m and hence $f \circ c$ and $f \circ c_1$ are tangent at $f(m)$. That is, $[f \circ c]_{f(m)} = [f \circ c_1]_{f(m)}$.

The basic properties of T are summarized in the following.

1.6.7 Theorem. *(i) (**C^1 composite mapping theorem**) Suppose $f: M \to N$ and $g: N \to K$ are C^1 maps of manifolds. Then $g \circ f: M \to K$ is of class C^1 and*

$$T(g \circ f) = Tg \circ Tf$$

(ii) If $h: M \to M$ is the identity map, then $Th: TM \to TM$ is the identity map.
(iii) If $f: M \to N$ is a diffeomorphism, then $Tf: TM \to TN$ is a bijection and $(Tf)^{-1} = T(f^{-1})$.

Proof. (i) Let (U, φ), (V, ψ), (W, ρ) be charts of M, N, K, with $f(U) \subset V$ and $g(V) \subset W$. Then we have, for the local representatives,

$$\begin{aligned}(g \circ f)_{\varphi\rho} &= \rho \circ g \circ f \circ \varphi^{-1} \\ &= \rho \circ g \circ \psi^{-1} \circ \psi \circ f \circ \varphi^{-1} \\ &= g_{\psi\rho} \circ f_{\varphi\psi}\end{aligned}$$

By the C^1 composite mapping theorem this, and hence $g \circ f$, is of class C^1. Moreover, $T(g \circ f)[c]_m = [g \circ f \circ c]_{g \circ f(m)}$ and $Tg \circ Tf[c]_m = Tg([f \circ c]_{f(m)}) = [g \circ f \circ c]_{g \circ f(m)}$. Hence $T(g \circ f) = Tg \circ Tf$.

Part (ii) is an immediate consequence of the definition of T. For (iii), f and f^{-1} are diffeomorphisms with $f \circ f^{-1}$ the identity on N, while $f^{-1} \circ f$ is the identity on M. But then using (i) and (ii), $Tf \circ Tf^{-1}$ is the identity on TN while $Tf^{-1} \circ Tf$ is the identity on TM. Thus (iii) follows. (See Exercise 1.1A.) ■

(As in the case of local manifolds, these properties signify that T is a functor.) Next let us show that in the case of local manifolds, Tf as defined in Sect. 1.3, which we denote $T_3 f$, coincides with Tf as defined here, which we denote $T_6 f$, when we identify $T_3 U$ with $T_6 U$.

1.6.8 Proposition. *Let $U \subset E$ and $V \subset F$ be local manifolds (open subsets) and $f: U \to V$ be of class C^1. Let $i: T_3(U) \to T_6(U)$ be the map defined in 1.6.4.*

ISBN 0-8053-0102-X

Then the diagram

$$\begin{array}{ccc} T_3(U) & \xrightarrow{T_3 f} & T_3(V) \\ \Big\downarrow{\scriptstyle i} & & \Big\downarrow{\scriptstyle i} \\ T_6(U) & \xrightarrow[T_6 f]{} & T_6(V) \end{array}$$

commutes, that is, $T_6 f \circ i = i \circ T_3 f$.

Proof. For $(u,e) \in T_3(U) = U \times E$, we have

$$\begin{aligned}(T_6 f \circ i)(u,e) &= T_6 f \cdot [c_{u,e}]_u \\ &= [f \circ c_{u,e}]_{f(u)}\end{aligned}$$

Also, $(i \circ T_3 f)(u,e) = i(f(u), Df(u)\cdot e) = [c_{f(u), Df(u)\cdot e}]_{f(u)}$. These will be equal provided the curves $t \mapsto f(u+te)$ and $t \mapsto f(u) + t(Df(u)\cdot e)$ are tangent at $t=0$. But this is clear from the definition of the derivative and the composite mapping theorem. ■

This theorem states that if we identify $T_3(U)$ and $T_6(U)$ by means of i then we can identify $T_3 f$ and $T_6 f$. Thus we will just write T, and will suppress the identification.

1.6.9 Proposition. *If $f: U \subset E \to V \subset F$ is a diffeomorphism, then $Tf: U \times E \to V \times F$ is a local vector bundle isomorphism.*

Proof. Since $Tf(u,e) = (f(u), Df(u)\cdot e)$, Tf is a local vector bundle mapping. But as f is a diffeomorphism, $(Tf)^{-1} = T(f^{-1})$ is also a local vector bundle mapping, and hence Tf is a vector bundle isomorphism. ■

For a chart (U, φ) on a manifold M, we can construct $T\varphi: TU \to TU'$. By 1.6.7, $T\varphi$ is a bijection, since φ is a diffeomorphism (Exercise 1.4C). Hence, on TM we can regard $(TU, T\varphi)$ as a local vector bundle chart. In the target of $T\varphi$ note that we have a special local vector bundle, where the fibers have the same dimension as the base.

1.6.10 Theorem. *Let M be a manifold and $\mathscr{Q}$ an atlas of admissible charts. Then $T\mathscr{Q} = \{(TU, T\varphi) | (U,\varphi) \in \mathscr{Q}\}$ is a vector bundle atlas of TM called a* ***natural atlas****.*

Proof. Since the union of chart domains of $\mathscr{Q}$ is M, the union of the corresponding TU is TM. Thus we must verify VBA2. Hence, suppose we have $TU_i \cap TU_j \neq \varnothing$. Then $U_i \cap U_j \neq \varnothing$ and we can form the overlap map

ISBN 0-8053-0102-X

$\varphi_i \circ \varphi_j^{-1}$, by restriction of φ_j^{-1} to $\varphi_j(U_i \cap U_j)$. Then we must verify $T\varphi_i \circ (T\varphi_j)^{-1} = T(\varphi_i \circ \varphi_j^{-1})$ is a local vector bundle isomorphism. But this is guaranteed by 1.6.9. ■

Hence TM has a natural vector bundle structure induced by the differentiable structure of M. If M is n-dimensional, Hausdorff, and second countable, TM will be $2n$-dimensional, Hausdorff, and second countable.

We shall now reconcile the bundle projection τ_M of 1.6.3 and the one given in Sect. 1.5 for an arbitrary vector bundle.

1.6.11 Proposition. *If $m \in M$, then $\tau_M^{-1}(m) = T_m M$ is a fiber of TM and $\tau_M|(TM)_0$: $(TM)_0 \to M$ is a diffeomorphism.*

Proof. Let (U, φ) be a local chart at $m \in M$, with φ: $U \to U' \subset \mathbf{E}$ and $\varphi(m) = u'$. Then $T\varphi$: $TM|U \to U' \times \mathbf{E}$ is a natural chart of TM, $T\varphi^{-1}(\{u'\} \times \mathbf{E}) = T\varphi^{-1}\{[c_{u',e}]_{u'} | e \in \mathbf{E}\}$ by definition of $T\varphi$, and this is exactly $T_m M$, by 1.6.7(iii). For the second assertion, $\tau_M|(TM)_0$ is obviously a bijection, and its local representative with respect to $T\varphi$ and φ is the natural identification $U' \times \{\mathbf{0}\} \to U'$, a local diffeomorphism. ■

We will often identify M with the zero section of TM, and τ_M with the bundle projection onto the zero section.

1.6.12 Proposition. *Let M and N be manifolds, and let f: $M \to N$ be of class C^{r+1}. Then Tf: $TM \to TN$ is a vector bundle mapping of class C^r.*

Proof. By Proposition 1.4.7 it is enough to check Tf using the natural atlas. For $u \in M$ choose charts (U, φ) and (V, ψ) on M and N so $u \in U$, $f(u) \subset V$ and $f_{\varphi\psi} = \psi \circ f \circ \varphi^{-1}$ is of class C^{r+1}. Then using $(TU, T\varphi)$ for TM and $(TV, T\psi)$ for TN, we must verify that $(Tf)_{T\varphi, T\psi}$ is a local vector bundle map of class C^r. But we have $(Tf)_{T\varphi, T\psi} = T\psi \circ Tf \circ T\varphi^{-1} = T(f_{\varphi\psi})$, and $Tf_{\varphi\psi}(u', e) = (f_{\varphi\psi}(u'), Df_{\varphi\psi}(u') \cdot e)$, which is a local vector bundle map of class C^r. ■

Now that TM has a manifold structure we can form higher tangents. For mappings f: $M \to N$ of class C^r, we define $T^r f$: $T^r M \to T^r N$ inductively to be the tangent of $T^{r-1} f$: $T^{r-1} M \to T^{r-1} N$. Induction readily yields the following.

1.6.13 Theorem (C^r composite mapping theorem). *Suppose f: $M \to N$ and g: $N \to K$ are C^r mappings of manifolds. Then $g \circ f$ is of class C^r and $T^r(g \circ f) = T^r g \circ T^r f$.*

The behavior of T under products is given by the following.

1.6.14 Proposition. *Let M and N be manifolds and $M \times N$ the product manifold. Then $T(M \times N)$ is related to $TM \times TN$ by a vector bundle isomorphism. If f: $K \to M$ and g: $K \to N$ are smooth mappings, then $T(f \times g) =$*

ISBN 0-8053-0102-X

$Tf \times Tg$, if we identify $T(M \times N)$ and $TM \times TN$. (The functor T is natural with respect to products.)

Proof. Let $[c]_{(m,n)} \in T(M \times N)$. Then $c: I \to M \times N$, $c(0) = (m,n)$ has the form $c(t) = (c_M(t), c_N(t))$, where c_M is a curve in M at m and c_N is a curve in N at n. Consider the map $[c]_{(m,n)} \mapsto ([c_M]_m, [c_N]_n)$: $T(M \times N) \to TM \times TN$. From the local representative we see that the mapping is well defined. Also, since it has an obvious inverse it is a bijection. Moreover, in local representation, it is merely the mapping $(u,v,e,f) \mapsto ((u,e),(v,f))$, which is a vector bundle isomorphism. The second part follows easily from definition 1.6.6. ■

As was the case for local manifolds, for $f: M \to N$ of class C^1, we have the commutative diagram

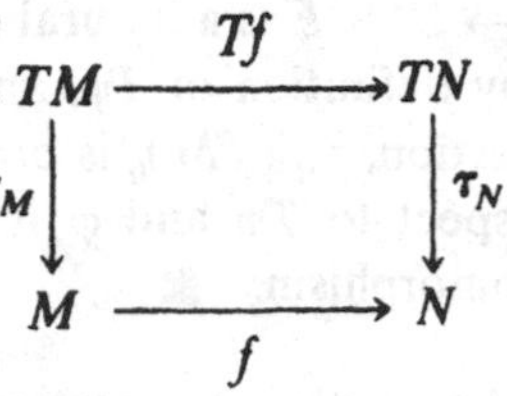

We sometimes write $\tau_f = (Tf, f): \tau_M \to \tau_N$ for this diagram. A special case is the *dual tangent rhombic*, with $f = \tau_M$:

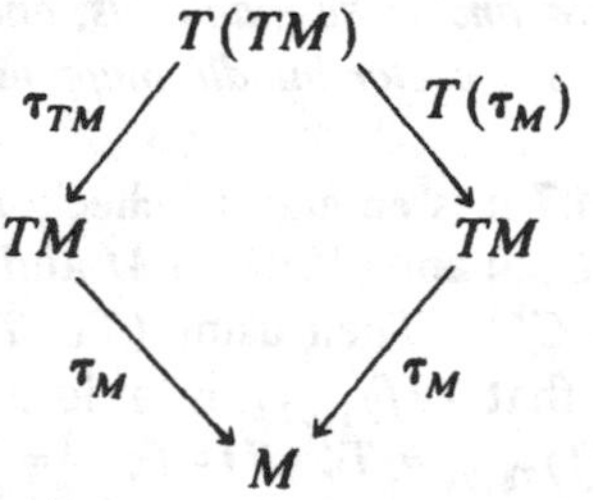

As it may be shown that the set $T(TM)$ has a vector bundle structure, in addition to the usual one, in which $T\tau_M$ is a projection, it is confusing to write $T(TM)$ alone for $(T(TM), T\tau_M)$. But "the bundle τ_{TM}: $T(TM) \to TM$" is adequate to indicate which vector bundle structure is implied.

The inverse mapping theorem takes the following form.

1.6.15 Theorem. *Let M and N be manifolds and $f: M \to N$ be of class C^r. Suppose Tf is an isomorphism on the fiber over $m \in M$. Then f is a C^r diffeomorphism from some neighborhood of m onto some neighborhood of $f(m)$.*

This follows easily from 1.3.12. Likewise the implicit function theorem (1.3.16) yields the following.

1.6.16 Theorem. *Let M and N be manifolds and $f: M \to N$ be of class C^∞. Suppose Tf restricted to the fiber T_mM is surjective to $T_{f(m)}N$. Then there are*

ISBN 0-8053-0102-X

charts (U,φ) *and* (V,ψ) *with* $m\in U$, $f(U)\subset V$, $\varphi\colon U\to U'\times V'$, $\varphi(m)=(0,0)$, $\psi\colon V\to V'$ *and* $f_{\varphi\psi}\colon U'\times V'\to V'$ *is the projection onto the second factor.*

The condition that $T_m f = Tf|T_m M$ be surjective means, for $f\colon R^k\to R^n$, with $k\geqslant n$, that the rank of the Jacobian matrix of f at $m\in M$ should be n.

1.6.17 Definition. *Suppose M and N are manifolds with $f\colon M\to N$ of class C^1. A point $n\in N$ is called a* ***regular value*** *of f if for each $m\in f^{-1}(\{n\})$, $T_m f$ is surjective. Let R_f denote the set of regular values of $f\colon M\to N$; note $N\setminus f(M)\subset R_f\subset N$. If, for each m in a set S, $T_m f$ is surjective, we say f is a* ***submersion*** *on S. (Thus $n\in R_f$ iff f is a submersion on $f^{-1}(\{n\})$.*

The next result will be important when we consider energy surfaces.

1.6.18 Proposition. *Suppose $f\colon M\to N$ is of class C^∞ and $n\in R_f$. Then $f^{-1}(n)=\{m|m\in M, f(m)=n\}$ is a submanifold of M.*

Proof. If $f^{-1}(n)=\varnothing$ the theorem is satisfied. Otherwise, for $m\in f^{-1}(n)$ we find charts (U,φ), (V,ψ) as described in the implicit mapping theorem.

Then it must be shown that the chart (U,φ) has the submanifold property. But $\varphi(U\cap f^{-1}(n))=f_{\varphi\psi}^{-1}(0)=U'\times\{0\}$. This is exactly the submanifold property. (See Fig. 1.6-2.) ■

This proof shows that $T_m f^{-1}(n)=\ker T_m f$.

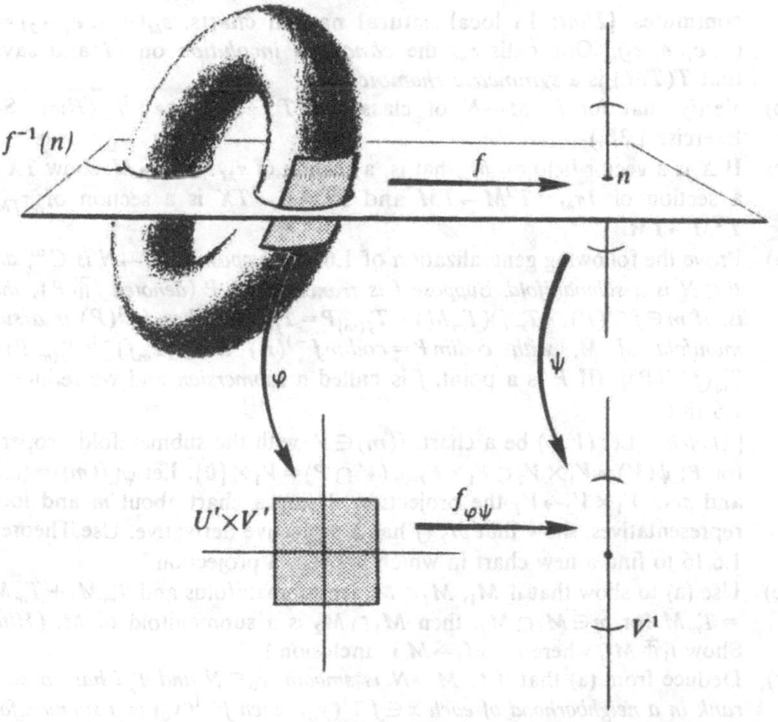

Figure 1.6-2

ISBN 0-8053-0102-X

Further versions and consequences of the implicit function theorem that will be used later in the book are developed in the exercises.

The following relatively deep result is a form of *Sard's theorem*. See, for example, Abraham–Robbin [1967, Chapter III].

1.6.19 Theorem (Sard). *Suppose $f: M \to N$ is of class C^∞. Then R_f is dense in N.*

EXERCISES

1.6A. Establish 1.6.15 and 1.6.16 from the corresponding local statements in Sect. 1.3. Deduce each of 1.6.15 and 1.6.16 from the other.

1.6B. Use 1.6.18 to given an alternative proof that S^n is a submanifold of $\boldsymbol{R}^{n+1}$.

1.6C. A *vector field* X on a local manifold U is a C^∞ section of $TU = U \times \boldsymbol{E}$. Interpret geometrically and examine the condition that a curve $c: I \to U$ is tangent to X, $Tc(t)\cdot 1 = X(c(t))$. Show also that $c'(t) = Tc(t)\cdot 1$ coincides with the usual derivative.

1.6D. (a) Show that there is a map $s_M: T(TM) \to T(TM)$ such that $s_M \circ s_M =$ identity and

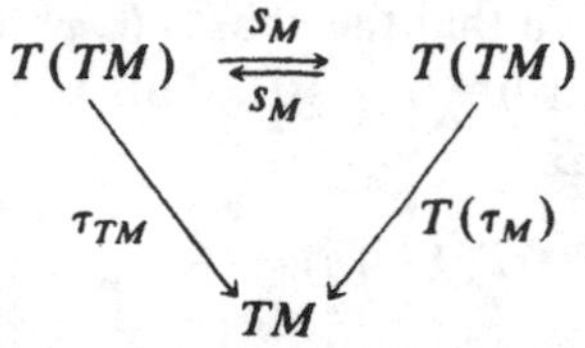

commutes. [*Hint*: In local natural natural charts, $s_M(u, e, e_1, e_2) = (u, e_1, e, e_2)$.] One calls s_M the *canonical involution* on M and says that $T(TM)$ is a *symmetric rhombic*.

(b) Verify that for $f: M \to N$ of class C^2, $T^2 f \circ s_M = s_N \circ T^2 f$. (*Hint*: See Exercise 1.3B.)

(c) If X is a vector field on M, that is, a section of $\tau_M: TM \to M$, show TX is a section of $T\tau_M: T^2M \to TM$ and $X^1 = s_M \circ TX$ is a section of $\tau_{TM}: T^2M \to TM$.

1.6E. (a) Prove the following generalization of 1.6.18: *Suppose $f: M \to N$ is C^∞, and $P \subset N$ is a submanifold. Suppose f is **tranversal** to P (denoted $f \pitchfork P$), that is, if $m \in f^{-1}(P)$, $(T_m f)(T_m M) + T_{f(m)}P = T_{f(m)}N$. Then $f^{-1}(P)$ is a submanifold of M, with $\operatorname{codim} P = \operatorname{codim} f^{-1}(P)$ and $(T_m f)^{-1}(T_{f(m)}P) = T_m(f^{-1}(P))$.* (If P is a point, f is called a *submersion* and we reduce to 1.6.18.)

[*Sketch*. Let (V, ψ) be a chart, $f(m) \in V$, with the submanifold property for P; $\psi(V) = V_1 \times V_2 \subset F_1 \times F_2$, $\psi(V \cap P) = V_1 \times \{0\}$. Let $\psi(f(m)) = (0, 0)$ and $pr_2: V_1 \times V_2 \to V_2$ the projection. Using a chart about m and local representatives, show that $pr_2 \circ f$ has a surjective derivative. Use Theorem 1.6.16 to find a new chart in which $pr_2 \circ f$ is a projection.]

(b) Use (a) to show that if $M_1, M_2 \subset M$ are submanifolds and $T_m M_1 + T_m M_2 = T_m M$ for $m \in M_1 \cap M_2$, then $M_1 \cap M_2$ is a submanifold of M. (*Hint*: Show $i_1 \pitchfork M_2$, where $i_1: M_1 \to M$ is inclusion.)

(c) Deduce from (a) that *if $f: M \to N$ is smooth, $y_0 \in N$ and $T_x f$ has constant rank in a neighborhood of each $x \in f^{-1}(y_0)$, then $f^{-1}(y_0)$ is a submanifold*

ISBN 0-8053-0102-X

with

$$T_x f^{-1}(y_0) = Ker\, T_x f$$

1.6F. (a) **If $f: M \to N$ is an *immersion*, that is, $T_m f$ is one-to-one at each $m \in M$, use the implicit function theorem to show that for each $m \in M$, there is a neighborhood U of m such that $f(U) \subset N$ is a submanifold. Show that if f is $1-1$ and is an open map onto its image, then $f(M)$ is a submanifold.** [*Sketch.* If $f: U \subset \boldsymbol{E} \to \boldsymbol{F}_0 \oplus \boldsymbol{F}$, where $\boldsymbol{F}_0$ is the range $Df(\mathbf{0})$ and $Df(\mathbf{0})$ is one to one, $D\Phi(\mathbf{0},\mathbf{0})$ is an isomorphism, where $\Phi: U \times \boldsymbol{F}_1 \to \boldsymbol{F}_0 \times \boldsymbol{F}_1$, $(\boldsymbol{u}, \boldsymbol{e}_1) \mapsto f(\boldsymbol{u}) + (\mathbf{0}, \boldsymbol{e}_1)$. Apply the inverse function theorem.]

(b) Prove the following generalization of (a): *Let $f: N \to P$ be a C^∞ map of manifolds such that $\ker Tf = \cup_m$ (kernel $T_m f$) $\subset TN$ is a subbundle of TN; let $d = \dim(\ker T_m f)$. Then $f(N)$ is a locally immersed submanifold of P and if f is open onto its image, $f(N) \subset P$ is a submanifold of dimension $\dim N - d$. (This formulation was suggested by J. Guckenheimer.)* [*Sketch.* Assume $P \subset \boldsymbol{F}$ is open; $\mathbf{0} \in N$, $f(\mathbf{0}) = \mathbf{0}$. Write $\boldsymbol{F}_1$ is the range of $T_{\mathbf{0}} f$ and $\boldsymbol{F} = \boldsymbol{F}_1 \oplus \boldsymbol{F}_2$. Let $\pi_1: \boldsymbol{F} \to \boldsymbol{F}_1$ be the projection. Use 1.6.18 to write $N = \boldsymbol{F}_1 \oplus \boldsymbol{F}_2$, where $\pi_1 \circ f$ is constant in the second variable. Let $i: \boldsymbol{E}_1 \times \{\mathbf{0}\} \to N$ be the injection. Show $\pi_1 \circ f \circ i$ is a local diffeomorphism and the image Q of $f \circ i$ is locally a submanifold. Use the fact that *ker Tf* is a subbundle to show Q is locally the image of f as well.]

(c) Let

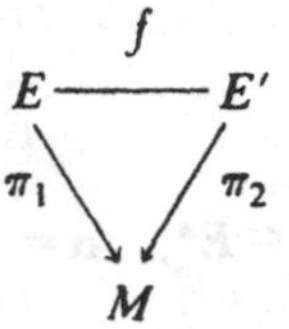

be a smooth vector bundle mapping. We say that f has *constant rank k* if for each $x \in M$ the linear map $f_x: E_x \to E'_x$ has rank k. Let

$$Ker f = \{ v \in E \mid f_x(v) = 0 \in E'_x,\ x = \pi_1(v) \}$$

and

$$Im f = \{ v' \in E' \mid \text{there exists } v \in E \text{ such that } f_x(v) = v',$$
$$\text{where } x = \pi_1(v) = \pi_2(v') \}$$

Show that *if f has constant rank, Ker f and Im f are subbundles of E and E′, respectively.*

1.6G. (Fibration theorem). Use 1.6E and 1.6F to prove the following: Suppose that $f: M \to N$ is smooth and, locally, $T_x f$ has constant rank. Let $Z \subset M$ be a submanifold, $x \in M$, $y \in N$ fixed, and assume that $T_x M = T_x Z + T_x f^{-1}(y)$. Then there are neighborhoods U of x, V of y such that $f(U)$ is a submanifold of N and f induces a diffeomorphism of $f^{-1}(V) \cap Z \cap U$ to $f(U) \cap V$.

1.6H. If N is a submanifold of M, then for each $n \in N$ there is an open neighborhood $U \subset M$ with $n \in U$ and a submersion $f: U \subset M \to \boldsymbol{R}^{m-n}$, where $m = \dim M$ and $n = \dim N$ such that $N \cap U = f^{-1}(0)$. (*Hint*: Choose a submanifold chart and compose it with an appropriate projection.)

ISBN 0-8053-0102-X

1.6I. Let $\boldsymbol{RP}^n$ be real projective n-space (see Sect. 1.1). Show $\boldsymbol{RP}^n$ is a manifold and that $\boldsymbol{RP}^1$ is a submanifold of $\boldsymbol{RP}^2$, which is not the level set of any submersion of $\boldsymbol{RP}^2$ into $\boldsymbol{RP}^1$; in fact, there are no such submersions. (*Hint*: $\boldsymbol{RP}^1$ is one-sided in $\boldsymbol{RP}^2$, that is, there is no continuous choice of normal direction to $\boldsymbol{RP}^1$ in $\boldsymbol{RP}^2$.)

1.7 TENSORS

Given a vector space $\boldsymbol{E}$ we can form a new vector space consisting of tensors on $\boldsymbol{E}$. When this is done on each fiber of a vector bundle, we obtain a new vector bundle structure. An important case occurs when this is applied to the tangent bundle of a manifold. Thus we begin with a review of tensors on vector spaces.

As in Sect. 1.2, $L^k(\boldsymbol{E}_1,\ldots,\boldsymbol{E}_k;\ \boldsymbol{F})$ denotes the vector space of multilinear maps from $\boldsymbol{E}_1\times\cdots\times\boldsymbol{E}_k$ into $\boldsymbol{F}$. The special case $L(\boldsymbol{E},\boldsymbol{R})$ is denoted $\boldsymbol{E}^*$, the **dual space** of $\boldsymbol{E}$. If $\hat{e}=(e_1,\ldots,e_n)$ is an ordered basis of $\boldsymbol{E}$, there is a unique ordered basis of $\boldsymbol{E}^*$, the **dual basis** $\hat{e}^*=(\alpha^1,\ldots,\alpha^n)$, such that $\alpha^j(e_i)=\delta_i^j$, where $\delta_i^j=1$ if $j=i$ and 0 otherwise. Furthermore, for each

$$e\in\boldsymbol{E},\quad e=\sum_{i=1}^{n}\alpha^i(e)e_i$$

and for each

$$\alpha\in\boldsymbol{E}^*,\quad \alpha=\sum_{i=1}^{n}\alpha(e_i)\alpha^i$$

Employing the **summation convention** whereby summation is implied when an index is repeated on upper and lower levels, these expressions become $e=\alpha^i(e)e_i$ and $\alpha=\alpha(e_i)\alpha^i$.

We may map $\boldsymbol{E}$ into $\boldsymbol{E}^{**}=L(\boldsymbol{E}^*,\boldsymbol{R})$ by associating with each $e\in\boldsymbol{E}$, $e^{**}\in\boldsymbol{E}^{**}$, given by $e^{**}(\alpha)=\alpha(e)$ for all $\alpha\in\boldsymbol{E}^*$. Because $\boldsymbol{E}$ has finite dimension, the map $e\mapsto e^{**}$ is an isomorphism (Exercise 1.2B).

1.7.1 Definition. *For a vector space* $\boldsymbol{E}$ *we put* $T_s^r(\boldsymbol{E})=L^{r+s}(\boldsymbol{E}^*,\ldots,\boldsymbol{E}^*,\boldsymbol{E},\ldots,\boldsymbol{E};\ \boldsymbol{R})$ *(r copies of* $\boldsymbol{E}^*$ *and s copies of* $\boldsymbol{E}$*). Elements of* $T_s^r(\boldsymbol{E})$ *are called* ***tensors on E, contravariant*** *of order r and* ***covariant*** *of order s; or simply,* ***of type*** $\binom{r}{s}$. *(This classical terminology is functorally backwards).*

Given $t_1\in T_{s_1}^{r_1}(\boldsymbol{E})$ *and* $t_2\in T_{s_2}^{r_2}(\boldsymbol{E})$ *the* ***tensor product*** *of* t_1 *and* t_2 *is the tensor* $t_1\otimes t_2\in T_{s_1+s_2}^{r_1+r_2}(\boldsymbol{E})$ *defined by*

$$t_1\otimes t_2(\beta^1,\ldots,\beta^{r_1},\gamma^1,\ldots,\gamma^{r_2},f_1,\ldots,f_{s_1},g_1,\ldots,g_{s_2})$$
$$=t_1(\beta^1,\ldots,\beta^{r_1},f_1,\ldots,f_{s_1})t_2(\gamma^1,\ldots,\gamma^{r_2},g_1,\ldots,g_{s_2})$$

where $\beta^j,\gamma^j\in\boldsymbol{E}^*$ *and* $f_j,g_j\in\boldsymbol{E}$.

ISBN 0-8053-0102-X

Because of the identifications discussed in Sect. 1.2, we have $T_0^1(E) \approx E$, $T_1^0(E) = E^*$, $T_2^0(E \approx L(E; E^*)$, and $T_1^1(E) \approx L(E; E)$. It is easy to see that $\otimes$ is associative and bilinear.

1.7.2 Proposition. *If $\dim E = n$, then $T_s^r(E)$ has the structure of a real vector space of dimension n^{r+s}. Indeed, for an ordered basis $\hat{e} = (e_1, \ldots, e_n)$ of E, a basis of $T_s^r(E)$ is given by*

$$\left\{ e_{i_1} \otimes \cdots \otimes e_{i_r} \otimes \alpha^{j_1} \otimes \cdots \otimes \alpha^{j_s} \,|\, i_k, j_k = 1, \ldots, n \right\}$$

where $\hat{e}^* = (\alpha^1, \ldots, \alpha^n)$ is the dual basis of $\hat{e}$.

Proof. We must show that the elements of $e_{i_1} \otimes \cdots \otimes e_{i_r} \otimes \alpha^{j_1} \otimes \cdots \otimes \alpha^{j_s}$ of $T_s^r(E)$ are linearly independent and span $T_s^r(E)$. Suppose $t^{i_1 \ldots i_r}_{j_1 \ldots j_s} e_{i_1} \otimes \cdots \otimes e_{i_r} \otimes \alpha^{j_1} \otimes \cdots \otimes \alpha^{j_s} = 0$. Then apply this to $(\alpha^{k_1}, \ldots, \alpha^{k_r}, e_{l_1}, \ldots, e_{l_s})$ using the identification $e_i(\alpha^j) = \alpha^j(e_i)$ to give $t^{k_1 \ldots k_r}_{l_1 \ldots l_s} = 0$. Next, we easily check that for $t \in T_s^r(E)$ we have

$$t = t\left(\alpha^{i_1}, \ldots, \alpha^{i_r}, e_{j_1}, \ldots, e_{j_s}\right) e_{i_1} \otimes \cdots \otimes e_{i_r} \otimes \alpha^{j_1} \otimes \cdots \otimes \alpha^{j_s}. \qquad \blacksquare$$

The coefficients $t^{i_1 \ldots i_r}_{j_1 \ldots j_s} = t(\alpha^{i_1}, \ldots, \alpha^{i_r}, e_{j_1}, \ldots, e_{j_s})$ are called the **components of t relative to $\hat{e}$.**

The classical operations of tensor algebra can be defined invariantly in this context. For example, the **Kronecker delta** is the tensor $\delta \in T_1^1(E)$ associated to the identity $I \in L(E, E)$ under the canonical isomorphism $T_1^1(E) \approx L(E, E)$, that is, $\delta(\alpha, e) = \alpha(e)$ for all $\alpha \in E^*$, $e \in E$. Relative to any basis, the components of δ are the usual δ_j^i. An example of an **interior product** is the mapping i_e: $T_1^1(E) \to T_0^1(E)$, where $e \in E$, defined by $i_e(t)(\alpha) = t(\alpha, e)$. An example of a **contraction** is the mapping tr: $T_1^1(E) \to R$, where $\mathrm{tr}(t)$ is the trace of the linear mapping associated to t by the isomorphism $T_1^1(E) \approx L(E, E)$. More general contractions may be defined by composing a sequence of inner products with this contraction. A reader with experience in the classical tensor calculus may wish to translate other operations into this language, but we will not need the full machinery in the sequel.

If $\varphi \in L(E, F)$, we can consider $\varphi \in L(T_0^1(E), T_0^1(F))$. Then we define the transpose of φ by $\varphi^* \in L(F^*, E^*) = L(T_1^0(F), T_1^0(E))$, $\varphi^*(\beta) \cdot e = \beta(\varphi(e))$, where $\beta \in F^*$ and $e \in E$. Unfortunately φ^* maps in the wrong direction for some purposes, but this may be remedied if φ is an isomorphism.

1.7.3 Definition. *If $\varphi \in L(E, F)$ is an isomorphism, let $T_s^r\varphi = \varphi_s^r \in L(T_s^r(E), T_s^r(F))$ be defined by*

$$\varphi_s^r t(\beta^1, \ldots, \beta^r, f_1, \ldots, f_s) = t(\varphi^*(\beta^1), \ldots, \varphi^*(\beta^r), \varphi^{-1}(f_1), \ldots, \varphi^{-1}(f_s))$$

where $t \in T_s^r(E)$, $\beta^1, \ldots, \beta^r \in F^*$, and $f_1, \ldots, f_s \in F$.

ISBN 0-8053-0102-X

Note $\varphi_1^0 = (\varphi^{-1})^*$, which maps "forward" like φ, and we identify φ with φ_0^1. The next proposition asserts essentially that T_s^r is a covariant functor.

1.7.4 Proposition. *Let $\varphi: E \to F$ and $\psi: F \to G$ be isomorphisms. Then*

(i) $(\psi \circ \varphi)_s^r = \psi_s^r \circ \varphi_s^r$;
(ii) *If $i: E \to E$ is the identity, then so is $i_s^r: T_s^r(E) \to T_s^r(E)$;*
(iii) $\varphi_s^r: T_s^r(E) \to T_s^r(F)$ *is an isomorphism, and* $(\varphi_s^r)^{-1} = (\varphi^{-1})_s^r$.

Proof. For (i),

$$\begin{aligned}
&\psi_s^r(\varphi_s^r t)(\gamma^1,\ldots,\gamma^r, g_1,\ldots,g_s)\\
&= \varphi_s^r t\big(\psi^*(\gamma^1),\ldots,\psi^*(\gamma^r), \psi^{-1}(g_1),\ldots,\psi^{-1}(g_s)\big)\\
&= t\big(\varphi^*\psi^*(\gamma^1),\ldots,\varphi^*\psi(\gamma^r), \varphi^{-1}\psi^{-1}(g_1),\ldots,\varphi^{-1}\psi^{-1}(g_s)\big)\\
&= t\big((\psi\circ\varphi)^*(\gamma^1),\ldots,(\psi\circ\varphi)^*(\gamma^r), (\psi\circ\varphi)^{-1}(g_1),\ldots,(\psi\circ\varphi)^{-1}(g_s)\big)\\
&= (\psi\circ\varphi)_s^r t(\gamma^1,\ldots,\gamma^r, g_1,\ldots,g_s)
\end{aligned}$$

where $\gamma^1,\ldots,\gamma^r \in G^*, g_1,\ldots,g_s \in G$, and $t \in T_s^r(E)$. We have used the fact that $(\psi\circ\varphi)^* = \varphi^*\circ\psi^*$ and $(\psi\circ\varphi)^{-1} = \varphi^{-1}\circ\psi^{-1}$, which the reader can easily check. Part (ii) is an immediate consequence of the definition and the fact that $i^* = i$ and $i^{-1} = i$. Finally, for (iii) we have $\varphi_s^r \circ (\varphi^{-1})_s^r = i_s^r$, the identity on $T_s^r(F)$, by (i) and (ii). Similarly, $(\varphi^{-1})_s^r \circ \varphi_s^r = i_s^r$ the identity on $T_s^r(E)$. Hence (iii) follows. ∎

The next proposition gives a connection with component notation. The proof is left as an easy exercise for the reader.

1.7.5 Proposition. *Let $\varphi \in L(E,F)$ be an isomorphism. For ordered bases $\hat{e} = (e_1,\ldots,e_n)$ of E and $\hat{f} = (f_1,\ldots,f_n)$ of F, suppose $\varphi(e_i) = A_i^j f_j$ and $(\varphi^{-1})^*(\alpha^i) = B_j^i \beta^j$. Then $B_j^i A_j^k = A_i^j B_j^k = \delta_i^k$, or the inverse matrix of (A_i^j) is (B_i^j), and for $t \in T_s^r(E)$ with components $t_{j_1\ldots j_s}^{i_1\ldots i_r}$ relative to $\hat{e}$, the components of $\varphi_s^r t$ relative to $\hat{f}$ are given by*

$$'t_{j_1'\ldots j_s'}^{i_1'\ldots i_r'} = t_{j_1\ldots j_s}^{i_1\ldots i_r} A_{i_1}^{i_1'}\cdots A_{i_r}^{i_r'} B_{j_1'}^{j_1}\cdots B_{j_s'}^{j_s}$$

In particular, if $E = F$ and φ is the identity, the above describes the components of t relative to the "new basis" $\hat{f}$.

We extend the tensor algebra next to local vector bundles, and finally to vector bundles. For $U \subset E$ (open) recall that $U \times F$ is a local vector bundle. Then $U \times T_s^r(F)$ is also a local vector bundle in view of 1.7.2. Suppose

ISBN 0-8053-0102-X

$\varphi: U \times F \to U' \times F'$ is a local vector bundle mapping and is an *isomorphism on each fiber*; that is, $\varphi_u = \varphi | \{u\} \times F \in L(F, F')$ is an isomorphism. Also, let φ_0 denote the restriction of φ to the zero section.

Then φ induces a mapping of the local tensor bundles as follows.

1.7.6 Definition. *If $\varphi: U \times F \to U' \times F'$ is a local vector bundle mapping such that for each $u \in U$, φ_u is an isomorphism, let $\varphi^r_s: U \times T^r_s(F) \to U' \times T^r_s(F')$ be defined by*

$$\varphi^r_s(u, t) = (\varphi_0(u), (\varphi_u)^r_s t)$$

where $t \in T^r_s(F)$.

Before proceeding we shall pause to recall some useful facts concerning linear isomorphisms.

1.7.7 Proposition. *Let $GL(E, F)$ denote the set of linear isomorphisms from E to F. Then $GL(E, F) \subset L(E, F)$ is open.*

This was proved in 1.3.14. Let us also recall 1.3.15.

1.7.8 Proposition. *Let $\mathcal{I}^*: L(E, F) \to L(F^*, E^*)$; $\varphi \mapsto \varphi^*$ and $\mathcal{I}^{-1}: GL(E, F) \to GL(F, E)$; $\varphi \to \varphi^{-1}$. Then $\mathcal{I}^*$ and $\mathcal{I}^{-1}$ are of class C^∞ and*

$$D\mathcal{I}^{-1}(\varphi) \cdot \psi = -\varphi^{-1} \cdot \psi \cdot \varphi^{-1}$$

Smoothness of $\mathcal{I}^*$ is clear since it is linear.

1.7.9 Proposition. *If $\varphi: U \times F \to U' \times F'$ is a local vector bundle map and φ_u is an isomorphism for all $u \in U$, then $\varphi^r_s: U \times T^r_s(F) \to U' \times T^r_s(F')$ is a local vector bundle map and $(\varphi_u)^r_s = (\varphi^r_s)_u$ is an isomorphism for all $u \in U$. Moreover, if φ is a local vector bundle isomorphism then so is φ^r_s.*

Proof. That φ^r_s is an isomorphism on fibers follows from the functorial property of φ^r_s and the last assertion follows from the former. By 1.7.6 we need only establish that $(\varphi_u)^r_s = (\varphi^r_s)_u$ is of class C^∞. Now, φ_u is a smooth function of u, and, by 1.7.8, φ^*_u and φ_u^{-1} are smooth functions of u. The Cartesian product of smooth functions is easily seen to be smooth and $(\varphi_u)^r_s$ is a multilinear mapping on a Cartesian product of smooth functions (this is not linearity in φ). Hence from the product rule $(\varphi_u)^r_s$ is smooth. ■

This smoothness can be verified also by using the standard bases in the tensor spaces as local bundle charts, and proving that the components $(\varphi^r_s t)^{i_1 \cdots i_r}_{j_1 \cdots j_s}$ are C^∞ functions.

ISBN 0-8053-0102-X

We have the following commutative diagram which says that φ_s^r preserves fibers:

$$\begin{array}{ccc} U \times T_s^r(F) & \xrightarrow{\varphi_s^r} & U' \times T_s^r(F') \\ \big\downarrow \pi & & \big\downarrow \pi' \\ U & \xrightarrow[\varphi_0]{} & U' \end{array}$$

1.7.10 Definition. *Let $\pi: E \to B$ be a vector bundle with $E_b = \pi^{-1}(b)$ the fiber over $b \in B$. Define $T_s^r(E) = \bigcup_{b \in B} T_s^r(E_b)$ and $\pi_s^r: T_s^r(E) \to B$ by $\pi_s^r(e) = b$ iff $e \in T_s^r(E_b)$. Furthermore, for a subset A of B, we put $T_s^r(E)|A = \bigcup_{b \in A} T_s^r(E_b)$. If $\pi': E' \to B'$ is another vector bundle and $(\varphi, \varphi_0): \pi \to \pi'$ is a vector bundle mapping with $\varphi_b = \varphi|E_b$ an isomorphism for all $b \in B$, let $\varphi_s^r: T_s^r(E) \to T_s^r(E')$ be defined by $\varphi_s^r|T_s^r(E_b) = (\varphi_b)_s^r$.*

Now suppose $(E|U, \varphi)$ is an admissible local bundle chart of π, where $U \subset B$ is an open set. Then the mapping $\varphi_s^r|[T_s^r(E)|U]$ is obviously a bijection onto a local bundle, and thus is a local bundle chart. Further, $(\varphi_s^r)_b = (\varphi_b)_s^r$ is a linear isomorphism, so this chart preserves the linear structure of each fiber, which in this case is given in advance. We shall call such a chart a **natural chart** of $T_s^r(E)$.

1.7.11 Theorem. *If $\pi: E \to B$ is a vector bundle, then the set of all natural charts of $\pi_s^r: T_s^r(E) \to B$ is a vector bundle atlas.*

Proof. Axiom (VBA1) is obvious. For (VBA2), suppose we have two overlapping natural charts, φ_s^r and ψ_s^r. For simplicity, let them have the same domain. Then $\alpha = \psi \circ \varphi^{-1}$ is a local vector bundle isomorphism, and by 1.7.4, $\psi_s^r \circ (\varphi_s^r)^{-1} = \alpha_s^r$, a local vector bundle isomorphism by 1.7.9. ■

This atlas of natural charts, the **natural atlas** of π_s^r, generates a vector bundle structure, and it is easily seen that the resulting vector bundle is Hausdorff, second countable, and has constant dimension. *Hereafter, π_s^r will denote all of this structure.*

1.7.12 Proposition. *If $f: E \to E'$ is a vector bundle map that is an isomorphism on each fiber, then $f_s^r: T_s^r(E) \to T_s^r(E')$ is also a vector bundle map that is an isomorphism on each fiber.*

Proof. Let (U, φ) be an admissible vector bundle chart of E, and let (V, ψ) be one of E' so that $f(U) \subset V$ and $f_{\varphi\psi} = \psi \circ f \circ \varphi^{-1}$ is a local vector bundle mapping. Then using the natural atlas, we see that $(f_s^r)_{\varphi_s^r, \psi_s^r} = (f_{\varphi\psi})_s^r$. ■

ISBN 0-8053-0102-X

1.7.13 Proposition. *Suppose $f: E \to E'$ and $g: E' \to E''$ are vector bundle maps that are isomorphisms on each fiber. Then so is $g \circ f$, and*

(i) $(g \circ f)^r_s = g^r_s \circ f^r_s$.
(ii) *If $i: E \to E$ is the identity, then $i^r_s: T^r_s(E) \to T^r_s(E)$ is the identity.*
(iii) *If $f: E \to E'$ is a vector bundle isomorphism, then so is f^r_s and $(f^r_s)^{-1} = (f^{-1})^r_s$.*

Proof. For (i) we examine representatives of $(g \circ f)^r_s$ and $g^r_s \circ f^r_s$. These representatives are the same in view of 1.7.4. Part (ii) is clear from the definition, and (iii) follows from (i) and (ii) by the same method as in 1.7.4. ■

This proposition asserts that T^r_s is a covariant functor on vector bundles and vector bundle mappings that are regular on fibers.

We now specialize to the important case where $\pi: E \to B$ is the tangent vector bundle of a manifold.

1.7.14 Definition. *Let M be a manifold and $\tau_M: TM \to M$ its tangent bundle. We call $T^r_s(M) = T^r_s(TM)$ the **vector bundle of tensors of contravariant order** r **and covariant order** s; or simply **of type** $\binom{r}{s}$. Also $T^0_1(M)$ is called the **cotangent bundle of** M and is denoted by $\tau^*_M: T^*M \to M$.*

Since E^{**} can be identified with E, we may identify $T^1_0(M)$ with TM, the tangent bundle. The zero section of $T^r_s(M)$ may be identified with M.

Recall that a section of a bundle assigns to each base point b a vector in the fiber over b. In the case of $T^r_s(M)$ these vectors are called **tensors**. Also, the addition and scalar multiplication of sections takes place within each fiber. The C^∞ sections of $\pi: E \to B$ were denoted $\Gamma^\infty(\pi)$, or $\Gamma^\infty(E)$.

1.7.15 Definition. *A **tensor field of type** $\binom{r}{s}$ on a manifold M is a C^∞ section of $T^r_s(M)$. We denote by $\mathfrak{T}^r_s(M)$ the set $\Gamma^\infty(T^r_s(M))$ together with its (infinite-dimensional) real vector space structure. Also we let $\mathfrak{F}(M)$ denote the set of mappings from M into $\mathbf{R}$ that are of class C^∞ (the standard local manifold structure being used on $\mathbf{R}$) together with its structure as a ring; namely, $f+g$, cf, fg for $f, g \in \mathfrak{F}(M)$, $c \in \mathbf{R}$ are given by $(f+g)(x) = f(x) + g(x)$, $(cf)(x) = c(f(x))$ and $(fg)(x) = f(x) \cdot g(x)$. A **vector field** on M is an element of $\mathfrak{X}(M) = \mathfrak{T}^1_0(M)$. A **covector field**, or **differential one-form**, is an element of $\mathfrak{X}^*(M) = \mathfrak{T}^0_1(M)$.*

Note that for the tangent bundle TM, a natural chart is obtained by taking $T\varphi$, where φ is an admissible chart of M. This in turn induces a natural chart $(T\varphi)^r_s = T^r_s\varphi$ for T^r_sM. We shall call these the **natural charts** of T^r_sM. If $t \in T^r_s(M)$ and $(T^r_sU, T^r_s\varphi)$ is a natural chart, we write ${}_\varphi t$ rather than $t_{T^r_s\varphi,\varphi}$ for the local representative. These local representatives lead to the classical notion of tensor fields if we start with a chart (U, φ) of M with range

ISBN 0-8053-0102-X

$\varphi(U) = U' \subset \mathbf{R}^n$. For if $\hat{e}$ is the standard ordered basis of $\mathbf{R}^n$, then

$${}_\varphi t(u') = {}_\varphi t^{i_1 \cdots i_r}_{j_1 \cdots j_s}(u') e_{i_1} \otimes \cdots \otimes \alpha^{j_s}$$

where ${}_\varphi t^{i_1 \cdots i_r}_{j_1 \cdots j_s}(u') = {}_\varphi t(u')(\alpha^{i_1}, \ldots, e_{j_s})$.

The algebraic operations on tensors, such as contraction and inner products, all carry over, fiberwise, to tensor fields. For example, if $\boldsymbol{\delta}_m \in T^1_1(T_m M)$ is the Kronecker delta, then $\delta\colon M \to T^1_1(M)$; $m \mapsto \boldsymbol{\delta}_m$ is obviously C^∞, and $\delta \in \mathfrak{T}^1_1(M)$ is also called the **Kronecker delta**. In addition, the mapping $T^r_s\varphi$ induces an action on tensor fields. We now treat the most important of these extensions explicitly.

1.7.16 Definition. *If $\varphi\colon M \to N$ is a diffeomorphism and $t \in T^r_s(M)$, let $\varphi_* t = (T\varphi)^r_s \circ t \circ \varphi^{-1}$, the* ***push-forward*** *of t by φ. If $t \in T^r_s(N)$, the* ***pull-back*** *of t by φ is given by $\varphi^* t = (\varphi^{-1})_* t$.*

1.7.17 Proposition. *If $\varphi\colon M \to N$ is a diffeomorphism, and $t \in \mathfrak{T}^r_s(M)$, then (i) $\varphi_* t \in \mathfrak{T}^r_s(N)$, (ii) $\varphi_*\colon \mathfrak{T}^r_s(M) \to \mathfrak{T}^r_s(N)$ is a linear isomorphism and (iii) $(\varphi \circ \psi)_* = \varphi_* \circ \psi_*$.*

Proof. (i) The differentiability is evident from the composite mapping theorem, together with 1.7.9. (ii, iii) follow from 1.7.4 and 1.7.13. ■

1.7.18 Definition. *If $f \in \mathfrak{F}(M)$ and $t \in \mathfrak{T}^r_s(M)$, let $ft\colon M \to T^r_s(M)\colon m \mapsto f(m)t(m)$. If also $X_i \in \mathfrak{X}(M)$, $i = 1, \ldots, s$, $\alpha^j \in \mathfrak{X}^*(M)$, $j = 1, \ldots, r$, let*

$$t(\alpha^1, \ldots, \alpha^r, X_1, \ldots, X_s)\colon M \to \mathbf{R};\ m \mapsto t(m)(\alpha^1(m), \ldots, X_s(m))$$

If also $t' \in \mathfrak{T}^{r'}_{s'}(M)$, let $t \otimes t'\colon M \to T^{r+r'}_{s+s'}(M)$; $m \mapsto t(m) \otimes t'(m)$.

1.7.19 Proposition. *With f, t, X_i, α^j, and t' as in 1.7.18, $ft \in \mathfrak{T}^r_s(M)$, $t(\alpha^1, \ldots, X_s) \in \mathfrak{F}(M)$, and $t \otimes t' \in \mathfrak{T}^{r+r'}_{s+s'}$.*

Proof. The differentiability is evident in each case from the product rule in local representation. ■

Finally, we describe an alternative approach to tensor fields. Suppose $\mathfrak{F}(M)$ is defined as above, and $\mathfrak{X}(M)$ either our way or some equivalent way. With the "scalar multiplication" $(f, X) \mapsto fX$ defined in 1.7.18, $\mathfrak{X}(M)$ becomes an $\mathfrak{F}(M)$-module. That is, $\mathfrak{X}(M)$ is essentially a vector space over $\mathfrak{F}(M)$, but the "scalars" $\mathfrak{F}(M)$ form only a commutative ring with identity, rather than a field, as $1/f$ may not exist, even if $f \neq 0$. We may thus define

$$L_{\mathfrak{F}(M)}(\mathfrak{X}(M), \mathfrak{F}(M)) = \mathfrak{X}^*(M)$$

ISBN 0-8053-0102-X

the $\mathcal{F}(M)$ linear mappings, and similarly

$$\mathfrak{T}^r_s(M) = L^{r+s}_{\mathcal{F}(M)}(\mathcal{X}^*(M), \ldots, \mathcal{X}(M); \mathcal{F}(M))$$

the $\mathcal{F}(M)$ multilinear mappings. From 1.7.17, we have a natural mapping $\mathcal{T}_s{}^r(M) \to \mathfrak{T}^r_s(M)$, which is an $\mathcal{F}(M)$ linear isomorphism.

The direct sum $\mathcal{T}(M)$ of the $\mathcal{T}_s{}^r(M)$, including $\mathcal{T}_0{}^0(M) = \mathcal{F}(M)$, is a real vector space with $\otimes$-product, including $f \otimes t = ft$, a "bigraded $\mathcal{F}(M)$-algebra," called the **tensor algebra** of M, and if $\varphi\colon M \to n$ is a diffeomorphism, φ_*. $\mathcal{T}(M) \to \mathcal{T}(N)$ is an algebra isomorphism.

EXERCISES

1.7A. Show in detail that $L(E, E) \approx T^1_1(E)$, and $L(E, L(E, E)) \approx T^1_2(E)$.

1.7B. Define, intrinsically, the contraction of $t \in T^r_s(M)$ between the ith contravariant and the jth covariant indices.

1.7C. Show that the local representative of $t \in T^r_s(M)$ with respect to a chart (U, φ) is $\varphi_*(t|U)$. Complete the details of 1.7.17 and 1.7.19.

1.7D. If $t \in \mathcal{T}_s{}^0(N)$ and $\varphi\colon M \to N$ is C^r, $r > 1$, show that $\varphi^* t$ is defined even if φ is not a diffeomorphism. Prove that $(\varphi \circ \psi)^* t = (\psi^* \circ \varphi^*) t$.

1.7E. Prove 1.7.5.

ISBN 0-8053-0102-X

CHAPTER 2

Calculus on Manifolds

Certainly a mathematical topic of wide practical use is calculus. For many applications, and this book is an example, it is essential to use calculus in the large. In this chapter we outline parts of the subject that are useful in mechanics. The reader may wish to skip the optional topics and technical proofs on first reading. The most important formulas we need are summarized on p. 121.

2.1 VECTOR FIELDS AS DYNAMICAL SYSTEMS

Recall that a vector field on a manifold M is a mapping from M to TM that assigns to each point $m \in M$ a vector in $T_m M$. A vector field may be interpreted alternatively as the right-hand side of a system of first-order ordinary differential equations in the large, that is, a dynamical system. In this section we develop this interpretation and discuss the basic existence and uniqueness theorems for the integral of the system.

The study of dynamical systems, also called *flows*, may be motivated as follows. Consider a physical system that is capable of assuming various "states" described by points in a set S. For example, S might be $\boldsymbol{R}^3 \times \boldsymbol{R}^3$ and a state might be the position and momentum $(\boldsymbol{q},\boldsymbol{p})$ of a particle.

Ralph Abraham and Jerrold E. Marsden, Foundation of Mechanics, Second Edition

ISBN 0-8053-0102-X

As time passes the state changes. If the state is s_1 at time t_1 and this changes to s_2 at a later time t_2, we set

$$F_{t_2,t_1}(s_1)=s_2$$

and call F_{t_2,t_1} the *evolution operator*; it maps a state at time t_1 to what the state would be after time t_2-t_1 has elapsed. "Determinism" is expressed by the law (sometimes called the Chapman–Kolmogorov law):

$$F_{t_3,t_2}\circ F_{t_2,t_1}=F_{t_3,t_1}, \qquad F_{t,t}=\text{identity}$$

By saying the evolution laws are time independent, we mean F_{t_2,t_1} depends only on t_2-t_1. Setting $F_{t_2-t_1}=F_{t_2,t_1}$, the above law becomes the *group property*:

$$F_t\circ F_s=F_{t+s}, \qquad F_0=\text{identity}$$

We call F_t a *flow* and F_{t_2,t_1} a *time-dependent flow*, or as above, an evolution operator. If the system is nonreversible, that is, defined only for $t_2>t_1$, we speak of a *semi-flow*.

In physics, it is usually not F_{t_2,t_1} that is given, but rather the *laws of motion*. In other words, some differential equations are given that we must solve in order to find the flow. These equations of motion have the form

$$\frac{ds}{dt}=X(s), \qquad s(0)=s_0$$

given where X is a (possibly time-dependent) vector field on S.

Let us now turn to the elaboration of these ideas when a vector field X is given on a manifold M.

Recall that a curve c at a point m of a manifold M is a C^1 map from an open interval I of $\mathbb{R}$ into M such that $0\in I$ and $c(0)=m$. For such a curve we may assign a tangent vector at each point $c(\lambda),\lambda\in I$, by $c'(\lambda)=Tc(\lambda,1)$.

2.1.1 Definition. *Let M be a manifold and $X\in\mathfrak{X}(M)$. An **integral curve** of X at $m\in M$ is a curve c at m such that $c'(\lambda)=X(c(\lambda))$ for each $\lambda\in I$.*

We now express the condition that c be an integral curve of X in terms of local representatives with respect to natural charts. Let (U,φ) be a chart of M and suppose the image of c is contained in U. Then the local representative of c with respect to the identity of $\mathbb{R}$ and (U,φ) is $c_\varphi=\varphi\circ c$, while the local representative of the curve c' with respect to the identity of $\mathbb{R}$ and the natural chart $(TM|U,T\varphi)$ is given by

$$\begin{aligned}(c')_\varphi(\lambda)&=T\varphi\circ c'(\lambda)=T\varphi\circ Tc(\lambda,1)\\&=T(\varphi\circ c)(\lambda,1)=(c_\varphi)'(\lambda)\end{aligned}$$

ISBN 0-8053-0102-X

by the composite mapping theorem. The local representative of $X \circ c$ with respect to the identity of $\boldsymbol{R}$ and the natural chart $T\varphi$ is

$$T\varphi \circ X \circ c = T\varphi \circ X \circ \varphi^{-1} \circ \varphi \circ c = X_\varphi \circ c_\varphi$$

where X_φ is the local representative of X. Thus c is an integral curve of X iff $c' = X \circ c$ iff $c'_\varphi = X_\varphi \circ c_\varphi$ iff c_φ is an integral curve of X_φ. This condition takes a simple form if $\varphi(U) \subset \boldsymbol{R}^n$. Then we have $X_\varphi(y) = (y; X_1(y), \ldots, X_n(y))$, where $y \in \varphi(U) \subset \boldsymbol{R}^n$, $\{X_i(y)\}$ are the components of X_φ, $c_\varphi(\lambda) = (c_1(\lambda), \ldots, c_n(\lambda))$, $c'_\varphi(\lambda) = (c_\varphi(\lambda); c'_1(\lambda), \ldots, c'_n(\lambda))$, and $c'_\varphi = X_\varphi \circ c_\varphi$ iff $c'_i(\lambda) = X_i(c_\varphi(\lambda))$ for $i = 1, \ldots, n$ and all $\lambda \in I$. Thus c is an integral curve of X iff the local representatives satisfy the system of first-order ordinary differential equations

$$\begin{cases} c'_1(\lambda) = X_1(c_1(\lambda), \ldots, c_n(\lambda)) \\ \vdots \\ c'_n(\lambda) = X_n(c_1(\lambda), \ldots, c_n(\lambda)) \end{cases}$$

Note that λ does not appear explicitly on the right. Such a system of equations (a local dynamical system) includes equations of higher order by their usual reduction to first-order systems and the Hamiltonian equations of motion as special cases.

For a system of ordinary differential equations there are well-known existence and uniqueness theorems. The form that we shall need is the following.

2.1.2 Theorem. *Let $X: U \subset \boldsymbol{R}^n \to \boldsymbol{R}^n$ be of class C^∞. For each $x_0 \in U$, there is a curve $c: I \to U$ at x_0 such that $c'(\lambda) = X(c(\lambda))$ for all $\lambda \in I$. Any two such curves are equal on the intersection of their domains. Moreover, there is a neighborhood U_0 of $x_0 \in U$, a real number $a > 0$, and a C^∞ mapping $F: U_0 \times I \to \boldsymbol{R}^n$, where $I = (-a, a)$ such that $c_u(\lambda): I \to \boldsymbol{R}^n$, $c_u(\lambda) = F(u, \lambda)$ is a curve at $u \in \boldsymbol{R}^n$ satisfying the differential equations $c'_u(\lambda) = X(c_u(\lambda))$ for all $\lambda \in I$.*

Because dynamical systems are such a central focus of the book, we shall include the proof. In fact, we shall prove it in Banach spaces with no added difficulty. The proof proceeds in several steps. We begin with the existence and uniqueness.

Lemma 1. *Let $\boldsymbol{E}$ be a Banach space, $U \subset \boldsymbol{E}$ an open set, and $X: U \subset \boldsymbol{E} \to \boldsymbol{E}$ a Lipschitz map; that is $\|X(x) - X(y)\| \leq K\|x - y\|$ for all $x, y \in U$ and a constant K. Let $x_0 \in U$ and suppose the ball of radius b, $B_b(x_0) = \{x \in \boldsymbol{E} \mid \|x - x_0\| \leq b\}$ lies in U, and $\|X(x)\| \leq M$ for all $x \in B_b(x_0)$. Let $\lambda_0 \in \boldsymbol{R}$ and let $\alpha = b/M$. Then there is a unique C^1 curve $x(\lambda)$, $\lambda \in [\lambda_0 - \alpha, \lambda_0 + \alpha]$ such that $x(\lambda) \in B_b(x_0)$ and*

$$\begin{cases} x'(\lambda) = X(x(\lambda)) \\ x(\lambda_0) = x_0 \end{cases}$$

ISBN 0-8053-0102-X

Proof. The conditions $x'(\lambda)=X(x(\lambda))$, $x(\lambda_0)=x_0$ are equivalent to the integral equation

$$x(\lambda)=x_0+\int_{\lambda_0}^{\lambda}X(x(s))\,ds$$

Put $x_0(\lambda)=x_0$ and define inductively $x_{n+1}(\lambda)=x_0+\int_{\lambda_0}^{\lambda}X(x_n(s))\,ds$. Clearly $x_n(\lambda)\in B_b(x_0)$ for all n and $\lambda\in[\lambda_0-\alpha,\lambda_0+\alpha]$ by definition of α. We also find by induction that

$$\|x_{n+1}(\lambda)-x_n(\lambda)\|\leqslant MK^n|\lambda-\lambda_0|^{n+1}/(n+1)!$$

Thus $x_n(\lambda)$ converges uniformly to a continuous curve $x(\lambda)$. Clearly $x(\lambda)$ satisfies the integral equation, and thus is the solution we sought.

For uniqueness, let $y(\lambda)$ be another solution. By induction we find that $\|x_n(\lambda)-y(\lambda)\|\leqslant MK^n|\lambda-\lambda_0|^{n+1}/(n+1)!$; thus, letting $n\to\infty$ gives $x(t)=y(t)$. ▼

Let us observe that exactly the same result holds if X depends explicitly on λ, is jointly continuous in (λ,x), and is Lipschitz in x uniformly in λ.

The next result is a basic estimate used to study the dependence on initial conditions. It is referred to as *Gronwall's inequality.*

Lemma 2. *Let $f,g:[a,b)\to\mathbf{R}$ be continuous and nonnegative. Suppose*

$$f(t)\leqslant A+\int_a^t f(s)\,g(s)\,ds,\qquad A\geqslant 0$$

Then

$$f(t)\leqslant A\exp\left(\int_a^t g(s)\,ds\right)\quad\textit{for}\quad t\in[a,b)$$

Proof. First suppose $A>0$. Let $h(t)=A+\int_a^t f(s)\,g(s)\,ds$; thus $h(t)>0$. Then $h'(t)=f(t)g(t)\leqslant h(t)g(t)$. Thus $h'(t)/h(t)\leqslant g(t)$. Integration gives $h(t)\leqslant A\exp(\int_a^t g(s)\,ds)$. This gives the result for $A>0$. If $A=0$, then we have the result replacing A by $\varepsilon>0$ for every $\varepsilon>0$, thus h and hence f is zero. ▼

Lemma 3. *Let X be as in Lemma 1. Let $F_\lambda(x_0)$ denote the solution (= integral curve) of $x'(\lambda)=X(x(\lambda)), x(0)=x_0$. Then there is a neighborhood V of x_0 and a number $\varepsilon>0$ such that for every $y\in V$ there is a unique integral curve $x(\lambda)=F_\lambda(y)$ satisfying $x'(\lambda)=X(x(\lambda))$ for $\lambda\in[-\varepsilon,\varepsilon]$ and $x(0)=y$. Moreover,*

$$\|F_\lambda(x)-F_\lambda(y)\|\leqslant e^{K|\lambda|}\|x-y\|$$

ISBN 0-8053-0102-X

Proof. The first part is clear from Lemma 1. For the second, let $f(t)=\|F_t(x)-F_t(y)\|$. Clearly

$$f(t)=\left\|\int_0^t X(F_s(x))-X(F_s(y))\,ds+x-y\right\|\leqslant\|x-y\|+K\int_0^t f(s)\,ds,$$

so the result follows from Lemma 2. ▼

This result shows that $F_\lambda(x)$ depends in a continuous, indeed Lipschitz, manner on the initial condition x and is jointly continuous in (λ,x). The next result shows that F_λ is C^k if X is, and completes the proof of 2.1.2.

Lemma 4. *Let X in Lemma 1 be of class C^k, $1\leqslant k\leqslant\infty$, and let $F_\lambda(x)$ be defined as above. Then locally in (λ,x), $F_\lambda(x)$ is of class C^k in x and is C^{k+1} in the λ-variable.*

Proof. We define $\psi(\lambda,x)\in L(E,E)$, the continuous linear maps of E to E, to be the solution of the "linearized" equations:

$$\frac{d}{d\lambda}\psi(\lambda,x)=DX(F_\lambda(x))\cdot\psi(\lambda,x)$$
$$\psi(0,x)=\mathit{identity}$$

where $DX(y)\colon E\to E$ is the derivative of X taken at the point y. By Gronwall's inequality it follows that $\psi(\lambda,x)$ is continuous in (λ,x) [using the norm topology on $L(E,E)$; see Exercise 1.2A].

We claim that $DF_\lambda(x)=\psi(\lambda,x)$. To show this, set $\theta(\lambda,h)=F_\lambda(x+h)-F_\lambda(x)$ and write

$$\begin{aligned}\theta(\lambda,h)-\psi(\lambda,x)\cdot h&=\int_0^\lambda\{X(F_s(x+h))-X(F_s(x))\}\,ds\\&\quad-\int_0^\lambda DX(F_s(x))\cdot\psi(s,x)\cdot h\,ds\\&=\int_0^\lambda DX(F_s(x))\cdot[\theta(s,h)-\psi(s,x)\cdot h]\,ds\\&\quad+\int_0^\lambda\{X(F_s(x+h))-X(F_s(x))\\&\quad-DX(F_s(x))\cdot[F_s(x+h)-F_s(x)]\}\,ds\end{aligned}$$

Since X is of Class C^1, given $\varepsilon>0$, there is a $\delta>0$ such that $\|h\|<\delta$ implies the second term is dominated in norm by $\int_0^\lambda\varepsilon\|F_s(x+h)-F_s(x)\|\,ds$, which is,

ISBN 0-8053-0102-X

in turn, smaller than $A\varepsilon\|h\|$ for a positive constant A. By Gronwall's inequality we obtain $\|\theta(\lambda, h) - \psi(\lambda, x)\cdot h\| \leqslant (\text{constant})\varepsilon\|h\|$. It follows that $DF_\lambda(x)\cdot h = \psi(\lambda, x)\cdot h$. Thus both partial derivatives of $F_\lambda(x)$ exist and are continuous; therefore, $F_\lambda(x)$ is of class C^1.

We prove $F_\lambda(x)$ is C^k by induction on k. Now

$$\frac{d}{d\lambda}F_\lambda(x) = X(F_\lambda(x))$$

so

$$\frac{d}{d\lambda}\frac{d}{d\lambda}F_\lambda(x) = DX(F_\lambda(x))\cdot X(F_\lambda(x))$$

and

$$\frac{d}{d\lambda}DF_\lambda(x) = DX(F_\lambda(x))\cdot DF_\lambda(x)$$

Since the right-hand sides are C^{k-1}, so are the solutions by induction. Thus F itself is C^k. ■

For another more "modern" proof of 2.1.2, see Robbin [1968] (this is reproduced in Lang [1970]). (Actually this proof referred to has a technical advantage: it works easily for other types of differentiability on X and F_λ, such as Hölder or Sobolev differentiability; see Ebin–Marsden [1970] for details.)

The mapping F gives a locally unique integral curve c_u for each $u \in U_0$, and for each $\lambda \in I$, $F_\lambda = F|(U_0 \times \{\lambda\})$ maps U_0 to some other set. It is convenient to think of each point u being allowed to "flow for time λ" along the integral curve c_u (see Fig. 2.1-1 and our opening motivation). This is a picture of a U_0 "flowing," and the system (U_0, a, F) is a local flow of X, or *flow box*. The analogous situation on a manifold is given by the following.

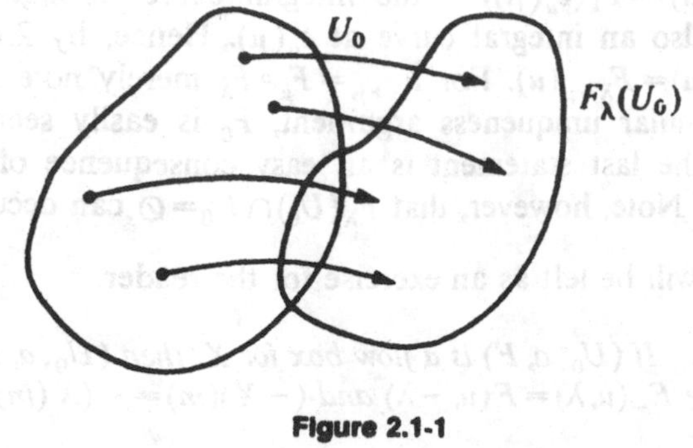

Figure 2.1-1

ISBN 0-8053-0102-X

2.1.3 Definition. *Let M be a manifold and X a vector field on M. A **flow box** of X **at** $m \in M$ is a triple (U_0, a, F), where*

(i) *$U_0 \subset M$ is open, $m \in U_0$, and $a \in \mathbf{R}$, $a > 0$ or $a = +\infty$;*
(ii) *$F\colon U_0 \times I_a \to M$ is of class C^∞, where $I_a = (-a, a)$;*
(iii) *for each $u \in U_0$, $c_u\colon I_a \to M$ defined by $c_u(\lambda) = F(u, \lambda)$ is an integral curve of X at u;*
(iv) *if $F_\lambda\colon U_0 \to M$ is defined by $F_\lambda(u) = F(u, \lambda)$, then for $\lambda \in I_a$, $F_\lambda(U_0)$ is open, and F_λ is a diffeomorphism onto its image.*

Before proving the existence of a flow box, it is convenient first to establish the following, which concerns uniqueness.

2.1.4 Proposition. *Suppose c_1 and c_2 are two integral curves of X at $m \in M$. Then $c_1 = c_2$ on the intersection of their domains.*

Proof. This does not follow at once from 2.1.2 for c_1 and c_2 may lie in different charts. (Indeed, if the manifold is not Hausdorff, examples show that this proposition is false.) Suppose $c_1\colon I_1 \to M$ and $c_2\colon I_2 \to M$. Let $I = I_1 \cap I_2$, and let $K = \{\lambda \mid \lambda \in I \text{ and } c_1(\lambda) = c_2(\lambda)\}$. From 1.1 B, K is closed since M is Hausdorff. We will now show that K is open. From 2.1.2, K contains some neighborhood of 0. For $\lambda \in K$ consider c_1^λ and c_2^λ, where $c^\lambda(t) = c(\lambda + t)$. Then c_1^λ and c_2^λ are integral curves at $c_1(\lambda) = c_2(\lambda)$. Again by 2.1.2 they agree on some neighborhood of 0. Thus some neighborhood of λ lies in K, and so K is open. Since I is connected, $K = I$. ∎

The next two propositions give elementary properties of flow boxes.

2.1.5 Proposition. *Suppose (U_0, a, F) is a triple satisfying (i), (ii), and (iii) of 2.1.3. Then for λ, μ, and $\lambda + \mu \in I_a$ we have $F_{\lambda+\mu} = F_\lambda \circ F_\mu = F_\mu \circ F_\lambda$, and F_0 is the identity map. Moreover, if $U_\lambda = F_\lambda(U_0)$ and $U_\lambda \cap U_0 \neq \varnothing$, then $F_\lambda | U_{-\lambda} \cap U_0\colon U_{-\lambda} \cap U_0 \to U_0 \cap U_\lambda$ is a diffeomorphism and its inverse is $F_{-\lambda} | U_0 \cap U_\lambda$.*

Proof. $F_{\lambda+\mu}(u) = c_u(\lambda + \mu)$, where c_u is the integral curve defined by F at u. But $d(\lambda) = F_\lambda(F_\mu(u)) = F_\lambda(c_u(\mu))$ is the integral curve through $c_u(\mu)$, and $f(t) = c_u(t + \mu)$ is also an integral curve at $c_u(\mu)$. Hence, by 2.1.4 we have $F_\lambda(F_\mu(u)) = c_u(\lambda + \mu) = F_{\lambda+\mu}(u)$. For $F_{\lambda+\mu} = F_\mu \circ F_\lambda$ merely note $F_{\lambda+\mu} = F_{\mu+\lambda} = F_\mu \circ F_\lambda$. By a similar uniqueness argument, F_0 is easily seen to be the identity. Finally, the last statement is an easy consequence of $F_\lambda \circ F_{-\lambda} = F_{-\lambda} \circ F_\lambda = \text{identity}$. Note, however, that $F_\lambda(U_0) \cap U_0 = \varnothing$ can occur. ∎

The following will be left as an exercise for the reader.

2.1.6 Proposition. *If (U_0, a, F) is a flow box for X, then (U_0, a, F_-) is a flow box for $-X$, where $F_-(u, \lambda) = F(u, -\lambda)$ and $(-X)(m) = -(X(m))$.*

ISBN 0-8053-0102-X

2.1.7 Theorem (Uniqueness of flow boxes). *Suppose (U_0, a, F), (U_0', a', F') are two flow boxes at $m \in M$. Then F and F' are equal on $(U_0 \cap U_0') \times (I_a \cap I_{a'})$.*

Proof. Again we emphasize that this does not follow at once from 2.1.2, for U_0, U_0' need not be chart domains. However, for each $u \in U_0 \cap U_0'$ we have $F|\{u\} \times I = F'|\{u\} \times I$, where $I = I_a \cap I_{a'}$. This follows from 2.1.4 and 2.1.3 (*iii*). Hence $F = F'$ on $(U_0 \cap U_0') \times I$. ■

Clearly uniqueness depends only on (i) and (iii) of 2.1.3.

2.1.8 Theorem (Existence of flow boxes). *Let X be a C^∞ vector field on a manifold M. For each $m \in M$ there is a flow box of X at m.*

Proof. Let (U, φ) be a chart in M with $m \in U$. It is enough to establish the result in $\varphi(U)$ by means of the local representation. That is, let (U_0', a, F') be a flow box of X_φ at $\varphi(m)$ as given by 2.1.2, with

$$U_0' \subset U' = \varphi(U) \quad \text{and} \quad F'(U_0' \times I_a) \subset U', \quad U_0 = \varphi^{-1}(U_0')$$

and let

$$F: U_0 \times I_a \to M; \qquad (u, \lambda) \mapsto \varphi^{-1}(F'(\varphi(u), \lambda))$$

Since F is continuous there is a $b \in (0, a) \subset \mathbf{R}$ and $V_0 \subset U_0$ open, $m \in V_0$, such that $F(V_0 \times I_b) \subset U_0$. We contend that (V_0, b, F) is a flow box at m (where F is understood as the restriction of F to $V_0 \times I_b$). Only (iv) need be established, that is, F_λ is a diffeomorphism. For $\lambda \in I_b$, F_λ has a C^∞ inverse, namely, $F_{-\lambda}$ as $V_\lambda \cap U_0 = V_\lambda$.

It follows that $F_\lambda(V_0)$ is open. And, since F_λ and $F_{-\lambda}$ are both of class C^∞, F_λ is a diffeomorphism. ■

The following result, called the "straightening out theorem," shows that near a point m that is not a critical point, that is, $X(m) \neq 0$, the flow can be modified by a change of variables so the integral curves become straight lines.

2.1.9 Theorem. *Let X be a vector field on a manifold M and suppose, for $m \in M$, $X(m) \neq 0$. Then there is a local chart (U, φ) with $m \in U$ so that*

(*i*) *$\varphi(U) = V \times I \subset \mathbf{R}^{n-1} \times \mathbf{R}$, $V \subset \mathbf{R}^{n-1}$ open, and $I = (-a, a) \subset \mathbf{R}$, $a > 0$;*

(*ii*) *$\varphi^{-1}|\{v\} \times I: I \to M$ is an integral curve of X at $\varphi^{-1}(v, 0)$, for all $v \in V$;*

(*iii*) *the local representative X_φ has the form $X_\varphi(y, \lambda) = (y, \lambda; \mathbf{0}, 1)$.*

ISBN 0-8053-0102-X

Proof. If $X(m)\neq 0$, there exists a local chart (U_0, ω) of M at m such that $X(m')\neq 0$ for all $m'\in U_0, \omega(U_0)=U_0'\subset \boldsymbol{R}^n$, and $\omega(m)=\mathbf{0}$. Let α be a linear isomorphism such that $\alpha(X_\omega(m))=(0,\ldots,0,1)$, where X_ω is the local representative of X relative to $(\omega, T\omega)$. Let $\psi=\alpha\circ\omega$. Then (U_0,ψ) is a local chart at $m\in M$ and $X_\psi(\mathbf{0})=(0,\ldots,0,1)$. Now let (U_1', b, F) be a flow box of X_ψ at $\mathbf{0}$, where $U_1'=V_0\times I_c\subset \boldsymbol{R}^{n-1}\times \boldsymbol{R}\approx \boldsymbol{R}^n, I_c=(-c,c), c>0$, and $F(U_1'\times I_b)\subset U_0'$. If $f_0=F|(V_0\times\{0\})\times I_b\colon V_0\times I_b\to U_0'$, we see that $Df_0(\mathbf{0},0)$ is a linear isomorphism because $X_\psi(\mathbf{0})=(0,\ldots,0,1)$. (The map f_0 substitutes the "time coordinate" λ for the last coordinate in $\boldsymbol{R}^n$.) By the inverse mapping theorem there is an open neighborhood $V\times I_a$ of $(\mathbf{0},0)\in V_0\times I_b$ such that $f=f_0|V\times I_a$ is a diffeomorphism onto an open set $U'\subset U_0'$. Let $U=\psi^{-1}(U')$ and $\varphi=f^{-1}\circ\psi$. Then (U,φ) is a chart at $m\in M$ with $\varphi(U)=V\times I_a\subset \boldsymbol{R}^{n-1}\times\boldsymbol{R}\approx\boldsymbol{R}^n$. By construction, $\varphi^{-1}=\psi^{-1}\circ f$, so $\varphi^{-1}|\{v\}\times I_a$ is an integral curve.

To prove (iii) from (ii), let $c=\varphi^{-1}|\{y\}\times I_a$ be the integral curve of X at $\varphi^{-1}(y,0)=m'\in U$. Then $X_\varphi(y,\lambda)=(T\varphi\circ X\circ\varphi^{-1})(y,\lambda)=T\varphi(X(\varphi^{-1}(y,\lambda)))$ $=T\varphi(c'(\lambda))=(\varphi\circ c)'(\lambda)=(y,\lambda;\mathbf{0},1)$ since $(\varphi\circ c)(t)=(\varphi\circ\varphi^{-1})(y,t)=(y,t)$. ■

At this point we may relate the flowbox idea to the classical notion of the complete solution of a dynamical system by means of a "complete set of integrals."

2.1.10 Definition. *Let X be a vector field on a manifold M. A* ***complete solution*** *of X is a triple (V, b, Ψ), where $V\subset M$ is an open set, $b\in\boldsymbol{R}$, $b>0$ or $b=+\infty$, $I_b=(-b,b)$, $n=\dim(M)$, and*

$$\Psi\colon V\times I_b\to\boldsymbol{R}^n$$

such that if $\Psi(u_0,0)=c\in\boldsymbol{R}^n$, then

$$\{u\in V|\Psi(u,t)=\mathbf{c}\}$$

is an integral curve of X at u_0. The component functions of a complete solution

$$\Psi(u,t)=(\psi_1(u,t),\ldots,\psi_n(u,t))$$

are known as a ***complete system of integrals*** *of X in the domain V.*

Note that the integral curves of X are defined by the n equations

$$\psi_i(u,t)=c_i,\qquad i=1,\ldots,n$$

and if M is a local manifold, $M\subset\boldsymbol{R}^n$, we may write

$$\psi_i(u_1,\ldots,u_n,t)=c_i,\qquad i=1,\ldots,n$$

which is the classical form.

ISBN 0-8053-0102-X

The existence of complete solutions is provided by the Flowbox Theorem 2.1.8.

2.1.11 Theorem. *Let X be a vector field on a manifold M, and $m \in M$. Then there is a complete solution of X, (V, b, Ψ), with $m \in V$.*

Proof. Let (U, φ) be a chart in M with $m \in U$, and let (U_0, a, F) be a flow box at m, $F_t = F|U_0 \times \{t\}$, and $U_t = F_t(U_0)$, chosen such that $U_0 \subset U$. As in the proof of 2.1.8, there is a $b \in \boldsymbol{R}$, $b > 0$, and an open neighborhood V of $m \in M$, such that

$$V \subset U_t \subset U \qquad \text{for all} \qquad t \in I_b$$

Now for each $t \in I_b$, we may define a modified coordinate chart on the common domain V by

$$\Psi_t \colon V \to V_t' \subset U' = \varphi(U) \subset \boldsymbol{R}^n$$

where

$$\Psi_t = \varphi \circ F_t^{-1} = \varphi \circ F_{-t}$$

Looking backwards from a point $c = \Psi_0(v)$ in $\boldsymbol{R}^n$, $t \mapsto \Psi_t^{-1}(c)$ is an integral curve at v, so (V, b, Ψ) is a complete solution of X, with $m \in X$, if we define

$$\Psi \colon V \times I_b \to \boldsymbol{R}^n; \qquad (v, t) \mapsto \Psi_t(v)$$

Note that Ψ_t is a diffeomorphism. ■

2.1.12 Corollary. *If (V, b, Ψ) is a complete solution for a vector field X on M, then*

$$\Psi_t \colon V \to V_t' \subset \boldsymbol{R}^n; \qquad v \mapsto \Psi(v, t)$$

is a diffeomorphism for each $t \in I_b$.

Now we shall turn our attention from local flows to global considerations. These ideas center on considering the flow of a vector field as a whole, extended as far as possible in the λ-variable.

2.1.13 Definition *Given a manifold M and a vector field X on M, let $\mathfrak{D}_X \subset M \times \boldsymbol{R}$ be the set of $(m, \lambda) \in M \times \boldsymbol{R}$ such that there is an integral curve $c \colon I \to M$ of X at m with $\lambda \in I$. The vector field X is* **complete** *if $\mathfrak{D}_X = M \times \boldsymbol{R}$. Also, a point $m \in M$ is called* **σ complete,** *where $\sigma = +$, $-$, or $\pm$, if $\mathfrak{D}_X \cap (\{m\} \times \boldsymbol{R})$ contains all (m, t) for $t > 0$, $t < 0$, or $t \in \boldsymbol{R}$, respectively.*

Thus, X is complete iff each integral curve can be extended so that its domain becomes $(-\infty, \infty)$.

ISBN 0-8053-0102-X

2.1.14 Examples. For $M = \boldsymbol{R}^2$, let X be the constant vector field, (0, 1). Then X is complete. On $\boldsymbol{R}^2 \setminus \{\boldsymbol{0}\}$, the same vector field X is not complete. For $M = \boldsymbol{R}$, let X be the vector field for which

$$c: (-\pi/2, \pi/2) \to \boldsymbol{R}: \theta \mapsto c(\theta) = \tan \theta$$

is the integral curve. Then X is not complete. [Here $X(x) = (1 + x^2)$.]

2.1.15 Proposition. *Let M be a manifold and $X \in \mathfrak{X}(M)$. Then*

(*i*) $\mathfrak{D}_X \supset M \times \{0\}$;
(*ii*) $\mathfrak{D}_X$ *is open in* $M \times \boldsymbol{R}$;
(*iii*) *there is a unique mapping* $F_X : \mathfrak{D}_X \to M$ *such that the mapping* $t \mapsto F_X(m, t)$ *is an integral curve at* m, *for all* $m \in M$.

Proof. Parts (i) and (ii) follow at once from the flow box existence theorem, and (iii) by the uniqueness of integral curves. ■

Thus, F_X is smooth, and for X complete, (M, ∞, F_X) is a flow box.

2.1.16 Definition. *Let M be a manifold and $X \in \mathfrak{X}(M)$. Then the mapping F_X is called the **integral** of X, and the curve $t \mapsto F_X(m, t)$ is called the **maximal integral curve** of X at m. In case X is complete, F_X is called the **flow** of X.*

Thus, if X is complete with flow F, then the set $\{F_t | t \in \boldsymbol{R}\}$ is a group of diffeomorphisms on M, sometimes called a **one-parameter group of diffeomorphisms.** The following is a useful criterion for completeness.

2.1.17 Proposition. *Suppose M is a compact manifold and X is a vector field on M. Then X is complete.*

In fact, we shall prove the following more general result.

2.1.18 Theorem. *Let X be C^k, $1 \leqslant k \leqslant \infty$. Let $c(\lambda)$ be a maximal integral curve of X such that for every open finite interval (a, b) in the domain of c, $c[(a, b)]$ lies in a compact subset of M. Then c is defined for all $t \in \boldsymbol{R}$.*

Proof. It suffices to show that $a \in I$, $b \in I$, where I is the interval of definition of c. Let $\lambda_n \in (a, b)$, $\lambda_n \to b$. By compactness we can assume some subsequence $c(\lambda_{n_k})$ converges, say, to a point x in M. Since the domain of the flow is open, it contains a neighborhood of $(x, 0)$. So there are $\varepsilon > 0$ and $\tau > 0$ such that integral curves starting at points [such as $c(\lambda_{n_k})$ for large k] closer than ε to x persist foı a time longer than τ. This serves to extend c to a time greater than b, so $b \in I$ since c is maximal. Similarly, $a \in I$. ■

2.1.19 Corollary. *A vector field with compact support on a manifold M is complete.*

ISBN 0-8053-0102-X

Completeness is often stressed in the literature since it corresponds to well-defined dynamics persisting eternally. In many circumstances (shock waves, singularities in elasticity, general relativity, etc.) one has to live with incompleteness. However, because of its importance we give two additional criteria for it. These results look ahead slightly in notation and can be returned to later. In the first we use the notation $X(f)=df\cdot X$ for the derivative of f in the direction X. The second uses ideas from Riemannian geometry, and may be deferred until Sect. 2.7 is read.

2.1.20 Proposition. *Suppose X is a C^k vector field on M, $k\geq 1$, and $f\colon M\to R$ is a C^1 **proper** map (that is, the inverse images of compact sets are compact). Suppose X admits the estimate*

$$|X(f)(m)|\leq A|f(m)|+B \quad \textit{for} \quad A,B\geq 0,\ m\in M$$

Then the flow of X is complete.

Proof. From the chain rule we have $(d/d\lambda)f(F_\lambda(m))=X(f)(F_\lambda(m))$ so that

$$f(F_\lambda(m))-f(m)=\int_0^\lambda X(f)(F_\tau(m))\,d\tau$$

Applying the hypothesis and Gronwall's inequality we see that $|f(F_\lambda(m))|$ is bounded on any finite λ-interval, so as f is proper, $F_\lambda(m)$ lies in a compact set. Hence 2.1.18 applies. ■

The next theorem gives a criterion in terms of a metric on M.

2.1.21 Proposition. *Let M be a complete Riemannian manifold and X a C^k vector field, $k\geq 1$. Let σ be any integral curve of X. Assume $\|X(\sigma(\lambda))\|_{\sigma(\lambda)}$ [the norm at the point $\sigma(\lambda)$] is bounded on finite λ-intervals. Then the flow of X is complete.*

Proof. Suppose $\|X(\sigma(\lambda))\|_{\sigma(\lambda)}\leq A$ for $\lambda\in(a,b)$. Let $\lambda_n\to b$ and let d be the metric induced on M from the Riemannian structure. For $\lambda_n<\lambda_m$ we have

$$d(\sigma(\lambda_n),\sigma(\lambda_m))\leq\int_{\lambda_n}^{\lambda_m}\|\sigma'(\lambda)\|\,d\lambda=\int_{\lambda_n}^{\lambda_m}\|X(\sigma(\lambda))\|\,d\lambda\leq A|\lambda_m-\lambda_n|$$

Hence $\sigma(\lambda_n)$ is a Cauchy sequence and therefore, converges. Now argue as in 2.1.18. ■

Later on in Chapter 7 we discuss stability questions rather extensively. However, it is convenient to include some of the more basic ideas and the useful spectral criterion of Liapunov here.

ISBN 0-8053-0102-X

2.1.22 Definition. *Let X be a C^1 vector field on a manifold M. A point m_0 is called a **critical point** (also called a **singular point** or an **equilibrium point**) of X if $X(m_0)=0$. The **linearization** of X at a critical point m_0 is the linear map*

$$X'(m_0): T_{m_0}M \to T_{m_0}M$$

defined by

$$X'(m_0)\cdot v = \frac{d}{d\lambda}(TF_\lambda(m_0)\cdot v)\Big|_{\lambda=0}$$

*where F is the flow of X. The eigenvalues of $X'(m_0)$ are called the **characteristic exponents** of X at m_0.*

Some remarks will clarify this definition. First of all, F_λ leaves m_0 fixed: $F_\lambda(m_0)=m_0$ since $c(\lambda)\equiv m_0$ is the unique integral curve through m_0. Conversely, it is obvious that if $F_\lambda(m_0)=m_0$ for all λ, then m_0 is a critical point. Thus $TF_\lambda(m_0)$ is a linear map of $T_{m_0}M$ to itself and so its λ derivative at 0, producing another linear map of $T_{m_0}M$ to itself, makes sense.

Computationally, the definition above is not so convenient. The following is useful.

2.1.23 Proposition. *Let m_0 be a critical point of X and let (U,φ) be a chart on M with $\varphi(m_0)=x_0\in \mathbf{R}^n$. Let $x=(x^1,\ldots,x^n)$ denote coordinates in $\mathbf{R}^n$ and $X^1(x^1,\ldots,x^n),\ldots,X^n(x^1,\ldots,x^n)$ the components of the local representative of X. Then the matrix of $X'(m_0)$ in these coordinates is*

$$\left(\frac{\partial X^i}{\partial x^j}\right)_{x=x_0}$$

Proof. This follows at once from the equations

$$X^i(F_\lambda^j(x)) = \frac{d}{d\lambda}F_\lambda^i(x)$$

after differentiating in x and setting $x=x_0$, $\lambda=0$. (See also the linearized equations in Lemma 4 of 2.1.2.) ■

One can also define $X'(m_0)$ directly without reference to the flow by noting that if $0\in T_{m_0}M$ represents the zero vector, then $T_0(TM)$ splits into $T_{m_0}M\oplus T_{m_0}M$, a horizontal and vertical space. Then $X'(m_0)$ is the vertical projection of $TX(m_0)$. The reader can work out the details of this case himself.

ISBN 0-8053-0102-X

The name characteristic exponent arises as follows. We have the linear differential equation

$$\frac{d}{d\lambda} T_{m_0}F_\lambda = X'(m_0) \circ T_{m_0}F_\lambda$$

and so

$$T_{m_0}F_\lambda = e^{\lambda X'(m_0)}$$

Here the exponential is defined, for example, by a power series. The actual computation of these exponentials is learned in differential equations courses, using the Jordan canonical form. (See Hirsch–Smale [1974], for instance.) In particular, if $\mu_1,\ldots,\mu_n$ are the characteristic exponents of X at m_0, the eigenvalues of $T_{m_0}F_\lambda$ are $e^{\lambda\mu_1},\ldots,e^{\lambda\mu_n}$.

2.1.24 Definition. *Let m_0 be a critical point of X. Then;*

(*i*) *m_0 is **stable** (or Liapunov stable) if for any neighborhood U of m_0, there is a neighborhood V of m_0 such that if $m \in V$, then m is + complete and $F_\lambda(m) \in U$ for all $\lambda \geq 0$. [See Fig. 2.1-2(a).]*

(*ii*) *m_0 is **asymptotically stable** if there is a neighborhood V of m_0 such that if $m \in V$, then m is + complete, $F_t(V) \subset F_s(V)$ if $t > s$ and*

$$\lim_{t\to+\infty} F_t(V) = \{m_0\}$$

[i.e., for any neighborhood U of m_0, there is a T such that $F_t(V) \subset U$ if $t \geq T$]. [See Fig. 2.1-2(b).]

It is obvious that asymptotic stability implies stability. The harmonic oscillator

$$\ddot{x} = -x$$

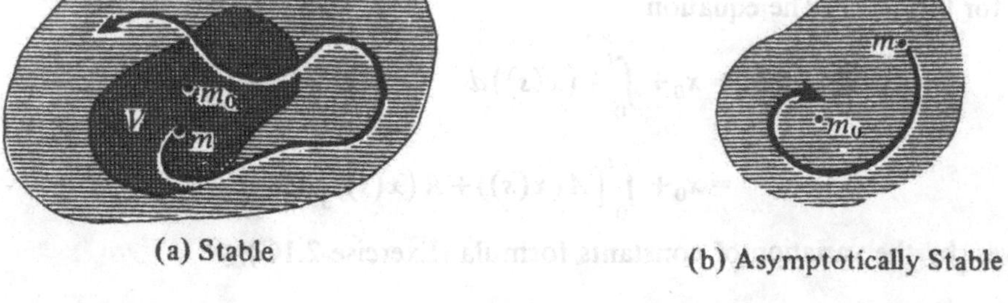

(a) Stable (b) Asymptotically Stable

Figure 2.1-2 (a) Stable. (b) Asymptotically stable.

ISBN 0-8053-0102-X

giving a flow in the plane shows that stability need not imply asymptotic stability.

The following result of Liapunov is basic.

2.1.25 Theorem. *Suppose X is C^1 and m_0 is a critical point of X. Assume the characteristic exponents of m_0 have strictly negative real parts. Then m_0 is asymptotically stable. [In a similar way, if $Re(\mu_i)>0$, m_0 is asymptotically unstable; i.e., asymptotically stable as $t\to-\infty$.]*

Proof. We can assume $M=E$ is a linear space and that $m_0=\mathbf{0}$. Let $-\varepsilon>r=\max(Re\lambda_1,\ldots,Re\lambda_n)$. Then we claim there is a norm $\|\cdot\|$ on E in which

$$\|e^{tX'(\mathbf{0})}\|\leqslant e^{-\varepsilon t}$$

If $X'(\mathbf{0})$ is diagonalizable (e.g., has distinct eigenvalues) as a complex matrix, this is easy, for we can let $\|\quad\|$ be the sup norm associated with a basis of eigenvectors. If $X'(\mathbf{0})$ is not diagonalizable, we can approximate it by one and get the same conclusion. (If the reader is familiar with the spectral radius formula, choose

$$\|x\|=\sup_{\substack{n>0\\ t>0}}\frac{|||e^{ntX'(0)}(x)|||}{e^{rnt}}$$

where $|||\cdot|||$ is any norm.) Write $A=X'(\mathbf{0})=DX(\mathbf{0})$.

From the local existence theory given in Theorem 2.1.2, there is a r-ball about $\mathbf{0}$ for which the time of existence is uniform if the initial condition x_0 lies in this ball. Let

$$R(x)=X(x)-DX(\mathbf{0})\cdot x$$

Find $r_2\leqslant r$ such that $\|x\|\leqslant r_2$ implies $\|R(x)\|\leqslant\alpha\|x\|$, where $\alpha=\varepsilon/2$.

Let B be the open $r_2/2$ ball about $\mathbf{0}$. We shall show that if $x_0\in B$, then the integral curve starting at x_0 remains in B and $\to\mathbf{0}$ exponentially as $t\to+\infty$. This will prove the result.

Let $x(t)$ be the integral curve of X starting at x_0. Suppose $x(t)$ remains in B for $0\leqslant t<T$. The equation

$$x(t)=x_0+\int_0^t X(x(s))\,ds$$

$$=x_0+\int_0^t\left[A(x(s))+R(x(s))\right]ds$$

gives, by the variation of constants formula (Exercise 2.1G),

$$x(t)=e^{tA}x_0+\int_0^t e^{(t-s)A}R(x(s))\,ds$$

ISBN 0-8053-0102-X

and so

$$\|x(t)\| \leqslant e^{-t\varepsilon}\|x_0\| + \alpha \int_0^t e^{-(t-s)\varepsilon}\|x(s)\|\, ds$$

Letting $f(t) = e^{t\varepsilon}\|x(t)\|$,

$$f(t) \leqslant \|x_0\| + \alpha \int_0^t f(s)\, ds$$

and so, by Gronwall's inequality,

$$f(t) \leqslant \|x_0\| e^{\alpha t}$$

Thus

$$\|x(t)\| \leqslant \|x_0\| e^{(\alpha-\varepsilon)t} = \|x_0\| e^{-\varepsilon t/2}$$

Hence $x(t) \in B$, $0 \leqslant t < T$, so $x(t)$ may be indefinitely extended in t and the above estimate holds. ∎

A critical point m_0 is called *hyperbolic* or *elementary* if none of its characteristic exponents has zero real part. Generalizations of Liapunov's theorem show that near a hyperbolic critical point the flow looks like that of its linearization. (See Sect. 7.2, Hartman [1973] and Nelson [1969], for proofs and for discussions.) In the plane, the possible hyperbolic flows near a critical point are shown in Fig. 2.1-3. (Remember that for real systems, the characteristic exponents occur in complex conjugate pairs.) For Hamiltonian systems, the nonhyperbolic case is common, so we shall have to eventually refine these ideas.

If m is a hyperbolic critical point of a vector field X, the number of eigenvalues (counting multiplicities) with negative real part is called the *index* of m. It is denoted $I(X,m)$. The Poincaré–Hopf index theorem states that if M is compact and X only has (isolated) hyperbolic critical points, then

$$\sum_{\substack{m \text{ is a} \\ \text{critical point of } X}} (-1)^{I(X,m)} = \chi(M)$$

where $\chi(M)$ is the Euler characteristic of M. For a proof and discussion, see Guillemin and Pollack [1974]. We do not really use this result, but it is one the reader should be aware of.

Another result of some importance in both theoretical and numerical work concerns writing a flow in terms of iterates of a known mapping. Let $X \in \mathfrak{X}(M)$ with flow F_t (maximally extended).

ISBN 0-8053-0102-X

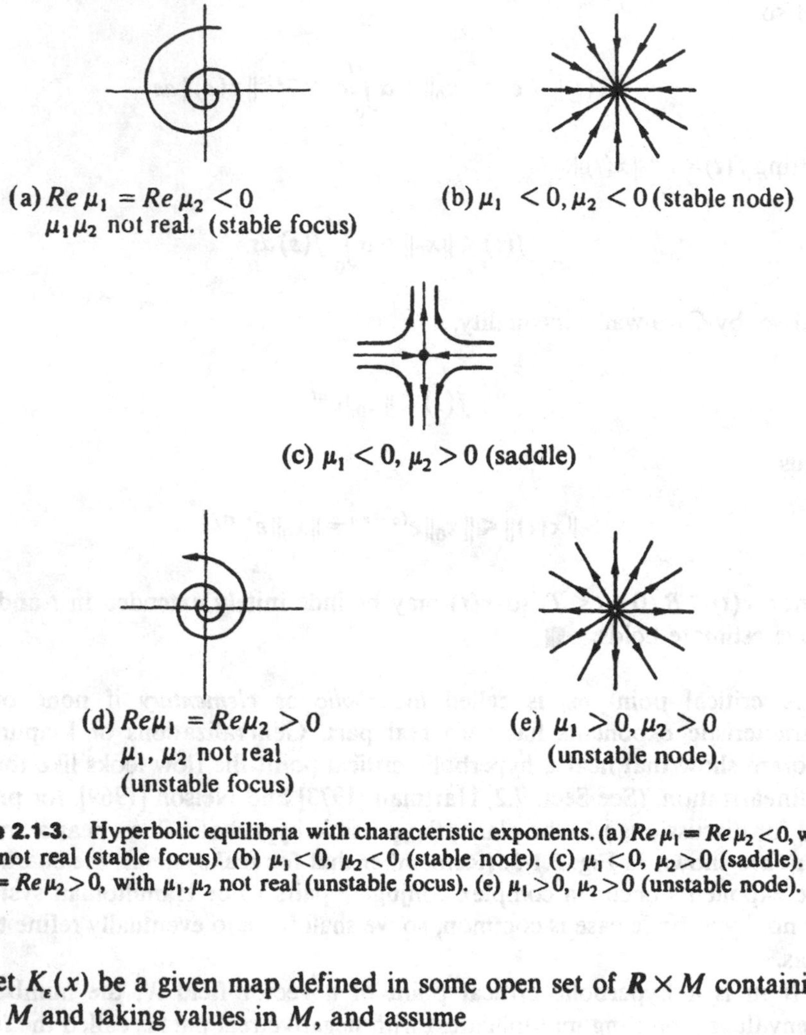

Figure 2.1-3. Hyperbolic equilibria with characteristic exponents. (a) $Re\,\mu_1 = Re\,\mu_2 < 0$, with μ_1, μ_2 not real (stable focus). (b) $\mu_1 < 0$, $\mu_2 < 0$ (stable node). (c) $\mu_1 < 0$, $\mu_2 > 0$ (saddle). (d) $Re\,\mu_1 = Re\,\mu_2 > 0$, with μ_1, μ_2 not real (unstable focus). (e) $\mu_1 > 0$, $\mu_2 > 0$ (unstable node).

Let $K_\varepsilon(x)$ be a given map defined in some open set of $\mathbf{R} \times M$ containing $\{0\} \times M$ and taking values in M, and assume

(i) $K_0(x) = x$
(ii) $K_\varepsilon(x)$ is C^1 in ε with derivative continuous in (ε, x).

We call K the "algorithm."

2.1.26 Theorem. *Assume that the algorithm $K_\varepsilon(x)$ is **consistent** with X in the sense that*

$$X(x) = \frac{d}{d\varepsilon} K_\varepsilon(x)\Big|_{\varepsilon=0}$$

Then, if (t, x) is in the domain of $F_t(x)$, $K^n_{t/n}(x)$ is defined for n sufficiently large and converges to $F_t(x)$ as $n \to \infty$. Conversely, if $K^n_{t/n}(x)$ is defined and converges for $0 \leq t \leq T$, then (T, x) is in the domain of F and the limit is $F_t(x)$.

ISBN 0-8053-0102-X

Proof. First, we prove that convergence holds locally. We begin by showing that for any x_0, the iterates $K^n_{t/n}(x_o)$ are defined if t is sufficiently small. Indeed, on a neighborhood of x_0, $K_\varepsilon(x)=x+O(\varepsilon)$, so if $K^j_{t/j}(x)$ is defined for x in a neighborhood of x_0, $j=1,\ldots,n-1$,

$$\begin{aligned}K^n_{t/n}(x)-x &= \left(K^n_{t/n}x-K^{n-1}_{t/n}x\right)+\left(K^{n-1}_{t/n}-K^{n-2}_{t/n}x\right)\\ &\quad+\cdots+\left(K_{t/n}(x)-x\right)\\ &=O(t/n)+\cdots+O(t/n)\\ &=O(t)\end{aligned}$$

This is small, independent of n for t sufficiently small; so, inductively, $K^n_{t/n}(x)$ is defined and remains in a neighborhood of x_0 for x near x_0.

Let β be a local Lipschitz constant for X so that $\|F_t(x)-F_t(y)\|\leqslant e^{\beta|t|}\times\|x-y\|$. Now write

$$\begin{aligned}F_t(x)-K^n_{t/n}(x) &= F^n_{t/n}(x)-K^n_{t/n}(x)\\ &=F^{n-1}_{t/n}F_{t/n}(x)-F^{n-1}_{t/n}K_{t/n}(x)\\ &\quad+F^{n-2}_{t/n}F_{t/n}(y_1)-F^{n-2}_{t/n}K_{t/n}(y_1)\\ &\quad+\cdots+F^{n-k}_{t/n}F_{t/n}(y_{k-1})-F^{n-k}_{t/n}K_{t/n}(y_{k-1})\\ &\quad+\cdots+F_{t/n}(y_{n-1})-K_{t/n}(y_{n-1})\end{aligned}$$

where $y_k=K^k_{t/n}(x)$. Thus

$$\begin{aligned}\|F_t(x)-K^n_{t/n}(x)\| &\leqslant \sum_{k=1}^{n} e^{\beta|t|(n-k)/n}\|F_{t/n}(y_{n-k-1})-K_{t/n}(y_{n-k-1})\|\\ &\leqslant ne^{\beta|t|}o(t/n)\to 0 \quad\text{as}\quad n\to\infty\end{aligned}$$

since $F_\varepsilon(y)-K_\varepsilon(y)=o(\varepsilon)$ by the consistency hypothesis.

Now suppose $F_t(x)$ is defined for $0\leqslant t\leqslant T$. We shall show $K^n_{t/n}(x)$ converges to $F_t(x)$. By the above proof and compactness, if N is large enough, $F_{t/N}=\lim_{n\to\infty}K^n_{t/nN}$ uniformly on a neighborhood of the curve $t\mapsto F_t(x)$. Thus, for $0\leqslant t\leqslant T$,

$$F_t(x)=F^N_{t/N}(x)=\lim_{n\to\infty}\left(K^n_{t/nN}\right)^N(x)$$

By uniformity in t,

$$F_T(x)=\lim_{j\to\infty}K^j_{T/j}(x)$$

Conversely, let $K^n_{t/n}(x)$ converge to a curve $c(t)$, $0\leqslant t\leqslant T$. Let $S=\{t\,|\,F_t(x)$ is defined and $c(t)=F_t(x)\}$. From the local result, S is a nonempty open set.

ISBN 0-8053-0102-X

Let $t_k \in S$, $t_k \to t$. Thus $F_{t_k}(x)$ converges to $c(t)$, so by local existence theory, $F_t(x)$ is defined, and by continuity, $F_t(x) = c(t)$. Hence $S = [0, T]$ and the proof is complete. ■

2.1.27 Corollary. *Let $X, Y \in \mathfrak{X}(M)$ with flows F_t and G_t. Let H_t be the flow of $X+Y$. Then*

$$H_t(x) = \lim_{n\to\infty} (F_{t/n} \circ G_{t/n})^n(x)$$

(Each side is defined if and only if the other is.)

Proof. Let $K_\varepsilon(x) = F_\varepsilon \circ G_\varepsilon(x)$ and use 2.1.26. ■

These results had their historical origins in Lie Group theory (see Exercise 4.1J). The above proofs were inspired by Nelson [1969] and Chorin et al [1978].

EXERCISES

2.1A. Use the example of Exercise 1.1D to show that the Hausdorff assumption in 2.1.4 cannot be dropped.

2.1B Let M and N be manifolds and $\varphi: M \to N$ a diffeomorphism; suppose $X \in \mathfrak{X}(M)$ and $c: I \to M$ is an integral curve of X at $m \in M$. Then show $\varphi \circ c$ is an integral curve of $\varphi_* X$ at $\varphi(m)$. Show that the flow of $\varphi_* X$ is $\varphi \circ F_\lambda \circ \varphi^{-1}$.

2.1C Let $F(m,\lambda)$ be a C^∞ mapping of $\boldsymbol{R} \times M$ to M such that $F_{t+s} = F_t \circ F_s$ and $F_0 =$ identity [where $F_t(m) = F(m,t)$]. Show that there is a unique C^∞ vector field X whose flow is F.

2.1D Let $\sigma(\lambda)$ be an integral curve of a vector field X and let $g: M \to \boldsymbol{R}$. Let $\tau(\lambda)$ satisfy $\tau'(\lambda) = g(\sigma(\tau(\lambda)))$. Then show $\lambda \mapsto \sigma(\tau(\lambda))$ is an integral curve of gX. Show by example that even if X is complete, gX need not be.

2.1E (i) (Gradient Flows) Let $f: \boldsymbol{R}^n \to \boldsymbol{R}$ be C^1 and let $X = (\partial f/\partial x^1, \ldots, \partial f/\partial x^n)$ be the gradient of f. Let F be the flow of X. Show that $f(F_t(x)) \geqslant f(F_s(x))$ if $t \geqslant s$.

(ii) Use (i) to find a vector field X on $\boldsymbol{R}^n$ such that $X(\mathbf{0}) = \mathbf{0}$, $X'(\mathbf{0}) = \mathbf{0}$, yet $\mathbf{0}$ is globally attracting.

2.1F If M is a manifold and ∇ is *any* connection on M, show that at a critical point of X, $X'(m_0) = \nabla X(m_0)$. (For the definition of a connection see Sect. 2.7.)

2.1G. (Variation of Constants Formula) Let $F_t = e^{tX}$ be the flow of a linear vector field X on E. Show that the solution of the equation

$$\dot{x} = X(x) + f(x)$$

with initial condition x_0 satisfies the integral equation

$$x(t) = e^{tX}x_0 + \int_0^t e^{(t-s)X} f(x(s))\, ds$$

2.2 VECTOR FIELDS AS DIFFERENTIAL OPERATORS

In this section we shall show how a vector field X on a manifold induces a differential operator L_X on the full tensor algebra $\mathfrak{T}(M)$, called the *Lie derivative*. Our development of this aspect of vector fields departs from the

ISBN 0-8053-0102-X

spirit of the previous sections in that it is special to the finite-dimensional case. Our definition, inspired by a theorem of Willmore, is adopted for reasons of efficiency. For the definition and the treatment of the infinite-dimensional case we refer the interested reader to Lang [1972]. At the end of this section, however, we show that the two definitions coincide. The Lie derivative seems to have first been introduced in connection with mechanics by Slebodzinski [1931].

We shall begin by defining L_X on $\mathfrak{F}(M)$ and $\mathfrak{X}(M)$, and then use a unique extension theorem to define L_X on $\mathfrak{T}(M)$.

2.2.1 Definition. *Let $f \in \mathfrak{F}(M)$ so that*

$$Tf: TM \to TR = R \times R$$

and

$$T_m f = Tf | T_m M \in L\big(T_m M, \{f(m)\} \times R\big)$$

*We then define $df: M \to T^*(M)$ by $df(m) = P_2 \circ T_m f$, where P_2 denotes the projection onto the second factor. We call df the **differential** of f.*

*For $X \in \mathfrak{X}(M)$, define $L_X f: M \to R$ by $L_X f(m) = df(m)[X(m)]$. We call $L_X f$ the **Lie derivative of f with respect to X**.*

2.2.2 Proposition. *(i) For $f \in \mathfrak{F}(M)$, $df \in \mathfrak{X}^*(M)$, and for $X \in \mathfrak{X}(M)$, $df(X) = P_2 \circ Tf \circ X$; that is, $df(X)(m) = P_2 \circ T_m f(X(m))$.*

(ii) For $f \in \mathfrak{F}(M)$ and $X \in \mathfrak{X}(M)$ we have $L_X f \in \mathfrak{F}(M)$.

Proof. For (i), we need only to show df is smooth. Let (U, φ) be an (admissible) chart on M so that the local representative of df in the natural charts is $(df)_\varphi = (T\varphi)^0_1 \circ df \circ \varphi^{-1}$: $U' \to U' \times E^*$, where φ: $U \subset M \to U' \subset E$. Then

$$(df)_\varphi(u') \cdot e = (T_u \varphi)^0_1 \circ df(u) \cdot e = (T_u \varphi)^0_1 \circ P_2 \circ T_u f \cdot e = P_2 \circ T_u f\big(T_{u'} \varphi^{-1} \cdot e\big)$$

$$= P_2 \circ T_{u'}\big(f \circ \varphi^{-1}\big) \cdot e = D\big(f \circ \varphi^{-1}\big)(u') \cdot e,$$

by the composite mapping theorem. Hence $(df)_\varphi$ is of class C^∞ and (i) is established. Then (ii) follows at once for $L_X f = df(X)$. ■

2.2.3 Proposition. *(i) Suppose φ: $M \to N$ is a diffeomorphism. Then L_X is **natural with respect to push-forward** by φ. That is, for each $f \in \mathfrak{F}(M)$,*

ISBN 0-8053-0102-X

$L_{\varphi_* X}(\varphi_* f) = \varphi_* L_X f$; or the following diagram commutes:

$$\begin{array}{ccc} \mathfrak{F}(M) & \xrightarrow{\varphi_*} & \mathfrak{F}(N) \\ {\scriptstyle L_X}\downarrow & & \downarrow{\scriptstyle L_{\varphi_* X}} \\ \mathfrak{F}(M) & \xrightarrow[\varphi_*]{} & \mathfrak{F}(N) \end{array}$$

*(ii) L_X is **natural with respect to restrictions**. That is, for U open in M and $f \in \mathfrak{F}(M)$, $L_{X|U}(f|U) = (L_X f)|U$; or, if $|U: \mathfrak{F}(M) \to \mathfrak{F}(U)$ denotes restriction to U, the following diagram commutes:*

$$\begin{array}{ccc} \mathfrak{F}(M) & \xrightarrow{|U} & \mathfrak{F}(U) \\ {\scriptstyle L_X}\downarrow & & \downarrow{\scriptstyle L_{X|U}} \\ \mathfrak{F}(M) & \xrightarrow[|U]{} & \mathfrak{F}(U) \end{array}$$

Proof. For (i), let $n = f(m)$. Then

$$L_{\varphi_* X}(\varphi_* f)(n) = d(f \circ \varphi^{-1}) \cdot \varphi_* X(n) = P_2 \circ T_n(f \circ \varphi^{-1}) \circ T_m \varphi \circ X \circ \varphi^{-1}(n)$$

$$= P_2(T_m f) \cdot X \cdot \varphi^{-1}(n) = \varphi_* L_X f(n)$$

Then (ii) follows from the fact that $d(f|U) = (df)|U$, which is clear from the definition of d. ■

Next we show that L_X has the "Leibniz rule" of derivatives.

2.2.4 Proposition. *(i) $L_X: \mathfrak{F}(M) \to \mathfrak{F}(M)$ is a **derivation** on the algebra $\mathfrak{F}(M)$. That is, L_X is $\mathbf{R}$ linear and for $f, g \in \mathfrak{F}(M)$, $L_X(fg) = (L_X f)g + f(L_X g)$.*

(ii) If c is a constant function, $L_X c = 0$.

Proof. By 2.2.3 (ii) it is enough to verify (i) in a chart (U, φ). Then the local representative of $L_X(fg)$ is

$$L_X(fg) \circ \varphi^{-1}(u') = (d(fg)X) \circ \varphi^{-1}(u')$$

$$= P_2 T_u(fg) T_{u'}(\varphi^{-1}) \cdot T_u \varphi X \varphi^{-1}(u')$$

$$= D_{u'}((fg) \circ \varphi^{-1}) \cdot X\varphi(u')$$

by the proof of 2.2.2(i). But $(fg) \circ \varphi^{-1} = (f \circ \varphi^{-1})(g \circ \varphi^{-1})$ and the result

ISBN 0-8053-0102-X

follows at once from 1.3.9. The result (ii) is a general property of derivations. Let 1 be the constant function with value 1. Then $L_X(1) = L_X(1^2) = 1 \cdot L_X 1 + 1 \cdot L_X 1$. Hence $L_X(1) = 0$. Then $L_X(c) = L_X(c \cdot 1) = cL_X(1) = 0$ by $\boldsymbol{R}$ linearity of L_X. ■

2.2.5 Corollary. *For $f, g \in \mathcal{F}(M)$ we have $d(fg) = (df)g + f(dg)$, and if c is constant, $dc = 0$.*

We saw in Sect. 1.7 that the tensor product has a natural extension to $\mathcal{T}(M)$. Then 2.2.4 and 2.2.5 become

$$L_X(f \otimes g) = L_X f \otimes g + f \otimes L_X g \in \mathcal{F}(M)$$

$$d(f \otimes g) = df \otimes g + f \otimes dg \in \mathfrak{X}^*(M)$$

2.2.6 Proposition. *If $\alpha_m \in T_m^* M$, there is an $f \in \mathcal{F}(M)$ such that $df(m) = \alpha_m$.*

Proof. If $M = \boldsymbol{R}^n$, so $T_m \boldsymbol{R}^n \cong \boldsymbol{R}^n$, let $f(x) = \alpha_m(x)$, a linear function on $\boldsymbol{R}^n$. Then df is constant and equals α_m.

The general case can be reduced to $\boldsymbol{R}^n$ using a local chart and a *bump function*; the latter is described as follows:

2.2.7 Lemma. *In $\boldsymbol{R}^n$, let U_1 be an open ball of radius r_1 about x_0 and U_2 an open ball of radius r_2, $r_1 < r_2$. Then there is a C^∞ function $h: \boldsymbol{R}^n \to \boldsymbol{R}$ such that h is one on U_1 and zero outside U_2. We call h a **bump function**. [In 2.5.3 we will prove more generally that on a manifold M, if U_1 and U_2 are two open sets with $cl(U_1) \subset U_2$, there is an $h \in \mathcal{F}(M)$ such that h is one on U_1 and is zero outside U_2.]*

Proof. By a scaling and translation, we can assume U_1 and U_2 are of radii 1 and 3 and centered at the origin. Let $\theta: \boldsymbol{R} \to \boldsymbol{R}$ be given by

$$\theta(x) = \begin{cases} exp(-1/(1-|x|^2)), & |x| < 1 \\ 0, & |x| \geqslant 1 \end{cases}$$

(See the remarks following 1.3.8.) Now set

$$\theta_1(s) = \frac{\int_{-\infty}^{s} \theta(t)\, dt}{\int_{-\infty}^{\infty} \theta(t)\, dt}$$

so $\theta_1(s)$ is a C^∞ function, 0 if $s < -1$, and 1 if $s > 1$. Let

$$\theta_2(s) = \theta_1(s-2)$$

ISBN 0-8053-0102-X

so θ_2 is a C^∞ function that is 1 if $s<1$ and 0 if $s>3$. Finally, let

$$h(\mathbf{x})=\theta_2(\|\mathbf{x}\|) \quad ▼$$

To complete 2.2.6, let $\varphi: U \to U' \subset \boldsymbol{R}^n$ be a local chart at m with $\varphi(m)=\mathbf{0}$ and such that U' contains the ball of radius 3. Let $\tilde{\alpha}_m$ be the local representative of α_m and let h be a bump function 1 on the ball of radius 1 and zero outside the ball of radius 2. Let $\tilde{f}(x)=\tilde{\alpha}_m(x)$ and let

$$f=\begin{cases}(h\tilde{f}\,)\circ\varphi, & \text{on } U\\ 0, & \text{on } M\setminus U\end{cases}$$

It is easily verified that f is C^∞ and $df(m)=\alpha_m$. ■

We saw in Sect. 1.7 that tensor fields can be regarded as $\mathfrak{F}(M)$ multilinear maps of $\mathfrak{X}^*(M)$, $\mathfrak{X}(M)$ into $\mathfrak{F}(M)$. Actually, this association is an isomorphism, according to the following.

2.2.8 Theorem. *$\mathfrak{T}^r_s(M)$ is isomorphic to the $\mathfrak{F}(M)$ multilinear maps from $\mathfrak{X}^*(M)\times\cdots\times\mathfrak{X}(M)$ into $\mathfrak{F}(M)$, regarded as $\mathfrak{F}(M)$ modules or as real vector spaces.*

Proof. We consider then the map $\mathfrak{T}^r_s(M)\to L^{r+s}_{\mathfrak{F}(M)}(\mathfrak{X}^*(M),\ldots,\mathfrak{X}(M);\mathfrak{F}(M))$ given by $l(\alpha^1,\ldots,\alpha^r,X_1,\ldots,X_s)(m)=l(m)(\alpha^1(m),\ldots,X_s(m))$. This map is clearly $\mathfrak{F}(M)$ linear. To show it is an isomorphism, given such a multilinear map l, define t by $t(m)(\alpha^1(m),\ldots,X_s(m))=l(\alpha^1,\ldots,X_s)(m)$. To show this is well defined we must show that, for each $v_0\in T_m(M)$, there is an $X\in\mathfrak{X}(M)$ such that $X(m)=v_0$, and similarly for dual vectors. Let (U,φ) be a chart at m and let $T_m\varphi(v_0)=(\varphi(m),v_0')$. Define $Y\in\mathfrak{X}(U')$ by $Y(u')=(u',v_0')$ on a neighborhood V_1 of $\varphi(m)$. Extend Y to U' so Y is zero outside V_2, where $cl(V_1)\subset V_2$, $cl(V_2)\subset U'$, by means of a bump function. Then X is defined by $X_\varphi=Y$ on U, and $X=0$ outside U. Then $X(m)=v_0$. The construction is similar for dual vectors. Also, $t(m)$ so defined is C^∞; indeed, using the chart φ we have

$$t(m)=t(m)\big(\underline{\alpha}^{i_1}(m),\ldots,\underline{e}_{j_s}(m)\big)\underline{e}_{i_1}(m)\otimes\cdots\otimes\underline{\alpha}^{j_s}(m)$$

where $\{e_1,\ldots,e_n\}$ is a basis of $\boldsymbol{R}^n\supset\varphi(U)$, $\underline{e}_i(m)=(T_m\varphi)^{-1}(\varphi(m),e_i)$ and $\underline{\alpha}^i$ is dual to $\underline{e}_i$. Since $t(m)(\underline{\alpha}^{i_1}(m),\ldots,\underline{e}_{j_s}(m))$, $\underline{e}_i(m)$, and $\underline{\alpha}^j(m)$ are smooth in m, so is t itself. ■

Returning to the Lie derivative, we have the following property, which is often taken as an alternative definition for $\mathfrak{X}(M)$; see our remarks at the beginning of Sect. 1.6.

ISBN 0-8053-0102-X

2.2.9 Proposition. *The collection of operators L_X on $\mathfrak{F}(M)$ forms a real vector space and $\mathfrak{F}(M)$ module, with $(fL_X)(g)=f(L_Xg)$, and is isomorphic to $\mathfrak{X}(M)$ as a real vector space and as an $\mathfrak{F}(M)$ module. In particular, $L_X=0$ iff $X=0$; and $L_{fX}=fL_X$.*

Proof. Consider the map $\sigma: X\mapsto L_X$. It is obviously $\mathbf{R}$ and $\mathfrak{F}(M)$ linear, as

$$L_{X_1+fX_2}=L_{X_1}+fL_{X_2}$$

To show that it is one-to-one, we must show $L_X=0$ implies $X=0$. But if $L_Xf(m)=0$, then $df(m)X(m)=0$ for all f. Hence, $\alpha_m X(m)=0$ for all $\alpha_m\in T_m^*(M)$. Thus $X(m)=0$ ■

2.2.10 Theorem. *The collection of all ($\mathbf{R}$ linear) derivations on $\mathfrak{F}(M)$ form a real vector space isomorphic to $\mathfrak{X}(M)$ as a real vector space. In particular, for each derivation $\boldsymbol{\theta}$ there is a unique $X\in\mathfrak{X}(M)$ such that $\boldsymbol{\theta}=L_X$.*

Proof. It suffices to prove the last assertion. First of all, we note that $\boldsymbol{\theta}$ is a *local operator*; that is, if $h\in\mathfrak{F}(M)$ vanishes on a neighborhood V of m, then $\boldsymbol{\theta}(h)(m)=0$. Indeed, let g be a bump function equal to one on a neighborhood of m and zero outside V. Thus $h=(1-g)h$ and so

$$\boldsymbol{\theta}(h)(m)=\boldsymbol{\theta}(1-g)(m)\cdot h(m)+\boldsymbol{\theta}(h)(m)(1-g(m))=0$$

If U is an open set in M, and $f\in\mathfrak{F}(U)$, define $(\boldsymbol{\theta}|U)(f)(m)=\boldsymbol{\theta}(gf)(m)$, where g is a bump function equal to one on a neighborhood of m and zero outside U. By the previous remark, $(\boldsymbol{\theta}|U)(f)(m)$ is independent of g, so $\boldsymbol{\theta}|U$ is well defined. For convenience we write $\boldsymbol{\theta}=\boldsymbol{\theta}|U$.

Let (U,φ) be a chart on M, $m\in U$, and $f\in\mathfrak{F}(M)$, where $\varphi: U\to U'\subset\mathbf{R}^n$; we can write, for $\mathbf{x}\in U'$ and $\mathbf{a}=\varphi(m)$,

$$(\varphi_*f)(\mathbf{x})=(\varphi_*f)(\mathbf{a})+\int_0^1\frac{\partial}{\partial t}(\varphi_*f)[\mathbf{a}+t(\mathbf{x}-\mathbf{a})]\,dt$$

$$=(\varphi_*f)(\mathbf{a})+\sum_{j=1}^{n}(x^j-a^j)\int_0^1(\varphi_*f)_{,j}[\mathbf{a}+t(\mathbf{x}-\mathbf{a})]\,dt$$

where

$$\mathbf{x}=(x^1,\ldots,x^n),\quad \mathbf{a}=(a^1,\ldots,a^n),\quad\text{and}\quad (\varphi_*f)_{,j}=\frac{\partial(\varphi_*f)}{\partial x^j}$$

This formula holds in some neighborhood $\varphi(V)$ of $\mathbf{a}$. Hence, for $u\in V$ we have

$$f(u)=f(m)+\sum_{i=1}^{n}(\varphi^i(u)-a^i)\,g_i(u)$$

ISBN 0-8053-0102-X

where $g_i \in \mathcal{F}(V)$ and

$$g_i(m) = \left.\frac{\partial(\varphi_* f)}{\partial x^i}\right|_a$$

Hence

$$\begin{aligned}\theta f(m) &= \sum_{i=1}^{n} g_i(m)\theta(\varphi^i)(m)\\ &= \sum_{i=1}^{n} \frac{\partial}{\partial x^i}(\varphi_* f)(a)\theta(\varphi^i)(m)\end{aligned}$$

and this is independent of the chart. Now define X on U by its local representative

$$X_\varphi(x) = (x, \theta(\varphi^1)(u), \ldots, \theta(\varphi^n)(u))$$

where $x = \varphi(u) \in U'$. We leave, as an exercise, that $X|U$ is independent of the chart φ and hence $X \in \mathfrak{X}(M)$. Then, for $f \in \mathcal{F}(M)$, the local representative of $L_X f$ is

$$D(f\circ\varphi^{-1})(x)\cdot X_\varphi(x) = \sum_{i=1}^{n} \frac{\partial}{\partial x^i}(f\circ\varphi^{-1})(x)\theta(\varphi^i)(u) = \theta f(u)$$

Hence $L_X = \theta$. Finally, uniqueness follows from 2.2.6. ■

We may say that the differential operators $\partial/\partial x^i$ in any chart (U,φ) form a basis of the space of derivations at a point m. Hence any vector field can be uniquely represented by

$$L_X f(m) = \sum_{i=1}^{n} \left\{ \left.\frac{\partial(f\circ\varphi^{-1})}{\partial x^i}\right|_{\varphi(m)} \right\} L_X(\varphi^i)(m)$$

2.2.11 Proposition. *If X and Y are vector fields on M, then $[L_X, L_Y] = L_X \circ L_Y - L_Y \circ L_X$ is an ($\mathbf{R}$ linear) derivation on $\mathcal{F}(M)$.*

Proof. More generally, let θ_1 and θ_2 be two derivations on an algebra $\mathcal{F}$. Clearly $[\theta_1, \theta_2] = \theta_1 \circ \theta_2 - \theta_2 \circ \theta_1$ is linear. Also

$$\begin{aligned}[\theta_1,\theta_2](fg) &= \theta_1\circ\theta_2(fg) - \theta_2\circ\theta_1(fg)\\ &= \theta_1((\theta_2 f)g + f(\theta_2 g)) - \theta_2[(\theta_1 f)g + f(\theta_1 g)]\\ &= \theta_1(\theta_2 f)g + (\theta_2 f)(\theta_1 g) + (\theta_1 f)(\theta_2 g) + f\theta_1(\theta_2 g)\\ &\quad - \theta_2(\theta_1 f)g - (\theta_1 f)(\theta_2 g) - (\theta_2 f)(\theta_1 g) - f\theta_2(\theta_1 g)\\ &= ([\theta_1,\theta_2]f)g + f([\theta_1,\theta_2]g)\end{aligned}$$

■

Because of 2.2.10, we can state the following.

ISBN 0-8053-0102-X

2.2.12 Definition. *$[X, Y] = L_X Y$ is the unique vector field such that $L_{[X,Y]} = [L_X, L_Y]$. We call $L_X Y$ the **Lie derivative** of Y **with respect to** X, or **the Lie bracket** of X and Y.*

2.2.13 Proposition. *The composition $[X, Y]$ on $\mathfrak{X}(M)$, together with the real vector space structure of $\mathfrak{X}(M)$, form a **Lie algebra**. That is,*

(*i*) *$[\ ,\ ]$ is **R** bilinear;*
(*ii*) *$[X, X] = 0$ for all $X \in \mathfrak{X}(M)$;*
(*iii*) *$[X, [Y, Z]] + [Y, [Z, X]] + [Z, [X, Y]] = 0$ for all $X, Y, Z \in \mathfrak{X}(M)$.*

Proof. More generally, the derivations on an algebra $\mathfrak{F}$ form a Lie algebra. For them (i), (ii), and (iii) are easily verified by direct computation. The special case 2.2.13 results from 2.2.10 and the definition 2.2.12. ■

Note that (i) and (ii) of 2.2.13 imply that $[X, Y] = -[Y, X]$, for

$$[X+Y, X+Y] = 0 = [X, X] + [X, Y] + [Y, X] + [Y, Y]$$

Also, (iii) may be written in the following suggestive way:

$$L_X[Y, Z] = [L_X Y, Z] + [Y, L_X Z]$$

or, *L_X is a Lie bracket derivation.*

From 2.2.10 it is easy to see in local representation,

$$X_\varphi(x) = (x, X^1(x), \ldots, X^n(x))$$

$$[X, Y]_\varphi = DY_\varphi \cdot X_\varphi - DX_\varphi \cdot Y_\varphi$$

In components,

$$[X, Y]^j_\varphi = X^i \frac{\partial Y^j}{\partial x^i} - Y^i \frac{\partial X^j}{\partial x^i}$$

Strictly speaking we should not use the same symbol L_X for both definitions 2.2.1 and 2.2.12. However, the meaning is generally clear from the context. The analog of 2.2.3 on the vector field level is the following.

2.2.14 Proposition. (*i*) *Let $\varphi: M \to N$ be a diffeomorphism and $X \in \mathfrak{X}(M)$. Then $L_X: \mathfrak{X}(M) \to \mathfrak{X}(M)$ is **natural with respect to push-forward** by φ. That is, $L_{\varphi_* X} \varphi_* Y = \varphi_* L_X Y$, or $[\varphi_* X, \varphi_* Y] = \varphi_* [X, Y]$, or the following diagram commutes:*

$$\begin{array}{ccc} \mathfrak{X}(M) & \xrightarrow{\varphi_*} & \mathfrak{X}(N) \\ {\scriptstyle L_X}\downarrow & & \downarrow{\scriptstyle L_{\varphi_* X}} \\ \mathfrak{X}(M) & \xrightarrow[\varphi_*]{} & \mathfrak{X}(N) \end{array}$$

ISBN 0-8053-0102-X

*(ii) L_X is **natural with respect to restrictions**. That is, for $U \subset M$ open, we have $[X|U, Y|U] = [X, Y]|U$; or the following diagram commutes:*

$$\begin{array}{ccc} \mathfrak{X}(M) & \xrightarrow{\;|U\;} & \mathfrak{X}(U) \\ {\scriptstyle L_X}\downarrow & & \downarrow{\scriptstyle L_{X|U}} \\ \mathfrak{X}(M) & \xrightarrow[\;|U\;]{} & \mathfrak{X}(U) \end{array}$$

Proof. For (i), let $f \in \mathfrak{F}(N)$ and $\varphi(m) = n \in N$. Then

$$\begin{aligned}
[L_{\varphi_*X}, L_{\varphi_*Y}]f(n) &= L_{\varphi_*X}\, df(n)T_m\varphi \circ Y \circ \varphi^{-1}(n) - L_{\varphi_*Y}\, df(n)T_m\varphi \circ X \circ \varphi^{-1}(n) \\
&= L_{\varphi_*X} P_2 T_m(f\circ\varphi)\circ Y \circ \varphi^{-1}(n) \\
&\quad - L_{\varphi_*Y} P_2 T_m(f\circ\varphi)\circ X \circ \varphi^{-1}(n) \\
&= P_2 T_n\big(T_m(f\circ\varphi)\circ Y\circ\varphi^{-1}\big)T_m\varphi\circ X\circ\varphi^{-1}(n) \\
&\quad - P_2 T_n\big(T_m(f\circ\varphi)\circ X\circ\varphi^{-1}\big)T_m\varphi\circ Y\circ\varphi^{-1}(n) \\
&= P_2 T_m\big(T_m(f\circ\varphi)\circ Y\big)\circ X\circ\varphi^{-1}(n) \\
&\quad - P_2 T_m\big(T_m(f\circ\varphi)\circ X\big)\circ Y\circ\varphi^{-1}(n) \\
&= L_{[X,Y]}(f\circ\varphi)(m) = L_{\varphi_*[X,Y]}f(n).
\end{aligned}$$

(ii) follows from the fact that $d(f|U) = df|U$. ■

2.2.15 Proposition. *For $X \in \mathfrak{X}(M)$, L_X is a derivation on $(\mathfrak{F}(M), \mathfrak{X}(M))$. That is, L_X is $\mathbf{R}$ linear on each, and $L_X(f \otimes Y) = L_X f \otimes Y + f \otimes L_X Y$.*

Proof. For $g \in \mathfrak{F}(M)$, we have

$$\begin{aligned}
[X, fY]\,g &= L_X(L_{fY}g) - L_{fY}L_Xg \\
&= L_X(fL_Yg) - fL_YL_Xg \\
&= (L_Xf)L_Yg + fL_XL_Yg - fL_YL_Xg
\end{aligned}$$

by 2.2.4 and 2.2.9, so $[X, fY] = (L_Xf)L_Y + f[X, Y]$ by 2.2.10. ■

Next, we shall develop machinery for extending L_X to the full tensor algebra.

ISBN 0-8053-0102-X

2.2.16 Definition. *A* ***differential operator*** *on the full tensor algebra $\mathfrak{T}(M)$ of a manifold M is a collection $\boldsymbol{D}^r_s(U)$ of maps of $\mathfrak{T}^r_s(U)$ into itself for each r, $s \geqslant 0$ ($\mathfrak{T}^0_0(U) = \mathfrak{F}(U)$) and each open set $U \subset M$, which we denote merely $\boldsymbol{D}$ (the r, s and U to be inferred from the context), such that*

(*DO* 1) *$\boldsymbol{D}$ is a* ***tensor derivation****; that is, $\boldsymbol{D}$ is $\boldsymbol{R}$ linear and for $t_1 \in \mathfrak{T}^{r_1}_{s_1}(M)$, $t_2 \in \mathfrak{T}^{r_2}_{s_2}(M)$: $\boldsymbol{D}(t_1 \otimes t_2) = \boldsymbol{D}t_1 \otimes t_2 + t_1 \otimes \boldsymbol{D}t_2$.*

(*DO* 2) *$\boldsymbol{D}$ is local, or is* ***natural with respect to restrictions****. That is, for $U \subset V \subset M$ open sets, and $t \in \mathfrak{T}^r_s(V)$*

$$(\boldsymbol{D}t)|U = \boldsymbol{D}(t|U) \in \mathfrak{T}^r_s(U)$$

or the following diagram commutes:

$$\begin{array}{ccc} \mathfrak{T}^r_s(V) & \xrightarrow{\;|U\;} & \mathfrak{T}^r_s(U) \\ \boldsymbol{D}\downarrow & & \downarrow\boldsymbol{D} \\ \mathfrak{T}^r_s(V) & \xrightarrow[\;|U\;]{} & \mathfrak{T}^r_s(U) \end{array}$$

(*DO* 3) *$\boldsymbol{D}\delta = 0$, where $\delta \in \mathfrak{T}^1_1(U)$ is Kronecker's delta.*

Note that we do not demand that $\boldsymbol{D}$ be natural with respect to push-forward by diffeomorphisms. The reason is that it is not needed for the following unique extension theorem, and indeed, the latter can be used to extend the covariant derivative, which is not natural with respect to diffeomorphisms; see Sect. 2.6 for details.

2.2.17 Theorem (Willmore). *Suppose for each $U \subset M$, open, we have maps $\boldsymbol{E}_U$: $\mathfrak{F}(U) \to \mathfrak{F}(U)$ and $\boldsymbol{F}_U$: $\mathfrak{X}(U) \to \mathfrak{X}(U)$, which are ($\boldsymbol{R}$ linear) tensor derivations and natural with respect to restrictions. That is*

(*i*) $\boldsymbol{E}_U(f \otimes g) = (\boldsymbol{E}_U f) \otimes g + f \otimes \boldsymbol{E}_U g \qquad f, g \in \mathfrak{F}(U)$;

(*ii*) *For $f \in \mathfrak{F}(M)$, $\boldsymbol{E}_U(f|U) = (\boldsymbol{E}_M f)|U$;*

(*iii*) $\boldsymbol{F}_U(f \otimes X) = (\boldsymbol{E}_U f) \otimes X + f \otimes \boldsymbol{F}_U X$;

(*iv*) *For $X \in \mathfrak{X}(M)$, $\boldsymbol{F}_U(X|U) = (\boldsymbol{F}_M X)|U$.*

Then there is a unique differential operator $\boldsymbol{D}$ on $\mathfrak{T}(M)$ that coincides with $\boldsymbol{E}_U$ on $\mathfrak{F}(U)$ and with $\boldsymbol{F}_U$ on $\mathfrak{X}(U)$.

Proof. Suppose that such a $\boldsymbol{D}$ exists. Let φ: $U \to U' \subset \boldsymbol{R}^n$ be a coordinate chart. By (DO 2) and (ii), (iv) above we may restrict attention to the chart (U, φ). Let $\boldsymbol{e}_i$ denote the standard basis of $\boldsymbol{R}^n$ and let

$$\underset{\sim}{e}_i(u) = T_u\varphi^{-1}(u', \boldsymbol{e}_i)$$

ISBN 0-8053-0102-X

for all $u \in U$, with $u' = \varphi(u)$. These are a basis of $T_u(M)$. Let $\underline{\alpha}^j(u)$ denote the dual basis. Note that the local representatives of $\underline{e}_i(u)$ and $\underline{\alpha}^j(u)$ appear as constant sections. We may write

$$t(u) = t^{i_1 \cdots i_r}_{j_1 \cdots j_s}(u) \underline{e}_{i_1} \otimes \cdots \otimes \underline{e}_{i_r} \otimes \underline{\alpha}^{j_1} \otimes \cdots \otimes \underline{\alpha}^{j_s}(u)$$

where

$$t^{i_1 \cdots i_r}_{j_1 \cdots j_s} \in \mathfrak{F}(U)$$

Then using $\boldsymbol{R}$ linearity and the derivation property of $\boldsymbol{D}$ we obtain a sum of terms all of which can be immediately expressed in terms of $\boldsymbol{E}_U$, $\boldsymbol{F}_U$ except for $\boldsymbol{D}\underline{\alpha}^j(u)$. However, by (DO 3),

$$\boldsymbol{D}\delta = 0 = \boldsymbol{D}\left(\underline{e}_j(u) \otimes \underline{\alpha}^j(u)\right) = \left(\boldsymbol{D}\underline{e}_j(u)\right) \otimes \underline{\alpha}^j(u) + e_j(u) \otimes \boldsymbol{D}\underline{\alpha}^j(u)$$

Applying this to $(\underline{\alpha}^i(u), \underline{e}_j(u))$ gives $0 = \underline{\alpha}^j(u)(\boldsymbol{F}_U(\underline{e}_i(u))) + \boldsymbol{D}\underline{\alpha}^j(u)(\underline{e}_i(u))$. Hence $\boldsymbol{D}\underline{\alpha}^j(u)$ is determined. Hence, such a $\boldsymbol{D}$, if it exists, is unique. For existence, we define $\boldsymbol{D}$ as obtained in the foregoing uniqueness argument. We leave it to the reader to check that the resulting $\boldsymbol{D}$ is well defined and satisfies (DO 1), (DO 2), and (DO 3). ■

There is an invariant way to write the computation just done:

(DO 4) Let $t \in \mathfrak{T}^r_s(M)$, $\alpha_1, \ldots, \alpha_r \in \Omega^1(M)$ and $X_1, \ldots, X_s \in \mathfrak{X}(M)$. Then

$$\boldsymbol{D}(t(\alpha_1, \ldots, \alpha_r, X_1, \ldots, X_s)) = (\boldsymbol{D}t)(\alpha_1, \ldots, \alpha_r, X_1, \ldots, X_s) + \sum_{j=1}^{r} t(\alpha_1, \ldots, \boldsymbol{D}\alpha_j, \ldots, \alpha_r, X_1, \ldots, X_s) + \sum_{k=1}^{s} t(\alpha_1, \ldots, \alpha_r, X_1, \ldots, \boldsymbol{D}X_k, \ldots, X_s).$$

We sometimes refer to this by saying that $\boldsymbol{D}$ *commutes with contractions.* Given $\boldsymbol{D}$ on functions, one forms, and vector fields, this formula determines $\boldsymbol{D}$.

The equivalence of (DO 3) and (DO 4) [under the assumption of (DO 1) and (DO 2)] may be proved as follows. Assuming (DO 1), (DO 2), and (DO 3), we prove (DO 4) by writing $t(\alpha_1, \ldots, \alpha_r, X_1, \ldots, X_s)$ in local coordinates as in the proof just given. Conversely, if (DO 4) holds, then (DO 3) follows by applying (DO 4) to the identity

$$\delta(\alpha, X) = \alpha(X)$$

Taking $\boldsymbol{E}_U$ and $\boldsymbol{F}_U$ to be $\boldsymbol{L}_{X|U}$ we see that the hypotheses of Willmore's theorem are satisfied. Hence we can define a differential operator as follows.

2.2.18 Definition. *If $X \in \mathfrak{X}(M)$, we let $\boldsymbol{L}_X$ be the unique differential operator on $\mathfrak{T}(M)$, called the* ***Lie derivative with respect to*** *X, such that $\boldsymbol{L}_X$ coincides with $\boldsymbol{L}_X$ as given in* 2.2.1 *and* 2.2.12.

ISBN 0-8053-0102-X

It may be instructive for the reader to examine Lie derivatives of higher-order tensors in component notation; see Exercise 2.2D.

2.2.19 Proposition. *Let $\varphi: M \to N$ be a diffeomorphism and X a vector field on M. Then L_X is natural with respect to push-forward by φ; that is, $L_{\varphi_* X}\varphi_* t = \varphi_* L_X t$ for $t \in \mathfrak{T}^r_s(M)$, or the following diagram commutes:*

$$\begin{array}{ccc} \mathfrak{T}^r_s(M) & \xrightarrow{\varphi_*} & \mathfrak{T}^r_s(N) \\ {\scriptstyle L_X}\downarrow & & \downarrow{\scriptstyle L_{\varphi_* X}} \\ \mathfrak{T}^r_s(M) & \xrightarrow[\varphi_*]{} & \mathfrak{T}^r_s(N) \end{array}$$

Proof. For an open set $U \subset M$ define $D: \mathfrak{T}^r_s(U) \to \mathfrak{T}^r_s(U)$ by $Dt = \varphi^* L_{\varphi_* X|U}(\varphi_* t)$, where we use the same symbol φ for $\varphi|U$. From 2.2.3(i) and 2.2.14(i), D coincides with $L_{X|U}$ on $\mathfrak{F}(U)$ and $\mathfrak{X}(U)$. Next, we show that D is a differential operator. For (DO 1) we use the fact that $\varphi_*(t_1 \otimes t_2) = \varphi_* t_1 \otimes \varphi_* t_2$, which follows from the definitions. Then

$$\begin{aligned} D(t_1 \otimes t_2) &= (\varphi_*)^{-1} \circ L_{\varphi_* X}\varphi_*(t_1 \otimes t_2) \\ &= (\varphi_*)^{-1} L_{\varphi_* X}(\varphi_* t_1 \otimes \varphi_* t_2) \\ &= (\varphi_*)^{-1}\left[(L_{\varphi_* X}\varphi_* t_1) \otimes \varphi_* t_2\right] + (\varphi_*)^{-1}\left[\varphi_* t_1 \otimes L_{\varphi_* X}\varphi_* t_2\right] \\ &= Dt_1 \otimes t_2 + t_1 \otimes Dt_2 . \end{aligned}$$

For (DO 2) we have, if $t \in \mathfrak{T}^r_s(M)$,

$$\begin{aligned} Dt|U &= \left[(\varphi_*)^{-1} L_{\varphi_* X}\varphi_* t\right]|U \\ &= (\varphi_*)^{-1} L_{\varphi_* X}\varphi_* t|U \\ &= (\varphi_*)^{-1} L_{\varphi_* X|U}\varphi_* t|U \qquad [\text{by (DO 2) for } L_X] \\ &= D(t|U) . \end{aligned}$$

Finally, (DO 3) follows from the fact that $\varphi_* \delta = \delta$, which the reader can easily check. Then we have

$$D\delta = (\varphi_*)^{-1} L_{\varphi_* X}\varphi_* \delta = (\varphi_*)^{-1} L_{\varphi_* X}\delta = 0$$

by (DO 3) for L_X. The result follows by Willmore's theorem. ■

ISBN 0-8053-0102-X

The reader can check, by using the same reasoning, that a differential operator that is natural with respect to diffeomorphisms on functions and vector fields is natural on all tensors.

We now turn to an alternative interpretation of the Lie derivative. For $t \in \mathfrak{T}^r_s(M)$ and $X \in \mathfrak{X}(M)$, we can find a curve at $t(m)$ in the fiber over m by using the flow of X. The derivative of this curve is the Lie derivative. In spirit, the flow plays the same role as parallel translation in covariant differentiation. (See Fig. 2.1-1 and Sect. 2.6.)

More precisely, for $m \in M$ and a vector field X on M let (U,a,F) be a flow box at m. For each $\lambda \in I_a = (-a,a)$ we can form the diffeomorphism $F_\lambda = F|U \times \{\lambda\}: U \to U_\lambda = F_\lambda(U)$. Now let $t \in \mathfrak{T}^r_s(M)$ and define $t_\lambda = F^*_\lambda(t|U_\lambda) = (F_\lambda^{-1})_*(t|U_\lambda) \in \mathfrak{T}^r_s(U)$. Define the map $t_\sharp(m): I_a \to T^r_s T_m(M): \lambda \mapsto t_\lambda(m)$. Using this notation we have the following.

2.2.20 Theorem. *$t_\sharp(m)$ is a curve in $T^r_s T_m(M)$ at $t(m)$ and $\mathbf{L}_X t(m) = t_\sharp(m)'(0)$.*

Proof. For smoothness of $t_\sharp(m)$, we have $t_\lambda(m) = (TF_\lambda^{-1}(m))^r_s t(F_\lambda(m))$. Hence we need verify only that $(TF_\lambda^{-1}(m))^r_s$ is a smooth function of λ. Consider $F: U \times I_a \to M$, which is smooth in (u, λ). Then $TF: TU \times (I_a \times \boldsymbol{R}) \to M$ is also smooth. For smoothness of TF_λ, note $TF_\lambda = TF|TM \times \{\lambda\} \times \{0\}$. Then from 1.7.9 we see that $(TF_\lambda^{-1})^r_s$ is smooth in λ. Since F_0 is the identity, it is clear that $t_\sharp$ is a curve at $t(m)$.

Now define $\boldsymbol{\theta}_X: \mathfrak{T}(M) \to \mathfrak{T}(M)$ by $\boldsymbol{\theta}_X t(m) = t_\sharp(m)'|_{\lambda=0}$. From the flow box existence and uniqueness theorem it is clear that $\boldsymbol{\theta}_X$ is well defined, and from smoothness of $t_\lambda(m)$, $\boldsymbol{\theta}_X t(m) = Tt_\sharp(m) \cdot (0, 1) \in \mathfrak{T}(M)$. Let us apply Willmore's theorem 2.2.17. First, $\boldsymbol{\theta}_X$ is $\boldsymbol{R}$ linear and is a derivation. This follows easily using the local representatives. Also, $\boldsymbol{\theta}_X$ is natural with respect to restrictions because it is defined locally. Moreover,

$$\boldsymbol{\theta}_X \delta = \frac{d}{d\lambda}(F_\lambda)^* \delta \bigg|_{\lambda=0} = \frac{d}{d\lambda}\delta \bigg|_{\lambda=0} = 0$$

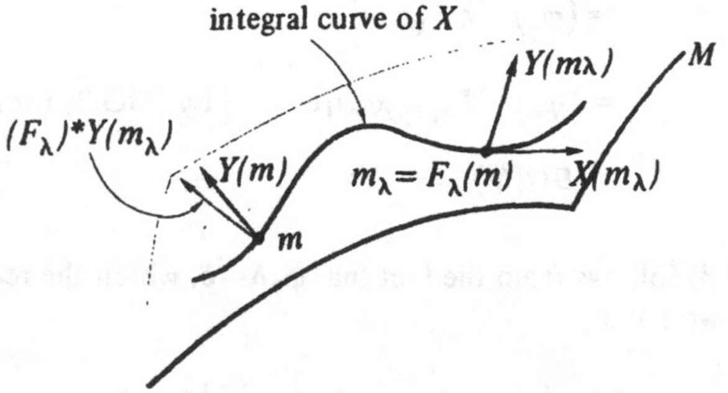

Figure 2.2-1

ISBN 0-8053-0102-X

Hence $\boldsymbol{\theta}_X$ is a differential operator. It remains to show that $\boldsymbol{\theta}_X$ coincides with L_X on $\mathfrak{F}(M)$ and $\mathfrak{X}(M)$. For $f \in \mathfrak{F}(M)$, we have $f_\lambda(m) = f \circ F_\lambda(m)$ and so

$$\frac{d}{d\lambda} f_\lambda(m)\Big|_{\lambda=0} = P_2 T_m f \circ T_\lambda F_\lambda(m)(0,1) = df \circ X(m) = L_X f(m)$$

Hence $\boldsymbol{\theta}_X f = L_X f$. Finally, we must verify that $\boldsymbol{\theta}_X Y = [X, Y] = L_X Y$, or equivalently $L_{\boldsymbol{\theta}_X Y} = [L_X, L_Y]$. First let us note that for $f \in \mathfrak{F}(M)$ there is a function $g_\lambda(u)$ so $f \circ F_\lambda = f + \lambda g_\lambda$ and $g_0 = L_X f$; namely, take

$$\lambda g_\lambda(u) = \int_0^1 \frac{\partial}{\partial t} f \circ F(t\lambda, u)\, dt$$

Then

$$\begin{aligned}
L_{\boldsymbol{\theta}_X Y} f(m) &= df(m) \frac{d}{d\lambda}\left[T_m F_{-\lambda} \circ Y \circ F_\lambda(m)\right]\Big|_{\lambda=0} \\
&= \frac{d}{d\lambda}\left[T_m(f \circ F_{-\lambda}) \circ Y \circ F_\lambda(m)\right]\Big|_{\lambda=0} \\
&= \frac{d}{d\lambda}\left[T_m(f - \lambda g_{-\lambda}) \circ Y \circ F_\lambda(m)\right]\Big|_{\lambda=0} \\
&= \frac{d}{d\lambda} T_m f \circ Y \circ F_\lambda(m)\Big|_{\lambda=0} - \frac{d}{d\lambda}[\lambda]\left[T_m g_{-\lambda} \circ Y \circ F_\lambda(m)\right]\Big|_{\lambda=0} \\
&= d(T_m f \circ Y) \circ F'_\lambda(m)\Big|_{\lambda=0} - T_m g_0 \circ Y \circ F_0(m) \\
&= (L_X L_Y f - L_Y L_X f)(m).
\end{aligned}$$

Hence from 2.2.11, $\boldsymbol{\theta}_X Y = [X, Y]$.

Finally, by Willmore's theorem, $\boldsymbol{\theta}_X = L_X$ for all $t \in \mathfrak{T}(M)$. ■

2.2.21 Corollary. *If $t \in \mathfrak{T}(M)$, $L_X t = 0$ iff t is constant along the flow of X. That is, $t = F_\lambda^* t$.*

We can write 2.2.20 in the following way:

$$\boxed{\frac{d}{d\lambda} F_\lambda^* t = F_\lambda^* L_X t}$$

This key identity between flows and Lie derivatives is sometimes taken as the *definition* of the Lie derivative.

As an application of the Lie derivative, we consider a partial differential equation on $\boldsymbol{R}^n$ of the form

$$(P)\begin{cases} \dfrac{\partial f}{\partial t}(x,t) = \displaystyle\sum_{i=1}^n X^i(x) \frac{\partial f}{\partial x^i}(x,t) \\ f(x,0) = g(x) \end{cases}$$

ISBN 0-8053-0102-X

for given smooth functions $X^i(x)$, $i=1,\ldots,n$, $g(x)$, and a scalar unknown $f(x,t)$.

2.2.22 Proposition. *Suppose $X=(X^1,\ldots,X^n)$ has a complete flow F_t. Then*

$$f(x,t)=g(F_t(x))$$

is a solution of the above problem (P). (See Exercise 2.2I for uniqueness.)

Proof.

$$\frac{\partial f}{\partial t}=\frac{d}{dt}F_t^*g=F_t^*L_Xg=L_X(F_t^*g)=X(f) \quad ■$$

Thus one can solve this *scalar* equation by computing the orbits of X. These trajectories are called *characteristics.*

These results on Lie derivatives extend to nonautonomous (i.e., time-dependent) systems as follows.

2.2.23 Definition. *A time-dependent vector field is a map $X: \mathbf{R}\times M\to TM$ such that $X(t,m)\in T_mM$. (As we mentioned earlier, the local existence theory also goes through in this case.) One defines $F_{t,s}(m)$ to be the integral curve of X_t; that is, $(d/dt)F_{t,s}(m)=X_t(F_{t,s}(m))$ such that $F_{s,s}(m)=m$. In other words, $F_{t,s}(m)$ is the solution curve starting at m at time $t=s$. In the time-independent, or autonomous case, $F_{t,s}=F_{t-s}$. We call $F_{t,s}$ the* **(time-dependent) flow** *of X_t or the* **evolution operator** *for X_t.*

We have $F_{t,s}\circ F_{s,r}=F_{t,r}$ replacing the flow property $F_{t+s}=F_t\circ F_s$, and $F_{t,t}=$identity. We often write $F_t=F_{t,0}$. In general, $F_t^*X_t\neq X_t$; however we do have:

2.2.24 Theorem. *Let X_t be of class C^k, $k\geqslant 1$ for each t and $X(t,m)$ be continuous in t,m. Then $F_{t,s}$ is of class C^k and for α a tensor on M,*

$$\frac{d}{dt}F_{t,s}^*\alpha=F_{t,s}^*(L_{X_t}\alpha)$$

The proof goes as in 2.1.2 and 2.2.20. Note that it is *not* true for time-dependent systems that the right-hand side is $L_{X_t}(F_{t,s}^*\alpha)$.

The three main pillars supporting differential topology and calculus on manifolds are the implicit function theorem, the existence theorem for ordinary differential equations, and Frobenius' theorem,* which we discuss briefly here. First some definitions:

2.2.25 Definition. *Let M be a manifold and let $E\subset TM$ be a subbundle of its tangent bundle. (So E is a vector bundle over M and is a submanifold of TM by means of a vector bundle chart of E.) We call E a* **distribution** *on M.*

*According to Lawson [1977], the theorem of Frobenius is due to A. Clebsch and F. Deahna.

ISBN 0-8053-0102-X

(i) *We say E is **integrable** if for any two vector fields X and Y defined on open sets of M and which take values in E, $[X, Y]$ takes values in E as well.*

(ii) *We say E **arises from a regular foliation** of M if for any $m_0 \in M$ there is a (local) submanifold $N \subset M$ called a leaf of the foliation containing m_0 whose tangent bundle is exactly E restricted to N.*

Frobenius' theorem asserts that these two conditions are equivalent.

2.2.26 Frobenius' Theorem. *A subbundle E of TM is integruble if and only if it arises from a regular foliation.*

Proof. The "if" part is obvious since the bracket of two vector fields on a manifold N is again a vector field on N.

If E has one-dimensional fibers, Frobenius' theorem amounts to the existence theorem for differential equations. The proof of the "only if" part actually proceeds along these lines.

By choosing a vector bundle chart, one is reduced to this local situation: $\mathbf{E}$ is a model space for the fibers of E, $\mathbf{F}$ is a complementary space, and $U \times V \subset \mathbf{E} \times \mathbf{F}$ is an open neighborhood of $(0,0)$. We have a map

$$f: U \times V \to L(\mathbf{E}, \mathbf{F})$$

such that the fiber of E over (x,y) is

$$E_{(x,y)} = \{(u, f(x,y)\cdot u) \mid u \in \mathbf{E}\} \subset \mathbf{E} \times \mathbf{F}$$

and we can assume we are working near $(0,0)$ and $f(0,0) = \textit{identity}$.

The condition of integrability is trivial to work out. One computes the bracket of two vector fields of the form

$$X(x,y) = (a(x,y), f(x,y)\cdot a(x,y))$$

$$Y(x,y) = (b(x,y), f(x,y)\cdot b(x,y))$$

Using the local formula following 2.2.13, one finds that E is integrable iff at a point (x,y)

$$Df\cdot(a, f\cdot a)\cdot b = Df\cdot(b, f\cdot b)\cdot a \tag{1}$$

for all $a, b \in \mathbf{E}$.

Consider these two time-dependent vector fields on $U \times V$:

$$X_t(x,y) = (0, f(tx,y)\cdot x)$$

and

$$Y_t(x,y) = (x, f(tx,y)\cdot tx)$$

ISBN 0-8053-0102-X

These are chosen with the form of $E_{(x,y)}$ in mind, so Y_0 is simple, but especially so that

$$[X_t, Y_t] + \frac{dY_t}{dt} = 0 \tag{2}$$

Condition (2) is easily verified [it does not use (1)]. From it, we get via 2.2.24

$$\frac{d}{dt} F_t^* Y_t = 0$$

where F_t is the flow of X_t, $F_0 = identity$. Since $X_t(0,0) = 0$, we can assume F_t is defined for $0 \leqslant t \leqslant 1$. Thus

$$F_1^* Y_1 = Y_0$$

Since $Y_1(x,y) = (x, f(x,y)x)$ and $Y_0(x,y) = (x, 0)$, we expect F_1 to be a local diffeomorphism of $U \times V$ such that F_1^* maps $E_{(x,y)}$ to $E \times \{0\}$ and be the required coordinate change to complete the proof.

Let $N = F_1(U \times \{y_0\})$. We shall complete the proof by showing that if $(x,y) \in N$,

$$T_{(x,y)}N = E_{(x,y)}$$

Clearly, since the first component of X_t is zero,

$$F_t(x,y) = (x, \varphi(t,x,y))$$

for a map $\varphi(t,x,y) \in F$. A typical tangent vector to $F_t(U \times \{y_0\})$ is clearly

$$(v, D_x\varphi(t,x,y)\cdot v), \qquad v \in E$$

Thus the proof is complete if we establish this identity:

$$D_x\varphi(t,x,y)\cdot v = tf(tx, \varphi(t,x,y))\cdot v$$

Indeed, they are equal at $t = 0$ and

$$\frac{d}{dt}\varphi(t,x,y) = f(tx, \varphi(t,x,y))\cdot x$$

so

$$\begin{aligned}\frac{d}{dt} D_x\varphi(t,x,y)\cdot v = {} & tD_x f(tx,\varphi(t,x,y))\cdot v\cdot x \\ & + D_y f(tx,\varphi(t,x,y))\cdot(D_x\varphi(t,x,y)\cdot v)\cdot x \\ & + f(tx,\varphi(t,x,y))\cdot v \end{aligned} \tag{3}$$

ISBN 0-8053-0102-X

Also,

$$\frac{d}{dt} tf(tx, \varphi(t,x,y))\cdot v = f(tx, \varphi(t,x,y))\cdot v$$
$$+ tD_x f(tx, \varphi(t,x,y))\cdot x \cdot v$$
$$+ tD_y f(tx, \varphi(t,x,y))\cdot \big(f(tx, \varphi(t,x,y))\cdot x\big)\cdot v \qquad (4)$$

For x, y, v fixed, let

$$\psi(t) = D_x\varphi(t, x, y)\cdot v - tf(tx, \varphi(t, x, y))\cdot v$$

so that $\psi(0) = 0$. Subtracting (3) and (4) and using (1) we find that

$$\frac{d}{dt}\psi(t) = D_y f(tx, \varphi(t, x, y))\cdot \psi(t)\cdot x$$

Hence by Gronwall's inequality, or by uniqueness of solutions of differential equations, $\psi(t) = 0$, and the result follows. ■

Remarks. (1) The method of using the time one map of a time-dependent flow to provide the appropriate coordinate change will be used several times throughout the book. (Sometimes these are called "Lie transforms".)

(2) For a direct proof from the implicit function theorem using manifolds of maps, see Penot [1970b]

(3) In Chapters 3 and 4 the concept of foliation will be needed again so we shall give the formal definition. For further information the reader is referred to Lawson [1977].

A *foliation* $\mathfrak{F}$ of class C^r, $0 \leqslant r$ and of dimension p on the m-dimensional manifold M is a decomposition of M into disjoint connected subsets $\mathfrak{F} = \{\mathfrak{L}_\alpha\}_{\alpha\in A}$ called the *leaves of the foliation*, such that each point of M has a neighborhood U and a system of C^r coordinates (x,y): $U \to R^p \times R^{m-p}$ such that for each leaf $\mathfrak{L}_\alpha$, the components of $U \cap \mathfrak{L}_\alpha$ are described by the equations

$$y_1 = constant, \ldots, y_{m-p} = constant$$

These coordinates are said to be *distinguished* by the foliation $\mathfrak{F}$.

Note that each leaf $\mathfrak{L}_\alpha$ is a connected *immersed* submanifold of dimension p. In general, this immersion is *not* an embedding, that is, the induced topology on $\mathfrak{L}_\alpha$ from M does not necessarily coincide with the topology of $\mathfrak{L}_\alpha$ (the leaf $\mathfrak{L}_\alpha$ may accumulate on itself, for example). The topology on $\mathfrak{L}_\alpha$ is given by the basis formed by the sets $\{x \in V \mid y_1(x) = constant, \ldots, y_{m-p}(x) = constant$, V open in U and (x,y): $U \to R^p \times R^{m-p}$, a distinguished chart of $\mathfrak{F}\}$. See Fig. 2.2-2.

ISBN 0-8053-0102-X

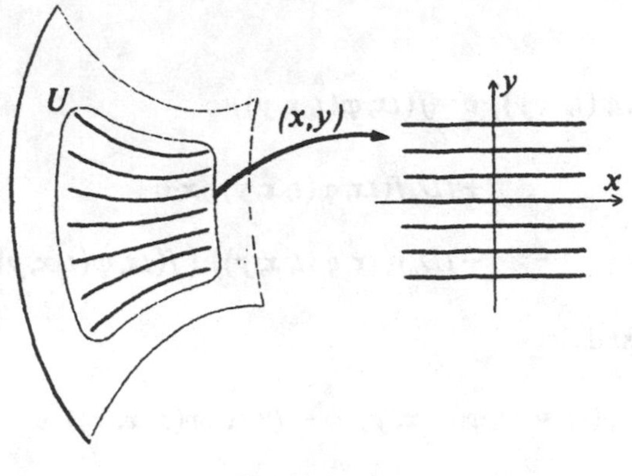

Figure 2.2-2

Globalizing the Frobenius theorem, one can show that an integrable distribution determines a foliation in the sense defined above (2.2.26 was local); see Exercise 2.2K.

Here are some simple examples of foliations.

(1) $\mathfrak{F}=\{M\}$ is the only dimension m foliation; it has one leaf, the whole manifold M.

(2) Let $f: M \to N$ be a submersion. Let $\mathfrak{F}$ be the collection of connected components of the manifolds $f^{-1}(y)$, $y \in N$. This is a dimension $m-n$ foliation of M.

(3) The orbits of a nowhere zero vector field on M define a dimension 1 foliation. (This result can be generalized to Lie group actions with the condition that the isotropy group has constant dimension on M.)

Foliations are by no means the only way to decompose a topological space into connected submanifolds. *Stratifications*, a concept generalizing foliations, turn out to be the natural tool to describe the topology of orbit spaces of compact Lie group actions (see, for instance, Levine [1959], Bredon [1972], Burghelea, Albu, Ratiu [1975], and Fischer [1970].)

Finally in this section we consider two miscellaneous (optional) topics; the first concerns *convergence of flows*. Often it is useful to know that the flows of a sequence of vector fields converge if the vector fields themselves converge.

2.2.27. Proposition. *Let X_α be locally Lipschitz vector fields on M for α in some topological space. Suppose the Lipschitz constants of X_α are locally bounded as $\alpha \to \alpha_0$ and $X_\alpha \to X_{\alpha_0}$ locally uniformly. Let $c(t)$ be an integral curve of X_{α_0}, $0 \leqslant t < T$ and $\varepsilon > 0$. Then the integral curves $c_\alpha(t)$ of X_α with $c_\alpha(0) = c(0)$ are defined for $t \in [0, T-\varepsilon]$ for α sufficiently close to α_0 and $c_\alpha(t) \to c(t)$ uniformly in $t \in [0, T-\varepsilon]$ as $\alpha \to \alpha_0$. If the flows are complete, $F_t^\alpha \to F_t$ locally uniformly. (The vector fields may be time dependent if the estimates are locally t-uniform.)*

ISBN 0-8053-0102-X

Proof. We have, in charts,

$$\|c_\alpha(t)-c(t)\| = \left\| \int_0^t X_\alpha(c_\alpha(\tau)) - X_{\alpha_0}(c(\tau))\,d\tau \right\|$$

$$\leqslant \int_0^t \|X_\alpha(c_\alpha(\tau)) - X_\alpha(c(\tau))\|\,d\tau + \int_0^t \|X_\alpha(c(\tau)) - X_{\alpha_0}(c(\tau))\|\,d\tau$$

$$\leqslant K\int_0^t \|c_\alpha(\tau)-c(\tau)\|\,d\tau + \rho_{\alpha,\alpha_0}$$

where $\rho_{\alpha,\alpha_0}\to 0$. From Gronwall's inequality $c_\alpha(t)\to c(t)$, as $\alpha\to\alpha_0$. This estimate shows that $c_\alpha(t)$ exists as long as $c(t)$ does on any compact subinterval of $[0,T)$. The result follows. ■

Next we consider invariant sets. If X is a smooth vector field on M and $N\subset M$ is a submanifold, the flow of X will leave N invariant (as a set) iff X is tangent to N. If N is not a submanifold (e.g., N is an open subset together with a smooth boundary) the situation is not so simple; however, for this there is a nice criterion going back to Nagumo [1942]. Our proof follows Brezis [1970].

2.2.28 Theorem. *Let X be locally Lipschitz on an open set $U\subset \boldsymbol{E}$, $\boldsymbol{E}$ a Banach space. Let $G\subset U$ be relatively closed and set $d(x,G)=\inf\{\|x-y\| \mid y\in G\}$. The following are equivalent:*

(1) *$\lim_{h\downarrow 0}(d(x+hX(x),\ G)/h)=0$ locally uniformly in $x\in G$ (or just pointwise if $\boldsymbol{E}=\boldsymbol{R}^n$); and*
(2) *if $x(t)$ is an integral curve of X starting in G, then $x(t)\in G$ for all $t\geqslant 0$ in the domain of $x(\cdot)$.*

Note (i) $x(t)$ need not lie in G for $t\leqslant 0$; so G is + invariant. (ii) If X is only continuous the theorem fails.

We give the proof assuming $\boldsymbol{E}=\boldsymbol{R}^n$ for simplicity.

Proof of 2.2.28. Assume (2) holds. Setting $x(t)=F_t(x)$, where F_t is the flow of X, we get

$$d(x+hX(x),G)\leqslant \|x(h)-x-hX(x)\| = |h|\left\|\frac{x(h)-x}{h}-X(x)\right\|$$

from which (1) follows.

Now assume (1). It suffices to show $x(t)\in G$ for small t. Near $x=x(0)$, say on a ball of radius r, we have $\|X(x_1)-X(x_2)\|\leqslant K\|x_1-x_2\|$ and $\|F_t(x_1)-F_t(x_2)\|\leqslant e^{Kt}\|x_1-x_2\|$. We can assume $\|F_t(x)-x\|<r/2$. Set $\phi(t)=d(F_tx,G)$. Since G is relatively closed, and $\boldsymbol{E}=\boldsymbol{R}^n$, $d(F_tx,G)=\|F_tx-y_t\|$ for

ISBN 0-8053-0102-X

some $y_t \in G$. (In the general case some approximation argument is needed here.) Thus, $\|y_t - x\| < r$. For small h, $\|F_h y_t - x\| < r$, so

$$\phi(t+h) \leqslant \|F_{t+h}(x) - F_h y_t\| + \|F_h(y_t) - y_t - hX(y_t)\| + d(y_t + hX(y_t), G)$$

$$\leqslant e^{Kh}\|y_t - F_t x\| + \|F_h(y_t) - y_t - hX(y_t)\| + d(y_t + hX(y_t), G)$$

or

$$\frac{\phi(t+h) - \phi(t)}{h} \leqslant \left(\frac{e^{Kh} - 1}{h}\right)\phi(t) + \left\|\frac{F_h(y_t) - y_t}{h} - X(y_t)\right\|$$

$$+ d(y_t + hX(y_t), G)/h$$

Hence

$$\limsup_{h \downarrow 0} \frac{\phi(t+h) - \phi(t)}{h} \leqslant K\phi(t)$$

As in Gronwall's inequality, we may conclude that

$$\phi(t) \leqslant e^{Kt}\phi(0)$$

so $\phi(t) = 0$. ■

For example, let X be a smooth vector field on $\boldsymbol{R}^n$, let $g: \boldsymbol{R}^n \to \boldsymbol{R}$ be smooth, and let $\lambda \in \boldsymbol{R}$ be a regular value for g, so $g^{-1}(\lambda)$ is a submanifold. Let $G = g^{-1}((-\infty, \lambda])$ and suppose that on $g^{-1}(\lambda)$,

$$\langle X, \operatorname{grad} g \rangle \leqslant 0$$

Then G is + invariant under F_t as may be seen by using 2.2.28. This result has been generalized to the case where ∂G might not be smooth by Bony [1969]. See also Redheffer [1972] and Martin [1973]. Related references are Yorke [1967], Hartman [1972], and Crandall [1972].

EXERCISES

2.2A. If each point $m \in M$ has a neighborhood U such that for all $f \in \mathfrak{F}(U)$, $L_X f = L_Y f$ (on U), show $X = Y$.

2.2B. Prove that φ_* is a tensor algebra homomorphism, as indicated in the proof of 2.2.19, and that $\varphi_* \delta = \delta$.

2.2C. On $\boldsymbol{R}$, let $X(x,y) = (x,y; y, -x)$. Find the flow of X and interpret 2.2.21 geometrically.

ISBN 0-8053-0102-X

2.2D. Show that, in coordinates,

$$[X, Y]^i = \sum_j \left(X^j \frac{\partial Y^i}{\partial x^j} - Y^j \frac{\partial X^i}{\partial x^j} \right)$$

and more generally,

$$(L_X t)^{a_1 \cdots a_r}_{b_1 \cdots b_s} = \sum_c X^c \frac{\partial t^{a_1 \cdots a_r}_{b_1 \cdots b_s}}{\partial x^c} - \sum_{c,\alpha} t^{a_1 \cdots a_{\alpha-1} c a_{\alpha+1} \cdots a_r}_{b_1 \cdots b_s} \frac{\partial X^{a_\alpha}}{\partial x^c}$$

$$+ \sum_{c,\alpha} t^{a_1 \cdots a_r}_{b_1 \cdots b_{\alpha-1} c b_{\alpha+1} \cdots b_s} \frac{\partial X^c}{\partial x^{b_\alpha}}$$

2.2E. (For readers with a knowledge of Riemannian geometry. See Sect 2.7.)

(i) Let M be a manifold with a torsion-free connection, denoted ∇, so that $\nabla_X Y$ is the covariant derivative of Y in the direction X. Show that $L_X Y = \nabla_X Y - \nabla_Y X$. Note, however, that L_X does not require a connection or metric for its definition.

(ii) Show that the formula in 2.2D remains valid if partial derivatives are replaced by covariant derivatives.

2.2F. Solve the following for $f(t,x,y)$:

$$\frac{\partial f}{\partial t} = y \frac{\partial f}{\partial x} - x \frac{\partial f}{\partial y}$$

if $f(0,x,y) = y(\sin x)$.

2.2G. Let X and Y be vector fields on M with complete flows F_t and G_t, respectively. Show that the following are equivalent:

(i) $[X, Y] = 0$

(ii) $F_t^* Y = Y$

(iii) $F_t \circ G_t = G_t \circ F_t$

If $[X, Y] = 0$, show that $X + Y$ has flow $H_t = F_t \circ G_t$.

2.2H. (i) Let $X = y^2 \partial/\partial x$ and $Y = x^2 \partial/\partial y$. Show that X and Y are complete on $\mathbb{R}^2$ but $X + Y$ is not.

(ii) Prove the following:

THEOREM *Let H be a Hilbert space and let X and Y be locally Lipschitz vector fields that satisfy the following:*

(a) *X and Y are bounded and Lipschitz on bounded sets;*

(b) *there is a constant $\beta \geq 0$ such that*

$$\langle Y(x), x \rangle \leq \beta \|x\|^2 \qquad \text{for all} \quad x \in H$$

(c) *there is a locally Lipschitz monotone increasing function $c(t) > 0$, $t \geq 0$ such that $\int^\infty \frac{dx}{c(x)} = +\infty$ and if $x(t)$ is an integral curve of X,*

$$\frac{d}{dt} \|x(t)\| \leq c(\|x(t)\|)$$

Then X, Y are $X + Y$, and are positively complete.

ISBN 0-8053-0102-X

Note: One may assume $\|X(x_0)\| \leqslant c(\|x_0\|)$ in (c) instead of $(d/dt)\|x(t)\| \leqslant c(\|x(t)\|)$. (*Hint.* Find a differential inequality for $\frac{1}{2}(d/dt)\|u(t)\|^2$, where $u(t)$ is an integral curve of $X+Y$, and use the following *comparison lemma*:

Suppose $r'(t)=c(r(t))$ and $r_0 \geqslant 0$. Then $r(t) \geqslant 0$ is defined for all $t \geqslant 0$. Suppose $f(t) \geqslant 0$, is continuous, and

$$f(t) \leqslant r_0 + \int_0^t c(f(s))\,ds, \qquad t \in [0,T)$$

Then $f(t) \leqslant r(t)$ for $t \in [0,T)$.)

2.2I. Show that, under suitable hypotheses, the solution $f(x,t)=g(F_t(x))$ of (P) given in 2.2.22 is unique. (*Hint.* Consider the function $E(t)=\int_{R_n}|f_1(x,t)-f_2(x,t)|^2\,dx$, where f_1 and f_2 are two solutions. Show that $dE/dt \leqslant \alpha E$ for a suitable constant α and conclude by Gronwall's inequality that $E=0$. The "suitable hypotheses" are conditions that enable integration by parts to be performed in the computation of dE/dt.)

2.2J. Use the method of proof of 2.2.26, especially (2), to prove the straightening-out theorem for flows.

2.2K. By analogy with the maximal extendability of integral curves of vector fields, formulate and prove a global version of Frobenius theorem.

2.2L. Using 2.1.26, prove that the flow H_t of $[X,Y]$ is given by

$$H_t(x) = \lim_{n\to\infty}\left(G_{-\sqrt{t/n}} \circ F_{-\sqrt{t/n}} \circ G_{\sqrt{t/n}} \circ F_{\sqrt{t/n}}\right)^n(x), \qquad t \geqslant 0$$

where $X, Y \in \mathfrak{X}(M)$ have flows F_t and G_t, respectively.

2.2M. (D. Burghelea) Let $f\colon M \to N$ be a smooth injective map, $\dim M = m$, $\dim N = n$. Show that $m \leqslant n$ and the set $P=\{x \in M \mid T_x f \text{ is injective}\}$ is dense in M.

(*Sketch*: The assertion $m \leqslant n$ follows by the implicit function theorem. For the second part it suffices to work in a local chart. Use induction on k to show that $\mathrm{cl}\, U_{i_1\cdots i_k} \supset V$, where

$$U_{i_1\cdots i_k} = \left\{ x \in V \,\middle|\, T_x f\left(\frac{\partial}{\partial x^{i_1}}\right)_x, \ldots, T_x f\left(\frac{\partial}{\partial x^{i_k}}\right)_x \textit{ linearly independent} \right\}$$

The case $k=n$ gives then the statement of the theorem. For example, the proof for $k=2$ goes as follows. We know that

$$U_i = \left\{ x \in V \,\middle|\, T_x f\left(\frac{\partial}{\partial x^i}\right) \neq 0 \right\}$$

is open and $cl\, U_i \supset V$. Consider the open set

$$U_{ij} = \left\{ x \in V \,\middle|\, T_x f\left(\frac{\partial}{\partial x^i}\right)_x, \; T_x f\left(\frac{\partial}{\partial x^j}\right)_x \textit{ linearly independent} \right\}$$

and define

$$U_j' = \left\{ x \in U_i \,\middle|\, T_x f\left(\frac{\partial}{\partial x^j}\right)_x \neq 0 \right\}$$

ISBN 0-8053-0102-X

U_i being open, U_j' is dense in U_i and hence in V. Let

$$U_{ij}' = \left\{ x \in U_j' \mid T_x f\left(\frac{\partial}{\partial x^i}\right)_x, T_x f\left(\frac{\partial}{\partial x^j}\right)_x \text{ linearly independent} \right\};$$

we show that U_{ij}' is dense in V. If this were not the case there is an open set $W \subset V$ such that

$$\lambda(x) T_x f\left(\frac{\partial}{\partial x^i}\right)_x + T_x f\left(\frac{\partial}{\partial x^j}\right)_x = 0$$

for a certain smooth function λ nonzero on W. Let $c: (-\varepsilon, \varepsilon) \to W$ be an integral curve of the vector field $\lambda(\partial/\partial x^i) - (\partial/\partial x^j)$. Then $(f \circ c)'(t) = T_{c(t)} f(c'(t)) = 0$ so that $f \circ c$ is constant on $(-\varepsilon, \varepsilon)$ contradicting injectivity of f.)

2.3 EXTERIOR ALGEBRA

The calculus of Cartan concerns exterior differential forms, which are sections of a vector bundle of linear exterior forms on the tangent spaces of a manifold. We begin with the exterior algebra of a vector space and extend this fiberwise to a vector bundle. As with tensor fields, the most important case is the tangent bundle of a manifold, which is considered in the next section.

2.3.1 Definition. *Let E be a finite-dimensional real vector space. Let* $\Omega^0(E) = R$. $\Omega^1(E) = E^*$, *and, in general,* $\Omega^k(E) = L_a^k(E, R)$, *the vector space of skew symmetric k multilinear maps or* ***exterior k-forms*** *on E.*

We leave as an easy exercise the fact that $\Omega^k(E)$ is a vector subspace of $T_k^0(E)$.

Recall that the permutation group on k elements, denoted S_k, consists of all bijections $\varphi: \{1, \ldots, k\} \to \{1, \ldots, k\}$ together with the structure of a group under composition. Clearly, S_k has order $k!$. Letting $(\tilde{R}, \times)$ denote $R \backslash \{0\}$ with the multiplicative group structure, we have a homomorphism *sign*: $S_k \to (\tilde{R}, \times)$. That is, for $\sigma, \tau \in S_k$, $sign(\sigma \circ \tau) = (sign\, \sigma)(sign\, \tau)$. The image of *sign* is the subgroup $\{-1, 1\}$, while its kernel consists of the subgroup of even permutations. One other fact we shall need is the following, which the reader can easily check: If G is a group and $g_0 \in G$, the map $R_{g_0}: G \to G: g \mapsto g g_0$ is a bijection.

2.3.2 Definition. ***The alternation mapping*** $A: T_k^0(E) \to T_k^0(E)$ *(as before, we do not index the A) is defined by*

$$At(e_1, \ldots, e_k) = \frac{1}{k!} \sum_{\sigma \in S_k} (sign\, \sigma) t(e_{\sigma(1)}, \ldots, e_{\sigma(k)})$$

where the sum is over all k! elements of S_k.

ISBN 0-8053-0102-X

2.3.3 Proposition. *A is a linear mapping onto $\Omega^k(E)$, $A|\Omega^k(E)$ is the identity, and $A\circ A=A$.*

Proof. Linearity of A follows at once. If $t\in\Omega^k(E)$, then

$$\begin{aligned}At(e_1,\dots,e_k)&=\frac{1}{k!}\sum_{\sigma\in S_k}(\operatorname{sign}\sigma)t(e_{\sigma(1)},\dots,e_{\sigma(k)})\\&=\frac{1}{k!}\sum_{\sigma\in S_k}t(e_1,\dots,e_k)\\&=t(e_1,\dots,e_k)\end{aligned}$$

since $(\operatorname{sign}\sigma)^2=1$ and S_k has order $k!$. Second, for $t\in T_k^0(E)$ we have

$$\begin{aligned}At(e_1,\dots,e_k)&=\frac{1}{k!}\sum_{\sigma\in S_k}(\operatorname{sign}\sigma)t(e_{\sigma(1)},\dots,e_{\sigma(k)})\\&=\frac{1}{k!}\sum_{\sigma\in S_k}(\operatorname{sign}\sigma\tau)t(e_{\sigma\tau(1)},\dots,e_{\sigma\tau(k)})\\&=(\operatorname{sign}\tau)At(e_{\tau(1)},\dots,e_{\tau(k)})\end{aligned}$$

since $\sigma\mapsto\sigma\tau$ is a bijection and *sign* is a homomorphism. This proves the first two assertions, and the last follows from them. ■

Then we may define the *exterior product* as follows.

2.3.4 Definition. *If $\alpha\in T_k^0(E)$ and $\beta\in T_l^0(E)$, define $\alpha\wedge\beta\in\Omega^{k+l}(E)$ by $\alpha\wedge\beta=(k+l)!/k!l!\ A(\alpha\otimes\beta)$. (Again, we do not index $\wedge$.) In particular, for $\alpha\in T_0^0(E)=R$, we put $\alpha\wedge\beta=\beta\wedge\alpha=\alpha\beta$.*

There are several possible conventions for defining the wedge product $\wedge$. The one here conforms to Spivak [1965], and Bourbaki [1971] but not to Kobayashi–Nomizu [1963] or Guillemin–Pollack [1976]. See Robbin [1974] for a lively discussion of what conventions are possible.

Our definition of $\alpha\wedge\beta$ is the one that eliminates the largest number of constants later. The reader should prove that, for exterior forms,

$$(\alpha\wedge\beta)(e_1,\dots,e_{k+l})=\Sigma'(\operatorname{sign}\sigma)\alpha(e_{\sigma(1)},\dots,e_{\sigma(k)})\beta(e_{\sigma(k+1)},\dots,e_{\sigma(k+l)})$$

where Σ' denotes the sum over all *shuffles*; that is, permutations σ of $\{1,2,\dots,k+l\}$ such that $\sigma(1)<\cdots<\sigma(k)$ and $\sigma(k+1)<\cdots<\sigma(k+l)$. The basic properties of the operation $\wedge$ are given in the following.

2.3.5 Proposition. *For $\alpha\in T_k^0(E)$, $\beta\in T_l^0(E)$, and $\gamma\in T_m^0(E)$, we have*

(*i*) $\alpha\wedge\beta=A\alpha\wedge\beta=\alpha\wedge A\beta$;

(*ii*) *$\wedge$ is bilinear;*

ISBN 0-8053-0102-X

(*iii*) $\alpha\wedge\beta=(-1)^{kl}\beta\wedge\alpha$;
(*iv*) $\alpha\wedge(\beta\wedge\gamma)=(\alpha\wedge\beta)\wedge\gamma$.

Proof. For (i), first note that if $\sigma\in S_k$ and $\sigma t(e_1,\ldots,e_k)=t(e_{\sigma(1)},\ldots,e_{\sigma(k)})$, then $A(\sigma t)=(\text{sign}\,\sigma)At$ for

$$A(\sigma t)(e_1,\ldots,e_k)=\frac{1}{k!}\sum_{\rho\in S_k}(\text{sign}\,\rho)t(e_{\rho\sigma(1)},\ldots,e_{\rho\sigma(k)})$$

$$=\frac{1}{k!}\sum_{\rho\in S_k}(\text{sign}\,\sigma)(\text{sign}\,\rho\sigma)t(e_{\rho\sigma(1)},\ldots,e_{\rho\sigma(k)})$$

$$=(\text{sign}\,\sigma)At(e_1,\ldots,e_k)$$

since $\rho\mapsto\rho\sigma$ is a bijection. Then,

$$A(A\alpha\otimes\beta)(e_1,\ldots,e_k,\ldots,e_{k+l})=A(A\alpha(e_1,\ldots,e_k)\beta(e_{k+1},\ldots,e_{k+l}))$$

$$=A\left(\frac{1}{k!}\sum_{\tau\in S_k}\text{sign}\,\tau\alpha(e_{\tau(1)},\ldots,e_{\tau(k)})\beta(e_{k+1},\ldots,e_{k+l})\right)$$

$$=A\frac{1}{k!}\sum_{\tau\in S_k}(\text{sign}\,\tau)(\tau\alpha\otimes\beta)(e_1,\ldots,e_k,\ldots,e_{k+l})$$

$$=\frac{1}{k!}\sum_{\tau\in S_k}(\text{sign}\,\tau)A(\tau\alpha\otimes\beta)(e_1,\ldots,e_{k+l})\quad(\text{linearity of }A)$$

$$=\frac{1}{k!}\sum_{\tau\in S_k}(\text{sign}\,\tau')A\tau'(\alpha\otimes\beta)(e_1,\ldots,e_{k+l})$$

where $\tau'\in S_{k+l}$,

$$\tau'(1,\ldots,k,\ldots,k+l)=(\tau(1),\ldots,\tau(k),k+1,\ldots,k+l)$$

so $\text{sign}\,\tau=\text{sign}\,\tau'$ and $\tau\alpha\otimes\beta=\tau'(\alpha\otimes\beta)$. Thus the above becomes

$$\frac{1}{k!}\sum_{\tau\in S_k}(\text{sign}\,\tau')(\text{sign}\,\tau')A(\alpha\otimes\beta)(e_1,\ldots,e_{k+l})$$

$$=A(\alpha\otimes\beta)(e_1,\ldots,e_{k+l})\frac{1}{k!}\sum_{\tau\in S_k}1$$

$$=A(\alpha\otimes\beta)(e_1,\ldots,e_{k+l})$$

Thus $A(A\alpha\otimes\beta)=A(\alpha\otimes\beta)$; that is, $(A\alpha)\wedge\beta=\alpha\wedge\beta$.

The other equality in (i) is similar.

Now (ii) is clear since $\otimes$ is bilinear and A is linear.

ISBN 0-8053-0102-X

For (iii), let $\sigma_0 \in S_{k+l}$ be given by $\sigma_0(1, \ldots, k+l) = (k+1, \ldots, k+l, 1, \ldots, k)$. Then $\alpha \otimes \beta(e_1, \ldots, e_{k+l}) = \beta \otimes \alpha(e_{\sigma_0(1)}, \ldots, e_{\sigma_0(k+l)})$. Hence, by the proof of (i), $A(\alpha \otimes \beta) = (sign\ \sigma_0)A(\beta \otimes \alpha)$. But $sign\ \sigma_0 = (-1)^{kl}$. Finally, (iv) follows from (i). ■

2.3.6 Definition. *The direct sum of the spaces* $\Omega^k(E)$ $(k = 0, 1, 2 \ldots)$ *together with its structure as a real vector space and multiplication induced by* $\wedge$, *is called the* ***exterior algebra*** *of* E, *or the* ***Grassmann algebra*** *of* E.

Using 2.3.5 and a simple induction argument, it follows that if α_i, $i = 1, \ldots, k$ are one-forms, then

$$(\alpha_1 \wedge \cdots \wedge \alpha_k)(e_1, \ldots, e_k) = \sum_\sigma (sign\,\sigma)\alpha_1(e_{\sigma(1)}) \cdots \alpha_k(e_{\sigma(k)})$$

We can now find a basis for $\Omega^k(E)$.

2.3.7 Proposition. *Let* $n = \dim E$. *Then for* $k > n$, $\Omega^k(E) = \{0\}$, *while for* $0 < k \leqslant n$, $\Omega^k(E)$ *has dimension* $\binom{n}{k}$. *The exterior algebra over* E *has dimension* 2^n. *Indeed, if* $\hat{e} = (e_1, \ldots, e_n)$ *is an ordered basis of* E *and* $\hat{e}^* = (\alpha^1, \ldots, \alpha^n)$ *its dual basis, a basis of* $\Omega^k(E)$ *is*

$$\{\alpha^{i_1} \wedge \cdots \wedge \alpha^{i_k} \mid 1 \leqslant i_1 < i_2 < \cdots < i_k \leqslant n\}$$

Proof. First we show that the indicated wedge products span $\Omega^k(E)$. If $t \in \Omega^k(E)$, then from 1.7.2 we know that

$$t = t(e_{i_1}, \ldots, e_{i_k})\alpha^{i_1} \otimes \cdots \otimes \alpha^{i_k}$$

where the summation convention indicates that this should be summed over all choices of $i_1, \ldots, i_k$ between 1 and n, not just the ordered ones of the proposition. Now if the linear operator A is applied to this sum, we have, since $t \in \Omega^k(E)$,

$$t = At = t(e_{i_1}, \ldots, e_{i_k})A(\alpha^{i_1} \otimes \cdots \otimes \alpha^{i_k})$$

so that

$$t(f_1, \ldots, f_k) = t(e_{i_1}, \ldots, e_{i_k})\frac{1}{k!}\sum_{\sigma \in S_k}(sign\,\sigma)(\alpha^{i_1} \otimes \cdots \otimes \alpha^{i_k})(f_{\sigma(1)}, \ldots, f_{\sigma(k)})$$

$$= t(e_{i_1}, \ldots, e_{i_k})\frac{1}{k!}(\alpha^{i_1} \wedge \cdots \wedge \alpha^{i_k})(f_1, \ldots, f_k)$$

ISBN 0-8053-0102-X

by the above remark. Therefore,

$$t=t(e_{i_1},\dots,e_{i_k})\frac{1}{k!}\alpha^{i_1}\wedge\cdots\wedge\alpha^{i_k}$$

The sum still runs over all choices of the $i_1,\dots,i_k$ and we want only distinct, ordered ones. However, since t is skew symmetric, the coefficient $t(e_{i_1},\dots,e_{i_k})$ is 0 if $i_1,\dots,i_k$ are not distinct. If they are distinct and $\sigma\in S_k$, then

$$t(e_{i_1},\dots,e_{i_k})\alpha^{i_1}\wedge\cdots\wedge\alpha^{i_k}=t(e_{\sigma(i_1)},\dots,e_{\sigma(i_k)})\alpha^{\sigma(i_1)}\wedge\cdots\wedge\alpha^{\sigma(i_k)}$$

since both t and the wedge product change by a factor of *sign* σ. [Use 2.3.5(iii), where α and β are one-forms.] Since there are $k!$ of these rearrangements, we are left with

$$t=\sum_{i_1<\cdots<i_k} t(e_{i_1},\dots,e_{i_k})\alpha^{i_1}\wedge\cdots\wedge\alpha^{i_k}$$

Secondly, we show

$$\{\alpha^{i_1}\wedge\cdots\wedge\alpha^{i_k}|i_1<\cdots<i_k\}$$

are linearly independent. Suppose that

$$\sum_{i_1<\cdots<i_k} t_{i_1\dots i_k}\alpha^{i_1}\wedge\cdots\wedge\alpha^{i_k}=0$$

For fixed $i'_1,\dots,i'_k$, let $j'_{k+1},\dots,j'_n$ denote the complementary set of indices, $j'_{k+1}<\cdots<j'_n$. Then

$$\sum_{i_1<\cdots<i_k} t_{i_1\dots i_k}\alpha^{i_1}\wedge\cdots\wedge\alpha^{i_k}\wedge\alpha^{j'_{k+1}}\wedge\cdots\wedge\alpha^{j'_n}=0$$

However, this reduces to

$$t_{i'_1\dots i'_k}\alpha^1\wedge\cdots\wedge\alpha^n=0$$

But $\alpha^1\wedge\cdots\wedge\alpha^n\neq0$, as $\alpha^1\wedge\cdots\wedge\alpha^n(e_1,\dots,e_n)=1$. Hence

$$t_{i'_1\dots i'_k}=0$$

The proposition now follows. ■

2.3.8 Definition. *The nonzero elements of the one-dimensional space $\Omega^n(E)$ are called **volume elements**. If ω_1 and ω_2 are volume elements, we say ω_1 and ω_2 are **equivalent** iff there is a $c>0$ such that $\omega_1=c\omega_2$. An equivalence class of volume elements on E is called an **orientation** on E.*

ISBN 0-8053-0102-X

We shall see shortly the close relationship between volume elements and determinants.

2.3.9 Proposition. *Let $\alpha_1,\ldots,\alpha_k\in E^*$. Then $\alpha_1,\ldots,\alpha_k$ are linearly dependent iff $\alpha_1\wedge\cdots\wedge\alpha_k=0$.*

Proof. If $\alpha_1,\ldots,\alpha_k$ are linearly dependent, then

$$\alpha_i=\sum_{j\neq i}c_j\alpha_j$$

for some i. Then, since $\alpha\wedge\alpha=0$, we see $\alpha_1\wedge\cdots\wedge\alpha_k=0$. Conversely, if $\alpha_1,\ldots,\alpha_k$ are linearly independent, extend to a basis $\alpha_1,\ldots,\alpha_n$. Then $\alpha_1\wedge\cdots\wedge\alpha_n\neq 0$, by 2.3.7 and hence $\alpha_1\wedge\cdots\wedge\alpha_k\neq 0$. ■

2.3.10 Proposition. *Let $\dim(E)=n$ and $\varphi\in L(E,E)$. Then there is a unique constant $\det\varphi$, called the **determinant** of φ, such that $\varphi^*:\Omega^n(E)\to\Omega^n(E)$, defined by $\varphi^*\omega(e_1,\ldots,e_n)=\omega(\varphi(e_1),\ldots,\varphi(e_n))$ satisfies $\varphi^*\omega=(\det\varphi)\omega$ for all $\omega\in\Omega^n(E)$.*

Proof. Clearly $\varphi^*:\Omega^n(E)\to\Omega^n(E)$ is a linear mapping. But, from 2.3.7, $\Omega^n(E)$ is one-dimensional so that if ω_0 is a basis and $\omega=c\omega_0$, $\varphi^*\omega=c\varphi^*\omega_0=b\omega$ for some constant b, clearly unique. ■

It is easy to see that this definition of determinant is the usual one (Exercise 2.3B.) However, it has the advantage of suggesting the proper global definition (Sect. 2.5), as well as making its basic properties trivial, as follows.

2.3.11 Proposition. *Let $\varphi,\psi\in L(E,E)$. Then*

(*i*) *$\det(\varphi\circ\psi)=(\det\varphi)(\det\psi)$;*
(*ii*) *if φ is the identity, $\det\varphi=1$;*
(*iii*) *φ is an isomorphism iff $\det\varphi\neq 0$, and in this case $\det(\varphi^{-1})=(\det\varphi)^{-1}$.*

Proof. For (i), $(\varphi\circ\psi)^*\omega=\det(\varphi\circ\psi)\omega$, but $(\varphi\circ\psi)^*\omega=\psi^*\circ\varphi^*\omega$ as we see from the definitions as in 1.7.17. Hence, $(\varphi\circ\psi)^*\omega=\psi^*(\det\varphi)\omega=(\det\psi)(\det\varphi)\omega$ and (i) follows. (ii) follows at once from the definition. For (iii), suppose φ is an isomorphism with inverse φ^{-1}. Then, by (i) and (iii), $1=\det(\varphi\circ\varphi^{-1})=(\det\varphi)(\det\varphi^{-1})$, and, in particular, $\det\varphi\neq 0$. Conversely, if φ is not an isomorphism there is an $e_1\neq 0$ so $\varphi(e_1)=0$ (Exercise 1.2B). Extend to a basis $e_1,e_2,\ldots,e_n$. Then for all n-forms ω, $\varphi^*\omega(e_1,\ldots,e_n)=\omega(0,\varphi(e_2),\ldots,\varphi(e_n))=0$. Hence, $\det\varphi=0$. ■

Recall from Chapter 1 that there is a unique vector space topology on $L(E,E)$ since it is finite-dimensional. One convenient norm giving this

ISBN 0-8053-0102-X

topology, which was used earlier in 1.7.7, is the following operator norm:

$$\|\varphi\| = sup\{\|\varphi(e)\| \,|\, \|e\|=1\} = sup\left\{\frac{\|\varphi(e)\|}{\|e\|}\,\middle|\, e\neq 0\right\}$$

where $\|e\|$ is a norm on E. (See Exercise 1.2A). Hence, for any $e\in E$,

$$\|\varphi(e)\| \leqslant \|\varphi\|\|e\|$$

2.3.12 Proposition. *$det: L(E,E)\to R$ is continuous.*

Proof. Note that

$$\|\omega\| = sup\{|\omega(e_1,\ldots,e_n)| \,|\, \|e_1\| = \cdots = \|e_n\| = 1\}$$
$$= sup\{|\omega(e_1,\ldots,e_n)|/\|e_1\|\cdots\|e_n\| \,|\, e_1,\ldots,e_n\neq 0\}$$

is a norm on $\Omega^n(E)$ and $|\omega(e_1,\ldots,e_n)| \leqslant \|\omega\|\,\|e_1\|\cdots\|e_n\|$. Then, for $\varphi,\psi\in L(E,E)$,

$$|det\,\varphi - det\,\psi|\|\omega\| = \|\varphi^*\omega - \psi^*\omega\|$$
$$= sup\{|\omega(\varphi(e_1),\ldots,\varphi(e_n)) - \omega(\psi(e_1),\ldots,\psi(e_n))| \,|\, \|e_1\| = \cdots = \|e_n\| = 1\}$$
$$\leqslant sup\{|\omega(\varphi(e_1)-\psi(e_1),\varphi(e_2),\ldots,\varphi(e_n)| + \cdots$$
$$+ |\omega(\psi(e_1),\psi(e_2),\ldots,\varphi(e_n)-\psi(e_n))| \,|\, \|e_1\| = \cdots = \|e_n\| = 1\}$$
$$\leqslant \|\omega\|\|\varphi-\psi\|\{\|\varphi\|^{n-1} + \|\varphi\|^{n-2}\|\psi\| + \cdots + \|\psi\|^{n-1}\}$$
$$\leqslant \|\omega\|\|\varphi-\psi\|(\|\varphi\|+\|\psi\|)^{n-1}$$

Consequently, $|det\,\varphi - det\,\psi| \leqslant \|\varphi-\psi\|(\|\varphi\|+\|\psi\|)^{n-1}$ and the result follows. ■

In 1.3.14 and 1.7.7 we saw that the isomorphisms are an open subset of $L(E,F)$. Using the determinant, we can give a simpler proof in the finite-dimensional case.

2.3.13 Proposition. *Suppose E and F are finite-dimensional and let $GL(E,F)$ denote those $\varphi\in L(E,F)$ that are isomorphisms. Then $GL(E,F)$ is an open subset of $L(E,F)$.*

ISBN 0-8053-0102-X

Proof. If $GL(E, F)=\varnothing$, the conclusion is true. If not, there is an isomorphism $\psi \in GL(E, F)$. A map φ in $L(E, F)$ is an isomorphism if and only if $\psi^{-1}\varphi$ is also. This happens precisely when $det(\psi^{-1}\varphi) \neq 0$. Therefore, $GL(E, F)$ is the inverse image of $\mathbf{R}\backslash\{0\}$ under the map taking φ to $det(\psi^{-1}\varphi)$. Since this is continuous and $\mathbf{R}\backslash\{0\}$ is open, $GL(E, F)$ is also open. ■

In order to define pull-back $\varphi^* t$ or push-forward $\varphi_* t$ of a general tensor t by a map φ, φ needs to be a diffeomorphism. For covariant tensors, however, pull-back makes sense if φ is merely a C^1 map. On the vector space level, this goes as follows.

2.3.14 Definition. *Let $\varphi \in L(E, F)$. For $\alpha \in T_k^0(F)$ define the **pull-back** of α by φ; $\varphi^*\alpha \in T_k^0(E)$ by $\varphi^*\alpha(e_1, \ldots, e_k) = \alpha(\varphi(e_1), \ldots, \varphi(e_k))$. If $\varphi \in GL(E, F)$, we denote by φ_* the **push-forward** map defined in* 1.7.3.

2.3.15 Proposition. *Let $\varphi \in L(E, F)$, $\psi \in L(F, G)$. Then*

(*i*) *$\varphi^*: T_k^0(F) \to T_k^0(E)$ is linear, and $\varphi^*(\Omega^k(F)) \subset \Omega^k(E)$;*
(*ii*) *$(\psi \circ \varphi)^* = \varphi^* \circ \psi^*$;*
(*iii*) *If φ is the identity, so is φ^*;*
(*iv*) *If $\varphi \in GL(E, F)$, then $\varphi^* \in GL(T_k^0(F), T_k^0(E))$, $(\varphi^*)^{-1} = (\varphi^{-1})^*$ and $\varphi^*\Omega^k(F) = \Omega^k(E)$;*
(*v*) *If $\varphi \in GL(E, F)$, then $\varphi_* \in GL(T_k^0(E), T_k^0(F))$, $(\varphi^{-1})^* = \varphi_*$, and $(\varphi_*)^{-1} = (\varphi^{-1})_*$; if $\psi \in GL(F, G)$, $(\psi \circ \varphi)_* = \psi_* \circ \varphi_*$;*
(*vi*) *If $\alpha \in \Omega^k(F)$, $\beta \in \Omega^l(F)$, then $\varphi^*(\alpha \wedge \beta) = \varphi^*\alpha \wedge \varphi^*\beta$.*

Proof. It is evident that (i) follows at once from the definition. For (ii),

$$\begin{aligned}(\psi \circ \varphi)^*\alpha(e_1,\ldots,e_k) &= \alpha(\psi \circ \varphi(e_1),\ldots,\psi \circ \varphi(e_k)) \\ &= \psi^*\alpha(\varphi(e_1),\ldots,\varphi(e_k)) \\ &= \varphi^* \circ \psi^*\alpha(e_1,\ldots,e_k)\end{aligned}$$

Then (iii) is clear and (iv) follows from (ii) and (iii). For (v), $\varphi_*\beta(f_1, \ldots, f_k) = \beta(\varphi^{-1}f_1, \ldots, \varphi^{-1}f_k) = (\varphi^{-1})^*\beta(f_1, \ldots, f_k)$ and $(\varphi_*)^{-1} = (\varphi^{-1})^{*-1} = \varphi^* = (\varphi^{-1})_*$. Finally, $\varphi^*(\alpha \wedge \beta)(e_1, \ldots, e_{k+l}) = \alpha \wedge \beta(\varphi e_1, \ldots, \varphi e_{k+l}) = \varphi^*\alpha \wedge \varphi^*\beta(e_1, \ldots, e_{k+l})$. ■

As in Sect. 1.7, we can consider the exterior algebra on the fibers of a vector bundle as follows.

2.3.16 Definition. *Let $\varphi: U \times F \to U' \times F'$ be a local vector bundle map that is an isomorphism on each fiber. Then define $\varphi_*: U \times \Omega^k(F) \to U' \times \Omega^k(F')$ by $(u, \omega) \mapsto (\varphi(u), \varphi_{u*}\omega)$, where φ_u is the second factor of φ (an isomorphism for each u).*

ISBN 0-8053-0102-X

2.3.17 Proposition. *If $\varphi: U\times F\to U'\times F'$ is a local vector bundle map that is an isomorphism on each fiber, then so is φ_*. Moreover, if φ is a local vector bundle isomorphism, so is φ_*.*

Proof. This is a special case of 1.7.9. ∎

2.3.18 Definition. *Suppose $\pi: E\to B$ is a vector bundle. Define*

$$\omega^k(E)|A=\bigcup_{b\in A}\Omega^k(E_b)$$

where A is a subset of B and $E_b=\pi^{-1}(b)$ is the fiber over $b\in B$. Let $\omega^k(E)|B=\omega^k(E)$ and define $\omega^k(\pi): \omega^k(E)\to B$ by $\omega^k(\pi)(t)=b$ if $t\in\Omega^k(E_b)$.

2.3.19 Theorem. *Suppose $\{E|U_i, \varphi_i\}$ is a vector bundle atlas of π, where φ_i: $E|U_i\to U_i'\times F_i'$. Then $\{\omega^k(E)|U_i, \varphi_{i_*}\}$ is a vector bundle atlas of $\omega^k(\pi)$: $\omega^k(E)\to B$, where $\varphi_{i_*}: \omega^k(E)|U_i\to U_i'\times\Omega^k(F_i')$ is defined by $\varphi_{i_*}|E_b=(\varphi_i|E_b)_*$ (as in 2.3.16).*

Proof. We must verify (VBA 1) and (VBA 2) of 1.5.2: (VBA 1) is clear; for (VBA 2) let φ_i, φ_j be two charts on π, so that $\varphi_i\circ\varphi_j^{-1}$ is a local vector bundle isomorphism. (We may assume $U_i=U_j$.) But then from 2.3.15, $\varphi_{i_*}\circ\varphi_{j_*}^{-1}=(\varphi_i\circ\varphi_j^{-1})_*$, which is a local vector bundle isomorphism by 2.3.17. ∎

Because of this theorem, the vector bundle structure of $\pi: E\to B$ induces naturally a vector bundle structure on $\omega^k(\pi)$: $\omega^k(E)\to B$, which is also Hausdorff, second countable, and of constant dimension. Hereafter $\omega^k(\pi)$ will denote this vector bundle.

EXERCISES

2.3A. If $k!$ is omitted in the definition of A (2.3.2), show that $\wedge$ fails to be associative.

2.3B. Show that, in terms of components, our definition of the determinant is the usual one.

2.3C. If α is a two-form and β is a one-form, show that

$$(\alpha\wedge\beta)(e_1,e_2,e_3)=\alpha(e_1,e_2)\beta(e_3)-\alpha(e_1,e_3)\beta(e_2)+\alpha(e_2,e_3)\beta(e_1)$$

2.3D. Show that if $e_1,\ldots,e_n$ is a basis of E and $\alpha^1,\ldots,\alpha^n$ is the dual basis, then $(\alpha^1\wedge\cdots\wedge\alpha^n)(e_1,\ldots,e_n)=1$.

2.4 CARTAN'S CALCULUS OF DIFFERENTIAL FORMS

We now specialize the exterior algebra of the preceding section to tangent bundles and develop a differential calculus that is special to this case. This is basic to the dual integral calculus of Sect. 2.6 and to the Hamiltonian mechanics of Chapter 3.

ISBN 0-8053-0102-X

If $\tau_M : TM \to M$ is the tangent bundle of a manifold M, let $\omega^k(M) = \omega^k(TM)$, and $\omega_M^k = \omega^k(\tau_M)$, so $\omega_M^k : \omega^k(M) \to M$ is the vector bundle of exterior k forms on the tangent spaces of M. Also, let $\Omega^0(M) = \mathfrak{F}(M)$, $\Omega^1(M) = \mathfrak{T}_1^0(M)$, and $\Omega^k(M) = \Gamma^\infty(\omega_M^k), k = 2, 3, \ldots$.

2.4.1 Proposition. *Regarding $\mathfrak{T}_k^0(M)$ as an $\mathfrak{F}(M)$ module, $\Omega^k(M)$ is an $\mathfrak{F}(M)$ submodule.*

Proof. If $t_1, t_2 \in \Omega^k(M)$ and $f \in \mathfrak{F}(M)$, we must show $f \otimes t_1 + t_2 \in \Omega^k(M)$. From 1.7.19, we have $f \otimes t_1 + t_2 \in \mathfrak{T}_k^0(M)$. But, by 2.3.1, $f \otimes t_1(m) + t_2(m) \in \Omega^k(T_m M)$ and the result follows. ∎

2.4.2 Proposition. *If $\alpha \in \Omega^k(M)$ and $\beta \in \Omega^l(M), k, l = 0, 1, \ldots, n$, define $\alpha \wedge \beta : M \to \omega^{k+l}(M)$ by $(\alpha \wedge \beta)(m) = \alpha(m) \wedge \beta(m)$. Then $\alpha \wedge \beta \in \Omega^{k+l}(M)$, and $\wedge$ is bilinear and associative.*

Proof. First, $\wedge$ is bilinear and associative by 2.3.5. To show $\alpha \wedge \beta$ is of class C^∞, consider the local representative of $\alpha \wedge \beta$ in natural charts. This is a map of the form $(\alpha \wedge \beta)_\varphi = \boldsymbol{B} \circ (\alpha_\varphi \times \beta_\varphi)$, with $\alpha_\varphi, \beta_\varphi, C^\infty$ and $\boldsymbol{B} = \wedge$, which is bilinear. Thus $(\alpha \wedge \beta)_\varphi$ is C^∞ by Leibniz' rule. ∎

2.4.3 Definition. *Let $\Omega(M)$ denote the direct sum of $\Omega^k(M)$, $k = 0, 1, \ldots, n$, together with its structure as an (infinite-dimensional) real vector space and with the multiplication $\wedge$ extended componentwise to $\Omega(M)$. We call $\Omega(M)$ the **algebra of exterior differential forms** on M. Elements of $\Omega^k(M)$ are called **k-forms**. In particular, elements of $\mathfrak{X}^*(M)$ are called **one-forms**.*

Note that we generally regard $\Omega(M)$ as a real vector space rather than an $\mathfrak{F}(M)$ module [as with $\mathfrak{T}(M)$]. The reason is that $\mathfrak{F}(M) = \Omega^0(M)$ is included in the direct sum, and $f \wedge \alpha = f \otimes \alpha = f\alpha$.

2.4.4 Notation. *Let (U, φ) be a chart on a manifold M with $U' = \varphi(U) \subset \boldsymbol{R}^n$. Let $\boldsymbol{e}_i$ denote the standard basis of $\boldsymbol{R}^n$ and let $\underline{e}_i(u) = T_{\varphi(u)}\varphi^{-1}(\varphi(u), \boldsymbol{e}_i)$. Similarly let $\boldsymbol{\alpha}^i$ denote the dual basis of $\boldsymbol{e}_i$ and $\underline{\alpha}^i(u) = (T_u\varphi)^*(\varphi(u), \boldsymbol{\alpha}^i)$. [Thus, for each $u \in U, \underline{e}_i(u)$ and $\underline{\alpha}^i(u)$ are dual bases of the fiber $T_u M$.] Then if $\varphi(u) = (x^1(u), \ldots, x^n(u)) \in \boldsymbol{R}^n$, we define*

$$\frac{\partial f}{\partial x^i} = L_{\underline{e}_i} f = \frac{\partial f_\varphi}{\partial y^i} \circ \varphi$$

at points $u \in U$.

With these notations, we see $dx^i(u) = \underline{\alpha}^i(u)$, for

$$dx^i(u)\big(\underline{e}_j(u)\big) = P_2 T_u x^i \circ T_{\varphi(u)}\varphi^{-1}(\varphi(u), \boldsymbol{e}_j) = P_2 T_u \big(x^i \circ \varphi^{-1}\big)(\varphi(u), \boldsymbol{e}_j)$$

$$= D\big(x^i \circ \varphi^{-1}\big)(\varphi(u)) \cdot \boldsymbol{e}_j = \delta_j^i$$

ISBN 0-8053-0102-X

Hence,

$$df(u)=df(\underline{e}_i)\underline{\alpha}^i(u)=\frac{\partial f}{\partial x^i}(u)\,dx^i(u)$$

Thus the components of the differential df are the partial derivatives $\partial f/\partial x^i$.

Also, for each $t\in\mathfrak{T}^r_s(U)$ we have

$$t(u)=t^{i_1\cdots i_r}_{j_1\cdots j_s}(u)\underline{e}_{i_1}\otimes\cdots\otimes\underline{e}_{i_r}\otimes dx^{j_1}\otimes\cdots\otimes dx^{j_s}$$

and for each $\omega\in\Omega^k(U)$

$$\omega(u)=\sum_{i_1<\cdots<i_k}\omega_{i_1\cdots i_k}(u)\,dx^{i_1}\wedge\cdots\wedge dx^{i_k}(u)$$

where

$$t^{i_1\cdots i_r}_{j_1\cdots j_s}=t\left(dx^{i_1},\ldots,dx^{i_r},\underline{e}_{j_1},\ldots,\underline{e}_{j_s}\right)$$

and

$$\omega_{i_1\cdots i_k}=\omega(\underline{e}_{i_1},\ldots,\underline{e}_{i_k})$$

The extension of d to $\Omega^k(M)$ is given by the following.

2.4.5 Theorem. *Let M be a manifold. Then there is a unique family of mappings $d^k(U)\colon\Omega^k(U)\to\Omega^{k+1}(U)$ ($k=0,1,2,\ldots,n$, and U is open in M), which we merely denote by d, called the* ***exterior derivative*** *on M, such that*

(i) *d is a $\wedge$* ***antiderivation.*** *That is, d is $\mathbb{R}$ linear and for $\alpha\in\Omega^k(U)$, $\beta\in\Omega^l(U)$,*

$$d(\alpha\wedge\beta)=d\alpha\wedge\beta+(-1)^k\alpha\wedge d\beta$$

(ii) *If $f\in\mathfrak{F}(U)$, $df=df$ (as defined in 2.2.1);*

(iii) *$d\circ d=0$ (that is, $d^{k+1}(U)\circ d^k(U)=0$);*

(iv) *d is* ***natural with respect to restrictions;*** *that is, if $U\subset V\subset M$ are open and $\alpha\in\Omega^k(V)$, then $d(\alpha|U)=(d\alpha)|U$, or the following diagram commutes:*

$$\begin{array}{ccc}\Omega^k(V) & \xrightarrow{\;|U\;} & \Omega^k(U)\\ {\scriptstyle d}\downarrow & & \downarrow{\scriptstyle d}\\ \Omega^{k+1}(V) & \xrightarrow[\;|U\;]{} & \Omega^{k+1}(U)\end{array}$$

As in Sect. 2.2, condition (iv) means that d is a local operator.

ISBN 0-8053-0102-X

Proof. We first establish uniqueness. Using (iv) it is sufficient to consider the local case $\omega \in \Omega^k(U)$; $U \subset M$. By $\boldsymbol{R}$ linearity, it is sufficient to consider the case in which ω has the form $\omega = f_0\, df_1 \wedge \cdots \wedge df_k$, where $f_i \in \mathfrak{F}(U)$. Hence, from (i), (ii), and (iii), $d\omega = df_0 \wedge df_1 \wedge \cdots \wedge df_k$ and thus, $d\omega$ is uniquely determined.

For existence we may again suppose $\omega = f_0\, df_1 \wedge \cdots \wedge df_k$ in some chart, and define $d\omega = df_0 \wedge df_1 \wedge \cdots \wedge df_k$, which is independent of the chart (exercise). Then (ii) and (iv) are clear, as is $\boldsymbol{R}$ linearity. To prove (i), note that if $\rho = g_0\, dg_1 \wedge \cdots \wedge dg_l$, then

$$\begin{aligned} d(\omega \wedge \rho) &= d(f_0 g_0) \wedge df_1 \wedge \cdots \wedge df_k \wedge dg_1 \wedge \cdots \wedge dg_l \\ &= g_0 df_0 \wedge df_1 \wedge \cdots \wedge df_k \wedge dg_1 \wedge \cdots \wedge dg_l \\ &\quad + f_0 dg_0 \wedge df_1 \wedge \cdots \wedge df_k \wedge dg_1 \wedge \cdots \wedge dg_l \\ &= d\omega \wedge \rho + (-1)^k \omega \wedge d\rho \end{aligned}$$

Finally, for (iii), it is clearly sufficient to verify $d \circ df = 0$ for functions. But in a local chart $df(u) = Df(u) \cdot e_i\, dx^i$ so that

$$\begin{aligned} d \circ df(u) &= DDf(u) \cdot (e_i, e_j)\, dx^j \wedge dx^i \\ &= \frac{\partial^2 f}{\partial x^i \partial x^j}\, dx^j \wedge dx^i = 0 \end{aligned}$$

by symmetry of the mixed partial derivatives. ■

2.4.6 Corollary. *Let $\omega \in \Omega^k(U)$, where $U \subset E$ (open). Then*

$$d\omega(u)(e_0, \ldots, e_k) = \sum_{i=0}^{k} (-1)^i D\omega(u) \cdot e_i(e_0, \ldots, \hat{e}_i, \ldots, e_k)$$

where $\hat{e}_i$ denotes that e_i is deleted. Also, we denote elements (u, e) of TU merely by e, for brevity. [Note that $D\omega(u) \cdot e \in L^k(E, \boldsymbol{R})$.]

Proof. First note that d defined this way is a map $\Omega^k(U) \to \Omega^{k+1}(U)$. Then it is sufficient to verify (i)–(iv) of 2.4.5. But $\boldsymbol{R}$ linearity, (ii), and (iv) are clear, and as $\wedge$ is bilinear, $D(\omega \wedge \rho) = \omega \wedge D\rho + D\omega \wedge \rho$, from which (i) readily follows. Finally, (iii) follows as in 2.4.5. ■

2.4.7 Definition. *Suppose $F\colon M \to N$ is a C^∞ mapping of manifolds. For $\omega \in \Omega^k(N)$, define $F^*\omega\colon M \to \omega^k(M)$ by $F^*\omega(m) = (T_m F)^* \circ \omega \circ F(m)$ (see 2.3.14). We say $F^*\omega$ is the **pull-back** of ω by F.*

Especially, note if $g \in \Omega^0(N)$, $F^*g = g \circ F$.

ISBN 0-8053-0102-X

2.4.8 Proposition. *Let $F: M \to N$ and $G: N \to W$ be C^∞ mappings of manifolds. Then*

(i) $F^*: \Omega^k(N) \to \Omega^k(M)$;
(ii) $(G \circ F)^* = F^* \circ G^*$;
(iii) *if $H: M \to M$ is the identity, then $H^*: \Omega^k(M) \to \Omega^k(M)$ is the identity;*
(iv) *if F is a diffeomorphism, then F^* is a vector bundle isomorphism and $(F^*)^{-1} = (F^{-1})^*$.*

Proof. Choose charts (U, φ), (V, ψ) of M and N so that $F(U) \subset V$, then $F_{\varphi\psi} = \psi \circ F \circ \varphi^{-1}$ is of class C^∞, as is $\omega_\psi = (T\psi)_* \circ \omega \circ \psi^{-1}$. Then

$$(T_{u'}T_{\varphi\psi})^* = (T_{u'}\varphi^{-1})^* \circ (T_u F)^* \circ (T_{F(u)}\psi)^* \qquad \text{(by 2.3.15)}$$

$$= (T_u \varphi)_* \circ (T_u F)^* \circ (T_{F(u)}\psi)^*$$

Hence the local representative of $F^*\omega$ is

$$(F^*\omega)_\varphi(u') = (T\varphi)_* \circ F^*\omega \circ \varphi^{-1}(u')$$
$$= (T_{u'}F_{\varphi\psi})^* \circ \omega_\psi \circ F_{\varphi\psi}(u')$$

which is of class C^∞ by the composite mapping theorem; $\mathbf{R}$ linearity is clear.

For (ii), we merely note that it holds for the local representatives by 2.3.15; (iii) follows at once from the definition; and (iv) follows in the usual way from (ii) and (iii) ■

As $F^*: \Omega^k(N) \to \Omega^k(M)$ is $\mathbf{R}$ linear, it induces a mapping on the direct sums, $F^*: \Omega(N) \to \Omega(M)$, which are differential algebras with $\wedge$ and $\mathbf{d}$.

2.4.9 Theorem. *Let $F: M \to N$ be of class C^∞. Then $F^*: \Omega(N) \to \Omega(M)$ is a homomorphism of differential algebras; that is,*

(i) $F^*(\psi \wedge \omega) = F^*\psi \wedge F^*\omega$, *and*
(ii) ***d is natural with respect to mappings;*** *that is, $F^*(\mathbf{d}\omega) = \mathbf{d}(F^*\omega)$, or the following diagram commutes:*

$$\begin{array}{ccc} \Omega^k(N) & \xrightarrow{F^*} & \Omega^k(M) \\ \mathbf{d}\downarrow & & \downarrow\mathbf{d} \\ \Omega^{k+1}(N) & \xrightarrow[F^*]{} & \Omega^{k+1}(M) \end{array}$$

ISBN 0-8053-0102-X

Proof. We first consider $F^*(\psi\wedge\omega)$ when ψ is a function. Then

$$\begin{aligned} F^*(\psi\omega)(m) &= (T_m F)^* \circ \psi\omega \circ F(m) \\ &= (T_m F)^* \circ \left[(\psi \circ F)\cdot(\omega \circ F) \right](m) \\ &= \psi(F(m)) F^*\omega(m) \end{aligned}$$

or $F^*(\psi\wedge\omega)=F^*\psi\wedge F^*\omega$, as $F^*\psi=\psi\circ F$ if $\psi\in\Omega^0(N)$. Then (i) follows immediately from 2.3.15(vi). For (ii) we shall show in fact that if $m\in M$, there is a neighborhood U of $m\in M$ such that $d(F^*\omega|U)=(F^*d\omega)|U$, which is sufficient, as F^* and d are both natural with respect to restriction. Let (V,φ) be a local chart at $F(m)$ and U a neighborhood of $m\in M$ with $F(U)\subset V$. Then for $\omega\in\Omega^k(V)$, we can write

$$\omega=\omega_{i_1\cdots i_k}\, dx^{i_1}\wedge\cdots\wedge dx^{i_k}$$

$$d\omega=\partial_{i_0}\omega_{i_1\cdots i_k}\, dx^{i_0}\wedge\cdots\wedge dx^{i_k}, \qquad \partial_{i_0}=\frac{\partial}{\partial x^{i_0}}$$

and by (i) above

$$F^*\omega|U=(F^*\omega_{i_1\cdots i_k})F^*dx^{i_1}\wedge\cdots\wedge F^*dx^{i_k}$$

But if $\psi\in\Omega^0(N)$, $d(F^*\psi)=F^*d\psi$ by the composite mapping theorem, so

$$\begin{aligned} d(F^*\omega|U) &= F^*(d\omega_{i_1\cdots i_k})\wedge F^*dx^{i_1}\wedge\cdots\wedge F^*dx^{i_k} \\ &= F^*(d\omega)|U \end{aligned}$$

by (i) above. ■

2.4.10 Corollary. *The operator* d *is natural with respect to diffeomorphisms. That is, if* $F: M\to N$ *is a diffeomorphism, then* $F_*d\omega=dF_*\omega$, *or the following diagram commutes:*

$$\begin{array}{ccc} \Omega^k(M) & \xrightarrow{\;F_*\;} & \Omega^k(N) \\ \Big\downarrow d & & \Big\downarrow d \\ \Omega^{k+1}(M) & \xrightarrow[\;F_*\;]{} & \Omega^{k+1}(N) \end{array}$$

Proof. With F_* defined as $F_*=(F)^0_k$, we see that $F_*=(F^{-1})^*$. The result then follows from 2.4.9(ii). ■

ISBN 0-8053-0102-X

The next few propositions give some important relations between the Lie derivative and the exterior derivative.

2.4.11 Theorem. *Let $X \in \mathfrak{X}(M)$. Then $\boldsymbol{d}$ is natural with respect to $\boldsymbol{L}_X$. That is, for $\omega \in \Omega^k(M)$ we have $\boldsymbol{L}_X\omega \in \Omega^k(M)$ and $\boldsymbol{d}\boldsymbol{L}_X\omega = \boldsymbol{L}_X\,\boldsymbol{d}\omega$, or the following diagram commutes;*

$$\begin{array}{ccc} \Omega^k(M) & \xrightarrow{\ L_X\ } & \Omega^k(M) \\ d\downarrow & & \downarrow d \\ \Omega^{k+1}(M) & \xrightarrow[\ L_X\ ]{} & \Omega^{k+1}(M) \end{array}$$

Proof. If $\alpha^1, \ldots, \alpha^k \in \Omega^1(M)$ we have

$$L_X(\alpha^1 \wedge \cdots \wedge \alpha^k) = L_X\alpha^1 \wedge \alpha^2 \wedge \cdots \wedge \alpha^k + \cdots + \alpha^1 \wedge \cdots \wedge L_X\alpha^k$$

This follows from the fact that L_X is $\boldsymbol{R}$ linear and is a tensor derivation. Since locally $\omega \in \Omega^k(M)$ is a linear combination of such products, it readily follows that $L_X\omega \in \Omega^k(M)$. For the second part, let (U, a, F) be a flow box at $m \in M$, so that from 2.2.20,

$$L_X\omega(m) = \frac{d}{d\lambda}(F_\lambda^*\omega)(m)\Big|_{\lambda=0}$$

But from 2.4.10 we have $F_\lambda^*\, d\omega = d(F_\lambda^*\omega)$. Then, since d is $\boldsymbol{R}$ linear, it commutes with $d/d\lambda$ and so $dL_X\omega = L_X\, d\omega$. ■

The foregoing proof can also be carried out in terms of local representatives.

2.4.12 Definition. *Let M be a manifold, $X \in \mathfrak{X}(M)$, and $\omega \in \Omega^{k+1}(M)$. Then define $\boldsymbol{i}_X\omega \in \mathfrak{T}_k^0(M)$ by*

$$i_X\omega(X_1, \ldots, X_k) = \omega(X, X_1, \ldots, X_k)$$

*If $\omega \in \Omega^0(M)$, we put $\boldsymbol{i}_X\omega = 0$. We call $\boldsymbol{i}_X\omega$ the **inner product** of X and ω.*

2.4.13 Theorem. *We have $\boldsymbol{i}_X\colon \Omega^k(M) \to \Omega^{k-1}(M)$, $k = 1, \ldots, n$, and, for $\alpha \in \Omega^k(M)$, $\beta \in \Omega^l(M)$, $f \in \Omega^0(M)$,*

(i) $\boldsymbol{i}_X$ *is a* $\wedge$ *antiderivation. That is,* $\boldsymbol{i}_X$ *is* $\boldsymbol{R}$ *linear and* $i_X(\alpha \wedge \beta) = (i_X\alpha) \wedge \beta + (-1)^k \alpha \wedge (i_X\beta)$;

(ii) $i_{fX}\alpha = f i_X\alpha$;

(iii) $i_X\, df = L_X f$;

(iv) $L_X\alpha = i_X\, d\alpha + d i_X\alpha$;

(v) $L_{fX}\alpha = fL_X\alpha + df \wedge i_X\alpha$.

ISBN 0-8053-0102-X

Proof. That $i_X\alpha \in \Omega^{k-1}(M)$ follows at once from 2.2.8. For (i), $\mathbf{R}$ linearity is clear. For the second part of (i)

$$i_X(\alpha \wedge \beta)(X_2, X_3, \ldots, X_{k+l}) = (\alpha \wedge \beta)(X, X_2, \ldots, X_{k+l})$$

and

$$i_X\alpha \wedge \beta + (-1)^k \alpha \wedge i_X\beta = \frac{(k+l-1)!}{(k-1)!\,l!} A(i_X\alpha \otimes \beta)$$

$$+ (-1)^k \frac{(k+l-1)!}{k!\,(l-1)!} A(\alpha \otimes i_X\beta)$$

But the sum over all permutations in the last term can be replaced by the sum over $\sigma\sigma_0$, where σ_0 is the permutation $(2, 3, \ldots, k+1, 1, k+2, \ldots, k+l) \mapsto (1, 2, 3, \ldots, k+l)$ whose sign is $(-1)^k$. Hence (i) follows. For (ii), we merely note α_X is linear, and (iii) is just the definition of $L_X f$.

For (iv) we proceed by induction on k. First note that for $k=0$, (iv) reduces to (iii). Now assume that (iv) holds for k. Then a $k+1$ form may be written as $\Sigma df_i \wedge \omega_i$, where ω_i is a k form, in some neighborhood of $m \in M$. But $L_X(df \wedge \omega) = L_X\, df \wedge \omega + df \wedge L_X\omega$ and

$$\begin{aligned} i_X d(df \wedge \omega) + d i_X(df \wedge \omega) &= -i_X(df \wedge d\omega) + d(i_X df \wedge \omega - df \wedge i_X\omega) \\ &= -i_X df \wedge d\omega + df \wedge i_X d\omega + d i_X df \wedge \omega \\ &\quad + i_X df \wedge d\omega + df \wedge d i_X \omega \\ &= df \wedge L_X\omega + dL_X f \wedge \omega \end{aligned}$$

by our inductive assumption and (iii). Since $dL_X f = L_X df$, the result follows.

Finally for (v) we have

$$\begin{aligned} L_{fX}\omega = i_{fX} d\omega + d i_{fX}\omega &= f i_X d\omega + d(f i_X\omega) \\ &= f i_X d\omega + df \wedge i_X\omega + f d i_X\omega \\ &= f L_X\omega + df \wedge i_X\omega \quad \blacksquare \end{aligned}$$

The behavior of inner products under diffeomorphisms is given by the following.

2.4.14 Proposition. *Let M and N be manifolds and $f: M \to N$ a diffeomorphism. Then, if $\omega \in \Omega^k(N)$ and $X \in \mathfrak{X}(N)$, we have*

$$i_{f^* X} f^*\omega = f^* i_X\omega$$

ISBN 0-8053O102-X

that is, inner products are natural with respect to diffeomorphisms; that is, the following diagram commutes:

$$\begin{array}{ccc} \Omega^k(N) & \xrightarrow{\;f^*\;} & \Omega^k(M) \\ \downarrow{\scriptstyle i_X} & & \downarrow{\scriptstyle i_{f^*X}} \\ \Omega^{k-1}(N) & \xrightarrow[\;f^*\;]{} & \Omega^{k-1}(M) \end{array}$$

Similarly for $Y \in \mathfrak{X}(M)$ we have the following commutative diagram:

$$\begin{array}{ccc} \Omega^k(M) & \xrightarrow{\;f_*\;} & \Omega^k(N) \\ \downarrow{\scriptstyle i_Y} & & \downarrow{\scriptstyle i_{f_*Y}} \\ \Omega^{k-1}(M) & \xrightarrow[\;f_*\;]{} & \Omega^{k-1}(N) \end{array}$$

Proof. Let $v_1, \ldots, v_{k-1} \in T_m(M)$ and $n = f(m)$. Then by 2.4.12 and 2.4.7

$$\begin{aligned} & i_{f^*X} f^*\omega(m)\cdot(v_1,\ldots,v_{k-1}) \\ &= f^*\omega(m)\cdot\big(f^*X(m), v_1,\ldots,v_{k-1}\big) \\ &= f^*\omega(m)\cdot\big(Tf^{-1}\circ X(n), v_1,\ldots,v_{k-1}\big) \\ &= \omega(n)\cdot\big(Tf\circ Tf^{-1}X(n), Tfv_1,\ldots,Tfv_{k-1}\big) \\ &= i_X\omega(n)\cdot(Tfv_1,\ldots,Tfv_{k-1}) \\ &= f^* i_X\omega(m)\cdot(v_1,\ldots,v_{k-1}) \qquad \blacksquare \end{aligned}$$

The next proposition expresses d in terms of the Lie derivative (Palais [1963]).

2.4.15 Proposition. *Let $X_i \in \mathfrak{X}(M)$, $i = 0, \ldots, k$, and $\omega \in \Omega^k(M)$. Then we have*

(i) $$(L_{X_0}\omega)(X_1,\ldots,X_k) = L_{X_0}(\omega(X_1,\ldots,X_k)) - \sum_{i=1}^{k} \omega(X_1,\ldots,L_{X_0}X_i,\ldots,X_k)$$

(ii) $$d\omega(X_0, X_1,\ldots,X_k) = \sum_{i=0}^{k}(-1)^i L_{X_i}(\omega(X_0,\ldots,\hat{X}_i,\ldots,X_k)) + \sum_{0<i<j<k}(-1)^{i+j}\omega(L_{X_i}(X_j), X_0,\ldots,\hat{X}_i,\ldots,\hat{X}_j,\ldots,X_k)$$

where $\hat{X}_i$ denotes that X_i is deleted.

ISBN 0-8053-0102-X

Proof. Part (i) is exactly condition (DO 4) following 2.2.17. For (ii) we proceed by induction. For $k=0$, it is merely $d\omega(X_0)=L_{X_0}\omega$. Assume the formula for $k-1$. Then if $\omega\in\Omega^k(M)$, we have, by 2.4.13(iv),

$$\begin{aligned} d\omega(X_0, X_1, \ldots, X_k) &= (i_{X_0}\, d\omega)(X_1, \ldots, X_k) \\ &= (L_{X_0}\omega)(X_1, \ldots, X_k) - (d(i_{X_0}\omega))(X_1, \ldots, X_k) \\ &= L_{X_0}(\omega(X_1, \ldots, X_k)) \\ &\quad - \sum_1^k \omega(X_1, \ldots, L_{X_0}X_i, \ldots, X_k) \\ &\quad - (di_{X_0}\omega)(X_1, \ldots, X_k) \qquad \text{(by (i))} \end{aligned}$$

But $i_{X_0}\omega\in\Omega^{k-1}(M)$ and we may apply the induction assumption. This gives, after a simple permutation and 2.4.12,

$$\begin{aligned} (d(i_{X_0}\omega))(X_1, \ldots, X_k) &= \sum_{i=1}^k (-1)^{i-1} L_{X_i}(\omega(X_0, X_1, \ldots, \hat{X}_i, \ldots, X_k)) \\ &\quad - \sum_{1\leq i<j\leq k} (-1)^{i+j}\omega(L_{X_i}X_j, X_0, X_1, \ldots, \hat{X}_i, \ldots, \hat{X}_j, \ldots, X_k) \end{aligned}$$

Substituting this into the above easily yields the result. ■

2.4.16 Definition. *We call $\omega\in\Omega^k(M)$ **closed** if $d\omega=0$, and **exact** if there is an $\alpha\in\Omega^{k-1}(M)$ such that $\omega=d\alpha$.*

2.4.17 Theorem. (*i*) *Every exact form is closed.*

(*ii*) (***Poincaré lemma***). *If ω is closed, then for each $m\in M$, there is a neighborhood U of m for which $\omega|U\in\Omega^k(U)$ is exact.*

Proof. Part (i) is clear since $d\circ d=0$. Using a local chart and 2.4.9(ii) together with 2.4.5(iv), it is sufficient to consider the case $\omega\in\Omega^k(U)$, $U\subset E$ a disk about $0\in E$, to prove (ii). On U we construct an R linear mapping H: $\Omega^k(U)\to\Omega^{k-1}(U)$ such that $d\circ H+H\circ d$ is the identity on $\Omega^k(U)$. This will give the result, for $d\omega=0$ implies $d(H\omega)=\omega$.

For $e_1,\ldots,e_k\in E$ define

$$H\omega(u)(e_1,\ldots,e_{k-1})=\int_0^1 t^{k-1}\omega(tu)(u,e_1,\ldots,e_{k-1})\,dt$$

ISBN 0-8053-0102-X

Then, by 2.4.6,

$$dH\omega(u)\cdot(e_1,\dots,e_k)=\sum_{i=1}^{k}(-1)^{i+1}DH\omega(u)\cdot e_i(e_1,\dots,\hat{e}_i,\dots,e_k)$$

$$=\sum_{i=1}^{k}(-1)^{i+1}\int_0^1 t^{k-1}\omega(tu)(e_i,e_1,\dots,\hat{e}_i,\dots,e_k)\,dt$$

$$+\sum_{i=1}^{k}(-1)^{i+1}\int_0^1 t^k D\omega(tu)\cdot e_i(u,e_1,\dots,\hat{e}_i,\dots,e_k)\,dt$$

(The interchange of D and $\int$ is permissible, as ω is smooth and bounded over $t\in[0,1]$.) However, we also have, by 2.4.6,

$$H\,d\omega(u)\cdot(e_1,\dots,e_k)=\int_0^1 t^k\,d\omega(tu)(u,e_1,\dots,e_k)\,dt$$

$$=\int_0^1 t^k D\omega(tu)\cdot u(e_1,\dots,e_k)\,dt$$

$$+\sum_{i=1}^{k}(-1)^{i}\int_0^1 t^k D\omega(tu)\cdot e_i(u,e_1,\dots,\hat{e}_i,\dots,e_k)\,dt$$

Hence

$$[dH\omega(u)+H\,d\omega(u)](e_1,\dots,e_k)=\int_0^1 kt^{k-1}\omega(tu)\cdot(e_1,\dots,e_k)\,dt$$

$$+\int_0^1 t^k D\omega(tu)\cdot u(e_1,\dots,e_k)\,dt$$

$$=\int_0^1 \frac{d}{dt}[t^k\omega(tu)\cdot(e_1,\dots,e_k)]\,dt$$

$$=\omega(u)\cdot(e_1,\dots,e_k)$$

which proves the assertion. ■

There is another proof of the Poincaré lemma that is useful to understand. This proof will help the reader master the proof of Darboux' theorem in Sect. 3.2, and is similar in spirit to the proof of Frobenius' theorem (2.2.26).

Alternative Proof of the Poincaré Lemma. We again let U be a ball about $\mathbf{0}$ in E. Let, for $t>0$, $F_t(u)=tu$. Thus F_t is a diffeomorphism and, starting at

ISBN 0-8053-0102-X

$t=1$, is generated by the time-dependent vector field

$$X_t(u)=u/t$$

that is, $F_1(u)=u$ and $dF_t(u)/dt=X_t(F_t(u))$. Therefore, since ω is closed,

$$\begin{aligned}\frac{d}{dt}F_t^*\omega &= F_t^* L_{X_t}\omega \\ &= F_t^*(di_{X_t}\omega) \\ &= d(F_t^* i_{X_t}\omega)\end{aligned}$$

For $0<t_0<1$, we get

$$\omega - F_{t_0}^*\omega = d\int_{t_0}^1 F_t^* i_{X_t}\omega\, dt$$

Letting $t_0 \to 0$, we get $\omega = d\beta$, where

$$\beta = \int_0^1 F_t^* i_{X_t}\omega\, dt$$

Explicitly,

$$\beta_u(e_1,\ldots,e_{k-1}) = \int_0^1 t^{k-1}\omega_{tu}(u,e_1,\ldots,e_{k-1})\,dt$$

(Note that this β agrees with that in the previous proof.) ■

See Exercise 2.4E for a relative Poincaré lemma.

It is not true that closed forms are always exact (for example, on a sphere). In fact, the quotient groups of closed forms by exact forms (called the de Rham cohomology groups of M) shed light on the manifold topology. A discussion may be found in Flanders [1963], Singer and Thorpe [1967], and in de Rham [1955].

In differential geometry the use of vector valued forms is important; that is, one replaces multilinear maps into $\boldsymbol{R}$ by multilinear maps into a vector space V. One can utilize the exterior calculus by taking the components of the form. For applications to geometry, see Kobayashi–Nomizu [1963], Chern [1972], or Spivak [1974].

The following table summarizes some of the important algebraic identities involving differential forms that have been obtained.

ISBN 0-8053-0102-X

Table 2.4-1

1. Vector fields on M with the bracket $[X, Y]$ form a Lie algebra; that is, $[X, Y]$ is real bilinear, skew symmetric, and Jacobi's identity holds:
$$[[X,Y],Z]+[[Z,X],Y]+[[Y,Z],X]=0.$$
2. For a diffeomorphism f, $f_*[X,Y]=[f_*X,f_*Y]$ and $(f\circ g)_*X=f_*g_*X$.
3. The forms on a manifold are a real associative algebra with $\wedge$ as multiplication. Furthermore, $\alpha\wedge\beta=(-1)^{kl}\beta\wedge\alpha$ for k and l forms α and β, respectively.
4. If f is a map, $f^*(\alpha\wedge\beta)=f^*\alpha\wedge f^*\beta$, $(f\circ g)^*\alpha=g^*f^*\alpha$.
5. d is a real linear map on forms and:
$$dd\alpha=0,\ d(\alpha\wedge\beta)=d\alpha\wedge\beta+(-1)^k\alpha\wedge d\beta \text{ for } \alpha \text{ a } k\text{-form.}$$
6. For α a k-form and $X_0,\ldots,X_k$ vector fields:
$$d\alpha(X_0,\ldots,X_k)=\sum_{i=0}^{k}(-1)^i X_i\big(\alpha(X_0,\ldots,\hat{X}_i,\ldots,X_k)\big)$$
$$+\sum_{i<j}(-1)^{i+j}\alpha([X_i,X_j],X_0,\ldots,\hat{X}_i,\ldots,\hat{X}_j,\ldots,X_k)$$
7. For a map f, $f^*d\alpha=df^*\alpha$.
8. (Poincaré lemma) If $d\alpha=0$, then α is locally exact; that is, there is a neighborhood U about each point on which $\alpha=d\beta$.
9. $i_X\alpha$ is a real bilinear in X,α and for $h: M\to R$, $i_{hX}\alpha=hi_X\alpha=i_Xh\alpha$. Also $i_Xi_X\alpha=0$, and
$$i_X(\alpha\wedge\beta)=i_X\alpha\wedge\beta+(-1)^k\alpha\wedge i_X\beta.$$
10. For a diffeomorphism f, $f^*i_X\alpha=i_{f^*X}f^*\alpha$.
11. $L_X\alpha=di_X\alpha+i_Xd\alpha$.
12. $L_X\alpha$ is real bilinear in X,α and $L_X(\alpha\wedge\beta)=L_X\alpha\wedge\beta+\alpha\wedge L_X\beta$.
13. For a diffeomorphism f, $f^*L_X\alpha=L_{f^*X}f^*\alpha$.
14. $(L_X\alpha)(X_1,\ldots,X_k)=X(\alpha(X_1,\ldots,X_k))-\sum_{i=1}^k\alpha(X_1,\ldots,[X,X_i],\ldots,X_k)$.
15. Locally,
$$(L_X\alpha)_x\cdot(v_1,\ldots,v_k)=D\alpha_x\cdot X(x)\cdot(v_1,\ldots,v_k)+\textstyle\sum_{i=1}^k\alpha_x\cdot(v_1,\ldots,DX_x\cdot v_i,\ldots,v_k).$$
16. The following identities hold:
$$L_{fX}\alpha=fL_X\alpha+df\wedge i_X\alpha$$
$$L_{[X,Y]}\alpha=L_XL_Y\alpha-L_YL_X\alpha$$
$$i_{[X,Y]}\alpha=L_Xi_Y\alpha-i_YL_X\alpha$$
$$L_Xd\alpha=dL_X\alpha$$
$$L_Xi_X\alpha=i_XL_X\alpha$$

EXERCISES

2.4A. On S^1 find a closed one-form α that is not exact. What are the cohomology groups of S^1?

2.4B. Show that the following properties uniquely characterize i_X:
 (i) $i_X: \Omega^k(M)\to\Omega^{k-1}(M)$ is a $\wedge$ antiderivation;
 (ii) $i_Xf=0$; $f\in\mathcal{F}(M)$;
 (iii) $i_X\omega=\omega(X)$ for $\omega\in\Omega^1(M)$;
 (iv) i_X is natural with respect to restrictions.
 Hence show $i_{[X,Y]}=L_Xi_Y-i_YL_X$. Finally, show $i_X\circ i_X=0$.

2.4C. If $\omega\in\Omega^k(M)$, and if, for some $f\in\mathcal{F}(M)$, $f(m)\neq0$ for all $m\in M$ and $f\omega$ is

ISBN 0-8053-0102-X

exact, there is a $\theta \in \Omega^1(U)$ with $d\omega = \theta \wedge \omega$ and $d\omega \wedge \omega = 0$. Interpret as a necessary condition for integrability of a total differential equation. Such a function f is an *integrating factor* of ω. For a partial converse, see Flanders [1963, p. 94].

2.4D. Let $s: T^2M \to T^2M$ be the canonical involution of the second tangent bundle (see Exercise 1.6D).
(i) If X is a vector field on M, show that $s \circ TX$ is a vector field on TM.
(ii) If F_t is the flow of X, TF_t is a flow on TM generated by $s \circ TX$.
(iii) If μ is a one form on M, $\hat{\mu}: TM \to R$ the corresponding function, and $w \in T^2M$, then show that

$$d\hat{\mu}(sw) = d\mu(\tau_{TM}(w), T\tau_M(w)) + d\hat{\mu}(w)$$

2.4E. Prove the following *relative Poincaré lemma*: Let ω be a closed k-form on a manifold M and let $N \subset M$ be a closed submanifold. Assume that the pull-back of ω to N is zero. Then there is a $(k-1)$-form α on a neighborhood of N such that $d\alpha = \omega$ and α vanishes on N. If ω vanishes on N, then α can be chosen so that all its first partial derivatives vanish on N. (*Hint:* Let φ_t be a homotopy of a neighborhood of N to N and construct an H operator as in the Poincaré lemma using φ_t.)

2.4F. (Angular Variables). Let S^1 denote the circle, $S^1 \approx R/(2\pi) \approx \{z \in C \mid |z| = 1\}$. Let $\gamma: R \to S^1: x \mapsto e^{ix}$, be the exponential map. Show that γ induces an isomorphism $TS^1 \approx S^1 \times R$. Let M be a manifold and let ω be an "angular variable," that is, a smooth map $\omega: M \to S^1$. Define $d\omega$, a one form on M by taking the R-projection of $T\omega$. Show that (i) if $\omega: M \to S^1$, then $d^2\omega = 0$; and (ii) if $f: M \to N$ is smooth, then $f^*(d\omega) = d(f^*\omega)$, where $f^*\omega = \omega \circ f$.

2.5 ORIENTABLE MANIFOLDS

The purpose of this section is to globalize the definitions of orientation and determinant discussed in Sect. 2.3. This leads naturally to the definition of the divergence of a vector field. First, we discuss partitions of unity, which are used in some proofs of this section, and which are essential for the definition of the integral (Sect. 2.6).

2.5.1 Definitions. *If t is a tensor field on a manifold M, the* ***support*** *of t is the closure of the set of $m \in M$ for which $t(m) \neq 0$, and is denoted* supp *t. Also, we say t has* ***compact support*** *if* supp *t is compact in M.*

A collection of subsets $\{C_\alpha\}$ of a manifold M (or, more generally, a topological space) is called ***locally finite*** *if for each $m \in M$ there is a neighborhood U of m such that $U \cap C_\alpha = \varnothing$ except for finitely many indices α.*

2.5.2 Definition. *A* ***partition of unity*** *on a manifold M is a collection $\{(U_i, g_i)\}$, where*

(i) $\{(U_i)\}$ is a locally finite open covering of M;
(ii) $g_i \in \mathfrak{F}(M)$, $g_i(m) \geq 0$ for all $m \in M$, g_i has compact support, and supp *$g_i \subset U_i$ for all i;*
(iii) For each $m \in M$, $\Sigma_i g_i(m) = 1$.

[By (i), this is a finite sum.]

ISBN 0-8053-0102-X

If $\mathcal{Q}$ is an atlas on M, a ***partition of unity subordinate*** *to $\mathcal{Q}$ is partition of unity $\{(U_i, g_i)\}$ such that each open set U_i is a restriction of a chart of $\mathcal{Q}$ to an open subset of its domain.*

2.5.3 Theorem. *If $\mathcal{Q}$ is an atlas of M, there is a partition of unity subordinate to $\mathcal{Q}$.*

Proof. The proof of 1.1.21 shows the following. Let M be an n manifold and $\{W_a\}$ be an open covering. Then there is a locally finite refinement consisting of charts $\{V_i, \phi_i\}$ such that $\phi_i(V_i)$ is the disk of radius 3, and such that $\phi_i^{-1}(D_1(0))$ cover M, where $D_1(0)$ is the unit disk, centered at the origin in the model space. Now let $\mathcal{Q}$ be an atlas on M and let $\{V_i, \phi_i\}$ be a locally finite refinement with these properties. From 2.2.7 there is a nonzero function $h_i \in \mathcal{F}(M)$ whose support lies in V_i and $h_j \geqslant 0$. Let

$$g_i(u) = \frac{h_i(u)}{\Sigma_i h_i(u)}$$

(the sum is finite). These are the required functions. ■

Proof of the parenthetical statement in 2.2.7. More generally, we prove a smooth version of Urysohn's lemma (1.1.23). Let A and B be two closed sets. Since manifolds are normal (see 1.1.21, and 1.1.22), there is an atlas $\{U_\alpha, \phi_\alpha\}$ such that $U_\alpha \cap A \neq \varnothing$ implies $U_\alpha \cap B = \varnothing$. Let $\{V_i, g_i\}$ be a subordinate partition of unity and $h = \Sigma g_i$, where the sum is over those i for which $V_i \cap A \neq \varnothing$. Then h is one on A and zero on B. ■

2.5.4 Definition. *A* ***volume*** *on an n-manifold M is an n-form $\Omega \in \Omega^n(M)$ such that $\Omega(m) \neq 0$ for all $m \in M$; M is called* ***orientable*** *if there is a volume on M.*

Thus, Ω assigns an orientation, as defined in 2.3.8, to each fiber of TM.

2.5.5 Theorem. *Let M be a connected n-manifold. Then (i) M is orientable iff $\Omega^n(M)$, regarded as an $\mathcal{F}(M)$ module, is one-dimensional (has one generator);*

(ii) M is orientable iff M has an atlas $\{(U_i, \varphi_i)\}$, where $\varphi_i: U_i \to U_i' \subset \mathbf{R}^n$, such that the Jacobian determinant of the overlap maps is positive (the Jacobian determinant being the determinant of the derivative, a linear map from $\mathbf{R}^n$ into $\mathbf{R}^n$).

Proof. For (i) assume first that M is orientable, with a volume Ω. Let Ω' be any other element of $\Omega^n(M)$. Now each fiber of $\Omega^n(M)$ is one-dimensional, so we may define a map $f: M \to \mathbf{R}$ by

$$\Omega'(m) = f(m)\Omega(m)$$

We must show that $f \in \mathcal{F}(M)$. In local representation,

$$\Omega'(m) = \omega'(m)\, dx^{i_1} \wedge \cdots \wedge dx^{i_n}(m)$$

ISBN 0-8053-0102-X

and $\Omega(m)=\omega(m)\,dx^{i_1}\wedge\cdots\wedge dx^{i_n}(m)$. But $\omega(m)\neq 0$ for all $m\in M$. Hence $f(m)=\omega'(m)/\omega(m)$ is of class C^∞. Conversely, if $\Omega^n(M)$ is generated by Ω, then $\Omega(m)\neq 0$ for all $m\in M$ since each fiber is one-dimensional.

To prove (ii), let $\{(U_i,\varphi_i)\}$ be an atlas with $U_i'\cap \boldsymbol{R}^n$. Also, we may assume that all U_i' are connected by taking restrictions if necessary. Now $\varphi_{i*}\Omega=f_i dx^1\wedge\cdots\wedge dx^n=f_i\Omega_0$, where Ω_0 is the standard volume element on $\boldsymbol{R}^n$. By means of a reflection if necessary, we may assume that $f_i(u')>0$ ($f_i\neq 0$ since Ω is a volume). However, a continuous real valued function on a connected space which is not zero is always >0 or always <0. Hence, for overlap maps we have

$$(\varphi_i\circ\varphi_j^{-1})_* dx^1\wedge\cdots\wedge dx^n=\varphi_{i*}\circ\varphi_{j*}^{-1}dx^1\wedge\cdots\wedge dx^n$$

$$=\frac{f_i}{f_j\circ\varphi_j\circ\varphi_i^{-1}}dx^1\wedge\cdots\wedge dx^n$$

But,

$$\psi^*(u)(\alpha^1\wedge\cdots\wedge\alpha^n)=D\psi(u)^*\cdot\alpha^1\wedge D\psi(u)^*\cdot\alpha^2\wedge\cdots\wedge D\psi(u)^*\cdot\alpha^n$$

where $D\psi(u)^*\cdot\alpha^1(e)=\alpha^1(D\psi(u)\cdot e)$. Hence, by definition of determinant we have

$$det\left(D\left(\varphi_j\circ\varphi_i^{-1}\right)(u)\right)=\frac{f_i(u)}{f_j\left[\varphi_j\circ\varphi_i^{-1}(u)\right]}>0$$

We leave as an exercise for the reader that the canonical isomorphism $L(E;E)\approx L(E^*;E^*)$, used above, does not affect determinants.

For the converse of (ii), let $\{(V_\alpha,\psi_\alpha)\}$ be an atlas with the given property, and $\{(U_i,\varphi_i,g_i)\}$ a subordinate partition of unity. Let

$$\Omega_i=\varphi_i^*(dx^1\wedge\cdots\wedge dx^n)\in\Omega^n(U_i)$$

and let

$$\tilde{\Omega}_i(m)=\begin{cases} g_i(m)\Omega_i(m) & \text{if } m\in U_i\\ 0 & \text{if } m\notin U_i\end{cases}$$

Since $supp\, g_i\subset U_i$, $\tilde{\Omega}_i\in\Omega^n(M)$. Then let

$$\Omega=\sum_i\tilde{\Omega}_i$$

Since this sum is finite in some neighborhood of each point, it is clear from local representatives that $\Omega\in\Omega^n(M)$. Finally, as the overlap maps have

ISBN 0-8053-0102-X

positive Jacobian determinant, then on $U_i \cap U_j$, $\Omega_i \neq 0$ and

$$\Omega_j = \varphi_j^*(dx^1_\wedge \cdots {}_\wedge dx^n) = \varphi_i^*(\varphi_j \circ \varphi_i^{-1})^*(dx^1 \wedge \cdots \wedge dx^n)$$

$$= \left[\det D(\varphi_j \circ \varphi_i^{-1}) \circ \varphi_i \right] \varphi_i^*(dx^1 \wedge \cdots \wedge dx^n)$$

Since $\Sigma_j g_j = 1$, it is clear then that $\Omega(m) \neq 0$ for each $m \in M$. ■

Thus, if M is an orientable manifold, with volume Ω, 2.5.5(i) defines a map from $\Omega^n(M)$ into $\mathfrak{F}(M)$; namely, for each $\Omega' \in \Omega^n(M)$, there is a unique $f \in \mathfrak{F}(M)$ such that $\Omega' = f\Omega$.

2.5.6 Definition. *Let M be an orientable manifold. Two volumes Ω_1 and Ω_2 on M are called **equivalent** iff there is an $f \in \mathfrak{F}(M)$ with $f(m) > 0$ for all $m \in M$ such that $\Omega_1 = f\Omega_2$. (This is clearly an equivalence relation.) An **orientation** of M is an equivalence class $[\Omega]$ of volumes on M. An **oriented manifold**, $(M, [\Omega])$, is an orientable manifold M together with an orientation $[\Omega]$ on M.*

*If $[\Omega]$ is an orientation of M, then $[-\Omega]$ (which is clearly another orientation) is called the **reverse orientation**.*

The next proposition tells us when $[\Omega]$ and $[-\Omega]$ are the only two orientations.

2.5.7 Proposition. *Let M be an orientable manifold. Then M is connected iff M has exactly two orientations.*

Proof. Suppose M is connected, and Ω, Ω' are two volumes with $\Omega' = f\Omega$. Since M is connected, and $f(m) \neq 0$ for all $m \in M$, $f(m) > 0$ for all m or else $f(m) < 0$ for all m. Thus Ω' is equivalent to Ω or $-\Omega$. Conversely, if M is not connected, let $U \neq \emptyset$ or M be a subset that is both open and closed. If Ω is a volume on M, define Ω' by

$$\Omega'(m) = \begin{cases} \Omega(m) & m \in U \\ -\Omega(m) & m \notin U \end{cases}$$

Obviously Ω' is a volume on M, and $\Omega' \notin [\Omega] \cup [-\Omega]$. ■

A simple example of a nonorientable manifold is the Möbius band (see Fig. 2.5-1

ISBN 0-8053-0102-X

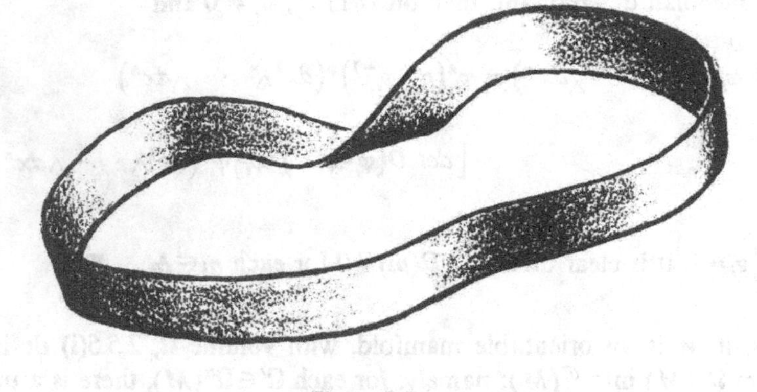

Figure 2.5-1

2.5.8 Proposition. *The equivalence relation in 2.5.6 is natural with respect to mappings and diffeomorphisms. That is, if $f: M \to N$ is of class C^∞, Ω_N and Ω'_N are equivalent volumes on N, and $f^*(\Omega_N)$ is a volume on M, then $f^*(\Omega'_N)$ is an equivalent volume. If f is a diffeomorphism and Ω_M and Ω'_M are equivalent volumes on M, then $f_*(\Omega_M)$ and $f_*(\Omega'_M)$ are equivalent volumes on N.*

Proof. This follows easily from the fact that

$$f^*(g\omega) = (g \circ f) f^*\omega$$

which implies

$$f_*(g\omega) = (g \circ f^{-1}) f_*\omega$$

when f is a diffeomorphism. ■

2.5.9 Definition. *Let M be an orientable manifold with orientation $[\Omega]$. A chart (U, φ) with $\varphi(U) = U' \subset \mathbf{R}^n$ is called **positively oriented** iff $\varphi_*(\Omega | U)$ is equivalent to the standard volume*

$$dx^1 \wedge \cdots \wedge dx^n \in \Omega^n(U')$$

From 2.5.8 we see that the above definition does not depend on the choice of the representative from $[\Omega]$.

If M is orientable, we can find an atlas in which every chart has positive orientation by choosing an atlas of connected charts and, if a chart has negative orientation, by composing it with a reflection. Thus, in 2.5.5(ii), the atlas consists of positively oriented charts.

2.5.10 Definition. *Let V be a submanifold of an n-manifold M. We say V has **codimension** k iff V, considered as a manifold, has dimension $n - k$.*

ISBN 0-8053-0102-X

Now since a curve in V is also a curve in M, we can say $T_v V \subset T_v M$, and it is clear from Sect. 1.6 that the submanifold V has codimension k iff $T_v V$ has dimension $n-k$ for each $v \in V$ iff for each $v \in V$ there is a vector space W_v of dimension k so that $T_v M = T_v V \oplus W_v$ (direct sum).

2.5.11 Proposition. *Suppose M is an orientable n-manifold and V is a submanifold of codimension k with trivial normal bundle. That is, there are C^∞ maps N_i: $V \to TM$, $i = 1, \ldots, k$ such that $N_i(v) \in T_v(M)$, and $N_i(v)$ span a subspace W_v such that $T_v M = T_v V \oplus W_v$ for all $v \in V$. Then V is orientable.*

Proof. Let Ω be a volume on M. Form $\Omega | V$: $V \to \Omega^n(M)$. Let us first note that $\Omega | V$ is a smooth mapping of manifolds. (This was obvious earlier when we considered open submanifolds.) This follows at once by using charts with the submanifold property, where the local representative is a restriction to a subspace. Now define Ω_0: $V \to \Omega^{n-k}(V)$ as follows: for

$$X_1, \ldots, X_{n-k} \in \mathfrak{X}(V)$$

put

$$\Omega_0(v)(X_1(v), \ldots, X_{n-k}(v)) = \Omega(v)(N_1(v), \ldots, N_k(v), X_1(v), \ldots, X_{n-k}(v))$$

(analogous to an inner product; however N_i are not vector fields on M). It is clear that $\Omega_0(v) \neq 0$ for all v. It remains only to show that Ω_0 is smooth, but this follows from the fact that $\Omega | V$ is smooth. ■

For some of the following proofs it will be convenient to use a Riemannian metric.

2.5.12 Definition. *A **Riemannian metric** on a manifold M is a tensor $g \in \mathfrak{T}^0_2(M)$ such that for all $m \in M$, $g(m)$ is symmetric and positive-definite.*

2.5.13 Proposition. *On any manifold there exists a Riemannian metric.*

Proof. Let $\{(U_i, \varphi_i, h_i)\}$ be a partition of unity on M, with $U_i' = \varphi_i(U_i) \subset \mathbf{R}^n$. If H_i is the standard Riemannian metric on U_i',

$$H_i(u)(v, w) = \sum v^i w^i$$

let $g_i \in \mathfrak{T}^0_2(m)$ be defined by

$$g_i(m) = \begin{cases} h_i(m)(\varphi_i^{-1})_* H_i(m) & \text{if } m \in U_i \\ 0 & \text{if } m \notin U_i \end{cases}$$

Then $g = \Sigma_i g_i$ is a Riemannian metric on M. ■

ISBN 0-8053-0102-X

Recall that we include second countable in our definition of a manifold. It is interesting that a manifold which admits a Riemannian metric (or a connection) must be second countable (see Abraham [1963]).

Note that if $g \in \mathfrak{T}^0_2(M)$, we may identify g with an $\mathfrak{F}$ linear mapping $g^\flat \in L(\mathfrak{X}, \mathfrak{X}^*)$ and if g is a Riemannian metric, obviously $g^\flat$ is an isomorphism. In this case we write $g^\sharp = (g^\flat)^{-1}$, and the maps $g^\sharp$ and $g^\flat$ are called raising and lowering indices, respectively.

2.5.14 Definition. *Let M be a manifold with a Riemannian metric g. For $f \in \mathfrak{F}(M)$, $\operatorname{grad} f = g^\sharp(df)$ is called the **gradient** of f. Thus, $\operatorname{grad} f \in \mathfrak{X}(M)$. In local coordinates, if $g_{ij} = g(e_i, e_j)$ and g^{ij} is the inverse matrix, then one checks that*

$$(\operatorname{grad} f)^i = g^{ij}\frac{\partial f}{\partial x^j}$$

The above machinery allows us to obtain the following consequence of 2.5.11.

2.5.15 Theorem. *Suppose M is an orientable manifold, $H \in \mathfrak{F}(M)$ and $c \in \mathbf{R}$ is a regular value of H. Then $V = H^{-1}(c)$ is an orientable submanifold of M of codimension one, if it is nonempty.*

Proof. Suppose c is regular value of H and $H^{-1}(c) = V \neq \phi$. Then V is a submanifold of codimension one. Let g be a Riemannian metric on M and $N = \operatorname{grad}(H)|V$. Then $N(v) \notin T_vV$ for $v \in V$, because T_vV is the kernel of $dH(v)$, and $dH(v)[N(v)] = g(N, N)(v) > 0$ as $dH(v) \neq 0$ by hypothesis. Then 2.5.11 applies, and so V is orientable. ■

Thus if we interpret V as the "energy surface," we see that it is an oriented submanifold for "almost all" energy values (Sard's theorem).

Let us now examine the effect of volumes under maps more closely.

2.5.16 Definition. *Let M and N be two orientable n-manifolds with volumes Ω_M and Ω_N, respectively. Then we call a C^∞ map $f: M \to N$ **volume preserving** (with respect to Ω_M and Ω_N) if $f^*\Omega_N = \Omega_M$, and we call f **orientation preserving** if $f^*(\Omega_N) \in [\Omega_M]$, and **orientation reversing** if $f^*(\Omega_N) \in [-\Omega_M]$.*

From 2.5.8, $[f^*\Omega_N]$ depends only on $[\Omega_N]$. Thus the first part of the definition depends explicitly on Ω_M and Ω_N while the last two parts depend only on the orientations $[\Omega_M]$ and $[\Omega_N]$. Furthermore, we see from 2.5.8 that if f is volume preserving with respect to Ω_M, Ω_N, then f is volume preserving with respect to $h\Omega_M$, $g\Omega_N$ iff $h = g \circ f$. It is also clear that if f is volume preserving with respect to Ω_M, Ω_N, then f is orientation preserving with respect to $[\Omega_M]$, $[\Omega_N]$.

ISBN 0-8053-0102-X

2.5.17 Proposition. *Let M and N be n-manifolds with volumes Ω_M and Ω_N, respectively. Suppose $f: M \to N$ is of class C^∞. Then (i) $f^*(\Omega_N)$ is a volume iff f is a local diffeomorphism; that is, for each $m \in M$, there is a neighborhood V of m such that $f|V: V \to f(V)$ is a diffeomorphism. (ii) If M is connected, then f is a local diffeomorphism iff f is orientation preserving or orientation reversing.*

Proof. If f is a local diffeomorphism, then clearly $f^*(\Omega_N)(m) \neq 0$, by 2.4.9(ii). Conversely, if $f^*(\Omega_N)$ is a volume, then the determinant of the derivative of the local representative is not zero, and hence the derivative is an isomorphism. The result then follows by the inverse function theorem. (ii) follows at once from (i) and 2.5.7. ■

Next we consider the global analog of the determinant.

2.5.18 Definition. *Suppose M and N are orientable n-manifolds with volumes Ω_M and Ω_N, respectively. If $f: M \to N$ is of class C^∞, the unique C^∞ function $\det_{(\Omega_M,\Omega_N)} f \in \mathfrak{F}(M)$ such that $f^*\Omega_N = (\det_{(\Omega_M,\Omega_N)} f)\Omega_M$ is called the **determinant** of f (with respect to Ω_M and Ω_N). If $f: M \to M$, we write $\det_{\Omega_M} f = \det_{(\Omega_M,\Omega_M)} f$.*

The basic properties of determinants given in Sect. 2.3 also hold in the global case, as follows.

2.5.19 Proposition. *In the notation of 2.5.18, f is a local diffeomorphism iff $\det_{(\Omega_M,\Omega_N)} f(m) \neq 0$ for all $m \in M$.*

This follows at once from 2.5.17.

2.5.20 Proposition. *Let M be an orientable manifold with volume Ω. Then*

(i) if $f: M \to M$, $g: M \to M$ are of class C^∞, then $\det_\Omega(f \circ g) = [(\det_\Omega f) \circ g][\det_\Omega g]$;
(ii) if $h: M \to M$ is the identity, then $\det_\Omega h = 1$;
(iii) if $f: M \to M$ is a diffeomorphism, then

$$\det_\Omega(f^{-1}) = 1/[(\det_\Omega f) \circ f^{-1}]$$

Proof. For (i),

$$\begin{aligned}\det_\Omega(f \circ g)\Omega &= (f \circ g)^*\Omega = g^* \circ f^*\Omega \\ &= g^*(\det_\Omega f)\Omega = ((\det_\Omega f) \circ g)\, g^*\Omega \\ &= ((\det_\Omega f) \circ g)(\det_\Omega g)\Omega\end{aligned}$$

Part (ii) follows since, by 2.4.8 (iii), h^* is the identity. For (iii) we have

$$\det_\Omega(f \circ f^{-1}) = 1 = ((\det_\Omega f) \circ f^{-1})(\det_\Omega f^{-1}) \qquad ■$$

ISBN 0-8053-0102-X

If $f: U \subset E \to E$, then $det f$ is the Jacobian determinant of f [that reduces to the determinant of f if f is linear since $Df(u)=f$ if f is linear]. Then in this case, (i) above is the usual "chain rule" for Jacobian determinants. (See the proof of 2.5.5.)

2.5.21 Proposition. *Let $(M,[\Omega_M])$ and $(N,[\Omega_N])$ be oriented manifolds and $f: M \to N$ be of class C^∞. Then f is orientation preserving iff $det_{(\Omega_M,\Omega_N)}f(m)>0$ for all $m \in M$, and orientation reversing iff $det_{(\Omega_M,\Omega_n)}f(m)<0$ for all $m \in M$. Also, f is volume preserving with respect to Ω_M, Ω_N iff $det_{(\Omega_M,\Omega_N)}f=1$.*

This proposition follows at once from the definitions. Note that the first two assertions depend only on the orientations $[\Omega_M]$ and $[\Omega_N]$ since

$$det_{(h\Omega_M, g\Omega_N)}f = \left(\frac{g \circ f}{h}\right) det_{(\Omega_M,\Omega_N)}f$$

which the reader can easily check. Here $g \in \mathfrak{F}(N)$, $h \in \mathfrak{F}(M)$, $g(n) \neq 0$, and $h(m) \neq 0$ for all $n \in N$, $m \in M$.

Suppose that X is a vector field on $\boldsymbol{R}^n$ and $\Omega_0 = dx^1 \wedge \cdots \wedge dx^n$ is the standard volume on $\boldsymbol{R}^n$. Then $L_X \Omega_0 = L_X\, dx^1 \wedge dx^2 \wedge \cdots \wedge dx^n + \cdots + dx^1 \wedge \cdots \wedge L_X\, dx^n$ (since L_X is a derivation). But $L_X\, dx^i = dL_X x^i$ and $L_X x^i = dx^i(X) = X^i$, the components of X. Hence

$$L_X\, dx^i = dX^i = \left(\frac{\partial X^i}{\partial x^j}\right) dx^j \quad \text{and} \quad L_X \Omega_0 = \left(\frac{\partial X^i}{\partial x^i}\right)\Omega_0$$

since $dx^i \wedge dx^i = 0$. That is, $L_X \Omega_0 = (div\, X)\Omega_0$ where $div\, X$ is the usual divergence of a vector field on $\boldsymbol{R}^n$. The generalization of this is as follows.

2.5.22 Definition. *Let M be an orientable manifold with volume Ω, and X a vector field on M. Then the unique function $div_\Omega X \in \mathfrak{F}(M)$, such that $L_X \Omega = (div_\Omega X)\Omega$ is called the* ***divergence*** *of X. We say X is* ***incompressible*** *(with respect to Ω) iff $div_\Omega X = 0$.*

2.5.23 Proposition. *Let M be an orientable manifold with volume Ω, and X a vector field on M. Then:*

(i) if $f \in \mathfrak{F}(M)$ and $f(m) \neq 0$ for all $m \in M$, then

$$div_{f\Omega} X = div_\Omega X + \frac{L_X f}{f}$$

(ii) for $g \in \mathfrak{F}(M)$, $div_\Omega gX = g\, div_\Omega X + L_X g$.

Proof. Since L_X is a derivation, we have

$$L_X(f\Omega) = (L_X f)\Omega + f L_X \Omega$$

ISBN 0-8053-0102-X

As $f\Omega$ is a volume, $(div_{f\Omega}X)(f\Omega)=(L_Xf)\Omega+f(div_\Omega X)\Omega$. Then (i) follows. For (ii), we have, by 2.4.13, $L_{gX}\Omega=gL_X\Omega+dg\wedge i_X\Omega$. Now from the antiderivation property of i_X, $dg\wedge i_X\Omega=-i_X(dg\wedge\Omega)+i_X\,dg\wedge\Omega$. But $dg\wedge\Omega\in\Omega^{n+1}(M)$, and hence $dg\wedge\Omega=0$. Also, $i_X\,dg=L_Xg$ and so $L_{gX}\Omega=gL_X\Omega+(L_Xg)\Omega$. The result follows at once from this. ■

2.5.24 Proposition. *Let M be a manifold with volume Ω and X a vector field on M. Then X is incompressible (with respect to Ω) iff every flow box of X is volume preserving; that is, for the diffeomorphism F_λ: $U\to V$, F_λ is volume preserving with respect to $\Omega|U$ and $\Omega|V$.*

Proof. If X is incompressible, $L_X\Omega=0$, Ω is constant along integral curves of X; $\Omega(m)=(F_\lambda)^*\Omega(m)$. Hence F_λ is volume preserving. Conversely, if $(F_\lambda)^*\Omega(m)=\Omega(m)$, then $L_X\Omega=0$. ■

2.5.25 Corollary. *Let M be an orientable manifold with volume Ω, and X a complete vector field with flow F on M. Then X is incompressible iff $\det_\Omega F_\lambda=1$ for all $\lambda\in\mathbf{R}$.*

EXERCISES

2.5A. Let f: $\mathbf{R}^n\to\mathbf{R}^n$ be a diffeomorphism with positive Jacobian and $f(\mathbf{0})=\mathbf{0}$. Prove that there is a continuous curve f_t of diffeomorphisms joining f to the identity. [*Hint*: first join f to $Df(0)$ by $g_t(x)=f(tx)/t$.]

2.5B. If t is a tensor density of M, that is, $t=t'\otimes\mu$, where μ is a volume, show that

$$L_Xt=(L_Xt')\otimes\mu+(div_\mu X)t\otimes\mu$$

2.5C. (T. Hughes) A map $A: E\to E$ is said to be derived from a variational principle if there is a function $L: E\to\mathbf{R}$ such that

$$dL(x)\cdot v=\langle A(x),v\rangle$$

where $\langle,\rangle$ is an inner product on E. Prove Vainberg's theorem: *A comes from a variational principle if and only if DA(x) is a symmetric linear operator.* Do this by applying the Poincaré lemma to the one form $\alpha(x)\cdot v=\langle A(x),v\rangle$.

2.6 INTEGRATION ON MANIFOLDS

The aim of this section is to define the integral of an n-form on an n-manifold M. We begin with a summary of the basic results on $\mathbf{R}^n$.

Suppose f: $\mathbf{R}^n\to\mathbf{R}$ is continuous and has compact support. Then $\int f\,dx^1\cdots dx^n$ is defined as the Riemann integral over any rectangle containing the support of f (see Marsden [1974a, Chapter 9]).

2.6.1 Definition. *Let $U\subset\mathbf{R}^n$ be open and $\omega\in\Omega^n(U)$ have compact support.*

ISBN 0-8053-0102-X

If, relative to the standard basis of $\mathbf{R}^n$,

$$\omega(u)=\frac{1}{n!}\omega_{i_1\cdots i_n}(u)\,dx^{i_1}\wedge\cdots\wedge dx^{i_n}=\omega_{1\cdots n}(u)\,dx^1\wedge\cdots\wedge dx^n$$

where

$$\omega_{i_1\cdots i_n}(u)=\omega(u)(e_{i_1},\ldots,e_{i_n})$$

we define

$$\int\omega=\int\omega_{1\ldots n}(u)\,dx^1\cdots dx^n$$

Clearly, if we regard $\omega\in\Omega^n(\mathbf{R}^n)$, the integral is unchanged.

The change of variables rule takes the following form.

2.6.2 Theorem. *Let U, V be open subsets of $\mathbf{R}^n$ and suppose $f\colon U\to V$ is an orientation preserving diffeomorphism. Then if $\omega\in\Omega^n(V)$ has compact support, $f^*\omega\in\Omega^n(U)$ has compact support and $\int f^*\omega=\int\omega$, that is, the following diagram commutes:*

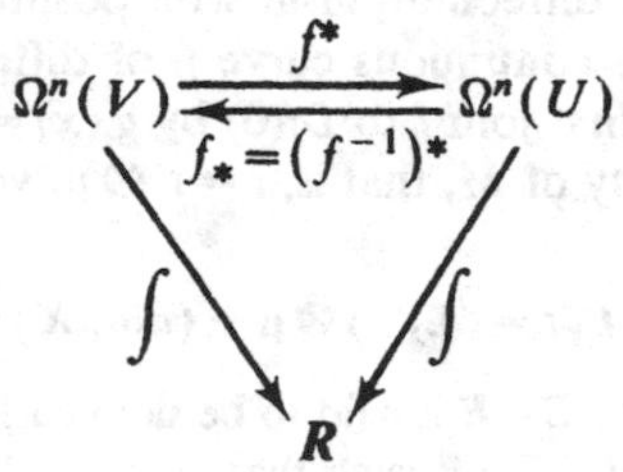

Proof. If $\omega=\omega_{1\cdots n}\,dx^1\wedge\cdots\wedge dx^n$, then $f^*\omega=(\omega_{1\cdots n}\circ f)(\det_{\Omega_0}f)\Omega_0$, where $\Omega_0=dx^1\wedge\cdots\wedge dx^n$ is the standard volume on $\mathbf{R}^n$. Since f is a diffeomorphism, the support of $f^*\omega$ is compact. Then

$$\int f^*\omega=\int(\omega_{1\cdots n}\circ f)(\det_{\Omega_0}f)\,dx^1\cdots dx^n$$

As was discussed in Sect. 2.5, $\det_{\Omega_0}f>0$ is the Jacobian determinant of f. Now by covering the support of ω by a finite number of disks, we see that the usual change of variables formula applies in this case (Marsden [1974a, Chapter 9]), namely,

$$\int\omega_{1\cdots n}\,dx^1\cdots dx^n=\int(\omega_{1\cdots n}\circ f)(\det_{\Omega}f)\,dx^1\cdots dx^n$$

which implies $\int f^*\omega=\int\omega$. ■

ISBN 0-8053-0102-X

Suppose that (U,φ) is a chart on a manifold M, and $\omega \in \Omega^n(M)$. Then if $supp\,\omega \subset U$, we may form $\omega|U$, which has the same support. Then $\varphi_*(\omega|U)$ has compact support, and we may state the following.

2.6.3 Definition. *Let M be an orientable n-manifold with orientation Ω. Suppose $\omega \in \Omega^n(M)$ has compact support $C \subset U$, where (U,φ) is a positively oriented chart. Then we define $\int_{(\varphi)}\omega = \int \varphi_*(\omega|U)$.*

2.6.4 Proposition. *Suppose $\omega \in \Omega^n(M)$ has compact support $C \subset U \cap V$, where (U,φ), (V,ψ) are two positively oriented charts on the oriented manifold M. Then*

$$\int_{(\varphi)} \omega = \int_{(\psi)} \omega$$

Proof. By 2.6.2, $\int \varphi_*(\omega|U) = \int (\psi \circ \varphi^{-1})_* \varphi_*(\omega|U)$. Hence $\int \varphi_*(\omega|U) = \int \psi_*(\omega|U)$. [Recall that for diffeomorphisms $f_* = (f^{-1})^*$ and $(f \circ g)_* = f_* \circ g_*$.] ■

Thus we merely define $\int \omega = \int_{(\varphi)} \omega$, where (U,φ) is any positively oriented chart containing the compact support of ω (if one exists).

More generally, we can define $\int \omega$ where ω has compact support as follows.

2.6.5 Definition. *Let M be an oriented manifold and $\mathcal{Q}$ an atlas of positively oriented charts. Let $P = \{(U_\alpha, \varphi_\alpha, g_\alpha)\}$ be a partition of unity subordinate to $\mathcal{Q}$. Define $\omega_\alpha = g_\alpha \omega$ (so ω_α has compact support in some U_i). Then define*

$$\int_P \omega = \sum_\alpha \int \omega_\alpha$$

2.6.6 Proposition. *(i) The above sum contains only a finite number of nonzero terms, and hence $\int_P \omega \in \mathbf{R}$.*

(ii) For any other atlas of positively oriented charts and subordinate partition of unity Q we have $\int_P \omega = \int_Q \omega$.

The common value is denoted $\int \omega$, the **integral** of $\omega \in \Omega^n(M)$.

Proof. For any $m \in M$, there is a neighborhood U such that only a finite number of g_α are nonzero on U. By compactness of $supp\,\omega$, a finite number of such neighborhoods cover $supp\,\omega$. Hence only a finite number of g_α are nonzero on the union of these U. For (ii), let $P = \{(U_\alpha, \varphi_\alpha, g_\alpha)\}$ and $Q = \{(V_\beta, \psi_\beta, h_\beta)\}$ be two partitions of unity with positively oriented charts. Then

ISBN 0-8053-0102-X

the functions $\{g_\alpha h_\beta\}$ have $g_\alpha h_\beta(m)=0$ except for a finite number of indices (α,β), and $\Sigma_\alpha\Sigma_\beta g_\alpha h_\beta(m)=1$, for all $m\in M$. Hence, since $\Sigma_\beta h_\beta=1$,

$$\int_P \omega=\sum_\alpha \int g_\alpha \omega$$

$$=\sum_\beta\sum_\alpha \int h_\beta g_\alpha \omega$$

$$=\sum_\alpha\sum_\beta \int g_\alpha h_\beta \omega=\int_Q \omega \qquad \blacksquare$$

The globalization of the change of variables formula is as follows.

2.6.7 Theorem. *Suppose M and N are oriented n-manifolds and $f\colon M\to N$ is an orientation preserving diffeomorphism. If $\omega\in\Omega^n(N)$ has compact support then $f^*\omega$ has compact support and $\int\omega=\int f^*\omega$.*

Proof. First, $\operatorname{supp} f^*\omega=f^{-1}(\operatorname{supp}\omega)$, which is compact. For the second part, let $\{U_i,\varphi_i\}$ be an atlas of positively oriented charts of M and let $P=\{g_i\}$ be a subordinate partition of unity. Then $\{f(U_i),\varphi_i\circ f^{-1}\}$ is an atlas of positively oriented charts of N and $Q=\{g_i\circ f^{-1}\}$ is a partition of unity subordinate to the covering $\{f(U_i)\}$. Then

$$\int f^*\omega=\sum_i\int g_i f^*\omega=\sum_i\int \varphi_{\alpha*}(g_i f^*\omega)$$

$$=\sum_i\int \varphi_{\alpha*}\cdot(f^{-1})_*(g_i\circ f^{-1})\omega$$

$$=\sum_i\int(\varphi_\alpha\circ f^{-1})_*(g_i\circ f^{-1})\omega$$

$$=\int\omega \qquad \blacksquare$$

As in 2.6.2, we have the following commutative diagram:

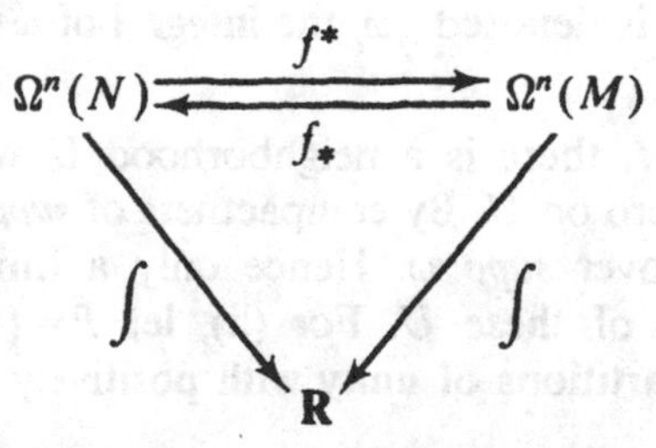

ISBN 0-8053-0102-X

We also can integrate functions of compact support as follows.

2.6.8 Definition. *Let M be an orientable manifold with volume Ω. Suppose $f \in \mathfrak{F}(M)$ and f has compact support. Then we define $\int_\Omega f = \int f\Omega$, the* ***integral*** *of f with respect to Ω.*

The reader can easily check that since the Riemann integral is $\boldsymbol{R}$ linear, so is the integral above.

The next theorem will show that the foregoing integral can be obtained in a unique way from a measure on M. (The reader unfamiliar with measure theory can find the necessary background in Royden [1963]. However, this will not be essential for future sections.) The integral we have described can clearly be extended to all continuous functions with compact support. Then we have the following.

2.6.9 Theorem (Riesz representation theorem). *Let M be an orientable manifold with volume Ω. Let $\mathfrak{B}$ denote the Borel sets of M, the σ algebra generated by the open (or closed, or compact) subsets of M. Then there is a unique measure μ_Ω on $\mathfrak{B}$ (and hence a completion $\bar{\mu}_\Omega$) such that for every continuous function of compact support, $\int f\,d\mu_\Omega = \int_\Omega f$.*

Proof. Existence of such a μ_Ω is proved in Royden [1963, p. 251]. For uniqueness, it is enough to consider bounded open sets (by the Hahn extension theorem). Thus, let U be open in M, and let C_U be its characteristic function. We can construct a sequence of C^∞ functions of compact support φ_n such that $\varphi_n \downarrow C_U$, pointwise. Hence from the monotone convergence theorem $\int_\Omega \varphi_n = \int \varphi_n\,d\mu_\Omega \to \int C_U\,d\mu_\Omega = \mu_\Omega(U)$. Thus, μ_Ω is unique. ■

Then one can define the space $L^p(M,\Omega)$, $p \in \boldsymbol{R}$, consisting of all measurable functions f such that $|f|^p$ is integrable. For $p>1$, the norm $\|f\|_p = (\int |f|^p\,d\mu_\Omega)^{1/p}$ makes $L^p(M,\Omega)$ into a Banach space (functions that differ only on a set of measure zero are identified).

The behavior of these spaces under mappings can give information about the manifold. In particular, the effect under flows is of importance in statistical mechanics. In this connection we have the following.

2.6.10 Proposition. *Let M be an orientable manifold with volume Ω. Suppose X is a complete vector field on M with flow F. Then X is incompressible iff μ_Ω is F invariant, that is, $\int f\,d\mu_\Omega = \int f \circ F_\lambda\,d\mu_\Omega$ for all λ, and $f \in L^1(M,\Omega)$.*

Proof. If X is incompressible, and f is continuous with compact support, then $\int (f \circ F_\lambda)\Omega = \int f \circ F_\lambda (F_\lambda)^*\Omega = \int (F_\lambda)^*(f\Omega) = \int f\Omega$. Hence, by uniqueness in 2.6.9, we have $\int f\,d\mu_\Omega = \int (f \circ F_\lambda)\,d\mu_\Omega$ for all integrable f. Conversely, if

ISBN 0-8053-0102-X

$\int(f\circ F_\lambda)\,d\mu_\Omega = \int f\,d\mu_\Omega$, then taking f continuous with compact support, we see

$$\int (f\circ F_\lambda)\Omega = \int (f\circ F_\lambda)F_\lambda{}_*\Omega$$

$$= \int (f\circ F_\lambda)(det_\Omega F_\lambda)\Omega$$

Thus, for every integrable f, $\int f\,d\mu_\Omega = \int (f\,det_\Omega F_\lambda)\,d\mu_\Omega$. Hence, $det_\Omega F_\lambda = 1$, which implies X is incompressible. ■

We now make a number of remarks and definitions preparatory to proving Stokes' theorem.

Let $\boldsymbol{R}^n_+ = \{x = (x_1, \ldots, x_n) \in \boldsymbol{R}^n | X_n > 0\}$ denote the upper *half-space* of $\boldsymbol{R}^n$ and let $U \subset \boldsymbol{R}^n_+$ be an open set (in the topology induced on $\boldsymbol{R}^n_+$ from $\boldsymbol{R}^n$). Call $Int\ U = U \cap \{x \in \boldsymbol{R}^n | x_n > 0\}$ the *interior* of U and $\partial U = U \cap (\boldsymbol{R}^{n-1} \times \{0\})$ the *boundary* of U. We clearly have $U = Int\ U \cup \partial U$, $Int\ U$ is open in U, ∂U closed *in* U (*not* in $\boldsymbol{R}^n$), and $\partial U \cap Int\ U = \varnothing$.

Let U, V be open sets in $\boldsymbol{R}^n_+$ and $f\colon U \to V$. We shall say that f is ***smooth*** if for each point $x \in U$ there exist open neighborhoods U_1 of x and V_1 of $f(x)$ in $\boldsymbol{R}^n$ and a smooth map $f_1\colon U_1 \to V_1$ such that $f|U \cap U_1 = f_1|U \cap U_1$. We then define $Df(x) = Df_1(x)$. We must prove that this definition is independent of the choice of f_1, that is, we have to show that if $\phi\colon W \to \boldsymbol{R}^n$ is a smooth map with W open in $\boldsymbol{R}^n$ such that $\phi|W \cap \boldsymbol{R}^n_+ = 0$, then $D\phi(x) = 0$ for all $x \in W \cap \boldsymbol{R}^n_+$. If $x \in Int(W \cap \boldsymbol{R}^n_+)$, there is nothing to prove. If $x \in \partial(W \cap \boldsymbol{R}^n_+)$, choose a sequence $x_n \in Int(W \cap \boldsymbol{R}^n_+)$ such that $x_n \to x$; but then $0 = D\phi(x_n) \to D\phi(x)$ and hence $D\phi(x) = 0$, which proves our claim.

Let $U \subset \boldsymbol{R}^n_+$ be open, $\phi\colon U \to \boldsymbol{R}^n_+$ be a smooth map, and assume that for some $x_0 \in Int\ U$, $\phi(x_0) \in \partial\boldsymbol{R}^n_+$. We claim that $D\phi(x_0)(\boldsymbol{R}^n) \subset \partial\boldsymbol{R}^n_+$. To see this, let $p_n\colon \boldsymbol{R}^n \to \boldsymbol{R}$ be the canonical projection onto the nth factor and notice that the relation

$$\phi(x_0 + tx) = \phi(x_0) + D\phi(x_0)\cdot tx + o(tx)$$

where $lim_{t\to 0}\, o(tx)/t = 0$, together with the hypothesis $(p_n \circ \phi)(y) \geq 0$ for all $y \in U$, implies $0 \leq (p_n \circ \phi)(x_0 + tx) = 0 + (p_n \circ D\phi)(x_0)\cdot tx + p_n(o(tx))$, whence for $t > 0$

$$0 \leq (p_n \circ D\phi)(x_0)\cdot x + p_n \circ \left(\frac{o(tx)}{t}\right)$$

Letting $t \to 0$, we get $(p_n \circ D\phi)(x_0)\cdot x \geq 0$ for all $x \in \boldsymbol{R}^n$. Similarly, for $t < 0$, letting $t \to 0$, we get $(p_n \circ D\phi)(x_0)\cdot x \leq 0$ for all $x \in \boldsymbol{R}^n$. The conclusion is

$$(D\phi)(x_0)(\boldsymbol{R}^n) \subset \boldsymbol{R}^{n-1} \times \{0\}$$

ISBN 0-8053-0102-X

We now prove the following assertion:

Lemma. *Let U, V be open sets in $\mathbf{R}^n_+$ and $f: U \to V$ a diffeomorphism. Then f induces diffeomorphisms $Int\, f: Int\, U \to Int\, V$ and $\partial f: \partial U \to \partial V$.*

Proof. Assume first that $\partial U = \varnothing$, that is, that $U \cap (\mathbf{R}^{n-1} \times \{0\}) = \varnothing$. We shall show that $\partial V = \varnothing$ and hence we take $Int\, f = f$. If $\partial V \neq \varnothing$, there exists $x \in U$ such that $f(x) \in \partial V$ and hence by definition of smoothness in $\mathbf{R}^n_+$, there are open neighborhoods in $\mathbf{R}^n$, $U_1 \subset U$, $x \in U_1$, $V_1 \subset \mathbf{R}^n$, $f(x) \in V_1$, and smooth maps $f_1: U_1 \to V_1$, $g_1: V_1 \to U_1$ such that $f|U_1 = f_1$, $g_1|V \cap V_1 = f^{-1}|V \cap V_1$. Let $x_n \in U_1$, $x_n \to x$, $y_n \in V_1 \backslash \partial V$, and $y_n = f(x_n)$. We have

$$\begin{aligned} Df(x) \circ Dg_1(f(x)) &= \lim_{y_n \to f(x)} (Df(g_1(y_n)) \circ Dg_1(y_n)) \\ &= \lim_{y_n \to f(x)} D(f \circ g_1)(y_n) = id_{\mathbf{R}^n} \end{aligned}$$

and similarly

$$Dg_1(f(x)) \circ Df(x) = id_{\mathbf{R}^n}$$

so that $Df(x)^{-1}$ exists and equals $Dg_1(f(x))$. But we saw above that $Df(x)(\mathbf{R}^n) \subset \mathbf{R}^{n-1} \times \{0\}$, which is impossible, $Df(x)$ being an isomorphism.

Assume that $\partial U \neq \varnothing$. If we assume $\partial V = \varnothing$, then, working with f^{-1} instead of f, the above argument leads to a contradiction. Hence $\partial V \neq \varnothing$. Let $x \in Int\, U$ so that x has a neighborhood $U_1 \subset U$, $U_1 \cap \partial U = \varnothing$, and hence $\partial U_1 = \varnothing$. Thus, by the above argument, $\partial f(U_1) = \varnothing$, and $f(U_1)$ is open in $V \backslash \partial V$. This shows that $f(Int\, U) \subset Int\, V$. Similarly, working with f^{-1}, we conclude $f(Int\, U) \supset Int\, V$ and hence $f: Int\, U \to Int\, V$ is a diffeomorphism. But then $f(\partial U) = \partial V$ and $f|\partial U: \partial U \to \partial V$ is a diffeomorphism as well. ■

Now we define a ***manifold with boundary*** exactly as in Sect. 1.4 with the following difference: if (U, ϕ) is a chart, we require that $\phi(U) \subset \mathbf{R}^n_+$. Let $\mathcal{A} = \{(U, \phi)\}$ be an atlas on the manifold with boundary M. Define $Int\, M = \cup_U \phi^{-1}(Int(\phi(U)))$ and $\partial M = \cup_U \phi^{-1}(\partial(\phi(U)))$ called, respectively, the ***interior*** and ***boundary*** of M. Their definition makes sense by the lemma above. $Int\, M$ is open in M and so is an n-dimensional manifold; ∂M is an $(n-1)$-dimensional manifold (possibly empty) without boundary.

If $\mathcal{A} = \{(U, \phi)\}$ is an atlas on M, then the atlas $\mathcal{B} = \{(\partial U, p_n \circ \partial\phi)\}$, $p_n \circ \partial\phi: \partial U \to \partial\phi(U) \subset \mathbf{R}^{n-1}$ defines the manifold structure on ∂M.

Summarizing, we have proved the following.

2.6.11 Proposition. *If M is an n-manifold with boundary, then its interior $Int\, M$ and its boundary ∂M are smooth manifolds without boundary of dimension n and $n-1$, respectively. Moreover, if $f: M \to N$ is a diffeomorphism, N being another n-manifold with boundary, then f induces, by restriction, two diffeomorphisms $Int\, f: Int\, M \to Int\, N$ and $\partial f: \partial M \to \partial N$.*

ISBN 0-8053-0102-X

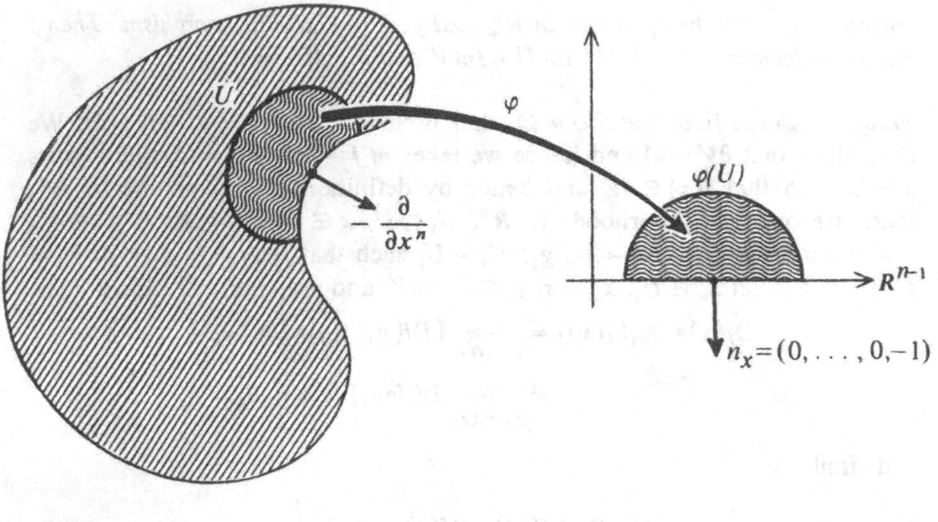

Figure 2.6-1.

Our next goal is Stokes' theorem, which deals with integration, so we have to define orientation on a manifold with boundary. A glance at the definition of orientability shows that the definition extends without difficulty to the case of manifolds with boundary. It is convenient to have in mind the following geometric interpretation of an orientation on M. An orientation on M is just a smooth choice of orientations of all the tangent spaces, "smooth" meaning that for all the charts of a certain atlas, the ***oriented charts***, the maps $D(\phi_j \circ \phi_i^{-1})(x)$: $\mathbf{R}^n \to \mathbf{R}^n$ are orientation preserving. With this picture in mind, we can define the ***boundary orientation*** of ∂M in the following way. At every $x \in \partial M$, $T_x(\partial M)$ has codimension one in $T_x(M)$ so that there are—in a chart on M intersecting ∂M—exactly two vectors perpendicular to $x_n = 0$: one points inward, the other outward. Our assertion preceding 2.6.11 assures us that a change of chart does not affect the quality of a vector being outward or inward. (See Fig. 2.6-1.)

We shall say that a basis $\{v_1, \ldots, v_{n-1}\}$ of $T_x(\partial M)$ is *positively oriented* if $\{-\partial/\partial x^n, v_1, \ldots, v_{n-1}\}$ is positively oriented in the orientation of M. This defines the *induced orientation on* ∂M.

2.6.12 Stokes' Theorem. *Let M be an oriented smooth n-manifold with boundary and $\alpha \in \Omega^{n-1}(M)$ have compact support. Let i: $\partial M \to M$ be the inclusion map so that $i^*\alpha \in \Omega^{n-1}(\partial M)$. Then*

$$\int_{\partial M} i^*\alpha = \int_M d\alpha$$

ISBN 0-8053-0102-X

or, for short,

$$\int_{\partial M} \alpha = \int_M d\alpha$$

Proof. Since integration was constructed with partitions of unity subordinate to an atlas and both sides of the equation to be proved are linear in α, we may assume without loss of generality that α is a form on $U \subset \mathbf{R}^n_+$ with compact support.

There are two cases: $\partial U = \varnothing$ and $\partial U \neq \varnothing$. Write

$$\alpha = \sum_{i=1}^{n} (-1)^{i-1} \alpha_i \, dx^1 \wedge \cdots \wedge \widehat{dx^i} \wedge \cdots \wedge dx^n$$

where $\wedge$ above a term means that it is deleted. Then

$$d\alpha = \sum_{i=1}^{n} \frac{\partial \alpha_i}{\partial x^i} dx^1 \wedge \cdots \wedge dx^n$$

and thus

$$\int_U d\alpha = \sum_{i=1}^{n} \int_{\mathbf{R}^n} \frac{\partial \alpha_i}{\partial x^i} dx^1 \cdots dx^n$$

If $\partial U = \varnothing$, we have $\int_{\partial U} \alpha = 0$. The integration of the ith term in the sum occurring in $\int_U d\alpha$ is

$$\int_{\mathbf{R}^{n-1}} \left(\int_{\mathbf{R}} \frac{\partial \alpha_i}{\partial x^i} dx^i \right) dx^1 \cdots \widehat{dx^i} \cdots dx^n \qquad \text{(no sum)}$$

and $\int_{-\infty}^{+\infty} \partial \alpha_i / \partial x^i \, dx^i = 0$ since α_i has compact support. Thus $\int_U d\alpha = 0$ as desired.

If $\partial U \neq \varnothing$, then we can do the same trick for each term except the last, which is

$$\int_{\mathbf{R}^{n-1}} \left(\int_0^\infty \frac{\partial \alpha_n}{\partial x^n} dx^n \right) dx^1 \cdots dx^{n-1} = -\int_{\mathbf{R}^{n-1}} \alpha_n(x^1, \ldots, x^{n-1}, 0) \, dx^1 \cdots dx^{n-1}$$

since α_n has compact support. Thus

$$\int_U d\alpha = -\int_{\mathbf{R}^{n-1}} \alpha_n(x^1, \ldots, x^{n-1}, 0) \, dx^1 \cdots dx^{n-1}$$

ISBN 0-8053-0102-X

On the other hand,

$$\int_{\partial U} \alpha = \int_{\partial R_+^n} \alpha = \int_{\partial R_+^n} (-1)^{n-1} \alpha_n(x^1, \ldots, x^{n-1}, 0)\, dx^1 \wedge \cdots \wedge dx^{n-1}$$

But $R^{n-1} = \partial R_+^n$ and the usual orientation on R^{n-1} is *not* the boundary orientation. The outward unit normal is $-e_n = (0, \ldots, 0, -1)$ and hence the boundary orientation has the sign of the ordered basis $\{-e_n, e_1, \ldots, e_{n-1}\}$, which is $(-1)^n$. Thus

$$\int_{\partial U} \alpha = \int_{\partial R_+^n} (-1)^{n-1} \alpha_n(x^1, \ldots, x^{n-1}, 0)\, dx^1 \wedge \cdots \wedge dx^{n-1}$$

$$= (-1)^{2n-1} \int_{R^{n-1}} \alpha_n(x^1, \ldots, x^{n-1}, 0)\, dx^1 \cdots dx^{n-1}$$

as desired. ■

This fundamental theorem reduces to the usual theorems of Stokes and Gauss in R^n (see Exercises 2.6C and 2.7B). (For forms without compact support, the best result is somewhat subtle. See Gaffney [1954] and Morrey [1966]. For manifolds with corners, see Lang [1972].)

We shall now discuss a topic called "Koopmanism" (after B. O. Koopman [1931]) and an important result in the subject due to Povzner [1968]. This material requires an acquaintance with functional analysis, specifically with Stone's theorem for self-adjoint operators, and may be omitted if desired. (Required background is obtained in almost any text on functional analysis, such as Reed and Simon [1975].)

Let M be a manifold and Ω a volume on M, with μ_Ω the corresponding measure. If F_t is a volume preserving flow on M, then F_t induces a *linear* one parameter group of isometries on $H = L^2(M, \mu_\Omega)$ by

$$U_t(f) = f \circ F_{-t}$$

The association of U_t with F_t replaces a nonlinear finite-dimensional problem with a linear infinite-dimensional one.

There have been several theorems that relate properties of F_t and U_t. The best known of these is the result of Koopman himself, which shows that U_t has one as a simple eigenvalue for all t if and only if F_t is ergodic. (If there are no other eigenvalues, F_t is weakly mixing.) A few basic results on ergodic theory are given in Sect. 3.7 below. We also refer the reader to the excellent texts of Halmos [1956], Arnold–Avez [1967], and Bowen [1975]. The spectral results may also be found, with further references, in Reed and Simon [1975].

ISBN 0-8053-0102-X

One can also attempt to use Koopmanism to study flows generated by vector fields that are in some sense singular. See, for example, Marsden [1968]. (See also Truesdell [1974].) For Hamiltonian systems, a satisfactory theory of the motion of a particle in a general potential with singularities (including collisions with walls and other particles) remains unsolved to this day. In this direction, however, there is an important result of Povzner that we shall present. A related result of note is that of Mackey [1963], which states that if U_t is a linear isometry on $L^2(M,\mu_\Omega)$, which is multiplicative $U_t(fg)=U_tf\cdot U_tg$, (where defined), then U_t is induced by some measure preserving flow F_t.

To present the result of Povzner, we need a lemma due to Nelson.

2.6.13 Lemma. *Let A be an (unbounded) self-adjoint operator on a Hilbert space H. Let $D_0\subset D(A)$ (the domain of A) be a dense linear subspace of H and suppose $U_t=e^{itA}$ (the unitary one-parameter group generated by A) leaves D_0 invariant. Then $A_0=(A$ restricted to $D_0)$ is essentially self-adjoint; that is, the closure of A_0 is A.*

Proof. Let $\bar{A}_0$ denote the closure of A_0. Since A is closed and extends A_0, A extends $\bar{A}_0$. We need to prove that $\bar{A}_0$ extends A.

For $\lambda>0$, $\lambda-iA$ is surjective with a bounded inverse. First of all, we prove $\lambda-iA_0$ has dense range. If not, there is a $v\in H$ such that

$$\langle v,\lambda x-iA_0x\rangle=0 \quad \text{for all } x\in D_0$$

In particular, since D_0 is U_t invariant,

$$\frac{d}{dt}\langle v,U_tx\rangle=\langle v,iAU_tx\rangle=\lambda\langle v,U_tx\rangle$$

so

$$\langle v,U_tx\rangle=e^{\lambda t}\langle v,x\rangle$$

Since D_0 is dense, this holds for all $x\in H$. Since $\|U_t\|=1$ and $\lambda>0$, we conclude that $v=0$.

Thus $(\lambda-iA_0)^{-1}$ makes sense and $(\lambda-iA)^{-1}$ is its closure. It follows that A is the closure of A_0. ■

2.6.14 Proposition. *Let X be a C^∞ divergence-free vector field on (M,Ω) with a complete flow F_t. Then iX is an essentially self-adjoint operator on $C_c^\infty=$ the C^∞ functions with compact support in the Hilbert space $L^2(M,\mu_\Omega)$.*

Proof. Let $U_tf=f\circ F_{-t}$ be the unitary one-parameter group induced from F_t. A straightforward convergence argument shows U_tf is continuous in t in

ISBN 0-8053-0102-X

$L^2(M, \mu_\Omega)$. In Lemma 2.6.13, choose $D_0 = C^\infty$ functions with compact support. This is clearly invariant under U_t. If $f \in D_0$,

$$\frac{d}{dt} U_t f \bigg|_{t=0} = \frac{d}{dt} f \circ F_{-t} \bigg|_{t=0} = -df \cdot X$$

so the generator of U_t is an extension of $-X$ (as a differential operator) on D_0. The corresponding essentially self-adjoint operator is therefore iX. ■

Now we prove the converse of 2.6.14. That is, if iX is essentially self-adjoint, then X has a complete flow. This is a functional-analytic characterization of completeness. Povzner's original proof was complicated and had some omissions. We give a simpler proof which was kindly communicated by E. Nelson.

2.6.15 Theorem. *Let M be a manifold with volume element Ω and let X be a C^∞ divergence-free vector field on M. Suppose that, as an operator on $L^2(M, \mu_\Omega)$, iX is essentially self-adjoint on the C^∞ functions with compact support. Then, except possibly for a set of points x of measure zero, the flow $F_t(x)$ of X is defined for all $t \in \mathbf{R}$.*

Actually we will prove more than this. Namely, if the defect index of iX is zero in the upper half-plane [i.e., if $(iX + i)(C_c^\infty)$ is dense in L^2], we shall show that the flow is defined, except for a set of measure zero, for all $t > 0$. Similarly, if the defect index of iX is zero in the lower half-plane, the flow is essentially complete for $t < 0$.

The converses of these more general results can be established along the lines of the proof of 2.6.13.

Proof. Suppose that there is a set E of finite positive measure such that if $x \in E$, $F_t(x)$ fails to be defined for t sufficiently large. Let E_T be the set of $x \in E$ for which $F_t(x)$ is undefined for $t \geq T$. Since $E = \cup_{T=1}^{\infty} E_T$, some E_T has positive measure. Replacing E by E_T, we may assume that all points of E "move to infinity" in a time $\leq T$.

If f is any function on M, we shall adopt the convention that $f(F_t(x)) = 0$ if $F_t(x)$ is undefined. For any $x \in M$, if $t < -T$, $F_t(x)$ must be either in the complement of E or undefined; otherwise it would be a point of E that did not move to infinity in time T. Hence we must have $\chi_E(F_t(x)) = 0$ for $t < -T$, where χ_E is the characteristic function of E. We now define a function on M by

$$g(x) = \int_{-\infty}^{\infty} e^{-\tau} \chi_E(F_\tau(x))\, d\tau$$

Note that the integral converges because the integrand vanishes for $t < -T$. In fact, we have $0 \leq g(x) \leq \int_{-T}^{\infty} e^{-\tau} d\tau = e^T$. Moreover, g is in L^2. Indeed,

ISBN 0-8053-0102-X

because F_t is measure-preserving, where defined, we have $\|\chi_E \circ F_\tau\|_2 \leq \|\chi_E\|_2$ so that

$$\|g\|_2 \leq \int_{-T}^{\infty} e^{-\tau}\|\chi_E \circ F_\tau\|_2 d\tau \leq \|\chi_E\|_2 e^T$$

The function g is nonzero because E has positive measure.

Fix a point $x \in M$. Then $F_t(x)$ is defined for t sufficiently small. It is easy to see that in this case $F_\tau(F_t(x))$ and $F_{\tau+t}(x)$ are defined or undefined together, and in the former case they are equal. Hence we have $\chi_E(F_\tau(F_t(x)) = \chi_E(F_{\tau+t}(x))$ for t sufficiently small. Therefore, for t sufficiently small

$$g(F_t(x)) = \int_{-\infty}^{\infty} e^{-\tau}\chi_E(F_{\tau+t}(x))\,d\tau$$

$$= \int_{-\infty}^{\infty} e^{t-\tau}\chi_E(F_\tau(x))\,d\tau$$

$$= e^t g(x)$$

Now if φ is C^∞ with compact support, we have

$$\int g(x)\cdot X\varphi(x)\,d\Omega = \lim_{t\to 0}\int g(x)\cdot\frac{\varphi(F_t(x))-\varphi(x)}{t}\,d\Omega$$

$$= \lim_{t\to 0}\int \frac{g(F_{-t}(x))-g(x)}{t}\varphi(x)\,d\Omega$$

$$= \lim_{t\to 0}\int \frac{e^{-t}-1}{t}\,g(x)\varphi(x)\,d\Omega$$

$$= -\int g(x)\varphi(x)\,d\Omega$$

(These equalities are justified because on the support of φ the flow F_t exists for sufficiently small t and is measure-preserving.)

Thus g is orthogonal to the range of $X+1$, and therefore the defect index of iX in the upper half-plane is nonzero.

The case of completeness for $t<0$ is similar. ■

EXERCISES

2.6A. Give the details for the construction of the sequence described in the proof of 2.6.9. Give the version of 2.6.10 that applies to vector fields that are not necessarily complete.

ISBN 0-8053-0102-X

2.6B. Suppose M is an orientable manifold and $X \in \mathfrak{X}(M)$ is incompressible. Let $A \subset M$ be a measurable set and $A_\lambda = F_\lambda(A)$, where F is the (local) flow of X. Then $\mu_\Omega(A) = \mu_\Omega(A_\lambda)$. Interpret physically. (*Hint:* Use 2.6.10.)

2.6C. Use Stokes' theorem to derive Gauss' theorem:

$$\int_{V_0} (div\, X)\,\Omega = \int_{\partial V_0} i_X \Omega$$

where X is a vector field on V_0 and Ω is a volume element.

2.6D. Verify from Gauss' theorem that if X is a divergence free vector field, then, as an operator, X is skew symmetric:

$$\langle X(f), g\rangle = \int X(f)\cdot g\Omega = -\int f X(g)\Omega = -\langle f, X(g)\rangle$$

2.6E. Prove the identity $\int_M L_X \alpha \wedge \beta = -\int_M \alpha \wedge L_X \beta$ if M is compact without boundary.

2.6F. Consider the flow in R^2 associated with a reflecting particle: for $t > 0$, set

$$F_t(q,p) = \begin{cases} q + tp & \text{if } q > 0,\ q + tp > 0 \\ -q - tp & \text{if } q > 0,\ q + tp < 0 \end{cases}$$

and set
$$F_t(-q,p) = -F_t(q,p), \quad F_{-t} = F_t^{-1}$$

Study the Koopmanism of this flow. Specifically, what is the exact generator of the induced unitary flow? Is it essentially self-adjoint on the C^∞ functions with compact support away from the line $q = 0$?

2.6G. Use the result of Mackey (see the remarks preceding 2.6.13) to give another proof of Povzner's theorem.

2.6H. Let M be orientable with a volume Ω and $X, Y \in \mathfrak{X}$. Prove that $div[X, Y] = X(div\, Y) - Y(div\, X)$.

2.7 SOME RIEMANNIAN GEOMETRY

In order to properly understand many important examples in mechanics, a knowledge of some Riemannian geometry is essential. In particular, this is needed if one wishes to understand geodesic motion as a Hamiltonian system. This material is treated rather quickly here. The reader who is anxious to get on to mechanics can merely use this section as a reference and come back to it as needed.

In 2.5.12 we defined a Riemannian metric g. We shall also consider pseudo-Riemannian metrics g, where we assume $g_m \equiv g(m)$ is nondegenerate in place of positive-definiteness.

In local coordinates, we write

$$g_m(v, w) = g_{ij}(x) v^i w^j$$

ISBN 0-80530102-X

where summation over i and j is implied and g_{ij} is a nonsingular symmetric matrix depending smoothly on x.

For example, a submanifold M of $\mathbf{R}^n$ is a Riemannian manifold, where for each $x \in M$, g_x is the inner product induced from that of $\mathbf{R}^n$. (The difficult theorem of J. Nash [1964] asserts that every Riemannian manifold is so obtained.)

In order to proceed most smoothly, it is convenient to first define the notion of affine connection. Below we shall see that any pseudo-Riemannian metric uniquely determines a connection satisfying certain conditions.

2.7.1 Definition. *An* ***(affine) connection*** *on a manifold M is a map that assigns to each pair of C^∞ vector fields X and Y on M (or on an open subset of M) another C^∞ vector field $\nabla_X Y$ such that*

(a) $\nabla_X Y$ is $\mathbf{R}$-bilinear in X and Y, and
(b) for $f: M \to \mathbf{R}$ smooth, $\nabla_{fX} Y = f\nabla_X Y$ and $\nabla_X fY = f\nabla_X Y + X(f)Y$.

We call $\nabla_X Y$ the ***covariant derivative*** *of Y along X.*

In a local chart on M with coordinates $x = (x^1, \ldots, x^n)$ we define the n^3 functions $\Gamma^i_{jk}(x)$ by

$$\nabla_{\partial/\partial x^j}\left(\frac{\partial}{\partial x^k}\right) = \Gamma^i_{jk}\frac{\partial}{\partial x^i}$$

The Γ^i_{jk} are called the ***Christoffel symbols*** *of the connection (in the given chart). The summation convention is in effect here and in what follows.*

The Christoffel symbols are not the components of a tensor on M. Rather, they are the components of an object on the second tangent bundle, namely, a "spray." This will be discussed in Sect. 3.7 below. If we change coordinates from $(x^1, \ldots, x^n)$ to $(\bar{x}^1, \ldots, \bar{x}^n)$, the transformation rule for the Γ's is easy to work out from the definition: substituting

$$\frac{\partial}{\partial \bar{x}^i} = \frac{\partial x^j}{\partial \bar{x}^i}\frac{\partial}{\partial x^j}$$

(the transformation rule for vectors) and using (a) and (b) and the chain rule, we find

$$\bar{\Gamma}^k_{ij} = \frac{\partial x^p}{\partial \bar{x}^i}\frac{\partial x^q}{\partial \bar{x}^j}\Gamma^r_{pq}\frac{\partial \bar{x}^k}{\partial x^r} + \frac{\partial \bar{x}^k}{\partial x^p}\frac{\partial^2 x^p}{\partial \bar{x}^i \partial \bar{x}^j}$$

Also from (a) and (b) we find that

$$(\nabla_X Y)^i = \frac{\partial Y^i}{\partial x^j}X^j + \Gamma^i_{jk}X^jY^k$$

ISBN 0-8053-0102-X

2.7.2 Example. If $M = R^n$ with the usual inner product, we choose, in Euclidean coordinates

$$(\nabla_X Y)^i = \frac{\partial Y^i}{\partial x^j} X^j$$

This is also commonly denoted $X \cdot \nabla Y$. Thus, the Euclidean connection is characterized by the fact that in standard coordinates, the Γ's are zero. In other coordinates (cylindrical, spherical, etc.) the Christoffel symbols are not zero. For instance, using the above transformation law to change from Euclidean (x,y,z) coordinates in R^3 to spherical (r,θ,φ) coordinates in R^3, $x = r\sin\theta\cos\varphi$, $y = r\sin\theta\sin\varphi$, $z = r\cos\theta$, we find that the nonzero Christoffel symbols are

$$\Gamma^r_{\theta\theta} = -r, \quad \Gamma^\theta_{r\theta} = \Gamma^\theta_{\theta r} = 1/r, \quad \Gamma^r_{\varphi\varphi} = -r\sin^2\theta$$

$$\Gamma^\theta_{\varphi\varphi} = -\sin\theta\cos\theta, \quad \Gamma^\varphi_{\varphi r} = \Gamma^\varphi_{r\varphi} = 1/r, \quad \Gamma^\varphi_{\varphi\theta} = \Gamma^\varphi_{\theta\varphi} = \cot\theta$$

Other examples will be easier to deal with after the relationship with a metric is introduced (see Exercise 2.7A).

2.7.3 Definition. *If $\gamma(t)$ is a curve and X is a vector field, define the* ***covariant derivative*** *of X along γ by*

$$\frac{DX}{dt} = \nabla_{\dot\gamma(t)} X$$

[This is well defined because, from the local formula for $(\nabla_X Y)^i$, we see that $(\nabla_X Y)(m)$ depends only on the value of X at m and not on its derivatives.]

We say X is ***autoparallel*** *along γ if $DX/dt = 0$. We call γ a* ***geodesic*** *if $\dot\gamma$ is autoparallel along γ.*

(The latter is well defined because, from the chain rule,

$$\left(\frac{DX}{dt}\right)^i = \frac{d}{dt}(X^i(\gamma(t))) + \Gamma^i_{jk}(\gamma(t))X^k(\gamma(t))\dot\gamma^j(t)$$

so DX/dt depends only on the values of X along γ.)

Thus $\gamma(t)$ is a geodesic if and only if, in any coordinate system,

$$\ddot\gamma^i + \Gamma^i_{jk}\dot\gamma^j\dot\gamma^k = 0$$

These are called the *geodesic equations.*

For example, in R^n it is obvious (using Euclidean coordinates) that the geodesics are straight lines.

From the existence theorem for ordinary differential equations (see Sect. 2.1), given $\gamma(0)$, $\dot\gamma(0)$, there is a unique geodesic $\gamma(t)$ defined on some

ISBN 0-8053-0102-X

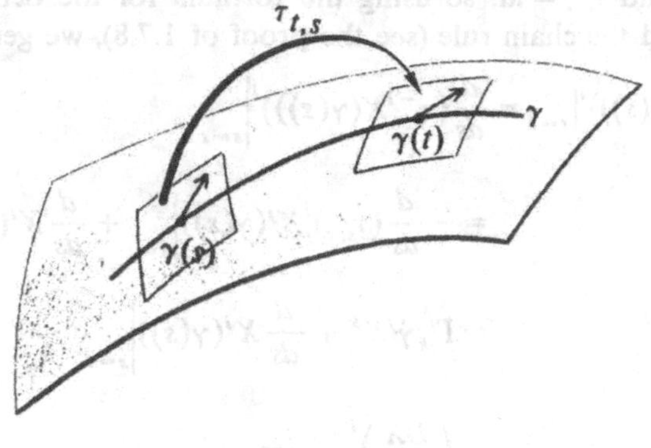

Figure 2.7-1

t-interval. Similarly, given a curve $\gamma(t)$ and $X_0 \in T_{\gamma(0)}M$, there is a unique autoparallel field $X(t)$ along γ with $X(t) \in T_{\gamma(t)}M$ and $X(0) = X_0$. There are no limitations on the time of existence for $X(t)$ because its differential equation is linear in X. Thus, for each curve γ and two points $\gamma(t)$ and $\gamma(s)$ on it, we get a linear isomorphism

$$\tau_{t,s}: T_{\gamma(s)}M \to T_{\gamma(t)}M$$

where $\tau_{t,s}(v) = v(t)$ is v at $\gamma(s)$ extended to be autoparallel along $\gamma(t)$. The map $\tau_{t,s}$ is called *parallel translation* along γ from $\gamma(s)$ to $\gamma(t)$. See Fig. 2.7-1 and compare with Fig. 2.2-1. From uniqueness of solutions of differential equations we see that the maps $\tau_{s,t}$ have the following flowlike property: $\tau_{s,t} \circ \tau_{t,r} = \tau_{s,r}$ and $\tau_{s,s} = \text{identity}$.

For example, in Euclidean space, parallel translation is exactly that in the Euclidean sense. On the other manifolds, the intuition is that if you were a creature living on the manifold (such as we are living on a sphere) and you carried an arrow around the manifold in a manner that seems parallel to you (*not* with respect to any containing space), then you are parallel transporting the vector.

There is a basic link between parallel translation and the covariant derivative much like that between flows and the Lie derivative, discussed in Sect. 2.2, as follows:

2.7.4 Theorem. *Let X be a vector field defined along γ. Then*

$$\frac{DX}{dt}(\gamma(t)) = \frac{d}{ds}(\tau_{t,s}X(\gamma(s)))\Big|_{s=t}$$

Proof. We work in a chart. By construction $v(t) = \tau_{t,s} \cdot v_0$, $v_0 \in T_{\gamma(s)}M$ satisfies $(d/dt)v^i + \Gamma^i_{jk} v^k \dot{\gamma}^j = 0$. If we write $(\tau_{t,s}v_0)^i = (\tau_{t,s})^i_j v_0^j$, we can conclude that

$$\frac{d}{dt}(\tau_{t,s})^i_k\Big|_{t=s} = -\Gamma^i_{jk}\dot{\gamma}^j$$

ISBN 0-8053-0102-X

Now $\tau_{t,s} = \tau_{s,t}^{-1}$ and $\tau_{s,s} = id$, so using the formula for the derivative of an inverse matrix and the chain rule (see the proof of 1.7.8), we get

$$\begin{aligned}\frac{d}{ds}(\tau_{t,s}X(\gamma(s)))^i\Big|_{s=t} &= \frac{d}{ds}(\tau_{s,t}^{-1}X(\gamma(s)))^i\Big|_{s=t}\\ &= -\frac{d}{ds}(\tau_{s,t})^i_j X^j(\gamma(s))\Big|_{s=t} + \frac{d}{ds}X^i(\gamma(s))\Big|_{s=t}\\ &= \Gamma^i_{jk}\dot{\gamma}^j X^k + \frac{d}{ds}X^i(\gamma(s))\Big|_{s=t}\\ &= \left(\frac{DX}{dt}\right)^i \quad \blacksquare\end{aligned}$$

From this formula and the property $\tau_{t,r}\circ\tau_{r,s} = \tau_{t,s}$, we get the more general formula

$$\frac{d}{ds}\tau_{t,s}X(\gamma(s)) = \tau_{t,s}\frac{DX}{ds}(\gamma(s))$$

Conversely, given a collection of parallel translation maps (satisfying the above flowlike property) one can use the formula in 2.7.4 to define a connection, just as one can define the Lie derivative in terms of a flow.

Given a connection ∇ and a vector field Y, ∇Y may be regarded as a (1, 1) tensor since $\nabla_X Y$ depends only on the point values of X. We write, in coordinates,

$$(\nabla Y)^i_j \equiv Y^i{}_{|j} = Y^i{}_{,j} + \Gamma^i_{jk}Y^k$$

where $Y^i{}_{,j} = \partial Y^i/\partial x^j$.

By Willmore's argument in Sect. 2.2, ∇ extends naturally to any tensor field. If t is a (p,q) tensor, ∇t is a $(p,q+1)$ tensor and is given in coordinates by

$$\begin{aligned}(\nabla t)^{i_1\cdots i_p}_{j_1\cdots j_q k} &\equiv t^{i_1\cdots i_p}_{j_1\cdots j_q|k}\\ &= t^{i_1\cdots i_p}_{j_1\cdots j_q,k} + t^{l i_2\cdots i_p}_{j_1\cdots j_q}\Gamma^{i_1}_{kl} + (\text{all upper indices})\\ &\quad - t^{i_1\cdots i_p}_{l j_2\cdots j_q}\Gamma^l_{kj_1} - (\text{all lower indices})\end{aligned}$$

The parallel translation map also extends in an analogous way to all tensors and the formula in 2.7.4 extends to all tensors by a proof just like we gave in Sect. 2.2 for the Lie derivative.

Given a connection ∇, we can define an exponential map as follows. By homogeneity of the equations, if $\gamma(t)$ is a geodesic and $\delta \in \boldsymbol{R}$, $\gamma(\delta t)$ is a geodesic as well, with initial velocity $\delta\dot{\gamma}(0)$. It follows that on some neighborhood V of $0 \in T_xM$, the corresponding geodesics are defined for $t \in [0,1]$.

ISBN 0-8053-0102-X

Then by definition exp_x maps $v \in V$ to the point $\gamma_v(1)$, where $\dot{\gamma}_v(0) = v$ and $\gamma_v(t)$ is a geodesic.

The homogeneity also gives $\gamma_{tv}(s) = \gamma_v(ts)$. Setting $s = 1$ gives $exp_x\,(tv) = \gamma_v(t)$. It follows also that $T exp_x(\mathbf{0}) = \text{identity}$, so by the implicit function theorem, exp_x is a diffeomorphism of a neighborhood of $\mathbf{0} \in T_x M$ to a neighborhood of $x \in M$. This then gives a local chart for M called *normal* or *Gaussian coordinates*. (The map *exp* is smooth because the solution of differential equations depends smoothly on the initial conditions; see Sect. 2.1.) These coordinates are often convenient for computations; for example, in them, rays through $\mathbf{0}$ are geodesics [from $exp_x\,(tv) = \gamma_v(t)$] and hence $\Gamma^i_{jk}(\mathbf{0}) = 0$.

If N is a submanifold of M, a tubular neighborhood of N is a neighborhood U of N that is diffeomorphic to a neighborhood of zero in a vector bundle over N.

2.7.5 Theorem. *If N is compact in a manifold M with a connection ∇, then N has a tubular neighborhood.*

Proof. The vector bundle E in question is any *normal bundle* of N; that is, for $x \in N$, E_x is such that $T_x N \oplus E_x = T_x M$. [For instance, if M has a Riemannian metric, $E_x = (T_x N)^{\perp}$ may be chosen.] The diffeomorphism is the exponential map. The calculation above shows *exp* has a derivative at $x \in N$ that is the identity on each factor $T_x N$ and E_x, and thus is a local diffeomorphism by the implicit function theorem. Since N is compact, and $exp_x\,(\mathbf{0}) = x$, it is a diffeomorphism on a neighborhood of N. ■

The theorem is also true if N is noncompact by using a partition of unity argument. See Lang [1972] for details.

Now we come to the question of how to obtain a specific connection from a Riemannian metric. The next result is sometimes called the "Fundamental Theorem of Riemannian Geometry."

2.7.6 Theorem. *Let M be a pseudo-Riemannian manifold. Then there is a unique connection ∇ on M such that*

(i) $\nabla_X Y - \nabla_Y X = [X, Y]$ *and*
(ii) parallel translation preserves the inner product (i.e., is an isometry).

Remark. Condition (i) is easily seen to be equivalent to the symmetry $\Gamma^i_{jk} = \Gamma^i_{kj}$ in any coordinate chart. For any connection, $\nabla_X Y - \nabla_Y X - [X, Y] = Tor(X, Y)$ defines a tensor called the ***torsion***. Thus (i) means ∇ is torsion free.

Proof of 2.7.6. Condition (ii) means that

$$\frac{d}{dt}\langle \tau_{t,s} X(\gamma(s)), \tau_{t,s} Y(\gamma(s))\rangle_{\gamma(t)} = 0$$

ISBN 0-8053-0102-X

From $\tau_{t,r} \circ \tau_{r,s} = \tau_{t,s}$, we see that this condition is equivalent to the same condition taken at $t = s$. Using our calculation in 2.7.4

$$\frac{d}{dt}\left(\tau_{t,s} X(\gamma(s))\right)\Big|_{t=s} = -\frac{DX}{dt}$$

(at first, the left-hand side makes sense only in a chart), the condition becomes

$$Z\langle X, Y\rangle = \langle \nabla_Z X, Y\rangle + \langle X, \nabla_Z Y\rangle$$

where $Z = \gamma'$. Thus (ii) is equivalent to this condition for all vector fields X, Y, Z By substitution of this identity and (i), we verify directly that

$$Z\langle X, Y\rangle + \langle Z, [X, Y]\rangle + Y\langle X, Z\rangle + \langle Y, [X, Z]\rangle$$

$$- X\langle Y, Z\rangle - \langle X, [Y, Z]\rangle = 2\langle X, \nabla_Z Y\rangle$$

This condition shows that $\nabla_Z Y$ is uniquely determined by the metric. Conversely, reversing the steps, if we define $\nabla_Z Y$ by this formula, it verifies the conditions of a connection, (i) and (ii). ■

In local coordinates, this formula relating the connection and metric is

$$2\Gamma_{ij}^k = g^{hk}\left\{\frac{\partial g_{hj}}{\partial x^i} + \frac{\partial g_{ih}}{\partial x^j} - \frac{\partial g_{ij}}{\partial x^h}\right\}$$

where $g^{hk}(x)$ is the inverse matrix of $g_{hk}(x)$.

Condition (ii) may also be written

$$\frac{d}{dt}\langle X, Y\rangle = \left\langle \frac{DX}{dt}, Y\right\rangle + \left\langle X, \frac{DY}{dt}\right\rangle$$

for vector fields X, Y defined along a curve γ. This shows, in particular, that if $\gamma(t)$ is a geodesic, the energy, $\langle\dot{\gamma}, \dot{\gamma}\rangle/2$ is constant in t.

A Riemannian manifold admits convex neighborhoods. Precisely, if $x \in M$, there is an open set U containing x such that any two points of U may be joined by one and only one geodesic lying in U; these geodesics are minimal in that if γ is the unique geodesic joining x_1 to x_2 in U and σ is any other curve from x_1 to x_2, then $l(\gamma) \leq l(\sigma)$, where

$$l(\sigma) = \int_a^b \|\dot{\sigma}(t)\|\, dt$$

is the *length* of σ. We shall refer to Milnor [1963] for a proof.

Let M be a connected, *finite-dimensional* Riemannian manifold. Set

$$d(x_1, x_2) = \inf\{l(\sigma) | \sigma \text{ is a } C^1 \text{ curve from } x_1 \text{ to } x_2\}$$

ISBN 0-8053-0102-X

One can show that d is a (distance) metric on M yielding the same topology as M was originally endowed with.

We shall leave it to the reader to fill in the proof of the next result (again, see Milnor [1963] for example):

2.7.7 Theorem. *The following are equivalent:*

(*i*) *M is geodesically complete.*
(*ii*) *d is a complete metric on M.*
(*iii*) *Closed and bounded subsets of M are compact.*

Here, geodesically complete means that every geodesic $\gamma(t)$ can be extended so as to be defined for all $t \in \mathbf{R}$. As in Sect. 2.1, one sees that compact Riemannian manifolds are complete. We shall return to these questions in Sect. 3.8.

Closely related to 2.7.7 is the Hopf–Rinow theorem, which asserts that for M complete, any two points x_1, x_2 can be joined by a minimal geodesic γ (not necessarily unique); that is, $l(\gamma) = d(x_1, x_2)$.

2.7.8 Definition. *A bijective map $f: M \to P$ between Riemannian (or pseudo-Riemannian) manifolds is called an* **isometry** *when f preserves the metric: $\langle v, w \rangle_x = \langle Tf \cdot v, Tf \cdot w \rangle_{f(x)}$ for $v, w \in T_x M$.*

It is not hard to see that this condition is the same as asking that f map geodesics to geodesics or that f is an isometry for the corresponding metric spaces.

Two isometries f_1, f_2 from a connected Riemannian manifold M to another P and which, together with their first derivatives agree at a point $x \in M$, are actually identical. This follows by letting $Q = \{ y \in M \mid T_y f_1 = T_y f_2 \}$ and observing that Q is both open and closed. For instance, it then follows that any isometry of $\mathbf{R}^n$ is a Euclidean motion, that is, a translation followed by an orthogonal transformation.

2.7.9 Definition. *A vector field X is called a Killing field if $L_X g = 0$.*

In components, $L_X g$ is easy to work out. (See Exercise 2.7F.) The most useful expression is

$$(L_X g)_{ij} = X_{i|j} + X_{j|i}$$

where $X_i = g_{ij} X^j$, the one form associated with X. (Likewise, g_{ij} and its inverse g^{kl} are used to associate contravariant with covariant tensors.)

2.7.10 Proposition. *A vector field is a Killing field if and only if its flow F_t consists of isometries.*

ISBN 0-8053-0102-X

Proof. This is a consequence of the formula

$$\frac{d}{dt}F_t^*g = F_t^*(L_Xg)$$

from Sect. 2.2. ■

We now define the volume element on a Riemannian manifold.

2.7.11 Definition. *Let M be an oriented Riemannian n-manifold. If $v_1,\dots,v_n \in T_xM$ are positively oriented, set*

$$\mu(v_1,\dots,v_n) = (det\langle v_i, v_j\rangle)^{1/2}$$

This is possible as $det\langle v_i, v_j\rangle$ is >0 for all $v_1,\dots,v_n \in T_xM$ since the metric is positive-definite. Now define μ on all n-tuples by skew symmetry.

Clearly μ is a volume form on M. Locally, $\mu = \sqrt{det\, g_{ij}}\; dx^1 \wedge \cdots \wedge dx^n$. The definition is motivated from the fact that the volume spanned by vectors $v_1,\dots,v_n \in R^n$ is $(det\; v_i \cdot v_j)^{1/2}$.

Since we have μ, we can use it to define the divergence of a vector field, $div\, X$. From the expression for μ above, and the definition $L_X\mu = (div\, X)\mu$ from Sect. 2.5, we find, locally, if $V = \sqrt{det\, g_{ij}}$,

$$div\, X = \frac{1}{V}\frac{\partial}{\partial x^i}(VX^i)$$

For $f: M \to R$, $grad f$ is the vector field defined by

$$\langle grad f(x), v_x\rangle = df(x)\cdot v_x \quad \text{for all} \quad v_x \in T_xM$$

In coordinates,

$$(grad f)^i = g^{ij}\frac{\partial f}{\partial x^j}$$

The *Laplace–Beltrami operator* on functions is defined by

$$\nabla^2 = div \cdot grad$$

so

$$\nabla^2 f = \frac{1}{V}\frac{\partial}{\partial x^k}\left(g^{ik}V\frac{\partial f}{\partial x^i}\right)$$

From Stokes' theorem we find that d and $-$div are adjoints and ∇^2 is

ISBN 0-8053-0102-X

symmetric:

$$\int_M df\cdot X\,d\mu = -\int_M f\,div X\,d\mu$$

$$\int_M f\nabla^2 g\,d\mu = -\int_M \langle grad f, grad g\rangle d\mu = \int_M g\nabla^2 f\,d\mu$$

for X, f, g having compact support.

Next we consider the *Laplace–de Rham Operator.*

2.7.12 Definition. *Let M be a Riemannian n-manifold and let β be a k-form. Define an $(n-k)$-form $*\beta$ by*

$$(*\beta)(v_{k+1},\ldots,v_n) = \beta(v_1,\ldots,v_k)$$

where $v_1,\ldots,v_n$ are oriented orthonormal vectors in T_xM. We call $*$ the ***Hodge star operator.***

For example, on $\boldsymbol{R}^3$, $*dx = dy\wedge dz$, $*dy = dx\wedge dz$, and so forth. One can then verify that

$$\langle\alpha,\beta\rangle_x\mu_x = \alpha_x\wedge*\beta_x$$

defines an inner product on k-forms.

2.7.13 Definition. *Set $(\alpha,\beta)=\int_M \alpha\wedge*\beta$, which gives an L^2 inner product on the sections of $\Omega^k(M)$. Also, define the* ***codifferential operator*** $\boldsymbol{\delta} = (-1)^{n(k+1)+1}*\boldsymbol{d}*$.

It is easily checked that $\boldsymbol{\delta}$ is adjoint of $\boldsymbol{d}$:

$$(\delta\gamma,\beta) = (\gamma, d\beta)$$

[Use the fact that $\int_M d(\beta\wedge*\gamma) = 0$ and $_{**}\beta = (-1)^{k(n-k)}\beta$.]

2.7.14 Definition. *The* ***Laplace–deRham*** *operator is defined by*

$$\Delta = d\delta + \delta d$$

The operator Δ is symmetric, and nonnegative

$$(\Delta\alpha,\beta) = (\alpha,\Delta\beta),\quad (\Delta\alpha,\alpha)\geqslant 0$$

A k-form α satisfying $\Delta\alpha = 0$ is called *harmonic*. On functions, Δ differs in sign from the Laplace–Beltrami operator ∇^2. (See, for instance, Nickerson-Spencer and Steenrod [1959].)

The operator Δ is at the basis of "Hodge–De Rham Theory." The central result states that on a compact manifold M without boundary, the kernel of Δ

ISBN 0-80530102-X

on the k-forms is isomorphic to the kth cohomology group of M. It is easy to see, however, in a formal way, that the kernel of Δ is the quotient of closed k-forms by exact ones. Indeed, first of all observe that if $d\omega=0$ and $\delta\omega=0$, then $\Delta\omega=0$. The converse is also true, since

$$(\Delta\omega,\omega)=(d\omega,d\omega)+(\delta\omega,\delta\omega)$$

from the fact that d and δ are adjoints. Now map $ker(\Delta)$ to the closed forms by injection. We claim that if $\omega\in ker(\Delta)$ is exact, then $\omega=0$. Indeed, let $\omega=d\alpha$; then $\delta d\alpha=0$, so $(\delta d\alpha,\alpha)=0$, so $(d\alpha,d\alpha)=0$, so $\omega=0$. Also, if ω is closed, there is a form α such that $\Delta(\omega-d\alpha)=0$. Indeed, formally, the orthogonal complement of the exact forms are the co-closed forms ($\delta=0$), so we orthogonally decompose $\omega=d\alpha+\beta$, where $\delta\beta=0$. Then, if $d\omega=0$, we get $\Delta\beta=0$.

More generally, one has the *Hodge decomposition*: for any k-form ω, ω splits uniquely into three L^2 orthogonal components

$$\omega=d\alpha+\delta\beta+\gamma$$

where $\Delta\gamma=0$. This is easy to see formally as well. Indeed, the exact forms and co-exact forms are orthogonal and their orthogonal complement consists of forms γ such that $d\gamma=0$ and $\delta\gamma=0$; that is, $\Delta\gamma=0$. To prove this rigorously requires elliptic operator theory. See Warner [1971] and, for the case of manifolds with boundary, Morrey [1966].

We conclude this section with a few basic facts about curvature.

2.7.15 Definition. *Let ∇ be a connection on M. The **curvature tensor** R of ∇ is a $(1,3)$ tensor (mapping $T_xM\times T_xM\times T_xM\to T_xM$ for $x\in M$) defined through the formula*

$$R(X,Y,Z)=\nabla_X\nabla_YZ-\nabla_Y\nabla_XZ-\nabla_{[X,Y]}Z$$

where X, Y, Z are vector fields on M.

Locally, writing

$$R\left(\frac{\partial}{\partial x^i},\frac{\partial}{\partial x^j},\frac{\partial}{\partial x^k}\right)=R_{ijk}{}^l\frac{\partial}{\partial x^l},$$

one computes that

$$R_{ijk}{}^l=\left(\frac{\partial\Gamma^l_{jk}}{\partial x^i}-\frac{\partial\Gamma^l_{ik}}{\partial x^j}\right)+(\Gamma^r_{jk}\Gamma^l_{ir}-\Gamma^r_{ik}\Gamma^l_{jr})$$

ISBN 0-8053-0102-X

Fixing Y and Z, $\boldsymbol{R}(\cdot, Y, Z)$ gives a linear map of T_xM to T_xM. Its trace is the Ricci tensor, *Ric*. In coordinates,

$$R_{ij} \equiv (Ric)_{ij} = R_{lij}{}^{l} \qquad \text{(summation on } l\text{)}$$

Some geometric insight may be gained from the fact that $\boldsymbol{R}=0$ if parallel translation from x to y is independent of the curve joining x to y. In general, if $S(x,y)$ maps $\boldsymbol{R}^2$ to M, the curvature measures the extent to which the covariant derivatives in the x and y directions on S fail to commute:

$$\frac{D}{dx}\frac{D}{dy}X - \frac{D}{dy}\frac{D}{dx}X = \boldsymbol{R}\left(\frac{\partial S}{\partial x}, \frac{\partial S}{\partial y}, X\right)$$

A manifold is *flat* if $\boldsymbol{R}=0$. In the Riemannian case, this is equivalent to the existence of coordinates in which g is constant. These coordinates are provided by the exponential map.

Given two orthonormal vectors $v_1, v_2 \in T_xM$ on a Riemannian manifold, the *sectional curvature* of their span Γ is defined by

$$K(\Gamma) = \langle v_1, \boldsymbol{R}(v_1, v_2, v_2)\rangle$$

If $K<0$, nearby geodesics starting out parallel tend to diverge, while if $K>0$, they tend to converge. Thus the sign of the curvature is relevant for the stability of a geodesic.

The *scalar curvature* on a Riemannian manifold is defined as the trace of the Ricci tensor. In coordinates:

$$R = \sum_{l,i} g^{li} R_{li}$$

($Ric: T_xM \times T_xM \to \boldsymbol{R}$, so taking its trace involves the metric).

Some identities satisfied by the curvature tensor are:

(i) $\boldsymbol{R}(X,Y,Z) = -\boldsymbol{R}(Y,X,Z)$ or $R_{ijk}{}^{l} = -R_{jik}{}^{l}$;
(ii) (Bianchi's identity) $\boldsymbol{R}(X,Y,Z)+\boldsymbol{R}(Z,X,Y)+\boldsymbol{R}(Y,Z,X)=0$ or $R_{ijk}{}^{l}$ is "cyclic" in i,j,k;
(iii) $\langle \boldsymbol{R}(X,Y,Z), W\rangle = \langle \boldsymbol{R}(Z,W,X), Y\rangle$ or $R_{ijkl} = R_{klij}$.

For our discussions on constrained systems later on in Sect. 3.7, the second fundamental form plays a central role.

Let M be an oriented (pseudo-) Riemannian n-manifold and $P \subset M$ and $(n-1)$-dimensional submanifold (i.e., a "hypersurface"). Therefore, we can, at least in the Riemannian case, and in the general case if P is spacelike, that is, $g_x > 0$ on T_xP, uniquely define a unit normal vector field N on P.

By 2.7.6, we have a connection on M and one on P as well, determined by the Riemannian structure. Let ∇ denote the connection on M, $\overline{\nabla}$ that on P.

For vector fields X, Y on P we can define $\nabla_X Y$; from its expression in coordinates, it depends only on their values on P. Also, we have *Gauss'*

ISBN 0-8053-0102-X

formula:

$$\nabla_X Y = \overline{\nabla}_X Y + S(X, Y)N$$

where S is a tensor on P, called the *second fundamental form*. In fact, a straightforward computation in coordinates shows that

$$S(X, Y) = -\langle \nabla_X N, Y\rangle = -\langle \nabla_Y N, X\rangle$$

(*Weingarten equation*). From these two equations and the definition of the curvature tensor, it is straightforward to show that

$$\langle \boldsymbol{R}(X, Y, Z)U\rangle = \langle \overline{\boldsymbol{R}}(X, Y, Z)U\rangle - \{S(X, U)S(Y, Z) - S(Y, U)S(X, Z)\}$$

(*Gauss' equation*), where $\overline{R}$ is the curvature on P and X, Y, Z, U are tangent to P, and that

$$\langle \boldsymbol{R}(X, Y, Z), N\rangle = \nabla_X S(Y, Z) - \nabla_Y S(X, Z)$$

(*Codazzi equation*).

EXERCISES

2.7A. Work out the metric components and Christoffel symbols using spherical coordinates on $M =$ the sphere of radius R in $\boldsymbol{R}^3$. Verify that the geodesics are great circles.

2.7B. (a) (*Classical Gauss' Theorem*) Let M be an oriented Riemannian manifold with boundary ∂M; let $\boldsymbol{\Omega}$ be the volume element of M and $\boldsymbol{\sigma}$ that on ∂M determined by the Riemannian metric. If $\boldsymbol{n}$ is the outward unit normal on ∂M and X is a vector field on M, show that $i_X\boldsymbol{\Omega} = (X\cdot\boldsymbol{n})\boldsymbol{\sigma}$ on ∂M. Prove Gauss' theorem:

$$\int_M (\operatorname{div} X)\boldsymbol{\Omega} = \int_{\partial M} (X\cdot\boldsymbol{n})\boldsymbol{\sigma}$$

(b) (*Classical Stokes' theorem*) Show that the curl on $\boldsymbol{R}^3$ is determined by

$$(\operatorname{curl} X)^{\sim} = *d\tilde{X}$$

where $\tilde{X}$ is the one form $\tilde{X} = X^1dx^1 + X^2dx^2 + X^3dx^3$ associated with the vector field $X = X^1e_1 + X^2e_2 + X^3e_3$. Also, show that $\tilde{X}$ pulled back to a curve in $\boldsymbol{R}^3$ is the tangential component of X. Use these results and (a) to derive the classical Stokes' theorem from the general Stokes' theorem 2.6.12. That is, prove that if S is an oriented smooth surface in $\boldsymbol{R}^3$, then

$$\int_S (\operatorname{curl} X)\cdot\boldsymbol{n}\, da = \int_{\partial S} X\cdot d\boldsymbol{s}$$

where $\boldsymbol{n}$ is the unit normal to S corresponding to the given orientation.

ISBN 0-8053-0102-X

2.7C. Define $L_X\nabla$ for a connection ∇ and show that $L_X\nabla = \nabla\nabla X + X\cdot R$, where R is the curvature tensor (See Yano [1957]).

2.7D. Let (M,g) be a Riemannian manifold and X a Killing vector field, that is, $L_Xg=0$. Let $x(t)$ be a geodesic and $T=\dot{x}$ its tangent vector. Prove that $\langle X,T\rangle$ is constant along the geodesic.

2.7E. If X is a vector field on a Riemannian manifold M, let $\tilde{X}$ be the associated one-form:

$$\tilde{X}(x)\cdot v = \langle X(x), v\rangle$$

Show that $\delta\tilde{X} = -\operatorname{div} X$.

2.7F. Show that $\nabla g=0$, where ∇ is the connection of g. Show that taking the Lie derivative of a covariant tensor does *not* commute with taking the contravariant form of the tensor via g, whereas this *is* true for ∇.

2.7G. Show that a hypersurface $P\subset M$ is totally geodesic (i.e., a geodesic in P starting tangent to P stays in P) if and only if the second fundamental form of P is zero.

2.7H. A hypersurface $P\subset M$ is called *minimal* if $\operatorname{tr} S$, the trace of the second fundamental form of P, vanishes. Show that the "variation" of the volume of hypersurfaces near P vanishes on P. Compute the second variation. (See Almgren [1967] or Choquet-Bruhat, Fischer, and Marsden [1978].)

ISBN 0-8053-0102-X

PART 2

ANALYTICAL DYNAMICS

This part develops most of the important theoretical topics in classical mechanics in the general setting of symplectic manifolds. Chapter 3 sets out the basic theory of Hamiltonian and Lagrangian mechanics. This is followed by a rather extensive chapter on systems with symmetry, including current accounts of reduction by algebras of integrals, and topology of invariant manifolds. The final chapter of this part has, as a focus, the Hamilton–Jacobi theory, with numerous related topics such as action angle variables and Lagrangian submanifolds, as well as offshoots to topics like quantization and the equations of mathematical physics as Hamiltonian systems.

The use of differential forms in mechanics and its eventual formulation in terms of symplectic manifolds has been slowly evolving since Cartan [1922].

The first modern exposition of Hamiltonian systems on symplectic manifolds seems to be due to Reeb [1952e]. An early version of Lagrangian systems in this context appears in Mackey [1963]. This formulation of mechanics was widely known in mathematical circles by 1962, and was explained in a letter by Richard Palais that circulated privately at about that time. The first systematic treatise concerning mechanics on Riemannian manifolds that we know of is Synge [1926]. The reader is referred to Whittaker [1959] for additional historical details.

ISBN 0-8053-0102-X

CHAPTER 3

Hamiltonian and Lagrangian Systems

This chapter begins our study of Hamiltonian mechanics. The basic structure of the classical theory will be given in the context of manifolds. We suggest that the reader have a good classical text available for comparison and additional insight, such as Whittaker [1959] or Goldstein [1950]. A re-reading of the Preview at this time may help motivate what follows. The treatment may seem unnecessarily abstract, but it is of ultimate benefit for a thorough understanding of a rigorous analysis of the applications in the later chapters.

3.1 SYMPLECTIC ALGEBRA

Symplectic manifolds constitute the arena for Hamiltonian mechanics. This section considers the linear case in preparation for the next section.

3.1.1 Definition *Let E be a finite-dimensional real vector space and $\omega \in L^2(E, R)$ a bilinear form on E, so $\omega: E \times E \to R$. We say that ω is* **nondegenerate** *if*

$$\omega(e_1, e_2) = 0 \quad \text{for all} \quad e_2 \in E \quad \text{implies} \quad e_1 = 0$$

Ralph Abraham and Jerrold E. Marsden, Foundation of Mechanics, Second Edition

ISBN 0-8053-0102-X

There are several equivalent ways of stating nondegeneracy. To give these, we need some notation:

(*a*) *If $\hat{e}=(e_i)$ is an (ordered) basis of $\mathbf{E}$ and (α^i) is the dual basis, $\omega_{ij}=\omega(e_i, e_j)$ is the* **matrix** *of ω; it is denoted $[\omega]_{\hat{e}}$. Then it is easy to see that*

$$\omega=\omega_{ij}\alpha^i\otimes\alpha^j \qquad \text{(summation understood)}$$

(*b*) *The* **transpose** *ω^t of ω is given by*

$$\omega^t(e_1, e_2)=\omega(e_2, e_1);$$

ω is **symmetric** *if $\omega^t=\omega$ and* **skew symmetric** *if $\omega^t=-\omega$.*

(*c*) *The linear map $\omega^\flat$: $\mathbf{E}\to\mathbf{E}^*$ is defined by*

$$\omega^\flat(e)\cdot e'=\omega(e, e')$$

Note that the matrix of $\omega^\flat$ relative to the bases e_i and α^j is exactly ω_{ij}; that is,

$$\omega^\flat(e_i)=\omega_{ij}\alpha^j$$

The rank of ω is the rank of the matrix ω_{ij}, that is, the dimension of $\omega^\flat(\mathbf{E})$.

If one uses (i) the fact that in finite dimensions a one-to-one linear map between spaces of the same dimension is an isomorphism and (ii) the observation that the definition of nondegeneracy is precisely that the linear map $\omega^\flat$ is one-to-one; that is, has trivial kernel, the following are easily seen to be equivalent:

(i) ω is nondegenerate;
(ii) ω^t is nondegenerate;
(iii) the matrix of ω is nonsingular;
(iv) $\omega^\flat$: $\mathbf{E}\to\mathbf{E}^*$ is an isomorphism.

In the infinite-dimensional case care must be taken with the type of nondegeneracy assumed. In that case ω is called *weakly nondegenerate* if $\omega^\flat$ is one-to-one and is called *nondegenerate* if $\omega^\flat$ is an isomorphism. Although these notions are equivalent in finite dimensions, this is not so in infinite dimensions and the distinction is important. Most important for a large number of examples such as the wave equation is the weakly nondegenerate case; this is for technical reasons which will become evident in Sect. 5.5.

In this book, however, with the exception of Sect. 5.5, we shall deal with the finite-dimensional case.

If $\hat{e}=(e_i)$ and $\hat{e}'=(e_i')$ are two (ordered) bases of $\mathbf{E}$ with $e_i'=A_i^je_j$, then the matrices of ω are related by congruence: $[\omega]_{\hat{e}'}=A^t[\omega]_{\hat{e}}A$. We have the following canonical form for symmetric and skew symmetric bilinear forms.

3.1.2 Proposition. *Let $\mathbf{E}$ be a p-dimensional real vector space. Then*

(*i*) *If ω is symmetric of rank s, there is an ordered basis $\hat{e}=(e_i)$ of $\mathbf{E}$ with dual ordered basis (α^i) such that*

$$\omega=\sum_{i=1}^{s}\eta_i\alpha^i\otimes\alpha^i \qquad \text{where} \quad \eta_i=\pm1, s\leqslant p$$

ISBN 0-8053-0102-X

or, equivalently, the matrix of ω *is*

$$\begin{bmatrix} \eta_1 & & & & & \\ & \ddots & & & 0 & \\ & & \eta_s & & & \\ 0 & & & 0 & & \\ & & & & \ddots & \\ & & & & & 0 \end{bmatrix}$$

(*ii*) *If* ω *is skew symmetric of rank* r, *then* $r=2n$ *for an integer* n *and there is an ordered basis* $\hat{e}=(e_i)$ *of* E *with dual ordered basis* (α^i) *such that*

$$\omega=\sum_{i=1}^{n}\alpha^i\wedge\alpha^{i+n}$$

or, equivalently, the matrix of ω *is*

$$\begin{bmatrix} 0 & I & 0 \\ -I & 0 & 0 \\ 0 & 0 & 0 \end{bmatrix}$$

where I *is the* $n\times n$ *identity matrix.*

Proof. (i) is just the assertion from linear algebra that any symmetric bilinear form can be brought into diagonal form.

We recall the argument. Since ω is symmetric we have the polarization identity

$$\omega(e,f)=\tfrac{1}{4}(\omega(e+f,e+f)-\omega(e-f,e-f))$$

Thus if $\omega\neq 0$, there is an $e_1\in E$ such that $\omega(e_1,e_1)\neq 0$. Rescaling, we can assume $\omega(e_1,e_1)=\eta_1=\pm 1$. Let E_1 be the span of e_1 and $E_2=\{e\in E\,|\,\omega(e,e_1)=0\}$. Clearly, $E_1\cap E_2=\{0\}$; also $E_1+E_2=E$ for if $z\in E$,

$$z-\eta_1\omega(z,e_1)e_1\in E_2$$

Now if $\omega\neq 0$ on E_2, there is an $e_2\in E_2$ such that $\omega(e_2,e_2)=\eta_2=\pm 1$. Continue inductively to complete the proof.

Remark. If $(V,\langle\,,\rangle)$ is an inner product space, this is the usual Gram–Schmidt argument showing the existence of an orthonormal basis.

(ii) Let $e_1\in E$, $e_{n+1}\in E$ be such that $\omega(e_1,e_{n+1})\neq 0$. This is possible if $\omega\neq 0$. Dividing e_1 by a constant we can assume $\omega(e_1,e_{n+1})=1$. Since $\omega(e_1,e_1)=\omega(e_{n+1},e_{n+1})=0$ and ω is skew, the matrix of ω in the plane P_1

ISBN 0-8053-0102-X

spanned by e_1, e_{n+1} is

$$\begin{pmatrix} 0 & 1 \\ -1 & 0 \end{pmatrix}$$

Let E_2 be the ω-orthogonal complement of P_1, that is,

$$E_2 = \{ z \mid \omega(z, z_1) = 0 \quad \textit{for all} \quad z_1 \in P_1 \}$$

Clearly $E_2 \cap P_1 = \{0\}$; also, $E = E_2 + P_1$, for if $z \in E$,

$$z - \omega(z, e_{n+1})e_1 + \omega(z, e_1)e_{n+1} \in E_2$$

as is easily checked. Thus $E = P_1 \oplus E_2$. Now we repeat the process on E_2, choosing e_2 and e_{n+2} such that $\omega(e_2, e_{n+2}) = 1$, and continue inductively.

This shows that ω has the stated matrix in the basis $e_1, \ldots, e_p$.

We conclude (ii) by showing $\Sigma_{i=1}^n \alpha^i \wedge \alpha^{i+n} = \rho$ is the same as ω in this basis. For any $\alpha \in E^*$, $\alpha = \alpha(e_j)\alpha^j$ (summation), so

$$\rho^\flat(e_i) = \sum_{j=1}^{p} \rho^\flat(e_i)(e_j)\alpha^j$$

But

$$\rho^\flat(e_i)(e_j) = \rho(e_i, e_j) = \sum_{k=1}^{n} \alpha^k \wedge \alpha^{k+n}(e_i, e_j)$$

$$= \sum_{k=1}^{n} (\delta_i^k \delta_j^{k+n} - \delta_i^{k+n} \delta_j^k) = (\delta_j^{i+n} - \delta_i^{j+n})$$

if $i, j \leqslant 2n$. Thus

$$\rho^\flat(e_i) = \begin{cases} \alpha^{i+n} & \text{if } i \leqslant n \\ -\alpha^{i-n} & \text{if } n < i \leqslant 2n \\ 0 & \text{if } 2n < i \end{cases}$$

In particular, the matrix of ρ is given by

$$\begin{bmatrix} 0 & I & 0 \\ -I & 0 & 0 \\ 0 & 0 & 0 \end{bmatrix}$$

so $\rho = \omega$. ■

With regard to (i) we prove for later use that if V is a finite-dimensional inner product space (with inner product $\langle , \rangle$), and ω is a symmetric bilinear

ISBN 0-8053-0102-X

form, then there is an orthonormal basis of V in which ω is diagonal. Indeed, consider the isomorphism $\langle,\rangle^\flat: V\to V^*$ associated with the inner product and its inverse $\langle,\rangle^\sharp$. Let $T=\langle,\rangle^\sharp\circ\omega\flat: V\to V$. Then one checks that T is symmetric, that is, $\langle Te_1,e_2\rangle=\langle e_1,Te_2\rangle$. Then from linear algebra there is an orthonormal basis of eigenvectors of T, that is, an orthonormal basis in which T, and hence ω is diagonal.

If we write a point $z\in E$ as

$$z=x_1e_1+\cdots+x_ne_n+y_1e_{n+1}+\cdots+y_ne_{2n}+\omega_{2n+1}e_{2n+1}+\cdots+\omega_pe_p$$

then we get a useful expression for ω in 3.1.2(ii) as a bilinear form:

$$\begin{aligned}\omega(z,z')&=\omega^\flat(z)\cdot z'\\&=\left(x_1\alpha^{1+n}+\cdots+x_n\alpha^{2n}-y_1\alpha^1-\cdots-y_n\alpha^n\right)\cdot z'\\&=x_1y_1'+\cdots+x_ny_n'-y_1x_1'-\cdots-y_nx_n'\end{aligned}$$

so

$$\omega(z,z')=\sum_{i=1}^n(x_iy_i'-y_ix_i')$$

A useful criterion for nondegeneracy is the following.

3.1.3. Proposition. *Let E be a finite-dimensional real vector space, and $\omega\in\Omega^2(E)$. Then ω is nondegenerate iff E has even dimension, say $2n$, and $\omega^n=\omega\wedge\cdots\wedge\omega$ is a volume on E.*

Proof. Suppose ω is nondegenerate. Choose a basis of E such that $\omega=\sum_{i=1}^n\alpha^i\wedge\alpha^{i+n}$. Since the rank of ω is the dimension of E, we have $\dim E=2n$. Then, by induction we easily verify that

$$\omega^k=\sum_{j_1,\ldots,j_k=1}^n\alpha^{j_1}\wedge\alpha^{j_1+n}\wedge\cdots\wedge\alpha^{j_k}\wedge\alpha^{j_k+n}$$

so

$$\omega^n=n!(-1)^{[n/2]}\alpha^1\wedge\cdots\wedge\alpha^{2n}$$

where $[n/2]$ is the largest integer in $n/2$. Thus ω^n is a volume. Conversely, if ω^n is a volume, the rank of ω is $2n$. ■

Some of this algebra carries over directly to manifolds as follows:

3.1.4 Definition. *Let M be a manifold, and $\omega\in\mathfrak{T}^0_2(M)$. Then ω is **nondegenerate** if $\omega(m)$ is nondegenerate for each $m\in M$.*

ISBN 0-8053-0102-X

Thus, from 3.1.3 we obtain the following.

3.1.5 Proposition. *Let M be a manifold and $\omega \in \Omega^2(M)$. Then ω is nondegenerate iff M is even-dimensional, say $2n$, and $\omega^n = \omega \wedge \cdots \wedge \omega$ is a volume on M.*

Thus, if $\omega \in \Omega^2(M)$ is nondegenerate, M is orientable. We shall use the standard volume

$$\Omega_\omega = \frac{(-1)^{[n/2]}}{n!}\omega^n$$

The globalization of the map $\omega^\flat$ is given as follows.

3.1.6 Definition. *Let M be a manifold and $\omega \in \mathcal{T}^0_2(M)$. Define $\omega^\flat : TM \to T^*M$ by $\omega^\flat(m) = \omega(m)^\flat$; that is, $\omega(m)^\flat(e)\cdot e' = \omega(m)(e,e')$, where $e,e' \in T_mM$. Also, for $X \in \mathfrak{X}(M)$, define $\omega^\flat X : M \to T^*M$ by $\omega^\flat X(m) = \omega^\flat(m)\cdot X(m)$.*

Note that the following proposition applies to Riemannian metrics, pseudo-Riemannian metrics and nondegenerate two forms.

3.1.7 Proposition. *Let M be a manifold and $\omega \in \mathcal{T}^0_2(M)$. Then*

(*i*) *$\omega^\flat : TM \to T^*M$ is a vector bundle mapping;*
(*ii*) *If $X \in \mathfrak{X}(M)$, then $\omega^\flat X \in \mathfrak{X}^*(M)$ and $\omega^\flat$ is $\mathcal{F}(M)$ linear as a mapping $\mathfrak{X}(M) \to \mathfrak{X}^*(M)$;*
(*iii*) *If ω is nondegenerate, then $\omega^\flat : TM \to T^*M$ is a vector bundle isomorphism (in particular, a diffeomorphism). In this case we write $\omega^\sharp = (\omega^\flat)^{-1}$.*

Proof. Let (U,φ) be a chart of M with $\varphi(U) = U' \subset E$. The local representative of $\omega^\flat$ with respect to the natural charts is

$$(T\varphi)^0_1 \circ \omega^\flat \circ (T\varphi^{-1}) : U' \times E \to U' \times E^* : (u',e) \mapsto (u', \omega_2^\flat(u)\cdot e)$$

where ω_2 is the second factor of the local representative of ω, so that $\omega_2^\flat(u)\cdot e(e') = \omega_2(u)(e,e')$. To prove (i) we show that $\omega_2^\flat : U' \to L(E,E^*)$ is smooth. We may use any convenient norm on the vector space $L(E,E^*)$ such as $\|\omega_2^\flat(u)\| = \max\{\|\omega_2^\flat(u)\cdot e\| \,|\, e \in E, \|e\| = 1\}$ for some norms on E and E^*. Now we can use the norm $\|\omega_2(u)\| = \|\omega_2^\flat(u)\|$ on $T^0_2(E)$. Since $\omega_2(u)$ is of class C^∞, $\omega_2^\flat(u)$ is also. Indeed, the derivative of $\omega_2^\flat$ is the following linear map (at $u' \in U'$): $e \mapsto L(e)$, where $L(e) \in L(E,E^*)$ is defined by $L(e)\cdot e_1(e_2) = D\omega_2(u')(e)\cdot(e_1,e_2)$. This can be readily checked. To prove (ii), note that $\mathcal{F}(M)$ linearity is clear from linearity of $\omega^\flat$ on each fiber, and $\omega^\flat X$ is of class C^∞ by (i) and 1.7.9 in local representation. Finally, for (iii), if ω is nondegenerate, $\omega^\flat$ is a bijection and an isomorphism on each fiber. We must show that $\omega^\sharp$ is smooth. The local representative of $\omega^\sharp$ is $(u',\alpha) \mapsto (u', \omega_2^\flat(u)^{-1}\cdot\alpha)$. From the proof of 1.7.8, however, $\omega_2^\flat(u)^{-1}$ is of class C^∞. ■

ISBN 0-8053-0102-X

If $t \in \mathfrak{T}^r_s(M)$, we can define $\omega^\flat t \in \mathfrak{T}^{r-1}_{s+1}(M)$ by

$$\omega^\flat t(\alpha^1,\ldots,\alpha^{r-1},X_1,\ldots,X_{s+1}) = t(\alpha^1,\ldots,\alpha^{r-1},\omega^\flat X_{s+1},X_1,\ldots,X_s)$$

to "lower the last contravariant index." Similarly other indices or groups of indices may be raised or lowered.

3.1.8 Definition. *A symplectic form on a vector space E is a nondegenerate two-form $\omega \in \Omega^2(E)$. The pair (E,ω) is called a **symplectic vector space**. If (E,ω) and (F,ρ) are symplectic vector spaces, a linear map $f\colon E \to F$ is **symplectic** iff $f^*\rho = \omega$.*

If (E,ω) is a symplectic vector space we have an orientation defined on E by

$$\Omega_\omega = \frac{(-1)^{[n/2]}}{n!}\omega^n$$

where E has dimension $2n$.

3.1.9 Proposition. *Let (E,ω) and (F,ρ) be symplectic vector spaces of dimension $2n$, and $f\colon E \to F$ be a symplectic mapping. Then f is volume preserving, is orientation preserving and $\det_{(\Omega_\omega,\Omega_\rho)} f = 1$. In particular, f is an isomorphism.*

Proof. We have

$$f^*\Omega_\rho = f^*\left[\frac{(-1)^{[n/2]}}{n!}\rho\wedge\cdots\wedge\rho\right] = \frac{(-1)^{[n/2]}}{n!}f^*\rho\wedge\cdots\wedge f^*\rho = \Omega_\omega$$

Hence f is volume preserving. The last statements follow from this. ■

3.1.10 Example. A volume preserving map need not be symplectic, for consider $\boldsymbol{R}^4$ with vectors denoted (x^1,x^2,y^1,y^2) and $\omega = \alpha^1\wedge\beta^1 + \alpha^2\wedge\beta^2$, where $(\alpha^1,\alpha^2,\beta^1,\beta^2)$ is a basis dual to the standard basis (e_1,e_2,f_1,f_2). Consider the map $(x^1,x^2,y^1,y^2) \mapsto (-x^1,-x^2,y^1,y^2)$. This preserves the volume $\Omega_\omega = \alpha^1\wedge\alpha^2\wedge\beta^1\wedge\beta^2$ but maps $\omega \mapsto -\omega$. On $\boldsymbol{R}^2$, however, every orientation and area preserving linear map is symplectic.

3.1.11 Proposition. *Let (E,ω) be a symplectic vector space. Then the set of all symplectic mappings $f\colon E \to E$ forms a group under composition, called the **symplectic group**, denoted by $Sp(E,\omega)$.*

Proof. Since $GL(E,E)$ forms a group, we need only show that if $f,g \in Sp(E,\omega)$, then $f\circ g$ and $f^{-1} \in Sp(E,\omega)$. But $(f\circ g)^*\omega = g^*\circ f^*\omega = g^*\omega = \omega$, and $(f^{-1})^*\omega = (f^*)^{-1}f^*\omega = \omega$. ■

Next we examine the condition that $f \in Sp(E,\omega)$ in matrix notation. As we saw in 3.1.2, there is an ordered basis of E such that the matrix of ω is

$$J = \begin{bmatrix} 0 & I \\ -I & 0 \end{bmatrix}$$

ISBN 0-8053-0102-X

Note that $J^{-1}=J^t=-J$, and $J^2=-I$. For $f\in L(E,E)$ with matrix $A=(A^i_j)$ relative to this basis, the condition $f^*\omega=\omega$, that is, $\omega(f(e),f(e'))=\omega(e,e')$ becomes

$$A^tJA=J$$

If $A=\begin{bmatrix} a & b \\ c & d\end{bmatrix}$, where a,b,c,d are $n\times n$ matrices, $f\in Sp(E,\omega)$ iff a^tc and b^td are symmetric and $a^td-c^tb=I$.

A condition on the eigenvalues of $f\in Sp(E,\omega)$ is given by the following.

3.1.12 Proposition. *Suppose (E,ω) is a symplectic vector space, $f\in Sp(E,\omega)$ and $\lambda\in C$ is an eigenvalue of f. Then $1/\lambda$, $\bar\lambda$, and $1/\bar\lambda$ are eigenvalues of f ($\bar\lambda$ denotes the complex conjugate of λ).*

Proof. Let $\hat e$ be an ordered basis of E such that $[\omega]_{\hat e}=J$ and $[f]_{\hat e}=A$. Then $A^tJA=J$, or $JAJ^{-1}=B$, where $B=(A^t)^{-1}=(A^{-1})^t$. Let $P(\lambda)=det(A-\lambda I)$, considered as a polynomial in the complex variable λ, with real coefficients. Then as $J^{-1}=-J$, the space is even dimensional, and $det A=1$, so

$$P(\lambda)=det(A-\lambda I)=det\left[J(A-\lambda I)J^{-1}\right]=det(B-\lambda I)=det(A^{-1}-\lambda I)$$

$$=det\left[A^{-1}(I-\lambda A)\right]=det(I-\lambda A)=det\left[\lambda\left(\frac{1}{\lambda}I-A\right)\right]$$

$$=\lambda^{2n}det\left(\frac{1}{\lambda}I-A\right)=\lambda^{2n}P\left(\frac{1}{\lambda}\right)\qquad(\lambda\neq 0)$$

if $2n=dim(E)$. As 0 is not an eigenvalue of A, $P(\lambda)=0$ iff $P(1/\lambda)=0$. As P has real coefficients, $P(\lambda)=0$ iff $P(\bar\lambda)=0$. ■

As a matter of fact, $Sp(E,\omega)\subset GL(E,E)=GL(E)$ is a submanifold, and composition is C^∞, so $Sp(E,\omega)$ is a Lie group (see Sect. 4.1). The final exposition of this section is a description of the Lie algebra of $Sp(E,\omega)$, denoted by $sp(E,\omega)\subset L(E,E)$. The reader should come back to this point after reading Sect. 4.1. The space $L(E,E)$ is a Lie algebra with the Lie bracket defined by $[u,v]=u\circ v-v\circ u$. (Lie algebras were defined in 2.2.13.) This algebra is associated to the group $GL(E)$ as follows. First of all, since $GL(E)$ is open in $L(E,E)$, $T_fGL(E)=\{f\}\times L(E,E)$. We identify $T_IGL(E)$ and $L(E,E)$. Secondly, let

$$F(t)=e^{-tu}e^{-tv}e^{tu}e^{tv}$$

where $e^{tu}=I+tu+t^2u^2/2+\cdots$ is a convergent power series. Then writing $e^{tu}=I+tu+(t^2/2)u^2+o(t^2)$ and expanding out $F(t)$ we find (after several lines of calculation) that

$$F(t)=I+t^2[u,v]+o(t^2)$$

ISBN 0-8053-0102-X

so that $F(0)=I$, $F'(0)=0$ and $\frac{1}{2}F''(0)=[u,v]$. (See Exercise 4.1J for the relationship with general Lie groups).

3.1.13 Definition. *A linear mapping $u \in L(E,E)$ is* ***infinitesimally symplectic*** *with respect to a symplectic form ω if $\omega(ue,e')+\omega(e,ue')=0$ for all $e,e' \in E$, that is, if u is ω-skew. Let $sp(E,\omega)$ denote the set of all linear mappings in $L(E,E)$ that are infinitesimally symplectic with respect to ω.*

3.1.14 Proposition. *The set $sp(E,\omega) \subset L(E,E)$ is a Lie subalgebra.*

The proof is a simple verification. The reader may also check (or wait until Sect. 4.1) that $u \in sp(E,\omega)$ iff $e^u \in Sp(E,\omega)$, which relates the Lie algebra to the corresponding Lie group.

In Sect. 3.3, we will refer to infinitesimally symplectic linear mappings as *Linear Hamiltonian* mappings. If we choose a basis in which the matrix of ω is

$$J=\begin{bmatrix} 0 & I \\ -I & 0 \end{bmatrix}$$

and if we write

$$u=\begin{bmatrix} A & B \\ C & D \end{bmatrix},$$

then u is infinitesimally symplectic iff $u^tJ+Ju=0$ iff $D=-A^t$ and C,B are symmetric.

If we proceed exactly as in 3.1.12, noting that $ue=\lambda e$ implies $\omega(e,\lambda e'+ue')=0$, we obtain the following.

3.1.15 Proposition. *If $u \in sp(E,\omega)$ and λ is an eigenvalue of u, so are $-\lambda$, $\bar{\lambda}$, and $-\bar{\lambda}$.*

The eigenvalue properties of 3.1.12 and 3.1.15 can be strengthened as follows.

3.1.16 Symplectic eigenvalue theorem. *Suppose (E,ω) is a symplectic vector space, $f \in Sp(E,\omega)$, and λ is an eigenvalue of f of multiplicity k. Then $1/\lambda$ is an eigenvalue of f of multiplicity k. Moreover, the multiplicities of the eigenvalues $+1$ and -1, if they occur, are even.*

Proof. We saw that if P is the characteristic polynomial of f, then $P(\lambda)=\lambda^{2n}P(1/\lambda)$, where $\dim E=2n$. Suppose λ_0 occurs with multiplicity k. Then $P(\lambda)=(\lambda-\lambda_0)^kQ(\lambda)$, so that

$$P\left(\frac{1}{\lambda}\right)\lambda^{2n}=(\lambda-\lambda_0)^kQ(\lambda)=(\lambda\lambda_0)^k\left(\frac{1}{\lambda_0}-\frac{1}{\lambda}\right)^kQ(\lambda)$$

ISBN 0-8053-0102-X

Now $(\lambda_0^k/\lambda^{2n-k})Q(\lambda)$ is a polynomial in $1/\lambda$, as Q is of degree $2n-k$ and $k<2n$. Hence $1/\lambda_0$ occurs with multiplicity $l \geqslant k$. Reversing the roles of $\lambda_0, 1/\lambda_0$, we see $k \geqslant l$, so $k=l$.

Note that $\lambda_0 = 1/\lambda_0$ iff λ_0 is $+1$ or -1. Thus, from the above, the multiplicity of the eigenvalues $+1$ and -1 is even. But, as $\det f = 1$ (the product of the eigenvalues) the number of each must be even. ■

In a similar way we can prove the following. (Note that if u is infinitesimally symplectic with characteristic polynomial P, then $P(\lambda)=P(-\lambda)$, so $tr(u)=0=$ sum of the eigenvalues of u).

3.1.17 Infinitesimally symplectic eigenvalue theorem. *Let (E, ω) be a symplectic vector space and $u \in sp(E, \omega)$. Then, if λ is an eigenvalue of multiplicity k, $-\lambda$ is an eigenvalue of multiplicity k. Moreover, 0, if it occurs, has even multiplicity.*

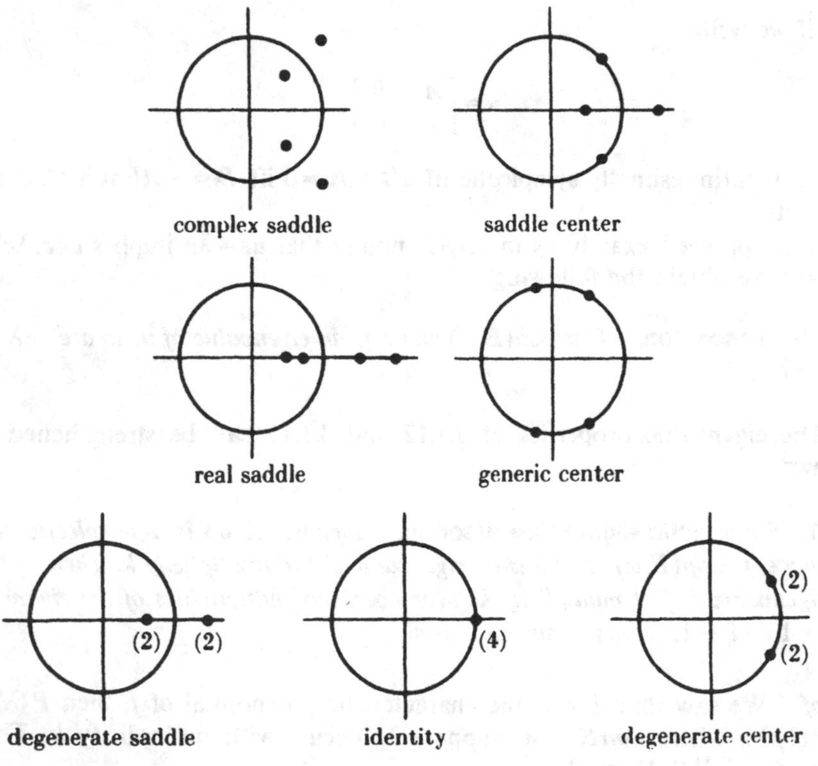

Figure 3.1-1

ISBN 0-8053-0102-X

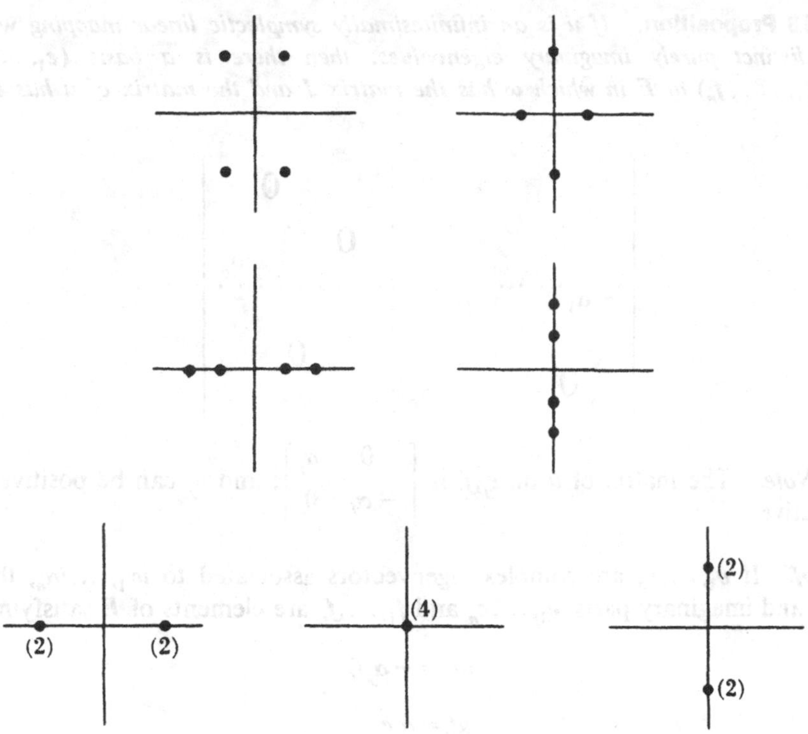

Figure 3.1-2

The possible eigenvalue configurations for a symplectic linear mapping $A \in Sp(\mathbf{R}^4, \omega_0)$, graphed with relation to the unit circle in the complex plane, are illustrated in Fig. 3.1-1.

The corresponding configurations for an infinitesimally symplectic mapping $u \in sp(\mathbf{R}^4, \omega_0)$ are illustrated in Figure 3.1-2.

These eigenvalue properties are basic to the qualitative theory and stability of Hamiltonian systems. Although $Sp(\mathbf{R}^{2n}, \omega_0)$ is the fundamental group underlying classical mechanics, very little application of its structure seems to have been made beyond these elementary eigenvalue properties. For additional properties of the symplectic group, see Sect. 4.1.

For information on how eigenvalues can move off the unit circle or imaginary axis see Krein [1950] and Arnold and Avez [1967], Appendix 9.

The question of canonical forms for infinitesimally symplectic mappings (and by exponentiation, symplectic mappings) is of some importance in mechanics. We give a simple version here and discuss the question further in Sect. 5.6.

ISBN 0-8053-0102-X

3.1.18 Proposition. *If u is an infinitesimally symplectic linear mapping with $2n$ distinct purely imaginary eigenvalues, then there is a basis $(e_1, \ldots, e_n, f_1, \ldots, f_n)$ of E in which ω has the matrix J and the matrix of u has the form*

$$
\left[\begin{array}{ccc:ccc}
 & & & \alpha_1 & & 0 \\
 & 0 & & & \ddots & \\
 & & & 0 & & \alpha_n \\
\hdashline
-\alpha_1 & & 0 & & & \\
 & \ddots & & & 0 & \\
0 & & -\alpha_n & & &
\end{array}\right]
$$

Note. The matrix of u on e_j,f_j is $\begin{bmatrix} 0 & \alpha_j \\ -\alpha_j & 0 \end{bmatrix}$; and α_j can be positive or negative.

Proof. If $v_1,\ldots,v_n$ are complex eigenvectors associated to $i\alpha_1,\ldots,i\alpha_n$, their real and imaginary parts, $e_1,\ldots,e_n$ and $f_1,\ldots,f_n$ are elements of E satisfying

$$ue_j = -\alpha_j f_j$$

$$uf_j = \alpha_j e_j$$

The complex eigenvectors associated with $-i\alpha_1,\ldots,-i\alpha_n$ are $\bar{v}_j = e_j - if_j$, as is easily checked.

Since the eigenvalues of u are distinct, $v_1,\ldots,v_n,\bar{v}_1,\ldots,\bar{v}_n$ are a basis of the complexification and hence $e_1,\ldots,e_n,f_1,\ldots,f_n$ are a basis of E itself.

Since u is ω-skew, the ω-orthogonal complement of $span(e_1,f_1)$ is u-invariant; therefore, it contains $e_2,\ldots,e_n,f_2,\ldots,f_n$. Thus $E_i = span(e_i,f_i)$ are ω-orthogonal and span E. Since ω is nondegenerate, it is nondegenerate on each E_i. Rescaling and relabeling the e_i,f_i if necessary, ω has the matrix $\begin{bmatrix} 0 & 1 \\ -1 & 0 \end{bmatrix}$ on E_i, so this is the required basis. ■

In the presence of multiple eigenvalues, 1's have to be inserted in off diagonal spots as in the Jordan canonical form (see Sect. 5.6).

For those interested in a basis free formulation of 3.1.2 and the infinite-dimensional analog, we include the following discussion. First some notation. Let E be a real vector space. By a *complex structure* on E we mean a linear map $J: E \to E$ such that $J^2 = -I$. By setting $ie = J(e)$, one gives E the structure of a complex vector space. We now show rather generally that a symplectic form is the imaginary part of a complex inner product. (This structure will come up again in our discussions of quantum mechanics.) The reader not familiar with Hilbert space theory can replace H by R^{2n} and derive the result from 3.1.2.

ISBN 0-8053-0102-X

3.1.19 Theorem. *Let H be a real Hilbert space and B a skew symmetric weakly nondegenerate bilinear form on H. Then there exists a complex structure J on H and a real inner product s such that*

$$s(x,y) = -B(Jx,y)$$

Setting

$$h(x,y) = s(x,y) - iB(x,y)$$

h is a hermitian inner product. Finally, h or s is complete on H iff B is nondegenerate.

Proof. Let $\langle , \rangle$ be the given complete inner product on H. By the Riesz theorem, $B(x,y) = \langle Ax, y\rangle$ for a bounded linear operator $A: H \to H$. Since B is skew, we find $A^* = -A$.

Since B is weakly nondegenerate, A is injective. Now $-A^2 > 0$, and from $A = -A^*$ we see that A^2 is injective. Let P be a symmetric nonnegative square root of $-A^2$. Hence P is injective. Since $P = P^*$, P has dense range. Thus P^{-1} is a well-defined (unbounded) operator. Set $J = AP^{-1}$, so that $A = JP$. From $A = -A^*$ and $P^2 = -A^2$, we find that $J^* = -J = J^{-1}$, J is orthogonal, and $J^2 = -I$. Thus J is bounded on the range of P, so extends to an orthogonal operator defined on all of H. Moreover, J is symplectic since $B(Jx, Jy) = B(x,y)$. Define $s(x,y) = -B(Jx,y) = \langle Px, y\rangle$. Thus s is an inner product on H. (Note that if $\langle Px, x\rangle = 0$, then, since $P = P^*$, $P > 0$, $\langle \sqrt{P}\,x, \sqrt{P}\,x\rangle = 0$, so $\sqrt{P}\,x = 0$, so $Px = 0$, so $x = 0$.) Finally, it is a straightforward check to see that h is a hermitian inner product. For example; $h(ix,y) = s(Jx,y) - iB(Jx,y) = B(x,y) + is(x,y) = ih(x,y)$. The theorem now follows. ■

In particular, this shows that any symplectic form is the negative imaginary part of some hermitian inner product.

If we identify C^n with R^{2n} and write

$$\begin{aligned} z &= (z_1, \ldots, z_n) \\ &= (x_1 + iy_1, \ldots, x_n + iy_n) \\ &= ((x_1, y_1), \ldots, (x_n, y_n)) \end{aligned}$$

then

$$\begin{aligned} -Im\langle z_1, \ldots, z_n, z_1', \ldots, z_n'\rangle &= -Im(z_1\bar{z}_1' + \cdots + z_n\bar{z}_n') \\ &= -(x_1'y_1 - x_1y_1' + \cdots + x_n'y_n - x_ny_n') \end{aligned}$$

ISBN 0-8053-0102-X

Thus, using the formula following 3.1.2, we have

$$\omega(z, z') = -Im\langle z, z'\rangle$$

The infinite dimensional analogue of 3.1.18 is a result due to Cook [1966] which is discussed in Sect. 5.6.

EXERCISES

3.1A. (i) In 3.1.1 show that ω is nondegenerate iff ω^b is an isomorphism. Deduce that ω is nondegenerate iff ω' is nondegenerate.
(ii) In 3.1.2(i) show that the number of -1's and 1's is independent of the diagonalization procedure by supplying intrinsic definitions.

3.1B. (i) Show algebraically that $A'JA = J$ implies $det\,A = 1$.
(ii) Show that the eigenvalue configurations shown for $A \in Sp(R^4, \omega_0)$ are the only ones possible (see Fig. 3.1-1).

3.1C. Let (E,ω) be a symplectic vector space. For $e \in E$ and $\lambda \in R$, let $\tau_{e,\lambda}: E \to E$

$$\tau_{e,\lambda}: x \mapsto x + \lambda\omega(x, e)e$$

(a) Prove $\tau_{e,\lambda} \in Sp(E,\omega)$. One calls $\tau_{e,\lambda}$ a *symplectic transvection*. (b) Show that $Sp(E,\omega)$ is generated by the symplectic transvections. (*Hint.* See Jacobson [1974]).

3.1D. Use 3.1C to show that the center of $Sp(E,\omega)$ is I and $-I$.

3.1E. Show that: $Sp(E,\omega)/\{I, -I\}$ is a simple group. (*Hint*: See Jacobson [1974].)

3.1F. Let E be a reflexive Banach space, i.e. the natural injection i of E into E^{**} is onto, and ω a weak symplectic form on E. Show that ω^b has closed range in E^* iff it is onto. [*Hint.* If $F \neq E^*$ is the closed range of ω^b, use the Hahn–Banach theorem to find $\phi \in E^{**}$, $\phi \neq 0$ such that $\phi(F) = 0$. If $\phi = i(v)$, show that $\omega(v, u) = 0$ for all $u \in E$.]

3.2 SYMPLECTIC GEOMETRY

The globalization of the simplectic algebra of Sect. 3.1 is symplectic geometry. Our first goal will be Darboux's theorem [1882], which states that for a nondegenerate, closed two-form ω on a manifold M (ω is called a *symplectic form*), the canonical form of 3.1.2(ii) can be extended to some chart about each $m \in M$. For the degenerate case, we refer the reader to 5.1.3.

3.2.1 Definition. *Let M be a manifold and $\omega \in \Omega^2(M)$ be nondegenerate. Then we define the map* $\flat: \mathfrak{X}(M) \to \mathfrak{X}^*(M): X \mapsto X^\flat = i_X\omega$ [*hence* $X^\flat = \omega^\flat(X)$] *and the map* $\sharp: \mathfrak{X}^*(M) \to \mathfrak{X}(M): \alpha \mapsto \alpha^\sharp = \omega^\sharp(\alpha)$.

Thus we see that $(X^\flat)^\sharp = X$ and $(\alpha^\sharp)^\flat = \alpha$.

The proof of Darboux's theorem we use is due to J. Moser [1965] and A. Weinstein [1977b]. This proof is considerably simpler than previous proofs

ISBN 0-8053-0102-X

and has several other applications (see 2.2.26, 3.2.3 below and Exercise 3.2B) and generalizations (see Sect. 5.3).

3.2.2 Theorem (Darboux). *Suppose ω is a nondegenerate two-form on a $2n$-manifold M. Then $d\omega=0$ iff there is a chart (U,φ) at each $m\in M$ such that $\varphi(m)=0$, and with $\varphi(u)=(x^1(u),\dots,x^n(u),y^1(u),\dots,y^n(u))$ we have*

$$\omega|U=\sum_{i=1}^{n}dx^i\wedge dy^i$$

Proof. It is obvious that $\sum_{i=1}^{n}dx^i\wedge dy^i$ is closed, so the "if" part is clear.

For the converse, it is sufficient by 3.1.2 to find a chart in which ω is constant.

For this purpose, we can assume that $M=\mathbf{E}$, a linear space, and $m=\mathbf{0}$.

Let ω_1 be the constant form equalling $\omega(\mathbf{0})$. Let $\tilde{\omega}=\omega_1-\omega$ and $\omega_t=\omega+t\tilde{\omega}$, $0\leqslant t\leqslant 1$. For each t, $\omega_t(\mathbf{0})=\omega(\mathbf{0})$ is nondegenerate. Hence by openness of the set of linear isomorphisms of $\mathbf{E}$ to $\mathbf{E}^*$, there is a neighborhood of $\mathbf{0}$ on which ω_t is nondegenerate for all $0\leqslant t\leqslant 1$. We can assume that this neighborhood is a ball. Thus, by the Poincaré lemma, $\tilde{\omega}=d\alpha$ for a one form α. We can suppose $\alpha(\mathbf{0})=0$.

Define a smooth vector field X_t by $i_{X_t}\omega_t=-\alpha$, which is possible since ω_t is nondegenerate. Moreover, since $X_t(\mathbf{0})=\mathbf{0}$, by the local existence theory, there is a ball about zero on which the "flow" of the time-dependent vector field X_t is defined for a time at least one; see Exercise 3.2C. Call this "flow" F_t (with initial condition $F_0=\text{Identity}$). Then by the basic link between flows and Lie derivatives,

$$\begin{aligned}\frac{d}{dt}(F_t^*\omega_t)&=F_t^*(L_{X_t}\omega_t)+F_t^*\frac{d}{dt}\omega_t\\&=F_t^*\,di_{X_t}\omega_t+F_t^*\tilde{\omega}\\&=F_t^*(-d\alpha+\tilde{\omega})=0\end{aligned}$$

Therefore, $F_1^*\omega_1=F_0^*\omega_0=\omega$, so F_1 provides the coordinate change transforming ω to the constant form ω_1. ■

Notice that this proof works in infinite dimensions; see Sect. 5.1 for additional comments.

This same argument can also be used to prove the important *Morse lemma* (Palais [1969]).

3.2.3 Morse Lemma. *Let $f: M\to\mathbf{R}$ be a smooth map with $m_0\in M$ a nondegenerate critical point; that is, $df(m_0)=0$ and $D^2f(m_0)$ is nondegenerate. Then there is a coordinate chart about m_0 in which m_0 is mapped to zero and the local*

ISBN 0-8053O102-X

representative of f satisfies

$$f(x)=f(0)+\tfrac{1}{2}D^2f(0)\cdot(x,x)$$

In particular, nondegenerate critical points of f are isolated.

Proof. We can assume that we are in $\boldsymbol{R}^n$ and $m_0=\mathbf{0}$, $f(m_0)=0$. Let $\omega_1=df$ and define the one form ω_2 by

$$\omega_2(x)\cdot h=D^2f(\mathbf{0})(h,x)$$

Let

$$\omega_t=t\omega_1+(1-t)\omega_2$$

Write

$$\omega_2=d\varphi, \qquad \varphi(x)=\tfrac{1}{2}D^2f(\mathbf{0})(x,x)$$

and define a vector field Z_t by

$$i_{Z_t}\omega_t+(f-\varphi)=0, \qquad Z_t(\mathbf{0})=\mathbf{0}$$

It is easy to see that Z_t exists near $\mathbf{0}$ by the nondegeneracy hypothesis. Let F_t be the flow of Z_t. Then

$$\frac{d}{dt}(F_t^*\omega_t)=F_t^*\left(di_{Z_t}\omega_t+i_{Z_t}d\omega_t+\omega_1-\omega_2\right)$$
$$=F_t^*d\left(i_{Z_t}\omega_t+f-\varphi\right)=\mathbf{0}$$

Thus $F_1^*\omega_1=\omega_2$, so F_1 gives (near $\mathbf{0}$), the coordinate change required. ■

In later chapters we will have occasion to use a small amount of Morse theory. We therefore supplement 3.2.3 with some additional results at the end of this section. We return now to our main topic of symplectic forms.

3.2.4 Definition. *A* ***symplectic form*** *(or a* ***symplectic structure****) on a manifold M is a nondegenerate, closed* two-form ω on M. A* ***symplectic manifold*** *(M, ω) is a manifold M together with a symplectic form ω on M. As in Sect. 3.1, we let* Ω_ω *denote the volume* $[(-1)^{[n/2]}/n!]\omega^n$*. The charts guaranteed by Darboux's theorem are called* ***symplectic charts*** *and the component functions* x^i, y^i *are called* ***canonical coordinates.***

Thus, in a symplectic chart,

$$\omega=\sum_{i=1}^{n}dx^i\wedge dy^i \quad\text{and}\quad \Omega_\omega=dx^1\wedge\cdots\wedge dx^n\wedge dy^1\wedge\cdots\wedge dy^n$$

*One can legitimately ask for the origin of the condition $d\omega=0$ and why it plays such a central role. One reason, looking ahead to Sect. 3.3, is that this condition is exactly the one needed to make Poisson brackets into a Lie algebra.

ISBN 0-8053-0102-X

From 3.2.2, we see that symplectic forms are much more flexible than Riemannian metrics. Indeed, the latter can be made constant in a local chart if and only if they are flat.

The global analog of a symplectic linear map is given as follows.

3.2.5 Definition. *Let (M,ω) and (N,ρ) be symplectic manifolds. A C^∞-mapping $F: M\to N$ is called* ***symplectic*** *or a* ***canonical transformation*** *if $F^*\rho=\omega$.*

From 2.4.9(i), 2.5.17, and 2.5.21 we obtain the following.

3.2.6 Proposition. *If (M,ω) and (N,ρ) are symplectic $2n$-manifolds and $F: M\to N$ is symplectic, then F is volume preserving, $\det_{(\Omega_\omega,\Omega_\rho)}F=1$, and F is a local diffeomorphism.*

It is clear that if (M,ω) is a symplectic manifold and $\varphi: M\to N$ is a diffeomorphism, then $(N,\varphi_*\omega)$ is a symplectic manifold and φ is a symplectic map.

3.2.7 Proposition. *Suppose (M,ω) and (N,ρ) are symplectic manifolds and $F: M\to N$ is of class C^∞. Suppose $\varphi: M\to M'$ and $\psi: N\to N'$ are diffeomorphisms. Then F is symplectic iff $\psi\circ F\circ\varphi^{-1}$ is a symplectic mapping of $(M',\varphi_*\omega)$ into $(N',\psi_*\rho)$. In particular, F is symplectic iff the local representatives of F are symplectic.*

Proof. If F is symplectic, then $(\psi\circ F\circ\varphi^{-1})^*\psi_*\rho=\varphi_*\circ F^*\circ\psi^*\circ\psi_*\rho=\varphi_*\circ F^*\rho=\varphi_*\omega$. Conversely, if $\psi\circ F\circ\varphi^{-1}$ is symplectic, then $F^*\rho=\varphi^*\circ\varphi_*\circ F^*\circ\psi^*\circ\psi_*\rho=\varphi^*\circ(\psi\circ F\circ\varphi^{-1})^*\circ\psi_*\rho=\varphi^*\circ\varphi_*\omega=\omega$. ■

3.2.8 Proposition. *Let (M,ω) and (N,ρ) be symplectic $2n$-manifolds and $f: M\to N$ a symplectic mapping. Then for each $m\in M$ there are symplectic charts (U,φ) at m and (V,ψ) at $f(m)$ such that $f(U)=V$, $\varphi(U)=\psi(V)$, and the local representative $f_{\varphi\psi}$ of f is the identity.*

Proof. Since f is a local diffeomorphism we can find neighborhoods U_1 of m and V_1 of $f(m)$ such that $f|U_1$: $U_1\to V_1$ is a diffeomorphism. Let (V,ψ) be a symplectic chart at $f(m)$ with $V\subset V_1$ (Darboux's theorem). Then let $U=(f|U_1)^{-1}(V)$ and $\varphi=\psi\circ f|U$. Clearly $f_{\varphi\psi}$ is the identity. Also, (U,φ) is a symplectic chart, for $\varphi_*\omega=\omega_\varphi=(\psi\circ f)_*\omega=\psi_*\circ f_*\omega=\psi_*\rho=\rho_\psi$ on $\varphi(U)=\psi(V)$. [Note that if $t\in\mathcal{T}^r_s(M)$, the local representative of t in the natural charts is φ_*t.] ■

The connection with Sect. 3.1 is given by the following.

3.2.9 Proposition. *Let (E,ω) and (F,ρ) be the symplectic vector spaces, which also may be regarded as symplectic manifolds (ω,ρ being constant sections).*

ISBN 0-8053-0102-X

Then a C^∞ map $F: U \subset E \to F$ is symplectic iff $DF(u) \in L(E, F)$ is symplectic for each $u \in U$.

Proof. This follows at once from the definition $F^*\rho = (TF)^* \circ \rho \circ F$ applied to the second factors. ■

In many mechanical problems, the basic symplectic manifold is the phase space of a configuration space. In fact, if the configuration space is a manifold Q, the momentum phase space is its cotangent bundle T^*Q, which has a standard symplectic form as follows.

3.2.10 Theorem. *Let Q be an n-manifold and $M = T^*Q$. Consider τ_Q^*: $M \to Q$ and $T\tau_Q^*$: $TM \to TQ$. Let $\alpha_q \in M$ $(q \in Q)$ denote a point of M and w_{α_q} a point of TM in the fiber over α_q. Define*

$$\theta_{\alpha_q}: T_{\alpha_q}M \to \mathbf{R}: w_{\alpha_q} \mapsto \alpha_q \cdot T\tau_Q^*(w_{\alpha_q}) \quad \text{and} \quad \theta_0: \alpha_q \mapsto \theta_{\alpha_q}$$

*Then $\theta_0 \in \mathfrak{X}^*(M)$, and $\omega_0 = -d\theta_0$ is a symplectic form on M; θ_0 and ω_0 are called **the canonical forms** on M.*

Proof. Let (U, φ) be a chart on Q with $\varphi(U) = U' \subset E$ and let

$$(T^*U, T^*\varphi),\ T^*\varphi: T^*U \to U' \times E^*,$$

$$(TT^*U, TT^*\varphi),\ TT^*\varphi: TT^*U \to U' \times E^* \times E \times E^*$$

$$(T^*T^*U, T^*T^*\varphi),\ T^*T^*\varphi: T^*T^*U \to U' \times E^* \times E^* \times E^{**}$$

be the corresponding charts on $T^*Q = M$, TM, and T^*M respectively. Denote $\varphi(q) = x$, $T_q^*\varphi(\alpha_q) = \alpha$, $T_{\alpha_q}T^*\varphi(w_{\alpha_q}) = (e, \beta)$.

Denoting by pr_1: $U' \times E^* \to U'$ the projection on the first factor, we get

$$\begin{aligned}
\left(T_{\alpha_q}^* T^*\varphi \circ \theta_0 \circ (T_q^*\varphi)^{-1}\right)(\alpha) \cdot (e, \beta) &= T_{\alpha_q}^* T^*\varphi(\theta_{\alpha_q}) \cdot (e, \beta) \\
&= \theta_{\alpha_q}\left((T_{\alpha_q}T^*\varphi)^{-1}(e, \beta)\right) \\
&= \theta_{\alpha_q}(w_{\alpha_q}) \\
&= \alpha_q\left(T_{\alpha_q}\tau_Q^*(w_{\alpha_q})\right) \\
&= \alpha_q\left(T_{\alpha_q}(\tau_Q^* \circ T^*\varphi^{-1})(e, \beta)\right) \\
&= \alpha_q\left(T_q(\varphi^{-1} \circ pr_1)(e, \beta)\right) \\
&= \alpha_q\left(T_q\varphi^{-1}(T_{\varphi(q)}pr_1(e, \beta))\right) \\
&= \alpha_q\left(T_q\varphi^{-1}(e)\right) \\
&= T_q^*\varphi(\alpha_q) \cdot e = \alpha(e)
\end{aligned}$$

ISBN 0-8053-0102-X

Therefore θ_0 is given locally by

$$(T^*T^*\varphi\circ\theta_0\circ T^*(\varphi^{-1}))(x,\alpha)\cdot(x,\alpha,e,\beta)=\alpha(e) \tag{1}$$

so that θ_0 is smooth and hence $\theta_0\in\mathfrak{X}^*(M)$. Since $\omega_0=-d\theta_0$, the above local expression for θ_0 shows that

$$\omega_0(x,\alpha)((x,\alpha,e_1,\beta_1),(x,\alpha,e_2,\beta_2))=\beta_2(e_1)-\beta_1(e_2) \tag{2}$$

Comparison with the formula for ω preceding 3.1.3 concludes the proof. ■

Note that formula (2) is independent of (x,α), reflecting the fact that the natural charts of T^*Q are symplectic charts.

In finite dimensions, denoting by $(q^1,\ldots,q^n)$ the coordinates on Q and by $(q^1,\ldots,q^n,p_1,\ldots,p_n)$ those on $T^*Q=M$, the above local formulas become

$$\boxed{\theta_0=\sum_{i=1}^{n}p_i\,dq^i}$$

and

$$\boxed{\omega_0=\sum_{i=1}^{n}dq^i\wedge dp_i}$$

As a mathematical curiosity, we note that the cotangent bundle of any manifold is orientable. Indeed, it carries a symplectic structure and hence a volume element.

The definition of the canonical one-form can be alternatively written as follows:

$$\langle\theta_0(\alpha_q),w_{\alpha_q}\rangle=\langle T\tau_Q^*w_{\alpha_q},\alpha_q\rangle$$

where $\langle\ ,\ \rangle$ denotes the natural pairing, or contraction, between vectors and one-forms.

The following proposition gives another description of θ_0, which will be of great utility later on.

3.2.11 Proposition. *The canonical one-form θ_0 on T^*Q is the unique one-form with the property that, for any one-form β on Q,*

$$\beta^*\theta_0=\beta$$

Here, regard $\beta\colon Q\to T^*Q$. Thus $\beta^*\omega_0=-d\beta$.

Proof. Let $v_q\in T_qQ$; then, by definition of pull-back,

$$\begin{aligned}\langle\beta^*\theta_0,v_q\rangle&=\langle\theta_0(\beta(q)),T(\beta)\cdot v_q\rangle\\&=\langle T\tau_Q^*\cdot T\beta\cdot v_q,\beta(q)\rangle\\&=\langle T(\tau_Q^*\circ\beta)\cdot v_q,\beta(q)\rangle\\&=\langle v_q,\beta(q)\rangle=\beta(q)\cdot v_q\end{aligned}$$

since $\tau_Q^*\circ\beta$ is the identity. Thus $\beta^*\theta_0=\beta$.

ISBN 0-8053-0102-X

This uniquely characterizes θ_0, since $\beta(q)$ and $T\beta(q)\cdot v_q$ span all of T_q^*Q and $T_{\beta(q)}(T^*Q)$ for variable β and v_q. ■

The coordinate proof of 3.2.11 may aid in seeing what is going on: $\theta_0 = \sum_{i=1}^n p_i\, dq^i$, and β maps q^i to $(q^i, p_i = \beta_i(q))$, so

$$\beta^*\theta_0 = \sum_{i=1}^n (\beta^* p_i)\, dq^i$$

$$= \sum_{i=1}^n \beta_i\, dq^i$$

$$= \beta$$

since $\beta^* p_i$ is the ith component of β.

A basic method for generating symplectic mappings on T^*Q from mappings on Q is given by the following.

3.2.12 Theorem. *Let Q be a manifold and $f\colon Q \to Q$ a diffeomorphism; define the lift of f by*

$$T^*f\colon T^*Q \to T^*Q;\ T^*f(\alpha_q)v = \alpha_q(Tf\cdot v),$$

where $q \in Q$ and $v \in T_{f^{-1}(q)}Q$. Then T^*f is symplectic and in fact $(T^*f)^*\theta_0 = \theta_0$, where θ_0 is the canonical one-form.

Proof. By definition, for $w \in T_{\alpha_q}(T^*Q)$

$$(T^*f)^*\theta_0(w) = \theta_0(TT^*f\cdot w)$$

$$= T^*f(\alpha_q)\cdot(T\tau_Q^* TT^*f\cdot w)$$

$$= T^*f(\alpha_q)\cdot\big(T(\tau_Q^* \circ T^*f)\cdot w\big)$$

$$= \alpha_q\cdot\big(Tf\cdot T(\tau_Q^* \circ T^*f)\cdot w\big)$$

$$= \alpha_q\cdot\big(T(f \circ \tau_Q^* \circ T^*f)\cdot w\big)$$

$$= \alpha_q\cdot(T\tau_Q^*\cdot w)$$

$$= \theta_0(\alpha_q)\cdot w$$

since, by construction, $f \circ \tau_Q^* \circ T^*f = \tau_Q^*$. ■

There is a similar theorem for diffeomorphisms $f\colon Q_1 \to Q_2$ and their lifts $T^*f\colon T^*Q_2 \to T^*Q_1$.

ISBN 0-8053-0102-X

In coordinates, if we write $f(q^1, \ldots, q^n) = (Q^1, \ldots, Q^n)$, then T^*f has the effect

$$(Q^1, \ldots, Q^n, P_1, \ldots, P_n) \mapsto (q^1, \ldots, q^n, p_1, \ldots, p_n)$$

where

$$p_j = \sum_{i=1}^{n} \frac{\partial Q^i}{\partial q^j} P_i$$

(evaluated at the correct points). That this transformation is always canonical and in fact preserves the canonical one-form may be verified directly:

$$\Sigma P_i \, dQ^i = \Sigma P_i \frac{\partial Q^i}{\partial q^k} dq^k = \Sigma p_k \, dq^k$$

Sometimes one refers to canonical transformations of this type as "point transformations" since they arise from general diffeomorphisms of Q to Q. One also speaks of a canonical transformation which preserves θ_0 as a *homogeneous canonical transformation* or according to Whittaker [1959, p. 301], a *Mathieu transformation*. A theorem of Robbin-Weinstein outlined in Exercise 3.2F shows that a canonical transformation defined on all of T^*Q is homogeneous if and only if it is a point transformation.

The point transformations clearly form a subgroup of the set of all canonical transformations. Those that are not point transformations are abundant and important (see Exercise 3.2E).

Notice that lifts of diffeomorphisms satisfy

$$f^{-1} \circ \tau_Q^* = \tau_Q^* \circ T^*f$$

that is, the following diagram commutes:

$$\begin{array}{ccc} T^*Q & \xrightarrow{\;T^*f\;} & T^*Q \\ {\scriptstyle \tau_Q^*}\downarrow & & \downarrow{\scriptstyle \tau_Q^*} \\ Q & \xleftarrow[\;f\;]{} & Q \end{array}$$

Notice also that

$$T^*(f \circ g) = T^*g \circ T^*f$$

and compare with

$$T(f \circ g) = Tf \circ Tg.$$

Next we consider symplectic forms induced by metrics. If $g = \langle \cdot, \cdot \rangle$ is a Riemannian (or pseudo-Riemannian) metric on Q, then from 3.1.7 the map

ISBN 0-8053-0102-X

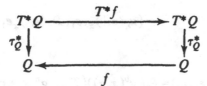

$g^\flat$: $TQ \to T^*Q$ defined by $g^\flat(u_q)\cdot w_q = \langle u_q, w_q\rangle_q$, $u_q, w_q \in T_qQ$ is a vector bundle isomorphism. Define

$$\Omega = (g^\flat)^*(\omega_0)$$

where $\omega_0 = -d\theta_0$ is the canonical two-form on T^*Q. Clearly $\Omega \equiv -d((g^\flat)^*\theta_0) = -d\Theta$, so Ω is exact.

3.2.13 Theorem. *Let $g = \langle\cdot,\cdot\rangle$ be a Riemannian (or pseudo-Riemannian) metric on Q. In a chart (U, φ) on Q we have*

(a) $\Theta(x, e)\cdot(x, e, e_1, e_2) = \langle e, e_1\rangle_x$, *that is,* $\Theta = \Sigma g_{ij}\dot q^i\, dq^j$, *where* $(q^1, \ldots, q^n, \dot q^1, \ldots, \dot q^n)$ *are coordinates for* TQ;

(b) $\Omega(x, e)((x, e, e_1, e_2), (x, e, e_3, e_4))$
$= D_x\langle e, e_1\rangle_x\cdot e_3 - D_x\langle e, e_3\rangle_x\cdot e_1 + \langle e_4, e_1\rangle_x - \langle e_2, e_3\rangle_x$
where D_x denotes the derivative with respect to x; that is,

$$\Omega = \Sigma g_{ij}\, dq^i \wedge d\dot q^j + \Sigma \frac{\partial g_{ij}}{\partial q^k}\dot q^i\, dq^j \wedge dq^k$$

Finally,

(c) Ω *is a symplectic form on* TQ.

Proof. (a) Locally $g^\flat$: $U\times E \to U'\times E^*$, $\varphi(U) = U' \subset E$ is given by $g^\flat(x,e) = (x, \langle e,\cdot\rangle_x)$, so that

$$Tg^\flat: U\times E\times E\times E \to U'\times E^*\times E\times E^*$$

is given by

$$\begin{aligned} Tg^\flat(x,e,e_1,e_2) &= \left(x, \langle e,\cdot\rangle_x, (Dg^\flat)_{(x,e)}(e_1,e_2)\right)\\ &= (x, \langle e,\cdot\rangle_x, e_1, D_x\langle e,\cdot\rangle_x e_1 + \langle e_2,\cdot\rangle_x) \end{aligned}$$

But then

$$\begin{aligned} (g^{\flat *}\theta_0)(x,e)\cdot(x,e,e_1,e_2) &= \theta_0(g^\flat(x,e))\left(T_{(x,e)}g^\flat(x,e,e_1,e_2)\right)\\ &= \theta_{(x,\langle e,\cdot\rangle_x)}(x, \langle e,\cdot\rangle_x, e_1, D_x\langle e,\cdot\rangle_x e_1 + \langle e_2,\cdot\rangle_x)\\ &= \langle e, e_1\rangle_x \end{aligned}$$

by the local formula of θ_0 given in Theorem 3.2.10.

(b) follows by taking the exterior derivative.

(c) Since everything is finite-dimensional, it suffices to prove weak nondegeneracy for $\Omega(x,e)$. Suppose that $\Omega(x,e)((x,e,e_1,e_2), (x,e,e_3,e_4)) = 0$

ISBN 0-8053-0102-X

for all (e_3, e_4). Setting $e_3=0$ and using the formula in (b), we find $\langle e_4, e_1\rangle_x=0$ for all e_4, whence $e_1=0$. Then we obtain $\langle e_2, e_3\rangle_x=0$ for all e_3 so that $e_2=0$, too. ∎

3.2.14 Corollary. *If Q_1 and Q_2 are Riemannian (or pseudo-Riemannian) manifolds and $f\colon Q_1\to Q_2$ is an isometry, then $Tf\colon TQ_1\to TQ_2$ is symplectic, and in fact preserves Θ.*

Proof. This follows from the formula

$$Tf=g_2^{\sharp}\circ(T^*f)^{-1}\circ g_1^{\flat}$$

All maps in the composition here are symplectic and hence so is Tf. ∎

We conclude this section with some remarks on Morse theory for later use. From 3.2.3, a function $f\colon M\to \mathbf{R}$ with a nondegenerate critical point at $x_0\in M$ can be written, in suitable local coordinates about x_0, as

$$f(x)=f(x_0)-\tfrac{1}{2}(x_1^2+\cdots+x_i^2)+\tfrac{1}{2}(x_{i+1}^2+\cdots+x_n^2)$$

The number i is called the *index* of f at x_0; it is the dimension of the largest subspace on which the Hessian $Hess\,f(x_0)=D^2f(x_0)$ is negative definite.

3.2.15 Definition (Bott [1954]). *Let $N\subset M$ be a submanifold and suppose each point in N is a critical point of f. We call N a nondegenerate* **critical submanifold** *if, in addition, for each $x_0\in N$*

$$\{v\in T_{x_0}M \mid Hess\,f(x_0)(v,w)=0 \text{ for all } w\in T_{x_0}M\}=T_{x_0}N$$

(Note that the inclusion $\supset$ is automatic.)

Equivalently, this means that on a subspace of $T_{x_0}M$, which is a complement to $T_{x_0}N$, $Hess\,f(x_0)$ is nondegenerate. On a tubular neighborhood about N, we can apply the Morse lemma to the variables transverse to N to get a canonical form for f parametrized by N; thus f has a well-defined *index* relative to N. (See Gromoll and Meyer [1969] for related results.)

Now we shall state two results from classical Morse theory that will be needed later.

3.2.16 Proposition. *Let M be a smooth, boundaryless manifold and $f\in \mathfrak{F}(M)$. Let $a, b\in \mathbf{R}$, $a<b$ and assume $f^{-1}([a,b])$ is compact and contains no critical points of f. Then the manifolds with boundary*

$$f^{-1}((-\infty,a]) \quad \text{and} \quad f^{-1}((-\infty,b])$$

are diffeomorphic and hence so are their boundaries, $f^{-1}(a)$ and $f^{-1}(b)$. Furthermore, $f^{-1}[a,b]$, $f^{-1}(a)\times[0,1]$ and $f^{-1}(b)\times[0,1]$ are all diffeomorphic.

ISBN 0-8053-0102-X

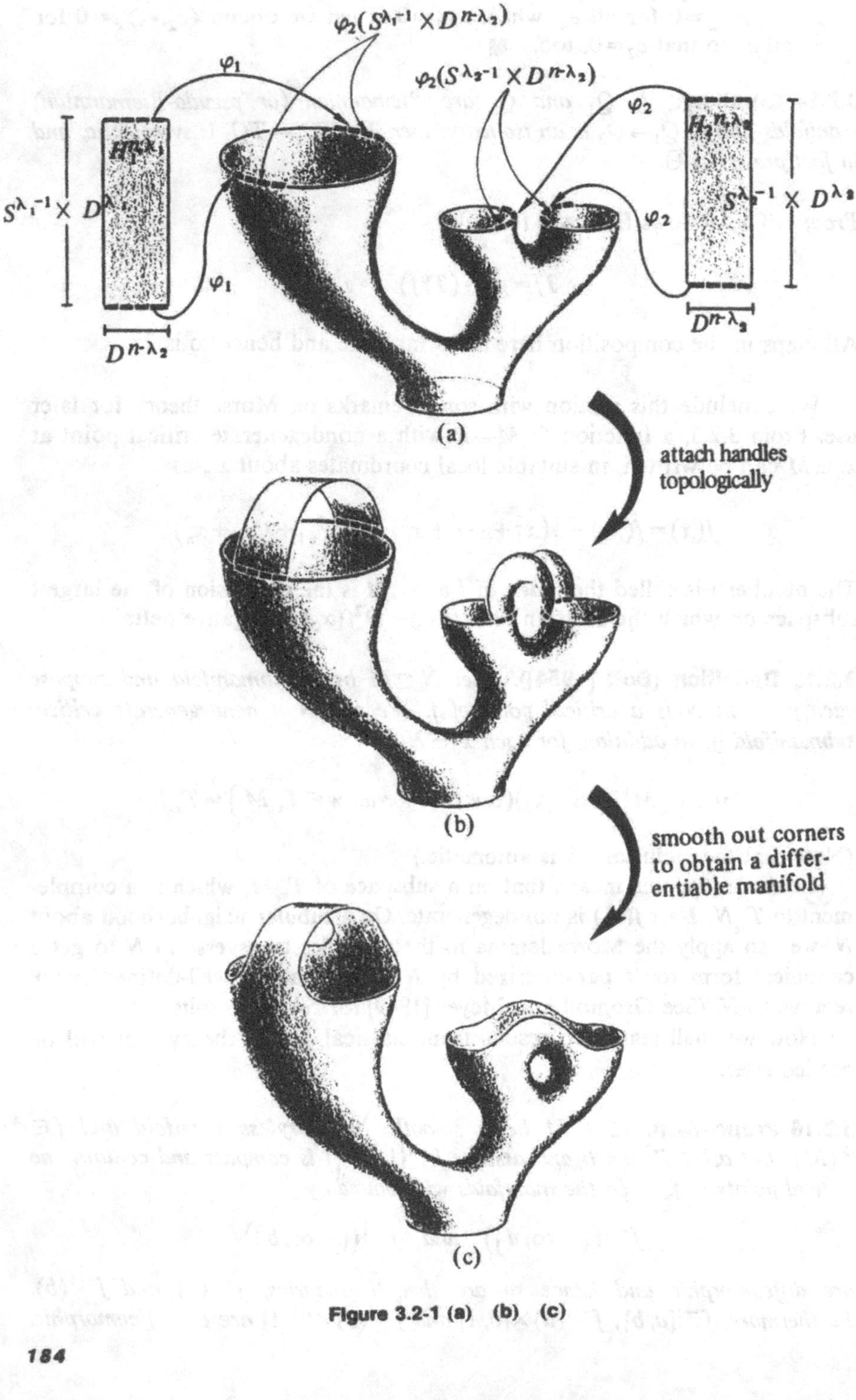

Figure 3.2-1 (a) (b) (c)

ISBN 0-8053-0102-X

The idea here is to construct a diffeomorphism by following the gradient curves of f (i.e., integral curves of $-\nabla f$); along such curves f decreases and since $f^{-1}([a,b])$ is compact with no critical points, such a curve starting in $f^{-1}((-\infty,b])$ must eventually reach $f^{-1}((-\infty,a])$. We refer the reader to Milnor [1963] and Hirsch [1976] for details.

Combining this with the Morse lemma one can deduce that if M is compact with boundary and $f\in\mathfrak{F}(M)$ has a nondegenerate minimum on M, has no other critical points and f is constant on ∂M, then M is diffeomorphic to a disk.

One can make more substantial topological deductions. To do so, we shall describe, without giving any proofs, the procedure of attachment of handles on a given manifold. The *standard n-dimensional handle of index* λ is $H^{n,\lambda}=D^\lambda\times D^{n-\lambda}$, where D^k denotes the unit ball in $\mathbf{R}^k$. If M is an n-dimensional manifold with boundary ∂M and $\phi: S^{\lambda-1}\times D^{n-\lambda}\to\partial M$ a smooth embedding, we can form the topological space $M\cup_\phi H^{n,\lambda}$ in the following way: Take the disjoint union $M\cup H^{n,\lambda}$ and identify $x\in H^{n,\lambda}$ with $\phi(x)\in\partial M$; the quotient space thus obtained is denoted by $M\cup_\phi H^{n,\lambda}$. It can be shown (see Milnor [1963]) that $M\cup_\phi H^{n,\lambda}$ admits a unique (up to diffeomorphism) smooth n-dimensional differentiable structure and in this way $M\cup_\phi H^{n,\lambda}$ becomes an n-dimensional manifold with boundary. (This differentiable structure depends only on the isotopy class of ϕ.) $M\cup_\phi H^{n,\lambda}$ is said to be obtained by the *attachment of the handle* $H^{n,\lambda}$ *to M via the embedding* ϕ. Similarly we define the manifold obtained by the simultaneous attachment of k n-dimensional handles $H_1^{n,\lambda_1},\ldots,H_k^{n,\lambda_k}$ of distinct indices (see Fig. 3.2-1).

The fundamental connection between critical points and the attachment of handles is given by the following theorem.

3.2.17 Theorem. *Let $f\in\mathfrak{F}(M)$, $\partial M=\varnothing$, and $a\in\mathbf{R}$. Assume that $\sigma(f)\cap f^{-1}(a)=\{x_1,\ldots,x_k\}$, where x_i is a critical point of index λ_i, $i=1,\ldots,k$, and $\sigma(f)$ is the set of critical points of f. Also assume that for $\varepsilon_0>0$, $f^{-1}([a-\varepsilon_0,a+\varepsilon_0])$ is a compact set* **not** *containing any other critical point of f except $x_1,\ldots,x_k$. Then for all ε satisfying $0<\varepsilon<\varepsilon_0$ the manifold $f^{-1}((-\infty,a+\varepsilon])$ is diffeomorphic to*

$$f^{-1}((-\infty,a-\varepsilon])\cup_{\phi_1}H_1^{n,\lambda_1}\cup_{\phi_2}H_2^{n,\lambda_2}\cup_{\phi_3}\cdots\cup_{\phi_k}H_k^{n,\lambda_k}$$

for some imbeddings $\phi_i: S^{\lambda-1}\times D^{n-\lambda}\to f^{-1}(a-\varepsilon)$.

EXERCISES

3.2A. Let (M,ω) be a symplectic manifold and $f: M\to M$ a local diffeomorphism. Prove that f is symplectic iff for every compact oriented two manifold B with boundary,

$$\int_{\partial B}\theta_0=\int_{f(\partial B)}\theta_0$$

ISBN 0-8053-0102-X

3.2B. (J. Moser) Use the method in 3.2.2 to prove that if M is a compact manifold and μ, ν are two volume elements with the same orientation and

$$\int \mu = \int \nu$$

then there is a diffeomorphism $f: M \to M$ with $f^* \nu = \mu$. [*Hint:* Since $\int \mu = \int \nu$, $\mu - \nu = d\alpha$ (this is a special case of de Rham's theorem). Put $\nu_t = t\nu + (1-t)\mu$ and $i_{X_t}\nu_t = \alpha$. Let φ_t be the flow of X_t and set $f = \varphi_1$.]

3.2C. Let X_t be a C^1 time-dependent vector field on E with $X_t(0) = 0$. Prove there is a ball about $\mathbf{0}$ on which the flow $F_{t,0}(x)$ of X_t is defined for $|t| \leq 1$.

3.2D. If θ_0 is the canonical one-form on T^*Q and $f: Q \to \mathbf{R}$ with $df: Q \to T^*Q$, show that for any vector field X on T^*Q, $\theta_0(X) \circ df = X(f \circ \tau_Q^*)$.

3.2E. Let β be a one-form on Q and let $f: T^*Q \to T^*Q$ be the map that is fiberwise translation by β. Prove that

$$f^*(\theta_0) = \theta_0 + (\tau_Q^*)^* \beta$$

so

$$f^*(\omega) = \omega \quad \text{if} \quad d\beta = 0$$

Show that such f's are examples of canonical transformations that are not lifts.

3.2F. (S. Lie, A. Weinstein and J. Robbin). Prove that a diffeomorphism φ of T^*Q is the lift of a diffeomorphism of Q iff φ preserves θ_0. [*Hint:* Suppose $\varphi^*\theta_0 = \theta_0$. We claim $\varphi = T^*f$ for a diffeomorphism $f: Q \to Q$. Let $X_0 = -\Sigma p_i \partial / \partial p^i$. Invariantly, $X_0 = \theta_0^\sharp$. Show that since φ preserves θ_0, $\varphi^* X_0 = X_0$, so φ preserves the integral curves of X_0. But X_0 is zero precisely on the zero section, so φ leaves the zero section invariant. This defines f. Show that $f \circ \tau_Q^* = \tau_Q^* \circ \varphi^{-1}$ using $\varphi^* X_0 = X_0$. Use this and $\varphi_* \theta_0 = \theta_0$ to conclude that $\varphi = T^*f$.]

3.2G. (Kähler Manifolds). Let M be a manifold and g a (pseudo-) Riemannian metric; let J be a complex structure on M; that is, J is an involution of TM with $J^2 = -I$, and J is g-orthogonal. M is called a *Kähler manifold* if $\nabla J = 0$, where ∇ is the connection of g (see Sect. 2.7) and J is regarded as a 1-1 tensor. Define, for vector fields X, Y on M,

$$\Omega(X, Y) = g(JX, Y)$$

(See 3.4.18.) Show that Ω is a symplectic structure on M (see Nelson [1967]).

3.2H. Show that Darboux' Theorem fails for weak symplectic forms as follows. Let H be a real Hilbert space. Let $S: H \to H$ be a compact operator with range a dense, but proper subset of H, which is selfadjoint and positive: $\langle Sx, x \rangle > 0$ for $0 \neq x \in H$. For example if $H = L^2(\mathbf{R})$, let $S = (1 - \Delta)^{-1}$ where Δ is the Laplacian; the range of S is $H^2(\mathbf{R})$.

Since S is positive, -1 is clearly not an eigenvalue. Thus, by the Fredholm alternative, $aI + S$ is onto for any real scalar $a > 0$. Define on H the weak metric $g(x)(e, f) = \langle A_x e, f \rangle$ where $A_x = S + \|x\|^2 I$. Clearly g is smooth in x, and is an inner product. Let Ω be the weak symplectic form on $H \times H = H_1$ induced by g, as in 3.2.13. Prove that *there is no coordinate chart about* $(0, 0) \in H_1$ *on which* Ω *is constant,* by showing that if there were such a chart,

ISBN 0-8053-0102-X

say $\phi: U \to \mathbf{H} \times \mathbf{H}$ where U is a neighborhood of (0,0), then in particular in this chart, the range F of Ω^b, as a map of $\mathbf{H}_1$ to $\mathbf{H}_1^*$, would be constant. (See Marsden [1972] and Tromba [1976].)

3.3. HAMILTONIAN VECTOR FIELDS AND POISSON BRACKETS

The Hamiltonian vector field of a function H on a symplectic manifold is formed in a manner analogous to the gradient of a function on a Riemannian manifold. However, the skew symmetry of the symplectic form leads to conservative properties for the Hamiltonian vector field whereas the symmetry of a Riemannian metric leads to dissipative properties for the gradient.

3.3.1 Definition. *Let (M, ω) be a symplectic manifold and $H: M \to \mathbf{R}$ a given C^r function. The vector field X_H determined by the condition*

$$\omega(X_H, Y) = dH \cdot Y \tag{1}$$

that is,

$$i_{X_H}\omega = dH \tag{2}$$

*is called the **Hamiltonian vector field** with energy function H. We call (M, ω, X_H) a **Hamiltonian system**. [Note that with our notations from 3.1.7, $X_H = \omega^\sharp(dH)$.]*

Nondegeneracy of ω guarantees that X_H exists. It is a C^{r-1} vector field. Clearly on a (connected) symplectic manifold any two Hamiltonians for the same X_H have the same differential by (2), so differ by a constant.

3.3.2 Proposition. *Let $(q^1,\ldots,q^n,p_1,\ldots,p_n)$ be canonical coordinates for ω, so $\omega = \Sigma dq^i \wedge dp_i$. Then in these coordinates (and dropping base points),*

$$X_H = \left(\frac{\partial H}{\partial p_i}, -\frac{\partial H}{\partial q^i}\right) = J \cdot dH \tag{3}$$

where $J = \begin{pmatrix} 0 & I \\ -I & 0 \end{pmatrix}$. Thus $(q(t), p(t))$ is an integral curve of X_H iff Hamilton's equations hold:

$$\boxed{\dot{q}^i = \frac{\partial H}{\partial p_i}, \quad \dot{p}_i = -\frac{\partial H}{\partial q^i}, \qquad i = 1, \ldots, n} \tag{4}^\dagger$$

Proof. Let X_H be defined by the formula (3). We then have to verify (2).

---

†As is discussed in Weinstein [1977a pp. 15, 16], Hamilton's equations were first discovered in linearized form by Lagrange in 1808. The reader may find a wealth of historical facts in Whittaker [1959].

ISBN 0-8053-0102-X

Now $i_{X_H} dq^i = \partial H/\partial p_i$, $i_{X_H} dp_i = -\partial H/\partial q^i$ by construction, so

$$i_{X_H}\omega = \sum i_{X_H}(dq^i \wedge dp_i) = \sum \left(i_{X_H} dq^i\right) \wedge dp_i - \sum dq^i \wedge i_{X_H} dp_i$$
$$= \sum \frac{\partial H}{\partial p_i} dp_i + \frac{\partial H}{\partial q^i} dq^i = dH \quad \blacksquare$$

Conservation of energy is easy to prove:

3.3.3 Proposition. *Let (M, ω, X_H) be a Hamiltonian system and let $c(t)$ be an integral curve for X_H. Then $H(c(t))$ is constant in t.*

Proof. By the chain rule and (1),

$$\frac{d}{dt} H(c(t)) = dH(c(t)) \cdot c'(t)$$
$$= dH(c(t)) \cdot X_H(c(t))$$
$$= \omega\left(X_H(c(t)), X_H(c(t))\right)$$
$$= 0$$

since ω is skew-symmetric. $\blacksquare$

The reader may also prove this using 3.3.2.

The next basic fact about Hamiltonian systems is that their flows consist of canonical transformations.

3.3.4 Proposition. *Let (M, ω, X_H) be a Hamiltonian system, and F_t be the flow of X_H. Then for each t, $F_t^*\omega = \omega$, that is, F_t is symplectic (on its domain). Thus F_t also preserves the phase volume Ω_ω (Liouville's Theorem).*

Proof. We have

$$\frac{d}{dt} F_t^*\omega = F_t^* L_{X_H}\omega$$
$$= F_t^*\left(i_{X_H} d\omega + d i_{X_H}\omega\right)$$
$$= F_t^*(0 + ddH) \quad (\textit{since } d\omega = 0 \textit{ and } i_{X_H}\omega = dH)$$
$$= 0$$

Thus $F_t^*\omega$ is constant in t. Since $F_0 = \textit{identity}$, the equation $F_t^*\omega = \omega$ results. $\blacksquare$

Notice that this is the first instance where we use the fact that ω is closed.

ISBN 0-8053-0102-X

3.3.5 Definition. *A vector field X on a symplectic manifold* (M, ω) *is called* ***locally Hamiltonian*** *if for every* $m \in M$ *there is a neighborhood U of m such that X restricted to U is Hamiltonian.*

3.3.6 Proposition. *(i) X is locally Hamiltonian iff* $\mathbf{i}_X\omega$ *is closed.*

(ii) X is locally Hamiltonian iff $\mathbf{L}_X\omega = 0$ *iff its flow consists of symplectic maps.*

(iii) X is locally Hamiltonian iff in a covering by symplectic charts (in which ω is constant; see Darboux's theorem), $DX(x)$ *is skew symmetric with respect to ω; that is,*

$$\omega(DX(m)\cdot e, f) = -\omega(e, DX(m)\cdot f)$$

Proof. (i) X is locally Hamiltonian iff $\mathbf{i}_X\omega$ is locally exact. By the Poincaré lemma, this is equivalent to $\mathbf{d}(\mathbf{i}_X\omega) = 0$.

(ii) As in (3.3.4) $(d/dt)F_t^*\omega = F_t^*(\mathbf{d}\mathbf{i}_X\omega)$ since $\mathbf{d}\omega = 0$. This vanishes iff $\mathbf{d}\mathbf{i}_X\omega = 0$, that is, X is locally Hamiltonian.

(iii) Let α be the one form $\mathbf{i}_X\omega$, so locally $\alpha_x \cdot e = \omega_x(X(x), e)$. Then X is locally Hamiltonian iff α is closed. But from the local formula for $\mathbf{d}$,

$$\begin{aligned}\mathbf{d}\alpha_x(e,f) &= D\alpha_x\cdot e\cdot f - D\alpha_x\cdot f\cdot e\\ &= \{\omega_x(DX(x)\cdot e, f) - \omega_x(DX(x)\cdot f, e)\\ &\quad + D\omega_x\cdot e(X(x), f) - D\omega_x\cdot f(X(x), e)\}\end{aligned}$$

The last two terms that, in a general chart, equal $D\omega_x\cdot X(x)\cdot(e,f)$ since ω is closed, vanish in a symplectic chart. ∎

Remarks (1) From $\mathbf{L}_{[X,Y]}\omega = \mathbf{L}_X\mathbf{L}_Y\omega - \mathbf{L}_Y\mathbf{L}_X\omega$ we see that the locally Hamiltonian vector fields, $\mathfrak{X}_{\mathcal{LH}}$ form a Lie subalgebra of $\mathfrak{X}$. Obviously a Hamiltonian vector field is locally Hamiltonian. The converse requires a topological condition sufficient to guarantee that closed one forms are exact, namely, the first cohomology group of M should vanish (see, e.g., Singer and Thorpe [1967]).

Here is an example of a locally Hamiltonian vector field that is not Hamiltonian. Consider the two torus T^2 with periodic coordinates x and y. Then $\omega = dx \wedge dy$ is a well-defined symplectic form on T^2. Identifying the tangent space of T^2 with $\mathbf{R}^2$, let, for any two constants a, b not both zero,

$$X(x,y) = (a,b)$$

Then

$$\begin{aligned}\mathbf{i}_X\omega &= (\mathbf{i}_X\,dx)\wedge dy - dx\wedge(\mathbf{i}_X\,dy)\\ &= a\,dy - b\,dx\end{aligned}$$

ISBN 0-8053-0102-X

which is closed. Thus X is locally Hamiltonian. But any locally Hamiltonian vector field that has no zeros on a *compact* symplectic manifold cannot be Hamiltonian. Indeed, if $X = X_H$ for some H, then since H has a critical point (a maximum or minimum point), X would correspondingly have a zero.

(2) There is a simple expression for H in terms of X_H and ω, namely, locally in a chart about $\mathbf{0}$,

$$H(x) = H(\mathbf{0}) + \int_0^1 \omega_{tx}(X_H(tx), x)\,dt$$

This follows from

$$\begin{aligned} H(x) - H(\mathbf{0}) &= \int_0^1 \frac{dH(tx)}{dt}\,dt \\ &= \int_0^1 dH(tx)\cdot x\,dt \\ &= \int_0^1 \omega_{tx}(X_H(tx), x)\,dt \end{aligned}$$

(3) If we specialize these results to linear vector fields X on a symplectic vector space $(\boldsymbol{E}, \omega)$, we obtain the following equivalent conditions:

(i) X is Hamiltonian with energy

$$H(e) = \tfrac{1}{2}\omega(X(e), e)$$

(ii) X is ω-skew

$$\omega(X(e), f) = -\omega(e, X(f))$$

(in the terminology of Sect. 3.1, X is infinitesimally symplectic);

(iii) the flow of X, that is, $F_t = e^{tX}$, preserves ω.

In 3.1.18 we showed that if X has distinct purely imaginary eigenvalues, then in symplectic coordinates $(x_1, \ldots, x_n, y_1, \ldots, y_n)$, X has matrix of the form

$$X = \left[\begin{array}{ccc|ccc} & & & \alpha_1 & & 0 \\ & 0 & & & \ddots & \\ & & & 0 & & \alpha_n \\ \hline -\alpha_1 & & & & & \\ & \ddots & 0 & & 0 & \\ 0 & & -\alpha_n & & & \end{array}\right]$$

ISBN 0-8053-0102-X

The corresponding energy is easily seen to be [by (i) above]

$$H(x,y)=\tfrac{1}{2}\sum_{i=1}^{n}\alpha_i\left(x_i^2+y_i^2\right)$$

that is, X is the sum of n noninteracting *harmonic oscillators* (If $\alpha_i<0$, the ith harmonic oscillator is running backwards).

For mechanics, one of the most important operations given by the symplectic structure of the phase space is that of the Poisson bracket. In fact, Jost [1964] has shown that a symplectic structure can be derived from the Poisson brackets (see Exercise 3.3F).

We shall first define Poisson brackets for one-forms and then for functions. If α is a one-form, let $\alpha^\#$ be the vector field corresponding to it via ω, that is, $i_{\alpha^\#}\omega=\alpha$, and for a vector field X, let $X^\flat=i_X\omega$.

3.3.7 Definition. *Suppose (M,ω) is a symplectic manifold and $\alpha,\beta\in\mathfrak{X}^*(M)$. The Poisson bracket of α and β is the one-form $\{\alpha,\beta\}=-[\alpha^\#,\beta^\#]^\flat$.*

Note that we have the following commutative diagram:

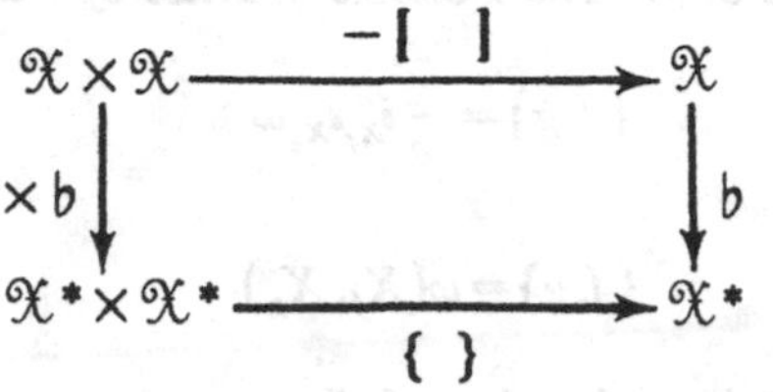

Since $[\ ,]$ makes $\mathfrak{X}$ a Lie algebra and $\flat$ is linear, then $\mathfrak{X}^*(M)$ as a real vector space, together with the composition $\{\ \}$, is a Lie algebra.

3.3.8 Proposition. *Let (M,ω) be a symplectic manifold and $\alpha,\beta\in\mathfrak{X}^*(M)$. Then $\{\alpha,\beta\}=-L_{\alpha^\#}\beta+L_{\beta^\#}\alpha+d(i_{\alpha^\#}i_{\beta^\#}\omega)$.*

Proof. From Table 2.4-1 we have the following formula for two-forms:

$$\begin{aligned}(d\omega)(X,Y,Z)=&L_X(\omega(Y,Z))+L_Y(\omega(Z,X))+L_Z(\omega(X,Y))\\&-\omega([X,Y],Z)-\omega([Y,Z],X)-\omega([Z,X],Y)\end{aligned}$$

Setting $X=\alpha^\#$, $Y=\beta^\#$ and observing that $\omega(\alpha^\#,Z)=\alpha(Z)$ yields

$$\begin{aligned}0=&L_{\alpha^\#}(\beta(Z))-L_{\beta^\#}(\alpha(Z))-L_Z(i_{\alpha^\#}i_{\beta^\#}\omega)\\&+\{\alpha,\beta\}(Z)+\alpha(L_{\beta^\#}Z)-\beta(L_{\alpha^\#}Z)\end{aligned}$$

ISBN 0-8053-0102-X

Then, using formulas (6) and (14) of Table 2.4-1, this becomes

$$0=(L_{\alpha^\sharp}\beta)(Z)-(L_{\beta^\sharp}\alpha)(Z)-d(i_{\alpha^\sharp}i_{\beta^\sharp}\omega)(Z)+\{\alpha,\beta\}(Z)$$

as required. ■

3.3.9 Proposition. *If $\alpha,\beta\in\mathfrak{X}^*(M)$ are closed, then $\{\alpha,\beta\}$ is exact.*

Proof. If ρ is closed then $L_X\rho=i_X d\rho+di_X\rho=di_X\rho$. Thus the result follows from 3.3.8. ■

3.3.10 Definition. *Let $\mathfrak{X}^*_{\mathcal{C}}(M)$ denote the set of closed one-forms, and $\mathfrak{X}^*_{\mathcal{E}}(M)$ the exact one-forms on a manifold M.*

Since d is $\mathbb{R}$ linear, it is clear that $\mathfrak{X}^*_{\mathcal{C}}(M)$ and $\mathfrak{X}^*_{\mathcal{E}}(M)$ are subspaces of $\mathfrak{X}^*(M)$ as real vector spaces. Also, if $\alpha,\beta\in\mathfrak{X}^*_{\mathcal{C}}(M)$, then $\{\alpha,\beta\}\in\mathfrak{X}^*_{\mathcal{E}}(M)\subset\mathfrak{X}^*_{\mathcal{C}}(M)$ by 3.3.9. It is also clear that if $\alpha,\beta\in\mathfrak{X}^*_{\mathcal{E}}(M)$, then $\{\alpha,\beta\}\in\mathfrak{X}^*_{\mathcal{E}}(M)$. Thus $\mathfrak{X}^*_{\mathcal{C}}$ and $\mathfrak{X}^*_{\mathcal{E}}$ are Lie subalgebras of $\mathfrak{X}^*$.

3.3.11 Definition. *Let (M,ω) be a symplectic manifold and $f,g\in\mathfrak{F}(M)$, with $X_f=(df)^\sharp\in\mathfrak{X}(M)$ as in 3.3.1. The **Poisson bracket** of f and g is the function*

$$\{f,g\}=-i_{X_f}i_{X_g}\omega$$

that is,

$$\{f,g\}=\omega(X_f,X_g)$$

Some properties of Poisson brackets follow.

3.3.12 Proposition. *Let (M,ω) be a symplectic manifold and $f,g\in\mathfrak{F}(M)$. Then*

$$\{f,g\}=-L_{X_f}g=L_{X_g}f$$

Proof. We have $L_{X_f}g=i_{X_f}dg=i_{X_f}i_{X_g}\omega$ since $dg=i_{X_g}\omega$. Since $i_Xi_Y\omega=-i_Yi_X\omega=\omega(Y,X)$, the last equality follows. ■

3.3.13 Corollary. *(i) For $f_0\in\mathfrak{F}(M)$, the map $g\mapsto\{f_0,g\}$ is a derivation.*

(ii) f is constant on the orbits of X_g iff $\{f,g\}=0$ iff g is constant on the orbits of X_f.

Proof. (i) $g\mapsto\{f_0,g\}=-L_{X_{f_0}}g$ is clearly a derivation. (ii) If F_t is the flow of X_f,

$$\frac{d}{dt}(g\circ F_t)=F_t^*L_{X_f}g=-F_t^*\{f,g\}$$

ISBN 0-8053-0102-X

which vanishes iff $\{f,g\}=0$. Since $\{f,g\}=-\{g, f\}$, this is equivalent to $(d/dt)(f\circ G_t)=0$, where G_t is the flow of X_g. ■

Note that the identity $\{H,H\}=0$ corresponds to conservation of energy by (ii).

3.3.14 Corollary. *In canonical coordinates (i.e., a symplectic chart) $(q^1,\ldots,q^n,p_1,\ldots,p_n)$, we have*

$$\{f,g\}=\sum_{i=1}^{n}\left(\frac{\partial f}{\partial q^i}\frac{\partial g}{\partial p_i}-\frac{\partial f}{\partial p_i}\frac{\partial g}{\partial q^i}\right)$$

(*Note:* $\{q^i,q^j\}=0, \{p_i,p_j\}=0, \{q^i,p_j\}=\delta^i_j$.)

Proof. $\{f,g\}=L_{X_g}f=df\cdot X_g$, so

$$\{f,g\}=\left(\frac{\partial f}{\partial q^i},\frac{\partial f}{\partial p_i}\right)\cdot\left(\frac{\partial g}{\partial p_i},-\frac{\partial g}{\partial q^i}\right)$$

$$=\sum_{i=1}^{n}\left(\frac{\partial f}{\partial q^i}\frac{\partial g}{\partial p_i}-\frac{\partial f}{\partial p_i}\frac{\partial g}{\partial q^i}\right) \quad ■$$

The following is often referred to as *the equations of motion in Poisson bracket notation.*

3.3.15 Corollary. *Let X_H be a Hamiltonian vector field on a symplectic manifold (M,ω) with Hamiltonian $H\in\mathfrak{F}(M)$ and flow F_t. Then for $f\in\mathfrak{F}(M)$ we have*

$$\frac{d}{dt}(f\circ F_t)=\{f\circ F_t,H\}$$

Proof.

$$\frac{d}{dt}(f\circ F_t)=\frac{d}{dt}F_t^*f=F_t^*L_{X_H}f$$

$$=L_{X_H}(f\circ F_t)=\{f\circ F_t,H\} \quad ■$$

The Poisson bracket of functions relates to the bracket of one forms as follows.

3.3.16 Proposition. *Let (M,ω) be a symplectic manifold and $f,g\in\mathfrak{F}(M)$. Then $d\{f,g\}=\{df,dg\}$.*

ISBN 0-8053-0102-X

Proof. This is a simple computation. Using 3.3.8,

$$\{df, dg\} = -L_{X_f} dg + L_{X_g} df + d(i_{X_f} i_{X_g} \omega) = -d(L_{X_f} g - L_{X_g} f - i_{X_f} i_{X_g} \omega)$$
$$\times = d\{f, g\} + d\{f, g\} - d\{f, g\} = d\{f, g\} \quad \blacksquare$$

3.3.17 Proposition. *The real vector space $\mathfrak{F}(M)$, together with the Poisson bracket $\{,\}$, forms a Lie algebra.*

Proof. Since d and $\omega^\sharp$ are R linear, the map $f \mapsto X_f$ is R linear. Hence $\{f,g\} = -i_{X_f} i_{X_g} \omega$ is R bilinear. It is also clear that $\{f,f\}=0$. For Jacobi's identity, we have

$$\{f, \{g, h\}\} = -L_{X_f}\{g, h\} = L_{X_f}(L_{X_g} h)$$
$$\{g, \{h, f\}\} = L_{X_g}(L_{X_h} f) = -L_{X_g}(L_{X_f} h)$$
$$\{h, \{f, g\}\} = L_{X_{\{f,g\}}} h$$

However, $X_{\{f,g\}} = (d\{f,g\})^\sharp = \{df, dg\}^\sharp = -[(df)^\sharp, (dg)^\sharp]$. Hence $X_{\{f,g\}} = -[X_f, X_g]$ and the result follows. $\blacksquare$

Jacobi's identity, restated, gives this corollary.

3.3.18 Corollary. $X_{\{f,g\}} = -[X_f, X_g]$. *In particular, the globally Hamiltonian vector fields $\mathfrak{X}_{\mathcal{H}}$ form a Lie algebra.*

From 3.3.9 and 3.3.16 one can show that $[\mathfrak{X}_{\mathcal{E}\mathcal{H}}, \mathfrak{X}_{\mathcal{E}\mathcal{H}}] \subset \mathfrak{X}_{\mathcal{H}}$. Actually, equality holds, a result of Arnold, Calabi, and Lichnerowicz (see Lichnerowicz [1973]).

A convenient criterion for symplectic diffeomorphisms is that they preserve the form of Hamilton's equations.

3.3.19 Theorem. (Jacobi [1837]) *Let (M, ω) and (N, ρ) be symplectic manifolds and $f: M \to N$ be a diffeomorphism. Then f is symplectic iff for all $h \in \mathfrak{F}(N)$, $f^* X_h = X_{h \circ f}$*

Proof. If f is symplectic, then $f^*(dh)^\sharp = (f^* dh)^\sharp = d(h \circ f)^\sharp = X_{h \circ f}$ (Exercise 3.3B). Conversely, if $f^* X_h = X_{h \circ f}$, then

$$d(h \circ f) = i_{X_{h \circ f}} \omega$$

On the other hand,

$$d(h \circ f) = f^* dh = f^* i_{X_h} \rho$$
$$= i_{f^* X_h} f^* \rho$$

ISBN 0-8053-0102-X

Therefore,

$$i_{X_{h\circ f}}\omega = i_{X_{h\circ f}} f^*\rho \quad \text{for all} \quad h\in\mathfrak{F}(N)$$

Every vector in T_mM has the form $X_{h\circ f}(m)$ for some h, so $\omega = f^*\rho$ and f is symplectic. ■

Preserving Poisson brackets also characterizes symplectic mappings as follows.

3.3.20 Proposition. *Let (M,ω) and (N,ρ) be symplectic manifolds and $F: M\to N$ a diffeomorphism. Then F is symplectic iff F preserves Poisson brackets of functions (resp. one-forms); that is, for all $f,g\in\mathfrak{F}(N)$, $\{F^*f, F^*g\} = F^*\{f,g\}$ (resp. for all $\alpha,\beta\in\Omega^1(N)$, $\{F^*\alpha, F^*\beta\} = F^*\{\alpha,\beta\}$); or F^* is a Lie algebra isomorphism on $\mathfrak{F}$ (resp. Ω^1).*

Proof. We have

$$F^*\{f,g\} = F^*L_{X_g}f = L_{F^*X_g}F^*f$$

Hence F preserves Poisson brackets iff

$$L_{F^*X_g}F^*f = L_{X_{F^*g}}F^*f$$

iff $F^*X_g = X_{F^*g}$ iff F is symplectic. We leave the second part as an exercise. ■

There is a useful characterization of symplectic charts in terms of coordinates as follows.

3.3.21 Proposition. *Let (M,ω) be a symplectic manifold, and (U,φ) a chart with $\varphi(u) = (q^1(u),\ldots,q^n(u),p_1(u),\ldots,p_n(u))$. Then (U,φ) is a symplectic chart, that is, $\omega = \sum dq^i\wedge dp_i$ iff $\{q^i,q^j\}=0$, $\{p_i,p_j\}=0$, and $\{q^i,p_j\}=\delta^i_j$ on U.*

Proof. If (U,φ) is a symplectic chart, the validity of these relations follows from 3.3.14.

Conversely, assume (U,φ) is a chart with $\{q^i,q^j\}=0$, $\{p_i,p_j\}=0$, and $\{q^i,p_j\}=\delta^i_j$. In this chart, let $\Omega=(\omega_{ij})$ be the $2n\times 2n$ matrix of ω (which equals the matrix of $\omega^\flat$) and let $A=(\alpha^{ij})$ be its inverse matrix (which equals the matrix of $\omega^\sharp$). Then

$$\{q^i,q^j\} = dq^i\cdot X_{q^j} = (X_{q^j})^i = \alpha^{ji}$$

Similarly,

$$\{q^i,p_j\} = \alpha^{j+n,i} = -\alpha^{i,j+n}, \qquad \{p_i,p_j\} = \alpha^{j+n,i+n}$$

ISBN 0-8053-0102-X

Thus, by assumption

$$A=\begin{bmatrix}0 & -I\\ I & 0\end{bmatrix}=J^{-1}$$

so $\Omega=J$ and the chart is symplectic. ■

The bracket expressions introduced by Lagrange (1808) to simplify the two-body problem are still important in celestial mechanics as we shall see in Part IV. They are closely related to the Poisson brackets (1809).

3.3.22 Definition. *If (M,ω) is a symplectic manifold and $X, Y\in\mathfrak{X}(M)$, the **Lagrange bracket** of the vector fields X and Y is the scalar function*

$$[\![X,Y]\!]=\omega(X,Y)$$

*If (U,φ) is a chart on M, the **Lagrange bracket of** φ is the matrix of functions on U given by*

$$[\![u^i,u^j]\!]=\left[\!\!\left[\frac{\partial}{\partial u^i},\frac{\partial}{\partial u^j}\right]\!\!\right]$$

where $\partial/\partial u^i$ are the standard basis vectors associated with the chart (U,φ), regarded as local vector fields on M.

Notice that $[\![X_f,X_g]\!]=\{f,g\}$, and that a diffeomorphism f is symplectic iff it preserves all Lagrange brackets; that is, $[\![f^*X,f^*Y]\!]=f^*[\![X,Y]\!]$.

3.3.23 Proposition. *Let (M,ω) be a 2n-dimensional symplectic manifold, (U,φ) a chart (not necessarily symplectic) and let $\varphi(m)=(u^1,\ldots,u^{2n})$. Then*

(i) $\omega|U=\frac{1}{2}\sum_{i,j}[\![u^i,u^j]\!]\,du^i\otimes du^j$;

(ii) *(U,φ) is a symplectic chart iff $J=(\omega_{ij})$, i.e., if $\omega_{ij}=[\![u^i,u^j]\!]$ is the matrix*

$$J=\begin{bmatrix}0 & I\\ -I & 0\end{bmatrix}$$

(iii) *If $\omega_\varphi=\varphi_*\omega$ is the push-forward of $\omega|U$ to $U'=\varphi(U)\subset\mathbb{R}^{2n}$, then*

$$[\![u^i,u^j]\!]\circ\varphi^{-1}=\omega_\varphi(e_i,e_j)$$

where e_i are the standard basis vectors in $\mathbb{R}^{2n}$;

(iv) *(Lagrange, 1808) if $f\colon M\to M$ is a diffeomorphism, (U,φ) and (V,ψ) are charts on M, $f(U)=V$, (U,φ) being a symplectic chart, and if we write*

$$\varphi(u)=(q^1,\ldots,q^n,p_1,\ldots,p_n)$$
$$\psi(v)=(Q^1,\ldots,Q^n,P_1,\ldots,P_n)$$

ISBN 0-8053-0102-X

$$(f_{\varphi\psi})^{-1} \circ (Q^1, \ldots, Q^n, P_1, \ldots, P_n) = (q^1, \ldots, q^n, p_1, \ldots, p_n)$$

then on V,

$$[Q, P] = \sum_{i=1}^{n} \left(\frac{\partial q^i}{\partial Q} \frac{\partial p_i}{\partial P} - \frac{\partial q^i}{\partial P} \frac{\partial p_i}{\partial Q} \right)$$

where Q and P are any of $(Q^1, \ldots, Q^n, P_1, \ldots, P_n)$.

(v) *Suppose that in (iv), f is a symplectic diffeomorphism and* (V, ψ) *a symplectic chart. Then*

$$[q, p] \circ f^{-1} = [Q, P]$$

Proof. (i) to (iii) are direct verifications and are left as an exercise to the reader. For (iv), recall that

$$\frac{\partial}{\partial Q} = \frac{\partial q^1}{\partial Q} \frac{\partial}{\partial q^1} + \cdots + \frac{\partial p_n}{\partial Q} \frac{\partial}{\partial p_n}$$

$$\frac{\partial}{\partial P} = \frac{\partial q^1}{\partial P} \frac{\partial}{\partial q^1} + \cdots + \frac{\partial p_n}{\partial P} \frac{\partial}{\partial p_n}$$

This combined with the local expression in U of ω, $\omega | U = \Sigma_{i=1}^{n} dq^i \wedge dp_i$ yields

$$[Q, P] = \omega\left(\frac{\partial}{\partial Q}, \frac{\partial}{\partial P} \right)$$

$$= \sum_{i=1}^{n} \left(\frac{\partial q^i}{\partial Q} \frac{\partial p_i}{\partial P} - \frac{\partial q^i}{\partial P} \frac{\partial p_i}{\partial Q} \right)$$

Suppose that in (iv), f is a *symplectic* diffeomorphism and (U, ψ) is a symplectic chart. Then

$$[\![q,p]\!] \circ f^{-1} = [\![Q,P]\!]$$

To see this, notice that

$$\frac{\partial}{\partial Q} = f_*\left(\frac{\partial}{\partial q} \right), \qquad \frac{\partial}{\partial P} = f_*\left(\frac{\partial}{\partial p} \right).$$

We have

$$([\![Q, P]\!] \circ f)(u) = \omega_{f(u)}\left(T_u f\left(\frac{\partial}{\partial q} \right)_u, T_u f\left(\frac{\partial}{\partial p} \right)_u \right)$$

$$= (f^*\omega)_u\left(\left(\frac{\partial}{\partial q} \right)_u, \left(\frac{\partial}{\partial p} \right)_u \right)$$

$$= \omega_u\left(\left(\frac{\partial}{\partial q} \right)_u, \left(\frac{\partial}{\partial p} \right)_u \right) = [\![q,p]\!](u) \quad \blacksquare$$

ISBN 0-8053-0102-X

Statement (iv) of the above proposition shows actually that the classical expressions of Lagrange provide an algorithm for computing the Lagrange brackets in the "new" coordinates (Q,P) from the "old" canonical coordinates (q,p) when these "old" coordinates are expressed as functions of the "new" ones. (This is one reason why the coordinate transformations in celestial mechanics are usually given backwards in classical texts.)

3.3.24 Proposition (*Lagrange* 1808). *Let X be a locally Hamiltonian vector field on the symplectic manifold (M,ω), (U,φ) a symplectic chart, and F_t the local flow of X on U. Write on $U_t'=F_t(U)$, $(\varphi\circ F_{-t})(u')=(Q_t^1,\ldots,P_{nt})$ and get, relative to the chart $(U_t',\varphi\circ F_{-t})$,*

$$[\![Q,P]\!]_t=\omega\left(\frac{\partial}{\partial Q_t},\frac{\partial}{\partial P_t}\right)$$

Then $[\![Q,P]\!]_t\circ F_t$ (defined on U) is independent of t.

Proof. Since X is locally Hamiltonian, F_t is a symplectic diffeomorphism, so that by (v) above,

$$[\![Q,P]\!]_t=[\![q,p]\!]\circ F_{-t} \quad \text{or} \quad [\![Q,P]\!]_t\circ F_t=[\![q,p]\!] \qquad \blacksquare$$

The independence of time of his brackets is Lagrange's celebrated result.

In terms of complete integrals (see Sect. 2.1), this result may be phrased as follows: let X be a locally Hamiltonian vector field on a symplectic manifold (M,ω) and (V,b,Ψ) a complete solution with the associated family of charts $\{\Psi_t\}$. Then if $[\![Q,P]\!]_t$ is the Lagrange bracket of Ψ_t, $[\![Q,P]\!]_t$ is independent of t.

We conclude this section with an important result called the "period-energy" relation. Let (P,ω) be a symplectic manifold and $H: P\to \mathbf{R}$ smooth. Let $F: D\subset P\times\mathbf{R}\to P$ be the flow of X_H and let $\tilde{F}$ be the graph of F, that is, $\tilde{F}(x,t)=(x,F_t(x))$, so $\tilde{F}: D\to P\times P$. Let $\Delta=\{(p,p)\,|\,p\in P\}$ be the diagonal in $P\times P$ and set

$$per_H=\left\{(x,t)\in\tilde{F}^{-1}(\Delta)\,|\,t\neq 0\right\}$$

which is just the collection of "periodic orbits" of X_H. We now show that the period and energy are always functionally related on surfaces of periodic orbits. This may be stated as follows:

3.3.25 Proposition. *$dt\wedge dH=0$ on any submanifold of $per_H\subset P\times\mathbf{R}$.*

Proof. (A. Weinstein). On $P\times P$, let π_i be the projection on the ith factor, $i=1,2$. Let $\Omega=\pi_1^*\omega-\pi_2^*\omega$. Then Ω is a symplectic form, as is easily checked.

ISBN 0-8053-0102-X

Furthermore, $\Delta \subset P \times P$ is a *canonical relation*, that is, $i^*\Omega = 0$, where $i: \Delta \to P \times P$ is inclusion. Now

$$T_{(x,t)}\tilde{F}\left(v, a\frac{d}{dt}\right) = \left(v, T_x F_t \cdot v + aX_H(F_t(x))\right)$$

from the definition of $\tilde{F}$, so if $dH(x) \neq 0$, which we can assume (otherwise the conclusion is trivial), $\tilde{F}$ will be an immersion; that is, $T_{(x,t)}\tilde{F}$ is 1-1. From this formula for $T\tilde{F}$, we see that $\tilde{F}^*\Omega = -dt \wedge dH$, since F_t is symplectic and $T_x F_t(X_H(x)) = X_H(F_t(x))$. (This computation will be left as an exercise.) Let $j: per_H \to P \times R$ be inclusion, and $\delta: P \times R \to \Delta: (x,t) \mapsto (x,x)$. Then $\tilde{F} \circ j = i \circ \delta$, so that $j^*(dt \wedge dH) = -j^*\tilde{F}^*\Omega = -(\tilde{F} \circ j)^*\Omega = -\delta^* i^*\Omega = 0$. ■

This proof is based on Gordon [1969]; see Moser [1970] for further comments and exercises 5.2G and 5.3I for a generalization.

EXERCISES

3.3A. Let (M,ω) be a symplectic manifold. Show that the collection of symplectic diffeomorphisms $\varphi: M \to M$ form a group under composition. Guess what the tangent space to this group at the identity (i.e., its Lie algebra) is.*

3.3B. Show that a diffeomorphism F between symplectic manifolds is symplectic iff $(F^*\alpha)^\sharp = F^*(\alpha^\sharp)$ for all one-forms α.

3.3C. Let (M,ω) be a symplectic manifold and (U,φ) a chart on M such that if $\varphi(u) = (x^1(u), x^2(u), y_1(u), y_2(u))$, then

$$\omega | U = dx^1 \wedge dy_1 + dx^2 \wedge dy_2 + f dx^1 \wedge dy_2$$

for some $f \in \mathfrak{F}(U)$ [so that (U,φ) is not a symplectic chart]. Then show, by determining the $\flat$ and $\sharp$ actions,

(i) $f = \{y^1, x_2\}$

(ii) If $H \in \mathfrak{F}(M)$, then in local representation,

$$X_H = -\frac{\partial H}{\partial x^1}e_3 - \frac{\partial H}{\partial x^2}e_4 + \frac{\partial H}{\partial y_1}e_1 + \frac{\partial H}{\partial y_2}e_2 + \left(\frac{\partial H}{\partial x^2}\right) f e_3 - \frac{\partial H}{\partial y_1} f e_2$$

where (e_1, e_2, e_3, e_4) is the standard basis;

*That this group really is a smooth infinite-dimensional manifold is proved in Ebin–Marsden [1970].

ISBN 0-8053-0102-X

(iii) A curve $c: I \to M$ is an integral curve of X_H iff, in local representation,

$$\frac{dx^1}{dt}(c(t)) = \frac{\partial H}{\partial y_1}(c(t))$$

$$\frac{dx^2}{dt}(c(t)) = \left(\frac{\partial H}{\partial y_2} - f\frac{\partial H}{\partial y_1}\right)(c(t))$$

$$\frac{dy_1}{dt}(c(t)) = \left(-\frac{\partial H}{\partial x^1} + f\frac{\partial H}{\partial x^2}\right)(c(t))$$

$$\frac{dy_2}{dt}(c(t)) = -\frac{\partial H}{\partial x_2}(c(t))$$

Compare with the Hamiltonian equations if the chart is symplectic ($f=0$). Note, however, that the integral curves are the same, irrespective of the chart used. That is, *the above equations are canonical even if they do not look it.*

3.3D. Consider the polar coordinate diffeomorphism ρ from the upper half of the cylinder $R \times S^1$ onto $R^2\backslash\{0\}$, defined by $(r,\theta) \mapsto (r\cos\theta, r\sin\theta)$ (note that θ is not defined globally on S^1, but $d\theta$ is). Show that $d(r^2/2) \wedge d\theta$ is a volume on $S^1 \times R$ and, relative to this volume and the standard one on R^2, ρ is symplectic. Compare with the statement: $dx\,dy = r\,dr\,d\theta$.

3.3E. (i) If $X \in \mathfrak{X}(M)$, define $P_X: T^*M \to R$ by $P_X(\alpha_m) = \alpha_m(X(m))$. Show that if $X, Y \in \mathfrak{X}(M)$, then $\{P_X, P_Y\} = -P_{[X,Y]}$ in the natural symplectic structure.

(ii) Let $X \in \mathfrak{X}(M)$ and F_t its flow. Let $G_t = T^*F_{-t}$. Show that G_t is the flow of X_{P_X}.

(iii) Suppose M is a Riemannian manifold and X is a Killing vector field (i.e., $L_X g = 0$, where g is the metric). Letting F_t be the flow of X, show that $G_t = TF_t$ is a Hamiltonian flow with $H(v) = \langle v, X\rangle$.

3.3F. (W. Pauli, R. Jost). Let $\{f,g\}$ be an R-bilinear bracket defined on $\mathfrak{F}(M) \times \mathfrak{F}(M)$ that makes $\mathfrak{F}(M)$ into a Lie algebra. Suppose $\{\,,\}$ is a derivation in each factor, and $\{f,g\} = 0$ for all g implies f is constant. Show that $\Lambda(df_x, dg_x) = \{f,g\}(x)$ defines a two tensor on M. Then show Λ is nondegenerate and that the corresponding two-form ω is a symplectic structure.

3.3G. (R. Jantzen). Show that a locally Hamiltonian vector field on a symplectic manifold (M,ω) is globally Hamiltonian if and only if as a derivation on $\mathfrak{F}(M)$ it is inner. (A Lie algebra derivation $h: \mathfrak{g} \to \mathfrak{g}$ is called *inner* if it is of the form $h(\xi) = [\xi, \xi_0]$ for some $\xi_0 \in \mathfrak{g}$.)

3.3H. Let θ_0 be the canonical one form on T^*Q and $f: T^*Q \to R$. What is $\{\theta_0, df\}$?

3.3I. (W. Tulczyjew). A *special symplectic manifold* is a quintuple $(P, M, \pi, \theta, \alpha)$, where $\pi: P \to M$ is a differentiable fibration [i.e., locally, $\pi^{-1}(U)$ is a product of U with a manifold], θ is a one-form on P and $\alpha: P \to T^*M$ is a diffeomorphism over M (i.e., $\tau_M^* \circ \alpha = \pi$) such that $\alpha^*\theta_M = \theta$, where θ_M is the canonical one-form on M. Clearly a special symplectic manifold is also symplectic.

(a) If (P,ω) is a symplectic manifold and $\alpha: TP \to T^*P$ is the map $\omega^\sharp$, let θ_T be given by

$$\theta_T(w) = \omega(\tau_{TP}(w), T\tau_P(w))$$

ISBN 0-8053-0102-X

Show that $(TP,P,\tau_P,\theta_T,\alpha)$ is a special symplectic manifold

(b) If $f: P\to P$ is symplectic, show that $Tf: TP\to TP$ is also symplectic, using the symplectic structure in (a).

(c) Let (P,M,π,θ,α) be a special symplectic manifold and γ a closed one-form on M. Then set $\beta=\alpha+\gamma\circ\pi$. Show that $(P,M,\pi,\alpha^*\theta_M+\pi^*\gamma,\beta)$ is a special symplectic manifold with the same symplectic structure.

(d) Show how $T(T^*Q)$ can be realized as a symplectic manifold in two ways, via two different special symplectic structures.

3.3J. Let (M,ω) be a symplectic manifold. If T^k denotes the k-dimensional torus with the symplectic manifold $(T(T^k), d\alpha_1\wedge d\dot{\alpha}_1+\cdots+d\alpha_k\wedge d\dot{\alpha}_k)$, then define an "angular chart" (U,ϕ), where $U\subset M$ is open and $\phi: U\to T(T^k)$ is a diffeomorphism onto an open set. Show:

(i) ϕ is symplectic iff the matrix defined by the Lagrange brackets is J;

(ii) the Lagrange brackets of ϕ are independent of time along the flow of a locally Hamiltonian vector field on M.

3.3K. (G. Marle). Let (P,ω) be a symplectic manifold, $H\in\mathfrak{F}(P)$ a Hamiltonian and Q a submanifold of P such that X_H is tangent to Q. Let (R,Ω) be another symplectic manifold and $\pi: Q\to R$ a submersion such that (i) $\pi^*\Omega=i^*\omega$, i: $Q\to P$ being the inclusion, and (ii) there exists $\overline{H}\in\mathfrak{F}(R)$ such that $\overline{H}\circ\pi=H\circ i$. Show that $X_H|Q$ and $X_{\overline{H}}\in\mathfrak{F}(R)$ are π-related, that is,

$$T\pi\circ X_H=X_{\overline{H}}\circ\pi \text{ on } Q$$

How are their flows related?

3.3L. (K. Meyer) Let (E,ω) be a symplectic vector space and $X\in L(E,E)$ a linear map. Show that X is Hamiltonian iff $(I+X)(I-X)^{-1}$ is symplectic (compare the Cayley transform in operator theory).

3.4. INTEGRAL INVARIANTS, ENERGY SURFACES, AND STABILITY

With the machinery of differential forms, Lie derivatives, and Cartan's calculus at hand, we can give a concise treatment of the integral invariants of Poincaré with emphasis on the symplectic context. Some other basic properties of Hamiltonian systems will be treated as well.

3.4.1 Definition. *Let M be a manifold and X a vector field on M. Let $\alpha\in\Omega^k(M)$. We call α an* **invariant** *k-form of X iff $L_X\alpha=0$.*

From the basic connection between flows and Lie derivatives, we obtain the following.

3.4.2 Proposition. *Let M be a manifold and $X\in\mathfrak{X}(M)$, $\alpha\in\Omega^k(M)$. Then α is an invariant k-form of X iff α is constant along the integral curves of X, that is, $(F_\lambda)^*\alpha$ is independent of λ, where F_λ is the flow of X.*

Thus, if we think of the integral curves of X as the motion of a system, α is a *constant of the motion*. The term *integral* k-form arises because of the following.

ISBN 0-8053-0102-X

3.4.3 Theorem (Poincaré–Cartan). *Let X be a complete vector field on a manifold M with flow F_λ, and let $\alpha \in \Omega^k(M)$. Then α is an invariant k-form of X iff for all oriented compact k-manifolds with boundary $(V, \partial V)$ and C^∞ mappings $\varphi: V \to M$, we have $\int_V (F_\lambda \circ \varphi)^* \alpha = \int_V \varphi^* \alpha$, independent of λ.*

Proof. If α is invariant (Poincaré), then $(F_\lambda)_* \alpha = \alpha$, $(F_\lambda)^* \alpha = \alpha$. Hence $\int_V (F_\lambda \circ \varphi)^* \alpha = \int_V \varphi^* \circ (F_\lambda)^* \alpha = \int_V \varphi^* \alpha$. Note that, since V is compact and orientable, the integral is well defined according to Sect. 2.6. Conversely (Cartan), if the integral is invariant under the flow, then for any closed k disk $(D, \partial D)$ embedded in V (a solid sphere in local representation in $\mathbf{R}^n$), we have $\int_D (F_\lambda \circ \varphi)^* \alpha = \int_D \varphi^* \alpha$, since D is compact. But the Lebesgue integral is a (signed) measure, and the disks above generate the Borel sets on V. Hence, over any measurable set A we have, by the Hahn extension theorem, $\int_A (F_\lambda \circ \varphi)^* \alpha = \int_A \varphi^* \alpha$. Thus $(F_\lambda \circ \varphi)^* \alpha = \varphi^* \alpha$. Then, by choosing V to be a portion of various subspaces in local representation, we see that $(F_\lambda \circ \varphi)^* \alpha = \varphi^* \circ F_\lambda^* \alpha = \varphi^* \alpha$ for all such φ implies $F_\lambda^* \alpha = \alpha$, so α is an invariant k-form of X. ∎

Note that X need not be complete; the statement of the theorem merely requires that the domain of F_t should contain $\varphi(V)$, $0 < t < \lambda$.

3.4.4 Proposition. *Let X be a vector field on a manifold M and α, β invariant forms of X. Then*

(i) *$i_X \alpha$ is an invariant form of X*
(ii) *$\mathbf{d}\alpha$ is an invariant form of X*
(iii) *$L_X \gamma$ is closed iff $\mathbf{d}\gamma$ is an invariant form, for any $\gamma \in \Omega^k(M)$;*
(iv) *$\alpha \wedge \beta$ is an invariant form of X.*

Proof. Note that $L_X i_X = i_X L_X$, since $L_X i_X = \mathbf{d} i_X i_X + i_X \mathbf{d} i_X = i_X \mathbf{d} i_X$ and $i_X L_X = i_X \mathbf{d} i_X + i_X i_X \mathbf{d} = i_X \mathbf{d} i_X$. This also follows from the relation $i_{[X,Y]} = L_X i_Y - i_Y L_X$. Thus $L_X i_X \alpha = i_X L_X \alpha = 0$ and (i) holds. For (ii), $L_X \mathbf{d}\alpha = \mathbf{d} L_X \alpha = 0$. This same relation $L_X \mathbf{d} = \mathbf{d} L_X$ proves (iii). Finally, (iv) follows since L_X is a tensor, and hence a $\wedge$ derivation; $L_X(\alpha \wedge \beta) = (L_X \alpha) \wedge \beta + \alpha \wedge L_X \beta$. ∎

Since L_X is $\mathbf{R}$ linear, we obtain the following.

3.4.5 Corollary. *Let $X \in \mathfrak{X}(M)$, and let $\mathcal{Q}_X$ denote the invariant forms of X. Then $\mathcal{Q}_X$ is a $\wedge$ subalgebra of $\Omega(M)$, which is closed under $\mathbf{d}$ and i_X.*

3.4.6 Definition. *Let X be a vector field on a manifold M and $\alpha \in \Omega^k(M)$. Then α is called a **relatively invariant k-form of X** iff $L_X \alpha$ is closed.*

Thus α is a relatively invariant k-form of X iff $\mathbf{d}\alpha$ is an invariant $(k+1)$-form of X.

ISBN 0-8053-0102-X

For the integral properties of relatively invariant forms (Whittaker [1959, p. 271]) we employ Stokes' theorem.

3.4.7 Theorem (Poincaré–Cartan). *Let X be a complete vector field with flow F_λ on a manifold M. Let $\alpha \in \Omega^{k-1}(M)$. Then α is a relatively invariant $(k-1)$-form of X iff for all oriented compact k manifolds with boundary $(V, \partial V)$ and C^∞ maps $\varphi: V \to M$ we have*

$$\int_{\partial V} (F_\lambda \circ \varphi \circ i)^* \alpha = \int_{\partial V} (\varphi \circ i)^* \alpha$$

that is, is independent of $\lambda \in R$ ($i: \partial V \to V$ is the inclusion map).

Proof. The form α is relatively invariant iff $d\alpha$ is an invariant form of X. But then, by Stokes' theorem and 3.4.3, we have

$$\int_{\partial V} (F_\lambda \circ \varphi \circ i)^* \alpha = \int_V (F_\lambda \circ \varphi)^* d\alpha = \int_V \varphi^* d\alpha = \int_{\partial V} i^* \circ \varphi^* \alpha$$

The converse may be proven as in 3.4.3. ■

We may now summarize the algebraic relationships between the invariant forms of a fixed vector field.

3.4.8 Definition. *If $X \in \mathfrak{X}(M)$, let $\mathcal{Q}_X$ be the set of all invariant forms of X, $\mathfrak{R}_X$ the set of all relatively invariant forms of X, $\mathcal{C}$ the set of all closed forms in $\Omega(M)$, and $\mathcal{E}$ the set of all exact forms in $\Omega(M)$.*

Note that $\mathcal{Q}_X \subset \mathfrak{R}_X$, and $\mathcal{E} \subset \mathcal{C} \subset \mathfrak{R}_X$. By 3.4.5, $\mathcal{Q}_X$ is a differential subalgebra of $\Omega(M)$, but $\mathfrak{R}_X$ is not. However, it is obviously an R subspace. Further relationships as R subspaces may be expressed in the convenient language of exact sequences of R linear mappings.

3.4.9 Definition. *Let E_i be a real vector space, and $\alpha_i: E_i \to E_{i+1}$ a linear mapping, $i \in Z$, the integers. Then the diagram*

$$\cdots \xrightarrow{\alpha_{i-1}} E_i \xrightarrow{\alpha_i} E_{i+1} \xrightarrow{\alpha_{i+1}} E_{i+2} \xrightarrow{\alpha_{i+2}} \cdots$$

*is an **exact sequence** iff for all i, $Im(\alpha_{i-1}) = Ker(\alpha_i)$.*

Especially, we say $E \xrightarrow{\alpha} F \longrightarrow 0$ is exact if α is surjective, and $0 \to E \xrightarrow{\alpha} F$ is exact if α is injective.

Note that for any linear mapping $\alpha: E \longrightarrow F$, we have the exact sequence

$$0 \longrightarrow Ker(\alpha) \xrightarrow{i} E \xrightarrow{\alpha} F \xrightarrow{\pi} F/Im(\alpha) \longrightarrow 0$$

where i is the inclusion map, and π the projection onto the quotient space.

ISBN 0-8053-0102-X

With these notations the following is a trivial restatement of definitions and elementary properties.

3.4.10 Proposition. *If $X \in \mathfrak{X}(M)$, the following sequences are exact:*

(i) $0 \longrightarrow \mathcal{Q}_X \xrightarrow{i} \Omega(M) \xrightarrow{L_X} \Omega(M) \xrightarrow{\pi} \Omega(M)/Im(L_X) \longrightarrow 0$

(ii) $0 \longrightarrow \mathcal{C} \xrightarrow{i} \mathcal{R}_X \xrightarrow{d} \mathcal{Q}_X \xrightarrow{\pi} \mathcal{Q}_X/\mathcal{E} \cap \mathcal{Q}_X \longrightarrow 0$

In addition,

(iii) $d(\mathcal{Q}_X) \subset \mathcal{Q}_X$ *and* $i_X(\mathcal{Q}_X) \subset \mathcal{Q}_X$.

The relevance of invariant forms for Hamiltonian systems is the following.

3.4.11 Proposition. *Let X be a locally Hamiltonian vector field on a symplectic $2n$-manifold (M, ω). Then $\omega, \omega^2, \ldots, \omega^n$ are invariant forms of X.*

This proposition follows at once from the fact that $L_X\omega = 0$ and L_X is a $\wedge$ derivation.

Next we shall discuss energy surfaces for Hamiltonian systems and obtain an invariant measure on them. This will then be generalized to level surfaces defined by several functions.

Let (M, ω, X_H) be a Hamiltonian system. Let $e \in R$ be a regular value of H, that is, $dH(m) \neq 0$ if $m \in H^{-1}(e)$, so that $H^{-1}(e)$ is a submanifold of M of codimension 1. We will write Σ_e for a connected component of $H^{-1}(e)$. It follows easily that Σ_e is a codimension one submanifold of M. Under these circumstances, we say Σ_e is a *regular energy surface*. By conservation of energy, integral curves of X_H starting in Σ_e stay in Σ_e. Thus X_H is tangent to Σ_e. We write $X_H|\Sigma_e$ for the vector field X_H restricted to points of Σ_e.

3.4.12 Theorem. *There is a volume element μ_e on Σ_e invariant under $X_H|\Sigma_e$.*

Proof. Work in a neighborhood of Σ_e on which $dH \neq 0$. Write Ω_ω for the phase volume, as usual. There is a form σ such that

$$\Omega_\omega = dH \wedge \sigma$$

In fact σ can be chosen in many ways; using a partition of unity, the question is local. Choosing local coordinates with H as one of the coordinates, possible since $dH \neq 0$, we can let σ be a function times the wedge of the remaining coordinates. Now

$$0 = L_{X_H}\Omega_\omega = L_{X_H}(dH \wedge \sigma) = dH \wedge L_{X_H}\sigma$$

Thus the expression of $L_{X_H}\sigma$ in a basis containing dH must contain dH in every term. Hence we can write

$$L_{X_H}\sigma = dH \wedge \tau$$

ISBN 0-8053-0102-X

Similarly if σ, σ' are two forms satisfying

$$\Omega_\omega = dH \wedge \sigma$$

then

$$\sigma - \sigma' = dH \wedge \rho$$

Now let $i: \Sigma_e \to M$ be inclusion and $\mu_e = i^*\sigma$. Then from $\sigma - \sigma' = dH \wedge \rho$, we see that

$$i^*\sigma = i^*\sigma' + i^*(dH \wedge \rho) = i^*\sigma'$$

since $i^*dH = 0$, so μ_e does not depend on which σ is chosen (so is unique in that sense).

Since $dH \wedge \sigma \neq 0$, μ_e is a volume. Finally, it is invariant since

$$L_{X_H}\mu_e = i^* L_{X_H}\sigma = i^*(dH \wedge \tau) = 0 \quad \blacksquare$$

By uniqueness, note that if $f: M \to M$ is a volume preserving diffeomorphism such that $H \circ f = f$, then $f|\Sigma_e$ preserves μ_e.

We shall now generalize this result. We shall replace the regular energy surface Σ_e by a manifold V that is the level surface of a family of constants of the motion (as opposed to just H) and seek an invariant measure on V.

3.4.13 Definition. *If $V \subset M$ is a submanifold and $X \in \mathfrak{X}(M)$, then V is an* ***invariant manifold*** *of X if for all $v \in V$, $X(v) \in T_vV \subset T_vM$.*

From this infinitesimal characterization of invariance follows immediately an integral characterization. For if V is invariant under X, then $X|V \in \mathfrak{X}(V)$. Thus, by the uniqueness of integral curves, we have the following.

3.4.14 Proposition. *If $V \subset M$ is an invariant manifold of $X \in \mathfrak{X}(M)$, $v \in V$, and $c: I \to M$ is an integral curve at v, then there is a neighborhood J of $0 \in I$ such that $c(J) \subset V$, and conversely.*

The invariant manifolds of interest to us are those defined by

$$\Sigma_c = F^{-1}(c)$$

where $F = \{f_1, \ldots, f_k\}$, $c \in R^k$, and $f_i: M \to R$ are constants of the motion for X_H.

3.4.15 Theorem. *Let (M, ω, X_H) be a Hamiltonian system and let $f_1, \ldots, f_k: M \to R$ be C^∞ constants of the motion for X_H; that is, $\{f_i, H\} = 0$. Let*

$$F = (f_1, \ldots, f_k): M \to R^k$$

ISBN 0-8053-0102-X

and let $c \in \mathbf{R}^k$ be a regular value of $\mathbf{F}$ [i.e., $df_1(x) \wedge \cdots \wedge df_k(x) \neq 0$ if $x \in \mathbf{F}^{-1}(c)$]. Let $\Sigma_c = \mathbf{F}^{-1}(c)$. Then Σ_c is an invariant manifold for X_H of codimension k and there is an invariant volume μ_c defined on Σ_c.*

Proof. That Σ_c is a codimension k invariant manifold (or is empty) is immediate from 1.6.16. As in 3.4.12, we can write, in a neighborhood of Σ_c,

$$\Omega_\omega = df_1 \wedge \cdots \wedge df_k \wedge \eta$$

From Liouville's theorem and $\{f_i, H\} = L_{X_H} f_i = 0$, it follows that

$$df_1 \wedge \cdots \wedge df_k \wedge L_{X_H} \eta = 0$$

Thus, since $df_1, \ldots, df_k$ may be completed to a basis, $L_{X_H}\eta$ must have at least one df_i in each term. Therefore,

$$L_{X_H}\eta = \sum_{i=1}^{k} df_i \wedge \rho_i$$

If η' is a second form with

$$\Omega_\omega = df_1 \wedge \cdots \wedge df_k \wedge \eta'$$

then

$$\eta - \eta' = \sum_{i=1}^{k} df_i \wedge \sigma_i$$

for similar reasons.

We now set

$$\mu_c = i^* \eta$$

where $i: \Sigma_c \to M$ is inclusion. Since each $f_i = constant$ on Σ_c, μ_c is independent of which η is chosen.

Also, $\mu_c \neq 0$ since $df_1 \wedge \cdots \wedge df_k \wedge \eta \neq 0$, so μ_c is a volume. Finally,

$$L_{X_H}\mu_c = i^* L_{X_H}\eta = i^*\left(\sum_{l=1}^{k} df_l \wedge \rho_l \right) = 0$$

so μ_c is invariant. ■

For another proof, see Exercise 4.3H.

The invariant measure on energy surfaces is important in statistical mechanics, for in some cases the flow there is ergodic. We shall discuss some

*Recall that by Sard's theorem the regular values for $\mathbf{F}$ form a dense set.

ISBN 0-8053-0102-X

elementary aspects of ergodicity in Sect. 3.7 in the context of the "Virial Theorem," but for now the reader is referred to Exercise 3.4F.

We conclude this section with a brief discussion of stability in the context of our work in Sect. 2.1. In such a discussion, energy surfaces play a key role. The first thing to observe is the following:

3.4.16 Proposition. *Let (M, ω, X_H) be a Hamiltonian system. Then a point $x_0 \in M$ is an equilibrium point if and only if x_0 is a critical point of H; that is, $dH(x_0)=0$.*

Proof. The point x_0 is an equilibrium if and only if $X_H(x_0)=0$. Since $X_H=(dH)^\sharp$ and ω is nondegenerate, this is equivalent to $dH(x_0)=0$. ■

The most elementary criterion for stability of Hamiltonian systems is as follows:

3.4.17 Theorem. *Let x_0 be an equilibrium point of X_H. Suppose that $D^2H(x_0)$ is positive- (or negative-) definite; that is, for all $v_0 \in T_{x_0}M$, $v_0 \neq 0$*

$$D^2H(x_0)(v_0, v_0) > 0$$

(or <0). Then x_0 is stable.

Proof. By the Morse lemma (actually Taylor's theorem suffices here) in a neighborhood of x_0, the level surfaces of H are diffeomorphic to concentric spheres. Since energy is conserved, any initial point near x_0 must remain on the associated energy sphere and hence remain in a neighborhood of x_0. ■

Under the circumstances of 3.4.17, all the characteristic exponents of X_H at x_0 [i.e., eigenvalues of $DX_H(x_0)$] must be purely imaginary. This can be seen algebraically by working in a symplectic chart. It is also consistent with 3.4.17, for if some eigenvalue was not purely imaginary, there would be at least one with negative and one with positive real part by the symplectic eigenvalue theorem in Sect. 3.1. Thus, along some directions we would have asymptotic stability and along others, asymptotic instability. The situation of 3.4.17 is called that of a pure *center*. These ideas will be taken up again later in a discussion of invariant manifolds and periodic orbits.

Definiteness of D^2H is not necessary for stability. For example on $\boldsymbol{R}^4$

$$H(x_1, x_2, y_1, y_2) = \tfrac{1}{2}(x_1^2+y_1^2) - \tfrac{1}{2}(x_2^2+y_2^2)$$

is not definite, but the flow is stable; it is two harmonic oscillators, one running backwards. (See Remark 3 following 3.3.6.)

In many delicate situations, D^2H is not definite, but criteria due to Kolmogorov, Arnold, and Moser are available. These are discussed in Chapter 8.

ISBN 0-8053-0102-X

EXERCISES

3.4A. Prove: Let X be locally Hamiltonian on a symplectic manifold (M,ω). Then the invariant one-forms of X form a Lie subalgebra of $\mathfrak{X}^*(M)$, and the invariant functions form a Lie subalgebra of $\mathfrak{F}(M)$.

3.4B. Let (M,ω) be a symplectic $2n$ manifold and X a locally Hamiltonian vector field on M. Suppose $\Lambda \in \Omega^{2n}(M)$ and $\Lambda = \rho\Omega_\omega$. Show Λ is an invariant form of X iff ρ is an invariant function.

3.4C. Consider the symplectic manifold T^*Q and let θ_0 be the canonical one-form. Prove that $X \in \mathfrak{X}(M)$ is locally Hamiltonian iff θ_0 is a relatively invariant one-form of X.

3.4D. Let H_1 and H_2 be two Hamiltonian functions on (P,ω) and suppose Σ is a regular energy surface for both of them. Prove that the integral curves of X_{H_1} and X_{H_2} are the same on Σ except possibly for a reparametrization. (*Hint:* For $x \in \Sigma$, show that $E_x = \{v \in T_x\Sigma \mid i_v\omega_x = 0\}$ is one dimensional.)

3.4E. Give an example of a Hamiltonian system on T^*R^2 that has an equilibrium at the origin, one direction is attracting, one is repelling and there is a two-dimensional manifold of closed orbits.

3.4F. (a) (Poincaré Recurrence Theorem). Let M be a compact manifold, X a smooth vector field on M with flow F_t, and Ω an X-invariant volume. For each open set U in M and $T > 0$, show that there is an $S > T$ such that $U \cap F_S(U) \neq \phi$. [*Hint:* Since $U, F_T(U), \ldots, F_{kT}(U)$ have the same measure, they cannot be disjoint if k is large enough.]

(b) What does (a) say in the context of 3.4.17?

3.5 LAGRANGIAN SYSTEMS

We saw in Sect. 3.2 that T^*Q has a natural symplectic structure. Therefore it is possible to study Hamiltonian vector fields on the *momentum phase space, T^*Q*. This section is concerned with an alternative description on the *velocity phase space, TQ*.

Roughly, the idea is as follows. We consider a function L on TQ and solutions to a certain second-order equation. From L we can derive an energy function E on TQ that, when translated to T^*Q by means of the "fiber derivative" $FL: TQ \rightarrow T^*Q$ (the derivative of L in each fiber of TQ), yields a suitable Hamiltonian. Then the solution curves in T^*Q (Hamiltonian equations) and in TQ (Lagrangian equations) will coincide when projected to Q. The following diagram may help to keep the locations in mind;

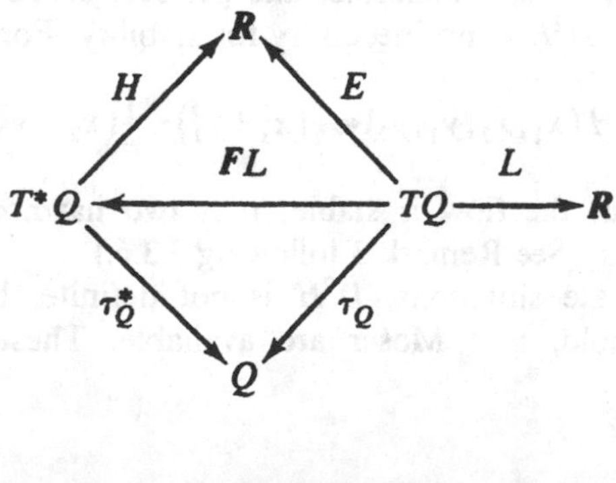

ISBN 0-8053-0102-X

We shall also see, in the next section, how the process may be reversed to allow passage from the Hamiltonian formulation to the Lagrangian.

Notice that the two formulations take place on different spaces, which in general cannot be canonically identified. Thus, the relation between H and L is not merely a change of variables.

We begin, then, with the fiber derivative in a slightly more general context.

3.5.1 Definition. *Let $\pi: E \to M$ and $\rho: F \to M$ be vector bundles over the common base space M, and let $f: E \to F$ be a C^∞ mapping (not necessarily a vector bundle mapping) that is fiber preserving and such that f_0 is the identity; that is, the following diagram commutes:*

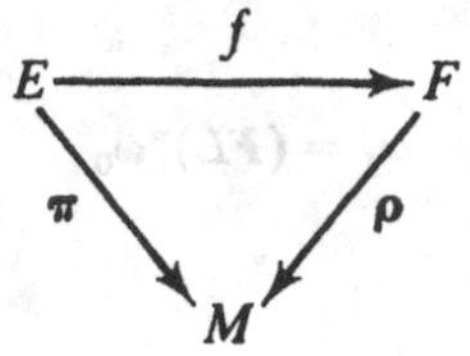

Let f_v denote $f|E_v$, where $E_v = \pi^{-1}(v)$ is the fiber over $v \in M$. Then the map

$$Ff: E \to \bigcup_{v\in M} L(E_v, F_v): e_v \mapsto Df_v(e_v) \in L(E_v, F_v)$$

*is called the **fiber derivative** of f.*

It may be easily shown, as in Sect. 1.7 that

$$L(E, F) = \bigcup_{v\in M} L(E_v, F_v)$$

is a vector bundle over M, with charts induced in a natural way from those of E and F. It is easy to see that $Ff: E \to L(E, F)$ is smooth and fiber preserving. Also notice that the association $f \mapsto Ff$ is linear. The situation that concerns us is $f: TQ \to \mathbf{R}$, so we will be content with the proofs in this case.

3.5.2 Definition. *Let Q be a manifold and $L \in \mathfrak{F}(TQ)$. Then the map $\mathbf{F}L$: $TQ \to T^*Q: w_q \mapsto DL_q(w_q) \in L(T_qQ, \mathbf{R}) = T_q^*Q$ is called the **fiber derivative** of L. Again L_q denotes the restriction of L to the fiber over $q \in Q$.*

We leave it to the reader to show that $\mathbf{F}L = \mathbf{F}\tilde{L}$ according to 3.5.1, where $\tilde{L}: TQ \to Q \times \mathbf{R}: w_q \mapsto (q, L(w_q))$ and $\tau_Q(w_q) = q$. Note that $\mathbf{F}L$ is not necessarily a vector bundle mapping. However, from our remarks following 3.5.1, we have:

3.5.3 Proposition. *Let Q be a manifold and $L \in \mathfrak{F}(TQ)$. Then $\mathbf{F}L$: $TQ \to T^*Q$ is a fiber preserving smooth mapping.*

ISBN 0-8053-0102-X

For computational purposes, we note that if $L: TQ \to \mathbf{R}$ is smooth and φ is a chart on M, then $(\mathbf{F}L)_\varphi = \mathbf{F}(L_\varphi)$.

3.5.4 Notation. *Let Q be a manifold and (U, φ) a chart on Q with $\varphi(u) = (q^1(u), \ldots, q^n(u)) \in \mathbf{R}^n$. Then we write*

$$T\varphi(v) = (q^1(v), \ldots, q^n(v), \dot{q}^1(v), \ldots, \dot{q}^n(v)) \in \mathbf{R}^{2n}$$

Then if $L \in \mathfrak{F}(TQ)$, we write $L_{\dot{q}} = D_2 L_\varphi \circ T\varphi^{-1}$ and $L_q = D_1 L_\varphi \circ T\varphi^{-1}$, with similar notation for higher derivatives. The components in the standard basis are denoted $L_{\dot{q}^i}$, and L_{q^i}. Thus $L_{\dot{q}^i}$, L_{q^i} represent the usual partial derivatives of L_φ.

3.5.5 Definition. *Let ω_0 be the canonical symplectic form on T^*Q and let $L \in \mathfrak{F}(TQ)$. Let*

$$\omega_L = (\mathbf{F}L)^* \omega_0$$

*called the **Lagrange two-form**.*

Clearly ω_L is a closed two-form on TQ since

$$d\omega_L = d(\mathbf{F}L)^* \omega_0 = (\mathbf{F}L)^* \, d\omega_0 = 0$$

The next proposition computes ω_L in coordinates. Let (U, φ) be a chart on M, with $\varphi(U) = U' \subset \mathbf{R}^n$. Then we have natural charts $T\varphi: TU \to U' \times \mathbf{R}^n \subset \mathbf{R}^n \times \mathbf{R}^n$ and $T^*\varphi: T^*U \to U' \times \mathbf{R}^{n^*} \subset \mathbf{R}^n \times \mathbf{R}^{n^*}$. Let the component functions of these two natural charts be $(q^1, \ldots, q^n, \dot{q}^1, \ldots, \dot{q}^n)$ and $(q^1, \ldots, q^n, p_1, \ldots, p_n)$, respectively. Then $\omega_L | TU$ is a linear combination of terms $dq^i \wedge d\dot{q}^j$, and so forth. In the notations of 3.5.4, we have $\mathbf{F}L: TU \to T^*U: (u, e) \mapsto L_{\dot{q}}(u, e)$. Also, let

$$L_{\dot{q}^i \dot{q}^j} = \frac{\partial^2 L}{\partial \dot{q}^i \partial \dot{q}^j}$$

and so forth. Thus in U,

$$dL_{\dot{q}^j} = L_{q^i \dot{q}^j} \, dq^i + L_{\dot{q}^i \dot{q}^j} \, d\dot{q}^i$$

3.5.6 Proposition. *If $L \in \mathfrak{F}(TM)$ and (U, φ) is a chart on M, then with the notations above*

$$\omega_L | TU = L_{\dot{q}^i q^j} dq^i \wedge dq^j + L_{\dot{q}^i \dot{q}^j} dq^i \wedge d\dot{q}^j$$

(summed on $i, j = 1, \ldots, n$).

ISBN 0-8053-0102-X

Proof. In terms of these coordinate functions, we have $\omega_0|T^*U = dq^i \wedge dp_i$. Thus

$$\begin{aligned}\omega_L|TU &= (FL^*\omega_0)|TU = FL^*(\omega_0|T^*U)\\ &= FL^*(dq^i \wedge dp_i) = (FL^*dq^i)\wedge(FL^*dp_i)\\ &= d(FL^*q^i)\wedge d(FL^*p_i)\\ &= d(q^i \circ FL)\wedge d(p_i \circ FL)\\ &= dq^i \wedge dL_{\dot{q}^i}\end{aligned}$$

from which the result follows. ■

In terms of bilinear forms, the expression for ω_L in 3.5.6 may be written this way:

$$\begin{aligned}\omega_L(u,e)((e_1,e_2),(e_3,e_4)) = {}& D_1D_2L(u,e)\cdot e_3\cdot e_1\\ &- D_1D_2L(u,e)\cdot e_1\cdot e_3 + D_2D_2L(u,e)\cdot e_4\cdot e_1\\ &- D_2D_2L(u,e)\cdot e_2\cdot e_3\end{aligned}$$

We leave it for the reader to check this, either from 3.5.6 or directly from the definition and formula (2) in the proof of 3.2.10.

3.5.7 Proposition. *Let $L: TQ \to R$ be a given smooth function and define the one form θ_L on TQ by*

$$\theta_L(w) = FL(\tau_{TQ}w)\cdot T\tau_Q(w)$$

where $w \in T(TQ)$. Then

$$\theta_L = (FL)^*\theta_0$$

where θ_0 is the canonical one-form on T^*Q,

$$\omega_L = -d\theta_L$$

and in a chart,

$$\theta_L = L_{\dot{q}^i}dq^i$$

Proof. For $w \in T(TQ)$ and $v = \tau_{TQ}(w)$, we have

$$\begin{aligned}(FL)^*\theta_0(w) &= \theta_0(T(FL)\cdot w)\\ &= \langle FL(v), T\tau_Q^*(T(FL)\cdot w)\rangle\end{aligned}$$

ISBN 0-8053-0102-X

from the definition of θ_0 in 3.2.10 and the fact that the base point of $T(FL)\cdot w$ is $FL(v)$. Thus, by the chain rule,

$$\begin{aligned}(FL)^*\theta_0(w) &= \langle FL(v), T(\tau_Q^* \circ FL)\cdot w\rangle \\ &= \langle FL(v), T\tau_Q \cdot w\rangle \\ &= \theta_L(w)\end{aligned}$$

This proves the first part. The second and third parts follow. Indeed,

$$\omega_L = (FL)^*\omega_0 = -(FL)^* d\theta_0 = -d(FL)^*\theta_0 = -d\theta_L$$

and

$$\begin{aligned}\theta_L &= (FL)^*(p_i\, dq^i) = (p_i \circ FL) d(q^i \circ FL) \\ &= L_{\dot q^i}\, dq^i \quad \blacksquare\end{aligned}$$

For most, but not all, of what follows we shall be interested in regular Lagrangians, or nondegenerate Lagrangians. They are defined as follows. (For additional results on the degenerate case see Proposition 3.7.19, Exercises 4.2B, 5.3L, Kunzle [1969] and Fischer–Marsden [1972]).

3.5.8 Definition. *Let Q be a manifold and $L \in \mathfrak{F}(TQ)$. We call L a* ***regular Lagrangian*** *if FL is regular (at all points) in the sense of* 1.6.17.

If one considers $L(x,y) = x$ on $T\mathbb{R}$, one sees that "regular Lagrangian" is not the same as "L has regular values."

3.5.9 Proposition. *Let $L \in \mathfrak{F}(TQ)$ for a manifold Q. Then L is a regular Lagrangian iff FL is a local diffeomorphism, iff $\omega_L = FL^*(\omega_0)$ is a symplectic form on TQ. (ω_0 denotes the canonical symplectic form on T^*Q.)*

Proof. From 3.2.6 it is sufficient to prove the first assertion. However, L is a regular Lagrangian iff $T_w FL$ is onto for each $w \in TQ$. Since the dimension of TQ and T^*Q are the same, $T_w FL$ is onto iff $T_w FL$ is an isomorphism. Thus the result follows at once from the inverse mapping theorem. $\blacksquare$

From the definition of ω_L, if L is a regular Lagrangian on Q, then FL: $TQ \to T^*Q$ is a symplectic mapping of the symplectic manifolds (TQ, ω_L), (T^*Q, ω_0).

Now FL: $TQ \to T^*Q$, so according to 3.5.1, $F^2L = F(FL)$: $TQ \to L(TQ, T^*Q) \approx T_2^0(Q)$. This object is used in the following.

3.5.10 Proposition. *Let L be a smooth function on the tangent bundle of a*

ISBN 0-8053-0102-X

manifold Q. Then $F^2L: TQ \to T^0_2(Q)$ is smooth and symmetric. Moreover, L is a regular Lagrangian iff F^2L is nondegenerate; that is, for each $w \in TQ$, $F^2L(w) \in L^2_s(TQ, \mathbf{R})$ is nondegenerate, or in charts, $(L_{\dot{q}^i\dot{q}^j})$ is a nondegenerate matrix.

Proof. It is sufficient to consider local vector bundles. Thus, assume $L: U \times E \to \mathbf{R}$ so that $FL: U \times E \to U \times E^*: (u,e) \mapsto (u, D_2L(u,e))$. Now L is a regular Lagrangian iff TFL is an isomorphism (in the fiber) at each point. This will be true iff DFL is an isomorphism at each point. A linear map between vector spaces of the same dimension is an isomorphism iff it is onto. However, $DFL(u,e): E \times E \to E \times E^*$ for each $(u,e) \in U \times E$, is given by $(e_1, e_2) \mapsto (e_1, DD_2L(u,e)\cdot(e_1,e_2))$. But

$$DD_2L(u,e)\cdot(e_1,e_2) = D_1D_2L(u,e)\cdot e_1 + D_2D_2L(u,e)\cdot e_2$$

where we identify $(e_1, 0)$ with e_1.

Now $DFL(u, e)$ is onto iff $D_2D_2L(u, e)$ is onto (for example, take $e_1 = 0$, etc.). However,

$$F^2L: U \times E \to U \times L^2_s(E, \mathbf{R}): (u,e) \mapsto (u, D_2D_2L(u,e))$$

and F^2L is nondegenerate iff $D_2D_2L(u, e)$ is an isomorphism for each (u, e). ■

This Proposition may also be proven from the expression for $\omega_L(u,e)$ following 3.5.6 which is easily seen to be nondegenerate iff $D_2D_2L(u,e)$ is nondegenerate.

Remark. In the infinite-dimensional case we say L is *weakly regular* if F^2L is weakly nondegenerate, which is equivalent to weak nondegeneracy of ω_L.

3.5.11 Definition. *Given $L: TQ \to \mathbf{R}$, define the **action** $A: TQ \to \mathbf{R}$ by $A(v_x) = FL(v_x)\cdot v_x$ and the **energy** by $E = A - L$. By a **Lagrangian vector field for** L we mean a vector field X_E on TQ such that $i_{X_E}\omega_L = dE$. If X_E exists we say that we can define **consistent equations of motion**.*

If L is regular, X_E exists; if L is weakly regular, there is at most one X_E. In general, X_E need not exist, nor need it be unique even in the finite-dimensional case.

Notice that in local coordinates, $A = \dot{q}^i L_{\dot{q}^i}$, so $E = \dot{q}^i L_{\dot{q}^i} - L$.

One of the main differences between the Hamiltonian and Lagrangian formulations is that *second-order equations* are possible on TQ, but not on T^*Q.

3.5.12 Definition. *A **second-order equation** on a manifold M is a vector field X on TM such that $T\tau_M \circ X$ is the identity on TM.*

ISBN 0-8053-0102-X

Thus, if X is a second-order equation on M we have the following commutative rhombic diagram

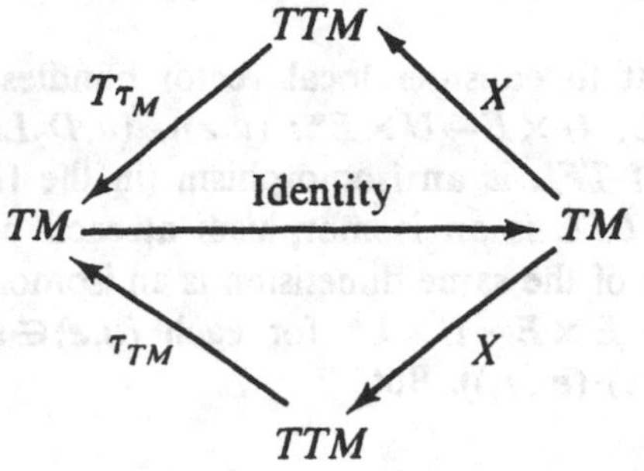

Second-order equations may be characterized in terms of their integral curves as follows.

3.5.13 Proposition. *Let X be a vector field on TM. Then X is a second-order equation on M iff for all integral curves $c: I \to TM$ of X, $(\tau_M \circ c)' = c$.*

Proof. From 2.1.2, for each $w_m \in TM$ there is a curve c at w_m such that $c'(t) = X(c(t))$ for $t \in I$. Then $T\tau_M \circ X$ is the identity iff $T\tau_M \circ c'(t) = c(t)$. But $T\tau_M \circ c'(t) = T\tau_M \circ Tc(t,1) = T(\tau_M \circ c)(t,1) = (\tau_M \circ c)'(t)$. ■

3.5.14 Definition. *If $c: I \to TM$ is an integral curve of a vector field X on TM, we call $\tau_M \circ c: I \to M$ a **base integral curve** of X. Similarly, if X is a vector field on T^*M and $c: I \to T^*M$ is an integral curve of X, $\tau_M^* \circ c: I \to M$ is called a **base integral curve** of X.*

Thus X is a second-order equation on M iff for every integral curve c of X, c equals the derivative of its base integral curve.

There is also a simple criterion for second-order equations in terms of local coordinates.

3.5.15 Proposition. *Let $X \in \mathfrak{X}(TM)$ and (U, φ) be a chart on M with $\varphi(U) = U' \subset E$. Suppose that the local representative of X has the form*

$$X_\varphi: U' \times E \to U' \times E \times E \times E;\ (u', e) \mapsto (u', e, X_1(u', e), X_2(u', e))$$

Then X is a second-order equation iff, for every chart, $X_1(u', e) = e$ for all $e \in E$.

Proof. Since $(\tau_M)_\varphi: U' \times E \to U'$, $(u', e) \mapsto u'$, and

$$T(\tau_{M\varphi}) = (T\tau_M)_\varphi: U' \times E \times E \times E \to U' \times E;\ (u', e, e_1, e_2) \mapsto (u', e_1)$$

we see that $(T\tau_M)_\varphi \circ X_\varphi = identity$ if $T\tau_M \circ X = identity$ iff $X_1(u', e) = e$. ■

ISBN 0-8053-0102-X

The usual notion of second-order equation is related to ours in the following way.

3.5.16 Proposition. *Let X be a second-order equation on M. Suppose (U,φ) is a chart on M with $\varphi(u)=(q^1(u),\dots,q^n(u))\in U'\subset \mathbf{R}^n$ and $T\varphi(v)=(q^1(v),\dots,q^n(v),\dot q^1(v),\dots,\dot q^n(v))\in U'\times\mathbf{R}^n\subset\mathbf{R}^n\times\mathbf{R}^n$. Suppose the local representative of X has the form $X_\varphi: U'\times\mathbf{R}^n\to U'\times\mathbf{R}^n\times\mathbf{R}^n\times\mathbf{R}^n$; $(u',e)\mapsto(u',e,e,X_2(u',e))$. Then $c: I\to M$ is a base integral curve of X iff*

$$\frac{d^2}{dt^2}q^i(c'(t)) = X_2^i(q(c'(t)),\dot q(c'(t)))$$

where X_2^i denotes the components of X_2, $i=1,\dots,n$, and $(q(c'(t)),\dot q(c'(t)))$ stands for $T\varphi(c'(t))$.

The proposition follows directly from 3.5.15.

Let us now return to the Lagrangian equations.

3.5.17 Theorem. *Let X_E be a Lagrangian vector field for $L: TQ\to\mathbf{R}$ (not necessarily regular). Assume X_E is a second-order equation. In a chart $U\times E$, if $(u(t),v(t))$ is an integral curve of X_E, it satisfies Lagrange's equations:*

$$\begin{cases} \dfrac{d}{dt}u(t)=v(t) \\ \dfrac{d}{dt}\{D_2L(u(t),v(t))\cdot w\} = D_1L(u(t),v(t))\cdot w \end{cases}$$

for all $w\in E$. In finite dimensions these are equivalent to the ***classical Euler–Lagrange equations:***

$$\boxed{\frac{d}{dt}\left(\frac{\partial L}{\partial \dot q^i}\right)=\frac{\partial L}{\partial q^i}}$$

Finally, if L is regular, then X_E is necessarily second order and always exists.

Proof. Write $X_E(x,e)=(Y(x,e),Z(x,e))$. Using the definition $A(x,e)=D_2L(x,e)\cdot e$, we get

$$\begin{aligned} DE(x,e)\cdot(e_1,e_2) &= D_1E(x,e)\cdot e_1 + D_2E(x,e)\cdot e_2 \\ &= D_1D_2L(x,e)\cdot e_1\cdot e - D_1L(x,e)\cdot e_1 + D_2D_2L(x,e)\cdot e_2\cdot e \end{aligned}$$

[a term $D_2L(x,e)\cdot e_2$ has cancelled out]. The condition $i_{X_E}\omega_L = dE$, that is, $\omega_L(x,e)((Y,Z),(e_1,e_2)) = DE(x,e)\cdot(e_1,e_2)$, using this formula for DE and

ISBN 0-8053-0102-X

the formula for ω_L following 3.5.6, becomes

$$D_1D_2L(x,e)\cdot e_1\cdot Y - D_1D_2L(x,e)\cdot Y\cdot e_1 + D_2D_2L(x,e)\cdot e_2\cdot Y - D_2D_2L(x,e)\cdot Z\cdot e_1$$
$$= D_1D_2L(x,e)\cdot e_1\cdot e - D_1L(x,e)\cdot e_1 + D_2D_2L(x,e)\cdot e_2\cdot e$$

If L is regular, setting $e_1=0$ we see $Y(x,e)=e$, so X_E is second order. In general, if we assume $Y(x,e)=e$, the condition reduces to

$$-D_1D_2L(x,e)\cdot e\cdot e_1 - D_2D_2L(x,e)\cdot Z\cdot e_1 = -D_1L(x,e)\cdot e_1$$

Setting $x=u(t)$, $e=v(t)=\dot{u}(t)$ we get

$$D_1D_2L(u,\dot{u})\cdot\dot{u}\cdot e_1 + D_2D_2L(u,\dot{u})\cdot\dot{v}\cdot e_1 = D_1L(u,\dot{u})\cdot e_1$$

which is Lagrange's equation. ■

Note that when L is regular the equations may be written

$$\ddot{u} = D_2D_2L(u,\dot{u})^{-1}\left[D_1L(u,\dot{u}) - D_1D_2L(u,\dot{u})\cdot\dot{u}\right]$$

where we regard $D_2D_2L(u,\dot{u})\colon E\to E^*$. In this case the coordinates $q=x$, $p=D_2L(x,e)$ are evidently canonical for ω_L. One calls p the *conjugate momentum.*

On a fixed symplectic manifold P, H and $\tilde{H}$ lead to the same equations of motion iff $H=\tilde{H}+$ *constant*. The situation for Lagrangian systems is as follows:

3.5.18 Proposition. *Let L and $\tilde{L}$ be regular Lagrangians on TQ and X_E, $X_{\tilde{E}}$ the corresponding Lagrangian vector fields. The following two assertions are equivalent:*

(i) $L=\tilde{L}+\alpha+$ *constant, where α is a closed one-form on Q, which we regard as a map $\alpha\colon TQ\to\mathbb{R}$;*
(ii) $X_E=X_{\tilde{E}}$ *and* $\omega_L=\omega_{\tilde{L}}$.

Proof. Assume (i). Clearly $FL=F\tilde{L}+\alpha\circ\tau_Q$ since α is linear on fibers. It follows at once that $E=\tilde{E}+$ *constant*. Also, $\theta_L=(FL)^*\theta_0=(F\tilde{L})^*\theta_0+\alpha^*\theta_0 = \theta_{\tilde{L}}+\alpha^*\theta_0$. Now $\alpha^*\omega_0=-d(\alpha^*\theta_0)$. From 3.2.11 $\alpha^*\theta_0=\alpha$, so as α is closed, $\alpha^*\omega_0=0$. Thus (ii) follows. If (ii) holds, $E=\tilde{E}+$ *constant*, so $L(v)=\tilde{L}(v)+[FL(v)-F\tilde{L}(v)]\cdot v+$ *constant*. Since $\omega_L=\omega_{\tilde{L}}$, $FL(v)-F\tilde{L}(v)$ is independent of $v\in T_xQ$. Thus $\alpha_x=FL(v)-F\tilde{L}(v)$, $v\in T_xQ$ is a well-defined one-form on Q. From $\alpha^*\omega_0=0$ we see, as above, that α is closed. ■

Because of this, one says that the closed one-forms on Q form the *gauge group* of Lagrangian mechanics. It can happen that for two Lagrangians,

ISBN 0-8053-0102-X

$E \neq \tilde{E} + constant$, yet $X_E = X_{\tilde{E}}$. For example, let $Q = \mathbf{R}$ and $L(x,y) = y^2$, $\tilde{L}(x,y) = 2y^2$. Both give $X_E(x,y) = X_{\tilde{E}}(x,y) = (y,0)$.

3.5.19 Proposition. *Let $L: TM \to \mathbf{R}$ be a Lagrangian with Lagrangian vector field X_E. Suppose $\Phi: M \to N$ is a diffeomorphism. Then a Lagrangian vector field for $\tilde{L} = L \circ T\Phi^{-1}$ is $(T\Phi)_* X_E$.*

Proof. From the definitions, $F\tilde{L} \circ T\Phi = T\Phi_* \circ FL$. It follows that $\omega_L = (T\Phi)^* \omega_{\tilde{L}}$ and $\tilde{E} = E \circ T\Phi^{-1}$. The result now follows from 3.3.19. ∎

EXERCISES

3.5A. Show that energy is conserved for degenerate Lagrangian systems.

3.5B. Show that a vector field X on TM is a second-order equation if and only if its flow F_t has this homogeneity property: $\tau_M(F_t(sv)) = \tau_M(F_{st}(v))$.

3.5C. On $T\mathbf{R}^n = \mathbf{R}^n \times \mathbf{R}^n$ suppose

$$L(q^1, \ldots, \dot{q}^n) = \sum_{i=1}^{n} m_i \frac{(\dot{q}^i)^2}{2} - V(q^1, \ldots, q^n)$$

where $m_i \in \mathbf{R}$ are constants and $V \in \mathfrak{F}(\mathbf{R}^n)$. Show L is regular iff $m_i \neq 0$ for all i, compute the action and energy of L, and write down Lagrange's equations.

3.5D. (a) Let $L: TQ \to \mathbf{R}$ be a Lagrangian, possibly degenerate, and let $f: Q \to Q$ be a diffeomorphism such that $L \circ Tf = L$. Show that $(Tf)^* \theta_L = \theta_L$; and, in particular, Tf is symplectic.

(b) Let F_t be the flow of a vector field X on Q, and suppose $L \circ TF_t = L$ for all t. Then prove that TF_t is generated by a vector field Y that is Hamiltonian with energy $\mu: v \mapsto FL(v) \cdot X$; that is, $i_Y \omega_L = d\mu$. (See Exercise 3.3E.)

(c) If $g: TQ \to TQ$ is a diffeomorphism and $L \circ g = L$, show that g need not preserve ω_L.

3.5E. Suppose ω is a closed two-form of constant rank on M. Show that $ker\,\omega = \{v_x | \omega(v_x, w_x) = 0 \text{ for all } w_x\}$ is an integrable subbundle, and for functions, f,g constant along its leaves, $\{f,g\}$ is unambiguously defined. (This is how one must define Poisson brackets...here called *Dirac brackets*...for degenerate Lagrangians. See Exercise 5.3L for further information.)

3.5F. Let $\pi: P \to M$ and $\eta: Q \to M$ be (smooth) submersions and $f: P \to Q$ a fibered map, that is, $f(\pi^{-1}(x)) \subset \eta^{-1}(x)$. Define $F_p f$ the fiber derivative at p to be the tangent of f restricted to $\pi^{-1}(\pi(p))$. Show that f is an immersion at p (resp. submersion, local diffeomorphism) if and only if $F_p f$ is injective (resp. surjective and isomorphism). [*Hint.* In a "fibered chart," f looks like $(x,y) \mapsto (x, \bar{f}(x,y))$. Now compute.]

3.5G. (J. P. Penot). Let $G: TM \to T^*M$ be a smooth fiber preserving map. We say G has *symmetric vertical derivative* if

$$\langle FG(u)v, w \rangle = \langle FG(u) \cdot w, v \rangle$$

for all $u, v, w \in TM$. Show that this occurs if $G = FL$ where $L: TM \to \mathbf{R}$.

ISBN 0-8053-0102-X

Prove the following proposition:

Proposition. *Let $F: TM \to T^*M$ have symmetric vertical derivative and let $V \in \mathfrak{F}(M)$. Let $\omega_F = F^*\omega_0$, where ω_0 is the canonical two-form on T^*M. Then:*

(i) *there is at most one $K \in \mathfrak{F}(TM)$ such that $K \circ \zeta_M = V$, where $\zeta_M: M \to TM$ is the zero section and $dK = i_X \omega_F$ for some second-order equation X on TM;*
(ii) *if $G = FL$, then $K(v) = -L(v) + FL(v)\cdot v + V(\tau_M(v)) + L(\zeta_M(\tau_M(v)))$;*
(iii) *if $FG(v) \in L(T_mM, T_m^*M)$ is injective for each $v \in T_mM$, then (TM, ω_F) is a weak symplectic manifold; if $FG(v)$ is an isomorphism for each $v \in T_mM$, there is a unique K and X satisfying (i).*

[*Hint.* For (i), proceed as in 3.5.17 and show that the conditions determine $K(x,0)$ and $D_2K(x,v)$ in charts, and hence K. It will be convenient to use Vainberg's theorem (see Exercise 2.4G) to relate existence of K to the symmetry assumption on G to do (ii).]

3.6. THE LEGENDRE TRANSFORMATION

Let us now give the relationship between the Lagrangian formulation on TQ and the Hamiltonian formulation on T^*Q. In fact, they are equivalent in the *hyperregular* case, and are transformed one into the other by the Legendre transformation.

3.6.1 Definition. *Let Q be a manifold and $L \in \mathfrak{F}(TQ)$. Then L is called a **hyperregular Lagrangian** if $FL: TQ \to T^*Q$ is a diffeomorphism.*

Recall that we define $\omega_L = (FL)^*\omega_0$, where ω_0 is the canonical two-form on T^*Q. Hence FL becomes a symplectic diffeomorphism, and thus preserves Poisson brackets. It is also clear that a hyperregular Lagrangian is regular. Notice that in coordinates, $FL: (q^1,\dots,q^n,\dot{q}^1,\dots,\dot{q}^n) \mapsto (q^1,\dots,q^n,p_1,\dots,p_n)$ where $p_i = \partial L/\partial \dot{q}^i$.

The transition from the Lagrangian formulation to the Hamiltonian is given by the following.

3.6.2 Theorem. *Let L be a hyperregular Lagrangian on Q and let $H = E \circ (FL)^{-1}: T^*Q \to R$, where E is the energy of L. Then X_E and X_H are FL related: $(FL)_*X_E = X_H$. The integral curves of X_E are mapped by FL onto integral curves of X_H. Furthermore, X_E and X_H have the same base integral curves.*

Proof. It suffices to prove that $(FL)_*X_E = X_H$ (see Exercise 2.1B). Note that $\tau_Q = \tau_Q^* \circ FL$, so once the integral curves are FL related, the base integral curves are deduced to be equal.

ISBN 0-8053-0102-X

Now writing $v^* = T_v(FL)(w)$ for $v \in TQ$, $w \in T_v TQ$, we get

$$\begin{aligned}\omega_0\big(TFL(X_E(v)), v^*\big) &= \omega_L(X_E(v), w)\\ &= dE(v)\cdot w\\ &= d(H \circ FL)(v)\cdot w\\ &= dH(FL(v))\cdot v^*\\ &= \omega_0\big(X_H(FL(v)), v^*\big)\end{aligned}$$

Since $T_v(FL)$ is an isomorphism, v^* is arbitrary, so

$$TFL(X_E(v)) = X_H(FL(v))$$

that is,

$$(FL)_* X_E = X_H \quad \blacksquare$$

The transformation $FL: TQ \to T^*Q$ thus maps the Lagrange equations into the Hamilton equations. In the literature FL itself is sometimes called the *Legendre transformation* (e.g., Sternberg [1964]), while classically the name is usually reserved for the map that takes

$$L(q^1,\ldots,q^n,\dot{q}^1,\ldots,\dot{q}^n) \text{ to } H(q^1,\ldots,q^n,p_1,\ldots,p_n) = \dot{q}^i p_i - L(q^1,\ldots,q^n,\dot{q}^1,\ldots,\dot{q}^n)$$

where $p_i = \partial L/\partial \dot{q}^i$ (e.g., Courant and Hilbert [1962, p. 34]). Using this coordinate notation the reader should verify 3.6.2, i.e., that the Legendre transformation converts Lagrange's equations to Hamilton's equations.*

For the reverse construction we need the following.

3.6.3 Proposition. *Let L by a hyperregular Lagrangian on Q and $H = E \circ (FL)^{-1}$, where E is the energy of L. Then $\theta_0(X_H) = A \circ (FL)^{-1}$, where A is the action of L, and θ_0 is the canonical one-form.*

---

*In the differential equations literature the Legendre transform is viewed in the following way. If $f: V \to \mathbf{R}$, and $Df: V \to V^*$ is a diffeomorphism, its Legendre transform $\tilde{f}: V^* \to \mathbf{R}$ is defined by $\tilde{f} = (\rho f - f) \circ (Df)^{-1}$ where ρ is the radial vector field $\rho = \Sigma x^i \partial/\partial x^i$ for coordinates x^i associated with a basis e_i of V. Geometrically, let $\varphi \in V^*$ and let P be the unique tangent hyperplane to the graph of f which is parallel to the graph of φ; $\tilde{f}(\varphi)$ is the height that the graph of φ has to be raised to give P. For a Lagrangian L the Hamiltonian H is just the Legendre transform of L performed fiber by fiber.

ISBN 0-8053-0102-X

Proof. We must show $\theta_0(X_H)\circ FL=A$. Let $w\in TQ$ and $\alpha=FL(w)$. Then

$$\begin{aligned}\theta_0(X_H)\circ FL(w)&=\theta_0(X_H)(\alpha)\\&=\langle\alpha, T\tau_Q^*X_H(\alpha)\rangle\\&=\langle\alpha, T\tau_Q^*FL_*X_E(\alpha)\rangle\\&=\langle\alpha, T\tau_Q^*\circ TFL\circ X_E(w)\rangle\\&=\langle\alpha, T(\tau_Q^*\circ FL)\circ X_E(w)\rangle\\&=\langle\alpha, T(\tau_Q)\circ X_E(w)\rangle\\&=\langle\alpha, w\rangle\end{aligned}$$

since X_E is a second-order equation. But $\langle\alpha,w\rangle=\langle FL(w),w\rangle=A(w)$ by definition of A. ■

The proof of 3.6.3 in coordinates goes as follows. From $\theta_0=p_i\,dq^i$, we get

$$\theta_0(X_H)=p_i\frac{\partial H}{\partial p_i}$$

We now change variables $q^i\mapsto q^i$, $\dot q^i\mapsto p_i=\partial L/\partial\dot q^i$ via the Legendre transform. Thus

$$E=\dot q^j\frac{\partial L}{\partial \dot q^j}-L=\dot q^j p_j-L$$

so

$$\frac{\partial E}{\partial p_i}=\dot q^i+\frac{\partial \dot q^j}{\partial p_i}p_j-\frac{\partial L}{\partial \dot q^j}\frac{\partial \dot q^j}{\partial p_i}=\dot q^i$$

Thus, since E as a function of q^i,p_i is just H, we get

$$\theta_0(X_H)=p_i\frac{\partial H}{\partial p_i}=\frac{\partial L}{\partial \dot q^i}\dot q^i=A$$

3.6.4 Corollary. *Let L be a hyperregular Lagrangian on Q and $\theta_L=FL^*\theta_0$ (so that $\omega_L=-d\theta_L$). Then $A=\theta_L(X_E)$, where E is the energy, and A the action of L.*

Proof. $A=\theta_0(X_H)\circ FL$. Let $w\in TQ$ and $\alpha=FL(w)$; then

$$\begin{aligned}A(w)=\theta_0(X_H)(\alpha)&=\langle\theta_0(\alpha),X_H(\alpha)\rangle=\langle\theta_0(\alpha),FL_*X_E(\alpha)\rangle\\&=\langle\theta_0(\alpha),TFL\circ X_E(w)\rangle=\langle\theta_L(w),X_E(w)\rangle=\theta_L(X_E)(w)\end{aligned}$$ ■

ISBN 0-8053-0102-X

This proposition tells us that we can recover L if we know $\mathbf{F}L$ and E. If $H\in\mathfrak{F}(T^*Q)$, note that $\mathbf{F}H: T^*Q\rightarrow T^{**}Q\approx TQ$. Then, as for L, $\mathbf{F}H$ is a smooth fiber preserving map, that is, $\tau_Q\circ\mathbf{F}H=\tau_Q^*$.

The following proposition is proven in the same way as the corresponding statement for L.

3.6.5 Proposition. *Let $H\in\mathfrak{F}(T^*Q)$. Then $\mathbf{F}H$ is a local diffeomorphism iff $\mathbf{F}^2H$ is nondegenerate. If this is the case, we call H a **regular Hamiltonian**.*

As was the case for L, we need the stronger condition of hyperregularity for transition to the Lagrangian formulation.

3.6.6 Definition. *If $H\in\mathfrak{F}(T^*Q)$, we call $G=\theta_0(X_H)$ the **action** of H. Also, H is called a **hyperregular Hamiltonian** if $\mathbf{F}H: T^*M\rightarrow TM$ is a diffeomorphism.*

3.6.7 Proposition. *Let H be a hyperregular Hamiltonian on T^*Q. Then define $E=H\circ(\mathbf{F}H)^{-1}$, $A=G\circ(\mathbf{F}H)^{-1}$, and $L=A-E$. Then L is a hyperregular Lagrangian on TQ, and in fact $\mathbf{F}L=(\mathbf{F}H)^{-1}$.*

Proof. The easiest proof we know of is done in coordinates. As in 3.6.3, $\theta_0=p_i\,dq^i$, so $G=p_i\partial H/\partial p_i$. Now $\mathbf{F}H$ is the map

$$(q^i,p_j)\mapsto\left(q^i,\frac{\partial H}{\partial p_j}\right)=(q^i,\dot q^j)$$

Thus

$$\begin{aligned}(L\circ\mathbf{F}H)(q^i,p_j)&=G(q^i,p_j)-H(q^i,p_j)\\&=p_k\frac{\partial H}{\partial p_k}-H(q^i,p_j)=p_k\dot q^k-H(q^i,p_j)\end{aligned}$$

Changing variables and using the chain rule,

$$\begin{aligned}\frac{\partial L}{\partial\dot q^i}&=\frac{\partial p_k}{\partial\dot q^i}\dot q^k+p_i-\frac{\partial H}{\partial p_k}\frac{\partial p_k}{\partial\dot q^i}\\&=p_i\end{aligned}$$

This shows that $\mathbf{F}L\circ\mathbf{F}H=id$. However, $\mathbf{F}H$ is assumed to be a diffeomorphism, so $\mathbf{F}L=\mathbf{F}L\circ(\mathbf{F}H\circ(\mathbf{F}H)^{-1})=(\mathbf{F}L\circ\mathbf{F}H)\circ(\mathbf{F}H)^{-1}=(\mathbf{F}H)^{-1}$, as required. ■

The reader can, as an exercise, translate this proof into one that works in infinite dimensions by writing

$$\begin{aligned}G(e,\alpha)&=\langle\alpha,D_2H(e,\alpha)\rangle\\(L\circ\mathbf{F}H)(e,\alpha)&=\langle\alpha,D_2H(e,\alpha)\rangle-H(e,\alpha)\quad\text{etc.}\end{aligned}$$

ISBN 0-8053-0102-X

Dual to 3.6.7 is:

3.6.8 Proposition. *Let L by a hyperregular Lagrangian on TQ and let $H=E\circ(FL)^{-1}$ be defined as in 3.6.2. Then H is a hyperregular Hamiltonian and $FH=(FL)^{-1}$.*

Proof. Consider the map FL: $(q^i,\dot q^j)\mapsto(q^i,\partial L/\partial\dot q^j)=(q^i,p_j)$. Then

$$
\begin{aligned}
H\circ FL &= E = A - L\\
&= \dot q^i\frac{\partial L}{\partial \dot q^i} - L = \dot q^i p_i - L
\end{aligned}
$$

Thus, by the chain rule, exactly as in the remarks following 3.6.3, we get

$$\frac{\partial H}{\partial p_i}=\dot q^i+\frac{\partial \dot q^k}{\partial p_i}p_k-\frac{\partial L}{\partial \dot q^k}\frac{\partial \dot q^k}{\partial p_i}=\dot q^i$$

Hence $FH\circ FL=id$, so as FL is a diffeomorphism, $FH=(FL)^{-1}$. ■

We are now ready to state the main result.

3.6.9 Theorem. *The hyperregular Lagrangians L on TQ and hyperregular Hamiltonians H on T^*Q correspond in a bijective manner: H is constructed from L by means of 3.6.2 and L from H by means of 3.6.7. The following diagrams commute:*

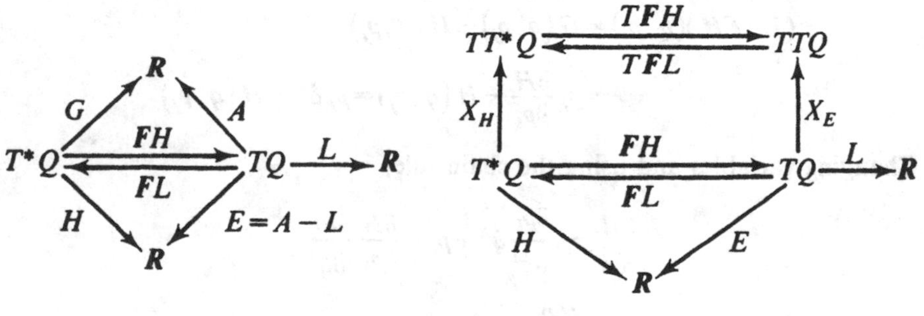

Proof. Let L be a given hyperregular Lagrangian and let H be constructed from it by 3.6.2. Thus, by 3.6.3 and 3.6.8,

$$H=E\circ(FL)^{-1}=(A-L)\circ FH=G-L\circ FH$$

From H we construct $\tilde L$ using 3.6.7:

$$
\begin{aligned}
\tilde L &= G\circ(FH)^{-1}-H\circ(FH)^{-1}\\
&= G\circ(FH)^{-1}-(G-L\circ FH)\circ(FH)^{-1}\\
&= L
\end{aligned}
$$

ISBN 0-8053-0102-X

Conversely, let H be a given hyperregular Hamiltonian and construct L by 3.6.7. From L we construct a Hamiltonian $\tilde{H}$ by 3.6.2.

We note that in 3.6.7, since $(FH)^{-1} = FL$,

$$A = G\circ(FH)^{-1} = \theta_0(X_H)\circ(FH)^{-1} = \theta_0(X_H)\circ FL$$
$$= \theta_0\left(X_{E\circ FL^{-1}}\right)\circ FL$$

so by 3.6.3, A is the action of L. Then $E = H\circ FL = A - L$ is the energy of L. Therefore, from the definition of $\tilde{H}$,

$$\tilde{H} = E\circ(FL)^{-1} = H$$

This shows that the map $L \mapsto H$ from hyperregular Lagrangians to hyperregular Hamiltonians given by 3.6.2 and 3.6.8 and the corresponding map $H \mapsto L$ given by 3.6.7 are inverses and hence each is a bijection. The commutativity of the diagrams now follows from what we have already proven. ■

The main examples will come in the next section, but for now we give this one:

3.6.10 Example. Let $\langle,\rangle$ be a pseudo-Riemannian metric on a manifold Q and let $L\colon TQ\to \mathbf{R}$ be given by $L(v) = \frac{1}{2}\langle v,v\rangle$. Then $FL\colon TQ \to T^*Q$ is $FL(w)\cdot v = \langle w,v\rangle$; that is, FL is the "flat map" associated to the metric. Thus L is a hyperregular Lagrangian. The action of L is $A(v) = FL(v)\cdot v = \langle v,v\rangle = 2L$, so the energy is $E = A - L = L$. The corresponding Hamiltonian $H\colon T^*Q \to \mathbf{R}$ is given by $H(\alpha) = \frac{1}{2}\langle \alpha^\sharp, \alpha^\sharp\rangle$, where $\alpha^\sharp$ is the vector corresponding to α by $(FL)^{-1}$; that is, $FL(\alpha^\sharp) = \alpha$. In coordinates $q^1,\ldots,q^n$,

$$L(v) = \tfrac{1}{2} g_{ij} v^i v^j$$

where g_{ij} is the matrix of the metric and $v^1,\ldots,v^n$ are the components of v. Also,

$$H(\alpha) = \tfrac{1}{2} g^{ij}\alpha_i\alpha_j$$

where g^{ij} is the inverse matrix of g_{ij}, as is readily checked.

The construction of Hamiltonian systems out of degenerate Lagrangian systems is more complicated. This theory, due to Dirac [1950] (in a formulation due to W. Tulczyjew), is given in Exercise 5.3L.

EXERCISES

3.6A. Let $\langle,\rangle$ be a pseudo-Riemannian metric on Q and let L be defined as in 3.6.10. Show that v is a critical point of E if and only if $v = 0$.

3.6B. Let $\langle,\rangle$ be a pseudo-Riemannian metric on Q, $V\colon Q \to \mathbf{R}$ and Y a vector field

ISBN 0-8053-0102-X

on Q. Let

$$L(v)=\tfrac{1}{2}\langle v,v\rangle+V(\tau_Q v)+\langle v,Y(\tau_Q v)\rangle$$

Show that L is hyperregular. What is the corresponding Hamiltonian? Write down Hamilton's equations in local representation.

3.6C. (The harmonic oscillator). Let $Q=\boldsymbol{R}$ and $H\in\mathfrak{F}(T^*Q)$ be given by $H(q,p)=\frac{1}{2}(p^2+\omega^2q^2)$. Find X_H, its flow and FH. Set up the corresponding Lagrangian and show that its solutions correspond to those of X_H via the Legendre transform.

3.6D. Show that the energy of a Lagrangian $L(q,\dot q)$ may be written

$$E(q,\dot q)=\tfrac{1}{2}A_{ij}(q,\dot q)\dot q^i\dot q^j+V(q)$$

where

$$A_{ij}(q,\dot q)=2\int_0^1 t\frac{\partial^2 L}{\partial\dot q^i\partial\dot q^j}(q,t\dot q)\,dt$$

and

$$V(q)=-L(q,0)$$

3.7 MECHANICS ON RIEMANNIAN MANIFOLDS

We begin this section with a simple, but rather basic result, namely, that geodesics are base integral curves of a Lagrangian (or Hamiltonian) system.

3.7.1 Theorem. *Let Q be a pseudo-Riemannian manifold and let $L\colon TQ\to\boldsymbol{R}$ be defined by $L(v)=\frac{1}{2}\langle v,v\rangle$ (see Example 3.6.10). Then $c_0(t)$ is a base integral curve of X_E, the Lagrangian vector field of L, if and only if $c_0(t)$ is a geodesic.*

Proof. In coordinates $(q^1,\ldots,q^n)$, we have $L(v)=\frac{1}{2}g_{ij}v^iv^j$, so Lagrange's equations for a base integral curve $c_0(t)=(q^1(t),\ldots,q^n(t))$, namely,

$$\frac{d}{dt}\frac{\partial L}{\partial\dot q^i}=\frac{\partial L}{\partial q^i},\qquad i=1,\ldots,n$$

become

$$\frac{d}{dt}(g_{ij}\dot q^j)=\frac{1}{2}\frac{\partial g_{jk}}{\partial q^i}\dot q^j\dot q^k$$

that is,

$$g_{ij}\ddot q^j=\left(\frac{1}{2}\frac{\partial g_{jk}}{\partial q^i}-\frac{\partial g_{ij}}{\partial q^k}\right)\dot q^j\dot q^k$$

ISBN 0-8053-0102-X

Relabeling indices and multiplying by g^{li} gives:

$$\ddot{q}^i = g^{li}\left(\frac{1}{2}\frac{\partial g_{jk}}{\partial q^l} - \frac{\partial g_{lj}}{\partial q^k}\right)\dot{q}^j\dot{q}^k$$

These are in fact equivalent to the geodesic equations

$$\ddot{q}^i + \Gamma^i_{jk}\dot{q}^j\dot{q}^k = 0$$

Indeed, from Sect. 2.7 we have the formula

$$\Gamma^i_{jk} = \tfrac{1}{2} g^{il}(g_{lk,j} + g_{lj,k} - g_{jk,l})$$

and so

$$\Gamma^i_{jk}\dot{q}^j\dot{q}^k = \tfrac{1}{2} g^{il}(g_{lk,j} + g_{lj,k} - g_{jk,l})\dot{q}^j\dot{q}^k$$

$$= g^{il}\left(g_{lj,k} - \tfrac{1}{2} g_{jk,l}\right)\dot{q}^j\dot{q}^k$$

since the first two terms give the same result in view of symmetry of $\dot{q}^j\dot{q}^k$. ■

The reader is invited to give a proof that works in infinite dimensions starting with the defining condition on $X_E(x,e) = (e, Y(x,e))$ in charts given in 3.5.17, namely

$$D_2D_2L(x,e)\cdot Y\cdot e_1 = D_1L(x,e)\cdot e_1 - D_1D_2L(x,e)\cdot e\cdot e_1$$

3.7.2 Definitions. *Let $\langle,\rangle$ be a pseudo-Riemannian metric on Q and let X_E be the associated Lagrangian vector field for $L(v) = \frac{1}{2}\langle v, v\rangle$. We call X_E the **geodesic spray** and its flow is called the **geodesic flow**.*

The characteristic property of a spray X on TQ is that it is a second-order equation quadratic in the velocities. Sprays are also characterized by this homogeneity property of their flows: $\tau_Q F_t(\lambda v) = \tau_Q F_{t\lambda}(v)$, as is easily seen. Earlier (2.7.6) we defined the exponential map of a metric. This can be done the same way for general sprays and the existence of tubular neighborhoods carries over unchanged (see Lang [1972] for details). One can also use sprays, specifically X_E, to reconstruct the connection of a metric. Indeed, in coordinates,

$$X_E(q^i, \dot{q}^j) = (\dot{q}^i, -\Gamma^j_{kl}\dot{q}^k\dot{q}^l)$$

so X_E carries the same information as the Christoffel symbols, since $\Gamma^j_{kl} = \Gamma^j_{lk}$. In other words, one can start with the Hamiltonian approach as basic and define the covariant derivative, and so forth, in terms of this structure.

ISBN 0-8053-0102-X

One other general remark: The vector field X_E shows where the Γ^i_{jk} properly live as geometric objects, namely, in T^2Q, since $X_E: TQ \to T^2Q$. If the reader will write down the vector field transformation property for X_E, the transformation rule for Christoffel symbols will result.

3.7.3 Proposition. *Let $\Sigma_e = \{v \in TQ \mid \frac{1}{2}\|v\|^2 = e\}$, where $e > 0$ and Q is a pseudo-Riemannian manifold. We call Σ_e a **sphere (or pseudosphere) bundle.** Then $\Sigma_e \subset TQ$ is a smooth submanifold and is invariant under the geodesic flow.*

Proof. Consider the energy function $E: TQ \to \mathbf{R}$, $E(v) = \frac{1}{2}\|v\|^2$. Then $dE(v) \cdot w = \langle v, w \rangle$, so if $v \neq 0$, $dE(v)$ is surjective. In particular, $e > 0$ is a regular value of E, so $\Sigma_e = E^{-1}(e)$ is a regular energy surface. Invariance of Σ_e is a restatement of conservation of energy. ■

For further information, see Sect. 3.4 and Problem 3.7B. We shall return to motion on energy surfaces in our discussion of the virial theorem below.

Next we consider particles moving on the pseudo-Riemannian manifold Q in the presence of a potential $V: Q \to \mathbf{R}$. We take $L: TQ \to \mathbf{R}$ of the form

$$L(v) = \tfrac{1}{2}\|v\|^2 - V(\tau_Q v)$$

Then L is a hyperregular Lagrangian. In fact, as in the geodesic case, $\mathbf{F}L(v) \cdot w = \langle v, w \rangle$. Thus the action is $A(v) = \|v\|^2$ and so $E = A - L$ is given by

$$E(v) = \tfrac{1}{2}\|v\|^2 + V(\tau_Q v)$$

3.7.4 Proposition. *With L as just defined, $c_0(t)$ is a base integral curve of X_E, that is, satisfies Lagrange's equations if and only if*

$$\nabla_{\dot{c}_0} \dot{c}_0 = -(\operatorname{grad} V)(c_0(t))$$

where ∇ is the covariant derivative. (It is understood that ∇ stands for the Levi–Cività connection of the metric, as in Sect. 2.7.)

Proof. The quickest proof is, again, a coordinate one. If we examine the proof of 3.7.1, the only change necessary is to add $-\partial V/\partial q^i$ to the geodesic equations. Thus the Lagrange equations become

$$\frac{d}{dt}\left(g_{ij}\dot{q}^j\right) = \frac{1}{2}\frac{\partial g_{jk}}{\partial q^i}\dot{q}^j\dot{q}^k - \frac{\partial V}{\partial q^i}$$

This simplifies, by the proof of 3.7.1 to

$$\ddot{q}^i + \Gamma^i_{jk}\dot{q}^j\dot{q}^k = -g^{ij}\frac{\partial V}{\partial q^j}$$

ISBN 0-8053-0102-X

This is exactly a coordinate statement of the conclusion of the proposition. ■

3.7.5 Definition. *Let $v, w \in T_xM$. Then the* **vertical lift** *of w with respect to v is defined by*

$$(w)_v^l = \frac{d}{dt}(v + tw)\Big|_{t=0} \in T_v(TM)$$

(Recall that the tangent vector to a curve at a point n in a manifold N is an element of T_nN.) The **horizontal part** *of a vector $z \in T_v(TM)$ is $T\tau_M \cdot z \in T_{\tau_M v}M$.*

In natural charts it is easy to see that if $v=(x,e)$ and $w=(x,e_1)$, then

$$(w)_v^l = ((x,e),(0,e_1))$$

The horizontal part of $z = ((x, e), (e_1, e_2))$ is given by

$$T\tau_M \cdot z = (x, e_1)$$

It is easy to check that a vector in TTM is the vertical lift of something if and only if its horizontal part is zero.

If we let S_Q denote the geodesic spray, then the result of 3.7.4 can be written

$$X_E = S_Q - (\mathrm{grad}\, V)^l$$

by which we mean

$$X_E(v) = S_Q(v) - (\mathrm{grad}\, V(\tau_Q v))_v^l$$

Remarks. In general there is no canonical way of taking the vertical part of a vector $z \in T_vTQ$. However, if we use a pseudo-Riemannian structure we can construct such a map. Namely, let $w(t)$ be a curve in TQ tangent to z at v. By parallel translation, we get a curve $\tilde{w}(t) \in T_xQ$, where $x = \tau_Q v$ and $\tilde{w}(0) = v$. Then $(d/dt)\tilde{w}(t)|_{t=0}$ is clearly vertical and is the vertical lift of the *vertical part* of z. In coordinates, if

$$z = (x, v^i, u^j, w^k)$$

then one computes from the results of Sect. 2.7,

$$(z_{vert})_v^l = (x, v^i, 0, \Gamma^i_{jk}u^jv^k + w^i)$$

The maps $z \mapsto z_{horiz} = T\tau_Q \cdot z$ and $z \mapsto z_{vert}$ then give an isomorphism

$$T_vTQ \approx T_xQ \oplus T_xQ$$

ISBN 0-8053-0102-X

The next result will be the important theorem of Jacobi, which states that motion in a potential is actually geodesic motion using a modified metric.

3.7.6 Definition. *Let $\langle , \rangle$ be a pseudo-Riemannian metric on Q and $V: Q \to \mathbf{R}$ be bounded above (if it is not, confine attention to a compact subset of Q). Let $e > V(x)$ for $x \in Q$. Define the **Jacobi metric** by*

$$g_e = (e - V)g$$

where $g(x)(\cdot, \cdot) = \langle \cdot, \cdot \rangle_x$ is the original metric.

3.7.7 Theorem. *The base integral curves of the Lagrangian $L(v) = \frac{1}{2}\langle v, v\rangle - V(\tau_Q v)$ with energy e are the same as geodesics of the Jacobi metric up to a reparametrization, with energy 1.*

We shall give a proof due to Godbillon [1969], simplified by A Weinstein. For a geometric proof starting from 3.7.4, see Ong [1975]. For Jacobi's proof using the calculus of variations, see Sect. 3.8.

Proof. It is more convenient to prove the result on T^*Q so that each system uses the same symplectic structure. On T^*Q, let $g^\#$ denote the metric corresponding to g (in coordinates, $g^\#$ has components g^{ij}, the inverse matrix to g_{ij}). Thus, the Hamiltonian on T^*Q corresponding to L is

$$H(\alpha) = \tfrac{1}{2} g^\#(\alpha, \alpha) + V(\tau_Q^* \alpha)$$

Similarly, the Hamiltonian for g_e is

$$H_e(\alpha) = \frac{1}{2} \frac{1}{\left[e - V(\tau_Q^* \alpha)\right]} g^\#(\alpha, \alpha)$$

The proof now consists of two remarks:

(a) $H^{-1}(e) = H_e^{-1}(1)$ (i.e., the two energy surfaces, regular by 3.7.3, coincide). Indeed, $\frac{1}{2} g^\#(\alpha, \alpha) + V(\tau_Q^* \alpha) = e$ is the same as

$$\frac{1}{2} \frac{1}{\left[e - V(\tau_Q^* \alpha)\right]} g^\#(\alpha, \alpha) = 1$$

(b) If two Hamiltonians H_1 and H_2 on a symplectic manifold (P, ω) have $H_1^{-1}(e_1) = H_2^{-1}(e_2) = \Sigma$, and Σ is a regular energy surface, then X_{H_1} and X_{H_2} have the same integral curves on Σ up to reparametrization.

Indeed, since Σ is of codimension one and ω is nondegenerate, $\{v \in T\Sigma \mid i_v \omega = 0\}$ is one dimensional at each point of Σ (see 3.1.2). But $i_{X_{H_1}} \omega = dH_1 = 0$ on Σ, and similarly $i_{X_{H_2}} \omega = 0$. Thus X_{H_1} and X_{H_2} are parallel and thus their integral curves are the same up to reparametrization. (See Exercise 3.4D.) ■

ISBN 0-8053-0102-X

Next we shall study constrained systems. (More properly, systems with "holonomic" constraints.) A typical constrained system consists of a collection of particles moving subject to given forces and subject to certain constraining relations. For example, in R^3 we might consider two particles connected by a (light) rod. In such cases, there arise forces of constraint such as centripetal forces. The constrained system par excellence is the rigid body, which is discussed in Sect. 4.4. Another one is the motion of an incompressible fluid in which the pressure gradient is the force of constraint—here the constraint is incompressibility; see Sect. 5.6.

From the variational point of view, the force of constraint appears as a Lagrange multiplier. We invite the reader to develop this method after reading Sect. 3.8 as an alternative to the present purely geometric approach.* Our set-up is as follows:

Let M be a Riemannian manifold and $N \subset M$ a submanifold. Let P: $TM|N \to TN$ be the bundle map defined by letting P_x: $T_xM \to T_xN$, $x \in N$ be the orthogonal projection of T_xM onto T_xN. Let V: $M \to R$ be given and let V_N be its restriction to N. From the definitions one checks that at points of N,

$$grad\, V_N = P \cdot (grad\, V)$$

Thus a particle moving on N according to the Lagrangian $L_N = L|N$, where $L(v) = \frac{1}{2}\langle v,v\rangle - V(\tau_M v)$ feels an external force that is just the projection of the force in T_xM onto T_xN.

Of more interest is the relationship between the geodesic sprays S_M on M and S_N on N. Although this relationship may be regarded as in the domain of Riemannian geometry, we can give a short proof using mechanics.

3.7.8 Theorem. *With the notations just described,*

$$S_N = TP \circ S_M$$

at points of TN.

Proof. Let E be the bundle TM restricted to N. Thus P: $E \to TN$, so TP: $TE \to T^2N$. Now S_M: $E \to T^2M$ and $T\tau \circ S_M = identity$, since S_M is a second-order equation. But $TE = \{w \in T^2M \mid T\tau(w) \in TN\}$ as is easy to see, so the composition $TP \circ S_M$ makes sense at points of TN.

Let θ_M and θ_N be the one-forms on TM and TN induced by their Riemannian structures [note that $\theta_M = (FL)^*\theta_0$, where θ_0 is the canonical one-form on T^*M and similarly for θ_N]. Thus $\theta_M(v) \cdot w = \langle v, T\tau \cdot w\rangle$ for $v \in TM$, $w \in T_vTM$. Let i: $E \to TM$ be inclusion and $\theta_E = i^*\theta_M$. We claim that

$$P^*\theta_N = \theta_E$$

*Constraints also arise naturally in degenerate Lagrangian systems. This theory (the Dirac theory of constraints) proceeds along somewhat different lines. See Exercise 5.3L.

ISBN 0-8053-0102-X

Indeed, for $v \in TM$, $w \in T_v E$, we have

$$
\begin{aligned}
(P^*\theta_N)_v \cdot w &= \langle Pv, T\tau \cdot TP \cdot w \rangle \\
&= \langle Pv, T(\tau \circ P) \cdot w \rangle \\
&= \langle v, PT\tau \cdot w \rangle \\
&= \langle v, T\tau \cdot w \rangle = \theta_E(v) \cdot w
\end{aligned}
$$

Here we have used $\tau \circ P = \tau$, the fact that P is symmetric and $T\tau \cdot w \in TN$ (since $\tau: E \to N$).

It follows that

$$P^*\omega_N = \omega_E$$

where $\omega_N = -d\theta_N$, $\omega_E = i^*\omega_M = -i^*d\theta_M = -d\theta_E$.

In particular, we have, for $v \in TN$, $w \in T_v E$, and $z \in T_v TN$,

$$\omega_N(TP \cdot w, z) = \omega_M(w, z)$$

since $TP \cdot z = z$. Since S_M and S_N are both Hamiltonian vector fields of $L(v) = \frac{1}{2}\langle v, v \rangle$, we have

$$dL \cdot z = \omega_M(S_M(v), z) = \omega_N(S_N(v), z)$$

Combining this and the previous formula gives the result. ■

3.7.9 Corollary. *For $v \in T_x N$,*

(a) $(S_M - S_N)(v)$ is the vertical lift of a vector $Z(v) \in T_x M$
(b) $Z(v)$ is orthogonal to $T_x N$
(c) $Z(v) = -\nabla_v v + P(\nabla_v v)$ (= normal component of $\nabla_v v$) where, in the right-hand side, v is extended to a vector field on M tangent N.

Proof. (a) This is clear since $T\tau \cdot S_M(v) = v = T\tau \cdot S_N(v)$.

(b) For $u \in T_x M$, we have $TP \cdot (u)_v^l = (Pu)_v^l$ since

$$(Pu)_v^l = \frac{d}{dt}(v + tPu)_{t=0} = \frac{d}{dt}P(v + tu)\bigg|_{t=0} = TP \cdot (u)_v^l$$

Thus

$$
\begin{aligned}
(PZ(v))_v^l &= TP \cdot (Z(v))_v^l = TP(S_M(v) - S_N(v)) \\
&= TP(S_M(v) - TPS_M(v)) = 0
\end{aligned}
$$

ISBN 0-8053-0102-X

since $P \circ P = P$. Hence $PZ(v) = 0$, so (b) holds.

(c) Let $v(t)$ be a curve of tangents to N; $v(t) = \dot{c}(t)$, $c(t) \in N$. Then, from 3.7.4,

$$S_M(v) = \frac{dv}{dt} - (\nabla_v v)_v'$$

so that by 3.7.8

$$S_N(v) = TP\frac{dv}{dt} - TP(\nabla_v v)_v'$$

$$= \frac{d}{dt}Pv - (P\nabla_v v)_v'$$

$$= \frac{dv}{dt} - (P\nabla_v v)_v'$$

Thus $S_M(v) - S_N(v) = -(\nabla_v v - P\nabla_v v)_v'$, so (c) holds. ■

This object $Z(v)$ is called the *force of constraint*. [It is also (the quadratic part of) the second fundamental form of N in M (if N is codimension 1)]. We may interpret $-Z(v)$ as the constraining force needed to keep particles in N. Clearly N is totally geodesic (i.e., geodesics in N are also geodesics in M) iff $Z = 0$.

For example, if $N \subset \mathbb{R}^2$ is a circle of radius $r > 0$, the force of constraint at v is the radial vector of length $\|v\|^2/r$, which is just the centrifugal force.

The result 3.7.9(b) is the famous principle of d'Alembert: *if a particle is constrained to move on a surface, the force of constraint is perpendicular to that surface.*

We now turn to questions of completeness for Lagrangian systems of the type $E = K + V$, where K is the kinetic energy, $K(v) = \frac{1}{2}\|v\|^2$, and $V(\tau_Q v)$ is the potential energy.

We shall begin by proving a result for Riemannian manifolds. Some discussion of this occurred in Sect. 2.7, but here we shall illustrate some methods using the Lagrangian point of view.

3.7.10 Definitions. (*i*) *A pseudo-Riemannian manifold Q is called **complete** if its geodesic spray S_Q is complete in the sense of a complete vector field (Sect. 2.1).*

(*ii*) *A pseudo-Riemannian manifold Q is called **homogeneous** if for $x, y \in Q$, there is an isometry $\Phi: Q \to Q$ such that $\Phi(x) = y$.*

3.7.11 Proposition. (*i*) *Any compact Riemannian manifold is complete.*

(*ii*) *Any homogeneous Riemannian manifold is complete.*

Remark. It is also true that a *compact homogeneous* pseudo-Riemannian manifold is complete, as we shall prove in Chapter 4.

ISBN 0-8053-0102-X

Proof of 3.7.11 (i) Let $e>0$ and $\Sigma_e=\{v\in TQ\,|\,\frac{1}{2}\|v\|^2=e\}$. It is a compact subset of TQ. By conservation of energy, any integral curve of the geodesic spray S_Q lies in such a set. Hence (i) follows from 2.1.18.

(ii) To prove this, it is convenient to have the following property of sprays:

3.7.12 Lemma *Let $S: TQ\to T^2Q$ be a spray. Let U be any bounded set in T_xQ. Then there is an $\varepsilon>0$ such that for any $v\in U$, the integral curve of S with initial condition v exists for a time $\geqslant\varepsilon$.*

Proof. There is a neighborhood V of 0 in T_xQ and a $\delta>0$ such that integral curves with initial data in V exist for a time $\geqslant\delta$. There is a constant $R>0$ such that $R^{-1}U\subset V$ since U is bounded. Thus by the homogeneity property of sprays, $\tau F_t(kv)=\tau F_{kt}(v)$, initial data in U are propagated for a time $\geqslant\delta/R$. ▼

3.7.13 Lemma *Let Φ: $Q\to Q$ be an isometry. Then Φ maps geodesics to geodesics.*

Proof. Since Φ is an isometry, $T\Phi$ preserves the one-form $\Theta=\theta_L$ on TQ associated with the metric. Thus $T\Phi$ is a canonical transformation, so by 3.3.19 $(T\Phi)_*X_E=X_{E\circ T\Phi^{-1}}=X_E$. But the integral curves of $(T\Phi)_*X_E$ are the images of integral curves of X_E, so $T\Phi$ maps the geodesic flow to itself and hence geodesics to geodesics. ▼

To prove (ii) of 3.7.11, let $v\in T_xQ$ and let U be an R-disk in T_xQ containing v. Choose ε as in 3.7.12. By assumption there is an isometry Φ mapping x to $x(\varepsilon)$, the base point of $v(\varepsilon)$, where $v(t)$ is the integral curve of the geodesic spray starting at v. But by 3.7.13, the geodesic starting at $x(\varepsilon)$ in direction $v(\varepsilon)$ is Φ applied to the geodesic at x in direction $T\Phi^{-1}\cdot v(\varepsilon)$. This lies in U, so the geodesic exists for time $\geqslant\varepsilon$. Thus $v(2\varepsilon)$ is defined. Repeating, $v(t)$ is defined for all $t\in\mathbf{R}$. ■

Now we consider the effect that adding a potential to the kinetic energy of a complete Riemannian manifold Q has on completeness. In many examples involving several particles, one removes points from Q corresponding to collisions, thereby making Q incomplete. Here we are instead concerned with the behavior of V at infinity that allows completeness.

First of all, if Q is complete Riemannian and V is bounded below, it is easy to see that if $E=K+V$, X_E is complete, at least if Q is finite dimensional. Indeed, let $c_0(t)$ be a base integral curve of X_E defined for $t\in(-T,T)$, $T<\infty$. Since E is constant along $c_0(t)$ and V is bounded below, $\|\dot{c}_0\|$ is bounded above, say, by α. Thus $c_0(t)$ lies in the ball $B_{T\alpha}$ of radius $T\alpha$ about $c_0(0)$, so $\dot{c}_0(t)\in\{v\in TM\,|\,\tau(v)\in B_{T\alpha}$ *and* $\|v\|\leqslant\alpha\}$. This is a compact set because Q is complete. Hence the base integral curve $c_0(t)$ must be extend-

ISBN 0-8053-0102-X

able in time beyond $(-T,T)$. Thus c_0 is defined for all time, by the maximal extendability of integral curves proved in Sect. 2.1.

We want to prove a generalization of this to allow V to be unbounded below.

3.7.14 Definition. *Let $V_0: [0,\infty)\to R$ be given. We call it* ***positively complete*** *if it is nonincreasing, C^2, and*

$$\int_0^\infty \frac{dx}{\sqrt{e-V_0(x)}}=+\infty$$

where $e>V_0(x)$ for all x. (This condition is easily seen to be independent of which e is chosen.)

For example, $V_0(x)=-x^\alpha$ is positively complete if $0\le\alpha\le 2$, as is $V_0(x)=-x[\log(x+1)]^\alpha$, $-x\log(x+1)[\log(\log(x+1)+1)]^\alpha$ etc.

3.7.15 Theorem. *Let Q be a complete Riemannian manifold, $V: Q\to R$ be C^2 and X_E the Lagrangian vector field for $L(v)=\frac{1}{2}\|v\|^2-V(\tau_Q v)$. Suppose there is a positively complete function V_0 such that for some $x_0\in Q$,*

$$V(x)\geq V_0(d(x,x_0))$$

and $d(x,x_0)$, the distance between x and x_0 is sufficiently large. Then X_E is complete.

Remark. This form of the theorem is due to Weinstein and Marsden [1970], which was, in turn, inspired by Gordon [1970b] and Ebin [1970b]. Related results are due to Lelong-Ferrand [1959] and Maslov [1965, p. 329]. We shall prove 3.7.15 for Q finite dimensional; for the general case, see Exercise 3.7C.

This theorem is the classical version of a quantum mechanical completeness theorem due to Ikebe and Kato [1962]. (Utilizing methods of Roeleke [1960], their proof extends from R^n to complete Riemannian manifolds.)

For example, on R^n if $V(x)\geq a-b\|x\|^2$, then X_E is complete. This in turn holds if $\|\mathrm{grad}\,V(x)\|\leq c\|x\|$. [Since $V(x)=V(x_0)+\int_\sigma \mathrm{grad}\,V\cdot\sigma'$, where σ is a path joining x to x_0.] The reader can formulate similar sufficient conditions on a Riemannian manifold Q without difficulty.

Proof of 3.7.15: Let $c(t)$ be an integral curve of X_E and $c_0=\tau_Q\circ c$ its base integral curve. Let $y=c_0(0)$ and let $f(t)$ be the solution of the differential equation

$$f''(t)=-\frac{dV_0}{ds}(f(t))$$

ISBN 0-8053-0102-X

with initial conditions $f(0)=d(x_0,y)$, $f'(0)=\sqrt{2(\beta-V_0(f(0)))}$, where $\beta=E(c(0))\geqslant V_0(f(0))$. Now β is also the energy of the curve $f(t)$: $\beta=\frac{1}{2}[f'(t)]^2+V_0(f(t))$. Actually, we can assume $\beta>V_0(f(0))$; that is, $\|\dot{c}(0)\|\neq 0$, for if $\dot{c}(t)$ is always zero, the problem is trivial; so start at t_0, where $\dot{c}(t_0)\neq 0$. The time it takes for $f(t)$ to increase from $f(0)$ to s is $\int_{f(0)}^{s} dx/\sqrt{2[\beta-V_0(s)]}$. By our assumption on V_0, it follows that $f(t)$ is defined for all $t\geqslant 0$.

Now let $g(t)=d(c_0(t),x_0)$ and $t\geqslant 0$. Then

$$
\begin{aligned}
g(t)&\leqslant d(c_0(t),y)+d(y,x_0)\\
&\leqslant d(x_0,y)+\int_0^t\|\dot{c}_0(s)\|\,ds\\
&\leqslant d(x_0,y)+\int_0^t\sqrt{2[\beta-V(c_0(s))]}\,ds
\end{aligned}
$$

by conservation of energy. Thus by our hypothesis,

$$
g(t)\leqslant d(x_0,y)+\int_0^t\sqrt{2[\beta-V_0(g(s))]}\,ds
$$

But

$$
f(t)=d(x_0,y)+\int_0^t\sqrt{2[\beta-V_0(f(s))]}\,ds
$$

Hence by the comparison lemma (Exercise 2.2H), $g(t)\leqslant f(t)$. Hence $c_0(t)$ remains in a compact set for finite t-intervals, so $c(t)$ does as well since V is bounded below on such a set. It follows that X_E is $+$complete. However, from reversibility, namely, $\tau_Q(F_{-t}(v))=\tau_Q(F_t(-v))$, it follows that X_E is $-$complete as well. ■

Next we shall define the concept of a dissipative system. (See also Exercise 3.7A. and Exercise 3.8F.)

3.7.16 Definition. *Let Q be a Riemannian manifold and K the kinetic energy function, $K(v)=\frac{1}{2}\|v\|^2$. A vector field Y on TQ is called **dissipative** if Y is vertical (i.e., $T\tau_Q\circ Y=0$) and $dK\cdot Y\leqslant 0$. By a **dissipative system** we mean a vector field X on TQ of the form*

$$
X=X_E+Y
$$

where $E(v)=K(v)+V(\tau_Q v)$ and Y is dissipative.

3.7.17 Proposition *Let $X=X_E+Y$ be a dissipative system on TQ. Then E is nonincreasing along integral curves of X.*

ISBN 0-8053-0102-X

Proof. Let $c(t)$ be an integral curve of X. Then

$$\begin{aligned}\frac{d}{dt}E(c(t)) &= dE(c(t))\cdot X(c(t))\\ &= dE(c(t))\cdot X_E(c(t)) + dE(c(t))\cdot Y(c(t))\\ &= dE(c(t))\cdot Y(c(t))\\ &= dK(c(t))\cdot Y(c(t)) \leqslant 0\end{aligned}$$

since $dE(c(t))\cdot X_E(c(t)) = \omega_L(X_E(c(t)), X_E(c(t))) = 0$, and $d(V\circ\tau_Q)\cdot Y = dV\cdot(T\tau_Q\circ Y) = 0$ as Y is vertical. ■

If E is strictly decreasing then closed orbits, common for Hamiltonian systems, cannot occur. If E is bounded below, orbits will generally converge to critical points of E, i.e. to equilibria of the associated Hamiltonian system.

3.7.18 Proposition. *Let the hypotheses of* 3.7.15 *hold and let Y be dissipative. Then* $X = X_E + Y$ *is positively complete.*

Proof. Note that X is a second-order equation since Y is vertical. Now we observe that, in view of 3.7.17, the proof of 3.7.15 carries over unchanged (omitting the last step on reversibility). ■

For example, frictional forces that depend linearly on the velocity are taken into account by choosing $Y(v) = -(\kappa v)_v^{\prime}$, where $\kappa > 0$ is a constant. The reader will see easily that Y is dissipative.

We now show that we can pass back and forth between configuration spaces Q and $Q\times\mathbf{R}$ by using (degenerate) homogeneous Lagrangians. This will be illustrated by considering the equations for a relativistic particle.

Let $L: TQ\to\mathbf{R}$ be a regular Lagrangian. Define $\bar{L}: T(Q\times\mathbf{R}) = TQ\times T\mathbf{R}\to\mathbf{R}$ by

$$\bar{L}(v_x,(t,\lambda)) = \lambda L(v_x/\lambda) \qquad (\lambda>0)$$

Thus $\bar{L}$ is homogeneous (of degree one), that is, $\bar{L}(sv_x,(t,s\lambda)) = s\bar{L}(v_x,(t,\lambda))$. From Euler's theorem on homogeneous functions, or directly, we see that $\bar{E} = 0$, where $\bar{E}$ is the energy of $\bar{L}$.

Let t: $\mathbf{R}\times\mathbf{R} = T\mathbf{R}\to\mathbf{R}$ be the projection onto the first factor, E be the energy of L, and $\theta_L = (\mathbf{F}L)^*\theta_0$, as usual.

3.7.19 Proposition. *We have (i)* $\theta_{\bar{L}}(v,t,\lambda) = \theta_L(v/\lambda) - E(v/\lambda)\,dt$.

(ii) The possible second-order Lagrangian vector fields $X_{\bar{E}}$ *for* $\bar{L}$ *are characterized as follows: Let* $\tau(t)$ *be a strictly increasing smooth function of t. Let* $(q(t),\dot{q}(t))$ *be an integral curve for* X_E*, the Lagrangian vector field for L. Then*

ISBN 0-8053-0102-X

$(q(\tau(t)), \dot{q}(\tau(t)), \tau(t), \dot{\tau}(t)) \in TQ \times TR$ is an integral curve for $X_{\bar{E}}$. Different choices of $\tau(t)$ yield the different possible choices of $X_{\bar{E}}$.

Proof of 3.7.19. The definition of $\bar{L}$ yields

$$F\bar{L}(v,(t,\lambda))\cdot(w,(t,s)) = FL(v/\lambda)\cdot w - sE(v/\lambda)$$

for $v, w \in T_xQ$. Now $\theta_{\bar{L}}(v, t, \lambda) = F\bar{L}(v, t, \lambda)\circ T(\tau_Q \times t)$ and combining this with the expression for FL we get (i).

To prove (ii) we use the fact that $X_{\bar{E}}$ is determined by Lagrange's equations, under the assumption that $X_{\bar{E}}$ is second order (see 3.5.17). The condition that $(q(\tau(t)), \dot{q}(\tau(t)), \tau(t), \dot{\tau}(t))$ be a solution to Lagrange's equations is

$$\text{(a)}\quad \frac{d}{dt}\left(\frac{\partial \bar{L}}{\partial \dot{q}}\right) = \frac{\partial \bar{L}}{\partial q} \quad\text{and}\quad \text{(b)}\quad \frac{d}{dt}\left(\frac{\partial \bar{L}}{\partial \dot{\tau}}\right) = \frac{\partial \bar{L}}{\partial \tau}$$

that is,

$$\frac{d}{dt}\left(\frac{\partial L}{\partial \dot{q}}\right) = \dot{\tau}\frac{\partial L}{\partial q} \quad\text{and}\quad \frac{d}{dt}\left(L - \frac{1}{\dot{\tau}}\frac{\partial L}{\partial \dot{q}}\dot{q}\right) = 0$$

The first equation expresses the fact that $q(t)$ is obtained from the Lagrange equations for L and the second equation expresses conservation of energy, that is, that E is the variable canonically conjugate to t for $\bar{L}$. The result follows. ■

Remark. We restricted our attention to $\lambda > 0$ and correspondingly to $\dot{\tau} > 0$ to ensure $\bar{L}$ be defined. One can consider the case $\lambda < 0$ and $\dot{\tau} < 0$ as well.

3.7.20 Example A free relativistic particle is described in terms of geodesics of the Lorentz metric on $\boldsymbol{R}^4$, that is,

$$\langle (x,t),(y,s)\rangle = x\cdot y - c^2ts$$

where $x\cdot y$ is the usual dot product on $\boldsymbol{R}^3$ and $c =$ *speed of light*. Because geodesics extremise arc length as well as energy, (see Sect. 3.8) it is possible to use the Lagrangian

$$\bar{L}((x,t),(\dot{x},\dot{t})) = -m_0c\sqrt{-\langle(\dot{x},\dot{t}),(\dot{x},\dot{t})\rangle}$$

to yield the same straightline geodesics in $\boldsymbol{R}^4$. Here m_0c ($m_0 =$ *mass*) is inserted for dimensional reasons. This Lagrangian is first-order homogeneous so we can use 3.7.19 to construct the corresponding three-dimensional L and recover the "physical" energy E of L. (Remember that $\bar{E} = 0$.) If we

ISBN 0-8053-0102-X

de-homogenize $\bar{L}$, we arrive at a Lagrangian on $T\mathbb{R}^3$:

$$L(x, \dot{x}) = -m_0 c\sqrt{c^2 - \|\dot{x}\|^2}$$

(Note that L is defined only on the open subset of $T\mathbb{R}^3$ corresponding to velocities $< c$). From L we compute the action and energy:

$$A(x, \dot{x}) = FL(x, \dot{x})\cdot(x, \dot{x}) = \frac{m_0 c\|\dot{x}\|^2}{\sqrt{c^2 - \|\dot{x}\|^2}}$$

$$E(x, \dot{x}) = A(x, \dot{x}) - L(x, \dot{x})$$

$$= \frac{m_0 c^2}{\sqrt{1 - \|\dot{x}\|^2/c^2}} = m_0 c^2 + \tfrac{1}{2} m_0 \|\dot{x}\|^2 + \cdots$$

We refer to Exercise 3.7F and books on relativity, such as Misner, Thorne and Wheeler [1973], for further details on these matters.

We shall conclude this section with some ideas that are important for the statistical theory of classical systems. The central notion is that of ergodicity which is intended to capture the idea that a flow is random or chaotic. In dealing with the motion of molecules, the founders of statistical mechanics, particularly Boltzmann and Gibbs, made such hypotheses at the outset. Our treatment will be brief and introductory. The reader should consult Arnold and Avez [1967], Ruelle [1969], and Markus and Meyer [1974] for further information.

One of the earliest precise definitions of randomness of a dynamical system was *minimality*: the orbit of almost every point is dense. In order to prove useful theorems, von Neumann and Birkhoff in the early 1930's required the stronger assumption of ergodicity, defined as follows.

3.7.21 Definition. *Let S be a measure space and F_t a (measurable) flow on S. We call F_t **ergodic** if the only invariant measurable sets are $\varnothing$ and all of S.*

Here, invariant means $F_t(A) = A$ for all $t \in \mathbb{R}$ and we agree to write $A = B$ if A and B differ by a set of measure zero. (It is not difficult to see that ergodicity implies minimality if we are on a second countable Borel space).

A measurable function $f\colon S \to \mathbb{R}$ will be called a *constant of the motion* iff $f \circ F_t = f$ a.e. for each $t \in \mathbb{R}$.

3.7.22 Proposition. *A flow F_t on S is ergodic iff the only constants of the motion are constant a.e.*

ISBN 0-8053-0102-X

Proof. If F_t is ergodic and f is a constant of the motion, the sets $\{x \in S \mid f(x) > a\}$, $\{x \in S \mid f(x) < a\}$ are invariant, it follows that f must be constant a.e. The converse follows by taking f to be a characteristic function. ■

One sometimes hopes that, by passing to the intersection of surfaces on which all obvious constants of the motion take specified values, one will end up with an ergodic flow. However, this is usually not the case, even if one can show there are no further *smooth* constants of the motion. See Sect. 4.3 and Markus and Meyer [1974]. In the Hamiltonian case, the first step is to pass to an energy surface. Verifying ergodicity can still be very difficult. For example a classical model for molecular motion is a collection of "hard spheres in a box" and showing one has ergodicity on an energy surface is the subject of work of Y. Sinai (see Arnold–Avez [1967] for references). In a spherical container, one would also, presumably, have to pass to a surface of constant total angular momentum.

Probably the most basic example where ergodicity can be verified is the following.

3.7.23 Theorem. (Hadamard) *Let M be compact, Riemannian and have negative sectional curvatures at each point. Then the geodesic flow on each sphere bundle* $(\{v \mid \|v\| = \text{constant} \neq 0\} \subset TM)$ *is ergodic.*

The proof may be found in Arnold–Avez [1967, Appendix 21]. Of course the sphere bundle is just an energy surface. The intuitive idea is this: negative sectional curvature implies that nearby geodesics tend to diverge apart, or to be "defocussed." Since one is on a compact manifold, this defocussing forces a randomness. Theorem 3.7.23 may be applicable to motion in a potential using Jacobi's theorem (3.7.7).

The first major step in ergodic theory was taken by J. von Neumann [1932] who proved the mean ergodic theorem. It remains as the most important basic theorem. The setting is in Hilbert space, but we shall see how it applies to classical systems shortly.

3.7.24 Theorem (Mean Ergodic Theorem). *Let H be a Hilbert space and $U_t: H \to H$ a strongly continuous one-parameter unitary group (i.e., U_t is unitary for each t, is a flow on H and for each $x \in H$, $t \mapsto U_t x$ is continuous).*

Let the closed subspace H_0 be defined by

$$H_0 = \{x \in H \mid U_t x = x \text{ for all } t \in \mathbf{R}\}$$

and let P be the orthogonal projection onto H_0. Then for any $x \in H$,

$$\lim_{t \to \pm\infty} \frac{1}{t} \int_0^t U_s(x)\, ds = P(x)$$

The limit in this result is called the *time average* of x and is customarily denoted $\bar{x}$.

ISBN 0-8053-0102-X

***Proof of 3.7.24* (*F. Riesz* [1944]).** We must show that

$$\lim_{t\to\infty}\left\|\frac{1}{t}\int_0^t U_s(x)\,ds - P(x)\right\| = 0$$

If $Px = x$, this means $x \in H_0$, so $U_s(x) = x$; the result is clearly true in this case. We can therefore suppose that $Px = 0$ by decomposing $x = Px + (x - Px)$.

We remark that

$$\{U_t y - y \mid y \in H,\ t \in R\}^{\perp} = H_0$$

where $\perp$ denotes the orthogonal complement. This is an easy verification using unitarity of U_t and $U_t^{-1} = U_{-t}$.

It follows from this remark that $\ker P$ is the closure of the space spanned by elements of the form $U_s y - y$. Indeed $\ker P = H_0^{\perp}$, and if A is any set in H, and $B = A^{\perp}$, then $B^{\perp}$ is the closure of the span of A.

Therefore, for any $\varepsilon > 0$, there exists $t_1, \ldots, t_n, x_1, \ldots, x_n$ such that

$$\left\|x - \sum_{j=1}^{n}\left(U_{t_j}x_j - x_j\right)\right\| < \varepsilon$$

It follows from this, again using unitarity of U_t, that it is enough to prove our assertion for x of the form $U_{t_0}y - y$. Thus we must establish that

$$\lim_{t\to\infty}\frac{1}{t}\int_0^t U_s\left(U_{t_0}y - y\right)ds = 0$$

For $t > t_0$ we may estimate this integral as follows:

$$\left\|\frac{1}{t}\int_0^t \left[U_s(U_{t_0}y) - U_s y\right]ds\right\| = \left\|-\frac{1}{t}\int_0^{t_0} U_s(y)\,ds + \frac{1}{t}\int_t^{t+t_0} U_s(y)\,ds\right\|$$

$$< \frac{1}{t}\int_0^{t_0}\|y\|\,ds + \frac{1}{t}\int_t^{t+t_0}\|y\|\,ds$$

$$= \frac{2t_0\|y\|}{t} \to 0$$

as $t \to \infty$. ∎

To apply 3.7.24 we use "Koopmanism" described in Sect. 2.6. Namely, given a measure-preserving flow F_t on S, we consider the unitary one parameter group $U_t(f) = f \circ F_{-t}$ on $L^2(S, \mu)$. We only require a minimal

ISBN 0-8053-0102-X

amount of continuity on F_t here, namely, we assume that if $t_n \to t$, $F_{t_n}(x) \to F_t(x)$ for a.e. $x \in S$. We shall also assume $\mu(S) < \infty$ for convenience. As in Sect. 2.6, under these hypotheses U_t is a strongly continuous unitary one-parameter group. Again we will leave the verification as an exercise in the use of the dominated convergence theorem.

3.7.25 Corollary. *To the hypotheses just described, add the condition that F_t be ergodic. Then for $f \in L^2(S)$,*

$$\lim_{t \to \pm\infty} \frac{1}{t} \int_0^t f \circ F_s \, ds = \frac{1}{\mu(S)} \int_S f \, d\mu$$

the limit being in the mean.

Proof. By 3.7.22, the space H_0 of 3.7.24 is one dimensional, consisting of constants. It is readily checked that

$$P(f) = \int_S f \, d\mu / \mu(S)$$

so 3.7.24 gives the result. ■

Thus, if F_t is ergodic, the time average $\bar{f}$ of a function is constant a.e. and equals its space average.

A refinement of 3.7.25 is the individual ergodic theorem of G. D. Birkhoff [1931], in which one obtains convergence almost everywhere. Also, if $\mu(S) = \infty$ but $f \in L^1(S) \cap L^2(S)$, one still concludes a.e. convergence of the time average. (If f is only L^2, mean convergence to zero is still assured by 3.7.24.)

One of the first things one does in statistical mechanics is introduce the notion of temperature. This is done by means of the *principle of equipartition of energy*, a consequence of the ergodic theorems. The result is as follows:

3.7.26 Proposition. *Let (P, ω, H) be a Hamiltonian system and e a regular value of H. Let $\Sigma_e = H^{-1}(e)$ and let μ_e be the invariant measure on Σ_e (see Sect. 3.4.) Assume*

(*i*) *Σ_e is compact (i.e., orbits of energy e are bounded),*
(*ii*) *F_t, the flow of X_H, is ergodic on Σ_e, and*
(*iii*) *$f_1, f_2 \colon \Sigma_e \to \mathbf{R}$ are continuous (or just L^2 will do) and $\phi \colon \Sigma_e \to \Sigma_e$ is an orientation preserving C^1 diffeomorphism such that $\phi^*(f_1 \mu_e) = f_2 \mu_e$.*

Then $\bar{f}_1 = \bar{f}_2$.

Proof. From (iii), (i), f_1, and f_2 have equal space averages. Hence from (ii) and 3.7.25, the time averages must be equal as well. ■

Here is how this result is applied.

ISBN 0-8053-0102-X

3.7.27 Example. On $P = R^{6N} = R^{3N} \times R^{3N}$ consider a Lagrangian of the form

$$L(q, \dot{q}) = \sum_{j=1}^{N} \tfrac{1}{2} m_j \|\dot{q}_j\|^2 - V(q_1, \ldots, q_n), \qquad \dot{q}_j, q_j \in R^3$$

where V is symmetric, with associated energy function

$$E(q, \dot{q}) = \sum_{j=1}^{N} \tfrac{1}{2} m_j \|\dot{q}_j\|^2 + V(q_1, \ldots, q_n)$$

Suppose Σ_e is compact and the flow is ergodic on Σ_e. Then the time averages of $\frac{1}{2} m_j \|\dot{q}_j\|^2, j = 1, \ldots, N$ are all equal.

Proof. Let $f_1 = \frac{1}{2} m_j \|\dot{q}_j\|^2$ and $f_2 = \frac{1}{2} m_k \|\dot{q}_k\|^2$ for given j, k. Define ϕ on P by interchanging q_j and q_k, likewise $\sqrt{m_j}\, \dot{q}_j$ and $\sqrt{m_k}\, \dot{q}_k$. Thus ϕ leaves E invariant, so maps Σ_e to Σ_e. The symplectic form here is

$$\omega_L = \Sigma m_i \, dq_i^{\alpha} \wedge d\dot{q}_i^{\alpha}$$

It follows easily that ϕ leaves the phase volume μ invariant (but not the symplectic form in general).

Since ϕ leaves μ and E invariant, it leaves μ_e invariant as well, and maps f_1 to f_2. The result therefore follows from 3.7.26. ■

Note that this example also works on $P = T(M \times \cdots \times M)$ where M is a Riemannian manifold.

The same argument can be used for the squares of individual components of the $\frac{1}{2} m_j \|q_j\|^2$. The time average of each component is assigned a value $kT/2$ (k is a constant—Boltzmann's constant—whose numerical value depends on the choice of units). Thus the total time average of the kinetic energy is

$$\tfrac{3}{2} NkT$$

which implicitly defines the systems' temperature T. One can treat the pressure in an analogous manner; see 3.7.32.

One word of caution. We are tacitly assuming V is a C^{∞} function. In the case that one wants to have particles enclosed in a container with hard walls, this smoothness question leads to nontrivial difficulties (c.f. Sect. 2.6).

Next we consider the classical virial theorem. Let Q be a (pseudo-) Riemannian manifold, K the kinetic energy function, and $V: Q \rightarrow R$ a given potential. Let

$$L(v) = K(v) - V(\tau_Q v)$$

be our usual Lagrangian and X_E the associated Lagrangian vector field on TQ.

ISBN 0-8053-0102-X

3.7.28 Definition. *Given a vector field X on Q, define the associated* ***momentum function*** *$P(X)$ by*

$$P(X): TQ \to R, \qquad P(X)(v) = \langle X(\tau_Q v), v\rangle$$

Define the ***virial function*** *by*

$$G(X) = \{E, P(X)\}$$

The momentum functions $P(X)$ will play an important role in our discussion of symmetry groups (see Chapter 4).

The following gives $G(X)$ in coordinates.

3.7.29 Proposition. *In local coordinates $q^i, \dot{q}^j$, we have*

$$G(X)(q,\dot{q}) = \sum_{i,j,k}\left\{\sum_l g_{il}X^i\Gamma^l_{jk}\dot{q}^j\dot{q}^k - \frac{\partial g_{ij}}{\partial q^k}\dot{q}^k\dot{q}^jX^i - g_{ij}\dot{q}^i\dot{q}^k\frac{\partial X^j}{\partial q^k}\right\} + \frac{\partial V}{\partial q^i}X^i$$

Proof. We have $G(X) = \{K, P(X)\} + \{V, P(X)\}$. Now $\{K, P(X)\} = -dP(X)\cdot X_K$. If v has components $\dot{q}^i$,

$$P(X)v = \Sigma \dot{q}^i X^j g_{ij}$$

Differentiating and applying dP to $X_K = (\dot{q}^i, -\Gamma^i_{jk}\dot{q}^j\dot{q}^k)$ gives the term $\{\ \}$ above. The term involving V is

$$\{V, P(X)\} = -dP(X)\cdot X_V = dV\cdot X$$

since $X_V = (0, -\operatorname{grad} V)$. ■

The virial theorem has the appearance of an ergodic theorem, but no ergodicity is assumed.

3.7.30 Virial Theorem. *Let Q, L be given as above and X a vector field on Q. Let e be a regular value of E and assume Σ_e is compact. Then the time and space averages of $G(X)$ on Σ_e are both zero.*

This results from the following.

ISBN 0-8053-0102-X

3.7.31 Lemma. *Let P be a symplectic manifold, $H: P \to \mathbf{R}$, and assume $\Sigma_e = H^{-1}(e)$ is compact. For any $f: P \to \mathbf{R}$, the time and space averages of $\{H, f\}$ are zero on Σ_e.*

Proof. Observe that the flow F_t of X_H is automatically complete on Σ_e as it is assumed to be compact. Now

$$\{f, H\} \circ F_t = \frac{d}{dt}(f \circ F_t),$$

so that

$$\frac{1}{t}\int_0^t \{f, H\} \circ F_s \, ds = \frac{1}{t}(f \circ F_t - f) \to 0 \quad \text{as} \quad t \to \infty$$

since f is bounded.

For the space averages, note that $L_{X_H}\mu_e = 0$ and hence

$$\{f, H\}\,\mu_e = L_{X_H}(f\mu_e) = d i_{X_H}(f\mu_e)$$

Thus

$$\int_{\Sigma_e} \{f, H\}\,\mu_e = 0$$

by Stokes' theorem. ■

If Σ_e is not compact but $G(X)$ is known to be bounded on Σ_e, then one can still conclude that $\overline{G(X)} = 0$ by the same argument.

3.7.32 Examples. (1) Let $Q = \mathbf{R}^n$ with the usual metric and $X = e_i$, a coordinate vector field. In this case, from 3.7.29, $G(X) = \partial V / \partial q^i$. Thus the force has the time and space average of zero on any compact energy surface. The same is true of the momenta since $P_i = \{q_i, E\}$.

(2) On $Q = \mathbf{R}^n$ again, let $X(x) = x$. Then we get

$$2\overline{K} = \overline{\frac{\partial V}{\partial q^i} q^i}$$

which for the special case of the gravitational potential $V(x) = -1/\|x\|$ on $\mathbf{R}^3 \backslash \{0\}$ becomes

$$2\overline{K} = -\overline{V}$$

a well-known property of central force motion.

ISBN 0-8053-0102-X

We can compare the result $2\overline{K} = \overline{(\partial V/\partial q^i)q^i}$ with the case in which the metric is not trivial. The terms

$$F_i = \sum_{j,k,l} g_{il}\Gamma^l_{jk}\dot{q}^j\dot{q}^k - \sum_{j,k} \frac{\partial g_{ij}}{\partial q^k}\dot{q}^k\dot{q}^j$$

$$= -\tfrac{1}{2}\sum_{j,k} g_{kj,i}\dot{q}^j\dot{q}^k$$

act like an additional force term to be added to $\partial V/\partial q^i$. In other words, the curvature of the space will have an effect on the average kinetic energy. For example, the virial theorem is used in astronomy to measure masses in distant galaxies, and the curvature of space might conceivably have an effect for dense clusters.

(3) On $\boldsymbol{R}^{6N}$ consider

$$E(q,\dot{q}) = \sum_{j=1}^{N} \tfrac{1}{2}m_j\|\dot{q}^j\|^2 + \sum_{j=1}^{N} V_j(q_j) + \sum_{j<k} V_{jk}(q_j - q_k)$$

Here V_j represents an external potential (such as a container) and V_{jk} are interaction potentials between the particles. Assume again Σ_e is compact. Letting $X(x) = x$ in the virial theorem, we conclude that the time and space averages of the following are zero:

$$-2\sum_{j=1}^{N} \tfrac{1}{2}m_j\|\dot{q}^j\|^2 + \sum_{j=1}^{N} \nabla V_j \cdot q_j + \sum_{j<k} \left[\nabla V_{jk}(q_j - q_k)\right]\cdot(q_j - q_k)$$

Let $3p|V|$ be the time average of the second term (p is the *pressure* and $|V|$ the volume of the "container"). Thus, if the flow is ergodic on Σ_e, we get the *equation of state*:

$$p|V| = NkT - \tfrac{1}{3}\sum_{j<k}\int_{\Sigma_e} \nabla V_{jk}(q_j - q_k)\mu_e$$

Note that on a curved space the pressure term would have to be modified by adding the "force" F_i to ∇V_i, as in Example 2.

EXERCISES

3.7A. Let $Y: TQ \to T^2Q$ be a dissipative vector field. Define a *generalized force* $\tilde{Y}: TQ \to TQ$ by

$$Y(K)(v) = \langle v, \tilde{Y}(v)\rangle$$

Conversely, given $\tilde{Y}$ define $Y(v) = [\tilde{Y}(v)]^l_v$ and show that it satisfies the same identity.

ISBN 0-8053-0102-X

If $\tilde{Y}(v) = -k(x)\cdot v$, where $k(x)$ is a nonnegative self-adjoint linear map on T_xQ, $x = \tau_Q v$, show that (i) Y is dissipative and (ii) (See Exercise 3.8F for a generalization.) $\tilde{Y}$ is the negative velocity gradient of the *Rayleigh dissipation function*

$$\mathcal{F}(v) = \tfrac{1}{2}\langle k(x)\cdot v, v\rangle$$

3.7B. Let Q be a Riemannian manifold. Put a Riemannian structure on TQ by writing

$$T_vTQ = T_xQ \oplus T_xQ, \quad x = \tau_Q v$$

the decomposition into horizontal and vertical subspaces, and declaring them to be orthogonal.

(i) Show that the Riemannian volume element on TQ is the same as the volume element induced by the symplectic form.

(ii) Show that the volume element on the sphere bundles $\Sigma_e \subset TQ$ induced by their Riemannian structure coincides with the natural Hamiltonian volume μ_e on energy surfaces.

3.7C. (a) If Q is a complete Riemannian manifold, show that TQ is as well by (i) a direct proof (see Ebin [1970b]), and by (ii) showing that geodesics on TQ are Jacobi fields along geodesics in Q. (See, e.g., Milnor [1963] for a discussion of Jacobi fields.)

(b) Observe that the result (a) holds in infinite dimensions.

(c) Give a proof of 3.7.15 valid for infinite-dimensional manifolds.

3.7D. Reformulate in terms of Riemannian manifolds and prove *Hadamard's theorem*: "If a particle is free to move on a surface which is everywhere regular and has no infinite sheets, the potential energy function being regular at all points of the surface and having only a finite number of maxima and minima on it, either the part of the orbit described in the attractive region is of length greater than any assignable quantity, or else the orbit tends asymptotically to one of the positions of unstable equilibrium." (See Whittaker [1959, p. 416].)

3.7E. Consider the motion of a particle on a line under a potential $V(x)$. Assume the line rotates in a plane with constant angular velocity ω. Show that the correct motion of the particle is that of a particle on the line in the *amended potential* $V_\omega(x) = V(x) - \frac{1}{2}m\omega^2/x^2$.

3.7F. Let M be a four-manifold with $\langle , \rangle$ a Lorentz metric, that is, a pseudo-Riemannian metric with diagonal form $(-1, +1, +1, +1)$. Let $H(q,p) = \frac{1}{2}g^{ij}p_ip_j$ be the Hamiltonian on T^*M. If $(q(t), p(t))$ is an integral curve of X_H, define its *mass* to be m_0, where m_0^2 is the constant energy $H(q(t), p(t))$ along the curve

(i) Show that two integral curves with initial data $(q_1(0), p_1(0))$ and $(q_2(0), p_2(0))$, which have $q_1(0) = q_2(0)$ and $p_1(0)$ parallel to $p_2(0)$, have the same base integral curve when parametrized by arc length.

(ii) Does this definition of m_0 agree with example 3.7.20 (use units with $c = 1$)? (*Hint.* Here we are on T^*M, not TM.)

(iii) Let F be a closed two-form on M (an electromagnetic field) and ω_0 the canonical two-form on T^*M. For a constant e, the charge, let

$$\omega_F = \omega_0 + e(\tau_M^*)^*F$$

Show that ω_F is a symplectic form on T^*M. Write down, in coordinates, the equations of motion for the Hamiltonian H relative to ω_F. Look up the

ISBN 0-8053-0102-X

equations of motion for a relativistic particle in the presence of an electromagnetic field in a physics text and see if they agree (modulo units) with what you just derived.

(iv) Let F be a closed two-form on a manifold M, ω_0 the canonical symplectic structure on T^*M, and $\omega_F = \omega_0 + (\tau_M^*)^*F$. Let $H \in \mathfrak{F}(T^*M)$. Show that the critical points of the two Hamiltonian vector fields formed using ω_0 and ω_F are the same.

3.7G. Holonomic constraints were described in 3.7.8. Sometimes constraints can be holonomic "without looking like it," an example being a hoop rolling without slipping on an inclined plane; see Goldstein [1950, p. 43]. Discuss and justify the following definition: "A holonomic constraint on TM is an integrable subbundle $E \subset TM$."

3.8 VARIATIONAL PRINCIPLES IN MECHANICS

Historically, variational principles have played a fundamental role in the evolution of mathematical models in mechanics, but in the last few sections we have obtained the bulk of classical mechanics without reference to the calculus of variations. In principle, we may envision two equivalent models for mechanics. In the first, we take the Hamiltonian or Lagrangian equations as an axiom and, if we wish, obtain variational principles as theorems. In the second, we may assume variational principles and derive the Hamiltonian and Lagrangian equations as theorems. We have taken the former course in this text because a complete account of the calculus of variations can be more complex. Others may prefer the second model, especially those metaphysicians who hold, with Maupertuis, that *nature always acts in the simplest way*.

In this section we summarize the basic ideas for the calculus of variations, but technicalities involving infinite-dimensional manifolds prevent us from presenting all the details. For these, we refer to, for example, Smale [1964], Palais [1963] and Klingenberg [1977]. For the classical geometrical theory without the modern infinite-dimensional framework, the reader should consult, for example, Bolza [1973], Whittaker [1959], Gelfand and Fomin [1963], or Hermann [1968].

3.8.1 Definition. *Let Q be a manifold and $L: TQ \to \mathbb{R}$ a regular Lagrangian. Fix two points q_1 and q_2 in Q and an interval $[a, b]$ and let*

$$\Omega(q_1, q_2, [a,b]) = \{c: [a,b] \to Q \mid c \text{ is a } C^2 \text{ curve}, c(a) = q_1 \text{ and } c(b) = q_2\}$$

called the ***path space*** *from q_1 to q_2. Define the map*

$$J: \Omega(q_1, q_2, [a,b]) \to \mathbb{R}$$

by

$$J(c) = \int_a^b L(c(t), c'(t))\, dt$$

ISBN 0-8053-0102-X

For certain purposes in the calculus of variations it is essential to be more delicate about the differentiability class chosen for the paths c; in particular, the Sobolev spaces H^s are often used. For our purposes, this will not be necessary.

What we shall *not* prove is that $\Omega(q_1,q_2,[a,b])$ is actually a C^∞ infinite-dimensional manifold. This is a special case of a general result in the topic of manifolds of mappings, wherein spaces of C^r (or H^s) maps from one manifold to another are shown to be smooth infinite dimensional manifolds. (See the above references and also Eells [1966], Eliasson [1967], Palais [1968] and Ebin–Marsden [1970].)

Granting this, the following is easy to see.

3.8.2 Proposition. *The tangent space to* $\Omega(q_1,q_2,[a,b])$ *at a curve* $c\in\Omega(q_1,q_2,[a,b])$ *is given as follows:*

$$T_c\Omega(q_1,q_2,[a,b])=\{v:[a,b]\to TQ \mid v \text{ is a } C^2$$

$$\text{map}, \tau_Q\circ v=c \text{ and } v(a)=0, v(b)=0\}$$

Idea of Proof. The tangent space to a manifold consists of tangents to curves in the manifold. Consider then a curve $c_\lambda\in\Omega(q_1,q_2,[a,b])$ with $c_0=c$. Thus a tangent vector is

$$v=\frac{d}{d\lambda}c_\lambda\Big|_{\lambda=0}$$

However $c_\lambda(t)$ for each t is a curve through $c_0(t)=c(t)$. (See Fig. 3.8-1.) Hence $(d/d\lambda)c_\lambda(t)|_{\lambda=0}$ is a tangent vector to Q based at $c(t)$. Hence $v(t)\in T_{c(t)}Q$, that is, $\tau_Q\circ v=c$. The restriction $c_\lambda(0)=q_1$ and $c_\lambda(1)=q_2$ leads to $v(0)=0$, $v(1)=0$, but otherwise v is an arbitrary C^2 function. ■

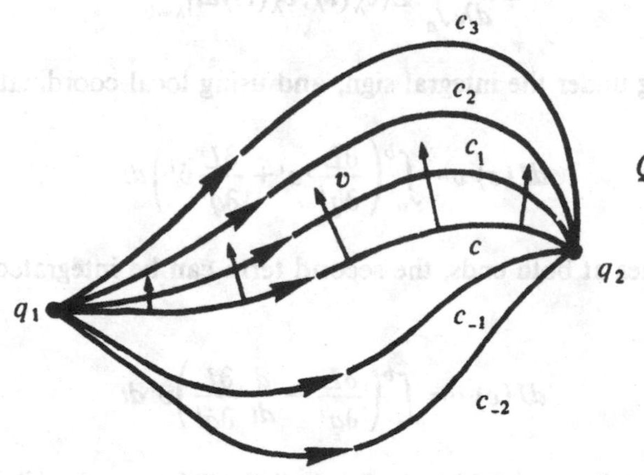

Figure 3.8-1

ISBN 0-8053-0102-X

One refers to v as an infinitesimal variation of the curve c subject to fixed end-points. Classically, the notation $v = \delta c$ is used.

Now we can state and sketch the proof of a main result in the calculus of variations.

3.8.3 Theorem. (*Variational Principle of Hamilton*). *Let L be a regular Lagrangian on TQ. A curve $c_0: [a,b] \to Q$ joining $q_1 = c_0(a)$ to $q_2 = c_0(b)$ is a base integral curve of X_E, that is, satisfies the Lagrange equations*

$$\frac{d}{dt}\frac{\partial L}{\partial \dot{q}^i} = \frac{\partial L}{\partial q^i}$$

if and only if c_0 is a critical point of the function $J: \Omega(q_1, q_2, [a,b]) \to \mathbf{R}$ defined above; that is, $dJ(c_0) = 0$.

Classically, the condition $dJ(c_0) = 0$ is denoted

$$\delta \int_a^b L(c_0(t), c_0'(t))\, dt = 0$$

that is, the integral is stationary when it is differentiated as if c were the independent variable.

Proof of 3.8.3. We first work out $dJ(c)\cdot v$ in the following way. Let $v = (d/d\lambda)c_\lambda|_{\lambda=0}$, that is, write v as the tangent to a curve in $\Omega(q_1, q_2, [a,b])$. By the chain rule,

$$dJ(c)\cdot v = \frac{d}{d\lambda} J(c_\lambda)\Big|_{\lambda=0}$$

$$= \frac{d}{d\lambda}\int_a^b L(c_\lambda(t), c_\lambda'(t))\, dt\Big|_{\lambda=0}$$

Differentiating under the integral sign, and using local coordinates we get

$$dJ(c)\cdot v = \int_a^b \left(\frac{\partial L}{\partial q^i}\cdot v^i + \frac{\partial L}{\partial \dot{q}^i}\dot{v}^i \right) dt$$

Since v vanishes at both ends, the second term can be integrated by parts to give

$$dJ(c)\cdot v = \int_a^b \left(\frac{\partial L}{\partial q^i} - \frac{d}{dt}\frac{\partial L}{\partial \dot{q}^i} \right) v^i\, dt$$

Now $dJ(c) = 0$ means $dJ(c)\cdot v = 0$ for all $v \in T_c\Omega(q_1, q_2, [a,b])$. This holds if

ISBN 0-8053-0102-X

and only if

$$\frac{\partial L}{\partial q^i} - \frac{d}{dt}\frac{\partial L}{\partial \dot{q}^i} = 0$$

since v is arbitrary, except $v=0$ at the ends, and the integrand is continuous.* ■

Next we discuss variational principles with the constraint of constant energy imposed. To balance this constraint, we let the interval $[a,b]$ be variable.

3.8.4 Definition. *Let L be a regular Lagrangian as above and Σ_e a regular energy surface for the energy E of L. Let $q_1, q_2 \in Q$ and let $[a,b]$ be a given interval. Set*

$$\begin{aligned}\Omega(q_1, q_2, [a,b], e) = \{(\tau, c) \mid \tau\colon [a,b] \to \mathbf{R} \text{ is } C^2, \tau' > 0, c\colon [\tau(a), \tau(b)] \to Q \\ \text{is a } C^2 \text{ curve}, c(\tau(a)) = q_1, c(\tau(b)) = q_2 \text{ and} \\ E(c(\tau(t)), \dot{c}(\tau(t))) = e \text{ for all } t \in [a,b]\}\end{aligned}$$

Arguing as above, differentiation of curves $(\tau(\lambda), c(\lambda))$ in $\Omega(q_1, q_2, [a,b], e)$ shows that the tangent space to $\Omega(q_1, q_2, [a,b], e)$ at (τ, c) consists of C^2 maps $\alpha\colon [a,b] \to \mathbf{R}$ and $v\colon [\tau(a), \tau(b)] \to TQ$ such that

$$v(t) \in T_{c(t)}Q,\ \dot{c}(\tau(a))\alpha(a) + v(\tau(a)) = 0,\ \dot{c}(\tau(b))\alpha(b) + v(\tau(b)) = 0$$

and

$$dE(c(\tau(t)), \dot{c}(\tau(t)))\cdot(\dot{c}(\tau(t))\alpha(t) + v(\tau(t)), \ddot{c}(\tau(t))\dot{\alpha}(t) + \dot{v}(\tau(t))) = 0$$

The Euler-Lagrange-Jacobi formulation of the *principle of least action* of Maupertuis is as follows.†

3.8.5 Theorem. *Let $c_0(t)$ be a base integral curve of X_E, $q_1 = c_0(a)$ and $q_2 = c_0(b)$. Let e be the energy of $c_0(t)$ and assume it is a regular value of E.*

*We use implicitly this easy lemma: if $f(t)$ is continuous on $[a,b]$, then $\int_a^b f(t)g(t)\,dt = 0$ for all C^r functions g vanishing at a and b if and only if $f = 0$.

†We thank M. Spivak for helping us formulate this theorem correctly. The authors like many others (we were happy to learn), were confused by the standard textbook statements. For instance the mysterious variation "Δ" in Goldstein [1950, p. 228] corresponds to our enlargement of the variables by $c \to (\tau, c)$.

ISBN 0-8053-0102-X

Define

$$I: \Omega(q_1, q_2, [a,b], e) \to \mathbf{R}$$

by

$$I(\tau, c) = \int_{\tau(a)}^{\tau(b)} A(c(t), \dot{c}(t))\, dt$$

where A is the action of L. Then

$$dI(Id, c_0) = 0, \text{ where } Id: [a,b] \to \mathbf{R};\ t \mapsto t.$$

Conversely, if (Id, c_0) is a critical point of I, and c_0 has energy e, a regular value of E, then c_0 is a solution of Lagrange's equations.

Proof. Since all curves have energy e,

$$I(\tau, c) = \int_{\tau(a)}^{\tau(b)} (L(c(t), \dot{c}(t)) + e)\, dt$$

Differentiating with respect to τ and c by the method of 3.8.3 gives

$$dI(Id, c_0)\cdot(\alpha, v) = \alpha(b)(L(c_0(b), \dot{c}_0(b)) + e) - \alpha(a)(L(c_0(a), \dot{c}_0(a)) + e)$$

$$+ \int_a^b \left(\frac{\partial L}{\partial q^i}(c_0(t), \dot{c}_0(t)) v^i(t) + \frac{\partial L}{\partial \dot{q}^i}(c_0(t), \dot{c}_0(t)) \dot{v}^i(t) \right) dt$$

Integrating by parts as in 3.8.3 gives

$$dI(Id, c_0)\cdot(\alpha, v) = \alpha(t)(L(c_0(t), \dot{c}_0(t)) + e)\Big|_a^b$$

$$+ \frac{\partial L}{\partial \dot{q}^i}(c_0(t), \dot{c}_0(t)) v^i(t)\Big|_a^b$$

$$+ \int_a^b \left(\frac{\partial L}{\partial q^i}(c_0(t), c_0(t)) - \frac{d}{dt}\frac{\partial L}{\partial \dot{q}^i}(c_0(t), \dot{c}_0(t)) \right) v^i(t)\, dt$$

Using the boundary conditions $v = -\dot{c}\alpha$, noted in the description of $T_{Id, c_0}\Omega(q_1, q_2, [a,b], e)$, and the energy constraint $(\partial L/\partial \dot{q}^i)\dot{c}^i - L = e$, the boundary terms cancel, leaving

$$dI(Id, c_0)\cdot(\alpha, v) = \int_a^b \left(\frac{\partial L}{\partial q^i} - \frac{d}{dt}\frac{\partial L}{\partial \dot{q}^i} \right) v^i\, dt$$

ISBN 0-8053-0102-X

However we can choose v arbitrarily; notice that the presence of α in the linearized energy constraint means that no restrictions are placed on the variations v' on the open set where $\dot{c}\neq 0$. The theorem therefore follows. ■

If $L=K-V$, where K is the kinetic energy of a Riemannian metric, then 3.8.5 states that a curve c_0 is a solution of Lagrange's equations if and only if

$$\delta_e\int_a^b 2K(c_0,\dot{c}_0)\,dt=0$$

where δ_e indicates a variation holding the energy and endpoints but *not* the parametrization fixed; i.e. symbolic notation for the precise statement in 3.8.5.

Since $K\geqslant 0$ in the Riemannian case (for timelike curves in the pseudo-Riemannian case consider $-K\geqslant 0$), this is the same as

$$\delta_e\int_a^b \sqrt{2K(c_0,\dot{c}_0)}\,dt=0.$$

that is, arc length is extremized (subject to constant energy). This is *Jacobi's form of the principle of least action* and represents a key to linking mechanics and geometrical optics, which was Hamilton's original motivation. In particular, geodesics are characterized as extremals of arc length.

We can see the link with Jacobi's theorem (3.7.7) as follows: let $L=K-V$ so $E=K+V$ and let c_0 be a solution of Lagrange's equations with energy e. Then along c_0

$$\delta_e\int_a^b 2K\,dt=0$$

so

$$\delta_e\int_a^b \sqrt{K}\,\sqrt{K}\,dt=0$$

that is,

$$\delta_e\int_a^b \sqrt{(e-V)K}\,dt=0$$

We have written the original variational principle in the form of that for the arc length of a geodesic in the metric $(e-V)g$, that is, in the Jacobi metric. This argument is essentially the original one given by Jacobi to prove 3.7.7.

Because geodesics extremize curvature as well as arc length, these ideas are closely related to the geometrical principles of Gauss and Hertz (see Whittaker [1959, Chapter IX]).

The reader who wishes to pursue the variational ideas should consult Klingenberg [1977], Weinstein [1978], and Duistermaat [1976a] in addition to references already given.

ISBN 0-8053-0102-X

EXERCISES

3.8A. (i) Give an example to show that J need not be minimized at a solution of Lagrange's equations. (*Hint*. Geodesics on the sphere.)
(ii) By considering second variations, show that, *locally*, J is minimized for geodesics of a Riemannian metric. (When minimization ceases one encounters conjugate points.)

3.8B. Suppose one begins with a variational formulation as in 3.8.3 as basic and then discovers the Lagrange equations. Show that by multiplying these equations by $\dot{q}^i$ and integrating you are led to the energy (Historically, this is a path to Hamilton's equations.)

3.8C. Let L be a Lagrangian depending on $q^i, \dot{q}^i, \ddot{q}^i, \ldots, \overset{(m)}{q}{}^i$.
(a) Derive the corresponding Euler–Lagrange equations by a variational argument.
(b) Show that these equations can be put into Hamiltonian form (see Whittaker [1959, pp. 265–267]).
(c) Formulate your results intrinsically on manifolds; you will need to find out about jet bundles. (See Sect. 5.5 and Rodrigues [1976].)

3.8D. Formulate and prove a principle of least action for (hyperregular) Hamiltonian systems on T^*Q.

3.8E. (A. Lichnerowicz, R. Jantzen and the authors). Let (P,ω) be a symplectic manifold and F_t the flow of a Hamiltonian vector field X_H. Let $G_t = TF_t$ be the tangent flow on TP and Y its generator. Show that Y is Hamiltonian with energy $\dot{H}(v) = -\omega(v, \tau_{TP}\cdot TX_H(v))$. (Use problems 1.6D and 3.3I). If (q^i, p_i) are canonical coordinates on P and (q^i, p_i, Q^i, P_i) are induced coordinates on TP show that

$$\dot{H}(q^i, p_i, Q^i, P_i) = Q^i \frac{\partial H}{\partial q^i} + P_j \frac{\partial H}{\partial p_j}.$$

(The system $X_{\dot{H}}$ on TP is called the *linearized Hamiltonian system of* X_H).

3.8F. (S. Shahshahani). Let $D \in \mathfrak{X}(TM)$ and define its *fiber differential* by $d_F D \in \mathfrak{X}^*(TM)$, $(d_F D)\cdot w_v = dD\cdot(T\tau_M\cdot w_v)^l_v$ (see 3.7.5). (a). In coordinates, show that $d_F D = \Sigma \frac{\partial D}{\partial \dot{q}^i} dq^i$. (b) Let ω be a symplectic form on TM which vanishes when pulled back to each fiber T_xM. Show that $\Delta = (d_F D)^\#$ is a vertical vector field. Let X_E be a second order Hamiltonian vector field on TM so that $Y = X_E + \Delta$ is also second order; Y is called a dissipative system with *Rayleigh dissipation function D*. (c) Show that this generalizes Exercise 3.7A(ii). (d) Let C denote the canonical vertical vector field on TM given by lifting vertically. Show that energy decreases along orbits of Y iff $C(D) > 0$. (e) Show that van der Pol's equation $\ddot{x} + \mu(x^2 - 1)\dot{x} + x = 0$ is a dissipative system in this sense with E the harmonic oscillator energy and $D(x,\dot{x}) = \frac{1}{2}\mu\dot{x}^2(x^2 - 1)$. Use (d) to study when energy is decreasing and verify by a direct calculation.

ISBN 0-8053-0102-X

CHAPTER **4**

Hamiltonian Systems with Symmetry

Associated with each one-parameter group of symmetries of a Hamiltonian system is a conserved quantity. For a group of symmetries we get thereby a vector-valued conserved quantity called the momentum. We shall discuss the properties of the momentum and how to construct it in Sect. 4.2, after summarizing the necessary topics from Lie group theory in Sect. 4.1. When symmetries are present the phase space can be reduced; that is, a number of variables eliminated. This topic is the subject of Sect. 4.3. Mechanical systems on Lie groups and the rigid body are discussed in Sect. 4.4. Smale's topological program for a mechanical system with symmetry is presented in Sect. 4.5 and this is applied to the rigid body problem in Sect. 4.6. A number of results presented in this chapter are new.

4.1. LIE GROUPS AND GROUP ACTIONS

In this section we develop the basic facts about Lie groups and actions of Lie groups on manifolds which we will need for applications to mechanics.

4.1.1 Definition. ***A Lie group is a finite-dimensional smooth manifold G that is a group and for which the group operations of multiplication,*** $\cdot: G\times G\to G$: $(g,h)\mapsto g\cdot h$, ***and inversion,*** $^{-1}: G\to G: g\mapsto g^{-1}$ ***are smooth. Let*** $e=$ ***identity.***

ISBN 0-8053-0102-X

Ralph Abraham and Jerrold E. Marsden, Foundation of Mechanics, Second Edition

4.1.2 Example. The group of linear isomorphisms of $\boldsymbol{R}^n$ to $\boldsymbol{R}^n$, denoted $Gl(n,\boldsymbol{R})$, is a Lie group of dimension n^2. It is a smooth manifold, being an open subset of $\boldsymbol{R}^{n^2}$ and the group operations are smooth since the formulas for the product and inverse of matrices are smooth in the matrix components.

For every $g \in G$ the maps $L_g: G \to G: h \mapsto gh$ and $R_g: G \to G: h \mapsto hg$ are, respectively, *left* and *right translation* by g. Since $L_g \circ L_h = L_{gh}$ and $R_h \circ R_g = R_{gh}$, $(L_g)^{-1} = L_{g^{-1}}$ and $(R_g)^{-1} = R_{g^{-1}}$. Thus both L_g and R_g are diffeomorphisms. Moreover, $L_g \circ R_h = R_h \circ L_g$.

Actually, smoothness of inversion follows *automatically* from smoothness of multiplication. This is easily seen by applying the inverse function theorem to the map $(g,h) \mapsto (g,gh)$ of $G \times G$ to $G \times G$.

A vector field X on G is called *left invariant* if for every $g \in G$, $(L_g)_* X = X$, that is,

$$T_h L_g X(h) = X(gh) \qquad \text{for every } h \in G$$

Let $\mathfrak{X}_L(G)$ be the set of left-invariant vector fields on G; then the maps $\rho_1: \mathfrak{X}_L(G) \to T_eG: X \mapsto X(e)$ and $\rho_2: T_eG \to \mathfrak{X}_L(G): \xi \mapsto \{g \mapsto X_\xi(g) = T_eL_g\xi\}$ satisfy $\rho_1 \circ \rho_2 = id_{T_eG}$ and $\rho_2 \circ \rho_1 = id_{\mathfrak{X}_L(G)}$. Therefore, $\mathfrak{X}_L(G)$ and T_eG are isomorphic as vector spaces. Actually $\mathfrak{X}_L(G)$ is a Lie subalgebra of the set of all vector fields on G because if $X, Y \in \mathfrak{X}_L(G)$, then for every $g \in G$,

$$\begin{aligned} L_{g*}[X,Y] &= [L_{g*}X, L_{g*}Y] \\ &= [X,Y] \end{aligned}$$

Defining a Lie bracket in T_eG by

$$[\xi,\eta] = [X_\xi, X_\eta](e) \qquad \text{for } \xi,\eta \in T_eG$$

makes T_eG into a Lie algebra (see 2.2.13). Note that $[X_\xi, X_\eta] = X_{[\xi,\eta]}$.

4.1.3 Definition. *The vector space T_eG with this Lie algebra structure is called the **Lie algebra** of G and is denoted by $\mathfrak{g}$ or if there is danger of confusion, by $\mathfrak{L}(G)$ or $\mathfrak{g}_G$.*

4.1.4 Example. For every $A \in L(\boldsymbol{R}^n,\boldsymbol{R}^n)$, $X_A: Gl(n,\boldsymbol{R}) \to L(\boldsymbol{R}^n,\boldsymbol{R}^n): Y \mapsto YA$ is a left-invariant vector field on $Gl(n,\boldsymbol{R})$ because for every $Z \in Gl(n,\boldsymbol{R})$, $X_A(L_Z Y) = ZYA = T_Y L_Z X_A(Y)$ and $L_Z: Gl(n,\boldsymbol{R}) \to Gl(n,\boldsymbol{R}): Y \mapsto ZY$ is a linear mapping.

Therefore, by the local formula $[X,Y](x) = DY(x)\cdot X(x) - DX(x)\cdot Y(x)$,

$$\begin{aligned} [A,B] &= [X_A, X_B](I) \\ &= DX_B(I)\cdot X_A(I) - DX_A(I)\cdot X_B(I) \end{aligned}$$

ISBN 0-8053-0102-X

But $X_B(Z) = ZB$ is linear in Z, so $DX_B(I)\cdot Z = ZB$. Hence $DX_B(I)\cdot X_A(I) = DX_B(I)\cdot A = AB$ and similarly $DX_A(I)\cdot X_B(I) = BA$. Thus, *$L(R^n, R^n)$ is the Lie algebra of $Gl(n, R)$ with Lie bracket given by*

$$[A, B] = AB - BA$$

4.1.5 Proposition. *Let H and G be Lie groups and $f: H \to G$ a smooth homomorphism. Then $T_ef: \mathfrak{L}(H) \to \mathfrak{L}(G)$ is a Lie algebra homomorphism.*

Proof. Since f is a homomorphism, $L_{f(h)} \circ f = f \circ L_h$ for every $h \in H$. Differentiation of this relation in h yields

$$X_{T_ef\cdot\xi} \circ f = Tf \circ X_\xi.$$

Even though f is not a diffeomorphism, we write this as

$$f_* X_\xi = X_{T_ef\cdot\xi}$$

Therefore,

$$T_ef\cdot[\xi, \eta] = T_ef\cdot[X_\xi, X_\eta](e) \qquad (e = e_H)$$

$$= f_*[X_\xi, X_\eta](e) \qquad (e = e_G)$$

$$= [f_*X_\xi, f_*X_\eta](e)$$

$$= [X_{T_ef\cdot\xi}, X_{T_ef\cdot\eta}](e)$$

$$= [T_ef\cdot\xi, T_ef\cdot\eta] \quad \blacksquare$$

For every $\xi \in T_eG$ let $\phi_\xi: R \to G: t \mapsto \exp t\xi$ denote the integral curve of X_ξ passing through e at $t = 0$. Because X_ξ is left invariant, its flow is complete. Indeed, the time of existence of the integral curve of X_ξ with initial condition g is the same as that with initial condition e since if $c(t)$ is an integral curve at e, $g\cdot c(t)$ is an integral curve at g; see Exercise 2.1B. Therefore, ϕ_ξ is defined for all $t \in R$. The following argument shows that

$$\phi_\xi(s + t) = \exp(s + t)\xi = \exp s\xi \exp t\xi = \phi_\xi(s)\phi_\xi(t) \tag{2}$$

for all $s, t \in R$; that is, ϕ_ξ is a smooth homomorphism of the (additive) group R into G and is therefore called a *one-parameter subgroup* of G. Fix $s \in R$ and define $\psi: R \to G: t \mapsto \phi_\xi(s)\phi_\xi(t) = L_{\phi_\xi(s)}\phi_\xi(t)$; then ψ is an integral curve of X_ξ passing through $\phi_\xi(s)$ at $t = 0$ by left invariance of X_ξ. Also $\theta: R \to G: t \mapsto \phi_\xi(s + t)$ is an integral curve of X_ξ passing through $\phi_\xi(s)$ at $t = 0$ because $\theta(0) = \phi_\xi(s)$ and

$$\frac{d\theta}{dt}(s + t) = \frac{d\theta}{d(s + t)}(s + t) = X_\xi(\theta(s + t))$$

Therefore, $\theta = \psi$ because the integral curve of X_ξ passing through $\psi_\xi(s)$ at

ISBN 0-8053-0102-X

$t=0$ is unique. Consequently (2) holds. Notice that we have proven that the flow of X_ξ is $F_t^\xi(g)=g\exp t\xi$.

If $\phi: \mathbf{R}\to G$ is a one-parameter subgroup of G, then $\phi=\phi_\xi$, where $\xi=d\phi/dt|_{t=0}$ because $\phi(t+s)=\phi(t)\phi(s)=L_{\phi(t)}\phi(s)$, which implies

$$\begin{aligned}\frac{d}{dt}\phi(t)&=\frac{d}{ds}\phi(t+s)\bigg|_{s=0}\\&=T_eL_{\phi(t)}\left(\frac{d\phi}{dt}\bigg|_{t=0}\right)\\&=T_eL_{\phi(t)}\xi=X_\xi(\phi(t))\end{aligned}$$

Thus ϕ is an integral curve of X_ξ passing through $e=\phi(0)$. But so is ϕ_ξ. Hence $\phi=\phi_\xi$ by uniqueness.

4.1.6 Definition. *The function* $\exp: T_eG\to G: \xi\mapsto\phi_\xi(1)$ *is called the* ***exponential mapping*** *of the Lie algebra of G into G. If there is any danger of confusion we will write* $\exp_G$.

The map $\exp$ is C^∞. Indeed, let Z be the vector field on $G\times\mathfrak{g}$ defined by $Z(g,\xi)=(X_\xi(g),0)$. Its flow is readily verified to be $F_t(g,\xi)=(g\exp t\xi,\xi)$. From its definition, $X_\xi(g)=T_eL_g\xi$ is smooth in ξ and g and thus Z and hence F_1 are smooth maps. In particular $F_1(e,\xi)=(\exp\xi,\xi)$ is smooth in ξ and so $\exp$ is smooth.

We have $T_0\exp=id_{T_eG}$ because

$$T_0(\exp)\xi=\frac{d}{dt}\exp t\xi\bigg|_{t=0}=\frac{d}{dt}\phi_\xi\bigg|_{t=0}=X_\xi(\phi_\xi(0))=\xi$$

Therefore, by the inverse function theorem, $\exp$ is a local diffeomorphism. In general, $\exp$ is not a diffeomorphism onto G as is shown in Example 4.1.9(c) below. Before giving these examples we note a basic property of the exponential map.

4.1.7 Proposition. *If* $f: H\to G$ *is a smooth homomorphism of Lie groups, then for all* $\eta\in\mathfrak{L}(H)$, $f(\exp_H\eta)=\exp_G(T_ef\cdot\eta)$.

Proof. The mapping $\phi: \mathbf{R}\to G: t\mapsto f(\exp_H t\eta)$ is a one-parameter subgroup of G. Therefore, $\phi(t)=\exp_G t\xi$, where $\xi=(d/dt)\phi|_{t=0}=T_ef\cdot\eta$, which implies $f(\exp_H\eta)=\phi(1)=\phi_\xi(1)=\exp_G\xi=\exp_G(T_ef\cdot\eta)$. ■

For every $g\in G$, let $I_g: G\to G: h\mapsto ghg^{-1}=R_{g^{-1}}L_gh$ be the inner automorphism associated with g. I_g is smooth and is a homomorphism because

$$I_g(hk)=ghkg^{-1}=ghg^{-1}gkg^{-1}=I_g(h)I_g(k)$$

ISBN 0-8053-0102-X

Let $Ad_g = T_e I_g = T_e(R_{g^{-1}} L_g)$: $T_eG \to T_eG$, called the *adjoint mapping* associated with g. Using proposition 4.1.7 we have:

4.1.8 Corollary. *$exp(Ad_g \xi) = I_g\, exp\, \xi = g(exp\, \xi)g^{-1}$ for every $\xi \in T_eG$ and every $g \in G$.*

4.1.9 Examples. (a) Consider $\boldsymbol{R}^n$ with the additive structure as a Lie group. Then the Lie algebra of $\boldsymbol{R}^n$ is $\boldsymbol{R}^n$ and $exp: \boldsymbol{R}^n \to \boldsymbol{R}^n$ is the identity.

(b) For every $A \in L(\boldsymbol{R}^n, \boldsymbol{R}^n)$, $\phi_A: \boldsymbol{R} \to Gl(n, \boldsymbol{R}) \subset \boldsymbol{R}^{n^2}: t \mapsto \Sigma_{n=0}^{\infty} t^n A^n / n!$ is a one-parameter subgroup because $\phi_A(0) = I$ and

$$\frac{d}{dt}\phi_A(t) = \sum_{n=1}^{\infty} \frac{t^{n-1}}{(n-1)!} A^n = \phi_A(t) A$$

which shows that ϕ_A is an integral curve of the left-invariant vector field X_A. Therefore, the exponential mapping is given by $exp: L(\boldsymbol{R}^n, \boldsymbol{R}^n) \to Gl(n, \boldsymbol{R})$: $A \mapsto \phi_A(1) = \Sigma_{n=0}^{\infty} A^n / n!$. For $P \in Gl(n, \boldsymbol{R})$, $P\phi_A P^{-1}$: $\boldsymbol{R} \to Gl(n, \boldsymbol{R})$: $t \mapsto P\phi_A(t)P^{-1}$ is an integral curve of $X_{PAP^{-1}}$ passing through I because $P\phi_A P^{-1}(0) = I$ and

$$\begin{aligned}\frac{d}{dt}\left(P\phi_A P^{-1}\right)(t) &= P\frac{d\phi_A(t)}{dt}P^{-1} \\ &= P\phi_A(t)AP^{-1} = \left(P\phi_A P^{-1}\right)(t) PAP^{-1}\end{aligned}$$

But so is $\phi_{PAP^{-1}}$. Therefore, $\phi_{PAP^{-1}} = P\phi_A P^{-1}$, that is,

$$exp\, PAP^{-1} = P(exp\, A)P^{-1}$$

(c) Consider $exp: L(\boldsymbol{R}^2, \boldsymbol{R}^2) \to Gl(2, \boldsymbol{R})$. The following argument shows that

$$B = \begin{pmatrix} -2 & 0 \\ 0 & -1 \end{pmatrix}$$

is not in the image of exp. By the real canonical form from linear algebra (see, e.g., Hirsch–Smale [1974], p. 129, Theorem 2) for every $A \in Gl(2, \boldsymbol{R})$, there is a $P \in Gl(2, \boldsymbol{R})$ such that PAP^{-1} is either

(i) $\begin{pmatrix} \lambda & 0 \\ 0 & \mu \end{pmatrix}$ for some $\mu, \lambda \in \boldsymbol{R}$,

(ii) $\begin{pmatrix} \alpha & -\beta \\ \beta & \alpha \end{pmatrix}$ for some $\alpha \in \boldsymbol{R}$, $\beta \in \boldsymbol{R} \setminus \{0\}$, or

(iii) $\begin{pmatrix} \lambda & 0 \\ 1 & \lambda \end{pmatrix}$ for some $\lambda \in \boldsymbol{R}$.

Therefore, $exp\, PAP^{-1}$ is either

ISBN 0-8053-0102-X

(i) $\begin{pmatrix} e^{\lambda} & 0 \\ 0 & e^{\mu} \end{pmatrix}$,

(ii) $e^{\alpha}\begin{pmatrix} \cos\beta & -\sin\beta \\ \sin\beta & \cos\beta \end{pmatrix}$, or

(iii) $\begin{pmatrix} e^{\lambda} & 0 \\ 1 & e^{\lambda} \end{pmatrix}$.

Suppose that $B = \exp A$ for some $A \in L(\mathbf{R}^2, \mathbf{R}^2)$; then for any $P \in Gl(2, \mathbf{R})$, $PBP^{-1} = P(\exp A)P^{-1} = \exp(PAP^{-1})$. Now cases (i) and (iii) cannot occur since $\operatorname{trace} B = -3$ and

$$\operatorname{trace} B = \operatorname{trace}(PBP^{-1}) = \operatorname{trace}\exp(PAP^{-1})$$

which is >0 in these cases. Also, B cannot have the canonical form (ii) on p. 257 since that form has eigenvalues $\lambda = \alpha \pm i\beta$ which can never equal $-2, -1$.

Of course, whenever G is not connected, $\exp$ cannot be onto because $\exp(\mathfrak{g})$ is connected. Any matrix with negative determinant is not in the component of the identity of $Gl(n, \mathbf{R})$ (since $\det > 0$ on the component of the identity) and hence is not in the image of $\exp$. However, in the example just given, B *is* in the component of the identity; it is joined to I by the curve

$$\begin{pmatrix} (1+\theta/\pi)\cos\theta & -\sin\theta \\ \sin\theta & \cos\theta \end{pmatrix} \text{ for } 0 \leqslant \theta \leqslant \pi$$

Thus, we *cannot* conclude that $\exp$ is onto the component of the identity. However, if G admits a bi-invariant Riemannian metric (e.g., if G is compact) then this is true. See Exercise 4.4D. [It follows that $Gl(2, \mathbf{R})$ does not admit a bi-invariant metric.]

4.1.10 Definition. *A Lie subgroup H of a Lie group G is a subgroup of G for which the inclusion mapping $i: H \to G$ is an immersion, that is, $i(H)$ is an immersed submanifold of G.*

The next example shows that the manifold topology on $i(H)$ need not be the topology induced from G. In other words, i need not be an embedding.

4.1.11 Example. Let $\alpha \in [0,1) \backslash \mathbf{Q}$ and define $\phi: \mathbf{R} \to T^2 = S^1 \times S^1 \subset C^2: t \mapsto (e^{2\pi i t}, e^{2\pi i \alpha t})$. Then ϕ is a one-parameter subgroup of T^2. Moreover, ϕ is injective, for $(e^{2\pi i t}, e^{2\pi i \alpha t}) = (e^{2\pi i s}, e^{2\pi i \alpha s})$ if and only if for some $m, n \in \mathbf{Z}$, $t = s + n$ and $\alpha t = \alpha s + m$; if $m \neq 0$ and $n \neq 0$, then $m = \alpha(t-s) = \alpha n$, which contradicts $\alpha \notin \mathbf{Q}$; hence either $m = 0$ or $n = 0$, which implies $t = s$. A similar argument shows that $T_t\phi(1) = d\phi/dt = 2\pi i(e^{2\pi i t}, \alpha e^{2\pi i \alpha t})$ is injective. Therefore,

ISBN 0-8053-0102-X

$\phi(\boldsymbol{R})$ is an injectively immersed submanifold of T^2. The following argument shows that $cl(\phi(\boldsymbol{R})) = T^2$, that is, $\phi(\boldsymbol{R})$ is *dense* in T^2. Let $p = (e^{2\pi i x}, e^{2\pi i y}) \in T^2$, then for all $m \in \boldsymbol{Z}$,

$$|\phi(x+m) - p| = |(0, e^{2\pi i \alpha x}(e^{2\pi i m \alpha} - e^{2\pi i z})|$$

where $y = \alpha x + z$. It suffices to show that $C = \{e^{2\pi i m \alpha} \in S^1 | m \in \boldsymbol{Z}\}$ is dense in S^1 because then there is a sequence $m_k \in \boldsymbol{Z}$ such that $e^{2\pi i m_k \alpha}$ converges to $e^{2\pi i z}$. Hence, $\phi(x + m_k)$ converges to p. If for each $k \in \boldsymbol{Z}_+$ we divide S^1 into k arcs of length $2\pi/k$, then, because $\{e^{2\pi i m \alpha} \in S^1 | m = 1, 2, \ldots, k+1\}$ are distinct, for some $1 \leqslant n_k < m_k \leqslant k+1$, $e^{2\pi i m_k \alpha}$ and $e^{2\pi i n_k \alpha}$ belong to the same arc. Therefore, $|e^{2\pi i m_k \alpha} - e^{2\pi i n_k \alpha}| < 2\pi/k$, which implies $|e^{2\pi i p_k \alpha} - 1| < 2\pi/k$, where $p_k = m_k - n_k$. Because

$$\bigcup_{j \in \boldsymbol{Z}_+} \{e^{2\pi i \alpha s} \in S^1 | s \in [jp_k, (j+1)p_k]\} = S^1$$

every arc of length less than $2\pi/k$ contains some $e^{2\pi i j p_k}$, which proves $cl(C) = S^1$. $\phi(\boldsymbol{R})$ is not an embedded submanifold of T^2 because it is not locally closed in the topology of T^2.

The difficulty in the above example is the fact that the subgroup is not closed.

4.1.12 Proposition. *If H is a closed subgroup of a Lie group G, then H is a submanifold of G and in particular is a Lie subgroup.*

Proof. (Adams [1969]). Put a norm $\|\cdot\|$ on $\mathfrak{g}_G = T_e G$. Let $\xi_n \in \mathfrak{g}_G, \xi_n \neq 0$ be such that $\exp \xi_n \in H$, $\xi_n \to 0$ and $\xi_n / \|\xi_n\| \to \xi \in \mathfrak{g}_G$. We will show that $\exp t\xi \in H$ for every $t \in \boldsymbol{R}$. Since $\xi_n \to 0$, for any $t \in \boldsymbol{R}$ there is a sequence of integers m_n such that $m_n \|\xi_n\| \to t$ as $n \to \infty$. Thus $\exp(m_n \xi_n) \to \exp t\xi$. But $\exp m_n \xi_n = (\exp \xi_n)^{m_n} \in H$ and H is closed, so $\exp t\xi \in H$.

Next, let $\mathfrak{g}_H = \{\xi \in \mathfrak{g}_G | \exp t\xi \in H \text{ for all } t \in \boldsymbol{R}\}$. We claim $\mathfrak{g}_H$ is a vector subspace of $\mathfrak{g}_G$. Clearly $\mathfrak{g}_H$ is closed under scalar multiplication. We need to show it is closed under addition. Let $\xi_1, \xi_2 \in \mathfrak{g}_H$ and suppose $\xi_1 + \xi_2 \neq 0$. For t sufficiently small, we can write $\exp t\xi_1 \cdot \exp t\xi_2 = \exp(f(t))$ since $\exp$ is a diffeomorphism of a neighborhood of e. Since

$$\xi_1 + \xi_2 = \frac{d}{dt} \exp t\xi_1 \exp t\xi_2 |_{t=0}$$

$(1/t)f(t) \to \xi_1 + \xi_2$ as $t \to 0$. Therefore, letting $\xi_n = f(1/n)$ and $\xi = (\xi_1 + \xi_2)/\|\xi_1 + \xi_2\|$, we see that $\exp t\xi \in H$, that is, $\xi_1 + \xi_2 \in \mathfrak{g}_H$.

Write $\mathfrak{g}_G = \mathfrak{g}_H \oplus \mathfrak{g}'$ and consider the diffeomorphism $\phi(\xi, \xi') = \exp \xi \cdot \exp \xi'$ between a neighborhood of 0 in $\mathfrak{g}_G$ and a neighborhood of e in G. (It is a local diffeomorphism by the implicit function theorem.) We use this map to show

ISBN 0-8053-0102-X

that *exp* maps a neighborhood of 0 in $\mathfrak{g}_H$ to a neighborhood of e in H. If not, there would be a sequence $(\xi_n, \xi'_n) \in \mathfrak{g}_H \oplus \mathfrak{g}'$ with $exp\,\xi_n \cdot exp\,\xi'_n \in H$, $exp\,\xi_n \cdot exp\,\xi'_n \to e$, and $\xi'_n \neq 0$. But $exp\,\xi_n \in H$, so $exp\,\xi'_n \in H$. There is a subsequence ξ'_{n_k} such that $\xi'_{n_k}/\|\xi'_{n_k}\| \to \xi \in \mathfrak{g}'$ (by compactness). But this would imply $\xi \in \mathfrak{g}_H$, which is impossible.

It follows from this that *exp* gives a submanifold chart for H near e modelled on $\mathfrak{g}_H$. By left translation, H is a submanifold around each of its points. ■

4.1.13 Proposition. *Let $i: H \to G$ be a Lie subgroup of G. Then for $\xi \in \mathfrak{L}(G)$, $\xi \in T_e i(\mathfrak{L}(H))$ if and only if $exp_G t\xi \in i(H)$ for all $t \in \mathbf{R}$.*

Proof. (Note that this was shown in the previous proof for closed subgroups.) If $\xi \in T_e i(\mathfrak{L}(H))$, then 4.1.7 yields at once that $exp_G t\xi \in i(H)$. Conversely, if $exp_G t\xi \in i(H)$ for all t, write $exp_G t\xi = i\phi(t)$ for $\phi(t) \in H$. Since i is an injective immersion, ϕ is a smooth one-parameter subgroup of H and so $\phi(t) = exp_H t\eta$ for $\eta \in \mathfrak{L}(H)$. Thus $\xi = T_e i \cdot \eta$. ■

Proposition 4.1.12 is just one of a number of "automatic smoothness" results for Lie groups. (See Varadarajan [1974] for further information.)

One can characterize Lebesgue measure up to a multiplicative constant on $\mathbf{R}^n$ by its invariance under translations. Similarly, on a locally compact group there is a unique left-invariant measure, called *Haar measure*. For Lie groups the existence of such measures is especially simple.

4.1.14 Proposition. *Let G be a Lie group. Then there is a volume form μ, unique up to nonzero multiplicative constants, which is left invariant. If G is compact, μ is right invariant as well.*

Proof. Pick any n-form μ_e on T_eG that is nonzero and define an n-form on T_gG by

$$\mu_g(v_1, \ldots, v_n) = \mu_e \cdot (TL_{g^{-1}} \cdot v_1, \ldots, TL_{g^{-1}} \cdot v_n)$$

Then μ_g is left invariant and smooth. For $n = \dim G$, μ_e is unique up to a scalar factor, so μ_g is as well.

Fix $g_0 \in G$ and consider $R^*_{g_0}\mu$. Since R_{g_0} and L_g commute, $R^*_{g_0}\mu$ is left invariant and hence $R^*_{g_0}\mu = c\mu$ for a constant c. Now if G is compact, this relationship may be integrated and by the change of variables formula, we deduce that $c = 1$. Hence μ is also right invariant. ■

This concludes our brief study of Lie groups as such, and we now turn to actions of groups on manifolds. Before proceeding the reader should be sure he understands some of the classical examples by consulting the exercises. For instance, later on, the fact that the Lie algebra of $SO(3)$ is $\mathbf{R}^3$ with the cross product as Lie bracket will be used without explicit mention.

ISBN 0-8053-0102-X

4.1.15 Definition. *Let M be a smooth manifold. An* ***action*** *of a Lie group G on M is a smooth mapping $\Phi: G\times M\rightarrow M$ such that (i) for all $x\in M$, $\Phi(e,x)=x$ and (ii) for every $g,h\in G$, $\Phi(g,\Phi(h,x))=\Phi(gh,x)$ for all $x\in M$.**

4.1.16 Examples. (a) A complete flow F on M is an action of $\mathbf{R}$ on M.

(b) If H is a subgroup of a Lie group G, then $\Phi: H\times G\rightarrow G: (h,g)\mapsto hg$ is an action of H on G.

For every $g\in G$ let $\Phi_g: M\rightarrow M: x\mapsto\Phi(g,x)$; then (i) becomes $\Phi_e=id_M$ while (ii) becomes $\Phi_{gh}=\Phi_g\circ\Phi_h$. Because $(\Phi_g)^{-1}=\Phi_{g^{-1}}$, Φ_g is a diffeomorphism. Definition 4.1.15 can be rephrased by saying that the mapping $g\mapsto\Phi_g$ is a homomorphism of G into the group of diffeomorphisms of M. If M is a vector space and each Φ_g is a linear transformation, the action of G on M is called a *representation* of G on M.

The following additional terminology regarding actions is useful.

4.1.17 Definitions. *Let Φ be an action of G on M. For $x\in M$, the* ***orbit*** *(or Φ-orbit) of x is given by*

$$G\cdot x=\{\Phi_g(x)\,|\,g\in G\}$$

An action is ***transitive*** *if there is just one orbit. It is* ***effective*** *(or* ***faithful****) if $\Phi_g=$ identity implies $g=e$; that is, $g\mapsto\Phi_g$ is one-to-one. An action is* ***free*** *if, for each $x\in M$, $g\mapsto\Phi_g(x)$ is one-to-one.*

The relation of belonging to the same Φ-orbit is an equivalence relation on M. We let M/G be the set of equivalence classes, that is, M/G is the set of Φ-orbits. Let $\pi: M\rightarrow M/G: x\mapsto[x]$, where $[x]$ is the Φ-orbit containing x. Give M/G the quotient topology, that is, $U\subset M/G$ is open if and only if $\pi^{-1}(U)$ is open in M.

4.1.18 Example. This example shows that the topology M/G need not be Hausdorff. Let $\Phi: \mathbf{R}\times\mathbf{R}\rightarrow\mathbf{R}, (t,x)\mapsto e^tx$, an action of the additive group $G=\mathbf{R}$ on $M=\mathbf{R}$. There are three orbits, $[-1]$, $[0]$, and $[1]$. It is readily checked that the open sets in M/G are the empty set, the whole space, $\{[-1]\}$, $\{[1]\}$, and $\{[-1],[1]\}$. In particular $\{[0]\}$ is not open, so the topology is not Hausdorff.

4.1.19 Proposition. *Let $\Phi: G\times M\rightarrow M$ be a smooth action and let $R=\{(m,\Phi_g m)\in M\times M\,|\,(g,m)\in G\times M\}$. If R is a closed subset of $M\times M$, then the quotient topology on M/G is Hausdorff.*

*Strictly speaking, this is a *left action*. A *right action* is a map $\Phi: M\times G\rightarrow M$ such that $\Phi(x,e)=x$ and $\Phi(\Phi(x,g),h)=\Phi(x,gh)$. For "automatic smoothness" results for group actions, see Bochner and Montgomery [1945] and Chernoff and Marsden [1970].

ISBN 0-8053-0102-X

Proof. Suppose M/G is not Hausdorff. Then there are distinct $[x], [y] \in M/G$ such that for any pair of neighborhoods U^x of $[x]$ and U^y of $[y]$, $U^x \cap U^y \neq \emptyset$.

Let V_i^x and V_i^y be nested bases of neighborhoods in M of x and y, $i=1,2,\ldots$. Let $W_i^x = \bigcup_{g\in G} \Phi_g V_i^x$ and $W_i^y = \bigcup_{g\in G} \Phi_g V_i^y$. Choosing $U_i^x = \pi(W_i^x)$ and $U_i^y = \pi(W_i^y)$, there must be $g_i, h_i \in G$ and $x_i \in V_i^x, y_i \in V_i^y$ such that

$$\Phi_{g_i} x_i = \Phi_{h_i} y_i, \quad \text{that is, } y_i = \Phi_{h_i^{-1} g_i} x_i$$

Now $y_i \to y$ and $x_i \to x$ as $i \to \infty$. Thus the points $(x_i, \Phi_{h_i^{-1}g_i} x_i) \in R$ converge, so the limit lies in R, as R is assumed closed. Thus $(x,y) \in R$; that is, $y = \Phi_g x$ for some $g \in G$ and so $[x]=[y]$. ■

The next theorem gives a necessary and sufficient condition for M/G to be a smooth manifold.

4.1.20 Theorem. *Let G act on M, and R be defined as in* 4.1.19. *Then R is a closed submanifold of* $M \times M$ *if and only if* M/G *has a smooth manifold structure such that* $\pi: M \to M/G$ *is a submersion.*

Proof. *Sufficiency.* Suppose that R is a closed submanifold of dimension r of $M \times M$ that has dimension $2n$. First we show that R is locally the graph of a smooth submersion of M into M, that is, for every $x \in M$ there is an open set $U \subset M$ with $x \in U$, a submanifold $N \subset M$ and a smooth submersion $\rho: U \subset M \to N \subset M$ such that for every $u \in U$, $\rho(u) \in N$ if and only if $(u, \rho(u)) \in R$. Since R is a submanifold of $M \times M$ and the map $R \to M: (x, \Phi_g(x)) \mapsto x$ is a submersion, by the local fibration theorem (Exercise 1.6G) find an open set $U_0 \subset M$ and a map $\eta: U_0 \times U_0 \subset M \times M \to \mathbf{R}^m$ such that $\eta^{-1}(\mathbf{0}) = (U_0 \times U_0) \cap R$ and $\eta_x = \eta_{1,x}: U_0 \subset M \to \mathbf{R}^m: y \mapsto \eta(x,y)$ is a submersion. This implies that $n \geqslant m$ and $\eta_x^{-1}(\mathbf{0})$ is a submanifold of U_0 with $T_x \eta_x^{-1}(\mathbf{0}) = E$, which has dimension $n-m$. Let F be a complement to E in $T_x M$. Shrinking U_0 if necessary, there is a submersion $\xi: U_0 \subset M \to \mathbf{R}^{n-m}$ such that $\xi^{-1}(\mathbf{0}) = N$ is a submanifold with $x \in N$ and $T_x \xi^{-1}(\mathbf{0}) = F$, which has dimension m. (See Exercise 1.6H.) Consider the mapping $\zeta: U_0 \times U_0 \subset M \times M \to \mathbf{R}^m \times \mathbf{R}^{n-m}$: $(y,z) \mapsto (\eta(y,z), \xi(z))$, then $\zeta(x,x) = (\mathbf{0},\mathbf{0})$ and the partial mapping $\zeta_x: U_0 \subset M \to \mathbf{R}_1^n = \mathbf{R}^m \times \mathbf{R}^{n-m}$: $z \mapsto (\eta_x(z), \xi(z))$; $T_x \zeta_x: T_x M \to T_{\zeta_x(x)} \mathbf{R}^n$ is bijective because $\ker T_x \zeta_x = \ker T_x \eta_x \cap \ker T_x \xi = E \cap F = \{\mathbf{0}\}$ and $\dim T_x M = n$. Therefore, the implicit function theorem applied to ζ gives an open set $U_1 \subset U_0$ with $x \in U_1$ and a smooth function $\rho: U_1 \subset M \to U_1 \subset M$ with $\rho(x) = x$ such that $\zeta^{-1}(\mathbf{0},\mathbf{0}) = \{(u, \rho(u)) \in U_1 \times U_1 \subset U \times U \mid u \in U_1\}$. For all $u \in U_1$, $\eta(u, \rho(u)) = \mathbf{0}$, which implies $(u, \rho(u)) \in (U_1 \times U_1) \cap R$ and $\xi(\rho(u)) = \mathbf{0}$, that is, $\rho(u) \in N$. It remains to show that ρ is a submersion near x. Differentiating $\eta(u, \rho(u)) = \mathbf{0}$ at $(x,x) = (x, \rho(x))$ gives

$$\mathbf{0} = T\eta_{2,x} + T\eta_{1,x} \circ T_x \rho$$

ISBN 0-8053-0102-X

where $\eta_{2,x}: U_1 \subset M \to R^m: z \mapsto \eta(z,x)$ and similarly $\eta_{1,x}(z) = \eta(x,z)$. Thus

$$\text{rank } T\eta_{2,x} \leqslant \text{rank } T_x\rho$$

But R is invariant under the diffeomorphism $j: M \times M \to M \times M: (y,z) \mapsto (z,y)$ and $\eta_{2,x} = (\eta \circ j)_{1,x}$. Therefore,

$$\text{rank } T\eta_{2,x} = \text{rank } T\eta_{1,x} = m$$

But $\dim N = m$, so $\text{rank } T_x\rho = m$, that is, ρ is a submersion near x.

Next we construct a chart at $[x]$ for M/G. Since R is closed, M/G is a Hausdorff space in the quotient topology. Since $T_x\rho(T_xM) = T_xN$ and $\rho(x) = x$, there is a chart (U,ϕ) at x of M such that $\phi \circ \rho \circ \phi^{-1}: \phi(U) = V \times W \subset R^m \times R^{n-m} \to V \times \{0\} \subset R^m \times \{0\}: (v,w) \mapsto (v,0)$. Therefore, for all $u \in U$, $u^1 = \rho(u)$ if and only if $\pi_1\phi(u) = \pi_1\phi(u^1)$, where $\pi_1: V \times W \to V: (v,w) \mapsto v$. Define $\omega: V \subset R^m \to \pi(U) \subset M/G: v \mapsto \pi \circ \phi^{-1}(v,0)$. Then ω is injective, for if $\pi(\phi^{-1}(v,0)) = \pi(\phi^{-1}(v^1,0))$, then $(\phi^{-1}(v,0), \phi^{-1}(v^1,0)) \in (U \times U) \cap R$; therefore, $\rho(\phi^{-1}(v,0)) = \phi^{-1}(v^1,0)$, which implies $v = \pi_1(\phi(\phi^{-1}(v,0))) = \pi_1(\phi(\phi^{-1}(v^1,0))) = v^1$. Since π and ϕ are continuous and open mappings, ω is a homeomorphism. Thus $(\pi(U), \omega^{-1})$ is a chart for M/G at $[x] = \pi(x)$.

Next we show that two charts $(\pi(U), \omega^{-1})$ and $(\pi(\tilde{U}), \tilde{\omega}^{-1})$ at $[x]$ are compatible. Let $Y = \pi^{-1}(\pi(U) \cap \pi(\tilde{U})) \cap U \subset U \cap \tilde{U}$ and $\tilde{Y} = \pi^{-1}(\pi(U) \cap \pi(\tilde{U})) \cap \tilde{U} \subset U \cap \tilde{U}$. The following argument shows that for every $v \in \pi_1\phi(Y) = V$ there is a unique $\tilde{v} \in \tilde{\pi}_1\tilde{\phi}(\tilde{Y}) = \tilde{V}$ such that

$$\pi(\phi^{-1}(v, W)) = \pi(\phi^{-1}(\tilde{v}, \tilde{W}))$$

where $W = \pi_2(\phi(Y))$, $\tilde{W} = \tilde{\pi}_2(\tilde{\phi}(\tilde{Y}))$ and $\pi_2: V \times W \to W: (v,w) \mapsto w$. For any $w, w' \in W$, $v = \pi_1(\phi\phi^{-1}(v,w)) = \pi_1(\phi\phi^{-1}(v,w^1))$, which implies $\pi(\phi^{-1}(v,w)) = \pi(\phi^{-1}(v,w^1))$. Therefore, for any $w \in W$, $\pi(\phi^{-1}(v,w)) = \pi\phi^{-1}(v,W)$. Similarly, for any $\tilde{w} \in \tilde{W}$, $\pi(\tilde{\phi}^{-1}(\tilde{v}^1,\tilde{w})) = \pi(\tilde{\phi}^{-1}(\tilde{v}^1,\tilde{W}))$. Since $Y = \phi^{-1}(V \times W)$, $\tilde{Y} = \tilde{\phi}^{-1}(\tilde{V} \times \tilde{W})$ and $\pi Y = \pi\tilde{Y}$, for every $(v,w) \in V \times W$ there is $(\tilde{v},\tilde{w}) \in \tilde{V} \times \tilde{W}$ such that

$$\pi(\phi^{-1}(v, W)) = \pi(\phi^{-1}(v,w)) = \pi(\tilde{\phi}^{-1}(\tilde{v},\tilde{w})) = \pi(\tilde{\phi}^{-1}(\tilde{v},\tilde{W}))$$

Suppose there is $(\tilde{v}^1,\tilde{w}^1) \in \tilde{V} \times \tilde{W}$ such that $\pi(\tilde{\phi}^{-1}(\tilde{v}^1,\tilde{w}^1)) = \pi(\tilde{\phi}^{-1}(\tilde{v},\tilde{w}))$, which implies $\tilde{v}^1 = \tilde{\pi}_1(\tilde{\phi}(\tilde{\phi}^{-1}(\tilde{v}^1,\tilde{w}^1)) = \tilde{\pi}_1(\tilde{\phi}(\tilde{\phi}^{-1}(\tilde{v},\tilde{w})) = \tilde{v}$. Thus $\tilde{v}$ is uniquely determined and so we can define a function $\psi: V \subset R^m \to \tilde{V} \subset R^m: v \mapsto \tilde{v}$. We now show that ψ is smooth. Since $\pi(\tilde{\phi}^{-1}(\tilde{v},\tilde{w})) = \pi(\phi^{-1}(v,w))$, $(\phi^{-1}(v,w), \tilde{\phi}^{-1}(\tilde{v},\tilde{w})) \in R$. Therefore, for some $g \in G$, $\Phi_g(\phi^{-1}(v,w)) = \tilde{\phi}^{-1}(\tilde{v},\tilde{w})$. Since Φ_g is a diffeomorphism, there is an open set $U \subset M$ with $x \in U \subset Y$ and $\Phi_g U \subset \tilde{Y}$. Therefore, the map

$$\tilde{\phi} \circ \Phi_g \circ \phi^{-1}: \phi(Y) \subset V \times W \subset R^m \times R^{n-m}$$
$$\to \tilde{\phi}(\tilde{Y}) \subset \tilde{V} \times \tilde{W}: (v,w) \mapsto (\psi(v), \theta(v,w))$$

ISBN 0-8053-0102-X

is smooth, which implies that ψ is smooth. Therefore, the charts $(\pi(U),\omega^{-1})$ and $(\pi(\tilde{U}),\tilde{\omega}^{-1})$ at $[x]$ are compatible, that is, M/G is a smooth manifold.

Let (U,ϕ) and $(\pi(U),\omega^{-1})$ be charts at $x\in M$ and $[x]\in M/G$, respectively, then $\omega^{-1}\circ\pi\circ\phi^{-1}$: $\phi(U)\subset V\times W\subset \boldsymbol{R}^m\times\boldsymbol{R}^n\to V\subset\boldsymbol{R}^m$: $(v,w)\mapsto v$ is a smooth submersion, which implies that π: $M\to M/G$ is a smooth submersion. This proves sufficiency.

Necessity. Since $\Delta_{M/G}=\{([x],[x])\in M/G\times M/G,\ [x]\in M/G\}$ is a closed submanifold of $M/G\times M/G$ and $\pi\times\pi$: $M\times M\to M/G\times M/G$: $(x,y)\mapsto(\pi(x),\pi(y))$ is a submersion, $(\pi\times\pi)^{-1}\Delta_{M/G}=R$ is a closed submanifold of $M\times M$. ■

We note the resemblance between the construction of charts on M/G and the proof of Exercise 1.6F.

A corollary of this argument, whose proof we leave to the reader, is the following useful technical remark: A map ϕ: $M/G\to N$ is smooth if and only if $\phi\circ\pi$: $M\to N$ is smooth.

This yields the following criterion of smoothness on quotient manifolds. Assume Φ: $G\times M\to M$, Ψ: $G\times N\to N$ are two smooth actions such that M/G, N/G are manifolds and π_M: $M\to M/G$, π_N: $N\to N/G$ are submersions. Let f: $M\to N$ be equivariant, that is, $f\circ\Phi_g=\Psi_g\circ f$ for all $g\in G$. This induces naturally a map $\hat{f}$: $M/G\to N/G$. Then smoothness of f implies smoothness of $\hat{f}$. This criterion is often called the *passage to quotients*.

The next result is a corollary of 4.1.20 concerning Lie groups themselves.

4.1.21 Corollary. *Let H be a closed subgroup of the Lie group G. If Φ: $H\times G\to G$: $(h,g)\mapsto hg$, then G/H is a smooth manifold and π: $G\to G/H$ is a submersion.*

Proof. Consider the mapping ξ: $G\times G\to G$: $(g,k)\mapsto g^{-1}k=m(i(g),k)$, where m: $G\times G\to G$: $(g,k)\mapsto gk$ and i: $G\to G$: $g\mapsto g^{-1}$. Since $T_{(g,k)}\xi(r,s)=T_k m_{i(g)}T_g i(r)+T_g m_k(s)$, where m_g: $G\to G$: $k\mapsto gk=L_g k$ and m_k: $G\to G$: $g\mapsto gk=R_k g$, $T_{(g,k)}\xi(r,s)=T_k L_{i(g)}\circ T_g i(r)+T_g R_k(s)$. Therefore, $T_{(g,k)}\xi(0,s)=T_g R_k(s)$. Thus $T_{(g,k)}\xi$ is surjective, because $T_g R_k$ is an isomorphism. Hence ξ is a submersion. Since H is a closed subgroup of G, H is a closed submanifold of G by 4.1.12. Hence $\xi^{-1}(H)$ is a closed submanifold of $G\times G$. But $(g,k)\in\xi^{-1}(H)$ if and only if $kg^{-1}\in H$, that is, if and only if $(g,k)\in R=\{(g,hg)\in G\times G\,|\,g\in G,\ h\in H\}$. Thus $\xi^{-1}(H)=R$ is closed submanifold of $G\times G$, which implies by Theorem 4.1.20 that G/H is a manifold and π: $G\to G/H$ is a submersion. ■

An action Φ: $G\times M\to M$ is called *proper* if and only if $\tilde{\Phi}$: $G\times M\to M\times M$ defined by $\tilde{\Phi}(g,x)=(x,\Phi(g,x))$ is a proper mapping, that is, if $K\subset M\times M$ is compact, then $\tilde{\Phi}^{-1}(K)$ is compact. Equivalently, if x_n converges in M and $\Phi_{g_n}x_n$ converges in M, then g_n has a convergent subsequence in G. For instance, if G is compact this condition is automatically satisfied. However, the $\boldsymbol{R}$ action in Example 4.1.18 is neither free nor proper, but the same action on $\boldsymbol{R}\backslash\{0\}$ is free and not proper.

ISBN 0-8053-0102-X

If $\Phi: G\times M\to M$ is a smooth action and $x\in M$, $G_x=\{g\in G|\Phi_g x=x\}$ is called the *isotropy group* of Φ at x. Since $G_x=\Phi_x^{-1}(x)$ and $\Phi_x: G\to M: g\mapsto\Phi(g,x)$ is continuous, G_x is a closed subgroup of G and hence by 4.1.12 a smooth submanifold. If the action is proper, G_x is compact. Because $\Phi_x(gh) = \Phi_g\circ\Phi_h x=\Phi_g x$ for every $h\in G_x$, Φ_x induces a mapping $\tilde{\Phi}_x: G/G_x\to G\cdot x\subset M: gG_x\mapsto\Phi_g x$. This map is injective because if $\Phi_g x=\Phi_h x$, then $g^{-1}h\in G_x$, that is, $gG_x=hG_x$.

4.1.22 Corollary. *If $\Phi: G\times M\to M$ is an action and $x\in M$, then $\tilde{\Phi}_x: G/G_x\to G\cdot x\subset M$ is an injective immersion. If Φ is proper, the orbit $G\cdot x$ is a closed submanifold of M and $\tilde{\Phi}_x$ is a diffeomorphism.*

Proof. First of all, $\tilde{\Phi}_x: G/G_x\to G\cdot x$ is smooth because $\tilde{\Phi}_x\circ\pi=\Phi_x$ is (see the Remark following 4.1.20). As we have already noted, $\tilde{\Phi}_x$ is one-to-one. To show it is an immersion, we show that $T\tilde{\Phi}_x([g])\cdot[\xi]$ is one-to-one. But $T\tilde{\Phi}_x([g])\cdot[\xi]=T\Phi_x(g)\cdot\xi$. Thus $T\tilde{\Phi}_x([g])$ will be one-to-one if we can show that $T_gG_x=\{\xi\in T_gG|T\Phi_x(g)\cdot\xi=0\}$. The inclusion $\subset$ is obvious. For the opposite inclusion, we first suppose $g=e$. Thus, let $\xi\in\mathfrak{g}$ satisfy $T\Phi_x(e)\cdot\xi=0$. Then

$$\frac{d}{dt}\Phi_x(exp\, t\xi)=T\Phi_x(exp\, t\xi)\cdot T(L_{exp\, t\xi})\cdot\xi$$

The defining property of an action may be written as

$$\Phi_x\circ L_g=\Phi_g\circ\Phi_x \quad \text{for all } g\in G,\ x\in M$$

Differentiating at e,

$$T\Phi_x(g)\cdot TL_g\cdot\xi=T\Phi_g(x)\cdot T\Phi_x(e)\cdot\xi$$

Taking $g=exp\, t\xi$,

$$\frac{d}{dt}\Phi_x(exp\, t\xi)=T\Phi_{exp\, t\xi}(x)\cdot T\Phi_x(e)\cdot\xi=0$$

Thus $\Phi_x(exp\, t\xi)=\Phi_x(e)=x$, so $exp\, t\xi\in G_x$, and thus $\xi\in T_eG_x$. This shows the inclusion $\supset$ for $g=e$. For the general case, note that the isomorphism $T_eL_g: T_eG\to T_gG$ satisfies $T_eL_g(T_eG_x)=T_gG_x$, $T_eL_g(\{\xi\in T_eG|T\Phi_x(e)\cdot\xi=0\})=\{\eta\in T_gG|T\Phi_x(g)\cdot\eta=0\}$ and the inclusion $\supset$ is proved. This completes the proof that $\tilde{\Phi}_x$ is an immersion.

If the action is proper, then $\tilde{\Phi}_x$ is a closed mapping and hence is a homeomorphism onto its image. ■

One can alternatively prove that the orbits $G\cdot x$ are immersed submanifolds by writing $G\cdot x=\Phi_x(G)$ and appealing directly to Exercise 1.6*F*(b). (Here, $ker\,\Phi_x$ is a subbundle of TG since it is $ker\, T\Phi_x(e)$ made left invariant.)

ISBN 0-8053-0102-X

Notice that if Φ is a transitive action of G on M, then for any $x \in M$, $G\cdot x = M$, so $M \approx G/G_x$. In this case M is called a *homogeneous space.* Conversely, if H is a subgroup of G and we set $M = G/H$ with G acting by left translation, M is a homogeneous space with $G_x = H$ for any $x \in M$. (See Exercises 4.1L, M.)

4.1.23 Proposition. *If $\Phi: G\times M \to M$ is a proper free smooth action, then M/G is a smooth manifold and $\pi: M \to M/G$ is a submersion.*

Proof. Let $\tilde{\Phi}: G\times M \to M\times M: (g,x) \mapsto (x, \Phi_g x)$, then the following argument shows that $\tilde{\Phi}$ has constant rank. Define the actions $\Lambda: G\times(M\times M) \to M\times M: (g,(x,y)) \mapsto (x,\Phi_g y)$ and $\Xi: G\times(G\times M) \to G\times M: (g,(h,x)) \mapsto (gh,x)$. Then $\Lambda_g: M\times M \to M\times M: (x,y) \mapsto (x,\Phi_g y)$ and $\Xi_g: G\times M \to G\times M: (h,x) \mapsto (gh,x)$ are diffeomorphisms for every $g \in G$. Also

$$\Lambda_g \circ \tilde{\Phi} = \tilde{\Phi} \circ \Xi_g \qquad \text{for every } g \in G \tag{1}$$

because

$$\Lambda_g \circ \tilde{\Phi}(h,x) = \Lambda_g(x,\Phi_h x) = (x, \Phi_g(\Phi_h x)) = (x, \Phi_{gh} x) = \tilde{\Phi}(gh,x) = \tilde{\Phi} \circ \Xi_g(h,x)$$

Taking the tangent of (1) at (e,x) gives

$$T_{\tilde{\Phi}(e,x)}\Lambda_g \circ T_{(e,x)}\tilde{\Phi} = T_{\Xi_g(e,x)}\tilde{\Phi} \circ T_{(e,x)}\Xi_g$$

which is equivalent to

$$T_{(x,x)}\Lambda_g \circ T_{(e,x)}\tilde{\Phi} = T_{(g,x)}\tilde{\Phi} \circ T_{(e,x)}\Xi_g$$

Therefore, the rank of $T_{(g,x)}\tilde{\Phi}$ equals the rank of $T_{(e,x)}\tilde{\Phi}$ since $T_{(x,x)}\Lambda_g$ and $T_{(e,x)}\Xi_g$ are isomorphisms. Thus the rank of $T_{(g,x)}\tilde{\Phi}$ is independent of $g \in G$. It remains hence to show that *rank* $T_{(e,x)}\tilde{\Phi}$ is independent of $x \in M$. We have

$$T_{(e,x)}\tilde{\Phi}: T_eG \times T_xM \to T_xM \times T_xM$$

$$T_{(e,x)}\tilde{\Phi}(\xi, v) = (v, T_e\Phi_x(\xi) + v)$$

and hence *rank* $T_{(e,x)}\tilde{\Phi} = n + \text{rank}\, T_e\Phi_x$, where $n = \dim M$. But by 4.1.22, the map $\Phi_x: G \mapsto G\cdot x$ is a diffeomorphism ($G\cdot x$ being an immersed submanifold in M) and hence

$$T_e\Phi_x(T_eG) = T_x(G\cdot x)$$

so that *rank* $T_e\Phi_x = \dim G$, which is independent of x. Since Φ is a free action, $\tilde{\Phi}$ is injective, for if $(x,\Phi_g x) = (y, \Phi_h y)$, then $x = y$ and $\Phi_g x = \Phi_h y = \Phi_h x$, which implies $x = \Phi_{h^{-1}g} x$; hence $h^{-1}g = e$. By the local fibration theorem

ISBN 0-8053-0102-X

(Exercise 1.6G) image $\tilde{\Phi} = R$ is an injectively immersed submanifold of $M \times M$. Because $\tilde{\Phi}$ is proper, it is a closed mapping. Thus R is closed and $\tilde{\Phi}^{-1}$ is continuous. Therefore, $\tilde{\Phi}$ is a homeomorphism, which implies that R is a submanifold of $M \times M$. By Theorem 4.1.20, M/G is a smooth manifold and $\pi: M \to M/G$ is a submersion. ■

Next we turn to the infinitesimal description of an action, which will be crucial for mechanics.

4.1.24 Definition. *Suppose $\Phi: G \times M \to M$ is a smooth action. If $\xi \in T_eG$, then $\Phi^\xi: \mathbf{R} \times M \to M: (t, x) \mapsto \Phi(\exp t\xi, x)$ is an* **R***-action on M, that is, Φ^ξ is a flow on M. The corresponding vector field on M given by*

$$\xi_M(x) = \frac{d}{dt}\Phi(\exp t\xi, x)\bigg|_{t=0}$$

is called the ***infinitesimal generator*** *of the action corresponding to ξ.*

Remark. From the proof of 4.1.22, we find that in the language of infinitesimal generators

$$T_x(G\cdot x) = \{\xi_M(x) | \xi \in \mathfrak{g}\}$$

4.1.25 Examples (a) Let $\Phi: G \times G \to G: (g, h) \mapsto gh = L_g h$, then Φ is a smooth action. If $\xi \in T_eG$, then $\Phi^\xi: \mathbf{R} \times G \to G: (t, h) \mapsto (\exp t\xi)h = R_h \exp t\xi$. Therefore, $\xi_G(g) = T_eR_g\xi$. Because $\Phi^\xi(t, R_gh) = (\exp t\xi)hg = R_g\Phi^\xi(t, h)$, ξ_G is *right* invariant and is therefore not equal to $X_\xi(g) = T_eL_g\xi$, which is *left* invariant, unless G is abelian.

(b) Let $\Phi: G \times T_eG \to T_eG: (g, \eta) \mapsto Ad_g\eta = T_e(R_{g^{-1}}L_g)\eta$, then Φ is a smooth action called the *adjoint action* of G on T_eG. If $\xi \in T_eG$, we claim that $\xi_{T_eG} = ad_\xi$, where $ad: T_eG \times T_eG \to T_eG: (\xi, \eta) \mapsto [\xi, \eta]$.

Indeed, let $\phi_t(g) = g \exp t\xi = R_{\exp t\xi}g$, the flow of X_ξ. Then

$$\begin{aligned}
[\xi, \eta] &= [X_\xi, X_\eta](e) \\
&= \frac{d}{dt}T_{\phi_t(e)}\phi_{-t}\cdot X_\eta(\phi_t(e))\Big|_{t=0} \\
&= \frac{d}{dt}T_{\exp t\xi}R_{\exp(-t\xi)}X_\eta(\exp t\xi)\Big|_{t=0} \\
&= \frac{d}{dt}T_{\exp t\xi}R_{\exp(-t\xi)}T_eL_{\exp t\xi}\eta\Big|_{t=0} \\
&= \frac{d}{dt}T_e(L_{\exp t\xi}R_{\exp(-t\xi)})\eta\Big|_{t=0}
\end{aligned}$$

ISBN 0-8053-0102-X

Therefore

$$\xi_{T_eG}(\eta)=\frac{d}{dt}Ad_{exp\,t\xi}\eta\Big|_{t=0}=[\xi,\eta]=ad_\xi\eta$$

(c) Rephrasing, the result (b) says that

$$\frac{d}{dt}Ad_{exp\,t\xi}\eta\Big|_{t=0}=[\xi,\eta]$$

that is, holding η fixed and differentiating,

$$T_e(Ad.\eta)\cdot\xi=[\xi,\eta]$$

where $Ad.\eta: G\to T_eG: g\mapsto Ad_g\eta$. More generally, from

$$Ad_{hg}\eta=Ad_h(Ad_g\eta)$$

we get, by differentiating in h,

$$T_g(Ad_g\eta)\cdot(T_eR_g\cdot\xi)=[\xi,Ad_g\eta],\quad \xi,\eta\in T_eG$$

that is,

$$T_g(Ad.\eta)\cdot\xi_g=[T_gR_{g^{-1}}\cdot\xi_g, Ad_g\eta],\quad \xi_g\in T_gG,\quad \eta\in T_eG$$

Therefore, if $x(t)$ is a smooth curve in G and $\xi\in T_eG$ and we let $\xi(t)=Ad_{x(t)}\xi$, the chain rule gives

$$\frac{d\xi(t)}{dt}=\left[T_{x(t)}R_{x(t)^{-1}}\frac{dx}{dt},\xi(t)\right]$$

a formula that we will need in Sect. 4.3. See (d) and Exercise 4.1H for the version on $\mathfrak{g}^*$.

(d) Let $\Phi: G\times(T_eG)^*\to(T_eG)^*: (g,\alpha)\mapsto Ad^*_{g^{-1}}\alpha$, where $Ad^*_g: (T_eG)^*\to(T_eG)^*: \beta\mapsto\{\eta\mapsto\beta(Ad_g\eta)\}$, then Φ is a smooth action called the *co-adjoint action* of G on $(T_eG)^*$. If $\xi\in T_eG$, then the following calculation shows that $\xi_{(T_eG)^*}=-ad^*_\xi$. Indeed, for $\eta\in T_eG$,

$$\begin{aligned}(\xi_{(T_eG)^*}(\alpha))\eta&=\frac{d}{dt}(Ad^*_{exp(-t\xi)}\alpha)(\eta)\Big|_{t=0}\\&=\frac{d}{dt}\alpha(Ad_{exp(-t\xi)}\eta)\Big|_{t=0}\\&=\alpha\left(\frac{d}{dt}Ad_{exp(-t\xi)}\eta\Big|_{t=0}\right)\\&=\alpha([-\xi,\eta])=-ad^*_\xi\alpha(\eta)\end{aligned}$$

ISBN 0-8053-0102-X

The next proposition gives the basic properties of infinitesimal generators.

4.1.26 Proposition. *Let $\Phi: G\times M\to M$ be a smooth action. For every $g\in G$ and $\xi,\eta\in T_eG$ we have*

(*i*) $(Ad_g\xi)_M=\Phi^*_{g^{-1}}\xi_M$ *and*
(*ii*) $[\xi_M,\eta_M]=-[\xi,\eta]_M$.

Proof. (i) For $x\in M$

$$(Ad_g\xi)_M(x)=\frac{d}{dt}\Phi(\exp tAd_g\xi,x)\bigg|_{t=0}\quad\text{(by Definition 4.1.24)}$$

$$=\frac{d}{dt}\Phi(g(\exp t\xi)g^{-1},x)\bigg|_{t=0}\quad\text{(by 4.1.8)}$$

$$=\frac{d}{dt}\Phi_g\circ\Phi(\exp t\xi,\Phi_{g^{-1}}(x))\bigg|_{t=0}\quad\text{(because }\Phi\text{ is an action)}$$

$$=T_{\Phi_{g^{-1}}(x)}\Phi_g\frac{d}{dt}\Phi(\exp t\xi,\Phi_{g^{-1}}(x))\bigg|_{t=0}\quad\text{(chain rule)}$$

$$=T_{\Phi_{g^{-1}}(x)}\Phi_g\xi_M(\Phi_{g^{-1}}(x))=(\Phi^*_{g^{-1}}\xi_M)(x)$$

(ii) Let $g=\exp t\eta$ in (i), so that

$$(Ad_{\exp t\eta}\xi)_M=\Phi^*_{\exp(-t\eta)}\xi_M$$

Now $\Phi_{\exp(-t\eta)}$ is the flow of $-\eta_M$, so differentiating in t at $t=0$, the right-hand side gives $[\xi_M,\eta_M]$. The derivative of the left-hand side at $t=0$ is, by 4.1.25(b), $([\eta,\xi])_M$. Thus (ii) follows. ■

Let $\tilde X_\xi(g)=T_eR_g\xi$. By 4.1.25(a), $\tilde X_\xi=\xi_G$ and so $[\tilde X_\xi,\tilde X_\eta]=-\tilde X_{[\xi,\eta]}$. The following ideas will play an important role in subsequent sections.

4.1.27 Definition. *Let M and N be manifolds and G a Lie group. Let Φ and Ψ be actions of G on M and N, respectively, and $f: M\to N$ a smooth map. We say f is **equivariant** with respect to these actions if for all $g\in G$,*

$$f\circ\Phi_g=\Psi_g\circ f$$

that is, the following diagram commutes.

$$\begin{array}{ccc} M & \xrightarrow{\ f\ } & N \\ {\scriptstyle\Phi_g}\downarrow & & \downarrow{\scriptstyle\Psi_g} \\ M & \xrightarrow{\ f\ } & N \end{array}$$

ISBN 0-8053-0102-X

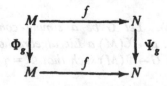

4.1.28 Proposition. *Let $f\colon M \to N$ be equivariant with respect to actions Φ and Ψ of G on M and N, respectively. Then for any $\xi \in \mathfrak{g}$,*

$$Tf \circ \xi_M = \xi_N \circ f$$

where ξ_M and ξ_N denote the infinitesimal generators on M and N, respectively, associated with ξ; in other words, the following diagram commutes:

$$\begin{array}{ccc} M & \xrightarrow{\quad f \quad} & N \\ {\scriptstyle \xi_M}\downarrow & & \downarrow{\scriptstyle \xi_N} \\ TM & \xrightarrow[\quad Tf \quad]{} & TN \end{array}$$

Proof. By equivariance,

$$f \circ \Phi_{\exp t\xi} = \Psi_{\exp t\xi} \circ f$$

Differentiating with respect to t at $t = 0$ and using the chain rule gives

$$Tf \circ \left(\frac{d}{dt} \Phi_{\exp t\xi} \bigg|_{t=0} \right) = \left(\frac{d}{dt} \Psi_{\exp t\xi} \bigg|_{t=0} \right) \circ f$$

that is, $Tf \circ \xi_M = \xi_N \circ f$. ■

We conclude this section with a few supplementary remarks on actions that help to unify ideas.

If we think of an action Φ of G on M as a homomorphism of G to $\mathfrak{D}(M)$, the diffeomorphism group of M, by $g \mapsto \Phi_g$, we can ask what the induced homomorphism of Lie algebras is [see 4.1.5 and Exercise 4.1G for a discussion of why $\mathfrak{X}$ is the Lie algebra of $\mathfrak{D}(M)$]. It is exactly the map

$$\Phi' = T_e\Phi\colon \mathfrak{g} \to \mathfrak{X}; \qquad \xi \mapsto \xi_M$$

[In 4.1.26(ii) we saw that Φ' is an *anti-homomorphism*; this is because the Lie algebra of $\mathfrak{D}(M)$ is $\mathfrak{X}(M)$ with bracket $-[X, Y]$; see Exercise 4.1G.] The action is called *essential* if this homomorphism $G \to \mathfrak{D}(M)\colon g \mapsto \Phi_g$ is injective.

It is not difficult to see that a homomorphism of connected Lie groups is determined by its tangent at the identity (Chevalley [1946, p. 113]). The analog of this for the maps $g \mapsto \Phi_g$ and Φ' is the following result of Palais [1957, Chapters II and III].

4.1.29 Theorem (Palais). *Let G be a simply connected Lie group, M a compact manifold, and $\varphi\colon \mathfrak{g} \to \mathfrak{X}(M)$ a Lie algebra homomorphism. Then there exists a unique action $\Phi\colon G \to \mathfrak{D}(M)$ such that $\Phi' = \varphi$.*

ISBN 0-8053-0102-X

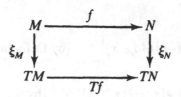

It follows that in the context of this theorem, the actions of G on M are in bijection with the potential infinitesimal generators, or homomorphisms of $\mathfrak{g}$ into $\mathfrak{X}(M)$. In our analogy $\Phi' = T_e\Phi$ above, it is clear that Φ should be essential iff Φ' is a monomorphism, and this is in fact made rigorous by the proof of 4.1.29 (omitted). Thus the essential actions of G on M are parametrized by isomorphisms of $\mathfrak{g}$ onto subalgebras of $\mathfrak{X}(M)$. These in turn may be parametrized as follows. Choose an ordered basis $(x_1, ,\ldots, x_k)$ for the real vector space $\mathfrak{g}$. The **constants of structure** $\{c^{\gamma}_{\alpha\beta}\}$ are defined by the **commutation relations**

$$[x_\alpha, x_\beta] = c^{\gamma}_{\alpha\beta} x_\gamma$$

(summed on $\gamma = 1, \ldots, k$). Then a monomorphism $\varphi: \mathfrak{g} \to \mathfrak{X}(M)$ is uniquely determined for any ordered linearly independent set $(Y_1, \ldots, Y_k) \subset \mathfrak{X}(M)$ satisfying the same commutation relations

$$[Y_\alpha, Y_\beta] = c^{\gamma}_{\alpha\beta} Y_\gamma$$

by the condition $\varphi(x_\alpha) = Y_\alpha$, and linear extension over $\mathfrak{g}$.

In this way we obtain a bijection between essential actions of G on M and k-tuples of vector fields on M that are linearly independent and satisfy a fixed system of commutation relations. In case G is not simply connected or M is not compact, the parametrization of essential actions is considerably more complicated. This aspect of the theory is not explicitly needed in the usual examples of actions in mechanics, but is useful for the intuition it provides. For more details, see Palais [1957] and Hermann [1966].

Some other investigations of Palais are also worth noting. Namely, he has a general result which asserts that the diffeomorphisms of a manifold which preserve a "geometric structure" form a Lie group. This generalizes the classic result of Myers and Steenrod [1939], which states that the isometries of a Riemannian manifold form a finite-dimensional Lie group. Of course the Euclidean group, $O(n, \boldsymbol{R}) \times \boldsymbol{R}^n$, the isometry group of $\boldsymbol{R}^n$, is a special case. See Kobayashi [1973] for the proof and discussion.

EXERCISES

4.1A. (i) Identify the Lie algebra of $SO(3, \boldsymbol{R}) = SO(3)$ with $\boldsymbol{R}^3$ as follows; define the map:

$$\boldsymbol{R}^3 \to T_e SO(3): \boldsymbol{x} = (x^1, x^2, x^3) \mapsto \hat{\boldsymbol{x}} = \begin{bmatrix} 0 & -x^3 & x^2 \\ x^3 & 0 & -x^1 \\ -x^2 & x^1 & 0 \end{bmatrix}$$

Show that $(\boldsymbol{x} \times \boldsymbol{y})\hat{} = [\hat{\boldsymbol{x}}, \hat{\boldsymbol{y}}]$, where $\times$ is the usual vector (cross) product on $\boldsymbol{R}^3$. Thus the Lie algebra of $SO(3)$ may be viewed as $\boldsymbol{R}^3$ with vector product as Lie bracket. Note that $\boldsymbol{x} \cdot \boldsymbol{y} = -\frac{1}{2}\, \mathrm{trace}\, (\hat{\boldsymbol{x}}\hat{\boldsymbol{y}})$.

ISBN 0-8053-0102-X

(ii) Suppose $\hat{x} \in T_eSO(3,R)$. Show that $\exp t\hat{x}$ is a rotation about the axis $x \in R^3$ through the angle $t\|x\|$, where $\|x\|$ is the Euclidean norm on R^3.

(iii) Consider the linear action $\Phi: SO(3)\times R^3 \to R^3: (A,x)\mapsto Ax$. For $\xi \in R^3$ let $\hat{\xi} \in T_eSO(3,R)$, where ˆ is given in (i), show that the infinitesimal generator is $\xi_{R^3}(x) = \xi \times x$.

(iv) Show that the adjoint action of $SO(3)$ on R^3 is the "usual" action described in (iii).

(v) Show that $S^2 \approx SO(3)/SO(2)$ is a homogeneous space, with the action of $SO(3)$ on S^2 being the "usual one."

4.1B. Let $R^\times$ be the multiplicative group of nonzero reals. Show that the homomorphism of Lie algebras corresponding to $det: GL(n; R) \to R^\times$ is the trace. Show this by showing that $T_e \det A = tr A$ directly from properties of the determinant. Deduce that $\det(e^A) = e^{trA}$ by a general fact about Lie groups.

4.1C. (i) Let E be a finite-dimensional vector space with a bilinear form $\langle , \rangle$. Let G be the group of isometries of E, that is, $G = \{F \in L(E) \mid F \text{ is an isomorphism of } E \text{ onto } E \text{ and } \langle Fe, Fe' \rangle = \langle e, e' \rangle \text{ for all } e, e' \in E\}$. Show that G is a subgroup and a closed submanifold of $Gl(E)$. Show that $\{K \in L(E) \mid \langle Ke, e' \rangle + \langle e, Ke' \rangle = 0 \text{ for all } e, e' \in E\}$ is the Lie algebra of G.

(ii) If $\langle , \rangle$ denotes the Minkowski metric on R^4; that is,

$$\langle x, y \rangle = \sum_{i=1}^{3} x^i y^i - x^4 y^4$$

then the group of linear isometries is called the *Lorentz group L*. Prove the following facts. The dimension of L is six and L has four connected components. If

$$S = \begin{pmatrix} I_3 & 0 \\ 0 & -1 \end{pmatrix} \in Gl(4,R)$$

then $L = \{A \in Gl(4,R) \mid A'SA = S\}$ and the Lie algebra of L is $\{A \in L(R^4, R^4) \mid SB + B'S = 0\}$. The identity component of L is $\{A \in L \mid \det A > 0 \text{ and } A_{44} > 0\} = L_\uparrow^+$; L and $L_\uparrow^+$ are not compact.

4.1D. This is a list of some important matrix Lie groups. Prove the facts stated.

(i) $Gl(n, C) = \{n \times n \text{ invertible complex matrices}\}$ has Lie algebra the set of all $n \times n$ matrices and has complex dimension n^2, that is, real dimension $2n^2$.

(ii) $Sl(n, C) = \{A \in Gl(n, C) \mid \det A = 1\}$ is a Lie subgroup of $Gl(n, C)$, has Lie algebra $\{B \in L(C^n, C^n) \mid tr B = 0\}$, and has complex dimension $n^2 - 1$, that is, real dimension $2(n^2-1)$.

(iii) $Gl(n, R) = \{n \times n \text{ invertible real matrices}\}$ has Lie algebra the set of all $n \times n$ real matrices and has real dimension n^2.

(iv) $Sl(n, R) = \{A \in Gl(n, R) \mid \det A = 1\}$ is a Lie subgroup of $Gl(n, R)$, has Lie algebra $\{B \in L(R^n, R^n) \mid tr B = 0\}$ and has real dimension $n^2 - 1$.

(v) $O(n)$, the group of orthogonal real matrices, $= \{A \in Gl(n, R) \mid \langle Ax, Ay \rangle = \langle x, y \rangle, \text{ where } \langle x, y \rangle = \sum_{i=1}^n x_i y_i \text{ and } x = (x_1, \ldots, x_n), y = (y_1, \ldots, y_n)\}$ is a Lie group with Lie algebra $\{B \in L(R^n, R^n) \mid \langle Bx, y \rangle + \langle x, By \rangle = 0\}$, has dimension $n(n-1)/2$ and is compact.

(vi) $SO(n)$, the special orthogonal group, $= \{A \in O(n) \mid \det A = 1\}$ is the connected component of $O(n)$ containing the identity I and is a compact Lie group of the same dimension as $O(n)$.

ISBN 0-8053-0102-X

(vii) $U(n)$, the unitary group, $=\{A \in Gl(n,C) | \langle Ax, Ay\rangle = \langle x,y\rangle$, where $\langle x,y\rangle = \sum_{i=1}^{n} x_i\bar{y}_i$, $x=(x_1,\ldots,x_n)\in C^n, y=(y_1,\ldots,y_n)\in C^n\}$ is a real Lie group with Lie algebra $\{B\in L(C^n,C^n)|\langle Bx,y\rangle+\langle x,By\rangle=0\}$, has real dimension n^2 and is compact.

(viii) $SU(n)=\{A\in U(n)| \det A=1\}$ is a Lie subgroup of $U(n)$ of real dimension n^2-1, has Lie algebra $\{B\in L(C^n,C^n)|\langle Bx,y\rangle+\langle x,By\rangle=0$, $\langle\ \rangle$ given in (vii) and $tr\,B=0\}$.

(ix) $Sp(2n,R)=\{A\in Gl(2n,R)|\sigma(Ax,Ay)=\sigma(x,y)$, where

$$\sigma(x,y)=x^t\begin{pmatrix}0 & -I_n\\ I_n & 0\end{pmatrix}y\Big\}$$

is the real symplectic group, has Lie algebra $\{B\in L(R^n,R^n)|\sigma(Bx,y)+\sigma(x,By)=0\}$, is of real dimension $2n^2+n$, and is noncompact.

4.1E. (a) (Lie's Theorem). Let G be a Lie group with Lie algebra $\mathfrak{g}$. Let $\mathfrak{h}\subset\mathfrak{g}$ be a Lie subalgebra. Show that $\mathfrak{h}$ is the Lie algebra of a connected Lie subgroup $H\subset G$. (*Hint:* Use Frobenius' theorem and take H to be the leaf of the foliation through e determined by the integrable subbundle defined by the left translates of $\mathfrak{h}$ in TG.)

(b) Show that a connected Lie group with an abelian Lie algebra is abelian. (This occurs only if G is a product of a torus with Euclidean space; see 4.1K.) Also show that a connected Lie subgroup H of a Lie group G is determined by its Lie algebra.

4.1F. (Requires a knowledge of covering spaces.)

(a) Let G be a Lie group with Lie algebra $\mathfrak{g}$ and identity e. Let $\tilde{G}$ be the universal covering space of G with projection $\pi\colon \tilde{G}\to G$. Show that $\tilde{G}$ becomes a Lie group (with an identity e' chosen in $\pi^{-1}(e)$).

(b) If $G=S^1\approx SO(2)$, show that $\tilde{G}=R$, $\pi(x)=e^{ix}$.

(c) Show that the universal covering group of $SO(3,R)$ is $SU(2,C)$ (the spin group) by considering the map $\pi\colon SU(2,C)\to SO(3,R)$ defined as follows.

Let $\sigma_1,\sigma_2,\sigma_3$ be the *Pauli spin matrices*, that is,

$$\sigma_1=\begin{pmatrix}0&1\\1&0\end{pmatrix},\quad \sigma_2=\begin{pmatrix}0&-i\\i&0\end{pmatrix},\quad \text{and}\quad \sigma_3=\begin{pmatrix}1&0\\0&-1\end{pmatrix}$$

and let

$$\sigma=(\sigma_1,\sigma_2,\sigma_3)$$

Then

$$x=(x^1,x^2,x^3)\leftrightarrow\begin{pmatrix}x^3 & x^1-ix^2\\ x^1+ix^2 & -x^3\end{pmatrix}$$

$$=x\cdot\sigma=x^1\sigma_1+x^2\sigma_2+x^3\sigma_3$$

sets up an isomorphism between R^3 and the 2×2 traceless Hermitian matrices. Note that

$$-\det(x\cdot\sigma)=\|x\|^2$$

ISBN 0-8053-0102-X

so that the orthogonal group $O(3)$ corresponds to the transformations that are determinant preserving.

Next show that $SU(2)$ is diffeomorphic to the three-sphere by

$$x=(x^1,\ldots,x^4)\leftrightarrow\begin{pmatrix} x^1+ix^2 & x^3+ix^4 \\ -x^3+ix^4 & x^1-ix^2\end{pmatrix}$$

In particular, deduce that $SU(2)$ is simply connected.

Now define $\pi: SU(2)\to 3\times 3$ matrices by

$$\pi(A)(x)\cdot\sigma=A(x\cdot\sigma)A^*=A(x\cdot\sigma)A^{-1}$$

Since $det(A(x\cdot\sigma)A^{-1})=det(x\cdot\sigma)$,

$$\pi(SU(2))\subset O(3)$$

But $\pi(SU(2))$ is connected, so

$$\pi(SU(2))\subset SO(3)$$

Also show that $\pi(A)=\pi(B)$ iff $A=\pm B$.

Finally show that π is onto and is a local diffeomorphism. To show it is a local diffeomorphism, use the inverse function theorem. [*Hint:* $T_e\pi$: $\alpha\mapsto\tilde{\alpha}$, where $\tilde{\alpha}(x\cdot\sigma)=(x\cdot\sigma)\alpha^*+\alpha(x\cdot\sigma)$. Use the fact that an open subgroup is also closed (its complement is a union of open cosets), and connectivity of $SO(3)$ to obtain $\pi(SU(2))=SO(3)$.]

4.1G. Let M be a manifold and let $\mathfrak{D}$ be the group of C^∞ diffeomorphisms of M to M. Ignoring all questions of smoothness,* show that:

(i) The tangent space to $\mathfrak{D}$ at η may be identified with maps $X: M\to TM$ such that $\tau_M\circ X=\eta$. (*Hint.* See 3.8.2.) In particular, the Lie algebra of $\mathfrak{D}$ is, as a vector space, the space of vector fields $\mathfrak{X}$ on M.

(ii) For $X\in\mathfrak{X}$, show that $exp(tX)$ is the flow of X. (*Warning: exp* is not onto a neighborhood of the identity.)

(iii) For $X\in\mathfrak{X}$, show that the *right*-invariant vector field $\tilde{X}$ corresponding to X is $\tilde{X}(\eta)=X\circ\eta$.

(iv) Show that

$$[\tilde{X},\tilde{Y}](e)=[X,Y]$$

the usual Lie bracket of vector fields. (*Hint:* Use the local formula for the bracket and $T\tilde{X}(e)\cdot Y=TX\circ Y$.)

(v) Conclude that the Lie bracket on $\mathfrak{X}$ as defined in the text (using *left*-invariant vector fields) is the negative of the usual one.

4.1H. Let G be a Lie group and $\mu\in\mathfrak{g}^*$. Let $x(t)$ be a curve in G and let $\xi_g\in T_gG$.

---

*These are answered in Leslie [1967], Omori [1970], Ebin [1970a], and Ebin–Marsden [1970] and references therein.

ISBN 0-8053-0102-X

Prove that

$$T_g\left(Ad^*_{g^{-1}}\mu\right)(\xi_g) = -ad\left(T_gR_{g^{-1}}(\xi_g)\right)Ad^*_{g^{-1}}\mu$$

and

$$T_g\left(Ad^*\mu\right)(\xi_g) = Ad^*_g ad\left(T_gR_{g^{-1}}(\xi_g)\right)^*\mu.$$

[*Hint:* This can be deduced from 4.1.25(c) by differentiating $\mu(t)\cdot\xi = \mu\cdot Ad_{x(t)^{-1}}\xi$ or by using 4.1.25(d) and the method of proof of 4.1.25(c).]

4.1I. Let G be a discrete subgroup of R^n (under addition). Show that there are linearly independent vectors $v_1,\ldots,v_l \in R^n$ such that

$$G = \{v \in R^n \mid v = \Sigma m_i v_i,\ m_i \in Z\}$$

4.1J. Let G be a Lie group and $\xi, \eta \in \mathfrak{g}$. Let

$$K_t(g) = g\,exp\,t\xi\,exp\,t\eta$$

and

$$F_t(g) = g\,exp\,t\xi\,exp\,t\eta\,exp(-t\xi)\,exp(-t\eta)$$

(a) Show that $K_0 = F_0 = Id$, $K_0'(e) = \xi + \eta$, $F_0'(e) = 0$, $\frac{1}{2}F_0''(e) = [\xi, \eta]$. (If ξ and η are replaced by $-\xi, -\eta$ in the formula for $F_0''(e)$, the Remark preceding 3.1.13 results.) (*Hint:* Use the fact that the flow of X_ξ, the left-invariant vector field coinciding with ξ at e, is $g \mapsto g\,exp\,t\xi$.)

(b) Use (a), 2.1.27, and Exercise 2.2L to prove the following product formulas of Lie:

(i) $$exp(\xi + \eta) = \lim_{n\to\infty}\left(exp\frac{\xi}{n}\,exp\frac{\eta}{n}\right)^n$$

(ii) $$exp[\xi, \eta] = \lim_{n\to\infty}\left(exp\frac{\xi}{\sqrt{n}}\,exp\frac{\eta}{\sqrt{n}}\,exp\left(-\frac{\xi}{\sqrt{n}}\right)exp\left(-\frac{\eta}{\sqrt{n}}\right)\right)^n$$

4.1K. (a) Use 4.1J(b) to show that if G is abelian then $exp: \mathfrak{g} \to G$ is a group homomorphism. Show that if G is connected, exp is onto.

(b) Show that if G is a connected abelian Lie group of dimension n, then G is isomorphic to $T^k \times R^{n-k}$. [*Hint:* By (a) $G \cong \mathfrak{g}/ker\,exp$ and $ker\,exp$ is discrete; then use 4.1I. Compare this with the proof of Arnold's invariant tori theorem 5.2.23.]

4.1L. Let H be a closed subgroup of the Lie group G and denote by $\pi: G \to G/H$ the canonical projection.

(a) Show that $\hat{L}: G \times G/H \to G/H: (g, \pi(g')) \mapsto \pi(gg')$ is a well-defined smooth action on G/H. (*Hint:* Use the passage to quotient criterion explained after 4.1.20.)

(b) Show that for each $[g] \in G/H$ there exists a neighborhood $\mathcal{U}$ of $[g]$ in G/H such that $\pi^{-1}(\mathcal{U})$ is diffeomorphic to $\mathcal{U} \times H$. (*Hint:* Show the existence of a section $\mathcal{U} \to \pi^{-1}(\mathcal{U})$ using $\mathcal{U}$ as chart at $[g]$ and the way

ISBN 0-8053-0102-X

the charts arise on G/H from those on G.) This proves that $\pi: G \to G/H$ is a principal fiber bundle with structural group H.

4.1M. (Requires knowledge of fiber bundles.)

(a) If G acts on M smoothly, properly, and freely, then $M \to M/G$ is a principal fiber bundle with structural group G. (*Hint:* Use the charts given in 4.1.20 to construct local sections as in 4.1L.)

(b) If G acts on M smoothly, properly, and such that all its isotropy groups are conjugate, that is, for all $x, y \in M$, there exists $g \in G$ such that $gG_xg^{-1} = G_y$, then M/G is a manifold and $\pi: M \to M/G$ is a locally trivial fiber bundle with structural group $N(H)/H$, $H = G_x$ and fiber $G \cdot x$, where $N(H)$ denotes the normalizer of H, that is, the smallest subgroup of G such that H is normal in it. (*Hint:* Modify the proof of 4.1.20 and (a) above; see also Hsiang [1966] and the slice theorem in, e.g., Palais [1957].)

4.2 THE MOMENTUM MAPPING

In this section we show how to obtain integrals (i.e., conserved quantities) for a Hamiltonian system with symmetries. Linear and angular momentum associated with translational and rotational invariance are the most common examples. However, topologically more complex examples such as the rigid body benefit from a rather abstract point of view. The final picture presented here is due primarily to J. M. Souriau [1970a], with important contributions by B. Kostant [1970], S. Smale [1970a] and J. Marsden [1968a].

4.2.1 Definition *Let (P, ω) be a connected symplectic manifold and $\Phi: G \times P \to P$ a **symplectic action** of the Lie group G on P; that is, for each $g \in G$, the map $\Phi_g: P \to P$; $x \mapsto \Phi(g, x)$ is symplectic. We say that a mapping*

$$J: P \to \mathfrak{g}^* \quad (\textit{the dual of the Lie algebra of } G)$$

*is a **momentum mapping** for the action provided that for every $\xi \in \mathfrak{g}$,*

$$d\hat{J}(\xi) = i_{\xi_P}\omega$$

where $\hat{J}(\xi): P \to \mathbf{R}$ is defined by $\hat{J}(\xi)(x) = J(x) \cdot \xi$ and ξ_P is the infinitesimal generator of the action corresponding to ξ. In other words, J is a momentum mapping provided

$$X_{\hat{J}(\xi)} = \xi_P$$

*for all $\xi \in \mathfrak{g}$. Sometimes (P, ω, Φ, J) is called a **Hamiltonian G-space.***

Shortly we shall see how to compute J for large classes of symplectic actions. Let us note, however, that not every symplectic action has a momentum mapping. The reason is precisely because not every locally Hamiltonian vector field is globally Hamiltonian; see Remark 1 following 3.3.6. If each ξ_P is globally Hamiltonian, there is a momentum mapping. Indeed, let $\xi_1, \ldots, \xi_k$ be a basis for $\mathfrak{g}$ and let $J_1, \ldots, J_k$ be the Hamiltonians for $(\xi_1)_P, \ldots, (\xi_k)_P$. Let

ISBN 0-8053-0102-X

$\hat{J}(\xi_i)=J_i$ and extend by linearity. Clearly if J and J' are two momentum mappings for the same action, there is a $\mu\in\mathfrak{g}^*$ such that $J(p)-J'(p)=\mu$ for all $p\in P$.

We want to develop a number of basic properties of momentum mappings as such, but before doing so we bring out their importance by means of the following fundamental conservation law.

4.2.2 Theorem. *Let Φ be a symplectic action of G on (P,ω) with a momentum mapping J. Suppose $H: P\to\mathbf{R}$ is invariant under the action, that is,*

$$H(x)=H(\Phi_g(x)) \quad \text{for all } x\in P, \quad g\in G$$

*Then J is an **integral** for X_H; that is, if F_t is the flow of X_H,*

$$J(F_t(x))=J(x)$$

Proof. For each $\xi\in\mathfrak{g}$ we have $H(\Phi_{\exp t\xi}x)=H(x)$ since H is invariant. Differentiating at $t=0$,

$$dH(x)\cdot\xi_P(x)=0$$

that is,

$$L_{X_{\hat{J}(\xi)}}H=0$$

that is,

$$\{H,\hat{J}(\xi)\}=0$$

which, by 3.3.13, proves that

$$\hat{J}(\xi)(F_t(x))=\hat{J}(\xi)(x)$$

for each ξ. But this is equivalent to our assertion. ■

We now turn to our study of momentum mappings. We begin with the following.

4.2.3 Proposition. *Let (P,ω,Φ,J) be a Hamiltonian G-space. Define, for $g\in G$ and $\xi\in\mathfrak{g}$,*

$$\psi_{g,\xi}: P\to\mathbf{R}: x\mapsto\hat{J}(\xi)(\Phi_g(x))-\hat{J}(Ad_{g^{-1}}\xi)(x)$$

*Then $\psi_{g,\xi}$ is constant on P. We let $\sigma: G\to\mathfrak{g}^*$ be defined by $\sigma(g)\cdot\xi=$ the constant value of $\psi_{g,\xi}$ and call it the **co-adjoint cocycle** associated to J. It satisfies the **cocycle identity**: $\sigma(gh)=\sigma(g)+Ad^*_{g^{-1}}\sigma(h)$.*

ISBN 0-8053-0102-X

Proof. We compute the derivative of $\psi_{g,\xi}$ as follows:

$$d\psi_{g,\xi}(x)=d\hat{J}\,(\xi)(\Phi_g(x))\cdot T_x\Phi_g(x)-d\hat{J}\,(Ad_{g^{-1}}\xi)(x)$$
$$=i_{\xi_P}\omega(\Phi_g(x))\cdot T_x\Phi_g(x)-i_{(Ad_{g^{-1}}\xi)_P}\omega(x)$$

by definition of the momentum mapping. As we saw in 4.1.26.(i)

$$(Ad_{g^{-1}}\xi)_P=\Phi_g^*\xi_P$$

Thus

$$d\psi_{g,\xi}(x)=\Phi_g^*(i_{\xi_P}\omega)(x)-i_{\Phi_g^*\xi_P}\omega(x)$$

But as Φ_g is symplectic,

$$\Phi_g^* i_{\xi_P}\omega=i_{\Phi_g^*\xi_P}\Phi_g^*\omega=i_{\Phi_g^*\xi_P}\omega$$

so the two terms cancel. Thus $d\psi_{g,\xi}=0$, and as P is connected, $\psi_{g,\xi}$ is constant.

The proof of the cocycle identity is as follows:

$$\sigma(gh)\cdot\xi=\hat{J}\,(\xi)(\Phi_{gh}(x))-\hat{J}\,(Ad_{(gh)^{-1}}\xi)(x)$$
$$=\hat{J}\,(\xi)(\Phi_g(\Phi_h(x)))-\hat{J}\,(Ad_{g^{-1}}\xi)(\Phi_h(x))$$
$$+\hat{J}\,(Ad_{g^{-1}}\xi)(\Phi_h(x))-\hat{J}\,(Ad_{h^{-1}}Ad_{g^{-1}}\xi)(x)$$
$$=\psi_{g,\xi}(\Phi_h(x))+\psi_{h,Ad_{g^{-1}}\xi}(x)$$
$$=\sigma(g)\cdot\xi+\sigma(h)\cdot Ad_{g^{-1}}\xi \quad \blacksquare$$

Notice that the definition of σ may be rewritten

$$\sigma(g)=J(\Phi_g(x))-Ad^*_{g^{-1}}J(x)$$

for any $x\in P$, and that $\sigma(e)=0$.

4.2.4 Definition. *Let G be a Lie group and $\mathfrak{g}$ its Lie algebra. A* ***(co-adjoint) cocycle*** *on G is a map*

$$\sigma: G\to\mathfrak{g}^*$$

such that the cocycle identity

$$\sigma(gh)=\sigma(g)+Ad^*_{g^{-1}}\sigma(h)$$

holds for all $g, h\in G$.

ISBN 0-8053-0102-X

*A cocycle Δ is called a **coboundary** if there is a $\mu \in \mathfrak{g}^*$ such that*

$$\Delta(g) = \mu - Ad^*_{g^{-1}}\mu$$

*The cocycles form a vector space and the coboundaries are a subspace. The quotient space; that is, equivalence classes $[\sigma]$ of cocycles mod coboundaries, is called the **cohomology** of G.*

Using these terms, we can make the following observation.

4.2.5 Proposition. *Let Φ be a symplectic action of G on P. If J_1 and J_2 are two momentum mappings with cocycles σ_1 and σ_2, then $[\sigma_1]=[\sigma_2]$. Thus to any symplectic action admitting a momentum mapping there is a well-defined cohomology class $[\sigma]$.*

Proof. We have

$$\sigma_1(g) - \sigma_2(g) = J_1(\Phi_g(x)) - J_2(\Phi_g(x)) - Ad^*_{g^{-1}}(J_1(x) - J_2(x))$$

However, since $d\hat{J}_1(\xi) = i_{\xi_P}\omega = d\hat{J}_2(\xi)$, $J_1 - J_2$ is a constant element of $\mathfrak{g}^*$, say μ. Thus

$$\sigma_1(g) - \sigma_2(g) = \mu - Ad^*_{g^{-1}}\mu$$

so $\sigma_1 - \sigma_2$ is a coboundary. ■

Specializing 4.1.27 we make:

4.2.6 Definition. *A momentum mapping J is called **Ad^*-equivariant** provided*

$$J(\Phi_g(x)) = Ad^*_{g^{-1}}J(x)$$

for every $g \in G$; that is, the following diagram commutes:

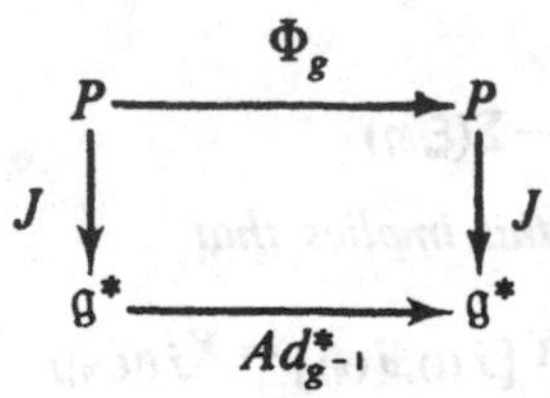

Thus the cocycle σ of a momentum mapping J measures its lack of equivariance. In particular, $\sigma = 0$ if and only if J is Ad^*-equivariant.

ISBN 0-8053-0102-X

4.2.7 Proposition. *Let J be a momentum mapping for the symplectic action Φ, with cocycle σ. Then:*

(i) the map $\Psi: (g,\mu)\to Ad^*_{g^{-1}}\mu+\sigma(g)$ is an action of G on $\mathfrak{g}^*$;
(ii) J is equivariant with respect to the action in (i).

Proof. Since $\sigma(e)=0$ and Ad^*_e is the identity, $\Psi(e,\mu)=\mu$. Also, using the cocycle identity,

$$\begin{aligned}\Psi(gh,\mu)&=Ad^*_{(gh)^{-1}}\mu+\sigma(gh)\\ &=Ad^*_{g^{-1}}(Ad^*_{h^{-1}}\mu)+\sigma(g)+Ad^*_{g^{-1}}\sigma(h)\\ &=Ad^*_{g^{-1}}(\Psi(h,\mu))+\sigma(g)\\ &=\Psi(g,\Psi(h,\mu))\end{aligned}$$

so Ψ is an action. This proves (i), and (ii) is obvious from the definition of σ and Ψ. ■

As we shall see, concrete momentum mappings one normally constructs are Ad^*-equivariant. However, in an "exotic" case where J is not Ad^*-equivariant, 4.2.7 shows that there is another action of G on $\mathfrak{g}^*$ with respect to which J is equivariant. See Exercise 4.2D for an example.

Next we shall discuss the commutation relations associated with a given momentum mapping J.

4.2.8 Theorem. *Let $\Phi: G\times P\to P$ be a symplectic action with a momentum mapping $J: P\to\mathfrak{g}^*$. Let $\sigma: G\to\mathfrak{g}^*$ be the cocycle of J and define*

$$\Sigma: \mathfrak{g}\times\mathfrak{g}\to\boldsymbol{R},\qquad \Sigma(\xi,\eta)=d\hat{\sigma}_\eta(e)\cdot\xi$$

where $\hat{\sigma}_\eta: G\to\boldsymbol{R}: g\mapsto\sigma(g)\cdot\eta$. Then:

(i) Σ is a skew symmetric bilinear form on $\mathfrak{g}$ satisfying Jacobi's identity

$$\Sigma(\xi,[\eta,\zeta])=\Sigma([\xi,\eta],\zeta)+\Sigma(\eta,[\xi,\zeta])$$

and

(ii) $\{\hat{J}(\xi),\hat{J}(\eta)\}=\hat{J}([\xi,\eta])-\Sigma(\xi,\eta)$

Since $\Sigma(\xi,\eta)$ is constant, this implies that

$$X_{\{\hat{J}(\xi),\hat{J}(\eta)\}}=X_{\hat{J}([\xi,\eta])}$$

Proof. Let us first show that

$$d\{\hat{J}(\xi),\hat{J}(\eta)\}=d\hat{J}([\xi,\eta])$$

ISBN 0-8053-0102-X

Indeed,

$$d\hat{J}([\xi,\eta]) = i_{[\xi,\eta]_P}\omega$$

$$= -i_{[\xi_P,\eta_P]}\omega \quad \text{(by 4.1.26)}$$

$$= -(L_{\xi_P} i_{\eta_P}\omega - i_{\eta_P} L_{\xi_P}\omega) \quad \text{(by Identity 16, Table 2.4.1)}$$

$$= -L_{\xi_P} i_{\eta_P}\omega \quad \text{(since } \xi_P = X_{\hat{J}(\xi)} \text{ and so } L_{\xi_P}\omega = 0)$$

$$= -L_{\xi_P} d\hat{J}(\eta)$$

$$= -dL_{X_{\hat{J}(\xi)}}\hat{J}(\eta)$$

$$= d\{\hat{J}(\xi), \hat{J}(\eta)\} \quad \text{(by 3.3.12)}$$

Thus $\rho(\xi,\eta) = \{\hat{J}(\xi), \hat{J}(\eta)\} - \hat{J}([\xi,\eta])$ is constant, so defines a skew symmetric map $\rho\colon \mathfrak{g}\times\mathfrak{g}\to \mathbf{R}$. Since $\{\ \}$ and $[\ ]$ satisfy Jacobi's identity, so does ρ. If we differentiate the relation

$$\hat{\sigma}_\eta(g) = \hat{J}(\eta)(\Phi_g(x)) - \hat{J}(Ad_{g^{-1}}\eta)(x)$$

in g at $g = e$, we get

$$\Sigma(\xi,\eta) = d\hat{\sigma}_\eta(e)\cdot\xi$$

$$= d\hat{J}(\eta)(x)\cdot\xi_P(x) + \hat{J}([\xi,\eta])(x) \quad \text{(by 4.1.25(b))}$$

$$= -\{\hat{J}(\xi), \hat{J}(\eta)\}(x) + \hat{J}([\xi,\eta])(x)$$

$$= -\rho(\xi,\eta) \quad \blacksquare$$

4.2.9 Corollary. *If J is an Ad*-equivariant momentum mapping, then*

$$\boxed{\{\hat{J}(\xi), \hat{J}(\eta)\} = \hat{J}([\xi,\eta])}$$

that is, $\hat{J}$ is a homomorphism of the Lie algebra $\mathfrak{g}$ to the Lie algebra of functions under the Poisson bracket.

If $\xi_{\mathfrak{g}^*}$ represents the infinitesimal generator of the action Ψ in 4.2.7, then by differentiating the equivariance relation

$$\Psi_g(J(x)) = J(\Phi_g(x))$$

ISBN 0-8053-0102-X

in g at $g=e$ (i.e., using 4.1.28) one finds that

$$\xi_{\mathfrak{g}^*}(J(x))\cdot\eta=\{\hat{J}(\xi),\hat{J}(\eta)\}(x)$$

which is a rephrasing of 4.2.8(ii). Likewise one may prove 4.2.9 by directly differentiating the condition that J be Ad^*-equivariant. Thus the *commutation relations represent the infinitesimal (or linearized) version of equivariance.*

The condition that Σ satisfies the Jacobi identity says that Σ defines a *two-cocycle on* $\mathfrak{g}$. A two-cocycle Σ is called *exact* if there is a $\mu\in\mathfrak{g}^*$ such that $\Sigma(\xi,\eta)=\mu([\xi,\eta])$. Thus, requiring any two-cocycle to be exact is a cohomology condition on the Lie algebra $\mathfrak{g}$.† (This is exactly the infinitesimal version of 4.2.8.) If Σ is exact and we replace $\hat{J}(\xi)$ by $\hat{J}(\xi)-\mu(\xi)$, we get a new momentum mapping satisfying 4.2.9 as is readily checked.

Next we turn to the important question of constructing momentum mappings. Many of the important results are derived from the following:

4.2.10 Theorem. *Let Φ be a symplectic action on P. Assume the symplectic form ω on P is exact, $\omega=-d\theta$, and that the action leaves θ invariant, that is, $\Phi_g^*\theta=\theta$ for all $g\in G$. Then $J: P\to\mathfrak{g}^*$ defined by*

$$\boxed{J(x)\cdot\xi=(i_{\xi_P}\theta)(x)}$$

is an Ad^*-equivariant momentum mapping for the action.

Proof. Since the action leaves θ invariant, we have $L_{\xi_P}\theta=0$, that is,

$$di_{\xi_P}\theta+i_{\xi_P}d\theta=0$$

that is,

$$d(i_{\xi_P}\theta)=i_{\xi_P}\omega$$

so $\hat{J}(\xi)=i_{\xi_P}\theta$ satisfies the definition of a momentum mapping.

For Ad^* equivariance, we must show that

$$\hat{J}(\xi)(\Phi_g(x))=\hat{J}(Ad_{g^{-1}}\xi)(x)$$

that is,

$$(i_{\xi_P}\theta)(\Phi_g(x))=i_{(Ad_{g^{-1}}\xi)_P}\theta(x)$$

However, this follows immediately from the identity $(Ad_{g^{-1}}\xi)_P=\Phi_g^*\xi_P$ proved in the previous section, together with invariance of θ under Φ_g. ■

†Whitehead's theorem asserts that if $\mathfrak{g}$ is a semisimple Lie algebra, then every two cocycle is exact.

ISBN 0-8053-0102-X

Consider now the important special case when $P=T^*Q$ with $\theta=\theta_0$, the canonical one-form. As we saw in Sect. 3.2, a diffeomorphism f of Q to Q lifts to a diffeomorphism T^*f of P that preserves θ_0 (and hence ω_0).

If we have an action Φ of G on Q we can lift it to obtain an action on T^*Q. Since $\Phi: G\times Q\to Q$, the notation $T^*\Phi$ could be misleading, so we shall write Φ^{T^*}; it is defined by lifting each Φ_g separately; that is,

$$\Phi^{T^*}(g,\alpha)=T^*\Phi_{g^{-1}}(\alpha)$$

or

$$\Phi_g^{T^*}=T^*\Phi_{g^{-1}}$$

We use g^{-1} so Φ^{T^*} is an action because from p. 181, $T^*(f\circ h)=T^*h\circ T^*f$ so $T^*(f\circ h)^{-1}=T^*f^{-1}\circ T^*h^{-1}$ and $\Phi_g^{-1}=\Phi_{g^{-1}}$. By 3.2.12 it is a symplectic action and in fact preserves the canonical one-form θ_0. Thus we can use 4.2.10 to obtain the following.

4.2.11 Corollary. *Let Φ be an action of G on Q and let Φ^{T^*} be the lifted action on $P=T^*Q$ as defined above. Then this action is symplectic and has an Ad^*-equivariant momentum mapping given by*

$$J: P\to\mathfrak{g}^*;\ \hat{J}(\xi)(\alpha_q)=\alpha_q\cdot\xi_Q(q)$$

where ξ_Q is the infinitesimal generator of Φ on Q. For a vector field X on Q we define

$$P(X): T^*Q\to\mathbf{R};\ \alpha_q\mapsto\alpha_q\cdot X(q)$$

*and call it the **momentum** corresponding to X. Thus $\hat{J}(\xi)=P(\xi_Q)$.*

Proof. By 3.2.12, $\tau_Q^*: T^*Q\to Q$ is equivariant, that is,

$$\tau_Q^*\circ\Phi_g^{T^*}=\Phi_g\circ\tau_Q^*$$

By 4.1.28, we find that

$$T\tau_Q^*\circ\xi_P=\xi_Q\circ\tau_Q^*$$

Now by definition of the canonical one-form from 3.2.10,

$$\begin{aligned} i_{\xi_P}\theta(\alpha_q)&=\langle T\tau_Q^*\circ\xi_P(\alpha_q),\alpha_q\rangle\\ &=\langle\xi_Q\circ\tau_Q^*(\alpha_q),\alpha_q\rangle\\ &=\langle\xi_Q(q),\alpha_q\rangle\\ &=P(\xi_Q)\cdot\alpha_q \quad\blacksquare\end{aligned}$$

ISBN 0-8053-0102-X

Notice that in coordinates $q^1,\ldots,q^n$ on Q and the corresponding induced coordinates $q^1,\ldots,q^n,p_1,\ldots,p_n$ on $P=T^*Q$, we have

$$P(X)(p,q)=p_iX^i(q)\quad\text{(summation understood)}$$

The general commutation relations proved in 4.2.9 may be specialized to the case of the momentum functions using 4.2.11. We can enlarge these relations by introducing, for any function $f\colon Q\to\mathbf{R}$, the corresponding *position function* $\tilde{f}=f\circ\tau_Q^*$. As we shall see in Sect. 5.4, these relations have played an important role in the relationship between classical and quantum mechanics.

4.2.12 Proposition. *For any two vector fields X and Y on Q and functions, $f,g\colon Q\to\mathbf{R}$, we have*

(*i*) $\{P(X),P(Y)\}=-P([X,Y]);$

(*ii*) $\{\tilde{f},\tilde{g}\}=0;$

(*iii*) $\{\tilde{f},P(X)\}=\widetilde{X(f)}.$

Instead of deducing this from 4.2.9, we shall prove it directly (see Exercises 4.1G and 4.2C for the "explanation" of the minus sign).

Proof. (i) From the coordinate formulas for { } and [ ] in Sects. 3.3 and 2.2,

$$\begin{aligned}\{P(X),P(Y)\}&=\{p_kX^k,p_jY^j\}\\&=\frac{\partial}{\partial q^i}(p_kX^k)\frac{\partial}{\partial p_i}(p_jY^j)-\frac{\partial}{\partial p_i}(p_kX^k)\frac{\partial}{\partial q^i}(p_jY^j)\\&=p_k\frac{\partial X^k}{\partial q^i}Y^i-X^ip_j\frac{\partial Y^j}{\partial q^i}\\&=-p_k\left(X^i\frac{\partial Y^k}{\partial q^i}-Y^i\frac{\partial X^k}{\partial q^i}\right)\\&=-p_k[X,Y]^k\\&=-P([X,Y])\end{aligned}$$

The assertion (ii) is clear since $\tilde{f}$ and $\tilde{g}$ are functions of q^i only. For (iii), note that

$$\begin{aligned}\{\tilde{f},P(X)\}&=\frac{\partial}{\partial q^i}f\cdot\frac{\partial}{\partial p_i}(p_jX^j)-\frac{\partial}{\partial q_i}(p_jX^j)\cdot\frac{\partial}{\partial p_i}f\\&=\frac{\partial f}{\partial q^i}\cdot X^i=X(f)\end{aligned}$$

which, as a function of q and p, is $\widetilde{X(f)}$. ■

ISBN 0-8053-0102-X

Another important special case of 4.2.10 that is proved the same way as 4.2.11 or that may be deduced from it, is as follows:

4.2.13 Corollary. *Let Q be a pseudo-Riemannian manifold and let a group G act on Q by isometries. Lift this action using 3.2.14 to a symplectic action on TQ. Its momentum mapping is given by*

$$\hat{J}(\xi)(v_q) = \langle v_q, \xi_Q(q) \rangle$$

This last corollary is actually a special case of a more general fact about Lagrangian systems. It is usually referred to as **Noether's theorem** in the finite-dimensional case. (For the result for continuous systems, see Sect. 5.5.)

4.2.14 Corollary. *Let G act on Q by $\Phi: G \times Q \to Q$ and let Φ^T denote the tangent action; $\Phi_g^T = T\Phi_g: TQ \to TQ$.*

Let L be a regular Lagrangian on TQ with $\theta_L = (FL)^*\theta_0$, as usual. Suppose L is invariant under the action Φ^T; that is, $L \circ \Phi_g^T = L$ for all $g \in G$. Then:

(*i*) $(\Phi_g^T)^*\theta_L = \theta_L$;

(*ii*) *the momentum for the action is*

$$\boxed{\hat{J}(\xi)(v_q) = FL(v_q)\cdot \xi_Q(q)}$$

and is Ad^*-equivariant;

(*iii*) *the momentum mapping J given by (ii) is an integral of Lagrange's equations X_E.*

Proof. (i) Differentiating $L = L \circ \Phi_g^T$ along fibers yields

$$FL(v)\cdot w = FL(\Phi_g^T(v))\cdot \Phi_g^T(w)$$

in other words, $\Phi_g^{T^*} \circ FL = FL \circ \Phi_g^T$, that is, $FL: TQ \to T^*Q$ is equivariant relative to the actions Φ^T of G on TQ and Φ^{T^*} of G on T^*Q.

By definition, $\theta_L = (FL)^*\theta_0$, and so

$$\begin{aligned}(\Phi_g^T)^*\theta_L &= (\Phi_g^T)^*(FL)^*\theta_0\\ &= (FL \circ \Phi_g^T)^*\theta_0\\ &= (\Phi_g^{T^*} \circ FL)^*\theta_0\\ &= (FL)^* \circ (\Phi_g^{T^*})^*\theta_0\\ &= (FL)^*\theta_0 \quad \text{(by 3.2.12)}\\ &= \theta_L\end{aligned}$$

ISBN 0-8053-0102-X

(ii) We compute as in 4.2.11. First note that $\tau_Q: TQ \to Q$ is equivariant, that is,

$$\tau_Q \circ \Phi_g^T = \Phi_g \circ \tau_Q$$

and hence

$$T\tau_Q \circ \xi_P = \xi_Q \circ \tau_Q$$

Therefore, using the definition of θ_L and letting $P = TQ$,

$$\begin{aligned}(i_{\xi_P}\theta_L)(v_q) &= \theta_L(v_q)\cdot \xi_P(v_q)\\ &= \theta_0\big(FL(v_q)\big)\cdot TFL\cdot \xi_P(v_q)\\ &= \langle T\tau_Q^* \circ TFL \circ \xi_P(v_q), FL(v_q)\rangle\end{aligned}$$

Since $T\tau_Q^* \circ TFL = T(\tau_Q^* \circ FL) = T\tau_Q$, we get

$$\begin{aligned}(i_{\xi_P}\theta_L)(v_q) &= \langle T\tau_Q \circ \xi_P(v_q), FL(v_q)\rangle\\ &= \langle \xi_Q(q), FL(v_q)\rangle\\ &= FL(v_q)\cdot \xi_Q(q)\end{aligned}$$

Thus (ii) follows from 4.2.10.

For (iii) we need only show that Φ_g^T leaves the energy E invariant, in view of 4.2.2. Because $FL(v)\cdot w = FL(\Phi_g^T(v))\cdot \Phi_g^T(w)$ (as noted in the proof of (i)), Φ_g^T leaves the action $A(v) = FL(v)\cdot v$ invariant. Hence it leaves $E = A - L$ invariant as well. ■

The relationship $FL(v) = FL(T\Phi_g \cdot v) \circ T\Phi_g$ proved in (i) is worth noting and is equivalent to commutivity of the diagram

$$\begin{array}{ccc} TQ & \xrightarrow{\ FL\ } & T^*Q \\ {\scriptstyle \Phi_g^T = T\Phi_g}\downarrow & & \downarrow{\scriptstyle T^*\Phi_{g^{-1}} = \Phi_g^{T^*}} \\ TQ & \xrightarrow[\ FL\ ]{} & T^*Q \end{array}$$

that is, to equivariance of FL. From 4.1.28, we deduce, for example,

$$TFL \circ \xi_{TQ} = \xi_{T^*Q} \circ FL.$$

Now we turn to some examples of momentum mappings. (Further examples are found in the exercises.)

ISBN 0-8053-0102-X

4.2.15 Examples. (i) Let $Q=\mathbf{R}^n$, $G=\mathbf{R}^n$, and let G act on $\mathbf{R}^n$ by translation: $\Psi\colon G\times Q\to Q\colon (s,q)\mapsto s+q$. The infinitesimal generator corresponding to $\xi\in\mathbf{R}^n$ is $\xi_Q(q)=\xi$. By 4.2.11 the momentum mapping on T^*Q is given by

$$\hat{J}(\xi)\cdot(q,p)=p\cdot\xi$$

In other words, $J(q,p)=p$, the *linear momentum.*

(ii) Let $Q=\mathbf{R}^n$ and let G be a Lie subgroup of $GL(n,\mathbf{R})$. Let G act on Q by $\Phi\colon G\times Q\to Q\colon (T,q)\mapsto Tq$. The infinitesimal generator corresponding to $B\in\mathfrak{g}\subset L(\mathbf{R}^n,\mathbf{R}^n)$ is $B_Q(q)=Bq$. Thus by 4.2.11 the momentum mapping on T^*Q is given by

$$\hat{J}(B)(q,p)=p(Bq)$$

Take the special case when $n=3$ and $G=SO(3,\mathbf{R})$. Then identifying the Lie algebra $so(3,\mathbf{R})$ of $SO(3,\mathbf{R})$ with $\mathbf{R}^3$ using $B\mapsto\tilde{B}$ (see Exercise 4.1A) and working on TQ, 4.2.13 and Exercise 4.1A gives

$$\begin{aligned}\hat{J}(B)(q,v)&=\langle v,Bq\rangle\\&=\langle\tilde{B}\times q,v\rangle=\det(\tilde{B},q,v)\\&=\langle q\times v,\tilde{B}\rangle\end{aligned}$$

Thus $J(q,v)=q\times v$ (under this identification of $so(3,\mathbf{R})$ with $\mathbf{R}^3$, and $\mathbf{R}^3$ with $\mathbf{R}^{3*}$ by $\langle,\rangle$) which is just the *angular momentum.*

(iii) Let G be a Lie group and consider the action $\Phi\colon G\times G\to G$; $(g,h)\mapsto gh$. The infinitesimal generator corresponding to $\xi\in\mathfrak{g}$ is the right-invariant vector field equal to ξ at e; that is, $\xi_G(g)=T_eR_g\cdot\xi$. [see Example 4.1.25(a).] Thus the corresponding momentum mapping on T^*G is

$$\hat{J}(\xi)\cdot\alpha_g=\alpha_g(T_eR_g\cdot\xi)=(T_e^*R_g\cdot\alpha_g)\xi$$

that is,

$$J(\alpha_g)=T_e^*R_g\,\alpha_g$$

(iv) Let $P=\mathbf{R}^n\times\mathbf{R}^n=\mathbf{R}^{2n}$ with the symplectic form associated with the Euclidean inner product as usual. Let $H(q,\dot{q})=\frac{1}{2}\|q\|^2+\frac{1}{2}\|\dot{q}\|^2$, the harmonic oscillator Hamiltonian. Using Example (ii) we can construct conserved quantities associated with $SO(n,\mathbf{R})$. We shall now construct a proper extension of this momentum mapping that gives rise to more conserved quantities.

Let

$$\begin{aligned}G&=Sp(2n,\mathbf{R})\cap SO(2n,\mathbf{R})\\&=\left\{\begin{pmatrix}a&-d\\d&a\end{pmatrix}\in Gl(2n,\mathbf{R})\,\middle|\,ad'=da'\text{ and }aa'+dd'=1\right\}\end{aligned}$$

ISBN 0-8053-0102-X

with Lie algebra

$$\mathfrak{g}=\left\{\begin{pmatrix} A & -D \\ D & A \end{pmatrix}\in gl(2n,\mathbf{R})\,\middle|\,A=-A',\ D=D'\right\}$$

Then G acts on P by $(c,Z)\mapsto cZ$; one checks that this is symplectic: $\omega(cZ,cW)=\omega(Z,W)$. The infinitesimal generator corresponding to $C\in\mathfrak{g}$ is

$$C_P(Z)=CZ$$

Since C is linear and ω-skew, it is a Hamiltonian vector field with energy $\frac{1}{2}\omega(CZ,Z)$. (See remarks following 3.3.6.) Thus the momentum mapping is

$$\begin{aligned}\hat{J}(C)(Z)&=\tfrac{1}{2}\omega(CZ,Z)\\ &=\langle Aq,\dot{q}\rangle-\tfrac{1}{2}\langle D\dot{q},\dot{q}\rangle-\tfrac{1}{2}\langle Dq,q\rangle\end{aligned}$$

where

$$C=\begin{pmatrix} A & -D \\ D & A \end{pmatrix}$$

The restriction of this momentum mapping to the subspace $D=0$ yields the momentum mapping associated with $SO(n,\mathbf{R})$. The mapping $\hat{J}$ is Ad^*-equivariant because $\hat{J}(C)(Z)$ vanishes when $Z=0$, so using 4.2.3, the cocycle must vanish as well. (This can also be seen directly. Using the fact that $\omega(cZ,cW)=\omega(Z,W)$, we get

$$\begin{aligned}\hat{J}(C)(cZ)-\hat{J}(c^{-1}Cc)Z&=\tfrac{1}{2}\{\omega(CcZ,cZ)-\omega(c^{-1}CcZ,Z)\}\\ &=0\end{aligned}$$

which is Ad^*-equivariance.)

We shall now reconsider the momentum mapping from a different point of view. The results can be used as an alternative characterization of the momentum mapping. The idea is that instead of considering $\Phi_g^*\theta$ for each $g\in G$, we consider $\Phi^*\theta$ for the *whole map* $\Phi: G\times P\to P$.

If J is a momentum mapping for a symplectic action Φ on (P,ω), define a one-form $\check{J}$ on $G\times P$ by

$$\check{J}(\xi_g,v_p)=J(p)\cdot TL_{g^{-1}}\cdot\xi_g$$

where $\xi_g\in T_gG$, $v_p\in T_pP$, and $p\in P$.

4.2.16 Theorem (Momentum Lemma). *Let the conditions of* 4.2.10 *hold and let J be as given there. Then*

$$\Phi^*\theta=\pi_2^*\theta+\check{J}$$

where $\pi_2: G\times P\to P$ is the projection onto the second factor.

ISBN 0-8053-0102-X

Notice that this result holds, in particular, in the context of 4.2.11, 4.2.13, and 4.2.14.

Proof. By definition of pull-back,

$$\Phi^*\theta(\xi_g, v_p) = \theta(\Phi(g,p))\cdot T\Phi(g,p)\cdot(\xi_g, v_p)$$

$$= \theta(\Phi_g(p))\cdot T\Phi_g(p)\cdot v_p + \theta(\Phi_g(p))\cdot T\Phi_p(g)\cdot \xi_g$$

$$= \theta(p)\cdot v_p + \theta(\Phi_g(p))\cdot T\Phi_p(g)\cdot \xi_g \qquad (1)$$

since $\Phi_g^*\theta = \theta$. Denote by $\xi = T_e L_{g^{-1}}(\xi_g)$. We have

$$T_g\Phi_p\xi_g = T_g\Phi_p \circ T_e L_g(\xi)$$

$$= T_e(\Phi_p \circ L_g)(\xi)$$

$$= T_e(\Phi_g \circ \Phi_p)(\xi)$$

$$= T_p\Phi_g \circ T_e\Phi_p(\xi)$$

and

$$T_e\Phi_p(\xi) = \frac{d}{dt}\Phi(\exp t\xi, p)\Big|_{t=0} = \xi_P(p)$$

so that

$$T_g\Phi_p\xi_g = T_p\Phi_g\cdot\xi_P(p) = T_p\Phi_g(T_e L_{g^{-1}}\xi_g)_P(p) \qquad (2)$$

Substituting (2) in (1), we get

$$\Phi^*\theta(\xi_g, v_p) = \theta(p)v_p + \theta(\Phi_g(p))\cdot T\Phi_g\cdot(TL_{g^{-1}}\xi_g)_P(p)$$

$$= \theta(p)v_p + \theta(p)\cdot(TL_{g^{-1}}\xi_g)_P(p)$$

since, again, $\Phi_g^*\theta = \theta$. But since $J(p)\xi = (i_{\xi_P}\theta)(p)$ we get $\Phi^*\theta(\xi_g, v_p) = \theta(p)\cdot v_p + J(p)\cdot(TL_{g^{-1}}\xi_g)$, and so the result follows. ■

There is another way to write $\check{J}$ that will be useful. Namely, we let $\pi_1\colon G\times P \to G$ be the projection on the first factor. Let ν be the Lie algebra valued one-form on G given by $\nu(\xi_g) = TL_{g^{-1}}\xi_g$. Then $\pi_1^*\nu$ is a Lie algebra valued one-form on $G\times P$:

$$\pi_1^*\nu(\xi_g, v_p) = TL_{g^{-1}}\cdot\xi_g$$

Thus we can write the definition of $\check{J}$ as

$$\check{J} = (J\circ\pi_2)\cdot\pi_1^*\nu$$

ISBN 0-8053-0102-X

where $\cdot$ signifies the natural pointwise pairing between maps with values in $\mathfrak{g}^*$ and $\mathfrak{g}$, respectively. Thus $\Phi^*\theta = \pi_2^*\theta + (J\circ\pi_2)\cdot\pi_1^*\nu$ and so applying $\boldsymbol{d}$, $\Phi^*\omega = \pi_2^*\omega + \boldsymbol{d}(J\circ\pi_2)\wedge\pi_1^*\nu + J\circ\pi_2\cdot \boldsymbol{d}\pi_1^*\nu$, with $\boldsymbol{d}$ and $\wedge$ appropriately defined.

The reader can similarly show that quite generally for a symplectic action with a momentum mapping J (not necessarily assuming ω is exact), we still have $\Phi^*\omega = \pi_2^*\omega + \boldsymbol{d}(J\circ\pi_2)\wedge\pi_1^*\nu + J\circ\pi_2\cdot\boldsymbol{d}\pi_1^*\nu$.

We now give an important reformulation of the momentum lemma.

4.2.17 Corollary (Whittaker's Lemma). *Under the same conditions as above, let $\Omega: P\to G$ be a smooth mapping and let*

$$B: P\to P, \quad B(p) = \Phi_{\Omega(p)}(p)$$

Then

$$B^*\theta = \theta + J\cdot\Omega^*\nu$$

where ν is the Lie algebra valued one form on G defined above, so that

$$\Omega^*\nu\cdot v_p = TL_{g^{-1}}(T_p\Omega\cdot v_p)$$

and $g = \Omega(p)$.

Proof. Factor B through $G\times P$ as follows:

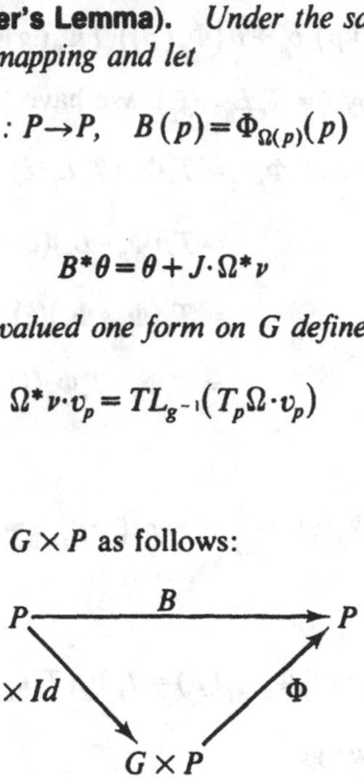

Thus

$$\begin{aligned}
B^*\theta &= (\Phi\circ(\Omega\times Id))^*\theta \\
&= (\Omega\times Id)^*\Phi^*\theta \\
&= (\Omega\times Id)^*\{\pi_2^*\theta + (J\circ\pi_2)\cdot\pi_1^*\nu\} \\
&= (\pi_2\circ(\Omega\times Id))^*\theta + J\circ(\pi_2\circ(\Omega\times Id))\cdot(\pi_1\circ(\Omega\times Id))^*\nu
\end{aligned}$$

But $\pi_2\circ(\Omega\times Id) = Id$ and $\pi_1\circ(\Omega\times Id) = \Omega$, so the result follows. ■

We refer to this basic result as Whittaker's lemma since he first discussed it in the context of $SO(2)$ and $SO(3)$ actions on $\boldsymbol{R}^2$ and $\boldsymbol{R}^3$, as we now explain.

We specialize the Momentum Lemma to the case of the usual $SO(2)\cong S^1$-action ρ on $\boldsymbol{R}^2$ by rotations. In this case $\overline{G}: T^*\boldsymbol{R}^2\to\boldsymbol{R}^*\cong\boldsymbol{R}$, $\overline{G}(q,p)=$

ISBN 0-8053-0102-X

$q^1p_2 - q^2p_1$ is an Ad^*-equivariant moment for the action

$$\rho^{T^*}: S^1 \times T^*\boldsymbol{R}^2 \to T^*\boldsymbol{R}^2: (e^{i\theta}, (\boldsymbol{q}, \boldsymbol{p})) \mapsto (e^{i\theta}\boldsymbol{q}, \boldsymbol{p}e^{-i\theta})$$

$$e^{i\theta}\boldsymbol{q} = (q^1 \cos\theta - q^2 \sin\theta, q^1 \sin\theta + q^2 \cos\theta),$$

$$\boldsymbol{p}e^{-i\theta} = (p_1 \cos\theta + p_2 \sin\theta, -p_1 \sin\theta + p_2 \cos\theta)$$

Ad^* equivariance means $\overline{G} \circ \rho_g^{T^*} = \overline{G}$ for all $g \in S^1$, since S^1 is abelian and hence $(Ad_{g^{-1}})^* = identity$.

Denote by $\pi_1: S^1 \times T^*\boldsymbol{R}^2 \to S^1$ and $\pi_2: S^1 \times T^*\boldsymbol{R}^2 \to T^*\boldsymbol{R}^2$ the canonical projections: $\overline{G} \circ \pi_2 \in \mathfrak{F}(S^1 \times T^*\boldsymbol{R}^2)$. If $\Omega: P \to S^1$ is an angular variable, we define $\boldsymbol{d}\Omega(p) \cdot v_p = TL_{g^{-1}}(T_p\Omega \cdot v_p)$ where $g = \Omega(p)$. (See Exercise 2.4F.) It follows from 4.2.17 that $\Omega^*\nu = \boldsymbol{d}\Omega$.

4.2.18 Corollary (Classical Momentum Lemma on the cotangent bundle.) *If ρ is the usual S^1-action on $\boldsymbol{R}^2$ by rotations and $\rho^{T^*}: S^1 \times T^*\boldsymbol{R}^2 \to T^*\boldsymbol{R}^2$ the induced action on the cotangent bundle,*

$$(\rho^{T^*})^*\theta_0 = \pi_2^*\theta_0 + (\overline{G} \circ \pi_2)\, \boldsymbol{d}\pi_1$$

$$(\rho^{T^*})^*\omega_0 = \pi_2^*\omega_0 - \boldsymbol{d}(\overline{G} \circ \pi_2) \wedge \boldsymbol{d}\pi_1$$

where $\pi_1: S^1 \times T^*\boldsymbol{R}^2 \to S^1$ and $\pi_2: S^1 \times T^*\boldsymbol{R}^2 \to T^*\boldsymbol{R}^2$ are the canonical projections.

With the remarks above, this follows from 4.2.16.

We now formulate a similar result on the tangent bundle. Assume that $L = K - V: T\boldsymbol{R}^2 \to \boldsymbol{R}$ is a Lagrangian with the kinetic energy $K(\boldsymbol{q}, \dot{\boldsymbol{q}}) = \frac{1}{2}\|\boldsymbol{q}\|^2$ and potential $V: \boldsymbol{R}^2 \to \boldsymbol{R}$. Then L is hyperregular and $\boldsymbol{F}L: (\boldsymbol{q}, \dot{\boldsymbol{q}}) \mapsto (\boldsymbol{q}, \boldsymbol{p} = \dot{\boldsymbol{q}})$. Hence $\theta_L = (\boldsymbol{F}L)^*\theta_0$, where θ_0 is the canonical one-form on $T^*\boldsymbol{R}^2$, has the form

$$\theta_L = \dot{q}^1 dq^1 + \dot{q}^2 dq^2$$

Then θ_L defines $\omega_L = -\boldsymbol{d}\theta_L = dq^1 \wedge d\dot{q}^1 + dq^2 \wedge d\dot{q}^2$, the standard, noncanonical symplectic form on $T\boldsymbol{R}^2$.

Let $S^1 \cong SO(2)$ act by rotations on $\boldsymbol{R}^2$ and call this action as before ρ. Then the induced action on the tangent bundle $\rho^T: S^1 \times T\boldsymbol{R}^2 \to T\boldsymbol{R}^2$ is (see 4.1) $\rho^T(e^{i\theta}, (\boldsymbol{q}, \dot{\boldsymbol{q}})) = (\rho(e^{i\theta}\boldsymbol{q}), \rho(e^{i\theta}\dot{\boldsymbol{q}}))$. An Ad^*-equivariant moment for this action is $G: T\boldsymbol{R}^2 \to \boldsymbol{R}^*: (\boldsymbol{q}, \dot{\boldsymbol{q}}) \mapsto q^1\dot{q}^2 - q^2\dot{q}^1$. Also notice that $G \circ \boldsymbol{F}L = \overline{G}$ and that L is invariant by ρ^T, assuming that V is by ρ. This happens, for example, if V depends only on $\|\boldsymbol{q}\|$, a case useful in the two- and three-body problems. As we saw in 4.2.14, invariance of L by ρ^T implies the equivariance of $\boldsymbol{F}L$, that is, $\boldsymbol{F}L \circ \rho_g^T = \rho_g^{T^*} \circ \boldsymbol{F}L$ for all $g \in S^1$. Denote as before by $\pi_1: S^1 \times T\boldsymbol{R}^2 \to S^1$ and $\pi_2: S^1 \times T\boldsymbol{R}^2 \to T\boldsymbol{R}^2$ the canonical projections. The following is easy to establish from 4.2.16 and 4.2.14.

ISBN 0-8053-0102-X

4.2.19 Corollary (Classical Momentum Lemma on the tangent bundle). *If ρ is the usual S^1-action on $\boldsymbol{R}^2$ by rotations and ρ^T: $S^1 \times T\boldsymbol{R}^2 \to T\boldsymbol{R}^2$ the induced action on the tangent bundle,*

$$(\rho^T)^*\theta_L = \pi_2^*\theta_L + (G \circ \pi_2)\, d\pi_1$$

$$(\rho^T)^*\omega_L = \pi_2^*\omega_L - d(G \circ \pi_2)\, d\pi_1$$

where L is a ρ^T-invariant Lagrangian on $T\boldsymbol{R}^2$.

The following special case of 4.2.17 and 4.2.18 will be useful in Sect. 9.5.

4.2.20 Corollary (Whittaker's Lemma in cotangent formulation.) *Let $U \subset T^*\boldsymbol{R}^2$ be an open set, Ω: $U \to S^1$ a smooth map, B: $U \to T^*\boldsymbol{R}^2$ the map defined by*

$$B(q,p) = (\rho_{\Omega(q,p)} q, p \circ \rho_{-\Omega(q,p)})$$

where ρ is the usual S^1-action on $\boldsymbol{R}^2$ using rotation, and $\omega_0 = -d\theta_0$ the canonical symplectic form on $T^*\boldsymbol{R}^2$. Then the pull-back forms are given by

$$B^*\theta_0 = \theta_0 + \bar{G}\cdot d\Omega$$

$$B^*\omega_0 = \omega_0 - d\bar{G} \wedge d\Omega$$

where $\bar{G}$ is the classical angular momentum on the cotangent bundle.

Similarly from 4.2.17 and 4.2.19, we get:

4.2.21 Corollary (Whittaker's Lemma in tangent formulation). *Let $U \subset T\boldsymbol{R}^2$ be an open set, Ω: $U \to S^1$ a smooth map, B: $U \to T\boldsymbol{R}^2$ the map defined by*

$$B(q,\dot{q}) = (\rho(\Omega(q,\dot{q}), q), \rho(\Omega(q,\dot{q}), \dot{q}))$$

where ρ is the S^1-action on $\boldsymbol{R}^2$ by rotations, and $\omega_L = -d\theta_L$, the standard (noncanonical) symplectic form on $T\boldsymbol{R}^2$; here L is any (regular) ρ^T-invariant Lagrangian on $T\boldsymbol{R}^2$. Then the pull-back forms are given by

$$B^*\theta_L = \theta_L + G\cdot d\Omega$$

$$B^*\omega_L = \omega_L - dG \wedge d\Omega$$

where G is the classical angular momentum on the tangent bundle.,

The original form of these lemmas, expressed in terms of coefficients of $B^*\omega_0$, $B^*\omega_L$—the classical Lagrange brackets—may be found in Whittaker [1959] or Brouer–Clemence [1961, p. 179]. In Whittaker's original paper, the

ISBN 0-8053-0102-X

standard action of $SO(3)$ on TR^3 was the goal. Using the general result 4.2.17, the reader may easily work out this case as well. (See Exercises 4.2E, G.)

Finally, in this section we give an application of the momentum mapping to prove a completeness theorem for pseudo-Riemmanian manifolds. We recall that a pseudo-Riemannian manifold is *homogeneous* if its isometry group acts transitively, that is, there is only one orbit. The following result should be compared with 3.7.11, and was proved by Hermann [1972b] for the special case of semi-simple groups. See also Mishchenko [1970].

4.2.22 Theorem (Marsden [1973a]). *Let M be a compact homogeneous pseudo-Riemannian manifold. Then M is geodesically complete.*

In this theorem, neither of the two assumptions of compactness nor homogeneity may be dropped. For example, there exists an incomplete Lorentz metric on the two torus T^2 (see Markus [1963, p. 189] and Wolf [1964]). For the proof of 4.2.22 we shall need the following three lemmas.

4.2.23 Lemma. *For each $m\in M$, $T_mM=\{\xi_M(m)|\xi\in\mathfrak{g}\}$, where $\mathfrak{g}$ is the Lie algebra of the isometry group G and ξ_M are the infinitesimal generators.*

Proof. According to Sect. 4.1 the set $\{\xi_M(m)|\xi\in\mathfrak{g}\}$ is the tangent space to the orbit through $m\in M$ (remember the orbit is an immersed submanifold). But the hypothesis of homogeneity means the orbit is all of M. ▼

4.2.24 Lemma. *Let Y and X be first countable Hausdorff spaces with X compact. Let $f: Y\to X$ be continuous and one-to-one. Suppose that if $x_n\in X$ is a convergent sequence, then $f^{-1}(x_n)=y_n\in Y$ has a convergent subsequence. Then f^{-1} is continuous and so Y is compact.*

Proof. If $x_n\to x$, and y_{n_k} is a subsequence of y_n that converges to, say, y, then $f(y)=x$ or $y=f^{-1}(x)$. Hence *any* subsequence of y_n has a subsequence that converges to the same point y. Thus the whole sequence y_n must converge to y. (If it did not we could find a neighborhood of y and a subsequence of y_n that did not enter this neighborhood.) Hence f^{-1} is continuous. ▼

4.2.25 Lemma. *Let $P: TM\to\mathfrak{g}^*$, $P(v)\cdot\xi=\langle v,\xi_M(m)\rangle$, the momentum for the action of G on TM. For $\alpha\in\mathfrak{g}^*$, set $S_\alpha=P^{-1}(\alpha)$. Then S_α is a compact subset of TM that is invariant under the geodesic flow.*

Proof. Invariance of S_α under the geodesic flow results from 4.2.13. We have to show S_α is compact. Certainly S_α is closed. Furthermore, the restriction of the projection $\tau_\alpha=\tau_M|S_\alpha$: $S_\alpha\to M$ is one-to-one because, from 4.2.23, S_α intersects each fiber in at most one point.

We claim $\tau_\alpha(S_\alpha)$ is closed and hence compact. Indeed $x\notin\tau_\alpha(S_\alpha)$ means that α is not in the range of the linear map obtained by restricting P to T_xM.

ISBN 0-8053-0102-X

Thus α is not in the range of $P|T_yM$ for y in a whole neighborhood of x. Hence $\tau_\alpha(S_\alpha)$ is closed.

To prove that S_α is compact, we shall use 4.2.24 to show that τ_α^{-1}: $\tau_\alpha(S_\alpha)\to S_\alpha$ is continuous. Let $\langle\langle,\rangle\rangle$ be a Riemannian metric on M with associated norm $|||\cdot|||$. Because our pseudo-Riemannian metric $\langle,\rangle$ is nondegenerate, there is an isomorphism $U_x\colon T_xM\to T_xM$, depending smoothly on x, such that for $u, v\in T_xM$,

$$\langle v, U_x u\rangle=\langle\langle v, u\rangle\rangle$$

Let $x_n=\tau_\alpha(v_n)$ and suppose $x_n\to x$. We want to show that v_n has a convergent subsequence. Now since $v_n\in S_\alpha$,

$$\langle v_n, \xi_M(x_n)\rangle=\alpha(\xi)\quad\text{for all } \xi$$

Choosing a Euclidean norm $\|\cdot\|$ on $\mathfrak{g}$, we can write

$$v_n=U_{x_n}^{-1}(\xi_n)_M(x_n)$$

for $\xi_n\in\mathfrak{g}$ and

$$\|\xi_n\|\leqslant C|||v_n|||$$

where C is a constant independent of n. [The constant C depends on the norm of U_{x_n} and on the norm of a right inverse for the surjective map $\xi\mapsto\xi_M(m)$.] Therefore,

$$\begin{aligned}|||v_n|||^2&=\langle\langle v_n, v_n\rangle\rangle\\&=\langle v_n, U_{x_n}v_n\rangle\\&=\langle v_n, (\xi_n)_M(x_n)\rangle\\&=\alpha(\xi_n)\\&\leqslant\|\alpha\|\,\|\xi_n\|\\&\leqslant\|\alpha\|C|||v_n|||\end{aligned}$$

Therefore, $|||v_n|||\leqslant\|\alpha\|C$, so v_n is uniformly bounded, and hence lies in a compact set; thus v_n has a convergent subsequence; since S_α is closed, it converges in S_α. The result now follows from 4.2.24. ▼

Proof of 4.2.22. Since $TM=\cup\{S_\alpha|\alpha\in\mathfrak{g}^*\}$, TM is a union of compact sets invariant under the geodesic flow. In particular, any integral curve of the geodesic flow remains a priori in a compact set. Therefore, by 2.1.18 the geodesic flow is complete. ■

Remark. If $\dim G=\dim M$, then S_α is a submanifold of TM since then $P\colon TM\to\mathfrak{g}^*$ is a submersion, as is easily checked.

ISBN 0-8053-0102-X

EXERCISES

4.2A. Let G act on Q and hence on T^*Q as usual. Let ω_0 be the canonical two-form, e a constant, and A be a one-form on Q. Set $F=dA$ and

$$\omega_F=\omega_0+e\left(\left(\tau_Q^*\right)^*F\right)$$

If the action of G on Q leaves A invariant, show that a momentum mapping for the action of G on (T^*Q,ω_F) is

$$\hat{J}(\xi)\cdot\alpha_q=(\alpha_q-eA(q))\cdot\xi_Q(q)$$

(This exercise is relevant for a particle moving in the electromagnetic field F; see Exercise 3.7F and 4.3.3.)

4.2B. (i) Show that 4.2.14 remains valid for degenerate Lagrangian systems.

(ii) Let $L\colon TQ\to R$ be a regular Lagrangian and $\tilde{L}\colon T(Q\times R)\to R$ the associated homogeneous Lagrangian (see 3.7.19). Apply 4.2.14 to this Lagrangian under time translations and conclude that the energy of the *original* Lagrangian is conserved.

4.2C. Let $\mathfrak{D}$ denote the group of diffeomorphisms of Q that acts on T^*Q by lifting. (See Exercise 4.1G for a description of the Lie algebra $\mathfrak{X}$ of $\mathfrak{D}$.)

(i) Compute the infinitesimal generator of $X\in\mathfrak{X}$ on T^*Q and show it is the Hamiltonian vector field $X_{P(X)}$.

(ii) Show that the action on T^*Q is Ad^*-equivariant.

(iii) Deduce from 4.2.9 that $\{P(X),P(Y)\}=-P([X,Y])$.

Let $\mathfrak{F}$ be the group of real valued functions f on Q under addition. Let $\mathfrak{F}$ act on T^*Q by translation by df along fibers.

(iv) Show that the infinitesimal generator corresponding to f is $X_{\tilde{f}}$.

(v) Show that the action is Ad^*-equivariant.

(vi) Deduce, from the fact that $\mathfrak{F}$ is abelian, that $\{\tilde{f},\tilde{g}\}=0$.

Let $G=\mathfrak{D}\times\mathfrak{F}$ with the semi-direct product structure:

$$(\eta,f)\cdot(\zeta,g)=(\eta\circ\zeta,f+g\circ\eta)$$

Let G act on T^*Q in the obvious way.

(vii) Show that the infinitesimal generator corresponding to (X,f) is $X_{J(X,f)}$, where $J(X,f)=\tilde{f}+P(X)$.

(viii) Show that the action is Ad^*-equivariant, and

(ix) the commutation relations for J reduce to those in 4.2.12.

4.2D. Let $G=R^2$ (coordinates η_1,η_2) under addition and $P=R^2$ (coordinates x_1,x_2) with the symplectic structure $\omega=dx_1\wedge dx_2$. Let G act on R^2 by translation:

$$\Phi_{(\eta_1,\eta_2)}(x_1,x_2)=(x_1+\eta_1,x_2+\eta_2)$$

(a) Show that a momentum mapping for this action is given by

$$J(x_1,x_2)\cdot(\xi_1,\xi_2)=\xi_1x_2-\xi_2x_1$$

and its cocycle is given by

$$\sigma(\xi_1,\xi_2)\cdot(\eta_1,\eta_2)=\xi_1\eta_2-\xi_2\eta_1$$

ISBN 0-8053-0102-X

(b) Show that $[\sigma]\neq 0$.
(c) Show that J is equivariant with respect to the action of G on $\mathfrak{g}^*$ given by

$$\Psi(\eta_1,\eta_2)\cdot(\mu_1,\mu_2)=(\mu_1+\eta_2,\mu_2-\eta_1)$$

(d) Verify the commutation relations

$$\{\hat{J}(\xi),\hat{J}(\eta)\}=\hat{J}([\xi,\eta])-\Sigma(\xi,\eta)$$

where $\Sigma(\xi,\eta)=\xi_1\eta_2-\xi_2\eta_1$.

4.2E. Write out Whittaker's lemma 4.2.17, computing explicitly a coordinate formula for $J\cdot\Omega^*\nu$ (and $dJ\wedge\Omega^*\nu$) for the standard action of $SO(3)$ on $T^*\mathbf{R}^3$. (You may find the discussion in Sternberg [1964, p. 234–236] helpful.)

4.2F. Let (V,ω) be a symplectic vector space and $Sp(\omega,\mathbf{R})=\{P\in L(V,V)|\omega(Pv,Pw)=\omega(v,w)$ *for all* $v,w\in V\}$ be the Lie group of linear symplectic mappings. If $H: V\to\mathbf{R}: v\mapsto\frac{1}{2}\tilde{H}(v,v)$, where $\tilde{H}$ is a symmetric bilinear form on V, show that:

(1) $G=\{Q\in Sp(\omega,\mathbf{R})|H(Qv)=H(v)$ *for all* $v\in V\}$ is a closed subgroup of $Sp(\omega,\mathbf{R})$ (and hence is a Lie group);
(2) the action $\Phi: G\times V\to V: (P,v)\mapsto Pv$ is symplectic and leaves H invariant;
(3) the momentum mapping of Φ is $J: V\to\mathfrak{g}^*$; where $\hat{J}(\xi)(v)=\frac{1}{2}\omega(\xi v,v)$.

Let ω be the standard symplectic form on $\mathbf{R}^{2n}$, so the matrix of ω is

$$J_{2n}=\begin{pmatrix}0 & I_n\\ -I_n & 0\end{pmatrix}$$

Compute G and the momentum map J for

$$H:\mathbf{R}^{2n}\to\mathbf{R}: x\mapsto\tfrac{1}{2}x'\begin{pmatrix}\alpha I_{p,q} & 0\\ 0 & \alpha I_{p,q}\end{pmatrix}x,$$

where

$$\alpha>0,\quad I_{p,q}=\begin{pmatrix}I_p & 0\\ 0 & -I_q\end{pmatrix},\quad \text{and } p+q=n$$

[*Hint*: Show that
(1) $Q\in G$ if and only if Q is symplectic and $QJ_{2n}\tilde{H}=J_{2n}\tilde{H}Q$, where

$$\tilde{H}=\begin{pmatrix}\alpha I_{p,q} & 0\\ 0 & \alpha I_{p,q}\end{pmatrix}$$

(2) In matrix terms,

$$G=\left\{\begin{pmatrix}a & b\\ -I_{p,q}bI_{p,q} & I_{p,q}aI_{p,q}\end{pmatrix}\in Gl(2n,\mathbf{R})\middle| a,b\in gl(n,\mathbf{R}),(I_{p,q}a)'(bI_{p,q})\right.$$

$$\left.=(bI_{p,q})'(I_{p,q}a) \text{ and } (I_{p,q}a)(I_{p,g}a)'+(I_{p,q}b)(I_{p,q})'=I\right\}$$

ISBN 0-8053-0102-X

(3) The map

$$\varphi: G \to U(p,q) = \{ S \in Gl(2n, C) | \bar{S}^t I_{p,q} S = I_{p,q} \}:$$

$$\begin{pmatrix} a & b \\ -I_{p,q} b I_{p,q} & I_{p,q} a I_{p,q} \end{pmatrix} \mapsto a + ib I_{p,q}$$

is an isomorphism of real Lie groups.] See Cushman [1974] for further information.

4.2G. (Spin Angular Momentum). Let $SO(n)$ be the rotation group on R^n and $Spin(n)$ be its universal (two-fold) covering group. In Exercise 4.1F we showed that $Spin(3)$ is isomorphic to $SU(2, C)$ (and that this is diffeomorphic to S^3).

Let $Spin(n)$ act irreducibly on a complex vector space S_n (if $n = 2k$ or $2k+1$, S_n has complex dimension 2^k; see Palais [1965]).

Let M be an oriented Riemannian manifold. To define a spin bundle over M we first define a local spin bundle and then globalize in the spirit of Sect. 1.4.

A vector bundle $\pi: E \to M$ with fiber S_n is a *spin bundle* if there is a covering U_α of M and bundle charts $\phi_\alpha: TU_\alpha \subset TM \to U_\alpha \times R^n$ of TM and $\phi_\alpha^*: \pi^{-1}(U_\alpha) \to U_\alpha \times S_n$ of E such that (i) ϕ_α preserves the metric and orientation and (ii) the overlap maps $\phi_\beta^* \circ \phi_\alpha^{*-1}$ have the form $(x,s) \mapsto (x, g_{\alpha\beta}(x) \cdot s)$, where $g_{\alpha\beta}: U_\alpha \cap U_\beta \to Spin(n)$ and $\rho(g_{\alpha\beta}(x)) = \phi_\beta \circ \phi_\alpha^{-1}$ (restricted to x), where $\rho: Spin(n) \to SO(n)$ is the canonical projection.

Thus a *spin bundle* over M is a (vector) bundle $\pi: E \to M$ whose local charts are local spin bundles and transition maps are local spin bundle isomorphisms. Roughly, when we have a coordinate change, the fibers "transform like" spinors rather than vectors; that is, according to $Spin(n)$ rather than $SO(n)$.

A *classical Hamiltonian system with spin** is, by definition, a Hamiltonian system on T^*E.

Let G be a Lie group acting on M by maps $\phi_g: M \to M$ and suppose this action *lifts* to E by maps $\psi_g: E \to E$, that is, (i) $\pi \circ \psi_g = \phi_g \circ \pi$ and (ii) there are chart coverings $\phi_\alpha, \phi_\alpha^*$ as above, such that over $x \in A$, $\phi_\beta^* \circ \psi_g \circ \phi_\alpha^{*-1} \in Spin(n)$ and $\rho(\phi_\beta^* \circ \psi_g \circ \phi_\alpha^{*-1}) = \phi_\beta \circ (T\psi_g) \circ \phi_\alpha^{-1}$. (Such a lifted action always exists locally and is unique. For the global problem, see Chichilnisky [1972].)

Let ξ_M be an infinitesimal generator of ϕ on M and ξ_E the corresponding one for ψ on E. Then show that locally $\xi_E = \xi_M + \xi_s$, where ξ_s at each point lies in the Lie algebra of $Spin(n)$, that is, $T_e Spin(n)$. If the ϕ_α above have the form Tf_α for charts f_α on M, then $T_e\rho(\xi_s) = T\xi_M$ (in the chart).

Deduce that if $H: T^*E \to R$ is invariant under the action ψ, then

$$\hat{J}(\xi) = P(\xi_M) \circ \pi + P(\xi_s)$$

("orbital" + "spin" angular momentum) defines a momentum mapping J that is a constant of the motion for X_H.

For $M = R^3$, $E = R^3 \times C^2$, and $G = SO(3)$, show that

$$J(x, p_x, \alpha, p_\alpha) = x \times p + \tfrac{1}{2} p_\alpha \cdot \sigma \cdot \alpha$$

where the components of σ are the Pauli spin matrices. Discuss Whittaker's lemma for this case following 4.2.17.

*A quantum mechanical system with spin is a quantum mechanical system on the Hilbert space of L^2 sections of E; see Sect. 5.4 and 5.5.

ISBN 0-8053-0102-X

4.3 REDUCTION OF PHASE SPACES WITH SYMMETRY

It is a classical theorem going back to Jacobi and Liouville that if one has k first integrals in involution (i.e., their Poisson brackets all vanish), then it is possible to reduce Hamilton's equations to a set of Hamiltonian equations in which $2k$ variables have been eliminated. Similarly, in the n-body problem, rotational invariance allows one to eliminate four variables (Jacobi's "elimination of the node"). We now give a procedure, called *reduction*, that includes both of these as special cases.

The general setting of reduction (going back to E. Cartan) is the following. Suppose M is a manifold and ω is a *closed* two-form on M; let $E_\omega = \{v \in TM \mid i_v\omega = 0\}$ the *characteristic distribution* of ω and call ω *regular* if E_ω is a subbundle of TM. (This will hold if ω has constant rank; see 5.1.2). In the regular case, we note that E_ω is an involutive distribution, that is, if X, Y are sections of E_ω, then so is $[X, Y]$. To see this, it is sufficient to recall the formula

$$i_{[X,Y]} = L_X i_Y - i_Y L_X, \quad L_X = i_X d + d i_X$$

to obtain $i_{[X,Y]}\omega = 0$. By Frobenius' theorem E_ω is integrable and hence defines a foliation $\mathfrak{F}$ on M. Form the quotient space $M/\mathfrak{F}$ by identification of all points on a leaf. Assume now that $M/\mathfrak{F}$ is a manifold with the canonical projection $M \to M/\mathfrak{F}$ a submersion. Then as in 4.1.20, the tangent space at a point $[x]$ is isomorphic to $T_xM/E_\omega(x)$ and hence ω will project on a well-defined, closed, *nondegenerate* two-form on $M/\mathfrak{F}$. In other words, $M/\mathfrak{F}$ is a *symplectic manifold, the process by which it was obtained being called* ***reduction***. We shall apply this general result to the case of submanifolds defined by an Ad^*-equivariant momentum mapping of a given symplectic action. (Some generalizations are given in the exercises.) Our formulation follows Marsden and Weinstein [1974], which in turn was inspired by Smale [1970a], Souriau [1970a], and Robbin [1973]. Related papers are Neboroshev [1972], Meyer [1973], Fong and Meyer [1975], and Marle [1976]. A discussion of these questions also occurs in Arnold [1978]. We shall return to a number of these topics in Sect. 5.3.

First we shall summarize the notation from the previous section that will be used. We let (P, ω) be a symplectic manifold and $\Phi: G \times P \to P$ a symplectic action. Assume that this action has an Ad^*-equivariant momentum mapping $J: P \to \mathfrak{g}^*$. Denote by G_μ the isotropy subgroup of G under the co-adjoint action Ad^*, that is, $G_\mu = \{g \in G \mid Ad^*_{g^{-1}}\mu = \mu\}$. Since J is Ad^*-equivariant under G_μ, the orbit space $J^{-1}(\mu)/G_\mu$ is well defined. This space $P_\mu = J^{-1}(\mu)/G_\mu$ is called the *reduced phase space*.

We impose two conditions that guarantee that P_μ is a manifold. Note that G_μ is a Lie group, being a closed subgroup of G. First we assume μ to be a regular value of J (which by Sard's theorem takes place for "almost all" μ). Then $J^{-1}(\mu)$ is a submanifold of P (see Sect. 1.5). Second, suppose G_μ acts

ISBN 0-8053-0102-X

freely and properly on $J^{-1}(\mu)$. Then we know* (see 4.1.23) that $J^{-1}(\mu)/G_\mu$ is a manifold with the canonical projection $\pi_\mu: J^{-1}(\mu) \to P_\mu = J^{-1}(\mu)/G_\mu$ a submersion.

4.3.1 Theorem. *Let (P, ω) be a (weak) symplectic manifold on which the Lie group G acts symplectically and let $J: P \to \mathfrak{g}^*$ be an Ad^*-equivariant momentum mapping for this action. Assume $\mu \in \mathfrak{g}^*$ is a regular value of J and that the isotropy group G_μ under the Ad^* action on $\mathfrak{g}^*$ acts freely and properly† on $J^{-1}(\mu)$. Then $P_\mu = J^{-1}(\mu)/G_\mu$ has a unique (weak) symplectic form ω_μ with the property*

$$\pi_\mu^* \omega_\mu = i_\mu^* \omega$$

where $\pi_\mu: J^{-1}(\mu) \to P_\mu$ is the canonical projection and $i_\mu: J^{-1}(\mu) \to P$ is the inclusion.

4.3.2 Lemma. *For $p \in J^{-1}(\mu)$ we have*

(i) $T_p(G_\mu \cdot p) = T_p(G \cdot p) \cap T_p(J^{-1}(\mu))$, *and*
(ii) $T_p(J^{-1}(\mu))$ *is the ω-orthogonal complement of $T_p(G \cdot p)$. Here, $G \cdot p$ denotes the orbit of p under the action of G, that is, $G \cdot p = \{\Phi(g, p) | g \in G\}$.*

Proof. Let $\xi \in \mathfrak{g}$ so that $\xi_P(p) \in T_p(G \cdot p)$ (see the remark following 4.1.24). Assertion (i) is that $\xi_P(p) \in T_p(J^{-1}(\mu))$ iff $\xi \in \mathfrak{g}_\mu$, the Lie algebra of G_μ. By equivariance, $T_p J(\xi_P(p)) = \xi_{\mathfrak{g}^*}(\mu)$ (see 4.1.28), so $\xi_P(p) \in T_p(J^{-1}(\mu)) = \ker T_p J$ iff $\xi_{\mathfrak{g}^*}(\mu) = 0$, which is equivalent to μ being a fixed point of $Ad^*_{\exp(-t\xi)}$, that is, $\exp \xi \in G_\mu$, or by the characterization of Lie algebras of subgroups (see 4.1.13), $\xi \in \mathfrak{g}_\mu$.

For (ii), if $\xi \in \mathfrak{g}$ and $v \in T_p P$, we have $\omega(\xi_P(p), v) = (d\hat{J}(\xi))_p(v) = (T_p J \cdot v)(\xi)$ since J is a momentum mapping. Thus $v \in T_p(J^{-1}(\mu)) = \ker T_p J$ iff $\omega(\xi_P(p), v) = 0$ for all $\xi \in \mathfrak{g}$, that is, v is in the ω-orthogonal complement of $T_p(G \cdot p) = \{\xi_P(p) | \xi \in \mathfrak{g}\}$. ▼

From (ii) it follows that $T_p(J^{-1}(\mu))$ and $T_p(G \cdot p)$ are ω-orthogonal complements of each other. This is needed in the following proof. (For the infinite-dimensional case, use Exercise 3.1F.)

Proof of 4.3.1. For $v \in T_p(J^{-1}(\mu))$, let $[v] = T\pi_\mu(v)$ denote the corresponding equivalence class in $T_p J^{-1}(\mu)/T_p(G_\mu \cdot p)$. The assertion $\pi_\mu^* \omega_\mu = i_\mu^* \omega$ is

$$\omega_\mu([v], [w]) = \omega(v, w), \quad v, w \in T_p J^{-1}(\mu)$$

Since π_μ and $T\pi_\mu$ are surjective, ω_μ is unique.

*In the infinite-dimensional case, add the condition that the map from the group to each orbit is an immersion.

†Add the conditions in the footnote above and the condition that the model space is reflexive, and $\omega_p^\flat$ has closed range in the infinite-dimensional case.

ISBN 0-8053-0102-X

Now from Lemma 4.3.2(ii), it follows that ω_μ is a well-defined two-form. It is smooth since $\pi_\mu^*\omega_\mu$ is smooth.

Next we prove that ω_μ is closed. Indeed, $d\pi_\mu^*\omega_\mu = di_\mu^*\omega = i_\mu^* d\omega = 0$, so $\pi_\mu^*(d\omega_\mu) = 0$. Since π_μ and $T_p\pi_\mu$ are surjective, $d\omega_\mu = 0$.

For (weak) nondegeneracy of ω_μ, suppose $\omega_\mu([v],[w]) = 0$ for all $w \in T_pJ^{-1}(\mu)$, hence $\omega(v,w) = 0$ for all $w \in T_pJ^{-1}(\mu)$. Thus by Lemma 4.3.2(ii), $v \in T_p(G \cdot p)$ and hence by (i), $v \in T_p(G_\mu \cdot p)$, that is, $[v] = 0$ and so ω_μ is a (weak) symplectic form.

Finally (in the infinite-dimensional case), if ω is a strong symplectic form, so is ω_μ. Indeed, let $\alpha \in T^*_{\pi_\mu(p)}P_\mu$ so α is a linear map from $T_pJ^{-1}(\mu)$ to $\boldsymbol{R}$ that vanishes on the closed subspace $T_p(G_\mu \cdot p)$. If $\omega(v,w) = \alpha(w)$ for all $w \in T_pJ^{-1}(\mu)$, we see that $\omega_\mu([v],[w]) = \alpha([w])$ for all $w \in T_pJ^{-1}(\mu)$. ■

We remark that even if $\omega = d\theta$ is exact and the action leaves θ invariant, ω_μ need *not* be exact.

The manifold P_μ is always even dimensional, being a symplectic manifold. Also $\dim P_\mu = \dim J^{-1}(\mu) - \dim G_\mu = \dim P - \dim G - \dim G_\mu$.

Important Remark. If μ is a regular value of J, the action of G_μ is locally free. Even if the action is not globally free and proper (and this occurs in a number of examples) this construction can be done locally. This is important because in a number of results later on, we only assume μ is regular; this is possible because we will be implicitly invoking the local version of 4.3.1. In fact, μ only needs to be *weakly regular*; i.e., $J^{-1}(\mu)$ is a submanifold with $T_pJ^{-1}(\mu) = \ker T_pJ$.

Next we describe a way, which will be important for later work, to explicitly realize P_μ in the case of the cotangent bundle. Let G act on Q and hence on T^*Q and let $J\colon T^*Q \to \mathfrak{g}^*$, $\hat{J}(\xi)(\alpha_q) = \alpha_q(\xi_Q(q))$ be the usual momentum mapping. Suppose μ is a regular value for J and the other conditions of 4.3.1 hold. Also, suppose G_μ acts freely and properly on Q so that we can form $Q/G_\mu = Q_\mu$.

4.3.3 Theorem. *In addition to these conditions, assume there is a G_μ-equivariant one-form α_μ on Q with values in $J^{-1}(\mu)$.*

Put on T^*Q the symplectic form $\Omega_\mu = \omega_0 + (\tau_Q^*)^* d\alpha_\mu$ (where ω_0 is the canonical symplectic form) and let T^*Q_μ be given the corresponding induced symplectic form (see Exercise 3.7F).

Then there is an induced symplectic embedding

$$\phi_\mu\colon P_\mu \to T^*Q_\mu$$

onto a subbundle over Q_μ. The map ϕ_μ is a diffeomorphism **onto** *T^*Q_μ if and only if $\mathfrak{g} = \mathfrak{g}_\mu$.*

ISBN 0-8053-0102-X

Remark. An appropriate one-form α_μ will be constructed in Sect. 4.5 in terms of a Riemannian metric. Note that if $\mu=0$ we can take $\alpha_\mu=0$, and if G is abelian $\mathfrak{g}=\mathfrak{g}_\mu$, so $P_\mu \approx T^*Q_\mu$. This special case is due to Satzer [1977].

Proof. We have

$$T^*Q_\mu = \{\alpha_q \in T^*Q \mid \alpha_q \cdot \xi_Q(q) = 0 \textit{ for all } \xi \in \mathfrak{g}_\mu\}/G_\mu$$

$$= F_\mu/G_\mu \quad (\text{this defines } F_\mu)$$

(Recall that if G acts freely on M,

$$T^*_{[x]}(M/G) \cong (T_x M / T_x(G\cdot x))^*$$

so
$$T^*(M/G) \cong \{\alpha_x \in T^*M \mid \alpha_x(T_x(G\cdot x)) = 0\}^\dagger$$

Also, $J^{-1}(\mu) = \{\alpha_q \in T^*Q \mid \alpha_q \cdot \xi_Q(q) = \mu(\xi) \textit{ for all } \xi \in \mathfrak{g}\}$. Let $\psi_\mu : J^{-1}(\mu) \to F_\mu$; $\alpha_q \mapsto \alpha_q - \alpha_\mu(q)$. Since this is translation on the fibers, it follows as in 3.2.11, or by a coordinate computation, that ψ_μ is symplectic, that is, $\psi_\mu^*(\Omega_\mu | F_\mu) = \omega_0 | J^{-1}(\mu)$. Also, ψ_μ is clearly an embedding (and onto iff $\mathfrak{g}=\mathfrak{g}_\mu$). Since α_μ is G_μ equivariant, ψ_μ passes to the quotient, defining ϕ_μ. By definition of the symplectic structure on P_μ, ϕ_μ is symplectic. ■

4.3.4 Examples. (i) We begin with the Jacobi–Liouville theorem to which we alluded in the introduction. Let (P,ω) be a symplectic $2n$-manifold and let $K_1,\ldots,K_k$ be k functions in involution: $\{K_i, K_j\}=0$, $i,j=1,\ldots,k$. Because the flows of X_{K_i} and X_{K_j} commute, we can use them to define a symplectic action of $G=\mathbf{R}^k$ on P. Here $\mu \in \mathbf{R}^k$ is in the range space of $K_1 \times \cdots \times K_k$ and $J = K_1 \times \cdots \times K_k$ is the momentum mapping of this action. Assume the dK_i are independent at each point, so μ is a regular value for J. Since G is abelian, $G_\mu = G$, so we get a symplectic manifold $J^{-1}(\mu)/G$ of dimension $2n-2k$.

This example will be completed below in 4.3.5, where we show that any invariant Hamiltonian system on P induces, canonically, one on P_μ. If $k=n$, the system is called *completely integrable*; see Sect. 5.2 and Exercises 5.2H and 5.2I.

(ii) Let us consider the above example from a different point of view. Let X_H be a Hamiltonian vector field on P so its flow gives a symplectic action of $\mathbf{R}$ on P. The momentum mapping is H itself. Hence, if e is a regular value of H, we get a symplectic structure on $H^{-1}(e)/\mathbf{R}$, assuming it is a regular quotient. In this quotient we consider each orbit, or solution, to Hamilton's equation to be a point. This space is called the *manifold of solutions of constant energy*. It has dimension *dim P*-2.

---

† For this to work on tangent bundles rather than cotangent bundles, one must use a metric on Q invariant under G and put a Riemannian structure on Q_μ that makes the projection $Q \to Q_\mu$ a Riemannian submersion, that is, an isometry on the orthogonal complements to its level sets.

ISBN 0-8053-0102-X

(iii) If $G=SO(3)$, then the adjoint action of G on $\mathfrak{g}=\boldsymbol{R}^3$ is the usual one [Exercise 4.1A(iv)]. For $\mu\in\boldsymbol{R}^3$, $\mu\neq 0$, $G_\mu=S^1$ corresponding to rotations about the axis μ. A moment for this action may be called angular momentum (since that is what it is if $P=T^*\boldsymbol{R}^3$), and the reduction from P to $J^{-1}(\mu)/S^1$ is a generalization of the procedure in celestial mechanics called "elimination of the nodes." It goes back to Jacobi. Note that $\dim P_\mu=\dim P-3-1=\dim P-4$.

(iv) Example (ii) above provides a method of endowing quotient manifolds with "canonical" symplectic structures, if one chooses the Hamiltonian carefully. For example, $\boldsymbol{CP}^{n-1}$, the complex $(n-1)$-dimensional projective space is a symplectic manifold. To see this, choose $P=\boldsymbol{R}^{2n}=T^*\boldsymbol{R}^n$ with the canonical symplectic form $\Sigma_{i=1}^n dq^i\wedge dp_i$ and consider the (harmonic oscillator) Hamiltonian $H(\boldsymbol{q},\boldsymbol{p})=\frac{1}{2}\Sigma_{i=1}^n((q^i)^2+p_i^2)$. Then

$$X_H(\boldsymbol{q},\boldsymbol{p})=\sum_{i=1}^n\left(p_i\frac{\partial}{\partial q^i}-q^i\frac{\partial}{\partial p_i}\right)$$

and the flow of X_H is given by

$$F_t:(\boldsymbol{q},\boldsymbol{p})\mapsto(\boldsymbol{q}\cos t+\boldsymbol{p}\sin t,\boldsymbol{p}\cos t-\boldsymbol{q}\sin t)$$

Since F_t is periodic with period 2π, it defines a symplectic action of S^1 on P. Since S^1 is compact, it is proper and it is obviously free. The value $\frac{1}{2}$ is clearly a regular value for H and $H^{-1}(\frac{1}{2})=S^{2n-1}$. By example (ii) above $H^{-1}(\frac{1}{2})/\boldsymbol{R}=H^{-1}(\frac{1}{2})/S^1=S^{2n-1}/S^1=\boldsymbol{CP}^{n-1}$ is a symplectic manifold of real dimension $2(n-1)$.

(v) Let G be a Lie group and denote by $\Lambda: G\times G\to G$ the action of G on itself by left translations, that is, $\Lambda_g=L_g$ for all $g\in G$. Consider the induced action Λ^{T^*} on T^*G. By 4.2.11, the momentum mapping of this action is $J: T^*G\to\mathfrak{g}^*$, $J(\alpha_g)(\xi)=\alpha_g(\xi_G(g))=\alpha_g(T_eR_g(\xi))=(T_eR_g)^*\alpha_g(\xi)$, since by 4.1.25(a), $\xi_G(g)=T_eR_g(\xi)$. Hence $J(\alpha_g)=(T_eR_g)^*\alpha_g$. Each $\mu\in\mathfrak{g}^*$ is a regular value for J and $J^{-1}(\mu)=\{\alpha_g\in T^*G\,|\,\alpha_g(T_eR_g\cdot\xi)=\mu(\xi)\ \textit{for all}\ \xi\in\mathfrak{g}\}$, which is the graph of the right-invariant one-form α_μ whose value at e is μ, that is, $\alpha_\mu(g)=T_g^*R_{g^{-1}}(\mu)=\mu\circ T_gR_{g^{-1}}$. It is easy to see that $G_\mu=\{g\in G\,|\,L_g^*\alpha_\mu=\alpha_\mu\}$ using the definitions, so G_μ acts on $J^{-1}(\mu)$ by left translation of the base point. Thus $J^{-1}(\mu)/G_\mu\approx G/G_\mu\approx G\cdot\mu\subset\mathfrak{g}^*$, the diffeomorphism being given by φ: $\pi_\mu(\alpha_\mu(g))\mapsto Ad_g^*\mu$ for any $g\in G$. Hence the reduced phase space, in this case, is naturally identifiable with the orbit of μ in $\mathfrak{g}^*$ under the coadjoint representation, that is, $G\cdot\mu$ is a symplectic manifold. This is the statement of the *Kirillov–Kostant–Souriau theorem*.

We now compute explicitly the symplectic form on $G\cdot\mu$. Let this form be denoted ω_μ. Let ζ: $J^{-1}(\mu)\to G\cdot\mu$ be given by $\alpha_\mu(g)\mapsto Ad_g^*\mu$ so that from 4.3.1, $\zeta^*\omega_\mu=i_\mu^*\omega$, where ω is the canonical two-form on T^*G. To work out ω_μ from this, we need a few steps.

ISBN 0-8053-0102-X

(1) Since $J^{-1}(\mu)$ is the graph of α_μ, $T_{\alpha_\mu(g)}J^{-1}(\mu)=\{T\alpha_\mu\cdot TR_g\xi\,|\,\xi\in\mathfrak{g}\}$

(2) $(i_\mu^*\omega)(T\alpha_\mu\cdot TR_g\xi, T\alpha_\mu\cdot TR_g\eta)=(\alpha_\mu^*\omega)(TR_g\xi, TR_g\eta)$

$$=-d\alpha_\mu(TR_g\xi, TR_g\eta);\ \text{(see 3.2.11)}$$

$$=-d\alpha_\mu(\tilde{X}_\xi, \tilde{X}_\eta)(g)$$

where $\tilde{X}_\xi$ is the *right*-invariant vector field corresponding to ξ. By Formula 6 in Table 2.4.1, this becomes

$$-\left\{\tilde{X}_\xi(\alpha_\mu(\tilde{X}_\eta))-\tilde{X}_\eta(\alpha_\mu(\tilde{X}_\xi))-\alpha_\mu([\tilde{X}_\xi,\tilde{X}_\eta])\right\}(g)$$

$$=\alpha_\mu([\tilde{X}_\xi,\tilde{X}_\eta])(g)\quad[\text{since } \alpha_\mu(\tilde{X}_\eta)(g)=\mu(\eta) \text{ is constant in } g]$$

$$=-\mu([\xi,\eta])$$

(it is *minus* since we are dealing with *right*-invariant vector fields).

(3) $(\zeta^*\omega_\mu)(\alpha_\mu(g))(T_g\alpha_\mu\cdot T_eR_g\xi, T_g\alpha_\mu\cdot T_eR_g\eta)=\omega_\mu(Ad_g^*\mu)(T_{\alpha_\mu(g)}\zeta\cdot T_g\alpha_\mu\cdot T_eR_g\xi,$ $T_{\alpha_\mu(g)}\xi\cdot T_g\alpha_\mu\cdot T_eR_g\eta)$. But $\zeta(\alpha_\mu(g))=Ad_g^*\mu$ by definition, so we get

$$\omega_\mu(Ad_g^*\mu)\left(T_g(Ad_\cdot^*\mu)\cdot T_eR_g\xi, T_g(Ad_\cdot^*\mu)\cdot T_eR_g\eta\right).$$

(4) By exercise 4.1H and Example 4.1.25(d) we have

$$T_g(Ad_\cdot^*\mu)\cdot T_eR_g\xi=Ad_g^*(ad\xi)^*\mu=-Ad_g^*(\xi_{\mathfrak{g}^*}(\mu))$$
$$=-\left(Ad_{g^{-1}}^*\xi_{\mathfrak{g}^*}\right)(Ad_g^*\mu)$$

By 4.1.26(i) applied to the coadjoint action,

$$Ad_{g^{-1}}^*\xi_{\mathfrak{g}^*}=(Ad_{g^{-1}}\xi)_{\mathfrak{g}^*}.$$

Putting these steps together we get the following formula:

$$\omega_\mu(Ad_g^*\mu)\left((Ad_{g^{-1}}\xi)_{\mathfrak{g}^*}(Ad_g^*\mu),(Ad_{g^{-1}}\eta)_{\mathfrak{g}^*}(Ad_g^*\mu)\right)=-\mu\cdot[\xi,\eta]$$

for any $g\in G$. Replacing here $Ad_{g^{-1}}\xi$ by ξ, $Ad_{g^{-1}}\eta$ by η and letting $\nu=Ad_g^*\mu$ be an arbitrary point on the orbit we get

$$\boxed{\omega_\mu(\nu)(\xi_{\mathfrak{g}^*}(\nu),\eta_{\mathfrak{g}^*}(\nu))=-\nu\cdot[\xi,\eta].}$$

From 4.1.26(i) and this formula for ω_μ, it also follows that ω_μ is $Ad_{g^{-1}}^*$-invariant. That is, $G\cdot\mu$ is a *homogeneous Hamiltonian G-space*. If (P,ω,Φ,J) is a

ISBN 0-8053-0102-X

homogeneous Ad^* equivariant Hamiltonian G-space, it is easy to see that $J: P \to \mathfrak{g}^*$ gives a symplectic covering map onto $G \cdot \mu$, a result of Kostant. (See Wallach [1977] and Guillemin and Sternberg [1977] for more information.) For examples using $G \cdot \mu$, see 4.4.7, 4.6G and 5.5L.

So far, the symplectic manifold alone has been reduced. We want now to induce Hamiltonian systems on the reduced phase space.

4.3.5 Theorem. *Under the assumptions of* 4.3.1, *let* $H: P \to R$ *be invariant under the action of G. Then the flow* F_t *of* X_H *leaves* $J^{-1}(\mu)$ *invariant and commutes with the action of* G_μ *on* $J^{-1}(\mu)$, *so it induces canonically a flow* H_t *on* P_μ *satisfying* $\pi_\mu \circ F_t = H_t \circ \pi_\mu$. *This flow is a Hamiltonian flow on* P_μ *with a Hamiltonian* H_μ *satisfying* $H_\mu \circ \pi_\mu = H \circ i_\mu$. H_μ *is called the* ***reduced Hamiltonian.***

Proof. From 4.2.2, J is an integral for X_H. It follows that $J^{-1}(\mu)$ is invariant under the flow and that we get a well-defined flow H_t induced on P_μ. If F_t denotes the flow of X_H on P and $J^{-1}(\mu)$, we clearly have $\pi_\mu \circ F_t = H_t \circ \pi_\mu$, so $\pi_\mu^* H_t^* \omega_\mu = F_t^* \pi_\mu^* \omega_\mu = F_t^* i_\mu^* \omega = i_\mu^* \omega = \pi_\mu^* \omega_\mu$, the third equality holding since F_t is symplectic and leaves $J^{-1}(\mu)$ invariant. But since π_μ is a surjective submersion, we conclude $H_t^* \omega_\mu = \omega_\mu$, so the flow H_t on P_μ is Hamiltonian.

The relation $H_\mu \circ \pi_\mu = H \circ i_\mu$ plus invariance of H under the action of G defines H_μ uniquely. Hence, if $[v] = T\pi_\mu(v) \in TP_\mu$, we have

$$dH_\mu[v] = i_\mu^* dH(v) = i_\mu^* \omega(X_H, v)$$

But from the construction of H_t, its generator Y satisfies $T\pi_\mu \circ X_H = Y \circ \pi_\mu$, so $dH_\mu[v] = i_\mu^* \omega(X_H, v) = \pi_\mu^* \omega_\mu(X_H, v) = \omega_\mu(Y, [v])$, that is, Y has energy H_μ. ■

In celestial mechanics where one has a rotational invariance, we can "eliminate the nodes" by passing to $P_\mu = J^{-1}(\mu)/S^1$ and we get a well-defined Hamiltonian system on the reduced phase space. It eliminates extra rotational freedom and, in effect, passes to a rotating coordinate system. (We shall see this sort of thing explicitly in Sect. 9.1 and 9.8.)

Again, in the Jacobi–Liouville theorem [4.3.4(i)], if we have a given Hamiltonian system X_H and k first integrals in involution and pass to P_μ, we still obtain equations in Hamiltonian form. Under the conditions of 4.3.4(v), the reduced Hamiltonian H_μ on $G \cdot \mu \subset \mathfrak{g}^*$ is given by $H_\mu(Ad_{g^{-1}}^* \mu) = H(T_g^* R_{g^{-1}}(\mu))$, where $H: T^*G \to R$ is a given Hamiltonian satisfying $H \circ T^* L_g = H$ for all $g \in G$.

The Hamiltonian system induced on the reduced phase space represents, in a sense, the "essential" dynamics; the explicitly known dynamics is factored out in the reduction process.

If we know the flow H_t *of the reduced system on* P_μ, *then we can find the flow of* F_t *on* $J^{-1}(\mu)$ *as follows.* Let $p_0 \in J^{-1}(\mu)$ and let $c(t)$ and $[c(t)]$ be the integral curves of X_H and X_{H_μ} with $c(0) = p_0$. We want to express $c(t)$ in terms

ISBN 0-8053-0102-X

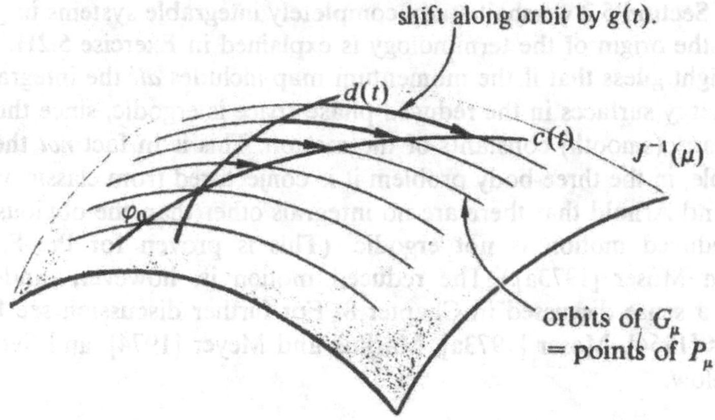

Figure 4.3-1

of $[c(t)]$. To do so, pick a smooth curve $d(t)\in J^{-1}(\mu)$ with $d(0)=p_0$ and $[d(t)]=[c(t)]$. Write $c(t)=\Phi_{g(t)}(d(t))$, for $g(t)\in G_\mu$. We then try to find $g(t)$ (see Fig. 4.3-1). Now [see Eq. (2) in the proof of 4.2.16]

$$\begin{aligned} X_H(c(t)) &= c'(t) \\ &= T\Phi_{g(t)}(d(t))\cdot d'(t) \\ &\quad + T\Phi_{g(t)}(d(t))\cdot\left(TL_{g(t)^{-1}}\cdot g'(t)\right)_P(d(t)) \end{aligned}$$

Using $\Phi_{g(t)}$ invariance of X_H, we get the equation

$$X_H(d(t)) = d'(t) + \left(TL_{g(t)^{-1}}g'(t)\right)_P(d(t))$$

This is an equation for $g(t)$ written only in terms of $d(t)$. We solve it by first solving the algebraic problem

$$\xi_P(d(t)) = X_H(d(t)) - d'(t) \tag{1}$$

for $\xi(t)\in\mathfrak{g}$ and then solving

$$g'(t) = TL_{g(t)}\xi(t) \tag{2}$$

for $g(t)$. (In examples, one will often get an answer "in quadratures.") Then finally the solution $c(t)$ sought is

$$c(t) = \Phi_{g(t)}d(t) \tag{3}$$

If the reduced phase space is a point, we say that the system is *completely integrable*, although this terminology is often reserved for integrals in involution. By the above, the problem of finding the flow is reduced to solving

ISBN 0-8053-0102-X

(1)–(3). In Section 5.2 we shall study completely integrable systems in greater detail and the origin of the terminology is explained in Exercise 5.2H.

One might guess that if the momentum map includes *all* the integrals, the flow on energy surfaces in the reduced phase space is ergodic, since there are no longer any (smooth) constants of the motion. This is in fact *not* the case. For example, in the three-body problem it is conjectured from classic work of Poincaré and Arnold that there are no integrals other than the obvious ones, yet the reduced motion is not ergodic. (This is proven for the Sitnikov example in Moser [1973a].) The reduced motion is, however, random or chaotic in a sense discussed in Chapter 8. For further discussion see Henon and Heiles [1964], Moser [1973a], Markus and Meyer [1974], and Sects. 8.3 and 8.8 below.

A circular orbit in a rotationally invariant system appears to be an equilibrium relative to a rotating observer. This idea leads to the following.

4.3.6 Definition. *Under the conditions of* 4.3.1 *and* 4.3.5, *a point* $p \in P$ *is called a* ***relative equilibrium*** *if* $\pi_\mu(p) \in P_\mu$ *is a fixed point for the induced Hamiltonian system* X_{H_μ} *on* P_μ, *where* $\mu = J(p)$.

Since the symplectic form on P_μ is nondegenerate, this condition is equivalent to $dH_\mu(\pi_\mu(p)) = 0$, that is, $\pi_\mu(p)$ is a critical point of H_μ.

Similarly, one defines a *relative periodic point* to be a point p such that $\pi_\mu(p) \in P_\mu$ lies on a periodic orbit of X_{H_μ}.

The next result characterizes relative equilibria and relative periodic points in terms of the original flow.

4.3.7 Proposition. *Let the conditions of* 4.3.1 *and* 4.3.5 *hold, and* $p \in J^{-1}(\mu)$. *Denote by* $\Phi: G \times P \to P$ *the symplectic action of G on P and by* F_t *the flow of the Hamiltonian vector field* $X_H \in \mathfrak{X}(P)$.

(*a*) *The following are equivalent:*
- (*i*) $p \in P$ *is a relative equilibrium;*
- (*ii*) *there exists a one-parameter subgroup* $g(t)$ *of G such that for all* $t \in \mathbb{R}$, $F_t(p) = \Phi(g(t), p)$;
- (*iii*) *there exists* $\xi \in \mathfrak{g}$ *such that* $X_H(p) = \xi_P(p)$.

(*b*) *The following are equivalent:*
- (*i*) $p \in P$ *is a relative periodic point;*
- (*ii*) *there exists* $g \in G$ *and* $\tau > 0$ *such that*

$$F_{t+\tau}(p) = \Phi(g, F_t(p)) \text{ for all } t \in \mathbb{R}.$$

Proof. (a) p is a relative equilibrium iff $\pi_\mu(F_t(p)) = \pi_\mu(p)$. If this holds, there is a unique curve $g(t) \in G_\mu$ such that $F_t(p) = \Phi(g(t), p)$ since the action of G_μ on $J^{-1}(\mu)$ is assumed to be free. The flow property $F_{t+s} = F_t \circ F_s$ shows $g(t+s) = g(t)g(s)$, so $g(t)$ is a one-parameter subgroup.

ISBN 0-8053-0102-X

Conversely, if $F_t(p) = \Phi(g(t), p)$, where $g(t)$ is a one-parameter subgroup of G, then, from Ad^* equivariance of J,

$$\Phi(g,p) \in J^{-1}(\mu) \quad \text{iff } g \in G_\mu$$

This, together with the invariance of J under F_t [hence $F_t(p) \in J^{-1}(\mu)$ for all $t \in \mathbf{R}$], shows that $g(t) \in G_\mu$ for all $t \in \mathbf{R}$, so $\pi_\mu(F_t(p)) = \pi_\mu(p)$ for all $t \in \mathbf{R}$, that is, p is a relative equilibrium.

This establishes the equivalence of (i) and (ii).

The equivalence of (ii) and (iii) follows from the fact that any one-parameter subgroup of G is of the form $\exp t\xi$ for some $\xi \in \mathfrak{g}$ [for (ii)⇒(iii)] and by uniqueness of integral curves [for (iii)⇒(ii)].

Part (b) is similar and is left as an exercise for the reader. ■

For example, in celestial mechanics, a relative equilibrium is a configuration of the bodies which is such that the bodies subsequently move in circles.

The following result gives a useful criterion for relative equilibria involving only the functions H, J on P.

4.3.8 Proposition. (*Souriau–Smale–Robbin*) *Let the conditions of* 4.3.1 *and* 4.3.5 *hold. Then* $p \in J^{-1}(\mu)$ *is a relative equilibrium iff p is a critical point of* $H \times J: P \to \mathbf{R} \times \mathfrak{g}^*$.

For the proof we need the following.

4.3.9 Lemma (Lagrange Multiplier Theorem). *Let* $T: E \to \mathbf{R}$, $A: E \to F$ *be linear maps, where A is surjective and E, F are finite-dimensional vector spaces.* Then T is surjective on* $\ker A$ *iff* $T \times A: E \to \mathbf{R} \times F$ *is surjective.*

Proof. Assume T is surjective on $\ker A$ and write $E = \ker A \oplus E_1$. Let $(t, f) \in \mathbf{R} \times F$ and find by surjectivity of A, $e_1 \in E_1$ such that $A(e_1) = f$. Since T is surjective on $\ker A$, there exists $k \in \ker A$ such that $T(k + e_1) = t$. Hence, if $e = e_1 + k$, $(T \times A)(e) = (t, f)$.

Conversely, if $T \times A$ is surjective, then for $(t, 0)$, $t \neq 0$, find $e \in E$ such that $(T \times A)(e) = (Te, Ae) = (t, 0)$. But this shows that T is surjective on $\ker A$. ■

Proof of 4.3.8. We saw before that p is a relative equilibrium iff $\pi_\mu(p)$ is a critical point of $H_\mu: P_\mu \to \mathbf{R}$. Since $H_\mu \circ \pi_\mu = H \circ i_\mu$ and π_μ is a surjective submersion, this is equivalent to p being a critical point of $H|J^{-1}(\mu)$, that is, $dH(p) = 0$ on $\ker T_pJ = T_p(J^{-1}(\mu))$. By the Lagrange Multiplier Theorem, this in turn is equivalent to $dH(p) \times T_pJ: T_pP \to \mathbf{R} \times \mathfrak{g}^*$ not surjective [if we assume $dH(p) \neq 0$] and hence $dH(p)$ surjective, which says that p is a critical

*In the infinite-dimensional case when E, F are supposed to be Banach spaces, add the condition that $\ker A$ has a closed complement.

ISBN 0-8053-0102-X

point of $H \times J: P \to \mathbf{R} \times \mathfrak{g}^*$. If $dH(p)=0$, then clearly p will be a critical point of $H \times J: P \to \mathbf{R} \times \mathfrak{g}^*$, so the theorem holds in this case, too. ■

We now investigate a notion of stability appropriate for relative equilibria, called *relative stability*. Recall that stability of a point (see Sect. 2.1) means intuitively that nearby trajectories stay nearby to the trajectory of the point for all time. Precisely, a fixed point $p \in P$ of a flow F_t is *stable* when for any neighborhood U of p, there is a neighbrohood $V \subset U$ of p such that $x \in V$ implies $F_t(x)$ is defined and remains in U for all $t \in \mathbf{R}$.

4.3.10 Definition. *Let (P, ω) be a symplectic manifold and G a Lie group acting symplectically on P and leaving a Hamiltonian $H \in \mathfrak{F}(P)$ invariant. Assume that the hypotheses of* 4.3.1 *and* 4.3.5 *hold. A relative equilibrium $p \in P$ is* ***relatively stable*** *if $\pi_\mu(p)$ is stable for the induced dynamical system X_{H_μ} on P_μ, where $\pi_\mu(p)$ appears as a fixed point of X_{H_μ}.*

The stability of relative equilibria in celestial mechanics is subtle, depending on deep properties of Hamiltonian systems as has been shown by Kolmogorov, Arnold, and Moser (see Chapter 10). However, for "simple" situations such as the rigid body, we have the following generalization of a result of V. Arnold. The specific case considered by Arnold is given in Sect. 4.4 below.

4.3.11 Theorem. *Let the conditions of* 4.3.1 *and* 4.3.5 *hold with $p \in J^{-1}(\mu)$ a relative equilibrium. Suppose the Hessian $(\mathrm{Hess}\, H_\mu)(\pi_\mu(p))$ is positive (or negative) definite. Then p is relatively stable.*

The proof is an immediate consequence of Definition 4.3.10 and of the general stability criterion for Hamiltonian systems 3.4.17.

In the above, we investigated the invariance of a Hamiltonian system under a one-parameter group, or a whole Lie group, of symmetries. Discrete symmetries can also play a profound role, such as reflection symmetries. Now we consider the simplest type of "symmetry" of this kind, namely, time-reversal symmetry.

4.3.12 Definition. *Let (P, ω) be a symplectic manifold. A map $\mu: P \to P$ is* ***antisymplectic*** *if $\mu^*\omega = -\omega$. A Hamiltonian system is called* ***reversible*** *if there is an antisymplectic involution μ (i.e., $\mu^2 = id$) such that $H \circ \mu = H$.*

4.3.13 Proposition. *Let H be reversible and let $c(t)$ be an integral curve of X_H. Then $\mu \circ c(-t)$ is also an integral curve of X_H. Thus $F_{-t}(x) = \mu F_t(\mu(x))$, where F_t is the flow of X_H.*

Proof. As in 3.3.19, one can show that $\mu^* X_H = -X_H$ or $T\mu \circ X_H = -X_H \circ \mu$ from which the conclusion is immediate. ■

ISBN 0-8053-0102-X

In such a situation, if the flow F_t of X_H is initially defined only for $t>0$, 4.3.13 assures us that F_t may be defined for $t<0$ as well. This is typical for many Hamiltonian systems. For example, if H is kinetic plus potential energy on a Riemannian manifold M, we can take $\mu\colon P=TM\to P$, $\mu(v)=-v$. Similarly, as we shall see later, quantum mechanical systems are always reversible as their Hamiltonians are quadratic forms. If f is a lift of an involution $\rho\colon M\to M$ to TM or T^*M, so f is symplectic, we call $f\circ\mu$ a "parity symmetry." Clearly $f\circ\mu$ is antisymplectic and is an involution iff $f\circ\mu=\mu\circ f$. Typically we might choose ρ to be reflection in the origin.

EXERCISES

4.3A. Under the hypotheses of 4.3.1 and 4.3.5, show that X_H is tangent to $J^{-1}(\mu)$ and that $X_H|J^{-1}(\mu)$ and X_{H_μ} are π_μ-related, that is, $T\pi_\mu\circ X_H=X_{H_\mu}\circ\pi_\mu$. For G invariant functions f,g on P, show that $\{f,g\}_\mu=\{f_\mu,g_\mu\}$. Discuss for 4.3.4(v).

4.3B. Let the hypotheses of 4.3.1 hold and suppose $\bar{G}$ is another Lie group acting symplectically on (P,ω) with an Ad^*-equivariant momentum mapping $\bar{J}$. If the actions of G and $\bar{G}$ commute and $\bar{J}$ is G-invariant, then

(i) $\bar{G}$ leaves J invariant;

(ii) the $\bar{G}$-action induces canonically a unique symplectic action on P_μ and the induced momentum mapping $\bar{J}_\mu\colon P_\mu\to\mathfrak{g}^*$ satisfies $\bar{J}_\mu\circ\pi_\mu=\bar{J}\circ i_\mu$. (*Hint*: Replace in the proof of 4.3.5 everywhere "flow" with $\bar{G}$-action; for the momentum computation on P_μ use the fact that $\xi_P|J^{-1}(\mu)$ and ξ_{P_μ} are π_μ-related.)

Note that 4.3.5 is a particular case of this statement with $\bar{G}=\mathbb{R}$, $\bar{J}=H$. Taking in 4.3.4(v) $\bar{G}=G$ acting on T^*G by right translations, we conclude that $G\cdot\mu\subset\mathfrak{g}^*$ is acted on symplectically by G. Thus, show that $G\cdot\mu$ is a "homogeneous Hamiltonian G-space." What is the momentum mapping for the action of G on $G\cdot\mu$? The reduced phase space?

4.3C. (a) Let S^1 act on $T^*\mathbb{R}^2$ by rotations. Construct the reduced phase space explicitly and identify it with "rotating coordinates."

(b) Describe the process of fixing the center of mass and angular momentum of a system of n particles in $\mathbb{R}^3$ in terms of reduction (see Robinson [1975c] and Section 10.4).

4.3D. This exercise shows how the reduction is done if the momentum mapping J is *not* Ad^*-equivariant (see also Exercise 4.3.E). Let (P,ω) be a symplectic manifold, G acting symplectically on P, and let $J\colon P\to\mathfrak{g}^*$ be a momentum mapping of this action.

(i) Recall that the one-cocycle of G for the co-adjoint representation is defined by $\sigma(g)=J(\Phi(g,p))-Ad^*_{g^{-1}}(J(p))$, $g\in G$, $p\in P$ and it does not depend on p (see Sect. 4.2). It also satisfies $\sigma(gh)=\sigma(g)+Ad^*_{g^{-1}}(\sigma(h))$. Let $\Psi\colon G\times\mathfrak{g}^*\to\mathfrak{g}^*$, $(g,\mu)\mapsto Ad^*_{g^{-1}}\mu+\sigma(g)$. By 4.2.7, J is equivariant with respect to this action, that is, $\Psi(g,J(p))=J(\Phi(g,p))$. Conclude that if $\xi_{\mathfrak{g}^*}$ denotes the infinitesimal generator for the action Ψ, that is, $\xi_{\mathfrak{g}^*}(\mu)=\frac{d}{dt}\Psi(\exp t\xi,\mu)|_{t=0}$, $\xi\in\mathfrak{g}$ we have $T_pJ(\xi_P(p))=\xi_{\mathfrak{g}^*}(J(p))$. (*Hint*: See 4.1.28.)

(ii) Show that Lemma 4.3.2 still holds, where G_μ denotes the isotropy group of μ under the action Ψ.

ISBN 0-8053-0102-X

(iii) Formulate and prove the analog of Theorem 4.3.1, dropping the assumption of Ad^* equivariance of J and letting G_μ be the isotropy subgroup of $\mu \in \mathfrak{g}^*$ under the action Ψ.

(iv) Prove the analog of Theorem 4.3.5, dropping the assumption of Ad^* equivariance of J.

(v) Show that if the action Ψ of G on $\mathfrak{g}^*$ is used in place of the co-adjoint action, the orbit $G\cdot\mu$ is symplectic with

$$\omega_\mu(\xi_{\mathfrak{g}^*}(\nu), \eta_{\mathfrak{g}^*}(\nu)) = -\mu[\xi\,\eta] + \Sigma(\xi\,\eta)$$

4.3E. (A. Weinstein)
Let $\pi: P \to M$ be a principal G-bundle and J_P the natural momentum mapping on T^*P. Let Q be a Hamiltonian G-space with momentum mapping J_Q. Show that $T^*P \times Q$ reduced by $J_P + J_Q$ at $\mu = 0$ is symplectically diffeomorphic to the space $(\tau_M^* P) \times_G Q$ constructed by Sternberg (Proc. Nat. Acad., Dec. 1977). Give the isomorphism explicitly for Exercise 3.7F.

4.3F. (Fong and Meyer)

(a) Let $\mathcal{Q} \subset \mathcal{F}(P)$ be a Poisson subalgebra on a symplectic manifold (P, ω). Let $S^0 = \{X \in \mathfrak{X}(P) | X(f) = 0 \textit{ for all } f \in \mathcal{Q}\}$ and $S^\# = \{X_f | f \in \mathcal{Q}\}$. Show that S^0, $S^\#$ and $S^0 \cap S^\#$ are involutive (and assume the pointwise objects are bundles). Let $p \in P$ and N be the leaf of the S^0 foliation through p. Let $p \sim p'$ if p and p' lie on the same $S^0 \cap S^\#$-leaf in N. Let $B = N/\sim$. Show that B is a symplectic manifold.

(b) Let G act symplectically on (P, ω) with a moment $J: P \to \mathfrak{g}^*$ and let $\mathcal{Q} = \{\hat{J}(\xi) | \xi \in \mathfrak{g}\}$. Show that $N = J^{-1}(\mu)$ if $J(p) = \mu$ and that the $S^0 \cap S^\#$-leaf in N is the orbit of p under $G_N = \{g \in G | \Phi_g N = N\}$, a Lie subgroup of G. If J is Ad^*-equivariant, show that this reproduces 4.3.1, while otherwise it reproduces the result of Exercise 4.3.D. (See Fong and Meyer [1975].)

4.3G. (Marle) Let the hypotheses of 4.3.1 hold and let G also act on $\mathfrak{g}^*$ by the coadjoint action (or the action Ψ if J is not Ad^*-equivariant; see Exercises 4.3D, F), so that $G\cdot\mu$ is a symplectic manifold 4.3.4(v). Let $N = J^{-1}(G\cdot\mu)$. On N, say $p \sim p'$ if p and p' lie in the same G-orbit and if $J(p) = J(p')$. Let $Q = N/\sim$ and $\tilde{J}: Q \to G\cdot\mu$ be the map induced from J. Show:

(a) Q is a smooth manifold for which the projection $\tilde{\pi}: N \to Q$ and $\tilde{J}$ are smooth submersions and G acts smoothly on Q.

(b) Q is a symplectic manifold and G acts symplectically on Q with momentum mapping $\tilde{J}$.

(c) The symplectic structure on Q can be obtained from the result of Fong and Meyer (Exercise 4.3F) by choosing $\mathcal{Q} = \{\hat{J}(\xi) | \xi \in \mathfrak{g} \text{ and } \xi_{\mathfrak{g}^*}(\mu) = 0\}$.

(d) Locally, Q is $P_\mu \times G\cdot\mu$.

(e) Q reduced by the action of G on it is again P_μ.

(The reader is referred to Marle [1976] and Kazhdan, Kostant, and Sternberg [1978] for a number of quite interesting applications of these results.)

4.3H. (Arms, Fischer and Marsden) Let J be an Ad^*-equivariant moment for a symplectic action Φ of G on (P, ω). Let P and G carry Riemannian structures.

ISBN 0-8053-0102-X

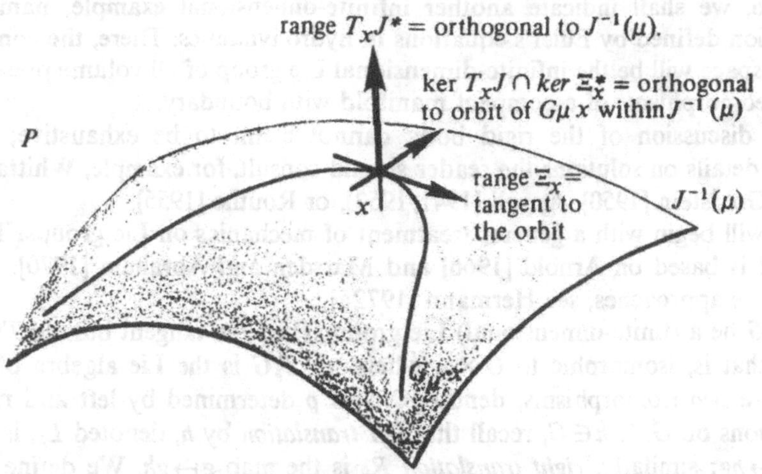

Figure 4.3-2

Then show that T_xP admits the orthogonal decomposition

$$T_xP = range\,\Xi_x + (ker\,T_xJ \cap ker\,\Xi_x^*) + range(T_xJ)^*$$
$$\approx T_x(G_\mu \cdot x) + ker\,T_xJ / T_x(G_\mu \cdot x) + T_x(G \cdot x)$$

where $\Xi_x\colon \mathfrak{g}_\mu \to T_xP\colon \xi \mapsto \xi_P(x)$ and $*$ is the adjoint relative to the inner products involved. [*Hint.* Write $T_xP = ker\,T_xJ + range(T_xJ)^* = range\,\Xi_x + ker\,\Xi_x^*$ and prove $range\,\Xi_x \subset ker\,T_xJ$.] See Fig. 4.3-2. (For applications of this decomposition, see Arms, Fischer, and Marsden [1975].)

4.3I. Prove theorem 3.4.15 in the following context. Let the conditions of 4.3.1 hold. Show that $I_\mu = J^{-1}(\mu)$ has a unique volume form δ_μ invariant under G_μ that projects to the volume form determined by the symplectic form ω_μ on P_μ. Deduce that if H is G-invariant, X_H leaves δ_μ invariant.

4.3J. (The Routhian) Let L be a Lagrangian on TQ and let G be a group acting on Q and hence on TQ, which leaves L invariant. Form from X_E the corresponding reduced Hamiltonian system on $P_\mu = J^{-1}(\mu)/G_\mu$, where J is the momentum map (see 4.2.14). The reduced Hamiltonian is called the *Routhian* and is denoted R_μ. (It depends parametrically on $\mu \in \mathfrak{g}^*$.) Show that this procedure generalizes the classical description of the Routhian for cyclic (or ignorable) coordinates (cf. Goldstein [1950], p. 218–220).

4.3K. Use reduction by the energy integral (4.3.4(ii)) to give another proof of the principle of least action 3.8.5.

4.4 HAMILTONIAN SYSTEMS ON LIE GROUPS AND THE RIGID BODY

This section studies some general features of Hamiltonian systems whose configuration space is a Lie group. The main example we consider is the rigid body; the configuration space here is the rotation group $SO(3, \boldsymbol{R})$. Later, in

ISBN 0-8053-0102-X

Sect. 5.6, we shall indicate another infinite-dimensional example, namely, the motion defined by Euler's equations in hydrodynamics. There, the configuration space will be the infinite-dimensional Lie group of all volume preserving diffeomorphisms of a compact manifold with boundary.

Our discussion of the rigid body cannot claim to be exhaustive; for specific details on solutions the reader should consult, for example, Whittaker [1959], Goldstein [1950], Appell [1941, 1953], or Routhe [1955].

We will begin with a general treatment of mechanics on Lie groups. This material is based on Arnold [1966] and Marsden and Abraham [1970]. For alternative approaches, see Hermann [1972a].

Let G be a (finite-dimensional) Lie group. Then the tangent bundle TG is trivial, that is, isomorphic to $G\times\mathfrak{g}$, where $\mathfrak{g}=T_eG$ is the Lie algebra of G. There are *two* isomorphisms, denoted λ and ρ determined by left and right translations on G. If $h\in G$, recall that *left translation* by h, denoted L_h, is the map $g\mapsto hg$; similarly, *right translation* R_h is the map $g\mapsto gh$. We define the isomorphism of vector bundles $\lambda\colon TG\to G\times\mathfrak{g}$ by

$$\lambda(v)=\left(g,T_gL_g^{-1}(v)\right)$$

where $g=\tau_G(v)$, $\tau_G\colon TG\to G$ being the natural projection. Similarly, we define the vector bundle isomorphism $\rho\colon TG\to G\times\mathfrak{g}$ by

$$\rho(v)=\left(g,T_gR_g^{-1}(v)\right)$$

We shall sometimes refer to λ as defining *body coordinates* [i.e., $\lambda(v)$ represents v in body coordinates] and ρ as defining *space coordinates*; the reason for this rather obscure terminology will be explained when we discuss the rigid body.

The transition from body to space coordinates is as follows:

$$\begin{aligned}(\rho\circ\lambda^{-1})(g,\xi)&=\rho\left(g,T_gL_g(\xi)\right)\\&=\left(g,T_eR_g^{-1}\circ T_eL_g(\xi)\right)\\&=(g,Ad_g\xi)\end{aligned}$$

where $g\mapsto Ad_g$ is the adjoint representation of G on $\mathfrak{g}$.

We now relate time derivatives in space and body coordinates. Thus, let $x(t)$ be a curve in G, and let $v_0(t)$ be a curve in TG with $v_0(t)\in T_{x(t)}G$. Let $\xi(t)=T_eL_{x(t)^{-1}}v_0(t)$, that is, $\xi(t)$ is the vector field over $x(t)$ in *body* coordinates so that the corresponding vector field in *space* coordinates is $\tilde{\xi}(t)=Ad_{x(t)}(\xi(t))$. Now recall the formula

$$\frac{d}{dt}\eta(t)=\left[T_eR_{x(t)^{-1}}\frac{dx(t)}{dt},\eta(t)\right]$$

ISBN 0-8053-0102-X

where $\eta(t) = Ad_{x(t)}(\eta)$, $\eta \in \mathfrak{g}$, proved in Example 4.1.25(c). Thus Leibniz' formula for derivatives yields

$$\frac{d\tilde{\xi}(t)}{dt} = Ad_{x(t)}\left(\frac{d\xi(t)}{dt}\right) + \left[v_s(t), \tilde{\xi}(t)\right]$$

that is,

$$\frac{d\tilde{\xi}}{dt} = \left(\frac{d\xi}{dt}\right)^{\sim} + \left[v_s(t), \tilde{\xi}(t)\right]$$

where $v_s(t) = T_e R_{x(t)^{-1}} \cdot dx/dt$ is the "velocity" in space coordinates. This formula becomes, in the special case of $SO(3, \boldsymbol{R})$ and using a slight abuse of notation,

$$\frac{d\xi_{space}}{dt} = \left(\frac{d\xi_{body}}{dt}\right)_{space} + v_s \times \xi_{space}$$

Finally, we compute the action of the left translation maps $T_e L_g$ in both space and body coordinates. This will be useful below in connection with our discussion of various left-invariant quantities. In *body* coordinates, left translation does not act on the vector component. Indeed, we have

$$\begin{aligned}
(\lambda \circ TL_g \circ \lambda^{-1})(h, \xi) &= (\lambda \circ TL_g)(h, TL_h(\xi)) \\
&= \lambda\left(gh, (TL_g \circ TL_h)(\xi)\right) \\
&= \left(gh, (TL_{gh}^{-1} \circ TL_{gh})(\xi)\right) \\
&= (gh, \xi)
\end{aligned}$$

On the other hand, a similar calculation shows that in *space* coordinates, the left translation map becomes

$$(\rho \circ TL_g \circ \rho^{-1})(h, \xi) = \left(gh, Ad_g(\xi)\right)$$

These results mirror the intuitive notion that the vector ξ is "attached" to the "body," and so does *not* vary as the body moves, relative to an observer fixed in the body; but ξ of course does vary relative to an observer fixed in "space."

To obtain only the equations of motion and the conservation laws the tangent bundle is the simplest and is all that is really needed. Indeed, on a Riemannian manifold, the geodesic flow appears most natural on the tangent bundle. However, in the literature on topology and mechanics, it is most common (although not essential) to use the cotangent bundle. In anticipation of this for later sections, we shall also describe the parallel situation on T^*G.

ISBN 0-8053-0102-X

The cotangent bundle T^*G is isomorphic in *two* ways, by $\bar{\lambda}$ and $\bar{\rho}$, to $G\times\mathfrak{g}^*$. These two isomorphisms of vector bundles are defined by

$$\bar{\lambda}(\alpha)=\left(g,\alpha\circ T_eL_g\right)=\left(g,(T_eL_g)^*\alpha\right)\in G\times\mathfrak{g}^*$$

$$\bar{\rho}(\alpha)=\left(g,\alpha\circ T_eR_g\right)=\left(g,(T_eR_g)^*\alpha\right)\in G\times\mathfrak{g}^*$$

where $g=\tau_G^*(\alpha)$, $\tau_G^*: T^*G\to G$ being the canonical projection. As before, we shall say that $\bar{\lambda}(\alpha)$ represents α in *body coordinates* and $\bar{\rho}(\alpha)$ represents α in *space coordinates*. The transition from body to space cooordinates is given by

$$\left(\bar{\rho}\circ\bar{\lambda}^{-1}\right)(g,\mu)=\left(g,Ad^*_{g^{-1}}(\mu)\right)$$

where $\mu\in\mathfrak{g}^*$.

As before, we want to relate time derivatives in space and body coordinates. Thus, let $x(t)$ be a curve in G and let $\alpha(t)$ be a curve in T^*G over $x(t)$, that is, for each t, $\alpha(t)\in T^*_{x(t)}G$. Let $\mu(t)=\alpha(t)T_eL_{x(t)}$, that is, $\mu(t)$ is the curve of one-forms over $x(t)$ in *body* coordinates. We just saw that the corresponding one-form in *space* coordinates is $\tilde{\mu}(t)=Ad^*_{x(t)^{-1}}(\mu(t))$. Now we have the formula

$$\frac{d\nu(t)}{dt}=-ad^*\left(T_eL_{x(t)^{-1}}\frac{dx}{dt}\right)\cdot\nu(t)\in\mathfrak{g}^*$$

where $\nu(t)=Ad^*_{x(t)^{-1}}(\nu)$ and $\nu\in\mathfrak{g}^*$; see Exercise 4.1.H. Then Leibniz' formula for derivatives yields

$$\frac{d\tilde{\mu}(t)}{dt}=Ad^*_{x(t)^{-1}}\frac{d\mu}{dt}-ad^*(v_b(t))\cdot\tilde{\mu}(t)$$

that is,

$$\frac{d\tilde{\mu}}{dt}=\left(\frac{d\mu}{dt}\right)^{\sim}-ad^*(v_b(t))\cdot\tilde{\mu}(t)$$

where $v_b(t)=T_eL_{x(t)^{-1}}\cdot dx/dt$ is the "velocity" in body coordinates.

It is easy to compute the action of the left translation maps $(T_eL_g)^*$ in both space and body coordinates:

$$\left(\bar{\lambda}\circ(TL_g)^*\circ\bar{\lambda}^{-1}\right)(h,\mu)=\left(g^{-1}h,\mu\right)$$

$$\left(\bar{\rho}\circ(TL_g)^*\circ\bar{\rho}^{-1}\right)(h,\mu)=\left(g^{-1}h,Ad^*_g(\mu)\right)$$

We now discuss Hamiltonian systems on T^*G and TG. We begin by finding the expressions for the canonical one- and two-forms θ_0 and ω_0 on T^*G, in body coordinates, that is, we want to determine $\theta_B=\bar{\lambda}_*\theta_0\in\Omega^1(G\times\mathfrak{g}^*)$ and $\omega_B=\bar{\lambda}_*\omega_0\in\Omega^2(G\times\mathfrak{g}^*)$.

ISBN 0-8053-0102-X

4.4.1 Proposition (Cushman [1977]) *Let* $(g,\mu)\in G\times\mathfrak{g}^*$ *and* (v,ρ), $(w,\sigma)\in T_{(g,\mu)}(G\times\mathfrak{g}^*)=T_gG\times\mathfrak{g}^*$. *Then*

(i) $\theta_B(g,\mu)\cdot(v,\rho)=\mu(T_gL_{g^{-1}}\cdot v)$

(ii) $\omega_B(g,\mu)((v,\rho),(w,\sigma))=-\rho(T_gL_{g^{-1}}w)+\sigma(T_gL_{g^{-1}}v)+\mu([T_gL_{g^{-1}}v,T_gL_{g^{-1}}w])$

Proof. (i)

$$\begin{aligned}\theta_B(g,\mu)\cdot(v,\rho)&=\bar{\lambda}_*\theta_0(g,\mu)\cdot(v,\rho)=\theta_0(\bar{\lambda}^{-1}(g,\mu))\cdot T\bar{\lambda}^{-1}(v,\rho)\\&=\bar{\lambda}^{-1}(g,\mu)\cdot(T\tau_G^*\circ T\bar{\lambda}^{-1})\cdot(v,\rho)\\&=(\mu\circ TL_{g^{-1}}\circ T(\tau_G^*\circ\bar{\lambda}^{-1}))\cdot(v,\rho)\\&=\mu(TL_{g^{-1}}\cdot v)\end{aligned}$$

(ii) To compute ω_B we shall use the formula

$$\begin{aligned}\omega_B(X,Y)&=-d\theta_B(X,Y)\\&=-L_X(\theta_B(Y))+L_Y(\theta_B(X))+\theta_B([X,Y])\end{aligned}$$

Let $X,Y\in\mathfrak{X}(G\times\mathfrak{g}^*)$, $X=(X^1,X^2)$, $Y=(Y^1,Y^2)$ be the vector fields on $G\times\mathfrak{g}^*$ such that X^2, Y^2 are constant and equal to ρ (respectively, σ) on $\mathfrak{g}^*$ and X^1, Y^1 are left-invariant vector fields on G whose value at $g\in G$ is v (respectively, w), that is, $X^1=X_\xi$, $Y^1=X_\eta$, where $\xi=T_gL_g^{-1}(v)$, $\eta=T_gL_g^{-1}(w)$. Denote by $\phi_t(h,\nu)=(\phi_t^1(h),\phi_t^2(\nu))$ the flow of X, that is, $\phi_t^1(h)$ is the flow of the left-invariant vector field X^1, and $\phi_t^2(\nu)$ the flow of the constant vector field X^2 on $\mathfrak{g}^*$ equal everywhere to ρ. Then

$$\begin{aligned}L_X(\theta_B(Y))(g,\mu)&=\frac{d}{dt}\theta_B(\phi_t^1(g),\phi_t^2(\mu))\big(Y^1(\phi_t^1(g)),Y^2(\phi_t^2(\mu))\big)\Big|_{t=0}\\&=\frac{d}{dt}\phi_t^2(\mu)\big(TL_{\phi_t^1(g)^{-1}}Y^1(\phi_t^1(g))\big)\Big|_{t=0}\quad\text{(by (i))}\\&=\rho\big(Y^1(e)\big)\ \text{(by left invariance of } Y^1 \text{ and the definition of } \phi^2)\\&=\rho\big(T_gL_g^{-1}\cdot w\big)\end{aligned}$$

A similar computation shows that

$$L_Y(\theta_B(X))(g,\mu)=\sigma\big(T_gL_{g^{-1}}\cdot v\big)$$

Since

$$\phi_t(h,\nu)=(\phi_t^1(h),\phi_t^2(\nu))$$

$$[X,Y](g,\mu)=([X^1,Y^1](g),[X^2,Y^2](\mu))$$

ISBN 0-8053-0102-X

so that

$$\begin{aligned}\theta_B([X,Y])(g,\mu)&=\theta_B(g,\mu)([X^1,Y^1](g),[X^2,Y^2](\mu))\\ &=\mu(T_gL_{g^{-1}}[X^1,Y^1](g)) \quad \text{(by (i))}\\ &=\mu([X^1(e),Y^1(e)]) \quad \text{(by left invariance)}\\ &=\mu([T_gL_{g^{-1}}(v),T_gL_{g^{-1}}(w)])\end{aligned}$$

Addition of these three equalities gives the desired formula. ■

There are similar formulas for θ_0 and ω_0 in space coordinates (replace L by R).

Assume that the Lie group G has a *left-invariant* metric or pseudometric $\langle,\rangle$, that is, for all v, $w\in T_hG$ and all $g\in G$,

$$\langle T_hL_gv, T_hL_gw\rangle_{gh}=\langle v,w\rangle_h$$

Left invariance amounts to requiring that in *body* coordinates $\langle,\rangle_g$ is independent of g.

The Riemannian (or pseudo-Riemannian) metric pulls back the natural symplectic structure of T^*G to TG as described in Sect. 3.2. Recall that the one-form Θ obtained in this way is given by

$$\Theta(v)\cdot w_v=\langle T\tau_G(w_v),v\rangle_{\tau_G(v)}$$

where $v\in TG$, $w_v\in T_vTG$. (See Theorem 3.2.13.). The symplectic form on TG is then $\Omega=-d\Theta$. We can now determine expressions for Θ and Ω in *body* coordinates, that is, for

$$\Theta_B=\lambda_*\Theta\in\Omega^1(G\times\mathfrak{g})$$

$$\Omega_B=\lambda_*\Omega\in\Omega^2(G\times\mathfrak{g})$$

4.4.2 Proposition. *Let* $(g,\xi)\in G\times\mathfrak{g}$ *and* $(v,\zeta),(w,\eta)\in T_{(g,\xi)}(G\times\mathfrak{g})=T_gG\times\mathfrak{g}$. *Then*

(*i*) $\Theta_B(g,\xi)\cdot(v,\zeta)=\langle T_gL_{g^{-1}}(v),\xi\rangle_e$

(*ii*) $\Omega_B(g,\xi)((v,\zeta),(w,\eta))$

$$=-\langle\zeta,T_gL_{g^{-1}}(w)\rangle_e+\langle\eta,T_gL_{g^{-1}}(v)\rangle_e$$
$$+\langle\xi,[T_gL_{g^{-1}}(v),T_gL_{g^{-1}}(w)]\rangle_e$$

Proof. Let ϕ: $TG\to T^*G$ be the diffeomorphism determined by the metric and ψ: $G\times\mathfrak{g}\to G\times\mathfrak{g}^*$ be given by $\psi(g,v)=(g,\langle v,\cdot\rangle_e)$. Then clearly $\psi\circ\lambda=$

ISBN 0-8053-0102-X

$\bar{\lambda}\circ\phi$, that is, the following diagram commutes:

$$\begin{array}{ccc} TG & \xrightarrow{\ \phi\ } & T^*G \\ \big\downarrow \lambda & & \big\downarrow \bar{\lambda} \\ G\times\mathfrak{g} & \xrightarrow[\ \psi\ ]{} & G\times\mathfrak{g}^* \end{array}$$

Thus,

$$\begin{aligned}\Theta_B=\lambda_*\Theta&=\psi^*\cdot\bar{\lambda}_*\cdot\phi_*\cdot\Theta\\ &=\psi^*\bar{\lambda}_*\theta_0\\ &=\psi^*\theta_B\end{aligned}$$

But from the expression for θ_B in 4.4.1 and the definition of ψ, $\psi^*\theta_B$ is given by (i). The proof of (ii) is similar. ■

Denote by $\Lambda: G\times G\to G$ the action of G on itself by left translation, that is, $\Lambda_g=L_g$ for all $g\in G$. Then 4.2.2 and 4.2.11 yield the following.

4.4.3 Theorem. *(i) The Ad*-equivariant momentum mapping $\bar{J}$ of the action Λ^{T^*} on T^*G is given by*

$$\bar{J}: T^*G\to\mathfrak{g}^*;\ \bar{J}(\alpha_g)(\xi)=\alpha_g(T_eR_g(\xi))$$

If the Hamiltonian $H: T^*G\to\mathbf{R}$ is left invariant, that is, $H\circ T^*L_g=H$ for all g, then $\hat{\bar{J}}(\xi)$ is constant on the orbits of X_H for all $\xi\in\mathfrak{g}$ and X_H is a left-invariant vector field.

(ii) If the Lie group G has a left-invariant metric $\langle,\rangle$, then the Ad-equivariant momentum mapping J of the action Λ^T on TG is given by*

$$J: TG\to\mathfrak{g}^*;\ J(v_g)(\xi)=\langle v_g, T_eR_g(\xi)\rangle_g$$

If $E: TG\to\mathbf{R}$ is left invariant, that is, $E\circ TL_g=E$ for all $g\in G$, then $\hat{J}(\xi)$ is constant on the orbits of the left-invariant vector field X_E for all $\xi\in\mathfrak{g}$. This holds in particular when $E=K$, where $K(v_g)=\frac{1}{2}\langle v,v\rangle_g$ is the kinetic energy of the metric.

(iii) The action Λ^{T^*} in body coordinates is given by

$$\Lambda_B^{T^*}: G\times G\times\mathfrak{g}^*\to G\times\mathfrak{g}^*;\ (g,(h,\mu))\mapsto(gh,\mu)$$

The Ad-equivariant momentum mapping of this action on $G\times\mathfrak{g}^*$, $\bar{J}_B: G\times\mathfrak{g}^*\to\mathfrak{g}^*$ is given by*

$$\hat{J}_B(\xi)=\hat{\bar{J}}(\xi)\circ\bar{\lambda}^{-1}\quad \textit{for all } \xi\in\mathfrak{g}$$

ISBN 0-8053-0102-X

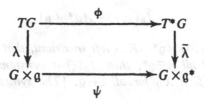

that is,

$$\bar{J}_B(g,\mu)=Ad^*_{g^{-1}}(\mu)$$

If the Hamiltonian $H\colon G\times\mathfrak{g}^*\to\mathbf{R}$ is left invariant, that is, $H(gh,\mu)=H(h,\mu)$ for all $g,\ h\in G$ and all $\mu\in\mathfrak{g}^*$, then $\hat{\bar{J}}_B(\xi)$ is constant on the orbits of the left-invariant vector field X_H for all $\xi\in\mathfrak{g}$. [The symplectic manifold here is $(G\times\mathfrak{g}^*,\omega_B)$.]

(iv) *Assume again that G has a left-invariant metric $\langle,\rangle$. The action Λ^T in body coordinates is given by*

$$\Lambda^T_B\colon G\times G\times\mathfrak{g}\to G\times\mathfrak{g};\ (g,(h,\xi))\mapsto(gh,\xi)$$

The Ad^*-equivariant momentum mapping of this action on $G\times\mathfrak{g}$ is the map $J_B\colon G\times\mathfrak{g}\to\mathfrak{g}^*$ given by

$$\hat{J}_B(\xi)=\hat{J}(\xi)\circ\lambda^{-1}\quad \textit{for all } \xi\in\mathfrak{g}$$

that is,

$$J_B(g,\eta)\cdot\xi=\langle\eta,Ad_{g^{-1}}\cdot\xi\rangle_e$$

If $E\colon G\times\mathfrak{g}\to\mathbf{R}$ is left invariant, that is, $E(gh,\eta)=E(h,\eta)$ for all $g,\ h\in G$, $\eta\in\mathfrak{g}$, then $\hat{J}_B(\xi)$ is constant on the orbits of X_E for all $\xi\in\mathfrak{g}$; the symplectic manifold here is $(G\times\mathfrak{g},\Omega_B)$. In particular, this holds if $E(g,\xi)=K$, where $K(\xi)=\frac{1}{2}\langle\xi,\xi\rangle_e$ is the kinetic energy of the metric.

(v) *The action Λ^{T^*} in space coordinates is given by*

$$\Lambda^{T^*}_S\colon G\times G\times\mathfrak{g}^*\to G\times\mathfrak{g}^*;\ (g,(h,\mu))\mapsto(gh,Ad^*_{g^{-1}}(\mu))$$

The Ad^*-equivariant momentum mapping of this action on $G\times\mathfrak{g}^*$ is the map $\bar{J}_S\colon G\times\mathfrak{g}^*\to\mathfrak{g}^*$ given by

$$\hat{\bar{J}}_S(\xi)=\hat{\bar{J}}(\xi)\circ\bar{\rho}^{-1}\quad \textit{for all } \xi\in\mathfrak{g}$$

that is,

$$\bar{J}_S(g,\mu)=\mu$$

On the symplectic manifold $(G\times\mathfrak{g}^*,\ \omega_s=\omega_0)$, if the Hamiltonian $H\colon G\times\mathfrak{g}^*\to\mathbf{R}$ is left invariant, that is, $H(gh,Ad^*_{g^{-1}}(\mu))=H(h,\mu)$ for all $g,\ h\in G,\ \mu\in\mathfrak{g}^*$, then $J_S(\xi)$ is constant on the orbits of the left-invariant vector field X_H for all $\xi\in\mathfrak{g}$.

(vi) *Again assume G has a left-invariant metric $\langle,\rangle$. The action Λ^T in*

ISBN 0-8053-0102-X

space coordinates is given by

$$\Lambda_S^T: G\times G\times \mathfrak{g}\to G\times\mathfrak{g};\ (g,(h,\xi))\mapsto(gh, Ad_g(\xi))$$

The Ad^*-equivariant momentum mapping of this action is given by

$$J_S: G\times\mathfrak{g}\to\mathfrak{g}^*,\ \hat{J}_S(\xi)=\hat{J}\,(\xi)\circ\rho^{-1}\quad \textit{for all } \xi\in\mathfrak{g}$$

that is,

$$J_S(g,\eta)\cdot\xi=\langle Ad_{g^{-1}}(\eta), Ad_{g^{-1}}(\xi)\rangle_e$$

If $E: G\times\mathfrak{g}\to\mathbf{R}$ is left invariant, that is, $E(gh, Ad_g(\eta))=E(h,\eta)$ for all $g, h\in G$, $\eta\in\mathfrak{g}$, then $\hat{J}_S(\xi)$ is constant on the orbits of the left-invariant vector field X_E for all $\xi\in\mathfrak{g}$; the symplectic manifold here is $(G\times\mathfrak{g}, \Omega_S=\rho_*\Omega)$. In particular, this holds if $E(g,\xi)=\frac{1}{2}\langle T_eR_g(\xi), T_eR_g(\xi)\rangle_g$.

Proof. All conservation statements are particular cases of the general theorem on conservation of momentum; see 4.2.2. The expressions for Λ^{T^*} and Λ^T in body and space coordinates have been established at the beginning of this section. (i) and (ii) are merely rephrasings of 4.2.11 and 4.2.13. The left invariance of the Hamiltonian vector fields in question follows from the fact that all actions are symplectic and from 3.3.19

(iii) $\bar{J}_B(g,\mu)\cdot\xi=\hat{\bar{J}}_B(\xi)(g,\mu)=\hat{\bar{J}}(\xi)(\bar{\lambda}^{-1}(g,\mu))$

$$=\hat{\bar{J}}(\xi)(\mu\circ T_eL_g^{-1})=(\mu\circ T_eL_g^{-1}\circ T_eR_g)(\xi)$$

$$=(\mu\circ Ad_{g^{-1}})(\xi)=(Ad^*_{g^{-1}}\mu)(\xi)$$

(iv) $J_B(g,\eta)\cdot\xi=\hat{J}_B(\xi)(g,\eta)=\hat{J}\,(\xi)(\lambda^{-1}(g,\eta))=\hat{J}\,(\xi)(T_eL_g(\eta))$

$$=\langle T_eL_g\eta, T_eR_g\xi\rangle_g=\langle\eta, Ad_{g^{-1}}(\xi)\rangle_e$$

(v) $\bar{J}_S(g,\mu)\cdot\xi=\hat{\bar{J}}_S(\xi)(g,\mu)=\hat{\bar{J}}(\xi)(\bar{\rho}^{-1}(g,\mu))=\hat{\bar{J}}(\xi)(\mu\circ T_eR_g^{-1})$

$$=(\mu\circ T_eR_g^{-1}\circ T_eR_g)(\xi)=\mu(\xi)$$

(vi) $J_S(g,\eta)\cdot\xi=\hat{J}_S(\xi)(g,\eta)=\hat{J}\,(\xi)(\rho^{-1}(g,\eta))=\hat{J}\,(\xi)(T_eR_g(\eta))$

$$=\langle T_eR_g(\eta), T_eR_g(\xi)\rangle_g=\langle Ad_{g^{-1}}(\eta), Ad_{g^{-1}}(\xi)\rangle_e \quad ■$$

The conservation laws stated above are the six possible formulations of the so-called *Euler conservation law.*

In the tangent bundle formulation in body coordinates the surfaces $\langle\xi,\xi\rangle_e=$ *constant* are called *inertia ellipsoids*. Thus, if $E=K=$ *kinetic energy,*

ISBN 0-8053-0102-X

these surfaces will be preserved in *body* coordinates by conservation of energy.

In 4.4.3(vi), if we write $\langle x,y\rangle_e = A(y)\cdot x$, which defines the linear map $A: \mathfrak{g}\to\mathfrak{g}^*$, then we can write

$$J_S(g,\eta)\cdot\xi = A(Ad_{g^{-1}}\xi)\cdot Ad_{g^{-1}}\eta$$
$$= \big((Ad^*_{g^{-1}}\circ A\circ Ad_{g^{-1}})\xi\big)\eta$$

and hence

$$\hat{J}_S(\xi) = (Ad^*_{g^{-1}}\circ A\circ Ad_{g^{-1}})(\xi)$$

Thus, Euler's conservation law amounts to conservation (in *space* coordinates) of the "vector" quantity

$$L_{\xi,g} = \big(Ad^*_{g^{-1}}\circ A\circ Ad_{g^{-1}}\big)\xi\in\mathfrak{g}^*$$

for each $\xi\in\mathfrak{g}$. In the case of $G = SO(3)$, this conserved quantity is simply the *angular momentum*. The plane $\mathcal{G}_{\xi,g} = \{\eta\in\mathfrak{g} \mid L_\xi(\eta) = 0\}$ is called the *invariable plane* for a given initial condition $\xi\in\mathfrak{g}$.

Poinsot, back in 1834, discovered a geometric way to visualize rigid body motion. Namely, he proved—for $SO(3)$—that the inertia ellipsoids move in space coordinates so as to always "roll" on (a translate of) the invariable plane. This rolling exactly characterizes the motion of the body since the inertia ellipsoid is attached to the body (see Fig. 4.4-1). These features are quite general for any Lie group as we now show.

4.4.4 Theorem. *Consider the situation of* 4.4.3(*vi*) *with* $E = K$, *that is*, $E(g,\xi) = \frac{1}{2}\langle T_eR_g\xi, T_eR_g\xi\rangle_g = \frac{1}{2}\langle Ad_{g^{-1}}\xi, Ad_{g^{-1}}\xi\rangle_e$. *Let* $w(t)$ *be an integral curve of* X_E,

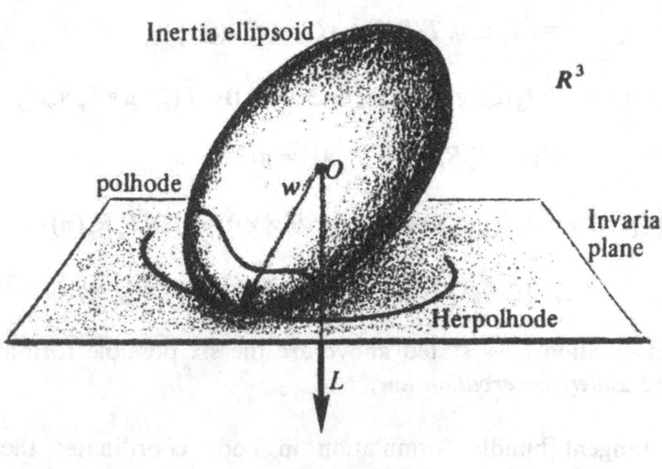

Figure 4.4-1

ISBN 0-8053-0102-X

in space coordinates. Let $E_0=\frac{1}{2}\langle w(0),w(0)\rangle_e$, *and let* $S(t)$ *denote the image of the inertia ellipsoid* $\langle\xi,\xi\rangle_e=2E_0$ *after time* t, *in space coordinates. That is,*

$$S(t)=\{\xi\in\mathfrak{g}\,|\,\langle Ad_{x(t)^{-1}}\xi, Ad_{x(t)^{-1}}\xi\rangle=2E_0\}$$

where $x(t)$ *is the geodesic in* G *obtained by projection of* $w(t)$ *onto* G.

Then letting $\mathcal{I}_{w(t),x(t)}$ *denote the invariable plane with initial condition* $w(t)\in\mathfrak{g}$ *we have:*

(*i*) $\mathcal{I}_{w(t),x(t)}$ *is tangent to* $S(t)$ *at* $w(t)$, *and*
(*ii*) $\mathcal{I}_{w(t),x(t)}$ *is independent of* t.

Proof. (i) Let $f(\xi)=\langle Ad_{x(t)^{-1}}\xi, Ad_{x(t)^{-1}}\xi\rangle_e$ and notice that by nondegeneracy of $\langle,\rangle_e$, $f^{-1}(2E_0)=S(t)$ is a codimension one submanifold of $\mathfrak{g}$. Moreover, $w(t)\in S(t)$ since, by left invariance of the metric $\langle\cdot,\cdot\rangle$,

$$\langle Ad_{x(t)^{-1}}w(t), Ad_{x(t)^{-1}}w(t)\rangle_e=\langle T_eR_{x(t)}w(t), T_eR_{x(t)}w(t)\rangle_{x(t)}$$

which is twice the value of the energy at $(x(t),w(t))$ by 4.4.3(vi). By conservation of energy, this equals $2E_0$. The tangent space to $S(t)$ at $w(t)$ is

$$T_{w(t)}S(t)=\ker df(w(t))=\mathcal{I}_{w(t),x(t)}$$

since

$$\mathcal{I}_{w(t),x(t)}=\{\xi\in\mathfrak{g}\,|\,\langle Ad_{x(t)^{-1}}w(t), Ad_{x(t)^{-1}}\xi\rangle_e=0\}$$

by definition.

(ii) By 4.4.3(vi), $J_S(x(t),w(t))\cdot\xi=\langle Ad_{x(t)^{-1}}w(t), Ad_{x(t)^{-1}}\xi\rangle_e$ is independent of t and thus the kernel of $J_S(x(t),w(t))$ is time independent. But this kernel is just $\mathcal{I}_{w(t),x(t)}$. ■

On the symplectic manifold (T^*G,ω_0) consider the left-invariant Hamiltonian $H\colon T^*G\to\boldsymbol{R}$ and let $H_B=H\circ\bar{\lambda}^{-1}\colon G\times\mathfrak{g}^*\to\boldsymbol{R}$ be its expression in *body* coordinates. Since by the definition of ω_B, $\bar{\lambda}\colon(T^*G,\omega_0)\to(G\times\mathfrak{g}^*,\omega_B)$ is a symplectic diffeomorphism, $\bar{\lambda}_*X_H=X_{H\circ\bar{\lambda}^{-1}}=X_{H_B}$ (see 3.3.19) is the Hamiltonian vector field $X_H\in\mathfrak{X}(T^*G)$ expressed in body coordinates. Hence

$$X_{H_B}\colon G\times\mathfrak{g}^*\to TG\times(\mathfrak{g}^*\times\mathfrak{g}^*), X_{H_B}(g,\mu)=\big(\overline{X}(g,\mu),\mu,\overline{Y}(g,\mu)\big)$$

where

$$\overline{X}\colon G\times\mathfrak{g}^*\to TG$$

$$\overline{Y}\colon G\times\mathfrak{g}^*\to\mathfrak{g}^*$$

$$\tau_G\circ\overline{X}=id_G$$

ISBN 0-8053-0102-X

This means, in particular, that $\bar{X}^\mu: g \mapsto \bar{X}(g,\mu)$ is a vector field on G. Since by definition of the action $\Lambda_B^{T^*}$ [see 4.4.3(iii)], $\Lambda_{Bg}^{T^*} = \bar{\lambda} \circ T^* L_{g^{-1}} \circ \bar{\lambda}^{-1}$, left invariance of H, that is, $H \circ T^* L_{g^{-1}} = H$, implies

$$\begin{aligned} H_B \circ \Lambda_{Bg}^{T^*} &= H \circ \bar{\lambda}^{-1} \circ \bar{\lambda} \circ T^* L_{g^{-1}} \circ \bar{\lambda}^{-1} \\ &= H \circ T^* L_{g^{-1}} \circ \bar{\lambda}^{-1} \\ &= H \circ \bar{\lambda}^{-1} = H_B \end{aligned}$$

that is, H_B is left invariant, too. Therefore by 4.4.3(iii), X_{H_B} is left invariant. But

$$\Lambda_{Bg}^{T^*} = (L_g, id_{\mathfrak{g}^*}) : G \times \mathfrak{g}^* \to G \times \mathfrak{g}^*,$$

which implies

$$T\Lambda_{Bg}^{T^*} = (TL_g, id_{\mathfrak{g}^*}, id_{\mathfrak{g}^*}) : TG \times (\mathfrak{g}^* \times \mathfrak{g}^*) \to TG \times (\mathfrak{g}^* \times \mathfrak{g}^*)$$

and hence

$$(\Lambda_{Bg}^{T^*})_* X_{H_B} = T\Lambda_{Bg}^{T^*} \circ X_{H_B} \circ (\Lambda_{Bg}^{T^*})^{-1}$$

Thus $\bar{X}^\mu$ is a left-invariant vector field on G for each $\mu \in \mathfrak{g}^*$ and $\bar{Y}(g^{-1}h, \mu) = \bar{Y}(g,\mu)$ for all $g, h \in G$ and $\mu \in \mathfrak{g}^*$, that is, $\bar{Y}(g,\mu) = \bar{Y}(e,\mu)$, and hence $\bar{Y}$ *is independent of* g.

Thus we have the complete description of the Hamiltonian vector field X_H in *body* coordinates: $\bar{\lambda}_* X_H = X_{H_B}$, $X_{H_B}(g,\mu) = (\bar{X}^\mu(g), \mu, \bar{Y}(\mu))$ with $\bar{X}^\mu \in \mathfrak{X}_L(G)$, the left-invariant vector fields depending smoothly on $\mu \in \mathfrak{g}^*$ and $\bar{Y}: \mathfrak{g}^* \to \mathfrak{g}^*$. We refer to $\bar{Y}$ as the *Euler vector field* or the *Euler equations in cotangent formulation.*

We want to determine the flow of $\bar{Y} \in \mathfrak{X}(\mathfrak{g}^*)$. Since $\bar{\lambda}_* X_H = X_{H_B}$, if $F_t: T^*G \to T^*G$ denotes the flow of X_H, $\bar{\lambda} \circ F_t \circ \bar{\lambda}^{-1}: G \times \mathfrak{g}^* \to G \times \mathfrak{g}^*$ is the flow of X_{H_B}. Hence, if $P_2: G \times \mathfrak{g}^* \to \mathfrak{g}^*$ is the projection,

$$\begin{aligned} \frac{d}{dt}(P_2 \circ \bar{\lambda} \circ F_t \circ \bar{\lambda}^{-1})(g,\mu) &= TP_2 \frac{d}{dt}(\bar{\lambda} \circ F_t \circ \bar{\lambda}^{-1})(g,\mu) \\ &= TP_2 X_{H_B}(\bar{\lambda} \circ F_t \circ \bar{\lambda}^{-1}(g,\mu)) \\ &= \left(P_2 \circ \bar{\lambda} \circ F_t \circ \bar{\lambda}^{-1}(g,\mu), \bar{Y}\left(\bar{\lambda} \circ F_t \circ \bar{\lambda}^{-1}(g,\mu)\right)\right) \\ &= \left(P_2 \circ \bar{\lambda} \circ F_t \circ \bar{\lambda}^{-1}(g,\mu), \bar{Y}\left(P_2 \circ \bar{\lambda} \circ F_t \circ \bar{\lambda}^{-1}(g,\mu)\right)\right) \end{aligned}$$

ISBN 0-8053-0102-X

the last equality holding since $\bar{Y}$ is independent of $g \in G$. We now compute $P_2 \circ \bar{\lambda} \circ F_t \circ \bar{\lambda}^{-1}$. Denote by $x(t) = \tau_G^*(F_t(\mu))$ for $\mu \in \mathfrak{g}^*$ fixed. Then by left invariance $T^*L_g \circ F_t = F_t \circ T^*L_g$, and hence

$$
\begin{aligned}
\left(P_2 \circ \bar{\lambda} \circ F_t \circ \bar{\lambda}^{-1}\right)(g,\mu) &= \left(P_2 \circ \bar{\lambda} \circ F_t\right)\left(T^*L_{g^{-1}}(\mu)\right)\\
&= \left(P_2 \circ \bar{\lambda} \circ T^*L_{g^{-1}} \circ F_t\right)(\mu)\\
&= P_2\left(gx(t), \left(T^*L_{gx(t)} \circ T^*L_{g^{-1}} \circ F_t\right)(\mu)\right)\\
&= \left(T^*L_{g^{-1}gx(t)} \circ F_t\right)(\mu)\\
&= \left(T^*L_{x(t)} \circ F_t\right)(\mu)\\
&= F_t(\mu) \circ T_e L_{x(t)}
\end{aligned}
$$

Thus $\bar{H}_t(\mu) = F_t(\mu) \circ T_e L_{x(t)}$ is the flow of $\bar{Y}$. Note that in the discussion above only left invariance of X_{H_B} was needed. We have proved the following.

4.4.5 Proposition. *(i) Let $X \in \mathfrak{X}(T^*G)$ be left invariant and denote by $X^B = \bar{\lambda}_* X$ its expression in body coordinates; then $X^B(g,\mu) = (\bar{X}(g,\mu), \mu, \bar{Y}(\mu))$, where $\bar{Y}: \mathfrak{g}^* \to \mathfrak{g}^*$ and $\bar{X}^\mu: g \mapsto \bar{X}(g,\mu)$ is a family of left-invariant vector fields on G depending smoothly on $\mu \in \mathfrak{g}^*$. The flow of $\bar{Y}$, denoted by $\bar{H}_t$, is given by*

$$\bar{H}_t(\nu) = (T_e^* L_{x(t)} \circ F_t)(\nu) = F_t(\nu) \circ T_e L_{x(t)}$$

where $x(t) = \tau_G^*(F_t(\nu))$. $\bar{Y}$ is called the ***cotangent Euler vector field.***

In particular, this holds for any Hamiltonian vector field X_H with left-invariant Hamiltonian H, in which case $X_H^B = X_{H_B}$, where $H_B = H \circ \bar{\lambda}^{-1}$ is the expression of the Hamiltonian H in body coordinates.

(ii) Assume G has a left-invariant metric $\langle , \rangle$. Let $X \in \mathfrak{X}(TG)$ be left invariant and denote by $X^B = \lambda_* X$ its expression in body coordinates; then $X^B(g,\xi) = (X^\xi(g), \xi, Y(\xi))$, where $Y: \mathfrak{g} \to \mathfrak{g}$ and X^ξ is a family of left-invariant vector fields on G depending smoothly on $\xi \in \mathfrak{g}$. The flow of Y, denoted by H_t, is given by

$$H_t(\xi) = T_{x(t)} L_{x(t)^{-1}}(F_t(\xi))$$

where $x(t) = \tau_G(F_t(\xi))$. Y is called the ***tangent Euler vector field.***

In particular, if $E: TG \to \mathbf{R}$ is left invariant, then the vector field X_E is left invariant and $X_E^B = X_{E_B}$, where $E_B = E \circ \lambda^{-1}$ is the expression of E in body coordinates.

Part (ii) of this proposition is proved along the same lines as part (i) and is left as an exercise to the reader.

ISBN 0-8053-0102-X

The derivation of the flow in body coordinates may be done slightly differently as follows. (We do the case of TG.) Let G be a Lie group, and let F_t be a flow on TG that is left invariant: $TL_g \circ F_t = F_t \circ TL_g$. Let us express F_t in body coordinates: we write $\lambda F_t \lambda^{-1}(g,v) = (A_t(g,v), B_t(g,v)) \in G \times \mathfrak{g}$. The condition of left invariance takes a particularly simple form in body coordinates since, as we have seen, $\lambda TL_g \lambda^{-1}(h,v) = (gh,v)$. It follows immediately that we must have

$$A_t(g,v) = gA_t(e,v)$$

and

$$B_t(g,v) = B_t(e,v)$$

Let us write $H_t(v)$ for $B_t(e,v)$, and $K_t(v)$ for $A_t(e,v)$. Thus H_t maps $\mathfrak{g} \to \mathfrak{g}$ and K_t maps $\mathfrak{g} \to G$. The condition $F_{s+t} = F_s \circ F_t$ implies that H_t and K_t obey certain identities. Indeed, we must have

$$(A_{s+t}(g,v), B_{s+t}(g,v)) = (A_s(A_t(g,v), B_t(g,v)), B_s(A_t(g,v), B_t(g,v)))$$

that is,

$$(gK_{s+t}(v), H_{s+t}(v)) = (gK_t(v)K_s(H_t(v)), H_s(H_t(v)))$$

so that

$$K_{s+t}(v) = K_t(v)K_s(H_t(v))$$

and

$$H_{s+t}(v) = H_s(H_t(v))$$

The latter condition says that H_t is a *flow* on $\mathfrak{g}$, which we may call the *flow in body coordinates*. The condition relating K and H may be less familiar; but such "cocycle identities" have been studied, for example, in connection with group representations.

4.4.6 Theorem. *(i) Let $X \in \mathfrak{X}(T^*G)$ be a left-invariant vector field with flow F_t. Let $\overline{Y}: \mathfrak{g}^* \to \mathfrak{g}^*$ be the corresponding cotangent Euler vector field with flow $\overline{H}_t$. Then we have the formula*

$$\overline{Y}(\mu)\cdot\eta = \left(d\hat{J}(\eta)\cdot\mu\right)(X(\mu)) + \mu\cdot\left[\left.\frac{dx(t)}{dt}\right|_{t=0}, \eta\right]$$

where $x(t) = \tau_G^*(F_t(\mu))$ for all $\mu \in \mathfrak{g}^*$, $\eta \in \mathfrak{g}$.

ISBN 0-8053-0102-X

In particular, if F_t is the flow of X_H, where H is left invariant, then we have

$$\bar{Y}(\mu)\cdot\eta=\mu\cdot\left[\left.\frac{dx}{dt}\right|_{t=0},\eta\right]$$

for all $\mu\in\mathfrak{g}^*$, $\eta\in\mathfrak{g}$.

(ii) Let G be a Lie group with left-invariant metric $\langle,\rangle$ and $X\in\mathfrak{X}(TG)$ a left-invariant second-order equation with flow F_t. Let $Y:\mathfrak{g}\to\mathfrak{g}$ be the corresponding tangent Euler vector field with flow H_t. Then we have the formula

$$\begin{aligned}\langle Y(\xi),\eta\rangle&=\langle[\xi,\eta],\xi\rangle+\frac{d}{dt}\{\hat{J}(\eta)(F_t(\xi))\}\Big|_{t=0}\\&=\langle[\xi,\eta],\xi\rangle+(d\hat{J}(\eta)\cdot\xi)(X(\xi))\end{aligned}$$

for all $\xi,\eta\in\mathfrak{g}$.

In particular, if F_t is the flow of X_E, where E: $TG\to\mathbf{R}$ is left invariant, the symplectic structure associated with the metric is used on TG, and X_E is a second order equation, then

$$\boxed{\langle Y(\xi),\eta\rangle_e=\langle[\xi,\eta],\xi\rangle_e}$$

for all $\xi,\eta\in\mathfrak{g}$.
Note: X is second order when $X^{\xi}(g)=T_eL_g(\xi)$; See Ex. 4.4.C.

Remark. This last formula obviously determines Y uniquely and is independent of the left-invariant energy function E. Taking then $E=K$, the kinetic energy defined by the metric, we can conclude that the geodesic flow is the unique *left-invariant* Hamiltonian vector field on TG that arises from a second-order equation. Recall that the flow is complete if the metric is Riemannian, and is complete in the pseudo-Riemannian case if G is compact [see 3.7.20(ii) and 4.2.22].

Proof. (i) Fix $\mu\in\mathfrak{g}^*$ and let $x(t)=\tau_G^*(F_t(\mu))$. Then

$$\begin{aligned}(d\hat{J}(\eta)\cdot\mu)(X(\mu))&=\frac{d}{dt}\{\hat{J}(\eta)(F_t(\mu))\}\Big|_{t=0}\\&=\frac{d}{dt}F_t(\mu)(T_eR_{x(t)}(\eta))\Big|_{t=0}\quad\text{by 4.4.3}(i)\\&=\frac{d}{dt}\{F_t(\mu)\circ T_eL_{x(t)}\circ T_{x(t)}L_{x(t)^{-1}}\circ T_eR_{x(t)}(\eta)\}\Big|_{t=0}\\&=\frac{d}{dt}\{\bar{H}_t(\mu)Ad_{x(t)^{-1}}(\eta)\}\Big|_{t=0}\\&=\bar{Y}(\mu)\cdot\eta+\mu\cdot\frac{d}{dt}Ad_{x(t)^{-1}}(\eta)\Big|_{t=0}\end{aligned}$$

ISBN 0-8053-0102-X

But $x(t)^{-1}x(t)=e$ implies that

$$\left.\frac{dx(t)^{-1}}{dt}\right|_{t=0}=-\left.\frac{dx(t)}{dt}\right|_{t=0}$$

and by Example 4.1.25(b), (c),

$$\begin{aligned}\frac{d}{dt}Ad_{x(t)^{-1}}(\eta)\big|_{t=0}&=\left[\left.\frac{dx(t)^{-1}}{dt}\right|_{t=0},\eta\right]\\&=-\left[\left.\frac{dx(t)}{dt}\right|_{t=0},\eta\right]\end{aligned}$$

This gives then the desired formula.

(ii) Fix $\xi\in\mathfrak{g}$ and let $x(t)=\tau_G(F_t(\xi))$. Then

$$\begin{aligned}(d\hat{J}(\eta)\cdot\xi)(X(\xi))&=\frac{d}{dt}\{\hat{J}(\eta)(F_t(\xi))\}\big|_{t=0}\\&=\frac{d}{dt}\langle T_eR_{x(t)}\eta,F_t(\xi)\rangle\big|_{t=0}\quad\text{by 4.4.3}(ii)\\&=\frac{d}{dt}\langle Ad_{x(t)^{-1}}\eta,\overline{H}_t(\xi)\rangle\big|_{t=0}\quad\text{(by left invariance of the metric)}\\&=\left\langle\frac{d}{dt}Ad_{x(t)^{-1}}\eta,\xi\right\rangle\bigg|_{t=0}+\langle\eta,Y(\xi)\rangle\\&=-\left\langle\left[\left.\frac{dx(t)}{dt}\right|_{t=0},\eta\right],\xi\right\rangle+\langle\eta,Y(\xi)\rangle\\&=-\langle[\xi,\eta],\xi\rangle+\langle\eta,Y(\xi)\rangle\end{aligned}$$

since $x(t)$ is a base integral curve of a second-order equation and hence

$$\left.\frac{dx(t)}{dt}\right|_{t=0}=F_t(\xi)\big|_{t=0}=\xi\quad\text{by 3.5.13}$$

The final assertions of (i) and (ii) follow from conservation of momentum in the case it concerns. ■

In the above derivation, the flow property of F_t was *not* essential. What was relevant was that we were dealing with a curve $u(t)$ in TG that "corresponds to a physical motion" in that $u(t)=(d/dt)x(t)$, if $x(t)=\tau_G\circ u(t)$. The

ISBN 0-8053-0102-X

Euler equations in tangent formulation are, in this sense, purely geometrical or "kinematical." Note also the discrepancy between the cotangent and tangent formulation in the above theorem.

4.4.7 Theorem. *(i) Let $H: T^*G \to R$ be a left-invariant Hamiltonian (i.e., $H \circ T^*L_g = H$ for all $g \in G$) and denote by $G\cdot\mu \subset \mathfrak{g}^*$ the orbit of G under the co-adjoint representation, $G\cdot\mu = \{Ad^*_{g^{-1}}\mu \mid g \in G\}$. Then, on the symplectic manifold $(G\cdot\mu, \omega_\mu)$, where*

$$\omega_\mu(Ad^*_{g^{-1}}\mu)\big(\xi_{\mathfrak{g}^*}(Ad^*_{g^{-1}}\mu), \eta_{\mathfrak{g}^*}(Ad^*_{g^{-1}}\mu)\big) = -(Ad^*_{g^{-1}}\mu)([\xi,\eta])$$

$$= -(ad\,\xi)^* Ad^*_{g^{-1}}\mu\cdot\eta = (ad\,\eta)^* Ad^*_{g^{-1}}\mu\cdot\xi$$

$\bar{Y}|G\cdot\mu$ is a Hamiltonian vector field with Hamiltonian $H_\mu: G\cdot\mu \to R$ given by

$$H_\mu\big(Ad^*_{g^{-1}}\mu\big) = H\big(T_g^*R_{g^{-1}}(\mu)\big) = H\big(Ad^*_{g^{-1}}\mu\big).$$

(ii) Let G have a bi-invariant metric $(,)$. Denote by $G\cdot\xi$ the orbit of G under the adjoint representation, that is, $G\cdot\xi = \{Ad_g\xi \mid g \in G\}$. Then on the symplectic manifold $(G\cdot\xi, \omega_\xi)$, where

$$\omega_\xi(Ad_g\xi)\big(\eta_{\mathfrak{g}}(Ad_g\xi), \zeta_{\mathfrak{g}}(Ad_g\xi)\big) = -([\eta,\zeta], Ad_g\xi)_e$$

$$= ([\eta, Ad_g\xi], \zeta)_e = ([Ad_g\xi, \zeta], \eta)_e,$$

$Y|G\cdot\xi$ is a Hamiltonian vector field with Hamiltonian $H_\xi: G\cdot\xi \to R$ given by

$$H_\xi(Ad_g\xi) = E(T_eR_g(\xi)) = E(Ad_g\xi)$$

where E is a left invariant energy function on TG.

Proof. (i) The expression for ω_μ was already derived in 4.3.4(v). That proof set up a symplectic diffeomorphism $\phi: P_\mu \to G\cdot\mu$; $\pi_\mu(\alpha_\mu(g)) \mapsto Ad^*_g\mu$. By 4.3.4, the original Hamiltonian system X_H induces a well-defined flow F_t^μ and hence vector field X_H^μ on P_μ; the vector field is Hamiltonian with the energy function induced naturally from H. We have $F_t^\mu(\pi_\mu(\alpha_\mu(g))) = \pi_\mu(F_t(\alpha_\mu(g)))$. The corresponding flow on $G\cdot\mu$ induced by ϕ is given as follows. For $\nu = Ad^*_{g^{-1}}\mu$, $x(t) = \tau_G^*(F_t(\nu))$,

$$
\begin{aligned}
H_t(\nu) &= \phi\circ F_t^\mu\circ\phi^{-1}\big(Ad^*_{g^{-1}}\mu\big) = \phi\circ F_t^\mu\circ\pi_\mu\alpha_\mu(g)\\
&= \phi\circ\pi_\mu\circ F_t\circ T^*_{g^{-1}}R_g\mu \quad \text{(by 4.3.4}(v))\\
&= \phi\circ\pi_\mu\circ F_t\circ T^*_{g^{-1}}L_g\nu\\
&= \phi\circ\pi_\mu\circ T^*_{g^{-1}x(t)}L_g\circ F_t\nu \quad \text{(since } F_t \text{ is left-invariant).}
\end{aligned}
$$

ISBN 0-8053-0102-X

Because J is an integral of X_H, $T^*_{g^{-1}x(t)}L_g \circ F_t(\nu) = F_t \circ T^*_{g^{-1}}L_g(\nu) \in J^{-1}(\mu)$. Since $J^{-1}(\mu)$ is the graph of the one-form α_μ, $\alpha_\mu(h) = T^*_{g^{-1}x(t)}L_g \circ F_t(\nu)$ for some $h \in G$. Applying τ_G^*, $h = g^{-1}\tau_G^*(F_t\nu) = g^{-1}x(t)$. Therefore,

$$H_t(\nu) = \phi \circ \pi_\mu \circ \alpha_\mu(g^{-1}x(t)) = Ad^*_{g^{-1}x(t)}\mu$$

$$= T_e^*L_{g^{-1}x(t)}T^*_{g^{-1}x(t)}R_{(g^{-1}x(t))^{-1}}\mu$$

$$= T_e^*L_{g^{-1}x(t)}T^*_{g^{-1}x(t)}L_gF_t(\nu) \quad \left(\text{since } \alpha_\mu(h) = T_h^*R_{h^{-1}}\mu = T^*_{g^{-1}x(t)}L_g \circ F_t(\nu)\right)$$

$$= T_e^*L_{x(t)}F_t(\nu)$$

The result (i) therefore follows. The assertion (ii) follows from (i) using the symplectic diffeomorphism between $\mathfrak{g}$ and $\mathfrak{g}^*$ induced by (,) and lemma 4.4.11. ■

From this argument and 4.3.6 we also see that relative equilibria of X_K on TG correspond to equilibria for the Euler vector field Y on $G\cdot\mu$. (Since $G\cdot\mu$ is invariant under Y, these are the zeros of Y itself.)

The stability criteria of equilibrium points for the tangent Euler vector field is given by the following.

4.4.8 Theorem (Arnold). *Let $Y(\xi)=0$, where $\xi \in \mathfrak{g}$ and Y is the tangent Euler vector field. Then if the bilinear form Q on $\mathfrak{g}$ given by*

$$Q(\eta,\zeta) = \langle A^{-1}(ad\eta)^*A\xi, A^{-1}(ad\zeta)^*A\xi\rangle + \langle \xi, A^{-1}(ad\zeta)^*(ad\eta)^*A\xi\rangle_e$$

is positive- or negative-definite, then ξ is a stable stationary (equilibrium) point of the vector field $Y|G\cdot\xi$. (A is defined on page 320.)

Proof. By 4.4.7(ii) $Y|G\cdot\xi$ is a Hamiltonian vector field on the symplectic manifold $(G\cdot\xi, \omega_\xi)$ and the Hamiltonian is $H(\eta) = \frac{1}{2}\langle\eta,\eta\rangle$ for $\eta \in G\cdot\xi$. Hence we can apply the stability criterion 3.4.17, that is, $D^2H(\xi)$ should be positive- or negative-definite. If this is established, ξ will be stable for $Y|G\cdot\xi$.

Recall that if M is an arbitrary manifold and $f\colon M \to \mathbb{R}$ is a smooth function with $m \in M$ a critical point, that is, $df(m)=0$, then

$$Hess\,f\colon T_mM \times T_mM \to \mathbb{R}$$

is a symmetric bilinear form which, in any chart, is given by the matrix of second partials. Intrinsically, if v, $w \in T_mM$ are extended to vector fields V, W; then,

$$D^2f_m(v,w) = (Hess\,f)_m(v,w) = V(Wf)(m) = W(Vf)(m)$$

ISBN 0-8053-0102-X

In our case, extend η, ζ to vector fields on $G\cdot\xi$ as follows:

$$\tilde{\eta}(v) = -A^{-1}(ad\eta)^*Av$$

$$\tilde{\zeta}(v) = -A^{-1}(ad\zeta)^*Av$$

where $v \in G\cdot\xi$. Clearly,

$$\tilde{\zeta}H_\xi(v) = \langle v, \tilde{\zeta}(v)\rangle_e$$

so that

$$\tilde{\eta}(\tilde{\zeta}H_\xi)(\xi) = \langle \tilde{\eta}(\xi), \tilde{\zeta}(\xi)\rangle_e$$

$$+\langle \xi, A^{-1}(ad\eta)^*A, A^{-1}(ad\zeta)^*A\xi\rangle_e$$

which is exactly the bilinear form given in the statement of the theorem. ■

For the case of the rigid body, this criterion reduces to a classical one: a rigid body spins stably about its longest and shortest principal axes, but unstably about its middle one (these terms will be defined below).

Another important point is the question of stability of whole geodesics on G, or of integral curves of X_H on TG. One way to approach this problem is by means of curvature. If one can show that a geodesic on G remains a priori in a region of positive (sectional) curvature, then it is stable. Unfortunately, the calculation of curvature of left-invariant metrics on G is not trivial. Arnold [1966] has done so for cases relevant to fluid mechanics (see Sect. 5.5 and Arnold [1978]). See also Milnor [1976] for an excellent review of what is currently known about curvatures of left-invariant metrics.

So far, all Hamiltonians on T^*G and TG have been supposed to be left invariant. We shall analyze now an important special case when this does *not* happen, left invariance being destroyed by a potential function.*

Let G be a Lie group with a left-invariant metric $\langle , \rangle$ and $V: G \to \mathbf{R}$ an aribtrary smooth function. Consider the energy function $E = K + V\circ\tau_G$ on TG, where $K(v_g) = \frac{1}{2}\langle v_g, v_g\rangle_g$ is the kinetic energy. If V is not left invariant, E will not be either. The associated vector field X_E will remain, of course, a second-order equation on TG by the last statement of 3.5.17. Let us define for $\eta \in \mathfrak{g}$

$$H_t(\eta) = T_{x(t)}L_{x(t)^{-1}}F_t(\eta)$$

Here F_t is the flow of X_H and $x(t) = x_\eta(t) = \tau_G(F_t(\eta))$. This will not define a

*Another case in which this happens is for a nonhomogeneous fluid; see Sect. 5.5 and Marsden [1976].

ISBN 0-8053-0102-X

flow on $\mathfrak{g}$ because F_t is not left invariant. However, we can define a time-dependent vector field at the points $H_t(\xi)$ by setting

$$Y_t(H_t(\xi)) = \frac{d}{dt}H_t(\xi)$$

The motion is then described as follows.

4.4.9 Proposition. *Using the above notation, we have*

$$Y_t(H_t(\xi)) = Y(H_t(\xi)) - T_{x(t)}L_{x(t)^{-1}}\operatorname{grad} V(x(t))$$

Here Y is the Euler vector field for G, $\langle,\rangle$ [see 4.4.6(ii)] and "grad" is the gradient with respect to the left-invariant metric $\langle,\rangle$.

Proof. Our argument is an extension of that of 4.4.3(ii). We have by 4.4.3(ii) and left invariance of the metric

$$\begin{aligned}\frac{d}{dt}\hat{J}(\eta)(F_t(\xi)) &= \frac{d}{dt}\langle T_eR_{x(t)}\eta, F_t(\xi)\rangle \\ &= \frac{d}{dt}\left\langle Ad\left(x(t)^{-1}\right)\eta, H_t(\xi)\right\rangle \\ &= \left\langle Ad\left(x(t)^{-1}\right)\eta, Y_t(H_t(\xi))\right\rangle + \left\langle \frac{d}{dt}Ad\left(x(t)^{-1}\right)\eta, H_t(\xi)\right\rangle\end{aligned}$$

Now, if we set $\eta(t) = Ad(x(t)^{-1})\eta$, we have

$$\eta = Ad(x(t))\eta(t)$$

Applying the formula giving the transition between time derivatives in space and body coordinates proved at the beginning of this section,

$$0 = Ad(x(t))\eta'(t) + \left[TR_{x(t)^{-1}}\frac{dx}{dt}, \eta\right]$$

Note that $dx/dt = F_t(\xi)$ because F_t arises from a second-order equation. Thus

$$\begin{aligned}\eta'(t) &= \left[Ad\left(x(t)^{-1}\right)\eta, Ad\left(x(t)^{-1}\right)TR_{x(t)^{-1}}F_t(\xi)\right] \\ &= \left[Ad\left(x(t)^{-1}\right)\eta, H_t(\xi)\right]\end{aligned}$$

Accordingly,

$$\begin{aligned}\frac{d}{dt}\hat{J}(\eta)F_t(\xi) &= \left\langle Ad\left(x(t)^{-1}\right)\eta, Y_t(H_t(\xi))\right\rangle \\ &\quad + \left\langle\left[Ad\left(x(t)^{-1}\right)\eta, H_t(\xi)\right], H_t(\xi)\right\rangle\end{aligned}$$

ISBN 0-8053-0102-X

Now

$$\frac{d}{dt}\hat{J}(\eta)F_t(\xi)=\left(L_{X_E}\hat{J}(\eta)F_t(\xi)\right)$$
$$=\{\hat{J}(\eta),E\}(F_t(\xi))$$
$$=\{\hat{J}(\eta),V\}(F_t(\xi))$$

since by 4.4.3(ii) $\{\hat{J}(\eta),K\}=0$. By 4.2.12(iii) and making all suitable identifications using the metric,

$$\{\hat{J}(\eta),V\}(F_t(\xi))=-dV(x(t))\cdot T_eR_{x(t)}\eta$$

Thus, if we let $w=Ad(x(t)^{-1})\cdot\eta$ and use 4.4.3(ii),

$$\langle Y_t(H_t(\xi)),w\rangle-\langle Y(H_t(\xi)),w\rangle=\langle Y_t(H_t(\xi)),w\rangle+\langle[w,H_t(\xi)],H_t(\xi)\rangle$$
$$=-dV(x(t))\cdot T_eR_{x(t)}Ad(x(t))w$$
$$=-dV(x(t))\cdot T_eL_{x(t)}w$$
$$=-\langle \operatorname{grad} V(x(t)),T_eL_{x(t)}w\rangle$$

for all $w\in\mathfrak{g}$. The result follows from this together with the left invariance of the metric $\langle,\rangle$. ■

In the case of $SO(3)$ the extra term in Y_t arising from V is called the "torque." If one has forces that are nonconservative, these lead to nonHamiltonian vector fields on TG, but the equations of motion retain the form given above, with the obvious replacement of $-\operatorname{grad} V$ by a general force term.

Previous completeness theorems apply, in particular, to the present context. The following consequence of 3.7.15 is illustrative.

4.4.10 Proposition. *Let G be a Lie group with a left-invariant metric $\langle,\rangle$. Let $K(v)=\frac{1}{2}\langle v,v\rangle$ on TG and let $V: G\to\mathbf{R}$ be smooth and bounded below (or, more generally, bounded below by a positively complete function). Then the flow of the Hamiltonian $E=K+V\circ\tau_G$ is complete.*

When one is using an "accelerating" coordinate system, fictitious forces are introduced. This happens, of course, when one uses coordinates fixed on a rigid body such as the Earth. The general set up is as follows. Let X_H be a given Hamiltonian vector field on a symplectic manifold P. Let ζ_t be a given flow on P corresponding to a "motion" of P. Let $c(t)$ be an integral curve of X_H and set $\tilde{c}(t)=\zeta_t(c(t))$. Then

$$\tilde{c}'(t)=(\zeta_t)_*X_H(\tilde{c}(t))+Y(\tilde{c}(t))$$

ISBN 0-8053-0102-X

where Y is the generator of ζ_t. Here $(\zeta_t)_* X_H$ is just X_H in the "new coordinate system" defined by ζ_t and the extra term Y is the "fictitious" or Coriolis forces.

Of course, using left- or right-invariant metrics is mostly a matter of convention. We saw that the flow in body coordinates is described by the Euler vector field Y. It is easy to see that the motion v_s in space coordinates may, dually, be regarded as an integral curve of the Euler vector field obtained by extending the metric to be right rather than left invariant. Thus changing "left" to "right" interchanges v_b with v_s. Both of them satisfy the same "Euler equations." When we study fluid mechanics as an example of Hamiltonian systems on Lie groups in Sect. 5.5 we will see an instance where it is more natural to use right invariance.

There is a simplification of the Euler equations on G if its Lie algebra $\mathfrak{g}$ carries a nondegenerate symmetric bilinear form $(,)$ that is invariant under the adjoint maps $Ad(g)$:

$$(Ad(g)\xi, Ad(g)\eta) = (\xi, \eta)$$

Such a bilinear form then defines a pseudo-Riemannian metric on G that is *both* left and right invariant. [Examples of Lie groups with this property are: Abelian Lie groups—for which $Ad(g) \equiv Id$; semi-simple Lie groups—for which the "Killing form" $(v,w) = \frac{1}{2} Trace(adv \circ adw)$ is nondegenerate; compact Lie groups—for which a *positive-definite* invariant form on $\mathfrak{g}$ may be obtained by "averaging" any given positive-definite form with respect to the invariant Haar measure on G. The group $SO(3)$ falls under the two latter categories; later we shall explicitly calculate the invariant inner product on its Lie algebra.]

Suppose that $(,)$ is a nondegenerate symmetric bilinear form on $\mathfrak{g}$. Then we can write $\langle \xi, \eta \rangle = (I\xi, \eta)$ where $I: \mathfrak{g} \to \mathfrak{g}$ is linear and symmetric with respect to $(,)$. The Euler equations for the geodesic flow of $\langle , \rangle$ then read

$$(IY(\xi), \eta) = (I\xi, [\xi, \eta]) = (I\xi, ad(\xi)\eta)$$

4.4.11 Lemma. *Suppose that $(,)$ is invariant under $Ad(g)$ for all $g \in G$. Then for each $\xi \in \mathfrak{g}$, $ad(\xi)$ is skew-symmetric with respect to $(,)$.*

Proof. We have

$$(ad(\xi)\eta, \zeta) = \frac{d}{dt}(Ad(\exp t\xi)\eta, \zeta)\Big|_{t=0}$$

$$= \frac{d}{dt}\left(\eta, Ad(\exp t\xi)^{-1}\zeta\right)\Big|_{t=0}$$

$$= \frac{d}{dt}\left(\eta, Ad(\exp(-t\xi))\zeta\right)\Big|_{t=0}$$

$$= (\eta, ad(-\xi)\zeta) = -(\eta, ad(\xi)\zeta) \qquad \blacksquare$$

ISBN 0-8053-0102-X

Hence, assuming that (,) is invariant, we have

$$(I\xi, ad(\xi)\eta) = -(ad(\xi)I(\xi), \eta)$$
$$= ([I\xi, \xi], \eta)$$

so that *the Euler vector field is given by*

$$IY(\xi) = [I\xi, \xi] \quad \text{that is,} \quad Y(\xi) = I^{-1}[I\xi, \xi]$$

The proof of Lemma 4.4.11 shows that $ad(\eta)$ is skew-symmetric with respect to a form provided that it is invariant under $Ad(exp(t\eta))$ for all t. We can apply this observation to deduce additional conservation laws for the geodesic flow of $\langle , \rangle$ when it has such invariance. (Applications to the case of the rigid body will be given below.)

4.4.12 Proposition. *Suppose that $\langle , \rangle$ is invariant under $\{Ad(exp\, t\eta): t \in \mathbf{R}\}$ for some $\eta \in \mathfrak{g}$. Then the function $\mu_\eta: \mathfrak{g} \to \mathbf{R}$ defined by*

$$\mu_\eta(\xi) = \langle \eta, \xi \rangle$$

is a constant of the motion for H_t, that is, $\mu_\eta \circ H_t =$ constant for all t. (Here H_t is, as above, the flow on $\mathfrak{g}$ in body coordinates corresponding to the geodesic flow of $\langle , \rangle$.) In particular, $\mu_\eta(Y(\xi)) = 0$ for all $\xi \in \mathfrak{g}$.

Proof. By our remarks above we have

$$\mu_\eta(Y(\xi)) = \langle \eta, Y(\xi) \rangle = \langle [\xi, \eta], \xi \rangle = -\langle ad(\eta)\xi, \xi \rangle$$
$$= \langle \xi, ad(\eta)\xi \rangle = \langle ad(\eta)\xi, \xi \rangle = -\mu_\eta(Y(\xi))$$

so that $\mu_\eta(Y(\xi))$ is 0 for all $\xi \in \mathfrak{g}$. Hence

$$\frac{d}{dt}\mu_\eta(H_t(\xi)) = \mu_\eta(Y(H_t(\xi))) = 0 \quad \blacksquare$$

Alternative Proof. On TG we have invariance of $\langle , \rangle$ under right translation by $exp(t\eta)$ so as in 4.4.3(ii) $\xi \mapsto \langle T_e L_g \eta, \xi \rangle$ is a constant of the motion. We get the result by passing to body coordinates. $\blacksquare$

4.4.13 Corollary. *Suppose that $\langle , \rangle$ is invariant under the adjoint representation of G. Then the corresponding Euler vector field vanishes identically, and therefore the geodesic flow F_t expressed in body coordinates is given by the exponential map, namely,*

$$\lambda \circ F_t \circ \lambda^{-1}(g, \xi) = ((exp\, t\xi)\, g, \xi)$$

ISBN 0-8053-0102-X

In other words, the two exponential maps in the Riemannian and Lie group sense coincide for a bi-invariant metric on a Lie group, and hence the geodesic curves through e are in this case just $\exp t\xi$ for $\xi \in \mathfrak{g}$.

Proof. If $\langle , \rangle$ is invariant under $Ad(g)$ for all g, it follows from the proof of 4.4.12 that $\langle \eta, Y(\xi)\rangle = 0$ for all $\eta \in \mathfrak{g}$. Consequently, $Y(\xi) = 0$ for all $\xi \in \mathfrak{g}$. But then $H_t(\xi) = TL_{x(t)^{-1}}F_t(\xi) = \xi$ for all $\xi \in \mathfrak{g}$ and hence $F_t(\xi) = T_eL_{x(t)}\xi$. The geodesic flow in body coordinates is

$$\begin{aligned}\lambda F_t \lambda^{-1}(g,\xi) &= \lambda F_t(T_eL_g\xi) \\ &= \lambda(T_eL_{x(t)g}\xi) = (x(t)g,\xi)\end{aligned}$$

Note that

$$\begin{aligned}(x(t_1+t_2)g,\xi) &= \lambda F_{t_1+t_2}\lambda^{-1}(g,\xi) \\ &= (\lambda F_{t_1}\lambda^{-1})(\lambda F_{t_2}\lambda^{-1}(g,\xi)) = (x(t_1)x(t_2)g,\xi)\end{aligned}$$

so that $x(t_1+t_2) = x(t_1)x(t_2)$, that is, the curve $x: \boldsymbol{R} \to G$ defines a smooth one-parameter subgroup of G and hence equals $\exp t\eta$, where $\eta = dx/dt|_{t=0}$. Since $X_K \in \mathfrak{X}(TG)$ is a second-order equation on G, $K(v) = \frac{1}{2}\langle v,v\rangle_g$ being the kinetic energy of the metric, $x: \boldsymbol{R} \to G$ is a geodesic. ■

The rotational motion of a rigid body may be described by the geodesic flow of a given left-invariant metric on $SO(3)$. This metric is determined by the body's mass distribution; it is called the *inertia tensor*. [The use of left rather than right invariance is due to the convention by which we represent the action of $SO(3)$ on $\boldsymbol{R}^3$.] The following physical remarks are intended to justify this mathematical model.

Consider a rigid body that is free to rotate about a fixed point, which, for convenience, we take to be the origin of $\boldsymbol{R}^3$. The notion of rigidity entails that the distances between points of the body are unchanged when it moves, so that, if $x(t,\xi)$ is the position at time t of the particle that was at ξ at time 0, we have

$$x(t,\xi) = A(t)\xi$$

where $A(t)$ is an orthogonal matrix. If the motion is continuous and starts out at the identity, we must have $A(t) \in SO(3)$, the identity component of the orthogonal group. The initial mass distribution of the body is described by a positive measure μ on $\boldsymbol{R}^3$; for $A(t)$ to be uniquely determined we must insist that the support of μ not be concentrated on a one-dimensional subspace of $\boldsymbol{R}^3$. Assuming this, the configuration space of the rigid body can be identified with the group $SO(3)$.

ISBN 0-8053-0102-X

The kinetic energy of the body at time t is given by

$$K(t)=\tfrac{1}{2}\int \|\dot{x}(t,\xi)\|^2 \, d\mu(\xi)$$
$$=\tfrac{1}{2}\int \|\dot{A}(t)\xi\|^2 \, d\mu(\xi)$$

Now $\dot{A}(0)\in T_eSO(3)$ is a 3×3 skew-symmetric matrix ω; as explained in Sect. 4.1, there is a corresponding vector $\tilde{\omega}\in \mathbf{R}^3$ such that $\omega(x)=\tilde{\omega}\times x$. The mapping $\omega\mapsto\tilde{\omega}$ identifies the Lie algebra $T_eSO(3)$ with $\mathbf{R}^3$; under this identification the Lie bracket corresponds to the cross product in $\mathbf{R}^3$.

At an arbitrary time t we have

$$\dot{x}(t,\xi)=\frac{d}{d\tau}A(t+\tau)\xi\Big|_{\tau=0}$$
$$=\frac{d}{d\tau}A(t+\tau)A(t)^{-1}x(t,\xi)\Big|_{\tau=0}$$
$$=\tilde{\omega}(t)\times x(t,\xi)$$

where

$$\omega(t)=\frac{d}{d\tau}A(t+\tau)A(t)^{-1}\Big|_{\tau=0}$$
$$=TR_{A(t)^{-1}}\dot{A}(t)=\rho(\dot{A}(t))$$

The vector representative $\tilde{\omega}(t)=\omega_S(t)\in\mathbf{R}^3$ is called the *angular velocity* (with respect to our fixed spatial coordinate system); note that the occurrence of the map ρ here is in agreement with (and indeed justifies) our general notion of "space coordinates." The corresponding vector as viewed from a coordinate system fixed in the body (and originally coinciding with the system fixed in space) is clearly $\omega_B(t)=A(t)^{-1}\omega_S(t)$. If our general notion of "body coordinates" is valid, this should correspond to the action of $Ad(A(t)^{-1})$ on $\omega_S(t)$. We verify this in the following lemma.

4.4.14 Lemma. *(i) The adjoint action of $SO(3)$ on $T_eSO(3)\approx\mathbf{R}^3$ is the usual action.*

(ii) The usual inner product on $\mathbf{R}^3$ is invariant under the adjoint action (and therefore extends to a bi-invariant metric on $SO(3)$).

Proof. The second assertion is an immediate consequence of the first; let us prove the first assertion. Let $\omega\in T_eSO(3)$ and let $\tilde{\omega}\in\mathbf{R}^3$ be the corresponding

ISBN 0-8053-0102-X

vector. Then, if $A \in SO(3)$,

$$Ad(A)\omega = \frac{d}{dt} A \exp(t\omega) A^{-1}\Big|_{t=0}$$

$$= A\omega A^{-1}$$

Therefore,

$$\widetilde{Ad(A)\omega} \times x = A\omega A^{-1}x$$

$$= A(\tilde{\omega} \times A^{-1}x)$$

$$= (A\tilde{\omega}) \times x$$

Since for $A \in SO(3)$, $A(y \times w) = Ay \times Aw$. We have thus shown that

$$\widetilde{Ad(A)\omega} = A\tilde{\omega}$$

which is our assertion. ■

The usual metric on R^3 therefore corresponds to a perfectly spherical rigid body whose motions are just one-parameter subgroups of rotations.

Returning to our expression for the kinetic energy, we see that

$$K(t) = \tfrac{1}{2} \int \|\omega_S(t) \times A(t)\xi\|^2 \, d\mu(\xi)$$

$$= \tfrac{1}{2} \int \|A(t)^{-1}\omega_S(t) \times \xi\|^2 \, d\mu(\xi)$$

$$= \tfrac{1}{2} \int \|\omega_B(t) \times \xi\|^2 \, d\mu(\xi)$$

$$= \tfrac{1}{2} \langle \omega_B(t), \omega_B(t) \rangle_e$$

where $\langle u, v \rangle_e = \int (u \times \xi, v \times \xi) \, d\mu(\xi)$. Thus $K(t)$ is the kinetic energy corresponding to the left-invariant extension of the form $\langle , \rangle_e$. The physical motion should conserve and "extremize" the energy K; thus we are led to the geodesic flow of $\langle , \rangle$, as promised.

The form $\langle u, v \rangle_e$ can be written (Iu, v) for some positive symmetric matrix I, where $(\cdot,\cdot)$ denotes the usual inner product in R^3. As noted above, following the remarks after 4.4.11, since the inner product $(,)$ is invariant under the adjoint representation, the Euler equations take on the simple form

$$I\dot{\omega}_B = I\omega_B \times \omega_B$$

where we have identified the Lie algebra $so(3)$ with R^3. We can choose the

ISBN 0-8053-0102-X

original (orthonormal) spatial coordinates so that I is diagonal: $Ix=(I_1x_1,I_2x_2,I_3x_3)$ for constants $I_1,I_2,I_3>0$. [The axes of this coordinate system are called *principal axes*, and the constants I_1,I_2,I_3 (the eigenvalues of I) are called the *principal moments of inertia*.] With respect to this coordinate system the Euler equations become

$$I_1\dot{\omega}_B^1-(I_2-I_3)\omega_B^2\omega_B^3=0$$

$$I_2\dot{\omega}_B^2-(I_3-I_1)\omega_B^3\omega_B^1=0$$

$$I_3\dot{\omega}_B^3-(I_1-I_2)\omega_B^1\omega_B^2=0$$

If torques are present, they are added to the right-hand side of the Euler equations, as indicated in the general case (but note that the torques depend on $A(t)$).

The conservation laws are as follows: Conservation of energy says that $\langle\omega_B,\omega_B\rangle$ is constant. Conservation of angular momentum says that

$$A(t)IA(t)^{-1}\omega_S(t)=A(t)I\omega_B(t)$$

is constant since

$$\begin{aligned}\hat{J}_B(\xi)(A(t),\omega_B(t))&=\langle\omega_B(t),A^{-1}(t)\xi\rangle_e\\&=(I\omega_B(t),A^{-1}(t)\xi)\\&=(A(t)I\omega_B(t),\xi)\end{aligned}$$

and $\hat{J}_B(\xi)$ is constant on the orbits $(A(t),\omega_B(t))$. Proposition 4.4.12 implies that if $\langle,\rangle_e$ is invariant under rotation about the vector $\tilde{v}$ (i.e., if the mass distribution has this symmetry), then $\langle\omega_B(t),\tilde{v}\rangle_e$ is constant. It follows that $(\omega_B(t),I\tilde{v})$ is constant. But the assumed symmetry property implies that $\tilde{v}$ is an eigenvalue of I. To see this note that a rotation $A\in SO(3)$ about $\tilde{v}$ is characterized by $A\tilde{v}=\tilde{v}$, that is, $Ad(A)v=v$ by Lemma 4.4.14 so that for any $u\in T_eSO(3)$

$$\begin{aligned}\langle u,v\rangle_e&=\langle Ad(A)u,Ad(A)v\rangle_e\\&=\langle Ad(A)u,v\rangle_e\end{aligned}$$

or $(\tilde{u},I\tilde{v})=(A\tilde{u},I\tilde{v})$ for any $\tilde{u}\in R^3$. It then follows that $AI\tilde{v}=I\tilde{v}$ and hence $\tilde{v}$ and $I\tilde{v}$ lie in the eigenspace of A corresponding to the eigenvalue 1. But since $A\in SO(3)$, it is an easy matter to see that 1 is either a simple eigenvalue, in which case $I\tilde{v}=t\tilde{v}$ for some $t\in R$, or it is a triple eigenvalue, in which case $A=e$. Thus we deduce the well-known fact (about a spinning top for instance) that the component of $\omega_B(t)$ along the direction $\tilde{v}$, that is, $(\omega_B(t),\tilde{v})$, is constant. This is also easy to see directly from the Euler equations.

ISBN 0-8053-0102-X

EXERCISES

4.4A. Determine $\theta_S=\bar{\rho}_*\theta_0\in\Omega^1(G\times\mathfrak{g}^*)$, $\omega_S=-d\theta_S$, $\Theta_S=\rho_*\Theta_0\in\Omega^1(G\times\mathfrak{g})$, and $\Omega_S=-d\Theta_S$.

4.4B. Formulate and prove the cotangent version of Theorem 4.4.4.

4.4C. In 4.4.6(ii) show that $X^\xi(g)=T_eL_g(\xi)$. [*Hint:* Show that $(P_1\circ T\lambda)(w_v)=TL_g(v)$, where $v\in T_gG$, $w_v\in T_vTG$, and P_1: $TG\times\mathfrak{g}\times\mathfrak{g}\to TG$ is the canonical projection.]

4.4D. (a) Prove that if G is a connected Lie group and G has a bi-invariant Riemannian metric, then $exp:\mathfrak{g}\to G$ is onto. (*Hint:* Use the fact that $C_\xi(t)=exp\, t\xi$ are geodesics and the Hopf–Rinow theorem.)

Remark. The most general simply connected group admitting a bi-invariant Riemannian metric is a compact group $\times\, \boldsymbol{R}^n$; see Milnor [1963, p. 115]; see also Sternberg [1964, p. 234].

(b) Show that the result in (a) is false if the metric is merely pseudo-Riemannian. (*Hint:* Use the Killing form on $Sl(2,\boldsymbol{R})$).

4.4E. In the notations of proposition 4.4.9 and the remarks preceding it, show that

(i) $X^B_{K+V\circ\tau_G}(g,\xi)=(T_eL_g(\xi),\xi,Y(\xi)-T_gL_{g^{-1}}(grad\,V(g)))$ by using the formula for Ω_B in 4.4.2(ii) and comparing with Y_t.

(ii) Let $(g,\xi)\mapsto(A_t(g,\xi),B_t(g,\xi))$ denote the flow of $X^B_{K+V\circ\tau_G}$. Show that

$$A_t(e,\xi)=x_\xi(t)=\tau_G F_t(\xi),\quad \text{and } H_t(\xi)=B_t(e,\xi).$$

(iii) Prove the version on T^*G of 4.4.9 and (i), (ii).

4.4F. (i) Let G be a Lie group with $\langle,\rangle$ a left-invariant metric and let $\tilde{G}$ be the universal covering group of G with a left-invariant metric $\langle\langle,\rangle\rangle$ coinciding with $\langle,\rangle$ at e. Show that geodesics on $\tilde{G}$ are lifts of geodesics on G.

(ii) Let U: $\tilde{G}\times\mathcal{H}\to\mathcal{H}$ be a unitary representation of $\tilde{G}$ on a Hilbert space $\mathcal{H}$. Define a metric (,) on $E=\tilde{G}\times\mathcal{H}$ by

$$((v_1,\phi_1),(v_2,\phi_2))=\langle v_1,v_2\rangle+(U_{g^{-1}}\phi_1)\cdot(U_{g^{-1}}\phi_2)$$

where g is the base point of v_1 and v_2 and $\cdot$ is the inner product in $\mathcal{H}$. Calculate the momentum mapping J for the action of $\tilde{G}$ on TE induced by the action of $\tilde{G}$ on $E=G\times\mathcal{H}$ given by $\Phi_g(h,\phi)=(gh,U_g(\phi))$.

(iii) For $G=SO(3)$, $\tilde{G}=SU(2)$ (see Exercise 4.1H) and $\mathcal{H}=\boldsymbol{C}^2$, and $\langle,\rangle$ a given moment of inertia tensor, the construction in (ii) gives a *rigid body with spin*. Show that

$$\begin{aligned} J(g,v,c,u) &= \text{rigid body angular momentum} \\ &\quad + \text{spin angular momentum} \\ &= A(t)I\omega_B(t)-\tfrac{1}{2}u\cdot i\sigma\cdot c \end{aligned}$$

where σ is the "vector" of Pauli spin matrices.

4.4G. Discuss the complete integrability of the rigid body using formulas (1)–(3) preceding 4.3.6. Consult Iacob [1973] and Mishchenko [1970], Dikii [1972], Manakov [1976], Mishchenko and Fomenko [1978].

4.5 THE TOPOLOGY OF SIMPLE MECHANICAL SYSTEMS

In Smale [1970a] a topological program for studying Hamiltonian systems with symmetry was laid out. This may be described as follows. Let H be a Hamiltonian on a symplectic manifold (P,ω) and let G be a (connected) Lie group acting on P, leaving H invariant, and having a momentum mapping

ISBN 0-8053-0102-X

$J: P \to \mathfrak{g}^*$. Form the *energy momentum mapping*

$$H \times J: P \to \mathbf{R} \times \mathfrak{g}^*, (H \times J)(p) = (H(p), J(p))$$

so that for each $c = (e, \mu) \in \mathbf{R} \times \mathfrak{g}^*$, the set

$$I_c = (H \times J)^{-1}(c)$$

is invariant under the flow of X_H. (Recall from the invariant volume Theorem 3.4.15 that if c is a regular value of $H \times J$, then I_c carries an invariant volume element.) To describe the topological nature of the flow of X_H one should determine:

(i) the topological type of I_c for all c;
(ii) the bifurcation set $\Sigma_{H \times J}$ of $H \times J$ (defined below);
(iii) the flow of X_H on each I_c (if the group action is explicitly known, this is equivalent to finding the flow of the reduced system on P_μ, see Sect. 4.3); and
(iv) how the sets I_c topologically fit together, as μ varies, to make up $H^{-1}(e)$.

The *bifurcation set* is defined as follows. A smooth map $f: M \to N$ is *locally trivial at the value* y_0 in its range if there is a neighborhood U of y_0 such that $f^{-1}(y)$ is a smooth submanifold of M for each $y \in U$ and there is a smooth map $h: f^{-1}(U) \to f^{-1}(y_0)$ such that $f \times h$ is a diffeomorphism from $f^{-1}(U)$ to $U \times f^{-1}(y_0)$ [hence $h | f^{-1}(y): f^{-1}(y) \to f^{-1}(y_0)$ is a diffeomorphism for each $y \in U$]. See Fig. 4.5.1. The *bifurcation set* of f is

$$\Sigma_f = \{ y_0 \in N \mid f \text{ fails to be locally trivial at } y_0 \}$$

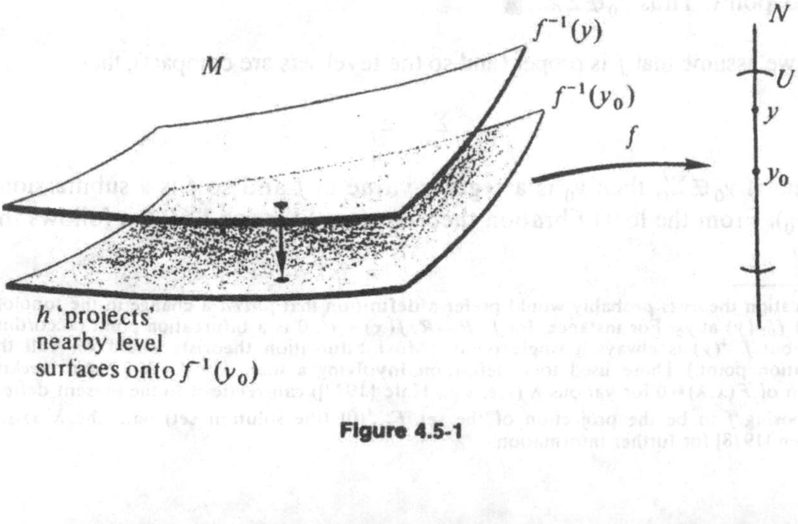

Figure 4.5-1

ISBN 0-8053-0102-X

The idea of the bifurcation set is that as y passes through Σ_f, $f^{-1}(y)$ *may* change topological type.*

Before studying mechanical systems, we give a general property of the bifurcation set.

Let $\sigma(f)$ denote the set of the critical points of f;

$$\sigma(f)=\{x\in M \mid Tf(x): T_xM\to T_{f(x)}N \text{ is not surjective}\}$$

which is a closed set in M. By Sard's theorem $f(\sigma(f))$, the *critical values*, has measure zero in N. Put

$$\Sigma'_f=f(\sigma(f))$$

We prove now that *critical values must be bifurcation points.*

4.5.1 Proposition. $\Sigma'_f\subset\Sigma_f$.

Proof. Suppose that $y_0\notin\Sigma_f$, so that f is locally trivial at the value y_0. By definition, $f\times h$ is a diffeomorphism of $f^{-1}(U)$ to $U\times f^{-1}(y_0)$. In particular, the derivative at $x_0\in f^{-1}(y_0)$,

$$T(f\times h)(x_0)=Tf(x_0)\times Th(x_0)$$

is an isomorphism. Writing $T_{x_0}M=E_{x_0}\oplus T_{x_0}(f^{-1}(y_0))$ and using the fact that $T_{x_0}f^{-1}(y_0)\subset \ker Tf(x_0)$, this derivative becomes the matrix

$$\begin{bmatrix} Tf(x_0)|E_{x_0} & 0 \\ Th(x_0)|E_{x_0} & Th(x_0)|T_{x_0}f^{-1}(y_0)\end{bmatrix}$$

If this is onto, then $Tf(x_0)|E_{x_0}$ is onto [and $T_{x_0}f^{-1}(y_0)=\ker Tf(x_0)$], so x_0 is a regular point. Thus $y_0\notin\Sigma'_f$. ■

If we assume that f is proper (and so the level sets are compact), then

$$\Sigma'_f=\Sigma_f$$

Indeed, if $y_0\notin\Sigma'_f$, then y_0 is a regular value of f and so f is a submersion on $f^{-1}(y_0)$. From the local fibration theorem (see Exercise 1.6G), it follows that f

---

*Bifurcation theorists probably would prefer a definition that *forced* a change in the topological type of $f^{-1}(y)$ at y_0. For instance, for $f: \mathbb{R}\to\mathbb{R}$, $f(x)=x^3$, 0 is a bifurcation point (according to 4.5.1), but $f^{-1}(y)$ is always a single point. (Most bifurcation theorists would *not* call this a bifurcation point.) Those used to a definition involving a map $F: X\times\Lambda\to Y$ and seeking a solution of $F(x,\lambda)=0$ for various λ (see, e.g., Hale [1977]) can relate it to the present definition by choosing f to be the projection of the set $F^{-1}(0)$ (the solution set) onto the λ axis. See Marsden [1978] for further information.

ISBN 0-8053-0102-X

is trivial near each point of $f^{-1}(y_0)$. Since $f^{-1}(y_0)$ is compact, it follows readily that f is trivial over a whole neighborhood of $f^{-1}(y_0)$.

Unfortunately, in many examples, f will not have compact level sets. [For such cases one needs to make sure $Df(x_0)$ has a uniformly bounded right inverse on all of $f^{-1}(y_0)$ to ensure $\Sigma'_f=\Sigma_f$. In examples in Sect. 9.8, $\Sigma'_f\neq\Sigma_f$ will actually occur.]

Since we are concerned with systems with symmetry, we are not studying the "generic" case. However, these cases are the interesting ones. The idea is that they are the hub around which qualitatively different classes of Hamiltonian vector fields hover; that is, they are the "bifurcation points" in the set of all Hamiltonian vector fields, the different classes being reached by breaking the symmetry in various ways. (See Chapter 8 for additional information.) Furthermore, most of the classical examples (n-body problem, rigid body, etc.) possess symmetries.

We will now carry out part of Smale's program, for the case of Hamiltonian systems of the form kinetic plus potential energy. For these a number of issues simplify. By using the methods of Sect. 4.3, a number of proofs become simpler than those existing in the literature, and 4.5.6 seems to be new. Most of the other results are due to Smale [1970a] (see also Iacob [1971]).

We make the following definition.

4.5.2 Definition. *A (simple)* ***mechanical system with symmetry*** *is a quadruple* (M,K,V,G), *where:*

(*i*) *M is a Riemannian manifold with metric $\gamma=\langle,\rangle$; M is called the* ***configuration space*** *and the cotangent bundle T^*M, with its canonical symplectic structure $\omega_0=-d\theta_0$, the* ***phase space*** *of the system;*

(*ii*) *$K\in\mathfrak{F}(T^*M)$ is the* ***kinetic energy*** *of the system defined by*

$$K(\alpha)=\tfrac{1}{2}\langle\alpha,\alpha\rangle_{\tau^*_M(\alpha)}$$

where we denote (by abuse of notation) $\langle,\rangle_x$ the metric on T^*_xM given by $\langle\alpha,\beta\rangle_x=\langle\gamma^\sharp(x)(\alpha),\gamma^\sharp(x)(\beta)\rangle_x$ for $\alpha,\beta\in T^*_xM$ and $\gamma^\sharp: T^*M\to TM$ is the usual isomorphism of vector bundles, $\gamma^\sharp=(\gamma^\flat)^{-1}$, $\gamma^\flat(v_x)=\langle\cdot,v_x\rangle_x$;

(*iii*) *$V\in\mathfrak{F}(M)$ is the* ***potential energy*** *of the system;*

(*iv*) *G is a connected Lie group acting on M by an action $\Phi: G\times M\to M$ under which the metric is invariant (i.e., Φ is an action by isometries) and V is invariant; these conditions mean that*

$$K\circ\Phi_g^{T^*}=K \quad\text{and}\quad V\circ\Phi_g=V$$

for all $g\in G$; G is called the ***symmetry group*** *of the system; finally,*

(*v*) *$H\in\mathfrak{F}(T^*M)$ defined by*

$$H=K+V\circ\tau^*_M$$

is the ***Hamiltonian*** *of the system.*

ISBN 0-8053-0102-X

Note that the above conditions imply $H\circ\Phi_g^{T^*}=H$ for all $g\in G$ and hence J: $T^*M\to\mathfrak{g}^*$, $J(\alpha_m)(\xi)=\alpha_m(\xi_M(m))$ is an Ad^*-equivariant momentum mapping, constant on the orbits of the Hamiltonian vector field X_H; that is, J is a constant of the motion (see Sect. 4.2).

First, we shall be concerned with the sets $I_{h,\mu}=(H\times J)^{-1}(h,\mu)$, which are, for "most" values, $(h,\mu)\in\boldsymbol{R}\times\mathfrak{g}^*$, manifolds. They will be called the *invariant manifolds*, even though for some values $(h,\mu)\in\Sigma_{H\times J}$ (the bifurcation set), they will not have a manifold structure. We know from Sect. 4.3 that the isotropy group $G_\mu=\{g\in G|Ad^*_{g^{-1}}\mu=\mu\}$ of the Ad^*-action acts on $J^{-1}(\mu)$. But H is G invariant, too, so that G_μ will act on $H^{-1}(h)$. Thus G_μ acts on $I_{h,\mu}=H^{-1}(h)\cap J^{-1}(\mu)$. Now we can apply the reduction procedure explained in Sect. 4.3 to obtain the submanifolds $\hat{I}_{h,\mu}=I_{h,\mu}/G_\mu$ of the symplectic manifolds $J^{-1}(\mu)/G_\mu$. On $J^{-1}(\mu)/G_\mu$ we have the Hamiltonian vector fields $(\pi_\mu)_*(X_H|J^{-1}(\mu))=X_{H_\mu}$ and on $\hat{I}_{h,\mu}$ the vector fields $(\pi_{h,\mu})_*(X_H|I_{h,\mu})$, where $\pi_{h,\mu}$: $I_{h,\mu}\to\hat{I}_{h,\mu}$ is the canonical projection. In fact, *$\hat{I}_{h,\mu}$ is just the energy surface $H_\mu^{-1}(h)$ for the reduced Hamiltonian system.*

But there may be cases where the rather restrictive conditions of 4.3.1 and 4.3.5 are not fulfilled and hence we do not know that $\hat{I}_{h,\mu}$ are manifolds. Even if $\hat{I}_{h,\mu}$ are not manifolds, but $X_H|I_{h,\mu}$ is complete, invariance under the symplectic group action assures that

$$\left(\Phi_g^{T^*}|I_{h,\mu}\right)\circ F_t=F_t\circ\left(\Phi_g^{T^*}|I_{h,\mu}\right)$$

where F_t is the flow of X_H. We can define a flow $\hat{F}_t$ on $\hat{I}_{h,\mu}$ by

$$\hat{F}_t([\alpha])=[F_t\alpha]$$

where $[\alpha]=\pi_{h,\mu}(\alpha)$.

Thus a version of the topological program for the energy-momentum mapping would be the study of the topological type of the sets $\hat{I}_{h,\mu}$ and the flows induced on them by X_H. If the conditions of 4.3.1 and 4.3.5 are satisfied, the dimensions of $I_{h,\mu}$ and $\hat{I}_{h,\mu}$ would be

$$\dim I_{h,\mu}=2\dim M-\dim G-1$$

$$\dim \hat{I}_{h,\mu}=2\dim M-\dim G-1-\dim G_\mu$$

It is clear now that working with the reduced "manifolds" $\hat{I}_{h,\mu}$ simplifies the problem considerably.

In what follows we want to construct these "invariant manifolds" and the bifurcation set. The first step is to characterize $\Sigma'_{H\times J}=(H\times J)(\sigma(H\times J))\subset\Sigma_{H\times J}$ (see 4.5.1).

4.5.3 Proposition.

$$\sigma(H\times J)=\sigma(J)\cup\left(\bigcup_{\mu\in\mathfrak{g}^*\setminus J(\sigma(J))}\sigma(H|J^{-1}(\mu))\right)$$

In other words, $\alpha\in T^*M$ is a critical point of $H\times J$ if and only if

ISBN 0-8053-0102-X

$T_\alpha J: T_\alpha(T^*M) \to \mathfrak{g}^*$ *is not surjective or* $\alpha \in J^{-1}(\mu)$ *is a critical point of* $H|J^{-1}(\mu): J^{-1}(\mu) \to \mathbf{R}$ *for some* $\mu \in \mathfrak{g}^* \backslash J(\sigma(J))$.

Proof. α is a critical point of $H \times J$ iff $T_\alpha(H \times J) = dH(\alpha) \times T_\alpha J: T_\alpha(T^*M) \to \mathbf{R} \times \mathfrak{g}^*$ is not surjective. Now $T_\alpha J$ is either surjective or not. If it is surjective then $\mu = J(\alpha)$ is a regular value of J, that is, $\mu \in \mathfrak{g}^* \backslash J(\sigma(J))$. By the Lagrange Multiplier Theorem (see 4.3.9), $T_\alpha(H \times J)$ not surjective is equivalent to $dH(\alpha) = 0$ on $\ker T_\alpha J = T_\alpha(J^{-1}(\mu))$, that is, α is a critical point of $H|J^{-1}(\mu)$. ■

Recall that from 4.3.8, critical points of $H \times J$ are exactly the relative equilibria. We prepare the following for later use.

4.5.4 Lemma. *(i) Let* $\Lambda = \{x \in M \mid J_x = J|T_x^*M: T_x^*M \to \mathfrak{g}^*$ *is not surjective*$\}$. *Then* $\Lambda = \{x \in M \mid \Xi_x: \mathfrak{g} \to T_xM,\ \xi \mapsto \xi_M(x)$ *is not injective*$\} = \{x \in M \mid \dim G_x \geqslant 1\}$, *where* $G_x = \{g \in G \mid \Phi(g,x) = x\}$ *is the isotropy group of the action of G on M;*

(ii) Λ *is closed and G-invariant.*

This result has been essentially given in the proof of 4.2.22, but we give it here for completeness.

Proof. (i) Recall that the Ad^*-equivariant momentum mapping $J: T^*M \to \mathfrak{g}^*$ is given by $J(\alpha_x)(\xi) = \alpha_x(\xi_M(x))$ for $x \in M$. Thus $\Xi_x^* = J_x$ and hence J_x not surjective is by linear algebra equivalent to Ξ_x not injective.

Now $\Xi_x: \mathfrak{g} \to T_xM$ not injective means that there exists $\xi \in \mathfrak{g} \backslash \{0\}$ such that $\Xi_x(\xi) = \xi_M(x) = 0$, that is, x is a fixed point of the flow of ξ_M (that is of $\Phi_{\exp t\xi}$) and hence this is equivalent to $\exp t\xi \in G_x$ for all $t \in \mathbf{R}$. Thus Ξ_x not injective is equivalent to the existence of a one-parameter subgroup of G_x, that is, to $\dim G_x \geqslant 1$.

(ii) Let $x \in \Lambda$ and $g \in G$. We want to prove that $y = \Phi(g,x) \in \Lambda$. By (i), $x \in \Lambda$ is equivalent to the existence of $\xi \in \mathfrak{g}$ with $\exp t\xi \in G_x$ for $t \in \mathbf{R}$, but then $\Phi(g(\exp t\xi)g^{-1}, y) = y$. Since $g(\exp t\xi)g^{-1} = \exp(tAd_g(\xi))$, we conclude that $\exp(tAd_g(\xi)) \in G_y$ for all $t \in \mathbf{R}$, that is, $y \in \Lambda$ and Λ is hence G-invariant.

Since J_x varies continuously with x, if J_{x_0} is surjective it will stay so in a neighborhood of x_0 which proves that $M \backslash \Lambda$ is open. ■

Because of this proposition $\Lambda \supset \tau_M^*(\sigma(J))$, that is, J has only regular values if Λ is excluded. We usually deal with $M \backslash \Lambda$ and Λ separately since in examples Λ can be readily worked out.

Recall from 4.3.3 that the reduced phase spaces for simple mechanical systems are symplectically embedded $P_\mu \subsetneq T^*M_\mu$, where $M_\mu = M/G_\mu$, if we can find a G_μ-invariant one-form α_μ on M with values in $J^{-1}(\mu)$. We shall now construct such an α_μ. Once we do this we shall examine the Hamiltonian induced on T^*M_μ.

ISBN 0-8053-0102-X

For $\mu \in \mathfrak{g}^*$, let the one-form $\alpha_\mu \in \Omega^1(M\setminus\Lambda)$ be defined by the following two conditions:

(1) $\alpha_\mu(x) \in J_x^{-1}(\mu) = J^{-1}(\mu) \cap T_x^*M$
(2) $K(\alpha_\mu(x)) = \inf_{\alpha \in J_x^{-1}(\mu)} K(\alpha)$, that is, $\alpha_\mu(x)$ is the minimum of $K|J_x^{-1}(\mu)$.

The existence and uniqueness of such an $\alpha_\mu(x)$ in $J_x^{-1}(\mu) = \{\alpha \in T_x^*M \mid \alpha(\xi_M(x)) = \mu(\xi)$ *for all* $\xi \in \mathfrak{g}\}$ follows from the well-known theorem guaranteeing the existence and uniqueness of elements of minimum norm in closed, convex sets of Hilbert space.*

The next result gives some basic properties of α_μ.

---

*For the convenience of the reader we shall give this standard proof.

Denote by

$$\varepsilon = \inf_{\alpha \in J_x^{-1}(\mu)} K(\alpha) = \tfrac{1}{2} \inf_{\alpha \in J_x^{-1}(\mu)} \|\alpha\|^2$$

where $\|\alpha\|^2 = \langle \alpha, \alpha \rangle_x$. Then for each $n \in N$ there exists $\alpha_\mu \in J_x^{-1}(\mu)$ such that

$$\varepsilon \leq K(\alpha_n) < \varepsilon + 1/n$$

and we can conclude that $\lim_{n\to\infty} K(\alpha_n) = \varepsilon$. We shall prove now that the sequence $\{\alpha_n\}$ constructed in this way is Cauchy and hence convergent to an element of $J_x^{-1}(\mu)$ since $J_x^{-1}(\mu)$ is closed. Since $\alpha_n, \alpha_m \in J_x^{-1}(\mu)$, then

$$\frac{\alpha_n + \alpha_m}{2} \in J_x^{-1}(\mu)$$

so that

$$\left\| \frac{\alpha_n + \alpha_m}{2} \right\|^2 \geq 2\varepsilon$$

We have

$$\left\| \frac{\alpha_n - \alpha_m}{2} \right\|^2 = 2\left\| \frac{\alpha_n}{2} \right\|^2 + 2\left\| \frac{\alpha_m}{2} \right\|^2 - \left\| \frac{\alpha_n + \alpha_m}{2} \right\|^2$$

$$\leq K(\alpha_n) + K(\alpha_m) - 2\varepsilon$$

$$< \varepsilon + \frac{1}{n} + \varepsilon + \frac{1}{m} - 2\varepsilon = \frac{1}{n} + \frac{1}{m}$$

and we conclude that the sequence $\{\alpha_n\} \subset J_x^{-1}(\mu)$ is Cauchy. Let $\alpha_\mu(x) = \lim_{n\to\infty} \alpha_n$. Then clearly $K(\alpha_\mu(x)) = \lim_{n\to\infty} K(\alpha_n) = \varepsilon$ and the existence of α_μ is proved. Finally, α_μ is unique. Indeed, if there were another $\alpha'_\mu(x)$ minimizing $K|J_x^{-1}(\mu)$, then

$$\left\| \frac{\alpha_\mu(x) - \alpha'_\mu(x)}{2} \right\|^2 = K(\alpha_\mu(x)) + K(\alpha'_\mu(x)) - \left\| \frac{\alpha_\mu(x) + \alpha'_\mu(x)}{2} \right\|^2$$

$$\leq 2\varepsilon - 2\varepsilon = 0$$

so that $\alpha_\mu(x) = \alpha'_\mu(x)$. ∎

In our case $J_x^{-1}(\mu)$ is an affine space. If $\beta_1, \beta_2 \in J_x^{-1}(\mu)$, we have $\langle \alpha_x(\mu), \beta_1 - \beta_2 \rangle = 0$. Indeed by the minimization property,

$$\langle \alpha_x(\mu), \alpha_x(\mu) \rangle \leq \langle \alpha_x(\mu) + t(\beta_1 - \beta_2), \alpha_x(\mu) + t(\beta_1 - \beta_2) \rangle$$

from which

$$2t\langle \alpha_x(\mu), \beta_1 - \beta_2 \rangle \leq t^2 \|\beta_1 - \beta_2\|^2 \qquad \text{for all } t \in R$$

which implies $\langle \alpha_x(\mu), \beta_1 - \beta_2 \rangle = 0$.

ISBN 0-8053-0102-X

4.5.5 Proposition. *(i)* $\alpha_\mu \in \Omega^1(M\backslash\Lambda)$; *that is,* α_μ *is a smooth one-form on* $M\backslash\Lambda$; $\alpha_\mu(x)$ *is the unique critical point of* $K|J_x^{-1}(\mu)$;

(ii) $\alpha_\mu(x)$ *is orthogonal to the subspace* $\ker J_x = J_x^{-1}(0)$ *with respect to the Riemannian metric* $\langle,\rangle_x$ *of* T_x^*M;

(iii) if $G_\mu = \{g\in G | Ad^*_{g^{-1}}\mu = \mu\}$ *is the isotropy subgroup of the co-adjoint action of G on* $\mathfrak{g}^*$, *then* α_μ *is* G_μ*-equivariant, that is,* $T^*\Phi_{g^{-1}}\circ\alpha_\mu = \alpha_\mu\circ\Phi_g$, *for all* $g\in G_\mu$.

Proof. (i) α_μ is the minimum of $K|J_x^{-1}(\mu)$, that is, it is the minimum of K subject to the constraint $J_x(\alpha_\mu(x)) = \mu$. Hence, using the Lagrange Multiplier Theorem (4.3.9) this minimum will be found among all solutions of the system:

$$\begin{cases} D(K+\xi\cdot J_x)(\alpha_\mu(x))\cdot\beta = 0 & \text{for all } \beta\in T_x^*M \\ J_x(\alpha_\mu(x)) = \mu \end{cases}$$

where $\xi\in\mathfrak{g}^{**}=\mathfrak{g}$, or, since $J_x\colon T_x^*M\to\mathfrak{g}^*$ is linear,

$$\begin{cases} \langle\alpha_\mu(x),\beta\rangle_x + \xi\cdot J_x(\beta) = 0 \\ J_x(\alpha_\mu(x)) = \mu \end{cases}$$

Hence by definition of the isomorphism $\gamma^\flat(x)\colon T_xM\to T_x^*M$, $\gamma^\flat(x)(v) = \langle\cdot,v\rangle_x$, we have

$$\begin{aligned} 0 &= \langle\alpha_\mu(x),\beta\rangle_x + \xi\cdot J_x(\beta) \\ &= \langle\gamma^\sharp(x)(\alpha_\mu(x)),\gamma^\sharp(x)(\beta)\rangle_x + (J_x^*\xi)\cdot\beta \\ &= \beta\cdot\gamma^\sharp(x)(\alpha_\mu(x)) + \beta\cdot(\Xi_x\xi) \quad \text{(notation of 4.5.4)} \end{aligned}$$

for all $\beta\in T_x^*M$. Thus

$$\alpha_\mu(x) = -(\gamma^\flat(x)\cdot\Xi_x)(\xi)$$

But $\mu = J_x(\alpha_\mu(x)) = -(J_x\circ\gamma^\flat(x)\circ\Xi_x)(\xi)$. We show now that $J_x\circ\gamma^\flat(x)\circ\Xi_x\colon \mathfrak{g}\to\mathfrak{g}^*$ is injective. Assume that $(J_x\circ\gamma^\flat(x)\circ\Xi_x)(\eta) = 0$. Then, in particular,

$$\begin{aligned} 0 &= (J_x\circ\gamma^\flat(x)\circ\Xi_x)(\eta)\cdot\eta \\ &= (J_x\gamma^\flat(x))(\eta_M(x))\cdot\eta \\ &= \gamma^\flat(x)(\eta_M(x))\cdot\eta_M(x) = \langle\eta_M(x),\eta_M(x)\rangle_x \end{aligned}$$

so that $\eta_M(x) = \Xi_x(\eta) = 0$. But since $x\in M\backslash\Lambda$, Ξ_x is injective by 4.5.4 and

ISBN 0-8053-0102-X

hence $\eta=0$. Since $\dim \mathfrak{g}^*=\dim \mathfrak{g}$ and $J_x\circ\gamma^\flat(x)\circ\Xi_x\colon \mathfrak{g}\to\mathfrak{g}^*$ is an injective linear map, it is an isomorphism, and so we can write

$$\xi=-\left(J_x\circ\gamma^\flat(x)\circ\Xi_x\right)^{-1}(\mu)$$

Thus $\alpha_\mu(x)$ is the *unique* critical point of $K|J_x^{-1}(\mu)$ and is given by

$$\alpha_\mu(x)=\left[\gamma^\flat(x)\circ\Xi_x\circ\left(J_x\circ\gamma^\flat(x)\circ\Xi_x\right)^{-1}\right](\mu)$$

Since all the maps on the right-hand side are clearly smooth in x, we conclude that α_μ is smooth.

(ii) Clearly $\ker J_x=J_x^{-1}(0)=J_x^{-1}(\mu)-\alpha_\mu(x)$, so that any $\beta\in\ker J_x$ can be written in the form $\beta=-\alpha_\mu(x)+\gamma$ for $\gamma\in J_x^{-1}(\mu)$. Thus,

$$\begin{aligned}\langle\alpha_\mu(x),\beta\rangle_x&=\langle\alpha_\mu(x),\gamma-\alpha_\mu(x)\rangle_x\\&=\langle\alpha_\mu(x),\gamma\rangle_x-\|\alpha_\mu(x)\|^2\geqslant 0\end{aligned}$$

because $\alpha_\mu(x)$ is the minimum of $K|J_x^{-1}(\mu)$. Hence $\langle\alpha_\mu(x),\beta\rangle_x\geqslant 0$ for all $\beta\in\ker J_x$. Reversing the sign of β, we must have $\langle\alpha_\mu(x),\beta\rangle_x=0$ for all $\beta\in\ker J_x$ (or use the last statement in the footnote on p. 344.)

(iii) For all $g\in G_\mu$, $T^*\Phi_{g^{-1}}(J_x^{-1}(\mu))\subset J_{\Phi_g(x)}^{-1}(\mu)$ because if $\beta_x\in J_x^{-1}(\mu)$, by Ad^* equivariance of J,

$$\begin{aligned}(J_{\Phi_g(x)}\circ T^*\Phi_{g^{-1}})(\beta_x)&=(Ad^*_{g^{-1}}\circ J_x)(\beta_x)\\&=Ad^*_{g^{-1}}(\mu)=\mu\end{aligned}$$

Thus $T^*\Phi_{g^{-1}}\colon J_x^{-1}(\mu)\to J^{-1}\Phi_g(x)(\mu)$ is a diffeomorphism with inverse $T^*\Phi_g$.

Since K is invariant under the action Φ^{T^*} and $\alpha_\mu(x)$ is a critical point of $K|J_x^{-1}(\mu)$, $T^*\Phi_{g^{-1}}(\alpha_\mu(x))$ is a critical point of $K|J_{\Phi_g(x)}^{-1}(\mu)$. But $\alpha_\mu(\Phi_g(x))$ is the *unique* critical point of $K|J_{\Phi_g(x)}^{-1}(\mu)$. Thus

$$T^*\Phi_{g^{-1}}\circ\alpha_\mu=\alpha_\mu\circ\Phi_g \qquad \blacksquare$$

Now we consider the symplectic embedding $\phi_\mu\colon P_\mu\to T^*M_\mu$ given by 4.3.3. (Recall that if μ is only assumed to be a regular value of J without the free and proper assumptions, this construction can only be done locally.) We know that the Hamiltonian H induces H_μ on P_μ where $H=H_\mu\circ\pi_\mu$ (see 4.3.5), and X_H induces X_{H_μ} on P_μ. In T^*M_μ we have $\hat{H}_\mu=H_\mu\circ\phi_\mu^{-1}$ induced on $\phi_\mu(P_\mu)$. Points in T^*M_μ are G_μ orbits of covectors α_x vanishing on $T_x(G_\mu\cdot x)$

ISBN 0-8053-0102-X

(see 4.3.2), so we can denote them by $G_\mu \cdot \alpha_x$. Then, by definition of ϕ_μ,

$$\begin{aligned}\hat{H}_\mu(G_\mu \cdot \alpha_x) &= H(\alpha_x + \alpha_\mu(x))\\ &= K(\alpha_x + \alpha_\mu(x)) + V(x)\\ &= K(\alpha_x) + 2\langle \alpha_x, \alpha_\mu(x)\rangle\\ &\quad + K(\alpha_\mu(x)) + V(x)\end{aligned}$$

By 4.5.5(ii), $\langle \alpha_x, \alpha_\mu(x)\rangle = 0$, so

$$\hat{H}_\mu(G_\mu \cdot \alpha_x) = K(\alpha_x) + K(\alpha_\mu(x)) + V(x)$$

Thus, it is natural to let

$$V_\mu(x) = K(\alpha_\mu(x)) + V(x)$$

so that $\hat{H}_\mu$ is the projection of $K + V_\mu$. Since $\phi_\mu(P_\mu) \subset T^*M_\mu$, (with equality if $G = G_\mu$), it makes sense to define a simple mechanical system on $\phi_\mu(P_\mu)$ as one obtained by restricting a simple mechanical Hamiltonian $K + V_\mu \circ \tau^*_{M_\mu}$ on T^*M_μ to $\phi_\mu(P_\mu)$. We shall say that we have a ***simple mechanical system in*** T^*M_μ. The following theorem of Satzer [1977] and Marsden now results.

4.5.6 Theorem *The reduction of a simple mechanical system on T^*M with $\Lambda = \varnothing$ is a simple mechanical system in T^*M_μ, where $M_\mu = M/G_\mu$. The kinetic energy of the reduced system is the function $\hat{K}$ induced from the G_μ-invariant kinetic energy function K on T^*M and the potential energy $\hat{V}_\mu$ is induced from the G_μ-invariant function on M defined by*

$$V_\mu(x) = V(x) + K(\alpha_\mu(x)) = H(\alpha_\mu(x))$$

which is called the ***effective potential,*** *or the* ***amended potential.***

Warning. The symplectic structure used on T^*M_μ is not necessarily the canonical one; see 4.3.3[†].

Since $\Lambda = \varnothing$ we are implicitly assuming that μ is a regular value of J and either the actions of G_μ on $J^{-1}(\mu)$ and M are free and proper or else we replace the statements by local ones. If $\Lambda \neq \varnothing$, we replace M by $M \backslash \Lambda$ and determine Λ separately.

[†] The motion in T^*M_μ is that of a particle in the potential V_μ plus an "electromagnetic" potential α_μ (see Exercise 3.7F). In several examples, (such as planar central force problems; Exercise 4.5C) α_μ is closed.

ISBN 0-8053-0102-X

For the version of this theorem suitable for the tangent bundle, see Exercise 4.5D.

For purposes of computation, notice that by definition of α_μ

$$V_\mu : M\backslash\Lambda \to \mathbf{R}$$

$$V_\mu(x) = V(x) + \inf_{\alpha \in J_x^{-1}(\mu)} K(\alpha)$$

$$= \inf_{\alpha \in J_x^{-1}(\mu)} H(\alpha)$$

At this point we recommend that the reader work Exercise 4.5C to get a feel for V_μ.

In T^*M_μ and in the subbundle $\phi(P_\mu)$ over M_μ, equilibrium points are points in the zero section that are critical points of $\hat{V}_\mu$. [The critical points of a Hamiltonian vector field on T^*M_μ are not affected by its noncanonical symplectic structure; see Exercise 3.7F(iii).] These critical points are obviously in one-to-one correspondence with critical orbits of V_μ. We note for later use that if a critical point of $\hat{V}_\mu$ is nondegenerate, the indices of $\hat{V}_\mu$ and V_μ along the corresponding nondegenerate critical manifold are equal (this follows easily from the definitions following 3.2.3).

By Definition 4.3.6, critical points of V_μ are in one-to-one correspondence (using the diffeomorphism induced by α_μ) with relative equilibria; that is, to critical points of $H \times J$ on $J^{-1}(\mu)$ or equivalently to critical points of $H|J^{-1}(\mu)$ (see 4.3.8).

We summarize the results obtained in the following.

4.5.7 Corollary. *(i) $V_\mu \circ \Phi_g = V_\mu$ for $g \in G_\mu$ and the set of critical points, $\sigma(V_\mu)$ is G_μ-invariant.*

(ii) $\sigma(H|J^{-1}(\mu)) = \alpha_\mu(\sigma(V_\mu))$ and is G_μ- and X_H-invariant.

The next basic theorem of Smale will characterize the set of critical values $\Sigma'_{H\times J}$ of the energy-momentum mapping in terms of the effective potential only.

4.5.8 Theorem. *Let μ be a regular value of J. Then*

$$\Sigma'_{H\times J|T^*(M\backslash\Lambda)} = \left\{(h,\mu) \in \mathbf{R} \times \mathfrak{g}^* \,|\, h \in V_\mu(\sigma(V_\mu))\right\}.$$

Proof. Recall that by 4.5.3,

$$\sigma(H \times J) = \sigma(J) \cup \bigcup_{\mu \in \mathfrak{g}^* \backslash J(\sigma(J))} \sigma(H|J^{-1}(\mu))$$

ISBN 0-8053-0102-X

By 4.5.7(ii), $\sigma(H|J^{-1}(\mu))=\alpha_\mu(\sigma(V_\mu))$ so that

$$(H\times J)\big(\sigma(H|J^{-1}(\mu))\big)=H\big(\sigma(H|J^{-1}(\mu))\big)\times\{\mu\}$$
$$=(H\alpha_\mu)\big(\sigma(V_\mu)\big)\times\{\mu\}$$
$$=V_\mu\big(\sigma(V_\mu)\big)\times\{\mu\}$$

Thus

$$(H\times J)\left(\bigcup_{\mu\in\mathfrak{g}^*\setminus J(\sigma(J))}\sigma(H|J^{-1}(\mu))\right)=\bigcup_{\mu\in\mathfrak{g}^*\setminus J(\sigma(J))}\big(V_\mu\big(\sigma(V_\mu)\big)\times\{\mu\}\big)$$
$$=\big\{(h,\mu)\in R\times\mathfrak{g}^*\setminus J(\sigma(J))\,|\,h\in V_\mu\big(\sigma(V_\mu)\big)\big\}$$

The statement now follows from the fact that $\tau_M^*(\sigma(J))\subset\Lambda$, observed in 4.5.4. ■

We now want to "construct" the invariant manifolds and the reduced invariant manifolds for the mechanical system with symmetry, up to a diffeomorphism. To do this, we shall need some constructions with vector bundles.

Let $p: E\to B$ be a vector bundle over a manifold B possibly with boundary ∂B, and $\langle,\rangle$ a Riemannian vector bundle metric, that is, a smooth map $b\mapsto\langle,\rangle_b$ that associates to each $b\in B$ a scalar product on the fiber E_b. $D_1(E)=\{v\in E\,|\,\|v\|\leq 1\}$ is the associated *unit disk bundle* and $S_1(E)=\{v\in E\,|\,\|v\|=1\}$ the associated *unit sphere bundle* of E. Clearly, if $\partial B=\varnothing$, $\partial D_1(E)=S_1(E)$. In $D_1(E)$ make the following construction: For each $x\in\partial B$, identify the whole fiber $p^{-1}(x)\cap D_1(E)$ with the point x. The space $\alpha(E)$ thus obtained is called the *reduced disk bundle* of E. Performing the identical construction with $S_1(E)$ we get the *reduced sphere bundle* $\beta(E)$ of E. Note that if $\partial B=\varnothing$, $\alpha(E)=D_1(E)$ and $\beta(E)=S_1(E)$. We shall prove in 4.5.14 that $\alpha(E)$ and $\beta(E)$ can be given smooth manifold structures, and that $\partial\alpha(E)=\beta(E)$; in particular, $\beta(E)$ is boundaryless.

If $M\times R^k\to M$ is the trivial bundle, we shall denote by $\alpha_k(M)=\alpha(M\times R^k)$ and $\beta_k(M)=\beta(M\times R^k)=\partial\alpha_k(M)$. We shall give some elementary examples that will be of use to us later in Sect. 10.6 when treating the topology of the three-body problem. Let

$$D^k=\{x\in R^k\,|\,\|x\|\leq 1\}\ \textit{and}\ S^{k-1}=\{x\in R^k\,|\,\|x\|=1\}.$$

(a) If $\partial M=\varnothing$, then $\alpha_k(M)\approx M\times D^k$, $\beta_k(M)\approx M\times S^{k-1}$.
(b) If $\partial M\neq\varnothing$, then $\beta_1(M)$ is the double of M, that is, the manifold obtained "glueing" together M to itself along ∂M, one copy of M having the reverse orientation, so that after glueing the resulting manifold is oriented and boundaryless.

ISBN 0-8053-0102-X

(c) $\beta_k(D^m) \approx S^{k+m-1}$.
(d) If $\partial M_1 = \varnothing$, then $\alpha_k(M_1 \times M_2) \approx M_1 \times \alpha_k(M_2)$, $\beta_k(M_1 \times M_2) \approx M_1 \times \beta_k(M_2)$.
(e) $\beta(TM)$ is the unit sphere bundle of M if M is boundaryless.

Further information on the α, β construction is given at the end of this section.

Let (M,K,V,G) be a mechanical system with symmetry with $\Lambda = \varnothing$. For $(h,\mu) \in \mathbf{R} \times \mathfrak{g}^*$, define

$$M_{h,\mu} = \{x \in M \mid V_\mu(x) \leqslant h\} = V_\mu^{-1}((-\infty, h])$$

If h is a regular value for V_μ, then $M_{h,\mu}$ is a smooth manifold with boundary $\partial M_{h,\mu} = V_\mu^{-1}(h)$. Let

$$E_{h,\mu} = \{\alpha \in T^*M \mid J(\alpha) = \mu,\ H(\alpha) \leqslant h\} = (H|J^{-1}(\mu))^{-1}((-\infty, h])$$

If h is a regular value of $V_\mu = H \circ \alpha_\mu$, then it is also a regular value of $H|J^{-1}(\mu)$, since $\sigma(H|J^{-1}(\mu)) = \alpha_\mu(\sigma(V_\mu))$. Hence $E_{h,\mu}$ is a submanifold (with boundary) of T^*M. We clearly have

$$\partial E_{h,\mu} = \{\alpha \in J^{-1}(\mu) \mid H(\alpha) = h\} = (H \times J)^{-1}(h,\mu) = I_{h,\mu}$$

4.5.9 Theorem (Invariant Manifold Theorem of Smale). *Given a mechanical system with symmetry (M, K, V, G) with $\Lambda = \varnothing$ and h a regular value of V_μ, the following hold:*

(i) $E_{h,\mu} = \{\alpha_x \in J^{-1}(\mu) \mid K(\alpha_x) - K(\alpha_\mu(x)) \leqslant h - V_\mu(x)\}$

$\partial E_{h,\mu} = I_{h,\mu}$

$\tau_M^*(E_{h,\mu}) \subset M_{h,\mu}$

[The last statement says that any orbit of momentum μ and energy $\leqslant h$ projects to $M_{h,\mu}$ and cannot cross the boundary $\partial M_{h,\mu} = V_\mu^{-1}(h)$.]

(ii) If $F = J^{-1}(0)$, then $\tau_M^*|F$: $F \to M$ is a vector subbundle of T^*M. Denote by $F|M_{h,\mu}$ its restriction to the submanifold $M_{h,\mu} \subset M$. Then if $(h,\mu) \notin \Sigma'_{H\times J}$, $E_{h,\mu}$ is diffeomorphic to $\alpha(F|M_{h,\mu})$; more precisely, there exists a G_μ-invariant diffeomorphism of manifolds with boundary $\phi_{h,\mu}$: $\alpha(F|M_{h,\mu}) \to E_{h,\mu}$.

(iii) The induced diffeomorphism on the boundaries

$$\partial\phi_{h,\mu} = \phi_{h,\mu}|\partial\alpha(F|M_{h,\mu}) = \phi_{h,\mu}|\beta(F|M_{h,\mu}) : \beta(F|M_{h,\mu}) \to I_{h,\mu}$$

is G_μ-equivariant.

(iv) If C is a nondegenerate critical manifold of $V_\mu|M_{h,\mu}$ of index λ, then $\alpha_\mu(C)$ is a nondegenerate critical submanifold of $H|E_{h,\mu}$ of the same index (see remarks following 4.5.6).

ISBN 0-8053-0102-X

Proof. (i) Since $V_\mu(x)=(H\circ\alpha_\mu)(x)=K(\alpha_\mu(x))+V(x)$, we have

$$H(\alpha_x)=K(\alpha_x)+V(x)=K(\alpha_x)-K(\alpha_\mu(x))+V_\mu(x)$$

so that $H(\alpha_x)\leqslant h$ is equivalent to

$$K(\alpha_x)-K(\alpha_\mu(x))\leqslant h-V_\mu(x)$$

and the first formula is proved. The second was shown above.

Let $\alpha_x\in T_x^*M\cap E_{h,\mu}$. To prove that $V_\mu(x)\leqslant h$, first note that $\alpha_x-\alpha_\mu(x)\in J_x^{-1}(0)$, so that by 4.5.5(ii) $\langle\alpha_x-\alpha_\mu(x),\alpha_\mu(x)\rangle=0$. Thus

$$\begin{aligned}K(\alpha_x)&=K(\alpha_x-\alpha_\mu(x)+\alpha_\mu(x))\\&=\tfrac{1}{2}\|\alpha_x-\alpha_\mu(x)\|^2+\tfrac{1}{2}\|\alpha_\mu(x)\|^2\\&=K(\alpha_x-\alpha_\mu(x))+K(\alpha_\mu(x))\end{aligned}$$

Then

$$\begin{aligned}V_\mu(x)&=V(x)+K(\alpha_\mu(x))\\&\leqslant V(x)+K(\alpha_\mu(x))+K(\alpha_x-\alpha_\mu(x))\\&=V(x)+K(\alpha_x)=H(\alpha_x)\leqslant h\end{aligned}$$

(ii) The map $J\times\tau_M^*: T^*M\to\mathfrak{g}^*\times M$ is a morphism of vector bundles of constant rank, since by hypothesis $\Lambda=\varnothing$, that is, J_x is surjective for all $x\in M$. Thus $J^{-1}(0)=ker(J\times\tau_M^*)$ is a subbundle of T^*M.

Let $D_1(F|M_{h,\mu})$ be the unit disk bundle associated to $M_{h,\mu}$. Define the map

$$\psi_{h,\mu}: D_1(F|M_{h,\mu})\to E_{h,\mu}$$

by

$$\psi_{h,\mu}(\alpha_x)=\sqrt{2(h-V_\mu(x))}\cdot\alpha_x+\alpha_\mu(x)$$

for $\alpha_x\in J_x^{-1}(0)$, $\|\alpha_x\|\leqslant 1$, $x\in M_{h,\mu}$. The first task is to prove that the definition of $\psi_{h,\mu}$ makes sense, that is, that $J_x(\psi_{h,\mu}(\alpha_x))=\mu$, $H(\psi_{h,\mu}(\alpha_x))\leqslant h$. We have

$$J_x(\psi_{h,\mu}(\alpha_x))=\sqrt{2(h-V_\mu(x))}\cdot J_x(\alpha_x)+J_x(\alpha_\mu(x))=\mu$$

by definition of $\alpha_\mu(x)$ and the fact that $\alpha_x\in J_x^{-1}(0)$. Now

$$\begin{aligned}H(\psi_{h,\mu}(\alpha_x))&=K(\psi_{h,\mu}(\alpha_x))+V(x)\\&=2(h-V_\mu(x))\cdot\tfrac{1}{2}\|\alpha_x\|^2+K(\alpha_\mu(x))+V(x)\\&\leqslant h-V_\mu(x)+V_\mu(x)\quad \textit{using } \|\alpha_x\|\leqslant 1\\&=h\end{aligned}$$

ISBN 0-8053-0102-X

If $\alpha_x \in D_1(F|M_{h,\mu})$ with $x \in \partial M_{h,\mu} = V_\mu^{-1}(h)$, then $\psi_{h,\mu}(\alpha_x) = \alpha_\mu(x)$. In other words, $\psi_{h,\mu}$ sends the whole fiber of $D_1(F|M_{h,\mu})$ over $x \in \partial M_{h,\mu} = V_\mu^{-1}(h)$ to the single point $\alpha_\mu(x) \in \partial E_{h,\mu} = I_{h,\mu}$. Denoting by $\rho_{h,\mu}: D_1(F|M_{h,\mu}) \to \alpha(F|M_{h,\mu})$ the canonical projection, we have defined a map

$$\phi_{h,\mu}: \alpha(F|M_{h,\mu}) \to E_{h,\mu}$$

by $\phi_{h,\mu} \circ \rho_{h,\mu} = \psi_{h,\mu}$. Since $\psi_{h,\mu}$ and $\rho_{h,\mu}$ are smooth and $\alpha(F|M_{h,\mu})$ is a quotient manifold, $\phi_{h,\mu}$ is smooth, too. The inverse of $\psi_{h,\mu}$ is given by

$$\phi_{h,\mu}^{-1}(\alpha_x) = \begin{cases} \dfrac{\alpha_x - \alpha_\mu(x)}{\sqrt{2(h - V_\mu(x))}} & \text{if } x \notin M_{h,\mu} = V_\mu^{-1}(h) \\ \rho_{h,\mu}(0_x) & \text{if } x \in \partial M_{h,\mu} \end{cases}$$

and is smooth.

Since J is Ad^*-equivariant and G acts by isometries, $E_{h,\mu}$ and $D_1(F|M_{h,\mu})$ are G_μ-invariant, where $G_\mu = \{ g \in G | Ad^*_{g^{-1}}\mu = \mu \}$. By 4.5.5(iii) and 4.5.7(ii), we can write:

$$\begin{aligned} \psi_{h,\mu}(T^*\Phi_{g^{-1}}(\alpha_x)) &= \sqrt{2(h - V_\mu(\Phi_g(x)))} \cdot T^*\Phi_{g^{-1}}(\alpha_x) + \alpha_\mu(\Phi_g(x)) \\ &= \sqrt{2(h - V_\mu(x))} \cdot T^*\Phi_{g^{-1}}(\alpha_x) + T^*\Phi_{g^{-1}}(\alpha_\mu(x)) \\ &= T^*\Phi_{g^{-1}}\left(\sqrt{2(h - V_\mu(x))} \cdot \alpha_x + \alpha_\mu(x)\right) \\ &= T^*\Phi_{g^{-1}}(\psi_{h,\mu}(\alpha_x)) \end{aligned}$$

that is, $\psi_{h,\mu}$ is equivariant with respect to the action Φ^{T^*} of G_μ. We want to show that this action induces one on $\alpha(F|M_{h,\mu})$. For this purpose, note that if $g \in G_\mu$ and $\alpha_x \in D_1(F|M_{h,\mu})$ with $x \in \partial M_{h,\mu} = V_\mu^{-1}(h)$, then $\Phi_g^{T^*}(\alpha_x) = T^*\Phi_{g^{-1}}(\alpha_x) \in T^*_{\Phi(g,x)}M$ and by 4.5.7(i), $\Phi(g,x) \in \partial M_{h,\mu} = V_\mu^{-1}(h)$. In other words, the action Φ^{T^*} sends the fiber over $x \in \partial M_{h,\mu}$ to the fiber over $\Phi(g,x) \in \partial M_{h,\mu}$. We can then define the action Ψ of G_μ on $\alpha(F|M_{h,\mu})$ by the formula

$$\Psi_g \circ \rho_{h,\mu} = \rho_{h,\mu} \circ \Phi_g^{T^*} = \rho_{h,\mu} \circ T^*\Phi_{g^{-1}}$$

Clearly we have

$$\Phi_g^{T^*} \circ \phi_{h,\mu} = \phi_{h,\mu} \circ \Psi_g \quad \text{for all } g \in G_\mu$$

and the equivariance statement is proved.

ISBN 0-8053-0102-X

(iii) $\partial E_{h,\mu} = I_{h,\mu}$ is G_μ-invariant since J is Ad^*-equivariant and $H \circ \Phi_g^{T^*} = H$ for all $g \in G$.

Since G acts by isometries, we conclude that $\Phi_g^{T^*}(S_1(F|M_{h,\mu})) \subset S_1(F|M_{h,\mu})$ so that $\beta(F|M_{h,\mu})$ is invariant under the action Ψ of G_μ on $\alpha(F|M_{h,\mu})$. Thus $\partial\phi_{h,\mu}: \beta(F|M_{h,\mu}) \to I_{h,\mu}$ is a G_μ-equivariant diffeomorphism.

(iv) If C is a nondegenerate critical submanifold of $V_\mu|M_{h,\mu}$, then since $\alpha_\mu: M \to \operatorname{graph}\alpha_\mu \subset T^*M$ is a diffeomorphism, $\alpha_\mu(C)$ will be a submanifold of T^*M and clearly $\alpha_\mu(C) \subset E_{h,\mu}$. Since $\partial C = \varnothing$, $\partial\alpha_\mu(C) = \varnothing$, and since $C \cap \partial M_{h,\mu} = \varnothing$, $\alpha_\mu(C) \cap \alpha_\mu(\partial M_{h,\mu}) = \alpha_\mu(C) \cap \partial E_{h,\mu} = \varnothing$; $\partial C = \varnothing$ and $C \cap \partial M_{h,\mu} = \varnothing$ are part of the definition of C as a critical submanifold. The rest of the statement is clear by 4.5.7(ii) since α_μ is a diffeomorphism onto its image and $V_\mu = H \circ \alpha_\mu$. ■

It was stressed throughout this section that all defined objects are G_μ-invariant so that a priori reduction is possible. Since by reducing, the dimensions of the manifolds become lower, it would be interesting to have a "topological" theorem similar to the above for the reduced invariant manifolds. We shall write $\hat{M}_{h,\mu} = M_{h,\mu}/G_\mu$, $\hat{E}_{h,\mu} = E_{h,\mu}/G_\mu$ and let $\hat{\tau}_M^*|E_{h,\mu}$ be the map for which the diagram

$$\begin{array}{ccc} E_{h,\mu} & \longrightarrow & \hat{E}_{h,\mu} \\ {\scriptstyle \tau_M^*|E_{h,\mu}}\downarrow & & \downarrow{\scriptstyle \hat{\tau}_M^*|\hat{E}_{h,\mu}} \\ M_{h,\mu} & \longrightarrow & \hat{M}_{h,\mu} \end{array}$$

commutes; the horizontal arrows are the canonical projections. Define by passing to the quotients: $\hat{H}: \hat{E}_{h,\mu} \to \mathbf{R}$, $\hat{\alpha}_\mu: \hat{M} = M/G_\mu \to \widehat{T^*M} = T^*M/G_\mu$, and $\hat{V}_\mu = \hat{H} \circ \hat{\alpha}_\mu: \hat{M}_{h,\mu} \to \mathbf{R}$. Note that $\partial\hat{E}_{h,\mu} = \hat{I}_{h,\mu}$.

4.5.10 Theorem (Reduced Invariant manifold theorem of Smale). *Assume G_μ acts freely and properly on $M_{h,\mu}$ and $E_{h,\mu}$. Then $\hat{M}_{h,\mu}$ and $\hat{E}_{h,\mu}$ are manifolds. Assume that $\hat{V}_\mu: \hat{M}_{h,\mu} \to \mathbf{R}$ has nondegenerate critical points; then:*

(i) If $\hat{x} \in \hat{M}_{h,\mu}$ is a nondegenerate critical point of $\hat{V}_\mu$, then $\hat{\alpha}_\mu(\hat{x})$ will be a nondegenerate critical point of $\hat{H}|\hat{E}_{h,\mu}$ of the same index. This index equals the index of V_μ on the nondegenerate critical manifold $\pi_\mu^{-1}(\hat{x}) \subset M_{h,\mu}$, where $\pi_\mu: M_{h,\mu} \to M_{h,\mu}/G_\mu$ is the projection.

(ii) If the vector bundle $J^{-1}(0)|M_{h,\mu}$ is trivial (i.e., isomorphic to $M_{h,\mu} \times \mathbf{R}^S$), then

$$I_{h,\mu} = \beta_S(M_{h,\mu}) \quad \text{and} \quad \hat{I}_{h,\mu} = \beta_S(\hat{M}_{h,\mu})$$

The proof is straightforward from what we have done and will be left as an exercise.

We conclude our study of the topology of the invariant and reduced invariant manifolds with an example that completely describes them in a simple case.

ISBN 0-8053-0102-X

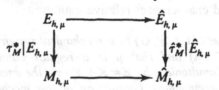

4.5.11 Example. Assume G_μ acts freely and properly on $M_{h,\mu}$ and that $\hat{M}_{h,\mu}$ is a compact manifold on which $\hat{V}_\mu|\hat{M}_{h,\mu}$ has a unique, nondegenerate critical point $\hat{x}_0$, which is a minimum, satisfying $\hat{V}_\mu(\hat{x}_0)<h$. Then $\hat{M}_{h,\mu}\approx D^m$, $M_{h,\mu}\approx G_\mu\times D^m$, $\hat{E}_{h,\mu}\approx D^j$, $E_{h,\mu}\approx G_\mu\times D^j$, and hence $I_{h,\mu}\approx G_\mu\times S^{j-1}$, $\hat{I}_{h,\mu}\approx S^{j-1}$, where $m=\dim M-\dim G_\mu$ and $j=2\dim M-\dim G-\dim G_\mu$.

This follows from the simple result in Morse theory which states that if M is a compact manifold with boundary on which f has a single nondegenerate interior minimum and f is constant on ∂M, then M is a disk. (See Exercise 4.5B.)

The next theorem gives additional characterizations of relative equilibria for simple mechanical systems with symmetry. Recall that under the hypotheses of 4.3.1 the set of relative equilibria coincides with the set $\sigma(H\times J)$ of critical points of the energy-momentum mapping $H\times J:\ T^*M\to\mathbf{R}\times\mathfrak{g}^*$, where $H=K+V\circ\tau_M^*$ is the Hamiltonian of the mechanical system with symmetry (M,K,V,G) and $J:\ T^*M\to\mathfrak{g}^*$ is the momentum mapping of the induced cotangent action of G on T^*M, that is, $J(\alpha_x)(\xi)=\alpha(\xi_M(x))$. Thus the part $\Sigma'_{H\times J}$ of the bifurcation set $\Sigma_{H\times J}$ is determined by the set $\sigma(H\times J)$ of relative equilibria. The next theorem is essentially due to S. Smale [1970a] and J. Robbin [1973], and characterizes relative equilibria.

4.5.12 Theorem. *Let (M,K,V,G) be a mechanical system with symmetry and assume that $\alpha_{x_0}\in J^{-1}(\mu)$ and that μ is a regular value of J. Denote by $H=K+V\circ\tau_M^*$ the Hamiltonian, by $E=K+V\circ\tau_M$ the energy function, and by $L=K-V\circ\tau_M$ the corresponding Lagrangian of this mechanical system. $\gamma^b=FL:\ TM\to T^*M$ denotes the Legendre transformation so that $H\circ FL=E$. The following are equivalent:*

(i) *α_{x_0} is a relative equilibrium;*
(ii) *there exists $\xi\in\mathfrak{g}$ satisfying $(J\circ\gamma^b_{x_0}\circ\xi_M)(x_0)=\mu$ such that $X_E(\xi_M(x_0))=\xi_{TM}(\xi_M(x_0))$;*
(iii) *there exists $\xi\in\mathfrak{g}$ satisfying $(J\circ\gamma^b_{x_0}\circ\xi_M)(x_0)=\mu$ such that x_0 is a critical point of $L\circ\xi_M:\ M\to\mathbf{R}$;*
(iv) *there exists $\xi\in\mathfrak{g}$ satisfying $(J\circ\gamma^b_{x_0}\circ\xi_M)(x_0)=\mu$ such that x_0 is a critical point of $V-K\circ\gamma^b\circ\xi_M:\ M\to\mathbf{R}$;*
(v) *x_0 is a critical point of the amended potential V_μ and $\alpha_{x_0}=\alpha_\mu(x_0)$.*

Proof. Equivalence of (i) and (ii) is clear since the Legendre transformation is a symplectic diffeomorphism. In detail, according to Proposition 4.3.7, α_{x_0} is a relative equilibrium iff there exists $\xi\in\mathfrak{g}$ such that $X_H(\alpha_{x_0})=\xi_{T^*M}(\alpha_{x_0})$. But since $L=K+V\circ\tau_M$, $FL=\gamma^b$ and since G acts by isometries $\Phi_g^{T^*}\circ\gamma^b=\gamma^b\circ\Phi_g^T$ as is easily checked; here $\Phi:\ G\times M\to M$ denotes the action of G on M. By 3.6.2, $X_H\circ FL=TFL\circ X_E$ and by 4.1.26, $\xi_{T^*M}\circ\gamma^b=T\gamma^b\circ\xi_{TM}$. Let $\alpha_{x_0}=\gamma^b(v_{x_0})$. Then

$$X_H(\alpha_{x_0})=(X_H\circ FL)(v_{x_0})=(TFL\circ X_E)(v_{x_0})$$

ISBN 0-8053-0102-X

and

$$\xi_{T^*M}(\alpha_{x_0})=\xi_{T^*M}(\gamma^\flat(v_{x_0}))=(T\gamma^\flat\circ\xi_{TM})(v_{x_0})=(TFL\circ\xi_{TM})(v_{x_0})$$

Since FL is in this case an isomorphism of vector bundles, $X_H(\alpha_{x_0})=\xi_{T^*M}(\alpha_{x_0})$ is equivalent to

$$X_E(v_{x_0})=\xi_{TM}(v_{x_0}),\quad \alpha_{x_0}=\gamma^\flat(v_{x_0})$$

But the last equality implies $v_{x_0}=\xi_M(x_0)$ because X_E is a second-order equation so that $T\tau_M\circ X_E=id_{TM}$ and hence

$$v_{x_0}=(T\tau_M\circ X_E)(v_{x_0})=(T\tau_M\circ\xi_{TM})(v_{x_0})=\xi_M(x_0)$$

Thus (i) and (ii) are equivalent.

We proved the equivalence of (i) and (v) in 4.5.7(iii), and the equivalence of (iii) and (iv) is clear. (Recall that the kinetic energy on T^*M is actually $K\circ\gamma^\sharp$, even though we denoted it always simply by K as in the tangent bundle.)

Finally, we show that (ii) and (iv) are equivalent. (This argument is due to A. Weinstein; cf. Robbin [1973] for another proof.) Condition (ii) is equivalent to

$$(X_E-X_{\hat{J}(\xi)})(\xi_M(x_0))=0$$

where J is the momentum mapping for the tangent bundle. This is, in turn, equivalent to

$$d(E-\hat{J}(\xi))(\xi_M(x_0))=0$$

that is, $d(E-\hat{J}(\xi))(\xi_M(x_0))\cdot w_v=0$ for all $w_v\in T_vTM$. Now $E-\hat{J}(\xi)=V\circ\tau_M+K-\hat{J}(\xi)$ and clearly $d(V\circ\tau_M)(\xi_M(x_0))w_v=dV(x_0)v$. However, we also have

$$d(K-\hat{J}(\xi))(\xi_M(x_0))w_v=-d(K\circ\xi_M)(x_0)v$$

where $w_v\in T_vTM$. [This identity is easily verified in coordinates: use $K(v)=\frac{1}{2}\gamma_{ij}v^iv^j$ and $\hat{J}(\xi)(v)=\gamma_{ij}\xi^i_M(x)v^j$.] Thus (ii) is equivalent to $dV(x_0)\cdot v-d(K\circ\xi_M)(x_0)\cdot v=0$ for all v, that is, (iv). ■

We conclude this section with some additional information on the α,β constructions (see remarks preceding 4.5.9). These results will be used in Sect. 10.4.

4.5.13 Proposition. *Let M be a manifold without boundary and $\pi: E\to M$ a vector bundle with a Riemannian bundle metric $\langle,\rangle$, that is, for each $x\in M$, $\langle,\rangle_x$ is inner product on the vector space $E_x=\pi^{-1}(x)$ depending smoothly on*

ISBN 0-8053-0102-X

$x \in M$. Let $f \in \mathfrak{F}(M)$ and $c \in \mathbf{R}$ be a regular value for f. Define $g \in \mathfrak{F}(E)$ by $g(v) = \langle v, v \rangle + (f \circ \pi)(v)$, $v \in E$. Then:

(*i*) *c is a regular value for g if and only if c is a regular value for f; $g^{-1}((-\infty, c])$ is a smooth manifold with boundary $g^{-1}(c)$;*

(*ii*) *$g^{-1}((-\infty, c])$ is homeomorphic to $\alpha(E|f^{-1}((-\infty, c]))$, where $E|f^{-1}((-\infty, c])$ is the restriction of the vector bundle E to the submanifold $f^{-1}((-\infty, c])$ of M; in this way $\alpha(E|f^{-1}((-\infty, c]))$ becomes a smooth manifold making the homeomorphism above into a diffeomorphism;*

(*iii*) *if $\pi_1: E_1 \to M$ is another Riemannian vector bundle over M and $f_1 \in \mathfrak{F}(M)$ is another smooth map such that c is a regular value for both f and f_1, $f_1^{-1}((-\infty, c]) = f^{-1}((-\infty, c])$, and $E_1|f_1^{-1}((-\infty, c])$ and $E|f^{-1}((-\infty, c])$ are isomorphic as vector bundles, then $\alpha(E_1|f_1^{-1}((-\infty, c]))$ is diffeomorphic to $\alpha(E|f^{-1}((-\infty, c]))$.*

Proof. (i) In a vector bundle chart of E with model space $\mathbf{E}$ for M and fiber $\mathbf{F}$, write the metric as

$$\langle , \rangle : U \subset \mathbf{E} \to L^2_{sym}(\mathbf{F}; \mathbf{R}); \; x \mapsto \langle , \rangle_x$$

(the local representative of $\langle , \rangle$), that is, $\langle , \rangle$ is a map that associates to each $x \in U$ an inner product in $\mathbf{F}$. In this chart, $g \in \mathfrak{F}(E)$ is written as

$$g(x, v) = \langle v, v \rangle_x + f(x)$$

so that $(x_0, v_0) \in U \times \mathbf{F}$ is a critical point for g iff

$$(D_1 g)_{(x_0, v_0)} \cdot w = (D\langle , \rangle)_{x_0}(v_0, v_0) \cdot w + (Df)(x_0) \cdot w = 0$$

$$(D_2 g)_{(x_0, v_0)} \cdot w = 2\langle v_0, w \rangle_{x_0} = 0$$

for all $w \in \mathbf{F}$. The second equation implies $v_0 = 0$ and then the first says that $(Df)(x_0) = 0$. Thus $\sigma(g) = \pi^{-1}(\sigma(f)) \subset M$, M being regarded here as the zero section of $\pi: E \to M$. Thus $c \in \mathbf{R}$ is a regular value for g iff c is a regular value for f. The rest of the statement is clear.

(ii) Let $D_1|f^{-1}((-\infty, c])$ be the unit disk bundle of E restricted to $f^{-1}((-\infty, c])$. Define the map

$$h: D_1|f^{-1}((-\infty, c]) \to g^{-1}((-\infty, c])$$

$$h(v) = (c - (f \circ \pi)(v))^{1/2} \cdot v$$

[Note that $c - (f \circ \pi)(v) \in \mathbf{R}$ so that h is essentially a rescaling.] If $\pi(v) \in f^{-1}(c) = \partial(f^{-1}((-\infty, c]))$, then $h(v) = 0$. Thus h induces a continuous map $\hat{h}$: $\alpha(E|f^{-1}((-\infty, c])) \to g^{-1}((-\infty, c])$ with inverse $\hat{h}^{-1}$: $g^{-1}((-\infty, c]) \to$

ISBN 0-8053-0102-X

$\alpha(E|f^{-1}((-\infty,c]))$ given by

$$\hat{h}^{-1}(u)=\begin{cases}(c-(f\circ\pi)u)^{-1/2}u & \text{if } \pi(u)\notin f^{-1}(c)\\ \hat{0}_{\pi(u)} & \text{if } \pi(u)\in f^{-1}(c)\end{cases}$$

where $\hat{0}_{\pi(u)}$ denotes the equivalence class of any $v\in E_{\pi(u)}$, $\|v\|<1$, $(f\circ\pi)(u)=c$. It is easily verified that $\hat{h}^{-1}$ is continuous, so that $\hat{h}$ is a homeomorphism.

(iii) It is clear that if $v\in g^{-1}((-\infty,c])$, then $g(v)=\|v\|^2+(f\circ\pi)(v)<c$ so that $\pi(v)\in f^{-1}((-\infty,c])$. If $k: E|f^{-1}((-\infty,c])\to E_1|f_1^{-1}((-\infty,c])$ denotes the isomorphism of vector bundles in the hypothesis, change the inner product in the second bundle so that k becomes an isometry. $k|g^{-1}((-\infty,c])$ is then a diffeomorphism between $g^{-1}((-\infty,c])$ and $g_1^{-1}((-\infty,c])$ (g_1 corresponding to E_1 with inner product changed, and to f_1) so that by (ii) the result follows. ■

4.5.14 Corollary. *For any Riemannian vector bundle $\pi: E\to M$, the reduced disk bundle $\alpha(E)$ and the reduced sphere bundle $\beta(E)$ have a natural smooth manifold structure and, $\partial\alpha(E)=\beta(E)$.*

Proof. We distinguish two cases: $\partial M=\varnothing$ and $\partial M\neq\varnothing$. If $\partial M=\varnothing$, then $\alpha(E)=D_1(E)$ and there is nothing to prove. If $\partial M\neq\varnothing$, a partition of unity argument shows that there exists $f\in\mathfrak{F}(M)$ such that $0\in\mathbf{R}$ is a regular value for f, $f^{-1}(0)=\partial M$ and $f(x)<0$ for all $x\in M$. Thus $M=f^{-1}((-\infty,0])$. Now take M_1 to be the double of M, that is, the manifold obtained by "glueing" together two copies of M along their boundaries, and define a vector bundle $\pi_1: E_1\to M_1$ such that $E_1|M=E$ in each copy of M in M_1 and identifying the fibers of E over ∂M (for this construction as well as a rigorous definition of the double of a manifold, see Hirsch [1976]). A choice of a Riemannian metric on E_1 and the previous proposition shows that $\alpha(E)$ is a manifold. ■

It will be useful in Sect. 10.4 to have an alternative description of $\alpha(E)$ using another function g. This is given by the following.

4.5.15 Proposition (Smale). *Let M, $\pi: E\to M$, f, c be as in Proposition* 4.5.13. *Assume there exists a map $h\in\mathfrak{F}(E)$ satisfying:*

(*i*) *for each $x\in M$, $h_x=h|E_x: E_x\to\mathbf{R}$, $E_x=\pi^{-1}(x)$ is a proper map and has a unique nondegenerate minimum at the origin $0_x\in E_x$;*
(*ii*) *$f(x)=\min\{h(v)|v\in E_x\}=h(0_x)$ for all $x\in M$.*

Then c is a regular value for h and $\alpha(E|f^{-1}((-\infty,c])$ is diffeomorphic to $h^{-1}((-\infty,c])$.

Proof. As before, working in a vector bundle chart with E a model for M and F a model for the fiber, $(x_0,v_0)\in U\times F\subset E\times F$ is a critical point for h iff

$$D_1h(x_0,v_0)\cdot w=D(h|U\times\{v_0\})(x_0)w=0$$

ISBN 0-8053-0102-X

and

$$D_2h(x_0, v_0)\cdot w = Dh_{x_0}(v_0)\cdot w = 0$$

for all $w \in F$. By (i), h_{x_0} has a *unique* critical point that is a nondegenerate minimum and this unique critical point is 0_{x_0}. Thus $h|U\times\{0_{x_0}\}=f$ and the first condition reads $Df(x_0)=0$. Thus $\sigma(h)=\pi^{-1}(\sigma(f))\cap M$, M regarded as the zero section of E. Thus, if c is a regular value for f, it is also a regular value for h.

From (i) we conclude that $h_x^{-1}((-\infty,c])$ is diffeomorphic to a disk [or is empty for $c<h(0_x)$] (see Exercise 4.5B). Also note that $\pi(h^{-1}((-\infty,c])=f^{-1}((-\infty,c])$. Let now $g\in\mathfrak{F}(E)$ be as in Proposition 4.5.13, that is, $g(v)=\|v\|^2+f(x)$, $x=\pi(v)$. The proposition will be proved if we show that $g^{-1}((-\infty,c])$ is diffeomorphic to $h^{-1}((-\infty,c])$. Denote by $r(x)$ the radius of the disk $h_x^{-1}((-\infty,c])$; $r(x)$ is a continuous function of x [and smooth where $h(x)<c$]. Recall that $g_x^{-1}((-\infty,c])$ is a disk of radius $\sqrt{c-f(x)}$. Define

$\varphi: g^{-1}((-\infty,c])\to h^{-1}((-\infty,c])$

$$\varphi(v)=\begin{cases}\dfrac{r(x)}{\sqrt{c-f(x)}}v & \text{if } \pi(v)=x\notin f^{-1}(c)\\ 0 & \text{if } \pi(v)=x\in f^{-1}(c) \text{ (in which case } v=0)\end{cases}$$

The definition of φ makes sense because

$$\left\|\frac{r(x)}{\sqrt{c-f(x)}}v\right\|^2=\frac{r^2(x)}{c-f(x)}\|v\|^2<\frac{r^2(x)}{c-f(x)}(c-f(x))=r^2(x)$$

implies that

$$\frac{r(x)}{\sqrt{c-f(x)}}v$$

belongs to the disk of radius $r(x)$ over $x\in f^{-1}((-\infty,c])$, hence $h(\varphi(0))<c$. The map φ is smooth since it is a fiber rescaling (this requires a short argument we leave to the reader). It clearly has an inverse

$$\varphi^{-1}(v)=\begin{cases}\dfrac{\sqrt{c-f(x)}}{r(x)}v & \text{if } \pi(v)\notin f^{-1}(c)\\ 0 & \text{if } \pi(v)\in f^{-1}(c) \text{ in which case } v=0\end{cases}$$

which is smooth. ■

ISBN 0-8053-0102-X

EXERCISES

4.5A. Complete the proof of 4.5.10.

4.5B. Let M be compact with boundary $f: M \to R$ have a single nondegenerate interior minimum and be constant on ∂M. Prove M is a disk. (*Hint:* Let F_t be the flow of $-\nabla f$; show that F_t for t large gives a diffeomorphism between M and a disk about the critical point. Alternatively, use 3.2.17.)

4.5C. Let S^1 act on $R^2\backslash\{0\}$ by rotations and let V be a potential on R^2 invariant under S^1, so V is a function of the radial coordinate r. If $K(v)=\frac{1}{2}m\|v\|^2$, show that the amended potential is given by

$$V_l(r)=V(r)+\frac{1}{2}\frac{l^2}{mr^2}$$

Verify 4.5.6 by direct computation, that is, show that motion in the plane under the potential V is governed by motion on the half line $r>0$ under the potential V_l. (Hint: See Goldstein [1950], Sect. 3.3.)

4.5D. Prove the analogs of 4.3.3 and 4.5.6 for simple mechanical systems on tangent bundles. Use a Riemannian structure on M_μ that makes the projection $M \to M_\mu$ a Riemannian submersion (see the footnote to 4.3.3).

4.5E. Determine the topology for the Hamiltonian on T^*R^2 given by

$$H(x,y,p_x,p_y)=\tfrac{1}{2}\left(p_x^2+p_y^2\right)+\tfrac{1}{2}\left(x^2+y^2\right)$$

for the usual S^1 action (by rotations) on R^2 with

$$J(x,y,p_x,p_y)=xp_y-yp_x$$

In particular, show that the critical manifolds of J on $H^{-1}(h)$ for $h>0$ are the curves

$$\gamma_{h,+}=\{(x,y,-y,x)\,|\,x^2+y^2=h^2\}$$

and

$$\gamma_{h,-}=\{(x,y,y,-x)\,|\,x^2+y^2=h^2\}$$

(periodic orbits of X_H). (The linking number of $\gamma_{h,+}$ with $\gamma_{h,-}$ in $H^{-1}(h)$ may be shown to be -1.)

4.5F. (Spherical Pendulum) Carry out the topological program for

$M=S^2$, with the standard Euclidean metric

$V=$ height function on S^2

$G=S^1=SO(2)$, acting on M by rotations about the z-axis

Use spherical coordinates (θ,ϕ) on S^2 and show that the effective potential is

$$V_\mu(\theta,\phi)=1+\cos\theta+\frac{\mu^2}{2\sin^2\theta}$$

ISBN 0-8053-0102-X

Show that there is exactly one relative equilibrium for the reduced system. Let

$$\tilde{V}_\mu(z) = 1 + z + \frac{\mu^2}{2(1-z^2)}, \quad -1 < z < 1$$

and let $h(\mu)$ be its minimum. Show

(i) if $h < h(\mu)$, $M_{h,\mu} = \varnothing$, $I_{h,\mu} = \varnothing$;

(ii) if $h = h(\mu)$, $M_{h,\mu} = S^1$, $I_{h,\mu} = S^1$;

(iii) if $h > h(\mu)$, $M_{h,\mu} = S^1 \times [a,b]$, $I_{h,\mu} = T^2$.

Also, show

(iv) $\Sigma' = \{(h,\mu) | h = h(\mu)\}$; and

(v) $\Sigma = \Sigma' \cup \{(h,\mu) | h = 0, h = 2\} \cup \{(h,\mu) | \mu = 0\}$.

Using the equations of motion in V_μ for the reduced system, express the solutions of the full system in terms of those of the reduced system; see the remarks following 4.3.5. (See Iacob [1973] and Cushman [1975] for further information.)

4.5G. Show that nondegenerate maxima or minima of the *reduced* amended potential give stable relative equilibria.

4.6 THE TOPOLOGY OF THE RIGID BODY

We now carry out pieces of the topological program for the rigid body. For a slightly easier example to start with, the reader may wish to first consult Sect. 9.8 on the topology of the two-body problem or Exercise 4.5E. The results of the present section are extensions of those due primarily to Iacob [1971, 1975] and Katok [1972] in a formulation due to Cushman [1977].

We recall that the motion of a rigid body is described by the geodesic flow on $SO(3)$ [$= SO(3,\boldsymbol{R})$] relative to a given left-invariant Riemannian metric $\langle,\rangle$ on $SO(3)$, the moment of inertia tensor. It will be convenient to work in body coordinates; that is, we work with the Hamiltonian system on $SO(3) \times so(3)^*$ with symplectic form given by 4.4.1 and with the energy momentum map given by

$$H \times J : SO(3) \times so(3)^* \to \boldsymbol{R} \times so(3)^*$$

$$(O, \mu) \mapsto (K(\mu), Ad^*_O \mu)$$

where K is the kinetic energy associated with the given inner product $\langle,\rangle$, that is, $K(\mu) = \frac{1}{2}\langle \mu, \mu \rangle$. We want to study the topology of this system following the procedures of Sect. 4.5. (Actually only nondegeneracy of $\langle,\rangle$ need be assumed).

We begin by summarizing some facts about the rotation group $SO(3)$ and its Lie algebra $so(3)$ which we shall need.

ISBN 0-8053-0102-X

4.6.1 Summary of SO(3) and notation (All of the facts below were proved or outlined in exercises in Sect. 4.1.)

(i) $(\,,\,)$ and $|\cdot|$ denote the Euclidean inner product and norm on $\boldsymbol{R}^3$, e_1, e_2, e_3 is the standard basis, and

$$SO(3)=\left\{O\in L(\boldsymbol{R}^3,\boldsymbol{R}^3)\,|\,(Ox,Oy)=(x,y) \text{ for all } x,y\in\boldsymbol{R}^3 \text{ and } \det O=1\right\}$$

which, using the standard basis, we can identify with

$$\left\{O\in GL(3,\boldsymbol{R})\,|\,OO'=I=O'O \text{ and } \det O=1\right\}$$

$SO(3)$ is a three-dimensional Lie group.

(ii) The Lie algebra of $SO(3)$ is

$$\begin{aligned} so(3)&=\left\{X\in L(\boldsymbol{R}^3,\boldsymbol{R}^3)\,|\,(Xx,y)+(x,Xy)=0 \text{ for all } x,y\in\boldsymbol{R}^3\right\}\\ &=\left\{X\in gl(3,\boldsymbol{R})\,|\,X+X'=0\right\}\end{aligned}$$

with $[X,Y]=X\circ Y-Y\circ X$. Let

$$E_1=\begin{bmatrix}0&0&0\\0&0&-1\\0&1&0\end{bmatrix},\quad E_2=\begin{bmatrix}0&0&1\\0&0&0\\-1&0&0\end{bmatrix},\quad \text{and}\quad E_3=\begin{bmatrix}0&-1&0\\1&0&0\\0&0&0\end{bmatrix}$$

Then $\{E_1,E_2,E_3\}$ is a basis for $so(3)$ and the bracket relations are

$$\begin{cases}[E_1,E_2]=E_1E_2-E_2E_1=E_3=-[E_2,E_1]\\ [E_2,E_3]=E_1=-[E_3,E_2] \quad\text{and}\quad [E_1,E_1]=[E_2,E_2]=[E_3,E_3]=0\\ [E_3,E_1]=E_2=-[E_1,E_3]\end{cases}$$

The vector product on $\boldsymbol{R}^3$ satisfies $(x\times y,z)=\det(x,y,z)$ and makes $\boldsymbol{R}^3$ into a Lie algebra. The map $j\colon \boldsymbol{R}^3\to so(3)\colon\ x=x_1e_1+x_2e_2+x_3e_3\mapsto X=x_1E_1+x_2E_2+x_3E_3$ is an isomorphism of the Lie algebras $(\boldsymbol{R}^3,\times)$ and $(so(3),[\,,\,])$. We shall identify these Lie algebras.

(iii) We have $Ad_O=O$ (more properly, $j^{-1}Ad_Oj=O$) and $ad_xy=x\times y$ (more properly, $ad_{j(x)}j(y)=j(x\times y)$).

The standard inner product on $\boldsymbol{R}^3$ [or $-\frac{1}{2}\operatorname{trace}(XY)$ on $so(3)$] is Ad-invariant. Let $(\,,\,)$ denote this inner product on $\boldsymbol{R}^3$ as well as on $\boldsymbol{R}^{3*}$.

(iv) For $\mu\in\boldsymbol{R}^{3*}$, the co-adjoint orbit of μ [which, for $\mu\neq0$, is a symplectic manifold by 4.3.4(vi)] is $S^2_{|\mu|}$, the sphere in $\boldsymbol{R}^{3*}$ of radius $|\mu|$. This follows directly from (iii). (It also follows from the general formula for the symplectic structure on $G\cdot\mu$ (4.3.4(v)) that it is given by the standard area element on S^2 if $\mu\neq0$.)

ISBN 0-8053-0102-X

Recall that the effective potential represents the potential of the reduced system. However, in our case, we already know the Hamiltonian for the reduced system on $S^2_{|\mu|}$ so we have, in effect, automatically computed the effective potential (see Exercise 4.6A). We can proceed then with a direct analysis of the topology of the reduced system.

The reduced Hamiltonian H_μ on $S^2_{|\mu|}$ is given by

$$H_\mu(\nu)=\tfrac{1}{2}\langle \nu,\nu\rangle = K(\nu)$$

where $\langle , \rangle$ denotes the given symmetric bilinear form on $so(3)$; that is, on $\mathbf{R}^3$. We can thus think of $\langle , \rangle$ as a given moment of inertia tensor as was explained in Sect. 4.4. Let us write, $\langle x,y\rangle=(Ix,y)$ which defines the symmetric linear map I.

4.6.2 Proposition. *The set of critical points of H_μ on $S^2_{|\mu|}$ is*

$$\sigma(H_\mu)=\bigcup_\lambda \{S^2_{|\mu|}\cap V_\lambda \mid V_\lambda = \text{eigenspace of } I \text{ corresponding to the eigenvalue } \lambda=\ker(I-\lambda Id)\}$$

Proof. $dH_\mu(x)y=(Ix,y)$. Now at $x\in S^2_{|\mu|}$, the tangent space is the set of y orthogonal to x, and the normal space is the set of multiples of x. Thus $dH_\mu(x)=0$ if and only if Ix is a multiple of x, that is, x is an eigenvector for I. ∎

If $n=\dim V_\lambda$ is the multiplicity of the eigenvalue λ, so $n=1$, 2 or 3, then each $S^{n-1}_\lambda=S^2_{|\mu|}\cap V_\lambda$ is a sphere of dimension 0, 1, or 2. The set S^{n-1}_λ is a nondegenerate critical manifold of H_μ with index equal to the number of eigenvalues less than λ and the corresponding critical value is $\frac{1}{2}\lambda|\mu|^2$ because the Hessian of H_μ on the normal to S^{n-1}_λ at x is $I-\lambda$ restricted to $\oplus_{\nu\neq\lambda}V_\nu$.

To keep things simple, we will assume that there is no external potential (Euler–Poinsot case) and that the eigenvalues $\lambda_1,\lambda_2,\lambda_3$ of I are distinct and are ordered $\lambda_2>\lambda_1>\lambda_3$ (see the exercises for the other cases). Then, from 4.6.2, H_μ is a Morse function with nondegenerate critical points $\pm x_{\lambda_i}$, $i=1,2,3$, where $Ix_{\lambda_i}=\lambda_i x_{\lambda_i}$ and $(x_{\lambda_i},x_{\lambda_i})=|\mu|^2$; these points have indices 1,2,0, respectively.

Since I is symmetric, there is an $\hat{O}\in SO(3)$ such that $\hat{O}(|\mu|e_1)=x_{\lambda_1}$, $\hat{O}(|\mu|e_2)=x_{\lambda_2}$, and $\hat{O}(|\mu|e_3)=x_{\lambda_3}$. Thus

$$\hat{O}^{-1}I\hat{O}=\begin{bmatrix}\lambda_1 & 0 & 0\\ 0 & \lambda_2 & 0\\ 0 & 0 & \lambda_3\end{bmatrix}=\bar{I}$$

Let $\bar{H}_\mu$: $S^2_{|\mu|}\subset\mathbf{R}^3\to\mathbf{R}$: $x\to\frac{1}{2}(\bar{I}x,x)$; then $\bar{H}_\mu$ has the same topological behavior as H_μ because $(\hat{O})^*H_\mu=\bar{H}_\mu$. The level sets $\bar{H}_\mu^{-1}(h)$ are either the

ISBN 0-8053-0102-X

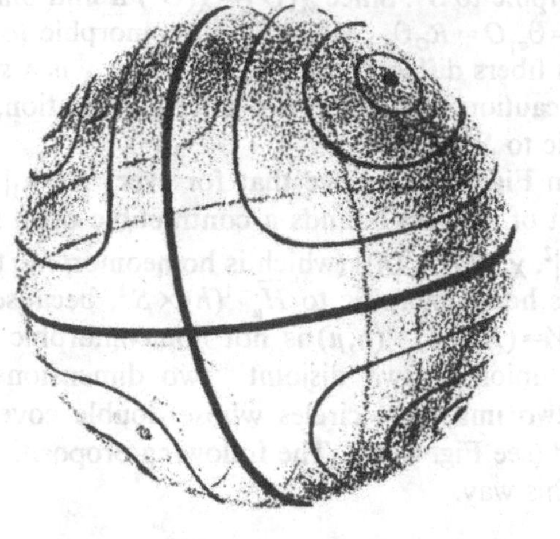

Figure 4.6-1

intersection of the ellipsoids $\frac{1}{2}(\lambda_1 x_1^2+\lambda_2 x_2^2+\lambda_3 x_3^2)=h$ with the sphere $x_1^2+x_2^2+x_3^2=|\mu|^2$ if $\lambda_2>\lambda_1>\lambda_3>0$ or the intersection of the hyperboloids $\frac{1}{2}(\lambda_1 x_1^2+\lambda_2 x_2^2-\lambda_3 x_3^2)=h$ with the sphere $x_1^2+x_2^2+x_3^2=|\mu|^2$ if $\lambda_2>\lambda_1>0$ and $\lambda_3>0$. In both cases, the topology of the level sets of $\overline{H}_\mu$ is the same and is given in Fig. 4.6-1 and Proposition 4.6.3 for the stated range of h.

4.6.3 Proposition. *The topology of the energy surfaces is given as follows.*

Level h	Topology of $H_\mu^{-1}(h)$
(1) $h=\frac{1}{2}\lambda_2\|\mu\|^2$	S^0, two points
(2) $\frac{1}{2}\lambda_2\|\mu\|^2<h<\frac{1}{2}\lambda_1\|\mu\|^2$	$S^0\times S^1$
(3) $h=\frac{1}{2}\lambda_1\|\mu\|^2$	W, two disjoint copies of S^1 with two distinct points identified
(4) $\frac{1}{2}\lambda_1\|\mu\|^2<h<\frac{1}{2}\lambda_3\|\mu\|^2$	$S^0\times S^1$
(5) $h=\frac{1}{2}\lambda_3\|\mu\|^2$	S^0

The mappings $\varphi: J^{-1}(\mu)\to SO(3)\cdot\mu: (O,Ad^*_{O^{-1}}\mu)\mapsto Ad^*_{O^{-1}}\mu$ and $\psi: SO(3)\to SO(3)\cdot\mu: O\mapsto Ad^*_{O^{-1}}\mu$ have the same topological behavior because $\psi=\varphi\circ\alpha_\mu$ where $\alpha_\mu: SO(3)\to J^{-1}(\mu): O\mapsto(O,Ad^*_{O^{-1}}\mu)$, which is a diffeomorphism. Define $\chi: SO(3)\to S^2_{|\mu|}\subset R^3: O\mapsto O(|\mu|e_1)$. Then χ and ψ have the same topological behavior because the linear action of $SO(3,R)$ on $S^2_{|\mu|}$ is transitive and equals the coadjoint action by 4.6.1(iii). Let $\mathbb{O}_{e_1}=\{O\in SO(3,R)|Oe_1=e_1\}$. Then

$$\mathbb{O}_{e_1}=\left\{\begin{bmatrix}1&0&0\\0&\cos\theta&-\sin\theta\\0&\sin\theta&\cos\theta\end{bmatrix}\middle|\,0\leqslant\theta\leqslant 2\pi\right\}$$

ISBN 0-8053-0102-X

which is diffeomorphic to S^1. Since $\chi(O)=\chi(O')$ if and only if $O'O^{-1}\in\mathcal{O}_{e_1}$, $\chi^{-1}(O^{-1}(|\mu|e_1))=\mathcal{O}_{e_1}O=R_O\mathcal{O}_{e_1}$, which is diffeomorphic to $\mathcal{O}_{e_1}$. Therefore, χ is a fibration with fibers diffeomorphic to S^1, since χ is a submersion.

The reader is cautioned that χ is not a trivial fibration, that is, $SO(3)$ is not homeomorphic to $S^2_{|\mu|}\times S^1$.

However, from Fig. 4.6-1 we see that for every $h\neq\lambda_1|\mu|^2/2$, every connected component of $H_\mu^{-1}(h)$ bounds a contractible open set in $S^2_{|\mu|}$. Therefore for $h\neq\frac{1}{2}\lambda_1|\mu|^2$, $\chi^{-1}(\overline{H}_\mu^{-1}(h))$ (which is homeomorphic to $\varphi^{-1}(H_\mu^{-1}(h))=(H\times J)^{-1}(h,\mu)$) is homeomorphic to $\overline{H}_\mu^{-1}(h)\times S^1$, because χ is a fibration. For $h=\frac{1}{2}\lambda_1|\mu|^2$, $V=(H\times J)^{-1}(h,\mu)$ is not homeomorphic to $W\times S^1$ but is topologically the union of two disjoint two dimensional tori which are identified along two imbedded circles whose double covering have linking number one in S^3 (see Fig. 4.6-4). The following proposition summarizes the results obtained this way.

4.6.4 Proposition. *The topological types for the level surfaces of the energy momentum mapping are:*

Level (h,μ), $u\neq 0$	Topological type of $(H\times J)^{-1}(h,\mu)=I_{h,\mu}$
(1) $h=\frac{1}{2}\lambda_2|\mu|^2$	$S^0\times S^1$
(2) $\frac{1}{2}\lambda_2|\mu|^2<h<\frac{1}{2}\lambda_1|\mu|^2$	$S^0\times T^2$; T^2, a two-dimensional torus
(3) $h=\frac{1}{2}\lambda_1|\mu|^2$	V; two disjoint tori T^2 identified along two embedded S^1 whose double coverings are linked once in S^3
(4) $\frac{1}{2}\lambda_1|\mu|^2<h<\frac{1}{2}\lambda_3|\mu|^2$	$S^0\times T^2$
(5) $h=\frac{1}{2}\lambda_3|\mu|^2$	$S^0\times S^1$

When $\mu=0$, the co-adjoint orbit is $\{0\}$. Therefore, $J^{-1}(\{0\})=SO(3)\times\{0\}$, which is precisely the set of critical points of $H: SO(3)\times so(3)^*\to\mathbf{R}: (O,\nu)\mapsto\frac{1}{2}\langle\nu,\nu\rangle$ since $\langle,\rangle$ is nondegenerate. The corresponding critical value is 0. Therefore:

4.6.5 Proposition. *The bifurcation set of $H\times J$ is the union of three paraboloids*

$$P_i=\left\{(h,\mu)\in\mathbf{R}\times so(3)^*\,\middle|\,\tfrac{1}{2}\lambda_i(\mu,\mu)=h\right\}$$

for $i=1,2,3$.

(See Fig. 4.6-2.)

In order to understand how the level sets $(H\times J)^{-1}(h,\mu)$ with h fixed fit together to form $J^{-1}(\mu)$ [which is homeomorphic to $SO(3)$], we use the solid ball model of $SO(3)$. In this model, $SO(3)$ is the ball $B^3_\pi=\{x\in\mathbf{R}^3|(x,x)<$

ISBN 0-8053-0102-X

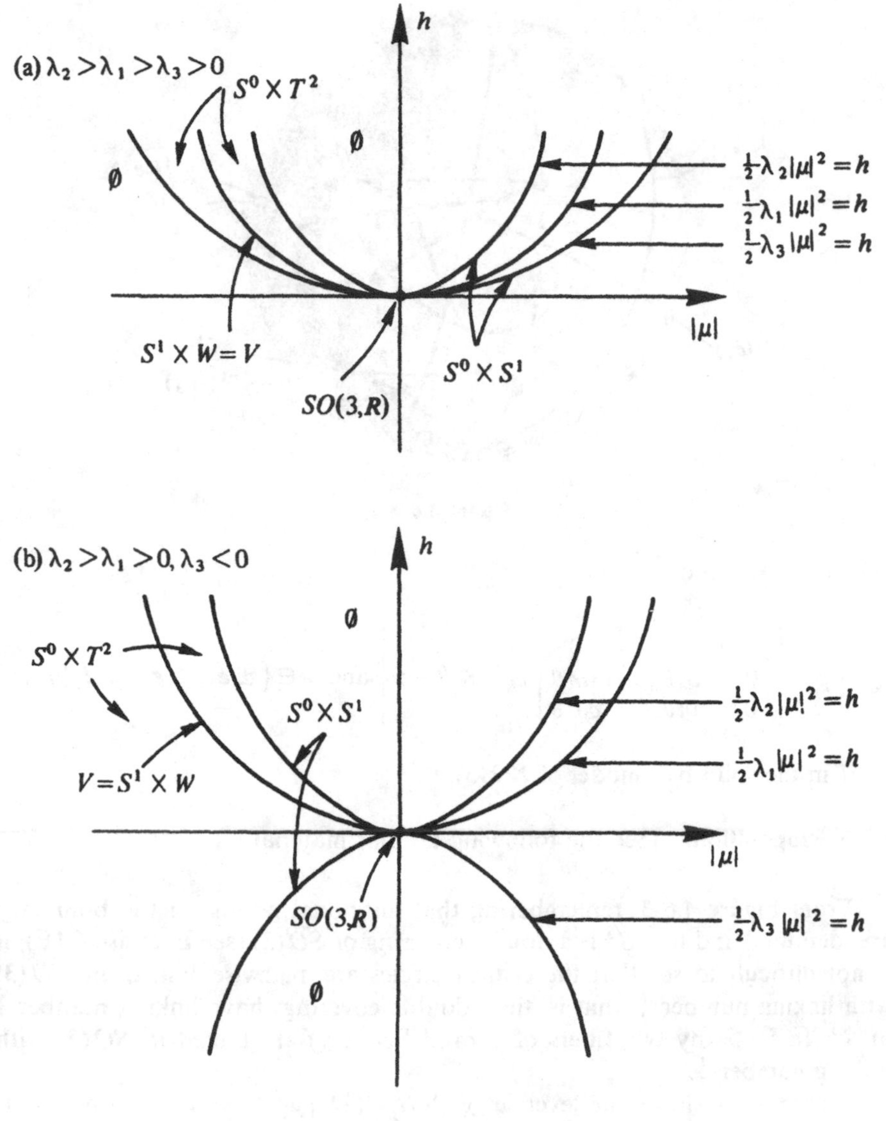

Figure 4.6-2

$\pi^2\}$ with diametrically opposite points y and $-y$ on its boundary $S_\pi^2 = \{x \in R^3 | (x,x) = \pi^2\}$ identified. B_π^3 is homeomorphic to $SO(3)$ under the homeomorphism that assigns to every $O \in SO(3)$ a vector $z \in R^3$ lying on the axis of rotation of O (which is the eigenspace spanned by the eigenvector of O corresponding to the eigenvalue 1) oriented so that the angle of rotation ϑ of O about that axis satisfies $0 \leq \vartheta \leq \pi$. The length of z is ϑ. Proposition 4.6.5 and Fig. 4.6-3 give the disposition of the six critical circles $\chi^{-1}(e) = \Theta_{e_1} O$

ISBN 0-8053-0102-X

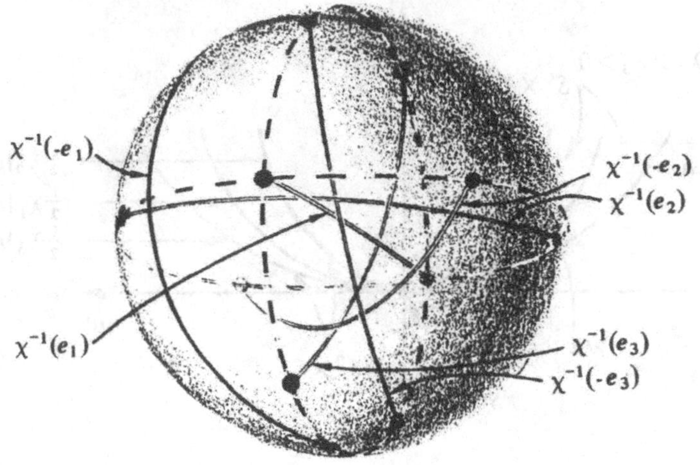

Figure 4.6-3

(where $Oe = e_1$ and

$$\mathcal{O}_{e_1} = \left\{ \begin{bmatrix} 1 & 0 & 0 \\ 0 & \cos\theta & -\sin\theta \\ 0 & \sin\theta & \cos\theta \end{bmatrix},\quad 0 \leqslant \theta \leqslant \pi \right\} \text{ and } e \in \{\pm e_1, \pm e_2, \pm e_3\})$$

of H in the solid ball model of $SO(3)$.

4.6.5 Proposition. (See the following tabular material.)

From Figure 4.6-3, remembering that antipodal points on the boundary are identified and that S^3 is a double covering of $SO(3)$ (see Exercise 4.1F), it is not difficult to see that the critical circles are pairwise linked in $SO(3)$ with linking number $\frac{1}{2}$, that is, their double coverings have linking number 1 in S^3. In fact, any two fibers of ψ (and hence χ) are linked in $SO(3)$ with linking number $\frac{1}{2}$.

Figure 4.6-4 shows the level set $\chi^{-1}(\overline{H}_\mu^{-1}(\frac{1}{2}\lambda_1|\mu|^2))$, which is homeomorphic to two disjoint tori $T^2_\pm$ joined along the two embedded linked circles $\chi^{-1}(e_1)$ and $\chi^{-1}(-e_1)$. From the figure it is clear that

$$\overline{M} = (H \times J)^{-1}\left(\left[\tfrac{1}{2}\lambda_3|\mu|^2, \tfrac{1}{2}\lambda_1|\mu|^2\right]\right)$$

is two disjoint solid tori $ST^2_\pm$ (which is diffeomorphic to $D^2_\pm \times T^2_\pm$) with $\chi^{-1}(\pm e_3)$ in its interior. It is also true, but is not clear from the figure, that $SO(3)\backslash\overline{M}$ is diffeomorphic to two disjoint solid tori $ST^2_\pm$ with $\chi^{-1}(\pm e_2)$ in its interior and $SO(3)\backslash\overline{M}$ is homeomorphic to $(H \times J)^{-1}([\frac{1}{2}\lambda_2|\mu|^2, \frac{1}{2}\lambda_1|\mu|^2])$. Note that the boundary of $\overline{M}$ is V.

ISBN 0-8053-0102-X

ISBN 0-8053-0102-X

4.6.5. Proposition. With this notation, we have:

Critical point	Critical circle	Axis of rotation	Angle of rotation ϑ	Comment
(1) e_1	$\Theta_{e_1}\begin{pmatrix} 1 & 0 & 0 \\ 0 & 1 & 0 \\ 0 & 0 & 1 \end{pmatrix}$	$(1,0,0),\ 0 \le \theta < 2\pi$	$\theta \bmod \pi$	Lies in plane $x_2 = x_3 = 0$
(2) $-e_1$	$\Theta_{e_1}\begin{pmatrix} -1 & 0 & 0 \\ 0 & 0 & 1 \\ 0 & 1 & 0 \end{pmatrix}$	$\begin{cases} (0, 1-\sin\theta, \cos\theta),\ 0 \le \theta < \pi/2 \\ (0,0,1),\ \theta = \pi/2 \\ (0, -1+\sin\theta, -\cos\theta),\ \pi/2 < \theta < 2\pi \end{cases}$	π	Lies in plane $x_1 = 0$
(3) e_2	$\Theta_{e_1}\begin{pmatrix} 0 & 1 & 0 \\ 1 & 0 & 0 \\ 0 & 0 & -1 \end{pmatrix}$	$\begin{cases} (\cos\theta+1, \cos\theta+1, \sin\theta),\ 0 \le \theta < \pi \\ (0,1,0),\ \theta = \pi \\ (-\cos\theta-1, -1-\cos\theta, -\sin\theta),\ \pi < \theta < 2\pi \end{cases}$	$\cos\vartheta = \frac{1}{2}(1+\cos\theta)$	Lies in plane $x_1 = x_2$
(4) $-e_2$	$\Theta_{e_1}\begin{pmatrix} 0 & -1 & 0 \\ 0 & 0 & -1 \\ 1 & 0 & 0 \end{pmatrix}$	$\begin{cases} (-\sin\theta-1, \sin\theta+1, -\cos\theta),\ 0 \le \theta < \pi/2 \\ (\sin\theta+1, -\sin\theta-1, \cos\theta),\ 3\pi/2 < \theta < 2\pi \\ (0,0,-1),\ \pi/2 < \theta \le 3/2 \end{cases}$	$\cos\vartheta = -\frac{1}{2}(1+\sin\theta)$	Lies in plane $x_1 = -x_2$
(5) e_3	$\Theta_{e_1}\begin{pmatrix} 0 & 0 & 1 \\ 0 & -1 & 0 \\ 1 & 0 & 0 \end{pmatrix}$	$\begin{cases} (-1+\sin\theta, -\sin\theta, -\sin\theta+1),\ 0 \le \theta < \pi/2 \\ (0,-1,0),\ \theta = \pi/2.\ 3\pi/2 \le \theta < 2\pi \\ (1-\sin\theta, \cos\theta, \sin\theta-1),\ \pi/2 < \theta < 3\pi/2 \end{cases}$	$\cos\vartheta = -\frac{1}{2}(1+\sin\theta)$	Lies in plane $x_1 = -x_3$
(6) $-e_3$	$\Theta_{e_1}\begin{pmatrix} 0 & 0 & -1 \\ 1 & 0 & 0 \\ 0 & -1 & 0 \end{pmatrix}$	$\begin{cases} (\cos\theta+1, -\sin\theta, \cos\theta+1),\ 0 \le \theta < \pi \\ (0,1,0),\ \theta = \pi \\ (-\cos\theta-1, \sin\theta, -\cos\theta-1),\ \pi < \theta < 2\pi \end{cases}$	$\cos\vartheta = -\frac{1}{2}(1+\cos\theta)$	Lies in plane $x_1 = x_3$

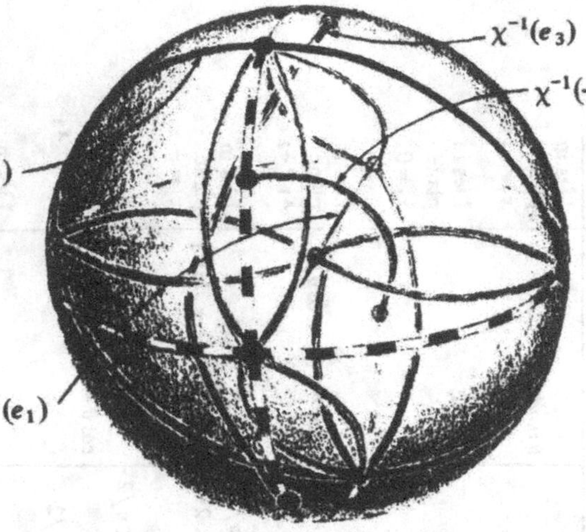

Figure 4.6-4

This completes the description of the topology of the energy-momentum mapping of the geodesic flow on $SO(3)$ for any nondegenerate left-invariant metric with distinct principal moments of inertia.

We did the analysis in this case by a direct examination of the Hamiltonian on the reduced phase space. For other cases, see the exercises, Iacob [1975], Katok [1972], and Tatarinov [1973].

EXERCISES

4.6A. Show that the effective potential, for a Lie group G with a left-invariant Riemannian metric $\langle , \rangle$ and potential V is

$$V_\mu(g) = V(g) + \tfrac{1}{2}\|Ad^*_{g^{-1}}\mu\|^2$$

(showing $\Lambda = \varnothing$). What is $\hat{V}_\mu$? Use the effective potential on $SO(3)$ to prove 4.6.4 and 4.6.5.

4.6B. Show that the flow on $I_{h,\mu}$ in cases 2 and 4 of 4.6.4 is quasi-periodic. (Cf. Arnold–Avez [1967, Appendix 3].)

4.6C. Study the topology of the rigid body for $\lambda_1 = \lambda_2 > \lambda_3$. Prove, in particular, that the generic invariant manifolds $I_{h,\mu}$ are tori T^3.

4.6D. Study the topology of the rigid body under gravity in the Lagrange–Poisson and Kovalevskaya case (see Iacob [1971], Katok [1972]).

4.6E. Use Exercise 4.5F and the effective potential in Exercise 4.6A to prove that the relative equilibria corresponding to λ_2 and λ_3 are stable, but λ_1 is unstable. (Assume $\lambda_2 > \lambda_1 > \lambda_3$ as in the text.)

4.6F. Show that $b: sl(2,\mathbf{R}) \times sl(2,\mathbf{R}) \to \mathbf{R}: x,y \mapsto \operatorname{trace} ad_x ad_y$, is a nondegenerate inner product on $sl(2,\mathbf{R})$ whose matrix with respect to the basis

$$E_1 = \frac{1}{2}\begin{pmatrix} 1 & 0 \\ 0 & -1 \end{pmatrix}, \quad E_2 = \frac{1}{\sqrt{2}}\begin{pmatrix} 0 & 1 \\ 0 & 0 \end{pmatrix}, \quad E_3 = \frac{1}{\sqrt{2}}\begin{pmatrix} 0 & 0 \\ -1 & 0 \end{pmatrix}$$

ISBN 0-8053-0102-X

is

$$\begin{pmatrix} 2 & -1 & 0 \\ -1 & 2 & 0 \\ 0 & 0 & 2 \end{pmatrix}$$

Clearly, b is invariant under $Ad: SL(2,\mathbf{R})\times sl(2,\mathbf{R})\to sl(2,\mathbf{R})$, $(A,X)\mapsto AXA^{-1}$. Conclude that the co-adjoint action orbits are diffeomorphic to the adjoint action orbits.

(i) Show that the cone $\{X\in sl(2,\mathbf{R})|b(X,X)=0\}$ consists of three adjoint orbits, those containing

$$\begin{pmatrix} 0 & 0 \\ 0 & 0 \end{pmatrix}, \begin{pmatrix} 0 & 1 \\ 0 & 0 \end{pmatrix}, \text{ and } \begin{pmatrix} 0 & -1 \\ 0 & 0 \end{pmatrix}$$

(ii) Show that the hyperboloid $\{X\in sl(2,\mathbf{R})|b(X,X)=-1\}$ contains two orbits.

(iii) Write the equations of each adjoint orbit as polynomial inequalities.

(iv) What is the Euler vector field for the dual metric $\langle,\rangle$ whose matrix with respect to the basis $\{b^{\#}(E_1),b^{\#}(E_2),b^{\#}(E_3)\}$ of $sl(2,\mathbf{R})^*$ has the form

$$\begin{bmatrix} \Gamma_1 & 0 & 0 \\ 0 & \Gamma_2 & 0 \\ 0 & 0 & \Gamma_3 \end{bmatrix}?$$

4.6G. (Van Moerbeke, M. Adler, B. Kostant, and R. Hermann)
Let G be the group of invertible lower triangular, $n\times n$ matrices. Let $\mathfrak{g}$ be identified with the lower triangular matrices and $\mathfrak{g}^*$ the upper triangular matrices. Show that the adjoint action of G on $\mathfrak{g}$ is by conjugation and on $\mathfrak{g}^*$ by conjugation followed by projection $A\mapsto A^+$ (taking the upper triangular part).

(i) Show that for $f\in\mathfrak{g}^*$,

$$T_h(G\cdot f)=\{[h,A]^+|A\in\mathfrak{g}\}$$

and that the sympletic structure on the coadjoint orbits is

$$\omega([h,A]^+,[h,B]^+)=-\operatorname{trace}(h[A,B])$$

(ii) Take the special case in which f has ones on the superdiagonal (the diagonal above the main diagonal) and zeros elsewhere. Show that $G\cdot f$, the orbit of f is parametrized by $(a_1,\ldots,a_{n-1}\ b_1,\ldots,b_{n-1})$ by associating to this $2(n-1)$ tuple, the upper-trianglular matrix C with $b_1,\ldots,b_{n-1}$, $b_n=-\sum_{i=1}^{n-1}b_i$, down the main diagonal and $a_1,\ldots,a_{n-1}$ down the superdiagonal. Show that the symplectic structure is

$$\omega=\sum_{j=1}^{n-1}db_j\wedge\sum_{i=j}^{n-1}\frac{da_i}{a_i}$$

and that $a_k=e^{(q_k-q_{k+1})}$ (with $\sum_{i=1}^{n}q_i=0$) $b_k=-p_k$, $k=1,\ldots,n-1$ brings ω to canonical form. This *Toda lattice* example is continued in Exercise 5.5K.

ISBN 0-8053-0102-X

CHAPTER 5

Hamilton–Jacobi Theory and Mathematical Physics

This chapter continues our development of the theory of Hamiltonian systems, giving a few more advanced topics. In addition to traditional topics related to canonical transformations and Hamilton–Jacobi theory, we present an introduction to some topics of more recent vintage including Lagrangian submanifolds, quantization and infinite-dimensional systems. The final section (5.6) is intended to provide a transition from this chapter to Part 3 on qualitative dynamics.

5.1 TIME-DEPENDENT SYSTEMS

Section 3.3 was concerned with Hamiltonian vector fields on a symplectic manifold(P,ω). If, instead, we are given a mapping $X\colon \boldsymbol{R}\times P\to TP$, (a *time-dependent vector field*), then the analysis of Sect. 3.3 is no longer appropriate. In fact, $\boldsymbol{R}\times P$ cannot be a symplectic manifold, as it has odd dimension; but it does have a *contact structure*. Contact manifolds will also be of importance for canonical transformations, which we study in the next section. (Historically, the terms *contact* and *canonical* have been used in a variety of ways; see Whittaker [1959, p. 290].)

Ralph Abraham and Jerrold E. Marsden, Foundation of Mechanics, Second Edition

ISBN 0-8053-0102-X

We shall begin with a study of closed two-forms that are not necessarily nondegenerate.

5.1.1 Definition. *Let ω be an exterior two-form on M. Then*

$$R_\omega = \{v \in TM \mid v^\flat = 0\}$$

is called the ***characteristic bundle*** *of ω. [Here $v^\flat$ is the one-form defined by $v^\flat(w) = \omega(v, w)$.] A* ***characteristic vector field*** *is a vector field X such that $i_X\omega = 0$, that is, $X(x) \in R_\omega$ for all $x \in M$.*

The following is a basic result that will be of use to us here and later. It is due to E. Cartan.

5.1.2 Proposition. *Let ω be a two-form on M of constant rank. Then R_ω is a subbundle of TM. If ω is closed, then R_ω is integrable as well.*

Proof. If ω is of constant rank, this implies the fibers of R_ω have constant dimension. We can choose a smoothly varying basis of R_ω in any chart by starting with a basis of vector fields for the tangent bundle and applying the algebraic procedure of 3.1.2(ii) to bring ω to canonical form. (The reader should write out the details.) Thus R_ω is a smooth subbundle.

The second part uses Frobenius' theorem and was already noted in the introduction to Sect. 4.3: Let X and Y be characteristic vector fields. Then

$$\begin{aligned} i_{[X,Y]}\omega &= L_X i_Y\omega - i_Y L_X\omega \\ &= -i_Y L_X \omega \\ &= -i_Y(i_X d\omega + d i_X\omega) = 0 \end{aligned}$$

so $[X, Y]$ is a characteristic vector field. ■

Remark. The reader can easily prove the converses to the two conclusions of 5.1.2.

Next we generalize Darboux' theorem proved in Sect. 3.2 to cover the case of closed two-forms with constant rank. (We shall explicitly give a finite-dimensional version, but our proof also works in infinite dimensions.) The classical proof may be found in Sternberg [1964].

5.1.3 Theorem (Darboux) *Let M be a $(2n+k)$-manifold and ω a closed two-form of constant rank 2n. For each $x_0 \in M$ there is a chart (U, φ) about x_0 such that, in this chart,*

$$\omega | U = \sum_{i=1}^{n} dx^i \wedge dy^i$$

where coordinates in the chart are written $(x^1, \ldots, x^n, y^1, \ldots, y^n, w^1, \ldots, w^k)$.

ISBN 0-8053-0102-X

Proof. Choose a chart (V,ψ) about x_0 and in it a ball B in a linear subspace E through x_0 such that $i^*\omega$ has rank $2n$ at x_0, where $i: B\to\psi(V)$ is inclusion. If we shrink B if necessary, we can find coordinates $x^1,\ldots,x^n,y^1,\ldots,y^n$ on B such that

$$i^*\omega = \sum_{i=1}^{n} dx^i\wedge dy^i$$

by Darboux's theorem. By 5.1.2, the characteristic subbundle R_ω is integrable. It is of dimension k and transverse to B. Thus a neighborhood of x_0 is diffeomorphic to $B\times N$, where the tangent space to N at $p\in B\times N$ coincides with $R_\omega(p)$. Here we may have to shrink B further and have used Frobenius' theorem. If we let $w^1,\ldots,w^k$ be arbitrarily chosen coordinates on N, then

$$\omega = \sum_{i=1}^{n} dx^i\wedge dy^i$$

since ω vanishes when applied to a pair of vectors, one of which is in R_ω. ■

We shall now specialize to the case of contact manifolds.

5.1.4 Definition. *A* ***contact manifold*** *is a pair (M,ω) consisting of an odd-dimensional manifold M and a closed two-form ω of maximal rank on M. An* ***exact contact manifold*** *(M,θ) consists of a $(2n+1)$-dimensional manifold M and a one-form θ on M such that $\theta\wedge(d\theta)^n$ is a volume on M.*

Note that the characteristic bundle R_ω of a contact form ω has one-dimensional fibers, so we sometimes call it the *characteristic line bundle*.

The next result gives the canonical form of ω and θ and shows that an exact contact manifold is a contact manifold. The converse can be proved locally using the Poincaré lemma.

5.1.5 Theorem. *Let (M,ω) be a contact manifold. Then for each $m\in M$ there is a chart (U,φ) at m with $\varphi(u)=(q^1(u),\ldots,q^n(u),p_1(u),\ldots,p_n(u),w(u))$ such that*

$$\omega|U = dq^i\wedge dp_i$$

Similarly, if (M,θ) is an exact contact manifold, there is a chart $(\tilde U,\tilde\varphi)$ at m such that

$$\theta|U = dw + p_i\, dq^i$$

Proof. The first statement follows at once from 5.1.3. For the second, let $\omega = d\theta$ and choose coordinates $(q^1,\ldots,q^n,p_1,\ldots,p_n,w)$ as in the first part. Then

$$d\left(\theta - p_i\, dq^i\right) = 0$$

ISBN 0-8053-0102-X

so locally

$$\theta - p_i\, dq^i = dw$$

for a function w. Since $\theta\wedge(d\theta)^n \neq 0$, the functions q^i, p_i and w are functionally independent, and thus define a local chart. ■

An illustrative characterization of exact contact manifolds is as follows:

5.1.6 Proposition. *Let θ be a nowhere zero one-form on a $(2n+1)$-manifold M and let $R_\theta = \{v \in TM \mid \theta(v) = 0\}$ be its characteristic bundle. Then (M, θ) is an exact contact manifold if and only if $d\theta$ is nondegenerate on the fibers of R_θ.*

Proof. R_θ is $2n$-dimensional, so by 3.1.3, $d\theta$ is nondegenerate on R_θ iff $(d\theta)^n \neq 0$. By definition of $\wedge$ and R_θ, this is so iff $\theta\wedge(d\theta)^n \neq 0$. ■

Here is an example of a contact structure:

5.1.7 Proposition. *Let (P, ω, H) be a Hamiltonian system and Σ_e a regular energy surface. Then $(\Sigma_e, i^*\omega)$ is a contact manifold, where $i: \Sigma_e \to P$ is inclusion. Moreover, $X_H|\Sigma_e$ is a characteristic vector field of $i^*\omega$ generating the characteristic line bundle of $i^*\omega$.*

Proof. Clearly $di^*\omega = i^*d\omega = 0$, so $i^*\omega$ is closed. Since ω is nondegenerate and Σ_e is of codimension one, ω has maximal rank on Σ_e. [$\omega | T_x\Sigma_e$ is of maximal rank iff the dimension of the ω-orthogonal complement to $T_x\Sigma_e$ in T_xP is one dimensional. But this is clear by 5.3.2(iii).]

Since X_H is tangent to Σ_e, $\omega(x)\cdot(X_H(x), v) = dH(x)\cdot v = 0$ for $x \in \Sigma_e$, $v \in T_x\Sigma_e \subset T_xP$. Thus $i_{X_H|\Sigma_e} i^*\omega = 0$, so $X_H|\Sigma_e$ is a characteristic vector field. Since e is a regular value, $X_H(x) \neq 0$ for $x \in \Sigma_e$, thus X_H generates the characteristic line bundle. ■

The argument given here shows that if ω is closed and its characteristic bundle is one dimensional, then ω is a contact form.

5.1.8 Example. If $P = T^*Q$, and $\omega = -d\theta_0$, where θ_0 is the canonical one-form and if $H = K + V\circ\tau_Q$, where K is the kinetic energy associated to a Riemannian metric and $V: Q \to R$, and if Σ_e is a regular energy surface that does not intersect the zero section of T^*Q, then $(\Sigma_e, i^*\theta_0)$ is an exact contact manifold. Indeed, by 5.1.7, $(\Sigma_e, i^*\omega) = (\Sigma_e, i^*(-d\theta_0))$ is a contact manifold, so $(d\theta_0)^n$ is nonzero on a complement to $R_{i^*\omega}$. Also, θ_0 is nonzero on $R_{i^*\omega}$ since $R_{i^*\omega}$ is generated by X_H and $\theta_0(X_H) = 2K$ is the action of H. Thus $\theta_0\wedge(d\theta_0)^n$ is nonzero, so we have an exact contact manifold.

The argument given here also shows that if $(M, d\theta)$ is a contact manifold and if θ is nonzero on the characteristic bundle R_ω, $\omega = d\theta$, then (M, θ) is an exact contact manifold.

ISBN 0-8053-0102-X

We now turn to a second main example of contact manifolds, namely, those associated with time-dependent Hamiltonian systems. We stress the Hamiltonian formulation, although the time-dependent Lagrangian formulation could be similarly described.

5.1.9 Proposition. *Let (P,ω) be a symplectic manifold, $\mathbf{R}\times P$ the product manifold of $\mathbf{R}$ and P and $\pi_2: \mathbf{R}\times P\to P$ the projection, $\pi_2(t,p)=p$. Let $\tilde{\omega}=\pi_2^*\omega$. Then $(\mathbf{R}\times P,\tilde{\omega})$ is a contact manifold.*

The characteristic line bundle of $\tilde{\omega}$ is generated by the vector field $\underset{\sim}{t}$ on $\mathbf{R}\times P$ given by

$$\underset{\sim}{t}(s,p)=((s,1),0)\in T_{(s,p)}(\mathbf{R}\times P)\approx T_s\mathbf{R}\times T_pP$$

If $\omega=\mathbf{d}\theta$ and $\tilde{\theta}=dt+\pi_2^*\theta$, where $t: \mathbf{R}\times P\to\mathbf{R}$ is the projection on the first factor, then $\tilde{\omega}=\mathbf{d}\tilde{\theta}$ and $(\mathbf{R}\times P,\theta)$ is an exact contact manifold.

Proof. Clearly $\mathbf{d}\tilde{\omega}=\pi_2^*\mathbf{d}\omega=0$, so $\tilde{\omega}$ is closed. As in 5.1.7, to show $\tilde{\omega}$ is of maximal rank, it suffices to show that its characteristic bundle is one dimensional. However, $((s,r),v_p)\in R_{\tilde{\omega}}$ means that

$$\tilde{\omega}_{(s,p)}\big(((s,r),v_p),((s,u),w_p)\big)=0$$

for all u,w_p. By definition of $\tilde{\omega}$, this is exactly the condition that

$$\omega_p(v_p,w_p)=0 \quad \text{for all } \omega_p$$

that is, $v_p=0$. Thus at $(s,p)\in\mathbf{R}\times P$,

$$R_{\tilde{\omega}}(s,p)=\{((s,r),0_p)|r\in\mathbf{R}\}$$

which is one dimensional, generated by $\underset{\sim}{t}(s,p)$.

The last part follows since $\mathbf{d}\tilde{\theta}=\tilde{\omega}$ and $\mathbf{d}t(\underset{\sim}{t})=1$, so $\tilde{\theta}$ is nonzero on $R_{\tilde{\omega}}$. ■

Recall from Sect. 2.2 that if M is a manifold, a smooth map $X: \mathbf{R}\times M\to TM$ is a *time-dependent vector field* on M if for each $t\in\mathbf{R}$, X is a vector field on M, that is, $(\tau_M\circ X)(t,m)=m$. We define $\tilde{X}: \mathbf{R}\times M\to T(\mathbf{R}\times M)\approx T\mathbf{R}\times TM$ by $(t,m)\mapsto((t,1),X(t,m))$ and observe that $\tilde{X}\in\mathfrak{X}(\mathbf{R}\times M)$, and $\tilde{X}=\underset{\sim}{t}+X$. Also, by definition, a curve $b: I\to M$ is an *integral curve* of X at m iff $b'(t)=X(t,b(t))$ for all $t\in I$, and $b(0)=m$. It follows that $c: I\to\mathbf{R}\times M$ is an integral curve of $\tilde{X}$ at $(0,m)$ iff $c(t)=(t,b(t))$, where $b: I\to M$ is an integral curve of X at m. Indeed, if $c(t)=(a(t),b(t))$, then $c(t)$ is an integral curve of $\tilde{X}$ if $c'(t)=(a'(t),b'(t))=\tilde{X}(c(t))$; that is, $a'(t)=1$ and $b'(t)=X(a(t),b(t))$. Since $a(0)=0$, $a(t)=t$. This proves the assertion. We call $\tilde{X}$ the *suspension* of X and note that its flow is $\tilde{F}_t(s,m)=(t+s,F_{t,s}(m))$ where $F_{t,s}$ is the flow of the time dependent vector field X.

ISBN 0-8053-0102-X

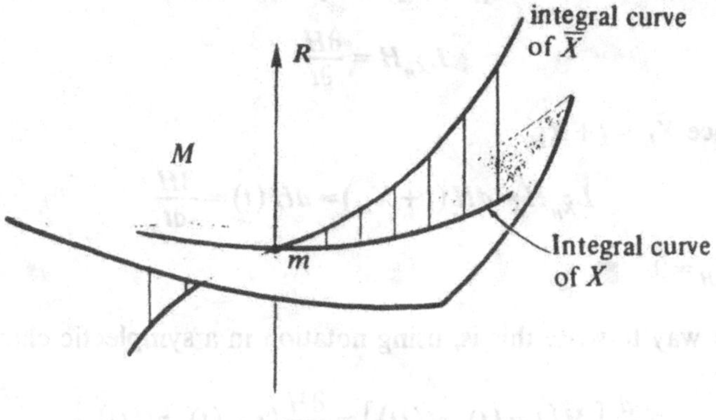

Figure 5.1-1

Thus, changing X to $\tilde{X}$ *suspends* the integral curves of X, as shown in Fig. 5.1-1.

5.1.10 Definition. *Let (P, ω) be a symplectic manifold and $H \in \mathfrak{F}(\mathbf{R} \times P)$. For each $t \in \mathbf{R}$ define $H_t: P \to \mathbf{R}$ by $H_t(p) = H(t,p)$. [Note that $H_t \in \mathfrak{F}(P)$ and $X_{H_t} = (dH_t)^\sharp \in \mathfrak{X}(P)$.] Then put $X_H: \mathbf{R} \times P \to TP: (t,p) \mapsto X_{H_t}(p)$ and define the suspension $\tilde{X}_H: \mathbf{R} \times P \to T(\mathbf{R} \times P)$ as above.*

Thus our time-dependent vector field X_{H_t} is obtained by merely freezing t and constructing the usual Hamiltonian vector field. Thus we obtain the following.

5.1.11 Proposition. *Let (U, φ) be a symplectic chart of P with*

$$\varphi(u) = \left(q^1(u), \ldots, q^n(u), p_1(u), \ldots, p_n(u)\right)$$

so $(\mathbf{R} \times U, t \times \varphi)$ is a chart of $\mathbf{R} \times P$, where $t: \mathbf{R} \times P \to \mathbf{R}$ is the projection onto the first factor. Then $c: I \to \mathbf{R} \times U: t \mapsto (t, b(t))$ is an integral curve of $\tilde{X}_H$ or, equivalently, $b: I \to U$ is an integral curve of X_H, iff

$$\frac{d}{dt}\left[q^i(b(t))\right] = \frac{\partial H}{\partial p_i}(t, b(t)), \qquad i = 1, \ldots, n$$

$$\frac{d}{dt}\left[p_i(b(t))\right] = -\frac{\partial H}{\partial q^i}(t, b(t)), \qquad i = 1, \ldots, n$$

In the time-independent case we saw that $L_{X_H} H = 0$, or H is constant along the integral curves of X_H. For the time-dependent case the Hamiltonian is not an integral invariant.

ISBN 0-8053-0102-X

5.1.12 Proposition. *With $\tilde{X}_H$ and $\underset{\sim}{t}$ as defined, we have*

$$L_{\tilde{X}_H} H = \frac{\partial H}{\partial t}$$

Proof. Since $\tilde{X}_H = \underset{\sim}{t} + X_H$,

$$L_{\tilde{X}_H} H = dH(\underset{\sim}{t} + X_H) = dH(\underset{\sim}{t}) = \frac{\partial H}{\partial t}$$

since $dH \cdot X_H = 0$. ■

Another way to write this is, using notation in a symplectic chart:

$$\frac{d}{dt}\left[H(t, q_i(t), p^j(t))\right] = \frac{\partial H}{\partial t}(t, q_i(t), p^j(t))$$

where $q_i(t), p^j(t)$ is an integral curve of Hamilton's equations.
Next we shall bring in the contact structure.

5.1.13 Theorem (Cartan). *Let (P, ω) be a symplectic manifold and $H \in \mathfrak{F}(\mathbf{R} \times P)$. Let $\tilde{\omega}$ be defined by* 5.1.9 *and set*

$$\omega_H = \tilde{\omega} + dH \wedge dt$$

Then

(*i*) *$(\mathbf{R} \times P, \omega_H)$ is a contact manifold;*
(*ii*) *$\tilde{X}_H$ generates the line bundle of ω_H; in fact, $\tilde{X}_H$ is the unique vector field satisfying*

$$i_{\tilde{X}_H}\omega_H = 0 \quad \textit{and} \quad i_{\tilde{X}_H} dt = 1;$$

moreover, if F is the flow of X_H, $F^*\omega = \tilde{\omega} - dH \wedge dt$; and
(*iii*) *if $\omega = d\theta$ and $\theta_H = \pi_2^*\theta + H\,dt$, then $\omega_H = d\theta_H$; if $H + (\theta \circ \pi_2)(X_H)$ is nowhere zero, then $(\mathbf{R} \times P, \theta_H)$ is an exact contact manifold.*

Proof. (i) We have $d\omega_H = d\tilde{\omega} + d(dH \wedge dt) = 0$, so ω_H is closed. To show ω_H is of maximal rank, observe that ω_H coincides with $\tilde{\omega}$ on "horizontal vectors" of the form $((s, 0), v_p) \in T_{(s,p)}(\mathbf{R} \times P)$ on which $\tilde{\omega}$ is nondegenerate by 5.1.9. (An alternative way to prove ω_H is of maximal rank is to observe that $dt \wedge \omega_H^n \neq 0$.)

(ii) For t fixed, let d_P denote exterior differentiation on P. Then

$$i_{\tilde{X}_H}\omega_H = i_{\tilde{X}_H}\tilde{\omega} + (i_{\tilde{X}_H} dH)\,dt - dH(i_{\tilde{X}_H} dt)$$

Since $T\pi_2 \cdot \tilde{X}_H(t, p) = X_{H_t}(p)$, we have

$$i_{\tilde{X}_H}\tilde{\omega}(Y) = \tilde{\omega}(\tilde{X}_H, Y) = \omega(T\pi_2 \circ \tilde{X}_H, T\pi_2 \circ Y)$$

$$= \omega(X_H, T\pi_2 \circ Y) = d_P H \cdot (T\pi_2 \circ Y)$$

ISBN 0-8053-0102-X

where Y is a vector field on $\mathbf{R}\times P$. Also, by 5.1.12,

$$i_{\tilde{X}_H}dH=dH(\underline{t})$$

and

$$i_{\tilde{X}_H}dt=i_{\underline{t}+X_H}dt=dt(\underline{t})=1$$

Thus

$$i_{\tilde{X}_H}\omega_H(Y)=d_PH\cdot(T\pi_2\circ Y)+\frac{\partial H}{\partial t}\cdot(Tt\circ Y)-dH\cdot Y=0$$

Since the characteristic bundle is one dimensional, $\tilde{X}_H$ is unique. The last statement follows from $F_{t,s}{}^*\omega=\omega$.

(iii) Clearly $d\theta_H=\omega_H$. Also, using the definition of θ_H,

$$\theta_H(\tilde{X}_H)=(\pi_2^*\theta)(X_H+\underline{t})+H\,dt(X_H+\underline{t})$$
$$=\theta(X_H)+H$$

so θ_H does not vanish on the characteristic bundle of ω. Thus $(\mathbf{R}\times P,\theta_H)$ is an exact contact manifold (see 5.1.8). ■

Notice that $H+(\theta\circ\pi_2)(X_H)$ is (minus) the Lagrangian associated to H according to Sect. 3.6.

Although energy is not conserved for time-dependent Hamiltonian systems, we do have some important integral invariants also due to Cartan.

5.1.14 Theorem (Cartan). *Let (P,ω) be a symplectic manifold, $H\in\mathfrak{F}(\mathbf{R}\times P)$ and ω_H be defined as in 5.1.13 (and θ_H if $\omega=d\theta$). Then:*

(*i*) *$\omega_H,\omega_H^2,\ldots,\omega_H^n$ are invariant forms of $\tilde{X}_H$ (and θ_H is relatively invariant if ω is exact);*
(*ii*) *$dt\wedge\omega_H^n=dt\wedge\tilde{\omega}^n$ is an invariant volume element for $\tilde{X}_H$.*

Proof. (i) $L_{\tilde{X}_H}\omega_H=i_{\tilde{X}_H}d\omega_H+di_{\tilde{X}_H}\omega_H=0$ since ω_H is closed and $\tilde{X}_H$ is a characteristic vector field of ω_H. Since $L_{\tilde{X}_H}$ is a $\wedge$ derivation, $L_{\tilde{X}_H}\omega_H^k=0$, $k=1,2,\ldots,n$.

(ii) Observe that from 5.1.9, $dt\wedge\tilde{\omega}^n$ is a volume element and clearly

$$dt\wedge\omega_H^n=dt\wedge\tilde{\omega}^n$$

Then, from (i),

$$L_{\tilde{X}_H}(dt\wedge\omega_H^n)=(L_{\tilde{X}_H}dt)\wedge\omega_H^n$$

But

$$L_{\tilde{X}_H}dt=d(L_{\tilde{X}_H}t)=d(dt\cdot\tilde{X}_H)=d(1)=0 \quad ■$$

ISBN 0-8053-0102-X

The integral invariants of Cartan just obtained illustrate the importance of suspension. For instance, the time-dependent flow of X_H itself need not be volume preserving, but the suspended flow is volume preserving.

For the interesting topic of adiabatic invariants for time dependent Hamiltonian systems by the method of averaging, we refer to Arnold [1978].

EXERCISES

5.1A. Adapt the Lagrangian formulation to the time-dependent case.

5.1B. Show that, for $f\in\mathfrak{F}(\boldsymbol{R}\times P)$,

$$L_{\tilde{X}_H}f=\frac{\partial f}{\partial t}+\{f,H\}$$

5.1C. (i) Let (M,θ) be an exact contact manifold. Show that there is a unique $Y\in\mathfrak{X}(M)$ such that $i_Y\theta=1$ and $i_Y d\theta=0$. Show that $L_Y d\theta=0$ and that there are local coordinates $(q^1,\ldots,q^n,p_1,\ldots,p_n,t)$ such that $Y=\partial/\partial t$ and $\theta=dt+p_i\,dq^i$.

(ii) Let (P,ω,H) be a Hamiltonian system, $\omega=d\theta$, Σ_e a regular energy surface such that $\theta(X_H)$ is never zero on Σ_e and $i:\Sigma_e\to P$ the inclusion. Show that $(\Sigma_e,i^*\theta)$ is an exact contact manifold. Show that the vector field defined by it, according to (i) is $(1/\theta(X_H))\cdot X_H$ restricted to Σ_e.

5.1D. Let M be a $2n+1$ manifold and ω a closed two-form on M. Then show (M,ω) is a contact manifold if and only if for each $m\in M$, there is an $\alpha_m\in T_m^*M$ such that $\alpha_m\wedge\omega_m^n\neq0$.

5.1E. (Reduction to the autonomous case) Let (P,ω) be a symplectic manifold, and $H\in\mathfrak{F}(\boldsymbol{R}\times P)$. Show that, on $\boldsymbol{R}\times(\boldsymbol{R}\times P)$ with first and second factor projections s and t, $ds\wedge dt+\tilde{\omega}$ is a symplectic form, as is $ds\wedge dt+\tilde{\omega}_H$ [$\tilde{\omega}=\pi_3^*\omega$, where π_3: $(s,t,x)\mapsto x$, etc.] Also, define $\tilde{H}(s,t,x)=H(t,x)-s$. Using the first symplectic form, show that the integral curves of $X_{\tilde{H}}$ are related to those of $\tilde{X}_H$ by projection. Moreover, show that, if F is the flow of $X_{\tilde{H}}$, then $s(F_t(x))=H(t,x)$. (*Hint:* Recall that $L_{\tilde{X}_H}H=\partial H/\partial t$.)

5.1F. Let (P,ω) be a symplectic manifold and $\Omega=dt\wedge\tilde{\omega}^n$, the volume on $\boldsymbol{R}\times P$. For $X\in\mathfrak{X}(\boldsymbol{R}\times P)$, write $X(t,x)=X_x(t)+X_t(x)$ as the vertical and horizontal components. Then show that

$$div_\Omega X(t,x)=\frac{dX_x}{dt}(t)+div_{\omega^n}X_t(x)$$

Use this to give another proof of 5.1.14.

5.1G. (i) Let M be a manifold and $X,Y\in\mathfrak{X}(M)$. Show that

$$L_X i_Y-i_X L_Y=d i_X i_Y-i_X i_Y d$$

(ii) Suppose $H\in\mathfrak{F}(\boldsymbol{R}\times T^*M)$ and $L_Y\theta_H=df$, where $\theta_H=\tilde{\theta}_0+H\,dt$. Then show $g=\theta_H(Y)-f$ is a constant of the motion (i.e., $L_{\tilde{X}_H}g=0$). Show that $dg=-i_Y\omega_H$, and hence that, in the circumstances of 5.1.14, we recover the same constant of the motion. (Under the weaker conditions here, g may depend on time.)

ISBN 0-8053-0102-X

5.2 CANONICAL TRANSFORMATIONS AND HAMILTON–JACOBI THEORY

If (P_1, ω_1) and (P_2, ω_2) are symplectic manifolds of the same dimension, we recall that a map $f: P_1 \to P_2$ is called *symplectic* when $f^*\omega_2 = \omega_1$. We shall begin this section by reformulating this definition in a way that facilitates the introduction of generating functions. Following this we generalize the concept of symplectic maps to the case of time-dependent systems. In either case we study the associated Hamilton–Jacobi equation that relates the generating function of the symplectic map or canonical transformation which trivializes the system, to the given Hamiltonian.

5.2.1 Proposition. *Let (P_1, ω_1) and (P_2, ω_2) be symplectic manifolds, π_i: $P_1 \times P_2 \to P_i$ the projection onto P_i, $i = 1, 2$ and*

$$\Omega = \pi_1^*\omega_1 - \pi_2^*\omega_2 \qquad (\textit{compare } 3.3.25)$$

Then:

(*i*) *Ω is a symplectic form on $P_1 \times P_2$;*
(*ii*) *a map $f: P_1 \to P_2$ is symplectic if and only if $i_f^*\Omega = 0$, where $i_f: \Gamma_f \to P_1 \times P_2$ is inclusion and Γ_f is the graph of f. (In the terminology of the next section, Γ_f is called a Lagrangian submanifold.)*

Proof. The first part is readily verified, as in 3.3.25. To prove (ii), notice that f induces a diffeomorphism of P_1 to Γ_f, so we can write

$$T_{(x, f(x))}\Gamma_f = \{(v, Tf \cdot v) | v \in T_x P_1\}$$

Therefore, by definition of Ω,

$$\begin{aligned}(i_f^*\Omega)((v_1, Tf\cdot v_1), (v_2, Tf \cdot v_2)) &= \omega_1(v_1, v_2) - \omega_2(Tf \cdot v_1, Tf \cdot v_2) \\ &= (\omega_1 - f^*\omega_2)(v_1, v_2)\end{aligned}$$

so (ii) is clear. ■

Suppose we write, locally, $\Omega = -d\Theta$. For example, $\Theta = \pi_1^*\theta_1 - \pi_2^*\theta_2$, where $-d\theta_i = \omega_i$ is a possible choice, but other choices exist as well. Thus $i_f^* d\Theta = di_f^*\Theta = 0$, that is, $i_f^*\Theta$ being closed is equivalent to f being symplectic. Locally, by the Poincaré lemma, $i_f^*\Theta = -dS$ for a function

$$S: \Gamma_f \to R$$

5.2.2 Definition. *We call such an S a **generating function** for the symplectic map f. (It depends on the choice of Θ and is locally defined.)*

At this point we recommend that the reader have a look at Exercises 5.2A and 5.2B in conjunction with Example 5.2.3.

ISBN 0-8053-0102-X

If $(q^1,\ldots,q^n,p_1,\ldots,p_n)$ are coordinates on P_2 and $(Q^1,\ldots,Q^n,P_1,\ldots,P_n)$ are coordinates on P_1, then Γ_f can be given a chart in several ways. For instance, S may appear as a function of $(q^1,\ldots,q^n,Q^1,\ldots,Q^n)$ or of $(q^1,\ldots,q^n,P_1,\ldots,P_n)$, and so forth.

The fact that S is only locally defined is important. Indeed global properties of Γ_f are important in problems of quantization using the Maslov index; see Maslov [1965] and Sect. 5.6. We also refer to Weinstein [1972] for a description of Poincaré's generating function in the context of 5.2.2. (see also Arnold [1978]).

5.2.3 Example. Let $f: P_1 \to P_2$ and write canonical coordinates on P_1 and P_2 as $(Q^1,\ldots,Q^n,P_1,\ldots,P_n)$ and $(q^1,\ldots,q^n,p_1,\ldots,p_n)$ and choose

$$\theta_2 = p_i\,dq^i, \qquad \theta_1 = P_i\,dQ^i$$

Then writing

$$f(Q^1,\ldots,Q^n,P_1,\ldots,P_n) = (q^1,\ldots,q^n,p_1,\ldots,p_n)$$

and regarding S as a function of $(q^1,\ldots,q^n,Q^1,\ldots,Q^n)$, the relationship $i_f^*\Theta = -dS$ reads

$$p_i = \frac{\partial S}{\partial q^i}, \qquad P_i = -\frac{\partial S}{\partial Q^i}$$

For instance, let

$$f: \mathbf{R}^2 \to \mathbf{R}^2$$

be given by

$$Q = \left(\frac{p}{\pi\omega}\right)^{1/2} \sin(2\pi q)$$

$$P = \left(\frac{p\omega}{\pi}\right)^{1/2} \cos(2\pi q)$$

Then f is easily seen to be symplectic away from $p=0$, that is, $dP \wedge dQ = dp \wedge dq$ and we can choose

$$S(q,Q) = -\tfrac{1}{2}\omega Q^2 \cot(2\pi q)$$

This canonical transformation maps the Hamiltonian of the harmonic oscillator $H(P,Q) = \frac{1}{2}(P^2 + \omega^2 Q^2)$ to $H(p,q) = (\omega/2\pi)p$; the latter has trivial integral curves.

ISBN 0-8053-0102-X

Recall from Sect. 3.6 that the canonical one-form $p_i dq^i$ is closely related to the action. Indeed, in the Lagrangian formulation, the action is

$$A = \frac{\partial L}{\partial \dot{q}^i}\dot{q}^i = p_i \dot{q}^i$$

so

$$A\,dt = p_i\,dq^i$$

Thus if we pull down the relationship $i_f^*\Theta = -dS$ to P_1 and integrate along a path, we see that S measures the change in the action induced by f. Therefore S itself is sometimes called the *action*.

We shall now discuss the time-independent Hamilton–Jacobi equation. The time-dependent case will be discussed below. The idea is rather simple: we seek a symplectic map f such that the new Hamiltonian will be totally in equilibrium, that is, $H \circ f = E = constant$, so Q^i and P_i can be treated as integration constants for $H \circ f$. The transformation of the original Hamiltonian then reads, by substituting $p_i = \partial S/\partial q^i$,

$$H\left(q^i, \frac{\partial S}{\partial q^i}\right) = E$$

which is the Hamilton–Jacobi equation.

The next result summarizes the situation.

5.2.4 Theorem (Hamilton–Jacobi). *Let $P = T^*Q$ with the canonical symplectic structure $\omega = -d\theta_0$. Let X_H be a given Hamiltonian vector field on P, and let $S: Q \to R$. Then the following two conditions are equivalent:*

(*i*) *for every curve $c(t)$ in Q satisfying*

$$c'(t) = T\tau_Q^* \cdot X_H(dS(c(t)))$$

the curve $t \mapsto dS(c(t))$ is an integral curve of X_H;

(*ii*) *S satisfies the Hamilton–Jacobi equation $H \circ dS = E$, a constant, that is,*

$$H\left(q^i, \frac{\partial S}{\partial q^i}\right) = E$$

Proof. Assume (ii) and let $p(t) = dS(c(t))$, where $c(t)$ satisfies the stated equation. Then, by the chain rule,

$$\begin{aligned} p'(t) &= T\,dS(c(t)) \cdot c'(t) \\ &= T\,dS(c(t)) \cdot T\tau_Q^* X_H(dS(c(t))) \\ &= T(dS \circ \tau_Q^*) \cdot X_H(dS(c(t))) \end{aligned}$$

ISBN 0-8053-0102-X

Now we use the following symplectic identity:

5.2.5 Lemma. *On T^*Q we have, for any function $S: Q \to \mathbf{R}$,*

$$\omega\big(T(dS\circ\tau_Q^*)\cdot v, w\big)=\omega\big(v, w-T(dS\circ\tau_Q^*)\cdot w\big)$$

Proof. From 3.2.11, $(dS)^*\omega=0$, so the identity in the lemma is equivalent to

$$\omega\big(v-T(dS\circ\tau_Q^*)v, w-T(dS\circ\tau_Q^*)w\big)=0$$

But this identity is obvious since $v-T(dS\circ\tau_Q^*)v$ is vertical ($T\tau_Q^*$ of it is zero) and from the local formula for ω (see formula 2 on p. 179), it vanishes on a pair of vertical vectors. ▼

Thus we get, for $w\in T_{p(t)}P$,

$$\begin{aligned}&\omega\big(T(dS\circ\tau_Q^*)\cdot X_H(p(t)), w\big)\\ &=\omega\big(X_H(p(t)), w\big)-\omega\big(X_H(p(t)), T(dS\circ\tau_Q^*)\cdot w\big)\\ &=\omega\big(X_H(p(t)), w\big)-dH(p(t))\cdot TdS(p(t))\cdot w\end{aligned}$$

But $dH(p(t))\cdot TdS(p(t))=d(H\circ dS)(p(t))$ vanishing is exactly the Hamilton–Jacobi equation. Thus, assuming (ii), we get

$$T(dS\circ\tau_Q^*)\cdot X_H(p(t))=X_H(p(t))$$

so (i) follows.

The proof that (i) implies (ii) follows from these arguments in the same way. ■

We encourage the reader to give a coordinate proof of this result, consulting a standard text, and then to reread the proof just given. Notice that this proof works in infinite dimensions. Future applications of the Hamilton–Jacobi equation in the infinite dimensional case may center around quantization problems for field theories (for applications in general relativity, see Misner, Thorne, and Wheeler [1973]).

Remarks. (1) Theorem 5.2.4 establishes a basic link between nonlinear *first-order scalar* partial differential equations and ordinary differential equations, often referred to as the method of characteristics (cf. 2.2.22 and textbooks such as Duff [1956] or John [1975]).

(2) In many examples, such as $H=K+V$, $T\tau_Q^*\cdot X_H$ is a metric, so condition (i) means $c(t)$ is a gradient line of S, so is orthogonal to level surfaces (or a "wave of action") of S. Additional results along these lines are given in Sect. 5.3.

ISBN 0-8053-0102-X

(3) S is related to the wave function ψ in quantum mechanics by being its phase:

$$\psi = e^{iS/\hbar}$$

where $\hbar$ is a constant (with the dimensions of action). For $H = K + V$, that is, $H(q,p) = (1/2m)p^2 + V(q)$, the Hamilton–Jacobi equation is easily seen to imply

$$-\frac{\hbar^2}{2m}\nabla^2\psi + V\psi = E\psi + \frac{\hbar}{2mi}\psi\nabla^2 S$$

If the last *nonlinear* term is omitted, this reduces to the Schrödinger equation.

(4) We explicate Remark 3 in a more general context. Suppose P is a linear partial differential equation on a manifold Q, written symbolically as

$$P\left(x, \frac{\partial}{\partial x}, \tau\right) u(x) = 0 \tag{1}$$

Assume that P is of order m and depends on a large parameter τ. We take P of the form

$$P\left(x, \frac{\partial}{\partial x}, \tau\right) = \sum_{j=0}^{m} P_{m-j}\left(x, \frac{\partial}{\partial x}\right)\tau^j$$

where P_{m-j} is a linear partial differential operator of order $m-j$. The *principal symbol* of P_{m-j} is the smooth real-valued function σ_{m-j} on T^*Q defined on $(x,\alpha) = \alpha_x \in T_x^*Q$ by replacing $\partial/\partial x^i$ by the coefficients α_i in the top order term of P_{m-j}. Let $H(x,\alpha)$ on T^*Q be defined by

$$H(x,\alpha) = \sum_{j=0}^{m} \sigma_{m-j}(x,\alpha)$$

Now we try to solve Eq. (1) by a solution of the form

$$u(x,\tau) = e^{i\tau S(x)} a(x,\tau) \tag{2}$$

If $a(x,\tau)$ is expanded in an asymptotic series in τ,

$$a(x,\tau) \sim \sum_{k=0}^{\infty} a_k(x)\tau^{\mu-k}$$

and Eq. (2) is substituted in Eq. (1), and coefficients of powers of τ are set equal to zero, the leading term is exactly the Hamilton–Jacobi equation:

$$H(x, dS(x)) = 0 \tag{3}$$

ISBN 0-8053-0102-X

(The constant E is built into H in this formulation.) Thus one expects that as $\tau \to \infty$, the solutions of Eq. (1) are approximated by $e^{i\tau S}$, where S solves Eq. (3), that is, by a Hamiltonian system. For the Schrodinger equation with $\tau = 1/\hbar$, this is the usual classical limit of quantum mechanics (by the "WKB" method; cf. Birkhoff [1933]). For the (reduced) wave equation $P = \nabla + \tau^2/f(x)^2$, Eq. (3) is the *eikonal equation* of geometrical optics.

Equations for other coefficients of powers of τ lead to *transport equations*; for geometrical optics, they govern the propagation of amplitudes normal to the wave fronts $S = constant$. These ideas form the basis of the fundamental generalized Huygens principle of Courant and Lax; see Courant–Hilbert [1962, p. 735].

The base integral curves of X_H in Q are called the *bicharacteristic curves* for the original operator P and play a key role in recent work of Egorov [1969] and Nirenberg and Treves [1970].

These ideas are related to the pioneering work of Keller [1958], Lax [1957], and Maslov [1965], and have led to and are related to the theory of Fourier Integral Operators and geometric quantization; see Hormander [1971], Duistermaat and Hormander [1972), Duistermaat [1974], Guckenheimer [1973c], Guillemin and Sternberg [1977], Sect. 5.4, and references therein. Some additional remarks are given at the end of Sect. 5.3.

Now we turn to time-dependent canonical transformations. These will map $\boldsymbol{R} \times P_1$ to $\boldsymbol{R} \times P_2$ and we wish them to preserve the form of the time-dependent Hamilton equations, as explained in the previous section.

As an example, consider a complete Hamiltonian vector field X_H on a symplectic manifold (P, ω). Suppose X_H has flow $F: \boldsymbol{R} \times P \to P$. Then define its suspension by

$$\tilde{F}: \boldsymbol{R} \times P \to \boldsymbol{R} \times P: (\lambda, x) \mapsto (\lambda, F(\lambda, x))$$

which becomes a diffeomorphism. In fact, $\tilde{F}^{-1}$ maps the integral curves of $\tilde{X}_H$ into integral curves of $\underset{\sim}{t}$, that is, fibers of the line bundle $\pi_2: \boldsymbol{R} \times P \to P$. We say that $\tilde{F}$ *transforms H to equilibrium*. (See Fig. 5.2-1.) We shall see shortly that $\tilde{F}$ is an example of a canonical transformation.

5.2.6 Definition. *Let (P_1, ω_1) and (P_2, ω_2) be symplectic manifolds and $(\boldsymbol{R} \times P_i, \tilde{\omega}_i)$ the corresponding contact manifolds. A smooth mapping $F: \boldsymbol{R} \times P_1 \to \boldsymbol{R} \times P_2$ is called a* ***canonical transformation*** *if each of the following holds:*

(C1) *F is a diffeomorphism;*

(C2) *F preserves time; that is, $F^* t = t$, or the following diagram commutes:*

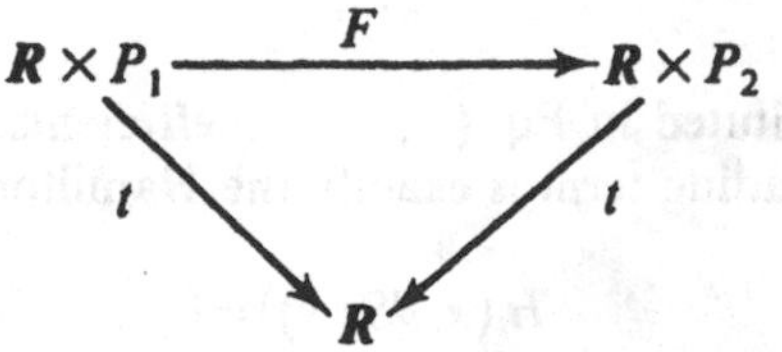

ISBN 0-8053-0102-X

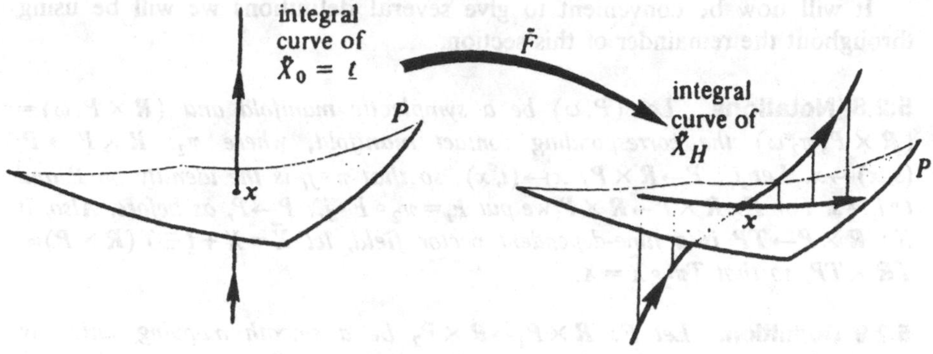

Figure 5.2-1

(C3) *There is a function $K_F \in \mathfrak{F}(R \times P_1)$ such that*

$$F^*\tilde{\omega}_2 = \omega_{K_F}$$

where $\omega_{K_F} = \tilde{\omega}_1 + dK_F \wedge dt$.

In the following, we will also allow the domain and range of F to be open subsets of $R \times P_1$ (or $R \times P_2$).

The function K_F will be related to the *generating function* of F in 5.2.15 below. We shall soon derive several more familiar conditions, equivalent to (C3). First, let us note the following.

5.2.7 Proposition. *The set of all canonical transformations on $(R \times P, \tilde{\omega})$ forms a group under composition.*

Proof. Let F and G be canonical transformations on $(R \times P, \tilde{\omega})$; that is, from $R \times P$ onto $R \times P$. We must show that $F \circ G^{-1}$ satisfies (C1), (C2), and (C3). However, (C1) and (C2) are clear, and for (C3),

$$(F \circ G^{-1})^*\tilde{\omega} = (G^{-1})^* \circ F^*\tilde{\omega} = (G^{-1})^*(\tilde{\omega} + dK_F \wedge dt)$$

But $G^*\tilde{\omega} = \tilde{\omega} + dK_G \wedge dt$, so $(G^{-1})^*\tilde{\omega} = \tilde{\omega} - (G^{-1})^*(dK_G \wedge dt)$ and hence

$$\begin{aligned}(F \circ G^{-1})^*\tilde{\omega} &= \tilde{\omega} - (G^{-1})^*(dK_G \wedge dt) + (G^{-1})^*(dK_F \wedge dt)\\ &= \tilde{\omega} - d(K_G \circ G^{-1}) \wedge d(t \circ G^{-1}) + d(K_F \circ G^{-1}) \wedge d(t \circ G^{-1})\\ &= \tilde{\omega} + d(K_F \circ G^{-1} - K_G \circ G^{-1}) \wedge dt \qquad \blacksquare\end{aligned}$$

As a corollary, we have

$$K_{F \circ G^{-1}} = K_F \circ G^{-1} - K_G \circ G^{-1}$$

ISBN 0-8053-0102-X

It will now be convenient to give several definitions we will be using throughout the remainder of this section.

5.2.8 Notations. *Let (P,ω) be a symplectic manifold and $(\boldsymbol{R}\times P,\tilde{\omega})=(\boldsymbol{R}\times P,\pi_2^*\omega)$ the corresponding contact manifold, where π_2: $\boldsymbol{R}\times P\to P$; $(s,x)\mapsto x$. Let j_t: $P\to\boldsymbol{R}\times P$; $x\mapsto(t,x)$, so that $\pi\circ j_t$ is the identity on P and $t\circ j_s=s$. For F: $\boldsymbol{R}\times P\to\boldsymbol{R}\times P$ we put $F_t=\pi_2\circ F\circ j_t$: $P\to P$, as before. Also, if X: $\boldsymbol{R}\times P\to TP$ is a time-dependent vector field, let $\tilde{X}=X+\underline{t}\in T(\boldsymbol{R}\times P)\approx T\boldsymbol{R}\times TP$, so that $T\pi_2\circ\tilde{X}=X$.*

5.2.9 Definition. *Let F: $\boldsymbol{R}\times P_1\to\boldsymbol{R}\times P_2$ be a smooth mapping satisfying* (C1). *Then F is said to have* **property** (S) *iff F_t: $P\to P$ is symplectic for each $t\in\boldsymbol{R}$.*

5.2.10 Proposition. *A mapping F: $\boldsymbol{R}\times P_1\to\boldsymbol{R}\times P_2$ has property (S) iff there is a one-form α on $\boldsymbol{R}\times P$ such that $F^*\tilde{\omega}_2=\tilde{\omega}_1+\alpha\wedge dt$.*

Proof. If $F^*\tilde{\omega}_2=\tilde{\omega}_1+\alpha\wedge dt$, then

$$\begin{aligned}F_t^*\omega_2&=(j_t^*\circ F^*\circ\pi_2^*)(\omega_2)=j_t^*\circ F^*\tilde{\omega}_2\\&=j_t^*(\tilde{\omega}_1+\alpha\wedge dt)=(\pi_2\circ j_t)^*\omega_1+j_t^*\alpha\wedge d(t\circ j_t)\\&=\omega_1\end{aligned}$$

since $\pi_2\circ j_t=$ *identity* and $t\circ j_t$ is constant.

Conversely, assume F_t is symplectic and let $\beta=F^*\tilde{\omega}_2-\tilde{\omega}_1$. Then

$$j_t^*\beta=j_t^*F^*\tilde{\omega}_2-j_t^*\tilde{\omega}_1=F_t^*\omega_2-\omega_1=0$$

Now we can write

$$\beta=\gamma+\alpha\wedge dt$$

where γ does not involve dt; that is, $\pi_2^*j_t^*\beta=\gamma$ at points of $\{t\}\times P$. Thus, since $j_t^*\beta=0$, $\gamma=0$. ■

Taking $\alpha=dK_F$ leads to the following.

5.2.11 Corollary. *Condition* (C3) *implies condition* (S).

In case the symplectic forms ω_i are exact, $\omega_i=-d\theta_i$, (C3) is clearly equivalent to:

(C4) *There is a K_F such that $F^*\tilde{\theta}_2-\theta_{K_F}$ is closed, where, as usual,*

$$\tilde{\theta}_i=dt+\pi_2^*\theta_i\quad\textit{and}\quad\theta_{K_F}=\tilde{\theta}_1-K_F dt$$

ISBN 0-8053-0102-X

5.2.12 Proposition. *Suppose $F: \mathbf{R}\times P_1 \to \mathbf{R}\times P_2$ satisfies* (C2). *Then* (C3) *is equivalent to the following.*

(C5) *For all $H\in\mathfrak{F}(\mathbf{R}\times P_2)$, there is a $K\in\mathfrak{F}(\mathbf{R}\times P_1)$ such that*

$$F^*\omega_H=\omega_K$$

Proof. If (C3) holds, let

$$K=H\circ F+K_F$$

Then

$$\begin{aligned} F^*\omega_H &= F^*(\tilde{\omega}_2+dH\wedge dt)\\ &= F^*\tilde{\omega}_2+d(H\circ F)\wedge d(t\circ F)\\ &= F^*\tilde{\omega}_2+d(H\circ F)\wedge dt\\ &=\tilde{\omega}_1+dK_F\wedge dt+d(H\circ F)\wedge dt\\ &=\tilde{\omega}_1+dK\wedge dt \end{aligned}$$

by (C2).

Conversely, let K_F denote the K determined by $H=0$. Then $F^*\omega_H=\omega_K$ reduces to (C3). ■

5.2.13 Proposition. *Let $F: \mathbf{R}\times P_1\to\mathbf{R}\times P_2$ satisfy* (C1) *and* (C2). *Then* (C3) *is equivalent to each of the following.*

(C6) *(S) holds and, for all $H\in\mathfrak{F}(\mathbf{R}\times P_2)$, there is a $K\in\mathfrak{F}(\mathbf{R}\times P_1)$ such that $F^*\tilde{X}_H=\tilde{X}_K$.*

(C7) *(S) holds, and there is a function $K_F\in\mathfrak{F}(\mathbf{R}\times P_1)$ such that $F^*\underline{t}=\tilde{X}_{K_F}$.*

Proof. Assume (C3) holds. Then by 5.2.11 (S) holds. Let K be given by $K=H\circ F+K_F$. From (C5), $F^*\omega_H=\omega_K$. Thus to prove (C6) it suffices to show

$$i_{F^*\tilde{X}_H}\omega_K=0 \quad\text{and}\quad i_{F^*\tilde{X}_H}dt=1$$

from 5.2.12. But

$$\begin{aligned} i_{F^*\tilde{X}_H}\omega_K &= i_{F^*\tilde{X}_H}F^*\omega_H\\ &= F^*i_{\tilde{X}_H}\omega_H=0 \end{aligned}$$

Similarly

$$\begin{aligned} i_{F^*\tilde{X}_H}dt &= i_{F^*\tilde{X}_H}F^*dt\\ &= F^*i_{\tilde{X}_H}dt\\ &= F^*\cdot 1=1 \end{aligned}$$

Second, (C6) implies (C7) by taking $H=0$.

ISBN 0-8053-0102-X

Finally, we must show (C7) implies (C3). From 5.2.10, $F^*\tilde{\omega}_2=\tilde{\omega}_1+\alpha\wedge dt$. Hence, writing $K_F=K(F)$,

$$i_{\tilde{X}_{K(F)}}F^*\tilde{\omega}_2=i_{\tilde{X}_{K(F)}}\tilde{\omega}_1+(i_{\tilde{X}_{K(F)}}\alpha)\wedge dt-\alpha\wedge i_{\tilde{X}_{K(F)}}dt$$

On the other hand,

$$\tilde{X}_{K_F}=F^*\underline{t} \quad \text{by (C7) so} \quad i_{\tilde{X}_{K(F)}}F^*\tilde{\omega}_2=F^*i_{\underline{t}}\tilde{\omega}_2=0$$

since $\tilde{\omega}_2=\pi_2^*\omega_2$, and $i_{\tilde{X}_{K(F)}}dt=1$ since $F^*\underline{t}=\underline{t}$ by (C2). Comparing the two expressions, and using $i_{\tilde{X}_{K(F)}}dt=1$ we have $\alpha=i_{F^*\underline{t}}\tilde{\omega}_1+(i_{F^*\underline{t}}\alpha)dt$. Thus $F^*\tilde{\omega}_2=\tilde{\omega}_1+(i_{F^*\underline{t}}\tilde{\omega}_1)\wedge dt$, and as

$$i_{F^*\underline{t}}\tilde{\omega}_1=i_{\tilde{X}_{K(F)}}\tilde{\omega}_1=dK_F-\frac{\partial K_F}{\partial t}dt,$$

we get $F^*\tilde{\omega}_2=\tilde{\omega}_1+dK_F\wedge dt$. ■

The statement $F^*\tilde{X}_H=\tilde{X}_K$ of (C6) is the precise meaning of the assertion: *F preserves the form of all time-dependent Hamiltonian equations.*

The classical form for a canonical transformation is essentially (C6) without the condition (S), as follows.

5.2.14 Theorem (Jacobi). *If $F: \mathbf{R}\times P_1\to\mathbf{R}\times P_2$ satisfies* (C1) *and* (C2), *then* (C3) *is equivalent to the following.*

(C8) *There is a function $K_F\in\mathfrak{F}(\mathbf{R}\times P_1)$ such that for all $H\in\mathfrak{F}(\mathbf{R}\times P_2)$, $F^*\tilde{X}_H=\tilde{X}_K$, where $K=H\circ F+K_F$.*

Proof. That (C3) implies (C8) was shown in 5.2.13. For the converse, taking $H=0$, we have $F^*\underline{t}=\tilde{X}_{K_F}$ and so for an arbitrary H we have

$$F^*\tilde{X}_H=F^*X_H+\tilde{X}_{K_F}$$

By (C8)

$$\tilde{X}_K=X_K+\underline{t}=X_{H\circ F}+X_{K_F}+\underline{t} \quad \text{and} \quad \tilde{X}_K=F^*\tilde{X}_H$$

so, combining the two expressions,

$$F^*X_H=X_{H\circ F}$$

Therefore,

$$j_t^*F^*\pi_2^*X_{H_t}=X_{H_t\circ F_t}$$

where

$$H_t=H\circ j_t=j_t^*H$$

ISBN 0-8053-0102-X

In other words,

$$F_t^* X_{H_t} = X_{H_t \circ F_t}$$

Since this holds for all $H \in \mathfrak{F}(R \times P_2)$, F_t is symplectic (see 3.3.19). Thus (C8) implies (C6) and hence (C3). ■

The (time-dependent) principal function of Hamilton is the analog of the generating function in the time-independent case and is derived from (C4).

5.2.15 Definition. *Let F be canonical and locally write $\omega_1 = -d\theta_1$, $\omega_2 = -d\theta_2$, and so forth as in (C4). Then if we locally write*

$$F^*\tilde{\theta}_2 - \theta_{K_F} = dW$$

*for $W: R \times P_1 \to R$, we call W a **generating function** for F.*

Notice that the local existence of W is *equivalent* to F being canonical. If we write down the definitions of $\tilde{\theta}_2$ and θ_{K_F}, the definition of W reads

$$F^*\left(dt + \pi_2^*\theta_2\right) - \left(dt + \pi_2^*\theta_1\right) + K_F\,dt = dW$$

that is,

$$F^*\pi_2^*\theta_2 - \pi_2^*\theta_1 + K_F\,dt = dW$$

since $F^*dt = dt$. Letting $\dot{F}$ be the coefficient of dt in $F^*\pi_2^*\theta_2$, we get

5.2.16 Proposition. *If F is canonical and has generating function W, then*

$$K_F = \partial W/\partial t - \dot{F}$$

and for a Hamiltonian H on $R \times P_2$,

$$F^*\tilde{X}_H = \tilde{X}_K \quad \textit{where } K = H \circ F + (\partial W/\partial t - \dot{F})$$

If F can be chosen so that K is constant, then we will have transformed the Hamiltonian system $\tilde{X}_H$ to one that is trivial; $\tilde{X}_K = \underset{\sim}{t}$. All points for $\tilde{X}_K$ are equilibrium (or fixed) points, so we have the following:

5.2.17 Definition. *Let $F: R \times P_1 \to R \times P_2$ be a canonical transformation and $H \in \mathfrak{F}(R \times P_2)$. We say that F **transforms** H to **equilibrium** if $K = H \circ F + K_F =$ constant.*

If we assume we can write W as a function

$$W(t, q^1, \ldots, q^n, Q^1, \ldots, Q^n)$$

ISBN 0-8053-0102-X

so that F is given by

$$p_i = \partial W/\partial q^i, \qquad P_i = -\partial W/\partial Q^i$$

as in 5.2.3, then from 5.2.16 and $(\partial W/\partial t)_{q,p=\text{const.}} = (\partial W/\partial t)_{q,Q=\text{const.}} + (\partial W/\partial Q^i)(\partial Q^i/\partial t)$, we see that F transforms H to equilibrium if and only if

$$H(t, q^i, \partial W/\partial q^i) + \partial W/\partial t = \text{constant}$$

This is the *time-dependent Hamilton–Jacobi equation*. Since the new Hamiltonian is in equilibrium, we can regard the Q^i as integration constants.

5.2.18 Theorem (Hamilton–Jacobi). *Let $P = T^*Q$ with the canonical symplectic structure. Let $\tilde{X}_H$ be a Hamiltonian system on $\mathbf{R} \times P$ and let W: $\mathbf{R} \times Q \to \mathbf{R}$. Then the following conditions are equivalent:*

(*i*) *for every curve $c(t)$ in Q satisfying*

$$c'(t) = T\tau_Q^* \cdot X_{H_t}(dW_t(c(t)))$$

the curve $t \mapsto dW_t(c(t))$ is an integral curve of X_H;

(*ii*) *W satisfies the Hamilton–Jacobi equation*

$$H_t \circ dW_t + \partial W/\partial t = \text{constant}$$

that is,

$$H(t, q^i, \partial W/\partial q^i) + \partial W/\partial t = \text{constant on } P$$

This is proved in the same way as 5.2.4, so may be omitted. The relationship between W and S, in case H is independent of time, is easily seen to be

$$W = S - (E + \text{constant})t$$

We refer to 5.3.33 for another formulation of the Hamilton–Jacobi theory. The remarks following 5.2.4 have analogs for the case of the time-dependent Hamilton–Jacobi equation; for instance, Remark 3 leads to the time-dependent Schrodinger equation.

We next give an application of the theory of canonical transformations just developed to show that, locally for a Hamiltonian system, coordinates can be found in which H and t appear as canonically conjugate variables.

The result is actually a special case of more general considerations treated in Chapter 4.* However, it is useful to treat it here from the present point of view. The result may be regarded as a refinement of the straightening out theorem proved in Sect. 2.1, to the Hamiltonian case.

---

*In Example 4.3.2(ii), it is shown that the manifold of solutions of constant energy is a symplectic manifold. If we choose coordinates H, t and symplectic coordinates on the reduced symplectic manifold (the remaining $2n-2$ coordinates), we obtain symplectic coordinates on all of P.

ISBN 0-8053-0102-X

5.2.19 Theorem (Hamiltonian flow box). *Let (P,ω) be a symplectic manifold, $H\in\mathfrak{F}(P)$, and suppose $dH(x_0)\neq 0$ for some $x_0\in P$. Then there is a symplectic chart (U,φ) at m with $\varphi(U)=I\times W\subset \mathbf{R}\times\mathbf{R}^{2n-1}$, $I=(-a,a)$, $\varphi(u)=(q^1(u),\ldots,q^n(u),p_1(u),\ldots,p_n(u))$ and $\varphi(x_0)=(0,\mathbf{0})$ such that $\varphi^{-1}|I\times\{w\}$ is an integral curve of X_H for all $w\in W$ (with parameter q^1) and $p_1(u)=H(u)-e$, where $e=H(x_0)$.*

Proof. First, we may assume that X_H is complete. For, choosing neighborhoods U_1, U_2 of x_0 with $U_1\subset U_2$ and U_2 compact, let $h\in\mathfrak{F}(P)$ be a bump function with $h|U_1=1$ and $h|P\setminus U_2=0$. Then by 2.1.19, X_{hH} is complete and coincides with X_H on U_1.

Thus suppose X_H complete, with flow F. Define $\tilde F: \mathbf{R}\times P\to\mathbf{R}\times P$ by $(\lambda,x)\mapsto(\lambda,F(\lambda,x))$. Clearly $\tilde F$ is a diffeomorphism and preserves time. Also, F_t is symplectic and $\tilde F^{-1}{}_*t=\tilde X_H$. Thus by (C7), $\tilde F$ is a canonical transformation with $K_{\tilde F^{-1}}=H$, and $K_{\tilde F}=-H\circ F$.

Next, let (U_0,ψ) be a symplectic chart such that $dH(u)\neq 0$ in U_0. Also, we may assume that at x_0, $X_H(x_0)=T\psi^{-1}(e_1)$; that is, X_H points in the direction of the first coordinate axis. This is possible because, in the proof of Darboux's theorem, the initial vector is arbitrary. If $\psi(u)=(q^1(u),\ldots,q^n(u),p_1(u),\ldots,p_n(u))$, let V_0 be the submanifold defined by $q^1=0$. From 2.1.9, there is an $\varepsilon>0$ and $V\subset V_0$ such that F maps $V_1=(-\varepsilon,\varepsilon)\times V$ diffeomorphically into U_0 (see Fig. 5.2-2.)

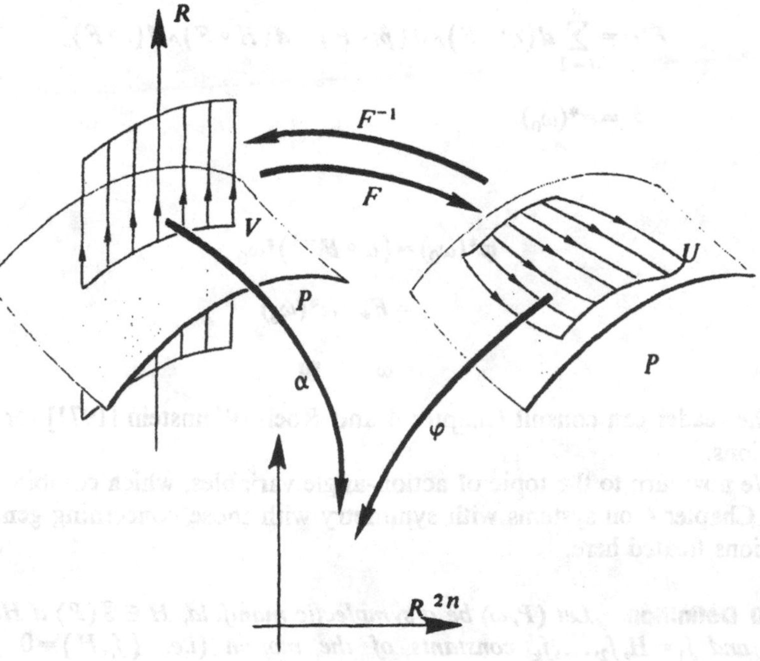

Figure 5.2-2

ISBN 0-8053-0102-X

Let $U=F(V_1)$ and F also stand for F restricted to V_1. On U we have defined, by means of the chart (U_0,ψ), $q^1,\ldots,q^n,p_1,\ldots,p_n\in\mathfrak{F}(U)$ and, in addition, $t=\pi_1\circ F^{-1}$, where π_1 is the projection onto the first factor. Define a mapping

$$\alpha:(-\varepsilon,\tau)\times V\to R^{2n}: v\mapsto(\pi_1(v),q^2\circ F(v),\ldots,q^n\circ F(v)$$

$$H\circ F(v)-e,p_2\circ F(v),\ldots,p_n\circ F(v))$$

and a mapping

$$\varphi: U\to R^{2n};\ u\mapsto(t(u),q^2(u),\ldots,q^n(u),H(u)-e,p_2(u),\ldots,p_n(u))$$

so the diagram indicated in Fig. 5.2-2 commutes. Let ω_0 be the canonical two-form on R^{2n}. We shall show that $\varphi^*(\omega_0)=\omega$.

Indeed,

$$\alpha^*(\omega_0)=d\pi_1\wedge d(H\circ F)+\sum_{i=2}^{n}d(q^i\circ F)\wedge d(p_i\circ F)$$

Also,

$$\tilde{F}^*\tilde{\omega}=\tilde{\omega}-d(H\circ F)\wedge d(t\circ F)$$

since $K_{\tilde{F}}=-H\circ F$. Hence, restricting F to V_1 $(q^1=0)$,

$$\begin{aligned}F^*\omega&=\sum_{i=2}^{n}d(q^i\circ F)\wedge d(p_i\circ F)-d(H\circ F)\wedge d(t\circ F)\\&=\alpha^*(\omega_0)\end{aligned}$$

Thus

$$\begin{aligned}\varphi^*(\omega_0)&=(\alpha\circ F^{-1})^*\omega_0\\&=F_*\circ\alpha^*(\omega_0)\\&=\omega\quad\blacksquare\end{aligned}$$

The reader can consult Chapter 4 and Roels–Weinstein [1971] for generalizations.

We now turn to the topic of action-angle variables, which combines ideas from Chapter 4 on systems with symmetry with those concerning generating functions treated here.

5.2.20 Definition. *Let (P,ω) be a symplectic manifold, $H\in\mathfrak{F}(P)$ a Hamiltonian, and $f_1=H,f_2,\ldots,f_k$ constants of the motion (i.e., $\{f_i,H\}=0$ for all $1<i<k$). The set $\{f_1,\ldots,f_k\}$ is said to be in **involution** if $\{f_i,f_j\}=0$ for all*

ISBN 0-8053-0102-X

$1 \leqslant i,j \leqslant k$. The set $\{f_1,\ldots,f_k\}$ is said to be ***independent*** *if the set of critical points of $F=f_1\times\cdots\times f_k$, $\sigma(F)=\{p\in P \mid df_1(p),\ldots,df_k(p)$ are linearly dependent$\}$ has measure zero in P. A set of constants of the motion that is in involution is called* ***integrable*** *(or* ***completely integrable****) if $k=n\equiv\frac{1}{2}\dim P$. (In infinite dimensions the set is called integrable if the set $\{X_{f_i}(p) \mid i=1,2,\ldots\}$ forms a basis for T_pP in the Hilbert or Banach space sense.)*

Recall that $\Sigma(F)$ denotes the *bifurcation set* of $F=f_1\times\cdots\times f_k: P\to \mathbf{R}^k$; that is, it is the set in $\mathbf{R}^k$ over which $F: P\to\mathbf{R}^k$ fails to be a locally trivial fibration (see 4.5.1).

5.2.21 Theorem (Arnold). *Let (P,ω) be a symplectic manifold, $H\in\mathfrak{F}(P)$ a Hamiltonian, $f_1=H, f_2,\ldots,f_n\in\mathfrak{F}(P)$ an independent, integrable system of constants of the motion for H, $n=\frac{1}{2}\dim(P)$. Denote by $F=f_1\times\cdots\times f_n: P\to\mathbf{R}^n$ and let $U\subset\mathbf{R}^n$ be an open set such that $F^{-1}(U)\cap\sigma(F)=\varnothing$.*

(i) If $F|F^{-1}(U): F^{-1}(U)\to U$ is a proper map, then each of $X_{f_i}|F^{-1}(U)$ is complete, $U\subset\mathbf{R}^n\backslash\Sigma(F)$ and the fibers of the locally trivial fibration $F|F^{-1}(U)$ are a disjoint union of manifolds diffeomorphic with the torus $\mathbf{T}^n$.

(ii) If $F|F^{-1}(U): F^{-1}(U)\to U$ is not proper but we assume that each of the Hamiltonian vector fields $X_{f_i}|F^{-1}(U)$, $1\leqslant i\leqslant n$ is complete and $U\subset\mathbf{R}^n\backslash\Sigma(F)$, then each fiber of $F|F^{-1}(U)$ is a disjoint union of manifolds diffeomorphic to the cylinders $\mathbf{R}^k\times\mathbf{T}^{n-k}$.

Proof. The proof of both assertions will be done together. For (i) we have an extra statement to prove: $U\subset\mathbf{R}^n\backslash\Sigma(F)$. By hypothesis, $\sigma(F)\cap F^{-1}(U)=\varnothing$ so that by the independence condition, $F|F^{-1}(U)$ is a *proper* submersion and hence a locally trivial fibration (see 4.5.1). Thus $U\subset\mathbf{R}^n\backslash\Sigma(F)$. Assume we have proved (ii). Then if $F|F^{-1}(U)$ is proper, the fiber $F^{-1}(c)$ is a compact manifold; but by (ii) it is diffeomorphic to $\mathbf{R}^k\times\mathbf{T}^{n-k}$ for some k so that it must be diffeomorphic to $\mathbf{T}^n$. Hence (i) follows from (ii).

Now we prove (ii). Since $U\subset\mathbf{R}^n\backslash\Sigma(F)$, $F|F^{-1}(U)$ is a locally trivial fibration. Let $c\in U$ and $I_c=F^{-1}(c)$ the fiber of $F|F^{-1}(U)$ over c and let I_c^0 be a connected component of I_c. By hypothesis $X_{f_i}|I_c$ is complete and hence so is $X_{f_i}|I_c^0$; note first that I_c is an invariant manifold for all $f_1,\ldots,f_n$, that is, any integral curve of X_{f_i} starting in I_c will stay completely in I_c, $i=1,\ldots,n$ by the involution condition $\{f_i,f_j\}=0$ for all $i,j=1,\ldots,n$. Let φ_t^i be the flow of X_{f_i}. Then φ_t^i acts on I_c^0 for all $i=1,\ldots,n$ and since $\{f_i,f_j\}=0$, $i,j=1,\ldots,n$, the flows φ_t^i and φ_t^j commute. Thus we can define an action

$$\Phi: \mathbf{R}^n\times I_c^0\to I_c^0$$

$$\Phi((t_1,\ldots,t_n),\alpha)=\left(\varphi_{t_1}^1\circ\varphi_{t_2}^2\circ\cdots\circ\varphi_{t_n}^n\right)(\alpha)$$

$\alpha\in I_c^0$, $(t_1,\ldots,t_n)\in\mathbf{R}^n$, $\mathbf{R}^n$ considered as a Lie group with its additive structure.

ISBN 0-8053-0102-X

We prove that this action is transitive. Clearly, $\Phi(\mathbf{R}^n,\alpha)$ is closed and connected in I_c^0. If we show it is open, it will follow that $\Phi(\mathbf{R}^n,\alpha)=I_c^0$ and hence Φ is a transitive action. To show that $\Phi(\mathbf{R}^n,\alpha)$ is open, we prove that $\Phi(\cdot,\alpha)\colon \mathbf{R}^n\to I_c^0$ is an open mapping. For this it is sufficient to show that it maps any open neighborhood of $0\in\mathbf{R}^n$ onto an open neighborhood of $\alpha\in I_c^0$. If $\mathbf{e}_1,\ldots,\mathbf{e}_n$ denotes the standard basis in $\mathbf{R}^n$, note that

$$T_0\Phi(\cdot,\alpha)\cdot \mathbf{e}_i=\frac{d}{dt}\varphi_t^i(\alpha)=X_{f_i}(\alpha)$$

so that by the independence assumption, $T_0\Phi(\cdot,\alpha)\colon \mathbf{R}^n\to T_\alpha I_c^0$ is an isomorphism ($\dim I_c^0=n$) and hence $\Phi(\cdot,\alpha)$ is a local diffeomorphism at $\mathbf{0}\in\mathbf{R}^n$. This proves our claim.

Since $\Phi\colon \mathbf{R}^n\times I_c^0\to I_c^0$ is transitive, the manifold I_c^0 is diffeomorphic to the homogeneous space $\mathbf{R}^n/H$, where H is the isotropy subgroup of an arbitrary element $\alpha_0\in I_c^0$, that is,

$$H=\{(t_1,\ldots,t_n)\in\mathbf{R}^n\,|\,\Phi((t_1,\ldots,t_n),\alpha_0)=\alpha_0\}$$

Since $\dim I_c^0=n$, we must have $\dim H=0$, that is, H is a discrete subgroup of $\mathbf{R}^n$ and is hence generated by $(n-k)$ linearly independent vectors over $\mathbf{R}$, $\mathbf{a}_{k+1},\ldots,\mathbf{a}_n$; that is,

$$H=\left\{v\in\mathbf{R}^n\,|\,v=\sum_{i=k+1}^{n} m_i\mathbf{a}_i,\ m_i\in\mathbf{Z}\right\}$$

(see Exercise 4.1I). Recall from Sect. 4.1 that we have a diffeomorphism $h\colon \mathbf{R}^n/H\to I_c^0$ given by the formula

$$h([t_1,\ldots,t_n])=\Phi((t_1,\ldots,t_n),\alpha_0)$$

where $[t_1,\ldots,t_n]\in\mathbf{R}^n/H$. Define the isomorphism $T\colon \mathbf{R}^n\to\mathbf{R}^n$ by $T\mathbf{a}_i=\mathbf{e}_i$ for $1\leqslant i\leqslant n$, where $\mathbf{a}_1,\ldots,\mathbf{a}_k$ is a completion to a basis for $\mathbf{R}^n$ of the linearly independent system of vectors $\mathbf{a}_{k+1},\ldots,\mathbf{a}_n$. T maps H isomorphically onto $\{(0,\ldots,0)\}\times\mathbf{Z}^{n-k}\subset\mathbf{R}^n$ and hence it induces a diffeomorphism

$$\hat{T}\colon \mathbf{R}^n/H\to\mathbf{R}^n/\{(0,\ldots,0)\}\times\mathbf{Z}^{n-k}=\mathbf{R}^k\times\mathbf{T}^{n-k}$$

Then $\hat{T}\circ h^{-1}\colon I_c^0\to\mathbf{R}^k\times\mathbf{T}^{n-k}$ is a diffeomorphism. ■

5.2.22 Definition. *Let $v\in\mathbf{R}^n$ be a fixed vector and consider the flow $F_t\colon \mathbf{R}^n\to\mathbf{R}^n$ given by $F_t(v)=v+tv$. Denote the canonical projection by $\pi\colon \mathbf{R}^n\to\mathbf{R}^k\times\mathbf{T}^{n-k}$ and let $\varphi_t\colon \mathbf{R}^k\times\mathbf{T}^{n-k}\to\mathbf{R}^k\times\mathbf{T}^{n-k}$ be the unique flow satisfying $\pi\circ F_t=\varphi_t\circ\pi$. φ_t is called a* ***translation-type flow.*** *(The definition equivalently reads*

$$\varphi_t(x_1,\ldots,x_k,\theta_{k+1},\ldots,\theta_n)$$
$$=(x_1+tv_1,\ldots,x_k+tv_k,\theta_{k+1}+tv_{k+1}(\mathrm{mod}\,1),\ldots,\theta_n+tv_n(\mathrm{mod}\,1))$$

ISBN 0-8053-0102-X

where $\nu=(\nu_1,\ldots,\nu_n)\in \boldsymbol{R}^n$, $x_1,\ldots,x_k\in \boldsymbol{R}$, *and* $\theta_{k+1},\ldots,\theta_n\in S^1$ *are the coordinates (angular variables) on the torus* $T^{n-k}=S^1\times\cdots\times S^1$ *[(n−k) times].*

The case $k=0$ deserves a special attention; the flow in this case is called *quasi-periodic*. Then $\varphi_t\colon T^n\to T^n$ and if $\theta_1,\ldots,\theta_n$ are the "coordinate functions" on T^n (angular variables defined *mod* 1), then the formula in the definition of φ_t shows that

$$\theta_i(\varphi_t(x))=\theta_i(x)+\nu_i t(\mathit{mod}\,1)$$

$x\in T^n$, $t\in \boldsymbol{R}$, $1\le i\le n$; $\nu_1,\ldots,\nu_n$ are called the *frequencies of the flow* and they determine completely its character as the next proposition shows.

5.2.23 Proposition. *Each orbit of* φ_t *is dense in* T^n *if and only if* $\sum_{i=1}^n k_i\nu_i=0$, $k_i\in \boldsymbol{Z}$ *implies* $k_i=0$ *for all* $1\le i\le n$, *that is,* ν_i *are independent over* $\boldsymbol{Z}$.

We refer the reader to Example 4.1.11 and Sell [1971] for a discussion of these toral automorphisms and the proof of this result.

5.2.24 Theorem (Arnold). *In the notations and hypotheses of Theorem 5.2.20, if* I_c^0 *denotes a connected component of* $I_c=F^{-1}(c)$ *and* $\varphi_t=\varphi_t^1$ *denotes the flow of* $X_H=X_{f_1}$, *then* $\varphi_t|I_c^0$ *is differentiably conjugate to a translation type flow on* $\boldsymbol{R}^k\times T^{n-k}$ *(for* $k=0$ *the flow is quasi-periodic).*

Proof. The diffeomorphism

$$h\circ\hat{T}^{-1}\colon \boldsymbol{R}^k\times T^{n-k}\to \boldsymbol{R}^k\times T^{n-k}$$

stated at the end of the proof of 5.2.20 defines the commutative diagram

$$\begin{array}{ccccc}
\boldsymbol{R}^k\times T^{n-k} & \xrightarrow{\hat{T}^{-1}} & \boldsymbol{R}^n/H & \xrightarrow{h} & I_c^0\\
\Big\downarrow{\scriptstyle \psi_t} & & \Big\downarrow{\scriptstyle \chi_t} & & \Big\downarrow{\scriptstyle \varphi_t|I_c^0}\\
\boldsymbol{R}^k\times T^{n-k} & \xrightarrow[\hat{T}^{-1}]{} & \boldsymbol{R}^n/H & \xrightarrow[h]{} & I_c^0
\end{array}$$

with the flows $\chi_t=h^{-1}\circ\varphi_t|I_c^0\circ h$, $\psi_t=\hat{T}\circ\varphi_t\circ\hat{T}^{-1}$, where

$$\begin{aligned}\chi_t([t_1,\ldots,t_n])&=(h^{-1}\circ\varphi_t)(\Phi((t_1,\ldots,t_n),\alpha_0))\\
&=h^{-1}\circ\Phi((t+t_1,t_2,\ldots,t_n),\alpha_0)\\
&=[t+t_1,t_2,\ldots,t_n]\end{aligned}$$

ISBN 0-8053-0102-X

Consider the flow $F_t\colon \boldsymbol{R}^n \to \boldsymbol{R}^n$, $F_t(v) = v + tTe_1$. We claim that ψ_t is the translation-type flow defined by F_t, that is, $\pi \circ F_t = \psi_t \circ \pi$, where $\pi\colon \boldsymbol{R}^n \to \boldsymbol{R}^k \times T^{n-k}$ is the canonical projection. Denote by $\pi_H\colon \boldsymbol{R}^n \to \boldsymbol{R}^n/H$ the canonical projection and note that by definition of $\hat{T}$ we have $\pi \circ T = \hat{T} \circ \pi_H$. If $(t_1,\ldots,t_n) \in \boldsymbol{R}^n$, denote $T^{-1}(t_1,\ldots,t_n) = (y_1,\ldots,y_n)$. Then we have

$$
\begin{aligned}
(\psi_t \circ \pi)(t_1,\ldots,t_n) &= (\hat{T} \circ \chi_t \circ \hat{T}^{-1} \circ \pi)(t_1,\ldots,t_n) \\
&= (\hat{T} \circ \chi_t \circ \pi_H \circ T^{-1})(t_1,\ldots,t_n) \\
&= (\hat{T} \circ \chi_t)[y_1,\ldots,y_n] \\
&= \hat{T} \circ \pi_H(t + y_1, y_2,\ldots,y_n) \\
&= \pi \circ T(te_1 + (y_1,\ldots,y_n)) \\
&= \pi(tTe_1 + (t_1,\ldots,t_n)) \\
&= (\pi \circ F_t)(t_1,\ldots,t_n) \qquad \blacksquare
\end{aligned}
$$

The two theorems of Arnold show us that complete integrability of a Hamiltonian system imposes serious restrictions on the topology of the manifold T^*M. This is the intuitive reason why integrable systems are *not generic*. (See Markus and Meyer [1974], Robinson [1970b].) The classical way to "integrate" a completely integrable system is given in Exercise 5.2H.

Recall that $F|F^{-1}(\boldsymbol{R}^n \backslash \Sigma_F)$ is a locally trivial fibration, that is, for each $c_0 \in \boldsymbol{R}^n$ there exists an open neighborhood U_0 of c_0 in $\boldsymbol{R}^n \backslash \Sigma_F$ and a smooth map $h_0\colon F^{-1}(U_0) \to I_{c_0} = F^{-1}(c_0)$ such that

$$
h = (F|F^{-1}(U_0)) \times h_0\colon F^{-1}(U_0) \to U_0 \times I_{c_0}
$$

is a diffeomorphism [and hence for any $c \in U_0$, $h_0|F^{-1}(c)\colon F^{-1}(c) = I_c \to I_{c_0}$ is a diffeomorphism]. If we have a Hamiltonian system that is completely integrable, by Arnold's theorem I_{c_0} is a disjoint union of "cylinders" $\boldsymbol{R}^k \times T^{n-k}$ so that we would expect the push-forward $h_*(X_H|F^{-1}(U_0))$ to have a simple form. The first question that naturally arises is if the vector field $h_*(X_H|F^{-1}(U_0))$ is Hamiltonian. The answer is in general "no." However, if I_{c_0} is compact, there is a local trivialization procedure that will make $h_*(X_H|F^{-1}(U_0))$ a Hamiltonian vector field on $U_0 \times T^k$. The variables that will be introduced for this purpose are called *action-angle coordinates*.

We begin with the *standard model for action-angle coordinates*. In the symplectic vector space $\boldsymbol{R}^{2n}$ with coordinates $(q^1,\ldots,q^n,p_1,\ldots,p_n)$ consider the equivalence relation which identifies two vectors $(\boldsymbol{q},\boldsymbol{p}),(\boldsymbol{q}',\boldsymbol{p}')$ if and only if $\boldsymbol{q} = \boldsymbol{q}'$ and $\boldsymbol{p} - \boldsymbol{p}' \in \boldsymbol{Z}^n$. The quotient space is clearly $\boldsymbol{R}^n \times T^n$, which inherits naturally a symplectic structure from $\boldsymbol{R}^{2n}$ via the canonical projection. Let $\boldsymbol{B}^n \subset \boldsymbol{R}^n$ be an open ball in $\boldsymbol{R}^n$ and consider the symplectic manifold $\boldsymbol{B}^n \times \boldsymbol{T}^n$

ISBN 0-8053-0102-X

in which we now introduce canonical coordinates $(I^1,\ldots,I^n,\varphi_1,\ldots,\varphi_n)$ by $I^i=q^i$ in B^n and $\varphi_i=p_i$ $(mod\,1)$, $i=1,\ldots,n$. Then we say that a Hamiltonian $H=H(I^1,\ldots,I^n)$ that does *not* depend on the variables $\varphi_1,\ldots,\varphi_n$ (i.e., these variables are ignorable) has *action-angle coordinates* (I,φ) in $B^n\times T^n$. Thus Hamilton's equations become

$$\dot{I}^i=0,\quad \dot{\varphi}_i=-\partial H/\partial I^i=\nu_i(I^1,\ldots,I^n)$$

and the maps $I^i\colon B^n\times T^n\to R$ become constants of the motion of X_H. Given initial conditions $I^i(0)=I_0^i$, $\varphi_i(0)=\varphi_i^0$, the solution is

$$\varphi_i(t)=\nu_i(I_0^1,\ldots,I_0^n)t+\varphi_0^i(mod\,1),\qquad I^i=I_0^i$$

Thus $I^i=I_0^i$, $i=1,\ldots,n$, are invariant n-tori and the motion on them is periodic or quasi-periodic with frequencies $\nu_i(I_0^1,\ldots,I_0^n)$.

5.2.25 Definition. *A Hamiltonian $H\in\mathfrak{F}(P)$ on a symplectic manifold (P,ω)* ***admits action-angle coordinates*** *(I,φ) in some open set $U\subset P$, if:*

(*i*) *there exists a symplectic diffeomorphism $\psi\colon U\to B^n\times T^n$;*
(*ii*) *$H\circ\psi^{-1}\in\mathfrak{F}(B^n\times T^n)$ admits the action-angle coordinates (I,φ) (as explained above) in $B^n\times T^n$, that is, the Hamiltonian vector field $\psi_*X_H=X_{H\circ\psi^{-1}}$ has the form*

$$\psi_*X_H=-\sum_{i=1}^{n}\frac{\partial(H\circ\varphi^{-1})}{\partial I^i}\frac{\partial}{\partial\varphi_i}$$

Thus Hamilton's equations can be written in this case as

$$\dot{I}^i=0$$

$$\dot{\varphi}_i=-\frac{\partial(H\circ\psi^{-1})}{\partial I^i}=\nu_i(I^1,\ldots,I^n)$$

and, according to what we said for the standard model, the flow admits invariant tori $I^i=I_0^i$, $i=1,\ldots,n$ on which it is quasi-periodic with frequencies $\nu_i(I^1,\ldots,I^n)$. The conclusion is that if we are able to write a Hamiltonian system in action-angle coordinates, we determine in one stroke the flow, the invariant tori, and the character of the flow on these invariant tori since it is given by the frequencies ν_i that are now explicitly computed.[†] For a nontrivial example of this, we refer to the Delaunay coordinates in the two-body problem (Sect 9.3). Another is the parametrization of the Korteweg–de Vries equation by scattering data (see Sect. 5.6 and Faddeev and Zakharov [1971]).

[†]In the terminology of the next section, these tori are Lagrangian submanifolds.

ISBN 0-8053-0102-X

We shall show now (following Arnold and Avez [1967]; see also Arnold [1978] and Iacob [1973]) how such action-angle coordinates are found on a manifold, where $I_c = T^n$ for all $c \in U \subset R^n \setminus \Sigma_F$. To be precise, assume from the beginning that we work in an open set of R^{2n} [instead of the symplectic manifold (P,ω)], the domain of a symplectic chart of (P,ω). Coordinates will be denoted here by $q^1,\ldots,q^n,p_1,\ldots,p_n$. We assume that we are given a Hamiltonian $H \in \mathfrak{F}(P)$ and n independent integrals in involution $f_1 = H$, $f_2,\ldots,f_n$, $2n = \dim P$. Let Σ_F denote the bifurcation set of F and assume that $U \subset R^n \setminus \Sigma_F$ and that $F^{-1}(U)$ is diffeomorphic to $U \times T^n$.

We shall construct the symplectic diffeomorphism $\psi: F^{-1}(U) \to B^n \times T^n$. Locally, the symplectic form $\omega = \Sigma_{i=1}^n dq^i \wedge dp_i$ is exact, $\omega = -d\theta$, where $\theta = \Sigma_{i=1}^n p_i\, dq^i$, $I_c \approx T^n = S^1 \times \cdots \times S^n$, and denote by $\gamma_1(c),\ldots,\gamma_n(c)$ the fundamental n cycles of I_c corresponding to the n factors S^1 [that is, $\gamma_1(c),\ldots,\gamma_n(c)$ forms a basis of the first homology group $H_1(T^n;\ R) \cong R^n$]. Define $\lambda = (\lambda_1,\ldots,\lambda_n): U \to R^n$ by

$$\lambda_i(c) = \oint_{\gamma_i(c)} i_c^*(\theta), \qquad 1 \leqslant i \leqslant n$$

where $i_c: I_c \to P$ is the canonical inclusion. $\lambda_i(c)$ is the integral of the one-form θ on the cycle $\gamma_i(c)$ [and the integrals depend only on the boundary class of $\gamma_i(c)$]. We shall assume, following Arnold, that λ *is a diffeomorphism onto its image*. Thus $\lambda \circ F: cf^{-1}(U) \to \lambda(U)$ and we shall assume, eventually shrinking U, that $\lambda(U) = B^n \subset R^n$. This is "half" of the desired diffeomorphism $\psi: F^{-1}(U) \to B^n \times T^n$.

We now search for a map $\Gamma: F^{-1}(U) \to T^n$ such that $(\lambda \circ F) \times \Gamma: F^{-1}(U) \to B^n \times T^n$ is a diffeomorphism; Γ will give us the angle coordinates. In the construction of Γ, the first step is to show that $i_c^*(\theta) \in \Omega^1(I_c)$ is a *closed* one-form. By hypothesis, the Hamiltonian vector fields $X_{f_1},\ldots,X_{f_n}$ are linearly independent at any point [since $U \subset R^n \setminus \Sigma_F$ implies $U \cap \sigma(F) = \varnothing$ and the set $f_1,\ldots,f_n$ is assumed to be independent]. Thus $X_{f_1}(q,p),\ldots,X_{f_n}(q,p)$ forms a basis of $T_{(q,p)}(I_c)$. Hence to show $d(i_c^*(\theta)) = 0$, it suffices to prove that

$$d(i_c^*(\theta))(X_{f_i}|I_c, X_{f_j}|I_c) = 0, \qquad 1 \leqslant i,j \leqslant n$$

But this is clear:

$$\begin{aligned} d(i_c^*(\theta))(X_{f_i}|I_c, X_{f_j}|I_c) &= i_c^*(d\theta)(X_{f_i}|I_c, X_{f_j}|I_c) \\ &= -i_c^*\omega(X_{f_i}|I_c, X_{f_j}|I_c) \\ &= -\omega(X_{f_i}, X_{f_j}) \circ i_c \\ &= -\{f_i, f_j\} \circ i_c = 0 \end{aligned}$$

ISBN 0-8053-0102-X

since the constants of the motion f_i, $i=1,\ldots,n$ are in involution. By the independence hypothesis, the matrix $(\partial f_i/\partial p_j)$ has nonzero determinant and hence fixing $(q_0^1,\ldots,q_0^n)$, the equation $F(q,p)-\lambda^{-1}(I)=0$ for I fixed can be solved for p in a neighborhood of q_0 (the implicit function theorem). Thus we get a function $p=p(q,I)$. Now define

$$S(q,I)=\int_{(q_0,p_0)}^{(q,p)} i^*_{\lambda^{-1}(I)}(\theta)$$

where the integral is taken over any path joining (q_0,p_0) to (q,p), the path lying in the torus $I_{\lambda^{-1}(I)}$. Since $i^*_{\lambda^{-1}(I)}(\theta)$ is closed, the integral does *not* depend on the path if (q,p) is close to (q_0,p_0). However, it should be noted that *globally* $S(q,I)$ does depend on the path of integration since T^n is not simply connected.

Define the map $\Gamma: F^{-1}(U)\to T^n$ by $\Gamma=(\Gamma_1,\ldots,\Gamma_n)$, where

$$\Gamma_i(q,p)=\left.\frac{\partial S(q,I)}{\partial I^i}\right|_{I=(\lambda\circ F)(q,p)}$$

Clearly, Γ_i are multi-valued functions, and this was to be expected since we want them to be angular variables. The variation of Γ_i on the cycle $\gamma_k(\lambda^{-1}(I))$ is given by

$$\begin{aligned}\oint_{\gamma_k(\lambda^{-1}(I))} d(\Gamma_i\circ i_{\lambda^{-1}(I)}) &= \oint_{\gamma_k(\lambda^{-1}(I))} d\left(\frac{\partial S}{\partial I^i}\circ i_{\lambda^{-1}(I)}\right)\\ &= \frac{\partial}{\partial I^i}\int_{\gamma_k(\lambda^{-1}(I))} dS\\ &= \frac{\partial}{\partial I^i}\int_{\gamma_k(\lambda^{-1}(I))} i^*_{\lambda^{-1}(I)}(\theta)\\ &= \frac{\partial I^k}{\partial I^i}=\delta_i^k\end{aligned}$$

so that *mod* 1, Γ_i are well defined and hence they determine angular coordinates on the torus, that is, $\Gamma: F^{-1}(U)\to T^n$.

Define now $\psi=(\lambda\circ F)\times\Gamma: F^{-1}(U)\to B^n\times T^n$ and *assume that it is bijective* (locally this is automatic by what we do below). Note that

$$\partial S/\partial q^i=p_i(q,I)$$

To see this, fix I and note that on the torus $I_{\lambda^{-1}(I)}$, the map $S(q,I)$ can be

ISBN 0-8053-0102-X

written as

$$S(q,I)=\int_{(q_0,p_0)}^{(q,p)}\sum_{i=1}^{n} p_i\,dq^i$$

$$=constant+\sum\int_{q_0}^{q} p_i(q,I)\,dq^i$$

by taking as path of integration the union of the two segments

$$\overline{(q_0,p_0),(q_0,p(q,I))}\quad\text{and}\quad\overline{(q_0,p(q,I)),(q,p)}$$

We have then the two relations

$$\Gamma_i(q,p)=\partial S/\partial I^i,\qquad p_i=\partial S/\partial q^i$$

so that S is a generating function of the map ψ: $(q,p)\mapsto(I,\varphi),\varphi=\Gamma$ and hence ψ is symplectic (see Example 5.2.3 and Proposition 5.2.1). As a symplectic map, it is a local diffeomorphism and since it is bijective, it is a global diffeomorphism. Thus (i) of Definition 5.2.24 is satisfied.

To show that the Hamiltonian $H=f_1$ is independent of the angle variables φ, we remark that

$$\frac{\partial(H\circ\psi^{-1})}{\partial\varphi_i}=dI^i\left(X_{H\circ\psi^{-1}}\right)$$

$$=d(\lambda_i\circ F)(X_H)\circ\psi^{-1}=(d\lambda_i\circ TF)(X_H)\circ\psi^{-1}$$

and that

$$TF(X_H)=(df_1(X_H),\ldots,df_n(X_H))$$

$$=-(\{H,f_1\},\ldots,\{H,f_n\})=0$$

which establishes condition (ii) of Definition 5.2.24.

It would be interesting to generalize this by introducing "action-angle variables" for any group (not just $\mathbf{R}^n$) acting on (P,ω) for which the reduced phase space (Sect. 4.3) is a point (i.e., a Hamiltonian system that is integrable in a generalized sense); see Exercise 5.2I.

EXERCISES

5.2A. Show that definition 5.2.2 reproduces all the generating functions listed in Goldstein [1950] and their associated identities.

5.2B. In 5.2.2, let $P_i=T^*Q_i$ and let $\Theta=\theta_1-\theta_2$, where θ_i is the canonical one-form. Show that a symplectic map f is the lift of a map of Q_1 to Q_2 if and only if the

ISBN 0-8053-0102-X

associated generating function is zero (or constant). (*Hint:* Consult Exercise 3.2F.)

5.2C. Show that if $\varphi: P_1 \to P_2$ is symplectic, its suspension $\tilde{\varphi}: R\times P_1 \to R\times P_2$; $(t,x)\mapsto(t,\varphi(x))$ is canonical with $K_{\tilde{\varphi}}=0$.

5.2D. Develop canonical transformations for Lagrangian systems up to and including the formula

$$L(Q,\dot{Q},t)=L(q,\dot{q},t)+\frac{d}{dt}W(q,Q,t)$$

It may help to consult one of the standard texts in mechanics.

5.2E. Check (i) directly using (C1), (C2), and (C4) and (ii) using the definition 5.2.15 of a generating function, that if E is a Banach space, then $W: R\times E\times E\to R$, and if the maps

$$F_1W: R\times E\times E\to R\times E\times E^*$$
$$(t,e_1,e_2)\mapsto(t,e_1,D_1W_t(e_1,e_2))$$

and

$$F_2W: R\times E\times E\to R\times E\times E^*$$
$$(t,e_1,e_2)\to(t,e_2,D_2W_t(e_1,e_2))$$

are diffeomorphisms, then

$$G=-F_1W\circ(F_2W)^{-1}: R\times T^*E\to R\times T^*E$$

is a canonical transformation with

$$K_G=-\frac{\partial W}{\partial t}\circ(F_2W)^{-1}$$

Use this to derive the Hamilton–Jacobi equation.

5.2F. A mapping $\chi: R\times P_1\to R\times P_2$ is a *locally canonical transformation* iff it satisfies (C1), (C2), and (S) (5.2.9.). Show that:

(i) For each $x\in P_1$ there is a neighborhood U of x such that $\chi|R\times U$ is canonical.

(ii) If P is contractible, then a locally canonical transformation is canonical.

(iii) Let $P=T^*S^1$. Then define $\chi: R\times P\to R\times P$ by $\chi(t,x)=(t,\varphi_t(x))$, where φ_t is the symplectic diffeomorphism induced by a rotation on S^1 by an angle t. Show that χ is locally canonical, but not canonical.

5.2G. (R. Hansen). Let Σ be a manifold of periodic orbits of a Hamiltonian system $(P,\omega=-d\theta,X_H)$ and for $p\in\Sigma$, let $J=\int_{c(p)}\theta$ where $c(p)$ is the periodic orbit, with period $\tau(p)$, through p. Prove that $dJ(p)=\tau(p)dH(p)$ and deduce the period-energy relation (see Sect. 3.3).

5.2H. (a) This exercise explains the term "completely integrable" and gives an algorithm for "integrating" completely integrable systems. Let M be a symplectic $2n$ manifold and let $P_1,P_2,\ldots,P_n$ be independent functions in involution; (one of these will normally be the energy of a given Hamiltonian system). In canonical coordinates $q^1,q^2,\ldots,q^n,p_1,p_2,\ldots,p_n$ on M, write the surfaces $P_i=$constant as $p_i=\theta_i(q^1,\ldots,q^n,P_1,\ldots,P_n)$. From the fact that the P_i are in involution, deduce from elementary exactness

ISBN 0-8053-0102-X

considerations that there is a function $S(q^1,\ldots,q^n,P_1,\ldots,P_n)$ such that $\theta_i=\frac{\partial S}{\partial q^i}$. Then show that $Q^i=-\frac{\partial S}{\partial P_i}$ and P_j define a symplectic chart. (In the language of the following section, the surfaces $P_i=$ constant define a Lagrangian foliation and S is the generating function.)

(b) Use (a) to integrate the harmonic oscillator.

5.2I. (The authors and A. Weinstein) generalize action-angle variables along the following lines. Let (P,ω) be a symplectic manifold, $\omega=-d\theta$, and Φ an action of G on P preserving θ (we have introduced θ so the action can be defined), with associated momentum mapping $J: P\to\mathfrak{g}^*$; $\hat{J}(\xi)=i_{\xi_P}\theta$. Fix a regular value μ of J and assume the reduced phase space is a point (generalized integrability; see Sect. 4.3), so $\dim P=\dim G+\dim G_\mu$. Consider the orbit $G\cdot\mu\subset\mathfrak{g}^*$, which is a symplectic manifold as shown in Sect. 4.3. Also, consider $T^*G_\mu\cong G_\mu\times\mathfrak{g}^*$. Show that the method of action-angle coordinates (in a generalized sense) produces a symplectic diffeomorphism between P and $G\cdot\mu\times T^*G_\mu$ (or $G\cdot\mu\times T^*G_\mu\times P_\mu$ if $\dim P_\mu>0$) (the latter with the product symplectic structure). The *angle* coordinates are in G_μ and the *action* coordinates are in $\mathfrak{g}^*$. Those in $G\cdot\mu$ are *residual.*

Show that this scheme has the right dimension count and reduces to standard action-angle coordinates if G is abelian. Prove the general case and work out the case of $SO(3)$ acting on T^*S^2 and $\mathbf{R}\times SO(3)$ acting on $T^*(\mathbf{R}^3\backslash\{0\})$ by the flow of the Kepler problem $\times SO(3)$. See Exercise 4.5H and Sect. 9.3; the results of Weinstein [1977b, Sect. 5] also seem relevant. (The action of $SO(n)$ on the Grassman manifolds provides another interesting example of generalized action angle variables).

5.3 LAGRANGIAN SUBMANIFOLDS

The last decade has seen a rapid development of the concept of a Lagrangian submanifold. Although the story is an ongoing one, we present some basic points in the theory here.

Virtually every physical system has a symplectic manifold associated with it, and the behavior of the system may be described in terms of Lagrangian submanifolds. This applies equally well to statics as to dynamics. As we shall explain, this approach may be regarded as basing mechanics on a *reciprocity principle*. Already we have seen some hints of Lagrangian submanifolds in our discussion of the period-energy relation (3.3.25) and canonical transformations (5.2.1). The reader should have another look at those topics after reading this section.

Some believe that the Lagrangian submanifold approach will give deeper insight into quantum theories than does the Poisson algebra approach. In any case, it gives deeper insight into classical mechanics and classical field theories.

For further information on the subject of this section, the reader should consult Weinstein [1977b], Guillemin and Sternberg [1977], and references given therein and below. Some use will be made of Lagrangian submanifolds in the next section.

ISBN 0-8053-0102-X

We shall begin by discussing the linear case, and then we shall globalize to manifolds. Following this we shall give some simple examples, explain the connections with reciprocity, and proceed to some more specialized topics.

5.3.1 Definitions. *Let (E,ω) be a symplectic vector space and $F\subset E$ a subspace. The **ω-orthogonal complement** of F is the subspace defined by*

$$F^{\perp}=\{e\in E\,|\,\omega(e,e')=0 \text{ for all } e'\in F\}$$

We say:

(*i*) *F is **isotropic** if $F\subset F^{\perp}$, that is, $\omega(e,e')=0$ for all $e,e'\in F$;*
(*ii*) *F is **co-isotropic** if $F\supset F^{\perp}$, that is, $\omega(e,e')=0$ for all $e'\in F$ implies $e\in F$;*
(*iii*) *F is **Lagrangian** if F is isotropic and has an isotropic complement, that is, $E=F\oplus F'$, where F' is isotropic.*
(*iv*) *F is **symplectic** if ω restricted to $F\times F$ is nondegenerate.*

Clearly each of these notions is invariant under symplectic isomorphisms.

The terminology "Lagrangian subspace" was apparently first used by Maslov [1965], although the ideas were in isolated use before that date.

Throughout this section we shall assume that our vector spaces and manifolds are *finite dimensional.* Of course many of the ideas work in infinite dimensions as well.

The following collects some properties related to definition 5.3.1.

5.3.2 Proposition.[†]

(*i*) *$F\subset G$ implies $G^{\perp}\subset F^{\perp}$.*
(*ii*) *$F^{\perp}\cap G^{\perp}=(F+G)^{\perp}$.*
(*iii*) *$\dim E=\dim F+\dim F^{\perp}$.*
(*iv*) *$F=F^{\perp\perp}$.*
(*v*) *$(F\cap G)^{\perp}=F^{\perp}+G^{\perp}$.*

Proof. The assertions (i) and (ii) are simple verifications. To prove (iii), consider the linear map $\omega^{b}\colon E\to E^{*}$. Now for $e\in F$, $\omega^{b}(e)$ annihilates $F^{\perp}$, so we get an induced linear map $\omega^{b}_{F}\colon F\to(E/F^{\perp})^{*}$. Since ω is nondegenerate, this map is injective, as is easily seen. Thus by linear algebra

$$\dim F\leqslant \dim(E/F^{\perp})^{*}=\dim E-\dim F^{\perp}$$

Next consider $\omega^{b}\colon E\to E^{*}\to F^{*}$. As a linear map $\hat{\omega}^{b}_{F}$ of E to F^{*}, this has kernel exactly $F^{\perp}$. Thus, by linear algebra again

$$\dim F\geqslant \dim \operatorname{range}\hat{\omega}^{b}_{F}=\dim E-\dim F^{\perp}$$

These two inequalities give (iii).

[†] For an infinite-dimensional version of (iv), see Exercise 3.1F, and for infinite-dimensional versions of many of the other results, see Weinstein [1971b].

ISBN 0-8053-0102-X

For (iv), notice that $F \subset F^{\perp\perp}$ is clear. From (iii) applied to F and to $F^{\perp}$ we get $\dim F = \dim F^{\perp\perp}$, so $F = F^{\perp\perp}$.

Finally, for (v), notice that, using (ii) and (iv),

$$(F \cap G)^{\perp} = (F^{\perp\perp} \cap G^{\perp\perp})^{\perp} = (F^{\perp} + G^{\perp})^{\perp\perp} = F^{\perp} + G^{\perp} \quad \blacksquare$$

The next result is often used to define Lagrangian subspaces.

5.3.3 Proposition. *Let (E, ω) be a symplectic vector space and $F \subset E$ a subspace. Then the following assertions are equivalent:*

(*i*) *F is Lagrangian.*
(*ii*) *$F = F^{\perp}$.*
(*iii*) *F is isotropic and $\dim F = \frac{1}{2} \dim E$.*

Proof. First we prove that (i) implies (ii). We have $F \subset F^{\perp}$ by definition. Conversely, let $e \in F^{\perp}$ and write $e = e_0 + e_1$, where $e_0 \in F$ and $e_1 \in F'$, where F' is given by Definition 5.3.1(iii). We shall show that $e_1 = 0$. Indeed, $e_1 \in F'^{\perp}$ by isotropy of F', and similarly $e_1 = e - e_0 \in F^{\perp}$. Thus $e_1 \in F'^{\perp} \cap F^{\perp} = (F' + F)^{\perp} = E^{\perp} = \{0\}$ by nondegeneracy of ω. Thus $e_1 = 0$, so $F^{\perp} \subset F$ and (ii) holds.

Secondly, (ii) implies (iii) follows at once from 5.3.2(iii).

Finally, we prove that (iii) implies (i). First, observe that (iii) implies that $\dim F = \dim F^{\perp}$ by 5.3.2(iii). Since $F \subset F^{\perp}$, we have $F = F^{\perp}$. Now we construct F' as follows. Choose arbitrarily $v_1 \notin F$ and let $V_1 = \mathrm{span}(v_1)$; since $F \cap V_1 = \{0\}$, $F + V_1^{\perp} = E$ by 5.3.2(v). Now pick $v_2 \in V_1^{\perp}$, $v_2 \notin F + V_1$, let $V_2 = V_1 + \mathrm{span}(v_2)$, and continue inductively until $F + V_k = E$. By construction, $F \cap V_k = \{0\}$, so $E = F \oplus V_k$. Also by construction,

$$\begin{aligned} V_2^{\perp} &= (V_1 + \mathrm{span}(v_2))^{\perp} \\ &= V_1^{\perp} \cap \mathrm{span}(v_2)^{\perp} \\ &\supset \mathrm{span}(v_1, v_2) = V_2 \end{aligned}$$

since $v_2 \in V_1^{\perp}$. Inductively, V_k is isotropic as well. Thus we can choose $F' = V_k$. $\blacksquare$

We can rephrase 5.3.3 by saying that Lagrangian subspaces are *maximal isotropic subspaces.*

5.3.4 Examples. (i) Any one-dimensional subspace of E is isotropic, so if E is two dimensional, any one-dimensional subspace is Lagrangian.

(ii) Let $E = R^2 \times R^2$ with elements denoted $v = (v_1, v_2)$ and with the usual symplectic structure

$$\omega((v_1, v_2), (w_1, w_2)) = \langle v_1, w_2 \rangle - \langle v_2, w_1 \rangle$$

ISBN 0-8053-0102-X

where $\langle , \rangle$ denotes the Euclidean inner product. Then the subspace spanned by linearly independent vectors v and w is Lagrangian if and only if

$$\langle v_1, w_2\rangle = \langle v_2, w_1\rangle$$

For instance $R^2\times\{0\}$ and $\{0\}\times R^2$ are Lagrangian subspaces, as is $span((1,1,1,1),(0,1,0,1))$, and so forth.

(iii) Let $E=V\oplus V^*$ with the canonical symplectic form

$$\omega_V((v_1,\alpha_1),(v_2,\alpha_2))=\alpha_2(v_1)-\alpha_1(v_2)$$

Then $V\oplus\{0\}\subset E$ and $\{0\}\oplus V^*$ are Lagrangian, since ω_V vanishes on them and they have half the dimension of E.

(iv) Let H be a complex inner product space (regarded as a real vector space) with the symplectic form

$$\omega(z,z')=-Im\langle z,z'\rangle$$

(see 3.1.18 and the ensuing discussion). Thus a subspace $V\subset H$ is isotropic if and only if all inner products of pairs of elements of V are *real*. Let $J: H\to H$ be multiplication by $i=\sqrt{-1}$. Then if V is isotropic, so is $V'=J\cdot V$. Also, $V\cap J\cdot V=\{0\}$, as is easily seen. Thus V is Lagrangian if and only if all pairs of inner products of elements of V are real and $V+JV=H$. This last decomposition of H identifies H with the complexification of a real inner product space V and within H, the "purely real" and "purely imaginary" subspaces are Lagrangian. Thus, this example merely rephrases Example (iii).

The next proposition shows that Example 5.3.4(iii) is, in a sense, the most general example.

5.3.5 Proposition. *Let (E,ω) be a symplectic vector space and $V\subset E$ a Lagrangian subspace. Then there is a symplectic isomorphism $A:(E,\omega)\to(V\oplus V^*,\omega_V)$ taking V to $V\oplus\{0\}$.*

Proof. Let $E=V\oplus V'$, where V' is isotropic, and consider the map

$$T: V'\to V^*;\ T(e_1)\cdot e=\omega(e,e_1)$$

We claim T is an isomorphism. Indeed, suppose that $T(e_1)=0$; then $\omega(e_1,e)=0$ for all $e\in V$ and hence—as V' is isotropic and $E=V\oplus V'$—for all $e\in E$. Since ω is nondegenerate, $e_1=0$. Hence T is one-to-one, and since $dim\,V=dim\,V'$, it is an isomorphism.

ISBN 0-8053-0102-X

Now let $A = Identity \oplus T$. It is now easy to verify that $A^*\omega_V = \omega$; indeed,

$$\begin{aligned}(A^*\omega_V)((e,e_1),(e',e_1')) &= \omega_V((e,Te_1),(e',Te_1'))\\ &= (Te_1')(e) - (Te_1)(e')\\ &= \omega(e,e_1') - \omega(e',e_1)\\ &= \omega(e+e_1, e'+e_1')\end{aligned}$$

since each of V and V' is isotropic. ■

The following definition will be convenient for products (see 5.2.1).

5.3.6 Definition. *Let (E_1,ω_1) and (E_2,ω_2) be symplectic vector spaces and π_i: $E_1 \times E_2 \to E_i$ the projection, $i = 1, 2$. Let $\omega_1 \ominus \omega_2 = \pi_1^*\omega_1 - \pi_2^*\omega_2$, a symplectic form on $E_1 \times E_2$.*

5.3.7 Proposition. *An isomorphism $A: E_1 \to E_2$ is symplectic if and only if its graph,*

$$\Gamma_A = \{(e_1, A(e_1)) | e_1 \in E_1\}$$

is a Lagrangian subspace of $(E_1 \times E_2, \omega_1 \ominus \omega_2)$.

Proof. Since Γ_A is half the dimension of $E_1 \times E_2$, Γ_A is Lagrangian if and only if it is isotropic, that is,

$$(\omega_1 \ominus \omega_2)((e_1, A(e_1)), (e_1', A(e_1'))) = 0$$

that is,

$$\omega_1(e_1, e_1') - \omega_2(A(e_1))\, A(e_1')) = 0$$

that is, A is symplectic. ■

We now introduce some additional terminology in connection with these products, generalizing the idea in 5.3.7.

5.3.8 Definition. *A Lagrangian subspace of $(E_1 \times E_2, \omega_1 \ominus \omega_2)$ is called a **linear canonical relation.***

If $S \subset E_1 \times E_2$ and $T \subset E_2 \times E_3$ are linear canonical relations, define the composition of S and T by

$$T \circ S = \{(e_1, e_3) \in E_1 \times E_3 | \text{there exists } e_2 \in E_2 \text{ such that } (e_1, e_2) \in S \text{ and } (e_2, e_3) \in T\}$$

ISBN 0-8053-0102-X

This definition of compositions is the usual one for relations, generalizing composition of functions. Specifically, if $S=\Gamma_A$ and $T=\Gamma_B$ for symplectic isomorphisms, then $T\circ S=\Gamma_{B\circ A}$, as is easily checked.

For linear canonical relations T and S, it is easy to see that $T\circ S$ is isotropic:

$$\begin{aligned}(\omega_1\ominus\omega_3)((e_1,e_3),(e_1',e_3'))&=\omega_1(e_1,e_1')-\omega_3(e_3,e_3')\\&=\left[\omega_1(e_1,e_1')-\omega_2(e_2,e_2')\right]\\&\quad+\left[\omega_2(e_2e_2')-\omega_3(e_3,e_3')\right]=0\end{aligned}$$

In fact, $T\circ S$ is actually Lagrangian. The proof will run most smoothly if we use the ideas about reduction from the beginning of Sect. 4.3.

5.3.9 Definition. *Let (E,ω) be a symplectic vector space and $F\subset E$ a subspace. On $F/F\cap F^\perp$, define*

$$\hat{\omega}(e+F\cap F^\perp,e'+F\cap F^\perp)=\omega(e,e')$$

*If F is co-isotropic, that is, $F\cap F^\perp=F^\perp$, we denote $F/F^\perp=E_F$ and call it the **reduced symplectic space**, or the space E **reduced by** F.*

To justify the terminology we shall check that $\hat{\omega}$ is well defined and is a symplectic form. Indeeed, let $f,f'\in F\cap F^\perp$; then

$$\begin{aligned}\omega(e+f,e'+f')&=\omega(e,e')+\omega(e+f,f')+\omega(f,e')\\&=\omega(e,e')\end{aligned}$$

The two terms involving f,f' vanish since $e+f\in F$, $f'\in F^\perp$ and since $e'\in F$, $f\in F^\perp$. Thus $\hat{\omega}$ is well defined. It is nondegenerate, for if $e\in F$ and if $\omega(e,e')=0$ for all $e'\in F$, then $e\in F^\perp$, so in $F/F\cap F^\perp$, e is zero.

5.3.10 Proposition. *Let $L\subset E$ be a Lagrangian subspace and let $F\subset E$ be co-isotropic. Then*

$$L_F=L\cap F/F^\perp\subset E_F$$

(the image of $L\cap F$ in E_F using projection $F\to F/F^\perp$) is Lagrangian in the reduced symplectic space.

Proof. From the definition of $L_F,\hat{\omega}$ and the fact that L is isotropic, it is clear that L_F is isotropic.

Next we will show that $\dim L_F=\frac{1}{2}\dim(F/F^\perp)$, which will complete the proof. From linear algebra,

$$\dim L_F=\dim(L\cap F)-\dim(L\cap F^\perp)$$

ISBN 0-8053-0102-X

From 5.3.2(v) and 5.3.2(iii), and $L = L^\perp$,

$$\dim(L\cap F^\perp) + \dim(L\cap F^\perp)^\perp = \dim E$$

that is,

$$\dim(L\cap F^\perp) + \dim(L+F) = \dim E$$

Thus

$$\begin{aligned}\dim L_F &= \dim(L\cap F) - \dim E + \dim(L+F)\\ &= \dim L + \dim F - \dim E\\ &= \dim F - \tfrac{1}{2}\dim E\end{aligned}$$

But

$$\begin{aligned}\dim(F/F^\perp) &= \dim F - \dim F^\perp\\ &= \dim F - (\dim E - \dim F)\\ &= 2\dim F - \dim E\end{aligned}$$

Thus $\dim L_F = \frac{1}{2}\dim(F/F^\perp)$. ■

The reader is referred to Weinstein [1977b] for another proof.

A little thought shows that the result 5.3.10 is actually somewhat remarkable. For example if we do not assume that F is co-isotropic and we replace E_F by the symplectic space $F/(F\cap F^\perp)$, then the result in 5.3.10 need not be true.

5.3.11 Example. Let (E,ω) be a symplectic space, let $A: E\to E$ be skew adjoint relative to ω, that is, $\omega(Ae_1,e_2) = -\omega(e_1,Ae_2)$, and let $H(e) = \frac{1}{2}\omega(A\cdot e,e)$. Then A is the linear Hamiltonian vector field with energy H, that is, $dH(e)\cdot e_1 = \omega(A(e),e_1)$. (See remarks following 3.3.6.) Fix $e\neq 0$ and let

$$F = \{e_1 | \omega(A(e),e_1) = 0\} = \ker dH(e)$$

Then F is co-isotropic; in fact, $F^\perp$ is the one-dimensional space

$$F^\perp = \operatorname{span}[A(e)] \subset F$$

Note that $\dim F/F^\perp = \dim E - 2$. If $L\subset E$ is Lagrangian, then the image of L in $F/F^\perp$ is Lagrangian, so has dimension $\frac{1}{2}\dim E - 1$. However, L may intersect F in dimension $n-1$ or n, where $2n = \dim E$. This may be seen by choosing $E = R^n\times R^n$ and $H(q_1,\ldots,q_n,p_1,\ldots,p_n) = \Sigma_{i=1}^n q_i^2 + \Sigma_{i=1}^n p_i^2$ and suitable (q,p).

ISBN 0-8053-0102-X

5.3.12 Proposition. *The composition of two linear canonical relations is again a linear canonical relation.*

Proof. Let T and S be linear canonical relations as above. We have seen that $T \circ S$ is isotropic. We want to show that it is Lagrangian.

Let $\Delta_{E_2} \subset E_2 \times E_2$ be the diagonal and notice $T \circ S$ is the projection on $E_1 \times E_3$ of

$$(S \times T) \cap E_1 \times \Delta_{E_2} \times E_3 \subset E_1 \times E_2 \times E_2 \times E_3$$

Let $(E, \omega) = (E_1 \times E_2 \times E_2 \times E_3, \omega_1 \ominus \omega_2 \oplus \omega_2 \ominus \omega_3)$. Now $F = E_1 \times \Delta_{E_2} \times E_3$ is co-isotropic. In fact, one checks that $F^{\perp} = \{0\} \times \Delta_{E_2} \times \{0\}$, so that E_F is isomorphic to $(E_1 \times E_3, \omega_1 \ominus \omega_3)$ by sending the class of (e_1, e_2, e_2, e_3) to (e_1, e_3).

Now $S \times T$ is Lagrangian in (E, ω), so by 5.3.10, its reduction by F is Lagrangian. Under the above isomorphism, this reduction is exactly $T \circ S$. Hence $T \circ S$ is Lagrangian. ■

This concludes our discussion of the linear theory. We turn now to the globalization of these ideas to symplectic manifolds.

5.3.13 Definitions. *Let (P, ω) be a symplectic manifold and $i: L \to P$ an immersion. We say L is an* ***isotropic (co-isotropic, symplectic) immersed submanifold*** *of (P, ω) if $(T_x i)(T_x L) \subset T_{i(x)} P$ is an isotropic (co-isotropic, symplectic) subspace for each $x \in L$. The same terminology is used for submanifolds of P and for subbundles of TP over submanifolds of P.*

A submanifold $L \subset P$ is called ***Lagrangian*** *if it is isotropic and there is an isotropic subbundle $E \subset TP|L$ such that $TP|L = TL \oplus E$.*

Notice that $i: L \to P$ is isotropic if and only if $i^*\omega = 0$. Also note, from the linear theory, that if $L \subset P$ is Lagrangian, $\dim L = \frac{1}{2} \dim P$ and $(T_x L)^{\perp} = T_x L$.

5.3.14 Proposition. *Let (P, ω) be a symplectic manifold and $L \subset P$ a submanifold. Then L is Lagrangian if and only if L is isotropic and $\dim L = \frac{1}{2} \dim P$.*

Proof. The preceding remark proves the "only if" part. For the "if" part, we know $T_x L$ has an isotropic complement E_x at each $x \in L$. The thing we need to check is that they can be chosen in a smooth manner. This can be done using the following device. Put a Riemannian metric $\langle , \rangle$ on P. Then by the argument in 3.1.18 we get a complex structure J on P and a complex inner product $h(v, w)$ such that $-\omega(v, w)$ is its imaginary part. Now $J: TP \to TP$ is a smooth involution and by Example 5.3.4(iv) we can take $E = J(TL)$, which is a smooth complementary isotropic bundle. ■

An important example of a Lagrangian submanifold is given in the next proposition.

ISBN 0-8053-0102-X

5.3.15 Proposition. *Let α be a one-form on Q and $L \subset T^*Q$ its graph. Then L is a Lagrangian submanifold if and only if α is closed.*

Proof. Clearly L is a submanifold with dimension $\frac{1}{2} \dim T^*Q$. However, from 3.2.11, $\alpha^*\theta_0 = \alpha$, so

$$d\alpha = \alpha^* d\theta_0 = -\alpha^*\omega_0$$

Thus α is closed if and only if $\alpha^*\omega_0 = 0$, that is, L is isotropic. ■

In particular, note that Q itself, being the zero section, is Lagrangian. The argument also shows that the Lagrangian submanifolds of T^*Q which project diffeomorphically onto Q are in one-to-one correspondence with the closed one-forms on Q. See also Exercise 5.3A.

In 5.3.15, since α is closed, locally $\alpha = dS$ for a function S.

5.3.16 Definition. *Let (P, ω) be a symplectic manifold, L a Lagrangian submanifold, and $i: L \to P$ the inclusion. If, locally, $\omega = -d\theta$, then $i^*\omega = -di^*\theta = 0$, so $i^*\theta = dS$ for a function $S: L \to \mathbf{R}$ (locally defined). We call S a* ***generating function*** *for L.*

If $L \subset T^*Q$ is the graph of dS, where $S: Q \to \mathbf{R}$, then L is Lagrangian and we can identify the generating function of L with S.

As we saw in Sect. 5.2, the idea of generating functions really goes back to Hamilton and Jacobi. However, the definition in a general context apparently is due to Sniatycki and Tulczyjew [1972]. (See also Maclane [1968] and Exercise 5.3K.) To link these ideas up with those in Sect. 5.2, we generalize products and canonical relations from the linear case.

5.3.17 Definition. *If (P_1, ω_1) and (P_2, ω_2) are symplectic manifolds, let $(P_1 \times P_2, \omega_1 \ominus \omega_2)$ be the symplectic manifold with $\omega_1 \ominus \omega_2 = \pi_1^*\omega_1 - \pi_2^*\omega_2$, where π_i: $P_1 \times P_2 \to P_i$ is the projection. A Lagrangian submanifold of $(P_1 \times P_2, \omega_1 \ominus \omega_2)$ is called a* ***canonical relation****. The* ***composition*** *of canonical relations is defined as in the linear case.*

As in 5.2.1 or 5.3.7 we see that a diffeomorphism $f: P_1 \to P_2$ is symplectic if and only if its graph is a canonical relation in $(P_1 \times P_2, \omega_1 \ominus \omega_2)$. Thus, *generating functions* of the graph of f reproduce the classical generating functions of canonical transformations given in Sect. 5.2.

A number of important phenomena in quantization theory arising from Maslov's work (see the following section) can be described in terms of Lagrangian submanifolds that are not necessarily graphs. Therefore, a study of them for their own sake is important.

We shall return to the study of canonical relations in the context of reduction, below. However, we shall first give a basic result of Weinstein that reduces one to the cotangent bundle case, generalizing 5.3.5.

ISBN 0-8053-0102-X

5.3.18 Theorem. *Let (P, ω) be a symplectic manifold and $L \subset P$ a Lagrangian submanifold. There is a neighborhood U of L in P, a neighborhood V of the zero section in T^*L, and a symplectic diffeomorphism $\varphi: U \to V$ such that $\varphi|L$ is the standard identification of L with the zero section of T^*L.*

Combined with 5.3.15, this shows that the Lagrangian submanifolds near L in P are in one-to-one correspondence with closed one-forms on L. ("Near" is in a sense strong enough to ensure that, using φ in 5.3.18, we get graphs. For example, if L is compact, a uniform C^1 topology will do.)

To prove 5.3.18 we use the following two lemmas.

5.3.19 Lemma. *There are neighborhoods U_1 of L in P and V_1 of L in T^*L and a diffeomorphism $f: U_1 \to V_1$ with $f|L = \text{identity}$ and $f_*\omega|L = \omega_0|L$, where ω_0 is the canonical symplectic form on T^*L.*

Proof. Let E be an isotropic complement to TL in TP. Define $\psi: E \to T^*L$ by $\psi(v)\cdot w = \omega(w, v)$. Then ψ is an isomorphism on fibers; indeed if $\psi(v) = 0$, then $v \in TL^\perp = TL$; but $TL \cap E = \{0\}$, so $v = 0$. Thus ψ is injective, so as $\dim E = \dim T^*L$, ψ is an isomorphism. Since E is a complementary bundle to TL, ψ will be a diffeomorphism on a tubular neighborhood defined by this bundle onto a neighborhood of L in T^*L. This restriction of ψ is f. Now on TL, Tf is the identity and on E, Tf is ψ since f is fiberwise linear. Thus, writing $v \in TP|L$ as $v = (v_1, v_2) \in TL \times E$ and splitting $TT^*L|L = TL \oplus T^*L$, we get, as in 5.3.5,

$$
\begin{aligned}
f^*\omega_0((v_1, v_2), (w_1, w_2)) &= \omega_0((v_1, \psi(v_2)), (w_1, \psi(w_2))) \\
&= \psi(w_2)\cdot v_1 - \psi(v_2)\cdot w_1 \\
&= \omega(v_1, w_2) - \omega(w_1, v_2) \\
&= \omega((v_1, v_2), (w_1, w_2)) \quad \blacktriangledown
\end{aligned}
$$

The next Lemma is a generalization of the Poincaré lemma.

5.3.20 Lemma. *Let $\pi: E \to N$ be a vector bundle and ω a closed k-form on E such that $i^*\omega = 0$, where $i: N \to E$ is the inclusion. Then, there is a $(k-1)$-form θ on E such that $\theta|N = 0$ and $\omega = d\theta$.*

Proof. Let $\pi_t: E \to E$ be fiber multiplication by $t \in [0, 1]$, and define the time-dependent vector field X_t on E by the requirement that π_t be its flow, that is, $X_t \circ \pi_t = d\pi_t/dt$. Then, as in the proof of the Poincaré lemma, there is no trouble at $t = 0$ and $(d/dt)(\pi_t^*\omega) = \pi_t^* L_{X_t}\omega$. Hence

$$\omega - \pi_0^*\omega = d\int_0^1 \pi_t^* i_{X_t}\omega \, dt$$

ISBN 0-8053-0102-X

or $\omega = \boldsymbol{d}\theta$, $\theta = \int_0^1 \pi_t^* i_{X_t}\omega\, dt$ since $i^*\omega = 0$ implies $\pi_0^*\omega = 0$. Clearly $\theta|N = 0$ since $X_t|N = 0$. ▼

Now we can complete the proof as in the proof of Darboux' theorem.

Proof of 5.3.18. By Lemma 5.3.19, we can define the pushforward ω_1 by f of ω in a neighborhood of L in T^*L. By Lemma 5.3.20, $\omega_1 = \boldsymbol{d}\theta_1$. Let $\omega_0 = -\boldsymbol{d}\theta_0$ be the canonical symplectic form on T^*L restricted to the same neighborhood. Recall that we have $\theta_1|L = \theta_0|L = 0$. Now we use the method of the proof of Darboux' theorem. Define $\omega_t = \omega_0 + t(\omega_1 - \omega_0)$ for $t \in [0,1]$ and notice that $\omega_t|L = \omega_1|L = \omega_0|L$ is nondegenerate. Since the condition that a closed skew-symmetric form be a symplectic structure is open, and since $[0,1]$ is compact, we can find for each $x \in L$ a neighborhood of x in T^*L on which all ω_t are symplectic. The union over L of all these neighborhoods gives a neighborhood of L in T^*L on which all ω_t are symplectic. Notice that $\omega_1 - \omega_0 = \boldsymbol{d}(\theta_1 + \theta_0)$. Define a smooth time-dependent vector field X_t on the above neighborhood by $i_{X_t}\omega_t = -(\theta_1 + \theta_0)$, which is possible since ω_t is symplectic. But on $L, X_t = 0$ and hence by the local existence theory, we can find for each $x \in L$ a neighborhood of x in T^*L on which the "flow" of X_t is defined for time at least one (see Exercise 3.2C). The union of these neighborhoods over all $x \in L$ gives a neighborhood of L in T^*L on which the "flow" F_t of X_t is defined for time at least one. Then we have:

$$\begin{aligned}\frac{d}{dt}(F_t^*\omega_t) &= F_t^*(L_{X_t}\omega_t) + F_t^*\frac{d}{dt}\omega_t\\ &= F_t^*(\boldsymbol{d}i_{X_t}\omega_t) + F_t^*(\omega_1 - \omega_0)\\ &= F_t^*(-\boldsymbol{d}(\theta_1 + \theta_0) + \omega_1 - \omega_0) = 0\end{aligned}$$

Therefore, $F_1^*\omega_1 = F_0^*\omega_0 = \omega_0$. Since $F_t|L = \textit{identity}$ (since $X_t|L = 0$), the composition $F_1 \circ f$ on the above appropriate neighborhood of L gives the result. ■

For additional motivation and insight into Lagrangian submanifolds, we turn to the ideas involved in reciprocity. First we consider an example,* namely, a 3 (or generally n) -port nonlinear DC electric network, schematically shown in Fig. 5.3-1. Let q^i denote voltages applied to each terminal and let p_i denote the currents flowing into the terminal (in specified directions, as in the figure). The applied voltages determine the currents, so we have relations

$$p_i = f_i(q^1, \ldots, q^n), \qquad i = 1, \ldots, n$$

*Many of the ideas here are due to W. Tulczyjew and G. Oster.

ISBN 0-8053-0102-X

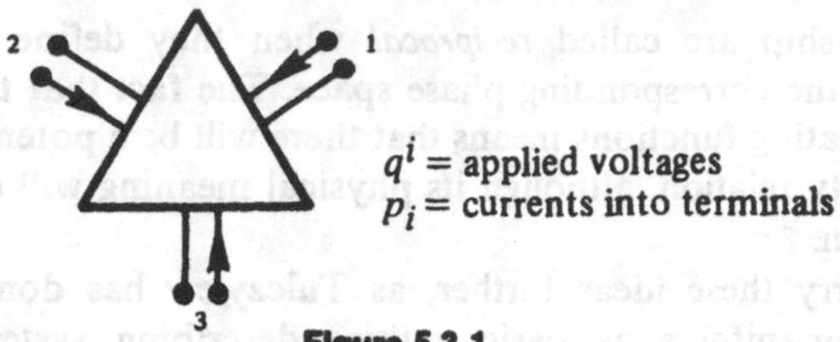

Figure 5.3-1

Thus small changes in the q's, Δq^i produce small changes Δp_i in the p's. Reciprocity means that

$$\frac{\Delta p_2}{\Delta q_1} = \frac{\Delta p_1}{\Delta q_2}$$

that is, the proportional change of current/voltage induces in port 2 by activating port 1 is the same if instead we activate port 1 and look at the current changed in port 2. Precisely, this means

$$\frac{\partial f_i}{\partial q^j} = \frac{\partial f_j}{\partial q^i}, \qquad i,j = 1,\ldots,n$$

Another way of saying this equality is that the manifold which is the graph of $f=(f_1,\ldots,f_n)$ in R^{2n} with its usual symplectic structure $\omega = \sum dq^i \wedge dp_i$ is isotropic and hence Lagrangian (since its dimension is n). This fits in with 5.3.15, for the above relations say exactly that f regarded as a one-form, $f = f_i\, dq^i$, is closed.

As we saw earlier, canonical transformations define Lagrangian submanifolds. However, reciprocity, that is, Lagrangian submanifolds, are more general. For instance, for a 2-port network, reciprocity (plus a nondegeneracy condition) is equivalent to having a function $F(q^1, q^2)$ such that $p_1 = -\partial F/\partial q^1$ and $p_2 = \partial F/\partial q^2$, that is, F is the generating function of a canonical transformation of (q^1, p_1) to (q^2, p_2). For three or higher ports, we still have generating function (5.3.16), but it is not associated with a canonical transformation.

Other physical examples are reciprocal relationships between generalized forces and displacements in an elastic suspension (see also Exercise 5.3G) and thermodynamic variables in thermostatics.* The generating functions for these examples are the "internal energy" and the "free energy." In the first example, $\theta = p_i\, dq^i$ is usually called the "virtual work." In thermostatics, the reciprocity relations are called the Maxwell relations, while in thermodynamics they are called the Onsager relations. Thus, pairs of variables in a

*See Tulczyjew [1977], Oster and Perelson [1974], and Exercise 5.5L.

ISBN 0-8053-0102-X

definite relationship are called *reciprocal* when they define a Lagrangian submanifold of the corresponding phase space. The fact that these submanifolds have generating functions means that there will be a potential associated to any reciprocity relation, although its physical meaning will depend on the particular system.

One can carry these ideas further, as Tulczyjew has done, and regard Lagrangian submanifolds as basic entities describing systems. We have already seen that graphs of canonical transformations are Lagrangian submanifolds.* What about Hamiltonian systems themselves? We have the following.

5.3.21 Proposition. *Let (P, ω) be a symplectic manifold and let ω_T be the symplectic form induced on TP (see Exercise 3.3I). Then a vector field X on P is locally Hamiltonian if and only if its graph in TP is Lagrangian.*

Proof. We claim that $X^*\theta_T = X^\flat$, where θ_T is the one-form on P given in Exercise 3.3I, namely,

$$\theta_T(w) = \omega(\tau_{TP}(w), T\tau_P(w))$$

In fact,

$$\begin{aligned} X^*\theta_T(v) &= \theta_T(TX\cdot v) \\ &= \omega(\tau_{TP}(TX\cdot v), T\tau_P\cdot TX\cdot v) \\ &= \omega(X(\tau_P v), v) \\ &= X^\flat \cdot v \end{aligned}$$

by definition of $X^\flat = i_X\omega$. Thus $X^*\omega_T = -dX^\flat$ is zero if and only if $X^\flat$ is closed, that is, X is locally Hamiltonian. ■

The generating function of this Lagrangian submanifold, in the sense of 5.3.16, is just $H: P \to \mathbf{R}$ (locally defined perhaps), the Hamiltonian for X, if we identify the graph of X with P using X.

There is also a neat way of describing the Legendre transformation using this idea, also due to Tulczyjew [1974]. Namely, let $P = T^*Q$. Then (see Exercise 3.3I), the symplectic structure on TP arises from two different diffeomorphisms:

$$i_h: TP \to T^*(T^*Q) \quad \text{and} \quad i_l: TP \to T^*(TQ)$$

*It is possible that some problems involving the dynamics of systems with singularities in classical mechanics may be describable in terms of Lagrangian submanifolds that are not necessarily graphs of canonical transformations.

ISBN 0-8053-0102-X

Thus if we have a Lagrangian submanifold $L \subset TP$, we get two Lagrangian submanifolds

$$L_h = i_h(L) \subset T^*(T^*Q) \quad \text{and} \quad L_l = i_l(L) \subset T^*(TQ)$$

as such, we get generating functions on T^*Q (the Hamiltonian) and on TQ (the Lagrangian).

The different ways of realizing this Lagrangian submanifold in TP as graphs of one-forms on T^*Q and TQ and the passage between them is, of course, exactly the Legendre transformation. In some problems of field theories, this formulation may have some technical advantages.

Next we turn to the topic of *reduction*, generalizing what we did in the linear case.*

5.3.22 Proposition. *Let (P, ω) be a symplectic manifold and $M \subset P$ a submanifold. Suppose $TM \cap (TM)^{\perp}$ is a subbundle of TP. Then:*

(i) *$TM / TM \cap (TM)^{\perp}$ is a symplectic vector bundle over M; that is, each fiber has an induced symplectic structure varying smoothly over M.*
(ii) *$TM \cap (TM)^{\perp}$ is an integrable subbundle of TP.*

Proof. (i) Follows from our remarks after 5.3.9. For (ii) let X_1, X_2 be vector fields on M that take values in $(TM)^{\perp}$. We have to prove that $[X_1, X_2]$ takes values in $(TM)^{\perp}$ as well. However, $\omega(X_i, Y) = 0$ for all vector fields Y on M. If this is substituted in the identity of 10.15(ii), that is,

$$\begin{aligned} 0 &= d\omega(X_1, X_2, Y) \\ &= X_1(\omega(X_2, Y)) - X_2(\omega(X_1, Y)) \\ &\quad + Y(\omega(X_1, X_2)) - \omega([X_1, X_2], Y) \\ &\quad + \omega([X_1, Y], X_2) - \omega([X_2, Y], X_1) \end{aligned}$$

we get $0 = \omega([X_1, X_2], Y)$. ■

By Frobenius' theorem there is a foliation $\mathfrak{F}$ of maximal integral manifolds of M of which $TM \cap (TM)^{\perp}$ is the tangent bundle. We can form the quotient space $M/\mathfrak{F}$. In general, this may not be a manifold, but we shall assume it is in what follows.† If we select a point on a leaf, the tangent space of $M/\mathfrak{F}$ can be identified with $TM / TM \cap (TM)^{\perp}$. See Fig. 5.3-2.

*The ideas here are taken from Weinstein [1977b]. They were in turn, partly inspired by Roels and Weinstein [1971] and Marsden and Weinstein [1974] (see Sect. 4.3).
†The reader may wish to look back to Sect. 4.1 where we constructed quotient manifolds for group actions. The constructions here are similar. Each point of $M/\mathfrak{F}$ is a leaf of the foliation.

ISBN 0-8053-0102-X

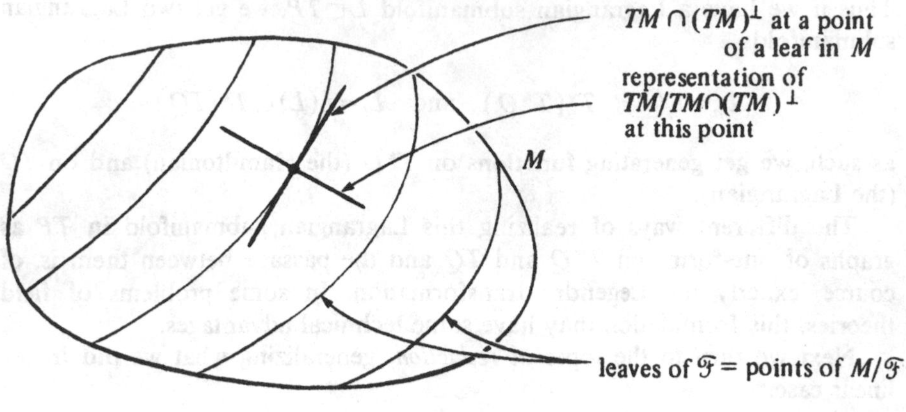

Figure 5.3-2

5.3.23 Theorem. *The quotient $M/\mathfrak{F}$ inherits the symplectic structure of $TM/TM\cap(TM)^{\perp}$ so that $M/\mathfrak{F}$ becomes a symplectic manifold. If M is co-isotropic, so $TM\cap TM^{\perp}=TM^{\perp}$, we write P_M for $M/\mathfrak{F}$ and call it the* ***reduction*** *of P by M.*

Proof. We know we get a nondegenerate two-form defined on each $TM/TM\cap TM^{\perp}$ at each point of M. We must show that this structure does not depend on the point on the leaf chosen. If X is tangent to a leaf, it is a section of $TM\cap TM^{\perp}$, and thus, on M, $i_X\omega=0$. Thus,

$$L_X\omega=di_X\omega=0$$

(precisely, $L_Xi^*\omega=di_Xi^*\omega=0$, where $i: M\to P$ is inclusion). Thus ω is constant along each leaf. The nondegenerate form ω_M on $M/\mathfrak{F}$ has the property that

$$\pi^*\omega_M=i^*\omega$$

where $\pi: M\to M/\mathfrak{F}$ is the projection and $i: M\to P$ is the inclusion. Thus $\pi^*d\omega_M=i^*d\omega=0$. Since π is a submersion, $d\omega_M=0$. Thus ω_M is a symplectic form. ■

Following the linear case 5.3.10, we are led to try projecting Lagrangian submanifolds of P to P_M. To do so we need to ensure $L\cap M$ is a manifold. If they intersect transversally, we know this is sufficient. But it is more than is actually required.

5.3.24 Definition (R. Bott [1954]). *If L and M are submanifolds of P, we say L and M have* ***clean intersection*** *if $L\cap M$ is a submanifold and $TL\cap TM=T(L\cap M)$.*

ISBN 0-8053-0102-X

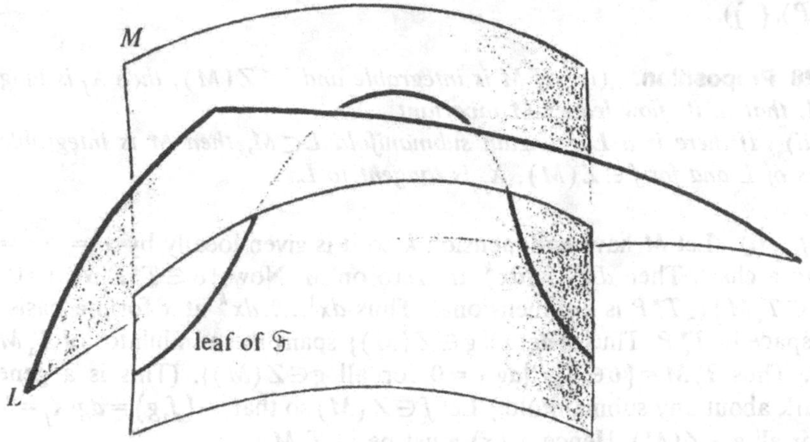

Figure 5.3-3

5.3.25 Proposition. *Suppose L is a Lagrangian submanifold, M is co-isotropic and L intersects M cleanly. Then the projection L_M of $L \cap M$ to P_M is an immersed Lagrangian submanifold.*

This is proved by reference to our calculations in the linear case and Exercise 5.3F(b). Note that the mapping of $L \cap M$ to P_M might not be one-to-one. See Fig. 5.3-3.

5.3.26 Proposition. *Let (P_i, ω_i), $i = 1, 2, 3$ be symplectic manifolds and let $S \subset P_1 \times P_2$ and $T \subset P_2 \times P_3$ be canonical relations. Suppose that $S \times T$ intersects the diagonal $\Delta \subset P_1 \times P_2 \times P_2 \times P_3$, $\Delta = P_1 \times \Delta_{P_2} \times P_3$ cleanly. Then $T \circ S$ is a canonical relation.*

This follows from 5.3.25 by the same procedure as in 5.3.12. Intersection theory and perturbations of Lagrangian submanifolds is an important topic we have omitted. See Weinstein [1973a] and, for applications, Moser [1978], and Arnold [1965].

The final topic in this section concerns the question of enlarging an isotropic submanifold $M \subset P$ to a Lagrangian submanifold. For this purpose, the following terminology is useful.

5.3.27 Definition. *Let (P, ω) be a symplectic manifold and $M \subset P$ a submanifold. Let*

$$Z(M) = \{ f \in \mathfrak{F}(P) \mid f|M = 0 \}$$

We say M is **integrable** *(resp. integrable* **at** *$x \in M$) if, for any $f, g \in Z(M)$, the Poisson bracket $\{f, g\}$ is zero on M (resp. at $x \in M$).*

ISBN 0-8053-0102-X

Another way of saying this is that $Z(M)$ is a Lie subalgebra of $(\mathfrak{F}(P),\{\ \})$.

5.3.28 Proposition. *(i) If M is integrable and $f\in Z(M)$, then X_f is tangent to M, that is, its flow leaves M invariant.*

(ii) If there is a Lagrangian submanifold $L\subset M$, then M is integrable at points of L and for $f\in Z(M)$, X_f is tangent to L.

Proof. (i) Let M have codimension k so it is given locally by $x^1=\cdots=x^k=0$ in a chart. Then $dx^1,\dots,dx^k$ are zero on M. Now $\{\alpha\in T_x^*P\,|\,\alpha(v)=0$ for all $v\in T_xM\}\subset T_x^*P$ is k-dimensional. Thus $dx^1,\dots,dx^k$ at x form a basis for this space in T_x^*P. Thus, $\{dg(x)\,|\,g\in Z(M)\}$ spans the annihilator of T_xM in T_x^*P. Thus $T_xM=\{v\in T_xP\,|\,dg\cdot v=0$ for all $g\in Z(M)\}$. (This is a general remark about any submanifold.) Let $f\in Z(M)$ so that $-\{f,g\}=dg\cdot X_f=0$ on M for all $g\in Z(M)$. Hence $X_f(x)$ must be in T_xM.

(ii) Let $i: M\to P$ and $j: L\to P$ be inclusions. Let $f\in Z(M)$. In particular, $j^*df=0$, so $df=i_{X_f}\omega$ will vanish on any vector tangent to L. Thus $X_f(x)\in(T_xL)^\perp=T_xL$, so X_f is tangent to L. If $f,g\in Z(M)$, then $-\{f,g\}=dg(X_f)$ is zero at all points of L since X_f is tangent to L and $g=0$ on L. ■

Notice that (i) also proves that for $x\in M$, $\{X_f(x)\,|\,f\in Z(M)\}$ is k-dimensional. Also note that much of what we have said is relevant for functions that are constant on M.

5.3.29 Lemma. *Let (P,ω) be a symplectic manifold, $H\in\mathfrak{F}(P)$ and suppose Σ_e is a regular energy surface. Let $M\subset\Sigma_e$ be isotropic and transverse to X_H (i.e., X_H is nowhere tangent to M). If $F: D\subset \mathbf{R}\times P\to P$ is the flow of X_H, then $F(\mathbf{R}\times M)\subset\Sigma_e$ is an isotropic immersion.*

Proof. Let $\tilde F$ be the graph of F, that is, $\tilde F(t,x)=(x,F_tx)$. By 3.3.25, $\tilde F^*\Omega=-dt\wedge dH$, where $\Omega=\pi_1^*\omega-\pi_2^*\omega$. Thus, as $F=\pi_2\circ\tilde F$ and $\pi_1\circ\tilde F: \mathbf{R}\times P\to P$ is the projection $(t,p)\mapsto p$, using the definition of $\tilde\omega$ in 5.1.9 we get $F^*\omega=\tilde F^*\pi_2^*\omega=\tilde F^*\pi_1^*\omega-\tilde F^*\Omega=\tilde\omega+dt\wedge dH$. This is also clear from 5.1; see 5.1.13. Thus if $i: \mathbf{R}\times M\to\mathbf{R}\times P$ is inclusion,

$$(F\circ i)^*\omega=i^*\tilde\omega+i^*(dt\wedge dH)$$

But $i^*\tilde\omega=0$ since M is isotropic and $i^*dH=0$ since $M\subset\Sigma_e$. That F is an immersion follows from the computations in the proof of 3.3.25 corroborated with the hypothesis that X_H is nowhere tangent to M. ■

Next we prove an extension theorem implicit in the paper of Roels and Weinstein [1971] as formulated by Guillemin and Sternberg [1977].

5.3.30 Theorem. *Let (P,ω) be a symplectic manifold of dimension $2n$. Let M be a submanifold of codimension $k\leqslant n$ that is integrable. Let $L_0\subset M$ be an*

ISBN 0-8053-0102-X

$(n-k)$-dimensional isotropic submanifold transversal to all X_f, $f \in Z(M)$. Then there is a unique Lagrangian submanifold L, $L_0 \subset L \subset M$.

Proof. By 5.3.28, locally there are $f_1, \ldots, f_k \in Z(M)$ such that $X_{f_1}, \ldots, X_{f_k}$ are a pointwise basis of $\{X_f | f \in Z(M)\}$. Let L_1 be the submanifold swept out by L_0 under the flow of X_{f_1}. By 5.3.29 it is isotropic and $\dim L_1 = \dim L_0 + 1$. Since M is integrable, $X_{f_2}, \ldots, X_{f_k}$ are not tangent to L_1. Repeating, we obtain finally $L_k = L$ of dimension $\dim L_0 + k = n$. Hence L is Lagrangian.

Doing this around each point of L_0 defines L locally. However, the pieces of L so obtained must coincide on overlaps. Indeed, by 5.3.28(ii), if L exists, all X_f, $f \in Z(M)$ are tangent to it. But the foliation determined by $\{X_f | f \in Z(M)\}$ must have leaves then coinciding with L; that is, L consists of all leaves passing through L_0. Thus if L exists, it is uniquely determined, so there is no ambiguity on overlaps. ■

5.3.31 Corollary (Lie). *Let (P, ω) be a symplectic manifold, $\dim P = 2n$. Let $f_1, \ldots, f_k \in \mathfrak{F}(M)$, $k \leq n$. Assume $\{f_i, f_j\} = 0$ and $df_1, \ldots, df_k$ are pointwise linearly independent. Then locally there exist canonical coordinates $(q^1, \ldots, q^n, p_1, \ldots, p_n)$ such that $f_1 = q^1, \ldots, f_k = q^k$.*

Proof. Pick, by Darboux' theorem, local coordinates $\bar{q}^1, \ldots, \bar{q}^n, \bar{p}_1, \ldots, \bar{p}_n$ such that $f_1, \ldots, f_k, \bar{p}_1, \ldots, \bar{p}_n$ are linearly independent. Let M be defined by $f_i = 0$ and L_0 by $f_i = 0$ and $\bar{p}_i = 0$. Then as $\{f_i, f_j\} = 0$, M is integrable and since $\{df_i\}$ are linearly independent, M has codimension k. Similarly L_0 is of dimension $n-k$. Since the space defined by $\bar{p}_i$ is isotropic and it contains L_0, L_0 is isotropic. Thus, by 5.3.30, there is a Lagrangian submanifold L, $L_0 \subset L \subset M$. Since a neighborhood of L in P is diffeomorphic to a neighborhood of zero in T^*L by 5.3.18, canonical coordinates on T^*L yield the desired coordinates. ■

This argument may be used to give another proof of Darboux' theorem (see Duistermaat [1973, p. 101] for details).

We now return to make a connection between the Hamilton–Jacobi equation and Lagrangian submanifolds. For orientation, we first make a simple observation, implicit in 5.3.30 and 5.3.31.

5.3.32 Proposition. *Let $L \subset P$ be Lagrangian and $H \in \mathfrak{F}(P)$.*

(i) If H is constant on L, then L is invariant under the flow of X_H.
(ii) If F_t is the flow of X_H, then $F_t(L) \subset P$ is Lagrangian.

Proof. (i) For $x \in L$, $T_xL = (T_xL)^\perp = \{v \in T_xP | \omega(v, w) = 0$ for all $w \in T_xL\}$. But $\omega(X_H(x), w) = dH(x) \cdot w = 0$ for $w \in T_xL$ since H is constant on L. Thus X_H is tangent to L.

(ii) This is obvious since F_t is a symplectic diffeomorphism. ■

ISBN 0-8053-0102-X

Now suppose that $L \subset T^*Q$ is the graph of dS, so S is a generating function for L, and $H \in \mathfrak{F}(M)$. Let F_t be the flow of X_H. Then, at least for a short time, $F_t(L)$ is the graph of dS_t for S_t a function S_t on Q. We can adjust the arbitrary constants in S_t so it depends smoothly on t and coincides with S (a given function) at $t=0$.

The Hamilton–Jacobi theorem can now be rephrased as follows:

5.3.33 Theorem. *The function $S(t,q)$ just described is the solution of the (time-dependent) Hamilton–Jacobi equation.*

This follows directly from 5.2.18; we leave the details to the reader.

For t's that are not small, $F_t(L)$ may eventually not be the graph of the differential of a function; $S(t,q)$ may in fact blow up in finite time. One then says that a *caustic* forms. Notice, however, that the picture of $F_t(L)$ as a Lagrangian submanifold remains valid; it just may "bend over" so that it fails to be a graph. This is the beginning of deep connections between symplectic geometry, geometric optics, partial differential equations, and Fourier integral operators. The reader is referred to the many excellent references on this topic, such as Duistermaat [1973, 1974], Guillemin and Sternberg [1977], Hörmander [1971], Maslov [1965], and Weinstein [1977b].

EXERCISES

5.3A. Show that the following result of Guckenheimer is a corollary of 5.3.15 and dimensional considerations: A submanifold $L \subset T^*Q$ is locally the image of a differential of a function on Q if and only if (i) L is isotropic and (ii) L is transverse to the fibers of T^*Q.

5.3B. Consider $\mathbf{R}^{2n} \approx \mathbf{R}^n \times \mathbf{R}^{n*}$ with its canonical symplectic structure ω, the Hamiltonian $H \in \mathfrak{F}(\mathbf{R}^{2n})$, $H(\mathbf{q},\mathbf{p}) = \|\mathbf{q}\|^2 + \|\mathbf{p}\|^2$ and its regular energy surface $H^{-1}(1) = S^{2n-1}$. Show that:

(i) X_H is ω-orthogonal to S^{2n-1} at each of its points (*Hint:*

$$T_{(q,p)}S^{2n-1} = \ker(dH)_{(q,p)}$$

$$= \left\{ \sum_{i=1}^{n} a_i \frac{\partial}{\partial q^i} + b_i \frac{\partial}{\partial p_i} \,\middle|\, \sum_{i=1}^{n} (a_i q^i + b_i p_i) = 0 \right\} \Bigg)$$

(ii) S^{2n-1} is a co-isotropic submanifold of $(\mathbf{R}^{2n}, \omega)$;

(iii) the leaves of the foliation $\mathfrak{F}$ defined by the co-isotropic submanifold S^{2n-1} in $\mathbf{R}^{2n}$ are the solution curves of X_H;

(iv) $S^{2n-1}/\mathfrak{F}$ is diffeomorphic to the complex projective plane $\mathbf{CP}^{n-1}$ [*Hint:* the solution curves in (iii) are all great circles in S^{2n-1}];

(v) $\mathbf{p}=\mathbf{0}$ is Lagrangian in $\mathbf{R}^{2n}$; call this manifold L;

(vi) $L \cap S^{2n-1} \to \mathbf{CP}^{n-1}$ is a Lagrangian immersion; it is a double cover of the image that is $\mathbf{RP}^{n-1}$, the real projective space; $\mathbf{RP}^{n-1}$ is hence imbedded as a Lagrangian submanifold in $\mathbf{CP}^{n-1}$.

ISBN 0-8053-0102-X

5.3C. (W. Tulczyjew). Let M be a co-isotropic submanifold of the finite-dimensional symplectic manifold (P,ω). If P_M denotes the reduction of P by M, show that $\{(m,S)|m\in M\cap S$ and S is a leaf of the foliation $\mathfrak{F}$ defined by $M^\perp$ in $M\}\subseteq P_M\times P$ is a canonical relation.

5.3D. (A. Weinstein). Let (P_i,ω_i), $i=1,2,3$ be symplectic manifolds, P_3 being the zero-dimensional manifold formed by one point. A canonical relation $L\subset P_2\times P_3$ is a Lagrangian submanifold of (P_2,ω_2). If $S\subset P_1\times P_2$ is a canonical relation, then $S\circ L=S(L)=\{p_1\in P_1|$ there exists $p_2\in L$ such that $(p_1,p_2)\in S\}$ is the image of $L\subset P_2$ by S in P_1. Deduce that under the assumption of clean intersection, canonical relations operate on Lagrangian submanifolds.

5.3E. (A. Weinstein). Let Q_1,Q_2 be two finite-dimensional manifolds and $(P_i,\omega_i)=(T^*Q_i,\omega_i)$ their cotangent bundles with their canonical symplectic structure. If $f: Q_1\to Q_2$ is a smooth map, notice that T^*f is symplectic, so its graph $\Gamma_{T^*f}=\{(\alpha,T^*f\alpha)|\alpha\in T^*Q_1=P_1\}$ is a canonical relation in $(P_1\times P_2,\omega_1\ominus\omega_2)$. Show that:

(i) if $f: Q_1\to Q_2$ is a submersion and L is a Lagrangian submanifold of $P_1=T^*Q_1$, then $(T^*f)(L)$ is a Lagrangian submanifold of $P_2=T^*Q_2$ [*Hint:* Work with the inverse relation $(T^*f)^{-1}$, show that $\Gamma_{(T^*f)^{-1}}(L)=(T^*f)(L)$ and prove that the intersection in question is clean; use Exercise 5.3D];

(ii) if $f: Q_1\to Q_2$ is an immersion and L is a Lagrangian submanifold of $P_2=T^*Q_2$, then $(T^*f)^{-1}(L)$ is a Lagrangian submanifold of $P_1=T^*Q_1$ [*Hint:* Apply 5.3D to the relation T^*f; show that $\Gamma_{T^*f}(L)=(T^*f)^{-1}(L)$];

(iii) if $L\subset P_1$ is generated by $S\in\mathfrak{F}(Q_1)$, then $(T^*f)(L)$ is generated by $S\circ f\in\mathfrak{F}(Q_2)$.

For those who know about generalized functions (distributions) on manifolds, the preceding exercise will suggest the idea that one can think of Lagrangian submanifolds of cotangent bundles as distributions. (i) and (ii) recapture two such properties: distributions can always be pulled back under submersions and pushed forward under immersions.

The analogy Lagrangian submanifolds/distributions is much deeper, both arising as "solutions" of partial differential equations, a subject not discussed in the book but touched upon in Lemma 5.3.29 and Theorem 5.3.30, which in fact gives the "method of characteristics" for first-order PDE's.

5.3F. (Guillemin and Sternberg [1977]). This concerns the Tricomi equation $u_{xx}+(x^2-1)u_{tt}=0$ and the method of characteristics. (See 5.3E.) Let $H\in\mathfrak{F}(T^*\mathbf{R}^2)$ be given by

$$H(q^1,q^2,p_1,p_2)=\tfrac{1}{2}\left[\left((q^2)^2-1\right)p_1+(p_2)^2\right]$$

(i) Show that

$$X_H(q^1,q^2,p_1,p_2)=\left((q^2)^2-1\right)p_1\frac{\partial}{\partial q^1}+p_2\frac{\partial}{\partial q^2}-q^2(p_1)^2\frac{\partial}{\partial p_2}$$

(ii) $H^{-1}(0)$ is integrable (*Hint:* As a codimension-one submanifold of $T^*\mathbf{R}^2$ it is integrable.)

(iii) Show that L_0 given by $(q^1,q^2,p_1,p_2)=(l,0,1,1)$ is a one-dimensional isotropic submanifold transversal to X_H.

(iv) Apply the method of characteristics to find a Lagrangian submanifold L such that $L_0\subset L\subset H^{-1}(0)$. Draw this submanifold in $\mathbf{R}^3$. [*Hint:* L is the cylinder $(q^2)^2+(p_2)^2=1$, $p_1=1$.]

ISBN 0-8053-0102-X

(v) Explain the reasons why in (iii) L_0 is a good choice for an isotropic submanifold lying in $H^{-1}(0)$ in the search of a Lagrangian submanifold using the method of characteristics.

(vi) Show that $\tau^*_{R^2}|L$ is not a diffeomorphism exactly on the lines $q^2 = \pm 1$ in the (q^1, q^2) plane.

5.3G. Show how to write Betti's reciprocal theorem in linear elastostatics and Graffi's reciprocal theorem in elastodynamics (see M. Gurtin [1972, p. 98, 218]) as statements about Lagrangian subspaces. (This exercise will require a fair amount of translation; the reader may wish to utilize the ideas in Sect. 5.6 below.)

5.3H. Show how, following Sniatycki and Tulczyjew [1971], to write the motion of a charged relativistic particle in terms of Lagrangian submanifolds.

5.3I. (A. Weinstein and the authors). Let G act symplectically on a symplectic manifold (P,ω) and have a momentum mapping $J: P \to \mathfrak{g}^*$ (see Sect. 4.2).

(a) Let T^*G be identified with $G \times \mathfrak{g}^*$ by left translations. Let

$$L = \{(g \cdot p, p, g, J(p)) \in P \times P \times T^*G\}$$

where $P \times P$ is given the symplectic structure $\omega \ominus \omega$. Show that L is Lagrangian (*Hint:* See the proof of the momentum lemma.) Call L the *bicharacteristic relation* of the original action.*

(b) Let $\Delta \subset P \times P$ be the diagonal; show that $\Delta \times T^*G$ is co-isotropic in $P \times P \times T^*G$.

(c) Reducing the bicharacteristic relation L by $\Delta \times T^*G$, show that, under a clean intersection hypothesis, the *character*

$$\{(g, J(p)) \in G \times \mathfrak{g}^* \mid g \cdot p = p\}$$

is Lagrangian.

(d) Show that the period energy relation (see Sect. 3.3 and problem 5.2G) is a special case of (c).

(e) Derive the results of Henon [1977] as a special case of (c).

5.3J. (A. Weinstein). A *local manifold* is an equivalence class of pairs (M,m), where M is a Banach manifold and $m \in M$ and $(M,m) \sim (M',m')$ if $m = m'$ and there is a manifold U containing $m = m'$ that is simultaneously an open submanifold of M and M'. Let the equivalence class be denoted $[M,m]$.

The usual differential geometric objects (vector fields, maps, etc.) can be localized in the obvious way. We work with representatives of the corresponding classes without explicit mention.

An *action* of a Lie algebra $\mathfrak{g}$ on $[M,m]$ is a homomorphism $\rho: \mathfrak{g} \to \mathfrak{X}[M,m]$. It induces a map $\theta_\rho: \mathfrak{g} \to T_m M$ by $\theta_\rho(\xi) = \rho(\xi)(m)$. The action is *free* if θ_ρ is injective. If $[M,m]$ is symplectic with form Ω, ρ is *symplectic* if the range of ρ consists of Hamiltonian vector fields and is *conformally symplectic* if, for $\xi \in \mathfrak{g}$, $\rho(\xi)$ is *conformally Hamiltonian*, that is, $L_{\rho(\xi)}\Omega = \phi(\xi)\Omega$ for a constant $\phi(\xi)$. We call ϕ the *multiplier* and $\theta^*_\rho\Omega$ (a form on $\mathfrak{g}$) the *associated form* of ρ.

(a) Show that two conformally symplectic free actions ρ and ρ' of $\mathfrak{g}$ on $[M,m]$ and $[M',m']$ are related by a [conformally] symplectic diffeomorphism if

*The name arises from the fact that L is identifiable with the character for a group representation whose generators are partial differential operators with bicharacteristic Hamiltonians that make up J (see Remark 4 following 5.2.4).

ISBN 0-8053-0102-X

and only if:

(i) there is a linear symplectic isomorphism from T_mM to $T_{m'}M'$;

(ii) ρ and ρ' have equal multipliers, and

(iii) ρ and ρ' have equal associated forms [forms that are scalar multiples of one another].

(*Hint:* Use the techniques of 5.3.18.)

(b) Deduce Darboux's theorem from (a) by taking $\mathfrak{g}=\{0\}$.

(c) Deduce Darboux's theorem for one-forms: if $\theta(m)\neq 0$, $\theta'(m')\neq 0$ and $d\theta=\Omega$, $d\theta'=\Omega'$, then locally $f^*\theta'=\theta$ for some f.

(d) Prove 5.3.31 using (a).

(e) Let $f_1,\ldots,f_k\in\mathfrak{F}([M,m])$ have independent differentials at m and $f_i(m)=a_i$, $\{f_i,f_j\}=\Sigma c_{ijk}f_k+b_{ij}$ for constants a_i, c_{ijk}, b_{ij}. Let f_i' be another such set of f's (with the same constants). Then there is a symplectic diffeomorphism transforming f_i to f_i'.

(f) If $f_1,\ldots,f_k\in\mathfrak{F}([M,m])$, $r<n$, $\dim M=2n$ have independent differentials at m and $\{f_i,f_j\}=b_{ij}$ for constants b_{ij}, then there exist symplectic coordinates in which

$$f_i=p_i-\tfrac{1}{2}\sum_{j=1}^{r}b_{ij}q_j,\qquad i=1,\ldots,r$$

(For the case $r=n$, see Roels and Weinstein [1971].)

.3K (Sniatycki and Tulczyjew [1972]). Let

$$\begin{array}{ccc} P & \xrightarrow{\ \alpha\ } & T^*M \\ \pi\big\downarrow & \swarrow{\scriptstyle \tau_M^*} & \\ M & & \end{array}$$

be a special symplectic manifold (Exercise 3.3I), $K\subset M$ a submanifold and $F\colon K\to\mathbb{R}$. Let

$$L=\{\,y\in P\,|\,\pi(y)\in K \quad\text{and}\quad \theta(v)=dF(T\pi\cdot v)$$
$$\text{for all } v\in T_yP \quad\text{such that}\quad T\pi\cdot v\in T_{\pi(y)}K\,\}.$$

Show that L is Lagrangian and F may be regarded as its generating function.

Show conversely that if $L\subset P$ is Lagrangian, $\pi|L$ has constant rank and π^{-1} (point) $\cap$ L is connected, then L is so represented (possibly with dF replaced by a closed one form γ).

5.3L (*Dirac Theory of Constraints*; Tulczyjew, Kunzle, Sniatycki, Weinstein and the authors) Let $L\colon TM\to\mathbb{R}$ be a given Lagrangian, possibly degenerate. Regarding $T(T^*M)$ as a special symplectic manifold over TM by

$$\begin{array}{ccc} T(T^*M) & \xrightarrow{\ \mathfrak{T}\ } & T^*(TM) \\ T\tau_M^*\big\downarrow & \swarrow{\scriptstyle \tau_{TM}^*} & \\ TM & & \end{array}$$

ISBN 0-8053-0102-X

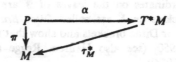

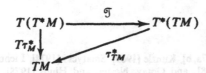

where $\mathfrak{T}: T^*(TM) \to T(T^*M)$ is the canonical symplectic diffeomorphism $((q,\dot{q},p,\dot{p}) \mapsto (q,\dot{p},\dot{q},p))$ and let Λ be the Lagrangian submanifold generated by L (Exercise 5.3K); Λ is defined by Lagrange's equations $p_i = \dfrac{\partial L}{\partial \dot{q}^i}, \dot{p}_i = \dfrac{\partial L}{\partial q^i}$. Realize $T(T^*M)$ as the special symplectic manifold

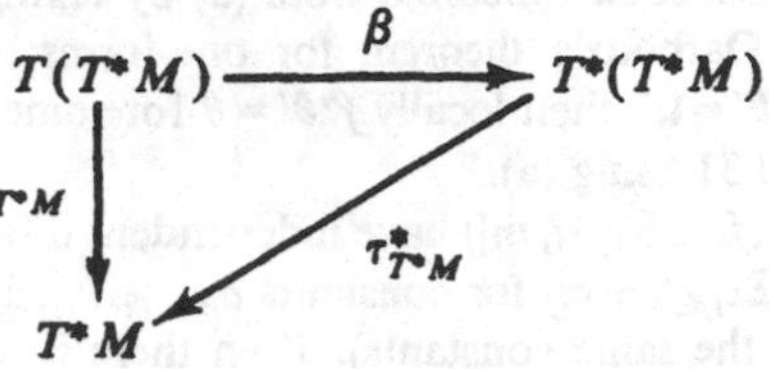

where β is induced by the symplectic structure on T^*M. If Λ is generated by $K \subset T^*M$ and $H: K \to R$, call K the *Hamiltonian constraint* and H the *Hamiltonian*. One says the constraint is *first class* if K is co-isotropic and otherwise is *second class*.

(a) If K is the zero set of a map $J: T^*M \to V^*$ (V a vector space), show that Λ can be regarded as being generated by $\overline{H} + \langle w, J\rangle$ where $w \in V$ is fixed and $\overline{H}: T^*M \to R$ is an extension of H (i.e. the dynamics is *generated* by the constraints).

(b) Describe K, H when L is the homogeneous Lagrangian for a relativistic particle.

(c) Let $\hat{K}$ be K reduced by the foliation $\mathfrak{F}$ determined by $TK \cap (TK)^\perp$ (see 5.3.23). Coordinates on the leaves of $\mathfrak{F}$ are called *gauge variables* and Poisson brackets on $\hat{K}$ are called *Dirac Brackets*. Work out coordinate expressions for Dirac brackets and show that they agree with the formulas of Dirac [1950] (see also Hansen, Regge and Teitelboim [1976] and Sudarshan [1974]).

(d) If $\tau_{T(T^*M)}$ maps TL to TK, (i.e., there are no secondary constraints) show that H is constant on the leaves of $\mathfrak{F}$ and hence induces a Hamiltonian $\hat{H}$ on $\hat{K}$. Show that this condition is not always satisfied (consider $L(q^1, q^2, \dot{q}^1, \dot{q}^2) = \frac{1}{2} f(q^1, q^2)(\dot{q}^1)^2$.)

(e) Show that Newton–Wigner coordinates for the relativistic top are canonical coordinates on K (determined by Darboux's theorem). (See Hansen, Regge and Teitelboim [1976, Chapter 3]).

(f) Let $\tilde{K}$ denote $TM/\mathfrak{L}$ where $\mathfrak{L}$ is the characteristic foliation determined by the Lagrange two form ω_L. Show that $\tilde{K}$ and $\hat{K}$ are symplectically diffeomorphic.

(g) Suppose there is a group action Φ of G on M leaving L invariant and whose orbits in TM are the degenerate directions (leaves of $\mathfrak{L}$). Let $J: T^*M \to \mathfrak{g}^*$ be a moment for the action. Show that (d) holds and we can take $K = J^{-1}(0)$ and $\hat{K}$ is the reduced phase space in the sense of 4.3.1.

references:

Tulczyjew [1977a, b], Kunzle [1969], Sniatycki [1974], Lichnerowicz [1975a], Menzio and Tulczyjew [1978], and Gotay, Nester, and Hinds [1978].

ISBN 0-80530102-X

5.4 QUANTIZATION*

Quantization refers to the process of forming a quantum mechanical system from a given classical system. This section provides merely an introduction and some background for this topic. More up to date and advanced work, along with references, may be found in Weinstein [1977b], Guillemin and Sternberg [1977], and Sniatycki [1978].

We shall concentrate on certain geometric problems associated with quantization, beginning with Dirac's quantization rule that associates quantum operators to the simplest classical observables, namely, the position and momentum functions. This leads naturally to the notion of quantization; that is, a map from classical observables to symmetric operators preserving the bracket relations and mapping 1 to the identity. The theorems of Gröenewald and van Hove are then established, which show that quantization is, in general, impossible if the q's and p's are to act with finite multiplicity; however, it is possible if they act with infinite multiplicity. The geometric construction of this using techniques of Souriau and Kostant is known as *pre-quantization.* The construction of the actual quantum mechanical Hilbert space uses the additional structure of a *polarization*.

The present view of quantization theory was developed by Souriau and Kostant, and our main goal is to give an introduction to it. However, it is important to keep it in context. We have therefore endeavored to include considerable background work of van Hove. We have also included a discussion of the Mackey–Wightman analysis at the end of the section. This shows, among other things, that Euclidean and Galilean invariance forces one to quantize in the usual way for free particles on $\boldsymbol{R}^n$. (See also Mackey [1963] and Varadarajan [1968].) The discussions are, however, incomplete because many key issues are omitted, such as the classical limit $\hbar \to 0$ (see Maslov [1965], Truman [1976]).

One should be warned of two things. First of all, the Souriau–Kostant theory has not yet reached a completely satisfactory state. Secondly, research is still active in the subject and it is rather complicated at the research frontier.

An analysis of Euclidean-invariant quantum systems may be summarized by a set of rules for assigning operators in Hilbert space to the simplest classical observables. For simplicity, let us consider the case of a spinless particle. Then the appropriate Hilbert space is $L^2(\boldsymbol{R}^3)$, the square integrable complex valued functions on $\boldsymbol{R}^3$. The classical coordinate observables x_j correspond to operators Q_j given by

$$(Q_j f)(x) = x_j f(x)$$

The classical momentum observables are the Hamiltonian generators of one-parameter groups of rigid motions; if X_α is the vector field on $\boldsymbol{R}^3$, which generates a one-parameter subgroup $t \mapsto \alpha(t)$ of rigid motions, then the

*The authors thank P. Chernoff for help with this section.

ISBN 0-8053-0102-X

corresponding quantum operator is $(1/i)X_\alpha$, regarded as a differential operator on $L^2(\boldsymbol{R}^3)$. In particular, the linear momentum operators are given by

$$P_k f = \frac{1}{i}\frac{\partial f}{\partial x_k}$$

(defined on a suitable domain). Corresponding to the classical energy function, $p^2/2m + V(x)$, in units with $\hbar = 1$, we have the Hamiltonian operator

$$Hf(x) = -\frac{1}{2m}\Delta f(x) + V(x)f(x)$$

Dirac observed that these operators obey commutation relations analogous to the Poisson bracket relations among the corresponding classical observables. He made this analogy the foundation of his approach to quantization. According to Dirac [1930, p. 87], "the problem of finding quantum conditions now reduces to the problem of determining Poisson brackets in quantum mechanics. The strong analogy between the quantum Poisson bracket (that is, $1/i$ times the commutator) and the classical Poisson bracket . . . leads us to make the assumption that the quantum Poisson brackets, or at any rate the simpler ones of them, have the same values as the classical Poisson brackets." We shall later examine the natural question of whether this "analogy" can be raised to the status of a *homomorphism* of Lie algebras. First of all, we shall show that Dirac's principle can be illustrated by a method of "canonical quantization" of any classical system whose configuration space is a finite-dimensional manifold. Special cases of this procedure appeared very early in the physics literature. Mathematical treatments have been given by Segal [1960] and Mackey [1963], whose outline we will follow. (See also Varadarajan [1968] and Marsden [1967].)

Our first task is to find the appropriate generalization of $L^2(\boldsymbol{R}^3)$. Unlike Euclidean space, an arbitrary finite-dimensional manifold Q does not carry a canonical measure. (Note, however, that there is a natural measure on a *Riemannian* manifold.) But Q does carry a natural *class* of measures, namely, those that are equivalent to Lebesgue in every coordinate chart of Q. We call such measures the *natural measures* on Q. We remind the reader that a measure μ is *absolutely continuous* with respect to another measure ν provided there is a density function ρ such that $\mu(E) = \int_E \rho\, d\nu$ for every Borel set E. The function ρ is often denoted by $d\mu/d\nu$, and is called the Radon–Nikodym derivative of μ with respect to ν. The measures μ and ν are said to be *equivalent* provided that each is absolutely continuous with respect to the other. In this case we have $d\nu/d\mu = (d\mu/d\nu)^{-1}$. Associated to this class of measures there is a Hilbert space, called the *intrinsic Hilbert space* of the manifold Q, and will be denoted by $\mathcal{H}(Q)$. Members of $\mathcal{H}$ will be called *half-densities*. Its formal definition follows.

ISBN 0-8053-0102-X

5.4.1 Definition *Consider the set of all pairs (f, μ), where μ is a natural measure and f is a complex (measurable) function such that $\int_Q |f|^2 d\mu < \infty$. Two pairs (f, μ) and (g, ν) will be called equivalent provided that $f\sqrt{d\mu/d\nu} = g$. [It is easy to show that this is indeed an equivalence relation by use of the formula $d\mu/d\lambda = (d\mu/d\nu)\cdot(d\nu/d\lambda)$] We denote the equivalence class of (f, μ) by $f\sqrt{d\mu}$. $\mathcal{H}(Q)$ is the set of all such equivalence classes. The Hilbert space structure of $\mathcal{H}(Q)$ can be defined as follows. Pick any natural measure μ. Then the map U_μ: $f \mapsto f\sqrt{d\mu}$ is a bijection from $L^2(Q, \mu)$ onto $\mathcal{H}(Q)$. We use U_μ to transfer the Hilbert space structure from $L^2(Q, \mu)$ to $\mathcal{H}(Q)$. (It is easy to check that the resulting structure is independent of the choice of μ. Indeed, if $\xi = f\sqrt{d\mu} = g\sqrt{d\nu}$ is a typical vector in $\mathcal{H}(Q)$, we have*

$$\|U_\mu^{-1}\xi\|^2 = \int |f|^2 d\mu = \int |f|^2 \frac{d\mu}{d\nu}\, d\nu$$

$$= \int |g|^2 d\nu = \|U_\nu^{-1}\xi\|^2$$

Thus the norm in $\mathcal{H}(Q)$ is independent of μ.)

Of course, "practical" computations in $\mathcal{H}(Q)$ must be made with the aid of one of the isomorphisms U_μ. The virtue of considering the intrinsic space $\mathcal{H}(Q)$ is that it puts all the measures μ on the same footing; choosing a particular $L^2(Q, \mu)$ would be akin to imposing a particular coordinate system on an abstract vector space.

We shall say that $\xi \in \mathcal{H}(Q)$ is C^∞ *with compact support* provided $\xi = f\sqrt{d\mu}$, where f is a C^∞ function with compact support on Q and μ is the measure associated with a smooth n-form Ω_μ on Q. These C^∞ vectors with compact support form a dense subspace $\mathfrak{D}_0^\infty$ of $\mathcal{H}(Q)$.

Let X be a smooth vector field on Q with local flow F_t. We shall show using Koopmanism (see Sect. 2.6) that there is a natural symmetric operator $\tilde{X}$ with domain $\mathfrak{D}_0^\infty$ associated to X. To see this, consider $f\sqrt{d\mu} \in \mathfrak{D}_0^\infty$. Then if t is sufficiently small,

$$U_t\left(f\sqrt{d\mu}\right) = f \circ F_t \sqrt{d(\mu \circ F_t)}$$

is a well-defined element of $\mathcal{H}(Q)$. We define

$$\tilde{X}\left(f\sqrt{d\mu}\right) = \frac{1}{i} \lim_{t \to 0} \frac{1}{t}\left[U_t\left(f\sqrt{d\mu}\right) - f\sqrt{d\mu} \right]$$

To see that this limit exists, note that

$$U_t\left(f\sqrt{d\mu}\right) = f \circ F_t \sqrt{\rho_t}\, \sqrt{d\mu}$$

ISBN 0-8053-0102-X

where $\rho_t = d(\mu \circ F_t)/d\mu$ is a smooth function. In terms of the n-form associated with μ, we have $F_t^*(\Omega_\mu) = \rho_t \Omega_\mu$. Moreover,

$$\lim_{t\to 0} \frac{\rho_t - 1}{t}\Omega_\mu = \lim_{t\to 0}\frac{1}{t}(F_t{}^*\Omega_\mu - \Omega_\mu) = L_X\Omega_\mu$$
$$= div_\mu X \cdot \Omega_\mu$$

where $div_\mu X$ is the divergence of X relative to Ω_μ. Hence we have

$$\lim_{t\to 0}\frac{1}{t}\left(f\circ F_t\sqrt{\rho_t} - f\right)\sqrt{d\mu} = \left(X\cdot f + \tfrac{1}{2}(div_\mu X)f\right)\sqrt{d\mu}$$

Thus we define

$$\tilde{X}\left(f\sqrt{d\mu}\,\right) = \frac{1}{i}\left(X\cdot f + \tfrac{1}{2}(div_\mu X)f\right)\sqrt{d\mu}$$

This is well-defined because it has been constructed by intrinsic methods; one could, of course, check directly that the right-hand side above depends only on the equivalence class of (f,μ).

5.4.2 Proposition *The operator $\tilde{X}$ is symmetric.*

Proof. For $\xi = f\sqrt{d\mu}\,,\ \eta = g\sqrt{d\mu} \in \mathfrak{D}_0^\infty$, we have

$$(U_t\xi, U_t\eta) = \int (f\circ F_t)\overline{(g\circ F_t)}\, d\mu\circ F_t$$
$$= \int f\bar{g}\, d\mu$$
$$= (\xi,\eta)$$

so $(\tilde{X}\xi,\eta) - (\xi,\tilde{X}\eta) = 0$ by differentiation of the above at $t=0$. ■

A fairly straightforward extension of Theorem 2.6.14 and its converse shows that $\tilde{X}$ is essentially self-adjoint on $\mathfrak{D}_0^\infty$ if and only if the flow F_t of X is complete except on a set of measure zero.

5.4.3 Definition. *Define the **classical momentum observable** $P(X)$ on T^*Q associated to the vector field X on Q by $P(X)(\alpha_x) = \alpha_x(X(x))$, $x\in Q$. We shall call the operator $\tilde{X}$ the corresponding **quantum momentum observable**.*

*If f is a C^∞ function on Q, that is, a **classical configuration observable**, we define the corresponding **quantum position observable** to be the operator*

$$Q_f \quad \text{on} \quad \mathcal{H}(Q) \quad \text{defined by}$$

$$Q_f\left(g\sqrt{d\mu}\,\right) = fg\sqrt{d\mu}$$

(i.e., Q_f is multiplication by f).

ISBN 0-8053-0102-X

In coordinates note that $P(X)=p_iX^i(q)$. In 4.2.12 we verified by a direct calculation in coordinates the classical Poisson bracket formulas:

$$\{P(X),P(Y)\}=-P([X,Y])$$

$$\{f,g\}=0$$

$$\{f,P(X)\}=X(f)$$

Here $\{f,g\}$ means $\{f\circ\tau_Q^*,g\circ\tau_Q^*\}$ and $\{f,P(X)\}$ means $\{f\circ\tau_Q^*,P(X)\}$.

The following theorem establishes the analogous operator commutation relations in accordance with Dirac's rule.

5.4.4 Theorem. *Let Q be a finite-dimensional manifold with intrinsic Hilbert space $\mathcal{H}(Q)$. Let X and Y be smooth vector fields on Q and let f,g be smooth real-valued functions on Q. Then we have the following commutation relations (valid on the domain $\mathfrak{D}_0^\infty\subset\mathcal{H}(Q)$):*

(i) $(1/i)[\tilde{X},\tilde{Y}]=-\widetilde{[X,Y]}$
(ii) $(1/i)[Q_f,Q_g]=0$
(iii) $(1/i)[Q_f,\tilde{X}]=Q_{X(f)}$

Proof. Pick a vector $\xi=\phi\sqrt{d\mu}$ in $\mathfrak{D}_0^\infty$. Then

$$\tilde{X}\xi=\frac{1}{i}\left[X\cdot\phi+\tfrac{1}{2}(\operatorname{div}_\mu X)\phi\right]\sqrt{d\mu}$$

From this, one computes that

$$\frac{1}{i}[\tilde{X},\tilde{Y}]\xi=i\left[(XY-YX)\phi+\tfrac{1}{2}(X\operatorname{div}_\mu Y-Y\operatorname{div}_\mu X)\phi\right]\sqrt{d\mu}$$

In terms of the volume form Ω_μ, we have

$$\begin{aligned}(\operatorname{div}_\mu[X,Y])\Omega_\mu&=L_{[X,Y]}\Omega_\mu\\&=L_XL_Y\Omega_\mu-L_YL_X\Omega_\mu\\&=L_X(\operatorname{div}_\mu Y\Omega_\mu)-L_Y(\operatorname{div}_\mu X\Omega_\mu)\\&=(X\cdot\operatorname{div}_\mu Y-Y\cdot\operatorname{div}_\mu X)\Omega_\mu\end{aligned}$$

Thus we have

$$\begin{aligned}\frac{1}{i}[\tilde{X},\tilde{Y}]\cdot\xi&=i\{[X,Y]\phi+\tfrac{1}{2}\operatorname{div}_\mu[X,Y]\phi\}\sqrt{d\mu}\\&=-\widetilde{[X,Y]}\cdot\xi\end{aligned}$$

ISBN 0-8053-0102-X

So (i) is true. The relation (ii) is trivial. As for (iii), we have

$$\frac{1}{i}[Q_f, \tilde{X}]\cdot\xi = -\left\{f\left(X\phi + \tfrac{1}{2}(div_\mu X)\phi\right) - X(f\phi) - \tfrac{1}{2}(div_\mu X)f\phi\right\}\sqrt{d\mu}$$

$$= \{X(f\phi) - fX\phi\}\sqrt{d\mu}$$

$$= X(f)\phi\sqrt{d\mu}$$

$$= Q_{X(f)}\cdot\xi \quad \blacksquare$$

We come next to the Hamiltonian operator. As we know, the classical kinetic energy arises from a Riemannian metric g on Q. The energy function on T^*Q associated with g is

$$K(w_x) = \tfrac{1}{2}\langle w_x, g_x^{-1}w_x\rangle$$

where $w_x \in T_x^*Q$ and $g_x\colon T_xQ \to T_x^*Q$ is the isomorphism induced by the Riemannian inner product on T_xQ.

The metric g induces a smooth measure Ω on Q, given by the volume form (see Sect. 2.7):

$$\Omega_x(v_1, v_2, \ldots, v_n) = \left[det\left(g_x(v_i, v_j)\right)\right]^{1/2}$$

The associated divergence operation on vector fields is the divergence relative to g. Thus, by definition, $L_X\Omega = div_g X\cdot\Omega$. Also, recall that if $\phi \in C^\infty(Q)$, $d\phi$ is a one-form, so $g^{-1}(d\phi) = grad_g\phi$ is a vector field, called the gradient of ϕ relative to g. We can now define the *Laplace–Beltrami operator* by

$$\Delta_g\phi = div_g\, grad_g\phi = \nabla^2\phi \text{ in §2.7}$$

The operator Δ_g is symmetric with respect to the inner product $\langle , \rangle$ in $L^2(Q,\Omega)$. Indeed, if ϕ and ψ are C^∞ functions with compact support, we have

$$\langle \Delta_g\phi, \psi\rangle = \int div\, grad\,\phi\bar{\psi}\Omega$$

$$= \int \bar{\psi} L_{grad\,\phi}\Omega$$

$$= \int L_{grad\,\phi}(\bar{\psi}\Omega) - \int\left(L_{grad\,\phi}\bar{\psi}\right)\Omega$$

$$= -\int\left(L_{grad\,\phi}\bar{\psi}\right)\Omega \quad \text{by Stokes' theorem}$$

ISBN 0-8053-0102-X

But

$$L_{grad\phi}\bar{\psi} = \langle d\bar{\psi}, g^{-1}(d\phi)\rangle$$

$$= \langle g^{-1}(d\bar{\psi}), d\phi\rangle \quad \text{since } g \text{ is symmetric}$$

$$= L_{grad\bar{\psi}}\phi$$

Thus

$$\langle \Delta_g\phi, \psi\rangle = -\int L_{grad\bar{\psi}}\phi\cdot\Omega$$

$$= \langle \Delta_g\bar{\psi}, \bar{\phi}\rangle = \langle \phi, \Delta_g\psi\rangle$$

Accordingly, using the canonical identification of $L^2(Q,\Omega)$ with $\mathcal{H}(Q)$, we can regard Δ_g as a symmetric operator with domain $\mathfrak{D}_0^\infty$. (It can be shown that Δ_g is essentially self-adjoint if Q is *complete* relative to the metric g; cf. Gaffney [1954], Roeleke [1960], Chernoff [1973].)

5.4.5 Definition. *If the classical potential energy is given by the function V on Q, we define the* ***Hamiltonian operator*** *to be*

$$H_{op} = -\tfrac{1}{2}\Delta_g + Q_V$$

on $\mathcal{H}(Q)$.

The question of essential self-adjointness of this operator is difficult in general. (See Reed and Simon [1975].) If we are willing to ignore the technical difficulties surrounding the self-adjointness question, then the rules that we have given suffice to quantize the basic observables of "any" classical system with a finite number of degrees of freedom.

Let us examine the relations between classical and quantum *velocity observables*. These are, by definition, the time derivatives of the configuration observables, relative to the flow generated by the Hamiltonian.

5.4.6 Proposition. *We have the classical formula*

$$\{K,f\} = -P(grad f)$$

and the analogous operator commutation relation

$$\frac{1}{i}\left[-\frac{1}{2}\Delta, Q_f\right] = -(grad f)^{\sim}$$

ISBN 0-8053-0102-X

Proof. It is straightforward to verify (e.g., by computing in local coordinates) the first classical formula. To establish the second relation we work in the space $L^2(Q,\Omega)$. If ϕ is C^∞ with compact support, then

$$\frac{1}{i}[-\Delta,Q_f]\phi=-\frac{1}{i}\Delta(f\phi)+\frac{1}{i}f\Delta\phi$$

Now $grad(f\phi)=f\,grad\,\phi+\phi\,grad\,f$. Moreover, for any smooth ψ and vector field X, we have $div(\psi X)=\psi\,div\,X+X\cdot\psi$. (Indeed,

$$\begin{aligned}div(\psi X)\Omega&=L_{\psi X}\Omega=di_{\psi X}\Omega\\&=d(\psi i_X\Omega)=\psi di_X\Omega+d\psi\wedge i_X\Omega\\&=\psi\,div\,X\,\Omega+d\psi\wedge i_X\Omega\end{aligned}$$

Now $d\psi\wedge\Omega=0$. Hence, since i_X is an antiderivation,

$$\begin{aligned}0=i_X(d\psi\wedge\Omega)&=(i_Xd\psi)\wedge\Omega-d\psi\wedge i_X\Omega\\&=(X\psi)\Omega-d\psi\wedge i_X\Omega\end{aligned}$$

Thus we have the stated formula.)

Hence

$$\begin{aligned}\frac{1}{i}[-\Delta,Q_f]\phi&=\frac{1}{i}f\Delta\phi-\frac{1}{i}[f\Delta\phi+(grad\,\phi)\cdot f+\phi\Delta f+(grad\,f)\cdot\phi]\\&=-\frac{1}{i}\phi\Delta f-\frac{2}{i}\langle grad\,f,grad\,\phi\rangle\\&=-\frac{1}{i}\{div(grad\,f)\phi+2\,grad\,f\cdot\phi\}\\&=-2(grad\,f)^{\tilde{}}\phi\qquad\blacksquare\end{aligned}$$

Unfortunately, the correspondence between classical and quantum brackets breaks down to some extent when we consider brackets involving the kinetic energy and momentum observables. In fact, computations reveal that the Poisson bracket $\{K,P(X)\}$ cannot be expressed in general as a finite linear combination of observables heretofore considered. The same holds true for the operator bracket. It is true, nevertheless, that $\{K,P(X)\}$ vanishes if and only if $[-\Delta,\tilde{X}]$ does; indeed, each of these conditions holds exactly when the flow of X consists of isometries.

5.4.7 Example (See Sect. 4.4). The *rigid body* provides an illustration of the Dirac quantization procedure. The classical configuration space is the group $G=SO(3)$. In this case there is a canonical measure on G, namely, Haar measure μ. Thus $\mathcal{H}=L^2(G,\mu)$ is our Hilbert space. For $v\in\mathfrak{g}$ (the Lie algebra

ISBN 0-8053-0102-X

of G), the right-invariant vector fields X_v on G correspond to generators of rotation in space coördinates, while the left-invariant vector fields X'_v correspond to the generators of rotation in body coördinates. Since Haar measure is bi-invariant, these vector fields are divergence-free, so the related momentum operators on $L^2(G,\mu)$ are given by

$$J_v = \frac{1}{i}\tilde{X}_v \quad \text{and} \quad J'_v = \frac{1}{i}\tilde{X}'_v$$

We have the commutation relations

$$\frac{1}{i}[J_v, J_w] = J_{[v,w]}$$

$$\frac{1}{i}[J'_v, J'_w] = -J'_{[v,w]}$$

$$[J_v, J'_w] = 0$$

for $v, w \in \mathfrak{g}$. These are the same as the corresponding classical brackets (See Sect. 4.3. and the relation $X_{\{f,g\}} = -[X_f, X_g]$.)

The kinetic energy operator is the Laplace–Beltrami operator associated with a given left-invariant Riemannian metric g on $SO(3)$. The metric g is determined by a symmetric positive operator I on the Lie algebra of $SO(3)$, which we identify with $\boldsymbol{R}^3$ in the usual way. If e_1, e_2, e_3 is an orthonormal basis that diagonalizes I, then the kinetic energy operator is

$$H = -\frac{1}{2}\Delta = \frac{1}{2A}(J'_{e_1})^2 + \frac{1}{2B}(J'_{e_2})^2 + \frac{1}{2C}(J'_{e_3})^2$$

where A, B, C are the principal moments of inertia.

According to the Peter–Weyl theorem (see, e.g. Yosida [1971, §XI 10]), the Hilbert space $L^2(G,\mu)$ is a direct sum of finite-dimensional subspaces that are invariant under both left and right translations. There is one such bi-invariant subspace for each irreducible representation $\mathfrak{D}$ of G, and its dimension is the square of the dimension of $\mathfrak{D}$. In the case of $SO(3)$ the dimensions of the irreducible representations are the successive odd integers, so $L^2(SO(3))$ is a direct sum of bi-invariant subspaces $\boldsymbol{H}_n$, $n=0,1,2,\ldots$, where the "spin n" subspace $\boldsymbol{H}_n$ has dimension $(2n+1)^2$. Since the spaces $\boldsymbol{H}_n$ are bi-invariant, they are invariant under all the operators J_v, J'_w and so also under the Hamiltonian H. Hence the determination of the eigenvalues and eigenfunctions of H reduces to a series of calculations with finite matrices. For further details and additional references see Wigner [1959], Chapter 19.

We turn now to the question of quantization. We are first going to prove that a quantization of all the classical observables is, in general, impossible. Therefore, let us work in the concrete setting of $Q = \boldsymbol{R}^n$ and introduce

ISBN 0-8053-0102-X

temporarily, the term "full quantization" for one which includes all the classical variables.

5.4.8 Definition.* *A **full quantization** of Q is a map taking classical observables f (i.e., continuous functions of $(q,p) \in T^*\mathbf{R}^n$) to self-adjoint operators $\hat{f}$ on Hilbert space $\mathcal{H}$ such that:*

(*i*) $(f+g)\hat{} = \hat{f}+\hat{g}$
(*ii*) $(\lambda f)\hat{} = \lambda \hat{f},\ \lambda \in \mathbf{R}$
(*iii*) $\{f,g\}\hat{} = (1/i)[\hat{f},\hat{g}]$
(*iv*) $\hat{1} = I$ *(1 = constant function, I = identity)*
(*v*) $\hat{q}^i$ *and* $\hat{p}_j$ *act irreducibly on* $\mathcal{H}$.

By the Stone–von Neumann theorem (stated on p. 452), condition (v) really means we can take $\mathcal{H} = L^2(\mathbf{R}^n)$ and that $\hat{q}_i$ and $\hat{p}_j$ are given by $\hat{q}_i = Q_{q_i}$ and $\hat{p}_j = (1/i)\partial/\partial q_j$; that is, the *Schrödinger representation*.

If we had just insisted on (i)–(iii) we could let $f = (1/i)X_f$ on $L^2(\mathbf{R}^{2n})$, but then (iv) [and(v)] would fail.

To allow for spin, one should relax (v) to

(*v'*) *The position and momentum operators are represented by a direct sum of **finitely many** copies of the Schrödinger representation. More precisely, we are asking that $\mathcal{H}$ be realized as the space of L^2 functions from $\mathbf{R}^n$ to $\mathfrak{h}_d$ (= d-dimensional Hilbert space, $d < \infty$) so that*

$$\hat{q}_i\phi(x) = q_i\phi(x)$$

$$\hat{p}_j\phi(x) = \frac{1}{i}\frac{\partial\phi}{\partial x_j}(x)$$

One may now ask whether there exists a full quantization satisfying (i)–(v) or (i)–(v'). The answer is "no." This has been in the literature since Gröenewald's paper [1946]; van Hove [1951] gave a rigorous proof of a result sharper than Gróenwold's (he excluded many classical observables as being a priori bad because they did not have complete classical flows, while any self-adjoint operator generates a complete quantum flow = one-parameter unitary group). Van Hove also showed that there *is* a quantization satisfying (i)–(iv), but the "multiplicity" of the representation of the p's and q's is infinite in this representation. We shall discuss van Hove's work below.

We shall now show that no full quantization satisfying (i)–(v') is possible. The fundamental reason for this seems to be the fact that the group $Sl(2, \mathbf{R})$ has no nontrivial finite-dimensional unitary representations.

Our exposition is taken from notes of P. Chernoff; see also A. Joseph [1970].

*Technical difficulties with addition and bracketing of unbounded self-adjoint operators are ignored here. They will not be important for what follows.

ISBN 0-8053-0102-X

5.4.9 Theorem. *Let $\mathfrak{A}$ be the Lie algebra of real-valued polynomials on $\boldsymbol{R}^{2n}$, where the bracket is the Poisson bracket. Let $\boldsymbol{H}=L^2(\boldsymbol{R}^n, \mathfrak{h}_d)$. Then there is* **no** *map $f \mapsto \hat{f}$ from $\mathfrak{A}$ to the self-adjoint operators on $\boldsymbol{H}$ that has the following properties:*

(0) *For each finite subset $S \subset \mathfrak{A}$ there is a dense subspace $\mathfrak{D}_s \subset \boldsymbol{H}$ such that for all $f \in S$, $\mathfrak{D}_s \subset \mathfrak{D}_{\hat{f}}$ and $\hat{f}\mathfrak{D}_s \subset \mathfrak{D}_s$.**

(*i*) *$(f+g)\hat{} = \hat{f}+\hat{g}$ pointwise on $\mathfrak{D}_s$ if $f, g \in S$*

(*ii*) *$(\lambda f)\hat{} = \lambda \hat{f}$ for $\lambda \in \boldsymbol{R}$*

(*iii*) *$\{f,g\}\hat{} = (1/i)[\hat{f}\hat{g} - \hat{g}\hat{f}]$ on $\mathfrak{D}_s$; and, more precisely, if $\{f,g\}=0$, then $\hat{f}, \hat{g}$ commute in the strong sense that their spectral resolutions commute.*

(*iv*) *$\hat{1} = I$*

(*v*) *$\hat{q}_i =$ multiplication by q_i and $\hat{p}_i = (1/i)\partial/\partial q_i$.*

For simplicity in what follows we shall take $n=1$, that is, one degree of freedom; but everything goes through in the general case.

We begin by noting some facts concerning $Sl(2,\boldsymbol{R})$ and its Lie algebra $sl(2,\boldsymbol{R})$.

$Sl(2,\boldsymbol{R})$ is the group of real 2×2 matrices with determinant one. Its Lie algebra, denoted $sl(2,\boldsymbol{R})$, is the algebra of real traceless 2×2 matrices $\begin{pmatrix} a & b \\ c & -a \end{pmatrix}$. It is three dimensional, with the standard basis being

$$H=\begin{pmatrix} 1 & 0 \\ 0 & -1 \end{pmatrix}, \quad E^+=\begin{pmatrix} 0 & 1 \\ 0 & 0 \end{pmatrix}, \quad E^-=\begin{pmatrix} 0 & 0 \\ 1 & 0 \end{pmatrix}$$

These satisfy the commutation relations

$$[H, E^+] = 2E^+$$
$$[H, E^-] = -2E^-$$
$$[E^+, E^-] = H$$

It is a well-known fact, which we shall use, that this algebra has no nonzero finite-dimensional representation by skew-symmetric operators. (This is a special case of a much more general result about semi-simple Lie algebras; cf. Varadarajan [1968].) Note that if $f=p^rq^s$ and $g=p^mq^n$, we have

$$\{f,g\} = (sm-rn)p^{r+m-1}q^{s+n-1}$$

---

*We call $\mathfrak{D}_s$ an *admissible common domain for S*.

ISBN 0-8053-0102-X

In particular, we have

$$\{q,p\}=1$$

$$\{q^2,q\}=0 \qquad \{q^2,p\}=2q$$

$$\{p^2,q\}=-2p \qquad \{p^2,p\}=0$$

$$\{qp,q\}=-q \qquad \{qp,p\}=p$$

$$\{q^2,p^2\}=4qp$$

$$\{q^2,qp\}=2q^2$$

$$\{p^2,qp\}=-2p^2$$

Hence q^2,qp,p^2 span a three-dimensional subalgebra of the polynomial algebra under $\{\ \}$. Moreover, this algebra is isomorphic to $sl(2,\boldsymbol{R})$, as can be seen by taking

$$\begin{cases} h=qp \\ e_+=p^2/2 \\ e_-=-q^2/2 \end{cases}$$

Finally, note that

$$\{q^3,q\}=0, \quad \{q^3,p\}=3q^2$$

$$\{q^3,p^2\}=6q^2p$$

$$\{p^3,q\}=-3p^2$$

$$\{p^3,q^2\}=-6p^2q$$

$$\{q^3,p^3\}=9q^2p^2$$

$$\{q^2p,qp^2\}=3q^2p^2$$

We have $\hat{q}$ = multiplication by $q\cdot I_d$ on $L^2(\boldsymbol{R},\mathfrak{h}_d)$. Suppose that $F(q)$ is a $d\times d$ matrix-valued function. Then the operator in $L^2(\boldsymbol{R},\mathfrak{h}_d)$, which is given by multiplication with $F(q)$, commutes with $\hat{q}$. It is a standard result that the *converse* is true: any operator $\mathfrak{F}$ commuting with $\hat{q}$ (in the strong sense if the operator is unbounded) must be of this form. For example, use the spectral theorem to bring $\mathfrak{F}$ and $\hat{q}$ simultaneously into diagonal form.

ISBN 0-8053-0102-X

Through the Fourier transform, we see that an analogous result is true for operators commuting with $\hat{p}$; and any operator commuting with both $\hat{q}$ and $\hat{p}$ must be just multiplication by a constant matrix.

We now have in hand the tools we need to prove Theorem 5.4.9.

Proof of 5.4.9 To begin with, note that every admissible domain $\mathfrak{D}$ for $\hat{p},\hat{q}$ consists of rapidly decreasing C^∞ functions (with values in $\mathfrak{h}_d$). Hence if a matrix-valued function $F(q)$ maps $\mathfrak{D}$ to $\mathfrak{D}$, $F(q)$ must be C^∞. (Through the Fourier transform we have a dual result for operators commuting with $\hat{p}$.)

For convenience, we abbreviate $\hat{p}=(1/i)\partial/\partial x$ by ∂. If $F(x)$ is a smooth matrix valued function, we have $[\partial F(x)]=(1/i)F'(x)$ (on a suitable domain $\mathfrak{D}$).

We shall now show what the operators $\widehat{q^2},\widehat{p^2},\widehat{pq}$ have to be.

Because $\{q^2,q\}=0$, $\widehat{q^2}$ and $\hat{q}$ have to commute in the strong sense [hypothesis (iii)]. Hence $\widehat{q^2}$ is multiplication by some $A(x)$, a $d\times d$ matrix-valued function. Moreover, restricting to an admissible domain for $\hat{q},\widehat{q^2},\hat{p},\ldots$, we see that $A(x)$ is C^∞. Now

$$\{p,q^2\}=-2q$$

so on this domain

$$\frac{1}{i}[\partial,A(x)]=-2x\cdot I$$

that is,

$$\frac{1}{i}\cdot\frac{1}{i}A'(x)=-2x\cdot I$$

Thus

$$A(x)=x^2\cdot I+\Lambda$$

where Λ is a constant self-adjoint matrix.

Similarly (using the Fourier transform)

$$\widehat{p^2}=\partial^2\cdot I+Q$$

where Q is a constant matrix.

Next, we consider the relation

$$4pq=\{q^2,p^2\}$$

ISBN 0-8053-0102-X

This gives us

$$4\widehat{pq}=\frac{1}{i}\left[\widehat{q^2},\widehat{p^2}\right] \quad \text{(on a suitable } \mathfrak{D}\text{)}$$

$$=\frac{1}{i}\left[x^2\cdot I+\Lambda, \partial^2\cdot I+Q\right]$$

$$=\frac{1}{i}\left[x^2,\partial^2\right]\cdot I+\frac{1}{i}\left[\Lambda,Q\right]$$

which simplifies to

$$\widehat{pq}=\tfrac{1}{2}(x\partial+\partial x)\cdot I+N$$

where

$$N=\frac{1}{4i}\left[\Lambda,Q\right] \quad \text{(a self-adjoint } d\times d \text{ matrix)}$$

Now, consider the relations

$$\{pq,p^2\}=2p^2 \quad \text{and} \quad \{pq,q^2\}=-2q^2$$

An easy computation shows that

$$\frac{1}{i}\left[N,Q\right]=2Q \quad \text{and} \quad \frac{1}{i}\left[N,\Lambda\right]=-2\Lambda$$

Hence

$$q^2\mapsto\frac{1}{i}\Lambda, \quad p^2\mapsto\frac{1}{i}Q, \quad pq\mapsto\frac{1}{i}N$$

generates a representation of $sl(2,\boldsymbol{R})$ by $d\times d$ skew-adjoint matrices. But since this must be trivial, we conclude that $\Lambda=Q=N=0$.

In other words, $\widehat{p^2}=\partial^2\cdot I$, $\widehat{q^2}=x^2\cdot I$,

$$\widehat{pq}=\tfrac{1}{2}(x\partial+\partial x)\cdot I$$

So far nothing surprising has happened. However, we now go on to higher things: cubic polynomials. As we shall see, a contradiction is arrived at.

First of all, what is $\widehat{q^3}$? Since $\widehat{q^3}$ commutes with $\hat{q}$, $\widehat{q^3}$ = multiplication by some C^∞ matrix function $B(x)$. The relation

$$\frac{1}{i}\left[\widehat{q^3},\hat{p}\right]=3\widehat{q^2}$$

gives us

$$B'(x)=3x^2\cdot I$$

ISBN 0-8053-0102-X

so

$$\widehat{q^3} = B(x) = x^3 \cdot I + B$$

B being a constant matrix.

The relation $\{q^3, pq\} = 3q^3$ clearly implies that $B=0$, so $\widehat{q^3} = x^3 \cdot I$ and, likewise,

$$\widehat{p^3} = \partial^3 \cdot I$$

Now we have

$$\frac{1}{i}[\widehat{q^2}, \widehat{p^3}] = 6\widehat{qp^2}$$

so (by calculation)

$$\widehat{qp^2} = \tfrac{1}{2}(x\partial^2 + \partial^2 x) \cdot I$$

Similarly

$$\widehat{q^2p} = \tfrac{1}{2}(x^2\partial + \partial x^2) \cdot I$$

Finally, observe that we must have

$$\frac{1}{i}[\widehat{q^3}, \widehat{p^3}] = 9\widehat{q^2p^2}$$

and

$$\frac{1}{i}[\widehat{q^2p}, \widehat{qp^2}] = 3\widehat{q^2p^2}$$

This leads us to

$$[\widehat{q^3}, \widehat{p^3}] = 3[\widehat{q^2p}, \widehat{qp^2}]$$

that is

$$x^3\partial^3\phi - \partial^3 x^3 \phi = \tfrac{3}{4}[(x^2\partial + \partial x^2)(x\partial^2 + \partial^2 x)\phi - (x\partial^2 + \partial^2 x)(x^2\partial + \partial x^2)\phi]$$

for all smooth ϕ. But applying the left-hand side to $\phi \equiv 1$ one gets $-6i$, while the right-hand side gives $-3i$. Hence the two operators are *not* equal, and we have reached the desired contradiction. ■

Now we shall turn to what is called *pre-quantization*; that is, a quantization satisfying (i)–(iv) but not (v) or (v′). Thus the position and momentum operators must act with infinite multiplicity. In order to make the construction we need a few definitions.

ISBN 0-8053-0102-X

Fix a manifold M. By a *principal circle bundle* over M we mean a fiber bundle $\pi: Q \to M$ with structural group $S^1 = T^1 = \{e^{is} | s \in \boldsymbol{R}\}$. Basically this means that each fiber $\pi^{-1}(m)$ is a circle and there is a consistent action

$$S^1 \times Q \to Q$$

which is just multiplication on each fiber, that is, $M = Q/S^1$, (see Sect. 4.1) with S^1 acting freely on Q.

5.4.10 Definition. *Let (P, ω) be a symplectic manifold. We say that (P, ω) is **quantizable** iff there is a principal circle bundle $\pi: Q \to P$ over P and a one-form α on Q such that*

(*i*) *α is invariant under the action of S^1 and*
(*ii*) $\pi^* \omega = d\alpha$.

As is standard in bundle theory (Kobayashi-Nomizu [1963]), one calls α a connection and ω its curvature; Q is the *quantizing manifold*. On fibers, write $\alpha | \pi^{-1}(m) = \hbar ds$, where $\hbar = h/2\pi$ is a constant, ultimately to be identified with Planck's constant. Let us fix this constant in the discussion.

The following result is a fairly easy exercise in fiber bundle theory whose proof is left to the reader versed in such matters (cf. Steenrod [1951]).

5.4.11 Theorem. *(P, ω) is quantizable if and only if $\omega/h \in H^2(P, \boldsymbol{Z})$, that is, ω/h is an integral cohomology class.*

Furthermore, the inequivalent quantizing manifolds are classified by $H^1(P, \boldsymbol{Z})$.

Basically, $\omega/h \in H^2(P, \boldsymbol{Z})$ means that when ω/h is integrated over a compact two-manifold $K \dot{\subset} P$ without boundary, we get an integer.

5.4.12 Examples. (1) If ω is exact, then P is quantizable (e.g., $P = T^*M$). If P is simply connected, then Q is unique. (If $\omega = d\theta$, we can let $Q = P \times S^1$ and let $\alpha = \theta + \hbar ds$.)

(2) (This example is relevant for the hydrogen atom; cf. Souriau [1970a], p. 324.) Consider S^2 the two sphere in $\boldsymbol{R}^3$ with radius mK with the following symplectic form (m = mass; K = attractive cst.; e = energy).

$$\omega = \frac{1}{\sqrt{-8me}} \cdot \frac{dx_1 \wedge dx_2}{x_3}$$

Then S^2 is quantizable if and only if $e = -2\pi^2 mK^2/N^2$ for an integer $N > 1$.

Since we have a principal bundle and S^1 acts on $\boldsymbol{C}$, we can construct from Q a *complex line bundle* L over P (by general principles of fiber bundle theory; each fiber S^1 is replaced by the space $\boldsymbol{C}$ on which it acts).

ISBN 0-8053-0102-X

The one-form α on Q gives us a connection ∇ on the line bundle L. (In differential geometry language, α is the connection form.) The condition $\pi^*\omega = d\alpha$ means that ω is the curvature form: $\nabla_X\nabla_Y - \nabla_Y\nabla_X - \nabla_{[X,Y]} = i\omega(X,Y)$. (In coordinates, $(\nabla_X s)^a = \dfrac{\partial s^a}{\partial x^i}X^i + \alpha^a_{ib}s^bX^i$).

5.4.13 Theorem (van Hove, Sourlau, Kostant). *Let (P,ω) be a quantizable symplectic manifold. Then there is a quantization map $f \mapsto \hat{f}$, where $\hat{f}$ is an operator on the space S of sections of L (the Hilbert space is then the L^2 sections of this bundle) satisfying (i)–(iv) of 5.4.8.*

Indeed, we can write down $\hat{f}$ explicitly:

$$\hat{f}: s \mapsto \frac{1}{i}(\nabla_{X_f})s + fs$$

(The $\hbar$ here is built into the connection.)

If $\omega = -d\theta$ is exact, so locally $\theta = \sum p_i\, dq^i$ and $Q = P \times S^1$, the above becomes

$$\hat{f}\cdot s = \frac{\hbar}{i}(X_f s) + \left(f - \sum p_i \frac{\partial f}{\partial p_i}\right)s$$

where $s: P \to \mathbf{C}$. This formula has been used by van Hove and Segal. Notice that for all f, $\hat{f}$ is a first-order operator and that our functions are on P; here $H = L^2(P, \omega^n)$.

Proof of 5.4.13. Because X_f preserves the phase volume, one sees that $\hat{f}$ is a symmetric operator. Conditions (i), (ii), and (iv) are trivial. For (iii):

$$\begin{aligned}
[\hat{f},\hat{g}]s &= \frac{1}{i}(\nabla_{X_f}(\hat{g}s) - \nabla_{X_g}(\hat{f}s)) + f\hat{g}(s) - g\hat{f}(s)\\
&= \frac{1}{i}\left\{\nabla_{X_f}\left(\frac{1}{i}\nabla_{X_g}s + gs\right) - \nabla_{X_g}\left(\frac{1}{i}\nabla_{X_f}s + fs\right)\right\}\\
&\quad + f\frac{1}{i}\nabla_{X_g}s + fgs - g\frac{1}{i}\nabla_{X_f}s - gfs\\
&= -(\nabla_{X_f}\nabla_{X_g} - \nabla_{X_g}\nabla_{X_f})(s) + \frac{1}{i}(X_f(g)s - X_g(f)s)
\end{aligned}$$

(here we used $\nabla_X(gs) = (Xg)s + g\nabla_X s$ and a similar identity with f and g interchanged.)

$$= -\nabla_{[X_f,X_g]}s + \frac{1}{i}\omega(X_f,X_g)s + \frac{1}{i}(\{g,f\} - \{f,g\})s$$

Since $[X_f, X_g] = -X_{\{f,g\}}$, we get

$$\begin{aligned}
[\hat{f},\hat{g}]s &= \nabla_{X_{\{f,g\}}}s - \frac{1}{i}\{f,g\}s\\
&= \frac{-1}{i}\widehat{\{f,g\}}s
\end{aligned}$$

This proves the above theorem. ■

ISBN 0-8053-0102-X

The condition that the "energy surfaces", $H^{-1}(e)/\boldsymbol{R}$ be quantizable amounts essentially to the Bohr–Sommerfeld quantization conditions.*

Since the above constructions are natural, it is fairly clear that any canonical transformations of P will lift to a transformation on Q—at least if the quantification is unique. Maps preserving α on Q are called *quantomorphisms* (after Souriau) and Souriau discusses these lifting problems (see p. 338–339 of Souriau [1970a], Weinstein [1977b] and Chichilnisky [1972].)

The Hilbert space for the quantization constructed above is not yet 'correct'. Elements of it are complex functions of both q and p. To make them just functions of q (or p), we must cut the space down. To do so, we introduce the notion of a polarization (="feuilletage de Planck").

5.4.14 Definition. *A real* ***polarization*** *of the symplectic manifold* (P,ω) *is a foliation* F *of* P *by Lagrangian submanifolds (as leaves). (It is important to allow complex tangent planes here, but we shall assume they are real for simplicity.)*

Recall that $L\subset P$ is a Lagrangian submanifold when L is isotropic (i.e., ω vanishes on $T_pL\times T_pL$) and is maximal (i.e., $\dim L=\frac{1}{2}\dim P$).

We now describe the quantization procedure of Segal, Kirillov, Kostant, and Souriau as follows:

5.4.15 Definition. *Let* (P,ω) *be a quantizable symplectic manifold and let* F *be a polarization. Let* $\mathbf{L}$ *be the line bundle obtained from the quantizing manifold. Then the* ***quantizing Hilbert space*** *is the space of* L^2 *sections of* $\mathbf{L}$ *that are constant on the leaves of* F. *(Assume the leaves are compact.)*

The term "Hilbert space" refers to "intrinsic Hilbert space" defined earlier.

5.4.16 Example. Let $P=T^*Q$, so $\boldsymbol{L}=P\times \boldsymbol{C}$, and sections of $\boldsymbol{L}$ are just complex valued functions. Let the leaves of F be the linear spaces T_x^*Q. This is a polarization and in this case

$$\mathcal{H}\approx L^2(Q) \quad \text{(intrinsic Hilbert space)}$$

Thus $\psi\in\mathcal{H}$ is just a function of the q's. If Q has a flat metric we can likewise obtain a horizontal Lagrangian foliation, so $\psi\in\mathcal{H}$ would be just a function of p's. Moreover, intermediate polarizations are also possible and have been introduced by physicists from time to time.

Kostant [1970a] investigated the relations between different polarizations. If $\mathcal{H}_{F_1}$ and $\mathcal{H}_{F_2}$ are the Hilbert spaces of two polarizations, Auslander and Kostant show that for invariant polarizations of orbits of certain solvable

*See Guillemin and Sternberg [1977] and Weinstein [1977b] for details.

ISBN 0-8053-0102-X

groups there is an intertwining map from $\mathcal{H}_{F_1}$ to $\mathcal{H}_{F_2}$; such maps are related to the *Fourier integral operators* (studied extensively by Hörmander, Maslov, Leray, and others). Kostant then studies how all these fit together—one must be able to consistently form half-forms from volume elements. This leads to the notion of a *metaplectic structure*. In general, however, different polarizations lead to different quantum systems that are equivalent only in the semi-classical limit. See Simms and Woodhouse [1976] and Guillemin and Sternberg [1977] for more information.

What about dynamics? How do we get the correct energy operator to put in Schrödinger's equation? The abstract formulation of this seems not to be completely settled. For example, suppose we can find a Lagrangian foliation F corresponding to constants of the motion, as in Chapter 4. Then—perhaps modulo some cohomology conditions—the classical flow F_t on P will induce a flow of unitary operators on $\mathcal{H}_F$ and thus will give the quantum dynamics.

This, or something like it, seems to be the final step in quantization. It is a crucial problem that has not yet found a satisfactory answer. Souriau has applied it to free particles, both relativistic and nonrelativistic, to obtain for instance the Klein–Gordon and Dirac equations.

The hydrogen atom has proved to be more of a problem. In 1972, Onofri and Pauri proposed a condition of selecting the correct quantum dynamics, again supposing there is a maximal symmetry group. Their conditions seem to lead to the correct equations, but on the other hand they do not appear to face the polarization question squarely. The hydrogen atom has also been studied by Souriau [1974] and Elhadad [1974] in a similar spirit, but their results *seem* quite special and depend on a careful analysis of the classical problem.

A more modest goal might be to obtain the correct energy levels and their multiplicity. To this end we consider, given (P,ω) and $H: P\to \mathbf{R}$, the following procedure: Fix $e\in \mathbf{R}$, consider $H^{-1}(e)$, and divide out by the flow to get $H^{-1}(e)/\mathbf{R}$ = space of all trajectories with energy e. For, modulo completeness problems (which seem to be real enough in the Kepler problem), we know from Chapter 4 that $\tilde{\Sigma}_e = H^{-1}(e)/\mathbf{R}$ is a symplectic manifold. We then try to determine those e for which Σ_e is quantizable and to construct $\mathcal{H}_{e_i}$ the Hilbert space for each such e_i and then write $\mathcal{H} = \Sigma \oplus \mathcal{H}_{e_i}$ as the quantized Hilbert space. The dimension of $\mathcal{H}_{e_i}$ is supposed to be the multiplicity of the spectrum with energy e_i. (One must work with complex polarizations here.)

Unfortunately this does not seem to work exactly. The crucial test case is the hydrogen atom.

We briefly summarize the results obtained by Simms [1968] for this case. Here

$$\Sigma_e \approx S^2\times S^2 = \{(x,y)\in \mathbf{R}^3\times \mathbf{R}^3 \mid \|x\|^2 = \|y\|^2\ x^2 = y^2 = m^2K^2\}$$

and

$$\omega = \frac{1}{\sqrt{-8me}}\left\{\frac{dx_1\wedge dx_2}{x_3} + \frac{dy_1\wedge dy_2}{y_3}\right\}$$

ISBN 0-8053-0102-X

This is quantizable if and only if $e = -2\pi^2 mK^2/N^2$, $N = 1, 2, \ldots$. Surprisingly, this seems to agree with the physics books if we set $\hbar = 1$ and use the right units—at least the N^2 is correct!

Simms uses the Riemann–Roch theorem to calculate the dimension of the quantized Hilbert space. It comes out to be $(N-1)^2$. This is too bad, because the physics books tell us N^2. A more careful analysis using the half-forms of Blattner mentioned above, however, apparently yields the correct answer.

A. Weinstein [1974] has done similar things for spheres, comparing the "quasi-classical" (i.e., quantized as above) spectrum with that for the Laplacian. He replaces the above conditions by the following procedure. He defines a *quasi-classical state* on $P = (T^*Q, d\theta)$ to be a Lagrangian submanifold $L \subset P$ such that, for any closed curve $\gamma \subset L$,

$$\frac{1}{2\pi\hbar}\int_\gamma \theta - \tfrac{1}{4} I_\gamma \quad \text{is an integer}$$

(quantization condition) where I_γ is the Maslov index—a generalization of the Morse index—of γ (cf. Arnold [1967]). The state L is an *eigenstate* if H is constant on it. Using these ideas, he calculates the quasi-classical spectrum (including multiplicities) for spheres. Again they do not agree exactly with (but do closely resemble) the exact spectrum of the Laplace Beltrami operator. Presumably this too can be corrected using half-forms.

For further information on these matters we again refer the reader to Weinstein [1977b] and Guillemin and Sternberg [1977].

We will now explain why the Hamiltonians of single particles in $\mathbb{R}^3$ (or in Minkowski space) for both classical and quantum mechanics must be chosen the way they are, at least if certain group invariance properties are assumed. The quantum mechanical case will be merely discussed, with references cited for detail expositions. The results leave little doubt that the quantization procedures for free particles are correct. The related Stone–von Neumann theorem shows that the Poisson bracket-commutator correspondence forces the quantum operators corresponding to q^i and p_j to be equivalent to the multiplication by q^i and $i(\partial/\partial q^j)$, respectively. Our constructions and proofs are done by "brute force." See Marle [1976] for a more geometric framework.

We shall begin by discussing what the Hamiltonians for Galilean and Lorentz invariant particles must be in the classical (i.e., nonquantum) case. To do so we must first understand what "invariance" means.

5.4.17 Definition. *Let (P, ω) be a symplectic manifold and let Φ_g be a symplectic action of a group G on P. Let X_H be a Hamiltonian vector field on P. We say the equations of motion are* ***invariant*** *if*

$$\Phi_g^* X_H = X_H \quad \textit{for all } g \in G$$

ISBN 0-8053-0102-X

This is equivalent to (see 3.3.19)

$$H\circ\Phi_g - H = c(g) = \textit{constant on } P$$

Since Φ_g is an action, we have

$$\begin{aligned} c(gh) &= H\circ\Phi_g\circ\Phi_h - H \\ &= (H\circ\Phi_g\circ\Phi_h - H\circ\Phi_h) + (H\circ\Phi_h - H) \\ &= c(g) + c(h) \end{aligned}$$

Thus $g\mapsto c(g)$ is a homomorphism of G to $\boldsymbol{R}$. If G is compact, we observe that $c\equiv 0$ [since $c(G)$ would be a compact subgroup of $\boldsymbol{R}$]. The study of these homomorphisms of G to $\boldsymbol{R}$ is basic to the study of invariant equations of motion and the determination of the structure of H. We also observe that if J is a momentum mapping for the action and if the equations of motion are invariant, then $\{\hat{J}(\xi), H\} = d(\xi) = \textit{constant}$, for $\xi\in\mathfrak{g}$.

We turn now to systems in $\boldsymbol{R}^3$ and Euclidean invariance. The Euclidean group on $\boldsymbol{R}^3$ is the group of orientation preserving isometries of $\boldsymbol{R}^3$; it is

$$\mathcal{E} = SO(3)\times\boldsymbol{R}^3$$

The group structure is the *semi-direct product* structure given by

$$(A,a)\cdot(B,b) = (AB, a+Ab)$$

5.4.18 Proposition. *Let X_H be a Hamiltonian system on $T^*\boldsymbol{R}^3$ that is invariant under the action of $\mathcal{E}$ on $T^*\boldsymbol{R}^3$. Then H itself is invariant and so H is a function of $\|\boldsymbol{p}\|$ alone.*

Proof. It suffices to show that any continuous homomorphism $c: \mathcal{E}\to\boldsymbol{R}$ is zero. Indeed since $SO(3)$ is compact and $SO(3)\times\{0\}$ is a subgroup, $c((A,0)) = 0$. Also, $c((I,a)) = \langle v,a\rangle$ for some $v\in\boldsymbol{R}^3$. Then the identity $(A,a)=(I,a)\cdot(A,0)$ yields $c(A,a)=\langle v,a\rangle$. But c is a homomorphism, so

$$c((A,a)\cdot(B,b)) = c(AB, a+Ab) = c(A,a) + c(B,b)$$

that is,

$$\langle v, a+Ab\rangle = \langle v,a\rangle + \langle v,b\rangle$$

from which it follows that $v=0$ [since the action of $SO(3)$ is transitive on the two-sphere]. ■

Since any H that is a function only of $\|\boldsymbol{p}\|$ is Euclidean invariant, this type of invariance is not enough to specify H completely. However, this end can be achieved if we enlarge the invariance group to the Galilean group.

ISBN 0-8053-0102-X

5.4.19 Definition. *The* ***Galilean group*** $\mathcal{G}$ *is the group of transformations on* $\mathbf{R}^3\times\mathbf{R}$ *generated by*

(*a*) *the Euclidean group* $\mathcal{E}$ *(on* $\mathbf{R}^3$*)*
(*b*) $L_v(x,t)=(x+tv,t)$ *for* $v\in\mathbf{R}^3$ *and*
(*c*) $T_\tau(x,t)=(x,t+\tau)$, $\tau\in\mathbf{R}$.

One calls L_v *a pure* ***Galilean transformation*** *and it is interpreted as the transformation to a frame of reference moving with velocity* v. T_τ *is a* ***time translation***.

Clearly $\mathcal{G}$ is a ten-dimensional Lie group.

For $a\in\mathbf{R}^3$, let $D_a(x,t)=(x+a,t)$ and for $A\in SO(3)$, let $R_A(x,t)=(Ax,t)$. The following commutation relations are easy to check:

$$D_aT_\tau=T_\tau D_a,\quad R_AT_\tau=T_\tau R_A,\quad T_\tau L_v=L_vD_{-\tau v}T_\tau$$

$$D_aR_A=R_AD_{A^{-1}a},\quad D_aL_v=L_vD_a,\quad R_AL_v=L_{Av}R_A$$

$$D_{a+b}=D_a\circ D_b,\quad R_{AB}=R_A\circ R_B,\quad T_{\tau+p}=T_\tau\circ T_p,\quad L_{v+w}=L_v\circ L_w$$

There are a number of ways to define Galilean invariance (used by various authors). We define it as follows.

5.4.20 Definition. *A Hamiltonian vector field* X_H *on* $T^*\mathbf{R}^3$ *is* ***Galilean invariant*** *if there is an action* W *of* $\mathcal{G}$ *on* $T^*\mathbf{R}^3$ *by symplectic diffeomorphisms such that*

$$W_a\equiv W_{D_a}=U_a;\qquad U_a(x,p)=(x+a,p),\quad a\in\mathbf{R}^3$$

$$W_A\equiv W_{R_A}=V_A;\qquad V_A(x,p)=(Ax,Ap),\quad A\in SO(3)$$

and

$$W_t\equiv W_{T_t}=F_t,\quad F_t=\textit{flow of } X_H$$

In other words, Galilean invariance means the condition that X_H is Euclidean invariant and that this Euclidean invariance together with time translations effected by F_t fit together to be part of a representation of $\mathcal{G}$. We make no demands on the structure of $W_{L_v}\equiv W_v$; its structure will follow.

A *free particle* of mass $m_0\neq 0$ is defined by $H(x,p)=\|p\|^2/2m_0+\textit{constant}$. This is Galilean invariant if we take $W_v(x,p)=(x,p-m_0v)$.

5.4.21 Theorem. *Let* X_H *be a Hamiltonian vector field on* $T^*\mathbf{R}^3$ *that is Galilean invariant. Then there exists a constant* $m_0\neq 0$ *such that* X_H *corresponds to a free particle of mass* m_0.

ISBN 0-8053-0102-X

Proof. Let X_{K_v} be the generator of $t \mapsto W_{tv}$. From the commutation relations $R_A L_v = L_{Av} R_A$ we get $V_A \circ W_v = W_{Av} \circ V_A$, so $V^*_{A^{-1}} X_{K_v} = X_{K_{Av}}$ and hence

$$K_v \circ V_{A^{-1}} = K_{Av} + \zeta(A, v) \tag{1}$$

where $\zeta(A,v)$ is a constant. Similarly from $D_a L_v = L_v D_a$,

$$K_v \circ U_a = K_v + \eta(a, v) \tag{2}$$

As in 5.4.18 η is linear in a, so $\eta(a,v) = \langle \eta(v), a \rangle$. Also, since $K_{v+w} = K_v + K_w + \textit{constant}$, $\eta(v)$ is linear in v.

From (1) and (2) we obtain

$$\begin{aligned}\langle \eta(Av), a \rangle &= K_{Av} \circ U_a - K_{Av} = K_v \circ V_{A^{-1}} \circ U_a - K_v \circ V_{A^{-1}} \\ &= K_v \circ U_{A^{-1}a} \circ V_{A^{-1}} - K_v \circ V_{A^{-1}} = (K_v \circ U_{A^{-1}a} - K_v) \circ V_{A^{-1}} \\ &= \langle \eta(v), A^{-1}a \rangle = \langle A\eta(v), a \rangle\end{aligned}$$

so $\eta(Av) = A\eta(v)$. It follows that $\eta(v) = m_0 v$ for a constant m_0.

Let $k_v(p) = K_v(0,p)$, so Eq. (2) becomes

$$K_v(q,p) = k_v(p) + m_0 \langle v, q \rangle \tag{3}$$

From $T_t L_v = L_v D_{-tv} T_t$ we get $F_t W_v = W_v U_{-tv} F_t$ and, in particular, for $t > 0$,

$$W_{-\sqrt{t}\, v} F_{-\sqrt{t}}\, W_{\sqrt{t}\, v} F_{\sqrt{t}} = U_{-tv}$$

and so (see Exercise 2.2L)

$$\{K_v, H\}(q,p) = \langle v, p \rangle + \gamma(v) \tag{4}$$

for a constant $\gamma(v)$. Using Eq. (1) and Euclidean invariance of H gives

$$\begin{aligned}\gamma(Av) &= \{K_{Av}, H\} - \langle Av, p \rangle \\ &= \{K_v \circ V_{A^{-1}}, H \circ V_{A^{-1}}\} - \langle Av, p \rangle \\ &= \{K_v, H\} \circ V_{A^{-1}} - \langle v, A^{-1}p \rangle \\ &= \gamma(v) \cdot V_{A^{-1}} = \gamma(v)\end{aligned}$$

Hence $\gamma(v) = 0$ in Eq. (4) since γ is linear in v and is rotationally invariant. On the other hand, from Eq. (3) and the fact that H depends only on $\|p\|$, we

ISBN 0-8053-0102-X

get

$$\{K_v, H\} = \{k_v, H\} + \{m_0\langle v, q\rangle, H\} = \{m_0\langle v, q\rangle, H\}$$
$$= m_0\langle v, \nabla H\rangle$$

where ∇H is the p-gradient of H. Comparing with Eq. (4),

$$m_0 \nabla H(\|p\|) = p$$

so $m_0 \neq 0$ and $H(\|p\|) = \|p\|^2/2m_0 + constant.$ ■

5.4.22 Corollary *The inequivalent actions on T^*R^3 satisfying* 5.4.20 *are precisely characterized by the mass $m_0 \neq 0$ ("equivalent" means up to a canonical transformation).*

Proof. From Eq. (3) $k_v(p)$ is linear in v and $k_v(p) = \langle v, k(p)\rangle + \sigma(v)$, with $\sigma(v)$ a constant. Let

$$\psi: T^*R^3 \to T^*R^3, \quad (q,p) \mapsto \left(q - \frac{k(p)}{m_0}, p\right)$$

Then ψ is symplectic and leaves H invariant, thus it commutes with F_t. Also, ψ commutes with U_a. From Eqs. (1) and (3), $k_v(Ap) = k_{A^{-1}v}(p)$, so k is rotationally invariant and hence ψ commutes with V_A. Note that $K_v \circ \psi(q,p) = m_0\langle v, q\rangle$, so

$$\begin{cases} W^0_{D_a} \equiv \psi \circ W_{D_a} \circ \psi^{-1} = U_a \\ W^0_{R_A} \equiv \psi \circ W_{R_A} \circ \psi^{-1} = V_A \\ W^0_{T_t} \equiv \psi \circ F_t \circ \psi^{-1} = F_t \\ W^0_v \equiv \psi \circ W_v \circ \psi^{-1}: (q,p) \mapsto (q, p - m_0 v) \end{cases}$$

is the standard mass m_0 representation. Hence any mass m_0 representation is canonically equivalent to the standard one. It is not hard to see that standard representations with different masses are symplectically inequivalent. ■

Some connections between Galilean transformations and the work on symmetries in Chapter 4 have been given by Marle [1976]. Also, in Souriau [1970a, Chapter III, §13], the mass is interpreted as the cohomology class of a Galilean group one-cocycle and the obstruction to equivariance.

Next we turn to Lorentz invariance.

ISBN 0-8053-0102-X

5.4.23 Definitions. *The **Poincaré group** (i.e., inhomogeneous Lorentz group) $\mathcal{P}$ differs from the Galilean group in that we now set*

$$L_v(x,t)=\left(\gamma(x-vt)-(\gamma-1)\left(x-v\frac{x\cdot v}{\|x\|^2}\right),\gamma(t-x\cdot v)\right)$$

where $\gamma=1/\sqrt{1-\|v\|^2}$ with D_a, R_A, T_τ as before.

*A Hamiltonian system X_H on $T^*\mathbf{R}^3$ is **Poincaré invariant** if there is a representation W of $\mathcal{P}$ on $T^*\mathbf{R}^3$ by canonical transformations such that*

$$W_{D_a}=U_a,\quad W_{R_A}=V_A \quad \text{and} \quad W_{T_\tau}=F_\tau=\text{flow of } X_H$$

5.4.24 Theorem. *Let X_H on $T^*\mathbf{R}^3$ be Poincaré invariant. Then there exists $m\in\mathbf{R}$ such that*

$$H(q,p)=\sqrt{m^2+\|p\|^2}+\text{constant}$$

Proof. In analyzing this situation, matters are complicated by the fact that $t\mapsto L_{vt}$ is no longer a one-parameter subgroup (addition of velocities in relativity is not "additive" as it is in the Galilean case). For this reason it is more expedient to work infinitesimally.

Let

$$P_i=\frac{d}{dt}D_{te_i}\bigg|_{t=0},\quad J_i=\frac{d}{dt}R_{\exp te_i}\bigg|_{t=0}$$

$$h=\frac{d}{dt}T_t\bigg|_{t=0},\quad K_i=\frac{d}{dt}L_{te_i}\bigg|_{t=0}$$

be the generators of the Lie algebra of $\mathcal{P}$; $e_i=i$th coordinate vector. Explicitly,

$$P_i(q,t)=(e_i,0)$$

$$J_i(q,t)=(e_i\times q,0)$$

$$h(q,t)=(0,1)$$

$$K_i(q,t)=(-e_it,-q_i)$$

The commutation relations shared with $\mathcal{G}$ are:

$$[P_i,P_j]=0,\quad [P_i,h]=0,\quad [J_i,h]=0,\quad [J_i,J_j]=\varepsilon_{ijk}J_k$$

$$[J_i,P_j]=\varepsilon_{ijk}P_k,\quad [J_i,K_j]=\varepsilon_{ijk}K_k,\quad [K_j,h]=P_j$$

ISBN 0-8053-0102-X

In the Galilean case we had $[K_i,K_j]=0$, $[K_i,P_j]=0$; here we have $[K_i,K_j]=-\varepsilon_{ijk}J_k$, $[K_i,P_j]=\delta_{ij}h$. [ε_{ijk} is the sign of the permutation $(1,2,3)\to(i,j,k)$, or zero if i,j,k are not distinct.]

To prove the result, we define K_v as in the Galilean case. As before,

$$K_v \circ V_{A^{-1}} = K_{Av} + \zeta(A,v) \tag{1}$$

and

$$\{K_v, H\} = \langle v,p\rangle \tag{2}$$

Also, by Euclidean invariance, H is a function of $\|p\|$.

Writing $K_i = K_{e_i}$ for the Hamiltonian function on $T^*\boldsymbol{R}^3$ (as well as the generator of $\mathcal{P}$), the relation $[K_i,P_j]=\delta_{ij}h$ gives us

$$\partial K_i/\partial q^j = constant, \quad i \neq j \tag{3}$$

$$\partial K_i/\partial q^i = H + constant \tag{4}$$

If one combines Eqs. (3) and (4) with Eq. (1), it is not hard to see that the constant in (3) is zero. The constant in Eq. (4) may be incorporated into H. Thus, Eq. (2) yields

$$H\,\partial H/\partial p_k = p_k$$

from which it follows that $H^2 = \|p\|^2 + constant$, or $H=\sqrt{m^2+\|p\|^2}$. ■

The literature on this type of result is extensive. See, for instance, Levy-Leblond [1969]. Some related papers are concerned with what interactions between particles are consistent with Lorentz invariance. (See, e.g., the papers of Currie, Jordan, Sudarshan, Foldy, Leutwyler, Arens, Babbitt, etc.)*

We next discuss, without proof, some of the corresponding ideas in the quantum mechanical case. Basic to this discussion is the Stone–von Neumann theorem. This theorem concerns the structure of self-adjoint operators $Q^1,\dots,Q^d,P_1,\dots,P_d$ on a Hilbert space $\mathcal{H}$ satisfying the *Heisenberg commutation relations*:

$$\left.\begin{aligned}[][Q^i,Q^j]&=0\\ [P_i,P_j]&=0\\ [P_j,Q^k]&=i\delta_j^k\end{aligned}\right\} \tag{1}$$

*The famous "no interaction theorem" states that in many cases a system of n particles governed by a Hamiltonian system which is Poincaré invariant is necessarily a system of *free* particles (i.e., in relativity, "action at a distance" does not work). References are given in the bibliography.

ISBN 0-8053-0102-X

We are interested in a quantum mechanical analog of Proposition 3.3.21. Technical problems are relieved if one considers the one-parameter groups

$$U_j(t)=e^{itP_j} \quad \text{and} \quad V^k(t)=e^{itQ^k}$$

generated by P_j and Q^k. The Heisenberg relations then become the *Weyl commutation relations*:

$$\left.\begin{aligned} &[V^k(t), V^j(s)]=0\\ &[U_j(t), U_k(s)]=0\\ &U_j(s)V^k(t)=e^{-i\delta_j^k st}V^k(t)U_j(s)\end{aligned}\right\} \tag{2}$$

If we set $t=(t_1,\ldots,t_d)\in \boldsymbol{R}^d$ and write

$$U(t)=U_1(t_1)\cdots U_d(t_d)$$

and

$$V(t)=V^1(t_1)\cdots V^d(t_d)$$

then the Weyl relations become

$$\left.\begin{aligned} &U(t+s)=U(t)U(s)\\ &V(t+s)=V(t)v(s)\\ \text{and}\\ &U(s)V(t)=e^{is\cdot t}V(t)U(s)\end{aligned}\right\} \tag{3}$$

where $s\cdot t=s_1t_1+\cdots+s_dt_d$.

The first two equations of (3) state that the maps $t\mapsto U(t)$, $t\mapsto V(t)$ are *representations* of $\boldsymbol{R}^d$ in Hilbert space. [Recall that a *representation* of a group G in $\mathcal{H}$ is a (continuous) action of G on $\mathcal{H}$ by bounded linear transformations.] If we let $U(t,s)=U(t)V(s)$, then the Weyl Relations become

$$U(t,s)U(t',s')=e^{i\omega((t,s),(t',s'))}U(t+t',s+s') \tag{3'}$$

where ω is the standard symplectic form on $T\boldsymbol{R}^d$. (See Segal [1963].)

The *Schrödinger representation* is, by definition, the representation of $\boldsymbol{R}^d$ on $L^2(\boldsymbol{R}^d)$ given by

$$\left.\begin{aligned} &(U(t)f)(x)=f(x-t)\\ &(V(t)f)(x)=e^{is\cdot x}f(x)\end{aligned}\right\} \tag{4}$$

ISBN 0-8053-0102-X

(corresponding to the usual choice of Q^i and P_j) and one sees that the Weyl relations (3) are satisfied. The Schrödinger representation is *irreducible*, that is, there is no closed subspace of $L^2(\boldsymbol{R}^d)$ [other than $\{\mathbf{0}\}$ and $L^2(\boldsymbol{R}^d)$] invariant under each $U(t)$ and $V(t)$. (This fact is part of the theorem below.) If we replace $L^2(\boldsymbol{R}^d)$ by $L^2(\boldsymbol{R}^d,\mathfrak{h})$, where $\mathfrak{h}$ is complex Hilbert space (finite or infinite dimensional), then we have card $\mathfrak{h}$ copies of the Schrödinger representation. The Stone–Von Neumann theorem states that this exhausts the possibilities.

5.4.25 Theorem (Stone [1932] and von Neumann [1932]). *Let $U(t)$ and $V(t)$ be (continuous) unitary representations of $\boldsymbol{R}^d$ on $\mathcal{H}$ satisfying the Weyl relations* (3). *Then there is a Hilbert space $\mathfrak{h}$ and a unitary map $T\colon \mathcal{H}\to L^2(\boldsymbol{R}^d,\mathfrak{h})$ that transforms $U(t)$ and $V(t)$ to the Schrödinger representation. The representation is irreducible if and only if $\mathfrak{h}$ is one dimensional.*

For systems with infinitely many degrees of freedom, the analog of the Schrödinger representation is called the *Fock representation* (see, for instance, Streater and Wightman [1964]). However, there are infinitely many other inequivalent irreducible representations as well (Gårding and Wightman [1954]) and according to a theorem of Haag (see Streater and Wightman [1964]) these cannot be avoided in nontrivial field theories. As mentioned earlier, the maps T implementing other representations of the Weyl relations are related to Fourier integral operators.

Mackey [1969] has given an important reformulation of the Stone–von Neumann theorem. One represents the position observables by orthogonal projections P_E in Hilbert space $\mathcal{H}$ for any (Borel) set $E\subset Q$, where Q represents position space. One requires $E\mapsto P_E$ to be a (projection-valued) measure. (For $Q=\boldsymbol{R}^3$, an example of these are the spectral projections associated with the usual position operators, i.e., with $\mathcal{H}=L^2(\boldsymbol{R}^3)$,

$$P_E\psi=\chi_E\cdot\psi$$

where χ_E is the characteristic function of $E\subset\boldsymbol{R}^3$.) If a group G acts on Q, the momentum observables will arise as a representation $U(g)$ of G on $\mathcal{H}$. (For example, if $G=\boldsymbol{R}^3=Q$, we obtain $U(g)$ as described earlier.) The position and momentum are linked by

$$U(g)P_EU(g)^{-1}=P_{g\cdot E} \tag{5}$$

where $g\cdot E$ is the translate of E under g in the given action. Equations (5) are an abstract form of the Weyl relations (3) [or the Heisenberg relations (1)]. One calls a projection-valued measure and a representation satisfying (5) a *system of imprimitivity*. Mackey then proves a general result of which the Stone–von Neumann theorem is a special case.

Besides $G=\boldsymbol{R}^d$, one wishes to take the Euclidean group for G and still impose (5). This leads to what is referred to as the *Mackey–Wightman*

ISBN 0-8053-0102-X

analysis. Since one should only work with expectation values, one should only require $U(g)$ to be a projective representation. As Bargmann has shown, we can then adjust things so that we have a true representation of the covering group $\tilde{\mathcal{E}} = \boldsymbol{R}^3 \times SU(2)$. Mackey and Wightman then use the generalized Stone–von Neumann theorem to show that if we have a system of imprimitivity based on $\boldsymbol{R}^3$ for $\tilde{\mathcal{E}}$, then it is unitarily equivalent to the system.

$$P_E f = \chi_E f$$

and

$$(U_{(a,A)} f)(x) = D_A \cdot f(A^{-1}(x-a))$$

on $L^2(\boldsymbol{R}^3, \mathfrak{h})$, where $a \in \boldsymbol{R}^3$, $A \in SU(2)$ (which by projection to $SO(3)$, acts on $\boldsymbol{R}^3$), and D_A is a unitary representation of $SU(2)$ on $\mathfrak{h}$.

Thus the unitary representations of $SU(2)$ classify Euclidean invariant systems. In quantum mechanics texts, the irreducible unitary representations of $SU(2)$ are shown to be of dimension n, $n = 1, 2, 3, \ldots$ and correspond to particles of spin $s = n/2$.

By analogy with the classical case, one can show that a quantum dynamical system with Hamiltonian operator H_{op} is Euclidean invariant on $\boldsymbol{R}^3$ when H_{op} is a function of the Laplacian; the relevant fact from operator theory is that every translation and rotational invariant operator on $\boldsymbol{R}^n$ is a function of the Laplacian.

We can go to the Galilei group and the Lorentz group as in the classical case. For the Galilei case we are again forced into $H_{op} = -(1/2m)\Delta$ acting on spin wave functions. For the case of the Lorentz group things are more interesting. Here H_{op} depends on the spin and one recovers, for example, the Klein–Gordon and Dirac operators, as Bargmann and Wigner have shown. Any such H_{op} satisfies

$$H_{op}^2 = m^2c^4 - c^2\Delta$$

the mass-energy relation, independent of spin. (Mass-zero particles, e.g.: the photon and neutrino are exceptional in that they are not localizable in the sense that their position operators have the form previously described, so this case is dealt with separately.) We refer the reader to Varadarajan [1968] for details of the aforementioned results and the appropriate references.

5.5 INTRODUCTION TO INFINITE-DIMENSIONAL HAMILTONIAN SYSTEMS

In this section we shall indicate by means of a number of examples how many of the ideas developed in this book for systems with finitely many degrees of freedom can be carried over to systems with infinitely many degrees of freedom. Because this topic is so vast, technicalities will be omitted and some of the examples will merely be sketched. For additional details and

ISBN 0-8053-0102-X

references, we refer the reader to Chernoff and Marsden [1974], Marsden [1974b], and references given below.

We shall begin with perhaps the most fundamental example, the wave equation. Then we shall discuss the Schrödinger and Korteweg–de Vries equations as Hamiltonian systems. We also discuss the equations of an ideal fluid and of general relativity as further examples and give a few results concerning field theory in general.

5.5.1 Example (The Wave Equation). The equation of motion governing small displacements from equilibrium of a homogeneous elastic medium is the wave equation

$$\frac{\partial^2\phi}{\partial t^2}=\Delta\phi=\sum_{i=1}^{s}\frac{\partial^2\phi}{\partial x_i^2} \tag{1}$$

Here $\phi(t,x_1,\ldots,x_s)$ is the "displacement" at $x\in\boldsymbol{R}^s$ at time t, taken to be scalar valued for simplicity. We have chosen units, as usual, so that the velocity of propagation is unity. In many physics books the above equation is derived by approximating the continuous medium by a discrete system of point masses interacting via "springs," that is, forces proportional to the displacements and acting against them. If one takes the limit of the corresponding kinetic and potential energies one finds (see Goldstein [1950])

$$\text{Kinetic energy}\quad K=\tfrac{1}{2}\int_{\boldsymbol{R}^s}(\dot{\phi})^2\,dx$$

and (2)

$$\text{Potential energy}\quad V=\tfrac{1}{2}\int_{\boldsymbol{R}^s}\|\nabla\phi\|^2\,dx$$

where

$$\|\nabla\phi\|^2=\nabla\phi\cdot\nabla\phi=\sum_{i=1}^{s}\frac{\partial\phi}{\partial x^i}\frac{\partial\phi}{\partial x^i}$$

More generally, we may consider possibly nonlinear restoring forces; a general class of potential energies is given by

$$V(\phi)=\int_{\boldsymbol{R}^s}\left\{\tfrac{1}{2}\|\nabla\phi\|^2+\tfrac{1}{2}m^2\phi^2+F(\phi)\right\}dx \tag{3}$$

Such potentials occur in the quantum theory of self-interacting mesons; the parameter m is related to the meson mass, while the function F governs the nonlinear part of the interaction.

ISBN 0-8053-0102-X

Another class of potential energies relevant for nonlinear elasticity has the form

$$V(\phi)=\int_{R^s}\left\{\tfrac{1}{2}\|\nabla\phi\|^2+W(\nabla\phi)\right\}dx \tag{4}$$

where now ϕ is R^s valued and

$$\|\nabla\phi\|^2=\sum_{i,j=1}^{s}\frac{\partial\phi^j}{\partial x^i}\frac{\partial\phi^j}{\partial x^i}$$

The particular type of elastic nonlinearity depends on the function W chosen. Notice that the arguments of W are matrices.

Of course in most cases of interest the fields ϕ will be defined not on all of R^s, but on some domain $\Omega\subset R^s$ with suitable boundary conditions imposed. We work with all of R^s for simplicity.

Configuration space is some space $\mathcal{C}$ of fields, that is, functions $\phi(x)$ on R^s. *At this point we will leave the precise structure of* $\mathcal{C}$ *unspecified;* later on we will make it precise. For now, let members of $\mathcal{C}$ be sufficiently differentiable to justify our manipulations below and let them form a linear space. We have the velocity space $T\mathcal{C}=\mathcal{C}\oplus\mathcal{C}$, and on $T\mathcal{C}$ we consider the Lagrangian

$$L(\phi,\dot{\phi})=K(\dot{\phi})-V(\phi)$$

where V is given by (2b), (3), or (4). Note that $K=\frac{1}{2}\langle\dot{\phi},\dot{\phi}\rangle$, where the brackets represent the usual L^2 inner product. We use the metric associated with K to pull back the canonical symplectic structure from $T^*\mathcal{C}=\mathcal{C}\oplus\mathcal{C}^*$ to $T\mathcal{C}$ in the usual way. On $T\mathcal{C}$ we have the canonical one-form given by 3.5.7:

$$\begin{aligned}\theta_L(x,e)\cdot(\alpha,\beta)&=\langle e,\alpha\rangle\\&=\int_{R^s}e(u)\alpha(u)\,du\end{aligned}$$

and the associated symplectic form $\omega_L=-d\theta_L$:

$$\omega_L(x,e)\cdot(\alpha,\beta;\alpha',\beta')=\langle\alpha,\beta'\rangle-\langle\alpha',\beta\rangle$$

Notice that ω_L is independent of the differentiability properties assumed for the members of $\mathcal{C}$. This is why ω_L is only *weakly* nondegenerate in general. Finally, we have the total energy

$$E(\phi,\dot{\phi})=K(\dot{\phi})+V(\phi)$$

ISBN 0-8053-0102-X

Next we find the equations of motion. Consider first the case (3). We seek a vector field X_E on $\mathcal{C}\oplus\mathcal{C}$ such that

$$dE = i_{X_E}\omega_L$$

From the formula for E we compute

$$dE(\phi,\dot{\phi})\cdot(\alpha,\beta) = \int \left\{\dot{\phi}\beta + \nabla\phi\cdot\nabla\alpha + m^2\phi\alpha + F'(\phi)\alpha\right\} dx$$

Notice that, by integration by parts,

$$\int \nabla\phi\cdot\nabla\alpha\, dx = -\int (\Delta\phi)\alpha\, dx$$

Thus, if we write $X_E(\phi,\dot{\phi}) = (Y(\phi,\dot{\phi}), Z(\phi,\dot{\phi}))$, then

$$dE(\phi,\dot{\phi})\cdot(\alpha,\beta) = \omega_L(\phi,\dot{\phi})\cdot(Y,Z;\alpha,\beta)$$

becomes

$$\langle\dot{\phi},\beta\rangle + \langle m^2\phi + F'(\phi) - \Delta\phi, \alpha\rangle = \langle Y,\beta\rangle - \langle Z,\alpha\rangle$$

Thus

$$\begin{cases} Y(\phi,\dot{\phi}) = \dot{\phi} \\ Z(\phi,\dot{\phi}) = \Delta\phi - m^2\phi - F'(\phi) \end{cases}$$

and so the equations of motion reduce to the *nonlinear wave equation*

$$\partial^2\phi/\partial t^2 = \Delta\phi - m^2\phi - F'(\phi)$$

that is,

$$\Box\phi = m^2\phi + F'(\phi) \tag{5}$$

where $\Box\phi = \Delta\phi - \partial^2\phi/\partial t^2$ is the d'Alembertian or wave operator.

The equations of motion for a general potential V are similarly given by

$$\partial^2\phi/\partial t^2 = -\,grad\,V(\phi) \tag{6}$$

where *grad* $V(\phi)$, the L^2 *gradient of* V is defined by

$$\int grad\,V(\phi)\cdot\psi\, dx = dV(\phi)\cdot\psi \tag{7}$$

ISBN 0-8053-0102-X

For the case of nonlinear elasticity with V given by (4), the equations become

$$\frac{\partial^2\phi}{\partial t^2}=\Delta\phi+div(T(\nabla\phi)) \tag{8}$$

where $T(\nabla\phi)=\partial W/\partial\nabla\phi$, the nonlinear part of the (Piola–Kirchhoff) stress tensor, and where

$$[div\,T]_i=\sum_{j=1}^{s}\frac{\partial T_i^j}{\partial x^j}$$

and

$$T_i^j=\frac{\partial W}{\partial(\partial\phi^i/\partial\phi^j)}$$

Thus, in coordinates on $\mathbf{R}^s$, (8) reads:

$$\frac{\partial^2\phi^i}{\partial t^2}=\sum_{j=1}^{s}\frac{\partial^2\phi^i}{(\partial x^j)^2}+\sum_{j=1}^{s}\frac{\partial}{\partial x^j}\left(\frac{\partial W}{\partial(\partial\phi^i/\partial x^j)}\right)$$

Formula (8) is derived as follows: First of all, by the chain rule, the derivative of

$$V(\phi)=\int_{\mathbf{R}^s}\left\{\tfrac{1}{2}\|\nabla\phi\|^2+W(\nabla\phi)\right\}dx$$

is

$$dV(\phi)\cdot\psi=\int_{\mathbf{R}^s}\nabla\phi\cdot\nabla\psi+\frac{\partial W}{\partial\nabla\phi}\cdot\nabla\psi\,dx$$

$$=\int_{\mathbf{R}^s}\left\{(-\Delta\phi)\psi-div\left(\frac{\partial W}{\partial\nabla\phi}\right)\psi\right\}dx$$

(integrating by parts). Thus, by (7) the L^2 gradient of V is

$$grad\,V(\phi)=-\Delta\phi-div(\partial W/\partial\nabla\phi)$$

and substitution of this in (6) gives (8). Equations of the form (8) [as contrasted to (5)] are also called nonlinear wave equations.

Now consider *conservation laws* for systems of the form (6). Naturally the *energy* $E(\phi,\dot\phi)$ is constant if ϕ is a solution of the equations of motion. (This is

ISBN 0-8053-0102-X

a *formal* consequence of general theory; it may be rigorously verified under the appropriate technical hypotheses. For Eq. (8), however, it is believed that shocks will generally develop after a finite time and conservation of energy now becomes a delicate issue.)

The symplectic form ω_L is, naturally, also a formal invariant. In this connection let us note that as $T\mathcal{C}=\mathcal{C}\oplus\mathcal{C}$ is a *linear* space and ω_L is constant in the natural chart, we may identify ω_L with a skew symmetric bilinear form $\tilde{\omega}_L$ on $\mathcal{C}\times\mathcal{C}$. In the case of the wave equation whose flow is given by *linear* operators F_t on $T\mathcal{C}$, this bilinear form is invariant. That is, if ϕ and ψ are two solutions of the wave equation, then

$$\langle\dot{\phi}(t),\psi(t)\rangle-\langle\dot{\psi}(t),\phi(t)\rangle=\int(\dot{\phi}\psi-\dot{\psi}\phi)\,dx$$

is time independent. The reason is that $DF_t=F_t$ by linearity; and so the invariance of ω_L as a two-form implies the invariance of the corresponding bilinear form.

Next, the group of motions of space $\boldsymbol{R}^s$ operates in a natural way on $\mathcal{C}$, at least if $\mathcal{C}$ is a suitable class of functions—one with an invariant norm. Thus for $\boldsymbol{v}\in\boldsymbol{R}^s$ and $\phi\in\mathcal{C}$ we consider translation by $\boldsymbol{v}$: $\phi\mapsto\phi_{\boldsymbol{v}}$, where $\phi_{\boldsymbol{v}}(x)=\phi(x+\boldsymbol{v})\in\mathcal{C}$, and similarly for rotations. Moreover, the Lagrangian L is clearly invariant under these operations. The theory of Sect. 4.2 gives us *momentum functions* that are formally conserved.

For example, translation in the $\boldsymbol{e}_i$ direction is given by the group

$$(U_\sigma^i\phi)(x)=\phi(x+\sigma e_i)$$

The corresponding vector field on $\mathcal{C}$, obtained by differentiation with respect to σ, is

$$X^i\phi=\partial\phi/\partial x_i$$

The corresponding momentum function is [see the formula for $P(X)$ in 4.2.11]

$$P(X^i)(\psi,\dot{\psi})=i_{X^i}\theta(\psi,\dot{\psi})=\langle\dot{\psi},X^i\psi\rangle$$

Written out in full,

$$P(X^i)(\psi,\dot{\psi})=\int\dot{\psi}\frac{\partial\psi}{\partial x_i}\,dx$$

A typical generator Y^{ij} of the rotation group yields the total angular momentum

$$P(Y^{ij})(\psi,\dot{\psi})=\int\dot{\psi}\left(x_j\frac{\partial\psi}{\partial x_i}-x_i\frac{\partial\psi}{\partial x_j}\right)dx$$

ISBN 0-8053-0102-X

One may verify by a direct formal calculation the invariance of these quantities if ϕ satisfies the equations of motion.

With these examples in mind we can formalize things for the linear case to indicate the general nature of the theory.*

5.5.2 Definition. *Let X be a Banach space and $\omega: X\times X\to \mathbf{R}$ a weak symplectic (bilinear) form. Let $Y\subset X$ be a Banach space densely and continuously included in X and let $A: Y\to X$ be a given continuous linear operator from*† *Y to X. We say A is* **Hamiltonian** *if there is a C^1 function $H: Y\to \mathbf{R}$ such that*

$$\omega(Au,v)=dH(u)\cdot v$$

for all $u, v\in Y$. (Note that H is automatically C^∞.)

Analogous to 3.3.6, we have:

5.5.3 Proposition. *(i) The operator A is Hamiltonian if and only if A is ω-skew; that is,*

$$\omega(Au,v)=-\omega(u,Av)$$

for all $u, v\in Y$

(ii) If A is Hamiltonian, we may choose as energy function, H_A defined by

$$H_A(u)=\tfrac{1}{2}\omega(Au,u)$$

Proof. (i) If A is Hamiltonian, we have

$$\omega(Au,v)=dH(u)\cdot v$$

Differentiating in u at 0:

$$\omega(Au,v)=D^2H(0)\cdot(u,v)$$

Thus $\omega(Au,v)$ is symmetric in u and v; that is, A is ω-skew.

Conversely, suppose A is ω-skew. Let $H_A(u)=\frac{1}{2}\omega(Au,u)$. Then $dH_A(u)\cdot v =\frac{1}{2}\omega(Au,v)+\frac{1}{2}\omega(Av,u)=\omega(Au,v)$, so A is Hamiltonian with energy H_A. This argument also proves (ii). ■

Normal forms for linear Hamiltonian systems in infinite dimensions are presented in the next section.

*For additional details on how to rigorously carry out the above manipulations for elasticity, see Marsden and Hughes [1978].

†Usually Y will be $D(A)$, the domain of A (with the graph norm if A is a closed operator).

ISBN 0-8053-0102-X

Flows of linear Hamiltonian operators are best approached by means of semi-group theory. For example, if A generates a semi-group e^{tA} in X, then one verifies that e^{tA} conserves energy and the form ω. For general existence theory in the Hamiltonian case, see Weiss [1967], Chernoff–Marsden [1974] and Marsden and Hughes [1978]. The proof of 2.6.13 shows that if A is a generator then it is ω-skew-*adjoint*. If A is ω-skew-adjoint and H_A is positive-definite, then a suitable modification of A is a generator. (See the aforementioned papers for details.) This covers the case of the linear wave equation for example. For that case, if we choose $X = L^2 \times L^2$ so ω is nondegenerate, then e^{tA} is not defined on X. To have it defined, we choose $X = H^1 \times L^2$ and thereby obtain only a weak symplectic form. (See, for instance Yosida [1974] or Marsden and Hughes [1978] for proofs of these facts from semi-group theory.

Some remarks on Poisson brackets in the linear case may be of some interest here. If we have two Hamiltonian operators A and B in X (not necessarily with the same domain), then we form H_A and H_B, their energy functions as in 5.5.3(ii). The following computes the Poisson bracket $\{H_A, H_B\}$.

5.5.4 Proposition. *We have the relationship*

$$\{H_A, H_B\} = H_{[A,B]}$$

where $[A,B] = AB - BA$, *on the domain of* $[A,B]$.

Proof. By definition, $H_A(x) = \frac{1}{2}\omega(Ax, x)$ and $H_B(x) = \frac{1}{2}\omega(Bx, x)$. Also, on $D(A) \cap D(B)$

$$\begin{aligned}\{H_A, H_B\}(x) &= \omega(Ax, Bx) \quad \text{(by definition of Poisson bracket)}\\ &= \tfrac{1}{2}\omega(Ax, Bx) - \tfrac{1}{2}\omega(Bx, Ax)\\ &= -\tfrac{1}{2}\omega(BAx, x) + \tfrac{1}{2}\omega(ABx, x)\\ &= \tfrac{1}{2}\omega([A,B]x, x)\\ &= H_{[A,B]}(x) \quad \blacksquare\end{aligned}$$

As we saw in 3.1.18 the symplectic form is the imaginary part of a complex inner product. Let us consider the complex linear case in more detail.

5.5.5 Proposition. *Let $\mathcal{H}$ be complex Hilbert space and $\omega(x,y) = -Im\langle x, y\rangle$. Then:*

(i) A (complex) linear operator A in $\mathcal{H}$ is Hamiltonian if and only if iA is symmetric.

ISBN 0-8053-0102-X

(ii) *The energy function associated with A is*

$$H_A(x) = \tfrac{1}{2}\langle iAx, x\rangle$$

(iii) *A bounded (complex) linear mapping $U: \mathcal{H} \to \mathcal{H}$ is symplectic if and only if it is unitary.*

This follows easily from 5.5.3, the relation $Re\langle ix,y\rangle = -Im\langle x,y\rangle$, and complex linearity, so the proof will be omitted. In this case the existence of a flow for A follows from Stone's theorem provided iA is self-adjoint, not merely symmetric (see, e.g., Reed and Simon [1975] for a proof of Stone's theorem).

5.5.6 Example (The Schrödinger Equation). An important class of complex linear Hamiltonian systems arises in *quantum mechanics*. The states of a quantum mechanical system are represented by unit vectors ψ in a complex Hilbert space $\mathcal{H}$, and the observables (physically measurable quantities) correspond to self-adjoint operators $\mathcal{O}$ on $\mathcal{H}$; $\langle \mathcal{O}\psi,\psi\rangle$ is interpreted physically as the expected value of the observable $\mathcal{O}$ when the state of the system is ψ. The time evolution is represented by a one-parameter group U_t that preserves the "transition probabilities" $|\langle\phi,\psi\rangle|^2$ and is therefore unitary. Hence U_t is symplectic with respect to the canonical skew form $\omega(x,y) = Im(x,y)$, and it is therefore given by $U_t = e^{itH_{op}}$, where H_{op} is self-adjoint. Accordingly H_{op} corresponds to an observable, namely, the *energy* (as in classical Hamiltonian mechanics). The Hamiltonian operator itself is $A = iH_{op}$.

For example, if we are dealing with a nonrelativistic particle of mass m moving in a force field derived from a potential $V(x)$, then the Hilbert space is $L^2(R^3; C)$ and the energy or Hamiltonian operator is

$$H_{op} = -\frac{\hbar^2}{2m}\left(\frac{\partial^2}{\partial x_1^2} + \frac{\partial^2}{\partial x_2^2} + \frac{\partial^2}{\partial x_3^2}\right) + V(x)$$

In Sect. 5.4 we saw that to a large extent one is forced into this choice. To begin with, this is a mere formal expression; it is important to derive conditions on V which guarantee that this expression corresponds to a unique self-adjoint operator—so that a well-determined dynamical group $U_t = e^{itH_{op}}$ exists. There has been a great deal of research in this area; the pioneer was Kato, who showed in 1949 that the usual Hamiltonians of nonrelativistic atomic and molecular physics are essentially self-adjoint. In other words, the corresponding Schrödinger equations can be integrated by virtue of Stone's theorem. For more recent work, consult Reed and Simon [1975].

The general theory of infinite-dimensional nonlinear Hamiltonian systems proceeds as in the finite-dimensional case. However, there are technical

ISBN 0-8053-0102-X

difficulties related to questions like the differentiability of the flow. These are outgrowths of the fact that the vector fields are only densely defined, since we are dealing with partial rather than ordinary differential equations. Once these are overcome,* the theory of symmetry groups and conservation laws, in general, may be carried through. See Chernoff and Marsden [1974] and Marsden and Hughes [1978] for details.

One of the most intriguing equations that has seen an explosion of study in the last decade is the Korteweg–de Vries (KdV) equation. It describes shallow water waves, but it is also of interest for its mathematical beauty. For background, see Witham [1974].

Our discussion of the KdV equation illustrates only a *few* of its aspects. The reader interested in this topic should consult one of the many excellent review articles on the subject, such as Scott, Chu, and McLaughlin [1973] and Miura [1976a], to see it in proper perspective. Although we discuss the higher order KdV equations, the reader should realize that these remarkable properties are shared by a whole family of them, including, for example, the sine-Gordon equation, $u_{tt} - u_{xx} = \sin u$. (This pun on the Klein–Gordon equation $u_{tt} - u_{xx} = m^2 u$ is due to Kruskal.) See Ablowitz, Kaup, Newell, and Segur [1974].

5.5.7 Example. (Korteweg–de Vries equation). The equation is†

$$u_t - 6uu_x + u_{xxx} = 0$$

where $u_t = \partial u/\partial t$, $u_x = \partial u/\partial x$, and so forth, $x \in \boldsymbol{R}$ or S^1 (the *periodic case*) and u is real valued. In a suitable space $\mathfrak{E}$ of fields u (e.g., any Sobolev space included in H^3), define the weak symplectic form

$$\omega(v_1, v_2) = \tfrac{1}{2}\int_{-\infty}^{+\infty}\left(\int_{-\infty}^{x}(v_2(x)v_1(y) - v_1(x)v_2(y))\,dy\right)dx$$

and the Hamiltonian

$$H(u) = \int_{-\infty}^{+\infty}\left(u^3 + \tfrac{1}{2}u_x^2\right)dx$$

(On S^1, replace integrals from $\int_{-\infty}^{\infty}$ with $\int_0^{2\pi}$.) Then we readily verify that the Hamiltonian vector field associated with H is

$$X_H(u) = \frac{\partial}{\partial x}(3u^2 - u_{xx})$$

that is, solutions of the KdV equation are integral curves of X_H.

---

*Passing to spaces of C^∞ functions so that the vector fields become everywhere defined does not seem to help much with existence and uniqueness questions.

†The most common other conventions are $u_t + uu_x + u_{xxx} = 0$ and $u_t - \frac{3}{2}uu_x + \frac{1}{4}u_{xxx} = 0$. Rescaling u, t, or x yields any set of conventions $u_t + auu_x + bu_{xxx} = 0$ desired.

ISBN 0-8053-0102-X

More generally, if

$$H(u)=\int_{-\infty}^{+\infty} f(u, u_x, u_{xx}, \ldots)\,dx,$$

then

$$X_H(u)=\frac{\partial}{\partial x}\left(\frac{\delta f}{\delta u}\right)$$

where

$$\frac{\delta f}{\delta u}=\frac{\partial f}{\partial u}-\frac{\partial}{\partial x}\left(\frac{\partial f}{\partial u_x}\right)+\frac{\partial^2}{\partial x^2}\left(\frac{\partial f}{\partial u_{xx}}\right)-\cdots$$

To prove this, the reader can verify by integration by parts that

$$\begin{aligned}(i_{X_H}\omega)_u(v)&=\omega(X_H(u), v)\\ &=\int_{-\infty}^{+\infty}\frac{\delta f}{\delta u}(x)v(x)\,dx=(dH)_u(v)\end{aligned}$$

The special case in which f is a (nonlinear) function of u alone gives the so-called *conservation laws*

$$u_t-\left(\frac{\partial f}{\partial u}\right)_x=0$$

much studied in the theory of shock waves; cf. Lax [1973].

A main interest of the KdV equation is that it possesses infinitely many integrals in involution; it is therefore completely integrable in some sense. These integrals were discovered by Gardner, Greene, Kruskal, and Miura [1967]. We will construct them algebraically; they are the Hamiltonians F_j for a hierarchy of equations $u_t = X_j(u)$, where

$$X_j(u)=\frac{\partial}{\partial x}\frac{\delta f_j}{\delta u}$$

is given recursively by the relation

$$\begin{aligned}X_j(u)&=(au+aDuD^{-1}+bD^2)X_{j-1}\\ &=(auD+aDu+bD^3)\frac{\delta f_{j-1}}{\delta u}\end{aligned}$$

where $X_1(u)=u_x$, $f_1(u)=\frac{1}{2}u^2$, and a,b are constants* and $D=\partial/\partial x$. The

*The constants a,b fix ones conventions in the KdV equation. In our conventions $a=2$, $b=1$.

ISBN 0-8053-0102-X

equations $u_t = X_j(u)$ are called *higher order KdV equations* and have Hamiltonians $F_j(u) = \int_{-\infty}^{+\infty} f_j(u)\,dx$ determined by the above recursion relation.* Note that for $a=2$, $b=-1$, $X_2(u) = 6uu_x - u_{xxx} = X_H(u)$ and $F_2 = H$.

We shall prove that $\{F_j, F_k\} = 0$ for all j, k which will show that all the F_j are first integrals of the KdV equation (since $F_2 = H$) and that they are in involution. We have, since $X_i = X_{F_i}$ for all i,

$$\begin{aligned}
\{F_j, F_k\} &= \omega(X_j, X_k) \\
&= \int_{-\infty}^{+\infty} \frac{\delta f_j}{\delta u} X_k(u)\,dx \\
&= \int_{-\infty}^{+\infty} \frac{\delta f_j}{\delta u}(auD + aDu + bD^3)\frac{\delta f_{k-1}}{\delta u}\,dx \\
&= -\int_{-\infty}^{+\infty} \frac{\delta f_{k-1}}{\delta u}(aDu + auD + bD^3)\frac{\delta f_j}{\delta u} \quad \text{(integration by parts)} \\
&= -\int_{-\infty}^{+\infty} \frac{\delta f_{k-1}}{\delta u} X_{j+1}(u) \\
&= -\omega(X_{k-1}, X_{j+1}) \\
&= \omega(X_{j+1}, X_{k-1}) = \{F_{j+1}, F_{k-1}\}
\end{aligned}$$

Successive application of the relation $\{F_j, F_k\} = \{F_{j+1}, F_{k-1}\}$ shows that if $j = 2i+1$, $k = 2l+1$,

$$\{F_j, F_k\} = \{F_{i+l+1}, F_{i+l+1}\} = 0$$

if $j = 2i$, $k = 2l$,

$$\{F_j, F_k\} = \{F_{i+l}, F_{i+l}\} = 0$$

*It is easily checked inductively that X_j is Hamiltonian; that is, $i_{X_j}\omega$ is closed; that is, $DX_j(u)$ is ω-skew. From

$$F_j(u) = \int_0^1 \omega(X_j(tu), u)\,dt$$

it follows inductively that F_j has the desired form:

$$F_j(u) = \int_{-\infty}^{\infty} f_j(u, u_x, u_{xx}, \ldots)\,dx$$

ISBN 0-8053-0102-X

if $j=2i+1$, $k=2l$,

$$\{F_j, F_k\}=\{F_{i+l}, F_{i+l+1}\}=\{F_{i+l+1}, F_{i+l}\}$$

and hence by antisymmetry of the Poisson bracket, $\{F_j, F_k\}=0$.

The fact that the KdV equation has such integrals in involution is believed to be closely related to the presence of *solitons*, that is, "the solitary waves which interact pairwise (2 body interactions) by passing through each other without changing shape," a remarkable property for a nonlinear equation. If the integrals were not in involution, it is believed that n-body interactions would occur. (See also Exercise 5.5K).

A second main point is that the KdV equations is related to the Schrödinger equation with potential u, that is, with the operator

$$H_{op}=-D^2+u$$

Notice that $H_{op}=A^*A$, where $A=D-v$ and $u=v_x+v^2$ (Riccati equation). Then if v satisfies the *modified KdV equation*, $v_t-6v^2v_x+v_{xxx}=0$, then u is a solution of the KdV equation. This is easily seen if we note that for $u=v_x+v^2$,

$$u_t-6uu_x+u_{xxx}=(2v+D)(v_t-6v^2v_x+v_{xxx})$$

Likewise one determines a hierarchy of modified KdV equations $v_t=Y_j(v)$.

Assume now that as t evolves, $u(t)$ changes subject to any of the conditions

$$u_t=X_j(u)=\frac{\partial}{\partial x}\left(\frac{\delta f_j}{\delta u}\right)$$

that is, $u(t)$ satisfies any higher-order KdV. We shall prove below that although the operator $H_{op}(t)=-D^2+u(t)$ changes (since u does), the spectrum of $H_{op}(t)$ is unchanged, that is, the evolution of u is *isospectral*. This is the tip of a deep connection between the KdV equation and the Schrödinger equation by means of the inverse scattering method which was discovered by Gardner, Greene, Kruskal and Miura [1967].

To show that the operators $H_{op}(t)$ are isospectral we follow the method of Lax [1968]. It is sufficient to show that they are all similar by unitary transformations in $L^2(\mathbf{R})$, that is, it is sufficient to find a (differentiable) family of unitary operators $U(t)$ such that

$$\frac{d}{dt}\left(U(t)^{-1}H_{op}(t)U(t)\right)=0 \tag{1}$$

Since $U(t)$ are unitary, $U(t)U(t)^*=I$; differentiation in t yields

$$\frac{dU(t)}{dt}U(t)^*+U(t)\left(\frac{dU(t)}{dt}\right)^*=0$$

ISBN 0-8053-0102-X

Let

$$B(t)=\frac{dU(t)}{dt}U(t)^*$$

and notice that the above relation implies

$$B(t)^*=\left(\frac{dU(t)}{dt}U(t)^*\right)^*$$

$$=(U(t)^*)^*\left(\frac{dU(t)}{dt}\right)^*$$

$$=U(t)\left(\frac{dU(t)}{dt}\right)^*$$

$$=-\frac{dU(t)}{dt}U(t)^*=-B(t)$$

that is, $B(t)$ is *skew-symmetric*. If the skew-symmetric operator $B(t)$ were known, $U(t)$ could be defined to be the solution of the linear initial value problem $dU(t)/dt=B(t)U(t)$, $U(0)=1$ (assuming the solution exists*). In order to find conditions on $B(t)$, use relation (1) and the fact that

$$\frac{d}{dt}U(t)^{-1}=-U(t)^{-1}\frac{dU(t)}{dt}U(t)^{-1}$$

The chain rule yields

$$-U^{-1}BH_{op}U+U^{-1}\frac{dH_{op}}{dt}U+U^{-1}H_{op}BU=0$$

that is,

$$\frac{dH_{op}}{dt}=BH_{op}-H_{op}B=[B(t),H_{op}(t)]$$

But since $H_{op}(t)=-D^2+u(t)$, $dH_{op}(t)/dt=(du/dt)I$. Hence the operator $B(t)$ has to be skew-symmetric and satisfy the condition that $[B(t),H_{op}(t)]=$ multiplication by a scalar function. A whole sequence of operators B_j satisfying these conditions is given by (cf. Lax [1975])

$$B_j=c_jD^{2j+1}+\sum_{i=1}^{j}\left(b_{ij}D^{2i-1}+D^{2i-1}b_{ij}\right)$$

*This is not trivial since $B(t)$ will be unbounded. However, below, B's independent of time are found and so Stone's theorem is applicable.

ISBN 0-8053-0102-X

where c_j are constants and b_{ij} are functions of u. For $j=1$ we can pick $c=4$ and $b=3u$ and find, by an easy calculation,

$$[B_1, H_{op}] = \text{multiplication by } X_2(u)$$

so the evolution of X_1 is isospectral. In general, one has the remarkable result that (with c_j and b_{ij} properly chosen) $[B_j, H_{op}]$ is multiplication by $X_{j+1}(u)$. The way to choose B_j so this occurs is by the following formula of McKean and van Moerbeke [1975]:

$$B_j = \sum_{k=1}^{j+1}\left(2\frac{\delta f_{k-1}}{\delta u}D - X_{k-1}(u)I\right)(4H_{op})^{j+1-k}$$

with the convention $f_0(u)=u$ so that $\delta f_0/\delta u=1$ and $X_0(u)=0$. To prove this, we note that B_j is clearly a differential operator of order $2j+1$ and is skew-symmetric. Using the recurrence relation for the X_j's, we get

$$\begin{aligned}
[B_j, H_{op}] &= \sum_{k=1}^{j+1}\left[\left(2\frac{\delta f_{k-1}}{\delta u}D - X_{k-1}(u)I\right)H_{op}\right.\\
&\qquad \left. - H_{op}\left(2\frac{\delta f_{k-1}}{\delta u}D - X_{k-1}(u)I\right)\right](4H_{op})^{j+1-k}\\
&= \sum_{k=1}^{j+1}\left(2u'\frac{\delta f_{k-1}}{\delta u} + 4\left(\frac{\delta f_{k-1}}{\delta u}\right)' - \left(\frac{\delta f_{k-1}}{\delta u}\right)'''\right)(4H_{op})^{j+1-k}\\
&= \sum_{k=1}^{j+1}\left[X_k(u) - X_{k-1}(u)(4H_{op})\right](4H_{op})^{j+1-k}\\
&= X_{j+1}(u)(4H_{op})^0 - X_0(u)(4H_{op})^{j+1}\\
&= X_{j+1}(u)I
\end{aligned}$$

This proves that all the X_j are isospectral. The procedure can be used as an alternative way to construct the hierarchy X_j.* We shall give below another proof that each X_j is isospectral (at least for the discrete part of the spectrum). First of all we shall prove, following Lax [1975], that not only are the F_j a family of integrals in involution, but so are the eigenvalues λ_i, regarded as functions of u. (If λ has multiplicity k at u, λ stands for $\lambda_1 + \cdots + \lambda_k$ at nearby u; this problem occurs only on S^1, not on $\boldsymbol{R}$; on S^1

*The integrals for the rigid body found by Mishchenko [1970] Manakov [1977] are related to the Lax procedure. In fact many completely integrable systems may be amenable to treatments like this using the fact that they are often systems on co-adjoint orbits; see Exercise 5.5K.

ISBN 0-8053-0102-X

eigenvalues can be double.†) In fact, F_j, λ_i are *all in involution*. That $\{F_j, \lambda_i\} = 0$ is another way of saying that X_j is isospectral; it is by this means that we shall give a second proof of the isospectrality of X_j. It should be noted that the λ_i's are not independent integrals; they are derivable from the F_j.

Let us first prove that $\{\lambda_i, \lambda_j\} = 0$ for all i,j. The first problem we face in this computation is the expression of the Poisson bracket of two functions that are *not* integrals of another function. Following Lax [1975], given a real valued function F of u, we let $G_F(u)$ be the L^2 gradient of dF_u, that is, $G_F(u)$ is the element in L^2 satisfying

$$dF_u(v) = \int_{-\infty}^{+\infty} G_F(u) v \, dx$$

for all $v \in L^2$. If dF_u is a differential operator, $G_F(u)$ can be constructed by integration by parts as in our earlier examples. Then the following formula for the Poisson bracket holds (Gardner [1971]):

$$\{F_1, F_2\} = \int_{-\infty}^{+\infty} G_{F_1}(u) \frac{\partial}{\partial x} G_{F_2}(u) \, dx$$

To prove it, notice first that by integration by parts

$$dF_u(v) = \omega\left(\frac{\partial}{\partial x} G_F(u), v\right)$$

and hence

$$X_F = \frac{\partial}{\partial x} G_F(u)$$

Then

$$\begin{aligned}\{F_1, F_2\} &= \omega(X_{F_1}, X_{F_2}) \\ &= \int_{-\infty}^{+\infty} G_{F_1}(u) X_{F_2}(u) \, dx \\ &= \int_{-\infty}^{+\infty} G_{F_1}(u) \frac{\partial}{\partial x} G_{F_2}(u) \, dx\end{aligned}$$

Thus

$$\{\lambda_i, \lambda_j\} = \int_{-\infty}^{+\infty} G_{\lambda_i} \frac{\partial}{\partial x} G_{\lambda_j} \, dx$$

and we are compelled to compute G_λ for an eigenvalue $\lambda(u)$.

†That is, on $\boldsymbol{R}$, $k=1$ while on S^1, $k \leq 2$; see Magnus and Winkler [1975].

ISBN 0-8053-0102-X

So let λ be an eigenvalue of H_{op} of multiplicity k at u and denote by $f_1,\ldots,f_k$ the corresponding L^2-orthogonal eigenfunctions, all of L^2-norm equal to one. We claim that

$$G_{\lambda(u)}=\frac{1}{k}(f_1^2+\cdots+f_k^2)$$

To prove this formula, start with the k equations $H_{op}(f_i)=\lambda f_i$, $i=1,\ldots,k$ defining λ. Replace u by $u+tv$ getting

$$(-D^2+(u+tv))f_i(u+tv)=\lambda(u+tv)f_i(u+tv)$$

and take $d/dt|_{t=0}$ of both sides; this gives

$$vf_i(u)+H_{op}\left(\left.\frac{df_i(u+tv)}{dt}\right|_{t=0}\right)=\left.\frac{d\lambda(u+tv)}{dt}\right|_{t=0}f_i(u)+\lambda(u)\left.\frac{df_i(u+tv)}{dt}\right|_{t=0}$$

Since H_{op} is symmetric, we have

$$\begin{aligned}\int_{-\infty}^{+\infty}H_{op}\left(\left.\frac{df_i(u+tv)}{dt}\right|_{t=0}\right)f_i(u)\,dx&=\int_{-\infty}^{+\infty}\left.\frac{df_i(u+tv)}{dt}\right|_{t=0}H_{op}(f_i)\,dx\\&=\int_{-\infty}^{+\infty}\left.\frac{df_i(u+tv)}{dt}\right|_{t=0}\lambda(u)f_i(u)\,dx\end{aligned}$$

so that taking the scalar product with $f_i(u)$ in the above relations and adding, we get

$$\begin{aligned}\int_{-\infty}^{+\infty}v(f_1^2+\cdots+f_k^2)\,dx&=\left.\frac{d\lambda(u+tv)}{dt}\right|_{t=0}\int_{-\infty}^{+\infty}(f_1^2+\cdots+f_k^2)\,dx\\&=\left.\frac{d\lambda(u+tv)}{dt}\right|_{t=0}k=kd\lambda_u(v)\end{aligned}$$

Thus

$$G_{\lambda(u)}=\frac{1}{k}(f_1^2+\cdots+f_k^2)$$

The following observation will yield the desired involution property of the λ_i: *If f is an eigenfunction of H_{op}, then f^2 satisfies the following third-order differential equation*

$$Tf^2=4\lambda\frac{d(f^2)}{dx}$$

ISBN 0-8053-0102-X

where T is the antisymmetric differential operator

$$T = -\frac{d^3}{dx^3} + 4u\frac{d}{dx} + 2u_x$$

To see this start with the equation

$$-f_{xx} + uf = \lambda f$$

Differentiating, we get

$$-f_{xxx} + u_x f + u f_x = \lambda f_x$$

We multiply the first by $6f_x$, the second by $2f$, and add.

Now we are ready to prove that $\{\lambda_i, \lambda_j\} = 0$. If λ_i is an eigenvalue of multiplicity k and λ_j is an eigenvalue of multiplicity l:

$$\begin{aligned}
\{\lambda_i, \lambda_j\} &= \int_{-\infty}^{+\infty} G_{\lambda_i} \frac{\partial}{\partial x} G_{\lambda_j}\, dx \\
&= \frac{1}{kl}\int_{-\infty}^{+\infty} (f_1^2 + \cdots + f_k^2)\frac{\partial}{\partial x}(g_1^2 + \cdots + g_l^2)\, dx \\
&= \frac{1}{4\lambda_j kl}\int_{-\infty}^{+\infty} (f_1^2 + \cdots + f_k^2) T (g_1^2 + \cdots + g_l^2)\, dx \\
&= -\frac{1}{4\lambda_j kl}\int_{-\infty}^{+\infty} T(f_1^2 + \cdots + f_k^2)(g_1^2 + \cdots + g_l^2)\, dx \\
&= -\frac{\lambda_i}{\lambda_j kl}\int_{-\infty}^{+\infty} \frac{\partial}{\partial x}(f_1^2 + \cdots + f_k^2)(g_1^2 + \cdots + g_l^2)\, dx \\
&= -\frac{\lambda_i}{\lambda_j}\int_{-\infty}^{+\infty} G_{\lambda_j} \frac{\partial}{\partial x} G_{\lambda_i}\, dx \\
&= -\frac{\lambda_i}{\lambda_j}\{\lambda_j, \lambda_i\} \\
&= \frac{\lambda_i}{\lambda_j}\{\lambda_i, \lambda_j\}
\end{aligned}$$

Since we assume λ_i and λ_j to be distinct eigenvalues we conclude $\{\lambda_i, \lambda_j\} = 0$ for all i, j.

Next we prove that $\{F_j, \lambda_i\} = 0$ for all i, j. For $j = 1$ this was proved above by construction of the appropriate operator B_1.

ISBN 0-8053-0102-X

We now note the following key fact: the recursion relation defining the X_j's may be written (in our conventions with $a=2$, $b=-1$)

$$X_{j+1}(u)=T\frac{\delta f_j}{\delta u}$$

that is, *the same operator T occurs in the recursion relation defining the higher-order KdV equations and in the equation satisfied by the squares of the eigenfunctions.*

Therefore, with the notation as above,

$$\begin{aligned}\{F_j,\lambda_i\}&=\int_{-\infty}^{\infty}\frac{\delta f_i}{\delta u}\frac{\partial}{\partial x}G_\lambda\,dx\\&=\frac{1}{k}\int_{-\infty}^{\infty}\frac{\delta f_j}{\delta u}\frac{\partial}{\partial x}(f_1^2+\cdots+f_k^2)\,dx\\&=\frac{1}{4\lambda_i k}\int_{-\infty}^{\infty}\frac{\delta f_i}{\delta u}\left(T(f_1^2+\cdots+f_k^2)\right)dx\\&=-\frac{1}{4\lambda_i k}\int_{-\infty}^{\infty}\left(T\frac{\delta f_j}{\delta u}\right)(f_1^2+\cdots+f_k^2)\,dx\\&=-\frac{1}{4\lambda_i k}\int_{-\infty}^{\infty}X_{j+1}G_\lambda\,dx\\&=\frac{1}{4\lambda_i k}\{F_{j+1},\lambda_i\}\end{aligned}$$

But $\{F_1,\lambda_i\}=0$, so $\{F_j,\lambda_i\}=0$ for all j.

Having found all these integrals in involution, we can suspect (very strongly suspect in view of theorems of Palais; see 4.1.29) that there must be an infinite-dimensional Abelian group G_{KdV} acting on the space $\mathcal{E}$ of fields $u(x)$ such that the above conserved quantities constitute its momentum mapping. Of course in a formal sense we can say that this group is the one generated by the flows of all the equations X_j, $j=1,2,\ldots$, but this is a posteriori and a little unsatisfactory. It appears that the corresponding group of contact transformations can be realized as the group of self-Backlund transformations (see Wadati, Sanuki, and Konno [1975] and Flaschka and McLaughlin [1976a]). If one is ever going to find interesting analogs of KdV in higher dimensions, the symmetry must be better understood. Some recent observations of Adler, Duistermaat, van Moerbeke, and the authors suggest that the proper setting for the KdV equation is as a Hamiltonian system on the group of invertible Fourier integral operators. M. Adler has already shown how to regard the KdV equation as a Hamiltonian system on a co-adjoint orbit in the Lie algebra of pseudodifferential operators (See also Exercise 5.5K.)

Finally, we should note that the presence of these integrals in involution allows one to introduce action-angle variables along the lines indicated in

ISBN 0-8053-0102-X

Sect. 5.2. (In particular, see Arnold's theorem 5.2.21.) The introduction of a symplectic transformation to action-angle variables is accomplished by using the scattering data; see Fadeev and Zakharoff [1970], McKean and van Moerbeke [1975], Flaschka and Newell [1975] and McKean and Trubowitz [1976] for further information.

Next we briefly describe how the Euler equations of a perfect incompressible fluid may be described as a Hamiltonian system. The discussion is based on the method of Sect. 4.4. (See Arnold [1966], Ebin–Marsden [1970], Marsden and Abraham [1970], and Marsden, Ebin, and Fischer [1972] for additional details.)

5.5.8 Example (Equations of a Perfect Fluid). The motion of a perfect fluid in a domain M, a smooth (oriented) Riemannian manifold with boundary, is governed by the Euler equations:

$$\frac{\partial u}{\partial t}+\nabla_u u=-\nabla p$$

$$\operatorname{div} u=0$$

and the boundary conditions $u \| \partial M$ (u is parallel to ∂M), that is, $u(x)\in T_x(\partial M)$ for each $x\in\partial M$. Here u is the velocity field of the fluid, a time-dependent vector field on M, and p is the pressure to be determined from the incompressibility condition $\operatorname{div} u=0$. Below we shall use the identity $(u\cdot n)a=i_u\mu$ on ∂M, where a is the area form on ∂M, μ is the volume form on M and n is the outward unit normal. (See Sect. 2.7.) Thus the boundary condition $u\|\partial M$ can be written $i_u\mu=0$ on ∂M. The expression $\nabla_u u=(u\cdot\nabla)u$ is the covariant derivative of u along u.

Let η_t be the flow of u. Then $\eta_t\in\mathfrak{D}_\mu$, the group of volume preserving diffeomorphisms* of M. The sense in which the equations are Hamiltonian is just this: η_t is a *geodesic* on $\mathfrak{D}_\mu$ if and only if u satisfies the Euler equations. The metric used on $\mathfrak{D}_\mu$ is the right-invariant weak Riemannian metric corresponding to the kinetic energy of the fluid, that is, $\frac{1}{2}\int\langle\dot u,\dot u\rangle\mu$. We will now prove this using the general Euler equations for geodesic motion on a Lie group derived in Sect. 4.4, namely,

$$\dot u=Y(u)$$

where

$$\langle Y(u),v\rangle=\langle[u,v],u\rangle$$

Here $\langle,\rangle$ stands for the (metric) inner product on the group $\mathfrak{D}_\mu$, that is, the L^2 inner product and $[u,v]$ is the Lie bracket, which is easily seen to be the

*One shows that $\mathfrak{D}_\mu$ is, in suitable function spaces, a C^∞ manifold and in an appropriate sense, a Lie group. See Exercise 4.1G and Ebin and Marsden [1970] for details.

ISBN 0-8053-0102-X

usual bracket of vector fields (Exercise 4.1G). We now compute: Let u^b be the one-form corresponding to u using the metric. Then

$$\langle [u,v], u\rangle = \int_M i_{[u,v]} u^b \mu = \int_M \left(i_{[u,v]} u^b - L_u i_v u^b\right)\mu$$

Letting

$$f = i_v u^b,$$

$$\int_M (L_u i_v u^b)\mu = \int_M L_u(f\mu) \quad (\text{since } \operatorname{div} u = 0)$$

$$= \int_M d i_u f\mu$$

$$= \int_{\partial M} f i_u \mu = 0$$

by Stokes's theorem and the boundary condition $u \| \partial M$; that is, $i_u\mu = 0$ on ∂M. But

$$i_{[u,v]} = L_u i_v - i_v L_u$$

so

$$\langle [u,v], u\rangle = -\int_M i_v L_u u^b \mu$$

$$= -\int_M (L_u u^b)\cdot(v)\mu = -\langle (L_u u^b)^\sharp, v\rangle$$

($\sharp$ means the metric sharp).

We cannot identify $Y(u)$ with $-(L_u u^b)^\sharp$ because we do not know $(L_u u^b)^\sharp$ is divergence free. The classical Hodge theorem states as a particular case that any vector field w on M can be uniquely decomposed as

$$w = w_0 + \nabla p$$

where w_0 is divergence free and parallel to ∂M.* The two summands are L^2 orthogonal, as is seen by integration by parts.

*The proof in this case is effected by solving the elliptic equation $\Delta p = \operatorname{div} w$,

$$\frac{\partial p}{\partial n} = w \cdot n \qquad (n = \textit{unit normal})$$

for p.

ISBN 0-8053-0102-X

Write $w_0 = Pw = w - \nabla p$ for the projection to the divergence-free part. Now we can write

$$Y(u) = -P\left(L_u u^\flat\right)^\sharp$$

Rewriting the Euler equations as $\partial u/\partial t = -P(\nabla_u u)$ and the identity

$$\left(L_u u^\flat\right)^\sharp = \nabla_u u + \tfrac{1}{2}\nabla\langle u, u\rangle$$

(whose proof we leave to the reader) shows that the equations $\partial u/\partial t = Y(u)$ are the Euler equations (with a redefinition of p). This completes the proof that geodesics on $\mathfrak{D}_\mu$ starting at the identity are in one-to-one correspondence with solutions of Euler's equations. (This can also be proved directly using the equations for a geodesic on $\mathfrak{D}_\mu$.)

A remarkable fact, discovered by Ebin and Marsden [1970], is that the geodesic spray on $\mathfrak{D}_\mu^s$ ($\mathfrak{D}_\mu$ completed in the Sobolev H^s topology) is a C^∞ map.* This means that the questions of existence (local in time) and uniqueness become simple, whereas for the Euler equations as they stand this is not so clear. In addition other facts (e.g., that one can uniquely join two nearby diffeomorphisms by a solution of the Euler equations and the convergence of certain numerical algorithms), can be proved rather easily this way.

Even though the Euler equations are the geodesic equations on a Lie group, we cannot apply Theorem 3.7.11 to conclude that the geodesic flow is complete. The reason is that the metric used for the geodesics is an L^2 metric and the topology on $\mathfrak{D}_\mu$ needed for the above analysis to work must be stronger than C^1. In fact, completeness (i.e., definability for all time) is known only in two-dimensional flows and may well be false in three or higher dimensions. We refer the reader to Ratiu and Bernard [1977], Marsden [1974b], and Marsden, Ebin, and Fischer [1972] for discussions of these and related issues and for further references.

After discussing some properties of field theories in general, we shall explain our final example, namely, how to write Einstein's equations of general relativity as a Hamiltonian system.

5.5.9 Example (Lagrangian Field Theory). The classical Lagrangian field theories possess a "local" structure in addition to the global structures that we have been discussing. Specifically, the Lagrangian is obtained by integrating a local "Lagrangian density." We shall indicate how to fit this sort of system into a general abstract framework. The literature on this subject is quite extensive, but we shall be content with a brief sketch.

*It is well known that the convective term $u\cdot\nabla u$ disappears in Lagrangian (or material) coordinates; that is, on $\mathfrak{D}_\mu$. However, it is less obvious that the terms corresponding to ∇p are well behaved under such a transformation.

ISBN 0-8053-0102-X

In the physics literature, one considers "fields" $\phi^i(x,t)$, which may have various transformation properties under coordinate changes. In current mathematical language, a field is a cross section of some vector (or fiber) bundle. A Lagrangian is just a real-valued function*

$$\mathcal{L}(\phi^i, \dot{\phi}^i, \partial\phi^i/\partial x^j, x^j)$$

which depends on the fields and their space–time derivatives. To formulate this in an invariant fashion we use the notion of jet bundles.

Let $\pi: E \to M$ be a vector bundle, and let $J^1(E)$ be the first jet bundle. At the point $x \in M$ we have, by definition,

$$J^1(E)_x \cong E_x \oplus L(T_xM, E_x)$$

We shall assume that M is endowed with a volume element μ, and that E is equipped with an inner product on each fiber as well as a connection. If ξ is a section of E, then its first jet is the section of $J^1(E)$ given by

$$j(\xi) = \xi \oplus D\xi$$

where $D\xi$ is the covariant derivative of ξ. In the language of jets, a Lagrangian density is simply a smooth map

$$\mathcal{L}: J^1(E) \oplus E \to \mathbf{R}$$

We can form the global Lagrangian, or action integral, as follows. Given two sections $\dot{\phi}, \phi$ of the bundle E, define

$$L(\phi, \dot{\phi}) = \int_M \mathcal{L}(\phi(x), \dot{\phi}(x), D\phi(x))\,\mu$$

Despite appearances, $\mathcal{L}$ can depend explicitly on the base point $x \in M$ because $\mathcal{L}$ is a map on $E \oplus J^1(E)$, and the fibers depend on the base point.

To set up the global machinery, one chooses for configuration space a suitable (Sobolev) class of sections $\mathcal{C}$ of E. With the appropriate choice one can prove that $L: T\mathcal{C} \to \mathbf{R}$ is a smooth function. Then using the chain rule one can establish formulas like the following (using the obvious notation):

$$\begin{aligned} DL(\phi, \dot{\phi})\cdot(h, \dot{h}) &= D_{\dot{\phi}}L(\phi, \dot{\phi})\cdot\dot{h} + D_{\phi}L(\phi, \dot{\phi})\cdot h \\ &= \int_M \partial_{\dot{\phi}}\mathcal{L}(\phi, \dot{\phi}, D\phi)\dot{h} \\ &\quad + \left(\int_M \partial_{\phi}\mathcal{L}(\phi, \dot{\phi}, D\phi)\cdot h\mu + \int_M \partial_{D\phi}\mathcal{L}(\phi, \dot{\phi}, D\phi)\cdot Dh\mu\right) \end{aligned}$$

*As we shall see in relativity below, it is sometimes more convenient to take $\mathcal{L}$ to be a density than a real-valued function.

ISBN 0-8053-0102-X

A common class of examples is the following. Let $\mathcal{V}: J^1(E)\to \boldsymbol{R}$ be a smooth function that we think of as a *potential energy density*. Define

$$\mathcal{L}(\phi(x),\dot\phi(x),D\phi(x))=\tfrac{1}{2}\langle\dot\phi(x),\dot\phi(x)\rangle_x-\mathcal{V}(\phi(x),D\phi(x))$$

where $\langle,\rangle_x$ is the inner product on the fiber E_x. A suitable domain for the global Lagrangian L might be the set of pairs $(\phi,\dot\phi)$ such that $\dot\phi$ is an L^2 section of E while ϕ lies in a Sobolev space on which the integral $V(\phi) = \int_M \mathcal{V}(\phi, D\phi)\,\mu$ is smooth. As a concrete example, consider the classical wave equation. Here $M=\boldsymbol{R}^3$, E is the trivial bundle $\boldsymbol{R}^3\times\boldsymbol{R}\to\boldsymbol{R}^3$, and $\mathcal{V}(\phi(x),D\phi(x))=\frac{1}{2}\nabla\phi\cdot\nabla\phi$. For wave equations with a nonlinear term of the form $\lambda\phi^p$, $\mathcal{V}$ has an additional term $\phi^{p+1}/(p+1)$; and if we have a "mass term," as in the Klein–Gordon equation, then $\mathcal{V}$ contains the term $\frac{1}{2}m\phi^2$.

Returning to our general Lagrangian field theory, consider *Lagrange's equations*, exactly as derived in Sect. 3.8:

$$\frac{d}{dt}D_{\dot\phi}L(\phi,\dot\phi)=D_\phi L(\phi,\dot\phi)$$

In our case this means that for any section h, the relation

$$\frac{d}{dt}\int\partial_{\dot\phi}\mathcal{L}(\phi,\dot\phi,D\phi)\cdot h\mu=\int\partial_\phi\mathcal{L}(\phi,\dot\phi,D\phi)\cdot h\mu$$

$$+\int\partial_{D\phi}\mathcal{L}(\phi,\phi,D\phi)\cdot Dh\mu$$

holds. Taking h to have compact support, we can integrate by parts the second integral on the right-hand side, obtaining

$$\int_M\left(\frac{\partial}{\partial t}\partial_{\dot\phi}\mathcal{L}\right)\cdot h\mu=\int_M(\partial_\phi\mathcal{L}-div\,\partial_{D\phi}\mathcal{L})\cdot h\mu$$

Since h is arbitrary, we must have the *Lagrangian density equation*

$$\frac{\partial}{\partial t}(\partial_{\dot\phi}\mathcal{L})=\partial_\phi\mathcal{L}-div\,\partial_{D\phi}\mathcal{L}$$

In coordinates, this reads

$$\frac{\partial}{\partial t}\frac{\partial\mathcal{L}}{\partial\dot\phi^i}=\frac{\partial\mathcal{L}}{\partial\phi^i}-\left(\frac{\partial\mathcal{L}}{\partial\phi^i_{|j}}\right)_{|j}$$

where the subscript $|j$ denotes the covariant derivative. The expression on the right-hand side of the last equation is often called the *functional derivative* of

ISBN 0-8053-0102-X

$\mathcal{L}$ and, as in the KdV example, is denoted

$$\frac{\delta \mathcal{L}}{\delta \phi} = \partial_\phi \mathcal{L} - div\, \partial_{D\phi} \mathcal{L}$$

This can be written a different way that will be convenient in our discussion of relativity below. (This remark is due to A. Fischer.) Namely, consider $\dot{\phi}$ as fixed and write

$$L(\phi, \dot{\phi}) = \int \mathcal{L}(\phi, \dot{\phi}, D\phi)\, \mu$$

Then the formula for $D_\phi L$ written above states that

$$D_\phi L(\phi, \dot{\phi})\cdot h = \int \frac{\delta \mathcal{L}}{\delta \phi} \cdot h\mu$$

Let $\hat{\mathcal{L}}(\phi, \dot{\phi})(x) = \mathcal{L}(\phi(x), \dot{\phi}(x), D\phi(x))$, so $\hat{\mathcal{L}}$ maps $T\mathcal{C}$, that is, pairs $\phi, \dot{\phi}$ to scalar functions. Then we have

$$\int D_\phi \hat{\mathcal{L}}(\phi, \dot{\phi})\cdot h\mu = \int \frac{\delta \mathcal{L}}{\delta \phi} \cdot h\mu$$

(The dot $\cdot$ on the left means application of a differential operator; the one on the right is a pointwise contraction of tensors.) If we use the L^2 pairing, this states no more than

$$\delta \mathcal{L} / \delta \phi = D_\phi \hat{\mathcal{L}}(\phi, \dot{\phi})^* \cdot 1$$

where * means the L^2 adjoint (of a linear operator) and 1 is the constant function. As we shall see in our relativity discussion, the presence of the 1 corresponds to the fact that in nonrelativistic mechanics we are dealing with a "universal time."

Returning to the Lagrange density equations, we know that the total energy is conserved under quite general circumstances. A simple computation establishes a local conservation law that is *formally* stronger:

Let $\mathcal{L}: J^1(E) \oplus E \to \mathbf{R}$ be a smooth Lagrangian density and let $\phi(t)$ be a differentiable curve of sections of E, that is, $\phi(t) \in \mathcal{C}$ such that the Lagrange density equation of motion holds:

$$\frac{\partial}{\partial t}(\partial_{\dot{\phi}} \mathcal{L}) = \partial_\phi \mathcal{L} - div\, \partial_{D\phi} \mathcal{L}$$

*Define the **energy density** by $\mathcal{E} = \dot{\phi} \partial_{\dot{\phi}} \mathcal{L} - \mathcal{L}$, where*

$$\dot{\phi} \frac{\partial \mathcal{L}}{\partial \dot{\phi}} = \frac{\partial \mathcal{L}}{\partial \dot{\phi}^i} \dot{\phi}^i$$

ISBN 0-8053-0102-X

is the contraction of tensors. Then $\mathcal{E}$ obeys the conservation equation ("continuity equation")

$$\frac{\partial \mathcal{E}}{\partial t} + div\left(\dot{\phi} \lrcorner \partial_{D\phi}\mathfrak{L}\right) = 0$$

that is,

$$\frac{\partial \mathcal{E}}{\partial t} + \left(\dot{\phi}^{i}\frac{\partial \mathfrak{L}}{\partial \phi^{i}_{|j}}\right)_{|j} = 0$$

Indeed, using the chain rule together with the equation of motion, we find

$$\begin{aligned}
\frac{\partial \mathcal{E}}{\partial t} &= \frac{\partial}{\partial t}\left(\dot{\phi}\partial_{\dot{\phi}}\mathfrak{L}\right) - \frac{\partial \mathfrak{L}}{\partial t} \\
&= \ddot{\phi}\partial_{\dot{\phi}}\mathfrak{L} + \dot{\phi}\frac{\partial}{\partial t}(\partial_{\dot{\phi}}\mathfrak{L}) - \ddot{\phi}\partial_{\dot{\phi}}\mathfrak{L} - \dot{\phi}\partial_{\phi}\mathfrak{L} - D\dot{\phi}\cdot\partial_{D\phi}\mathfrak{L} \\
&= \dot{\phi}\frac{\partial}{\partial t}(\partial_{\dot{\phi}}\mathfrak{L}) - \dot{\phi}\partial_{\phi}\mathfrak{L} - D\dot{\phi}\cdot\partial_{D\phi}\mathfrak{L} \\
&= \dot{\phi}\{\partial_{\phi}\mathfrak{L} - div\,\partial_{D\phi}\mathfrak{L}\} - \dot{\phi}\partial_{\phi}\mathfrak{L} - D\dot{\phi}\cdot\partial_{D\phi}\mathfrak{L} \\
&= -\dot{\phi}\,div\,\partial_{D\phi}\mathfrak{L} - D\dot{\phi}\cdot\partial_{D\phi}\mathfrak{L} \\
&= -div\left(\dot{\phi}\partial_{D\phi}\mathfrak{L}\right)
\end{aligned}$$

One can similarly localize the conservation laws associated with more general symmetry groups. This proceeds, briefly, as follows. Let Ψ_g be an action of a Lie group G on M and let $\tilde{\Psi}_g$ be an action of G on E, linear on fibers, covering it. This extends, naturally, to an action on $J^1(E)$, called say $\tilde{\tilde{\Psi}}_g$. (It is determined by: $\tilde{\tilde{\Psi}}_g \circ j(\phi) \circ \Psi_g^{-1} = j(\tilde{\Psi}_g \circ \phi \circ \Psi_g^{-1})$ for ϕ a smooth section of E.) If $\xi \in \mathfrak{g}$, let, as usual, ξ_M and ξ_E be the corresponding infinitesimal generators on M and E (see Sect. 4.1). Assume that $\mathfrak{L}$ is invariant in the sense that $\mathfrak{L}\,\mu$ is unchanged under pull-back:

$$\mathfrak{L} \circ \left(\tilde{\Psi}_g \oplus \tilde{\tilde{\Psi}}_g\right)\Psi_g^{*}\mu = \mathfrak{L}\mu$$

If ϕ is a solution of the Lagrange density equations, set

$$T_{\xi} = \mathfrak{L}\xi_M + \partial_{D\phi}\mathfrak{L}\cdot(\xi_E \circ \phi - T\phi \cdot \xi_M) \quad \text{(a vector field on } M\text{)}$$

or, in coordinates,

$$T^{i}_{\xi} = \mathfrak{L}\xi^{i}_{M} + \frac{\partial \mathfrak{L}}{\partial \phi^{j}_{|i}}\left(\xi^{j}_{E} \circ \phi - \phi^{j}_{,k}\xi^{k}_{M}\right) \quad \text{(sum on } j \text{ and } k \text{ understood)}$$

ISBN 0-8053-0102-X

and

$$\mathfrak{T}_\xi = \partial_{\dot\phi}\mathfrak{L}\cdot(\xi_E\circ\phi - T\phi\cdot\xi_M) \quad \text{(a scalar field on } M\text{)}$$

that is,

$$\mathfrak{T}_\xi = \frac{\partial\mathfrak{L}}{\partial\dot\phi^i}(\xi_E^i\circ\phi - \phi^i_{,k}\xi_M^k) \quad \text{(sum on } i \text{ and } k \text{ understood)}$$

where $\mathfrak{L}$ stands for $L(\phi,\dot\phi,D\phi)$, and so forth. Then *Noether's theorem* states that

$$\frac{\partial\mathfrak{T}_\xi}{\partial t} + \operatorname{div} T_\xi = 0$$

and it is readily proved along the lines indicated earlier. The connection with moments is this: the momentum mapping is the map of $T\mathcal{C}$ to $\mathfrak{g}^*$ given by

$$(\phi,\dot\phi)\mapsto\left(\xi\mapsto\int\mathfrak{T}_\xi(\phi,\dot\phi)\mu\right)$$

Thus Noether's theorem provides a localization of the conservation laws given by the general theory of symmetry groups.

5.5.10 Example (Einstein's Vacuum Field Equation of General Relativity). Let V_4 be a four-manifold with a Lorentzian metric ${}^{(4)}g$. The vacuum field equations state that the Ricci tensor vanishes: $Ric\,{}^{(4)}g=0$. These are the Euler Lagrange equations for the Lagrangian density

$$\mathfrak{L}(g) = R({}^{(4)}g)\,\mu({}^{(4)}g)$$

where $R({}^{(4)}g)$ denotes the scalar curvature of ${}^{(4)}g$ (defined in Sect. 2.7) and $\mu({}^{(4)}g)=\sqrt{-\det{}^{(4)}g}\,d^4x$ is its volume element. For the proof of this and additional background material, see Misner, Thorne, and Wheeler [1973].

In the pioneering work of Dirac [1959], Wheeler [1964], and Arnowitt, Deser, and Misner [1962], it was shown how to write these equations as Hamiltonian evolution equations. Such a procedure is basic to a study of, for example, the initial value problem, quantization, and stability. We present an account of this "ADM formalism" following the methods and conventions of Fischer and Marsden [1978]. It will be brief and the technical papers cited should be consulted for those who wish to pursue the matter.

The first thing we need to do is to describe how to split up V_4 into space and time. We wish a formalism that reflects the fact that such a splitting can be done in an arbitrary way, that is, there is no preferred space or time

ISBN 0-8053-0102-X

coordinates. We let M be a compact* oriented three-dimensional manifold, and let $i: M \to V_4$ be an embedding of M such that the embedded manifold $i(M) = \Sigma$ is spacelike, that is, the pull-back $i^*({}^{(4)}g) = g$ is a Riemannian metric on M. Let $E^\infty(M, V_4, {}^{(4)}g)$ denote the set of all such spacelike embeddings. As with $\mathfrak{D}$ in 5.5.8, this is a smooth manifold. Let k denote the second fundamental form of the embedding as defined in Sect. 2.7. In our present notation, at $m \in M$ and for $X, Y \in T_m M$, it is given by

$$k_m(X,Y) = -{}^{(4)}g \circ i(m)\Big((T_m i \circ Y),\, {}^{(4)}\nabla_{(T_m i X)}\, {}^{(4)}Z_\Sigma \circ i(m)\Big)$$

where ${}^{(4)}Z_\Sigma \circ i(m)$ is the forward pointing unit timelike normal to Σ at $i(m)$. Thus $k_{ij} = -Z_{i;\ j}$, where $;$ denotes covariant differentiation using ${}^{(4)}g$. Covariant differentiation using g is denoted with a vertical bar. Let $\pi = \pi' \otimes \mu(g)$ be a two-contravariant tensor density, whose tensor part π' is defined by $\pi' = ((trk)g - k)^\#$, where $\#$ indicates the contravariant form of a covariant tensor with indices raised by $g^\# = g^{ij}$; similarly $\flat$ denotes the covariant form of a contravariant tensor. In the Hamiltonian formulation of Arnowitt, Deser, and Misner, k plays the role of a velocity variable and π is its canonical momentum.

Now suppose we have a curve in $E^\infty(M, V_4, {}^{(4)}g)$; that is, a curve i_λ of spacelike embeddings of M into $(V_4, {}^{(4)}g)$. The λ-derivative of this curve defines a one-parameter family of vector fields ${}^{(4)}X_{\Sigma_\lambda}$ on the embedded hypersurfaces by the equation

$$di_\lambda / d\lambda = {}^{(4)}X_{\Sigma_\lambda} \circ i_\lambda : M \to TV_4$$

(see Fig. 5.5-1). The normal and tangential projections of ${}^{(4)}X_{\Sigma_\lambda}$ define a curve of functions $N_\lambda : M \to \mathbf{R}$ and vector fields $X_\lambda : M \to TM$ on M by the equation

$${}^{(4)}X_{\Sigma_\lambda} \circ i_\lambda(m) = N_\lambda(m)\, {}^{(4)}Z_{\Sigma_\lambda} \circ i_\lambda(m) - T_m i_\lambda \circ X_\lambda(m)$$

where ${}^{(4)}Z_{\Sigma_\lambda}$ is the forward pointing unit timelike normal to Σ_λ. If $N_\lambda > 0$, then the map

$$F: I \times M \to V_4;\ (\lambda, m) \mapsto i_\lambda(m)$$

is a diffeomorphism of $I \times M$ onto a tubular neighborhood of $i_0(M) = \Sigma_0$, if the interval $I = (-\beta, \beta)$ is chosen small enough (see 2.7.5). In this case we call either the curve i_λ or the embedded spacelike hypersurfaces $\Sigma_\lambda = i_\lambda(M)$ a *slicing* of V_4.

The functions N_λ and the vector fields X_λ are called the *lapse functions* and *shift vector fields*.

---

*Compactness of M is more than a technical convenience here. The noncompact case has a different flavor. See the articles of Choquet-Bruhat, Fischer, and Marsden [1978] and Hansen, Regge, and Teitelboim [1976] for the later case.

ISBN 0-8053-0102-X

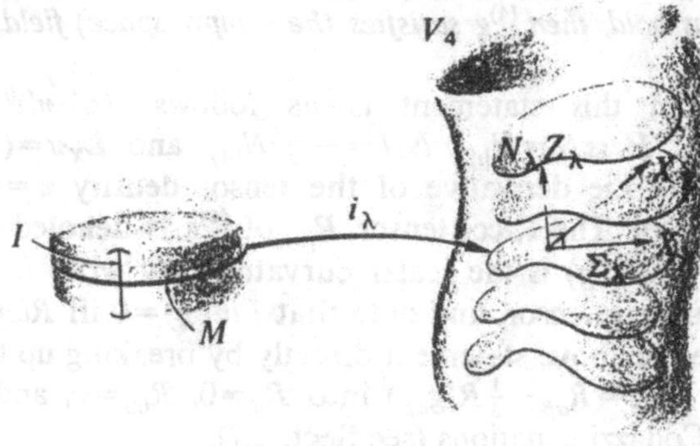

Figure 5.5-1

Using $F: I\times M\to V_4$ as a coordinate system for a tubular neighborhood of Σ_0 in V_4, coordinates (x^i), $i=1,2,3$ on M, and $(x^\alpha)=(\lambda,x^i)$, $\alpha=0,1,2,3$ as coordinates on $I\times M$, we can write the pulled back metric $F_\lambda^{*(4)}g$ as follows:

$$(F^{*(4)}g)_{\alpha\beta}dx^\alpha dx^\beta = -(N^2 - X_iX^i)d\lambda^2 + 2X_i dx^i d\lambda + g_{ij}dx^i dx^j$$

where $g_{ij}=(g_\lambda)_{ij}$ and $g_\lambda = i_\lambda^{*(4)}g$.

Let k_λ be the curve of second fundamental forms for the embedded hypersurfaces $\Sigma_\lambda = i_\lambda(M)$, and let π_λ be their associated canonical momenta.

The basic geometrodynamical equations are contained in the following statement; the notation is explained below.

Let the vacuum Einstein field equations $Ein({}^{(4)}g)=0$ hold on V_4. Then for each one-parameter family of spacelike embeddings $\{i_\lambda\}$ of V_4, the induced metrics g_λ and momentum π_λ on Σ_λ satisfy the following equations:

(*Evolution Equations*)

$$\partial g/\partial\lambda = 2N\left((\pi') - \tfrac{1}{2}g(tr\pi')\right) + L_Xg$$

$$\begin{aligned}\partial\pi/\partial\lambda = &-N\left(Ric(g) - \tfrac{1}{2}R(g)g\right)^\sharp \mu(g)\\ &+\tfrac{1}{2}Ng^\sharp\left(\pi' : \pi' - \tfrac{1}{2}(tr\pi')^2\right)\mu(g)\\ &-2N\left(\pi'\cdot\pi' - \tfrac{1}{2}(tr\pi')\pi'\right)\mu(g)\\ &+(Hess\,N + g\Delta N)^\sharp\ \mu(g) + L_X\pi\end{aligned}$$

and

(*Constraint Equations*)

$$\mathcal{H}(g,\pi) = \left(\pi' : \pi' - \tfrac{1}{2}(tr\pi')^2 - R(g)\right)\mu(g) = 0$$

$$\mathcal{J}(g,\pi) = -2(\delta_g\pi) = 2\pi_i^j{}_{|j} = 0$$

ISBN 0-8053-0102-X

Conversely, if i_λ is a slicing of $(V_4, {}^{(4)}g)$ such that the above evolution and constraint equations hold, then ${}^{(4)}g$ satisfies the (empty space) field equations.

Our notation in this statement is as follows: $(\pi'\cdot\pi')^{ij}=(\pi')^{ik}(\pi')^j_k$, $\pi':\pi'=(\pi')^{ij}(\pi')_{ij}$, $Hess\,N=N_{|i|j}$, $\Delta N=-g^{ij}N_{|i|j}$, and $L_X\pi=(L_X\pi')\mu(g)+\pi'(div\,X)\mu(g)$ is the Lie derivative of the tensor density $\pi=\pi'\mu(g)$ [note $L_X\mu(g)=(div\,X)\mu(g)$]. The Ricci tensor $R_{\mu\nu}$ of ${}^{(4)}g$ is denoted $Ric({}^{(4)}g)$ and that of g by $Ric(g)$. $R(g)$ is the scalar curvature. We write $Ein(g)=Ric(g)-\frac{1}{2}R(g)g$, the Einstein tensor, and note that $Ein(g)=0$ iff $Ric(g)=0$.

One can prove the above statement directly by breaking up the statement $R_{\alpha\beta}=0$ (or better $G_{\alpha\beta}=R_{\alpha\beta}-\frac{1}{2}R'g_{\alpha\beta}$) into $R_{ij}=0$, $R_{00}=0$, and $R_{0i}=0$ and using the Gauss–Codazzi equations (see Sect. 2.7).

Now in a sense we wish to make precise, the evolution equations are the Hamiltonian equations for the Hamiltonian density $N\mathcal{H}+X\cdot\mathcal{J}$. Also, in a sense we will not consider here, the momentum map corresponding to the symmetry group ${}^{(4)}\mathcal{D}$ = all spacetime diffeomorphisms is

$$\Psi(g,\pi)\cdot(N,X)=\int N\mathcal{H}+X\cdot\mathcal{J}$$

The constraints are then just the condition that this natural conserved quantity vanish identically (see Exercise 5.5G). Write the quadratic algebraic part of $\partial\pi/\partial\lambda$ as

$$S_g(\pi,\pi)=-2\left\{\pi'\cdot\pi'-\tfrac{1}{2}(tr\,\pi')\pi'\right\}\mu(g)$$
$$+\tfrac{1}{2}g^{\#}\left\{\pi':\pi'-\tfrac{1}{2}(tr\,\pi')^2\right\}\mu(g)$$

This is the spray of the DeWitt metric, that is, the terms quadratic in π'. The terms in the evolution equation for π may be interpreted as follows:

$\dfrac{\partial\pi}{\partial\lambda}=NS_g(\pi,\pi)$	geodesic spray of the DeWitt metric
$-N\,Ein(g)^{\#}\mu(g)$	force term of the scalar curvature potential, i.e., the term $R(g)$ in
$+(Hess\,N+g\Delta N)^{\#}\mu(g)$	"tilt" term due to nonconstancy of N
$+L_X\pi$	"shift" term due to a nonzero shift

ISBN 0-8053-0102-X

Our goal now is to rewrite the equations in a way that makes their dependence on the slicing (i.e., on N, X) and their Hamiltonian nature more explicit.

We consider the space $\mathfrak{M}$ of Riemannian metrics on M, and the diffeomorphism group $\mathfrak{D}$ of M.

Let $T\mathfrak{M} \approx \mathfrak{M} \times S_2$ denote the tangent bundle of $\mathfrak{M}$, where S_2 is the space of C^∞ two-covariant symmetric tensor fields on M. Let S_d^2 denote the space of C^∞ two-contravariant symmetric tensor densities on M. Define $T^*\mathfrak{M} \approx \mathfrak{M} \times S_d^2 = \{(g,\pi) | g \in \mathfrak{M}, \pi \in S_d^2\}$. We shall think of $T^*\mathfrak{M}$ as the "L^2-cotangent bundle to $\mathfrak{M}$." For $k \in T_g\mathfrak{M} \approx S_2$, $\pi \in T_g^*\mathfrak{M} \approx S_d^2$, there is a natural pairing

$$(\pi, k)_{L^2} = \int_M \pi : k$$

Thus $T^*\mathfrak{M}$ as defined is a subbundle of the "true" contangent bundle. Since $T^*\mathfrak{M}$ is open in $S_2 \times S_d^2$, the tangent space of $T^*\mathfrak{M}$ at $(g,\pi) \in T^*\mathfrak{M}$ is $T_{(g,\pi)}(T^*\mathfrak{M}) \approx S_2 \times S_d^2$.

On $T^*\mathfrak{M}$ we define the globally constant symplectic structure

$$\Omega = \Omega_{(g,\pi)} : T_{(g,\pi)}(T^*\mathfrak{M}) \times T_{(g,\pi)}(T^*\mathfrak{M}) \to \mathbf{R}$$

in the usual way : for $(h_1, \omega_1), (h_2, \omega_2) \in T_{(g,\pi)}(T^*\mathfrak{M}) = S_2 \times S_d^2$,

$$\Omega_{(g,\pi)}((h_1, \omega_1), (h_2, \omega_2)) = \int_M \omega_2 : h_1 - \omega_1 : h_2$$

Let

$$J = \begin{pmatrix} 0 & I \\ -I & 0 \end{pmatrix} : S_d^2 \times S_2 \to S_2 \times S_d^2$$

be defined by

$$(\omega, h) \mapsto J \begin{pmatrix} \omega \\ h \end{pmatrix} = \begin{pmatrix} h \\ -\omega \end{pmatrix}$$

so that

$$J^{-1} = \begin{pmatrix} 0 & -I \\ I & 0 \end{pmatrix} : S_2 \times S_d^2 \to S_d^2 \times S_2, (h, \omega) \mapsto (-\omega, h)$$

Then, as usual,

$$\Omega((h_1, \omega_1), (h_2, \omega_2)) = -\int_M (h_1, \omega_1) \cdot J^{-1} \begin{pmatrix} h_2 \\ \omega_2 \end{pmatrix}$$

ISBN 0-8053-0102-X

Let $C^\infty = C^\infty(M; R)$ denote the smooth real-valued functions on M,

$$C_d^\infty = \text{smooth scalar densities on } M$$

$$\mathfrak{X} = \text{smooth vector fields on } M$$

$$\Lambda_d^1 = \text{smooth one-form densities on } M$$

Consider the functions

$$\mathcal{H}: T^*\mathfrak{M} \to C_d^\infty: (g, \pi) \mapsto \mathcal{H}(g, \pi) = (\pi' : \pi' - \tfrac{1}{2}(tr\,\pi')^2 - R(g))\,\mu(g)$$

$$\mathcal{J} = 2\delta: T^*\mathfrak{M} \to \Lambda_d^1;\ (g, \pi) \mapsto 2(\delta_g \pi) = -2\pi^j_{i\,|j}$$

$$\Phi = (\mathcal{H}, \mathcal{J}): T^*\mathfrak{M} \to C_d^\infty \times \Lambda_d^1;\ (g, \pi) \mapsto (\mathcal{H}(g, \pi), \mathcal{J}(g, \pi))$$

Using functional derivative notation (see the previous examples), the evolution equations may be written as Hamilton's equations with Hamiltonian $N\mathcal{H} + X \cdot \mathcal{J}$, that is,

$$\frac{\partial g}{\partial \lambda} = \frac{\delta}{\delta \pi}(N\mathcal{H} + X \cdot \mathcal{J})$$

$$\frac{\partial \pi}{\partial \lambda} = -\frac{\delta}{\delta g}(N\mathcal{H} + X \cdot \mathcal{J})$$

(This is a long but straightforward calculation to show the equivalence.) If we take the lead mentioned above, we can write these equations concisely using adjoint notation (adjoints are taken relative to the L^2 pairing above):

The Einstein system, defined by the evolution equations and constraint equations above, can be written as

(*Evolution equations*)
$$\frac{\partial}{\partial \lambda}\begin{pmatrix} g \\ \pi \end{pmatrix} = J \circ (D\Phi(g, \pi))^* \cdot \begin{pmatrix} N \\ X \end{pmatrix}$$

(*Constraint equations*)
$$\Phi(g, \pi) = (\mathcal{H}(g, \pi), \mathcal{J}(g, \pi)) = 0$$

Notice that in this sense the constraints determine the dynamics. From a more general point of view, this situation is covered by the Dirac theory of constraints (see Exercise 5.3L).

A way to arrive at these equations is to start with the four-dimensional variational principle $\delta \int R({}^{(4)}g)\,\mu({}^{(4)}g) = 0$ and break it into space and time components and then take the corresponding variations.

The above adjoint form of the Einstein equations can be extended to include field theories coupled to gravity, that is, nonvacuum spacetimes. This

ISBN 0-8053-0102-X

extended form is at the basis of a covariant formulation of Hamiltonian systems (Kuchar [1976], and Fischer–Marsden [1976]). For example, the canonical formulation of the covariant scalar wave equation $\Box\phi = m^2\phi + F'(\phi)$ on a spacetime $V_4 = (I \times M, {}^{(4)}g)$ in terms of a general lapse and shift is as follows: Consider the Hamiltonian

$$\mathcal{H}(\phi, \pi_\phi) = \left\{ \tfrac{1}{2}\left((\pi'_\phi)^2 + |\nabla\phi|^2 + m^2\phi^2 \right) + F(\phi) \right\} \mu(g)$$

for the scalar field (the background metric is considered as implicitly given for this example). We construct the stress tensor, a two-contravariant symmetric tensor density $\mathcal{T}$ by varying $\mathcal{H}(\phi, \pi_\phi)$ with respect to g:

$$\mathcal{T} = -2D_g\mathcal{H}(\phi, \pi_\phi)^* \cdot 1$$

and a one-form density $\mathcal{J}(\phi, \pi_\phi)$ from the relationship

$$\int \langle X, \mathcal{J}(\phi, \pi_\phi) \rangle = -\int \langle \pi, L_X\phi \rangle$$

so that $\mathcal{J}(\phi, \pi_\phi) = -\pi_\phi \cdot d\phi$. This condition expresses $\mathcal{J}$ as the conserved quantity for the coordinate invariance group on M (Exercise 5.5G). If we set $\Phi = (\mathcal{H}, \mathcal{J})$, then the Hamiltonian equations of motion for ϕ in a general slicing of the spacetime with lapse N and shift X are

$$\frac{\partial}{\partial\lambda}\begin{pmatrix} \phi \\ \pi_\phi \end{pmatrix} = J \circ D\Phi(\phi, \pi_\phi)^* \begin{pmatrix} N \\ X \end{pmatrix}$$

exactly as for general relativity. A computation shows that this system is equivalent to the covariant scalar wave equation given above.

If we couple the scalar field with gravity by regarding the scalar field as a source, the equation for the gravitational momentum $\partial\pi/\partial\lambda$ is altered by the addition of the term $\frac{1}{2}N\mathcal{T}$, and the equation for $\partial g/\partial\lambda$ is unchanged. The constraint equations become $\mathcal{H}_{geom}(g, \pi) + \mathcal{H}_{scalar}(g, \phi, \pi_\phi) = 0$ and $\mathcal{H}_{geom}(g, \pi) + \mathcal{H}_{scalar}(\phi, \pi_\phi) = 0$. More generally, if one considers the total Hamiltonian $\mathcal{H}_T = \mathcal{H}_{geom} + \mathcal{H}_{fields}$ and a total universal flux tensor $\mathcal{J}_T = \mathcal{J}_{geom} + \mathcal{J}_{fields}$ (and if the nongravitational fields are nonderivatively coupled to the gravitational fields), the general form of the equations

$$\frac{\partial}{\partial\lambda}\begin{pmatrix} g, \phi_A \\ \pi, \pi^A \end{pmatrix} = J \circ \left(D\Phi_T(g, \pi), (\phi_A, \pi^A) \right)^* \cdot \begin{pmatrix} N \\ X \end{pmatrix}$$

$$\Phi_T(g, \pi, \phi, \pi_\phi) = 0$$

remains valid. (For the case of Yang–Mills fields, see Arms [1977, 1978].) Here, ϕ_A represents all nongravitational fields, π^A the conjugate momenta,

ISBN 0-8053-0102-X

and $\Phi_T = (\mathcal{H}_T, \mathcal{J}_T)$. These results provide a unified covariant Hamiltonian formulation of general relativity coupled to other Lagrangian field theories.

Many of the ideas from Chapter 4 can be used in general relativity. One example is given in Exercise 5.5.G. In fact, if the reduction procedure is carried out, there results an important new Hamiltonian system called the *space of gravitational degrees of freedom*. See Fischer and Marsden [1976a, b, 1978]. For approaches to the symplectic structure directly from the four-dimensional point of view, see Szczyryba [1977] (some background theory is given in Guillemin and Sternberg [1977]).

EXERCISES

5.5A. Let $\mathcal{H}$ be a real Hilbert space and A a generator of a one-parameter group U_t of isometries in $\mathcal{H}$. Assume A^{-1} is a bounded operator. Then show:

(i) A is Hamiltonian relative to $\omega(x,y) = \langle A^{-1}x, y\rangle$

(ii) $\mathcal{H}$ can be given a complex (linear) structure relative to which U_t is unitary.

Apply this result to the Klein–Gordon equation. (*Note*: This is related to results of Weiss [1967] and Cook [1966].)

5.5B. Consider the symmetric hyperbolic systems of Friedrichs:

$$\frac{\partial u}{\partial t} = \sum_{j=1}^{m} a_j(x)\frac{\partial u}{\partial x^j} + b(x)u(x)$$

where u takes values in $\boldsymbol{R}^N$ and $a_j(x), b(x)$ are $N \times N$ matrices with $a_j(x)$ symmetric.

(i) Letting

$$u = \left(\frac{\partial \phi}{\partial x^1}, \ldots, \frac{\partial \phi}{\partial x}, \phi\right)$$

show that the Klein–Gordon equation may be written as a symmetric hyperbolic system.

(ii) If there is a uniformly positive-definite symmetric matrix $c(x)$ such that $ca_j = 0$ and

$$cb + b^*c = -\left(b + b^* + 2\sum_{j=1}^{m} \frac{\partial a_j}{\partial x_j}\right)$$

then the energy

$$H(u) = \tfrac{1}{2}\{\langle u, u\rangle + \langle u, cu\rangle\}$$

is conserved. (See Exercise 5.5A.)

(iii) Write Maxwell's equations as a Hamiltonian system.

5.5C. Let $\mathcal{H}$ denote complex Hilbert space with $\omega(x,y) = Im\langle x, y\rangle$. Let $G = S^1 = \{z \in \boldsymbol{C} \,|\, |z| = 1\}$ act on $\mathcal{H}$ by $\Phi_z(x) = zx$. Show that:

(i) this action is symplectic;

(ii) a momentum mapping for the action is $\psi(x)\cdot z = \tfrac{1}{2}\|x\|^2 z$;

(iii) the corresponding reduced symplectic space (see Sect. 4.3) is projective Hilbert space, $\tilde{\mathcal{H}}$.

ISBN 0-8053-0102-X

(*Note*: The Bargmann–Wigner theorem states that any complex linear one-parameter group of Hamiltonian transformations of $\mathcal{H}$ are implemented by a unitary one-parameter group on $\mathcal{H}$; this is proved in Varadarajan [1968] using different terminology.)

5.5D. Show that the linear beam equation

$$u_{tt}+u_{xxxx}=0,\quad x\in[0,1],\quad u,u_x=0 \text{ at } 0,1$$

is Hamiltonian, as is the nonlinear equation

$$u_{tt}+u_{xxxx}+K\left(\int_0^1 (u_x)^2\,dx\right)u_{xx}=0$$

(K represents the stiffness).

5.5E. (a) In Example 5.5.8 write down the momentum map explicitly following the ideas in Sect. 4.2. (Note that the tangent space to $\mathfrak{D}_\mu$ at a point η consists of maps $v\circ\eta^{-1}$, where v is a divergence-free vector field parallel to the boundary.)

(b) Identify the co-adjoint action of $\mathfrak{D}_\mu$ on $T_e^*\mathfrak{D}_\mu\approx$ all one-forms on M that are divergence free and parallel to ∂M with pull-back of one-forms.

(c) Identify the reduced phase spaces for Euler's equations with subsets of the space of all divergence-free vector fields on M.

5.5F. Let X be a Banach space with weak symplectic form ω. Let $A: Y\to X$ be a linear Hamiltonian operator and let $B: Y\times Y\to X$ be a continuous symmetric bilinear map such that, for fixed x, $B(x,\cdot)$ is linear Hamiltonian. Then show that the nonlinear operator $G(x)=Ax+B(x,x)$ is Hamiltonian with $H(x)=H_A(x)+\frac{1}{3}\omega(B(x,x),x)$. Generalize.

5.5G. Let M be a compact manifold and $\mathfrak{M}$ the space of Riemannian metrics on M. Let the group of diffeomorphisms $\mathfrak{D}$ act on $\mathfrak{M}$ by pull-back. Let $T^*\mathfrak{M}$ be the set of pairs (g,π) of metrics and tensor densities as in the text. Show that $\mathfrak{D}$ induces a symplectic action on $T^*\mathfrak{M}$. Compute its momentum mapping (using the formulas of Chapter 4) to be

$$\Psi(g,\pi)\cdot X=\int_M\langle\pi,L_X g\rangle=-\int_M X\cdot\mathfrak{J}(g,\pi)$$

where $\mathfrak{J}(g,\pi)_i=2\pi_{i|j}^{j}$, twice the divergence of π. Generalize by replacing $\mathfrak{M}$ by a general space of tensors. (See Fischer and Marsden [1972]).

5.5H. (a) (Quantum Mechanical Systems with Spin). Let M be an oriented Riemannian manifold and $\pi: E\to M$ a spin bundle (see Exercise 4.2H). Let $\mathcal{H}$ be the Hilbert space of L^2 sections of E and let H_{op} be a self-adjoint operator on $\mathcal{H}$. Let ϕ be an action of a Lie group G on M that preserves the volume form μ on M, and let ψ be an action of G on E covering ϕ, which is an isometry on fibers. Let H_{op} be invariant under the action Ψ on $\mathcal{H}$ induced by ψ. Show that the expectation of $i(\xi_M+\xi_s)$ (defined in Exercise 4.2H) as a differential operator in $\mathcal{H}$ is a constant of the motion.

(b) For $M=\mathbb{R}^3$, $E=\mathbb{R}^3\times\mathbb{C}^2$, deduce that the function

$$P_{jk}=u\mapsto\sum_{\alpha=1}^{2}\int_{\mathbb{R}^3} i\bar{u}^\alpha\left(x_j\frac{\partial u_\alpha}{\partial x_k}-x_k\frac{\partial u_\alpha}{\partial x_j}\right)dx$$
$$+\tfrac{1}{2}\int_{\mathbb{R}^3}\bar{u}\cdot\sigma_l\cdot u\,dx$$

ISBN 0-8053-0102-X

(orbital + spin angular momentum), where (j,k,l) is an even permutation of (1,2,3) is a constant of the motion.

(c) Define a notion of spin bundle appropriate for the Dirac equation on a Lorentz manifold and prove a result like (b) in Minkowski space. (See, e.g., Hitchin [1974] and references therein.)

5.5I. (Bharucha–Reid, and Chernoff) The Heisenberg and Schrödinger Pictures. Let $\mathcal{H}$ be complex Hilbert space and F_t a one-parameter unitary group with generator $A = iH_{op}$. (This situation, discussed in the text, is usually called the Schrödinger picture.) Let $\mathcal{S}$ be the space of Hilbert–Schmidt operators on $\mathcal{H}$ with norm $\|T\|^2 = Trace(T^*T)$. Let $G_t(T) = F_tTF_{-t}$. Show that G_t is a one-parameter unitary group on $\mathcal{S}$ (the Heisenberg representation). Show that the generator Y of G_t is given as follows: $D(Y) = \{T \in \mathcal{S} \mid T\colon D(A) \to D(A)$ and $AT - TA$ is the restriction to $D(A)$ of an element $B \in \mathcal{S}\}$ and $Y(T) = B$.

5.5J. Relate the generating function defined by Gardner [1971] for canonical transformations of the KdV equation to the generating functions defined in Sect. 5.2.

5.5K. (Adler, Kostant, Hermann, van Moerbeke) The Toda lattice. Let $G, \mathfrak{g}$ be as in exercise 4.6G and (P, ω) the co-adjoint orbit of C described there. Let $I_k = trC^k/k$. Show that $\{I_j, I_k\} = 0$ and that $H = I_2 = \frac{1}{2}\Sigma p_k^2 + \frac{1}{4}\Sigma e^{2(q_k - q_{k+1})}$. Deduce that the Hamiltonian system (P, ω, H) is completely integrable. Find a Hamiltonian system on T^*G which, when reduced according to 4.3.4(v), yields the one here. This model is a discrete version of the KdV equation on $\mathbb{R}$; see Adler [1978, 9] and Ratiu [1979].

5.5L. (Tulczyjew) An elastic beam. Consider an elastic beam in Euclidean space. The equilibrium configuration of the beam with no external forces is a straight line l. Small deflections induced by external forces and torques can be represented by points of a plane M perpendicular to the line l. The distance measured along l from an arbitrary reference point is denoted by s. We select a section of the beam corresponding to an interval $[s_1, s_2]$ and assume that external forces and bending torques are applied only to the ends of the section. The configuration manifold Q of the section of the beam is the product $TM \times TM$ with coordinates $(q_2^i, q'^j_2, q_1^k, q'^l_1)$, $i,j,k,l = 1,2$. The force bundle $F = T^*TM \times T^*TM$ has coordinates $(q_2^i, q'^j_2, f^2{}_k, t^2{}_l, q_1{}^m, q_1{}^n, f^1{}_p, t^1{}_r)$. The coordinates $f^2{}_k$ and $t^2{}_l$ are components of the reaction force and the reaction torque respectively at q_2^i; $f^1{}_p$ and $t^1{}_r$ are components of the force and the torque applied to the end of the beam section at $q_1{}^m$. If $(\delta q_2^i, \delta q'^j_2, \delta f^2{}_k, \delta t^2{}_l, \delta q_1{}^m, \delta q'^n_1, \delta f^1{}_p, \delta t^1{}_r)$ are components of an infinitesimal displacement u in F at $(q_2^i, q'^j_2, f^2{}_k, t^2{}_l, q_1{}^m, q'^n_1, f^1{}_p, t^1{}_r)$ then the virtual work is

$$\begin{aligned} w &= f^1{}_i \delta q_1{}^i + t^1{}_i \delta q'^i_1 - f^2{}_i \delta q_2{}^i - t^2{}_i \delta q'^i_2 \\ &= -\langle u, \vartheta \rangle, \end{aligned}$$

where

$$\vartheta = f^2{}_i dq_2{}^i + t^2{}_i dq'^i_2 - f^1{}_i dq_1{}^i - t^1{}_i dq'^i_1$$

is the special symplectic form on F (see 3.3I).

In the limit $s_2 \to s_1$ the configuration manifold Q' is the bundle TTM with coordinates $(q^i, q'^j, \dot{q}^k, \dot{q}'^l)$ and the force bundle F' is the bundle TT^*TM with coordinates $(q^i, q'^j, f_k, t_l, \dot{q}^m, \dot{q}'^n, \dot{f}_p, \dot{t}_r)$. The special symplectic form becomes

$$\vartheta' = \dot{f}_i dq^i + f_i d\dot{q}^i + \dot{t}_i dq'^i + t_i d\dot{q}'^i.$$

ISBN 0-8053-0102-X

Equilibrium conditions are

$$\dot{f}_i = 0, \qquad f_i + \dot{t}_i = 0, \qquad t_i = k_{ij}\dot{q}'^j,$$

where k_{ij} is a tensor characterizing the elastic properties of the beam. These conditions express the vanishing of the total force and the total torque, and also Hooke's law. In addition to these conditions there is a constraint condition $\dot{q}^i = q'^i$, (i.e. the configuration of the bent beam is a differentiable curve.) This condition defines a constraint submanifold

$$C = \{w \in TTM \mid \tau_{TM}(w) = T\tau_m(w)\}.$$

We use on C coordinates $(q^i, \dot{q}^j, \ddot{q}^k)$ related to coordinates $(q^i, q'^j, \dot{q}^k, \dot{q}'^l)$ by $\dot{q}'^i = \ddot{q}^i$. Show that these conditions define a Lagrangian submanifold $S' \subset F'$ generated (in the sense of 5.3K) by $-L$, where

$$L(q^i, \dot{q}^j, \ddot{q}^k) = \frac{1}{2} k_{ij}\ddot{q}^i\ddot{q}^j$$

is the potential energy per unit length of the beam and is defined on C. Finally, show that the equivalent equations

$$\dot{q}^i = q'^i, \dot{q}'^i = k^{ij}t_j, \left(k^{ij}k_{jl} = \delta^i_l\right), \dot{f}_i = 0, \dot{t}_i = -f_i$$

define a Hamiltonian vector field X on F' and that the Lagrangian submanifold S' is the image of the field X.

5.6 INTRODUCTION TO NONLINEAR OSCILLATIONS

This section is intended to provide a transition between the more formal analytical dynamics we have been considering and the qualitative theory that constitutes the next three chapters (Part 3).

Nonlinear oscillations is a whole subject on its own and our intention is merely to highlight a few points relevant to Hamiltonian mechanics. For a more extensive discussion, see, for instance, Hale [1963], Minorsky [1974], Andronov, Witt, and Chaikin [1966], and Pliss [1966].

Let us begin by recalling the linear situation and the small oscillation approximation, a topic initiated by Lagrange, and now standard in virtually every elementary mechanics text (Goldstein [1950] is particularly readable).

5.6.1 Proposition. *Consider the Lagrangian system $L = K - V$, where*

$$K = \frac{1}{2}\sum_{i,j=1}^{n} m_{ij}\dot{q}^i\dot{q}^j$$

and

$$V = \frac{1}{2}\sum_{i,j=1}^{n} c_{ij}q^iq^j$$

ISBN 0-8053-0102-X

on $\boldsymbol{R}^{2n}$. Assume that the matrices m_{ij} and c_{ij} are symmetric (this is no loss of generality) and m_{ij} is positive-definite. Then there is a linear change of coordinates

$$Q^i = \sum_{j=1}^{n} \alpha^i_j q^j$$

$$\dot{Q}^i = \sum_{j=1}^{n} \alpha^i_j \dot{q}^j$$

such that the Lagrangian in the new coordinates $Q^i, \dot{Q}^i$ is

$$\bar{L} = \bar{K} - \bar{V}$$

where

$$\bar{K} = \tfrac{1}{2} \sum_{i=1}^{n} m_i (\dot{Q}^i)^2, \qquad m_i > 0$$

and

$$\bar{V} = \tfrac{1}{2} \sum_{i=1}^{n} c_i (Q^i)^2$$

The new coordinates $Q^1, \ldots, Q^n, \dot{Q}^1, \ldots, \dot{Q}^n$ are called **normal modes** *and Lagrange's equations become*

$$\ddot{Q}^i + \lambda_i^2 Q^i = 0, \qquad i = 1, \ldots, n$$

where

$$\lambda_i^2 = -c_i / m_i$$

If $c_i > 0$ (i.e., λ_i are purely imaginary, or equivalently, $E = K + V$ is positive-definite), there are n one-parameter families of periodic orbits. Setting $\omega_i = \sqrt{c_i / m_i}$, $\omega_i / 2\pi$ are called the ***fundamental frequencies.***

Proof. The result follows from a theorem in linear algebra which states that two quadratic forms, one of which is positive-definite, can be simultaneously diagonalized by a linear transformation. This, in fact, has already been proved in 3.1. Indeed, consider the inner product on $\boldsymbol{R}^n$ given by

$$\langle v, w \rangle = \sum_{i,j=1}^{n} m_{ij} v^i w^j$$

It is an inner product since m_{ij} is positive-definite. Then there is an orthonormal basis for this inner product that diagonalizes the quadratic form V. This change of coordinates defines the matrix α^i_j.

ISBN 0-8053-0102-X

This transformation on $q^i, \dot{q}^i$ is the lift of the transformation on q^i alone; thus it induces a symplectic transformation of the corresponding Lagrangian systems.

Note that $\lambda_1, \ldots, \lambda_n, -\lambda_1, \ldots, -\lambda_n$ are the characteristic roots of the linear Hamiltonian system X_E and so can be computed in the original coordinates. Thus, to compute the resonant frequencies, the change of coordinates need not be made.

In the new coordinates, the system decouples into n harmonic oscillators and so (if λ_i are purely imaginary) there are n one-parameter families of periodic orbits. There may be more periodic orbits depending on whether or not $\lambda_1, \ldots, \lambda_n$ are rationally related. If we start with general initial conditions $Q^i(0), \dot{Q}^i(0)$, the solution curve is

$$Q^i(t) = Q^i(0)\cos(\omega_i t) + \frac{\dot{Q}^i(0)}{\omega_i}\sin(\omega_i t)$$

so each component has period $2\pi/\omega_i$. If components with indices $i \in I \subset \{1, \ldots, n\}$ are excited [i.e., $Q^i(t)$ is non trivial, $i \in I$], we have a periodic orbit if and only if there is a number T and nonzero integers k_i such that

$$T = k_i \cdot 2\pi/\omega_i \quad \text{for all } i \in I$$

This is equivalent to $\omega_i = r_{ij}\omega_j$, $i, j \in I$, r_{ij} rational; that is, a *resonance* between the ith and jth normal modes. Thus there are no periodic orbits other than the obvious ones iff $\lambda_1, \ldots, \lambda_n$ are *pairwise independent over the integers* $\boldsymbol{Z}$, that is, no nontrivial relation of the form $m_i\lambda_i + m_j\lambda_j = 0$, $m_i, m_j \in \boldsymbol{Z}$ exists for any $i \neq j$. ■

It is natural to ask if any linear Hamiltonian system with purely imaginary eigenvalues can be brought into the form of n uncoupled equations of the form $\ddot{Q}^i + \lambda_i^2 Q^i = 0$, that is, if we can bring X_H to the following *real normal form*:

$$\left[\begin{array}{ccc|ccc} 0 & & & \alpha_1 & & 0 \\ & & & & \ddots & \\ & & & 0 & & \alpha_n \\ \hline -\alpha_1 & & 0 & & & \\ & \ddots & & & & \\ 0 & & -\alpha_n & & & 0 \end{array}\right] \quad [\text{real normal form (RNF)}]$$

The answer is no. The problem of bringing a general X_H to some normal form by a *real* canonical transformation was solved by Williamson [1936, 1937, 1939]. See also Wintner [1934], Laub and Meyer [1974], and Burgoyne and Cushman [1974]. For a statement of Williamson's result, see these references and Arnold [1978].

ISBN 0-8053-0102-X

However, there are three important cases where the above normal form (RNF) *is* valid:*

(a) When $\lambda_1,\ldots,\lambda_n$ are all distinct and are all pure imaginary (see Sect. 3.1).
(b) Under the hypotheses of 5.6.1. Here the λ_i's need not be distinct, but are assumed to be pure imaginary.
(c) If H is positive-definite. [This generalizes (b) since, under the conditions of 5.6.1, E is positive-definite if the λ_i are pure imaginary.]

In the last case, then, we have (at least) n periodic orbits on each energy surface.

The case (c) is the most satisfying since it corresponds to stability of the linear dynamical system, as discussed in Sect. 3.4. This result goes back to Krein [1950], Gelfand and Lidskii [1955], and Moser [1958].

We present a version due to J. Cook [1966] as it is valid in infinite dimensions as well. (See also Segal [1966].)

The infinite-dimensional version of (c) takes the following form:

5.6.2 Theorem *Let $\mathcal{H}$ be a real Hilbert space carrying a strongly nondegenerate symplectic form ω. Let F_t be a continuous one-parameter group of symplectic linear transformations. Suppose that the following* **stability condition** *holds: for each $x, y \in \mathcal{H}$, the function $t \mapsto \omega(F_t x, y)$ is bounded on* **R**. *Then $\mathcal{H}$ can be made into a complex Hilbert space such that ω is the imaginary part of the inner product, and such that the operators F_t are unitary.*

If $\mathcal{H}$ is a finite-dimensional symplectic space and H is a real-valued quadratic function on $\mathcal{H}$ that is positive-definite, we know that the orbits of the flow of X_H are bounded (by conservation of energy) and thus the stability condition above holds. With respect to a new Hermitian inner product, X_H is skew adjoint, so iX_H can be diagonalized by a unitary transformation. This change of coordinates is a complex change in $\mathcal{H}$, but since $\mathcal{H}$ has a complex structure, we can express it in real terms. (Recall that on a complex Hilbert space a *complex* unitary operator regarded as a *real* linear operator is symplectic.) Thus:

5.6.3 Corollary. *If E is a finite-dimensional real symplectic space and H is a positive-definite quadratic form on E, there is a set of real canonical coordinates relative to which X_H has the matrix (RNF) given above.*

We shall give another proof of this below that does not rely on complexifications.

Proof of 5.6.2. (Assumes some knowledge of functional analysis.) By Theorem 3.1.18, we can introduce a complex structure J and a Hermitian inner

*If we look for normal forms on a complexified symplectic space, things are somewhat simpler, but are not as useful as the *real* normal forms. See Siegel and Moser [1971] and for an elegant approach using Lagrangian subspaces, Duistermaat [1973].

ISBN 0-8053-0102-X

product on $\mathcal{H}$ such that ω is the imaginary part of the inner product. Accordingly, we may as well suppose that $\mathcal{H}$ is a complex Hilbert space to begin with, with $\omega(x,y) = -Im\langle x,y\rangle$. The group F_t consists of real-linear, symplectic transformations. Our goal is to construct a *new* complex structure, with respect to which the operators F_t are complex linear (hence unitary).

We first note the following consequence of the stability condition: for each $x,y \in \mathcal{H}$, $-Im\langle F_t x,y\rangle = \omega(F_t x,y)$ is bounded, $-\infty < t < \infty$. Hence, by the uniform boundedness theorem $\|F_t\|$ is bounded in t.

We want to renorm $\mathcal{H}$ so that the F_t's are not merely uniformly bounded, but isometric. The basic idea is to "average" the original inner product over the group $\mathbf{R}$. If $\mathbf{R}$ were compact, we could simply integrate over the group. Instead, we shall use the following:

5.6.4 Lemma. *Let $\mathcal{E} = BC(\mathbf{R})$ be the space of bounded continuous functions on $\mathbf{R}$, with the supremum norm. Then there is a functional $\mu \in \mathcal{E}^*$ such that $\mu(1) = 1$; μ is **positive**, that is, if $f \geq 0$, then $\mu(f) \geq 0$; and μ is **translation invariant**, that is, $\mu(f_a) = \mu(f)$, where $f_a(x) = f(x+a)$.*

Proof. Define $\mu_n \in \mathcal{E}^*$ by

$$\mu_n(f) = \frac{1}{2n}\int_{-n}^{n} f(x)\,dx$$

Because the unit ball of $\mathcal{E}^*$ is compact in the weak* topology, there is a subnet μ_{n_i} converging in this topology to $\mu \in \mathcal{E}^*$. That is, $\mu(f) = \lim \mu_{n_i}(f)$ for every f. It follows immediately that μ is positive with $\mu(1) = 1$. As for translation invariance, note that

$$|\mu_n(f_a - f)| = \frac{1}{2n}\left|\int_n^{n+a} f(x)\,dx - \int_{-n}^{-n+a} f(x)\,dx\right|$$

$$\leq \frac{|a|}{n}\|f\|_\infty$$

which vanishes as $n \to \infty$. Hence $\mu(f_a - f) = 0$. ▼

A more sophisticated argument (see Segal [1966]), using the Markov–Kakutani fixed point theorem, enables one to prove an analog of the lemma for "amenable" groups—in particular, abelian or solvable groups.

We can now finish the proof of 5.6.2. Define a real symmetric bilinear form on $\mathcal{H}$ by the relation

$$s(x,y) = \mu_t(Re\langle F_t x, F_t y\rangle)$$

Here the subscript t indicates our averaging of the bounded function $t \mapsto Re\langle F_t x, F_t y\rangle$. By the translation invariance of μ and the group property of F we have, obviously, $s(F_\tau x, F_\tau y) = s(x,y)$. Moreover, s is a symmetric

ISBN 0-8053-0102-X

nonnegative form. Finally, by the positivity of μ, we have

$$s(x,x)=\mu_t(Re\langle F_t x, F_t x\rangle)\leqslant M^2\|x\|^2$$

where $M=sup\|F_t\|$. On the other hand, we have $\|x\|^2=\|F_{-t}F_t x\|^2\leqslant M^2\|F_t x\|^2$; whence

$$\|x\|^2\leqslant M^2 s(x,x)$$

Thus $s(\cdot,\cdot)$ is a real inner product equivalent to the original inner product on $\mathcal{H}$.

Accordingly, by the proof of Theorem 3.1.18, there is an invertible bounded linear operator T_0, skew-adjoint relative to s, such that

$$\omega(x,y)\equiv s(T_0 x, y)$$

Since ω and s are invariant under F_t, it follows that F_t and T_0 commute. Hence F_t and J_0 commute, where J_0 is the isometric factor in the polar decomposition of T_0. The operator J_0 satisfies $J_0^2=-I$, hence gives a complex structure—which makes the F_t unitary. ■

Alternative Proof of 5.6.3 (Weinstein [1971].) Let $\tau>0$ and V_τ be the sum of the eigenspaces of X_H corresponding to eigenvalues of the form $2\pi ni/\tau$, where $n\in \mathbf{Z}$. Then clearly any orbit in V_τ is periodic with period dividing τ. Since H is positive-definite, $H^{-1}(1)\cap V_\tau$ is a compact submanifold of V_τ consisting of periodic orbits; moreover, $H^{-1}(\varepsilon)\cap V_\tau, \varepsilon>0$ is diffeomorphic to a sphere by definiteness of H. It must be odd dimensional since X_H is nowhere vanishing on it; equivalently, V_τ is even dimensional, a fact we know from the symplectic eigenvalue theorem. Now as X_H is linear, there is a two-dimensional subspace E_1 of V_τ containing a one-parameter family of these closed orbits (parametrized by ε). Since X_H is tangent to and nontrivial on E_1, E_1 is nondegenerate and so we can choose canonical coordinates q,p on E_1; X_H is then given in E_1 by $(\partial H/\partial p, -\partial H/\partial q)$. Let $H(q,p)=aq^2+bpq+cp^2$ on E_1; since H is positive-definite, $a>0$, $c>0$, and $4ac-b^2>0$. Write $H(q,p)=\alpha q^2+c\bar{p}^2$, where $\bar{p}=p+\gamma q$ and where $\gamma=b/2c$ and

$$\alpha=a-c\gamma^2=\frac{4ac-b^2}{4c^2}>0$$

Now $dq\wedge dp=dq\wedge d\bar{p}$, so $q,\bar{p}$ are canonical coordinates and in them

$$X_H=\left(\frac{\partial H}{\partial \bar{p}}, \frac{-\partial H}{\partial q}\right)=(2c\bar{p}, -2\alpha q)$$

so X_H has the matrix $\begin{bmatrix} 0 & 2c \\ -2\alpha & 0 \end{bmatrix}$. We get the required (RNF) on E_1 by changing to the coordinates $\bar{q}=\mu q$, $\bar{\bar{p}}=(1/\mu)\bar{p}$ for suitable μ.

ISBN 0-8053-0102-X

Now one chooses an ω-orthogonal complement to E_1 (as in 3.1.2), picks another two-dimensional E_2 containing a family of closed orbits, and repeats.

This brings X_H into (RNF) on V_τ and then one repeats for all the V_τ. Since the different (nontrivial) V_τ are ω-orthogonal (by 3.1.2), we get the desired (RNF) on all of E. ■

If we have a nonlinear system whose energy has the form

$$E=\tfrac{1}{2}\sum_{i,j=1}^{n} m_{ij}\dot{q}^i\dot{q}^j+V(q^i)$$

and if the origin is a fixed point that is, $\partial V/\partial q^i(0)=0$, then the (quadratic) term in the Taylor expansion of V about 0 gives a system of the form 5.6.1. Ignoring the nonlinear terms here is called the *small oscillation approximation*.

More generally, one can consider a Hamiltonian H on a symplectic manifold (P,ω). Suppose x_0 is a fixed point, so $dH(x_0)=0$. If $D^2H(x_0)$ is positive-definite, one might hope that the motion near x_0 is decribable by the small oscillation approximation, that is, by replacing H by its quadratic part. (Recall from Sect. 3.4 that the eigenvalues, of $DX_H(x_0)$, which we can label $\lambda_1,\ldots,\lambda_n,-\lambda_1,\ldots,-\lambda_n$, are then purely imaginary.) For instance, does the nonlinear system possess n periodic orbits on each energy surface near x_0 as the linearized system does? This question actually leads one into deeper waters.

Before describing this situation, a couple of elementary classical examples are relevant. The reader is asked to fill in the details for himself.

5.6.5 Examples (a) The simple pendulum: $\ddot{q}+k\sin q=0$, $k>0$. Here a study of the energy surfaces $\dot{q}^2/2-k\cos q=constant$ yields the phase portrait shown in Fig. 5.6-1. (The Morse Lemma will simplify your efforts; the global structure of the level sets $H^{-1}(\varepsilon)$ is a good example to illustrate the ideas of Morse theory.)

(b) The van der Pol or Liénard equation* (without friction or forcing): $\ddot{q}=\tfrac{1}{2}(1-q^2)$. Again, this is easiest to analyze by looking at the level surfaces of the energy

$$\frac{\dot{q}^2}{2}+\frac{q}{6}(q^2-3)=constant$$

See Fig. 5.6-2.

Now we will turn to a study of periodic orbits for nonlinear Hamiltonian systems in the small oscillation region. Our main goal is a proof of Liapunov's

*The names "van der Pol and Liénard equations" refers to equations of the form $\ddot{u}+f(u)\dot{u}+g(u)=0$. We have taken $f(u)=0$ and $g(u)=\tfrac{1}{2}(q^2-1)$. This form arises in several applications.

ISBN 0-8053-0102-X

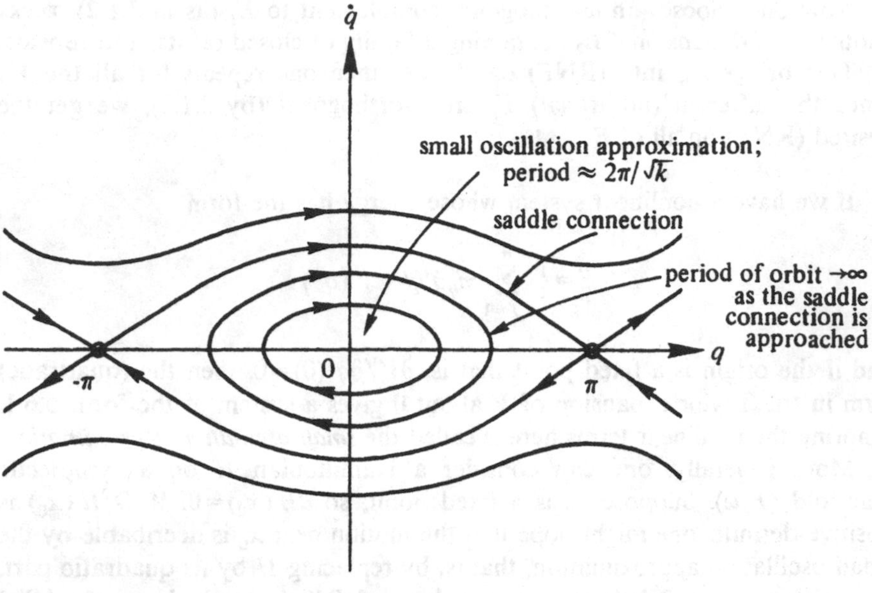

Figure 5.6-1. The simple pendulum; $\ddot{q} = -k \sin q$.

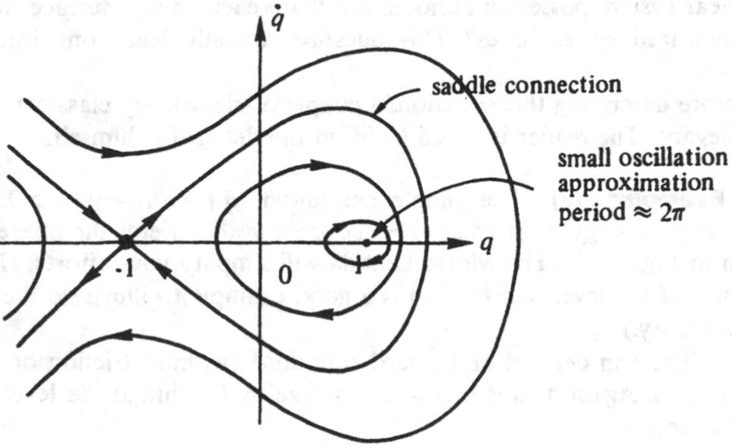

Figure 5.6-2. Phase portrait for the equation $\ddot{q} = \frac{1}{2}(1-q^2)$.

theorem that, in effect, justifies the small oscillation approximation, at least if certain technical conditions are met.

We shall begin with a rather general result on perturbations of periodic orbits for vector fields possessing an integral. Our exposition follows Duistermaat [1972], although this sort of result is found in several places, such as Hale [1969].

5.6.6 Theorem. *Let X_ε be a family of C^k vector fields on a manifold M, $0 < \varepsilon < \varepsilon_0$ depending in a C^k manner on ε, $k \geq 1$. Write $X = X_0$.*

ISBN 0-8053-0102-X

Let $H: [0, \varepsilon_0) \times M \to \mathbf{R}: (\varepsilon, x) \mapsto H_\varepsilon(x)$ be a C^k function on $[0, \varepsilon_0) \times M$ that is constant on the orbits of X_ε.

Let $x_0 \in M$ and let γ be a periodic orbit of X with period $T_0 > 0$ passing through x_0. Let γ lie on a regular energy surface

$$\Sigma_{e_0} = H^{-1}(e_0) \qquad e_0 = H_0(\gamma)$$

Let F_t^ε be the flow of X_ε and $F_t = F_t^0$. Assume:

(*i*) *$\ker(D_x F_{T_0}(x_0) - Id)$ is two dimensional;*
(*ii*) *$X(x_0) \notin \operatorname{Im}(D_x F_{T_0}(x_0) - Id)$.*

Let S_{x_0} be a transverse section at x_0. Then for ε, $|e - e_0|$ sufficiently small there is a unique $x(\varepsilon, e)$ near x_0 and in S_{x_0} and $T(\varepsilon, e)$ near T_0 (depending C^k on ε, e) such that the solution starting at $x(\varepsilon, e)$ is periodic for X_ε with period $T(\varepsilon, e)$, and energy e.

Proof. For e near e_0 and ε near 0, $\Sigma_{e,\varepsilon} = H_\varepsilon^{-1}(e)$ form C^k submanifolds depending C^k on ε, e. Let $x^1, \ldots, x^n$ be coordinates around x_0. We work in this chart without explicit mention.

Let

$$\Theta(x, T, \varepsilon) = F_T(x) - x \tag{1}$$

so that zeros of Θ are periodic orbits.

Let $Z = \ker DH(x_0)$ and let $\pi: \mathbf{R}^n \to Z$ be projection. Clearly we can write $\Sigma_{e,\varepsilon}$ as the graph of a map $\varphi(x, e, \varepsilon)$ defined on a neighborhood of 0 in Z.

Thus finding zeros of (1) is equivalent to finding zeros of

$$\Psi(x, T, e, \varepsilon) \equiv \pi(\Theta(\varphi(x, e, \varepsilon), T, \varepsilon))$$

called the *reduced periodicity equation.**

Now

$$D_x \Psi(x_0, T_0, e_0, 0) = \pi\left(D_x F_{T_0}(x_0) - Id\right)$$

and

$$D_T \Psi(x_0, T_0, e_0, 0) = \pi(X(x_0)) = X(x_0)$$

Notice that we have $D_x F_{T_0}(x_0) \cdot X(x_0) = X(x_0)$, that is, $X(x_0) \in \ker(D_x F_{T_0}(x_0) - Id) \cap Z$, and, by conservation of energy, $dH(x_0) \cdot X(x_0) = 0$. Also, from $H(F_{T_0}(x)) = H(x)$, we get $dH(x_0) \circ (DF_{T_0}(x_0) - Id) = 0$. By a dimension count, then (i) and (ii) imply that (i′) $\ker(D_x F_{T_0}(x_0) - Id) \cap \ker dH(x_0)$ is one dimensional spanned by $X(x_0) \neq 0$.

*This plays the same role as the bifurcation equation in bifurcation problems, where this procedure is called the Liapunov–Schmidt procedure (see, e.g., Hale [1977]).

ISBN 0-8053-0102-X

Therefore, as a map of $Z \times$(interval about T_0) to Z, condition (i′) says that $D_x\Psi$ is one-to-one on a transverse subspace to $X(x_0)$ and (ii) says that $D_T\Psi$ is independent of the range of $D_x\Psi$. Thus $D_{(x,T)}\Psi(x_0, T_0, e_0, 0)$ is an isomorphism on a transverse to $X(x_0)$ and so the result follows by the implicit function theorem. ■

We now turn to the application of this result to Hamiltonian systems. The proof will be done by the technique of "blowing up the singularity."

5.6.7 Theorem (Llapunov). *Let (P, ω) be a symplectic manifold and H be C^l, $l \geqslant 2$. Let x_0 be a critical point for H and let $X'_H(x_0)$ be the linearized Hamiltonian system. Let $X'_H(x_0)$ have characteristic exponents*

$$\lambda_1, \ldots, \lambda_n, -\lambda_1, \ldots, -\lambda_n$$

Assume

(i) $\lambda_1 = i\alpha_1$, $\alpha_1 > 0$, that is, λ_1, $-\lambda_1$ are pure imaginary,
and the non-resonance condition
(ii) no λ_j, $j = 2, \ldots, n$ is an integer multiple of λ_1.

Then there is a one-parameter family γ_ε defined for $0 < \varepsilon \leqslant \varepsilon_0$ of closed orbits of X_H that approach x_0 as $\varepsilon \to 0$ and whose periods approach $2\pi/\alpha_1$ (see Fig. 5.6-3).

Proof. We can work in $T^*\boldsymbol{R}^n$ and let $x_0 = 0$ and $H(x_0) = 0$. We blow up the singularity by letting

$$\mathcal{H}(x, \varepsilon) = \frac{1}{\varepsilon^2} H(\varepsilon x)$$
$$= H_2(x) + h(x, \varepsilon)$$

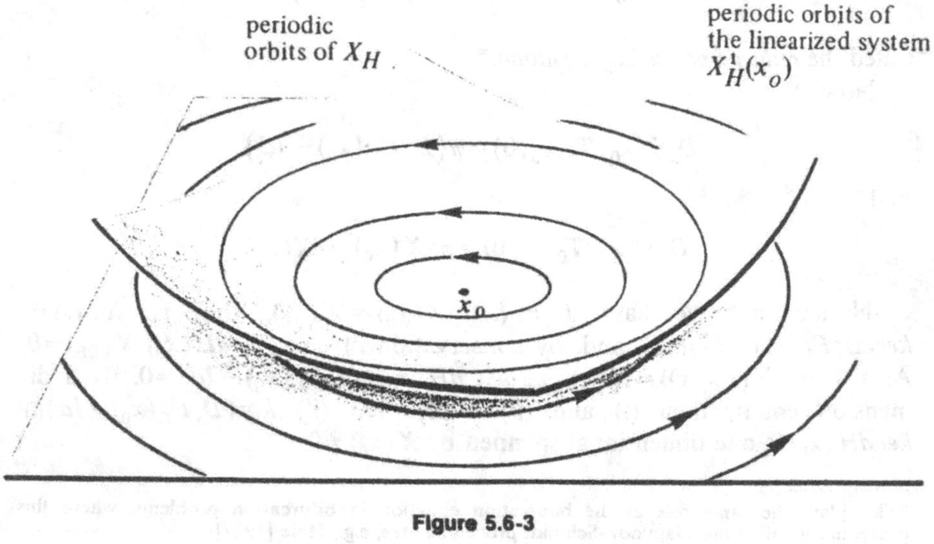

Figure 5.6-3

ISBN 0-8053-0102-X

where $H_2 = \frac{1}{2}D^2H(0)$, h is C^l, and h and its first two derivatives vanish at $x=0$; $h(x,0)=0$ and $h(x,\varepsilon)=0(\varepsilon)$ uniformly in x near x_0.

We can introduce a linear symplectic change of coordinates so that H_2 takes the form

$$H_2(q_1,\ldots,q_n,p_1,\ldots,p_n)=\tfrac{1}{2}\alpha_1(q_1^2+p_1^2)+\tilde{H}_2(q_2,\ldots,q_n,p_2,\ldots,p_n)$$

(see 3.1.2). The two corresponding subspaces are ω-orthogonal and invariant under $X'_H(x_0)=X_{H_2}$.

Let γ be the closed orbit $q_1^2+p_1^2=1$; it has period $T_0=2\pi/\alpha_1$. Choose x_0 arbitrarily on γ. We will verify conditions (i) and (ii) of 5.6.6 [where X_ε is the Hamiltonian vector field of $\mathcal{H}(x,\varepsilon)$]. For (i), note that $D_xF_{T_0}(x_0)=e^{AT_0}$, where $A=X_{H_2}(x_0)=X'_H(x_0)$. Now to prove (i) we must show that

$$\dim\ker(e^{T_0A}-Id)=2$$

We know that the eigenspace corresponding to eigenvalues $i\alpha_1$ and $-i\alpha_1$ is in $\ker(e^{T_0A}-Id)$. If $(q,p)\in\ker(e^{T_0A}-Id)$, then

$$e^{T_0A}(q,p)=(q,p)$$

so (q,p) would be on a periodic orbit for A, with period dividing T_0; that is, (q,p) would be an eigenvector of A with eigenvalue an integer multiple of α_1. [Thus $\ker(e^{tA_0}-Id)$ equals the eigenspace for $i\alpha_1$ and $-i\alpha_1$, which has dimension 2.]

As for (ii), x_0 lies in the subspace spanned by q_1,p_1, as does $Ax_0\neq 0$, since this space and its complement are invariant under A by construction; but since the space spanned by q_1,p_1 lies in the kernel of $e^{T_0A}-Id, Ax_0\in Im(e^{T_0A}-Id)$ is impossible.

Thus by 5.6.6, $\mathcal{H}(x,\varepsilon)$ has a periodic orbit for each ε sufficiently small. But a periodic orbit through $x(\varepsilon)$ for $\mathcal{H}(x,\varepsilon)$ yields a periodic orbit through $\varepsilon^2x(\varepsilon)$ for $H(x)$. (If $x(t)$ is a periodic orbit for $\mathcal{H}$, $y(t)=\varepsilon x(t)$ is one for H.) The result therefore follows. ■

This "blowing up" argument also allows one to show that the periodic orbits fill up a C^1 manifold tangent to the eigenspace corresponding to $\pm i\alpha_1$. This argument requires one to keep track of the differentiability and parametrize the manifold by polar coordinates (Duistermaat [1972]). The C^1 result is due to Kelley [1967*c*], and the analytic result is due to Siegel (see Siegel and Moser [1971]). The C^r case (if H is C^{r+2} the orbits lie on a C^r manifold) may be deduced by using the Birkhoff normal form (see below and also Moser [1976]).

The blowing up technique (also called rescaling) is a useful general tool in bifurcation theory, the theory of singularities (see below and, e.g., Takens [1971], Hale [1977], and Buchner, Marsden, and Schecter [1978]) and in singular perturbation theory.

ISBN 0-8053-0102-X

The Liapunov theorem has in it the important nonresonance condition that is related to the problem of "small divisors." (This terminology arises naturally when the power series approach is used, cf. Siegel and Moser [1971].) The blowing up technique enables one to deal with certain resonant cases, for example, the 2 : 1 resonance ($\lambda_2 = 2\lambda_1$). For a discussion and application, see Duistermaat [1972] and Cushman [1973]. The literature on these resonant cases is extensive; the reader interested may scour the bibliography; Roels [1971] and Henrard [1973] are representative.

For a study of those resonant cases the Birkhoff normal form has played a key role. We now give a brief informal discussion of it. For details and convergence questions, Siegel and Moser [1971] and Moser [1973a] should be consulted. See Deprit [1969] for an efficient algorithm.

Let $H = H_2 + H_3 + \cdots + H_n + \cdots$ be a formal power series on a symplectic vector space (V, ω), where H_n is a homogeneous polynomial of degree n. Generally, let $[f]_k$ be the homogeneous polynomial of degree k in the formal power series expansion of a function f; thus $[H]_n = H_n$.

Suppose that the linear Hamiltonian vector field X_{H_2} on (V, ω) has purely imaginary eigenvalues and can be put in the real normal form (RNF) discussed above. The formal power series H is said to be in *Birkhoff normal form* if $L_{X_{H(2)}} H = 0$ that is, $L_{X_{H(2)}} H_n = 0$ for all $n > 2$, where $H(2) = H_2$.

Formally, the flow of $X_H(x, y)$ is $exp\left(t\,ad_H\right)\begin{pmatrix} x \\ y \end{pmatrix}$, where (x, y) are symplectic co-ordinate functions on (V, ω), $ad_H f = \{f, H\}$ for $f \in C^\infty(V, \boldsymbol{R})$, and

$$exp(t\,ad_H) = \sum_{n=0}^{\infty} \frac{t^n}{n!} ad_H^n, \quad \text{where } ad_H^n = \underbrace{ad_H \circ \cdots \circ ad_H}_{n \text{ times}}$$

Note that since $ad_H x_i = \partial H/\partial y_i$ and $ad_H y_i = -\partial H/\partial x_i$, $ad_H(x,y) = X_H(x,y)$.

5.6.8 Normal form algorithm. Suppose the formal power series H is in normal form up to terms of degree n, that is, $L_{X_{H(2)}}[H]_i = 0$ for $2 < i < n-1$. We now find a homogeneous polynomial F_n of degree n such that the symplectic diffeomorphism

$$\varphi_{F_n}(x, y) = exp\left(adF_n\begin{pmatrix} x \\ y \end{pmatrix}\right) = (x, y) + X_{F_n}(x, y) + \cdots \tag{1}$$

has the property that $L_{X_{H(2)}}[\varphi_{F_n}^* H]_n = 0$, then the terms of degree n are in normal form while the terms of degree $i < n-1$ are unaffected because of (1) and hence remain in normal form. Now

$$\begin{aligned} [\varphi_{F_n}^* H]_n &= \left[H_2((x,y) + X_{F_n}(x,y))\right]_n + H_n(x,y) \\ &= dH_2(x,y) X_{F_n}(x,y) + H_n(x,y) \qquad (2) \\ &= L_{X_{H(2)}} F_n + H_n \end{aligned}$$

ISBN 0-8053-0102-X

Let $P_n = P_n(V,R)$ be the vector space of homogeneous polynomials of degree n on V and let $L_{X_{H(2)}}$ operate on P_n. Since X_{H_2} is in (RNF),

$$P_n = \ker L_{X_{H(2)}} \cap P_n \oplus \operatorname{im} L_{X_{H(2)}} \cap P_n$$

Write

$$H_n = H_n' + G_n \in \ker L_{X_{H(2)}} \cap P_n \oplus \operatorname{im} L_{X_{H(2)}} \cap P_n$$

then choose F_n so that $L_{X_{H(2)}} F_n = -G_n$. Thus

$$L_{X_{H(2)}} F_n + H_n = -G_n + H_n' + G_n = H_n' \in \ker L_{X_{H(2)}} \cap P_n$$

Hence by (2) $[\varphi_{F_n}^* H] \in \ker L_{X_{H(2)}}$. Thus $\varphi_{F_n}^* H$ is in normal form up to terms of degree n.

To see what all this means, write

$$L_{X_{H(2)}} = \sum_{i=1}^{n} \alpha_i \left(y_i \frac{\partial}{\partial x_i} - x_i \frac{\partial}{\partial y_i} \right)$$

and introduce complex conjugate coordinates $z_j = x_j + iy_j$, $\bar{z}_j = x_j - iy_j$, then

$$\frac{\partial}{\partial z_j} = \frac{1}{2} \left(\frac{\partial}{\partial x_j} - i \frac{\partial}{\partial y_j} \right)$$

and

$$\frac{\partial}{\partial \bar{z}_j} = \frac{1}{2} \left(\frac{\partial}{\partial x_j} + i \frac{\partial}{\partial y_j} \right)$$

Thus in terms of $z_j, \bar{z}_j$, $L_{X_{H(2)}}$ becomes

$$\tilde{L}_{X_{H(2)}} = \sum_{j=1}^{n} \alpha_j \left(z_j \frac{\partial}{\partial z_j} - \bar{z}_j \frac{\partial}{\partial \bar{z}_j} \right)$$

Every real homogeneous polynomial of degree m may be written

$$\sum_{m = j_1 + \cdots + j_n + l_1 + \cdots + l_n} C_{j_1 \cdots j_n l_1 \cdots l_n} z_1^{j_1} \cdots z_n^{j_n} \bar{z}_1^{l_1} \cdots \bar{z}_n^{l_n}$$

$$= \sum_{|j| + |l| = m} c_{jl} z^j \bar{z}^l$$

where $c_{jl} = \bar{c}_{lj}$. Since

$$L_{X_{H(2)}} z^k \bar{z}^l = \left(\sum_{j=1}^{n} \alpha_j (k_j - l_j) \right) z^k \bar{z}^l := \langle \alpha, k - l \rangle z^k \bar{z}^l$$

ISBN 0-8053-0102-X

$z^k\bar{z}^l \in \ker \tilde{L}_{X_{H(2)}} \cap P_m$ if and only if $\langle \alpha, k-l\rangle = 0$ and $|k|+|l| = m$. Thus $Re(z^k\bar{z}^l)$ and $Im(z^k\bar{z}^l)$ for $\langle \alpha, k-l\rangle = 0$ and $|k|+|l| = m$ span $\ker L_{X_{H(2)}} \cap P_m$.

5.6.9 Example. Take $\alpha = (1,2)$, $m = 3$, $k = (k_1, k_2)$, and $l = (l_1, l_2)$. We must find solutions of

$$\left.\begin{aligned} 0 &= \langle (1,2), (k_1 - l_1, k_2 - l_2)\rangle \\ k_1 + k_2 &+ l_1 + l_2 = 3 \end{aligned}\right\} \tag{3}$$

Since $0 \leq k_1, k_2, l_1, l_2 \leq 3$, $0 \leq |k_1 - l_1|$, $|k_2 - l_2| \leq 3$; thus,

$$k_1 - l_1 = 2, \qquad k_2 - l_2 = -1 \tag{4}$$

$$k_1 - l_1 = -2, \qquad k_2 - l_2 = 1 \tag{5}$$

are the only possible solutions of the first equation of (3). For (4) the only possibility is $k_1 = 2$, $l_1 = 0$, $k_2 = 0$, $l_2 = 1$ while for (5) the only possibility is $k_1 = 0$, $l_1 = 2$, $k_2 = 1$, $l_2 = 0$.

Thus $Re\, z_1^2\bar{z}_2 = x_2(x_1^2 - y_1^2) + 2x_1y_1y_2$ and $Im\, z_1^2\bar{z}_2 = 2x_1y_1x_2 - y_2(x_1^2 - y_1^2)$ span the kernel of $L_{X_{H(2)}}$ in P_3. This gives the following normal form on $T^*\boldsymbol{R}^2$.

$$H(x_1, x_2, y_1, y_2) = \tfrac{1}{2}(x_1^2 + y_1^2) + x_2^2 + y_2^2 + c_1x_2(x_1^2 - y_1^2)$$

$$+ 2c_1x_1y_1y_2 + 2c_2x_1y_1x_2 - c_2y_2(x_1^2 - y_1^2) + (\text{higher-order terms})$$

This is a 2 : 1 resonance and the cubic terms must be taken into account in the blowing up technique. For details, see Duistermaat [1972], and for a study of the topology of the problem, see Cushman [1975].

If $\alpha_1, \ldots, \alpha_n$ are independent over $\boldsymbol{Z}$, that is, if $a_1\alpha_1 + \cdots + a_n\alpha_n = 0$ for integers a_i implies $a_i = 0$, $i = 1, \ldots, n$, then arguing as above, we see that each H_k is a function only of the products $z_i\bar{z}_i$, $i = 1, \ldots, n$. This is the form for the Birkhoff normal form that is usually given. The condition that $\alpha_1, \ldots, \alpha_n$ are independent over $\boldsymbol{Z}$ clearly implies that there are no resonances (i.e., on each energy surface for the linearized problem there are exactly n periodic orbits).

The difficulties with the resonances might make one think that the situation in the presence of general resonances is hopelessly complicated. Actually this is not so. Already Seifert [1947] showed that for the motion of a particle about equilibrium in the presence of a potential with a nondegenerate minimum there (see 5.6.1), that there is at least one periodic orbit on each energy surface near equilibrium. In a sense this is a problem in geometry because one can reduce it to a problem in geodesics using the Jacobi metric (see Sect. 3.7). There are, naturally, a few complications. However, the key thing is that no non-resonance assumptions are made.

The general case is due to A. Weinstein.

ISBN 0-8053-0102-X

5.6.10 Theorem (A. Weinstein). *Let (P, ω) be a symplectic manifold and $H: P \to R$ be C^2. Let x_0 be a critical point of H and let $D^2H(x_0)$ be positive-definite there (so, by 5.6.3, the linearized system has at least n periodic orbits on each energy surface).*

Then on any energy surface $H^{-1}(H(x_0)+\varepsilon^2)$ for ε sufficiently small there are at least n periodic orbits of X_H whose periods are close to that of the linearized system.

The original proof of Weinstein [1973a, b] uses algebraic topology (Liusternick–Schnirelman category) to estimate the number of closed orbits. Moser [1976] gave a proof based on Hamilton's variational principle and bifurcation theory (blowing up!) and the category result. Finally, Weinstein [1978*a*] gave another yet simpler and more general proof using the variational argument more fully and ideas on perturbation of critical manifolds due to Bott [1954] and Reeken [1973]. See also Bottkol [1977] and Rabinowitz [1978b].

In general, examples (Duistermaat [1972], Schmidt and Sweet [1973], Moser [1976]) show that without the resonance condition, the periodic orbits do not lie on manifolds through x_0, but only on cones.

There is a global result due to Weinstein [1978*b*] that is closely related to 5.6.10. Namely, if $H: T^*R^n \to R$ and Σ_e is a regular energy surface of H that is the boundary of a compact convex region (or if Σ_e is close to such a surface), then there are n periodic orbits of X_H on Σ_e. This is related to 5.6.10 through the fact that for ε small, $H^{-1}(H(x_0)+\varepsilon^2)$ are approximate ellipsoids by the Morse lemma. Some related global results are found in Chow and Mallet–Paret [1977], Chow, Mallet–Paret, and Yorke [1977] and Rabinowitz [1978a].

We conclude this section with a brief account of nonlinear oscillations in a more general setting. One can imagine Hamiltonian systems as pivotal points in the family of all dynamical systems. We can perturb them in purely Hamiltonian ways (Chapter 8) or we can destroy their Hamiltonian character by forcing them or adding on dissipative terms. The latter moves the system into the context of Chapter 7.

We shall confine ourselves to a few remarks and examples on the aspect of nonlinear oscillations connected with bifurcation theory. Bifurcation theory for zeros of *maps* is a well-developed subject, but is not our concern here (see, e.g., Hale [1977]). We are concerned with dynamic bifurcation theory. We consider vector fields X_λ on a manifold M depending on a parameter $\lambda \in R^p$. As λ changes, the dynamical system X changes; if a qualitative change in the flow F_t^λ occurs at $\lambda = \lambda_0$, we call λ_0 a *bifurcation point*.

We begin by describing the simplest bifurcations for *one*-parameter systems, that is, $p=1$. In a certain sense these bifurcations are the generic local ones. (See Chapter 7 for details.) If one imposes a symmetry, however, what is generic may change, as we shall explain.

5.6.11 Saddle Node. This is a bifurcation of fixed points; a saddle and a sink come together and annihilate one another, as shown in Fig. 5.6-4. A

ISBN 0-8053-0102-X

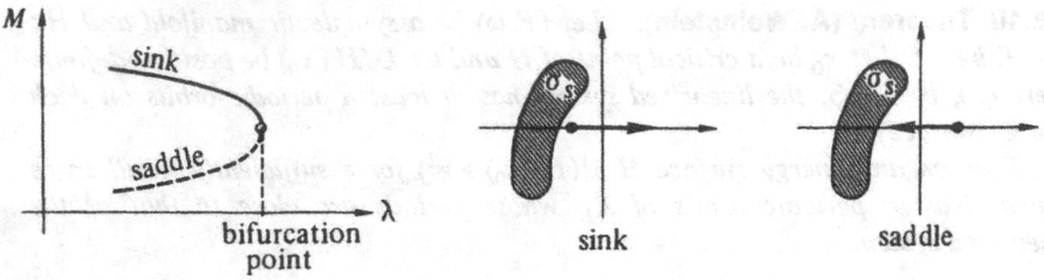

saddle-node bifurcation eigenvalue evolution

Figure 5.6-4

simple real eigenvalue of the sink crosses the imaginary axis at the moment of bifurcation; one for the saddle crosses in the opposite direction. The symmetric situation of a saddle source is also possible.

If an axis of symmetry is present, then a symmetric bifurcation can occur, as in Fig. 5.6-5. This occurs in Euler buckling for instance (Zeeman [1975]). A small asymmetric perturbation or imperfection "unfolds" this into a simple nonbifurcating path and a saddle node.

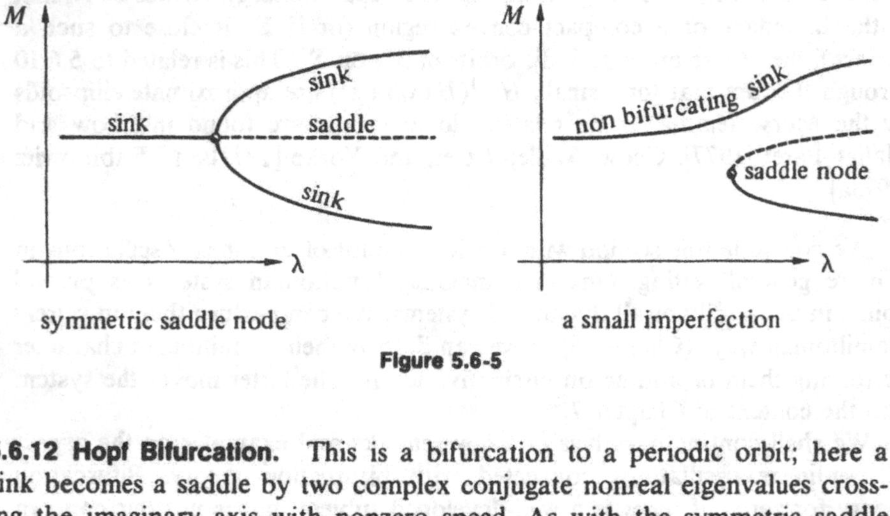

symmetric saddle node a small imperfection

Figure 5.6-5

5.6.12 Hopf Bifurcation. This is a bifurcation to a periodic orbit; here a sink becomes a saddle by two complex conjugate nonreal eigenvalues crossing the imaginary axis with nonzero speed. As with the symmetric saddle node, the bifurcation can be sub- (unstable closed orbits) or super- (stable closed orbits) critical. (See Marsden and McCracken [1976] for calculations to determine which one.) Figure 5.6-6 depicts the attracting case.

The precise statement and proof of the Hopf theorem may be given using the same blowing-up technique we used to prove Liapunov's theorem. (In fact, the latter may be deduced from Hopf's theorem; see Alexander and Yorke [1977] and Schmidt [1976].)

The physical idea behind the Hopf theorem is that as λ increases the system is excited, and beyond a certain critical point excitations and dissipation balance to enable self-sustaining oscillations to be supported.

ISBN 0-8053-0102-X

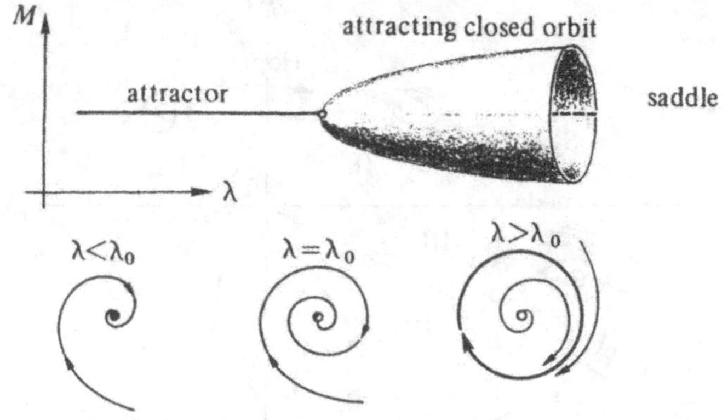

Figure 5.6-6. The Hopf bifurcation.

These two bifurcations are local in the sense that they can be detected by linearization about a fixed point. There are, however, some *global* bifurcations that are more difficult to detect. A *saddle connection* is shown in Fig. 5.6-7. In this example note that *at* bifurcation the system may be Hamiltonian, as in 5.6.5(b).

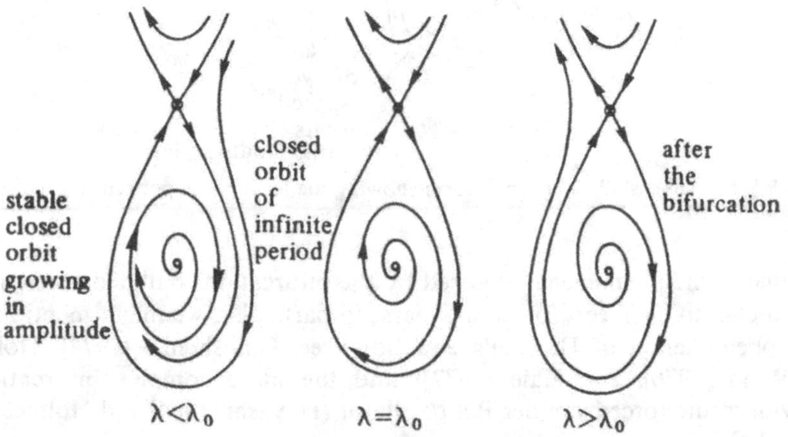

Figure 5.6-7. Bifurcation point.

These global bifurcations can occur as *part* of local bifurcations of systems with additional parameters. This approach has been developed by Takens [1973a] who has classified certain generic or stable bifurcations of *two-parameter* families of vector fields on the plane. This is an outgrowth of extensive work of the Russian school led by Andronov. An example of one of Takens' bifurcations with a symmetry imposed and depending on parameters λ_1, λ_2 is shown in Fig. 5.6-8.

For an example of a physical system (panel flutter) in which the bifurcation illustrated in Fig. 5.6-8 occurs, see Holmes [1977] and Holmes and Marsden [1977a, 1977b].

ISBN 0-8053-0102-X

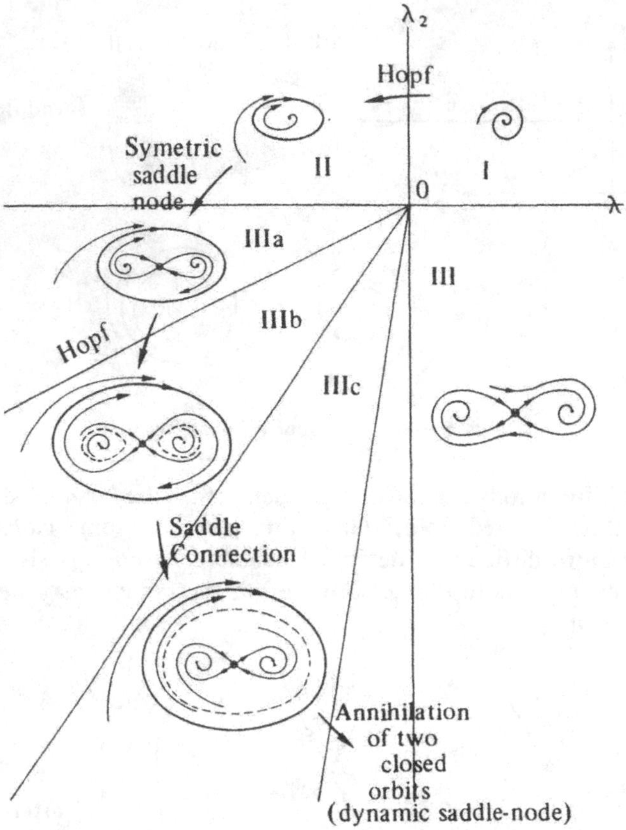

Figure 5.6-8. Takens' (2,—) normal form showing the local phase portrait in each region of parameter space.

Some of the phenomena captured by the bifurcations outlined above have been known to engineers for many years. In particular, we might mention the jump phenomenon of Duffing's equation (see Timoshenko [1974], Holmes and Rand [1976], and Hale [1977]) and the more complex bifurcational behavior of the forced van der Pol oscillator (Hayashi [1964] and Holmes and Rand [1977]).

ISBN 0-8053-0102-X

PART 3

AN OUTLINE OF QUALITATIVE DYNAMICS

The qualitative theory of flows has evolved into a vast subject since the early work of Poincaré, Liapunov, and Hadamard in the 1880s. Its offspring—topological dynamics, differentiable dynamics, and Hamiltonian dynamics—have become huge domains on their own, increasingly disjoint since the pioneering days of Birkhoff in the 1920s.

This part is concerned with the qualitative theory of dynamical systems, emphasizing the Hamiltonian case. There is a change in style from Part 2 necessitated by the volume of material covered. While Part 2 included most proofs and was reasonably complete, this part will often refer the reader to other sources for proofs. The qualitative theory culminates in the Hamiltonian theory where we describe some important recent results.

ISBN 0-8053-0102-X

PART 3

AN OUTLINE OF QUALITATIVE DYNAMICS

The qualitative theory of flows has evolved into a vast subject since the early works of Poincaré, Lyapunov, and Hadamard in the 1880s. Its offspring—topological dynamics, differentiable dynamics, and Hamiltonian systems—have become large domains in their own, increasingly distant since the pioneering days of Birkhoff in the 1920s.

This part is concerned with the qualitative theory of dynamical systems, emphasizing the Hamiltonian case. There is a change in style from Part 2 necessitated by the volume of material covered. While Part 2 included most proofs and was reasonably complete, this part will often refer the reader to other sources for proofs. The qualitative theory culminates in the Hamiltonian theory where we describe some important recent results.

CHAPTER 6

Topological Dynamics

This short chapter provides a backdrop for the next two chapters. We have deliberately set off this material so that the distinction between the topological and differentiable aspects of dynamical systems is clearly maintained. Here we collect only the small number of topics in the topological dynamics of continuous flows that are used in the sequel.

6.1 LIMIT AND MINIMAL SETS

Suppose X is a vector field on a manifold M. An *orbit* of X is the image of a maximal integral curve of X. The conception of M decomposed into orbits (perhaps oriented with arrowheads) is known as the *phase portrait* of X. In the topological theory the emphasis is on topological properties of the flow F and so one usually starts with a continuous flow on a (Hausdorff) topological space M. However, to keep in mind the fact that the flows we are concerned with usually come from vector fields X, we shall still regard X as given.

An original program in the qualitative theory of Poincaré is the classification of all phase portraits on a given manifold, perhaps up to equivalence under homeomorphisms of M preserving (oriented) orbits. Although some results have been obtained (see Kneser [1924], Whitney [1933], Kaplan [1940],

ISBN 0-8053-0102-X

Ralph Abraham and Jerrold E. Marsden, Foundation of Mechanics, Second Edition and Markus [1954]) in the two-dimensional case, it became clear rather early that this program was too ambitious.

A more modest program, initiated by Birkhoff in 1912, aimed at the classification of only the "final motions" in the phase portrait: the asymptotic behavior as time goes to infinity. The fundamental concepts of this program —limit and minimal sets, recurrent and almost-periodic motions— have been central to topological dynamics ever since. For further information in this area, Sell [1971] is strongly recommended.

In this section we give the basic properties of limit and minimal sets, and illustrate their role in the qualitative theory by the Poincaré–Bendixson–Schwartz theorem for the two-dimensional case.

6.1.1 Definition. *Suppose X is a vector field on the manifold M with integral $F\colon \mathfrak{D}_X \subset M \times \boldsymbol{R} \to M$. Then the λ^σ* ***limit set*** *of m, $\sigma = +, -$,* or $\pm$, *is defined by*

$$\lambda^+(m) = \bigcap_{n \in \boldsymbol{Z}} cl\left\{ F\left[(\{m\} \times (n, +\infty)) \cap \mathfrak{D}_X \right] \right\}$$

$$\lambda^-(m) = \bigcap_{n \in \boldsymbol{Z}} cl\left\{ F\left[(\{m\} \times (-\infty, n)) \cap \mathfrak{D}_X \right] \right\}$$

and

$$\lambda^\pm(m) = \lambda^+(m) \cup \lambda^-(m)$$

where $\boldsymbol{Z}$ denotes the integers. Also, let $\Lambda_X^\sigma = \cup \{\lambda^\sigma(m) | m \in M\}$, $\sigma = +, -, \pm$, and $\Lambda_X = \Lambda_X^\pm$.

The λ^+ (resp. λ^-) limit set is sometimes called the ω (resp. α) limit set.

6.1.2 Proposition. *If m is σ complete, the limit set $\lambda^\sigma(m)$ is the set of points $m_0 \in M$ for which there exists a sequence $\{t_n\}$ with $t_n \to \sigma\infty$ and $F(m, t_n) \to m_0$, for $\sigma = +$ or $\sigma = -$. If m is not σ complete, then $\lambda^\sigma(m)$ is empty for $\sigma = +$ or $\sigma = -$.*

Proof. Let $A_n = F[\{m\} \times (n, \infty)]$. Then by definition, $m_0 \in \lambda^+(m)$ iff $m_0 \in cl(A_n)$ for all $n \in \boldsymbol{Z}$. Let (U_n) be a sequence of open neighborhoods of m_0 such that $\cap_{n=1}^{\infty} U_n = \{m_0\}$. As $A_n \supset A_{n'}$ for $n < n'$, $m_0 \in cl(A_n)$ for all $n \in \boldsymbol{Z}$ iff $U_n \cap A_n = \phi$ for all $n \in \boldsymbol{Z}$. If $m_0 \in \lambda^+(m)$, choose $t_n \in (n, \infty)$ such that $F(m, t_n) \in U_n$. Then $t_n > n$, so t_n tends to ∞ as n goes to ∞, and $F(m, t_n)$ converges to m_0. The converse and the case $\sigma = -$ are similar. The second assertion is obvious. ■

For example, Fig. 6.1-1 illustrates a possible situation on $\boldsymbol{R}^2$.

6.1.3 Definition. *Let X be a complete vector field on a manifold M with flow F. A subset S of M is called* ***positively invariant*** *iff $F_t(S) \subset S$ for all $t > 0$ and*

ISBN 0-8053-0102-X

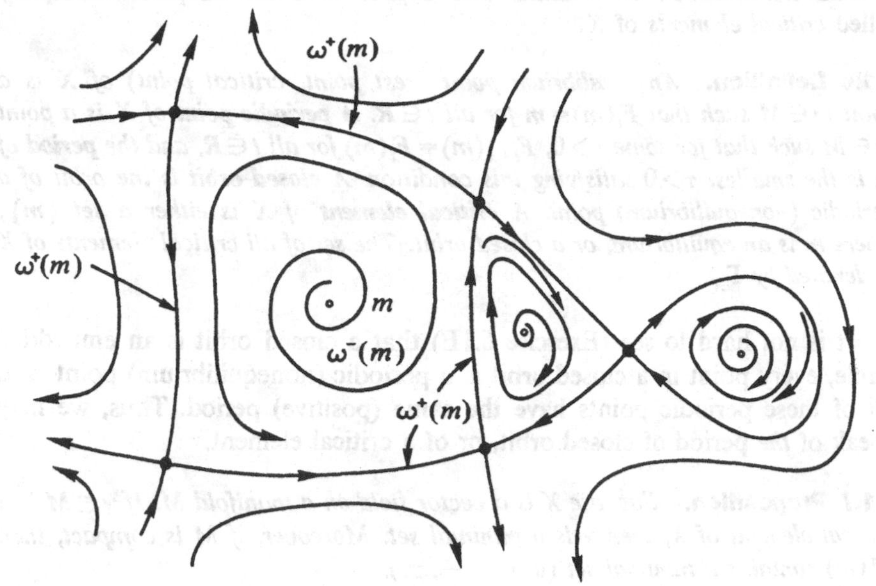

Figure 6.1-1

negatively invariant *iff $F_t(S) \subset S$ for all $t \leqslant 0$. S is called* ***invariant*** *if it is both positively and negatively invariant (cf. 2.2.28). A subset S of M is a* ***minimal set*** *iff S is closed, nonempty, invariant, and no proper subset of S has these properties.*

6.1.4 Proposition. *If $X \in \mathfrak{X}(M)$ is complete and S is a minimal set of X, then S is connected. If S and T are minimal sets, then either $S = T$ or $S \cap T = \varnothing$.*

Proof. Suppose S is not connected. Then components of S are closed, nonempty, and invariant, which contradicts minimality of S. If S and T are minimal sets and $S \cap T \neq \varnothing$, then $S \cap T$ is closed, nonempty, and invariant. By minimality, $S \cap T = S = T$. ■

6.1.5 Proposition. *Let X be a vector field on a compact manifold M. Then $\lambda^\sigma(m)$ is closed, nonempty, connected, and invariant ($\sigma = +$ or $\sigma = -$).*

Proof. It is clear that $\lambda^\sigma(m)$ is closed. To show that it is invariant, let $m_0 \in \lambda^\sigma(m)$ and $m_t = F_t(m_0)$. Suppose $F_{t_i}(m) \to m_0$. Then $F_{t_i+t}(m) \to m_t$. Thus $m_t \in \lambda^\sigma(m)$. Connectedness of $\lambda^\sigma(m)$ follows at once from the fact that in a compact Hausdorff space, a nested sequence of closed connected sets has connected intersection (see Sect. 1.1.7). ■

ISBN 0-8053-0102-X

The classification of minimal sets begins with the simplest examples, called *critical elements* of X.

6.1.6 Definition. *An **equilibrium point** (rest point, critical point) of X is a point $m \in M$ such that $F_t(m) = m$ for all $t \in \mathbf{R}$. A **periodic point** of X is a point $m \in M$ such that for some $\tau > 0$, $F_{t+\tau}(m) = F_t(m)$ for all $t \in \mathbf{R}$, and the **period** of m is the smallest $\tau > 0$ satisfying this condition. A **closed orbit** is the orbit of a periodic (nonequilibrium) point. A **critical element** of X is either a set $\{m\}$, where m is an equilibrium, or a closed orbit. The set of all critical elements of X is denoted by Γ_X.*

It is not hard to see (Exercise 6.1E) that a closed orbit is an embedded circle, every point in a closed orbit is a periodic (nonequilibrium) point, and all of these periodic points have the same (positive) period. Thus, we may speak of *the* period of closed orbit, or of a critical element.

6.1.7 Proposition. *Suppose X is a vector field on a manifold M. If $\gamma \subset M$ is a critical element of X, then γ is a minimal set. Moreover, if M is compact, then $\lambda^\sigma(m)$ contains a minimal set ($\sigma = +, -, \pm$).*

Proof. Obviously $\{m\}$ is minimal if m is a critical point. If γ is a closed orbit, then it is closed, invariant, and nonempty. In addition, γ is minimal, for if $m_1, m_2 \in \gamma$ there is a t such that $F_t(m_1) = m_2$. Also, if M is compact, $\lambda^\sigma(m)$ is nonempty, closed, and invariant by 6.1.5, hence contains a minimal set. ■

An important theorem on minimal sets in the two-dimensional case is the following.

6.1.8 Theorem (A. Schwartz). *Suppose M is a compact, two-dimensional manifold and $X \in \mathfrak{X}(M)$. Let A be a minimal set of X that is nowhere dense. Then $A \in \Gamma_X$; that is, A is a critical point or a closed orbit.*

For a proof see A. Schwartz [1963] or Hartman [1973, p. 185].

This result, and the next two, assume that the vector field is smooth (actually C^2), and belong to *differentiable* dynamics. However, the conclusions are topological. Moreover, Hajek [1968] shows that the closely related Poincaré–Bendixon theory in $\mathbf{R}^2$ follows using only C^0 hypotheses. We have, therefore, placed these C^2 results here. For a C^1 counterexample, see Denjoy [1932].

6.1.9 Corollary. *Let M be a compact, connected, two-dimensional manifold, $X \in \mathfrak{X}(M)$ and A a minimal set of X. Then either*

(*i*) *A is a critical point;*
(*ii*) *A is a closed orbit;*
(*iii*) *$A = M$ and $M = T^2 = S^1 \times S^1$.*

ISBN 0-8053-0102-X

Here is the method of proof. Suppose (i) and (ii) do not hold. Then, by 6.1.8, the interior of A is nonempty, that is, $int(A) \neq \varnothing$. Also $int(A)$ is invariant, as F_t is a homeomorphism. Thus, $bd(A)$ is closed and invariant. By minimality of A, $bd(A) = \varnothing$. Hence A is both open and closed and so $A = M$. As A is minimal, it contains no critical points or closed orbits (6.1.7) and, by a theorem of Kneser [1924], $M = T^2$.

The next result shows that, in two dimensions, limit sets are usually tori or closed orbits.

6.1.10 Theorem (Poincaré–Bendixson–Schwartz). *Let M be a compact, connected, orientable two manifold and $X \in \mathfrak{X}(M)$. For $m \in M$ suppose $\lambda^+(m)$ contains no critical points. Then either*

(i) $\lambda^+(m) = M = T^2$; *or*
(ii) $\lambda^+(m) = \gamma$ *is a closed orbit.*

The idea of the proof is as follows (see Schwartz [1963]). By 6.1.7 $\lambda^+(m)$ contains a minimal set, so by 6.1.9 either $\lambda^+(m) = M = T^2$ or $\lambda^+(m)$ contains a closed orbit. Then by a geometric argument special to the two-dimensional case, one finds that in fact $\lambda^+(m) = \gamma$.

For further details on Poincaré–Bendixson theory see Hartman [1973].

EXERCISES

6.1A Prove the converse in 6.1.2.

6.1B Construct an example of a vector field on $\mathbb{R}^2$ in which only one point is contained in a minimal set, and another in which every limit set is empty.

6.1C Discuss 6.1.8 in the case M is not orientable.

6.1D Prove that a closed orbit is an embedded circle, or periodic points of a common period.

6.2 RECURRENCE

Many different notions of recurrent or almost-periodic motion have been explored in topological dynamics. Here we collect some of those needed in the differentiable theory. One of these recurrence results, the Poincaré recurrence theorem, was already given in Exercise 3.4F and related ideas of ergodicity were discussed in Sects. 2.6 and 3.7.

6.2.1 Definition. *If $X \in \mathfrak{X}(M)$ with integral $F: \mathfrak{D}_X \subset M \times \mathbb{R} \to M$ and $m \in M$, then m is a **nonwandering point** of X iff $(m, t) \in \mathfrak{D}_X$ for all $t > 0$, and for all neighborhoods U of $m \in M$ and all $t_0 > 0$ there is a $t > t_0$ such that $U \cap F_t(U)$ is nonempty. Let Ω_X denote the set of all nonwandering points of $X \in \mathfrak{X}(M)$.*

ISBN 0-8053-0102-X

6.2.2 Proposition. *If $X \in \mathfrak{X}(M)$ and is + complete, then $\Omega_X \subset M$ is closed, invariant, and $cl(\Gamma_X) \subset cl(\Lambda_X) \subset \Omega_X$.*

Proof. Let $m_n \in \Omega_X$ be a sequence and $m_n \to m$ as $n \to \infty$. Let U be a neighborhood of m and $t_0 > 0$. Choose N so that $m_n \in U$ if $n > N$. Then since $m_N \in \Omega_X$ and $m_N \in U$, there is a t such that $U \cap F_t(U) \neq \varnothing$. Hence $m \in \Omega_X$. Thus Ω_X is closed.

Let $m \in \Omega_X$ and $t_1 \in \boldsymbol{R}$. To show that $m_1 = F_{t_1} m \in \Omega_X$, let U be a neighborhood of m_1 and $t_0 > 0$. Then $F_{-t_1}(U) = U_1$ is a neighborhood of m and so there is a $t > t_0$ such that $U_1 \cap F_t(U_1) \neq \varnothing$. Hence $\varnothing \neq F_{t_1}(U_1 \cap F_t(U_1)) = F_{t_1}(U_1) \cap F_t(F_{t_1}(U_1) = U \cap F_t(U)$. Thus $m_1 \in \Omega_X$ and so Ω_X is invariant.

It is obvious that $\Gamma_X \subset \Lambda_X$ and hence that $cl(\Gamma_X) \subset cl(\Lambda_X)$. It remains to show that $\Lambda_X^\sigma \subset \Omega_X$. Let $m_0 \in \Lambda_X^\sigma$ so $m_0 = lim_{n \to \infty} F_{t_n} m$ for some $m \in M$ and $t_n \to \sigma\infty$. Let U be a neighborhood of m_0 and $t_0 > 0$. But there is an N such that $F_{t_n} m \in U$ if $n \geqslant N$. Let $\sigma = +$. Since $t_n \to +\infty$, we can choose $n_1 > n_2 \geqslant N$ such that $t = t_{n_1} - t_{n_2} > t_0$. Then $F_{t_{n_1}} m \in U \cap F_t(U)$ since $F_{t_{n_1}} m = F_t F_{t_{n_2}} m$. The case $\sigma = -$ is similar. ■

6.2.3 Definition. *For $X \in \mathfrak{X}(M)$, recall that a point $m \in M$ is **complete** if it is both + and − complete (i.e., its orbit is defined for all time). If m is complete, the **hull** of m is the orbit closure, $H(m) = cl\{F_t(m) | t \in \boldsymbol{R}\}$. A point $m \in M$ is called **compact** if it is complete, and $H(m)$ is a compact set. We say $m \in M$ is a **recurrent point** of X if it is complete, and for all neighborhoods U of $m \in M$,*

$$h_U = \{t \in \boldsymbol{R} | F(m,t) \in U\}$$

*is **relatively dense** in $\boldsymbol{R}$: that is, there is a (large) $\tau > 0$ such that for all $\alpha \in \boldsymbol{R}$, $[\alpha, \alpha + \tau] \cap h_U \neq \varnothing$. Let R_X denote the set of all recurrent points.*

The connection between recurrence and minimal sets is this classical result of Birkhoff [1966, p. 199].

6.2.4 Birkhoff Recurrence Theorem. (*i*) *A set $H \subset M$ is a compact minimal set of $X \in \mathfrak{X}(M)$ iff $H = H(m)$ for some compact recurrent point $m \in M$.*

(*ii*) *If $H \subset M$ is a compact minimal set and $m \in H$, then m is a compact recurrent point.*

Proof. (i) Suppose H is compact and minimal. Let $m \in H$. We will show that m is compact, recurrent, and $H = H(m)$. Since H is compact and invariant, the orbit through m lies a priori in H and so is complete (see Sect. 2.1). By invariance of H, $H(m) \subset H$ and so $H(m)$ is compact. Since H is minimal, $H(m) = H$. It remains to show that m is recurrent. If not, there would be a neighborhood of m such that h_U is not relatively dense in $\boldsymbol{R}$. Then for any n there is an $\alpha_n \in \boldsymbol{R}$ such that $[\alpha_n, \alpha_n + n] \cap h_U = \varnothing$; that is, $F_t(m) \notin U$

ISBN 0-8053-0102-X

for all $\alpha_n \leq t \leq \alpha_n + n$. Now, by compactness, $F_{\alpha_n+n/2}(m) = m_n$ has a convergent subsequence, converging to, say, m_0. Clearly, $F_t(m_n) \notin U$ for $t \in [-n/4, n/4]$ and, letting $n \to \infty$, we see that the whole orbit of m_0 does not meet U. This contradicts the fact that H is minimal.

Conversely, suppose m is compact and recurrent. Then, clearly, $H = H(m)$ is compact and invariant. Suppose H were not minimal. Then there is a closed invariant set $B \subsetneq H$. If $m \in B$, the orbit of m would belong to B and so, as B is closed, $H(m) \subset B$, which is impossible if $B \neq H(m)$. Thus $m \notin B$. Let U, V be disjoint open neighborhoods of m and B, respectively. Since $B \subset H$, there exists a sequence $t_n \to \infty$ (or $-\infty$) such that $F_{t_n}(m) \to m_0 \in B$. Since the orbit of m_0 lies in B, for t_n sufficiently large $F_{t_n}(m) \in V$ and $F_t(F_{t_n}(m)) \in V$ for all $t \in [-\beta_n, \beta_n]$, where $\beta_n \to \infty$. But from recurrence, there is a $\tau > 0$ such that $F_t(F_{t_n}(m)) \in U$ for some t in any interval of length τ. This is a contradiction for n large enough.

Part (ii) is contained in the first part of the proof of (i). ■

This argument also yields the following, which can be taken as a characterization of minimal sets (cf. Gottshalk–Hedlund [1955]).

6.2.5 Corollary. *A nonempty set $H \subset M$ is minimal if and only if $H(m) = H$ for every $m \in H$.*

The different notions of recurrence are related as follows.

6.2.6 Proposition. *If $X \in \mathfrak{X}(M)$, $\Gamma_X \subset R_X \subset \Omega_X$.*

We leave the proof as an exercise.

The literature of topological dynamics is replete with further notions of recurrence, which fit between Γ_X and Ω_X in this scheme.

EXERCISES

6.2A. (a) If m is recurrent, show that $\lambda^+(m) = \lambda^-(m)$.
(b) If m is recurrent, show that it is *pseudorecurrent*:

$$m \in \lambda^+(m) \cap \lambda^-(m)$$

6.2B. Prove 6.2.6.

6.3 STABILITY

There are many different notions of stability of an orbit of a vector field. In this section we give a unified definition of several of these in terms of continuity of set valued mappings.

Throughout this section we suppose that for a manifold M we have chosen a metric ρ, and let $\bar{\rho}$ denote the Hausdorff pseudometric on 2^M induced by ρ (see Sect. 1.1).

ISBN 0-8053-0102-X

As the topology on the subset of compact subsets in 2^M induced in this way is independent of ρ, the definitions that follow are indifferent to the choice of ρ, if M is compact.

6.3.1 Notation. *Let M be a manifold and X a vector field on M, with integral $F\colon \mathfrak{D}_X \subset M \times \mathbf{R} \to M$. For $(m,t) \in \mathfrak{D}_X$ let $m_t = F(m,t)$. Then for each $m \in M$, let*

$$m_+ = \cup\{m_t | (m,t) \in \mathfrak{D}_X, t \geqslant 0\}, \quad \textit{the + orbit of } m$$

$$m_- = \cup\{m_t | (m,t) \in \mathfrak{D}_X, t \leqslant 0\}, \quad \textit{the − orbit of } m$$

$$m_\pm = m_+ \cup m_-, \quad \textit{the full orbit of } m$$

These will be denoted m_σ, where σ can be $+$, $-$, or $\pm$.

In a similar way, if $\Theta\colon U \subset M \to M$ is a diffeomorphism onto $\Theta(U)$, let $\mathfrak{D}_\Theta \subset U \times \mathbf{Z}$ be the set of points (u,n) such that $\Theta^n(u)$ is defined, where $\Theta^n(u) = \Theta \circ \cdots \circ \Theta(u)$ (n factors) for $n > 0$, Θ^0 the identity, and $\Theta^n = (\Theta^{-1})^{-n}$ if $n < 0$. Let $F_\Theta\colon \mathfrak{D}_\Theta \to M\colon (u,n) \to \Theta^n(u)$, and $u_n = F_\Theta(u,n)$ for $(u,n) \in \mathfrak{D}_\Theta$. Then define u_σ as above, and $\lambda^\sigma(u)$ analogous to 6.1.1.

In either case above, we define

$$S^\sigma(m) = \{m' \in M | \lambda^\sigma(m') \subset cl(m_\sigma)\}$$

and

$$A^\sigma(m) = \left\{m' \in M | m' \text{ is } \sigma \text{ complete and } \lim_{t \to \sigma\infty} \rho(m_t, m'_t) = 0\right\}$$

if m is σ complete, and $A^\sigma(m) = \{m\}$ otherwise. $S^+(m)$ and $S^-(m)$ are known as the ***inset and outset*** *of m, respectively.*

We also let 2_1^M denote 2^M (the set of all subsets of M) with the Hausdorff topology, and 2_0^M denote 2^M with the discrete topology.

Then eighteen notions of stability of orbits may be defined as follows.

6.3.2 Definition. *Let M be a manifold with $X \in \mathfrak{X}(M)$, or $\Theta\colon U \subset M \to M$ a diffeomorphism onto $\Theta(U)$. Then $m \in M$ is called* ***α^σ-stable with respect to X****, or Θ, where $\alpha = o$, a, or L and $\sigma = +$, $-$, or $\pm$, iff*

(*i*) *$\alpha = o$* ***(orbital stability of Birkhoff)***
$\sigma\colon M \to 2_1^M$; $m' \mapsto m'_\sigma$ is continuous at m;

(*ii*) *$\alpha = a$* ***(asymptotic stability of Poisson)***
$S^\sigma\colon M \to 2_0^M$; $m' \mapsto S^\sigma(m')$ is continuous at m;

(*iii*) *$\alpha = L$* ***(Liapounov stability)***
$A^\sigma\colon M \to 2_0^M$; $m' \mapsto A^\sigma(m')$ is continuous at m.

If m is not α^σ-stable with respect to X (or Θ) we say m is ***α^σ-unstable*** *with respect to X (or Θ).*

ISBN 0-8053-0102-X

Perhaps more familiar is the following equivalent form.

6.3.3 Proposition. *Let M be a manifold, $X \in \mathfrak{X}(M)$ [or Θ: $U \subset M \to M$ a diffeomorphism onto $\Theta(U)$] and $m \in M$ be σ complete. Then m is α^σ-stable iff for every $\varepsilon > 0$ there is a $\delta > 0$ so that $\rho(m', m) < \delta$ implies*

(i) $\alpha = o$; $\bar{\rho}(m_\sigma, m'_\sigma) < \varepsilon$ ($\bar{\rho}$ *is the Hausdorff metric*);

(ii) $\alpha = a$; $\lim_{t \to \sigma\infty} \rho(m_\sigma, m'_t) = 0$;

(iii) $\alpha = L$; $\lim_{t \to \sigma\infty} \rho(m_t, m'_t) = 0$.

This follows directly from the definitions above and the definition of continuity. For other forms of these definitions and additional types of stability of orbits, see Coppel [1965]. The three cases $\alpha = o, a, L$ for a vector field and $\sigma = +$ are illustrated in Fig. 6.3-1.

6.3.4 Proposition. *Under the conditions of* 6.3.3, *if m is L^σ-stable, then m is a^σ-stable.*

The first part is clear and the second follows easily by continuity of the flow of X (or Θ).

These conditions simplify if m is a rest point, and in this case m is L^σ-stable iff m is a^σ-stable, but of course o^σ-stable remains weaker. (For the eigenvalue conditions for L^σ stability, see 2.1.25.)

Another notion of stability is attraction.

6.3.5 Definition. *A subset $A \subset M$ is an **attractor** of a complete vector field $X \in \mathfrak{X}(M)$ if it is closed, invariant, and has an open neighborhood $U_0 \subset M$ that is (i) positively invariant, and (ii) for each open neighborhood V of A ($A \subset V \subset U_0 \subset M$) there is $\tau > 0$ such that $U_t = F_t(U_0) \subset V$ for all $t \geqslant \tau$. An attractor $A \subset M$ is **stable** if for every neighborhood U_0 of $A \subset M$ there is a neighborhood V of $A \subset M$ such that $V_t \subset U_0$ for all $t \geqslant 0$. If $A \subset M$ is an attractor, the **basin** of A is the union of all open neighborhoods of A satisfying (i) and (ii) above.*

6.3.6 Examples (a) Figure 6.3-2 illustrates a rest point that is an attractor in $\boldsymbol{R}^2$ but is not stable. Its basin is all of $\boldsymbol{R}^2$.

(b) Consider the flow of

$$\ddot{u} + \dot{u} - u + u^3 = 0$$

in $\boldsymbol{R}^2$. There are three rest points at $(0,0), (\pm 1, 0)$. The flow and basin of $(-1,0)$ are sketched in Fig. 6.3-3. The determination of the basins of attraction is actually not entirely trivial. The proof uses the function $V(u,\dot{u}) = \frac{1}{2}\dot{u}^2 - \frac{1}{2}u^2 + \frac{1}{4}u^4$, which decreases on orbits (a Liapunov function) (cf. Ball and Carr [1976] and references therein).

For the determination of the attractors or, more generally, the limit sets in the topological context, Liapunov functions are very useful. For general

ISBN 0-8053-0102-X

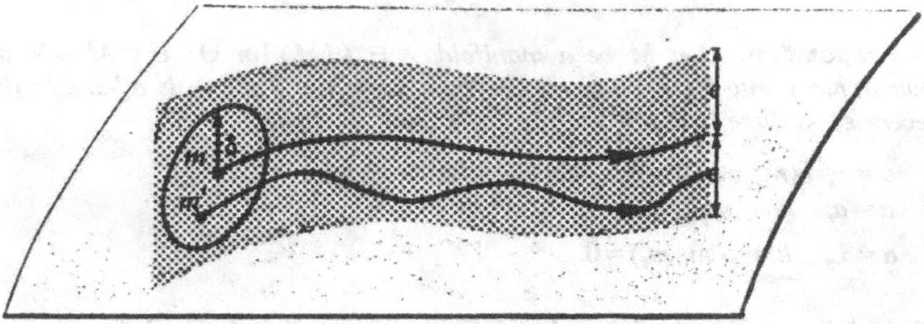

(i) $\alpha = o$ (orbital stability)

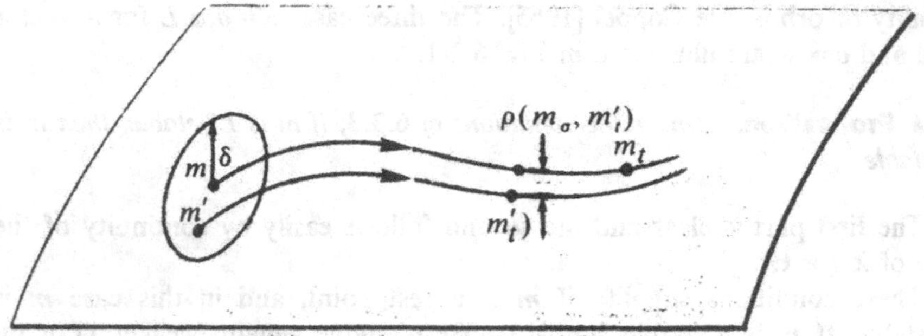

(ii) $\alpha = a$ (asymptotic stability)

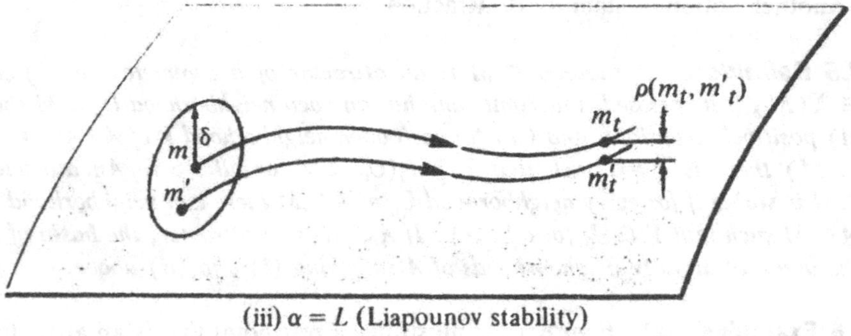

(iii) $\alpha = L$ (Liapounov stability)

Figure 6.3-1

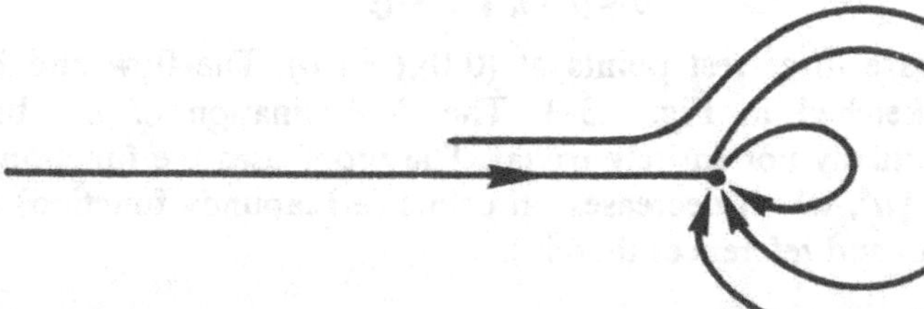

Figure 6.3-2

ISBN 0-8053-0102-X

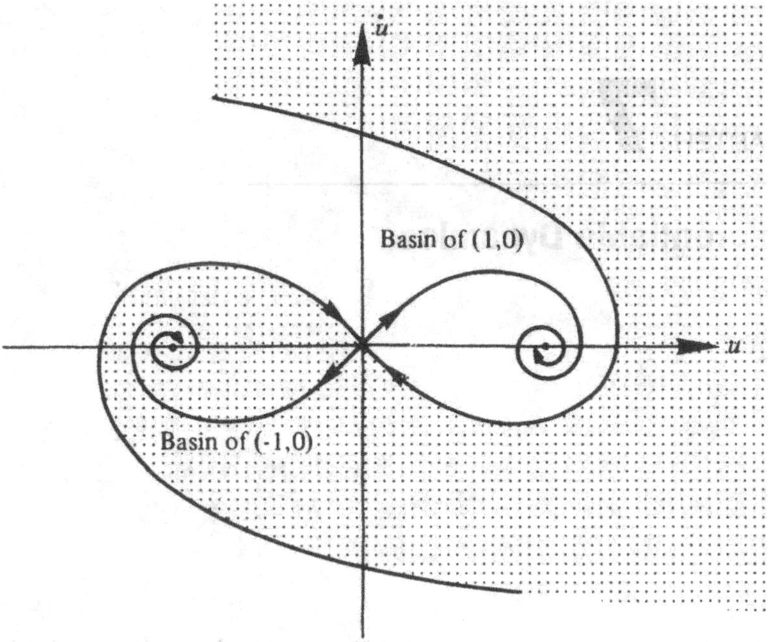

Figure 6.3-3

information, see LaSalle [1976], and for a substantial application, see, for example, Ball [1974].

EXERCISES

6.3A. Prove 6.3.3 and 6.3.4.

6.3B. Find a vector field on $\boldsymbol{R}^2$ and $m \in \boldsymbol{R}^2$ that is o^+-stable using the standard metric, but that is o^+-unstable using some other equivalent metric.

6.3C. Find a vector field on $\boldsymbol{R}^2$ and $m \in \boldsymbol{R}^2$ that is a^+-stable, but not o^+-stable.

6.3D. Relate α^+-stability of an equilibrium or closed orbit to attractor stability, for $\alpha = o, a, L$.

ISBN 0-8053-0102-X

CHAPTER 7

Differentiable Dynamics

In this chapter we give a survey of differentiable dynamics as a backdrop for the contrasting Hamiltonian picture of the next chapter. This area, revived by Lefshetz, Peixoto, Reeb, Smale, and Thom in the late 1950s, is still advancing rapidly.

7.1 CRITICAL ELEMENTS

One of the main goals of differentiable dynamics is to determine the location of critical elements (i.e., fixed points and closed orbits) in the phase portrait and the asymptotic behavior of nearby orbits. The latter is revealed by a linear map derived from the flow, characterized by the *characteristic exponents* or *multipliers* of the critical element. For equilibrium points these have been defined and discussed in Sect. 2.1. Special reference is made to Liapunov's theorem 2.1.25. Details on the sense in which the linearization $X'(m_0)$ of a vector field X at a equilibrium m_0 approximates the flow of X near m_0 in cases not covered by 2.1.25 are discussed in the next section; see also Hartman [1973] and Nelson [1969].

The basic tool in the investigation of the asymptotic behavior of orbits close to a closed orbit is the Poincaré map on a local transversal section, defined as follows.

Ralph Abraham and Jerrold E. Marsden, Foundation of Mechanics, Second Edition

ISBN 0-8053-0102-X

7.1.1 Definition. *Let X be a vector field on M. A* ***local transversal section*** *of X at $m \in M$ is a submanifold $S \subset M$ of codimension one with $m \in S$ and for all $s \in S$, $X(s)$ is not contained in T_sS.*

Let X be a vector field on a manifold M with integral $F: \mathfrak{D}_X \subset M \times \mathbf{R} \to M$, γ a closed orbit of X with period τ, and S a local transversal section of X at $m \in \gamma$. A ***Poincaré map*** *of γ is a mapping $\Theta: W_0 \to W_1$ where:*

(PM 1) *$W_0, W_1 \subset S$ are open neighborhoods of $m \in S$, and Θ is a diffeomorphism;*

(PM 2) *there is a continuous function $\delta: W_0 \to \mathbf{R}$ such that for all $s \in W_0$, $(s, \tau - \delta(s)) \in \mathfrak{D}_X$, and $\Theta(s) = F(s, \tau - \delta(s))$; and finally,*

(PM 3) *if $t \in (0, \tau - \delta(s))$, then $F(s,t) \notin W_0$ (see Fig. 7.1-1).*

7.1.2 Theorem (Existence and uniqueness of Poincaré maps). *(i) If X is a vector field on M, and γ is a closed orbit of X, then there exists a Poincaré map of γ.*

(ii) If $\Theta: W_0 \to W_1$ is a Poincaré map of γ (in a local transversal section S at $m \in \gamma$) and Θ' also (in S' at $m' \in \gamma$), then Θ and Θ' are locally conjugate. That is, there are open neighborhoods W_2 of $m \in S$, W_2' of $m' \in S'$, and a diffeomorphism $H: W_2 \to W_2'$, such that $W_2 \subset W_0 \cap W_1$, $W_2' \subset W_0' \cap W_1'$ and the

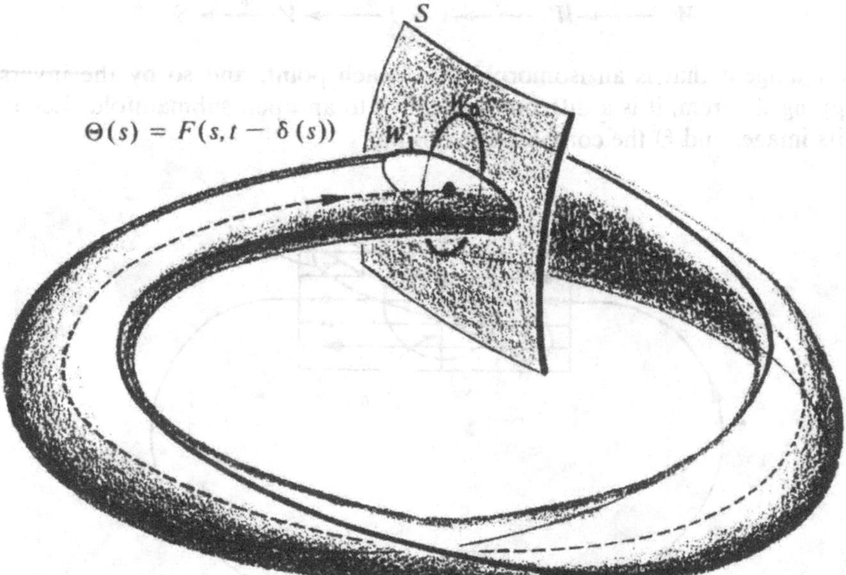

Figure 7.1-1

ISBN 0-8053-0102-X

diagram

$$\begin{array}{ccc} \Theta^{-1}(W_2)\cap W_2 & \xrightarrow{\quad\Theta\quad} & W_2\cap\Theta(W_2) \\ {\scriptstyle H}\big\downarrow & & \big\downarrow{\scriptstyle H} \\ W_2' & \xrightarrow{\quad\Theta'\quad} & S' \end{array}$$

commutes.

Proof. (i) At any point $m\in\gamma$ we have $X(m)\neq 0$, so there exists a flow box chart (U,φ) at m with $\varphi(U)=V\times I\subset \mathbf{R}^{n-1}\times\mathbf{R}$ (2.1.9). Then $S=\varphi^{-1}(V\times\{0\})$ is a local transversal section at m. If $F\colon \mathfrak{D}_X\subset M\times\mathbf{R}\to M$ is the integral of X, $\mathfrak{D}_X$ is open, so we may suppose $U\times[-\tau,\tau]\subset\mathfrak{D}_X$, where τ is the period of γ. As $F_\tau(m)=m\in U$ and F_τ is a homeomorphism, $U_0=F_\tau^{-1}U\cap U$ is an open neighborhood of $m\in M$ with $F_\tau U_0\subset U$. Let $W_0=S\cap U_0$ and $W_2=F_\tau W_0$. Then W_2 is a local transversal section at $m\in M$ and $F_\tau\colon W_0\to W_2$ is a diffeomorphism (see Fig. 7.1-2).

Now if $U_2=F_\tau U_0$, then we may regard U_0, U_2 as open submanifolds of the vector bundle $V\times\mathbf{R}$ (by identification using φ) and then $F_\tau\colon U_0\to U_2$ is a diffeomorphism mapping fibers into fibers, as φ identifies orbits with fibers, and F_τ preserves orbits. Thus W_2 is a section of an open subbundle. More precisely, if $\pi\colon V\times I\to V$ and $\rho\colon V\times I\to I$ are the projection maps, then the composite mapping

$$W_0\xrightarrow{F_\tau}W_2\xrightarrow{\varphi}V\times I\xrightarrow{\pi}V\xrightarrow{\varphi^{-1}}S$$

has a tangent that is an isomorphism at each point, and so by the inverse mapping theorem, it is a diffeomorphism onto an open submanifold. Let W_1 be its image, and Θ the composite mapping.

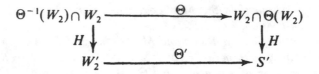

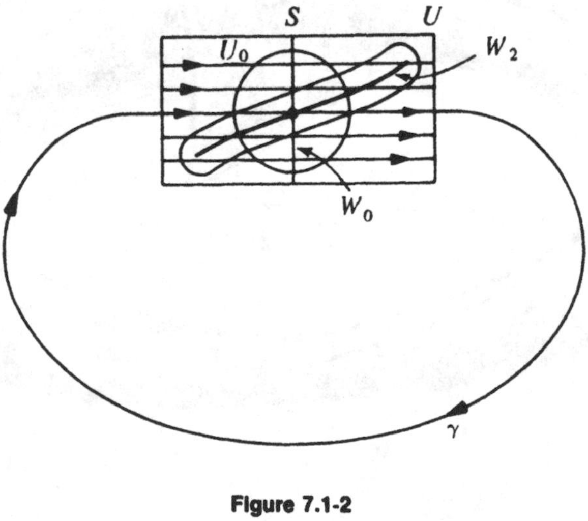

Figure 7.1-2

ISBN 0-8053-0102-X

We now show that $\Theta: W_0 \to W_1$ is a Poincaré map. Obviously (PM 1) is satisfied. For (PM 2), we identify U and $V \times I$ by means of φ to simplify notations. Then $\pi: W_2 \to W_1$ is a diffeomorphism, and its inverse $(\pi | W_2)^{-1}$: $W_1 \to W_2 \subset W_1 \times \boldsymbol{R}$ is a section corresponding to a smooth function $\sigma: W_1 \to \boldsymbol{R}$. In fact, σ is defined implicitly by

$$F_\tau(w_0) = (\pi F_\tau w_0, \rho F_\tau w_0) = (\pi F_\tau w_0, \sigma \pi F_\tau w_0)$$

or $\rho F_\tau w_0 = \sigma \pi F_\tau w_0$. Now let $\delta: W_0 \to \boldsymbol{R}: w_0 \mapsto \sigma \pi F_\tau w_0$. Then we have

$$\begin{aligned} F_{\tau - \delta(w_0)}(w_0) &= F_{-\delta(w_0)} \circ F_\tau(w_0) \\ &= (\pi F_\tau w_0, \rho F_\tau w_0 - \delta(w_0)) \\ &= (\pi F_\tau w_0, 0) \\ &= \Theta(w_0) \end{aligned}$$

Finally, (PM 3) is obvious as (U, φ) is a flow box.

(ii) The proof is burdensome because of the definition of local conjugacy, so we will be satisfied to prove this uniqueness under additional simplifying hypotheses that lead to global conjugacy (identified by italics). The general case will be left to the reader.

We consider first the special case $m = m'$. Then we choose a flow box chart (U, φ) at m, and *assume $S \cup S' \subset U$, and that S and S' intersect each orbit arc in U at most once, and that they intersect exactly the same sets of orbits.* (These three conditions may always be obtained by shrinking S and S'.) Then let $W_2 = S$, $W_2' = S'$, and $H: W_2 \to W_2'$ the bijection given by the orbits in U. As in (i), this is easily seen to be a diffeomorphism, and $H \circ \Theta = \Theta' \circ H$.

Finally, suppose $m \neq m'$. Then $F_a(m) = m'$ for some $a \in (0, \tau)$, and as $\mathfrak{D}_X$ is open there is a neighborhood U of m such that $U \times \{a\} \subset \mathfrak{D}_X$. Then $F_a(U \cap S) = S''$ is a local transversal section of X at $m' \in \gamma$, and $H = F_a$ effects a conjugacy between Θ and $\Theta'' = F_a \circ \Theta \circ F_a^{-1}$ on S''. By the preceding paragraph, Θ'' and Θ' are locally conjugate, but conjugacy is an equivalence relation. This completes the argument. ■

If γ is a closed orbit of $X \in \mathfrak{X}(M)$ and $m \in \gamma$, the behavior of nearby orbits is given by a Poincaré map Θ on a local transversal section S at m. Clearly $T_m\Theta \in L(T_m S, T_m S)$ is a linear approximation to Θ at m. By uniqueness of Θ up to local conjugacy, $T_{m'}\Theta'$ is similar to $T_m\Theta$, for any other Poincaré map Θ' on a local transversal section at $m' \in \gamma$. Therefore, the eigenvalues of $T_m\Theta$ are independent of $m \in \gamma$ and the particular section S at m.

7.1.3 Definition. *If γ is a closed orbit of $X \in \mathfrak{X}(M)$, the* ***characteristic multipliers*** *of X at γ are the eigenvalues of $T_m\Theta$, for any Poincaré map Θ at any $m \in \gamma$.*

ISBN 0-8053-0102-X

Another linear approximation to the flow near γ is given by $T_m F_\tau \in L(T_m M, T_m M)$ if $m \in \gamma$ and τ is the period of γ. Note that $F_\tau^*(X(m)) = X(m)$, so $T_m F_\tau$ always has an eigenvalue 1 corresponding to the eigenvector $X(m)$. The $(n-1)$ remaining eigenvalues (if $dim(M) = n$) are in fact the characteristic multipliers of X at γ.

7.1.4 Proposition. *If γ is a closed orbit of $X \in \mathfrak{X}(M)$ of period τ and c_γ is the set of characteristic multipliers of X at γ, then $c_\gamma \cup \{1\}$ is the set of eigenvalues of $T_m F_\tau$, for any $m \in \gamma$.*

Proof. We can work in a chart modelled on $\boldsymbol{R}^n$ and assume $m = \boldsymbol{0}$. Let V be the span of $X(m)$ so $\boldsymbol{R}^n = T_m S \oplus V$. Write the flow $F_t(x,y) = (F_t^1(x,y), F_t^2(x,y))$. By definition, we have

$$D_1 F^1(m) = T_m \Theta$$

and

$$D_2 F_\tau(m) X(m) = X(m)$$

Thus the matrix of $T_m F_\tau$ is of the form

$$\begin{pmatrix} T_m \Theta & 0 \\ A & 1 \end{pmatrix}$$

where $A = D_1 F_\tau^2(m)$. From this it follows that the spectrum of $T_m F_\tau$ is $\{1\} \cup c_\gamma$. ■

If the characteristic exponents of an equilibrium lie (strictly) in the left half-plane, we know from 2.1.25 that the equilibrium is stable. Likewise, we have:

7.1.5 Proposition. *Let γ be a closed orbit of $X \in \mathfrak{X}(M)$ and let the characteristic multipliers of γ lie strictly inside the unit circle. Then γ is asymptotically stable.*

We can sharpen this statement a little using the following.

7.1.6 Definition. *Let X be a vector field on a manifold M and γ a closed orbit of X. An orbit $F_t(m_0)$ is said to **wind toward** γ if m_0 is + complete and for any transversal S to X at $m \in \gamma$ there is a t_0 such that $F_{t_0}(m_0) \in S$ and successive applications of the Poincaré map yield a sequence of points that converges to m.*

We also use the term "wind toward" for equilibria m to mean merely a + complete orbit $F_t(m_0)$ that converges to m as $t \to +\infty$. To prove that an orbit winds toward a closed orbit it is sufficient to use any Poincaré map, by 7.1.2(ii). Then the following implies 7.1.5:

ISBN 0-8053-0102-X

7.1.5′ Proposition. *Let γ be a closed orbit of $X \in \mathfrak{X}(M)$ and let the characteristic multipliers of γ lie strictly inside the unit circle. Then there is a neighborhood U of γ such that for any $m_0 \in U$, the orbit $F_t(m_0)$ winds toward γ.*

This follows easily from:

7.1.7 Lemma. *Let $f: S \to S$ be a smooth map on a manifold S with $f(s_0) = s_0$ for some s_0. Let the spectrum of $T_{s_0}f$ lie inside the unit circle. Then there is a neighborhood U of s_0 such that if $s \in U$, $f(s) \in U$ and $f^n(s) \to s_0$ as $n \to \infty$, where $f^n = f \circ f \circ \cdots \circ f$ (n times).*

The lemma is proved in the same way as 2.1.25.

7.1.8 Definition. *If $X \in \mathfrak{X}(M)$ and γ is a critical element of X, γ is an* ***elementary*** *or* ***hyperbolic*** *critical element iff none of the characteristic multipliers of X at γ has modulus one.*

The local qualitative behavior near an elementary critical element is especially simple. Also, elementary critical elements are isolated (see Abraham–Robbin [1967, Chapter V]).

Nowadays, *hyperbolic* is frequently used in place of *elementary*.

EXERCISES

7.1A. Show that every (topologically) closed orbit is a point on a one-dimensional embedded submanifold. Find an example of an orbit that is not a submanifold. (*Hint:* Consider a vector field on the torus with irrational slope.)

7.1B. Let $X \in \mathfrak{X}(M)$, $\varphi: M \to N$ be a diffeomorphism and $Y = \varphi_* X$. Then:

(i) $m \in M$ is a critical point of X iff $\varphi(m)$ is a critical point of Y and the characteristic multipliers are the same for each.

(ii) $\gamma \subset M$ is a closed orbit of X iff $\varphi(\gamma)$ is a closed orbit of Y and the characteristic multipliers are the same.

7.1C. Complete the proof of 7.1.2.

7.1D. Prove Lemma 7.1.7.

7.1E. Let γ be a closed orbit of $X \in \mathfrak{X}(M)$. For $m, m' \in \gamma$, show that m is α°-stable if and only if m' is. Show that it is also equivalent to α°-stability with respect to any Poincaré map on a transversal section. (In such a case we say γ is α°-stable.) Show that the hypotheses of 7.1.5 imply γ is α°-stable.

7.1F. Derive a formula for the derivative of the Poincaré map at the fixed point, as the integral of a linearized equation along the closed orbit.

7.2 STABLE MANIFOLDS.

A key feature of differentiable dynamics is the smooth structure of the insets and outsets of an elementary critical element—thus in this context they are renamed: stable and unstable manifolds.

ISBN 0-8053-0102-X

7.2.1 Definition. *If $X \in \mathfrak{X}(M)$ and $\Lambda \subset M$ is a closed invariant set, let*

$$S^{\sigma}(\Lambda) = \{ m \in M \mid \lambda^{\sigma}(m) \subset \Lambda \}$$

*for $\sigma = +, -$, or $\pm$, where $\lambda^{\sigma}(m)$ is the λ^{σ} limit set of m. If γ is an elementary critical element, $S^{+}(\gamma)$ is called the **stable manifold** of γ, and $S^{-}(\gamma)$ the **unstable manifold** of γ.*

Note that $S^{+}(\gamma)$ is the union of orbits that wind toward γ (with increasing time), and $S^{-}(\gamma)$ the union of orbits that wind away from γ (wind toward γ with decreasing time). Compare with 6.2.1.

The following theorem, which is basic to the qualitative behavior near a critical element, has a long history going back to Poincaré. For the proof, see the appendix of A. Kelley in Abraham–Robbin [1967], Hartman [1973], Robbin [1971] or Hirsch, Pugh, and Shub [1977].

7.2.2 Theorem (Local center-stable manifolds). *If $\gamma \subset M$ is a critical element of $X \in \mathfrak{X}(M)$, there exist submanifolds $S_l^{+}, CS_l^{+}, C_l, CS_l^{-}, S_l^{-}$ of M such that:*

(*i*) *Each is invariant under X and contains γ.*
(*ii*) *For $m \in \gamma$, $T_m(S_l^{+})$ [resp. $T_m(CS_l^{+}), T_m(C_l), T_m(CS_l^{-}), T(S_l^{-})$] is the sum of the eigenspace in $T_m M$ of characteristic multipliers of modulus <1 [resp. $\leqslant 1, =1, \geqslant 1, >1$], and the subspace $T_m\gamma$.*
(*iii*) *If $m \in S_l^{\sigma}$ then $\lambda^{\sigma}(m) = \gamma(\sigma = +$ or $-)$.*
(*iv*) *S_l^{+} and S_l^{-} are locally unique.*

Note that the configuration of these manifolds is slightly different in the two cases covered: $\gamma = \{m\}$, a critical point, in which case $T_m\gamma = \{0\}$, or γ is a closed orbit, in which case $T_m\gamma$ is the subspace generated by $X(m)$. The two

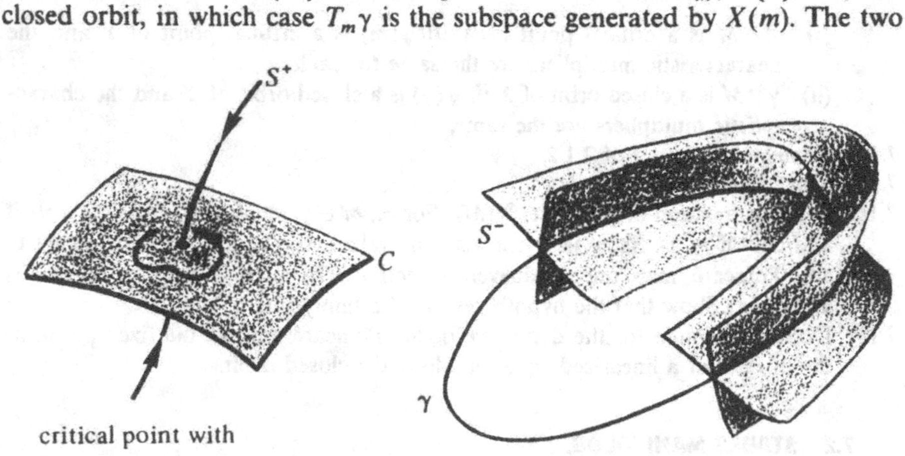

critical point with
$|\lambda_1| < 1, |\lambda_2| = |\lambda_3| = 1$

closed orbit with
$|\lambda_1| < 1, |\lambda_2| > 1$

Figure 7.2-1

ISBN 0-8053-0102-X

cases are illustrated in R^3 in Fig. 7.2-1. The theorem says, in addition, that if γ is elementary, then the nearby orbits behave qualitatively like the linear case.

These manifolds are called respectively the *local stable* (S_l^+), *local center-stable* (CS_l^+), *local center* (C_l), *local center-unstable* (CS_l^-), and *local unstable* (S_l^-) *manifold* of γ. Compare 7.2.1.

In the case of an elementary critical element γ, we have only the locally unique manifolds $S_l^\sigma(\gamma)$, ($\sigma = +$ or $-$). These are easily extended to globally unique manifolds by expanding the local manifolds by means of the integral of X. Recall from Sect. 1.5 that A subset $S \subset M$ is an **immersed submanifold** if it is the image of a mapping $f: V \to M$ that is injective and locally a diffeomorphism onto a submanifold of M.

7.2.3 Corollary (Global stable manifold theorem of Smale). *If γ is an elementary critical element of $X \in \mathfrak{X}(M)$, then $S^+(\gamma)$ and $S^-(\gamma)$ are immersed submanifolds. Also, $\gamma \subset S^+(\gamma) \cap S^-(\gamma)$, and for $m \in \gamma$, $T_m S^+(\gamma)$ and $T_m S^-(\gamma)$ generate $T_m M$. If n_+ is the number of characteristic multipliers of γ of modulus greater than one, and n_- the number of modulus less than one, then the dimension of $S^\sigma(\gamma)$ is $n_{-\sigma}$ (if γ is a critical point) or $n_{-\sigma}+1$ (if γ is a closed orbit), for $\sigma = +$ or $-$.*

In the case of an elementary critical point on a two-dimensional manifold, there are from the stable manifold point of view, three possible local phase portraits (see Fig. 7.2-2).

The stable and unstable manifolds of all critical elements are special features of the phase portrait that are second in importance only to the critical elements.

In the previous section we obtained the basic stability criterion for closed orbits in terms of characteristic multipliers.

A deeper stability theorem is the following.

7.2.4 Theorem (Pliss–Kelley). *Let X be a vector field on a manifold M and $\sigma = +$ or $-$. Let $\gamma \subset M$ be a critical element of X, and S^+, S^-, C the stable,*

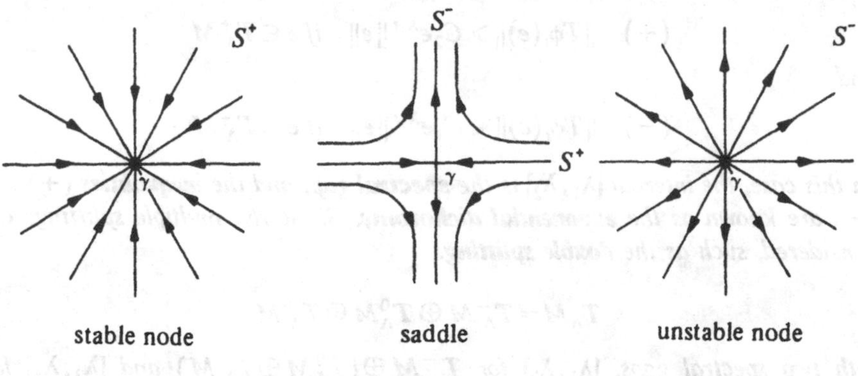

Figure 7.2-2

ISBN 0-8053-0102-X

unstable, and center manifolds of γ, respectively. Then γ is α^o-stable with respect to X on C iff m is α^o-stable with respect to X on CS^o.

For the proof, see Kelley [1967b] or Duistermaat [1976b]. For example, if C and S^+ only occur, and o^+-stability on C is established, then it holds in a neighborhood of $m \in M$.

From the local center-stable manifold theorem we also obtain a condition for instability.

7.2.5 Proposition. *Let $X \in \mathfrak{X}(M)$ and γ be a critical element of X. Then if γ has a characteristic multiplier of modulus greater than one, γ is α^+-unstable.*

This completes the basic ideas of stable manifold theory, for critical elements—equilibria and closed orbits—that are the simplest minimal (or nonwandering) sets. One of the greatest breakthroughs of differentiable dynamics was the discovery—by Smale in the early 1960s—of the generalization to arbitrary nonwandering sets. The obstacle here is the lack of a spectrum (analogous to the characteristic multipliers) to use in the definition of the hyperbolic (elementary) property. Here is a snapshot of the general theory of stable manifolds. For full details, see Hirsch, Pugh, and Shub [1977], Duistermaat [1976b] and Fenichel [1977].

7.2.6 Definition. *Let $\Lambda \subset M$ be a closed set (not necessarily—and not usually—a submanifold) invariant under the flow $\{\phi_t\}$ of a complete vector field $X \in \mathfrak{X}(M)$. Let $T_\Lambda M$ denote the restriction of the tangent bundle of M to Λ. Then a **spectral splitting** of $T_\Lambda M$ **with respect to** X is a splitting*

$$T_\Lambda M = T_\Lambda^- M \oplus T_\Lambda^+ M$$

as a Whitney sum of C^0 vector bundles on Λ, invariant under the derived flow $\{T\phi_t\}$ on TM, such that there exists a Riemannian metric on M, with associated norm $\|\cdot\|$, constants $C_1, C_2 > 0$ and constants λ_1, λ_2 with $\lambda_1 < \lambda_2$ such that, for all $t \geq 0$,

$$(+) \quad \|T\phi_t(e)\| \geq C_2 e^{\lambda_2 t}\|e\| \quad \text{if } e \in T_\Lambda^+ M$$

and

$$(-) \quad \|T\phi_t(e)\| \leq C_1 e^{\lambda_1 t}\|e\| \quad \text{if } e \in T_\Lambda^- M$$

*In this case, the interval $[\lambda_1, \lambda_2]$ is the **spectral gap**, and the **inequalities** $(+)$ and $(-)$ are known as the **exponential dichotomy**. Similarly, multiple splittings are considered, such as the **double splitting**:*

$$T_\Lambda M = T_\Lambda^- M \oplus T_\Lambda^0 M \oplus T_\Lambda^+ M$$

with two spectral gaps, $[\lambda_1, \lambda_2]$ for $T^- M \oplus \{T_\Lambda^0 M \oplus T_\Lambda^+ M\}$ and $[\lambda_3, \lambda_4]$ for $\{T_\Lambda^- M \oplus T_\Lambda^0 M\} \oplus T_\Lambda^+ M$ with $\lambda_2 \leq \lambda_3$.

ISBN 0-8053-0102-X

Thus, for $e \in T^{\sigma}_{\Lambda}M$ ($\sigma = +, 0, -$) there is an ***exponential trichotomy:***

(+) $C_4 e^{\lambda_4 t}\|e\| \leq \|T\phi_t(e)\|$
(0) $C_2 e^{\lambda_2 t}\|e\| \leq \|T\phi_t(e)\| \leq C_3 e^{\lambda_3 t}\|e\|$
(−) $\|T\phi_t(e)\| \leq C_1 e^{\lambda_1 t}\|e\|$

For example, the case $\lambda_2 = \lambda_3 = 0$ occurs as follows.

7.2.7 Definition. *Let $\Lambda \subset M$ be a closed set, invariant under the flow $\{\phi_t\}$ of a complete vector field $X \in \mathfrak{X}(M)$. Then Λ is a* ***hyperbolic set*** *of X iff there is a double splitting*

$$T_\Lambda M = T^-_\Lambda M \oplus T^0_\Lambda M \oplus T^+_\Lambda M$$

with spectral gaps $[\lambda_1, \lambda_2]$, $[\lambda_3, \lambda_4]$ satisfying

$$-\lambda = \lambda_1 < \lambda_2 = 0 = \lambda_3 < \lambda_4 = \lambda$$

and $T^0_\Lambda M$ is a real line bundle or zero.

Note that either $X|\Lambda \equiv 0$ or else X has no equilibrium in Λ (i.e., $X|\Lambda$ is nowhere zero) and $T^0_\Lambda M = \langle X(m) \rangle$ the span of $X(m)$ for $m \in \Lambda$. Thus $T^-_\Lambda M \oplus T^+_\Lambda M$ is a bundle of hyperplanes transversal to the orbits of X in Λ. It is known (e.g., Duistermaat [1976b]) that the invariance of these subbundles under $\{T\phi_t\}$ need not be assumed in 7.2.6, and, further, that the splitting (for any given gap) is unique.

There is a corresponding concept for discrete flows. Throughout this chapter, we will ignore this parallel theory to minimize confusion.

Finally, most of the results of generalized stable manifold theory for smooth flows can be obtained from the following master theorem. Note that compactness of Λ is assumed and everything is C^∞ here.

7.2.8 Stable Manifold Master Theorem. *Suppose $\Lambda \subset M$ is a compact set, invariant under the flow $\{\phi_t\}$ of a complete vector field $X \in \mathfrak{X}(M)$, and*

$$T_\Lambda M = T^-_\Lambda M \oplus T^+_\Lambda M$$

is a spectral splitting with gap $[\lambda^-, \lambda^+]$ and Riemannian metric g.

(*i*) *Then there is a g-uniform neighborhood of the zero section in $T^-_\Lambda M$*

$$N^-_\delta = \{e \in T^-_\Lambda M \mid \|e\|_g < \delta\}$$

for some $\delta > 0$ and a C^0 embedding

$$\sigma : N^-_\delta \to M$$

ISBN 0-8053-0102-X

such that for all $x \in \Lambda$, $\sigma(0_x) = x$,

$$S_\delta(x) = \sigma(N_\delta^- | x)$$

(*where* $N_\delta^- | x = N_\delta^- \cap T_x M$) *is a submanifold of M tangent to* $T_x^- M$ *at x.*

(*ii*) *The disks* $\{S_\delta(x)\}$ *depend continuously on* $x \in \Lambda$ *and are locally invariant under* $\{\phi_t\}$.

(*iii*) *Suppose further* $\lambda^- < 0$ *and* $\lambda \in (\lambda^-, 0] \cap (\lambda^-, \lambda^+)$. *Let* $d(x,y)$ *denote the (geodesic) metric on M induced by g. For* $x_0 \in \Lambda$, *let*

$$\omega^\rho(x_0) = \{ y_0 \in M \,|\, d(y_t, x_t) \leqslant \delta e^{\rho t} \text{ for all } t \geqslant 0 \}$$

called a ***local isochron****. Then there is a* $\delta' \in (0, \delta)$ *such that*

$$S_{\delta'}(x) \subset \omega^\rho(x) \subset S_\delta(x)$$

for all $x \in \Lambda$. *In addition, the* ***global isochrons***

$$W_\rho^+(x_0) = \{ y_0 \in M \,|\, d(y_t, x_t) = O(e^{\rho t}) \text{ for } t \to \infty, \rho < 0 \}$$

are immersed k^- *cells* (k^- *is the fiber dimension of* $T\Lambda^-$) *for all* $x_0 \in \Lambda$, *are invariant under* $\{\phi_t\}$, *and are independent of* $\rho < 0$.

(*iv*) *If, on the other hand,* $\lambda^- \geqslant 0$, *then* $y_0 \in S_\delta(x_0)$ *if* $d(y_t, x_t) \leqslant \delta$ *for all* $t \geqslant 0$.

This is taken nearly verbatim from Duistermaat [1976b], who gives much credit to Perron.

In case Λ is a critical element, the results cited previously on center-stable manifolds are easily obtained from this more general result. Similarly, if Λ is an elementary (= hyperbolic) critical element, the stable manifold theorem is obtained. These are good exercises.

Suppose now that Λ is more complicated, a general connected, compact, invariant set of $X \in \mathfrak{X}(M)$. Probably there will be no uniform trichotomy, or spectral gaps, to create double splittings

$$T_\Lambda M = T_\Lambda^- M \oplus T_\Lambda^0 M \oplus T_\Lambda^+ M$$

But if there are such, with gaps $[\lambda_1, 0]$ and $[0, \lambda_4]$, then Λ will have center-stable manifolds. If, further, the central subbundle $T_\Lambda^0 M$ is one-dimensional, then Λ is hyperbolic and has a global stable manifold,

$$S^+(\Lambda) = \bigcup_{x \in \Lambda} W^+(x)$$

and a global unstable manifold $S^-(\Lambda)$, which are continuous families of smoothly immersed cells.

We summarize the implications for the hyperbolic case, for future reference. Recall the distinction between insets and stable manifolds (6.3.1).

ISBN 0-8053-0102-X

7.2.9 Stable Manifold Theorem for Hyperbolic Sets (Smale [1967]). *Suppose $\Lambda \subset M$ is a compact set, invariant under the flow $\{\phi_t\}$ of a complete vector field $X \in \mathfrak{X}(M)$, and hyperbolic with Riemannian metric g and distance d. Then for any initial point $x \in \Lambda$ the stable manifold (isochron) $W^+(x)$ is an injectively immersed open k^- cell. If y is a point of the inset of Λ and $y \in S^+(\Lambda)$, or equivalently $\lambda^+(y) \subset \Lambda$, then $y \in W^+(x)$ for some unique $x \in \Lambda$. Also, the inset is the union of stable manifolds,*

$$S^+(\Lambda) = \bigcup_{x \in \Lambda} W^+(x)$$

and this union is a continuous bundle over Λ, invariant under $\{\phi_t\}$.

An identical statement applies to the outset, and unstable manifolds,

$$S^-(\Lambda) = \bigcup_{x \in \Lambda} W^-(x).$$

EXERCISES

7.2A. Deduce 7.2.2 from 7.2.8.
7.2B. Deduce 7.2.3 from 7.2.2.
7.2C. Deduce 7.2.9 from 7.2.8.
7.2D. Deduce 7.2.3 from 7.2.9.
7.2E. Prove 7.2.5.

7.3 GENERIC PROPERTIES

At one point in the history of differentiable dynamics, it was hoped that even though an arbitrary phase portrait was too complicated to classify, at least the typical garden variety would be manageable. And so began the search for generic properties of vector fields. An early result, property (G2) described below, was found by Markus [1960]. Since then, most properties proposed eventually fell to a counterexample, and the ever-widening moat between the generic and the classifiable became known as the *yin–yang problem*. In any case, this section—a survey of the known generic properties of differentiable dynamics—is understandably brief. And be warned: none of these results is applicable to Hamiltonian dynamics.

We wish now to express the fact that a property is actually *generic*, and we begin by making this word precise. The main requirement is a topology on $\mathfrak{X}(M)$. As the definition of the topology is somewhat involved, we give here just an outline. For details, see Abraham–Robbin [1967, Chapter II].

If $U \subset \boldsymbol{E}$ is an open subset of a vector space and $\boldsymbol{F}$ is a vector space, we may put on the vector space $\mathcal{B}^r(U,\boldsymbol{F})$ of C^r mappings f from U to $\boldsymbol{F}$ $(r < \infty)$ with $D^kf\colon U \to L_s^k(\boldsymbol{E},\boldsymbol{F})$ bounded $(k = 0,\ldots,r;\ D^0f = f)$ a structure of Banach

ISBN 0-8053-0102-X

space by defining the following C^r norm:

$$\|f\|_r = sup\left\{ \sum_{k=0}^{r} \|D^k f(u)\| \,|\, u \in U \right\}$$

Let $\mathfrak{X}^r(M)$ be the set of C^r vector fields on M. If (U,φ) is a chart on $M, \varphi(U) = U' \subset E$, $(TU, T\varphi)$ is a natural chart on TM, and for $X \in \mathfrak{B}^r(M)$ we have a local representative $X_\varphi \in C^\infty(U', E)$. If U' is contained in the image of a larger chart and is bounded in E, then $X_\varphi \in \mathfrak{B}^r(U', E)$. Thus there exists an atlas $\mathfrak{A} = \{(U_i, \varphi_i)\}$ on M such that $X_{\varphi_i} \in \mathfrak{B}^r(U_i', E_i)$ for every i, and the $\mathfrak{A}^r$ *uniform topology* on $\mathfrak{X}(M)$ is the smallest topology such that the mapping $X \to X_{\varphi_i}$ is continuous for all i. If M is compact, it may be shown that this topology is independent of the atlas $\mathfrak{A}$; it is then called the *C^r topology*. If M is not compact, we may take the union of $\mathfrak{A}^r$ topologies for all admissible atlases $\mathfrak{A}$ on M having the boundedness property above, which is the *Whitney C^r topology* on $\mathfrak{X}^r(M)$.

We may now say precisely what we mean by a generic property of vector fields.

7.3.1 Definition. *The space of C^r vector fields on M with the Whitney C^r topology is denoted by $\mathfrak{X}^r(M)$. A* ***property of vector fields*** *in $\mathfrak{X}^r(M)$ is a proposition $P(X)$ with a variable $X \in \mathfrak{X}^r(M)$. A property $P(X)$ in $\mathfrak{X}^r(M)$ is* ***generic*** *if the subspace $\{X \in \mathfrak{X}^r(M) | P(X)\} \subset \mathfrak{X}^r(M)$ contains a residual set (see Sect. 1.1). A property $P(X)$ in $\mathfrak{X}(M)$ is* ***C^r generic*** *if $\{X \in \mathfrak{X}(M) | P(X)\}$ is the intersection of $\mathfrak{X}(M)$ with a subspace of $\mathfrak{X}^r(M)$ containing a residual set.*

Note that a property generic in $\mathfrak{X}^r(M)$ is C^r generic in $\mathfrak{X}(M)$ with the relative Whitney C^r topology. For it can be shown that the C^∞ vector fields $\mathfrak{X}(M) \subset \mathfrak{X}^r(M)$ is a residual set. Also, $\mathfrak{X}^r(M)$ is a Banach space if M is compact, and is thus a Baire space. For noncompact M, $\mathfrak{X}^r(M)$ is still a Baire space with the Whitney topology (see Morlet [1963]). Thus in either case, residual sets are dense, so if $P(X)$ is a generic property, every vector field can be approximated as closely as we wish by one with the generic property, see Hirsch [1977].

We may now state very easily the main results.

7.3.2 Definition. *A vector field $X \in \mathfrak{X}(M)$ has* ***property (G2)*** *if every critical element is hyperbolic.*

This property (defined in 7.1.8) implies that the critical elements are isolated and have stable and unstable manifolds, denoted $S^+(\gamma)$ and $S^-(\gamma)$. We do not define a separate property (G1), but (G2) begins a sequence of generic properties (Gi).

Another property of vector fields concerns the position of the intersections of these stable and unstable manifolds.

ISBN 0-8053-0102-X

7.3.3 Definition. *A vector field $X \in \mathfrak{X}(M)$ has* ***property (G3)*** *if X has property (G2) and if $m \in S^+(\gamma) \cap S^-(\delta)$ for any critical elements γ and δ, then $T_m S^+(\gamma)$ and $T_m S^-(\delta)$ generate $T_m M$, that is, $S^+(\gamma)$ and $S^-(\delta)$ intersect transversally at m.*

7.3.4 Theorem (Kupka–Smale). *If M is compact and $r > 1$, the property (G3) on $\mathfrak{X}(M)$ is C^r generic.*

This theorem was proved independently by Kupka [1963] and Smale [1963a] in the case of a compact manifold M. The proof consists of a long sequence of careful approximations which, when written out in full detail, fill thirty pages (Abraham–Robbin [1967]). The theorem has been extended to the noncompact case by Peixoto [1967b]. See also Palis and deMelo [1975].

Two additional properties have been shown to be C^1 generic: (G4) and (G5).

7.3.5 Definition. *For a vector field $X \in \mathfrak{X}(M)$ let C_X denote the set of initial points $m \in M$ such that the orbit $o(m)$ is complete and has compact closure, and recall that Ω_X is the set of nonwandering points of X. Then*

$$\Omega_X^C = cl[C_X \cap \Omega_X]$$

is the ***o-compact-nonwandering set*** *of X. The vector field X has* ***property (G4)*** *if*

$$cl(\Gamma_X) = \Omega_X^C$$

where Γ_X is the union of all critical elements of X.

Note that if M is compact, $\Omega_X^C = \Omega_X$. Also, as $cl(\Gamma_X) \subset \Omega_X^C$ always, (G4) is equivalent to $\Omega_X^C \subset cl(\Gamma_X)$. Compare (G3).

7.3.6 Theorem. *Property (G4) is C^1 generic.*

This was established by Pugh [1967] for compact M using his closing lemma, and Robinson [1970a] for the general case.

The genericity of (G5) also follows from this very difficult lemma.

7.3.7 Definition. *A* ***regular first integral*** *of a vector field $X \in \mathfrak{X}(M)$ is a proper function $f: M \to R$ of class C^1 such that f is not constant on any open subset of M, and $L_X f = 0$. A vector field $X \in \mathfrak{X}(M)$ has* ***property(G5)*** *if X has no regular first integral.*

7.3.8 Theorem. *Property (G5) is C^1 generic.*

For a proof, see Peixoto [1959] and Robinson [1970a].

ISBN 0-8053-0102-X

It is this theorem which suggests that, generically, a Hamiltonian vector field X_H has no regular first integral other than H itself. Let $\mathcal{G}_3 = \{X \in \mathfrak{X}(M) | X$ has property (G3)$\}$ with the C^1 Whitney topology. A different type of stability has been introduced by Pugh.

7.3.9 Definition. *A vector field $X \in \mathcal{G}_3$ is **critically stable or has property (G6)** if the mapping $cl(\Gamma): \mathcal{G}_3 \to 2_1^M : X \mapsto cl(\Gamma_X)$ is continuous at X.*

7.3.10 Theorem (Pugh). *If M is compact, then property (G6) is C^1 generic in $\mathfrak{X}(M)$.*

This is in fact a fairly easy corollary of the Kupka–Smale (G3) density theorem. This was proved very ingeniously by Pugh [1967] using properties of set valued mappings with the Hausdorff metric, and used by him with the closing lemma to prove the (G4) density theorem. See also Robinson [1970a].

EXERCISES

7.3A. Prove that if M is compact and $X \in \mathfrak{X}(M)$, then $C_X = M$ and $\Omega_X^c = \Omega_X$.

7.3B. Find an example $X \in \mathfrak{X}(\mathbf{R}^2)$ with $\Omega_X^c \neq \Omega_X$.

7.3C. If $X \in \mathfrak{X}(\mathbf{R}^2)$ has a regular first integral f, show that the critical points of f include the equilibria of X.

7.3D. Is the vector field X_0 of the Hopf bifurcation (Sect. 5.6) critically stable?

7.4 STRUCTURAL STABILITY

Structural stability was an early candidate for a generic property of vector fields. Although it turned out to be generic only in the two-dimensional case, it may be important in applications. Some weaker notion of stability may be found to be generic eventually. In any case, we outline now some of the results of this program, initiated by Andronov and Pontriagin [1937].

7.4.1 Definition. *Let X be a vector field on M. Then X is **C^r structurally stable** if there is a neighborhood $\mathcal{O}$ of $X \in \mathfrak{X}(M)$ in the Whitney C^r topology, such that $Y \in \mathcal{O}$ implies X and Y are **topologically conjugate**, that is, they have equivalent phase portraits: there is a homeomorphism $h: M \to M$ carrying oriented orbits of X to oriented orbits of Y. The set of C^r structurally stable vector fields on M is denoted by $\Sigma_S^r(M)$.*

7.4.2 Definition. *A vector field $X \in \mathfrak{X}(M)$ is said to be a **Morse–Smale system** if it has properties (G3), (G4), and (F): it has a finite number of critical elements.*

For the two-dimensional case note that (G3) is equivalent to: No orbit connects two saddle points; that is, there are no *saddle connections*.

ISBN 0-8053-0102-X

7.4.3 Theorem (Peixoto). *If M is compact, orientable, and two dimensional, and $1 \leqslant r < \infty$, then:*

(*i*) $X \in \Sigma_S^r(M)$ *iff X is a Morse–Smale system;*
(*ii*) Σ_S^r *is an open, dense subset of $\mathfrak{X}(M)$ in the C^r topology.*

For the proof see Peixoto [1962].

For higher dimensions, the foregoing approach fails because of the following results.

7.4.4 Theorem (Smale [1964a]). *For every $n > 2$, there exists a compact manifold M, a vector field X_0 on M, and an open neighborhood U of $X_0 \in \mathfrak{X}(M)$ in the C^1 topology, such that every $X \in U$ has property (G4) but not property (F).*

7.4.5 Theorem (Anosov [1967]). *For each $n > 2$, there is a compact n-manifold M and $X \in \mathfrak{X}(M)$ such that X is C^1 structurally stable but does not have property (F).*

7.4.6 Theorem (Smale [1966]). *There is a manifold of dimension four such that in the C^1 topology, Σ_S^1 is not dense in $\mathfrak{X}^1(M)$.*

Concerning the relations between properties ($G\alpha$) and structural stability there are the following results.

7.4.7 Theorem. *Let M be a compact manifold. If $X \in \mathfrak{X}(M)$ is C^1 structurally stable, then X has:*

(*i*) *property (G3) (Markus [1961]);*
(*ii*) *property (G4) (Pugh [1967]);*
(*iii*) *property (G5) (Thom, see Peixoto [1967a]).*

When the nongenericity of structural stability was accepted, weaker concepts of stability were proposed. Here is an important one, due to Smale [1970b].

7.4.8 Definition. *If $X, Y \in \mathfrak{X}(M)$, with nonwandering sets Ω_X, Ω_Y and limit sets Λ_X, Λ_Y, then X and Y are* **Λ-conjugate** *(resp.,* **Ω-conjugate***) if there is a homeomorphism $h: \Lambda_X \to \Lambda_Y$ (resp. $h: \Omega_X \to \Omega_Y$) preserving oriented orbits. A vector field $X \in \mathfrak{X}(M)$ is* **Λ-stable** *(resp.* **Ω-stable***) if it has a C^1 neighborhood $\mathfrak{U}$ in $\mathfrak{X}(M)$ such that $Y \in \mathfrak{U}$ implies Y is Λ-conjugate (resp. Ω-conjugate) to X.*

As $\Lambda_X \subset \Omega_X \subset M$, it is obvious that Λ-stability is weaker than Ω-stability, which is weaker than S-stability. Within days of the proposal of these weaker

ISBN 0-8053-0102-X

notions of stability, the first of the counterexamples was constructed (Abraham–Smale [1970]), killing hopes that they might be generic properties of vector fields.

Concerning stability stronger than structural, here is a key result.

7.4.9 Theorem [Palis–Smale (1970)]. *Morse–Smale implies structural stability.*

Let $\Sigma_M \subset \mathfrak{X}(M)$ denote the Morse–Smale vector fields (7.4.2) and $\Sigma_\alpha \subset \mathfrak{X}(M)$ the α-stable vector field ($\alpha = S$, Λ, or Ω).

Then we have, in summary, the *tower of stability*

$$\Sigma_M \subset \Sigma_S \subset \Sigma_\Omega \subset\subset \mathcal{G}_4 \subset \mathfrak{X}(M)$$

where $\mathcal{G}_4$ is the set of vector fields satisfying generic properties (G2) to (G4). Here the double inclusion sign $\subset\subset$ (informally) signifies inclusion *and* a large gap. Finally, recent results (Zeeman [1975]) allow replacement—in the tower of stability—of the smallest set (Σ_M = Morse–Smale vector fields) by a much larger one (Σ_Z = Smale–Zeeman vector fields).

Parallel to the tower of stability is another:

$$\Sigma_M \subset \Sigma_{AS} \subset \Sigma_{A\Omega} \subset\subset \mathcal{G}_4 \subset \mathfrak{X}(M)$$

the *tower of absolute stability*. This parallel tower, and its geometric description, is set out in the next section.

EXERCISES

7.4A. Prove that in the two-dimensional case property (G3) is equivalent to: No orbit connects two saddle points, and (G2).

7.4B. Characterize α-stability in terms of continuity of a set mapping, $\alpha = S, \Lambda, \Omega$.

7.5 ABSOLUTE STABILITY AND AXIOM A

We may think of the generic vector fields $\mathcal{G}_4$ as heaven (yin) and the Morse–Smale systems Σ_M as earth (yang). Differentiable dynamics attempts to build a tower from earth to heaven, in spite of warnings from Puritan poets and others (Abraham [1971]).

The literature is full of examples demonstrating the largeness of the gap between the top of the tower of stability Σ_Λ and heaven. And so, in 1970, Smale proposed a new tower, the *A-tower*, based on his "Axiom A" of 1966:

$$\Sigma_M \subset \mathcal{A}_{ST} \subset \mathcal{A}_{NC} \subset\subset \mathcal{G}_4$$

Although it subsequently turned out to be a little shorter than the tower of stability, it has wonderful features: geometric descriptions, statistical

ISBN 0-8053-0102-X

mechanics, algebraic classification, reasonable bounds for numbers of critical elements, and so forth. These features are still being elaborated; for details, see Takens [1973a], Duistermaat [1976b], Bowen [1975], Manning [1975], Robbin [1972a], and Hirsch–Pugh–Shub [1977], and Bowen [1978].

The definitions of $\mathcal{Q}_{ST}$ and $\mathcal{Q}_{NC}$ require a substantial diversion, as we shall see. And throughout this section, *let M be compact*, to keep things simple.

Recall that for $X \in \mathfrak{X}(M)$, $cl(\Gamma_X)$ is the closure of the critical elements of X in M, $cl(\Lambda_X)$ is the closure of the α and ω limit sets, Ω_X is the set of nonwandering points, R_X is the set of recurrent points, and

$$\begin{array}{ccccc} & & cl(\Lambda_X) & & \\ & \subset & & \subset & \\ cl(\Gamma_X) & & & & \Omega_X \\ & \subset & & \subset & \\ & & cl(R_X) & & \end{array}$$

always (6.3.2 and 6.3.5).

7.5.1 Definition. *A vector field $X \in \mathfrak{X}(M)$ satisfies* **Axiom A** *[or has* **property** *(A)] if Ω_X is hyperbolic, and (G4), $cl(\Gamma_X) = \Omega_X$. Let $\mathcal{Q}(M) \subset \mathfrak{X}(M)$ denote the set of vector fields satisfying Axiom A.*

We know (G4) is generic, but (A) is not (Abraham–Smale, [1970]). The geometric characterization of Axiom A vector fields is based on the following.

7.5.2 Spectral Decomposition Theorem (Smale, 1967). *If $X \in \mathcal{Q}(M)$ (with M compact), then Ω_X is a finite union of pairwise disjoint closed connected invariant sets*

$$\Omega_X = \Omega_1 \dot{\cup} \cdots \cup \Omega_s$$

which are topologically transitive: there is a dense orbit in each.

These components, Ω_i, are called the **basic sets** of $X \in \mathcal{Q}$. They play the same fundamental role in differentiable dynamics that minimal sets play in the topological theory. Yet, basic sets are rarely minimal. It is known that the flow of X within a basic set is not only topologically transitive, but even may be mixing (in the sense of ergodic theory) (see Bowen [1975]).

There are three sorts of basic sets: an equilibrium point, a closed orbit, and a chaotic set. A **chaotic set** means simply any basic set other than a critical element.

If $\lambda^\sigma(m)$ is in the basic set Ω^σ ($\sigma = +, -$) we say the orbit of m *goes from Ω^- to Ω^+*.

ISBN 0-8053-0102-X

7.5.3 Definition. *If $X \in \mathcal{Q}(M)$ and Ω_i, Ω_j are basic sets, then we write $\Omega_i \to \Omega_j$ if there is an orbit which goes from Ω_i to Ω_j.*

This is simple to visualize if X is Morse–Smale, and we may pretend that any Axiom A flow is a Morse–Smale system with some of its critical elements "blown up" into chaotic sets.

A directed graph (or **quiver**) may be drawn to symbolize the relation $\to$ on basic sets. This will have a finite number of vertices (if M is compact) corresponding to the basic sets, and directed edges (arrows) for the relation. This is the **Smale quiver** of $X \in \mathcal{Q}(M)$. If, in addition, each vertex is labelled with its inset and outset dimensions and the dynamical system $(\Omega_i, X|\Omega_i)$, and each directed edge labelled with its intersection topology, we could hope that the flow of X could be recovered from the labelled quiver. At present this is unknown, but see Robinson and Williams [1977].

Now we are ready to return to the A-tower of Smale.

7.5.4 Definition. *A vector field $X \in \mathcal{Q}(M)$ satisfies **property (NC)**, or has the **No Cycle property**, if there are no cycles:*

$$\Omega_{i_1} \to \cdots \to \Omega_{i_s} \to \Omega_{i_1}$$

(where $s > 1$ and $\Omega_{i_j} \neq \Omega_{i_k}$ for $j \neq k$) in its Smale quiver. Let $\mathcal{Q}_{NC} \subset \mathcal{Q}$ denote the set of $X \in \mathcal{Q}$ satisfying property (NC).

Note that for $X \in \mathcal{Q}_{NC}$ we may orient the Smale quiver in a descending "order," the source vertices at the top and sinks at the bottom, and saddle-type basic set vertices in between. However, $\to$ is not an order relation, but just a partial ordering.

The first justification for the definition of $\mathcal{Q}_{NC}$ as a rung of the A-tower was the celebrated Ω-stability theorem of Smale [1970a]: for discrete flows, $\mathcal{Q}_{NC} \subset \Sigma_\Omega$. Shortly afterwards, this inclusion was also established for continuous flows, by Pugh and Shub [1970a].

7.5.5 Ω-Stability Theorem. *For vector fields on a compact manifold, $\mathcal{Q}_{NC} \subset \Sigma_\Omega$.*

For the remaining set in the A-tower, $\mathcal{Q}_{ST}$, the stable and unstable manifolds must be reconsidered.

Let $X \in \mathcal{Q}(M)$ and Ω be a basic set of X. Then Ω is a compact connected invariant set with a hyperbolic structure. So according to the stable manifold theorem the inset of Ω is a continuous cell bundle of stable manifolds of points of Ω,

$$S^+(\Omega) = \bigcup_{x \in \Omega} W^+(x)$$

ISBN 0-8053-0102-X

and similarly, for the outset,

$$S^-(\Omega) = \bigcup_{x \in \Omega} W^-(x)$$

Further, the flow on Ω, $\{\phi_t|\Omega\}$, has a dense orbit. Suppose $x \in \Omega$ has $\lambda^+(x) = \Omega$. Then $S^\sigma(\Omega) = S^\sigma(x)$, as defined earlier (6.3.1). For any $m \in M$, $\lambda^+(m)$ is in one of the basic sets, say Ω_i. Thus $m \in W^+(x)$ for some unique $x \in \Omega_i$. Likewise, $m \in W^-(y)$ for a unique $y \in \Omega_j$, where $\Omega_j \to \Omega_i$. These two injectively immersed cells intersect at m,

$$m \in W^-(y) \cap W^+(x)$$

In addition, the inset and outset meet,

$$m \in S^-(\Omega_j) \cap S^+(\Omega_i)$$

Note that if m' is a point on the orbit of m, $m \neq m'$, then $m \in W^+(x)$, $m' \in W^+(x')$ implies $x' \in S^+(\Omega_i)$, but $x' \notin W^+(x)$. To unite the distinct isochrons through distinct points on the same orbit, we introduce yet another version of stable manifold.

7.5.6 Definition. *For $X \in \mathcal{X}(M)$, complete, and $m_0 \in M$, let $\tilde{W}^\sigma(m_0) = \bigcup_{t \in R} W^\sigma(m_t)$, $\sigma = +, -$. If $\gamma \subset M$ is an orbit, let $\tilde{W}^\sigma(\gamma) = \tilde{W}^\sigma(m_0)$ for any $m_0 \in \gamma$. These are the* ***suspended stable*** *($\sigma = +$) and* ***unstable*** *($\sigma = -$)* ***manifolds*** *of γ.*

Note that if γ is an equilibrium point, $\tilde{W}^\sigma(\gamma) = W^\sigma(\gamma)$. If γ is a closed orbit, then $\tilde{W}^\sigma(\gamma) = S^\sigma(\gamma)$. If γ is in a chaotic hyperbolic set Ω, then $\tilde{W}^\sigma(\gamma) \subset S^\sigma(\Omega)$.

The complete theory of differentiable dynamics has two parallel, distinct branches: discrete and continuous flows. Each is necessary for the other, and they are nearly identical. In the research literature, the discrete case is usually treated explicitly and the continuous case is sometimes disposed of with a wave of the hand. One reason is that the distinction between stable manifolds and their suspensions is a major annoyance. Throughout this chapter, we try to treat the continuous case exclusively, for expository reasons. But many of the references to the literature cross to the discrete track.

We return now to the A-tower of Smale, in the continuous version.

7.5.7 Definition. *A vector field $X \in \mathcal{X}(M)$ satisfies* ***property (ST)*** *or the* ***Strong Transversality property*** *if for all $m \in M$, the suspended stable and unstable manifolds meet transversally,*

$$\tilde{W}^+(m) \pitchfork \tilde{W}^-(m)$$

Let $\mathcal{X}_{ST} \subset \mathcal{X}$ denote the set of vector fields satisfying this property.

ISBN 0-8053-0102-X

This completes the definition of all levels of the A-tower. It is easy to see, directly from the definitions, that $\Sigma_M \subset \mathcal{Q}_{ST}$. If the definition of Σ_Z were at hand, it would be equally evident that $\Sigma_M \subset \Sigma_Z \subset \mathcal{Q}_{ST}$, and that this short tower is a very natural progression. So, to complete this tower, we lack only

$$\mathcal{Q}_{ST} \subset \mathcal{Q}_{NC}$$

That is, Strong Transversality implies No Cycles. To prove this, it is necessary to bridge over to the third tower, of absolute stability. The construction of this bridge, a tour de force of global analysis, is well described in Robbin [1972a]. We will finally obtain:

$$\mathcal{Q}_{ST} = \Sigma_{AS} \subset \Sigma_{A\Omega} \subset \mathcal{Q}_{NC}$$

(see Fig. 7.5-1) to complete the A-tower. Meanwhile, the inclusions $\Sigma_M \subset \Sigma_{AS} \subset \Sigma_{A\Omega}$ comprises the third tower, which we will now establish.

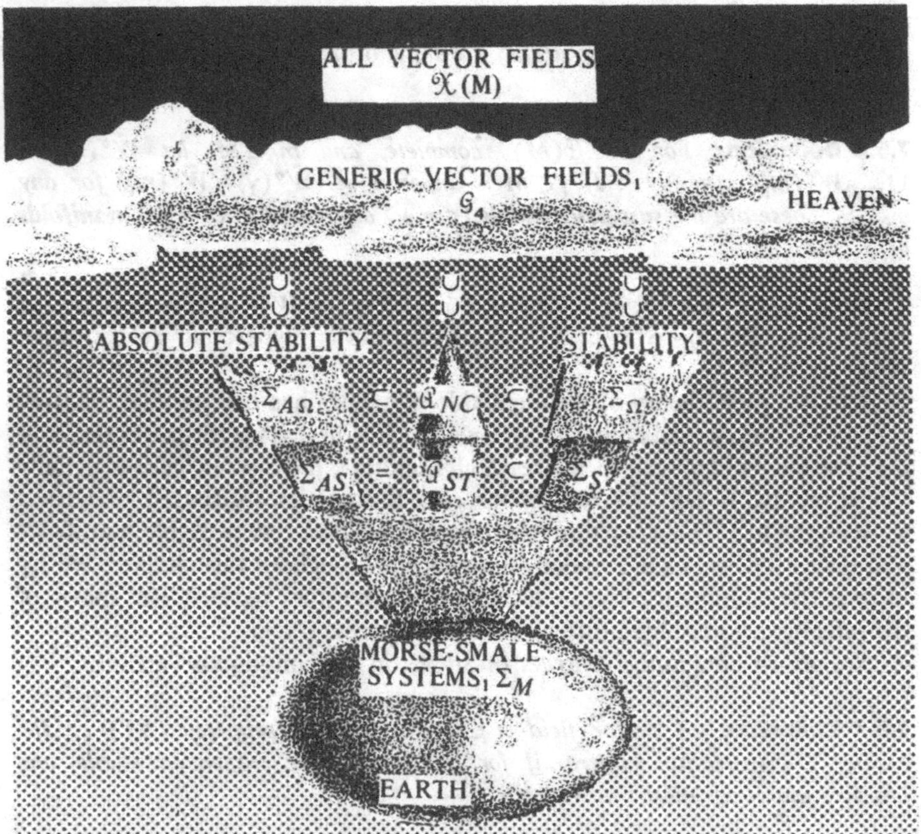

Figure 7.5-1. The three towers of differentiable dynamics.

ISBN 0-8053-0102-X

Let $C^0(M,M)$ denote continuous maps with g a Riemannian metric on M and d_0 the induced metric on $C^0(M,M)$. We endow the vector fields $\mathfrak{X}(M)$ with the C^∞ Whitney topology, and let $\| \ \|_0$ denote the C^0 sup norm induced by g.

7.5.8 Definition. *A vector field $X \in \mathfrak{X}(M)$ is* ***absolutely structurally stable*** *if there is a neighborhood $\mathfrak{N}$ of $X \in \mathfrak{X}(M)$, a function Φ: $\mathfrak{N} \to C^0(M,M)$, and a real number $K>0$ such that for all $Y \in \mathfrak{N}$:*

(i) $d_0(Id_M, \Phi(Y)) < K\|Y-X\|_0$;
(ii) *$\Phi(Y)$ carries oriented orbits of Y to oriented orbits of X.*

Let Σ_{AS} denote the set of absolutely structurally stable vector fields in $\mathfrak{X}(M)$.

From condition (ii) above, it is evident that $\Sigma_{AS} \subset \Sigma_S$. The added condition (i) is a sort of Lipschitz condition. While not obviously motivated, it simply grew out of the attempts to characterize $\mathcal{Q}_{ST}$; see Robbin [1972a] for full details. Here is the result due to Robbin [1971a], Franks [1972], Robinson [1975a], and Mañé [1975a, 1975b].

7.5.9 Theorem. *For compact manifolds, $\mathcal{Q}_{ST} = \Sigma_{AS}$.*

The proof in the context of continuous flows is found in Duistermaat [1976b].

Finally, we come to absolute Ω-stability, a concept due to Guckenheimer [1972]. We suppose, for a start, that we have $X \in \mathfrak{X}(M)$ with nonwandering set Ω_X, and use the notations of 7.5.8. Let $I_X = Id|\Omega_X \in C^0(\Omega_X, M)$.

7.5.10 Definition. *A vector field $X \in \mathfrak{X}(M)$ is* ***absolutely Ω-stable*** *if there is a neighborhood $\mathfrak{N}$ of $X \in \mathfrak{X}(M)$, a function Φ: $\mathfrak{N} \to C^0(\Omega_X, M)$, and a real number $K>0$ such that for all $Y \in \mathfrak{N}$:*

(i) $d_0(I_X, \Phi(Y)) < K\|Y-X\|_0$;
(ii) $Im[\Phi(Y)] = \Omega_Y$;
(iii) *$\Phi(Y)$ is a homeomorphism that carries oriented orbits of X in Ω_X to oriented orbits of Y in Ω_Y.*

This compares with 7.4.8, as 7.5.8 compares with 7.4.1. And as $\Sigma_S \subset \Sigma_\Omega$, so also $\Sigma_{AS} \subset \Sigma_{A\Omega}$. This inclusion of the tower of absolute stability is trivial. The final inclusion is the following, due to Franks [1972] and Guckenheimer [1972] in the case of discrete flows. Knowing of no explicit proof in the literature for continuous flows, we append a question mark to "theorem." But the techniques of Duistermaat [1976b] should be easily adapted to prove this.

7.5.11 Theorem (?) *For compact manifolds, $\Sigma_{A\Omega} \subset \mathcal{Q}_{NC}$.*

ISBN 0-8053-0102-X

Referring again to the tower picture (Fig. 7.5-1), all of the sets are defined and all the inclusions explained. This final step completes the A-tower by

$$\mathcal{Q}_{ST} = \Sigma_{AS} \subset \mathcal{Q}_{A\Omega} \subset \mathcal{Q}_{NC}$$

but a direct proof of $\mathcal{Q}_{ST} \subset \mathcal{Q}_{NC}$ should be possible, using the homoclinic technique of Abraham–Smale [1970].

We have been vague about the topmost inclusions—$\mathcal{Q} \subset \mathcal{G}_4$, for example—as the literature is incomplete, and they are not too important. Clearly, $\mathcal{Q} \subset \mathcal{G}_4$. But $\mathcal{Q} \subset \mathcal{G}_5$ is questionable, and $\mathcal{Q} \subset \mathcal{G}_6$ is probably false. So we may consider

$$\Sigma_M \subset \Sigma_Z \subset \mathcal{Q}_{ST} \subset \mathcal{Q}_{NC} \subset \mathcal{Q} \subset\subset \mathcal{G}_4$$

as the current state-of-the-art A-tower. Except for Σ_Z (see Zeeman [1975]), this was described in Smale [1970a].

What of the future? The gap $\mathcal{Q} \subset\subset \mathcal{G}_4$ still prevents pilgrims from climbing to heaven by performing good works. In these heights, however, is a glimmer. Recent work on Lorenz' equation suggests a promising way to weaken hyperbolicity in Axiom A. See Guckenheimer [1976a], Rössler [1976], Williams [1977], Ratiu and Bernard [1977] and Shaw [1978] for a description of the strange nonwandering set of this vector field on $\boldsymbol{R}^3$.

We have not mentioned an important consequence of Axiom A: The equilibria are isolated points in Ω. Thus Ω consists of isolated equilibria, isolated closed orbits, and distinct chaotic basic sets. In these latter sets, closed orbits are dense and there are no equilibria. It is this feature that is violated by the Lorenz system. Hopefully, it will lead to a new rung, or a new tower.

Finally, we should point out that large parts of the theory of $\mathcal{Q}$ have not been mentioned—especially symbolic dynamics, entropy, and homology—and the interested reader should go to the bibliography. The review articles such as Markus [1975] and Robbin [1972a], and basic books such as Manning [1975], Bowen [1975], Hirsch–Pugh–Shub [1977], Bowen [1978], and most of all Smale [1967] should be consulted.

EXERCISES

7.5A. Identify the basic sets and draw the Smale quivers for the vector field $X = -\textit{grad}(f)$, where $f\colon \boldsymbol{T}^2 \to \boldsymbol{R}$ is any Morse function: all critical points are nondegenerate (see 3.2.3). Does X satisfy Axiom A?

7.5B. Find a vector field $X \in \mathfrak{X}(\boldsymbol{R}^3)$ having a cycle of hyperbolic sets.

7.5C. Prove $\Sigma_M \subset \mathcal{Q}_{ST}$.

7.5D. Prove $\mathcal{Q}_{ST} \subset \mathcal{Q}_{NC}$ directly.

7.5.E. Prove (and publish) 7.5.11.

7.5.F. Is $\mathcal{Q} \subset \mathcal{G}_6$?

ISBN 0-8053-0102-X

7.6 BIFURCATIONS OF GENERIC ARCS

Bifurcation of vector fields refers to an instability within a parametrized family of vector fields. Let M be a paracompact manifold, so $\mathfrak{X}(M)$ is a Fréchet space. We suppose that the differential calculus has been developed in this context, as described in Chapter 1. Let C be a finite-dimensional manifold.

7.6.1 Definition. *A C^r **controlled vector field** is a C^r map $\mu: C \to \mathfrak{X}(M)$. The **control space** of μ is C and M is the **phase space**. In case C is an open disk at $0 \in R^n$ and μ is an embedding, then μ is called an **n-dimensional family** of vector fields, or an **n-parameter perturbation** of $X_0 = \mu(0)$, the **focus** of μ. A point $c \in C$ is a **robust point** of μ if there is an open neighborhood U of $c \in C$ such that for all $u \in U$, $\mu(u)$ and $\mu(c)$ are topologically conjugate vector fields. A point $c \in C$ is a **bifurcation point** of μ if it is not a robust point. Let $C_B(\mu)$ denote the set of bifurcation points of μ.*

As the control $c \in C$ is changed, the phase portrait of $X_c = \mu(c)$ is unchanged (up to topological conjugacy) until c crosses the bifurcation set C_B.

Note the similarity between ordinary points in C and structurally stable vector fields in $\mathfrak{X}(M)$. Let $\mathfrak{B}$ denote the *bad set* in $\mathfrak{X}$, that is, of nonstructurally-stable vector fields

$$\mathfrak{B} = \mathfrak{X} \backslash \Sigma_S$$

It looks as if $C_B(\mu) = \mu^{-1}(\mathfrak{B})$. This is not so, although $C_B(\mu) \subset \mu^{-1}(\mathfrak{B})$, or equivalently, $c \in C_B(\mu)$ implies $X_c \in \mathfrak{B}$. For the image $\mu[C]$ can hit $\mathfrak{B}$ a glancing blow, from one side, without actually entering a distinct component of Σ_S. Worse yet, $\mu[C] \subset \mathfrak{B}$ is possible.

Within differentiable dynamics, the study of bifurcations may be regarded as the experimental branch. A two-way street, it brings experimental results into the theory of structural stability, and brings applicable results from theory into the empirical domain.

The earliest work on bifurcations known to us is the famous experiments of Chladni in 1787. A contemporary of Beethoven in search of new musical instruments, he sprinkled fine sand on hand-held glass and copper plates, which he bowed with a cello bow, and discovered his famous nodal lines. The pressure of the bow is the control parameter.

Inspired by Chladni, Melde discovered analogous bifurcations with stretched strings, while Faraday (1831) and Matthiessen (1868) examined vibrating fluids, disagreeing over the results. The discovery of the vector field modelling these phenomena by Mathieu (1868) led to the initiation of a mathematical theory of bifurcation by Lord Rayliegh (1877). Among other things, he correctly explained (1883) the disagreement between Faraday and Matthiessen by interpolating a bifurcation between their control parameters. This explanation was rigorously justified much later by Benjamin and Ursell [1954].

ISBN 0-8053-0102-X

The origin of bifurcation theory, as a branch of differentiable dynamics, is generally attributed to Poincaré [1885], the father of dynamics. The current emphasis—on the relationship between bifurcations and structural stability—emerged in Andronov and Leontovitch (1938). For further history of this period, see Minorsky [1974, Chapter 7]. The theory assumed its modern form during the 1960s, especially in the work of Sotomayor [1974].

In this section, we give a brief introduction to the theory of generic bifurcations for arcs ($n=1$) and finally say a few words about planar families ($n=2$). This represents more-or-less the current frontier of the theory.

We wish to make clear right at the start that an exact theory does not yet exist. As long as there is a gap in the tower of stability (described in the previous section) bifurcation theory will be a house of cards. Why, then, carry on so long about this subject? There are two equal and opposite reasons. The theorists of differentiable dynamics see bifurcation theory as a tool to extend the tower of stability to heaven by dissecting the bad set $\mathfrak{B}=\mathfrak{X}\setminus\Sigma_S$. Their idea is to study a little slice C of $\mathfrak{X}$, through $\mathfrak{B}$, as if it were the whole thing.

Meanwhile, the applied dynamics community feels that attractors are the only observable features of a phase portrait, in experimental situations. For example, see Abraham [1972a] or Ruelle and Takens [1971]. In these situations, only a finite number of parameters can be varied. The universal experiment consists of a black box with n control knobs. A microcomputer inside models a dynamical system with n parameters evaluated by the knobs. When an initial point is fed into a slot at the top of the box, an ω-limit set plops out the bottom. After numerous repetitions with different initial points, most of the attractors are located. Then, one of the knobs is incremented a bit, and the process repeated. (Experimental dynamics is slow and tedious, yet a robot mathematician has been built to automate this process; see Stein and Ulam [1964].)

A theory useful to this group would provide an encyclopedia of generic bifurcations. Meanwhile, an experiment interesting to the theorists would draw pictures of all bifurcations the theory should eventually classify. (The experimentalists are ahead at the moment.) Both groups are inspired, to some extent, by the success of the unfolding technique for singularities of functions, described in Thom [1975].

Like vector fields, only generic families are interesting, or manageable. What is a generic family?

As for vector fields, a generic property of families is defined in terms of residual sets in $\mathcal{C}^r(C,\mathfrak{X}(M))$, the set of all C^r controlled vector fields with control space C and phase space M. But in what topology? There are two choices in use: The C^r Whitney topology and the C^r **graph topology**. The latter is defined as follows.

For $\mu\in\mathcal{C}^r(C,\mathfrak{X}(M))$, let $G_\mu\in\mathfrak{X}^r(C\times M)$ be defined by

$$G_\mu(c,m)=(0,\mu(c)(m))$$

Visualizing $C\times M$ with C horizontal and M vertical, G_μ is a vertical vector

ISBN 0-8053-0102-X

field. Thus

$$G: \mathcal{C}^r(C, \mathfrak{X}(M)) \to \mathfrak{X}^r(C \times M)$$

This is a standard construction in global analysis. The C^r graph topology in $\mathcal{C}^r(C, \mathfrak{X}(M))$ is defined by pulling back the C^r Whitney topology of $\mathfrak{X}^r(C \times M)$.

7.6.2 Definition. *Let $\mathcal{C}^r_F(C, \mathfrak{X}(M))$ denote $\mathcal{C}^r(C, \mathfrak{X}(M))$ with the C^r Whitney topology, and $\mathcal{C}^r_G(C, \mathfrak{X}(M))$ denote the same set with the C^r graph topology. A property $P(\mu)$ for $\mu \in \mathcal{C}^r(C, \mathfrak{X}(M))$ is **G^r-generic** (resp. **F^r-generic**) if*

$$\{\mu | P(\mu)\} \subset \mathcal{C}^r(C, \mathfrak{X}(M))$$

is residual in the C^r graph (resp. Whitney) topology.

7.6.3 Proposition. *The graph map*

$$G: \mathcal{C}^r_F(C, \mathfrak{X}(M)) \to \mathfrak{X}^r(C \times M)$$

is continuous.

Thus the graph topology is contained in the Whitney topology, and both are Baire, so residual sets are dense. A set residual in $\mathcal{C}^r_G$ is automatically residual in $\mathcal{C}^r_F$, so it is harder for a property to be generic in the graph topology.

This will make it possible for the reader to go to the literature without getting lost. In the sequel, just plain *generic* (for a property of control systems) will mean G^r-generic, and we will consider C^1 families.

So now, the known results of bifurcation may be described as generic properties of families of vector fields.

Most of these concern the one-parameter case $n = 1$. It is already known that planar perturbations are frighteningly complicated (Takens [1974a] and Arnold [1972]); some examples were mentioned in Sect. 5.6. But take courage, much is revealed by arcs.

The first result on generic arcs of vector fields, $\mu \in \mathcal{C}^1(I, \mathfrak{X}(M))$, $I = [-1, 1] \subset \boldsymbol{R}$, is special to the case $dim(M) = 2$ This is very important for the understanding of the general case.

Recall that the two-dimensional case is special, from the perspective of structural stability, because there

$$\Sigma_M = \Sigma_S$$

is generic, in fact open and dense (7.4.3), whereas for $dim(M) > 2$

$$\Sigma_M \subset \cdots \subset \mathcal{G}_4$$

ISBN 0-8053-0102-X

is a long reach. In fact, Morse–Smale ($X \in \Sigma_M$) means, in the two-dimensional case:

(F) finite number of critical elements, $\gamma_1, \ldots, \gamma_k$

(G3) all hyperbolic, and

(G4) $\Omega = \Lambda = \Gamma$: for all initial points $m \in M$, the limit sets $\alpha(m)$ and $\omega(m)$ must be critical elements, so

$$\gamma_- = \alpha(m) \to \omega(m) = \gamma_+$$

For the critical points there are three hyperbolic cases: source, saddle, and sink. For the closed orbits, there are two hyperbolic cases: source, and sink. The basins of the attracting critical elements are dense, the complement consists of the sources, saddles, and insets of saddles. This is the famous theorem of Peixoto [1959]. Note that the two-dimensional minimal sets (see 6.4.9) are not allowed. Neither are saddle connections: $\gamma_- = \alpha(m) \to \omega(m) = \gamma_+$, where both γ_- and γ_+ are equilibria of saddle type.

In this context, the first result of modern (generic dynamic) bifurcation theory gives a very satisfactory analysis of generic arcs.

7.6.4 Theorem (Sotomayor [1968]). *The following property is generic, for arcs of vector fields on two-dimensional manifolds;* $\mu \in C^1_G(I, \mathfrak{X}(M))$:

(*i*) *The bifurcation set* $I_B \subset I$ *is closed, and nowhere dense.*

(*ii*) $I_B = \mu^{-1}(\mathfrak{B})$, *that is, whenever* X_c *fails to be Morse–Smale for some* $c \in I$, *then c is actually a bifurcation point,* $c \in I_B$.

(*iii*) *Whenever c passes a bifurcation point* $b \in I_B$, *exactly one of the following four violations of the Morse–Smale conditions occurs:*

Q_1: *one of the equilibria is nonhyperbolic*

Q_2: *one of the closed orbits is nonhyperbolic*

Q_3: *two equilibria of saddle type (not necessarily distinct) have a saddle connection*

Q_4: *nontrivial recurrence,* $\Omega \neq cl(\Gamma)$, *so* X_b *has (G3) and (F), but not (G4).*

For the proof—an elegant application of transversality theory (and excellent diagrams)—see Sotomayor [1974]. The idea is to show that Q_1, Q_2, and Q_3 describe a submanifold Q in $\mathfrak{X}(M)$ of codimension one (plus bits of junk) then perturb an arc $\mu: I \to \mathfrak{X}(M)$ to be transversal to Q. The distinction between F-generic and G-generic must be treated carefully.

To make the transition to generic arcs on phase spaces of higher dimension, $dim(M) \geqslant 3$, we must first of all give up the condition (F) of finiteness of Ω. The reason for this is that, in case $dim(M) = 3$, for example, we can have a two-dimensional outset $Out(\gamma_-)$ meeting a two-dimensional inset $In(\gamma_+)$ *transversally*, in a one-dimensional orbit. Thus, saddle connections $\gamma_- \to \gamma_+$ cannot be perturbed away. And if there is a cycle,

$$\gamma_1 \to \gamma_2 \to \cdots \to \gamma_{k+1} = \gamma_1$$

ISBN 0-8053-0102-X

with all Inset–Outset intersections transversal, then it is known that $Out(\gamma_i) \cap In(\gamma_i)$ transversally for each $i=1,\ldots,k$. Further, each intersection $Out(\gamma_i) \cap In(\gamma_j)$ is contained in the nonwandering set Ω and even in $cl(\Gamma)$, the closure of the set of critical elements. This situation is called a *homoclinic tangle*, or a *generic cycle*. When, in an arc of vector fields, a saddle connection is created that completes a cycle, this tangle becomes a large-scale addition to the nonwandering set—an *Ω-explosion*—and the sets $cl(\Gamma) \subset \Omega$ are necessarily infinite.

Thus, for $dim(M)>2$, we start with the tower:

$$\Sigma_M \subset \mathcal{Q}_{ST} = \Sigma_{AS} \subset \Sigma_{A\Omega} \subset \mathcal{Q}_{NC} \subset \mathcal{Q} \subset \subset \mathcal{G}_4 \subset \mathcal{G}_3 \subset \mathcal{X}(M)$$

rather than the tower of structural stability. Recall that in this tower,

Σ_M = Morse–Smale
$\mathcal{Q}_{ST}$ = Axiom A + Strong Transversality
Σ_{AS} = Absolute Structural Stability
$\Sigma_{A\Omega}$ = Absolute Ω Stability
$\mathcal{Q}_{NC}$ = Axiom A + No Cycles
$\mathcal{Q}$ = Axiom A
$\mathcal{G}_4$ = Kupka–Smale + $[cl(\Gamma)=\Omega]$
$\mathcal{G}_3$ = Kupka–Smale
$\mathcal{X}(M)$ = All C^∞ vector fields

Corresponding to the approximation theorem of Sotomayor in the case $dim(M)=2$, which gives arcs that are nice with respect to the Morse–Smale set $\Sigma_M \subset \mathcal{X}(M)$, is the following, which yields arcs in general position with respect to the Kupka–Smale set $\mathcal{G}_3 \subset \mathcal{X}(M)$. This combines results of Brunovsky [1970], Sotomayor [1973b], and Newhouse and Palis [1973b].

7.6.5 Theorem of Generic Arcs. *The following property is generic, for arcs $\mu \in C^1(I, \mathcal{X}(M))$ of vector fields on finite-dimensional manifolds:*

(*i*) *The bifurcation set $I_B \subset I$ is closed.*
(*ii*) *$I_B = I_3 \cup I_4$, where*

$$I_3 = \mu^{-1}(\mathcal{X} \setminus \mathcal{G}_3)$$

$$I_4 = \mu^{-1}(\mathcal{X} \setminus \mathcal{G}_4)$$

are closed, I_3 is countable and nowhere dense, and $I_3 \cap int(I_4) = \varnothing$. Thus, whenever X_c fails to be Kupka-Smale, or has nontrivial recurrence, $\Omega \neq cl(\Gamma)$, then c is actually a bifurcation point.

(*iii*) *Whenever c passes a bifurcation point $b \in I_3$, exactly one of the following three violations of Kupka–Smale occurs:*
Q_1: one of the equilibria is quasi-hyperbolic
Q_2: one of the closed orbits is quasi-hyperbolic

ISBN 0-8053-0102-X

Q_3: two critical elements (not necessarily distinct) have a saddle connection which is quasi-transversal.

(iv) *Whenever a passes a bifurcation point $b \in I_4$, X_b has nontrivial recurrence $\Omega \neq cl(\Gamma)$.*

Here quasi-hyperbolic means, roughly, that the critical element fails to be hyperbolic through the passage of only one (or, a single complex conjugate pair) of the characteristic exponents across the imaginary axis (compare the Hopf bifurcation described in Sect. 5.6). Similarly, *quasi-transversal* means, more or less, that the two immersed manifolds cross nontransversally in the simplest way, by only one dimension too much. For the precise definitions, see Newhouse and Palis [1973b] (beware, the definition of bifurcation there uses Σ_{AS} in place of Σ_S—we could call this **absolute bifurcation**), where the Ω-explosion caused by creation of a homoclinic tangle is fully dissected.

At this point you may ask: What happens to the phase portrait of X_c as c passes $b \in I_B$ such that Q_i occurs, $i=1$, 2, or 3? We return to this question in the next section, at least for $i=1$ or 2, and give there some examples. Also, the diagrams in Sotomayor [1973b] are very instructive.

Before ending this section, it would be appropriate to describe the generic bifurcations for k-parameter families $k>1$. But unfortunately not only is there precious little known here, but worse yet, what is known is frighteningly complex. These discoveries we owe to Arnold and Takens—who ventured where others feared. Considering the two-parameter analogs of Q_1 and Q_2, Takens finds generic properties and normal forms for the simultaneous passage of two principal characteristic exponents across the imaginary axis. The resulting classification is not finite, due to "certain symmetry properties." For an excellent discussion, see Arnold [1972] and Takens [1973a]. Also, see Fig. 5.6-8.

EXERCISES

7.6A. Prove 7.6.3.

7.6B. Draw a microscopic portrait of a homoclinic tangle.

7.7 A ZOO OF STABLE BIFURCATIONS

To be interesting, a controlled vector field or family of vector fields should be more than generic—it should be stable. The motivation of stability and the corresponding definitions are similar to those for vector fields. As the theory of stable families is hardly begun, we will not present these definitions.

ISBN 0-8053-0102-X

Presumably, a tower will be constructed (for arcs)

$$\mathcal{S} \subset \Sigma \subset\subset \mathcal{G} \subset C^1(I, \mathfrak{X}(M))$$

where $\mathcal{S}$ stands for *simple arcs*, Σ for *stable arcs* (however defined), and $\mathcal{G}$ for generic arcs, having all known generic properties—including those of Brunovsky, Sotomayor, Newhouse, and Palis, described in the preceding section. The first proposal for Σ (Sotomayor [1973b]) was promptly defeated as a generic property (Guckenheimer [1973a]).

As this subject settles down, it will hopefully be extended to k-parameter families with $k > 1$. The concomitant maturation of stability theory for vector fields will aid this extension. In the meanwhile, one might say that the theory of stable bifurcations consists of a few prototypical examples, which ought to be stable according to any reasonable definition. That is, if one perturbs them, no qualitative features seem to change.

In this taxonomic section, we describe these prototypical bifurcations. In the descriptions, yielding to our softness toward the viewpoint of applications, the bifurcations of attractors will be especially emphasized. Let M have a probability measure σ, that is, a measure with $\sigma(M) = 1$. The probability of putting $m \in M$ into the top of the empirical black box, and having a limit set $\Omega_i = \omega(m)$ plop out the bottom, is $P_i = \sigma[In(\Omega_i)]$. So if Ω_i is an attractor, the probability is the volume of its basin $B_i = In(\Omega_i)$, and otherwise we expect $P_i = 0$.

We consider single, isolated, stable bifurcations of arcs. Thus $\mu: I = [-1, 1] \to \mathfrak{X}(M)$ and $I_B = \{0\}$. For simplicity, we imagine $\mu[-1, 0) \subset \Sigma_M$, and μ is "stable," hence generic, that is, $X_0 = \mu(0)$ is a bifurcation characterized by Q_1, Q_2, or Q_3 in the theorem on generic arcs (7.6.5). We consider a taxonomy of five types:

I. *Local*: near a Q_1 equilibrium
II. *Local*: near a Q_2 closed orbit
III. *Global*: caused by a Q_3 saddle connection
IV. *Global*: caused by a local bifurcation
V. *Global*: other types—especially those caused by chaotic sets

Without further ado, here is the zoo of prototypical bifurcations. Each illustration consists of three parallel movies, using the following conventions:

———	track of an attractor
- - - - - -	track of saddle or repellor
○[○$^{S^1}$]	a repellant equilibrium (resp. closed orbit)
◐	a quasi-hyperbolic set
●[●$^{S^1}$]	an attracting equilibrium (resp. closed orbit)

ISBN 0-8053-0102-X

INDEX OF ILLUSTRATIONS

Figures 7.7-1–7.7-10

Index of Illustrations

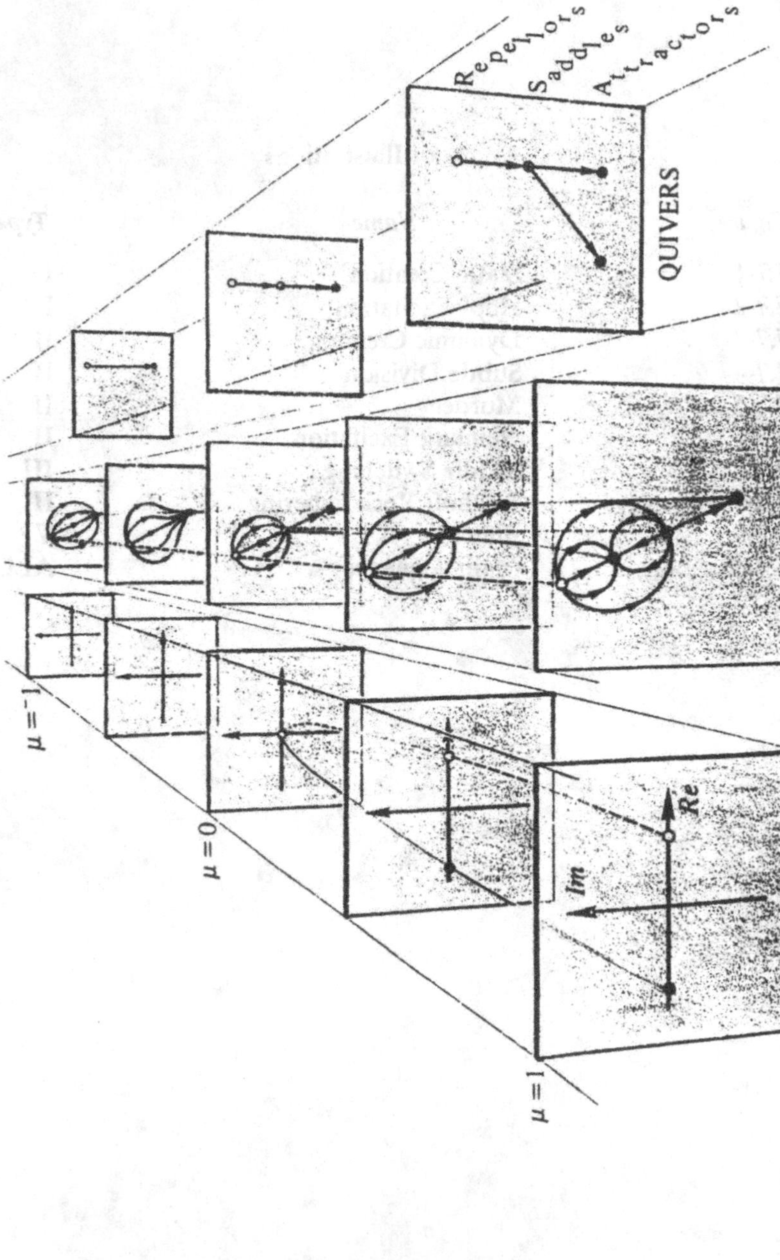

ISBN 0-8053-0102-X

ISBN 0-8053-0102-X

Figure 7.7-1. Static Creation.
Type: I—Local Q_1
Other names: True bifurcation; Fork; Saddle-node
History: Poincaré [1885]
Modern reference: Takens [1973a] and Sotomayor [1974]
Minimum dimension: One
Description: The vector field contracts near a point $p \in M$. All changes are inside a small ball B at p. The solid cylinder of orbits through B are pinched like a ponytail. The cylindrical surface (the husk) is pinched to a goblet at p. This goblet becomes the inset $In(p)$ of the new equilibrium point. This becomes a separatrix, confining the basin of a new attractor within the goblet.
Variants: Static annihilation (replace μ by $-\mu$).

The new separatrix can receive outsets from several critical elements.

Instead of an attractor and a saddle, any two equilibria p,q with $dim\, In(p) = dim\, In(q) + 1$ can be created.

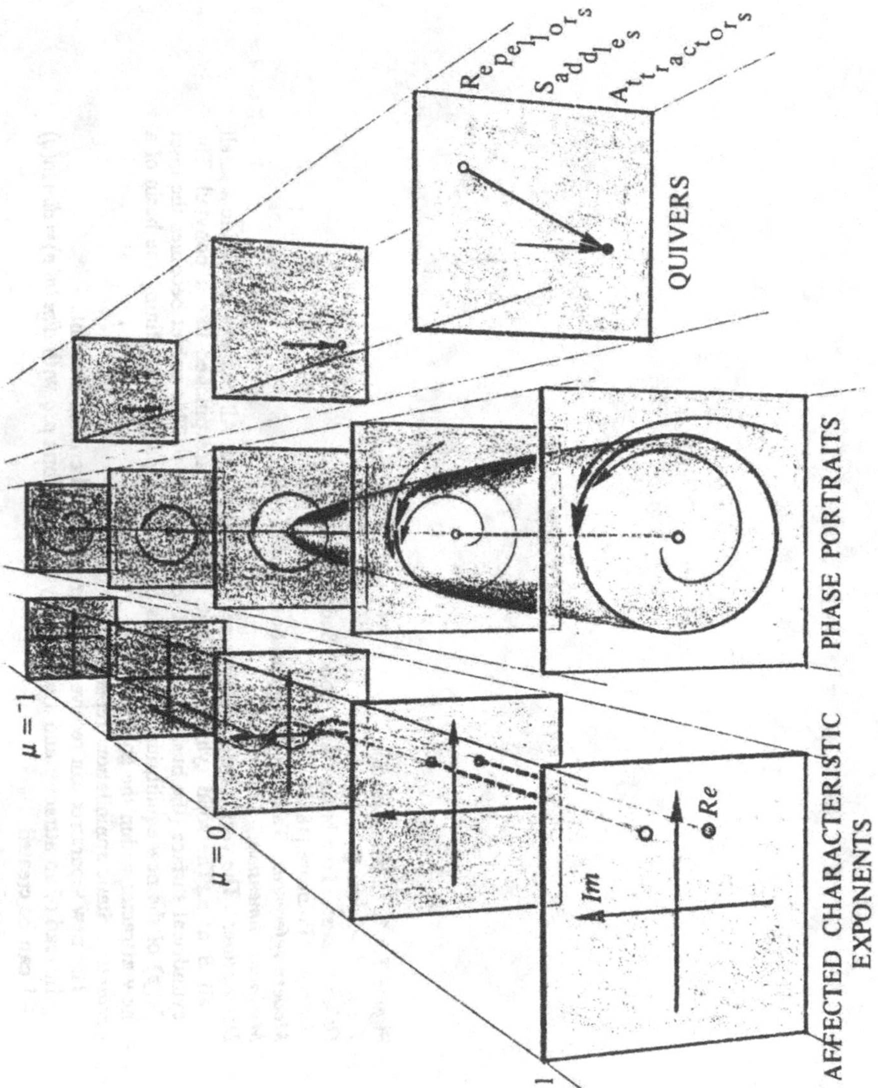

ISBN 0-8053-0102-X

ISBN 0-8053-0102-X

Figure 7.7-2. Hopf Excitation.
Type: I—Local Q_1
Other names: Hopf catastrophe or bifurcation; Soft self-excitation
History: Poincaré [1885], Andronov and Witt [1930], Hopf [1942]
Modern reference: Takens [1973a], Kopell–Howard [1974], Marsden–McCracken [1976], Alexander and Yorke [1978]
Minimum dimension: Two
Description: Suddenly, a point attractor is replaced by a tiny attractive closed orbit, which then grows in amplitude. A pair of conjugate characteristic exponents crosses through the imaginary axis.
Variants: Death (replace μ by $-\mu$).

Instead of an attractor, a saddle p may become a closed orbit γ, with $\dim In(p) = \dim In(\gamma)$.

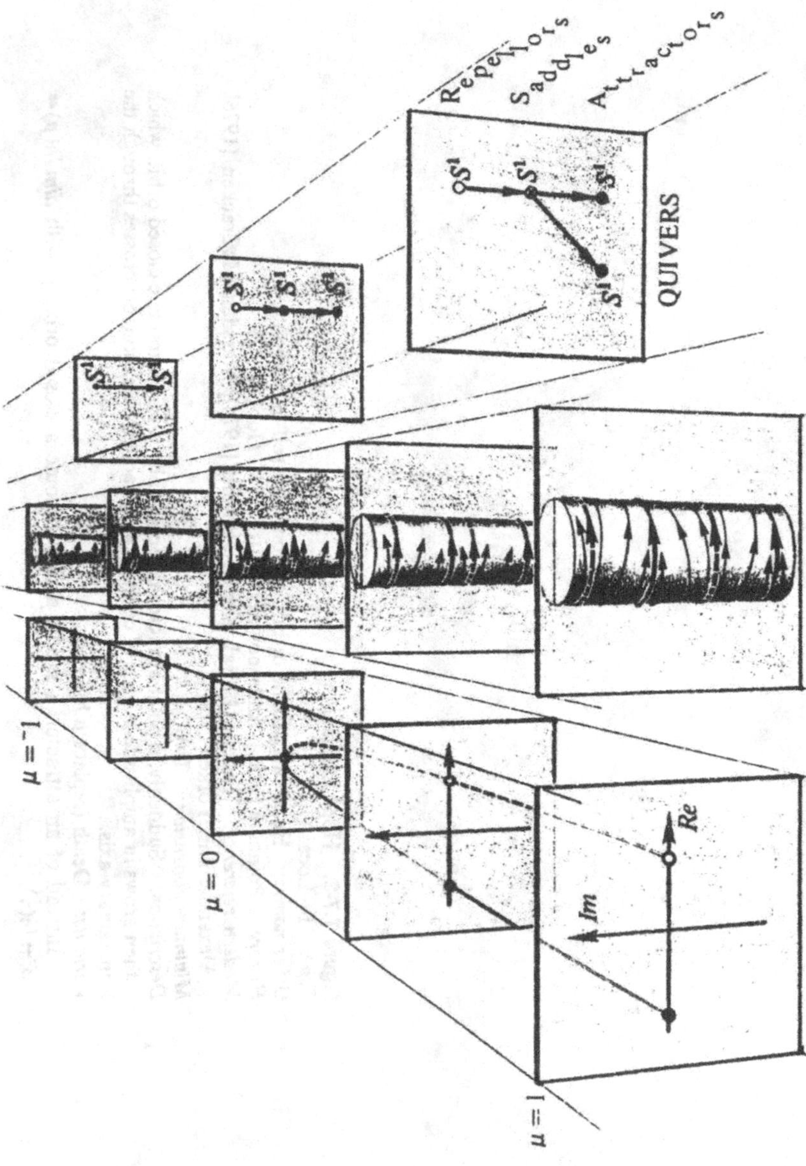

ISBN 0-8053-0102-X

ISBN 0-8053-0102-X

Figure 7.7-3. Dynamic Creation.
Type: II—Local Q_2
Other names: True bifurcation; Hard self-excitation
History: Poincaré [1885]
Modern reference: Brunovsky [1971a], Takens [1973]
Minimum dimension: Two
Description: Suddenly, an attractive closed orbit appears, with a small basin. Nearby, a new closed orbit of saddle type—the phantom dual—has a hypersurface shaped like a jello aspic mold for an inset. This surface is a new separatrix, delimiting the new basin. The new characteristic exponents diverge from zero.
Variants: Dual suicide (replace μ by $-\mu$).
The new closed orbits need not be an attractor and an adjacent saddle.
Any pair of closed orbits of adjacent type will do:

$$\dim In(\gamma_1) = \dim In(\gamma_2) + 1$$

The separatrix (inset of new saddle) can receive outsets (saddle connections) from several saddles or repellors.

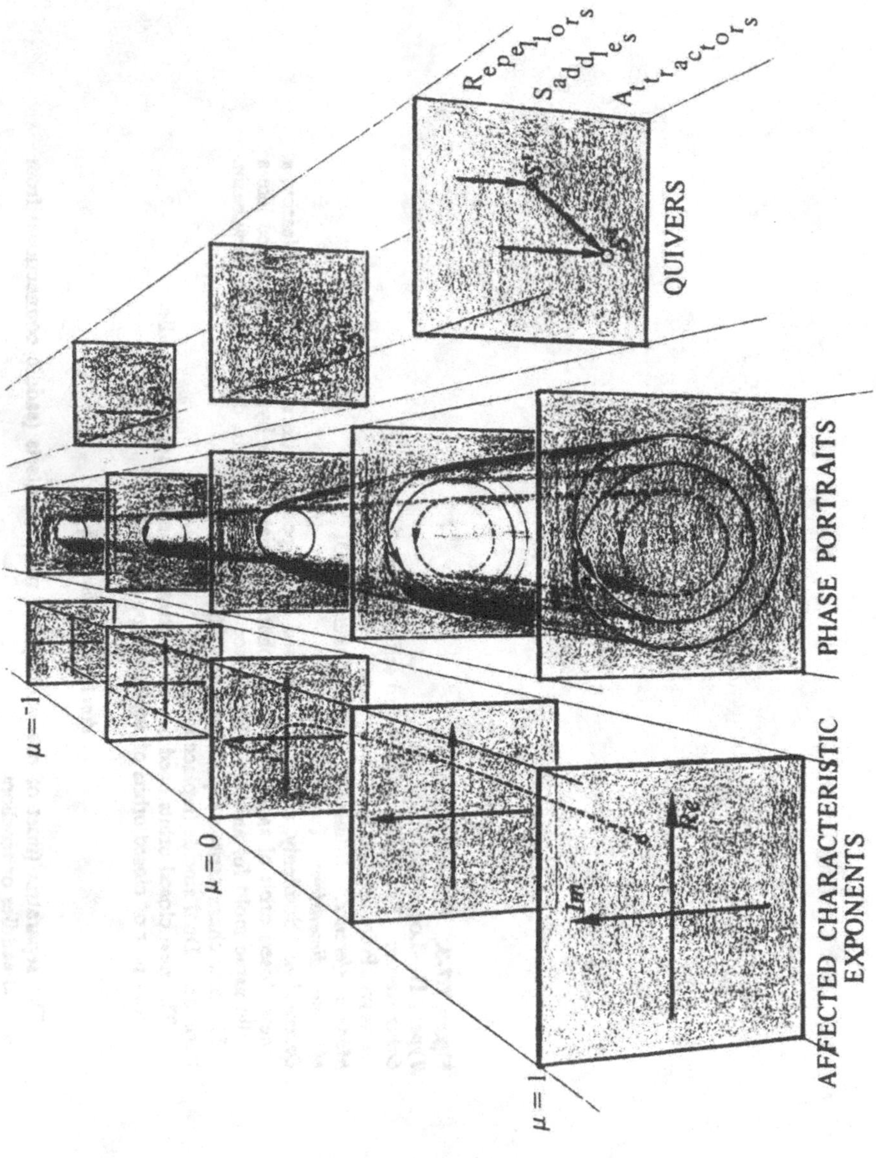

ISBN 0-8053-0102-X

ISBN 0-8053-0102-X

Figure 7.7-4. Subtle Division.
Type: II—Local Q_2
Other names: Subharmonic resonance, flip
History: Steinmetz [1933].
Modern reference: Brunovsky [1971], Iooss and Joseph [1977], Newhouse and Palis [1976].
Minimum dimension: Two
Description: An attractive closed orbit γ becomes a thick [$dim\, In(\gamma) = dim\, M - 1$] saddle, as one of its characteristic multipliers passes through -1 [or, equivalently, a characteristic exponent passes through $(2k+1)i$]. A new attractive closed orbit, of twice the period, is emitted.
Variants: Murder (see next figure)

The closed orbit affected could be a saddle, which loses one dimension of thickness. In this case, the emitted subharmonic saddle cycle is as thick as the original.

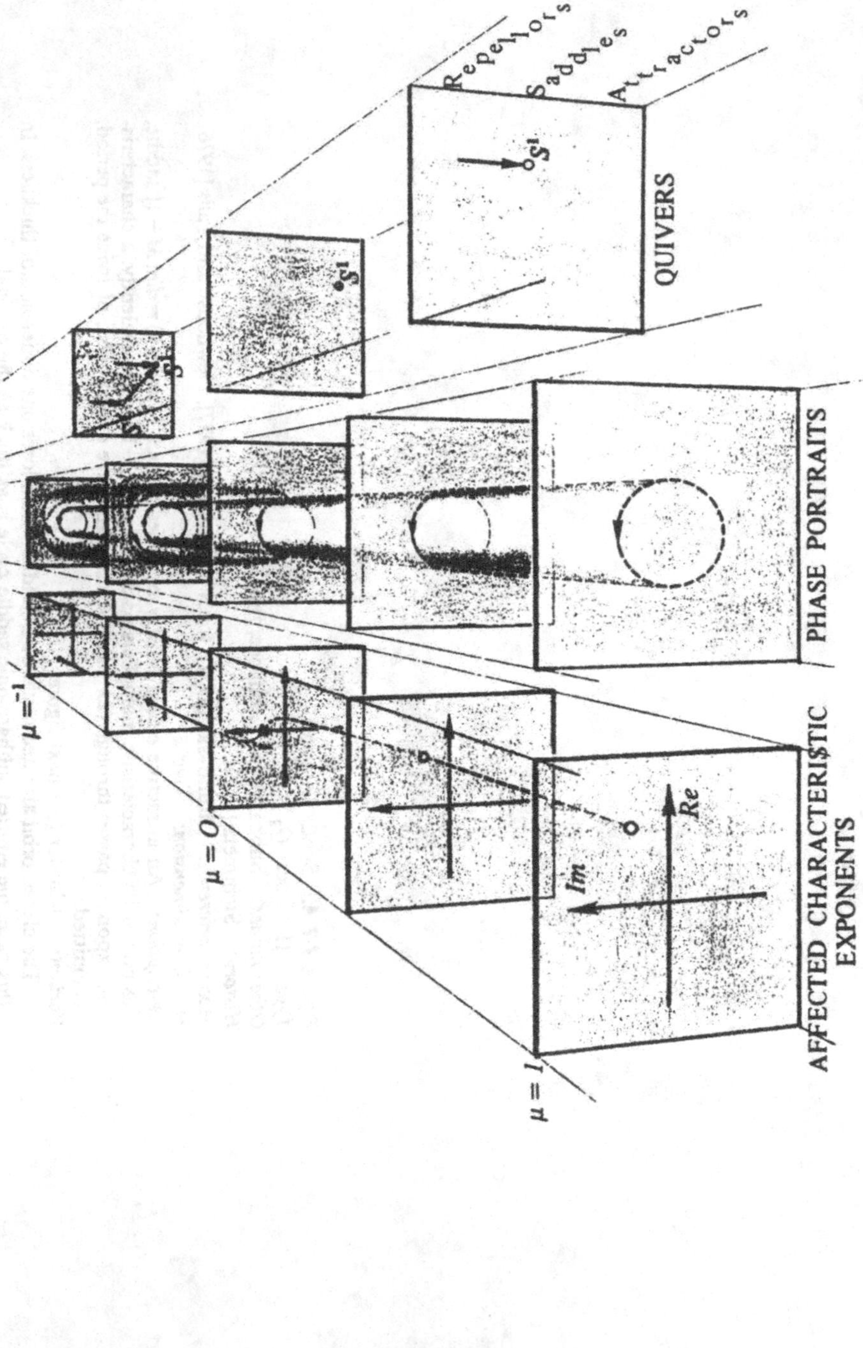

ISBN 0-8053-0102-X

ISBN 0-8053-0102-X

Figure 7.7-5. Murder.
Type: II—Local Q_2
Other names: Destabilization; Hard self-excitation (with μ reversed)
History: None?
Modern reference: Brunovsky [1971], Iooss and Joseph [1977]
Minimum dimension: Two
Description: Exactly like subtle division, except that a thin harmonic is absorbed, rather than an attractive subharmonic emitted.
Variants: The attractive closed orbit killed could be a saddle. Then a subharmonic saddle cycle one-dimension thinner arrives, and the original saddle cycle loses one dimension in thickness.

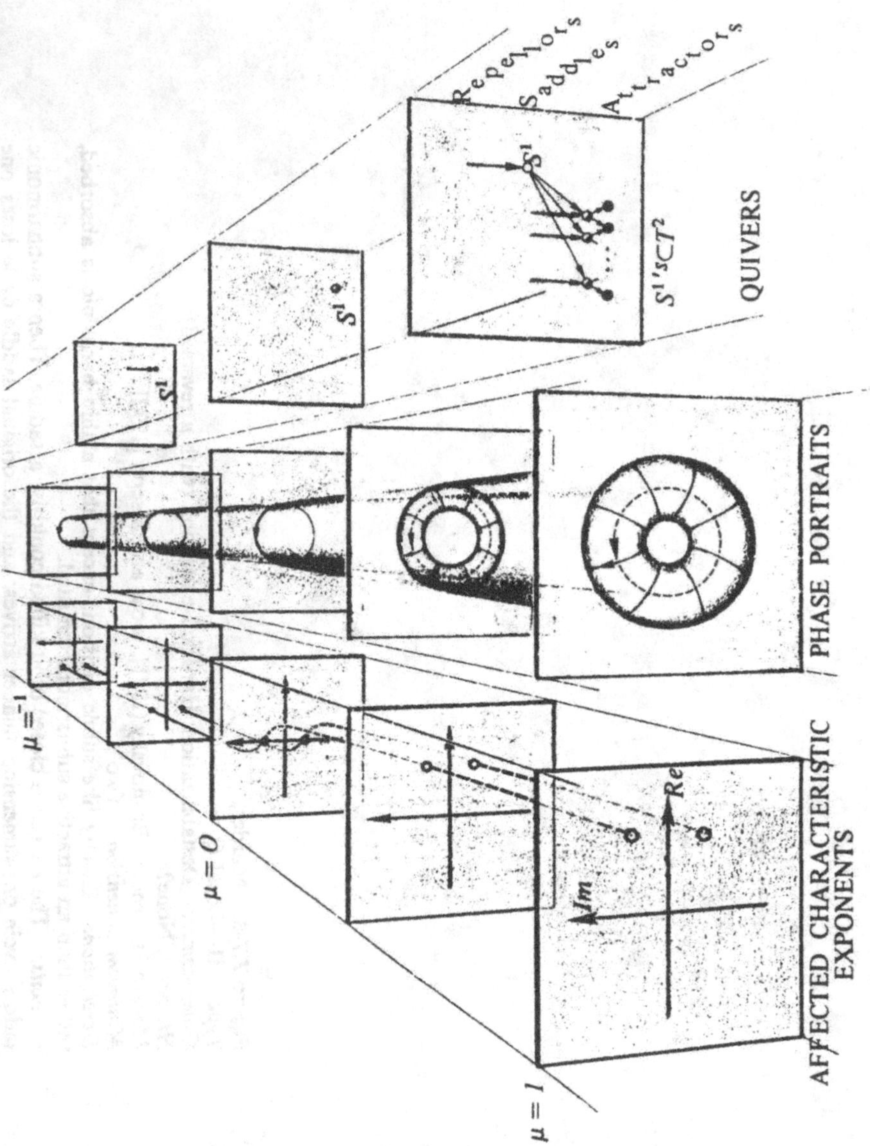

ISBN 0-8053-0102-X

ISBN 0-8053-0102-X

Figure 7.7-6. Naimark Excitation.

Type: II—Local Q_2

Other names: Hopf bifurcation; Naimark bifurcation

History: Naimark [1959]

Modern reference: Ruelle–Takens [1971], Marsden–McCracken [1976], Iooss [1975–1976]

Minimum dimension: Two

Description: Similar to Hopf excitation, but starts with an attractive closed orbit. A pair of conjugate characteristic multipliers traverse the unit circle outwards (or exponents cross the imaginary axis rightwards) and an attractive invariant torus T^2 is created surrounding the closed orbit—now a thick saddle.

Warning: The torus is not a basic set, but will contain a finite number of closed orbits, some attracting.

Variants: The attractive closed orbit might be a saddle.

An invariant torus of higher dimension T^k can become excited to a T^{k+1} (*Takens excitation*; cf. Iooss and Chencinere [1977]).

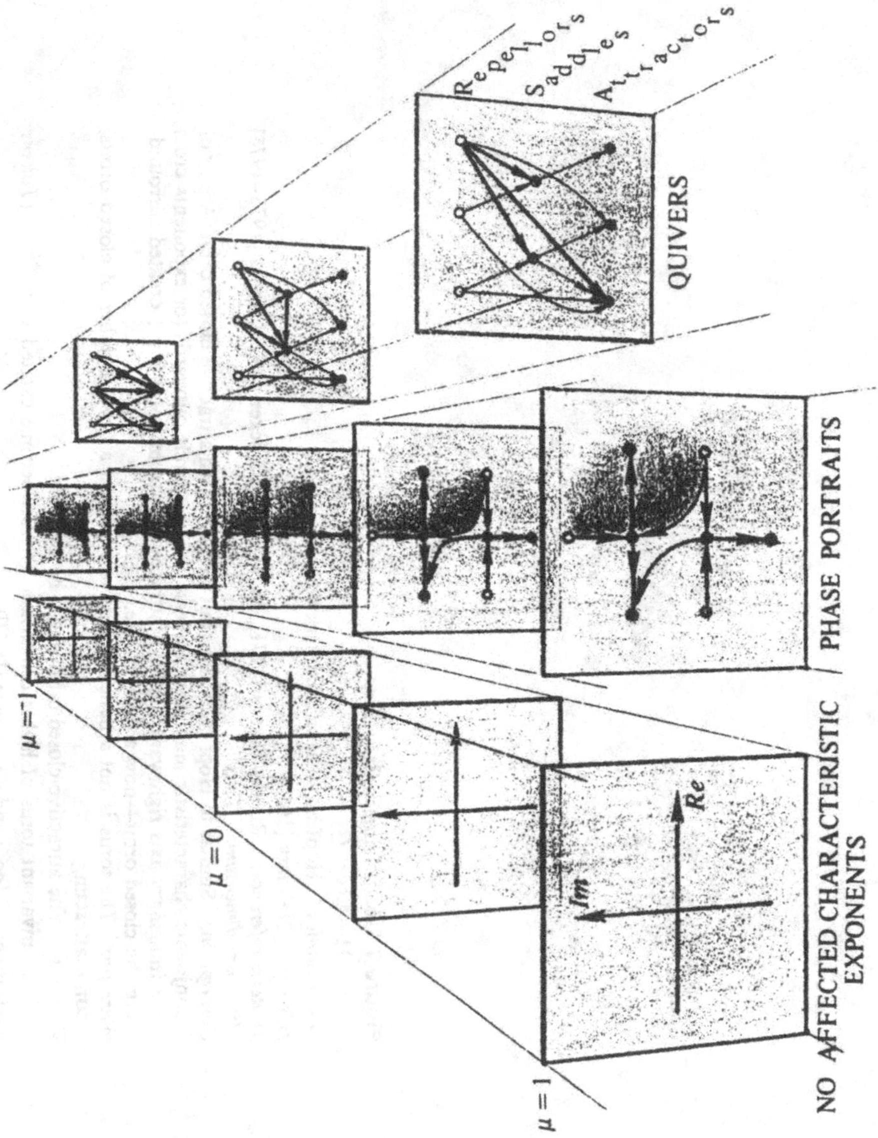

ISBN 0-8053-0102-X

ISBN 0-8053-0102-X

Figure 7.7-7. Saddle Switching.

Type: III—Equilibrium saddle to equilibrium saddle

Other names: Basin bifurcation

History: ?

Modern reference: Sotomayor [1974]

Minimum dimension: Two

Description: The outset of a saddle is moved from one attractor to another. Enroute, it must pass the separatrix of these basins, which is the (thick) inset of another saddle. En passant, there is an illegal (but quasi-hyperbolic) saddle connection (touché). In these phase portraits, one basin is shaded. Only the basins are changed in this bifurcation—possibly in topology as well as volume.

Variants: The inset of any equilibrium of saddle type can be moved, but here we are interested in the thick insets (hypersurfaces) which comprise the separatrices between the basins of attraction. The passage of a thick inset through any outset may be quasi-transversal—in higher dimensions—even though that outset be completely engulfed (touché) at contact.

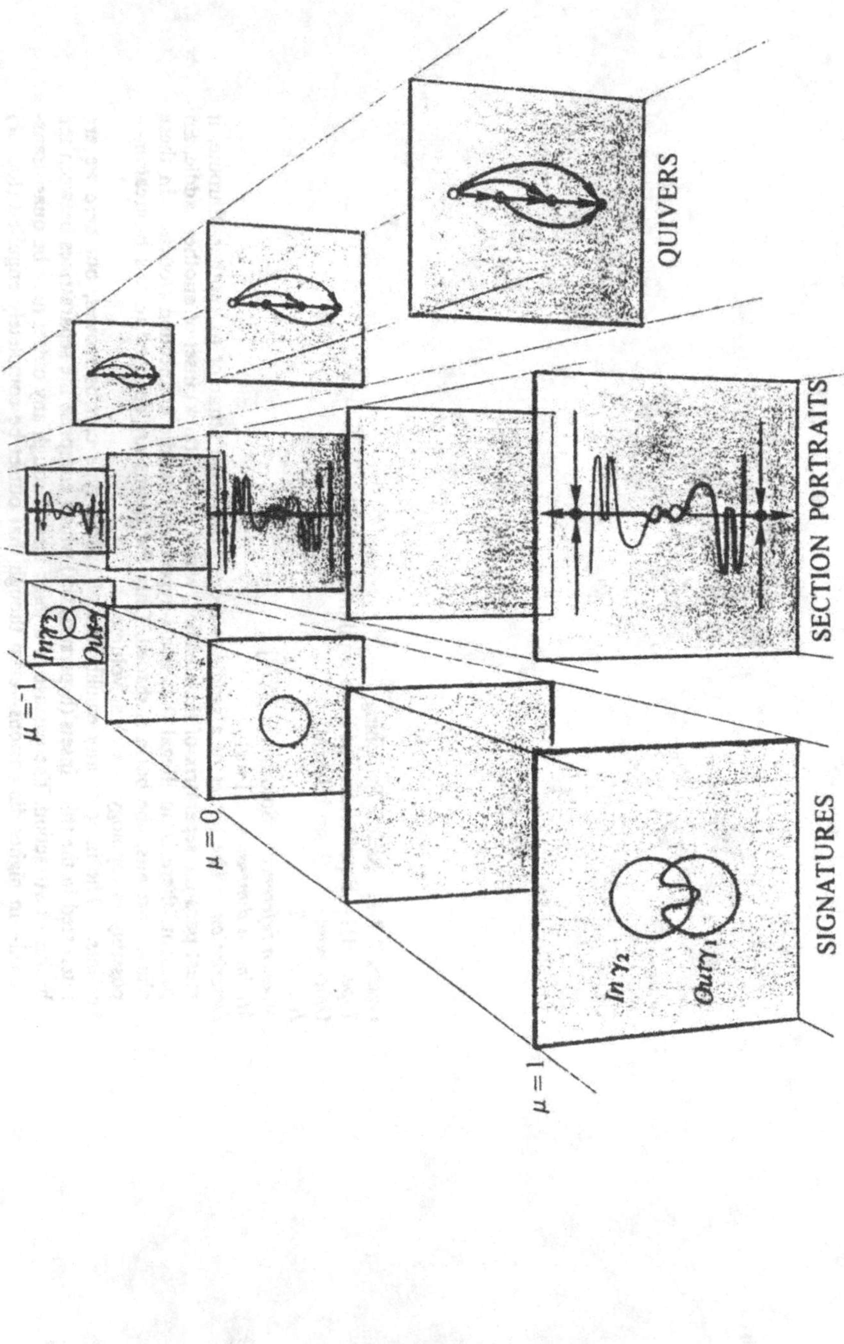

ISBN 0-8053-0102-X

ISBN 0-8053-0102-X

Figure 7.7-8. Birkhoff rechambering.
Type: III
Other names: Signature bifurcation
History: Birkhoff [1935]
Modern reference: Newhouse–Palis [1973a]
Minimum dimension: Three
Description: This particular example starts with the gradient vector field of the usual height function on T^2, then multiplies by S^1 to get a flow on T^3 with a global section.

There are four closed orbits: a repellor at the top, two thick saddles (γ_1, γ_2) ($dim\,In = dim\,Out = 2$), and an attractor at the bottom. This is then perturbed so $Out(\gamma_1) \cap In(\gamma_2)$ is transversal. We illustrate the crossection T^2 flattened a bit. The action in this movie has $Out(\gamma_1)$ fixed, while $In(\gamma_2)$ rotates. At the bifurcation point $c = 0$ the two cylinders are coincident, $Out(\gamma_1) = In(\gamma_2)$. This is a quasi-transversal intersection, in three dimensions. After rotating through $Out(\gamma_1)$, $In(\gamma_2)$ looks much as before. But the topology of the intersection (called the *signature* by Birkhoff) is changed. The two infinitely crossing cylinders in T^3 look like a Chambered Nautilus shell, hence the name, *rechambering*.

Variants: This is characteristic of saddle connections involving closed orbits, in dimensions greater than two.

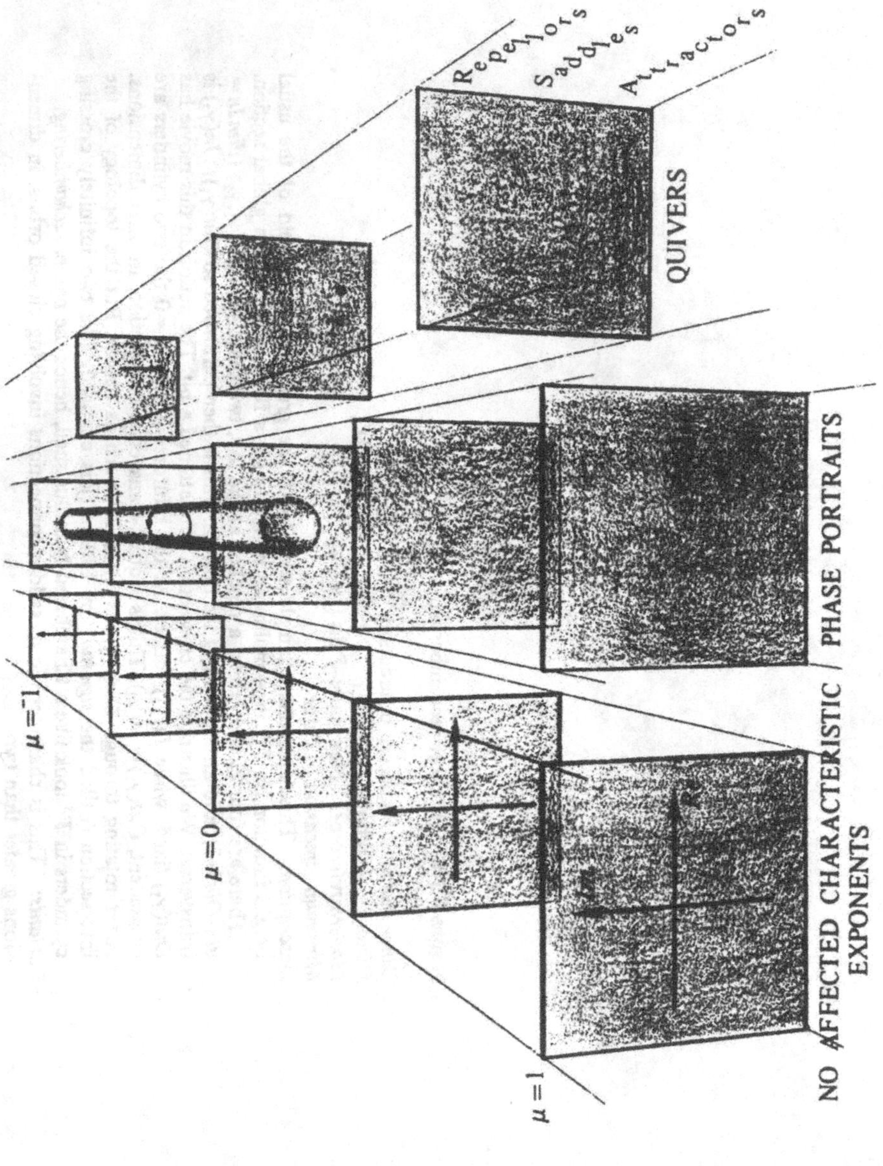

ISBN 0-8053-0102-X

Figure 7.7-9. Blue sky catastrophe.
Type: V—Other
Other names: Disappearance into the blue
History: Ruelle–Takens [1971]
Modern reference: Alexander and Yorke [1978]; Devaney [1978a]
Minimum dimension: Two?
Description: As μ approaches 0 from the left, the period of the attracting closed orbit tends to infinity.
Variants: Closed orbit need not be attracting.
Out of the blue: reverse μ.
There is a chaotic basic set, and the closed orbit disappears into it.
The closed orbit may disappear into a cyclic saddle connection (see Fig. 5.6-1).

ISBN 0-8053-0102-X

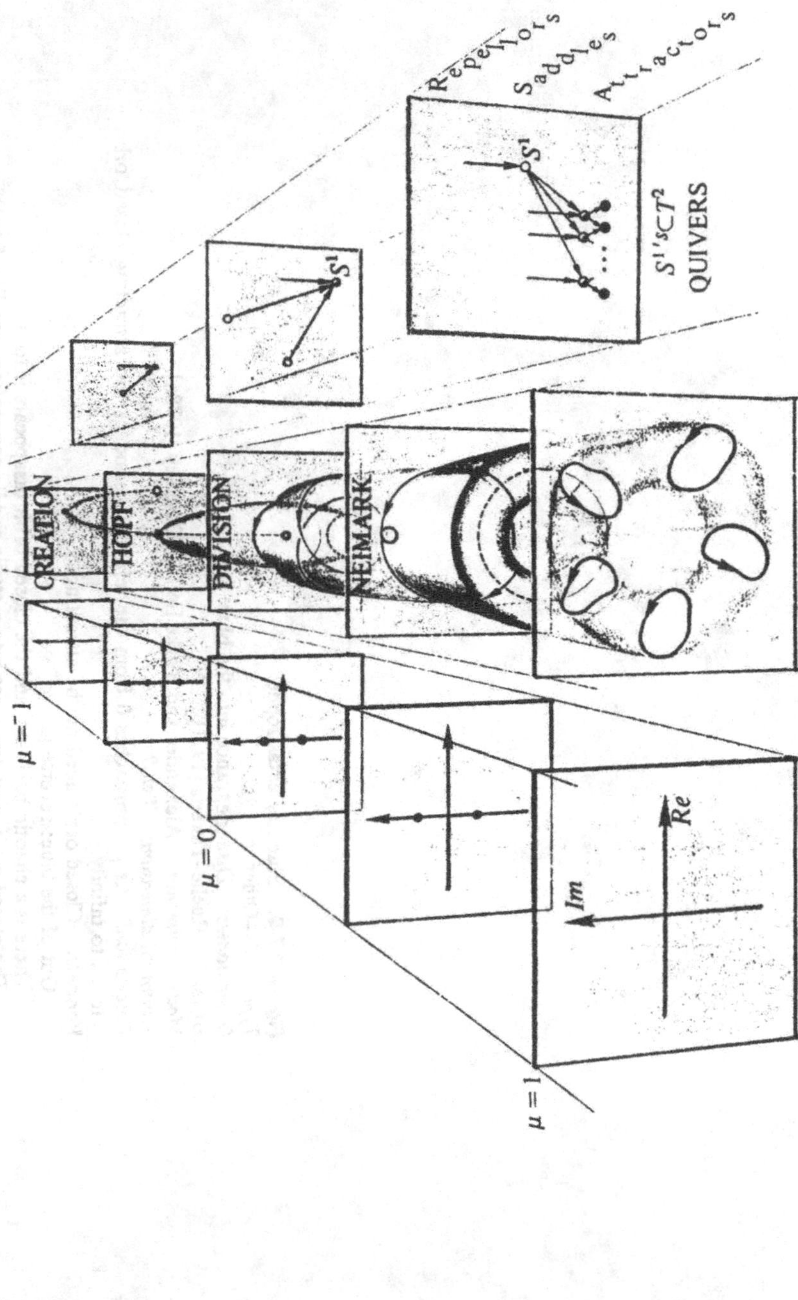

ISBN 0-8053-0102-X

Figure 7.7-10. The Main Sequence.
Other names: Generic evolution
History: Abraham [1972a]
Modern reference: Takens [1973a]
Minimum dimension: Three
Description: A vector field with no attractors moves along a generic arc, passing a sequence of stable bifurcations. In the first ($\mu = -1$, stable creation) an attracting equilibrium is born. Thereafter, the other nonfatal, increasing bifurcations occur. A large family is produced. One generation is illustrated here.
Variants: Countable.

ISBN 0-8053-0102-X

7.8 EXPERIMENTAL DYNAMICS

The prior section suggests that, whenever a vector field is perturbed, a bifurcation is imminent. And there are so many possibilities—according to the nascent theory—especially if there is more than one dimension of perturbation. (One example: The *Takens bifurcation* was illustrated in Sect. 5.6. See Fig. 5.6-8.)

One wonders: If a particular vector field is taken in hand, its portrait drawn, then varied, does all this happen? So there comes a time when a theorist might turn to experimental work.

In this section, we will say just a few introductory words about the history and prospects for this field—experimental dynamics.

A. The special device period: 1787 – 1918. We have already had occasion to speak of the pioneering work of the musicians Chladni and Melde in the era of Beethoven, the consequent work of Faraday, Lord Rayleigh, and so on. We may not fairly distinguish this line of inquiry (which continues in the present day) from experimental physics, hydrodynamics, continuum mechanics, and so on.

B. The radio period: 1918–1953. Vacuum tube oscillators and radio frequency circuits were used to draw phase portraits for forced oscillations of vector fields in the plane—that is, two-parameter families. Subharmonic resonance was thoroughly studied.

The works of van der Pol, Appleton, Lienard, McCrumm, Roelle, Duffing, Strutt, Pederson, and many others are systematically explained in Hayashi [1953a], the outstanding experimentalist in this period, and Stoker [1950].

C. The analog period: 1953–1962. As modular architecture evolved in the electronic industry, it became possible to model most classical vector fields (with polynomial or sinusoidal coefficients). The oscillator was replaced by the analog computer. High-speed and graphical output are the outstanding characteristics of these machines, still widely used. See Shaw [1978] for an outstanding example.

D. The digital period: 1962 on. Mathematicians lost no time in adapting the general purpose digital computer to the problem of phase portraiture of dynamical systems. The pioneering works of Lorenz [1962] and Stein and Ulam [1964] are still studied. The current trend is toward utilization of new developments in computer graphics for phase portrait output, and special

ISBN 0-8053-0102-X

architecture to implement faster portrait algorithms, see Abraham [1978]. It seems likely that this field will expand quickly, along with the electronic revolution.

To date, the leading accomplishments of the experimentalists are the discoveries of Hopf bifurcation by van der Pol, of subharmonic resonance by Duffing, and the onset of turbulence by Lorenz, Stein and Ulam.

ISBN 0-8053-0102-X

CHAPTER 8

Hamiltonian Dynamics

In this chapter we outline the recent developments of this very specialized qualitative theory. The typical picture of a Hamiltonian system that emerges is extremely complex, poorly understood, and still evolving. The main results, which differ rather sharply from the differentiable case of the preceding chapter, are relevant for a number of applications including the rigid body and the n-body problem. The latter will be briefly discussed in the next chapters.

8.1 CRITICAL ELEMENTS

In this section we consider the characteristic multipliers for critical elements of Hamiltonian vector fields and explain why such a critical element cannot be expected to be elementary in general.

First we take up the case of a critical point. Suppose (M,ω) is a symplectic manifold, $H \in \mathfrak{F}(M)$, and X_H is the Hamiltonian vector field with Hamiltonian H. Recall that $m \in M$ is a critical point of X_H iff $X_H(m)=0$ and obviously this occurs iff $dH(m)=0$, that is, m is a critical point of H. The characteristic exponents of X_H at m are defined as the eigenvalues of the linear mapping $X_H'(m) \in L(T_mM, T_mM)$ and T_mM is symplectic with the form $\omega(m)$. The main restriction on the characteristic exponents in the Hamiltonian

Ralph Abraham and Jerrold E. Marsden, Foundation of Mechanics, Second Edition

ISBN 0-8053-0102-X

case results from the fact that $X'_H(m)$ is infinitesimally symplectic, that is, is a linear Hamiltonian system. See Sect. 3.3.

From the infinitesimally symplectic eigenvalue theorem it follows that *the characteristic exponents of X_H at a critical point $m \in M$ occur in pairs $(\lambda, -\lambda)$ of the same multiplicity. Thus if λ is a characteristic exponent, so are $\bar{\lambda}, -\lambda, -\bar{\lambda}$, all of these having the same multiplicity. The exponent zero always has even multiplicity.*

We see now why a Hamiltonian critical point cannot be elementary in general. For m is elementary iff there are no characteristic exponents on the imaginary axis. However, if there are exponents $\pm i\beta$ of multiplicity one, small perturbations in H, thus X_H and $X'_H(m)$, perturb the exponents only slightly, and the exponents $\pm i\beta$ are trapped on the imaginary axis. Moreover, it follows that the stable and unstable manifolds of the critical point $m \in M$ have the same dimension, and the center manifold is even dimensional. The center manifold cannot in general be removed by a small perturbation of H alone, although its dimension may be reduced by four if there is a purely imaginary exponent of multiplicity two.

In the remainder of the section we consider analogously the case of a closed orbit $\gamma \subset M$ of the Hamiltonian vector field X_H. The characteristic multipliers of X_H at γ are the eigenvalues of the tangent $T_m\Theta$, where $m \in \gamma$ and Θ is a Poincaré map on a local transversal section. Alternatively, the characteristic multipliers are the eigenvalues (omitting one $+1$) of the tangent T_mF_τ, where $m \in \gamma$, F is a flow box around γ, and τ is the period of γ. As F_τ is a symplectic diffeomorphism and $F_\tau(m) = m$, we get the following restriction on the characteristic multipliers.

8.1.1 Proposition. *The characteristic multipliers of X_H at a closed orbit $\gamma \subset M$ occur in pairs (λ, λ^{-1}) of the same multiplicity. Thus if λ is a characteristic multiplier, so are $\bar{\lambda}, \lambda^{-1}, \bar{\lambda}^{-1}$, all having the same multiplicity. The multiplier one always occurs with odd multiplicity at least one.*

This is an immediate consequence of the symplectic eigenvalue theorem. We see that γ can never be elementary, as there is always at least one multiplier equal to one, and thus of modulus one.

Even restricted to an energy surface Σ_e, the Poincaré mapping (see 8.1.3) must satisfy the symplectic eigenvalue theorem. Thus, as in the case of a critical point we still expect in M (resp. in Σ_e) stable and unstable manifolds of the same dimension, and possibly a center manifold of even (resp. odd) dimension that cannot be eliminated by small perturbations of the Hamiltonian function.

Following Robinson [1970b] we strike out the redundant characteristic multipliers (CM's) as follows.

8.1.2 Definition. *The **principal characteristic multipliers**, or PCM's of X_H at γ are defined as follows: the CM 1 of multiplicity $2k+1$ is a PCM of multiplicity*

ISBN 0-8053-0102-X

k, the CM -1 *of multiplicity 2k is a PCM of multiplicity k. To a unimodular CM pair* $(\lambda, \bar{\lambda};\ |\lambda|=1,\ Im(\lambda)>0)$ *of multiplicity k corresponds the single PCM* λ *with multiplicity k. To a real CM pair* $(\lambda, \lambda^{-1};\ |Re(\lambda)|>1)$ *of multiplicity k corresponds the single PCM* λ *with multiplicity k. And finally, to a CM quadruplet* $(\lambda, \bar{\lambda}, \lambda^{-1}, \bar{\lambda}^{-1};\ |\lambda|>1,\ Im(\lambda)>0)$ *of multiplicity k corresponds a pair of PCM's* $(\lambda, \bar{\lambda})$ *of multiplicity k. Counting multiplicities, the* $2n-1=2(n-1)+1$ *CM's have been replaced by* $(n-1)$ *PCM's (see Fig. 8.1-1). If* λ *is a unimodular PCM of* γ, $\lambda=\exp(2\pi i\alpha)$ *with* $\alpha\in[0,\frac{1}{2}]$, *then* α *is a* ***transverse frequency*** *of* γ.

The following proposition gives some additional information about the Poincaré map in the Hamiltonian case. Note that if γ is a closed orbit of X_H, then we may assume γ lies in some regular energy surface Σ_e since near γ, dH must be nonzero.

8.1.3 Proposition. *Let* (M, ω) *be a symplectic manifold,* $H\in\mathfrak{F}(M)$ *and* γ *a closed orbit of* X_H *lying in a regular energy surface* Σ_e. *Then there exists a local transversal section S at* $m\in\gamma$ *and a Poincaré map* $\Theta: W_0\to W_1$ *on S, such that the following hold:*

(*i*) (W_0, ω_0) *and* (W_1, ω_1) *are contact manifolds, where* $\omega_i=i_j^*\omega$, $i_j: W_j\to M$ *being the inclusion,* $j=0, 1$;

(*ii*) Θ *is a canonical transformation; that is,* Θ *preserves H, and there is a function* $\delta\in\mathfrak{F}(W_0)$ *such that* $\Theta^*\omega_1=\omega_0-d\delta\wedge dH$; *moreover,* δ *is the period shift of the Poincaré map described in 7.1.2, p. 523;*

(*iii*) *There exists* $\varepsilon>0$ *and regular energy surfaces* $\Sigma_{e'}$ *for* $e'\in(e-\varepsilon, e+\varepsilon)$, *such that* $(S_{e'}, \omega_{e'})$ *is a symplectic submanifold of codimension two and* $\Theta|W_0\cap S_{e'}$ *is a symplectic diffeomorphism onto* $W_1\cap S_{e'}$, *where* $S_{e'}=S\cap\Sigma_{e'}$, $i: S_{e'}\to M$ *is the inclusion mapping, and* $\omega_{e'}=i^*\omega$.

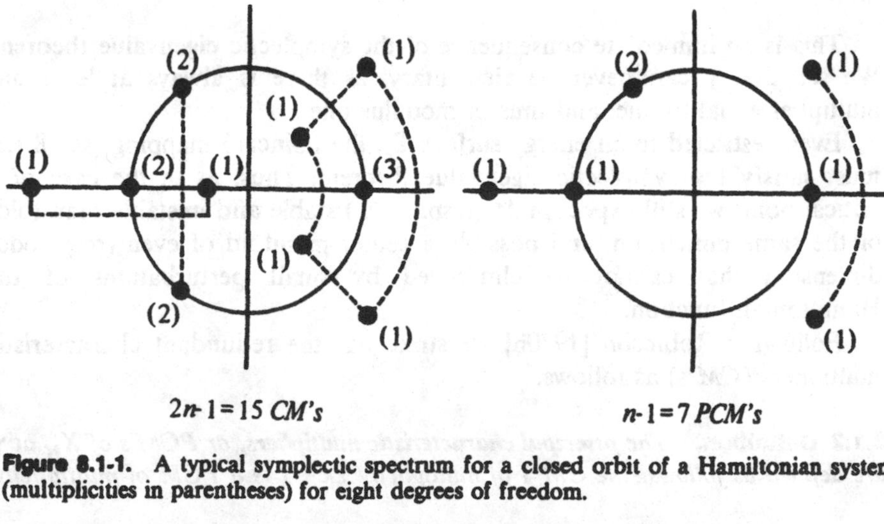

Figure 8.1-1. A typical symplectic spectrum for a closed orbit of a Hamiltonian system (multiplicities in parentheses) for eight degrees of freedom.

ISBN 0-8053-0102-X

Proof. Let (U,φ) be a Hamiltonian flow box chart at $m\in\gamma$ and S be defined by $t=0$. Then, if $i: S\to M$ is inclusion,

$$i^*\omega=i^*\left(dH\wedge dt+\sum_{i=2}^{n}dq^i\wedge dp_i\right)=\sum_{i=2}^{n}dq^i\wedge dp_i$$

as $t\circ i=0$. Hence (i) is clear. Also, since γ is compact, there is an open neighborhood V of γ on which $dH\neq 0$. Hence $\Sigma_{e'}=V\cap H^{-1}(e')$ is a regular energy surface for e' in some interval $(e-\varepsilon,e+\varepsilon)$, and, restricted to $S\cap\Sigma_{e'}$, ω becomes $\Sigma_{i=2}^{n}dq^i\wedge dp_i$, so the first part of (iii) is clear. For (ii), a simple computation shows that for $s\in S$.

$$T\Theta(s)\cdot Y=T_1F(s,\tau-\delta(s))\cdot Y-(d\delta(s)\cdot Y)X_H$$

where $\Theta(s)=F(s,\tau-\delta(s))$ as in 7.1.1. Also, as $F_{\tau-\delta(s)}$ is symplectic and $H\circ F_{\tau-\delta(s)}=H$, we have

$$\Theta^*\omega_1(s)(X,Y)=\omega(\Theta(s))(T\Theta(s)\cdot X,T\Theta(s)\cdot Y)$$

$$=\omega(s)(X,Y)-\tfrac{1}{2}X_H(Y)d\delta(X)(s)+\tfrac{1}{2}X_H(X)d\delta(Y)(s)$$

so that (ii) follows. Finally, (ii) implies (iii) by restricting to $\Sigma_{e'}$. ■

Thus, on $S_{e'}$, Θ preserves the volume element

$$i^*(\omega^{n-1})=dq^2\wedge\cdots\wedge dq^n\wedge dp_2\wedge\cdots\wedge dp_n$$

a classical and useful fact (see, for example, Pars [1965, p. 446]). In addition, the properties of 8.1.3 and this consequence hold for any transversal section S (sufficiently small). This follows from existence and local conjugacy. Recall that time t and energy H are canonically conjugate coordinates according to the Hamiltonian flow box theorem. According to the preceeding theorem, we have constructed a chart (U,ϕ) at $m\in\gamma$ so that

$$\phi: U\to \mathbf{R}\times\mathbf{R}^{n-1}\times\mathbf{R}\times\mathbf{R}^{n-1}$$

$$: u\mapsto(\delta(u),q,\varepsilon(u),p)$$

where $\varepsilon(u)=H(m)-H(u)$ and $\delta(u)$ is the time along the orbit through u, from the Poincaré section S, defined by $\delta(u)=0$.

We refer to this in future as a *power chart* (see Fig. 8.1-2)

ISBN 0-8053-0102-X

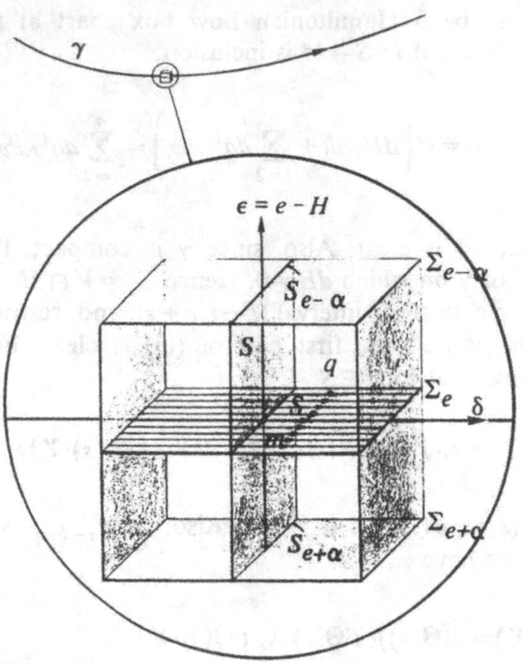

Figure 8.1-2. A power chart at a point in a closed orbit.

8.2 ORBIT CYLINDERS

Now we return to the fact that the closed orbit γ cannot be elementary, as there is always at least one *CM* equal to one, hence on the unit circle. The multiplier one corresponds to an eigenspace on which the Poincaré map is the identity in first approximation, suggesting the possibility of the existence of an entire cylinder of closed orbits $\{\gamma_s\}$ with a parameter s in which γ is an element $\gamma = \gamma_0$, say.

8.2.1 Definition. *An **orbit cylinder** of X_H is an embedding $\Gamma: S^1 \times (a,b) \to M$ such that for all $e \in (a,b)$, $\gamma_e = \Gamma[S^1 \times \{e\}]$ is a closed orbit of X_H. An orbit cylinder is **regular** if $H[\gamma_e] = e$, and Γ is transversal to every energy surface Σ_e. That is, $H \circ \Gamma$ has no critical point. See Fig.* 8.2-1.

Using the implicit mapping theorem, we now show the existence of an orbit cylinder if γ has the characteristic multiplier one with multiplicity one, or equivalently, if one is not a *PCM* of γ.

The proof is rather similar to the proof of 5.6.6 on the existence of periodic orbits for perturbed systems.

8.2.2 Regular orbit cylinder theorem. *If γ is a closed orbit of X_H, then it is contained in a regular orbit cylinder iff one is not a PCM of γ.*

ISBN 0-8053-0102-X

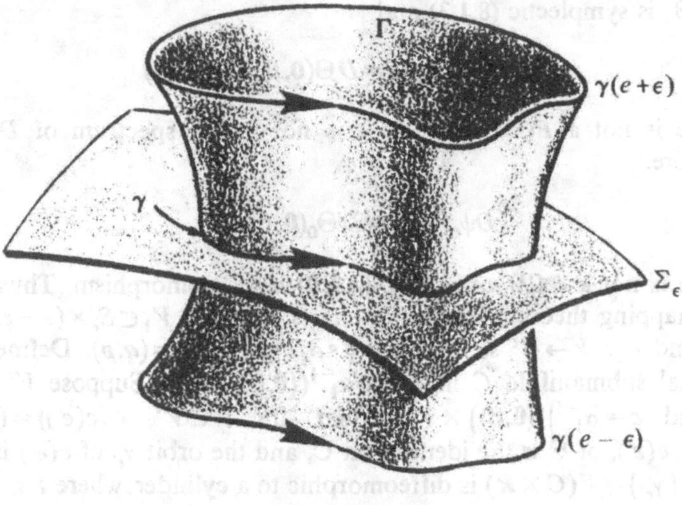

Figure 8.2-1

Proof. Let (U,ϕ) be a power chart for γ,

$$\phi: U \to \boldsymbol{R} \times \boldsymbol{R}^{n-1} \times \boldsymbol{R} \times \boldsymbol{R}^{n-1}$$
$$: u \mapsto (\delta, q, \varepsilon, p)$$

where $\delta(u)$ is the time since the Poincaré section S and $\varepsilon(u) = e - H(u)$. Let Θ denote the Poincaré map, represented in this chart. Thus, within $\delta = 0$,

$$\Theta(q, \varepsilon, p) = (Q, \varepsilon, P)$$

where P, Q: $W_0' \to \boldsymbol{R}^{n-1}$ are the nontrivial components of Θ. We now define auxiliary maps:

$$\psi: W_0' \to \boldsymbol{R}^{n-1} \times \boldsymbol{R}^{n-1}$$
$$: (q, \varepsilon, p) \mapsto (Q, P) - (q, p)$$

and

$$\Theta_0: W_{00}' \to \boldsymbol{R}^{n-1} \times \boldsymbol{R}^{n-1}$$
$$: (q, p) \mapsto (Q(q, 0, p), P(q, 0, p))$$

where

$$W_{00}' = \{(q, 0, p) \in W_0'\}$$

and finally define

$$\psi_0 = \Theta_0 - I$$

ISBN 0-8053-0102-X

Now Θ_0 is symplectic (8.1.3) and

$$D\Theta_0(0,0)(h,k)=D\Theta(0,0,0)(h,0,k)$$

Thus: one is not a *PCM* of γ iff it is not in the spectrum of $D\Theta_0(0,0)$. Furthermore,

$$D\psi_0(0,0)=D\Theta_0(0,0)-I$$

Thus: one is not a *PCM* of γ iff $D\psi_0(0,0)$ is an isomorphism. Thus, by the implicit mapping theorem, there are charts $\alpha_1\colon U_1\times V_1\subset S_e\times(e-\varepsilon,e+\varepsilon)\to U'\times V'$ and $\alpha_2\colon V\to U'$ so that $\alpha_2\circ\psi\circ\alpha_1^{-1}((q,p),e')=(q,p)$. Define a one-dimensional submanifold C in S by $\alpha_1^{-1}(\{(0,0)\}\times V')$. Suppose $V'=(e-\varepsilon', e+\varepsilon')$ and $c=\alpha_1^{-1}|\{(0,0)\}\times V'$. Then, for $e'\in V'$, $\psi(c(e'))=(0,0)$ or $\Theta(c(e'))=c(e')$, or Θ is the identity on C, and the orbit $\gamma_{e'}$ of $c(e')$ is closed. Clearly $\cup\{\gamma_{e'}\}=F(C\times R)$ is diffeomorphic to a cylinder, where F is the flow of X_H. ■

In general, this cylinder cannot be extended for all energies without encountering a singularity, that is, a critical point or a closed orbit for which the hypotheses of the theorem fail. In particular, as the transversal intersection of the orbit cylinder with energy surfaces is one of the defining properties for a regular orbit cylinder, the hypothesis (no $PCM=1$) may fail because the orbit cylinder becomes tangent to an energy surface.

This situation has been studied by Robinson [1970a] who found a nondegenerate way for tangency to occur so that the orbit cylinder may be continued past the nondegenerate tangency. We now describe this nondegeneracy condition.

Again, let (U,ϕ) be a power chart at $m\in\gamma$, a closed orbit of X_H. If $e=H(m)$, then $\gamma\subset\Sigma_e$ and $S_e=S\cap\Sigma_e$, where S is the Poincaré section ($\delta=0$). Let $\pi\colon S\to S_e$ be the map whose representative with respect to (U,ϕ) is the projection

$$\pi'\colon(0,q,\varepsilon,p)=(q,p)$$

Note that in the proof of the preceding theorem $\psi'=(\pi\circ\Theta)'-\pi'$, where the prime denotes local representation.

Using the auxiliary map π, we may now give the nondegeneracy condition of Robinson.

8.2.3 Definition. *A closed orbit γ of X_H is* **0-elementary** *if there exists a power chart at a point $m\in\gamma$ such that with its Poincaré section projection π, Poincaré map Θ, and section S, the linear map*

$$T_m(\pi\circ\Theta)-T_m\pi\in L(T_mS,T_mS)$$

is surjective.

ISBN 0-8053-0102-X

It is easy to show that this condition is independent of the point $m \in \gamma$ and the power chart.

Global analysts will recognize this as the definition of transversal intersection $\pi \pitchfork \pi \circ \Theta$ for the two maps

$$\pi, \pi \circ \Theta : S \to S_e$$

in a neighborhood of $m \in S_e$. Or, equivalently, $\Theta \pitchfork I$ (for the maps Θ, I: $S \to S) \bmod \pi$ (see Abraham and Robbin [1967]).

Note that the time δ in the second case of 8.2.4 is related to the period τ_δ of the closed orbit γ_δ by

$$\tau_\delta = \tau_0 - \delta$$

so the period τ_δ may be used as the cylinder parameter.

8.2.4 Orbit cylinder theorem (Robinson). *If a closed orbit γ of X_H is 0-elementary, it belongs to an orbit cylinder. Furthermore, either no PCM of γ is one and the cylinder is regular, or, exactly one PCM of γ is one (multiplicity one) and the period parametrizes the cylinder: $\tau(\gamma_\lambda) = \lambda$.*

The proof is a simple modification of the argument given above for the regular case.

EXERCISES

8.2A. Demonstrate 8.2.2 directly in the case of a Hamiltonian derived from a Riemannian metric.

8.2B. Show that the family of closed orbits produced in Liapunov's theorem (5.6.7) is a regular orbit cylinder.

8.2C. Show that 8.2.3 is independent of the power chart.

8.2D. Prove 8.2.4 in the irregular case.

8.3 STABILITY OF ORBITS

We consider now the question of orbital stability of critical elements in the Hamiltonian case. We have seen that if a critical element has a stable manifold, it also has an unstable manifold and is therefore α^+ unstable for all cases $\alpha = o, a, L$. Thus there is the possibility of stability only if all of the characteristic multipliers have modulus one. This case, in which the entire manifold is a center manifold for the critical element, is called an **elliptic element**, a **pure center**, or the **oscillatory case.** If the characteristic multipliers are expressed $(1, e^{\pm i\alpha_1}, \ldots, e^{\pm i\alpha_{n-1}})$ for a closed orbit, or $(e^{\pm i\alpha_1}, \ldots, e^{\pm i\alpha_n})$ for a critical point, $\alpha_i \in [0, 2\pi)$, the real numbers $\{\alpha_i / 2\pi\}$ are called the **frequencies** (or, in the case of a closed orbit, the **transverse frequencies**).

In the oscillatory case, the flow is a rotation in linear approximation, so asymptotic Liapounov stability is not to be expected. Orbital stability is

ISBN 0-8053-0102-X

natural, however, and always occurs in the case of a critical point in two dimensions. Thus the natural question for a Hamiltonian vector field is this: *When is a critical element of pure center type o^+-stable?* Certainly this has an obvious importance in celestial mechanics, for example, in Laplace's problem of the stability of the solar system.

In this section, we give some limited results on stability of oscillatory critical points and closed orbits.

We begin by rephrasing the results obtained in Sect. 5.6 on Liapunov's theorem.

8.3.1 Definition. *Let (M, ω) be a symplectic manifold and $H \in \mathfrak{F}(M)$. Then a critical point m of X_H is called* $\mathcal{H}$ ***elementary*** *if each of the following conditions hold:*

(*i*) *Zero is not a characteristic exponent*
(*ii*) *If λ is a characteristic exponent with real part zero, then λ has multiplicity one.*
(*iii*) *If λ and μ are characteristic exponents with real part zero and imaginary part positive, then λ and μ are independent over the integers; that is, if $n_1\lambda + n_2\mu = 0$ for $n_1, n_2 \in \mathbf{Z}$, then $n_1 = n_2 = 0$.*

Thus the critical element is $\mathcal{H}$ elementary iff the frequencies are independent over the rationals. The nonelementary case is sometimes called the **problem of small divisors** in celestial mechanics.

Of course $\mathcal{H}$ *elementary* is not as elementary as *elementary*, and the qualitative behavior of orbits close to an $\mathcal{H}$ elementary critical point can be much more complicated. In addition to the center manifold, which exists in any case, we get in the $\mathcal{H}$ elementary case an additional very important simplification in the behavior of nearby orbits, the splitting of the center manifold into the two-dimensional invariant *subcenter manifolds* discovered by Liapounov.

8.3.2 Theorem (Liapounov subcenter stability). *Let (M, ω) be a symplectic manifold. $H \in \mathfrak{F}(M)$, and $m \in M$ be an $\mathcal{H}$ elementary critical point of X_H. Then, if $i\beta$ is a characteristic exponent of m ($\beta \in \mathbf{R}$), there is a two-dimensional submanifold C_β with $m \in C_\beta$ such that:*

(*i*) *$T_m C_\beta$ is the eigenspace corresponding to the characteristic exponents $i\beta$ and $-i\beta$;*
(*ii*) *C_β is an invariant submanifold of X_H;*
(*iii*) *C_β is a union of closed orbits γ_r such that there is a diffeomorphism $\varphi: C_\beta \to D_1$ (D_1 is the disk of radius one in $\mathbf{R}^2$) with $\varphi(\gamma_r)$ a circle of radius r about $0 = \varphi(m)$. Moreover, if τ_r denotes the period of γ_r, $\lim_{r\to 0}\tau_r = 2\pi/\beta$.*

See Sect. 5.6 for the proof and discussion. (The hypothesis that the characteristic exponents are independent over $\mathbf{Z}$ is of course a little stronger

ISBN 0-8053-0102-X

than required; we only need to know that no other characteristic exponents are integer multiples of $i\beta$ for C_β to exist.)

Note that if all characteristic exponents are imaginary, then $T_mC_{\beta_1}\oplus\cdots\oplus T_mC_{\beta_k}=T_mC$ in the linear case.

For the case of an $\mathcal{H}$ elementary critical point, the subcenter stability theorem gives a splitting of the center into two-dimensional invariant manifolds that are o^+-stable. Thus a very important question is: Under what conditions does stability on all subcenter manifolds imply stability of the center? This is somewhat similar to the center-stable theorem, but at present we do not even have a plausible conjecture to offer.

In the case of a closed orbit of oscillatory type, the analogous questions are still important, and in addition we do not even know the existence of subcenter manifolds. (See Exercise 8.3E.) Stability in a given subcenter manifold is, however, the subject of Moser's theorem.

Consider the case of *two degrees of freedom*, $M=T^*W$, where W is a two-dimensional manifold. If Σ_e is a regular energy surface, $\gamma\subset\Sigma_e$ a closed orbit and S a local transversal section in Σ_e; then Σ_e is three dimensional, and S is two dimensional. A Poincaré map Θ on S can be considered a diffeomorphism in the plane $\mathbf{R}^2$ keeping the origin fixed. Then γ is a pure center iff $T_0\Theta$ is a rotation. In this case the entire three-manifold Σ_e is a center (or subcenter) manifold for $\gamma\subset\Sigma_e$. Moser's theorem gives a sufficient condition for the existence of a dense set of invariant circles in S, thus invariant tori in Σ_e, implying $o^\pm$-stability of γ. In the remainder of this section we describe Moser's results without proofs. These results are applied, in Sect. 10.3, to the restricted three-body problem.

To gain some perspective for the results we are going to describe we pause momentarily to consider the history of the problem.

Poincaré already realized how important the study of area preserving maps of the plane are for systems of two degrees of freedom. These maps model the Poincaré map on a transversal to the closed orbit γ within an energy surface as just described. Thus fixed points of this map give closed orbits near γ.

Motivated this way, Poincaré [1912] formulated his "last geometric theorem": *Any area preserving homeomorphism of an annulus in $\mathbf{R}^2$ to itself which shifts the two boundary circles in opposite directions has at least two fixed points.*

Poincaré did not claim a general proof; this was supplied by Birkhoff [1913]. In his book [1927] Birkhoff shows how this and related results apply to problems on closed geodesics and the three body problems.*

Next came the Birkhoff–Lewis theorem [1933] that allowed one to prove the existence of infinitely many periodic points of arbitrarily high period. This theorem remains of interest today (cf. the remarks following the proof of 10.3.7 below). For an elegant proof of this result, see Moser [1977]. The Birkhoff–Lewis theorem is of interest because it is not restricted to systems

*See Arnold [1978] for further discussion.

ISBN 0-8053-0102-X

with two degrees of freedom,* whereas the result of Moser which we now describe is so restricted; however, in this case it is also more powerful. Most of the key results were discovered first by Arnold [1963] in the analytic case.

To formulate the results we need the following terminology.

8.3.3 Definition. *Let $U \subset \mathbf{R}^2$ be an open neighborhood of the origin. A C^∞ mapping $F: U \to \mathbf{R}^2$ is an **(α, β)-normal form**, $\alpha \in [0, 2\pi)$ and $\beta \in \mathbf{R}$, if $F(\mathbf{u}) = \mathbf{u}e^{i(\alpha+\beta|\mathbf{u}|^2)} + R_4(\mathbf{u})$ (in complex notation, $\mathbf{R}^2$ identified with the complex plane), where for some $K>0$, $|R_4(\mathbf{u})| < K|\mathbf{u}|^4$ for all $\mathbf{u} \in U$. A C^∞ mapping $F: U \to \mathbf{R}^2$ is an **α-twist mapping** if $F(\mathbf{0}) = \mathbf{0}$, and $DF(\mathbf{0})$ has eigenvalues $e^{\pm i\alpha}$.*

We consider $\mathbf{R}^2$ a symplectic manifold with symplectic form $dx \wedge dy$, as usual.

The following is an outgrowth of the Birkhoff normal form discussed in Sect. 5.6.

8.3.4 Theorem (Birkhoff-Sternberg-Moser normal form). *If $F: U \subset \mathbf{R}^2 \to \mathbf{R}^2$ is an α-twist mapping with α not zero or an integral multiple of $\pi/2$ or $2\pi/3$, then there is a symplectic chart at $\mathbf{0} \in \mathbf{R}^2$ such that the local representative of F is an (α, β)-normal form, and sign$(\beta) = (+, 0$, or $-)$ for all symplectic charts having this property.*

For the proof, see Sternberg [1969] and Siegel and Moser [1971]. The excluded values of the eigenvalues of $DF(\mathbf{0})$ are illustrated in Fig. 8.3-1

8.3.5 Definition. *An α-twist mapping is an **elementary twist mapping** if α is not zero or an integral multiple of $\pi/2$ or $2\pi/3$, and the invariant β is not zero. A **cycle** in U is a homeomorphic image of the circle S^1.*

8.3.6 Theorem (Moser twist stability). *If $F: U \subset \mathbf{R}^2 \to \mathbf{R}^2$ is an elementary twist mapping, then:*

(i) In every neighborhood of $\mathbf{0} \in U$, there is an invariant cycle σ having no periodic points. That is, $F(\sigma) = \sigma$, and for all $\mathbf{u} \in \sigma$ and integers k, $F^k(\mathbf{u}) \neq \mathbf{u}$.

(ii) For all $\varepsilon > 0$ there is a $\delta > 0$ such that the set of invariant cycles in $D_\delta(\mathbf{0})$ has measure greater than $(1-\varepsilon)(2\pi\delta^2)$.

(iii) For every neighborhood V of $\mathbf{0} \in U$ and integer k there are points $\mathbf{v} \in V$ such that $F^k(\mathbf{v}) = \mathbf{v}$.

For the proof of (i) and (ii) see Siegel and Moser [1971]. For the proof of (iii), see Arnold and Avez [1967].

By applying this theorem to the Poincaré map Θ on a local transversal section S within the energy surface Σ_e, we obtain a condition for $o^\pm$-stability of $\gamma \subset \Sigma_e$ in the case of two degrees of freedom.

*In fact, with the right technical conditions, Moser's proof works for systems with infinitely many degrees of freedom.

ISBN 0-8053-0102-X

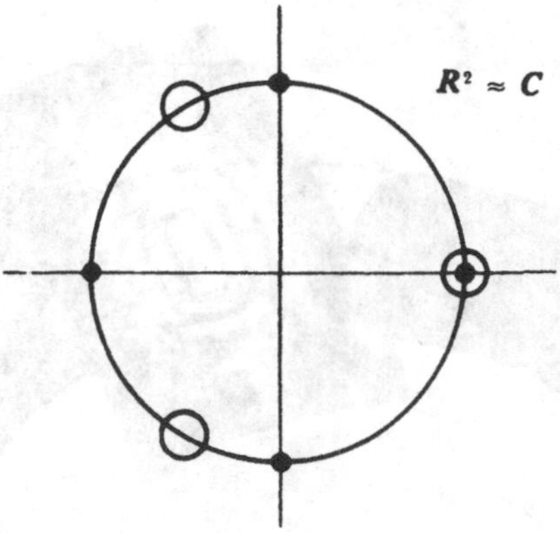

Figure 8.3-1

8.3.7 Corollary. *Suppose X_H is a Hamiltonian vector field on a symplectic four-manifold M, γ is a closed orbit of X_H in a regular energy surface Σ_e, and Θ is a Poincaré map of $X_H|\Sigma_e$ at $\gamma \subset \Sigma_e$. Then in the oscillatory case (characteristic multipliers $1, e^{\pm i\alpha}$), if Θ is an elementary twist mapping, γ is $o^{\pm}$-stable within Σ_e and within M.*

Proof. It follows at once from 8.3.6 that $\gamma \subset \Sigma_e$ is $o^{\pm}$-stable. To show that $\gamma \subset M$ is $o^{\pm}$-stable, we consider a local transversal section $\tilde{S}$ for $\gamma \subset M$. Let $\Gamma = \cup\{\gamma_{e'} | e-\varepsilon < e' < e+\varepsilon\}$ be a cylinder of closed orbits through $\gamma = \gamma_e$, $H(\gamma_{e'}) = e'$. Then if $\Sigma_{e'}$ is a regular energy surface containing $\gamma_{e'}$, $S_{e'} = \tilde{S} \cap \Sigma_{e'}$ is a local transversal section for $\gamma_{e'}$. We may suppose in addition that $\tilde{S}$ and Θ are constructed, so $\Theta|S_{e'} = \Theta_{e'}$ is a Poincaré map on $S_{e'}$. Then Θ_e is an elementary twist mapping by hypothesis, and the derivatives of $\Theta_{e'}$ of all orders are continuous functions of e'. So for e' sufficiently close to e, $\Theta_{e'}$ is an elementary twist mapping also. Thus for some $\varepsilon' > 0$, $\gamma_{e'}$ is $o^{\pm}$-stable for all $e' \in (e-\varepsilon', e+\varepsilon')$. Then it follows easily that $\gamma \subset M$ is $o^{\pm}$-stable, as γ is compact and H is invariant. ■

The conclusion of Moser's twist stability theorem is illustrated in Fig. 8.3-2. The nested concentric tori (which contain no closed orbits among themselves) delimit invariant regions which contain closed orbits of long periods, according to 8.3.6(iii). It is further known (see Arnold [1963b], Arnold and Avez [1967] and Zehnder [1971]) that these occur in pairs, elliptic and heteroclinic-hyperbolic. The complex picture that emerges, which Poincaré was reluctant to draw in 1899, is illustrated in Fig. 8.3-3. It is this picture that is proposed by Thom [1975, p. 27] as the Hamiltonian analog of the attractor in differentiable dynamics under the name *vague attractor*. We shall call this configuration *VAK*, for *Vague Attractor of Kolmogorov*, and for the goddess of vibration in the Rig Veda.

Note that within the VAK are smaller VAKs. If one of these is magnified, the same picture (with time dilated) is obtained; a solenoid.

ISBN 0-8053-0102-X

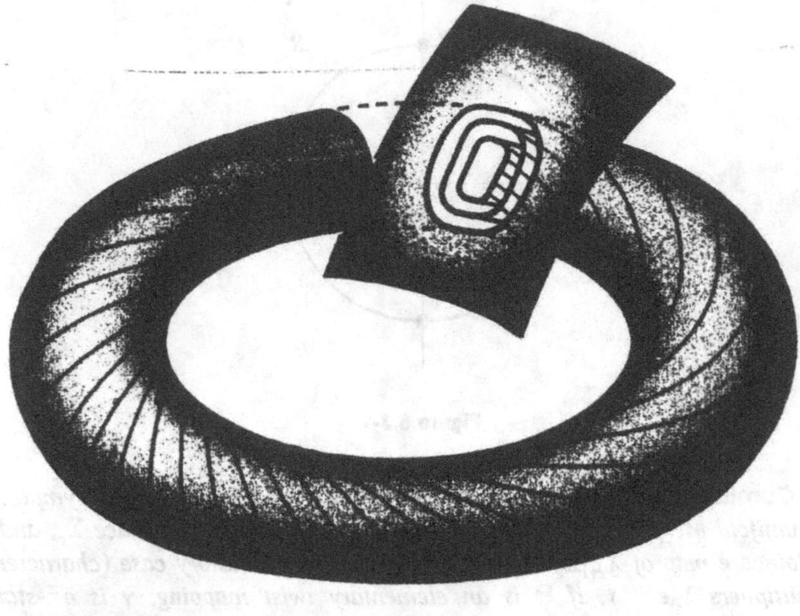

Figure 8.3-2. The VAK drawn in a three-dimensional energy surface.

This picture is central to much of the writings of Poincaré and Birkhoff on Hamiltonian dynamics. The classic of Birkhoff [1935] is still well worth reading for further understanding of it. Markus and Meyer [1974] trace the idea back to Lagrange (1762), while Whittaker [1959] gives some credit to D. Bernoulli (1753).

The Moser theorem admits a generalization to systems of three or more degrees of freedom (see Moser [1963a] and Arnold [1963a, 1963b]), but the generalization does not imply $o^{\pm}$-stability. The escape of orbits through the non-bounding concentric tori, known as **Arnold diffusion**, is important; see Arnold [1978].

For discussion of stability in the elliptic case, see Robinson [1970b] and Markus and Meyer [1974]. They propose a more rigorous condition of nondegeneracy for critical points of elliptic type, the *general elliptic* point. Combining suggestive results of Arnold [1963b] and Sternberg [1969], Markus and Meyer offer a very plausible conjecture, generalizing the twist theorem (8.3.6) for general elliptic points. Although the stability corollary (8.3.7) does not generalize, nonergodicity would follow.

Moser [1973a] states that the invariant tori are Lagrangian submanifolds (compare 5.3.32). This fact can probably be exploited, although to our knowledge it has not been.

By the Oxtoby–Ulam [1941] theorem, area preserving homeomorphisms of the annulus are C^0 generically ergodic. For C^r, $r \geqslant 2$, Moser's result shows that C^r area preserving diffeomorphisms of the annulus are *not* generically ergodic. For C^1, Takens [1971] has shown that Moser's theorem breaks down,

ISBN 0-8053-0102-X

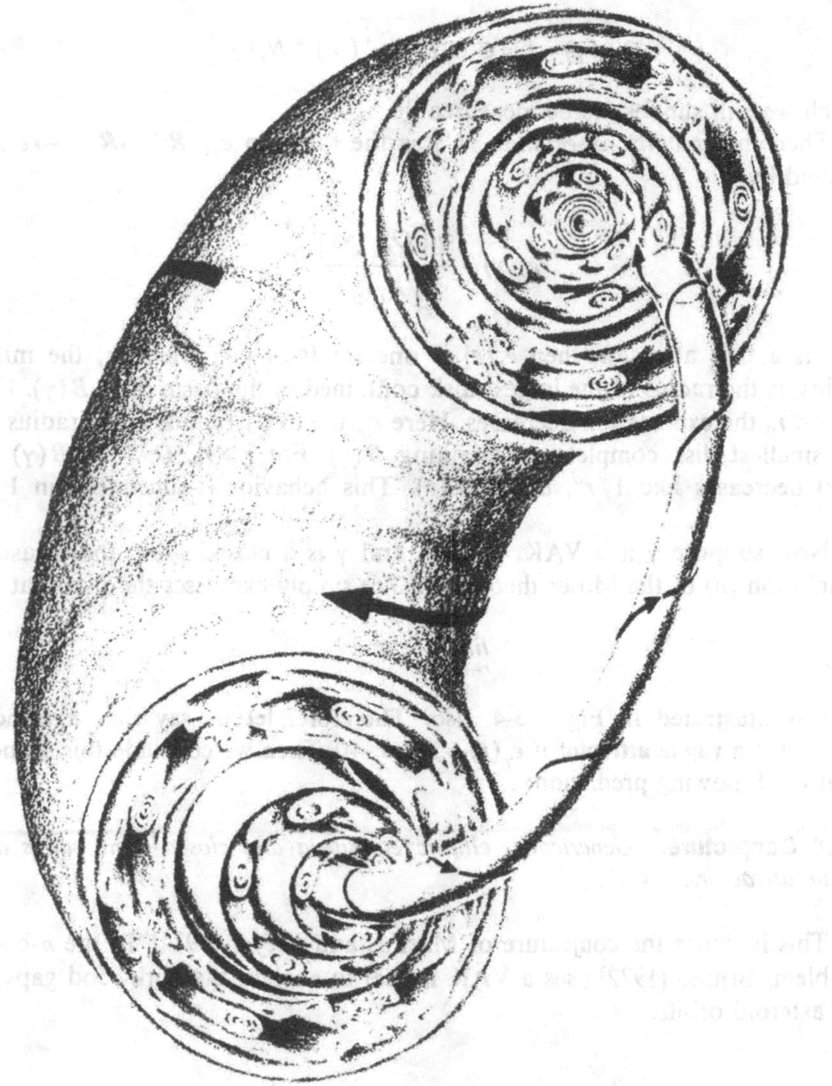

Figure 8.3-3. The VAK, according to Arnold [1963b].

yet Winkelnkemper [1977] shows by an "elementary" argument that even there C^1 area preserving diffeomorphisms are not generically ergodic.

The vague o^+-stability to be expected near a general elliptic point or closed orbit cylinder (VAK point or band) in higher dimensions can be expressed by means of an expectation function.

Let M have a Riemannian metric with derived distance function d, and measure μ. If γ is a critical element of X_H and $r>0$, let

$$N_r(\gamma)=\{m\in M|d(m,\gamma)<r\}$$

ISBN 0-8053-0102-X

and

$$O_r^+(\gamma) = \{m \in N_r(\gamma) | o^+(m) \subset N_r(\gamma)\}$$

which we will suppose to be measureable.

Then the *stability expectation* of γ is the function $e_\gamma: \boldsymbol{R}^+ \to \boldsymbol{R}: r \mapsto e_\gamma(r)$ defined by

$$e_\gamma(r) = \frac{\mu[O_r^+(\gamma)]}{\mu[N_r(\gamma)]}$$

If γ is a true attractor, then $e_\gamma(r)$ is one for $0 < r \leqslant r_i$, where r_i, the inner radius, is the radius of the largest disk contained in the basin of γ, $B(\gamma)$. For $r_i < r < r_0$ the expectation decreases. Here r_0, the outer radius, is the radius of the smallest disk completely containing $B(\gamma)$. For $r \geqslant r_0$, $O_r^+(\gamma) = B(\gamma)$ so $e_\gamma(r)$ decreases like $1/r^d$, $d = dim(M)$. This behavior is illustrated in Fig. 8.3-4.

Now suppose γ is a VAK. If $d=4$ and γ is a closed orbit, the measure conclusion (ii) of the Moser theorem (8.3.6) simply expresses the fact that

$$\lim_{r \to 0} e_\gamma(r) = 1$$

This is illustrated in Fig. 8.3-4 also. Therefore, let us say that a critical element is a **vague attractor** if $e_\gamma(r) \to 1$ as $r \to 0$. Then we conclude this section with the following prediction:

8.3.8 Conjecture. *Generically, elliptic equilibria and closed orbit bands are vague attractors.*

This includes the conjecture of Markus and Meyer [1974]. In the n-body problem, Brjuno [1972] uses a VAK model to explain the Kirkwood gaps in the asteroid orbits.

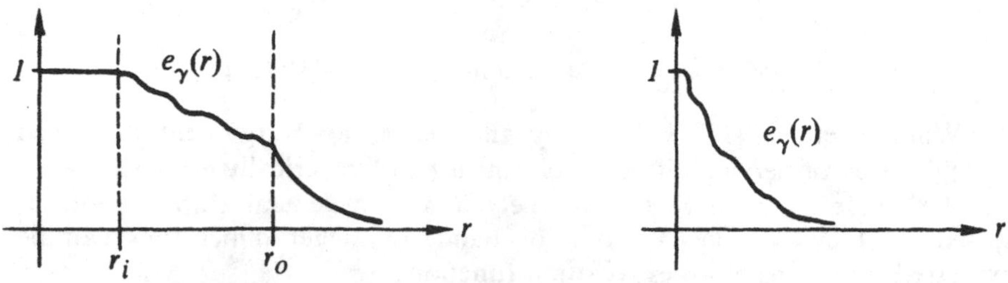

True attractor

Vague attractor

Figure 8.3-4. The stability expectation function.

ISBN 0-8053-0102-X

The VAK in infinite dimensions, including this vague stability property, has been proposed by Thom [1975] as a model for the stable states of quantum mechanics.

EXERCISES

8.3A. Compute Taylor's formula to order four for an α-twist mapping both in complex and in real notations.

8.3B. Find necessary conditions, on the second and third derivatives at the origin of a chart, that an α-twist mapping be changed to an (α,β)-normal form.

8.3C. Show that if Θ and Θ' are Poincaré maps on local transversal sections S and S' of dimension two, and Θ is an elementary twist mapping, then so is Θ'.

8.3D. Show that if X_H is a Hamiltonian vector field, $\tilde{S}$ is a local transversal section, Σ is a regular energy surface, and $S=\tilde{S}\cap\Sigma$, then S is a local transversal section of $X_H|\Sigma$.

8.3E. Find conditions on a symplectic diffeomorphism, at a fixed point of purely elliptic type, for the existence of invariant subcenter manifolds of dimension two.

8.4 GENERIC PROPERTIES

As critical elements in the Hamiltonian case are not generally elementary, we will now describe alternative notions of nondegeneracy for this context. Recently, fairly complete results on the genericity of these properties have appeared.

8.4.1 Definition. *A Hamiltonian $H\in\mathfrak{F}(M)$, or a Hamiltonian vector field $X_H\in\mathfrak{X}_{\mathcal{H}}(M)$ or $X\in\mathfrak{X}_{\mathcal{LH}}(M)$, has **property** (H1) iff every critical point is $\mathcal{H}$-elementary.*

The genericity of this property was established by Buchner [1970].

8.4.2 Theorem. *Property* (*H*1) *is C^r generic in $\mathfrak{X}_{\mathcal{H}}(M)$ for all $r>1$.*

The proof—a delicate exercise in transversality and matrix varieties—is characteristic of all these recent results. A good exposition is found in Robinson [1971a].

In the case of closed orbits, a condition on the transverse frequencies—corresponding to the defining property [8.3.1(iii)] for the oscillatory frequencies of an $\mathcal{H}$-elementary point—could be proposed as a generic condition. But unlike the equilibria, which are generically isolated points, the closed orbits must lie in orbit cylinders $\{\gamma_e\}$. As the cylinder parameter e varies, the phase portrait near γ_e may behave as in an arc of vector fields and violate the restraining relation on transverse frequencies, at least at exceptional (bifurcating) values of e. And so, the properties (H2) and (H3), which will be analogs of (G2) and (G3), must take orbit cylinders into account.

ISBN 0-8053-0102-X

To describe (H2) on the transverse frequencies of γ_e and their dependence on e, we make use of a power chart. Suppose γ is an 0-*elementary* closed orbit of X_H, $m \in \gamma$, and (U,ϕ) is a power chart at m,

$$\phi: U \to \boldsymbol{R} \times \boldsymbol{R}^{n-1} \times \boldsymbol{R} \times \boldsymbol{R}^{n-1}$$

$$: u \mapsto (\delta, \boldsymbol{q}, \varepsilon, \boldsymbol{p})$$

and let

$$\theta': W_0' \to W_1' \subset \boldsymbol{R}^{n-1} \times \boldsymbol{R} \times \boldsymbol{R}^{n-1}$$

$$: (\boldsymbol{q}, \varepsilon, \boldsymbol{p}) \mapsto (\boldsymbol{Q}, \varepsilon, \boldsymbol{P})$$

be the local representative of the Poincaré map on the transversal section defined by $\delta = 0$. Suppressing the preserved energy coordinate, let

$$\theta_\varepsilon: W_{00}' \to W_{1\varepsilon}' \subset \boldsymbol{R}^{n-1} \times \boldsymbol{R}^{n-1}$$

$$: (\boldsymbol{q}, \boldsymbol{p}) \mapsto (\boldsymbol{Q}(\boldsymbol{q}, \varepsilon, \boldsymbol{p}), \boldsymbol{P}(\boldsymbol{q}, \varepsilon, \boldsymbol{p}))$$

denote the reduced map, where $W_{00}' \in \boldsymbol{R}^{n-1} \times \boldsymbol{R}^{n-1}$ is an open disk at $(\boldsymbol{0},\boldsymbol{0})$ such that $W_{00}' \times (-a,a) \subset W_0'$. Note that for each $\varepsilon \in (-a,a)$, ϕ_ε is a symplectic diffeomorphism from W_{00}' to $W_{1\varepsilon}'$, and the closed orbit γ_ε of the orbit cylinder is represented by a fixed point $(\boldsymbol{q}_\varepsilon, \boldsymbol{p}_\varepsilon)$ of ϕ_ε. These fixed points lie on a smooth curve through $(\boldsymbol{0},\boldsymbol{0})$ parametrized by ε. Thus

$$D\theta_\varepsilon(\boldsymbol{q}_\varepsilon, \boldsymbol{p}_\varepsilon) \in Sp(\boldsymbol{R}^{n-1} \times \boldsymbol{R}^{n-1})$$

has the *PCM*'s of γ_ε (and their inverses) as its spectrum. Clearly, the choice of the power chart (U,ϕ) is unimportant. But we have created, for any 0-elementary closed orbit γ_0 an arc

$$D_\phi \theta_\varepsilon: (-a,a) \to Sp(\boldsymbol{R}^{2n-2})$$

corresponding to the tangent of the energy-restricted Poincaré section map θ_e of each closed orbit γ_e near γ_0 in its orbit cylinder. This is the essential construction of this section. We shall call this map the **tangent arc** of γ_0 with respect to (U,ϕ).

8.4.3 Definition. *A Hamiltonian vector field $X_H \in \mathfrak{X}_{\mathcal{H}}(M)$ has* ***property (H2)*** *if*

H2-0: all closed orbits are 0-elementary and for all integers $N>0$,
H2-N: if $\{\gamma_e : e \in (-a,a)\}$ is an orbit cylinder, and

$$D_\phi \theta_e: (-a,a) \to Sp(\boldsymbol{R}^{2n-2})$$

ISBN 0-8053-0102-X

its tangent arc with respect to any power chart, then the transverse frequencies $\{\alpha_i\}$ *are linearly independent over the integers* $\{-N,\ldots,N\}$ *for all but a finite set of points* $e\in(-a,a)$:

$$\left.\begin{array}{l} p_1,\ldots,p_k\in\{-N,\ldots,N\} \\ \Sigma p_i\alpha_i \text{ is an integer} \end{array}\right\} \Rightarrow p_i=0 \text{ all } i$$

(*no resonance of order* $\leqslant N$).

Of course, independence of α_i over $\mathbf{Z}$ is the same condition that is needed for the Birkhoff normal form (see Sect. 5.6). An additional condition for nondegeneracy of higher-order terms in the Birkhoff normal form, and implying the existence of invariant tori, could be added to H2. See Markus and Meyer [1974]. This definition, a triumph in itself, is due to Robinson [1970a]. It is good enough to render his density theorem only moderately difficult.

8.4.4 Theorem (Robinson [1970a, b.]). *Property* $(H2)$ *is* C^r *generic,* $r\geqslant 2$.

The proof requires making the tangent arcs transversal to bad sets $B_N\subset Sp(\mathbf{R}^{2n-2})$ where $(H2-N)$ is violated, by perturbation. As B_n contains submanifolds of codimension 1, transversal intersections in isolated points are inescapable. Hence, the result cannot be improved, and bifurcations along orbit cylinders—the famous resonances of celestial mechanics—are characteristic of generic Hamiltonians. We return to this problem in a future section, 8.6.

The situation for intersection of stable and unstable manifolds is similar in that unreasonable behavior is inescapable at isolated closed orbits within an energy cylinder. This situation has been thoroughly analyzed in Robinson [1970b].

Recall that for a closed orbit or equilibrium point of a Hamiltonian vector field, the stable manifold $W^+(\gamma)$ corresponds to the CM's strictly within the unit circle. The unstable manifold $W^-(\gamma)$, with the same dimension as $W^+(\gamma)$, corresponds to the CM's strictly outside the unit circle. And the center manifold (defined locally only) corresponds to the oscillatory CM's, which are even in number. Obviously, $W^+(\gamma)$ and $W^-(\gamma)$ cannot intersect transversally at γ if there is a center manifold. So we can only ask for

$$(W^+(\gamma)\backslash\gamma)\,\overline{\pitchfork}\,(W^-(\gamma)\backslash\gamma)$$

($\overline{\pitchfork}$ means intersect transversally) in a property (H3), analogous to (G3), for the Hamiltonian case. But even then, there remains a problem due to the conservation of energy: for $\gamma\subset\Sigma_e$, $W^\pm(\gamma)\subset\Sigma_e$ also. So we can ask, at best, for transversal intersection within Σ_e. Let us denote the transversal intersection of $W^+(\gamma)\backslash\gamma$ and $W^-(\gamma)\backslash\gamma$ *within* Σ_e $(e=H(\gamma))$ by $W^+(\gamma)\times W^-(\gamma)$. But even this condition is too much to ask for all γ.

ISBN 0-8053-0102-X

For suppose X_H has the generic property H2, so all closed orbits lie in orbit cylinders. Further, these cylinders are tangent to energy surfaces only at isolated closed orbits. In between two such critical (that is, 1 is a *PCM*) closed orbits is a regular orbit cylinder. And for a regular orbit cylinder $\{\gamma_e\}$, $e=H(\gamma_e)$, the dimensions of $W^+(\gamma_e)$ and $W^-(\gamma_e)$ are constant. But given two regular orbit cylinders, Γ and Δ, the condition

$$W^+(\gamma_e) \times W^-(\delta_e)$$

is bound to be violated for isolated energy values.

So if $\Gamma=\{\gamma_e | e\in(a,b)\}$ is a regular orbit cylinder, the sets $W^+(\Gamma)=\cup\{W^+(\gamma_e)|e\in(a,b)\}$ and $W^-(\Gamma)=\cup\{W^-(\gamma_e)|e\in(a,b)\}$ will be called the *stable* and *unstable ribbons* of the regular orbit cylinder Γ. And now, we could ask for the transversal intersection of stable and unstable ribbons as submanifolds of M. Taking into account, again, that a center manifold for γ_e implies $W^+(\Gamma)\cap W^-(\Gamma)$ at γ_e nontransversally, we write

$$W^+(\Gamma) \times W^-(\Delta)$$

to mean $W^+(\Gamma)\backslash\Gamma$ and $W^-(\Delta)\backslash\Delta$ intersect transversally within M.

So here, at last, is Robinson's definition.

8.4.5 Definition. *A Hamiltonian vector field $X_H \in \mathfrak{X}_{\mathcal{H}}(M)$ has* ***property (H3)*** *if:*

H3-0: it has property (H2);

H3-1: all equilibrium points lie on different energy surfaces;

H3-2: if m is an equilibrium point of X_H,

$$W^+(m) \times W^-(m)$$

H3-3: if m is an equilibrium and γ is a closed orbit in the same energy surface, then

$$W^+(m) \times W^-(\gamma) \quad \textit{and} \quad W^-(m) \times W^+(\gamma)$$

H3-4: if Γ and Δ are regular orbit cylinders, then

$$W^+(\Gamma) \times W^-(\Delta)$$

H3-5: for all but a countable set B of closed orbits,

$$W^+(\gamma) \times W^-(\delta)$$

H3-6: every bad closed orbit $\beta\in B$ is interior to a regular orbit cylinder (that is, 1 is not a PCM).

ISBN 0-8053-0102-X

Again, the definition is adapted to the transversality technique used in the proof of the density theorem.

8.4.6 Theorem (Robinson [1970b]). *Property* $(H3)$ *is* C^r*-generic,* $r \geq 2$.

The proof is described well in Robinson [1971a]. An additional condition for nondegeneracy of the purely oscillatory, or elliptic, closed orbits is established in Robinson [1970b] and also in Markus and Meyer [1969, 1974]. This could be added to (H3) and guarantees invariant tori, and nonergodicity.

The remaining generic properties are less trouble. The next, (G4) (closed orbits dense in the nonwandering set), was established as generic in $\mathfrak{X}_{\mathcal{H}}(m)$ in Pugh [1966]. But this result was improved by Takens [1972] who showed that the homoclinic hyperbolic closed orbits are dense in the nonwandering set.

Recall that C_X denotes the set of initial points $m \in M$ such that the orbit $o(m)$ is complete, and has compact closure, and that $\Omega_X^C = cl[C_X \cap \Omega_X]$. The property (G4) was expressed as $cl(\Gamma_X) \subset \Omega_X^C$. So now let h_X denote the union of all hyperbolic closed orbits γ of X of homoclinic type. Here *hyperbolic* means $dim\, W^+(\gamma) = dim\, W^-(\gamma) = \frac{1}{2} dim(M)$ and *homoclinic* means

$$[W^+(\gamma) \backslash \gamma] \cap [W^-(\gamma) \backslash \gamma] \neq \varnothing$$

8.4.7 Definition. *A Hamiltonian vector field* $X_H \in \mathfrak{X}_{\mathcal{H}}(M)$ ***has property (H4)*** *if*

H4-1: every hyperbolic closed orbit is homoclinic;

H4-2: homoclinic closed orbits are dense in the (compact) nonwandering set: $cl[h_{X_H}] = \Omega_{X_H}^C$.

This definition is due to Takens [1972], who proved it is a generic property, at least if M is compact. (This condition is not necessary, however.)

8.4.8 Theorem. *Property* $(H4)$ *is* C^r*-generic,* $r \geq 1$.

The next, property (G5), excludes regular first integrals. In the Hamiltonian case, we always have one—the Hamiltonian itself, so (G5) must be modified. This, also, is due to Robinson [1970b].

8.4.9 Definition. *A function* $f \in \mathfrak{X}(M)$ *is a* ***regular second integral*** *of a Hamiltonian vector field* $X_H \in \mathfrak{X}_{\mathcal{H}}(M)$ *iff* $\{f, H\} = 0$, *and f is not constant on any open set of any level surface* Σ_e *of H. A Hamiltonian vector field* $X_H \in \mathfrak{X}_{\mathcal{H}}(M)$ ***has property (H5)*** *iff either*

H5-1: X_H *has no regular second integral; or*

H5-2: $int\, \Omega_{X_H}^C = \varnothing$.

8.4.10 Theorem. *Property* $(H5)$ *is* C^r*-generic,* $r \geq 1$.

ISBN 0-8053-0102-X

In fact, Robinson [1970b] shows that (H3) and (G4) imply (H5).

For property (G6), critical stability, there is an analog for Hamiltonian systems. This follows from the Hamiltonian closing lemma of Pugh and Robinson. But as the normal situation for Hamiltonian vector field is $cl(\Gamma) = M$, critical stability will not be important.

This ends the current list of generic properties for the Hamiltonian case. All are easily adapted from the analogs of differentiable dynamics, except (H2) and (H3).

The theory of generic orbit cylinders may evolve further in the future. A new direction is indicated in Newhouse [1977a], which shows that the density of elliptic closed orbits may be generic. Furthermore, recent work of Markus and Meyer indicates that generically Hamiltonian systems contain solenoids of all types.

8.5 STRUCTURAL STABILITY

We have seen several reasons why a Hamiltonian vector field cannot be structurally stable. For example, compare the stability of closed orbits under perturbation of the energy (8.2.2) with the necessity of property (G2), or the necessity of (G5) (7.4.7), which is violated by the Hamiltonian function itself.

From the experimental point of view, we may substitute other versions of structural stability that are more appropriate to the Hamiltonian case. For if we assume that the mathematical model of a theory is a conservative Hamiltonian system, the uncertainty of the experimental domain is represented by perturbations of the Hamiltonian function. That is, we arbitrarily exclude non-Hamiltonian and nonautonomous perturbations. Then for stability of the phase portrait under perturbations within $\mathfrak{X}_{\mathcal{H}}(M)$, we get an appropriate analogue of the previous definition 7.4.1 by restricting it to the subspace $\mathfrak{X}_{\mathcal{H}}(M) \subset \mathfrak{X}(M)$, with the Whitney C^r topology.

8.5.1 Definition. *A Hamiltonian vector field X_H on a symplectic manifold (M, ω) is* $\mathcal{H}^r$ ***structurally stable*** *(or the Hamiltonian H is* $\mathcal{H}^{r+1}$ ***structurally stable****) if there is a neighborhood $\mathcal{O}$ of $X_H \in \mathfrak{X}_{\mathcal{H}}(M)$ in the Whitney C^r topology such that $X_K \in \mathcal{O}$ implies X_K and X_H have equivalent phase portraits.*

As this notion of stability is very strong, and is known to be nongeneric (Robinson [1970b], the construction is outlined below) we might seek a weaker one.

Perhaps an intermediate notion requiring stability of the phase portrait on a single energy surface Σ_e under perturbations of the Hamiltonian and the energy e is more appropriate.

8.5.2 Definition. *Suppose $X_H \in \mathfrak{X}_{\mathcal{H}}(M)$ and $\Sigma_e = H^{-1}(e)$ is a regular energy surface. Then $X_H|\Sigma_e$ is* Σ^r ***structurally stable*** *iff there is a neighborhood $\mathcal{O}$ of $X_H \in \mathfrak{X}_{\mathcal{H}}(M)$ in the Whitney C^r topology, and an $\varepsilon > 0$, such that if $X_K \in \mathcal{O}$ and $e' \in (e - \varepsilon, e + \varepsilon)$, then $K^{-1}(e')$ is a regular energy surface, and there is a*

ISBN 0-8053-0102-X

homeomorphism $h: H^{-1}(e) \to K^{-1}(e')$ *that maps orbits of* $X_H|H^{-1}(e)$ *into orbits of* $X_K|K^{-1}(e)$.

It seems plausible that $\mathcal{H}$ structural stability implies Σ^r structural stability on any regular energy surface. Perhaps this would actually be the case if in 8.5.1 we had required in addition that the phase portrait homeomorphism $h: M \to M$ preserve the energy surfaces, or in other words that there exists a function $h_0: \mathbf{R} \to \mathbf{R}$ such that the diagram

$$\begin{array}{ccc} M & \xrightarrow{\ h\ } & M \\ {\scriptstyle H}\downarrow & & \downarrow{\scriptstyle K} \\ \mathbf{R} & \xrightarrow{\ h_0\ } & \mathbf{R} \end{array}$$

commutes, and if in 8.5.2 we had permitted $\varepsilon = 0$. In any case, Σ^r structural stability is weaker in some sense, and close in spirit to the applications.

As generically X_H has property (H5), or $X_H|\Sigma_e$ has no first integrals, this avoids the conflict between structural stability and the existence of first integrals.

This version of stability is also nongeneric because of Robinson's example.

Another stability definition has been proposed by Thom [1975, p. 26]. This calls for a stronger type of equivalence of phase portraits. A symplectic diffeomorphism replaces the homeomorphism, and in addition, it is required to preserve parametrization of the integral curves, not just the orbits.

Recall that if $\varphi: (M, \omega_0) \to (N, \omega_1)$ is a symplectic diffeomorphism and $H \in \mathcal{F}(N)$, then $\varphi^* X_H = X_{\varphi^* H}$.

8.5.3 Definition. *A Hamiltonian vector field* $X_H \in \mathfrak{X}_{\mathcal{H}}(M)$ *is* T^r ***structurally stable*** *if there is a neighborhood* $\mathcal{O}$ *of* $X_H \in \mathfrak{X}_{\mathcal{H}}(M)$ *(in the Whitney* C^r *topology) such that* $X_K \in \mathcal{O}$ *implies there is a symplectic diffeomorphism* $\varphi: M \to M$ *such that* $\varphi^* X_H = X_K$.

If M is connected, the last condition is equivalent to the commutativity of the diagram:

$$\begin{array}{ccc} M & \xrightarrow{\ \varphi\ } & M \\ {\scriptstyle K}\downarrow & & \downarrow{\scriptstyle H} \\ \mathbf{R} & \xrightarrow{\ c\ } & \mathbf{R} \end{array}$$

where c is translation by a constant. This implies preservation of energy surfaces, a natural condition. This is not generic, either, despite Thom's conjecture [1975; p. 26].

Finally, we might consider the structural stability of a Hamiltonian vector field X_H restricted to an energy surface Σ_e. That is, the phase portrait in Σ_e

ISBN 0-8053-0102-X

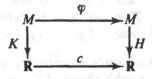

should be stable under perturbation by arbitrary vector fields tangent to Σ_e. This appeals to those who doubt that the universe is conservative. Of course, it implies that restricted to Σ_e the phase portrait has all the generic properties of differentiable dynamics; (Gi), $i=2,\ldots,6$. This is extremely unlikely.

In fact, the example of Robinson shows that none of these notions of stability is appropriate in the Hamiltonian context. Here is the idea of his construction.

Let X_H be a generic Hamilton vector field, with H3. Then there are numerous orbit cylinders, occasionally tangent to energy surfaces. Let Γ be a band of an orbit cylinder, between tangencies—a maximal regular orbit cylinder $\Gamma=\{\gamma_e\}$. Along Γ, the *PCM*'s of γ_e vary continuously with e. For some exceptional values of the energy, the dimension of the center manifold, $W^0(\gamma_e)\subset\Sigma_e$, will change. Deleting these exceptional orbits, Γ is broken into a finite number of subcylinders Γ_i, each with constant dimensions of $W^+(\Gamma_i)$, $W^-(\Gamma_i)$, and $W^0(\Gamma_i)$. Here $W^0(\gamma_e)\subset\Sigma_e$ and $W^0(\Gamma)=\cup\{W^0(\gamma_e)|\gamma_e\subset\Gamma\}$. Note for one of these, say Γ_0, if $\gamma_e\in\Gamma_0$ has u *PCM*'s (or, equivalently, $2u$ *CM*'s, note $u\leqslant n-1$) on the unit circle, and $dim(M)=2n$, then

$$dim\,W^0(\gamma_e)=2u+1$$

$$dim\,W^0(\Gamma_0)=2u+2$$

$$dim\,W^+(\gamma_e)=dim\,W^-(\gamma_e)=n-u$$

$$dim\,W^+(\Gamma_0)=dim\,W^-(\Gamma_0)=n-u+1$$

$$dim\,\Gamma_0=2$$

The purely hyperbolic bands ($u=0$) are unusual. So suppose Γ_0 is not purely hyperbolic, with $1\leqslant u\leqslant n-1$ transverse frequencies for each $\gamma_e\in\Gamma_0$ ($u=n-1$ is the purely elliptic case). For the moment, we will call this a *typical band*. Then the result of Robinson [1970b], which eliminates all the preceding notions of structural stability from the Hamiltonian context, is the following.

8.5.4 Critical instability theorem. *Let $X_H\in\mathfrak{X}_{\mathcal{K}}(M)$ be a Hamiltonian vector field with a typical band Δ, and $\mathcal{V}$ any neighborhood of $X\in\mathfrak{X}_{\mathcal{K}}(M)$ in the Whitney C^r topology, $r\geqslant 1$. Then there is a Hamiltonian vector field $X_K\in\mathcal{V}$ such that the sets of critical elements Γ_{X_H} and Γ_{X_K} are not homeomorphic.*

Nasty. The idea of Robinson's constructive proof is to introduce a perturbation confined to a neighborhood of a closed orbit δ_e in the typical band, so that the center manifold $W^0(\delta_e)$ is perturbed to a VAK (nest of invariant tori) with typical closed orbits denser than in the original phase portrait.

Contemplation of this situation shows that an appropriate notion of structural stability for Hamiltonian dynamics must be extremely vague and

ISBN 0-8053-0102-X

fuzzy, if not downright statistical. This is a real obstacle to a reasonable philosophy of stability in the Hamiltonian context, and no relief is visible on the horizon. Yet, it is possible that further study of generic arcs of symplectic diffeomorphisms will yield a sort of stability for vague attractors, excepting a countable (but perhaps dense) set of exceptional values of the parameter, where the VAK's change topological type.

8.6 A ZOO OF STABLE BIFURCATIONS

In Hamiltonian dynamics, orbit cylinders provide built-in one-parameter families. These arcs have been endowed with generic properties—analogous to Sect. 7.6 in the differentiable context—within property H3. The bifurcations that result—analogous to those drawn in Sect. 7.7—become part of the phase portrait of a single Hamiltonian vector field. Thus the restricted vector field $X_H|\Sigma_e$ is more or less analogous to an arc of vector fields with the energy e as parameter. However, in this case, the manifold Σ_e may bifurcate, as well as the phase portrait.

In this section, we illustrate the bifurcations and terminations of generic orbit cylinders, as discovered by Deprit and Henrard [1968], Meyer and Palmore [1970], and Meyer [1970, 1971a]. This is the beginning of a complete taxonomy of orbit cylinder pathology, as general techniques of Takens [1973a] point the way.

In these examples, we consider only $dim(M)=4$, and X_H with H3. The techniques of Meyer are based on the Poincaré generating function (see also Weinstein [1972] and Arnold [1978, Appendix 9]).

Discussion of Figure 8.6-1: The Burst. First, an orbit cylinder can originate or terminate at a critical point. This bifurcation, a metaphor for asexual creation, was known to Liapounov, and is described in the Liapounov theorem (Sect. 5.6). It is similar to the Hopf bifurcation in the context of one-parameter families of vector fields.

Let $m \in M$ be an $\mathcal{H}$-elementary equilibrium of X_H. Two cases arise (as $n=2$): either m is a saddle-center, with CM's $[exp(\pm 2\pi i\alpha), \mu, \mu^{-1}]$ with $\alpha \in (0, \frac{1}{2})$ and irrational, $\mu > 1$ real, or m is a pure center, with CM's $[exp(\pm 2\pi i\alpha), exp(\pm 2\pi i\beta)]$ with $\alpha, \beta \in (0, \frac{1}{2})$, and irrationally related. Take the saddle-center case first, and let $A \subset T_m M$ be the eigenspace of $exp(\pm 2\pi i\alpha)$. By the Liapounov theorem, there is a two-dimensional submanifold $C \subset M$ tangent to A at m consisting entirely of closed orbits of transverse frequency approximately α, the *center manifold* of m. The center manifold must therefore be an orbit cylinder Γ closed by the point m, as shown in Fig. 8.6-1. As X_H is assumed H2, this cylinder is H2-N for all N. Suppose this cylinder is parametrized as $\{\gamma_\lambda\} = \Gamma$, with γ_λ tending to m as $\lambda > 0$ tends to zero. The flow normal to γ_λ, for λ sufficiently small, is governed by the CM's (μ, μ^{-1}) of m, as $C = \Gamma \cup \{m\}$ is tangent to A at m. Thus the CM μ_λ of γ_λ approaches μ as λ tends to zero, and therefore is eventually real. Thus a disk in C around m

ISBN 0-8053-0102-X

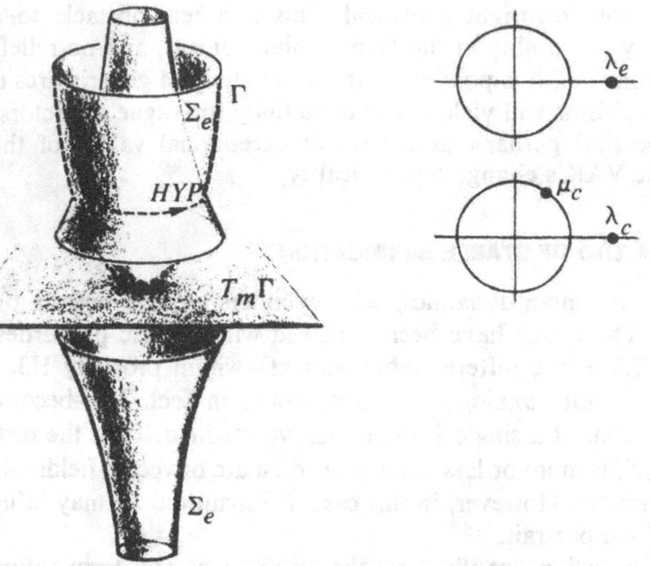

Figure 8.6-1(a). *Phantom burst.* The Liapounov bifurcation, in $n=2$ degrees of freedom, for a single hyperbolic orbit cylinder incident at a critical point of saddle-center type along the center. A three-dimensional energy surface is shown as a two-dimensional surface of rotation. The *PCM*'s of the critical point are shown below the *PCM* of the approaching closed orbits. Here Γ is shown dashed because it is hyperbolic, and therefore qualitatively invisible.

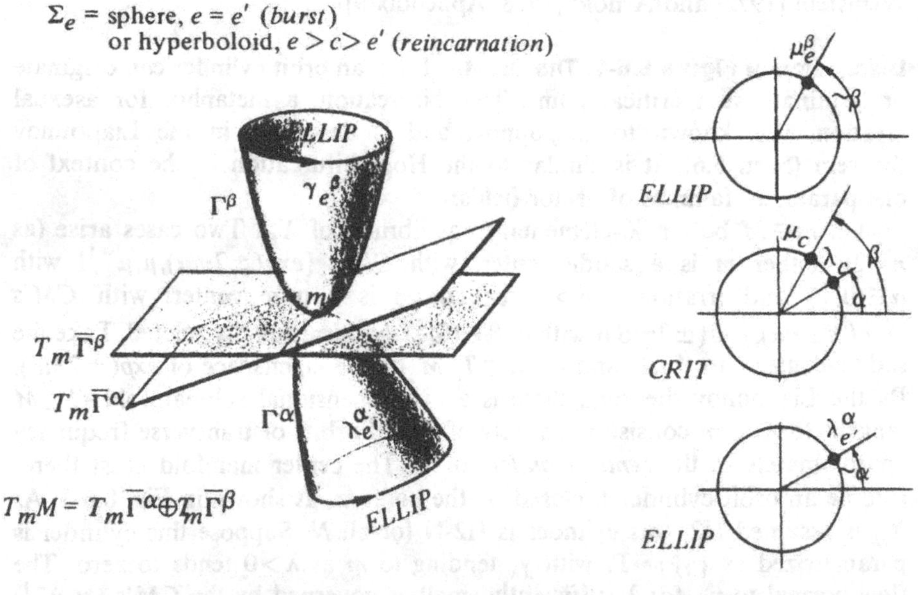

Figure 8.6-1(b). *Stable burst.* The Liapounov bifurcation in $n=2$ degrees of freedom for a pair of elliptic orbit cylinders incident at a critical point of pure-center type along the sub-centers. In four dimensions they do not intersect. The *PCM* of γ_e^β is shown above the *PCM*'s of m with the *PM* of γ_e^α at the bottom.

ISBN 0-8053-0102-X

consists entirely of closed orbits of hyperbolic type (real *PCM*) closed by m. In our qualitative view only elliptic orbit cylinders (unimodular *PCM*) are significant because of their generic orbital stability (8.4.4), so this case is of no qualitative significance.

Now consider the second possibility, in which m is an elliptic equilibrium. Let $A \subset T_m M$ be as before, and $B \subset T_m M$ be similarly the eigenspace of the *CM*'s $exp(\pm 2\pi i \beta)$. The Liapounov construction now applies to both A and B, so we have two orbit cylinders (not intersecting) closed by m, comprising the two subcenter manifolds of m, say

$$C_i = \Gamma_i \cup \{m\}, \quad \Gamma_i = \{\gamma_\lambda^i\}, \quad i = \alpha, \beta$$

Here the *PCM* of γ_λ^α approaches $exp(2\pi i \alpha)$ as λ tends to zero. That is, the transverse frequency of γ_λ^α approaches α. Similarly, the transverse frequency of γ_λ^β approaches β. If the period of a closed orbit γ is τ, then its orbital frequency is $2\pi/\tau$. The orbital and transverse frequencies must not be confused. In this case, the transverse frequency of γ_λ^α approaches the orbital frequency of γ_λ^β, and vice versa. Eventually, both γ_λ^α and γ_λ^β are of elliptic type, and therefore of qualitative significance. As $PCM \neq 1$ in either case, the parameter λ can be taken to be the energy. Suppose m is a local minimum of H. Then each energy surface Σ_e near m is a three-sphere, which contains two elliptic closed orbits γ_e^α and γ_e^β, collapsing to m as e approaches its minimum value $c = H(m)$, as shown in Fig. 8.6-1(b). This represents the simultaneous creation in vacuo of twin stable oscillations of (possibly) large orbital and transverse frequencies, with amplitude increasing from (or decreasing to) zero as e passes its critical value c at m, the relative extremum, the **stable burst** catastrophe.

In case m is a local maximum, the parameter is reversed.

In the saddle case, the energy surfaces are hyperboloidal, and each contains a closed orbit γ_e^α for $e > c$, γ_e^β for $e < c$. As e passes c, γ_e^β shrinks, dies, and is reborn as γ_e^α, **reincarnation.**

Discussion of Figure 8.6-2: Creation. In the case of $n=2$ degrees of freedom, a closed orbit γ has only one *PCM* μ, either real ($|\mu| > 1$, the hyperbolic case), unimodular ($\mu = exp(2\pi i \alpha), \alpha \in (0, \frac{1}{2})$, the elliptic case), or both ($\mu = \pm 1$, the degenerate cases). Thus for $n=2$, the bad set B_N corresponds to $\mu = exp(2\pi i \alpha)$ with $\alpha \in [0, \frac{1}{2}]$, and there is a nonzero integer $p \in [-N, N]$ such that $p\alpha$ is an integer, or $\alpha = q/p$ since α is nonnegative, we may assume without loss of generality that p is positive and q nonnegative, so μ is a pth root of unity. In other words, for an orbit cylinder $\{\gamma_\lambda\}$ in the case $n=2$, we have non-H2 behavior whenever the *PCM* μ_λ is a pth root of unity, $p = 1, 2, \ldots,$ and so forth. In the rest of this section we will consider these cases one at a time and describe the results of Meyer [1970] who classified all the generic phenomena that arise with $n=2$. He calls these the generic p-bifurcations, and in this section we consider the first case, $p=1$, which Meyer calls an *extremal closed orbit*. Thus we have a regular orbit cylinder $\{\gamma_\lambda\}$ with

ISBN 0-8053-0102-X

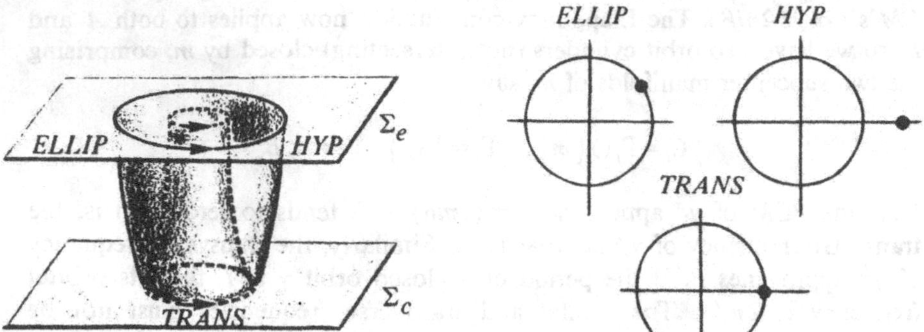

Figure 8.6-2. *Creation.* A hyperbolic to elliptic transition via tangency of an orbit cylinder (here shown as a surface of revolution, hyperbolic dashed, elliptic solid) to an energy surface (here, a horizontal plane). The *PCM* for the elliptic, transitional, and hyperbolic cases are shown alongside. The transitional orbit is unstable, and therefore belongs to the dashed portion of Γ.

PCM μ_λ and γ_0 is extremal, or $\mu_0 = 1$. By the Regular Orbit Cylinder Theorem, the orbit cylinder is tangent to the energy surface Σ_c, $c = H(\gamma_0)$, and on one side. Generically, Meyer shows that this occurs only when μ_λ changes transversally from real to unimodular values, passing through 1 at $\lambda = 0$, so γ_λ changes suddenly from hyperbolic to elliptic type (or vice versa) as λ increases through zero, as shown in Fig. 8.6-2. Also, he shows that γ_0 is orbitally unstable.

With the energy e as parameter, the vector field $X_H|\Sigma_e = X_e$ suddenly develops an unstable periodic orbit γ_0 for $e = c$ of large amplitude, presumably by a **Pugh catastrophe**: the closing of a recurrent orbit. For $e > c$, this extremal orbit γ_0 splits into two closed orbits γ_e^- and γ_e^+, where $\gamma_e^- = \gamma_\lambda$ for some $\lambda < 0$ and is hyperbolic, and $\gamma_e^+ = \gamma_\lambda$ for some $\lambda > 0$ and is elliptic. As only γ_e^+ is qualitatively "visible," a single elliptic closed orbit has suddenly made its appearance in the phase portrait of X_e, as e increased past $e = c$, and nearby is its phantom dual γ_e^-, which is qualitatively invisible. Alternatively, the process could be read in reverse, as the instantaneous annihilation of a

ISBN 0-8053-0102-X

large closed orbit, through cancellation by a phantom dual. We therefore call this phenomenon **creation** (or **annihilation**).

Discussion of Figure 8.6-3: Subtle Division. Next let $p=2$, the two-bifurcation or *transitional orbit* of Meyer [1970]. As 1 is not a *PCM* in this case (or in fact in any of the remaining cases) the orbit cylinder may be parametrized by the energy according to the regular orbit cylinder theorem. Thus $\Gamma=\{\gamma_e\}$, $\mu_e=PCM(\gamma_e)$, and $\mu_c=-1$. Generically, according to Meyer, the transitional orbit γ_c occurs only for a transversal change of μ_e from unimodular to real values through the common point -1, and γ_e undergoes "transition" from elliptic to hyperbolic type as e increases through c, or vice versa. This aspect is similar to the extremal orbit of creation, but μ_e moves through -1 instead of $+1$. But in this case energy is the parameter, and there is a further pathology in the incidence at $\gamma_c \subset \Gamma$ of another orbit cylinder $\Delta=\{\delta_e | e<c\}$. Two cases arise. In the first, δ_e is of elliptic type. As e approaches c from above, δ_e a tends to a double covering of γ_c and the orbital frequency of δ_e approaches half the orbital frequency of γ_c. Thus we may consider δ_e a sub-harmonic of γ_c, approaching resonance as shown in Fig. 8.6-3. Meyer has

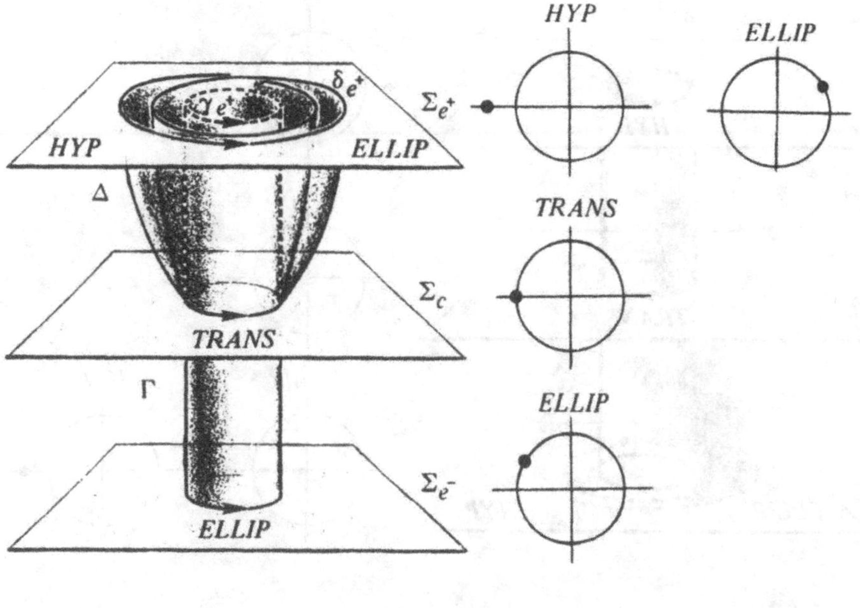

Figure 8.6-3. *Subtle division.* An elliptic to hyperbolic transition via crossing of *PCM* through -1 with emission of a subtly halved elliptic cylinder. The transitional orbit is stable.

ISBN 0-8053-0102-X

shown that the transitional orbit γ_c is orbitally stable. Thus as e increases through c, we have the significant orbit γ_e replaced by the qualitatively visible sub-harmonic δ_e, while γ_e itself becomes invisible. Qualitatively, the behavior is not changed very much, as δ_e is approximately a double covering of γ_c, so the orbital frequency and amplitude of the oscillation are not catastrophically changed. Only later, as e increases considerably, will it become apparent to an observer that δ_e has doubled its period because δ_e is no longer running twice around in a neighborhood of γ_e. Hence we call this phenomenon **subtle halving**. Read the other way, with energy decreasing with time, a visible oscillation doubles over itself and resonates with a phantom oscillation having twice its orbital frequency, a **subtle doubling**. These are two versions of the first of the two cases arising generically when the *PCM* is -1.

Discussion of Figure 8.6-4: Murder. In this case, the arriving sub-harmonic orbit cylinder $\Delta = \{\delta_e\}$ is of hyperbolic type and approaches from the other side of Σ_c, that is, along the elliptic part of Γ. Therefore, the configuration is identical to subtle division, with "elliptic" and "hyperbolic" interchanged everywhere in Fig. 8.6-4, and the energy parameter reversed. Thus γ_e changes

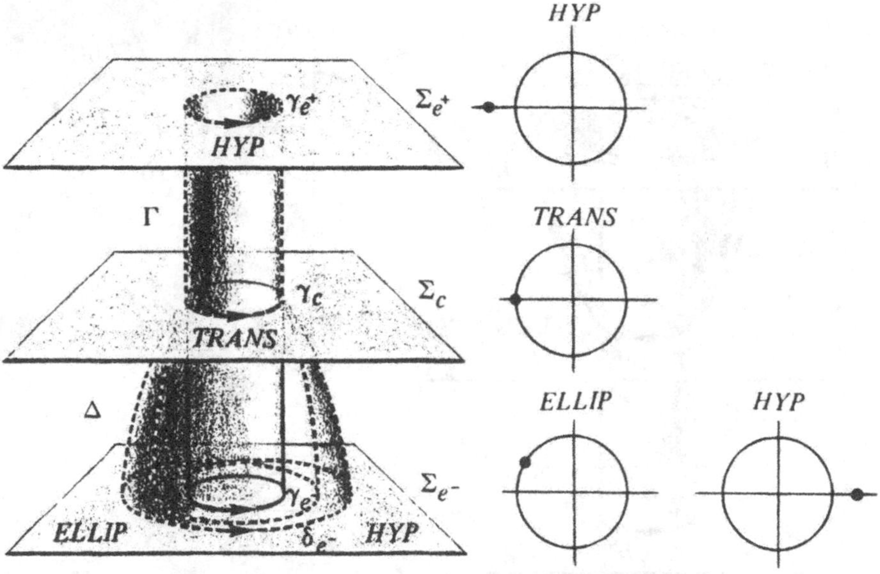

Figure 8.6-4. *Murder.* An elliptic to hyperbolic transition via crossing of *PCM* through -1 with absorbtion of a sub-harmonic hyperbolic cylinder. The transitional orbit is unstable.

ISBN 0-8053-0102-X

from elliptic type ($e<c$) to hyperbolic type ($e>c$) at γ_c, the transitional orbit, which in this case is orbitally unstable. The hyperbolic sub-harmonic δ_e ($e<c$) approaches a double covering of γ_e as e increases to c, and the orbit cylinder Δ terminates at γ_c, as shown in Fig. 8.6-4. Only γ_e, for $e<c$, is visible. The phantom killer δ_e approaches the stable orbit γ_e as half its orbital frequency. At $e=c$ **resonance** occurs, δ_e disappears, and γ_e dies. For $e>c$, γ_e persists only as a ghost, so we call it **murder**. In this case also, the transition can be interpreted with the parameter (energy) reversed. Thus a ghost (γ_e) suddenly materializes (becomes elliptic), and emits a phantom sub-harmonic (δ_e, hyperbolic) **materialization**.

Discussion of Figure 8.6-5: The Phantom Kisses. The cases already described cover transitions between hyperbolic and elliptic states ($PCM=\pm 1$) and the remaining p-bifurcations ($PCM=p$th root of unity, $p=3,4,\ldots$) all must occur along orbit cylinders, parametrized by the energy, which remain elliptic during the bifurcation. The case $p=3$ and one of the two cases with $p=4$ are very similar. The orbit cylinder $\Gamma=\{\gamma_e\}$ is elliptic, and γ_e has PCM $exp(2\pi i\alpha_e)$. For $e=c$, the transverse frequency is $\alpha_c=\frac{1}{3}$ [*resp.* $\frac{1}{4}$], and generically, α_e passes transversally through this value. Nearby are two other orbit cylinders of hyperbolic type:

$$\Delta^+=\{\delta_e: e\in(c-\varepsilon,c)\}$$

$$\Delta^-=\{\delta_e: e\in(c,c+\varepsilon)\}$$

As e approaches c from above or below, δ_e approaches a triple (resp. quadruple) covering of γ_e, as shown in Fig. 8.6-5. Both cylinders terminate at γ_c, which is orbitally unstable. The set of closed orbits

$$\Delta=\Delta^+\cup\gamma_c\cup\Delta^-=\{\delta_e|e\in(c-\varepsilon,c+\varepsilon)\}$$

with $\delta_c=\gamma_c$ may be considered an orbit "cylinder," degenerate at $e=c$. The phantom δ_e approaches γ_e, resonates to its third (resp. fourth) harmonic, kisses Γ at $\gamma_c=\delta_c$ (an excited state), undergoes *subtle frequency division* (falls to its ground state), and departs again. The actor γ_e loses his stability momentarily, but recovers immediately. An observer sees nothing but this momentary instability, if anything.

Discussion of Figure 8.6-6: Emission. The second case of four-bifurcation, and all cases of p-bifurcation for $p>4$ (there is generically only one phenomenon for each $p>4$, according to Meyer) are similar. Again $\Gamma=\{\gamma_e\}$ is an elliptic orbit cylinder, and the transverse frequency α_e passes transversally through $\alpha_c=\frac{1}{4}$ [resp. q/p, $p>4$, $1<q<(p/2)$]. There are two nearby orbit cylinders, an elliptic one $\Delta=\{\delta_e\}$ and a hyperbolic one $E=\{\varepsilon_e\}$, both defined only for $e>c$. As e approaches c from above, both δ_e and ε_e approach a p-fold covering of γ_e, Δ and E terminating at γ_c, as shown in Fig. 8.6-6. The

ISBN 0-8053-0102-X

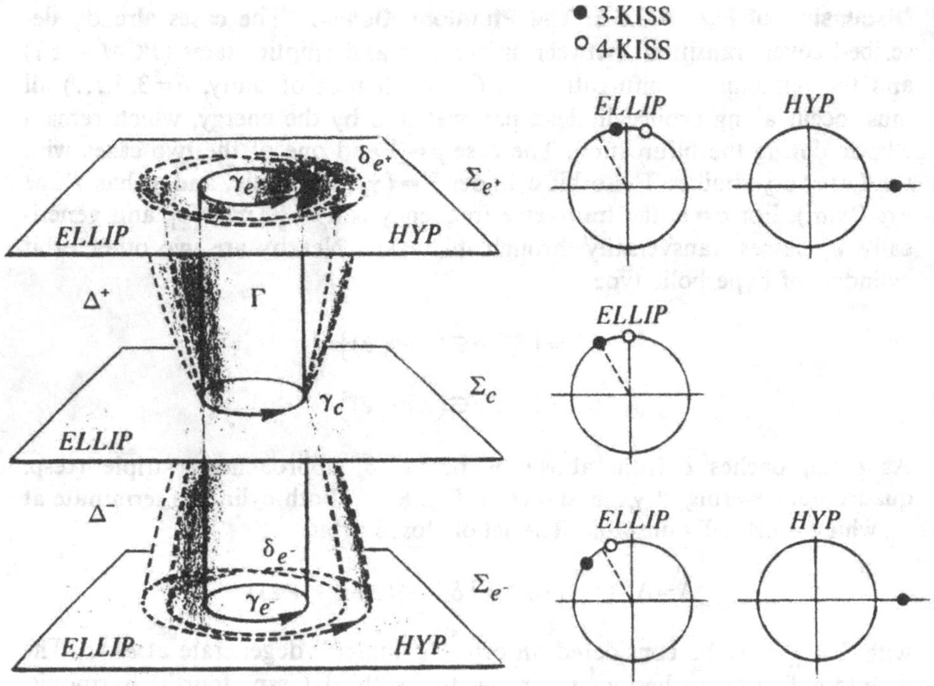

Figure 8.6-5. *Phantom kiss.* A crossing of a *PCM* past $exp(2\pi i\alpha)$, $\alpha = \frac{1}{3}$, a three-bifurcation, with a kiss by a hyperbolic sub-harmonic. The original elliptic cylinder is unperturbed. The 4-kiss, in which α passes $\frac{1}{4}$, is identical, except that the osculating sub-harmonic has one-fourth the orbital frequency instead of one-third, as shown, of the elliptic cylinder.

ISBN 0-8053-0102-X

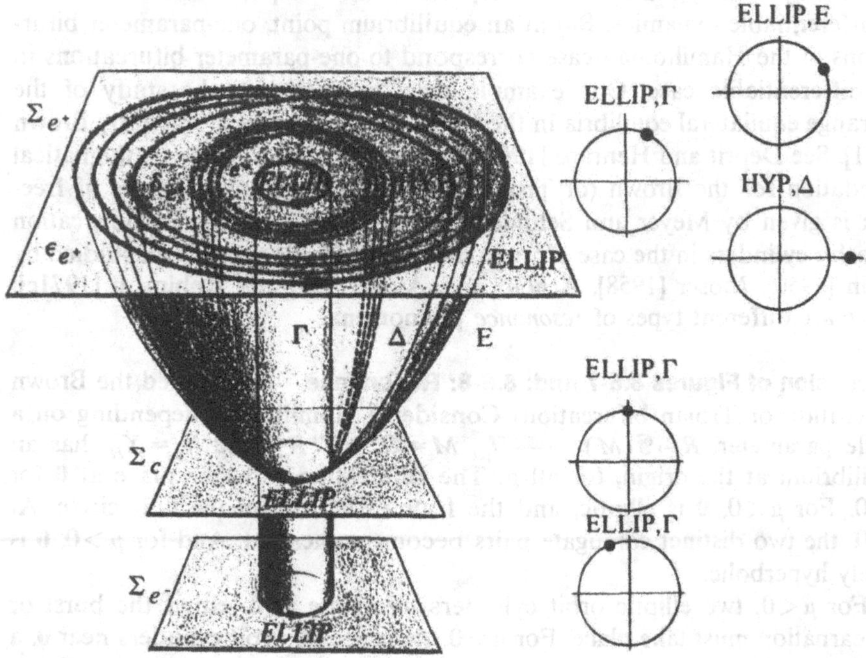

Figure 8.6-6. *Emission.* A crossing of a *PCM* past *exp*($2\pi i\alpha$), a *p*th root of unity, $p > 4$, with emission of elliptic and hyperbolic sub-harmonics of one-*p*th the orbital frequency of the original elliptic cylinder, shown here for $p = 4$.

ISBN 0-8053-0102-X

critical orbit γ_c is stable. Here something significant happens. As e increases through c, the stable orbit γ_e splits into two stable orbits, γ_e and δ_e, the latter at a subtly divided (sub-harmonic) orbital frequency. Although no change is observed in the principal actor γ_e, a new actor is emitted from γ_c, along with a phantom twin ε_e. We call this **emission**. Read in reverse, it is called **absorption**.

For these bifurcations of orbit cylinders in the case of two degrees of freedom, an elegant treatment is suggested in Takens [1973b].

Another bifurcation situation arises if the Hamiltonian vector field is perturbed by a k-parameter family, $\boldsymbol{R}^k \to \mathfrak{X}_{\mathcal{K}}(M)$. Even for $k=1$, the bifurcations along orbit cylinders will correspond to the two-parameter bifurcations of differentiable dynamics. But at an equilibrium point, one-parameter bifurcations in the Hamiltonian case correspond to one-parameter bifurcations in the differentiable case. One example was discovered in the study of the Lagrange equilateral equilibria in the restricted three-body problem by Brown [1911]. See Deprit and Henrard [1968] for the early history. The mathematical foundation for the Brown (or Trojan) bifurcation with two degrees of freedom is given by Meyer and Schmidt [1971]. Also, an analogous bifurcation for orbit cylinders in the case of three degrees of freedom has been studied by Krein [1950], Moser [1958], Arnold and Avez [1967] and Robinson [1971c]. These are different types of *resonance* phenomena.

Discussion of Figures 8.6-7 and: 8.6-8: Resonance. (also called the Brown bifurcation or Trojan bifurcation) Consider a Hamiltonian depending on a single parameter, $\boldsymbol{R} \to \mathfrak{F}(M)$: $\mu \mapsto H_\mu$, $M = \boldsymbol{R}^4 = T^*\boldsymbol{R}^2$, and $X_\mu = X_{H_\mu}$ has an equilibrium at the origin, for all μ. The bifurcation will take place at $\mathbf{0}$ for $\mu = 0$. For $\mu < 0$, $\mathbf{0}$ is elliptic, and the four CM's are on the unit circle. At $\mu = 0$, the two distinct conjugate pairs become coincident. And for $\mu > 0$, $\mathbf{0}$ is purely hyperbolic.

For $\mu < 0$, two elliptic orbit cylinders terminate at $\mathbf{0}$: either the burst or reincarnation must take place. For $\mu > 0$, there are no orbit cylinders near $\mathbf{0}$, a saddle point. Where did they go? Meyer and Schmidt prove that they collapse to a point (the **bubble**), or drift away as a single orbit cylinder (**liberation**). Both cases are illustrated in Fig. 8.6-7. It is the latter case that occurs in the restricted three-body problem for a mass ratio near Routh's critical value (exactly where is explained at the end of Sect. 10.2).

Finally, at an equilibrium of a Hamiltonian vector field perturbed by a one-parameter family, we may have a passage of CM's on the unit circle, as shown in Fig. 8.6-8 (compare Fig. 8.6-7). Also, for $|\alpha_1| < |\alpha_2|$, the passage (as μ increases through zero, or any other value) of $\alpha_2(\mu)$ through a multiple of $\alpha_1(\mu)$,

$$\alpha_2 = k\alpha_1, \qquad k \in \boldsymbol{Z}$$

creates a similar phenomenon, known as **resonance**. In this case the hypotheses of the Liapounov Sub-Center Theorem are violated, and bifurcations occur. Related bifurcations occur in a passage of μ past a critical value μ_0,

ISBN 0-8053-0102-X

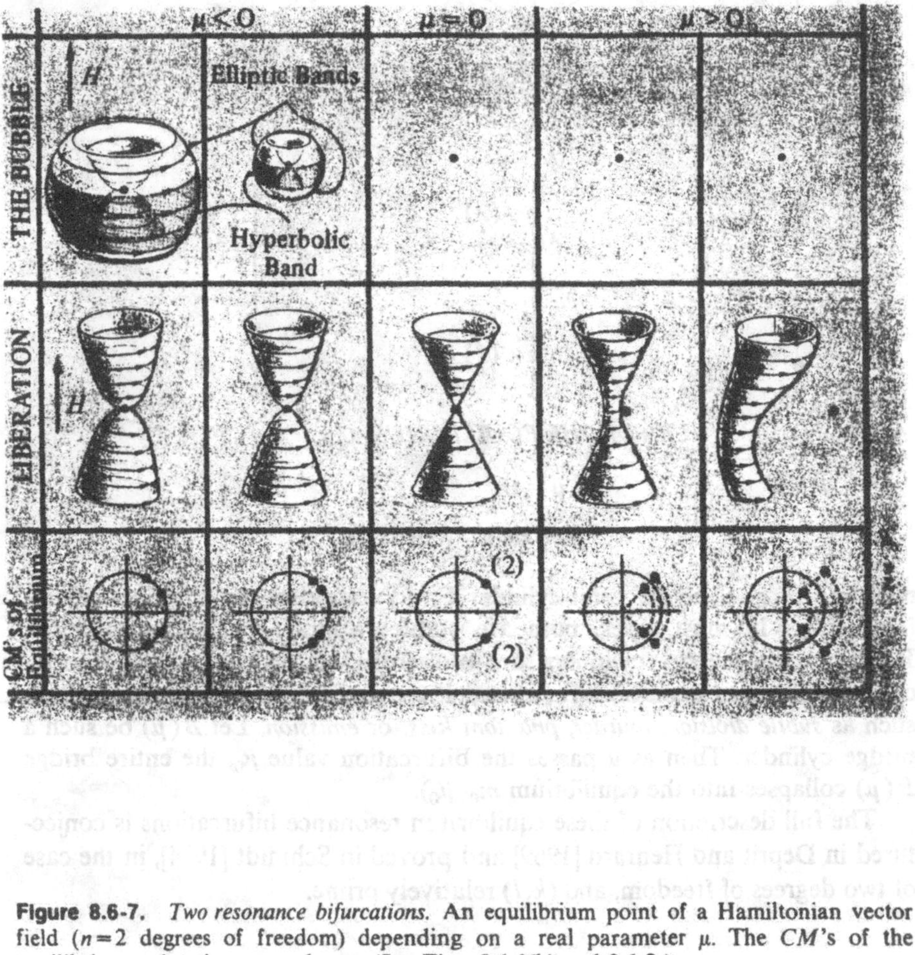

Figure 8.6-7. *Two resonance bifurcations.* An equilibrium point of a Hamiltonian vector field ($n=2$ degrees of freedom) depending on a real parameter μ. The *CM*'s of the equilibrium point change as shown. (See Figs. 8.6-1(b) and 8.6-2.)

where a relation

$$l\alpha_2(\mu_0)=k\alpha_1(\mu_0), \quad l,k\in \mathbf{Z}$$

is satisfied. These are the **subharmonic resonances** that *generally occur in this context, for almost every value of the parameter* μ. The general idea of these bifurcations is this: corresponding to the frequency α_1 is an orbit cylinder $\Gamma_l(\mu)$ incident at the equilibrium $m=m_0(\mu)$ and comprising the subcenter

ISBN 0-8053-0102-X

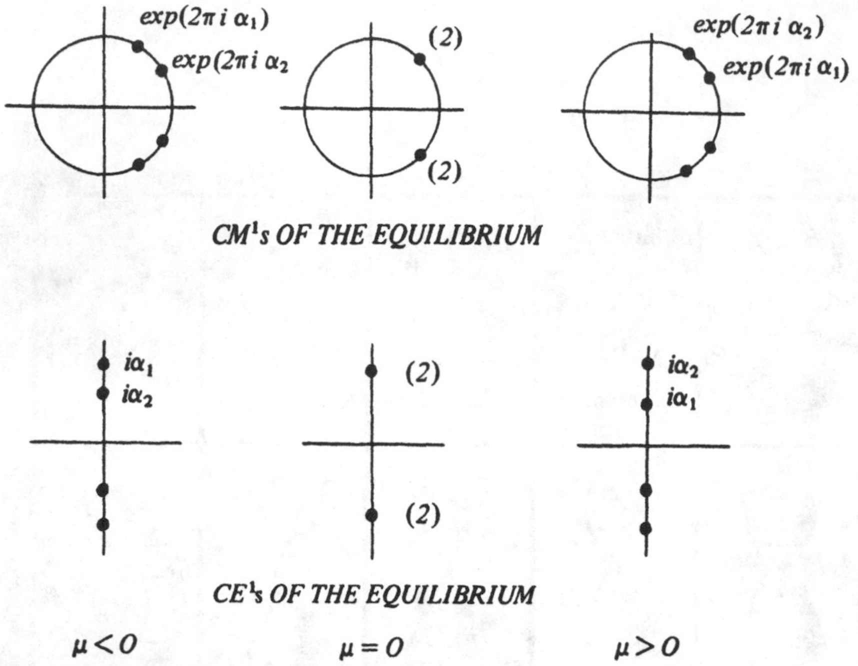

Figure 8.6-8. Resonance

manifold of α_1. Another orbit cylinder $\Gamma_s(\mu)$ comprises the subcenter manifold of α_2. In most cases, both are stable: a *stable burst* configuration. Between these energy cylinders, are numerous *bridges*, consisting of orbit cylinders incident upon both $\Gamma_s(\mu)$ and $\Gamma_l(\mu)$ in an orbit cylinder bifurcation such as *subtle division, murder, phantom kiss,* or *emission.* Let $B(\mu)$ be such a bridge cylinder. Then as μ passes the bifurcation value μ_0, the entire bridge $B(\mu)$ collapses into the equilibrium $m_0(\mu_0)$.

The full description of these equilibrium resonance bifurcations is conjectured in Deprit and Henrard [1969] and proved in Schmidt [1974], in the case of two degrees of freedom, and (k,l) relatively prime.

8.7 THE GENERAL PATHOLOGY

We now combine these results and the invariant tori of Kolmogorov, Arnol'd, and Moser in a single picture. Suppose M has dimension four, $H: M \to \boldsymbol{R}$ is C^r, r sufficiently large, and the Hamiltonian system X_H satisfies all of the generic conditions envisioned so far. Then X_H has a set C of isolated critical points that are of hyperbolic (real or complex), saddle-center, or elliptic type. Write $C = C_h \cup C_s \cup C_e$. Each point $m \in C$ has a different energy $H(m)$. The complement $M \setminus C$ is foliated by energy surfaces Σ_e of dimension three. A point $m \in C_h$ has two-dimensional insets and outsets, and no center set. A point $m \in C_s$ has one-dimensional insets and outsets, and a two-dimen-

ISBN 0-8053-0102-X

sional center, which is a hyperbolic orbit cylinder. A point $m \in C_e$ has two sub-centers of dimension two, each an elliptic orbit cylinder—either the burst or reincarnation. The remaining closed orbits comprise orbit cylinders that originate and terminate with these center cylinders, or on each other. Each orbit cylinder Γ may be tangent to an energy surface only at isolated values of its cylinder parameter—the *creation* and *annihilation* events. At each of these isolated tangencies, the cylinder changes from hyperbolic to elliptic type, or vice versa. The creation bifurcations of a given cylinder divide it into bands of closed orbits, parametrized by the energy. At isolated closed orbits in these bands, there may occur either *subtle division* (arrival of an elliptic cylinder of half the orbital frequency) or *murder* (arrival of a hyperbolic cylinder of half the orbital frequency). In either case, the original band changes from elliptic to hyperbolic type. Omitting the transitional orbits of all three types, the components of the rest of the original cylinder Γ is a union of connected orbit bands, each completely elliptic or completely hyperbolic. The hyperbolic bands have further bifurcations: *rechambering*—at isolated closed orbits where insets and outsets cross (quasi-transversally)—and *renesting*—where the critical instabilities occur (8.5.5).

The hyperbolic bands may only terminate in certain ways:

—by bursting at an equilibrium;
—on an elliptic band at a kiss, murder, or emission;
—or by transition to an elliptic band through creation, subtle division, or murder.

At isolated orbits in an elliptic band E, hyperbolic cylinders may touch E in one of the *kisses*, resonating to either their third or fourth octaves at contact. On the rest of E, the transverse frequency α_e of $\gamma_e = E \cap \Sigma_e$ passes continuously through rational and irrational values. At each rational value, the *emission* of a pair of sub-harmonic cylinders occurs, one elliptic and one hyperbolic. A few of these are illustrated in Fig. 8.7-1, within a three-dimensional submanifold $S \subset M$, which is transversal to the cylinder E ($S \cap E$ is a curve) and such that S is a transversal section for each orbit $\gamma_e \subset E$. The energy subsurfaces $S_e = S \cap \Sigma_e$ are drawn as horizontal subspaces, and the curve $S \cap E$ is a vertical line. Each S_e is a transversal section for γ_e within Σ_e, so closed orbits δ_e of an emitted cylinder Δ appear as periodic orbits of the Poincaré section map of S_e. As e varies, the periodic points corresponding to δ_e trace curves in S, which appear as ribs of an upside-down umbrella. At each e with α_e rational, such an umbrella originates with $2p$ ribs if α_e is a pth root of unity. The ribs alternate elliptic/hyperbolic and are tangent to S_e at $x_e = \gamma_e \cap S$, according to Meyer [1970]. Only a few umbrellas are drawn in Fig. 8.7-1, but in fact, S is practically full of them. See Figs. 8.3-3 and 8.7-2.

Now fix an elliptic closed orbit $\gamma_e \subset E$. As X_H satisfies property H2, the Poincaré section S_e has many concentric invariant circles around $x_e = \gamma_e \cap S$, corresponding to invariant tori around γ_e, on which X_H is an irrational rotation, according to the Moser stability theorem. The ribs of emitted

ISBN 0-8053-0102-X

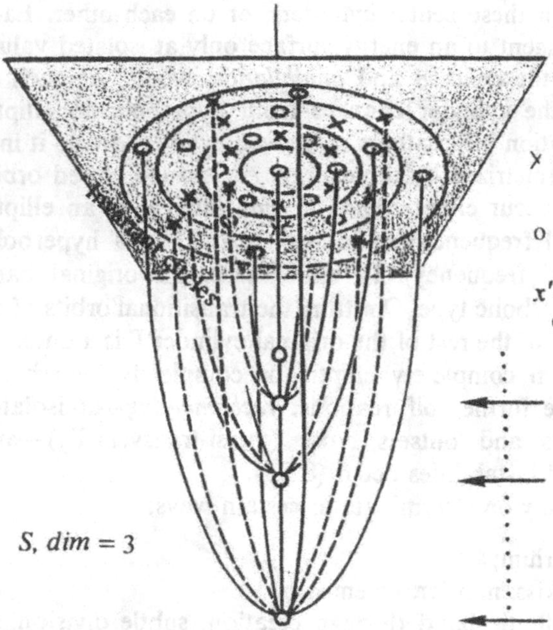

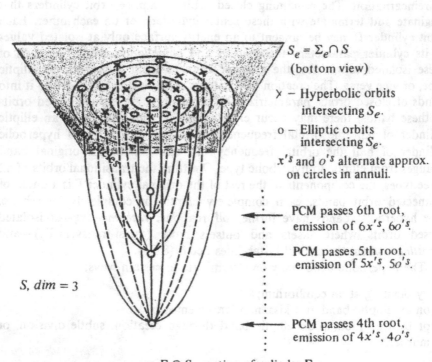

Figure 8.7-1 *Nested umbrellas.* A cross section S of an elliptic orbit cylinder Γ showing the loci of successive emitted sub-harmonic cylinders. The energy surfaces, $S_e = \Sigma_e \cap S$, shown as horizontal planes, are actually two dimensional. The intersection $S \cap \Gamma$, shown as a vertical line, is actually one dimensional. Pictorials of all the bifurcations, shown schematically (four dimensions represented as three) in the previous illustrations, could be made in this manner. Only a few ribs of three umbrellas are shown. There is a countable set of umbrellas.

umbrellas for $x_c = \gamma_c \cap S$ below x_e pierce the annuli bounded by the Moser circles. As x_c gets closer to x_e, the ribs emitted from x_c pierce progressively smaller annuli. For a given p, there will be a finite number of umbrellas of $2p$ ribs, so the ribs of the umbrella of x_c must get increasingly numerous as x_c approaches x_e.

Further, the VAK around γ_e and within Σ_e generates a "VAK cylinder" as e is changed: each invariant torus T^2 of recurrent motion generates a cylinder $T^2 \times I$, and each pair of (elliptic/hyperbolic) closed orbits of long period between two T^2's generate a pair of (elliptic/hyperbolic) orbit cylinders.

ISBN 0-8053-0102-X

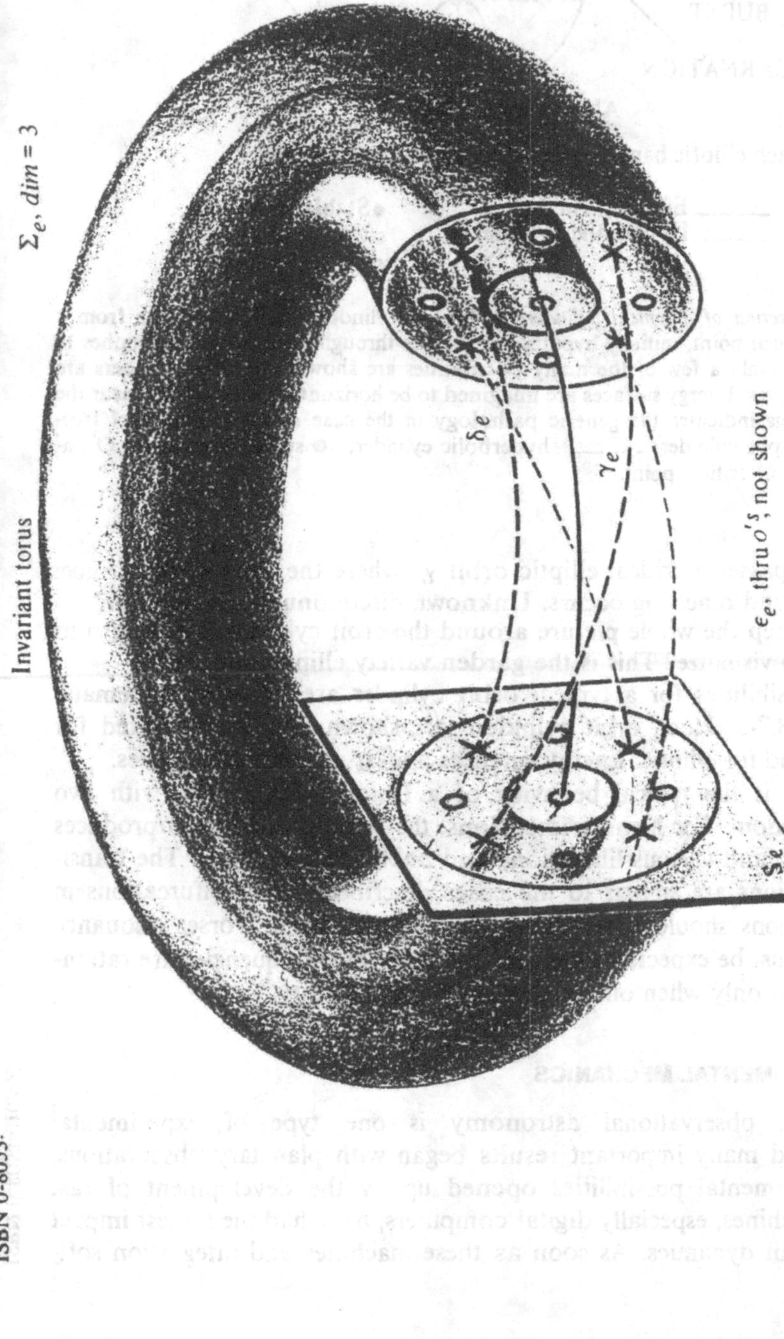

Figure 8.7.2. *The VAK-: nested invariant tori of an elliptic orbit* γ_e. Here a single three-dimensional energy surface Σ_e is shown. As *e* moves, this figure translates in four-space to generate the orbit cylinders Γ and those of the ribs of the umbrellas. For example, as *e* decreases, suppose γ_e is stationary within Σ_e, as ϵ_e approaches, faster and faster, a *p*-fold covering of γ_e.

ISBN 0-8053-

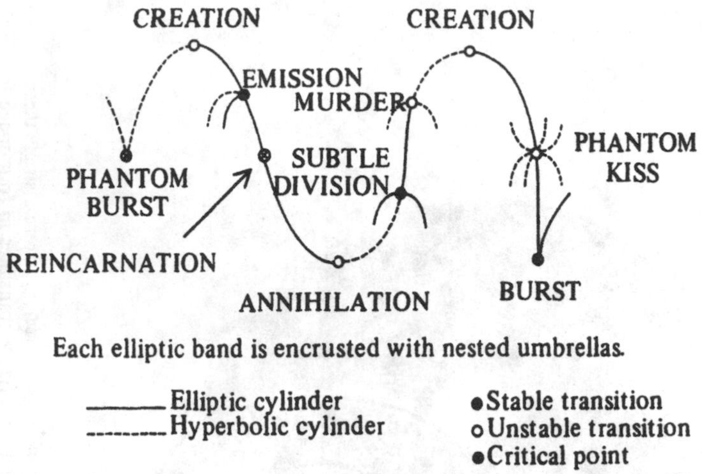

Figure 8.7-3. *Section of a typical cylinder.* A typical cylinder, from birth (here from a saddle-center critical point, initially hyperbolic) evolves through successive catastrophes to death at a burst. Only a few of the many possibilities are shown. All orbit cylinders are represented as curves. Energy surfaces are imagined to be horizontal planes, except near the critical point. This indicates the generic pathology in the case of $n=2$ degrees of freedom. ——— elliptic cylinder; - - - - - - hyperbolic cylinder; ● stable transition; O unstable transition; ⊗ critical point.

Eventually, e passes a critical elliptic orbit γ_c, where the twist invariant goes through zero, and renesting occurs. Unknown discontinuities may occur.

Finally, sweep the whole picture around the orbit cylinder E to return to M. Difficult to visualize? This is the garden variety elliptic orbit band.

A few possibilities for a typical orbit cylinder are shown in schematic form in Fig. 8.7-3. Here, orbit cylinders are shown as curves—dotted for hyperbolic, solid for elliptic. Energy surfaces appear as horizontal planes.

This, then, is the typical behavior of a Hamiltonian system with two degrees of freedom. For higher dimensions, the burst phenomenon produces more cylinders, and various liberations and bubbles are possible. The transitional bifurcations are similar to the cases described. The p-bifurcations in higher dimensions should admit many new catastrophes. Worse, resonance bifurcations must be expected whenever the transverse frequencies are rationally related, not only when one of them is a root of unity.

8.8 EXPERIMENTAL MECHANICS

Of course, observational astronomy is one type of experimental mechanics, and many important results began with planetary observations. But the experimental possibilities opened up by the development of fast numerical machines, especially digital computers, have had the largest impact on Hamiltonian dynamics. As soon as these machines and integration soft-

ISBN 0-8053-0102-X

ware were available, Hamiltonian systems were among the first studied. Thus, the historical remarks of Sect. 7.8 apply here as well.

The literature on experimental mechanics was dominated from the start by the restricted three-body problem, and this development is probably as essential as rocketry to the physical exploration of the solar system.

In this section, we describe three examples.

8.8.1 Example. An early classic is Henon and Heiles [1964], who studied the motion of a star in a cylindrical galaxy by numerical integration. This reduces to a Lagrangian system

$$L: T\boldsymbol{R}^2 \to \boldsymbol{R}^2$$
$$: (x, y, \dot{x}, \dot{y}) \mapsto \tfrac{1}{2}(\dot{x}^2 + \dot{y}^2) + U(x, y)$$

Fixing energy e, a global section is obtained with section map

$$\theta_e: V_e \subset \boldsymbol{R}^2 \to \boldsymbol{R}^2$$

where V_e is a neighborhood of (0,0) in the $(y, \dot{y})$ plane. This is constructed by projecting the energy surface Σ_e into $\boldsymbol{R}^3$,

$$\pi: \Sigma_e \subset T\boldsymbol{R}^2 \to \boldsymbol{R}^3$$
$$: (x, y, \dot{x}, \dot{y}) \to (x, y, \dot{y})$$

and taking the section S defined by $x = 0$. Figures 8.8-1(a)–8.8-1(c) show the

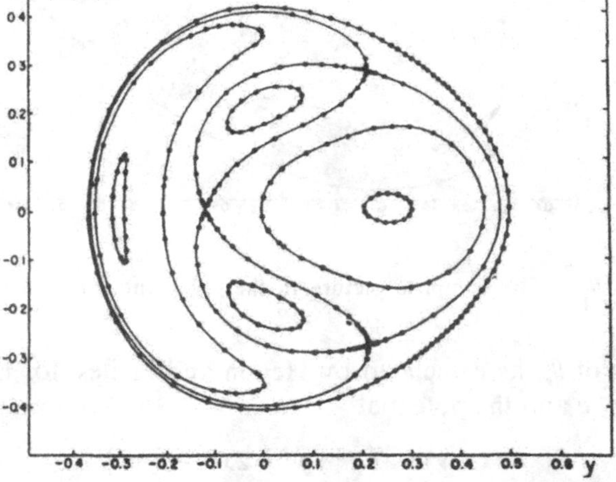

Figure 8.8-1(a). $\theta_{1/12}$. The complete picture in the $y, \dot{y}$ plane for a given value of the energy: $e = \frac{1}{12} = 0.08333$.

ISBN 0-8053-0102-X

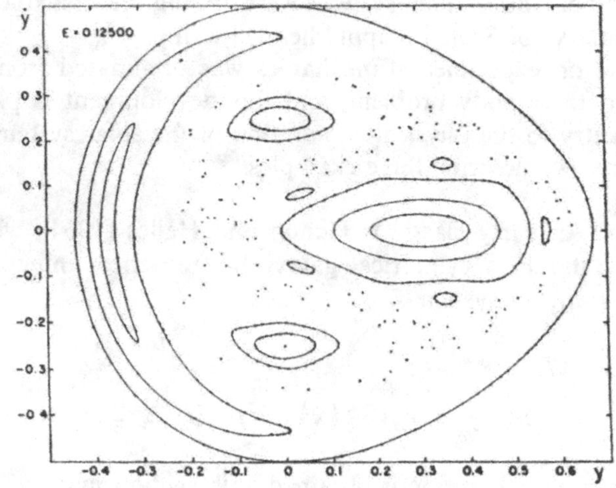

Figure 8.8-1(b). $\theta_{1/8}$. The complete picture in the $y,\dot{y}$ plane for higher energy: $e=0.12500$.

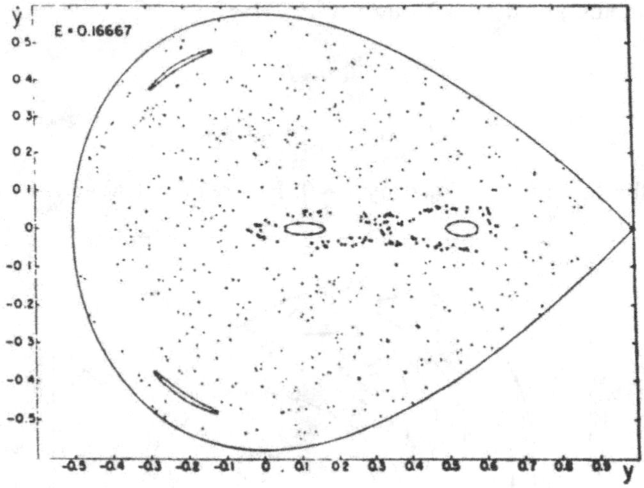

Figure 8.8-1(c). $\theta_{1/6}$. The complete picture in the $y,\dot{y}$ plane for higher energy: $e=\frac{1}{6}=$ 0.16667.

phase portraits of θ_e, as calculated by Henon and Heiles, for three values of the fixed energy e and the potential

$$U(x,y)=\tfrac{1}{2}\left(x^2+y^2+2x^2y-\tfrac{2}{3}y^3\right)$$

They found that some of the discrete orbits appear to lie on closed curves, which are rotated irrationally by θ_e, as expected because of the Moser twist theorem. These curves have been added to the phase portraits. The final

ISBN 0-8053-0102-X

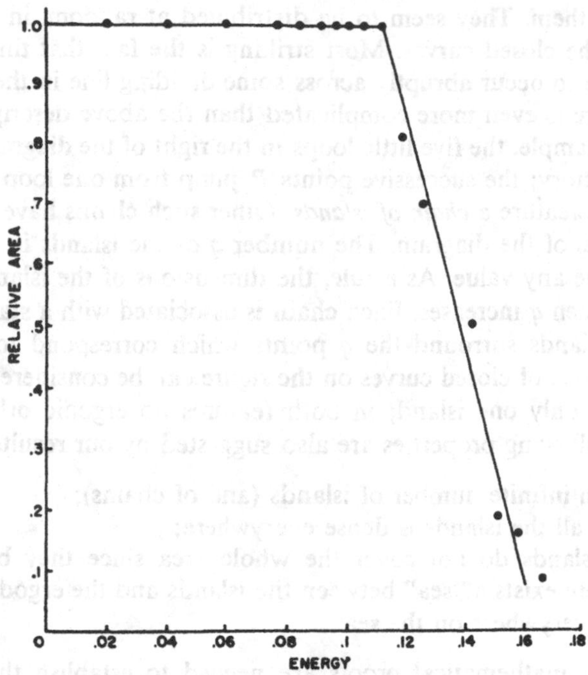

Figure 8.8-1(d). VAK density versus energy. Relative area covered by the curves as a function of energy.

drawing, Fig. 8.8-1(d), shows the density of these curves as a function of the energy *e*. This is similar to the expectation of stability for a single VAK shown in Fig. 8.3-4. The commentaries in Fig. 8.8-1 are quotations from Henon and Heiles [1964]. Compare Fig. 8.3-3! For further discussion, see Churchill, Pecelli and Rod [1978b].

Commentary on Fig. 8.8-1(a). Each set of points linked by a curve corresponds to one computed trajectory. In fact, more trajectories and more points on each have been computed than shown on this picture. It appears that in every case, the points seem to lie exactly on a curve. These curves form a one-parameter family that fills completely the available area, within the outer curve.

"In the middle of the four small loops are four invariant points of the mapping (not represented on the figure); they correspond to stable periodic orbits. The three intersections of curves are also invariant points, corresponding to unstable periodic orbits." (From Henon and Heiles [1964].)

Commentary on Fig. 8.8-1(b). Here we continue to have a set of closed curves around each stable invariant point. But these curves no longer fill the whole area. All the isolated points on the figure correspond to one and the same trajectory, just as the points on one of the closed curves; but they behave in a completely different way. It is clearly impossible to draw any

ISBN 0-8053-0102-X

curve through them. They seem to be distributed at random, in an area left free between the closed curves. Most striking is the fact that this change of behavior seems to occur abruptly across some dividing line in the plane.

"The picture is even more complicated than the above description would suggest. For example, the five little loops in the right of the diagram belong to the same trajectory; the successive points P_i jump from one loop to the next. Let us call this feature a *chain of islands*. Other such chains have been found in various parts of the diagram. The number q of the islands in a chain can apparently have any value. As a rule, the dimensions of the islands decrease very rapidly when q increases. Each chain is associated with a stable periodic orbit; the q islands surround the q points which correspond to that orbit. Note that each set of closed curves on the figure can be considered as a chain constituted by only one island; in both features no ergodic orbit seems to appear. The following properties are also suggested by our results:

(1) there is an infinite number of islands (and of chains);
(2) the set of all the islands is dense everywhere;
(3) but the islands do not cover the whole area since they become very small; there exists a "sea" between the islands and the ergodic trajectory is dense everywhere on the sea.

But, of course, mathematical proofs are needed to establish these points." (From Henon and Heiles [1964].)

Commentary on Fig. 8.8-1(c). With even higher energy the picture again changes drastically. All the isolated points correspond to one trajectory, and it is apparent that this "ergodic" trajectory covers almost the whole area within the outer line. Its random character is most strikingly seen when one plots the successive points; they jump from one part of the diagram to another without any apparent law. Two of the sets of closed curves, those on the $\dot{y}$ axis, have now disappeared, presumably because their central invariant point has become unstable. The two other sets of closed curves have degenerated, each one into a chain of two small islands, successive points P_i jumping from one to the other. No other chain of islands has been found; probably they still exist, but the dimensions of the islands are so small that finding them is difficult.

"The open circles in the middle of the diagram correspond to a trajectory of a new kind, intermediate between the closed curves and the ergodic behavior. They are approximately situated on an eight-shaped line, but with an important dispersion around it. The ultimate behavior of such an orbit is not known; perhaps the points will always remain in the vicinity of the same line, and fill an eight-shaped band; or perhaps they will after some time penetrate into the ergodic region. Some recent results, not shown here, are in favor of this last hypothesis." (From Henon and Heiles [1964].)

Commentary on Fig. 8.8-1(d). "A remarkable feature of Figs. 8.8-1(a)–8.8-1(c) is the complete change in the picture over a moderate interval of the

ISBN 0-8053-0102-X

energy e. For $e=0.08333$, the area is completely covered with curves; for twice that value, the curves are almost completely replaced by an ergodic region.

"In order to study this transition in more detail, we have computed, for a number of values of e, the proportion of the total allowable area in the $y,\dot{y}$ plane which is covered by curves.

"The figure shows the results. Up to a critical energy (about $e=0.11$) the curves cover the whole area; there is no ergodic orbit. For higher energies the area covered by curves shrinks very rapidly. Thus the situation could be very roughly described by saying that the second integral exists for orbits below a "critical energy," and does not exist for orbits above that energy. $e=\frac{1}{6}$ is the *energy of escape* in the potential; for $e>\frac{1}{6}$, the equipotential lines open and the star can eventually escape to infinity, if the orbit is ergodic. The area in the $y,\dot{y}$ plane becomes infinite and the relative area ceases to have meaning. No obvious connection exists between the critical energy and the energy of escape; in the present case the critical energy is less than the energy of escape. But results from computations with $U=\frac{1}{2}(x^2+y^2-x^2y^2)$, not shown here, indicate the opposite situation, as do the results of computations by Ollongren [1962] with an approximation to the Galactic potential." (From Henon and Heiles [1964].)

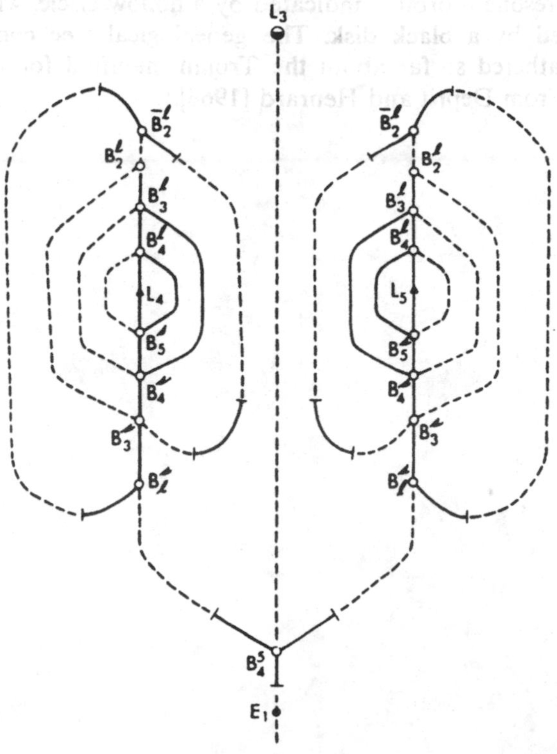

Figure 8.8-2. Genealogical tree of the Trojan manifold for Routh's mass ratio.

ISBN 0-8053-0102-X

8.8.2 Example. As early as 1965, a computer graphic film was made of "bus orbits" in the three-body problem by Arentsdorf. (See Fig. 10.3-2.) Computational techniques for such graphics are well described in Daniel and Moore [1970].

8.8.3 Example. The previous section described some outstanding features of the phase portrait of a typical Hamiltonian vector field in four dimensions. According to the theory, some wild behavior is possible: bifurcating layers of infinite umbrellas of VAK nests, and so on. One might wonder if any real system is this complex, in spite of Poincaré's warnings about the three-body problem. Deprit and Henrard [1968] describe how the liberation phenomenon was discovered by digital simulation. And furthermore, they have attempted a full experimental exploration and mapping of the phase portrait of the restricted three-body problem in the neighborhood of the point corresponding (roughly) to the Trojan planets in the Sun–Jupiter system (see Sect. 10.3). Their results are shown in Fig. 8.8-2. Compare with Fig. 8.7-3.

Commentary on Figure 8.8-2. "Each orbit is represented by a point; two orbits having the same Jacobi constant lie on the same horizontal line. A branch of periodic orbits is represented by a curve; it is a solid curve as long as the constituent orbits are stable; otherwise it is a dashed line. A bifurcation orbit or a self-resonant orbit is indicated by a hollow circle, while an ejection orbit is marked by a black disk. The genealogical tree condenses all the information gathered so far about the Trojan manifold for Routh's critical mass ratio." (From Deprit and Henrard [1968].)

ISBN 0-8053-0102-X

PART 4

CELESTIAL MECHANICS

Although many new and important applications of analytical dynamics bespeak its broad role in the theoretical sciences, the grandfather of the entire edifice is celestial mechanics, or the study of the *n*-body problem. In this Part we introduce the simplest cases. Far from being simple, we shall see that all of the preceding results must be used.

At this point, we return to our policy prior to Part III. Henceforward, we include complete proofs for the principal results.

ISBN 0-8053-0102-X

CHAPTER 9

The Two-Body Problem

In this chapter we establish the classical results on the two-body problem. These are expressed in a form dictated by the three-body problem, discussed in the next chapter.

9.1 MODELS FOR TWO BODIES

In this section we consider several mathematical models for the two-body problem and their interrelationships. The experimental domain might be the Earth and Sun, a binary star, and so forth. As the heuristic derivation of the models from Newton's gravitational theory is so well known, we shall not discuss it, its experimental domain, nor its interpretation.

9.1.1 Definition. *The first model for the two-body problem* (I) *is a system* (M, H^{μ}, m, μ), *where:*

(i) $M = T^*W$, $W = R^3 \times R^3 \backslash \Delta$, $\Delta = \{(q,q) | q \in R^3\}$, *where M has the canonical symplectic structure;*
(ii) $m \in M$ *(initial conditions);*
(iii) $\mu \in R, \mu > 0$ *(the mass ratio);*

Ralph Abraham and Jerrold E. Marsden, Foundation of Mechanics, Second Edition

ISBN 0-8053-0102-X

(*iv*) $H^\mu \in \mathfrak{F}(M)$ *defined by*

$$H^\mu(q,q',p,p') = \frac{\|p\|^2}{2\mu} + \frac{\|p'\|^2}{2} - \frac{1}{\|q-q'\|}$$

where $q, q' \in R^3$, $p, p' \in (R^3)^*$, *and* $\| \ \|$ *denotes the standard norm in* R^3.

The predictions of the model are: the maximal integral curve of the Hamiltonian vector field X_{H^μ} *through m, the orbits* m_σ *of m* ($\sigma = +, -, \pm$), *and limit sets* $\omega^\sigma(m)$.

Note that collision points, represented by the set Δ, are excluded from $R^3 \times R^3$ so that $H^\mu \in \mathfrak{F}(M)$. Also, X_{H^μ} is not complete as integral curves can "run off the manifold" in finite time due to collisions.

9.1.2 Proposition (Conservation of linear momentum). *In model I the components of* $p+p'$ *(relative to the standard dual basis) are constants of the motion.*

Proof. Consider the Lie group $G=(R^3, +)$ acting on W by translations: $\phi_r(q,q') = (q+r, q'+r)$ for $r \in G$. On $M = T^*W$ consider the induced action ϕ^{T^*} (see Sect. 4.2) given by

$$\phi_r^{T^*}(q,q',p,p') = (q+r, q'+r, p, p')$$

A computation analogous to the one done for the linear momentum in Sect. 4.2 shows that $J: M \to (R^3)^*$, $J(q,q',p,p') = p+p'$ is a moment for the action ϕ^{T^*}, and hence $\hat{J}(\xi)$ is a constant of the motion for all $\xi \in (R^3)^*$ since the Hamiltonian H^μ is clearly ϕ^{T^*}-invariant. Making ξ successively equal to $(1,0,0)$, $(0,1,0)$, and $(0,0,1)$, we find that the maps $(q,q',p,p') \mapsto p_i + p_i'$, $i = 1,2,3$ are constants of the motion. ■

The next proposition in effect changes us to *coordinates relative to the center of mass.*

9.1.3 Proposition. *There is a symplectic diffeomorphism* $F: M \to N$, *where M is as in* 9.1.1 *and* $N = T^*V$, $V = R^3 \times (R^3 \backslash \{0\})$ *such that*

$$H^\mu \circ F^{-1}(q,q',p,p') = \frac{\|p+p'\|^2}{2\mu} + \frac{\|p'\|^2}{2} - \frac{1}{\|q'\|}$$

where H^μ *is given in* 9.1.1.

Proof. Consider the diffeomorphism $f: W \to V$; $(q,q') \mapsto (q, q-q')$. The induced symplectic diffeomorphism $F = T^*f^{-1}$: $T^*W = M \to T^*V = N$ is easily

ISBN 0-8053-0102-X

seen to be $(q,q',p,p')\mapsto(q,q-q',p+p',-p')$ with inverse $F^{-1}: N\to M$ given by $(q,q',p,p')\mapsto(q,q-q',p+p',-p')$. This gives the result. ■

In 9.1.3, the components of p are constants of the motion for $'H^{\mu}=H^{\mu}\circ F^{-1}$. If $n\in N$ represents the initial conditions and $n=(q_0,q_0',p_0,p_0')$, then using the symplectic diffeomorphism $(q,q',p,p')\mapsto(q-q_0,q',p-p_0,p')$, we may assume $q_0=0$, $p_0=0$. Thus

$$'H^{\mu}(q,q',p,p')=\left(\frac{1}{\mu}+1\right)\frac{\|p'\|^2}{2}-\frac{1}{\|q'\|}$$

along the integral curve, where $(1/\mu+1)^{-1}$ is called the *reduced mass*.

This leads naturally to another model that is "equivalent" in some sense (the one-body problem). This is an example of *reduction by first integrals*, the general case of which was described in Sect. 4.3. (The approach of Sect. 4.3 is actually carried out in 10.4.2 for the planar n-body problem.)

9.1.4 Definition. *The second model for the two-body problem* (II) *is a system* (M,H,m), *where:*

(*i*) $M=T^*U$, $U=R^3\backslash\{0\}$;
(*ii*) $m\in M$ *(initial conditions); and*
(*iii*) $H\in\mathfrak{F}(M)$, *defined by* $H(q,p)=\|p\|^2/2-1/\|q\|$.

The predictions of the model are: the maximal integral curve of the Hamiltonian vector field X_H *through m, and the corresponding orbits and limit sets.*

Note that we have now rescaled so the mass is one.

9.1.5 Proposition (Conservation of angular momentum). *In model II the following quantities are constants of the motion,*

$$(q^2p_3-q^3p_2,\ q^3p_1-q^1p_3,\ q^1p_2-q^2p_1)=(G_1,G_2,G_3)$$

where

$$q=(q^1,q^2,q^3),\ p=(p_1,p_2,p_3)$$

Proof. Consider the Lie group $G=SO(3)$ acting on $U=R^3\backslash\{0\}$ through rotations: ϕ: $(A,q)\mapsto Aq$. On $M=T^*U$ consider the induced action $\phi^{T^*}(A,(q,p))=(Aq,p\circ A^{-1})$ and recall from 4.2.15(ii) that $J: M\to so(3)^*$, $J=(G_1,G_2,G_3)$ is a momentum mapping for this action, called angular momentum. Since the Hamiltonian H is clearly ϕ^{T^*}-invariant, $\hat{J}(\xi)$ is a constant of the motion for all $\xi\in so(3)$. Making ξ successively equal to $(1,0,0)$, $(0,1,0)$, and $(0,0,1)$, we find that the maps $(q,p)\mapsto G_i$, $i=1,2,3$ are constants of the motion. ■

ISBN 0-8053-0102-X

Thus 9.1.5 tells us that the orbit of (q,p) lies in the hyperplane perpendicular to $(G_1,G_2,G_3,G_1,G_2,G_3)$ at (q,p) in the standard metric. A rotation in $R^3\backslash\{0\}$ together with the same one in the phase variables is, as we have seen, a symplectic diffeomorphism (compare 9.1.3). Thus we may orient the above hyperplane to obtain the following "equivalent" model. Again this is a particular case of reduction.

9.1.6 Definition. *The third model for the two-body problem (III) is a system (M,H,m) where:*

(i) $M=T^*(R^2\backslash\{0\})$, with canonical symplectic structure;
(ii) $m\in M$ (initial conditions); and
(iii) $H\in\mathcal{F}(M)$ defined by $H(q,p)=\|p\|^2/2-1/\|q\|$.

The predictions of the model are: the maximal integral curve through m of the Hamiltonian vector field X_H, and the corresponding orbits and limit sets.

Again, $\|\ \|$ denotes the Euclidean norm.

From the comments preceding 3.7.4, the above Hamiltonian is hyperregular with Lagrangian $\mathcal{L}$ on $T(R^2\backslash\{0\})$ given by

$$\mathcal{L}(q,\dot{q})=\frac{\|\dot{q}\|^2}{2}+\frac{1}{\|q\|}$$

Thus, in this case, the Lagrangian equations become

$$\frac{d^2q^i(t)}{dt}=-\frac{q^i(t)}{\|q(t)\|^3},\qquad i=1,2$$

where $q\colon I\to R^2\backslash\{0\}$, $q(t)=(q^1(t),q^2(t))$ denotes the base integral curve (3.5.17). We also have conservation of energy:

$$E(q,\dot{q})=\tfrac{1}{2}\|\dot{q}\|^2-1/\|q\|$$

and angular momentum (area integral):

$$G(q,\dot{q})=q^1\dot{q}^2-q^2\dot{q}^1$$

The fiber derivative of the Hamiltonian H (or Lagrangian $\mathcal{L}$) relates X_H to X_E as we have seen in 3.6.9. Moreover, $FH\colon T^*(R^2\backslash\{0\})\to T(R^2\backslash\{0\})\colon (q,p)\mapsto(q,\dot{q})$ (relative to the standard bases) so we may freely pass between the two formulations.

The fourth model is just the Lagrangian version of the third model, living in velocity space rather than momentum phase space. This distinction is trivial in this context, as we have already transformed the masses away, by normalization. Thus FH is essentially the identity mapping, and so also is the Legendre transformation, $F\mathcal{L}=FH^{-1}$.

ISBN 0-8053-0102-X

9.1.7 Definition. *The fourth model for the two-body problem (IV) is a system (N, L, n), where:*

(*i*) $N=T(\boldsymbol{R}^2\backslash\{\boldsymbol{0}\})$ *with the symplectic structure* $\omega_{\mathfrak{L}}=(F\mathfrak{L})^*\omega_0$, ω_0 *the canonical symplectic form on* $T^*(\boldsymbol{R}^2\backslash\{\boldsymbol{0}\})$;

(*ii*) $\boldsymbol{n}\in N$ *(initial conditions); and*

(*iii*) $\mathfrak{L}\in\mathfrak{F}(N)$ *is the hyperregular Lagrangian defined by*

$$\mathfrak{L}(\dot{\boldsymbol{q}},\dot{\boldsymbol{q}})=\tfrac{1}{2}\|\dot{\boldsymbol{q}}\|^2+1/\|\boldsymbol{q}\|$$

The predictions of the model are: the maximal integral curve of the Lagrangian vector field X_E through n, and the corresponding orbits and limit sets. Here $E\in\mathfrak{F}(N)$ is the energy of $\mathfrak{L}$,

$$E(\boldsymbol{q},\dot{\boldsymbol{q}})=\tfrac{1}{2}\|\dot{\boldsymbol{q}}\|^2-1/\|\boldsymbol{q}\|$$

Note that the symplectic structure induced on N, $\omega_{\mathfrak{L}}=(F\mathfrak{L})^*\omega_0$ is the standard one,

$$\omega_{\mathfrak{L}}=dq^1\wedge d\dot{q}^1+dq^2\wedge d\dot{q}^2$$

where for $n\in N=T(\boldsymbol{R}^2\backslash\{\boldsymbol{0}\})$ we use the notations

$$\boldsymbol{n}=(\boldsymbol{q},\dot{\boldsymbol{q}})=(q^1,q^2,\dot{q}^1,\dot{q}^2)$$

for the component functions.

The fourth model has an associated angular momentum $G\in\mathfrak{F}(N)$ defined by

$$G(\boldsymbol{q},\dot{\boldsymbol{q}})=q^1\dot{q}^2-q^2\dot{q}^1$$

This is the familiar momentum of the usual action of $SO(2)$ on $\boldsymbol{R}^2$, in the tangent formulation, except we here identify the dual Lie algebra of $SO(2)$ with $\boldsymbol{R}$. Thus, the momentum is a real valued function on $T(\boldsymbol{R}^2\backslash\{\boldsymbol{0}\})$.

The Lagrangian vector field X_E in model IV is the classical second-order system of Newton for the two-body problem. The well-known solution to these equations (a base integral curve of X_E in model IV) is a conic section, possibly degenerate. In fact, if $G=0$, the solution is a straight line (degenerate case), whereas if $G\neq 0$, the path is an ellipse, parabola, or hyperbola, according as $E<0$, $E=0$, or $E>0$, the sense of rotation being determined by the sign of G. We shall see all this in detail in the next section for the elliptical case. It is not simple, however, to obtain a formula for the global flow explicitly. For the details and a discussion of the use of the Jacobi metric, see Wintner [1941, Chapter IV].

We shall be mainly interested in the case of closed orbits ($E<0$, $G\neq 0$), which we discuss in the next section.

ISBN 0-8053-0102-X

EXERCISES

9.1A. Work out the details that lead above to the conclusion that $G(\mathbf{q},\dot{\mathbf{q}})=q^1\dot{q}^2-q^2\dot{q}^1$ is a constant of the motion in model IV.

9.1B. Show in detail that the reduction procedure of Sect. 4.3 applied to the linear and angular momentum leads to models II and III, respectively.

9.2 ELLIPTIC ORBITS AND KEPLER ELEMENTS

In model IV for the two-body problem a large open subset of the velocity space is filled with closed orbits. In this section we study this region, primarily because it is basic to the study of closed orbits of the restricted three-body problem. Then we shall do the same thing for the phase space.

We begin by prescribing the domains of closed orbits, then introduce the classical parameters of the elliptic orbits known variously as anomalies, elliptic elements, Kepler elements, and so forth. The balance of this section is devoted to the classical properties of these parameters, discovered by Kepler (1627), Newton (1687), and Lagrange (1808).

Recall that for the planar two-body problem, in model IV (9.1.7) we found the following equations for the base integral curves (Lagrange's equations):

$$\frac{d^2q^i(t)}{dt^2}=-\frac{q^i(t)}{r(t)^3}, \qquad r=\|\mathbf{q}\|$$

with constants of the motion

$$\frac{1}{2}\left\|\frac{d\mathbf{q}(t)}{dt}\right\|^2-\frac{1}{r(t)}=E.$$

and

$$q^1(t)\frac{dq^2(t)}{dt}-q^2(t)\frac{dq^1(t)}{dt}=G$$

E represents the energy and G is the derivative of twice the area swept out by the radius arm; that is why G is often referred to as the "area integral." To see this, denote by (r,α) a polar coordinate system in the plane; hence $q^1(t)=r(t)\cos\alpha(t)$, $q^2(t)=r(t)\sin\alpha(t)$, and the area swept out by the radius arm when the angle α goes from ϕ_1 to ϕ_2 is $\frac{1}{2}\int_{\phi_1}^{\phi_2}r^2(\alpha)\,d\alpha=A$. We conclude then that $dA/d\alpha=\frac{1}{2}r^2(\alpha)$ and hence $dA/dt=\frac{1}{2}r^2(t)\dot{\alpha}(t)$. A straightforward computation shows that

$$\begin{aligned}\tfrac{1}{2}G&=\tfrac{1}{2}\left(q^1(t)\frac{dq^2(t)}{dt}-q^2(t)\frac{dq^1(t)}{dt}\right)\\&=\tfrac{1}{2}r^2(t)\dot{\alpha}(t)=\frac{dA}{dt}\end{aligned}$$

ISBN 0-8053-0102-X

The fact that G is constant along the integral curves of X_E means that $dA/dt = constant$ and hence that the radius arm *sweeps out equal areas in equal times*. This is Kepler's Second Law discovered in 1602 (published in 1609).

9.2.1 Proposition. *If, in the above model, $E<0$ and $G \neq 0$ on the base integral curve (computed from initial conditions $\mathbf{q}_0, \dot{\mathbf{q}}_0$), then the curve is an ellipse with eccentricity $b=(1+2EG^2)^{1/2}$ semimajor axis $a=(-2E)^{-1}$ and one focus at the origin (Kepler's First Law). Also, the rate of sweeping out an area by $\mathbf{q}_t$ is constant and the period of the closed orbit is $\tau = 2\pi a^{3/2}$ (Kepler's Third Law).*

Proof. It readily follows that

$$G\frac{d^2q^2(t)}{dt^2} = \frac{d}{dt}\left(\frac{q^1(t)}{r(t)}\right)$$

so that

$$G\frac{dq^2(t)}{dt} = \frac{q^1(t)}{r(t)} + A$$

Similarly,

$$G\frac{dq^1(t)}{dt} = -\frac{q^2(t)}{r(t)} - B$$

which gives $A^2+B^2 = 1+2EG^2$ and $G^2 = Aq^1(t)+Bq^2(t)+r(t)$, that is, in polar coordinates $q^1 = r\cos\alpha$, $q^2 = r\sin\alpha$,

$$r(\alpha) = \frac{G^2}{1+\sqrt{A^2+B^2}\cos(\alpha-\beta)}$$

where $\cos\beta = A/\sqrt{A^2+B^2}$, $\sin\beta = B/\sqrt{A^2+B^2}$, and β represents the *argument of the perihelion*, that is, the angle formed by $\overline{SP}$ and the q^1-axis (see Fig. 9.2-1 or 9.2-2). That β has indeed this interpretation follows from the fact that $r(\beta)$ is minimal. From the general equations of the ellipse in polar coordinates and since $1+2EG^2 \geqslant 0$, we see that the above equation represents an ellipse with eccentricity $b=\sqrt{A^2+B^2} = (1+EG^2)^{1/2} < 1$. Since the numerator in the expression for $r(\alpha)$ must be equal to $a(1-b^2)$, we conclude $a=(-2E)^{-1}$. The above equation thus represents an ellipse with one focus at the origin, a and b given in the statement and inclined by the angle β with respect to the q^1-axis. Finally, the rate of sweeping out area by the vector $\mathbf{q}_t$ is the constant $G/2$ as we saw earlier. Since the semi-minor axis of the ellipse is $a(1-b^2)^{1/2}$, the area of the ellipse is $\pi a^2(1-b^2)^{1/2}$, and hence the period is

$$\tau = \frac{\pi a^2(1-b^2)^{1/2}}{G/2} = 2\pi a^{3/2}$$

Note also the compatibility condition $0 \leqslant 1+2EG^2 < 1$. ■

ISBN 0-8053-0102-X

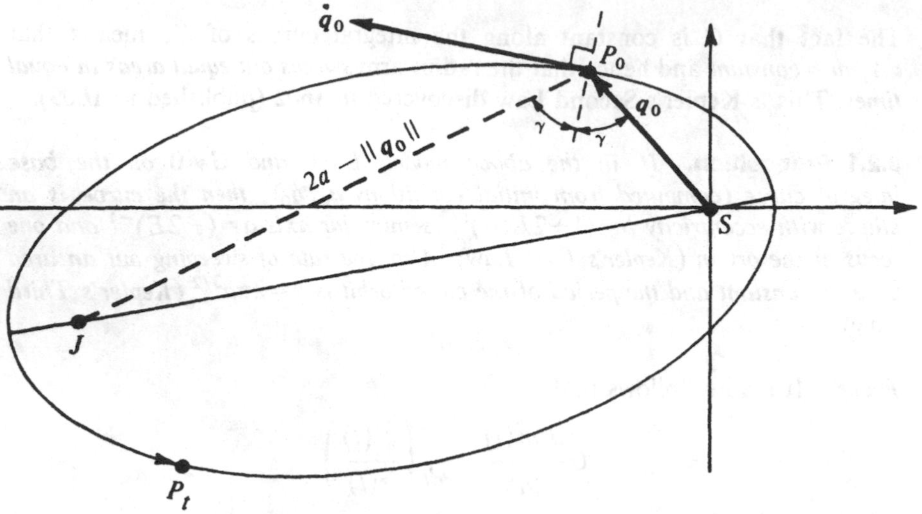

Figure 9.2-1. Construction of the apocenter J from initial conditions $(q_0, \dot{q}_0)$ with $E<0$ and $G>0$.

Given initial conditions $q_0, \dot{q}_0$ the orbit may be easily reconstructed by geometrical means as indicated in Fig. 9.2-1. Recall that the sum of the distances to the foci is always $2a$. (We assume $E<0$, $G\neq 0$.)

The position of $q=(q^1,q^2)$ on the elliptical orbit is described by the various "anomalies" or angular parameters in $[0,2\pi)\approx S^1$.

9.2.2 Definition. *Consider a point q_t moving on the ellipse described in 9.2.1 in accordance with the equations of motion. Then the following quantities are defined in Fig. 9.2-2, all in S^1:*

$\alpha(t)$, *the* ***polar angle;***

$u(t)$, *the* ***eccentric anomaly.***

We also define, in case the eccentricity $b\neq 0$,

β, *the* ***argument of the perihelion*** *(a constant);*

$f(t)=\alpha(t)-\beta$, *the* ***true anomaly,*** *and*

$l(t)$, *the* ***mean anomaly,*** *defined by*

$$l(t)=a^{-3/2}(t-T)=\frac{2\pi}{\tau}(t-T)=l_0+\frac{2\pi}{\tau}t$$

where $T\in(-\tau,0]$ is the ***time of perihelion passage,*** *that is, $q_T=P$, and thus the* ***initial mean anomaly*** *is $l_0=-(2\pi/\tau)T\in[0,2\pi)$. Note that $l_0=\alpha(0)-\beta$.*

That this diagram can be constructed follows at once from the equations of the ellipse.

ISBN 0-8053-0102-X

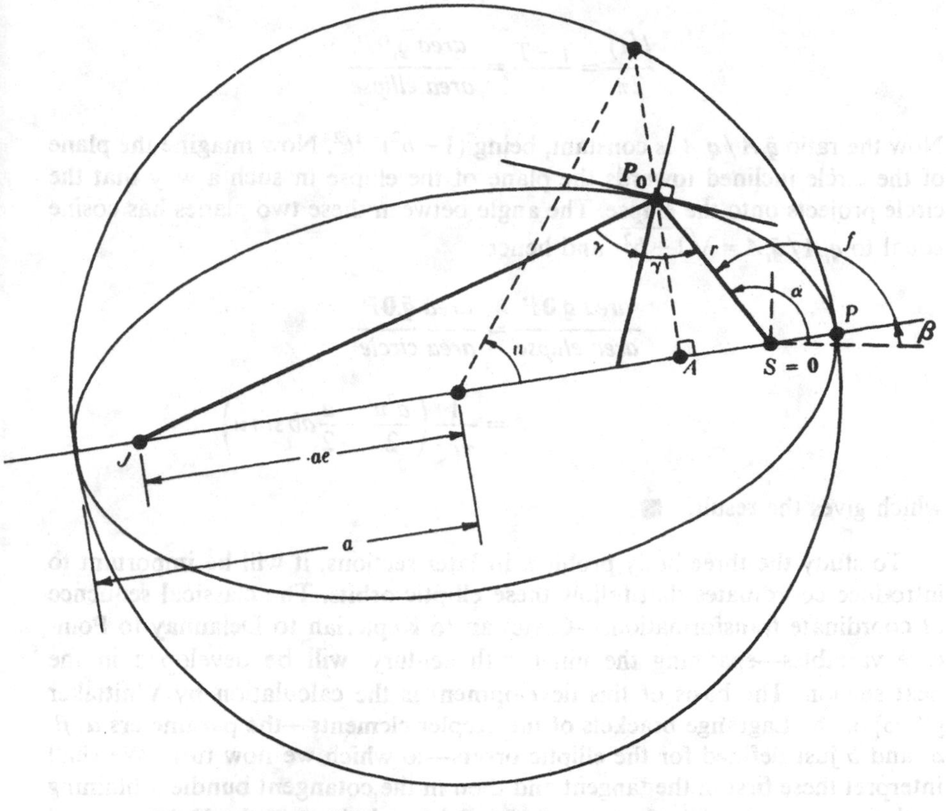

Figure 9.2-2. Definition of the anomalies, and related parameters.

9.2.3 Proposition (Kepler's Law). *In the situation described above, we have: for all $t \in [0, \tau)$,*

(i) $r(t) = a(1 - b\cos u(t))$, *and*
(ii) $l(t) = u(t) - b\sin u(t)$.

Proof. (i) In Fig. 9.2-2, if $q_tA = \|q_t - A\|$, and so forth,

$$\begin{aligned} r^2 &= (q_tA)^2 + (0A)^2 \\ &= \left(a\sqrt{1-b^2}\,\sin u\right)^2 + a^2(\cos u - b)^2 \\ &= a^2(1 - b\cos u)^2 \end{aligned}$$

since $q_tA = a\sqrt{1-b^2}\,\sin u$. This relation follows from the well-known property of ellipses which states that $\tilde{q}_tA / q_tA = (1-b^2)^{-1/2}$ and the fact that $\tilde{q}_tA = a\sin u$. As $r > 0$, the result follows. For (ii) we have, from the law of

ISBN 0-8053-0102-X

areas:

$$\frac{l(t)}{2\pi}=\frac{t-T}{\tau}=\frac{\text{area } \boldsymbol{q}_t \mathbf{0} P}{\text{area ellipse}}$$

Now the ratio $\tilde{\boldsymbol{q}}_t A / \boldsymbol{q}_t A$ is constant, being $(1-b^2)^{-1/2}$. Now imagine the plane of the circle inclined towards the plane of the ellipse in such a way that the circle projects onto the ellipse. The angle between these two planes has cosine equal to $\boldsymbol{q}_t A / \tilde{\boldsymbol{q}}_t A = \sqrt{1-b^2}$ and hence

$$\frac{\text{area } \boldsymbol{q}_t \mathbf{0} P}{\text{area ellipse}}=\frac{\text{area } \tilde{\boldsymbol{q}}_t \mathbf{0} P}{\text{area circle}}$$

$$=\frac{1}{\pi a^2}\left(\frac{a^2 u}{2}-\frac{a}{2} ab \sin u\right)$$

which gives the result. ■

To study the three-body problem in later sections, it will be important to introduce coordinates that follow these elliptic orbits. The classical sequence of coordinate transformations—Cartesian to Keplerian to Delaunay to Poincaré variables—spanning the nineteenth century, will be developed in the next section. The basis of this development is the calculation by Whittaker [1896] of the Lagrange brackets of the Kepler elements—the parameters α, β, a, and b just defined for the elliptic orbits—to which we now turn. We shall interpret these first in the tangent and then in the cotangent bundle, obtaining at the end two couples of (nonequivalent) formulations of the Delaunay and Poincaré models.

Recall from the closed orbit characterization of Proposition 9.2.1 that proper elliptical orbits were obtained for all initial conditions $(\boldsymbol{q},\dot{\boldsymbol{q}})\in N=T(\boldsymbol{R}^2\backslash\{\mathbf{0}\})$ such that

$$E(\boldsymbol{q},\dot{\boldsymbol{q}})<0$$

$$G(\boldsymbol{q},\dot{\boldsymbol{q}})\neq 0$$

$$b(\boldsymbol{q},\dot{\boldsymbol{q}})>0$$

where b, the eccentricity, satisfies $b^2=1+2EG^2$. Motion along the ellipse is direct (counterclockwise) if $G>0$, and retrograde (clockwise) if $G<0$, as the construction of Fig. 9.2-1 shows.

We restrict attention now to the direct elliptical orbits. To deal with the angular variables, we use the torus. Just as we interpreted the circle S^1 as the reals $mod 2\pi$, $S^1 \approx \boldsymbol{R}/(2\pi)$, we now identify the two-torus $\boldsymbol{T}^2$ with the plane $mod 2\pi$. This we denote by

$$\boldsymbol{T}^2=S^1\times S^1\approx \boldsymbol{R}^2/(2\pi)\times(2\pi)$$

ISBN 0-8053-0102-X

which is isomorphic (as a Lie group) to the usual torus (see Sect. 1.1). As a Lie group, this manifold has trivial tangent bundle, so we will identify

$$T(T^2) \approx T^2 \times R^2$$

for convenience in dealing with the angular variables.

Finally, we are ready to define the first change of variables, the *Kepler map*.

9.2.4 Definition. *In the context of model IV for the planar two-body problem, the **direct elliptical domain** of $N = T(R^2 \setminus \{0\})$ is the open subset*

$$\mathcal{E} = \{n \in N \mid E(n) < 0 < b(n) < 1, G(n) > 0\}$$

*The **Kepler domain** of $T(T^2)$ is the subset*

$$\mathcal{K} = T^2 \times S \subset T^2 \times R^2 \approx T(T^2)$$

where $S \subset R^2$ is the open half-strip defined by

$$S = \{(a, b) \in R^2 \mid a > 0, b \in (0, 1)\}$$

*The **Kepler elements** are the following functions:*

α: $\mathcal{E} \to S^1$, *the polar angle at epoch defined in* 9.2.2;
β: $\mathcal{E} \to S^1$, *the argument of the perihelion, defined in* 9.2.2;
a: $\mathcal{E} \to R^+$, *the semi-major axis, defined in* 9.2.1; *and*
b: $\mathcal{E} \to (0, 1)$, *the eccentricity, from* 9.2.1.

*The **Kepler map** is*

$$K: \mathcal{E} \to \mathcal{K} : n \mapsto (\alpha(n), \beta(n), a(n), b(n))$$

If this map is to serve as a change of variables, it ought to be a diffeomorphism.

9.2.5 Proposition. *The Kepler map is a C^∞ diffeomorphism.*

The proof, which belongs to analytic geometry, is left as an exercise. The implications are not only that the Kepler elements are new coordinates in the direct elliptical domain, but also that, topologically, this domain is a thickened two-torus.

The classical analog of the Kepler map is usually expressed in the full two-body problem of three degrees of freedom. In that six-dimensional context, it becomes

$$\tilde{K}: \tilde{\mathcal{E}} \subset T(R^2 \setminus \{0\}) \to \tilde{\mathcal{K}} \subset T^2 \times R^2 \times T^2$$

$$\tilde{K}: n \mapsto (\alpha, \beta, a, b, I, \Omega)$$

ISBN 0-8053-0102-X

and the additional Kepler elements—*inclination I* and *longitude of the ascending node* Ω—are known as the *orientation elements*. They orient the orbital plane in $\boldsymbol{R}^3$.

Note: In most texts of celestial mechanics, the Kepler variables we call (α,β,a,b) are denoted $(\varepsilon,\tilde{\omega},a,e)$.

Returning to the planar problem in the Lagrangian context, we have specialized the domain to the open set $\mathcal{E}\subset N=T(\boldsymbol{R}^2\backslash\{\boldsymbol{0}\})$, which is a symplectic manifold with the standard form

$$\omega_{\mathfrak{L}}=dq^1\wedge d\dot{q}^1+dq^2\wedge d\dot{q}^2$$

Then we introduced a classical change of coordinates

$$K:\mathcal{E}\rightarrow\mathcal{K}\subset T(T^2)$$

The new domain, also has a standard symplectic form ω_{l}, so it is natural to ask whether the Kepler map is symplectic ($K^*\omega_{\mathrm{l}}=\omega_{\mathfrak{L}}$ or $K_*\omega_{\mathfrak{L}}=\omega_{\mathrm{l}}$) or not. As Lagrange knew well, it is not. In fact, he calculated $K_*\omega_{\mathfrak{L}}$ by the method of Lagrange brackets, which he introduced for this purpose. We present this calculation in Sect. 9.4, after introducing the Delaunay variables in Sect. 9.3.

So far, the whole formulation has been on the tangent bundle. For reasons that will become clear in Chapter 10, we also need a formulation on the cotangent bundle.

Recall that models III and IV were equivalent by the Legendre transformation that in this case is just the identity, $q^i=q^i$, $\dot{q}^i=p_i$, $i=1,2$. Since the base integral curves of X_H (in model III) and X_E (in model IV) are the same, Proposition 9.2.1 remains unchanged with the exception of the expressions for a and b that in this case are

$$b=(1+2H\overline{G}^2)^{1/2},\qquad a=(-2H)^{-1}$$

where $\overline{G}(\boldsymbol{q},\boldsymbol{p})=q^1p_2-q^2p_1$. Proper elliptical orbits are obtained for all initial conditions $(\boldsymbol{q},\boldsymbol{p})\in M=T^*(\boldsymbol{R}^2\backslash\{\boldsymbol{0}\})$ such that $H(\boldsymbol{q},\boldsymbol{p})<0$, $\overline{G}(\boldsymbol{q},\boldsymbol{p})\neq0$, $b(\boldsymbol{q},\boldsymbol{p})>0$.

The *direct elliptical domain* in this cotangent formulation is

$$\mathcal{E}^*=\{m\in M\,|\,H(m)<0<b(m)<1,\overline{G}(m)>0\}$$

The *Kepler domain* of $T^*(T^2)\approx T^2\times\boldsymbol{R}^2$ is the subset

$$\mathcal{K}^*=T^2\times S\subset T^2\times\boldsymbol{R}^2\approx T^*(T^2)$$

where $S\subset\boldsymbol{R}^2$ is the open half-strip defined by

$$S=\{(a,b)\in\boldsymbol{R}^2\,|\,a>0,b\in(0,1)\}$$

ISBN 0-8053-0102-X

The *Kepler elements* (α, β, a, b) are the ones defined in 9.2.4, and the Kepler map in this cotangent formulation will be denoted by

$$\bar{K}: \mathcal{E}^* \to \mathcal{K}^*, \qquad m \mapsto (\alpha(m), \beta(m), a(m), b(m))$$

It is a C^∞-diffeomorphism (9.2.5): $\mathcal{E}^*$ and $\mathcal{K}^*$ are open subsets of M and $T^*(T^2)$, respectively, so they carry in a natural canonical way symplectic structures defined by the two-forms $dq^1 \wedge dp_1 + dq^2 \wedge dp_2$ on $\mathcal{E}^*$ and $d\alpha \wedge da + d\beta \wedge db$ on $\mathcal{K}^*$. Is $\bar{K}$ symplectic? As we shall see, it is not. Since the computations involved are identical with the ones for $K: \mathcal{E} \to \mathcal{K}$, we shall work out the details only for K. Notice that $\mathcal{K}$ and $\mathcal{K}^*$ are in fact one and the same space; only the interpretation differs.

9.3 THE DELAUNAY VARIABLES

As the Kepler elements turn out to be noncanonical coordinates for the direct elliptical domain, we proceed immediately to a further change of variables. These, suggested by Lagrange's calculation of $K_* \omega_{\mathcal{E}}$ by his bracket method, were introduced by Delaunay [1860].

We will use the "angular variable" idea or, more precisely, action-angle variables (see Sect. 5.2). Let S^1 denote $\boldsymbol{R} \operatorname{mod} 2\pi$, and

$$\mathcal{A}(M) = C^\infty(M, S^1)$$

for any manifold M. For an angular variable $\omega \in \mathcal{A}(M)$, its differential is a one-form on M, $d\omega \in \mathfrak{X}^*(M)$, see Ex. 2.4F.

Denote the coordinate functions of $T(T^2) \approx S^1 \times S^1 \times \boldsymbol{R} \times \boldsymbol{R}$ by $(\gamma, \delta, \dot{\gamma}, \dot{\delta})$. Note that $\gamma, \delta \in \mathcal{A}(T(T^2))$ are angular variables, not real-valued functions. Note, if $(\gamma, \delta, \dot{\gamma}, \dot{\delta}) \in \mathcal{K}$, then

$$\dot{\gamma} > 0 \quad \text{and} \quad 0 < \dot{\delta} < 1$$

For $K: \mathcal{E} \to \mathcal{K}: (q^1, q^2, \dot{q}^1, \dot{q}^2) \mapsto (\alpha, \beta, a, b)$ and $\mathcal{K} = K(\mathcal{E})$, so for some $n \in \mathcal{E}$,

$$\dot{\gamma} = a(n) \in \boldsymbol{R}^+, \quad \text{the semi-major axis}$$

$$\dot{\delta} = b(n) \in (0, 1), \quad \text{the eccentricity}$$

by 9.2.4.

Here are the Delaunay variables.

9.3.1 Definition. *Let $\pi, \lambda \in \mathcal{A}(\mathcal{K})$ be defined by*

$$\pi: \mathcal{K} \subset T(T^2) \to S^1: (\gamma, \delta, \dot{\gamma}, \dot{\delta}) \mapsto \delta$$

$$\lambda: \mathcal{K} \subset T(T^2) \to S^1: (\gamma, \delta, \dot{\gamma}, \dot{\delta}) \mapsto \gamma - \delta$$

ISBN 0-8053-0102-X

and Γ, $\Lambda \in \mathfrak{F}(K)$ *by*

$$\Gamma: \mathcal{K} \to \boldsymbol{R}: (\gamma, \delta, \dot{\gamma}, \dot{\delta}) \mapsto [\dot{\gamma}(1-\dot{\delta}^2)]^{1/2}$$

$$\Lambda: \mathcal{K} \to \boldsymbol{R}: (\gamma, \delta, \dot{\gamma}, \dot{\delta}) \mapsto \dot{\gamma}^{1/2}$$

Let $g, l \in \mathfrak{A}(\mathcal{E})$ *and* $G, L \in \mathfrak{F}(\mathcal{E})$ *be defined by composition with the Kepler map* $K: \mathcal{E} \to \mathcal{K}$, *that is,*

$$g = K^*\pi, \quad l = K^*\lambda, \quad G = K^*\Gamma, \quad L = K^*\Lambda$$

These are the ***Delaunay variables*** *of model IV.*

For reference, we collect here the relationship between the Delaunay variables and the anomalies defined previously in 9.2.2.

9.3.2 Proposition. *The angular variable* $g \in \mathfrak{A}(\mathcal{E})$ *is the argument of the perihelion of* 9.2.2, $g \equiv \beta$. *The variable* $l \in \mathfrak{A}(\mathcal{E})$ *defined in* 9.3.1 *is the same as the initial mean anomaly (or mean anomaly at epoch) of* 9.2.2, $l = l_0$. *The function* $G \in \mathfrak{F}(\mathcal{E})$ *defined in* 9.3.1 *is the angular momentum of model IV,* $G \equiv G$, *and the function* $L \in \mathfrak{F}(\mathcal{E})$ *is related to the energy of* 9.1.7 *by*

$$1 + 2L^2E = 0$$

or, equivalently, $L^2 = a$, *the semi-major axis function defined in* 9.2.1.

Putting together the Delaunay variables into a map and identifying $T(\boldsymbol{T}^2) \approx \boldsymbol{T}^2 \times \boldsymbol{R}^2$, we get

$$D: \mathcal{K} \subset T(\boldsymbol{T}^2) \to \mathcal{D} \subset T(\boldsymbol{T}^2), \qquad D = (\pi, \lambda, \Gamma, \Lambda)$$

where $\mathcal{D} = D(\mathcal{K})$. Further, let $\Delta = D \circ K: \mathcal{E} \subset T(\boldsymbol{R}^2) \to \mathcal{D} \subset T(\boldsymbol{T}^2)$; see the following commutative diagram:

$$\begin{array}{ccc} T(\boldsymbol{R}^2) \supset \mathcal{E} \in (q^1, q^2, \dot{q}^1, \dot{q}^2) & \xrightarrow{\quad K \quad} & (\alpha, \beta, a, b) \in \mathcal{K} \subset T(\boldsymbol{T}^2) \\ & \searrow^{\Delta} & \big\downarrow D = (\pi, \lambda, \Gamma, \Lambda) \\ & & (g, l, G, L) \in \mathcal{D} \subset T(\boldsymbol{T}^2) \end{array}$$

9.3.3 Definition. *The map* $\Delta: \mathcal{E} \to \mathcal{D}$ *is the* ***Delaunay map*** *of the direct elliptical domain* $\mathcal{E}$, *and* $\mathcal{D} = \Delta(\mathcal{E})$ *is the* ***Delaunay domain.***

9.3.4 Proposition. (*i*) *The Delaunay domain is*

$$\mathcal{D} = \boldsymbol{T}^2 \times \mathcal{O} \subset \boldsymbol{T}^2 \times \boldsymbol{R}^2 \approx T(\boldsymbol{T}^2)$$

ISBN 0-8053-0102-X

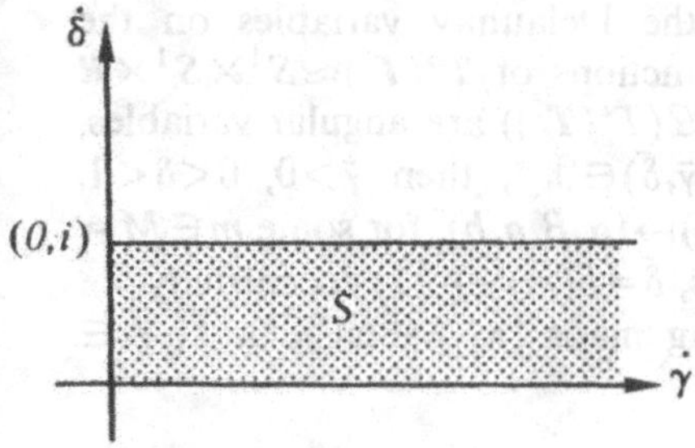

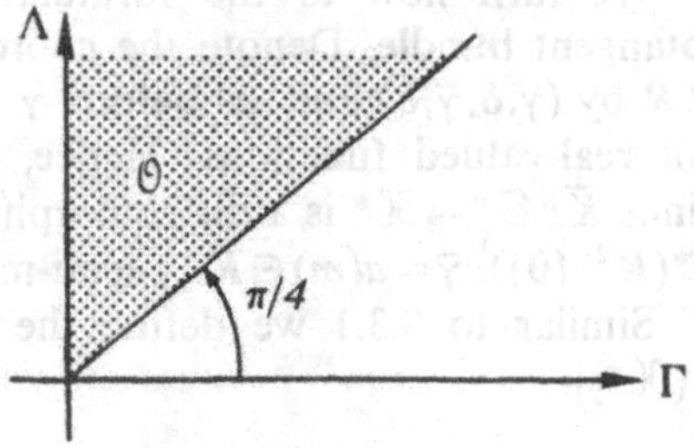

Figure 9.3-1

where $\Theta \subset \boldsymbol{R}^2$ *is the open set*

$$\Theta = \{(\Gamma, \Lambda) \in \boldsymbol{R}^2 | 0 < \Gamma < \Lambda\}$$

(*ii*) *The Delaunay map is a diffeomorphism.*

The open set $\Theta \subset \boldsymbol{R}^2$ is a triangle in the first quadrant illustrated in Fig. 9.3-1. Thus, the Delaunay domain $\mathcal{D} = \boldsymbol{T}^2 \times \Theta$ can be visualized as a "figure of double revolution," or thickened torus. Of course, this reveals the topology of the elliptical domain $\mathcal{E} \subset T(\boldsymbol{R}^2)$. But this was already established, by 9.2.5, as $\mathcal{E} \approx \boldsymbol{T}^2 \times S$, where $S \subset \boldsymbol{R}^2$ is an open half-strip. This is also illustrated in Fig. 9.3-1.

Proof. As $\mathcal{K} = \boldsymbol{T}^2 \times S$, by definition (9.3.3),

$$\mathcal{D} = \Delta(\mathcal{E}) = (D \circ K)(\mathcal{E}) = D(\boldsymbol{T}^2 \times S)$$

But $D\colon \mathcal{K} \to \mathcal{D}\colon (\gamma, \delta, \dot{\gamma}, \dot{\delta}) \mapsto (\delta, \gamma - \delta, \Gamma, \Lambda)$, where Γ, Λ depend only on the fiber coordinates $\dot{\gamma}, \dot{\delta}$. Thus D is fiber preserving. On the zero section, D is a toral isomorphism. On the fiber, the map $V\colon (\dot{\gamma}, \dot{\delta}) \mapsto (\Gamma, \Lambda)$ has Jacobian matrix

$$DV_{(\dot{\gamma}, \dot{\delta})} = \begin{bmatrix} \dfrac{1 - \dot{\delta}^2}{2} & \dfrac{1}{2\Lambda} \\ -\dfrac{\dot{\gamma}\dot{\delta}}{\Gamma} & 0 \end{bmatrix}$$

which is nonsingular on S. And trivially, V is a bijective map with $V(S) = \Theta$. ■

The Kepler map $K\colon \mathcal{E} \to \mathcal{K}$ is a diffeomorphism, but not symplectic. However, the Delaunay map $\Delta = D \circ K$ is a *symplectic diffeomorphism*. This will be established in the next section.

ISBN 0-8053-0102-X

We turn now to the formulation of the Delaunay variables on the cotangent bundle. Denote the coordinate functions of $T^*(T^2) \approx S^1 \times S^1 \times R \times R$ by $(\gamma, \delta, \bar{\gamma}, \bar{\delta})$ and, as before, γ and $\delta \in \mathcal{Q}(T^*(T^2))$ are angular variables, not real-valued functions. Hence, if $(\gamma, \delta, \bar{\gamma}, \bar{\delta}) \in \mathcal{K}^*$, then $\bar{\gamma} > 0$, $0 < \bar{\delta} < 1$. Since $\bar{K}: \mathcal{E}^* \to \mathcal{K}^*$ is a diffeomorphism $(q,p) \mapsto (\alpha, \beta, a, b)$, for some $m \in M = T^*(R^2 \backslash \{0\})$, $\bar{\gamma} = a(m) \in R^+$, semi-major axis, $\bar{\delta} = b(m) \in (0,1)$ eccentricity.

Similar to 9.3.1 we define the following maps $[\pi, \lambda \in \mathcal{Q}(\mathcal{K}^*)$, $\bar{\Gamma}, \bar{\Lambda} \in \mathcal{F}(\mathcal{K}^*)]$:

$$\pi: \mathcal{K}^* \subset T^*(T^2) \to S^1: (\gamma, \delta, \bar{\gamma}, \bar{\delta}) \mapsto \delta$$

$$\lambda: \mathcal{K}^* \subset T^*(T^2) \to S^1: (\gamma, \delta, \bar{\gamma}, \bar{\delta}) \mapsto \gamma - \delta$$

$$\bar{\Gamma}: \mathcal{K}^* \subset T^*(T^2) \to R^1: (\gamma, \delta, \bar{\gamma}, \bar{\delta}) \mapsto \left[\bar{\gamma}(1 - \bar{\delta}^2) \right]^{1/2}$$

$$\bar{\Lambda}: \mathcal{K}^* \subset T^*(T^2) \to R: (\gamma, \delta, \bar{\gamma}, \bar{\delta}) \mapsto \bar{\gamma}^{1/2}$$

Let $g, l \in \mathcal{Q}(\mathcal{E}^*)$ and $\bar{G}, \bar{L} \in \mathcal{F}(\mathcal{E}^*)$ be defined by

$$g = \bar{K}^* \pi, \quad l = \bar{K}^* \lambda, \quad \bar{G} = \bar{K}^* \bar{\Gamma}, \quad \bar{L} = \bar{K}^* \Lambda$$

These are the *Delaunay variables* on the cotangent bundle. As in 9.3.2, these variables have the following interpretation:

$g = \beta$, the argument of the perihelion (9.2.2);
l, the same as the initial mean anomaly l_0 (9.2.2);
$\bar{G}$, the angular momentum (9.2.6);
$\bar{L} = a^{1/2}$ or $1 + 2\bar{L}^2 H = 0$, where a is the semi-major axis and H the Hamiltonian.

Define the map

$$\bar{D}: \mathcal{K}^* \subset T^*(T^2) \to \mathcal{D}^* \subset T^*(T^2), \quad D = (\pi, \lambda, \bar{\Gamma}, \bar{\Lambda})$$

where $\mathcal{D}^* = \bar{D}(\mathcal{K}^*)$. Let $\bar{\Delta} = \bar{D} \circ \bar{K}: \mathcal{E}^* \subset T^*(R^2) \to \mathcal{D}^* \subset T^*(T^2)$ be the Delaunay map in the contangent formulation. $\mathcal{D}^* = \bar{\Delta}(\mathcal{E}^*) = \bar{D}(\mathcal{K}^*)$ is the *Delaunay domain* in the cotangent formulation. See the following commutative diagram:

$$T^*(R^2) \supset \mathcal{E}^* \ni (q^1, q^2, p_1, p_2) \xrightarrow{\quad \bar{K} \quad} (\alpha, \beta, a, b) \in \mathcal{K}^* \subset T^*(T^2)$$

$$\bar{\Delta} \searrow \qquad \downarrow \bar{D} = (\pi, \lambda, \bar{\Gamma}, \bar{\Lambda})$$

$$(g, l, \bar{G}, \bar{L}) \in \mathcal{D}^* \subset T^*(T^2)$$

ISBN 0-8053-0102-V

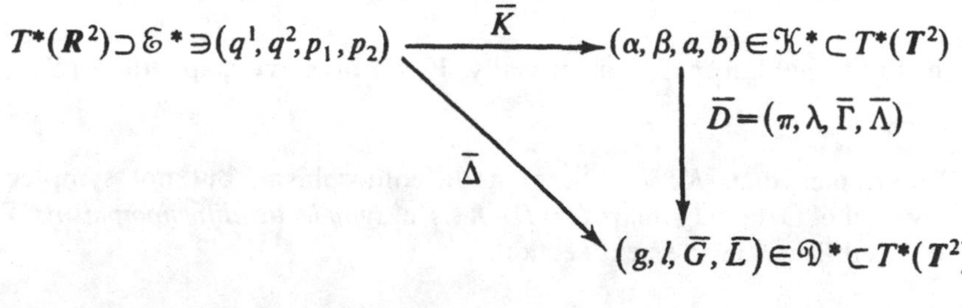

Redoing the proof of 9.3.4, we conclude that the Delaunay domain is

$$\mathfrak{D}^* = T^2 \times \mathcal{O}^* \subset T^2 \times (R^2)^* \approx T(T^2)$$

where $\mathcal{O}^*$ is the open set

$$\mathcal{O}^* = \left\{ (\bar{\Gamma}, \bar{\Lambda}) \in (\dot{R}^2)^* = R^2 | 0 < \bar{\Gamma} < \bar{\Lambda} \right\}$$

and that the Delaunay map $\bar{\Delta}$ is a diffeomorphism. We shall prove in the next section that $\bar{\Delta}$ is a symplectic diffeomorphism.

9.4 LAGRANGE BRACKETS OF KEPLER ELEMENTS

Recall that $\omega_{\mathfrak{L}}$ denotes the standard symplectic form

$$\omega_{\mathfrak{L}} = dq^1 \wedge d\dot{q}^1 + dq^2 \wedge d\dot{q}^2$$

on $\mathcal{E} \subset TR^2$, while ω_1 denotes the corresponding symplectic form

$$\omega_1 = d\alpha \wedge d\dot{\alpha} + d\beta \wedge d\dot{\beta}$$

on $\mathcal{K} \subset T(T^2)$ [or $\mathfrak{D} \subset T(T^2)$]. In Sect. 9.2 we posed the question: Is the Kepler map symplectic, $K_* \omega_{\mathfrak{L}} = \omega_1$? (It is not.) In Sect. 9.3 we asked: Is the Delaunay map symplectic, $\Delta_* \omega_{\mathfrak{L}} = \omega_1$? (It is.)

Here, finally, is Lagrange's result.

9.4.1 Theorem (Lagrange, 1808). *With $K: \mathcal{E} \to \mathcal{K}$ as above and $\omega_{\mathfrak{L}}$ the standard symplectic form on $\mathcal{E} \subset TR^2$, the push-forward of $\omega_{\mathfrak{L}}$ to $\mathcal{K}$ by the Kepler map is*

$$K_* \omega_{\mathfrak{L}} = d\pi \wedge d\Gamma + d\lambda \wedge d\Lambda$$

The proof, a somewhat intricate calculation, is set aside in the next section.

We can express this equivalently in the original domain by pulling back the Lagrange formula with the Kepler map.

9.4.2 Corollary. *With $K: \mathcal{E} \to \mathcal{K}$ as above and $\omega_{\mathfrak{L}}$ the standard symplectic form on $\mathcal{E}$,*

$$\omega_{\mathfrak{L}} = dg \wedge dG + dl \wedge dL$$

where $g, l \in \mathcal{Q}(\mathcal{E})$ and $G, L \in \mathcal{F}(\mathcal{E})$ are defined in 9.3.1 *and identified in* 9.3.2. *Equivalently, the Delaunay map $\Delta: \mathcal{E} \to \mathfrak{D}$ is a symplectic diffeomorphism.*

The corollary follows directly from Lagrange's theorem.

ISBN 0-8053-0102-X

This result establishes the equivalence of the following fifth model for the planar two-body problem—at least for the direct elliptical domain (9.2.4) of model IV (9.1.7).

9.4.3 Definition. *The* ***Delaunay model for the two-body problem (V)*** *is* $(\mathfrak{D}, K, d)$, *where* $\mathfrak{D} \subset T(T^2)$ *is the Delaunay domain* (9.3.3), *the energy function* $K(g, l, G, L) = -1/2L^2$, *and* $d \in \mathfrak{D}$ *(initial condition).*

Note that the **retrograde elliptical domain**

$$\mathcal{E}^- = \{n \in N \mid E(n) < 0 < b(n) < 1, G(n) < 0\}$$

could have been included in this model. Then the domain $\mathfrak{D} \cup \mathfrak{D}^-$ would consist of two components, each a thickened torus.

Of course, Lagrange would not recognize his result in the form

$$\omega_{\mathfrak{L}} = dg \wedge dG + dl \wedge dL$$

So now, we translate to Lagrange's bracket notation.

Recall from Sect. 3.3, that the Lagrange bracket is a matrix of functions defined by a chart on a symplectic manifold. Take $(\mathcal{E}, \omega_{\mathfrak{L}})$ as the symplectic manifold and $K: \mathcal{E} \to \mathcal{K} \subset T(T^2)$ as the chart. Of course this is not a coordinate chart in the formal sense, as the first two components are angular variables rather than real coordinates. The trivial extension of Lagrange bracket theory to this context is an exercise in Sect. 3.3 (Exercise 3.3J).

Thus the Lagrange bracket for

$$K: \mathcal{E} \subset TR^2 \to \mathcal{K} \subset T(T^2)$$

$$: (q^1, q^2, \dot{q}^1, \dot{q}^2) \mapsto (\alpha, \beta, a, b)$$

is the skew-symmetric matrix of functions on $\mathcal{E}$

$$\Omega_K = \begin{bmatrix} 0 & [\![\alpha, \beta]\!] & [\![\alpha, a]\!] & [\![\alpha, b]\!] \\ & 0 & [\![\beta, a]\!] & [\![\beta, b]\!] \\ & & 0 & [\![a, b]\!] \\ & & & 0 \end{bmatrix}$$

comprising the Lagrange brackets of the Kepler variables. By 3.3.23(iii)

$$[\![\alpha, \beta]\!] \circ K^{-1} = K_* \omega_{\mathfrak{L}}\left(\frac{\partial}{\partial \gamma}, \frac{\partial}{\partial \delta}\right)$$

As

$$\omega_{\mathfrak{L}} = [\![\alpha, \beta]\!] d\alpha \wedge d\beta + [\![\alpha, a]\!] d\alpha \wedge da + [\![\alpha, b]\!] d\alpha \wedge db$$
$$+ [\![\beta, a]\!] d\beta \wedge da + [\![\beta, b]\!] d\beta \wedge db + [\![a, b]\!] da \wedge db$$

ISBN 0-8053-0102-X

we may compare with 9.4.2

$$\omega_{\mathfrak{L}} = dg \wedge dG + dl \wedge dL$$

to obtain the classical form:

9.4.4 Proposition. *The Lagrange brackets of the Kepler elements are given by the skew symmetric matrix:*

$$\Omega_K = \begin{bmatrix} 0 & 0 & \dfrac{1}{2L} & 0 \\ & 0 & \dfrac{1-b^2}{2G} - \dfrac{1}{2L} & -\dfrac{ab}{G} \\ & & 0 & 0 \\ & & & 0 \end{bmatrix}$$

Proof. As functions on $\mathfrak{S}$ we have $g = \beta$, $l = \alpha - \beta$, $G = [a(1-b^2)]^{1/2}$, $L = a^{1/2}$ (see 9.3.2 and 9.2.1) and hence $dg = d\beta$, $dl = d\alpha - d\beta$,

$$dG = \frac{1}{2G}\left[(1-b^2)\,da - 2ab\,db\right]$$

and

$$dL = \frac{1}{2L}\,da$$

Thus

$$dg \wedge dG = \frac{1-b^2}{2G}\,d\beta \wedge da - \frac{ab}{G}\,d\beta \wedge db$$

$$dl \wedge dL = \frac{1}{2L}(d\alpha \wedge da - d\beta \wedge da)$$

Comparison of the two expressions of $\omega_{\mathfrak{L}}$ yields the result. ■

This is Lagrange's result, in its original form. Its proof will be completed in the next section, where we give Whittaker's proof of 9.4.1.

Lagrange's result in the contangent formulation is

$$\overline{K}_*(dq^1 \wedge dp_1 + dq^2 \wedge dp_2) = d\pi \wedge d\overline{\Gamma} + d\lambda \wedge d\overline{\Lambda}$$

where $\overline{K}$: $\mathfrak{S}^* \to \mathfrak{K}^*$ is the Kepler map in cotangent formulation. As a corollary, we conclude that the standard canonical two-form $dq^1 \wedge dp_1 + dq^2 \wedge dp_2$ on $\mathfrak{S}^*$ has the expression

$$dq^1 \wedge dp_1 + dq^2 \wedge dp_2 = dg \wedge d\overline{G} + dl \wedge d\overline{L}$$

ISBN 0-8053-0102-X

where $g, l \in \mathcal{Q}(\mathcal{E}^*)$ and $\bar{G}, \bar{L} \in \mathcal{F}(\mathcal{E}^*)$ are defined in 9.3.5. This statement is equivalent to the fact that $\bar{\Delta}: \mathcal{E}^* \to \mathcal{D}^*$ is a symplectic diffeomorphism. Redoing the same computations as in 9.4.4, we find that the Lagrange brackets of the Kepler elements in cotangent formulation are given by the skew symmetric matrix

$$\Omega_{\bar{K}} = \begin{bmatrix} 0 & 0 & \dfrac{1}{2\bar{L}} & 0 \\ & 0 & \dfrac{1-b^2}{2\bar{G}} - \dfrac{1}{2\bar{L}} & -\dfrac{ab}{\bar{G}} \\ & & 0 & 0 \\ & & & 0 \end{bmatrix}$$

As in 9.4.3 we can use this to formulate the Delaunay model on the cotangent bundle.

9.4.5 Definition. *The Delaunay model for the two-body problem (VI) in cotangent formulation is a triple $(\mathcal{D}^*, \bar{K}, d^*)$, where $\mathcal{D}^* \subset T^*(T^2)$ is the Delaunay domain, $d^* \in \mathcal{D}^*$ the initial condition, and $\bar{K}(g, l, \bar{G}, \bar{L}) = -1/2\bar{L}^2$ the Hamiltonian.*

By construction, both models V and VI are *Hamiltonian*, V on the tangent and VI on the cotangent bundle. However, note that both models have *degenerate Hamiltonians*, and thus are not equivalent via a Legendre transformation.

9.5 WHITTAKER'S METHOD

In this section we prove Lagrange's theorem (9.4.1) of 1808, using the method of Whittaker [1896]. (A modern reference is Brouwer and Clemence [1961].) In particular, this will complete the proof that the Delaunay variables are canonical, a result needed for the three-body problem.

Our proof consists of three steps. The first is the factorization of the Kepler map,

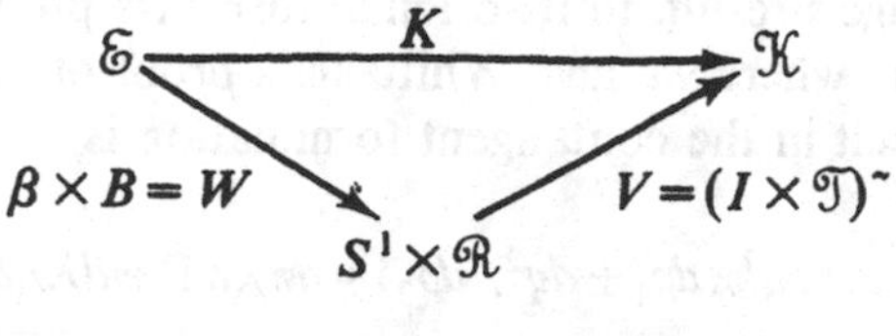

so $K_* \omega_{\mathcal{E}} = V_* W_* \omega_{\mathcal{E}}$; the second is the calculation of $(\beta \times B)_* \omega_{\mathcal{E}} = \omega'$ by means of Whittaker's lemma; the third step is the calculation of $(I \times \mathcal{T})^{\sim}{}_* \omega'$ by brute force.

The idea of the map B is to reduce the direct elliptical domain $\mathcal{E}$, considered as the union of elliptical orbits parametrized by the semi-major

ISBN 0-8053-0102-X

axis a, the eccentricity b, and the argument of the perihelion β, to the case of ellipses in standard position, that is, $\beta = 0$.

Using the function $\beta: \mathcal{E} \to S^1$ defined above, we will define a map

$$B: \mathcal{E} \to TR^2$$

by rotation in R^2 through the angle $-\beta$, which depends upon initial conditions $(q, \dot{q}) \in \mathcal{E}$. This map rotates velocities as well as positions. Thus, if $\rho: S^1 \times R^2 \to R^2$ is the usual action of $S^1 \cong SO(2)$ on R^2 by rotations, its induced action on the tangent bundle is

$$\rho^T: S^1 \times TR^2 \to TR^2$$
$$: (\theta, (q, \dot{q})) \mapsto (\rho(\theta, q), \rho(\theta, \dot{q}))$$

Now $-\beta: \mathcal{E} \to S^1$, where $\mathcal{E} \subset TR^2$, so the graph of $-\beta$, $gr(-\beta): \mathcal{E} \to S^1 \times TR^2$, can be composed with ρ^T.

9.5.1 Definition. *The **Whittaker map** is the composite*

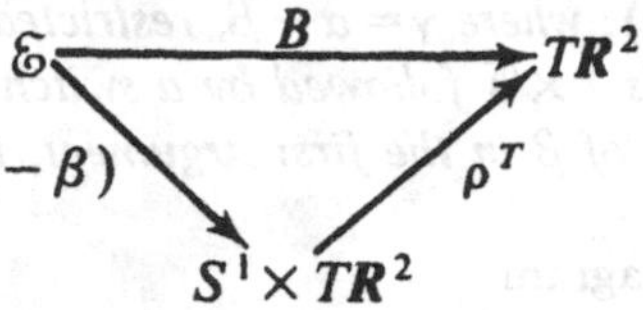

that is, $B(q, \dot{q}) = (\rho(-\beta(q, \dot{q}), q), \rho(-\beta(q, \dot{q}), \dot{q}))$. *The **elliptical ring** $\mathfrak{R} \subset TR^2$ is the image of the Whittaker map $\mathfrak{R} = B(\mathcal{E})$. The **suspended Whittaker map** is*

$$W = \beta \times B: \mathcal{E} \to S^1 \times \mathfrak{R}$$

$$: (q, \dot{q}) \mapsto (\beta(q, \dot{q}), \rho(-\beta(q, \dot{q}), q), \rho(-\beta(q, \dot{q})\dot{q}))$$

9.5.2 Proposition. *The suspended Whittaker map is a diffeomorphism, the elliptical ring is a submanifold of TR^2 of codimension one, and the Whittaker map is a submersion of corank one.*

Proof. $gr(-\beta)$ is a diffeomorphism onto its image and clearly ρ^T is a submersion of corank one. Hence $B = \rho^T \circ gr(-\beta)$ is a submersion of corank one. It follows that the Jacobian of $\beta \times B = W$ is nonsingular. Bijectivity of W then implies that W is a diffeomorphism. Hence $W(\mathcal{E}) = S^1 \times \mathfrak{R}$ is a four-dimensional submanifold of $S^1 \times TR^2$, that is, $\mathfrak{R}$ is a three-dimensional submanifold of TR^2. ■

To conclude our first step, we shall express the map $V = K \circ W^{-1}$ directly, so the factorization $K = V \circ W$ will be useful.

ISBN 0-8053-0102-X

Let us begin with a point $(\boldsymbol{q},\dot{\boldsymbol{q}})\in\mathcal{E}$. How do we calculate the Kepler elements $K(\boldsymbol{q},\dot{\boldsymbol{q}})=(\alpha,\beta,a,b)$ for this initial point? The construction of the elliptical base integral curve is illustrated in Fig. 9.2-1. The parameters (β,a,b) describe the ellipse—argument of the perihelion, semi-major axis, and eccentricity, respectively. These three correspond to the *radius* in polar coordinates. The remaining element α is the argument of the point $\boldsymbol{q}$ along the ellipse, and is analogous to the angle in polar coordinates. Note that $l=\alpha-\beta$, the true anomaly, is the angular distance from the perihelion to $\boldsymbol{q}$.

Now rotate $\boldsymbol{q}$ and $\dot{\boldsymbol{q}}$ by $-\beta(\boldsymbol{q},\dot{\boldsymbol{q}})$. We obtain the same elliptical diagram, but with the argument of the perihelion transformed to zero. Thus points in $\mathfrak{R}=B(\mathcal{E})\subset\mathcal{E}$ are parametrized by the elliptical elements a and b (semi-major axis and eccentricity) without change, and the angular variable has become $l=\alpha-\beta$. This establishes the following characterization of the factor V.

9.5.3 Proposition. *Let*

$$\mathfrak{T}:\mathfrak{R}\to S^1\times \boldsymbol{R}^2$$

be defined by $\mathfrak{T}=(\gamma,a,b)$, *where* $\gamma=\alpha-\beta$, *restricted to* $\mathfrak{R}\subset\mathcal{E}$. *Let* $I:S^1\to S^1$ *be the identity. Then* V *is* $I\times\mathfrak{T}$ *followed by a switch of the first two arguments, and then by the addition of* β *in the first argument, that is,* $V=(I\times\mathfrak{T})^{\sim}$.

Proof. Consider the diagram

$$\begin{array}{ccccc}
\mathcal{E} & \xrightarrow{\;W\;} & S^1\times\mathfrak{R} & \xrightarrow{\;V\;} & \mathcal{K}\subset T(T^2)\\
 & & & {\scriptstyle I\times\mathfrak{T}}\searrow & \uparrow{\scriptstyle \nu}\\
 & & & & S^1\times S^1\times\boldsymbol{R}^2
\end{array}$$

Here $\nu: S^1\times S^1\times\boldsymbol{R}^2\to\mathcal{K}\subset T(T^2)$: $(\beta,\gamma,a,b)\mapsto(\alpha=\gamma+\beta,\beta,a,b)$ and hence $V=(I\times\mathfrak{T})^{\sim}=\nu\circ(I\times\mathfrak{T})$. We have then:

$$\begin{array}{l}
V\circ W:(\boldsymbol{q},\dot{\boldsymbol{q}})\overset{W}{\longmapsto}(\beta,\rho_{-\beta}\boldsymbol{q},\rho_{-\beta}\dot{\boldsymbol{q}})\overset{I\times\mathfrak{T}}{\longmapsto}(\beta,\gamma=\alpha-\beta,a,b)\\
\qquad\qquad\qquad\qquad\qquad\qquad\qquad\qquad\qquad\downarrow{\scriptstyle\nu}\\
\qquad\qquad\qquad\qquad\qquad\qquad\qquad\qquad\quad(\alpha,\beta,a,b)
\end{array}$$

and so we see that $K=V\circ W$. The diagram above expresses V as a composition of diffeomorphisms (ignoring proper specification of the domain $(I\times\mathfrak{T})(S^1\times\mathfrak{R})\subset S^1\times S^1\times\boldsymbol{R}^2$ at the bottom of the diagram). ■

This completes the first step, the factorization. This may seem a bit tedious, but the geometric motivation is simple enough.

Next, the second step. We must calculate $W_*\omega_{\mathcal{E}}$. For this computation we use the composite definition $W=\beta\times(\rho^T\circ gr(-\beta))$.

ISBN 0-8053-0102-X

If $\sigma: S^1 \times \mathcal{R} \to \mathcal{R}$ is projection, and $i: \mathcal{R} \to TR^2$ is the inclusion map, let $\omega_1 = \sigma^* i^* \omega_{\mathcal{E}}$. Let $\Omega: S^1 \times TR^2 \to S^1$ be the projection on the first factor and let $d\Omega$ be its differential, as in Exercise 2.4F and the remarks preceding 4.2.18.

9.5.4 Proposition. *The push-forward of the symplectic form $\omega_{\mathcal{E}}$ with the Whittaker map is*

$$W_* \omega_{\mathcal{E}} = \omega_1 + d\Omega \wedge dG_1$$

where $G_1 = G \circ W^{-1} = W_* G$, and $G: \mathcal{E} \subset TR^2 \to R$ is the momentum of the $S^1 \cong SO(2)$ action ρ^T on $TR^2 = R^2 \times R^2$ interpreted as a real-valued function. (See the end of Sect. 9.1 and Exercise 9.1A.)

Proof. To obtain this from Whittaker's lemma (4.2.20) we reexpress the result we want in terms of the pull-back, that is,

$$W^* \omega_1 = \omega_{\mathcal{E}} - d\beta \wedge dG$$

where $\beta = W^* \Omega = \Omega \circ W: \mathcal{E} \to S^1$ is the argument of the perihelion. Here $\mathcal{E} \xrightarrow{W} S^1 \times \mathcal{R} \xrightarrow{id \times i} S^1 \times TR^2$ and $\omega_1 = \sigma^* i^* \omega_{\mathcal{E}}$. But $i \circ \sigma \circ W = B$ since $W = \beta \times B$, and hence $W^* \omega_1 = W^* \sigma^* i^* \omega_{\mathcal{E}} = B^* \omega_{\mathcal{E}}$, so we are reduced to showing that

$$B^* \omega_{\mathcal{E}} = \omega_{\mathcal{E}} - d\beta \wedge dG$$

where $B: \mathcal{E} \to TR^2$ is the Whittaker map. But this is exactly Whittaker's lemma (4.2.21) for the map $-\beta$. ■

This completes the second step. Note that $W_* \omega_{\mathcal{E}} = \omega_1 + d\Omega \wedge dG_1$ implies $K_* \omega_{\mathcal{E}} = V_* W_* \omega_{\mathcal{E}} = V_* \omega_1 + V_*(d\Omega \wedge dG_1) = V_* \omega_1 + d(\Omega \circ V^{-1}) \wedge d(G_1 \circ V^{-1})$. Recall from 9.5.3 that $V(\beta, \rho_{-\beta} q, \rho_{-\beta} \dot{q}) = (\alpha, \beta, a, b)$, that is, $V^{-1}(\alpha, \beta, a, b) = (\beta, \rho_{-\beta} q, \rho_{-\beta} \dot{q})$, so $(\Omega \circ V^{-1})(\alpha, \beta, a, b) = \beta$, that is, $\Omega \circ V^{-1} = \pi$, the map defined in 9.3.1. Also, $G_1 \circ V^{-1} = G \circ W^{-1} \circ V^{-1} = G \circ K^{-1} = \Gamma$ (by Definition 9.3.1). Thus we have found that

$$K_* \omega_{\mathcal{E}} = V_* \omega_1 + d\pi \wedge d\Gamma$$

and comparing with our goal—the Lagrange theorem (9.4.1)—it remains to show that $V_* \omega_1 = d\lambda \wedge d\Lambda$, where $\lambda(\gamma, \delta, \dot{\gamma}, \dot{\delta}) = \gamma - \delta$ and $\Lambda(\gamma, \delta, \dot{\gamma}, \dot{\delta}) = \dot{\gamma}^{1/2}$. This is the third and final step.

9.5.5 Proposition (Whittaker, 1896). *With $V: S^1 \times \mathcal{R} \to T(T^2)$ as in 9.5.3, $\omega_1 = \sigma^* i^* \omega$ as in 9.5.4 and $\lambda, \Lambda: T(T^2) \to R$ as in 9.3.1, the push-forward of ω_1 by V is given by*

$$V_* \omega_1 = d\lambda \wedge d\Lambda$$

SBN 0-8053-0102-X

Proof. Finally, we must do a brute force calculation in classical style, following Whittaker's original method. The components of $V_*\omega_1$ are given by the Lagrange brackets as functions on $\mathcal{K}\subset T(\mathbf{T}^2)$, the Kepler domain (see 3.3.23), and these functions are independent of time in the following sense (see 3.3.24). Let X_0 denote the original Hamiltonian vector field on $\mathcal{E}$, defining the two-body problem in the positive elliptic domain. Let X_1 denote the push-forward of this vector field to the suspended elliptical ring $S^1\times\mathfrak{R}$ and let $X_1=W_*X_0$, where $W:\mathcal{E}\to S^1\times\mathfrak{R}$ is the suspended Whittaker map. This is a globally Hamiltonian vector field with respect to the (unusual) symplectic form

$$W_*\omega_{\mathcal{E}}=\omega_1+d\Omega\wedge dG_1$$

and for this vector field X_1, a complete system of integrals is provided by the Keplerian elements. That is, the function

$$\Psi: S^1\times\mathfrak{R}\times\mathbf{R}\to T(\mathbf{T}^2)$$

$$:(\beta,x,\dot{x},t)\mapsto(\alpha+nt,\beta,a,b)$$

(here $n=a^{-3/2}$ is the mean motion) is a complete system of integrals for X_1 in the sense of 2.1.10. Thus the components of $V_*W_*\omega_{\mathcal{E}}=V_*\omega_1+d\pi\wedge d\Gamma$ in $T(\mathbf{T}^2)$ are independent of the first component γ by Lagrange's theorem 3.3.24. Since π (as a function on $\mathcal{K}$) and Γ are clearly independent of the first component γ, we see that $V_*\omega_1$ has components independent of γ and of time. Thus

$$V_*\omega_1=[\![\gamma,\delta]\!]\,d\gamma\wedge d\delta+\cdots+[\![c,d]\!]\,dc\wedge dd$$

and the Lagrange brackets $[\![\gamma,\delta]\!]$, and so forth depend only on the coordinates δ, c, and d. We may therefore calculate them at any convenient point on the integral ellipses, say at perihelion. As usual in the classical Lagrangian context, the computation is done in terms of the inverse chart V^{-1}, expressing the old coordinates $(\beta,x,\dot{x})$ in terms of the new (γ,δ,d,c). As

$$V(\beta,x,\dot{x})=(\alpha,\beta,a,b)$$

is defined by the Keplerian elements, we have

$$V^{-1}(\gamma,\delta,c,d)=(\delta,y,\dot{y})$$

where y is on the ellipse with semi-major axis c, eccentricity d, argument of the perihelion 0, and angular distance $\gamma-\delta$ from perihelion. Note that perihelion is described [in $T(\mathbf{T}^2)$] by $\gamma=\delta$. Then $\dot{y}$ is determined by tangency to this ellipse in standard position.

So we must express the components (X,Y) of y, and $(\dot{X},\dot{Y})$ of $\dot{y}$ and their partial derivatives as functions of (γ,δ,c,d), and substitute in the classical

ISBN 0-8053-0102-X

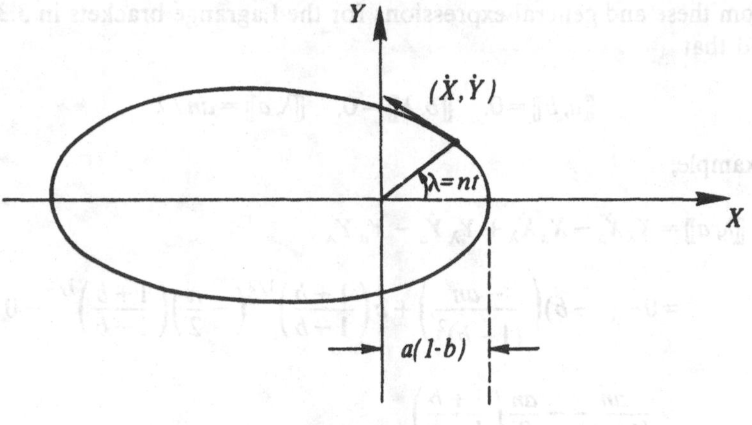

Figure 9.5-1

expressions for $[\![\gamma,\delta]\!]$, and so forth given in 3.3.23. The functions $X, Y, \dot{X}, \dot{Y}$ are obviously independent of δ, and the computations need only be done at perihelion $\gamma=\delta$. Introducing the auxiliary variable $\lambda=\gamma-\delta$, the mean anomaly, and writing a for c, b for d, the functions $X(\lambda,a,b), Y(\lambda,a,b)$ are provided by the explicit solution of the dynamical system X_0, as given in 9.2.1. Their change in time is given by $\lambda(t)=nt$, where $n=a^{-3/2}$, so (see Fig. 9.5-1)

$$\dot{X}(\lambda,a,b)=n\frac{\partial X}{\partial\lambda}, \qquad \dot{Y}(\lambda,a,b)=n\frac{\partial Y}{\partial\lambda}$$

At perihelion ($\lambda=0$) the component functions X, Y of V^{-1} may be developed as Taylor polynomials in λ, from the equations of the ellipses, as

$$X=X_0+\dot{X}_0t+\tfrac{1}{2}\ddot{X}_0t^2+R_Xt^2=a(1-b)-\frac{a\lambda^2}{2(1-b)^2}+\cdots$$

$$Y=Y_0+\dot{Y}_0t+\tfrac{1}{2}\ddot{Y}_0t^2+RY\,t^2=a\lambda\left(\frac{1+b}{1-b}\right)^{1/2}+\cdots$$

Thus, the partial derivatives with respect to a,b,λ at perihelion are given by

$$X_a=1-b \qquad X_b=-a \qquad X_\lambda=0$$

$$Y_a=0 \qquad Y_b=0 \qquad Y_\lambda=a\left(\frac{1+b}{1-b}\right)^{1/2}$$

$$\dot{X}_a=0 \qquad \dot{X}_b=0 \qquad \dot{X}_\lambda=\frac{-an}{(1-b)^2}$$

$$\dot{Y}_a=-\frac{n}{2}\left(\frac{1+b}{1-b}\right)^{1/2} \qquad \dot{Y}_b=an(1+b)^{-1/2}(1-b)^{-3/2} \qquad \dot{Y}_\lambda=0$$

ISBN 0-8053-0102-X

From these and general expressions for the Lagrange brackets in 3.3.23(iv) we find that

$$[\![a, b]\!] = 0, \quad [\![b, \lambda]\!] = 0, \quad [\![\lambda, a]\!] = an/2$$

For example,

$$[\![\lambda, a]\!] = X_\lambda \dot{X}_a - X_a \dot{X}_\lambda + Y_\lambda \dot{Y}_a - Y_a \dot{Y}_\lambda$$

$$= 0 - (1-b)\left(\frac{-an}{(1-b)^2}\right) + a\left(\frac{1+b}{1-b}\right)^{1/2}\left(-\frac{n}{2}\right)\left(\frac{1+b}{1-b}\right)^{1/2} - 0$$

$$= \frac{an}{(1-b)} - \frac{an}{2}\left(\frac{1+b}{1-b}\right)$$

$$= \frac{an}{2}$$

Thus, the only nonzero term in the expression for $V_* \omega_1$ in terms of Lagrange brackets comes from $[\![\lambda, a]\!]$, and so

$$V_* \omega_1 = \frac{an}{2} d\lambda \wedge da$$

On the other hand, since $\Lambda = a^{1/2}$ and $n = a^{-3/2}$ in the present notation, we have

$$d\lambda \wedge d\Lambda = \frac{1}{2a^{1/2}} d\lambda \wedge da$$

$$= \frac{an}{2} d\lambda \wedge da$$

and so $V_* \omega_1 = d\lambda \wedge d\Lambda$ as asserted. ∎

We now turn to the cotangent bundle formulation of Whittaker's result. We proceed as before in three steps and will only sketch the proofs, since they are entirely similar to the ones just given.

Step 1. Here we want a commutative diagram

$$
\begin{array}{ccc}
\mathcal{E}^* & \xrightarrow{\bar{K}} & \mathcal{K}^* \\
\beta \times \bar{B} = \bar{W} \searrow & & \nearrow \bar{V} = (\bar{I} \times \bar{\mathfrak{I}})^- \\
& S^1 \times \mathcal{R}^* &
\end{array}
$$

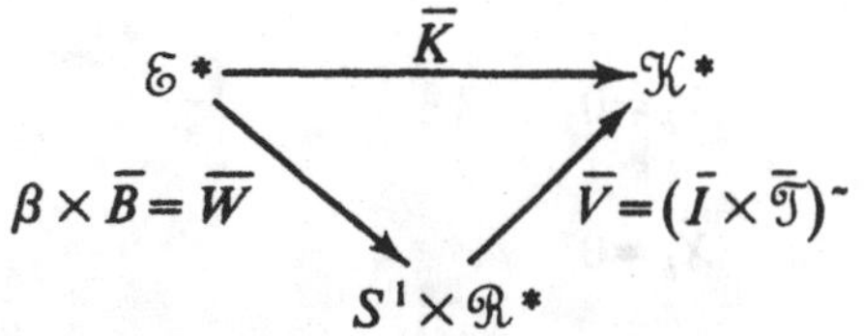

As before, the idea of the map $\bar{B}$ is to reduce the direct elliptical domain $\mathcal{E}^* \subset T^* R^2$, considered as the union of ellipses parametrized by semi-major

ISBN 0-8053-0102-X

axis a, eccentricity b, and argument of the perihelion β, to the case of ellipses in standard position only, that is, with $\beta=0$.

Using the function $\beta\colon \mathcal{E}^*\to S^1$, we will define a map $\bar{B}\colon \mathcal{E}^*\to T^*\mathbf{R}^2$ by rotation in $\mathbf{R}^2$ through the angle $-\beta$, which depends upon initial conditions $(q,p)\in\mathcal{E}^*$. Thus, if $\rho\colon S^1\times \mathbf{R}^2\to\mathbf{R}^2$ is the usual action by rotations in $\mathbf{R}^2$, its induced action on the cotangent bundle is

$$\rho^{T^*}\colon S^1\times T^*\mathbf{R}^2 \longrightarrow T^*\mathbf{R}^2$$

$$(e,^{i\theta}(q,p))\longmapsto(e^{i\theta}q, p\circ e^{-i\theta})$$

where

$$e^{i\theta}q=(q^1\cos\theta-q^2\sin\theta, q^1\sin\theta+q^2\cos\theta)$$

$$p\circ e^{-i\theta}=(p_1\cos\theta-p_2\sin\theta, +p_1\sin\theta+p_2\cos\theta)$$

Now $-\beta\colon \mathcal{E}^*\to S^1$, so that the graph of $-\beta$, $gr(-\beta)\colon \mathcal{E}^*\to S^1\times T^*\mathbf{R}^2$ can be composed with ρ^{T^*}. Then define the ***Whittaker map in cotangent formulation*** by the commutative diagram

$$\begin{array}{ccc}
\mathcal{E}^* & \xrightarrow{\quad\bar{B}\quad} & T^*\mathbf{R}^2 \\
 & {\scriptstyle gr(-\beta)}\searrow \qquad \nearrow{\scriptstyle \rho^{T^*}} & \\
 & S^1\times T^*\mathbf{R}^2 &
\end{array}$$

that is, $\bar{B}(q,p)=(\rho_{-\beta(q,p)}q, p\circ\rho_{\beta(q,p)})$. The ***elliptical ring*** $\mathcal{R}^*\subset T^*\mathbf{R}^2$ is the image of the Whittaker map $\mathcal{R}^*=\bar{B}(\mathcal{E}^*)$. The ***suspended Whittaker map*** is

$$\bar{W}=\beta\times\bar{B}\colon \mathcal{E}^*\longrightarrow S^1\times\mathcal{R}^*$$

$$(q,p)\longmapsto(\beta(q,p), \rho_{-\beta(q,p)}q, p\circ\rho_{\beta(q,p)})$$

Exactly as in 9.5.2 it can be proved that *$\bar{W}$ is a diffeomorphism, $\mathcal{R}^*$ is a codimension one submanifold of $T^*\mathbf{R}^2$, and $\bar{B}$ is a corank one submersion.*

Identical arguments as those preceding 9.5.3 lead us to define $\bar{V}=(\bar{I}\times\bar{\mathcal{T}})^{\sim}=\nu\circ(\bar{I}\times\bar{\mathcal{T}})$, where $\bar{I}\colon S^1\to S^1$ is the identity, $\bar{\mathcal{T}}\colon \mathcal{R}^*\to S^1\times\mathbf{R}^2$, $\bar{\mathcal{T}}=(\gamma,a,b)$, $\gamma=\alpha-\beta$, and $\nu\colon S^1\times S^1\times\mathbf{R}^2\to\mathcal{K}^*\subset T^*(T^2)$, $\nu(\beta,\gamma,a,b)=(\alpha=\gamma+\beta,\beta,a,b)$. Hence we have the diagram

$$\begin{array}{ccccc}
\mathcal{E}^* & \xrightarrow{\ \bar{W}\ } & S^1\times\mathcal{R}^* & \xrightarrow{\qquad\bar{V}\qquad} & \mathcal{K}^*\subset T^*(T^2) \\
 & & & {\scriptstyle \bar{I}\times\bar{\mathcal{T}}}\searrow & \uparrow{\scriptstyle \nu} \\
 & & & & S^1\times S^1\times\mathbf{R}^2
\end{array}$$

ISBN 0-8053-0102-X

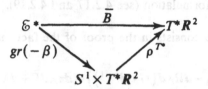

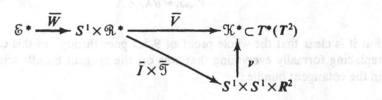

We have then

$$\bar{V}\circ\bar{W}:(q,p)\overset{\bar{W}}{\longmapsto}(\beta,\rho_{-\beta}q,p\circ\rho_\beta)\overset{\bar{I}\times\bar{\mathfrak{T}}}{\longmapsto}(\beta,\gamma=\alpha-\beta,a,b)\overset{\nu}{\longmapsto}(\alpha,\beta,a,b)$$

that is, $\bar{V}\circ\bar{W}=\bar{K}$, which completes the first step.

Step 2. This consists in the proof of

$$\bar{W}_*\omega_0=\omega_1+d\Omega\wedge d\left(\bar{G}\circ\bar{W}^{-1}\right)$$

where $\bar{G}$: $\mathcal{E}^*\subset T^*R^2\to R$ is the classical angular momentum in cotangent formulation (i.e., the momentum for the action ρ^{T^*} of S^1 on T^*R^2), Ω: $S^1\times T^*R^2\to S^1$ the canonical projection on the first factor, and $\omega_1=\sigma^*i^*\omega_0$, σ: $S^1\times\mathfrak{R}^*\to\mathfrak{R}^*$ being the projection and i: $\mathfrak{R}^*\hookrightarrow T^*R^2$ the inclusion. This proof is identical to 9.5.4 with the sole difference that one uses the Whittaker lemma in cotangent formulation (see 4.2.17 and 4.2.19).

Step 3. This step consists in the proof of the fact that

$$\bar{V}_*\left(\omega_1+d\Omega\wedge d\left(\bar{G}\circ W^{-1}\right)\right)=d\pi\wedge d\bar{\Gamma}+d\lambda\wedge d\bar{\Lambda}$$

As before, this can be reduced to a simpler statement involving only $\bar{V}$ by noticing that $\bar{V}_*(\bar{G}\circ\bar{W}^{-1})=G\circ\bar{W}^{-1}\circ\bar{V}^{-1}=G\circ\bar{K}^{-1}=\bar{\Gamma}$, $\Omega\circ\bar{V}^{-1}(\alpha,\beta,a,b)=\beta$ [since $\bar{V}(\beta,\rho_{-\beta}q,p\circ\rho_\beta)=(\alpha,\beta,a,b)$], that is, $\Omega\circ\bar{V}^{-1}=\pi$. We shall have then

$$\begin{aligned}\bar{K}_*\omega_0&=\bar{V}_*\left(\omega_1+d\Omega\wedge d\left(\bar{G}\circ\bar{W}^{-1}\right)\right)\\&=\bar{V}_*\omega_1+d\pi\wedge d\bar{\Gamma}\end{aligned}$$

so all we have to prove is

$$\bar{V}_*\omega_1=d\lambda\wedge d\bar{\Lambda}$$

But it is clear that the whole proof of 9.5.5 goes through in this case, just by replacing formally everything that was on the tangent bundle with variables in the cotangent bundle.

ISBN 0-8053-0102-X

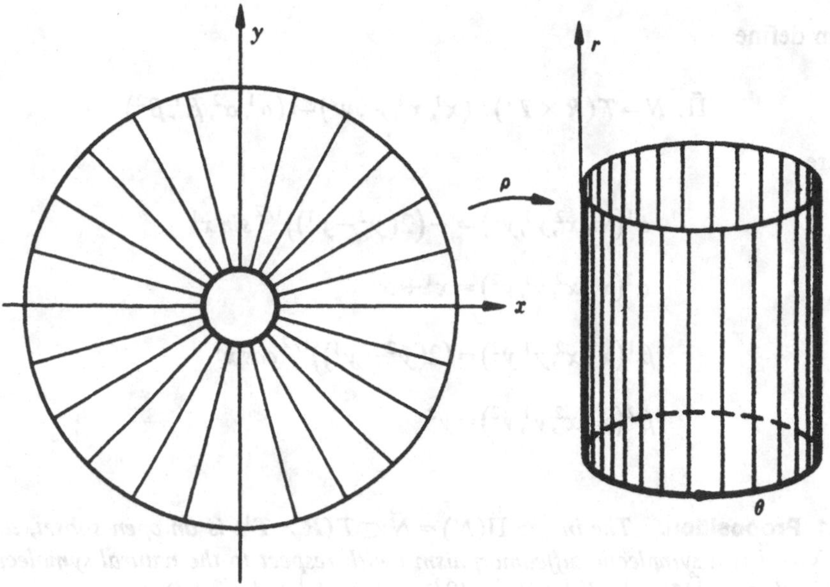

Figure 9.6-1

9.6 POINCARÉ VARIABLES

The main weakness of the Delaunay model is that it does not contain the circular Kepler orbits ($b=0$; see Fig. 9.3-2). The inverse $\Delta^{-1}: \mathfrak{D}\to\mathcal{E}$ of the Delaunay map can be extended so that its image contains the circular orbits, but the extension is no longer injective. The situation is analogous to that for polar coordinates (r,θ) in the plane. These coordinates are described by a map $\rho: \boldsymbol{R}^2\backslash\{0\}\to TS^1$ (see Fig. 9.6-1) for which ρ^{-1} can be extended by mapping the entire θ-"axis" ($r=0$) into the origin.

For the purpose of studying the circular orbits (in Chapter 10), the Poincaré model is helpful, and is obtained from the Delaunay model by an additional symplectic diffeomorphism Π. The mapping $P=\Pi\circ\Delta$ relating the new model to $\mathcal{E}$ in model IV (9.2.4) then extends to a diffeomorphism on $\mathcal{E}_1$, the domain of *counterclockwise elliptical and circular orbits* in model IV. This *Poincaré diffeomorphism* P_1 preserves integral curves, so the Poincare model is "equivalent" to model IV restricted to $\mathcal{E}_1$.

We begin by constructing Π on a larger domain $N\subset T(T^2)$ that contains $\mathfrak{D}$; we shall call this mapping $\tilde{\Pi}$. Let (x^1,x^2,y^1,y^2) be "coordinate functions" in $T(T^2)\approx \boldsymbol{T}^2\times\boldsymbol{R}^2$, with (x^1,x^2) angular variables. Then we define the open set $N\subset T(T^2)$ by $N=\{(x^1,x^2,y^1,y^2)|y^2-y^1>0\}$. Let $\boldsymbol{T}^1$ denote the circle S^1 parametrized by $[0,4\pi)$; that is, $\boldsymbol{T}^1$ is the quotient space of $\boldsymbol{R}$ *mod*4π, and in $T(\boldsymbol{R}\times\boldsymbol{T}^1)$ we use coordinate variables $(\alpha^1,\alpha^2,\beta^1,\beta^2)$ with α^2 cyclic (*mod*4π).

ISBN 0-8053-0102-X

Then define

$$\tilde{\Pi}: N \rightarrow T(\boldsymbol{R} \times \boldsymbol{T}^1): (x^1, x^2, y^1, y^2) \mapsto (\alpha^1, \alpha^2, \beta^1, \beta^2)$$

where

$$\alpha^1(x^1, x^2, y^1, y^2) = -(2(y^2 - y^1))^{1/2} \sin x^1$$

$$\alpha^2(x^1, x^2, y^1, y^2) = x^1 + x^2$$

$$\beta^1(x^1, x^2, y^1, y^2) = (2(y^2 - y^1))^{1/2} \cos x^1$$

$$\beta^2(x^1, x^2, y^1, y^2) = y^2$$

9.6.1 Proposition. *The image $\tilde{\Pi}(N) = \tilde{N} \subset T(\boldsymbol{R} \times \boldsymbol{T}^1)$ is an open subset, and $\tilde{\Pi}: N \rightarrow \tilde{N}$ is a symplectic diffeomorphism (with respect to the natural symplectic forms; that is, $\tilde{\Pi}^*(d\alpha^1 \wedge d\beta^1 + d\alpha^2 \wedge d\beta^2) = dx^1 \wedge dy^1 + dx^2 \wedge dy^2$).*

The proof is a simple computation and left as an exercise at the end of this section.

We may now apply this symplectic diffeomorphism to the Delaunay model $(\mathcal{D}, K, d)$ to obtain the Poincaré model $(\mathcal{P}_1, \Omega, p)$. We let $\mathcal{P} = \tilde{\Pi}(\mathcal{D}) \subset T(\boldsymbol{R} \times \boldsymbol{T}^1)$. Then, as $\Pi = \tilde{\Pi}|\mathcal{D}$ is a symplectic diffeomorphism, we have $\Pi_* X_K = X_{K \circ \Pi^{-1}}$ (see 3.3.19). But $K(x^1, x^2, y^1, y^2) = -1/2(y^2)^2$ (9.4.3) and $\beta^2 \circ \Pi = y^2$, so $y^2 \circ \Pi^{-1} = \beta^2$ and $K \circ \Pi^{-1} = -\frac{1}{2}(\beta^2)^2$. This new model is related to the fourth model $(\mathcal{E}, L, n)$ by the Delaunay mapping $\Delta: \mathcal{E} \rightarrow \mathcal{D}$ composed with the mapping $\Pi: \mathcal{D} \rightarrow \mathcal{P}$. The composite mapping $P = \Pi \circ \Delta: \mathcal{E} \rightarrow \mathcal{P}$ is the **Poincaré mapping**, and traditionally the component functions of this mapping are denoted by

$$P(q^1, q^2, \dot{q}^1, \dot{q}^2) = (\eta, \lambda, \xi, \Lambda)$$

Then since $P = \Pi \circ \Delta$ and $\Delta(q^1, q^2, \dot{q}^1, \dot{q}^2) = (g, l, G, L)$ (see Fig. 9.3-1), we may write

$$\eta = -(2(L - G))^{1/2} \sin g$$

$$\lambda = g + l$$

$$\xi = (2(L - G))^{1/2} \cos g$$

$$\Lambda = L$$

As these equations express the connection between the new model and model IV, and thus the experimental domain, it is suggestive to relabel the

ISBN 0-8053-0102-X

coordinates $(\alpha^1,\alpha^2,\beta^1,\beta^2)$ by $(\eta,\lambda,\xi,\Lambda)$. Then the new energy function becomes $\Omega=-1/2\Lambda^2$.

We now extend this model slightly to include the circular orbits. In model IV, the domain N includes all orbits, $\{n\in N|E(n)<0<b(n)<1,\ G(n)\neq 0\}$ contains elliptical orbits only ($b\neq 0$), and $\mathcal{E}$ contains the direct elliptical orbits only.

Since $b=(1+2EG^2)^{1/2}=(1-G^2/a)^{1/2}=(1-G^2/L^2)^{1/2}$, $\mathcal{E}$ is defined by the condition $0<G<L$. Thus the enlarged domain $\mathcal{E}_1\subset N$ defined by $0<G\leqslant L$ contains all elliptical and circular orbits. It does not follow immediately that $\mathcal{E}_1$ is an open subset of N; however, this seems plausible if we recall that a circular orbit may be completely surrounded by elliptical orbits, as illustrated in Fig. 9.6-2.

On the other hand, in the Poincaré model, (ξ,η) are polar coordinates in the plane with radius $r=(2(L-G))^{1/2}$. As $0<G<L$ in $\mathcal{E}$, $r=0$ is excluded from $\mathcal{P}=\Pi(\mathcal{D})$. Thus we may extend $\mathcal{P}$ to a larger set $\mathcal{P}_1\subset T(\boldsymbol{R}\times\boldsymbol{T}^1)$ by adding the points $\{(0,\lambda,0,\Lambda)|\lambda\in\boldsymbol{T}^1,\Lambda>0\}$. Obviously $\mathcal{P}_1$ is open. Call $\mathcal{P}_1$ the **Poincaré domain.**

We now extend the Poincaré mapping $P\colon\ \mathcal{E}\to\mathcal{P}$ to a mapping $P_1\colon\ \mathcal{E}_1\to\mathcal{P}_1$ to include the circular orbits. From Fig. 9.2-2 we see that for a circular orbit the polar coordinate α replaces the sum $g+l=\lambda$. We may therefore define

$$P_1(q^1,q^2,\dot{q}^1,\dot{q}^2)=(0,\lambda,0,\Lambda)$$

if

$$n=(q^1,q^2,\dot{q}^1,\dot{q}^2)\in\mathcal{E}_1$$

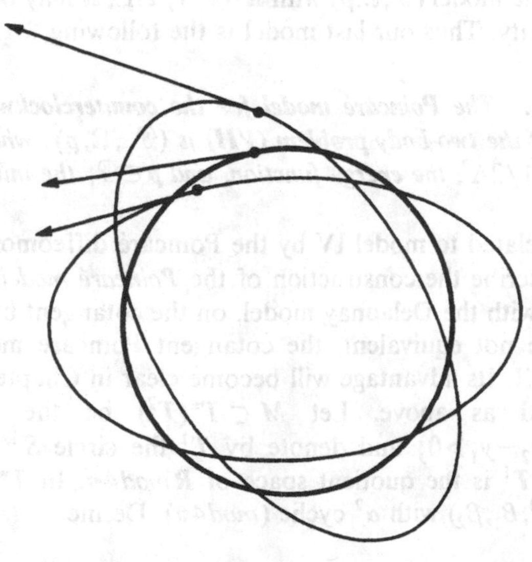

Figure 9.6-2

ISBN 0-8053-0102-X

with $G=L$ or $n\in\mathcal{E}_1\backslash\mathcal{E}$ by defining $\lambda=\alpha$ and $\Lambda=L=G$. This is clearly a bijection on $\mathcal{E}_1\backslash\mathcal{E}$. We now summarize the properties of the Poincaré mappings. Let ω_2 denote the standard symplectic form on $T(\boldsymbol{R}\times\boldsymbol{T}^1)$, that is,

$$\omega_2=d\eta\wedge d\xi+d\lambda\wedge d\Lambda$$

9.6.2 Theorem. *The subsets $\mathcal{E}_1\subset T(\boldsymbol{T}^2)$ and $\mathcal{P}_1\subset T(\boldsymbol{R}\times\boldsymbol{T}^1)$ are open. The Poincaré mapping $P_1:\mathcal{E}_1\to\mathcal{P}_1$ defined by $P_1(m)=(\eta,\lambda,\xi,\Lambda)$ for $m\in\mathcal{E}$ is a diffeomorphism, where*

$$\eta=-(2(L-G))^{1/2}\sin g$$

$$\lambda=g+l$$

$$\xi=(2(L-G))^{1/2}\cos g$$

$$\Lambda=L$$

and $P_1(m)=(0,\lambda,0,\Lambda)$ for $m\in\mathcal{E}_1\backslash\mathcal{E}$, where $\lambda=\alpha$, the polar angle of Fig. 9.6-2, *and $\Lambda=L$. Also,*

$$P_1^*\omega_2=dg\wedge dG+dl\wedge dL$$

The proof is straightforward and is left as an exercise. Note that P_1 is symplectic. In the model $(\mathcal{P},\Omega,p)$ with $\Omega=-1/2\Lambda^2$, Ω may be extended to $\mathcal{P}_1$ without singularity. Thus our last model is the following.

9.6.3 Definition. *The **Poincaré model for the counterclockwise elliptical and circular orbits of the two-body problem (VII)** is $(\mathcal{P}_1,\Omega,p)$, where $\mathcal{P}_1=P_1(\mathcal{E}_1)$, $\Omega(\eta,\lambda,\xi,\Lambda)=-1/2\Lambda^2$, the energy function, and $p\in\mathcal{P}_1$ the initial conditions.*

This model is related to model IV by the Poincaré diffeomorphism P_1.

We now describe the construction of the *Poincaré model on the cotangent bundle* starting with the Delaunay model, on the cotangent bundle (VI). Since V and VI were not equivalent, the cotangent Poincaré model will not be equivalent to VII. Its advantage will become clear in Chapter 10.

We proceed as above. Let $M\subset T^*(\boldsymbol{T}^2)$ be the open set $M=\{(x^1,x^2,y_1,y_2)|y_2-y_1>0\}$ and denote by $\boldsymbol{T}^1$ the circle S^1 parametrized by $[0,4\pi)$, that is, $\boldsymbol{T}^1$ is the quotient space of $\boldsymbol{R}\,mod\,4\pi$. In $T^*(\boldsymbol{R}\times\boldsymbol{T}^1)$ we use variables $(\alpha^1,\alpha^2,\beta_1,\beta_2)$ with α^2 cyclic $(mod\,4\pi)$. Define

$$\tilde{\bar{\Pi}}: M\to T^*(\boldsymbol{R}\times\boldsymbol{T}^1): (x^1,x^2,y_1,y_2)\mapsto(\alpha^1,\alpha^2,\beta_1,\beta_2)$$

ISBN 0-8053-0102-X

where

$$\alpha^1(x^1,x^2,y_1,y_2) = -(2(y_2-y_1))^{1/2} \sin x^1$$

$$\alpha^2(x^1,x^2,y_1,y_2) = x^1+x^2$$

$$\beta_1(x^1,x^2,y_1,y_2) = (2(y_2-y_1))^{1/2} \cos x^1$$

$$\beta_2(x^1,x^2,y_1,y_2) = y_2$$

Then it is easy to show (see 9.6.1) that $\tilde{\Pi}(M)=\tilde{M}\subset T^*(R\times T^1)$ is an open subset and $\tilde{\Pi}: M\to\tilde{M}$ is a symplectic diffeomorphism with respect to the canonical symplectic structures, that is,

$$\tilde{\Pi}(d\alpha^1\wedge d\beta_1 + d\alpha^2\wedge d\beta_2) = dx^1\wedge dy_1 + dx^2\wedge dy_2$$

We apply this symplectic diffeomorphism to the Delaunay model (VI) $(\tilde{\mathfrak{D}}^*,\overline{K},d^*)$. Let $\mathcal{P}^*=\tilde{\Pi}(\mathfrak{D}^*)\subset T^*(R\times T^1)$. Since $\overline{\Pi}=\tilde{\Pi}|\mathfrak{D}^*$ is symplectic, $\overline{\Pi}_* X_{\overline{K}} = X_{\overline{K}\circ\overline{\Pi}^{-1}}$ (3.3.19), and because $\overline{K}(x^1,x^2,y_1,y_2) = -1/2y_2^2$ (9.4.6), $\beta_2\circ\overline{\Pi} = y_2$, that is, $y_2\circ\overline{\Pi}^{-1}=\beta_2$. Thus $\overline{K}\circ\overline{\Pi}^{-1} = -1/2\beta_2^2$ is the new Hamiltonian on $\mathcal{P}^*$. Call $\overline{P}=\overline{\Pi}\circ\overline{\Delta}: \mathcal{E}^*\to\mathcal{P}^*$ the *Poincaré mapping in cotangent formulation*. We shall denote the component functions by

$$\overline{P}(q^1,q^2,p_1,p_2) = (\eta,\lambda,\xi,\Lambda)$$

Since $\overline{\Delta}(q^1,q^2,p_1,p_2) = (g,l,\overline{G},\overline{L})$ (see Fig. 9.3-3), we may write

$$\eta = -\left(2(\overline{L}-\overline{G})\right)^{1/2} \sin g$$

$$\lambda = g+l$$

$$\xi = \left(2(\overline{L}-\overline{G})\right)^{1/2} \cos g$$

$$\overline{\Lambda} = \overline{L}$$

The new Hamiltonian is $\overline{\Omega} = -1/2\overline{\Lambda}^2$.

We can extend this model to include the circular orbits. Recall that $\mathcal{E}^*$ is defined by the conditions $H<0<b<1$, $\overline{G}>0$. But since $b=(1+2H\overline{G}^2)^{1/2} = (1-\overline{G}^2/a)^{1/2} = (1-\overline{G}^2/\overline{L}^2)^{1/2}$, $\mathcal{E}^*$ can be defined equivalently by $0<\overline{G}<\overline{L}$. Define the *enlarged elliptical domain* $\mathcal{E}_1^*\subset M=T^*(R^2\backslash\{0\})$ by the condition $0<\overline{G}\leqslant\overline{L}$; it will contain all elliptical *and* circular orbits. The same reasoning as above shows us that $\mathcal{E}_1^*$ is open in M.

In the Poincaré model described above (ξ,η) are polar coordinates in the plane with radius $r=(2(\overline{L}-\overline{G}))^{1/2}$. Since $0<\overline{G}<\overline{L}$ in $\mathcal{E}^*$, $r=0$ is excluded

ISBN 0-8053-0102-X

from $\mathcal{P}^* = \overline{\Pi}(\mathcal{D}^*)$. We extend $\mathcal{P}^*$ to a larger domain $\mathcal{P}_1^* \subset T^*(\boldsymbol{R}\times T^1)$ by adding the points $\{(0,\lambda,0,\overline{\Lambda})|\lambda\in T^1, \overline{\Lambda}>0\}$. $\mathcal{P}_1^*$ is clearly open and $\overline{P}_1$: $\mathcal{E}_1^* \to \mathcal{P}_1^*$

$$\overline{P}_1(m) = \begin{cases} \overline{P}(m) & \text{for } m\in\mathcal{E}^* \\ (0, \lambda=\alpha, 0, \overline{\Lambda}=\overline{L}=\overline{G}) & \text{for } m\in\mathcal{E}_1^*\backslash\mathcal{E}^* \ (i.e.,\ \overline{L}=\overline{G}) \end{cases}$$

Here α is the polar coordinate, which in the case of circular orbits replaces the sum $\lambda = g+l$. $\mathcal{P}_1^*$ is called the *Poincaré domain in cotangent formulation* and $\overline{P}_1$ the *Poincaré mapping*. $\overline{P}_1$ is a symplectic diffeomorphism (see 9.6.2) and also

$$\overline{P}_1^*\left(d\eta\wedge d\xi + d\lambda\wedge d\overline{\Lambda}\right) = dg\wedge d\overline{G} + dl\wedge d\overline{L}$$

In the model $(\mathcal{P}^*, \overline{\Omega}, p^*)$ described above, $\overline{\Omega} = -1/2\overline{\Lambda}^2$, so that $\overline{\Omega}$ may be extended to $\mathcal{P}_1^*$ without singularity. Thus we obtain the following model.

9.6.5 Definition. ***The Poincaré model for the counterclockwise elliptical and circular orbits of the two-body problem in cotangent formulation (VIII) is*** *$(\mathcal{P}_1^*, \overline{\Omega}, p^*)$, where $\mathcal{P}_1^* = \overline{P}_1(\mathcal{E}_1^*)$ is the Poincaré domain, $p^*\in\mathcal{P}_1^*$, the initial condition, and $\overline{\Omega}(\eta,\lambda,\xi,\overline{\Lambda}) = -1/2\overline{\Lambda}^2$ the Hamiltonian.*

Note that this model has a degenerate Hamiltonian. It is completely analogous to VII, but not equivalent (via the Legendre transformation) to it.

EXERCISES

9.6A Prove Proposition 9.6.1.

9.6B Prove Theorem 9.6.2 (*Hint:* Use the fact that Δ and Π are symplectic and 9.4.2.)

9.7 SUMMARY OF MODELS

We now have eight models for the planar two-body problem. Some are restricted to small domains (such as direct elliptical orbits), and some are in Hamiltonian form and others in Lagrangian. Here we list for reference each of the models in both formats. (This will enlarge the number of models to 13.) The notation (2BIIH) will mean model II for the two-body problem in Hamiltonian format.

The Hamiltonian models are:

(2BIH) This model is a system (M, H^μ, m, μ), where:

(i) $M = T^*W$ with canonical symplectic structure,

$$W = \boldsymbol{R}^3\times\boldsymbol{R}^2\backslash\Delta, \quad \Delta = \{(q,q)|q\in\boldsymbol{R}^2\};$$

ISBN 0-8053-0102-X

(ii) $m \in M$ (initial conditions);
(iii) $\mu \in \boldsymbol{R}$, $\mu > 0$ (mass ratio);
(iv) $H^\mu \in \mathfrak{F}(M)$ given by

$$H^\mu(\boldsymbol{q},\boldsymbol{q}',\boldsymbol{p},\boldsymbol{p}') = \frac{\|\boldsymbol{p}\|^2}{2\mu} + \frac{\|\boldsymbol{p}'\|^2}{2} - \frac{1}{\|\boldsymbol{q}-\boldsymbol{q}'\|}$$

where $\boldsymbol{q},\boldsymbol{q}' \in \boldsymbol{R}^3$, $\boldsymbol{p},\boldsymbol{p}' \in (\boldsymbol{R}^3)^*$ and $\| \ \|$ denotes the Euclidean norm in $\boldsymbol{R}^3$.

(2BIIH) This model is a system (M,H,m), where:

(i) $M = T^*U$ with canonical symplectic structure, $U = \boldsymbol{R}^3 \backslash \{0\}$;
(ii) $m \in M$ (initial conditions);
(iii) $H \in \mathfrak{F}(M)$ given by $H(\boldsymbol{q},\boldsymbol{p}) = \|\boldsymbol{p}\|^2/2 - 1/\|\boldsymbol{q}\|$, where $\boldsymbol{q} \in \boldsymbol{R}^3$, $\boldsymbol{p} \in (\boldsymbol{R}^3)^*$, and $\| \ \|$ denotes the Euclidean norm in $\boldsymbol{R}^3$.

(2BIIIH) This model is a system (M,H,m), where:

(i) $M = T^*(\boldsymbol{R}^2 \backslash \{0\}$ with canonical symplectic structure;
(ii) $m \in M$ (initial conditions);
(iii) $H \in \mathfrak{F}(M)$ defined by $H(\boldsymbol{q},\boldsymbol{p}) = \|\boldsymbol{p}\|^2/2 - 1/\|\boldsymbol{q}\|$, where $\boldsymbol{q} \in \boldsymbol{R}^2, \boldsymbol{p} \in (\boldsymbol{R}^2)^*$, and $\| \ \|$ denotes the Euclidean norm in $\boldsymbol{R}^2$.

These are our three models from Sect. 9.1.

(2BIVH) This model is a system (M,H,m), where:

(i) $M = T^*((0,\infty) \times S^1)$ with canonical symplectic structure; $dr \wedge dp_r + d\theta \wedge dp_\theta$;
(ii) $m \in M$ (initial conditions);
(iii) $H \in (M)$ defined by

$$H(r,\theta,p_r,p_\theta) = \frac{1}{2}\left(p_r^2 + \frac{p_\theta^2}{r^2}\right) - \frac{1}{r}$$

where $(r,\theta) \in (0,\infty) \times S^1$, $(p_r,p_\theta) \in \boldsymbol{R}^2$.

Note that this model is obtained from (2BIIIH) from the symplectic diffeomorphism ϕ: $(0,\infty) \times S^1 \to \boldsymbol{R}^2 \backslash \{0\}$, $\phi(r,\theta) = (r\cos\theta, r\sin\theta)$ so that $p_r = \cos\theta p_1 + \sin\theta p_2$, $p_\theta = -r\sin\theta p_1 + r\cos\theta p_2$.

The two Delaunay models from Sect. 9.4 follow.

(2BVH) This model is a system $(\mathfrak{D},K,d)$, where:

(i) $\mathfrak{D} \subset T(T^2)$ is the Delaunay domain (9.3.3) with standard (noncanonical) symplectic form $dg \wedge dG + dl \wedge dL$;

ISBN 0-8053-0102-X

(ii) $d \in \mathfrak{D}$ (initial condition);
(iii) $K \in \mathfrak{F}(\mathfrak{D})$, the energy function, defined by

$$K(g,l,G,L) = -1/2L^2$$

This is a *Hamiltonian* model on the *tangent bundle*, the Hamiltonian being the energy function K. While all the previous models are hyperregular, this one has degenerate Hamiltonian.

(2BVIH) This model is a system $(\mathfrak{D}^*, \bar{K}, d^*)$, where:

ii(i) $\mathfrak{D}^* \subset T^*(T^2)$ is the Delaunay domain in cotangent formulation (9.3.5) with canonical symplectic form $dg \wedge d\bar{G} + dl \wedge d\bar{\Lambda}$;
(ii) $d^* \in \mathfrak{D}^*$ (initial condition);
(iii) $\bar{K} \in \mathfrak{F}(\mathfrak{D}^*)$, the Hamiltonian, defined by

$$K(g,l,\bar{G},\bar{L}) = -1/2\bar{L}^2$$

(Again, the Hamiltonian is degenerate.)

The last two Hamiltonian models are the Poincaré models from Sect. 9.6.

(2BVIH) This model is a system $(\mathfrak{P}_1, \Omega, p)$ where

(i) $\mathfrak{P}_1 \subset T(\boldsymbol{R} \times T^1)$ is the Poincaré domain (9.6.3) with standard (noncanonical) symplectic form $d\eta \wedge d\xi + d\lambda \wedge d\Lambda$;
(ii) $p \in \mathfrak{P}_1$ (initial condition);
(iii) $\Omega \in \mathfrak{F}(\mathfrak{P}_1)$, the Hamiltonian, defined by

$$\Omega(\eta, \lambda, \xi, \Lambda) = -1/2\Lambda^2$$

(This model again has a degenerate Hamiltonian.)

(2BVIIH) This model is a system $(\mathfrak{P}_1^*, \bar{\Omega}, p^*)$, where:

(i) $\mathfrak{P}_1^* \subset T^*(\boldsymbol{R} \times T^1)$ is the Poincaré domain in cotangent formulation (9.6.4) with canonical symplectic form $d\eta \wedge d\xi + d\lambda \wedge d\bar{\Lambda}$;
(ii) $p^* \in \mathfrak{P}_1^*$ (initial condition);
(iii) $\bar{\Omega} \in \mathfrak{F}(\mathfrak{P}_1^*)$, the Hamiltonian, defined by

$$\Omega(\eta, \lambda, \xi, \bar{\Lambda}) = -1/2\bar{\Lambda}^2$$

(Again a degenerate Hamiltonian.)

The Lagrangian models are the following.

(2BIL) This model is a system (N, L^μ, n, μ), where:
(i) $N = TW$, $W = \boldsymbol{R}^3 \times \boldsymbol{R}^3 \backslash \Delta$, $\Delta = \{(q,q) | q \in \boldsymbol{R}^3\}$, with standard (noncanonical) symplectic form

$$dq^1 \wedge d\dot{q}^1 + dq^2 \wedge d\dot{q}^2 + \cdots + dq'^3 \wedge d\dot{q}'^3$$

ISBN 0-8053-0102-X

(ii) $n \in N$ (initial conditions);
(iii) $\mu \in \mathbf{R}$, $\mu > 0$;
(iv) $L^{\mu} \in \mathfrak{F}(N)$, the Lagrangian, given by

$$L^{\mu}(q, q', q, q') = \frac{1}{2\mu}\|\dot{q}\|^2 + \frac{1}{2}\|\dot{q}'\|^2 + \frac{1}{\|q - q'\|}$$

where $q, q', \dot{q}, \dot{q}' \in \mathbf{R}^3$ and $\| \ \|$ denotes the norm in $\mathbf{R}^3$. The energy function is

$$E^{\mu}(q, q', \dot{q}, \dot{q}') = \frac{1}{2\mu}\|\dot{q}\|^2 + \frac{1}{2}\|\dot{q}'\|^2 - \frac{1}{\|q - q'\|}.$$

This model is the Lagrangian counterpart of (2BIH). The next one is the Lagrangian counterpart to (2BIIH).

(2BIIL) This model is a system (N, L, n), where:

(i) $N = T(\mathbf{R}^3 \backslash \{\mathbf{0}\}$ with the standard (noncanonical) symplectic structure defined by

$$dq^1 \wedge d\dot{q}^1 + dq^2 \wedge d\dot{q}^2 + dq^3 \wedge d\dot{q}^3$$

(ii) $n \in N$ (initial conditions);
(iii) $L \in \mathfrak{F}(N)$, the Lagrangian, given by

$$L(q, \dot{q}) = \frac{1}{2}\|\dot{q}\|^2 + \frac{1}{\|q\|}$$

where $q, \dot{q} \in \mathbf{R}^3$ and $\| \ \|$ denotes the Euclidean norm in $\mathbf{R}^3$. The energy function is

$$E(q, \dot{q}) = \tfrac{1}{2}\|\dot{q}\|^2 - 1/\|q\|$$

(2BIIIL) This model is a system (N, L, n), where:

(i) $N = T(\mathbf{R}^2 \backslash \{\mathbf{0}\})$ with the standard (noncanonical) symplectic structure given by

$$dq^1 \wedge d\dot{q}^1 + dq^2 \wedge d\dot{q}^2$$

(ii) $n \in N$ (initial conditions);
(iii) $L \in \mathfrak{F}(N)$, the Lagrangian, given by

$$L(q, \dot{q}) = \tfrac{1}{2}\|\dot{q}\|^2 + 1/\|q\|$$

where $q, q \in \mathbf{R}^2$ and $\| \ \|$ denotes the Euclidean norm on $\mathbf{R}^2$. The energy function is

$$E(q, \dot{q}) = \tfrac{1}{2}\|\dot{q}\|^2 - 1/\|q\|$$

ISBN 0-8053-0102-X

This model is the Lagrangian counterpart of (2BIIIH) and is IV in Sect. 9.1.

(2BIVL) This model is a system (N, L, n), where:

(i) $N = T((0,\infty)\times S^1)$ with the standard (noncanonical) symplectic structure given by $dr\wedge d\dot{r} + r^2\, d\theta\wedge d\dot{\theta}$;
(ii) $n\in N$ (initial conditions);
(iii) $L\in(N)$, the Lagrangian, given by

$$L(r,\theta,\dot{r},\dot{\theta}) = \tfrac{1}{2}(\dot{r}^2 + r^2\theta^2) + 1/r$$

where $(r,\theta)\in(0,\infty)\times S^1$, $(\dot{r},\dot{\theta})\in \boldsymbol{R}^2$. The energy function is

$$E(r,\theta,\dot{r},\dot{\theta}) = \tfrac{1}{2}(\dot{r}^2 + r^2\dot{\theta}^2) - 1/r$$

Note that this model is the Lagrangian counter part of (2BIVH) and can also be obtained from (2BIIIL) using the diffeomorphism

$$\phi\colon (0,\infty)\times S^1 \to \boldsymbol{R}^2\backslash\{0\},\ \phi(r,\theta) = (r\cos\theta, r\sin\theta).$$

The model of most use in the previous treatment was the restriction of model (2BIIIL) to $\mathfrak{E}$, the direct elliptical domain. We shall call this model (2BIIIeL). Notice that the models (2BVH) to (2BVIIH) do *not* have Lagrangian equivalent since their Hamiltonians are degenerate.

EXERCISE

9.7A. Establish all the equivalences stated in this section between the Lagrangian and Hamiltonian models.

9.8 TOPOLOGY OF THE TWO-BODY PROBLEM

This section will carry out the topological program for the two-body problem outlined in Sect. 4.5. It is convenient to work with model (2BIVH) of Sect. 9.7 as in A. Iacob [1973], even though this topological analysis of the invariant manifolds was first done by S. Smale [1970a] using model (2BIVL); noncanonicity of the symplectic form in (2BIVL) is the reason for our preferring the Hamiltonian formulation.

The mechanical system with symmetry under consideration will be (M, K, V, G), where

$M = (0,\infty)\times S^1$ considered as a Riemannian manifold with the metric

$$\langle (r_1,\theta_1,\dot{r}_1,\dot{\theta}_1),(r_2,\theta_2,\dot{r}_2,\dot{\theta}_2)\rangle = \dot{r}_1\dot{r}_2 + r_1 r_2\dot{\theta}_1\dot{\theta}_2$$

K is the kinetic energy of the metric above whose expression on T^*M is given

ISBN 0-8053-0102-X

by

$$K(r,\theta,p_r,p_\theta)=\tfrac{1}{2}(p_r^2+p_\theta^2/r^2)$$

V is the potential energy given by $V(r,\theta)=-1/r$; $G=SO(2)\cong S^1$ is the Lie group that acts on M by rotations, that is, if $R_\varphi\in SO(2)$ is the rotation through an angle φ,

$$\Phi:\left(R_\varphi,(r,\theta)\right)\mapsto(r,\theta+\varphi)$$

so that the induced actions on the tangent and cotangent bundles are given by

$$\Phi^T:\left(R_\varphi,(r,\theta,\dot r,\dot\theta)\right)\mapsto(r,\theta+\varphi,\dot r,\dot\theta)$$

$$\Phi^{T^*}:\left(R_\varphi,(r,\theta,p_r,p_\theta)\right)\mapsto(r,\theta+\varphi,p_r,p_\theta)$$

Clearly, G acts by isometries and leaves V invariant.

The Hamiltonian of the system is

$$H(r,\theta,p_r,p_\theta)=\tfrac{1}{2}(p_r^2+p_\theta^2/r^2)-1/r$$

The momentum mapping $J: T^*M\to R$ is given by $J(r,\theta,p_r,p_\theta)=p_\theta$ and is clearly invariant under the action of S^1 on T^*M.

Let us first determine the "unpleasant set" $\Lambda\subset M$, where $J_x: T_x^*M\to R$ is not surjective, $x=(r,\theta)$. The expression of $J_x: (p_r,p_\theta)\mapsto p_\theta$ shows that J_x is surjective for all $x\in M$, so that $\Lambda=\varnothing$. It should also be noted that $dJ(r,\theta,p_r,p_\theta)=dp_\theta$ so J has no critical points on T^*M, that is, $\sigma(J)=\varnothing$.

We next determine the effective potential. First, we compute the one-form α_μ defined by the conditions

$$\alpha_\mu(x)\in J^{-1}(\mu)\cap T_x^*M=J_x^{-1}(\mu)$$

$$K(\alpha_\mu(x))=\inf_{\alpha\in J_x^{-1}(\mu)}K(\alpha)$$

If $x=(r,\theta)$ is fixed, then

$$J_x^{-1}(\mu)=\{(r,\theta,p_r,\mu)\,|\,p_r\in R\}$$

so that $\alpha_\mu(x)$ is the minimum of $\tfrac{1}{2}(p_r^2+\mu^2/r^2)$ with respect to p_r, which is attained for $p_r=0$, that is,

$$\alpha_\mu(x)=\alpha_\mu(r,\theta)=(r,\theta,0,\mu)$$

ISBN 0-8053-0102-X

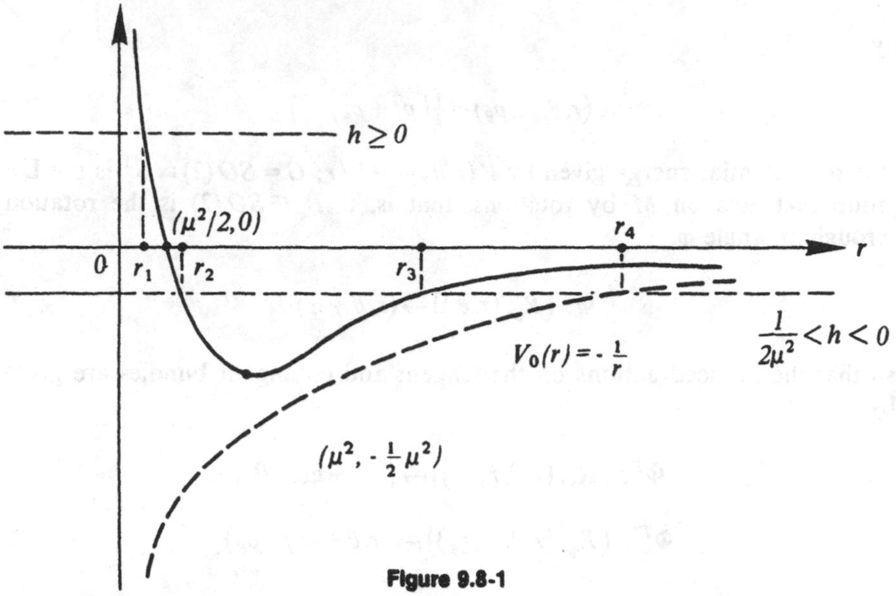

Figure 9.8-1

Hence the effective potential is given by

$$V_\mu(r,\theta)=(H\circ\alpha_\mu)(r,\theta)=\frac{\mu^2}{2r^2}+V(r)=\frac{\mu^2}{2r^2}-\frac{1}{r}$$

whose graph is shown in Fig. 9.8-1. For $\mu=0$, the effective potential is $V_0(r)=V(r)=-1/r$ whose graph is the dotted line in Fig. 9.8-1; it does not have critical points.

By Proposition 4.5.8, the set of critical points of the energy-momentum mapping is given by

$$\begin{aligned}\Sigma'_{H\times J}&=\{(h,\mu)\in \mathbf{R}^2 \mid h\in V_\mu(\sigma(V_\mu))\}\\&=\{(h,\mu)\in \mathbf{R}^2 \mid h=-1/2\mu^2\}\end{aligned}$$

which is pictured in Fig. 9.8-2. Note that this set is not closed, so the bifurcation set $\Sigma_{H\times J}$ may be bigger than $\Sigma'_{H\times J}$.

9.8.1 Theorem (Smale [1970a]). *(a) If $\mu\neq 0$, the invariant manifolds are given as follows:*

(i) *if $h\geqslant 0$, $I_{h,\mu}\approx S^1\times \mathbf{R}$ (a cylinder);*
(ii) *if $-1/2\mu^2<h<0$, $I_{h,\mu}\approx S^1\times S^1$ (a torus);*
(iii) *if $-1/2\mu^2=h$, $I_{h,\mu}\approx S^1$ (a circle);*
(iv) *if $h<-1/2\mu^2$, $I_{h,\mu}=\varnothing$.*

ISBN 0-8053-0102-X

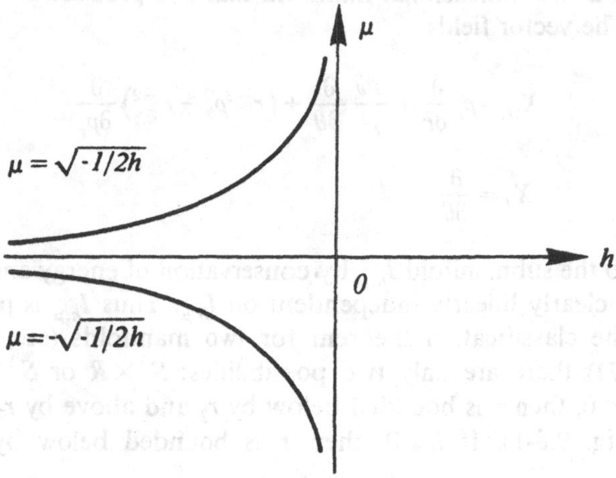

Figure 9.8-2

(b) If $\mu=0$, the invariant manifolds are:

(i) if $h<0$, $I_{h,0}\approx S^1\times R$;
(ii) if $h\geq 0$, $I_{h,0}\approx S^0\times S^1\times R$.

Proof. (After A. Iacob [1973].) (a) The first thing to notice is that on $I_{h,\mu}=(H\times J)^{-1}(h,\mu)$, $h\geq V_\mu(r,\theta)$ and hence, in particular, $h\geq -1/2\mu^2$; this will then prove (iv). Indeed, the defining equations of $I_{h,\mu}$ are

$$\tfrac{1}{2}(p_r^2+p_\theta^2/r^2)-1/r=h,\quad p_\theta=\mu$$

that is,

$$p_r=\pm\sqrt{2(h-V_\mu(r,\theta))}$$

so that $h\geq V_\mu(r,\theta)$.

Now, if $h=-1/2\mu^2$, the equation above, together with the formula for V_μ, yields $r=\mu^2$, $p_r=0$. Indeed, since $-1/2\mu^2$ is the minimum of $V_\mu(r,\theta)$ and $-1/2\mu^2-V_\mu(r,\theta)\geq 0$, we must have $V_\mu(r,\theta)=-1/2\mu^2$, hence $p_r=0$; this happens at $r=\mu^2$. Hence $I_{h,\mu}=\{(\mu^2,\theta,0,0)\in R_+\times S^1\times R\times R\}$, which is a circle. This proves (iii).

For (i) and (ii) note that (h,μ) will be a regular value of $H\times J$ and hence $I_{h,\mu}$ is a two-dimensional submanifold of T^*M. We have

$$I_{h,\mu}=\left\{(r,\theta,p_r,p_\theta)\in(0,\infty)\times S^1\times R^2\,|\,p_r=\pm\sqrt{2(h-V_\mu(r,\theta))}\ ,p_\theta=\mu\right\}$$

$$=S^1\times\left\{(r,p_r)\in(0,\infty)\times R\,|\,p_r=\pm\sqrt{2(h-V_\mu(r,\theta))}\ \right\}\times\{\mu\}$$

ISBN 0-8053-0102-X

so that $I_{h,\mu}$ is a two-dimensional manifold that is a product, S^1 being one of the factors. The vector fields

$$X_H = p_r \frac{\partial}{\partial r} + \frac{p_\theta}{r^2}\frac{\partial}{\partial \theta} + (r^{-3}p_\theta^2 - r^{-2})\frac{\partial}{\partial p_r}$$

$$X_J = \frac{\partial}{\partial \theta}$$

are tangent to the submanifold $I_{h,\mu}$ by conservation of energy and momentum and they are clearly linearly independent on $I_{h,\mu}$. Thus $I_{h,\mu}$ is parallelizable* and, using the classification theorem for two manifolds (see, e.g., Massey ([1967], p. 37)), there are only two possibilities: $S^1 \times \boldsymbol{R}$ or $S^1 \times S^1$. Now, if $-1/2\mu^2 < h < 0$, then r is bounded below by r_2 and above by r_3, so $I_{h,\mu} \approx S^1 \times S^1$ (see Fig. 9.8-1). If $h \geqslant 0$, then r is bounded below by r_1, so that $I_{h,\mu} \approx S^1 \times \boldsymbol{R}$.

(b) ($\mu = 0$) Here $(h,0)$ is a regular value for $H \times J$ so that $I_{h,\mu}$ will be a parallelizable two-manifold (since X_H and X_J are linearly independent on $I_{h,0}$). We have, as before,

$$I_{h,0} = S^1 \times \left\{ (r,p_r) \in (0,\infty) \times \boldsymbol{R} \,|\, p_r = \pm\sqrt{2(h+1/r)} \right\} \times \{0\}$$

If $h<0$, we must have $h + 1/r \geqslant 0$, that is, $r \leqslant -1/h$. See Fig. 9.8-3, the graph of the second factor of $I_{h,0}$, which shows that this factor is diffeomorphic to $\boldsymbol{R}$. If $h \geqslant 0$, the graph of the second factor (see Fig. 9.8-4) shows that this factor is diffeomorphic to two copies of $\boldsymbol{R}$, that is, to $S^0 \times \boldsymbol{R}$. ■

Looking at Figs. 9.8-3 and 9.8-4 it is clear how at $h=0$ a bifurcation takes place: the point $(-1/h, 0)$ is "pulled" to the right until the graph "breaks."

9.8.2 Corollary. *The bifurcation set* $\Sigma_{H\times J}$ *is*

$$\Sigma_{H\times J} = \Sigma'_{H\times J} \cup \{(h,0) \,|\, h \in \boldsymbol{R}\} \cup \{(0,\mu) \,|\, \mu \in \boldsymbol{R}\}$$

that is, $\Sigma_{H\times J}$ *is the graph in Fig.* 9.8-2 *together with the coordinate axes.*

Proof. The theorem above shows that when crossing the coordinate axes, the topological type of $I_{h,\mu}$ changes, so the given set is in the bifurcation set. For case ($\mu \neq 0$), a similar phenomenon occurs: the second factor, which is S^1 for $h<0$, has a point that is pulled to the right until the circle "breaks," becoming a line; this happens exactly at $h=0$.

$\Sigma_{H\times J}$ is clearly closed and it is easy to see that on its complement $H\times J$ is a locally trivial fibration. ■

*A manifold M of dimension n is parallelizable if $\mathfrak{X}(M)$ has elements $X_1, \ldots, X_n$ that are linearly independent at each point of M.

ISBN 0-8053-0102-X

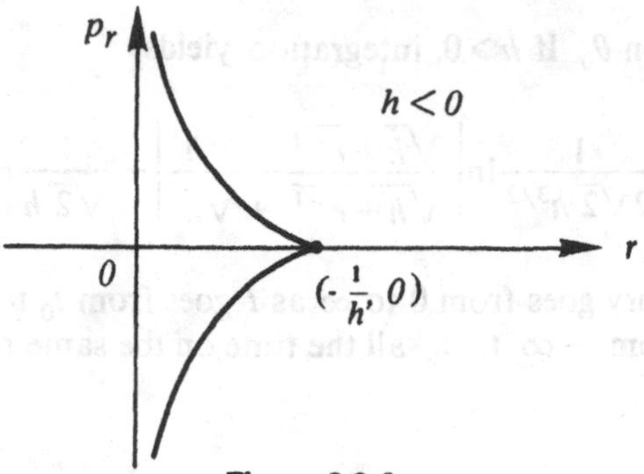

Figure 9.8-3

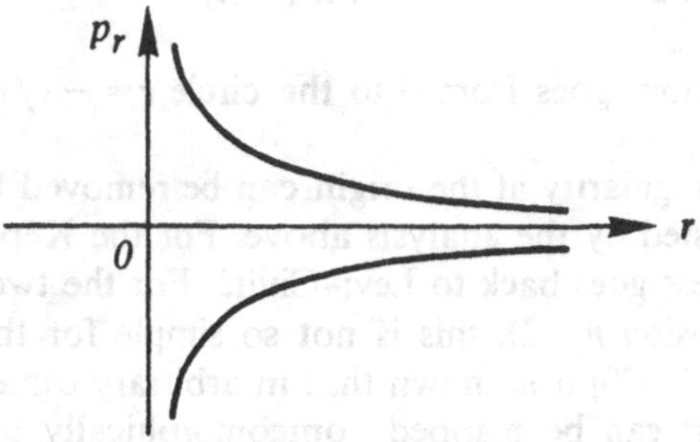

Figure 9.8-4

This completes the first two points of the topological program: the characterization of $I_{h,\mu}$ for all values of (h,μ) and the specification of the bifurcation set $\Sigma_{H\times J}$.

Regarding the characterization of the flow on each $I_{h,\mu}$ we resort to Arnold's theorem (see 5.2.23). On each two-dimensional $I_{h,\mu}$, this flow is a translation-type flow. On the circle obtained for $h=-1/2\mu^2$, the flow is periodic. Working in action-angle variables, the frequencies are computed and then it is easy to see that on the invariant tori obtained for $\mu \neq 0$, $-1/2\mu^2 < h < 0$, the flow is periodic. The orbits remain in the region between r_2 and r_3. Except for $h \geqslant 0$, $\mu=0$, the orbits of X_H are always bounded away from the origin by a circle.

Finally, if $\mu=0$, Hamiltonian's equations on $I_{h,\mu}$ take a particularly simple form:

$$\dot{r}=p_r, \quad \dot{\theta}=0, \quad \dot{p}_r=-1/r^2, \quad \dot{p}_\theta=0$$

that is,

$$p_\theta=0, \quad \theta=\theta_0, \quad p_r=\dot{r}, \quad \dot{r}=\pm\sqrt{2(h+1/r)}$$

This shows that these trajectories lie on a ray from the origin characterized by

ISBN 0-8053-0102-X

the initial condition θ_0. If $h \geqslant 0$, integration yields

$$t - t_0 = \pm\left[\frac{1}{2\sqrt{2}\,h^{3/2}}\ln\left|\frac{\sqrt{h+r^{-1}}-\sqrt{h}}{\sqrt{h+r^{-1}}+\sqrt{h}}\right| + \frac{1}{\sqrt{2}\,h} r\sqrt{h+r^{-1}}\right]$$

so that the trajectory goes from 0 to ∞ as t goes from t_0 to ∞ and from $-\infty$ to 0 as t goes from $-\infty$ to t_0 all the time on the same ray $\theta = \theta_0$. If $h<0$, integration yields

$$t - t_0 = \pm\left[\frac{r}{\sqrt{2}\,h}\sqrt{h+r^{-1}} - \frac{1}{\sqrt{2}\,(-h)^{3/2}}\arctan\frac{\sqrt{h+r^{-1}}}{\sqrt{-h}}\right]$$

and thus the trajectory goes from 0 to the circle $r = -1/h$ and returns to 0 along the ray $\theta = \theta_0$.

The apparent singularity at the origin can be removed by a compactification process suggested by the analysis above. For the Kepler problem in the plane, such a process goes back to Levi–Cività. For the two-body problem in space (or in dimension $n>2$), this is not so simple for the negative energy surfaces. In Moser [1970] it is shown that in arbitrary dimension n, for $h<0$, the surface $H^{-1}(h)$ can be mapped homeomorphically to the unit tangent bundle of S^n punctured at the North Pole (collision states) and the flow on $H^{-1}(h)$ is mapped to the geodesic flow after a change of independent variable. The proof is "elementary" and we recommend Moser [1970] for the proof of this and a number of important related issues.

ISBN 0-8053-0102-X

CHAPTER **10**

The Three-Body Problem

In the preceding chapters, we have presented only a few "nontrivial" examples illustrating the general theory of Hamiltonian systems and the important qualitative results. In this chapter we give a complete discussion of one of the most important systems to which the theory has been applied: the *restricted three-body problem*. The analysis requires the full power of the theory we have developed, and uses nearly every major result of the book.

In this program we follow the method of Barrar [1965a, 1965b] based on the *Poincaré variables*, with minor modifications, in Sects. 10.1 to 10.3. Then in 10.4 we take a look at the topology of the n-body problem following the ideas of Smale [1970a].

10.1 MODELS FOR THREE BODIES

We consider two bodies S and J moving in circles about their center of mass. We let μ be the reduced mass so that $0<\mu<1$, where

$$M_S = \text{mass of } S = 1-\mu$$

and

$$M_J = \text{mass of } J = \mu$$

Ralph Abraham and Jerrold E. Marsden, Foundation of Mechanics, Second Edition

ISBN 0-8053-0102-X

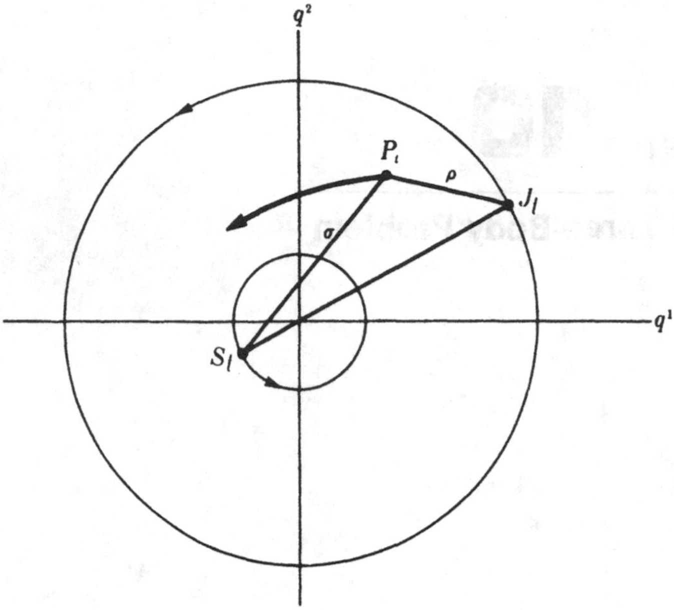

Figure 10.1-1

The problem is to determine the motion of a small third body P moving under the influence of the first two and lying in the same plane. For example, the experimental domain might be the Sun, Jupiter, and a small asteroid, or the Earth, the Moon, and a space vehicle, and so forth.

As in the two-body problem we omit those points where the potential is singular. Thus, we consider the following model (see Fig. 10.1-1).

10.1.1 Definition. *For $t \in \mathbf{R}$, $0 < \mu < 1$, let*

$$S_t = (-\mu \cos t, -\mu \sin t) \in \mathbf{R}^2$$

$$S_* = \cup \{(t, S_t) | t \in \mathbf{R}\}$$

$$J_t = ((1-\mu)\cos t, (1-\mu)\sin t) \in \mathbf{R}^2$$

and

$$J_* = \cup \{(t, J_t) | t \in \mathbf{R}\}$$

*The **first Hamiltonian model for the restricted three-body problem (3BIH)** is a system (M, H, m, μ), where:*

(i) *$M \subset \mathbf{R} \times T^*\mathbf{R}^2$ (phase space) defined by*

$$M = \mathbf{R} \times \left[\mathbf{R}^2 \times (\mathbf{R}^2)^* \setminus (S_* \cup J_*) \times (\mathbf{R}^2)^*\right]$$

with the standard contact structure;

ISBN 0-8053-0102-X

(ii) $m \in M$ (*initial conditions*);
(iii) $\mu \in \mathbf{R}$, $0 < \mu < 1$, *the reduced mass; and*
(iv) $H \in \mathfrak{F}(M)$, *the Hamiltonian, defined by*

$$H(t,q,p) = \frac{\|p\|^2}{2} - \frac{\mu}{\rho(t,q)} - \frac{1-\mu}{\sigma(t,q)}$$

where $\|\ \|$ *denotes the Euclidean norm,* $\rho(t,q) = \|q - J_t\|$, *and* $\sigma(t,q) = \|q - S_t\|$.

The prediction consists of the orbit of m in the time dependent Hamiltonian system $\tilde{X}_H$.

As in the two-body problem, to obtain a smooth Hamiltonian $H \in \mathfrak{F}(M)$ we sacrifice completeness of $\tilde{X}_H$ (collisions). The assumptions that the three bodies lie in one plane and that P does not contribute to the gravitational potential are made to simplify the problem (*restricted* problem of three bodies). Note that H is not invariant under rotations and so we do *not* have conservation of angular momentum.

Important Remark. Because the model is invariant with respect to the involution $(q^1, q^2, \mu) \mapsto (-q^1, -q^2, 1-\mu)$, we can replace our assumption $\mu \in (0,1)$ by $\mu \in (0, \frac{1}{2}]$, that is, we can assume $M_S \geqslant M_J$.

10.1.2 Proposition. *Let M and H be as in* 10.1.1. *Then there is a canonical transformation* $F: M \to \mathbf{R} \times T^*W$, $W = \mathbf{R}^2 \backslash \{(-\mu, 0), (1-\mu, 0)\}$ *such that* $H' = H \circ F^{-1} + K_{F^{-1}}$ *is given by*

$$H'(t,q,p) = \frac{\|p\|^2}{2} + q^1 p_2 - q^2 p_1 - \frac{\mu}{\rho(q)} - \frac{1-\mu}{\sigma(q)}$$

where $\rho(q) = \|q - (1-\mu, 0)\|$ *and* $\sigma(q) = \|q - (-\mu, 0)\|$.

Proof. Consider the clockwise rotation mapping

$$F: M \to \mathbf{R} \times T^*W = N: (t, q^1, q^2, p_1, p_2) \mapsto (t, x^1, x^2, y_1, y_2)$$

where

$$x^1 = (q^1 \cos t + q^2 \sin t), \qquad y_1 = (p_1 \cos t + p_2 \sin t)$$

$$x^2 = (-q^1 \sin t + q^2 \cos t), \qquad y_2 = (-p_1 \sin t + p_2 \cos t)$$

It is clear that F satisfies C1 and C2 of 5.2.6. For C3 we have, by 2.4.9,

$$F^*\tilde{\omega}_N = F^*(dx^1 \wedge dy_1 + dx^2 \wedge dy_2) = d(x^1 \circ F) \wedge d(y_1 \circ F) + d(x^2 \circ F) \wedge d(y_2 \circ F)$$

ISBN 0-8053-0102-X

and by direct computation we see that

$$F^*\tilde{\omega}_N = \tilde{\omega}_M + d(q^1p_2 - q^2p_1)\wedge dt$$

The proposition follows. ■

Thus $F_*\tilde{X}_H = \tilde{X}_{H'}$ by 5.2.14. Hence we obtain the following model.

10.1.3 Definition. *The second Hamiltonian model for the restricted three-body problem (3BIIH) is a system (M, H, m, μ), where:*

(i) $M \in T^*\boldsymbol{R}^2$ *(phase space) defined by*

$$M = T^*W, \quad W = \boldsymbol{R}^2 \backslash \{(-\mu, 0), (1-\mu, 0)\},$$

together with the natural symplectic structure;

(ii) $m \in M$ *(initial conditions);*

(iii) $0 < \mu < 1$, *the reduced mass; and*

(iv) $H \in \mathfrak{F}(M)$ *(the Hamiltonian) is defined by*

$$H(\boldsymbol{q}, \boldsymbol{p}) = \frac{\|\boldsymbol{p}\|^2}{2} + q^1p_2 - q^2p_1 - \frac{\mu}{\rho(\boldsymbol{q})} - \frac{1-\mu}{\sigma(\boldsymbol{q})}$$

The prediction of the model is the integral curve of X_H at m.

The extra term in H may be considered the rotational energy introduced by the rotating coordinate system.

10.1.4 Proposition. *The Hamiltonian H in model 3BIIH is hyperregular. The corresponding Lagrangian on TW is given by*

$$\mathcal{L}(\boldsymbol{q}, \dot{\boldsymbol{q}}) = \frac{\|\dot{\boldsymbol{q}}\|^2}{2} + \dot{q}^1q^2 - \dot{q}^2q^1 + \frac{\|\boldsymbol{q}\|^2}{2} + \frac{\mu}{\rho(\boldsymbol{q})} + \frac{1-\mu}{\sigma(\boldsymbol{q})}$$

with ρ, σ as in 10.1.2 *and notation as in Sect.* 3.5.

Proof. For $H: T^*W \to \boldsymbol{R}$, $FH: T^*W \to TW$ is given by $(q^1, q^2, p_1, p_2) \mapsto (q^1, q^2, p_1 - q^2, p_2 + q^1)$, which is a diffeomorphism.. Hence H is hyperregular with inverse

$$(FH)^{-1}: (q^1, q^2, \dot{q}^1, \dot{q}^2) \mapsto (q^1, q^2, \dot{q}^1 + q^2, \dot{q}^2 - q^1)$$

The action of H is (see Sect. 3.7)

$$G = p_1\frac{\partial H}{\partial p_1} + p_2\frac{\partial H}{\partial p_2} = \|\boldsymbol{p}\|^2 + q^1p_2 - q^2p_1$$

Hence $\mathcal{L} = A - E$, where $E = H \circ (FH)^{-1}$ and $A = G \circ (FH)^{-1}$, so that after

ISBN 0-8053-0102-X

simplification:

$$A(q,\dot{q}) = \|\dot{q}\|^2 + \dot{q}^1 q^2 - \dot{q}^2 q^1$$

and

$$E(q,\dot{q}) = \frac{\|\dot{q}\|^2}{2} - \frac{\mu}{\rho(q)} - \frac{1-\mu}{\sigma(q)} - \frac{\|q\|^2}{2}$$

This gives the desired form for $\mathfrak{L}$. ■

Formally, then, this transition to the Lagrangian formulation may be regarded as giving another "equivalent"model.

10.1.5. Definition. *The second Lagrangian model for the restricted three-body problem (3BIIL) is* $(TW, \mathfrak{L}, x, \mu)$, *where:*

(*i*) $W = \mathbf{R}^2 \backslash \{(-\mu, 0), (1-\mu, 0)\}$;
(*ii*) $x \in TW$;
(*iii*) $0 < \mu < 1$
(*iv*) $\mathfrak{L} \in \mathfrak{F}(TW)$ *the Lagrangian, is defined by*

$$\mathfrak{L}(q,\dot{q}) = \frac{\|\dot{q}\|^2}{2} + \dot{q}^1 q^2 - \dot{q}^2 q^2 + \frac{\mu}{\rho(q)} + \frac{1-\mu}{\sigma(q)} + \frac{\|q\|^2}{2}$$

The prediction of the model is the integral curve of X_E *at* x.

The symplectic structure on TW is given by the symplectic form

$$\omega_{\mathfrak{L}} = \mathfrak{L}_{\dot{q}^i q^j}\, dq^i \wedge dq^j + \mathfrak{L}_{\dot{q}^i \dot{q}^j}\, dq^i \wedge d\dot{q}^j$$

(see 3.5.6), that is,

$$\omega_{\mathfrak{L}} = dq^1 \wedge d\dot{q}^1 + dq^2 \wedge d\dot{q}^2 + 2\, dq^1 \wedge dq^2$$

The prediction is obtained from the Lagrangian equations that in this case become:

$$\frac{d^2 q^1(t)}{dt^2} + \frac{2\, dq^2(t)}{dt} = q^1(t) - (1-\mu)\frac{(q^1(t)+\mu)}{\sigma(q(t))^3} - \mu\frac{(q^1(t)-1+\mu)}{\rho(q(t))^3}$$

$$\frac{d^2 q^2(t)}{dt^2} - \frac{2\, dq^1(t)}{dt} = q^2(t) - (1-\mu)\frac{q^2(t)}{\sigma(q(t))^3} - \frac{\mu q^2(t)}{\rho(q(t))^3}$$

together with the energy integral

$$E = \frac{\|\dot{q}\|^2}{2} - \frac{\|q\|^2}{2} - \frac{\mu}{\rho(q)} - \frac{1-\mu}{\sigma(q)}$$

ISBN 0-8053-0102-X

We obtain additional models for the restricted three-body problem by applying the Delaunay and Poincaré mappings to model 3BIIL.

Recall that in model 2BIIIL the state space was $N=T(R^2\backslash\{0\})$ with standard (noncanonical) symplectic form $dq^1\wedge d\dot{q}^1+dq^2\wedge d\dot{q}^2$, Lagrangian $L(q,\dot{q})=\frac{1}{2}\|\dot{q}\|^2+1/\|q\|$, and energy function $E(q,\dot{q})=\frac{1}{2}\|\dot{q}\|^2-1/\|q\|$. We will denote this Hamiltonian E on the tangent bundle by E_{2B}. The Delaunay mapping is a symplectic diffeomorphism

$$\Delta: \mathcal{E}\subset T(R^2\backslash\{0\})\to\mathfrak{D}\subset T(T^2)$$
$$:(q^1,q^2,\dot{q}^1,\dot{q}^2)\mapsto(g,l,G,L)$$

where $\mathcal{E}$ is the direct elliptical domain (9.2.4).

In model 3BIIL the state space is $N^\mu=TW^\mu$, where $W^\mu=R^2\backslash\{(-\mu,0),(1-\mu,0)\}$, so Δ may be defined on

$$\mathcal{E}\backslash(\{(-\mu,0),(1-\mu,0)\}\times R^2)=\mathcal{E}\cap N^\mu\subset TR^2$$

Let $\mathfrak{D}^\mu=\Delta(\mathcal{E}\cap N^\mu)$, which is an open subset of the Delaunay domain $\mathfrak{D}$. We also have

$$E^\mu=\frac{\|\dot{q}\|^2}{2}-\frac{\mu}{\rho(q)}-\frac{1-\mu}{\sigma(q)}-\frac{\|q\|^2}{2}$$
$$=E_{2B}+\mu\left(\frac{1}{\sigma(q)}-\frac{1}{\rho(q)}\right)+\left(\frac{1}{\|q\|}-\frac{1}{\sigma(q)}-\tfrac{1}{2}\|q\|^2\right)$$
$$E^\mu=-1/2L^2+S_\mu$$

where

$$S_\mu=\mu\left(\frac{1}{\sigma(q)}-\frac{1}{\rho(q)}\right)+\left(\frac{1}{\|q\|}-\frac{1}{\sigma(q)}-\tfrac{1}{2}\|q\|^2\right)$$

and E_{2B} is the Hamiltonian in model 2BIIIL. Thus $E^\mu=E^0+S_\mu$, where $E^0=-1/2L^2$. If $K^\mu=\Delta_*E^\mu=E^\mu\circ\Delta^{-1}$, then $K^\mu=K^0+R_\mu$, where $K^0=\Delta_*E^0=-1/2L^2$ and $R_\mu=\Delta_*S_\mu$. Note that $S_0=-\frac{1}{2}\|q\|^2$ and

$$R_0=-\frac{G^4}{2}\left[1+\left(1-\frac{G^2}{L^2}\right)^{1/2}cos(l+g)\right]^{-2}$$

ISBN 0-8053-0102-X

10.1.6 Definition. *The Delaunay model for the restricted three-body problem in tangent bundle formulation is the system* $(\mathcal{D}^\mu, K^\mu, d, \mu)$, *where:*

(*i*) $\mathcal{D}^\mu = \Delta(\mathcal{E} \cap N^\mu)$ *with symplectic structure defined by*

$$\begin{aligned}\omega_D &= \Delta_*(\omega_{\mathcal{L}})\\ &= \Delta_*(dq^1 \wedge d\dot{q}^1 + dq^2 \wedge d\dot{q}^2 + 2\,dq^1 \wedge dq^2)\\ &= \left(1 - 2r\frac{\partial r}{\partial G}\right) dg \wedge dG + \left(1 - 2r\frac{\partial r}{\partial L}\right) dl \wedge dL\\ &\quad - 2r\frac{\partial r}{\partial L} dg \wedge dL - 2r\frac{\partial r}{\partial G} dl \wedge dG\end{aligned}$$

where

$$r = \frac{G^2}{1 + (1 - G^2/L^2)^{1/2}\cos(l+g)}$$

(*ii*) $d \in \mathcal{D}^\mu$ *(initial conditions);*
(*iii*) $0 < \mu < 1$; *and*
(*iv*) $K^\mu(g, l, G, L) = -(1/2L^2) + R_\mu(g, l, G, L)$, *the energy function.*

The prediction in this model is the integral curve of the Hamiltonian vector field X_{K^μ} *with initial condition d.*

As in the two-body problem, the Delaunay model is Hamiltonian on the tangent bundle.

Note that in this model the domain $\mathcal{D}^\mu$ and the Hamiltonian K^μ depend on the parameter μ. The model is "equivalent" to the second Lagrangian model 3BIIL restricted to an open subset (not necessarily invariant) of the state space, by a symplectic diffeomorphism preserving the Hamiltonian vector field and therefore the predictions. For $\mu = 0$, the model corresponds to the elliptical closed orbit domain of the two-body problem in a rotating coordinate system. The domain does not contain the circular orbits, nor can the model be extended in a straightforward fashion to include them.

From this Delaunay model one gets the Poincaré model in tangent bundle formulation. We just state here this model which we will not use later (it is too complicated) and leave its relationship to 3BIIL and the tangent Delaunay model as an exercise.

ISBN 0-8053-0102-X

10.1.7 Definition. *The* ***Poincaré model for the restricted three-body problem in tangent bundle formulation*** *is the system* $(\mathcal{P}_1^\mu, \Omega^\mu, p, \mu)$, *where:*

(*i*) $\mathcal{P}_1^\mu = P_1(\mathcal{E}_1 \cap N^\mu)$ *with the symplectic structure defined by*

$$\omega_P = P_{1*}(\omega_{\mathcal{E}})$$

$$= d\eta \wedge d\xi + \left(1 + r\frac{\partial r}{\partial G} + r\frac{\partial r}{\partial L}\right) \circ \Pi^{-1} d\lambda \wedge d\Lambda + \eta d\eta \wedge d\lambda - \xi d\lambda \wedge d\xi$$

where

$$r = \frac{G^2}{1 + (1 - G^2/L^2)^{1/2} \cos(l+g)} \circ \Pi^{-1}$$

(*ii*) $p \in \mathcal{P}_1^\mu$ (*initial conditions*);
(*iii*) $0 < \mu < 1$ (*reduced mass*);
(*iv*) $\Omega^\mu(g, l, G, L) = -1/2L^2 + P_{1*}S_\mu\ (g, l, G, L)$, *the energy function.*

The prediction in this model is the integral curve of the Hamiltonian vector field X_{Ω^μ}, *with initial conditions p.*

Here $P_1 : \mathcal{E}_1 \to \mathcal{P}_1$ is the symplectic diffeomorphism from the enlarged elliptical domain $\mathcal{E}_1$ onto the Poincaré domain $\mathcal{P}_1$ (see Sect. 9.6).

Exactly as in the Delaunay model above, or in the Poincaré model for the two-body problem, this model is Hamiltonian on the tangent bundle.

Since $S_0 = -\frac{1}{2}\|\boldsymbol{q}\|^2$, and

$$R_0 = -\frac{G^4}{2}\left[1 + \left(1 - \frac{G^2}{L^2}\right)^{1/2} \cos(l+g)\right]^{-2}$$

we have

$$P_{1*}S_0 = -\tfrac{1}{2}\left[\Lambda - \tfrac{1}{2}(\xi^2 + \eta^2)\right]^4 \left[1 + \Lambda^{-1}(\xi^2 + \eta^2)^{1/2}\left(\Lambda - \tfrac{1}{4}\xi^2 - \tfrac{1}{4}\eta^2\right)^{1/2} \cos\lambda\right]^{-2}$$

The construction of the model is summarized in the following diagram:

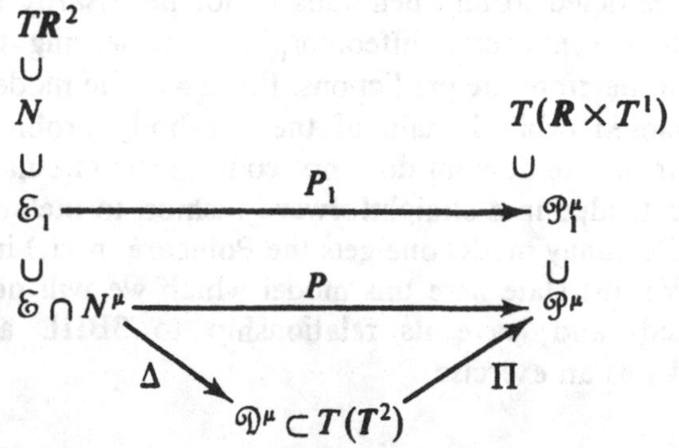

ISBN 0-8053-0102-X

The great disadvantage of both of the previous formulations on the tangent bundle is their use of nonstandard symplectic forms. The equations of motion are correspondingly complicated. We turn now to the formulations on the cotangent bundle that will provide us with more natural Delaunay and Poincaré models.

Recall that in model 2BIIIH the phase space was $M=T^*(R^2\backslash\{0\})$ with canonical symplectic structure and Hamiltonian $H(q,p)=\frac{1}{2}\|p\|^2-1/\|q\|$. We will denote this Hamiltonian by H_{2B}. The Delaunay mapping is a symplectic diffeomorphism

$$\bar{\Delta}: \mathcal{E}^*\to\mathcal{D}^*\subset T^*(T^2): (q^1,q^2,p_1,p_2)\mapsto(g,l,\bar{G},\bar{L})$$

where $\mathcal{E}^*\subset T^*R^2$ is the direct elliptical domain in cotangent formulation, both symplectic structures on $\mathcal{E}^*$ and $\mathcal{D}^*$ being the canonical ones.

In model 3BIIH, the phase space is $M^\mu=T^*W^\mu$, where $W^\mu=R^2\backslash\{(-\mu,0),(1-\mu,0)\}$, so $\bar{\Delta}$ may be defined on $\mathcal{E}^*\backslash(\{(-\mu,0),(1-\mu,0)\}\times R^2)=\mathcal{E}^*\cap M^\mu\subset T^*R^2$. Let $\mathcal{D}^{*\mu}=\bar{\Delta}(\mathcal{E}^*\cap M^\mu)$, which is an open subset of the Delaunay domain in cotangent formulation. We have

$$\begin{aligned}H^\mu(q,p)&=\frac{\|p\|^2}{2}+\bar{G}(q,p)-\frac{\mu}{\rho(q)}-\frac{1-\mu}{\sigma(q)}\\&=H_{2B}(q,p)+\bar{G}(q,p)+\mu\left(\frac{1}{\sigma(q)}-\frac{1}{\rho(q)}\right)+\left(\frac{1}{\|q\|}-\frac{1}{\sigma(q)}\right)\\&=-1/2\bar{L}^2+\bar{G}+\bar{S}_\mu\end{aligned}$$

where

$$\bar{S}_\mu=\mu\left(\frac{1}{\sigma(q)}-\frac{1}{\rho(q)}\right)+\left(\frac{1}{\|q\|}-\frac{1}{\sigma(q)}\right)$$

and $H_{2B}(q,p)=\|p\|^2/2-1/\|q\|$ is the Hamiltonian in model 2BIIIH. Thus $H^\mu=H^0+\bar{S}_\mu$, where $H^0=-1/2\bar{L}^2+\bar{G}$. If $K^\mu=\bar{\Delta}_*H^\mu=H^\mu\circ\bar{\Delta}^{-1}$, then $K^\mu=K^0+R_\mu$, where $K^0=\bar{\Delta}_*H^0=-1/2L^2$ and $R_\mu=\bar{\Delta}_*\bar{S}_\mu$. Note that $\bar{S}_0=0$ and $R_0=0$.

10.1.8 Definition. *The Delaunay model for the restricted three-body problem in cotangent formulation is the system* $(\mathcal{D}^{*\mu},\bar{K}^\mu,d^*,\mu)$, *where:*

(*i*) $\mathcal{D}^{*\mu}=\bar{\Delta}(\mathcal{E}^*\cap M^\mu)$ *with the symplectic structure defined by the canonical two-form*

$$dg\wedge d\bar{G}+dl\wedge d\bar{L}$$

(*ii*) $d^*\in\mathcal{D}^{*\mu}$ *(initial conditions);*

ISBN 0-8053-0102-X

(iii) $0<\mu<1$ *(reduced mass); and*
(iv) $\overline{K}^{\mu}\in\mathfrak{F}(\mathfrak{D}^{*\mu})$, *the Hamiltonian, defined by*

$$\overline{K}^{\mu}(g,l,\overline{G},\overline{L})=-1/2\overline{L}^{2}+\overline{G}+\overline{R}_{\mu}(g,l,\overline{G},\overline{L})$$

The prediction in this model is the integral curve of the Hamiltonian vector field $X_{\overline{K}^{\mu}}$ with initial condition d^{*}.

In the Delaunay model in cotangent formulation the equations of motion are

$$\frac{dg}{dt}=1+\frac{\partial\overline{R}_{\mu}}{\partial\overline{G}},\qquad \frac{dl}{dt}=\frac{1}{\overline{L}^{3}}+\frac{\partial\overline{R}_{\mu}}{\partial\overline{L}}$$

$$\frac{d\overline{G}}{dt}=-\frac{\partial\overline{R}_{\mu}}{\partial g},\qquad \frac{d\overline{L}}{dt}=-\frac{\partial\overline{R}_{\mu}}{\partial l}$$

Note that in this model the domain $\mathfrak{D}^{*\mu}$ and the Hamiltonian $\overline{K}^{\mu}$ depend on the parameter μ. The model is "equivalent" to the second Hamiltonian model 3BIIH restricted to an open subset (not necessarily invariant) of the phase space, by a symplectic diffeomorphism preserving the Hamiltonian vector field and therefore the predictions. For $\mu=0$, the model corresponds to the elliptical closed orbit domain of the two-body problem in a rotating coordinate system. The domain does not contain circular orbits, nor can the model be extended in a straight-forward fashion to include them.

For the study of the circular orbits in the case $\mu\cong 0$, the model obtained from the Poincaré model is more useful. As the derivation of this model is very similar to the case of the Delaunay model treated above, we will simply state the model and leave the relationship to 3BIIH and the cotangent Delaunay model as an exercise.

10.1.9 Definition. *The* ***Poincaré model for the restricted three-body problem in cotangent bundle formulation*** *is the system* $(\mathfrak{P}_{1}^{*\mu},\overline{\Omega}^{\mu},p^{*},\mu)$, *where:*

(i) $\mathfrak{P}_{1}^{*\mu}=\overline{P}_{1}(\mathcal{E}_{1}^{*}\cap M^{\mu})$ *with the canonical symplectic structure induced from* $T^{*}(\boldsymbol{R}\times\boldsymbol{T}^{1})$, *that is, given by the two-form*

$$d\eta\wedge d\bar{\xi}+d\lambda\wedge d\overline{\Lambda}$$

(ii) $p^{*}\in\mathfrak{P}_{1}^{*\mu}$ *(initial conditions);*
(iii) $0<\mu<1$ *(reduced mass); and*
(iv) $\overline{\Omega}^{\mu}\in\mathfrak{F}(\mathfrak{P}_{1}^{*\mu})$, *the Hamiltonian, defined by*

$$\overline{\Omega}^{\mu}=-1/2\overline{\Lambda}^{2}+\overline{\Lambda}-\tfrac{1}{2}(\bar{\xi}^{2}+\eta^{2})+\overline{P}_{1*}\overline{S}_{\mu}$$

ISBN 0-8053-0102-X

The prediction in this model is the integral curve of the Hamiltonian vector field $X_{\bar{\Omega}^\mu}$ with initial condition p^*.

The construction of this model is summarized in the following diagram:

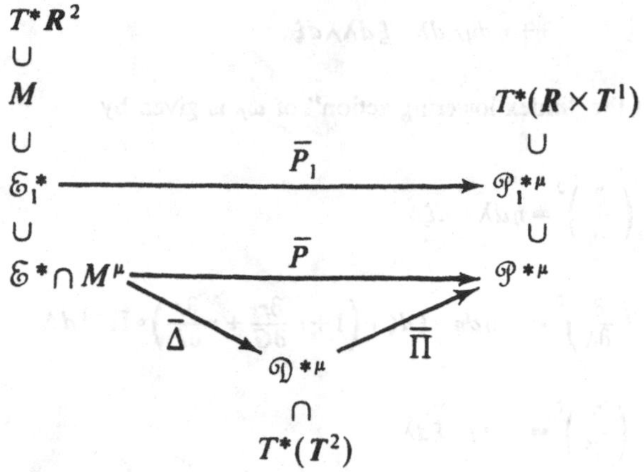

Since $\bar{P}_{1*}\bar{S}_0 = 0$, the equations of motion for $\mu = 0$ are:

$$\frac{d\eta}{dt} = -\bar{\xi}, \qquad \frac{d\lambda}{dt} = \frac{1}{\bar{\Lambda}^3} + 1$$

$$\frac{d\bar{\xi}}{dt} = \eta, \qquad \frac{d\bar{\Lambda}}{dt} = 0$$

The simplicity of the equations of motion is why the cotangent formulation is much more advantageous. (See Exercise 10.1C.)

EXERCISES

10.1A. Derive the Lagrangian stated after 10.1.5.

10.1B. Show that $\omega_D = \Delta_*(\omega_{\mathcal{E}})$ has the form asserted in 10.1.6.

10.1C. (a) Show that on $\mathcal{P}_1^\mu$ we have:

$$dg = -\frac{\xi}{\xi^2+\eta^2}\,d\eta + \frac{\eta}{\xi^2+\eta^2}\,d\xi$$

$$dl = \frac{\xi}{\xi^2+\eta^2}\,d\eta + d\lambda - \frac{\eta}{\xi^2+\eta^2}\,d\xi$$

$$dG = -\eta\, d\eta - \xi\, d\xi + d\Lambda$$

$$dL = d\Lambda$$

This expresses the differentials of the Delaunay variables viewed as defined on $\mathcal{P}_1^\mu$ by $\Pi^{-1}: \mathcal{P}_1^\mu \to \mathcal{D}^\mu$.

ISBN 0-8053-0102-X

(b) Conclude that

$$\omega_P = d\eta\wedge d\xi + \left(1 + r\frac{\partial r}{\partial G} + r\frac{\partial r}{\partial L}\right)\circ\Pi^{-1} d\lambda\wedge d\Lambda$$

$$+\eta\, d\eta\wedge d\lambda - \xi\, d\lambda\wedge d\xi$$

(c) Show that the "index lowering action" of ω_P is given by

$$\left(\frac{\partial}{\partial\eta}\right)^{\flat} = \eta\, d\lambda + d\xi$$

$$\left(\frac{\partial}{\partial\lambda}\right)^{\flat} = -\eta\, d\eta - \xi\, d\xi + \left(1 + r\frac{\partial r}{\partial G} + r\frac{\partial r}{\partial L}\right)\circ\Pi^{-1} d\Lambda$$

$$\left(\frac{\partial}{\partial\xi}\right)^{\flat} = -d\eta + \xi\, d\lambda$$

$$\left(\frac{\partial}{\partial\Lambda}\right)^{\flat} = \left(1 + r\frac{\partial r}{\partial G} + r\frac{\partial r}{\partial L}\right)\circ\Pi^{-1} d\lambda$$

(d) Show that the "index raising action" of ω_P is given by

$$(d\lambda)^{\#} = \frac{1}{\left(1 + r\frac{\partial r}{\partial G} + r\frac{\partial r}{\partial G}\right)\circ\Pi^{-1}}\frac{\partial}{\partial\Lambda}$$

$$(d\xi)^{\#} = \frac{\partial}{\partial\eta} - \frac{\eta}{\left(1 + r\frac{\partial r}{\partial G} + r\frac{\partial r}{\partial L}\right)\circ\Pi^{-1}}\frac{\partial}{\partial\Lambda}$$

$$(d\eta)^{\#} = -\frac{\partial}{\partial\xi} + \frac{\xi}{\left(1 + r\frac{\partial r}{\partial G} + r\frac{\partial r}{\partial L}\right)\circ\Pi^{-1}}\frac{\partial}{\partial\Lambda}$$

$$(d\Lambda)^{\#} = \frac{1}{\left(1 + r\frac{\partial r}{\partial G} + r\frac{\partial r}{\partial L}\right)\circ\Pi^{-1}}\left(\xi\frac{\partial}{\partial\eta} + \frac{\partial}{\partial\lambda} - \eta\frac{\partial}{\partial\xi}\right)$$

(e) For $\mu = 0$, the energy function becomes

$$\Omega^0 = -\frac{1}{2\Lambda^2} - \tfrac{1}{2}r^2$$

ISBN 0-8053-0102-X

Show that

$$\frac{\partial \Omega^0}{\partial \eta} = r\eta \frac{\partial r}{\partial G} \circ \Pi^{-1}$$

$$\frac{\partial \Omega^0}{\partial \lambda} = -r\frac{\partial r}{\partial l} \circ \Pi^{-1}$$

$$\frac{\partial \Omega^0}{\partial \xi} = r\xi \frac{\partial r}{\partial G} \circ \Pi^{-1}$$

$$\frac{\partial \Omega^0}{\partial \Lambda} = -r\left(\frac{\partial r}{\partial G} + \frac{\partial r}{\partial L}\right) \circ \Pi^{-1}$$

Compute from here $(d\Omega^0)^\# = X_{\Omega^0}$ and write down the equations of motion. Then compare this with the simple equations for $\mu = 0$ on the cotangent bundle.

10.1D. Derive the Poincaré model in cotangent formulation stated in 10.1.9.

10.2 CRITICAL POINTS IN THE RESTRICTED THREE BODY PROBLEM

In this section we examine the critical points in the restricted three-body problem using model 3BIIL (see 10.1.5). These correspond to periodic orbits of period 2π in the time dependent model 3BIH.

Recall that $m \in TW$ is a critical point iff $X_E(m) = 0$ iff $dE(m) = 0$. Carrying out the differentiation yields the following.

10.2.1 Proposition. *The point $(q^1, q^2, \dot{q}^1, \dot{q}^2) \in TW$ is a critical point of X_E iff $\dot{q} = 0$, and q is a critical point of the function $V \in \mathfrak{F}(W)$ defined by*

$$V(q^1, q^2) = -\tfrac{1}{2}\left[(q^1)^2 + (q^2)^2\right] - \frac{\mu}{\rho} - \frac{1-\mu}{\sigma}$$

or equivalently:

(i) $\dot{q}^1 = \dot{q}^2 = 0$

(ii) $-V_{q^1} = q^1 - \dfrac{\mu(q^1 - 1 + \mu)}{\rho^3} - \dfrac{(1-\mu)(q^1+\mu)}{\sigma^3} = 0$, *and*

(iii) $-V_{q^2} = q^2\left(1 - \dfrac{1-\mu}{\sigma^3} - \dfrac{\mu}{\rho^3}\right) = 0$

It turns out that there are five critical points; three with $q^2 = 0$, the *collinear solutions* (m_1, m_2, m_3) *of Euler* (1767) and the *two equilateral solutions* (m_4, m_5) *of Lagrange* (1773).

ISBN 0-8053-0102-X

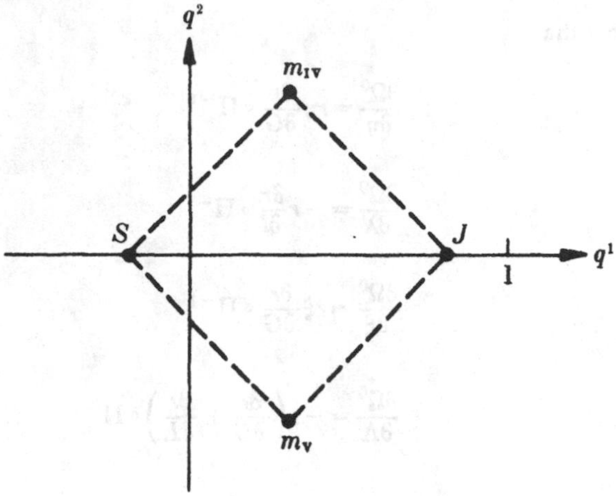

Figure 10.2-1. The equilateral triangle solutions of Lagrange.

The Lagrange points may be found explicitly:

$$m_4 = \left(\tfrac{1}{2} - \mu, \sqrt{3}/2\right)$$

$$m_5 = \left(\tfrac{1}{2} - \mu, -\sqrt{3}/2\right)$$

and for them, $\sigma = \rho = 1$. The reader may verify by direct substitution that these are critical points. (See Fig. 10.2-1.) To show the existence of the collinear critical points and to show that there are no others, we analyze the geometry a little further.

10.2.2 Definition. *Let $B: R^2 \to R^2 : (q^1, q^2) \mapsto (\rho, \sigma)$, the* ***bipolar map****, where*

$$\rho(q^1, q^2) = \sqrt{(q^1 + \mu - 1)^2 + (q^2)^2}$$

$$\sigma(q^1, q^2) = \sqrt{(q^1 + \mu)^2 + (q^2)^2}$$

as before, and $\mu \in (0, \tfrac{1}{2}]$. Let W_+ denote the open upper half-plane ($q^2 > 0$), W_- the open lower half-plane ($q^2 < 0$). Thus $W = R^2 \backslash \{S, J\} = W_+ \cup W_- \cup C_*$, where C_* is the q^1-axis with $S = (-\mu, 0)$ and $J = (1 - \mu, 0)$ deleted. Write $C_* = C_1 \cup C_2 \cup C_3$, where

$$C_1 = \{(q^1, 0) | q^1 < -\mu\}$$

$$C_2 = \{(q^1, 0) | -\mu < q^1 < 1 - \mu\}$$

$$C_3 = \{(q^1, 0) | 1 - \mu < q^1\}$$

and $C_i' = B(C_i)$, $i = *, 1, 2, 3$ and $S = B(W)$.

ISBN 0-8053-0102-X

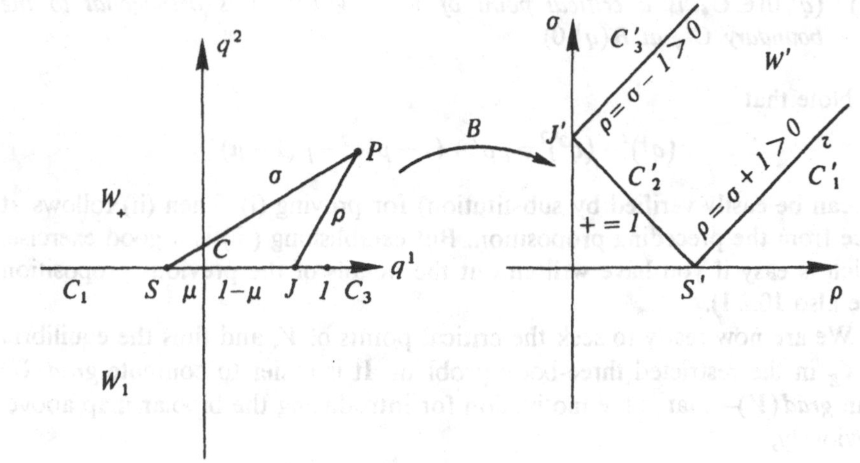

Figure 10.2-2

These notations are illustrated in Fig. 10.2-2: the bipolar map, folding the plane onto the strip W' with cusps at the primaries S and J.

10.2.3 Proposition. *The primary image points are* $S' = B(S) = (1,0)$ *and* $J' = B(J) = (0,1)$. *The image curves* C_1', C_2', C_3' *are straight line segments of slope* $+1, -1, +1$ *(resp.) through* S' *and* J', *as shown in Fig.* 10.2-1. *The image of* W, $W' = B(W)$ *is the closed half-strip bounded by* C'_* *with the corners* S', J' *removed. The restrictions* $B_\sigma = B|W_\sigma : W_\sigma \to W'$ $(\sigma = +, -)$ *are diffeomorphisms onto the open half-strip* $(W')^0$. *The restriction* $B_* = B|C_*$ *is a one-dimensional diffeomorphism.*

The proof is a straightforward verification. We shall use this map to locate the critical points of $V \in \mathfrak{F}(W)$ in 10.2.1. Note that to each value of the reduced mass μ there corresponds a different bipolar map B.

Note that V is symmetric with regard to the inversion $q^2 \mapsto -q^2$ in W. So also is the bipolar map B. Thus it is no surprise that there is a function $U \in \mathfrak{F}(W')$ such that $V = U \circ B$. In fact, the following is readily verified.

10.2.4 Proposition. *Let* $U \in \mathfrak{F}(W'')$ *be defined by*

$$U(\rho,\sigma) = -\tfrac{1}{2}\left[\mu\rho^2 + (1-\mu)\sigma^2 - \mu(1-\mu)\right] - \frac{\mu}{\rho} - \frac{1-\mu}{\sigma}$$

where $W'' = R^2 \setminus \{S', J'\}$. *Then*

(*i*) $V = U \circ B$

(*ii*) $(q^1, q^2) \in W_\sigma$ *is a critical point of* V *iff* $B(q^1, q^2)$ *is a critical point of* U $(\sigma = +, -)$

ISBN 0-8053-0102-X

(iii) $(q^1,0)\in C_*$ *is a critical point of* V *iff* $grad(U)$ *is orthogonal to the boundary* C'_* *at* $B(q^1,0)$.

Note that

$$(q^1)^2+(q^2)^2=\mu\rho^2+(1-\mu)\sigma^2-\mu(1-\mu)$$

(as can be easily verified by substitution) for proving (i). Then (ii) follows at once from the preceding proposition. But establishing (iii) is a good exercise, which is easy if you have written out the details of the previous proposition (use also 10.2.1).

We are now ready to seek the critical points of V, and thus the equilibria of X_E in the restricted three-body problem. It is easier to compute $grad(U)$ than $grad(V)$—that is the motivation for introducing the bipolar map above. Obviously,

$$\nabla U(\rho,\sigma)=\left(\frac{\mu}{\rho^2}[1-\rho^3],\frac{1-\mu}{\sigma^2}[1-\sigma^3]\right)$$

and, wonderfully, the variables separated!

The most interesting critical points may now be found at once, namely, the *equilateral triangle solutions* of Lagrange.

10.2.5 Proposition. *There is exactly one critical point of* U *in the interior of* W' : $\rho=\sigma=1$. *Thus there are exactly two critical points of* V *off the primary axis* C_*, $m_4=(\frac{1}{2}-\mu,\sqrt{3}/2)$ *and* $m_5=(\frac{1}{2}-\mu,-\sqrt{3}/2)$.

Now we must scour the boundary $C'_*=C'_1\cup C'_2\cup C'_3$ for points satisfying (iii) of 10.2.4. We will find one in each component: m_1, m_2, and m_3, the *collinear solutions* of Euler. We analyze the three cases separately.

Case I: Opposition. Corresponding to critical points of V on C_1: $q^1<-\mu$, $q^2=0$, we seek points on C'_1: $\rho=\sigma+1>1$, where ∇U is orthogonal, or $U_\rho=-U_\sigma$. Substituting $\rho-1$ for σ in the equation $U_\rho+U_\sigma=0$, we obtain

$$\frac{\mu}{\rho^2}[1-\rho^3]+\frac{1-\mu}{(\rho-1)^2}[1-(\rho-1)^3]=0$$

and as $\rho,\sigma\neq0$ in the domain of U, this is equivalent to

$$\text{(I)}\qquad f(\rho)=\rho^5-(3-\mu)\rho^4+(3-2\mu)\rho^3+(\mu-2)\rho^2+2\mu\rho-\mu=0$$

As $\mu\in(0,\frac{1}{2}]$, there is only one change of sign. By Descartes' rule of signs, there is one real root, ρ_1. From

$$\rho=\sqrt{(q^1+\mu-1)^2+(q^2)^2}=1-\mu-q^1$$

ISBN 0-8053-0102-X

on C_1 we obtain one critical point $m_1=(1-\mu-\rho_1,0)$. Since $f(1)<0$ and $f(+\infty)>0$, we conclude that $\rho_1>1$. Thus $(\rho_1,\rho_1-1)\in C_1'$ and $m_1\in C_1$.

Case II. Inferior conjunction. For critical points of V on C_2: $-\mu<q^1<1-\mu$, $q^2=0$, we examine U on C_2': $\rho+\sigma=1$, $0<\rho,\sigma<1$. Thus $\nabla U\perp C_2'$ if $U_\rho=U_\sigma$. Putting $1-\rho$ for σ in $U_\rho-U_\sigma=0$, we obtain

$$\frac{\mu}{\rho^2}\left[1-\rho^3\right]-\frac{1-\mu}{(1-\rho)^2}\left[1-(1-\rho)^3\right]=0$$

or, equivalently,

$$\text{(II)}\qquad \rho^5-(3-\mu)\rho^4+(3-2\mu)\rho^3-\mu\rho^2+2\mu\rho-\mu=0$$

This is identical to the quintic equation (I) above, except for one coefficient. This yields only one root ρ_2 that, since $f(0)<0$ and $f(1)>0$, lies in the desired range,

$$\rho_2\in(0,1)$$

providing $m_2=(1-\mu-\rho_1,0)\in C_2$.

Case III: Exterior conjunction. For the critical points of V on C_3: $1-\mu<q^1$, $q^2=0$, consider C_3': $\rho=\sigma-1>0$ and $U_\rho+U_\sigma=0$ as in Case I. Thus

$$\frac{\mu}{\rho^2}\left[1-\rho^3\right]+\frac{1-\mu}{(\rho+1)^2}\left[1-(\rho+1)^3\right]$$

or

$$\text{(III)}\qquad \rho^5+(3-\mu)\rho^4+(3-2\mu)\rho^3-\mu\rho^2-2\mu\rho-\mu=0$$

This also yields one positive real root ρ_3, hence one critical point $m_3=(1-\mu+\rho_3,0)\in C_3$. (Note that $q^1=1-\mu+\rho$ in this case.)

For comparison, we collect here the three quintic equations:

$$\text{(I)}\qquad \rho^5-(3-\mu)\rho^4+(3-2\mu)\rho^3+(\mu-2)\rho^2+2\mu\rho-\mu=0\quad(1<\rho_1)$$

$$\text{(II)}\qquad \rho^5-(3-\mu)\rho^4+(3-2\mu)\rho^3-\mu\rho^2+2\mu\rho-\mu=0\quad(0<\rho_2<1)$$

$$\text{(III)}\qquad \rho^5+(3-\mu)\rho^4+(3-2\mu)\rho^3-\mu\rho^2-2\mu\rho-\mu=0\quad(0<\rho_3)$$

and throughout, $0<\mu\leqslant\frac{1}{2}$. As we proved above, we have the estimates $1<\rho_1$ and $0<\rho_3$, so we have established the solutions of Euler. The situation is summarized in the following proposition.

ISBN 0-8053-0102-X

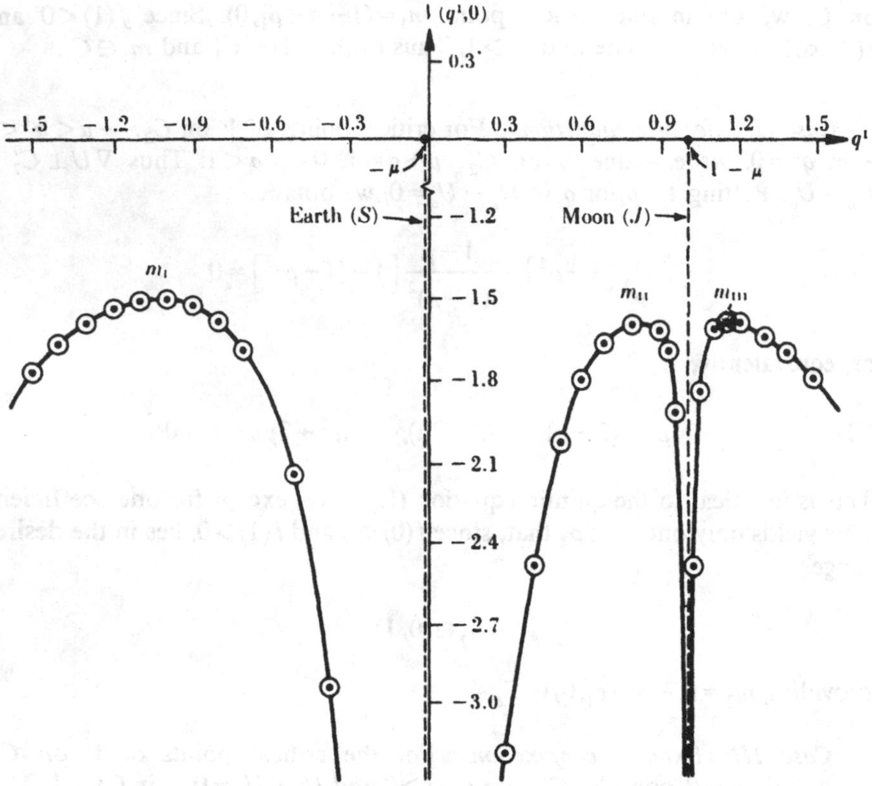

Figure 10.2-3. The potential V for $q^2=0$ showing the three collinear solutions of Euler in the Earth–Moon system: $\mu=0.012277471$.

10.2.6 Proposition. *On the fold curve C_* there are exactly three critical points of V:*

$$m_1=(1-\mu-\rho_1,0)\in C_1$$

$$m_2=(1-\mu-\rho_2,0)\in C_2$$

and

$$m_3=(1-\mu+\rho_3,0)\in C_3$$

where the ρ_i are determined by the ith quintic equation above, $i=I, II, III$.

A numerical calculation of the three collinear points yields the data in Figs. 10.2-3 and 10.2-4.

The relative positioning of all five critical points for the Earth-Moon system is shown in Fig. 10.2-5.

Next we turn to a study of the stability of these five critical points. Our first job is to study the characteristic exponents. This can be done in either

ISBN 0-8053-0102-X

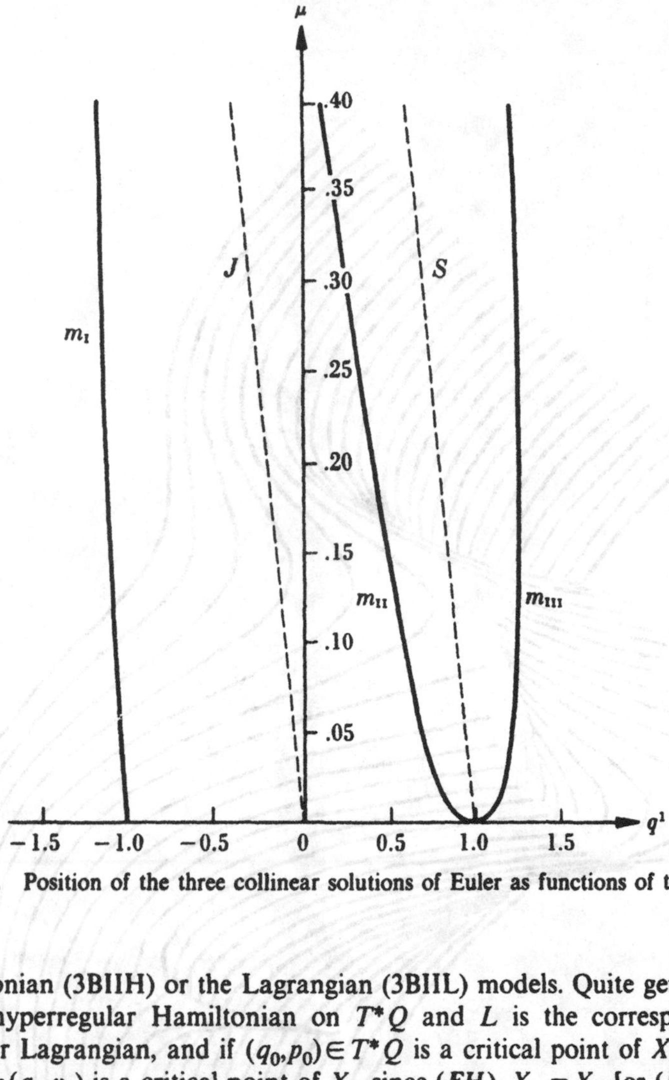

Figure 10.2-4. Position of the three collinear solutions of Euler as functions of the mass ratio.

the Hamiltonian (3BIIH) or the Lagrangian (3BIIL) models. Quite generally, if H is a hyperregular Hamiltonian on T^*Q and L is the corresponding hyperregular Lagrangian, and if $(q_0,p_0) \in T^*Q$ is a critical point of X_H, then $FH(q_0,p_0)=(q_0,v_0)$ is a critical point of X_E since $(FH)_*X_H=X_E$ [or $(FL)_*X_E = X_H$]. In fact, the linear maps $X_H'(q_0,p_0)$ and $X_E'(q_0,v_0)$ are similar and so have the same eigenvalues.

From the Lagrangian equations in model 3BIIL we see that the components of X_E relative to the natural chart are

$$X^1(q,\dot{q})=\dot{q}^1$$

$$X^2(q,\dot{q})=\dot{q}^2$$

$$X^3(q,\dot{q})=-2\dot{q}^2-V_1(q)$$

$$X^4(q,\dot{q})=2\dot{q}^1-V_2(q)$$

ISBN 0-8053-0102-X

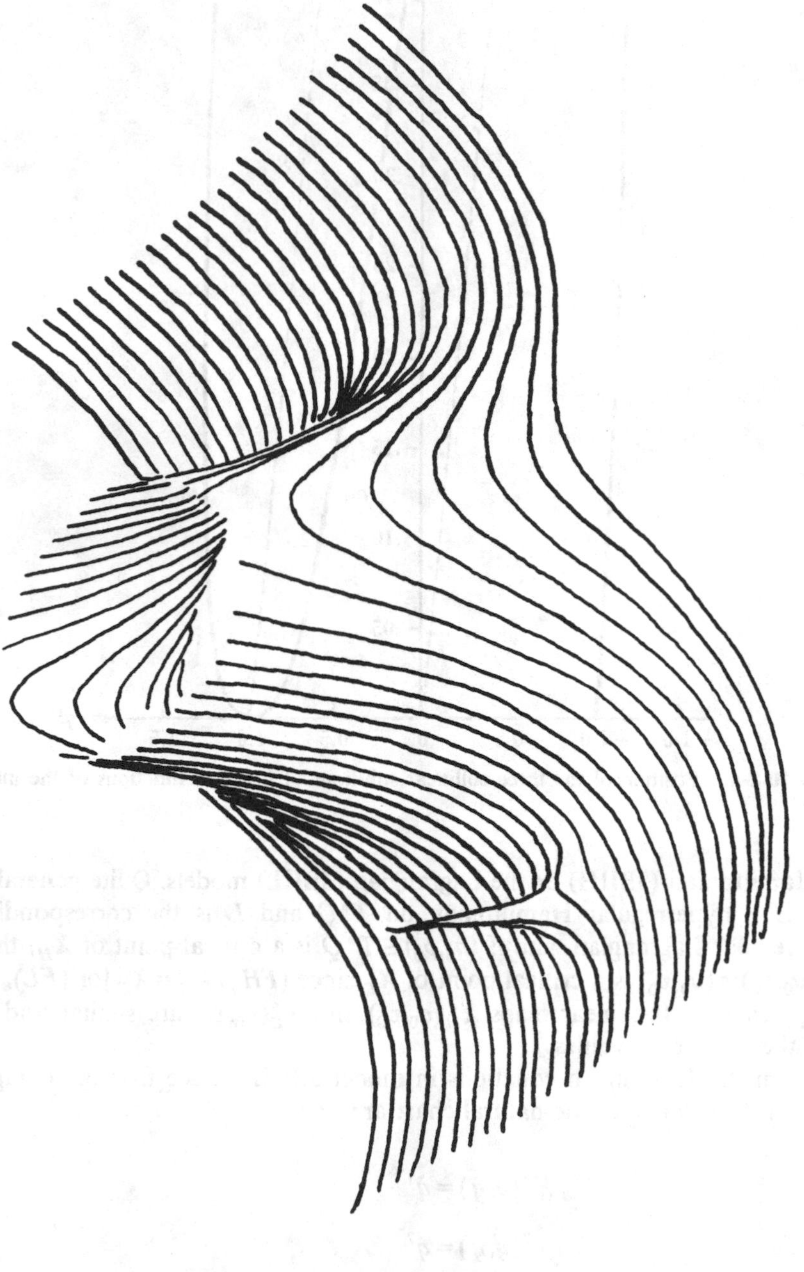

Figure 10.2-5. The graph of the potential $-V(q_1, q_2)$ of the Earth and Moon in rotating coordinates, $\mu = 0.012277471$, showing the five critical points.

ISBN 0-8053-0102-X

where

$$-V_1(q) = -V_{q^1}(q) = q^1 - \frac{\mu(q^1-1+\mu)}{[\rho(q)]^3} - \frac{(1-\mu)(q^1+\mu)}{[\sigma(q)]^3}$$

and

$$-V_2(q) = -V_{q^2}(q) = q^2 - \frac{(1-\mu)q^2}{[\sigma(q)]^3} - \frac{\mu q^2}{[\rho(q)]^3}$$

Thus the characteristic exponents are eigenvalues of the matrix representing $X_E'(m)$:

$$\begin{bmatrix} 0 & 0 & -V_{11} & -V_{12} \\ 0 & 0 & -V_{21} & -V_{22} \\ 1 & 0 & 0 & 2 \\ 0 & 1 & -2 & 0 \end{bmatrix}$$

where $V_{ij} = V_{q^i, q^j}$. The characteristic polynomial is then

$$P(\lambda) = \lambda^4 + (4 + V_{11} + V_{22})\lambda^2 + \left(V_{11}V_{22} - V_{12}^2\right)$$

This is quadratic in λ^2 so that solutions have the form $\alpha_1, \alpha_2, -\alpha_1, -\alpha_2$ as they must according to the general theory in Sect. 3.1.

10.2.7 Theorem (Plummer [1901]). *The Euler collinear points are unstable (i.e., not α^σ stable for any type α^σ).*

Proof. It suffices to prove that at these points $P(\lambda)$ does not have all its roots pure imaginary, for then (by the symplectic eigenvalue theorem) some eigenvalue must have positive real part and hence the fixed point has an unstable manifold. This assertion is the same as the claim that $\hat{P}(\xi) = \xi^2 + (4 + V_{11} + V_{22})\xi + (V_{11}V_{22} - V_{12}^2)$ does not have two negative real roots. We shall, in fact, prove that $\hat{P}$ has two real roots of opposite sign, that is, P has two real and two imaginary roots [case (b) in Fig. 3.1-2].

Direct calculation of the second derivatives gives

$$-V_{11} = 1 - \frac{\mu}{\rho^3} - \frac{1-\mu}{\sigma^3} + \frac{3\mu}{\rho^5}(q^1 - 1 + \mu)^2 + \frac{3(1-\mu)}{\sigma^5}(q^1+\mu)^2$$

$$-V_{12} = 3q^2\left\{\frac{\mu}{\rho^5}(q^1 - 1 + \mu) + \frac{(1-\mu)}{\sigma^5}(q^1+\mu)\right\}$$

$$-V_{22} = 1 - \frac{\mu}{\rho^3} - \frac{(1-\mu)}{\sigma^3} + 3(q^2)^2\left\{\frac{\mu}{\rho^5} + \frac{1-\mu}{\rho^5}\right\}$$

ISBN 0-8053-0102-X

For $q^2=0$, $V_{12}=0$ so $\hat{P}(\xi)=\xi^2+(4+V_{11}+V_{22})\xi+V_{11}V_{22}$ and $\rho=|q^1-1+\mu|, \sigma=|q^1+\mu|$, thus

$$-V_{11}=1+2\Delta \quad \text{and} \quad -V_{22}=1-\Delta$$

where $\Delta=\mu/\rho^3+(1-\mu)/\sigma^3$.

We note that $-V_{11}>0$ and

$$(4+V_{11}+V_{22})^2-4V_{11}V_{22}=\Delta(9\Delta-8)$$

Thus the roots of $\hat{P}$ are real if $\Delta>8/9$. We are going to in fact prove that $-V_{22}<0$, that is, $\Delta>1$ at the three collinear points. Since $-V_{11}>0$ and the product of the roots is $V_{11}V_{22}<0$, the roots will then be real and of opposite sign.

We now show that $-V_{22}<0$ at the three collinear points. For this purpose, we consider the function U (the potential in the variables ρ,σ) from 10.2.4:

$$-2U=\mu\left(\rho^2+\frac{2}{\rho}\right)+(1-\mu)\left(\sigma^2+\frac{2}{\sigma}\right)-\mu(1-\mu)$$

so

$$-\frac{\partial U}{\partial \rho}=\mu\left(\rho-\frac{1}{\rho^2}\right), \qquad -\frac{\partial U}{\partial \sigma}=(1-\mu)\left(\sigma-\frac{1}{\sigma^2}\right)$$

Write

$$V_2=\frac{\partial U}{\partial \rho}\frac{\partial \rho}{\partial q^2}+\frac{\partial U}{\partial \sigma}\frac{\partial \sigma}{\partial q^2}=q^2\left(\frac{1}{\rho}\frac{\partial U}{\partial \rho}+\frac{1}{\sigma}\frac{\partial U}{\partial \sigma}\right)$$

Thus, at $q^2=0$,

$$V_{22}=\frac{1}{\rho}\frac{\partial U}{\partial \rho}+\frac{1}{\sigma}\frac{\partial U}{\partial \sigma}$$

Now consider the two cases m_2 and m_1, m_3 separately. First of all, along the axis between the masses (i.e., on C_2) we have

$$\rho+\sigma=1 \text{ so } \frac{\partial \rho}{\partial q^1}+\frac{\partial \sigma}{\partial q^1}=0$$

By the chain rule,

$$V_1=\frac{\partial U}{\partial \rho}\frac{\partial \rho}{\partial q^1}+\frac{\partial U}{\partial \sigma}\frac{\partial \sigma}{\partial q^1}$$

ISBN 0-8053-0102-X

Hence if $V_1=0$, $\partial U/\partial\rho=\partial U/\partial\sigma$, that is,

$$\mu\left(\frac{1}{\rho^2}-\rho\right)=(1-\mu)\left(\frac{1}{\sigma^2}-\sigma\right)$$

(Since $\mu<1-\mu$, this implies that $\sigma>\rho$.) Thus

$$-V_{22}=-\left(\frac{1}{\rho}+\frac{1}{\sigma}\right)\frac{\partial U}{\partial\rho}$$

$$=\left(\frac{1}{\rho}+\frac{1}{\sigma}\right)\mu\left(\rho-\frac{1}{\rho^2}\right)$$

which is negative since $0<\rho<1$.

We turn to the case in which $\rho=\sigma+1$, that is, m_1. Here

$$\frac{\partial\rho}{\partial q^1}=\frac{\partial\sigma}{\partial q^1}$$

and so if $V_1=0$, we get, as above,

$$\frac{\partial U}{\partial\rho}=-\frac{\partial U}{\partial\sigma}$$

that is,

$$\mu\left(\frac{1}{\rho^2}-\rho\right)=(1-\mu)\left(\sigma-\frac{1}{\sigma^2}\right)$$

which implies that $\sigma<1<\rho$. Thus

$$-V_{22}=\left(\frac{1}{\rho}-\frac{1}{\sigma}\right)\mu\left(\rho-\frac{1}{\rho^2}\right)$$

is negative (the first factor is negative, the second positive).

This argument also holds for m_3 by interchanging σ and ρ. ■

10.2.8 Corollary. *Each collinear solution has a one-parameter family of closed orbits about it, together with a stable and unstable manifold.*

This follows from the above proof and Liapunov's theorem. The closed orbits surrounding the collinear solutions lie on the center manifold and, since one eigenvalue remaining is positive, these closed orbits are *unstable.*

Next we turn to the equilateral triangle solutions of Lagrange, m_4 and m_5. Here things are a good deal more subtle.

ISBN 0-8053-0102-X

Substituting into the general expressions for V_{12}, V_{11}, V_{22} in the proof of 10.2.7 we find that at

$$m_4=\left(\tfrac{1}{2}-\mu, \frac{\sqrt{3}}{2}\right): \quad V_{11}=-\tfrac{3}{4}$$
$$V_{22}=-\tfrac{9}{4}$$
$$V_{12}=\frac{3\sqrt{3}}{4}(2\mu-1)$$

and at

$$m_5=\left(\tfrac{1}{2}-\mu, -\frac{\sqrt{3}}{2}\right): \quad V_{11}=-\tfrac{3}{4}$$
$$V_{22}=-\tfrac{9}{4}$$
$$V_{12}=-\frac{3\sqrt{3}}{4}(2\mu-1)$$

[Notice that the matrix

$$\begin{pmatrix} V_{11} & V_{12} \\ V_{21} & V_{22} \end{pmatrix}$$

has $V_{11}<0$ and determinant $\frac{27}{4}\mu(1-\mu)>0$ so the potential has a maximum at the equilateral solutions. This does *not* mean these points are unstable. Although our energy function E is kinetic energy plus potential energy V, and hence E has a saddle point at the equilateral solutions as well (three "stable" and one "unstable" direction), this does not imply instability. This is, basically, why the Kolmogorov, Arnold, and Moser results play such a key role; they allow one to still prove stability in such situations.]

For the equilateral solutions the characteristic equation becomes

$$P(\lambda)=\lambda^4+\lambda^2+\tfrac{27}{4}\mu(1-\mu)=0$$

so

$$\lambda^2=-\tfrac{1}{2}\pm\tfrac{1}{2}[1-27\mu(1-\mu)]^{1/2}$$

The roots are purely imaginary if

$$1-27\mu(1-\mu)>0$$

that is, if $0<\mu<\mu_1=\frac{1}{2}-\sqrt{69}/18=.03852\ldots$ (*The Routh critical value.*) We recall that we restricted $0<\mu\leqslant\frac{1}{2}$; with only $0<\mu<1$, there would be another critical value at $\mu_2=\frac{1}{2}+\sqrt{69}/18$ above which the roots are purely imaginary.

The evolution of eigenvalues as μ crosses μ_1 is shown in Fig. 10.2-6. (The pictures reverse as μ crosses μ_2.)

ISBN 0-8053-0102-X

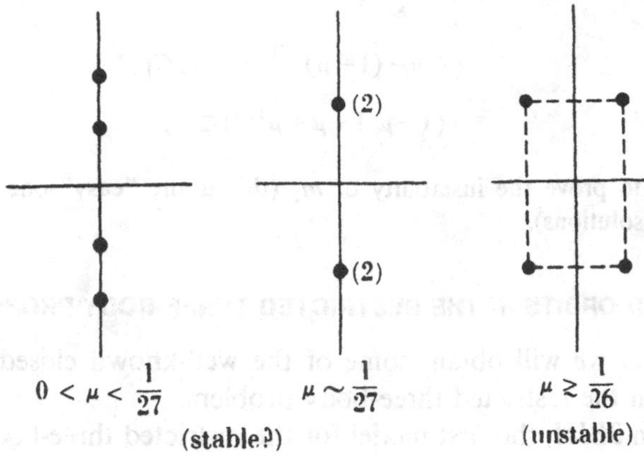

Figure 10.2-6

For $\mu_1 < \mu \leqslant \frac{1}{2}$, the equilateral solutions are clearly unstable. For $0 < \mu < \mu_1$, Leontovich [1962] proved (using a result of Arnold [1961] that gives invariant tori of quasi-periodic orbits) stability for all μ except possibly those in a set of measure zero. Then Deprit and Deprit–Bartholomé [1967] using Moser's improvement of Arnold's theorem showed stability for all but three values of μ in this range. This is consistent with experimental evidence; for the Sun-Jupiter system satellites (the Trojans) are observed near these positions (van de Kamp [1964, p. 114]).

The passage of μ past μ_1 is a Hamiltonian bifurcation involving a loss of stability. It corresponds to the *Trojan bifurcation* discussed in Chapter 8; see Meyer and Schmidt [1971] and Deprit and Henrard [1968].

EXERCISES

10.2A. Complete the details of 10.2.1.

10.2B. Show that the characteristic exponents for a periodic orbit of a hyperregular Hamiltonian on T^*Q are the same as those of the corresponding periodic orbit in the Lagrangian formulation.

10.2C. Regarding the proof of 10.2.7, let $\Delta_1 \subset W$ denote the set

$$\Delta_1 = \{(q^1, q^2) | \Delta(q^1, q^2) > 1\}$$

where

$$\Delta(q^1, q^2) = \frac{\mu}{[\rho(q^1, q^2)]^3} + \frac{1-\mu}{[\sigma(q^1, q^2)]^3}$$

Clearly Δ_1 contains the disk around S of radius $(1-\mu)^{1/3}$ and the disk around J of radius $\mu^{1/3}$ (excepting of course the points S and J themselves). As $(1-\mu)^{1/3} > 1-\mu$ and $\mu^{1/3} > \mu$, conclude that these disks contain C_2 and

ISBN 0-8053-0102-X

intervals

$$(-\mu-(1-\mu)^{1/3}, -\mu)\subset C_1$$

$$(1-\mu, 1-\mu+\mu^{1/3})\subset C_3$$

Use this to prove the instability of m_2 (this is the "easy" one of the three collinear solutions).

10.3 CLOSED ORBITS IN THE RESTRICTED THREE-BODY PROBLEM

In this section we will obtain some of the well-known closed orbits and periodic orbits in the restricted three-body problem.

Recall that in 3BIH, the first model for the restricted three-body problem, the two bodies rotate around the origin and the Hamiltonian is time dependent. The system is a vector field on $M_1\subset \boldsymbol{R}\times T^*\boldsymbol{R}^2$ with a constant upward vertical component, and the orbits are always rising. The "trajectory" in the phase space is obtained by projecting an integral curve into $T^*\boldsymbol{R}^2$. Thus an orbit in this model can never be closed (that is, a cycle), but the integral curve might be periodic in the sense that the projected trajectory is a closed curve (not necessarily simple, see Fig. 10.3-1); i.e., if the orbit is $t\mapsto(t,m_t)$, then for some $\tau>0$, $m_{t+\tau}=m_t$ for all t.

A *periodic orbit* in this sense, which spirals up covering a closed curve in phase space, is the analog of a closed orbit for the time-independent case. The projection into the phase space is called a *closed orbit*, but remember that it is not an orbit of an autonomous vector field, and may have self-intersections.

Recall also that 3BIIH, the second model, has the two bodies transformed to rest on the q^1-axis, and the Hamiltonian is conservative. The models are related by a canonical transformation $M_{\mathrm{I}}\to\boldsymbol{R}\times M_{\mathrm{II}}$, which preserves integral curves. Thus critical points in M_{II} become periodic orbits in M_{I} with period 2π, hence "closed orbits" in the phase space of M_{I}.

In the last section we showed the existence of exactly five critical points in the second model, the collinear solutions m_1, m_2, m_3 of Euler and the triangular solutions m_4 and m_5 of Lagrange. These critical points are mapped into periodic orbits $\gamma_1,\ldots,\gamma_5$ of period 2π in the first model by the canonical transformation relating the two models. As m_1, m_2, and m_3 are unstable critical points, γ_1, γ_2, and γ_3 are unstable periodic orbits. The stability of m_4 and m_5, so also γ_4 and γ_5, for $0<\mu<\mu_1$ was also discussed. In addition, we recall that m_1, m_2, and m_3 have one-dimensional stable and unstable manifolds and a two-dimensional center manifold. The critical points m_4 and m_5 have purely imaginary characteristic exponents for $\mu<\mu_1$, and for almost all such μ, the two exponents on the positive imaginary axis are independent over the integers. Applying the subcenter stability theorem (8.3.2) we have in these cases one-parameter families of closed orbits γ_{II}^s with periods τ_s depending continuously on s. For nearly all s we have $\tau_s=2p\pi/q$ for relatively prime integers p and q, and γ_{II}^s is mapped into a closed orbit γ_{I}^2 in 3BIH of period $2p\pi$. These are the closed orbits discovered by Liapounov [1947].

ISBN 0-8053-0102-X

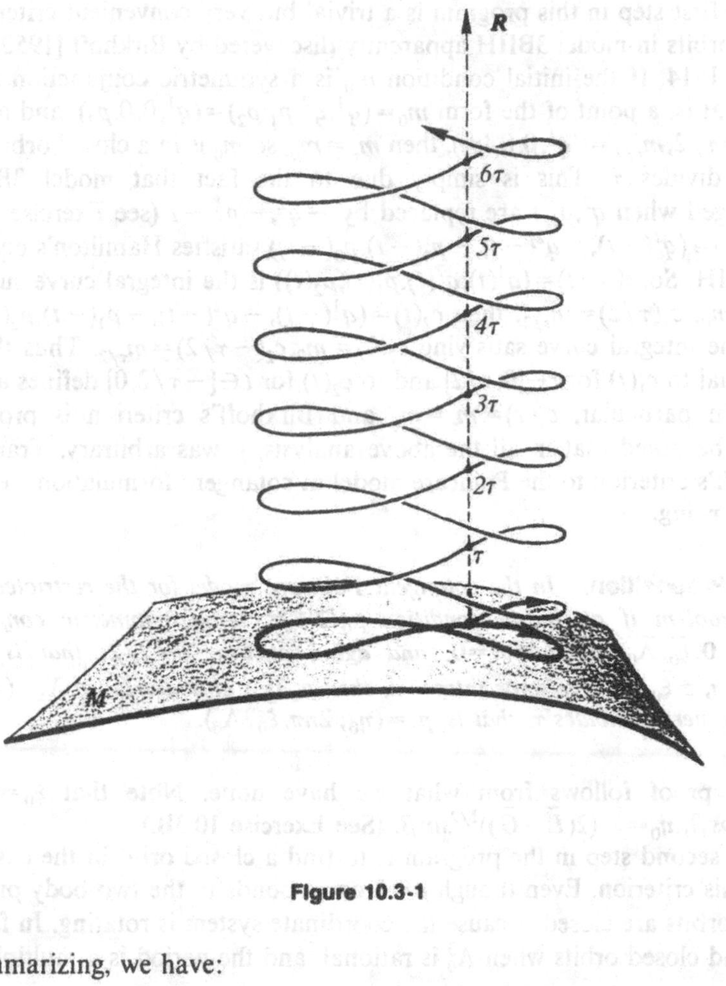

Figure 10.3-1

Summarizing, we have:

10.3.1 Theorem (Llapounov). *In M_{I} there are five closed orbits of period 2π corresponding to the critical points in M_{II}, and for almost all μ, $0<\mu<\mu_1$, every neighborhood of any of these contains infinitely many closed orbits of arbitrarily high period.*

As was noted in the last section, about m_1, m_2, m_3 there are no exceptional μ and about m_4 and m_5 there are just three.

Next we turn to the closed orbits that are obtained by "analytic continuation" from $\mu=0$. These are of two types. The *closed orbits of the first kind*, discovered by Poincaré [1892] are close to the circular Keplerian orbits in the second model. The *closed orbits of the second kind*, discovered by Arenstorf

ISBN 0-8053-0102-X

[1963a], are close to Keplerian orbits of arbitrary (positive) eccentricity. We shall treat both types simultaneously, using the cotangent Poincaré model, in a program suggested by Barrar [1965a].

The first step in this program is a trivial but very convenient criterion for closed orbits in model 3BIIH apparently discovered by Birkhoff [1950; v.I, p. 713] in 1914. If the initial condition m_0 is a symmetric conjunction at time $t=0$, that is, a point of the form $m_0=(q^1,q^2,p_1,p_2)=(q^1,0,0,p_2)$, and again at time $t=\tau/2, m_{\tau/2}=('q^1,0,0,'p_2)$, then $m_\tau=m_0$, so m_0 is in a closed orbit whose period divides τ. This is simply due to the fact that model 3BIIH is unchanged when q^2,p_1,t are replaced by $-q^2,-p_1,-t$ (see Exercise 10.3A). Hence $t\mapsto(q^1(-t),-q^2(-t),-p_1(-t),p_2(-t))$ satisfies Hamilton's equations for 3BIIH. So, if $c_1(t)=(q^1(t),q^2(t),p_1(t),p_2(t))$ is the integral curve such that $c_1(0)=m_0$, $c_1(\tau/2)=m_{\tau/2}$, then $c_2(t)=(q^1(-t),-q^2(-t),-p_1(-t),p_2(-t))$ is the same integral curve satisfying $c_2(0)=m_0, c_2(-\tau/2)=m_{\tau/2}$. Thus the map $c(t)$ equal to $c_1(t)$ for $t\in[0,\tau/2]$ and to $c_2(t)$ for $t\in[-\tau/2,0]$ defines a closed orbit. In particular, $c_1(\tau)=m_\tau=m_0$ and Birkhoff's criterion is proved. It should be noted that in all the above analysis, μ was arbitrary. Translating Birkhoff's criterion to the Poincaré model in cotangent formulation we obtain the following.

10.3.2 Proposition. *In the cotangent Poincaré model for the restricted three-body problem if an initial condition $p_0\in\mathcal{P}_1^{*\mu}$ is in symmetric conjunction $p_0=(\eta_0,0,\bar\xi_0,\bar\Lambda_0)$ at time $t=0$, and again at time $t=\tau/2$, that is $p_{\tau/2}=(\pm\eta_0,n\pi,\pm\bar\xi_0,\bar\Lambda_0)$ for some integer n, then p_0 is a closed orbit of $X_{\bar\Omega^\mu}$ (for any μ) whose period divides τ, that is, $p_\tau=(\eta_0,2n\pi,\bar\xi_0,\bar\Lambda_0)$.*

The proof follows from what we have done. Note that $\bar\xi_0=(2(\bar L-\bar G))^{1/2}\cos\beta$, $\eta_0=-(2(\bar L-\bar G))^{1/2}\sin\beta$. (See Exercise 10.3B.)

The second step in the program is to find a closed orbit in the case $\mu=0$ using this criterion. Even though $\mu=0$ corresponds to the two-body problem, not all orbits are closed because the coordinate system is rotating. In fact, we only find closed orbits when Λ_0^3 is rational, and the period is a multiple of π.

10.3.3 Corollary. *In the cotangent Poincaré model with $\mu=0$, initial conditions of the form $p_0=(\eta_0,0,\bar\xi_0,\bar\Lambda_0)$ with $\bar\Lambda_0^{-3}=m/k$ (m, k integers) are in closed orbits of $X_{\bar\Omega^0}$ whose period divides $2k\pi$. If m and k are relatively prime, the period equals $2k\pi$.*

Proof. From 10.1.9 we have $X_{\bar\Omega^0}=(-\bar\xi,\bar\Lambda^{-3}+1,\eta,0)$ so the flow is

$$\left(\eta_0,\lambda_0,\bar\xi_0,\bar\Lambda_0;\ t\right)\mapsto\left(\eta_0\cos t+\bar\xi_0\sin t,\bar\lambda_0+\left(\bar\Lambda_0^{-3}+1\right)t,\ -\eta_0\sin t+\bar\xi_0\cos t,\bar\Lambda_0\right)$$

ISBN 0-8053-0102-X

In the case of symmetric conjunction, this becomes

$$(p_0; t) = (\eta_0, 0, \xi_0, \overline{\Lambda}_0; t) \mapsto p_t$$
$$= \left(\eta_0 \cos t + \xi_0 \sin t, \left(\overline{\Lambda}_0^{-3} + 1\right)t, -\eta_0 \sin t + \xi_0 \cos t, \overline{\Lambda}_0\right)$$

so if $\overline{\Lambda}_0^{-3} = m/k$ and $t = k\pi$, we have $p_{k\pi} = (\pm\eta_0, (m+k)\pi, \pm\xi_0, \overline{\Lambda}_0)$, which satisfies the above criterion. The proof of the last statement is left as an exercise. ■

Note that in the above we obtain $+\eta_0, +\xi_0$ or $-\eta_0, -\xi_0$ in $p_{k\pi}$ according as k is even or odd. In any case the period is a rational multiple of 2π, and thus gives rise to a periodic orbit in model 3BIH.

The third step in the program consists of showing that these particular closed orbits are preserved under small perturbations of μ away from zero. This requires the consideration of the domain $\mathcal{P}_1^{*\mu}$, Hamiltonian $\overline{\Omega}^\mu$, vector field $X_{\overline{\Omega}^\mu}$, and integral F^μ, all depending on the parameter μ. It is important that all vary smoothly with μ. We let $\mathfrak{D}^\mu = \mathfrak{D}_{X_{\overline{\Omega}^\mu}} \subset \mathcal{P}_1^{*\mu} \times \mathbf{R}$ denote the domain of the integral F^μ of $X_{\overline{\Omega}^\mu}$ and we extend all of our models for the restricted three-body problem by allowing μ to take on negative values, thus any real value. For each μ we have $\mathcal{P}_1^{*\mu} \subset T^*(\mathbf{R} \times \mathbf{T}^1)$, and we let

$$\mathcal{P}^\# = \bigcup_{\mu \in \mathbf{R}} (\{\mu\} \times \mathcal{P}_1^{*\mu}) \subset \mathbf{R} \times T^*(\mathbf{R} \times \mathbf{T}^1)$$

and

$$\mathfrak{D}^\# = \bigcup_{\mu \in \mathbf{R}} (\{\mu\} \times \mathfrak{D}^\mu) \subset \mathcal{P}^\# \times \mathbf{R}$$

Note that we may consider the family of vector fields $\{X_{\overline{\Omega}^\mu} \mid \mu \in \mathbf{R}\} = X^\#$ itself as a vector field on $\mathcal{P}^\#$.

10.3.4 Lemma. *In the context above,*

(i) the function $\Omega^\# : \mathcal{P}^\# \to \mathbf{R}$ defined by $\Omega^\# | \mathcal{P}_1^{*\mu} = \overline{\Omega}^\mu$ is of class C^∞; $\mathcal{P}^\# \subset \mathbf{R} \times T^*(\mathbf{R} \times T^1)$ and $\mathfrak{D}^\# \subset \mathcal{P}^\# \times \mathbf{R}$ are open sets; the vector field $X^\#$ is C^∞; the mapping $F^\# : \mathfrak{D}^\# \to \mathcal{P}^\#$ defined by $F^\# | \mathfrak{D}^\mu = F^\mu$ (the integral of $X_{\overline{\Omega}^\mu}$) is none other than the integral of $X^\#$; and $F^\#$ is of class C^∞;

(ii) If γ is a closed orbit of $X_{\overline{\Omega}^{\mu_0}}$ in $\mathcal{P}_1^{*\mu_0}$ with period τ, then there is a neighborhood U of $\gamma \subset \mathcal{P}_1^{*\mu_0}$ and $\delta, \varepsilon > 0$ such that $U \subset \mathcal{P}_1^{*\mu}$ for all $\mu \in I = (\mu_0 - \delta, \mu_0 + \delta)$, $U \times J \subset \mathfrak{D}^\mu$ for all $\mu \in I$, where $J = (-\varepsilon, \tau + \varepsilon)$, and the mapping

$$F : I \times U \times J \to \mathcal{P}^\# : (\mu, p, t) \mapsto F^\mu(p, t)$$

is of class C^∞.

ISBN 0-8053-0102-X

The proof is a routine verification (see Exercise 10.3D).

To show the closed orbits of 10.3.3 are preserved under small perturbations of μ around zero, we will separate the two cases of elliptic ($\xi_0 \neq 0$) and circular ($\xi_0 = 0$) orbits. We consider first the circular case, to obtain the closed orbits of the first kind of Poincaré.

We begin with a circular closed orbit γ_0 of $X_{\bar{\Omega}^0}$ with initial condition $p_0 = (0,0,0,\bar{\Lambda}_0)$, $\bar{\Lambda}_0^{-3} = m/k$, and period $2k\pi$, m and k relatively prime. Thus in the notations of 10.3.4 we have the C^∞ local integral

$$F: I \times U \times J \to \mathcal{P}^\# : (\mu, p, t) \mapsto F^\mu(p, t)$$

and for $\mu = 0$ we have

$$F^0(0,0,0,\bar{\Lambda}_0;\, k\pi) = (0, n\pi, 0, \bar{\Lambda}_0)$$

where $n = k + m$, the order of γ_0. According to the criterion 10.3.2 we seek $(\mu, t, \bar{\Lambda})$ near $(0, k\pi, \bar{\Lambda}_0)$ such that

$$F^\mu(0,0,0,\bar{\Lambda};\, t) = (0, n\pi, 0, \bar{\Lambda})$$

as well, so that $p = (0,0,0,\bar{\Lambda})$ is in a closed orbit γ_μ of period $2(1+\sigma)k\pi$. Obviously we have a job for the implicit mapping theorem, which we may use because everything depends smoothly on μ as well as the other variables.

10.3.5 Theorem (Poincaré). *In the cotangent Poincaré model for the restricted three-body problem, the closed orbit γ_0 of $X_{\bar{\Omega}^0}$ containing the initial conditions $p_0 = (0,0,0,\bar{\Lambda}_0)$, with $\bar{\Lambda}_0^{-3} = m/k$ and period $2k\pi$, is preserved under perturbation of the mass ratio μ away from zero. That is, there is an $\varepsilon > 0$ and a C^∞ function $f: (-\varepsilon, \varepsilon) \to \mathbf{R}$ such that if $\mu \in (-\varepsilon, \varepsilon)$, then p^μ is in a closed orbit γ_μ of $X_{\bar{\Omega}^\mu}$ of period $2k\pi$, where $p^\mu = (0,0,0,f(\mu))$, and $f(0) = \bar{\Lambda}_0$.*

Proof. We consider the mapping

$$G: I \times K \to \mathbf{T}^1 : (\mu, \bar{\Lambda}) \mapsto G(\mu, \bar{\Lambda})$$

where $K \subset U$ is the intersection of U and the $\bar{\Lambda}$ axis, $K = \{(0,0,0,\bar{\Lambda}) \in U\}$, and $G(\mu, \bar{\Lambda})$ is the λ-component of $F^\mu(0,0,0,\bar{\Lambda};\, k\pi)$. By the discussion above, we seek to "solve" the implicit equation $G(\mu, \bar{\Lambda}) = n\pi$ for $\bar{\Lambda}$ as a function of μ, where $n = m + k$, and we have

$$G(0, \bar{\Lambda}_0) = n\pi$$

when $\bar{\Lambda}_0^{-3} = m/k$. As

$$\left(\frac{\partial G}{\partial \bar{\Lambda}}\right)_{(0,\bar{\Lambda}_0)} = \left(\frac{\partial}{\partial \bar{\Lambda}}((\bar{\Lambda}^{-3}+1)k\pi)\right)_{\bar{\Lambda}_0} = \frac{-3k\pi}{\bar{\Lambda}_0^4} \neq 0$$

ISBN 0-8053-0102-X

the implicit mapping theorem shows there is an $\varepsilon>0$ and a mapping $f\colon(-\varepsilon,\varepsilon)\to\boldsymbol{R}$ such that $G(\mu,f(\mu))=n\pi$ for all $\mu\in(-\varepsilon,\varepsilon)$. The result follows at once from the criterion 10.3.2. ■

Note that we also obtain the smoothness of the energy, or $\overline{\Lambda}(\gamma_\mu)=f(\mu)$, of the closed orbit γ_μ. In addition, we could have obtained by a similar argument a closed orbit γ'_μ of the same energy as γ_0, $\overline{\Lambda}(\gamma'_\mu)=\overline{\Lambda}_0$, for each μ, but then the period of γ'_μ depends on μ. Even more, there is a curve $s\mapsto\gamma^s_\mu$ of closed orbits for each μ, of which these are but two examples (see Exercise 10.3E).

We turn now to the second case, the elliptical orbits. We begin with an elliptical closed orbit γ_0 of $X_{\overline{\Omega}^0}$ of the type given by 10.3.3. Thus we have $\not{p}_0=(0,0,\bar{\xi}_0,\overline{\Lambda}_0)$ with $\overline{\Lambda}_0^{-3}=m/k$, where m and k are relatively prime integers, and $F^0(\not{p}_0,k\pi)=(0,n\pi,\pm\bar{\xi}_0,\overline{\Lambda}_0)$, where $n=m+k$, so $\not{p}_0\in\gamma_0$ has period $2k\pi$. For simplicity we suppose k even so that $(+\bar{\xi}_0)$ is obtained in the third component. Note that $\bar{\xi}_0\neq0$ in this case. As in the circular case ($\bar{\xi}_0=0$) we seek $(\mu,t,\overline{\Lambda})$ near $(0,k\pi,\overline{\Lambda}_0)$ such that $F^\mu(0,0,\bar{\xi}_0,\overline{\Lambda};t)=(0,n\pi,\bar{\xi}_0,\overline{\Lambda})$. That is, we keep $\bar{\xi}_0$ fixed (but not zero).

10.3.6 Theorem (Arenstorf [1963a]). *In the cotangent Poincaré model for the restricted three-body problem, the closed orbit γ_0 containing $\not{p}_0=(0,0,\bar{\xi}_0,\overline{\Lambda}_0)$ (where $\overline{\Lambda}_0^{-3}=m/k$, m and k being relatively prime integers, $\bar{\xi}_0\neq0$, and the period is $2k\pi$) is preserved under perturbation of the mass ratio μ away from zero. That is, there is an $\varepsilon>0$ and C^∞ functions $f,g\colon(-\varepsilon,\varepsilon)\to\boldsymbol{R}$ such that if $\mu\in(-\varepsilon,\varepsilon)$, then $\not{p}^\mu$ is in a closed orbit γ_μ of period $g(\mu)$, where $\not{p}^\mu=(0,0,\bar{\xi}_0,f(\mu))$, $f(0)=\overline{\Lambda}_0$, and $g(0)=2k\pi$.*

The proof is very analogous to the previous one (which was inspired by this one of Arenstorf and Barrar rather than the original of Poincaré) and so may be relegated to the exercises. (See Exercise 10.3F.)

A typical example of a closed orbit of the second kind of Arenstorf is illustrated in Fig. 10.3-2, both in the inertial frame of the Sun (similar to 3BIH) and rotating coordinates (configuration space of 3BIIH).

Here E and M replace S and J, as the orbit is computed for $\mu=0.012277471$, corresponding to the Earth and the Moon. The orbit shown was computed at the G. G. Marshall Space Flight Center of the National Aeronautics and Space Administration and is reproduced here through the courtesy of Dr. Arenstorf. The interest in these "bus orbits" is evident from the fact that they pass very close to the Earth and to the Moon.

Note that for the ε range in Theorems 10.3.5 and 10.3.6 no collisions occur. Through Lemma 10.3.4(ii) we have chosen a domain in which the orbits do not run off the manifold in the times involved, hence do not arrive at S or J in these times. The last theorem (elliptical case) can also be proved very similarly in the cotangent Delaunay model (see Barrar [1965a]), but the circular case cannot be attacked in that model because it does not contain circular orbits.

ISBN 0-8053-0102-X

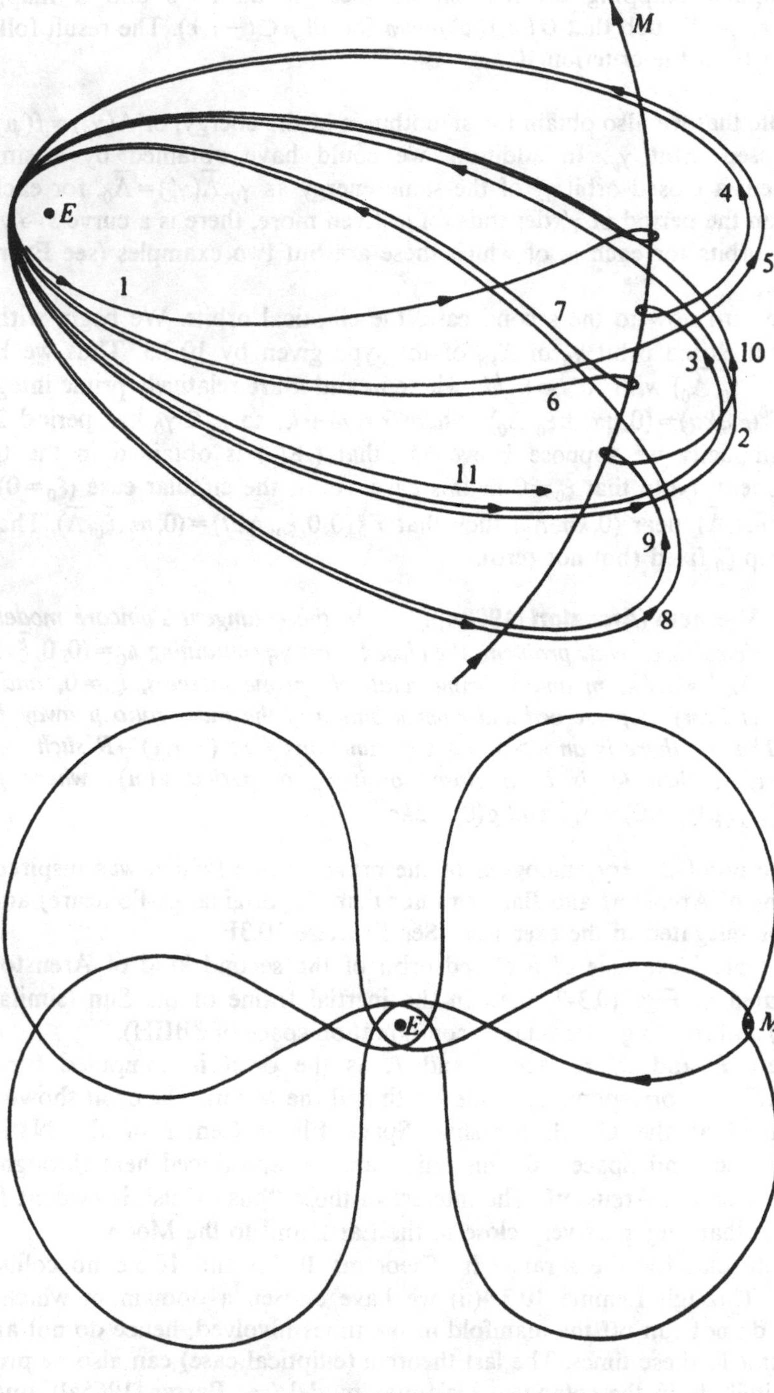

Figure 10.3-2. (a) Inertial frame. (b) Rotating frame.

ISBN 0-8053-0102-X

Finally, we shall show the existence of the *closed orbits of the second kind of Moser* and the o^+-stability of the closed orbits of the first kind of Poincaré. We obtain these two results of Moser simultaneously by applying the twist theorem to the closed orbits of the first kind. The program of the proof is quite analogous to a proof by Barrar [1965b] of the existence of the Moser orbits.

Let γ_0 be a circular closed orbit for $\mu=0$, as in Theorem 10.3.5, with $\overline{\Lambda}_0^{-3}=m/k$ and period $2k\pi$. By Lemma 10.3.4(ii) we may choose an open neighborhood U of $\gamma\subset\mathcal{P}_1^{*\mu_0}$ such that for $\mu\in(-\delta,\delta)$ we have $U\subset\mathcal{P}_1^{*\mu}$ as well, and $U\times J\subset\mathcal{D}^\mu$, $J=(-\varepsilon,2k\pi+\varepsilon)\subset R$. Thus the integral F^μ of $X_{\overline{\Omega}^\mu}$ is defined for points $(t,\not{p})\in J\times U$, or especially $F^\mu(\not{p},2k\pi)$ is defined for $\not{p}\in U$ and $|\mu|<\delta$. As above, there are no "collisions" in this range. Let

$$\not{p}^0=\left(0,0,0,\overline{\Lambda}_0\right)\in\gamma_0$$

so

$$F^0(\not{p}^0,2k\pi)=\left(0,2n\pi,0,\overline{\Lambda}_0\right)=\left(0,0,0,\overline{\Lambda}_0\right)=\not{p}^0$$

From 10.3.5, the Poincaré orbits γ_μ for sufficiently small μ are defined by initial conditions $\not{p}^\mu=(0,0,0,f(\mu))$, and we have $F^\mu(\not{p}^\mu,2k\pi)=\not{p}^\mu$.

To apply the Moser twist theorem, we must construct a local transversal section S^μ for γ_μ and Poincaré section map Θ^μ on S^μ such that S^μ, and its derivatives up to order four, at least, depend continuously on the parameter μ. We may then verify the elementary twist hypothesis for $\mu=0$, and assert it is satisfied also for $\mu\approx0$. Note that the local transversal section should be tangent to the energy surface $\Sigma^\mu_{e(\mu)}$ defined by $\overline{\Omega}^\mu=\overline{\Omega}^\mu(\not{p}^\mu)$, which depends on μ in two different ways. This construction is the heart of the proof of the following.

10.3.7 Theorem (Moser). *In the cotangent Poincaré model for the restricted three-body problem, let γ_0 be a closed orbit of $X_{\overline{\Omega}^0}$ containing $\not{p}^0=(0,0,0,\overline{\Lambda}_0)$, with $\overline{\Lambda}_0^{-3}=m/k$, m and k relatively prime integers. Let γ_μ denote the closed orbit of the first kind of $X_{\overline{\Omega}^\mu}$ containing $\not{p}^\mu=(0,0,0,\overline{\Lambda}_\mu)$ given by* 10.3.5 *($\overline{\Lambda}_\mu=f(\mu)$). Then if $k/(m+k)\neq p/q$ for $q=1,2,3,4$ and any integer p, there is an $\varepsilon>0$ such that*

(*i*) *if $|\mu|<\varepsilon$, γ_μ is $o^\pm$-stable, and*
(*ii*) *if $|\mu|<\varepsilon$, V is a neighborhood of $\gamma_\mu\subset\mathcal{P}_1^{*\mu}$, and N is a positive integer, there exists a closed orbit of $X_{\overline{\Omega}^\mu}$ in V with period greater than N.*

Proof. We first construct the Poincaré section map Θ^μ on the local transversal section S^μ. Let $e(\mu)=\overline{\Omega}^\mu(\not{p}^\mu)=\overline{\Omega}^\mu(\gamma_\mu)$ denote the energy of the closed orbits of the first kind, and $\Sigma_0^\mu=(\overline{\Omega}^\mu)^{-1}(e(\mu))$ be the corresponding energy surfaces. Restricting to the neighborhood U of $\gamma_0\subset\mathcal{P}^0$ of the preceding discussion, $F^\mu(\not{p},2k\pi)$ is defined for all $\not{p}\in U$. If U is taken sufficiently small, it contains none of the five critical points of $X_{\overline{\Omega}^\mu}$, so $\Sigma^\mu=\Sigma_0^\mu\cap U$ is a regular energy surface. Let $\tilde{S}$ denote the intersection of U and the coordinate

ISBN 0-8053-0102-X

hyperplane in $\mathcal{P}^0$ defined by $\lambda=0$ (in the Poincaré variables: η, λ, $\bar{\xi}$, and $\bar{\Lambda}$). Then for $\mu=0$, the second (λ) component of $X_{\bar{\Omega}^0}$ is not zero on $\tilde{S}$, so $\tilde{S}$ is a local transversal section of $X_{\bar{\Omega}^0}$. As $X_{\bar{\Omega}^\mu}$ depends continuously on μ, there is an $\varepsilon_1>0$ such that $\tilde{S}$ is a local transversal section of $X_{\bar{\Omega}^\mu}$ if $|\mu|<\varepsilon_1$. Hereafter we suppose $|\mu|<\varepsilon_1$ and $\varepsilon_1<\varepsilon_0$ (the ε of 10.3.5).

Note that the initial condition p^μ is in $\tilde{S}$ for all μ. Let $S^\mu=\tilde{S}\cap\Sigma^\mu$. As $\tilde{S}$ is a local transversal section and Σ^μ an energy surface, S^μ is a submanifold, necessarily of dimension two, and is a local transversal section of $X_{\bar{\Omega}^\mu}|\Sigma^\mu$. As $p^\mu\in S^\mu$, there is a (locally) unique Poincaré map Θ^μ of $X_{\bar{\Omega}^\mu}|\Sigma^\mu$ at $p^\mu\in S^\mu$.

This finishes the construction, and we now study the map Θ^0 corresponding to the unperturbed orbit γ_0 to establish the elementary twist hypothesis of Moser's theorem. Recall that

$$\bar{\Omega}^0(\eta,\lambda,\xi,\Lambda)=\frac{1}{2\bar{\Lambda}^2}+\bar{\xi}-\tfrac{1}{2}(\bar{\xi}^2+\eta^2)$$

so

$$\dot{e}(0)=\bar{\Omega}^0(p^0)=-\frac{1}{2\bar{\Lambda}_0^2}+\bar{\Lambda}_0$$

We may thus solve the equation $\bar{\Omega}^0(p)=e(0)$ (even explicitly if we wish), for $\bar{\Lambda}$ as a function of $(\bar{\xi}^2+\eta^2)$. That is, we have a C^∞ (in fact analytic) function ν: $U\to \boldsymbol{R}$ such that

$$\bar{\Omega}^0\big(\eta,\lambda,\bar{\xi},\nu(\bar{\xi}^2+\eta^2)\big)=e(0)$$

for all $(\eta,\lambda,\bar{\xi})$. Thus a point $p=(\eta,\lambda,\bar{\xi},\bar{\Lambda})$ is in the local transversal section Σ^0 iff $\lambda=0$ and $\bar{\Lambda}=\nu(\bar{\xi}^2+\eta^2)$. Consider the integral $F^0(p,t)$ of $X_{\bar{\Omega}^0}$ restricted to Σ^0. By integrating the equations $X_{\bar{\Omega}^0}$ explicitly, we find (see the proof of 10.3.3)

$$F^0(\eta,0,\bar{\xi},\bar{\Lambda};t)=\big(\eta\cos t-\bar{\xi}\sin t,(\bar{\Lambda}^{-3}+1)t,\eta\sin t+\bar{\xi}\cos t,\bar{\Lambda}\big)$$

As $p\in\Sigma^0$ we have $\bar{\Lambda}=\nu(\bar{\xi}^2+\eta^2)$, and we may choose for each $p\in\Sigma^0$ a time $t=\chi(p)$ so that the point $F^0(\eta,0,\bar{\xi},\mu;\chi)$ is again in Σ^0, as in the definition of the Poincaré section map. Namely, let $\chi(\xi,0,\eta,\nu)=2\pi(\nu^{-3}+1)^{-1}$. Then we have

$$F^0\big(\eta,0,\bar{\xi},\nu(\bar{\xi}^2+\eta^2);\chi(\bar{\xi}^2+\eta^2)\big)$$

$$=\big(\eta\cos\chi-\bar{\xi}\sin\chi,2k\pi,\eta\sin\chi+\bar{\xi}\sin\chi,\nu\big)$$

This is in fact the Poincaré map on the local transversal section S^0. By choosing $(\eta,\bar{\xi})$ as a coordinate chart in S^0, we obtain

$$\Theta^0(\eta,\xi)=\big(\eta\cos\chi-\bar{\xi}\sin\chi,\eta\sin\chi+\bar{\xi}\cos\chi\big)$$

where $\chi(\eta,\bar{\xi})=\chi(\bar{\xi}^2+\eta^2)$ is determined above. Looking back through the construction above, we have $\chi(\eta,\bar{\xi})=2\pi(\nu^{-3}+1)^{-1}$, where $\nu(\eta,\bar{\xi})$ is defined

ISBN 0-8053-0102-X

implicitly by

$$\frac{-1}{2\nu^2}+\nu-\tfrac{1}{2}(\xi^2+\eta^2)=\frac{-1}{2\overline{\Lambda}_0^2}+\overline{\Lambda}_0$$

and $\overline{\Lambda}_0^{-3}=m/k$. Luckily, *the section map is already in* (α,β)*-normal form.* For Θ^0, in complex notation, is $\Theta(v)=ve^{i\nu(|v|^2)}$. It only remains to expand $\nu(|v|^2)$ in a Taylor formula to order four, which we leave as Exercise 10.3H. The result is

$$\chi(|v|^2)=\alpha+\beta|v|^2+|v|^4\mathfrak{R}(v)$$

where $\mathfrak{R}(0)=0$,

$$\alpha=\chi(0)=\frac{2k\pi}{m+k}=\frac{k}{n}2\pi$$

and

$$\beta=\frac{-3k\pi\overline{\Lambda}_0^2}{\left(\overline{\Lambda}_0^{-3}+1\right)^3}=\frac{-3k^4\pi\overline{\Lambda}_0}{(m+k)^3}$$

As $\beta\neq0$, this is an elementary twist iff α is not zero or an integral multiple of $\pi/2$ or $3\pi/2$. But this is the case iff $k/n\neq p/q$ for $q=1,2,3,4$ or any integer p.

This completes the proof that Θ^0 is stable, and the result follows from the continuity of Θ^μ in μ. ■

Concerning this theorem, it is possible to prove the existence of the closed orbits of the second kind (ii) without the "order of resonance" assumption $k/n\neq p/q$ (see Moser [1953] or Barrar [1965b]) by using the Birkhoff fixed point theorem (Siegel and Moser [1971, ch. 22]) in place of the Moser twist theorem. For stability the assumption $k/n\neq p/q$ with $q=1,2,3$ is known to be necessary ("order of resonance $\leqslant 3$") but Moser [1963a] states that the assumption $k/n\neq p/q$ with $q=4$ can be removed.

Finally, it is clear that the proof actually demonstrates a most important fact: the twist configuration (and $o^{\pm}$-stability) of the circular γ_0 is *preserved under any Hamiltonian perturbation of* Ω^0 *that is symmetric with respect to the* (q^2,p_1) *plane.*

It is tempting to apply the method of this proof to the closed orbits of the second kind of Arenstorf to prove their stability. However, the Poincaré section map Θ^0 in that case $(\xi_0\neq0)$ is not in (α,β)-normal form, so one is faced with a difficult problem in computing the invariants α and β. However a rather different argument proves that they are not $o^{\pm}$-stable, as follows.

10.3.8 Proposition (E. Chorosoff*). *In the cotangent Poincaré model for the restricted three-body problem, the closed orbits of the second kind given by* 10.3.6 (*the closed orbits of Arenstorf*) *are not* $o^{\pm}$*-stable.*

Proof. (See Fig. 10.3-3). From Proposition 6.3.3 and Theorem 10.3.6 it follows that it is enough to show that the closed orbit obtained for $\mu=0$ and

*Private communication.

ISBN 0-8053-0102-X

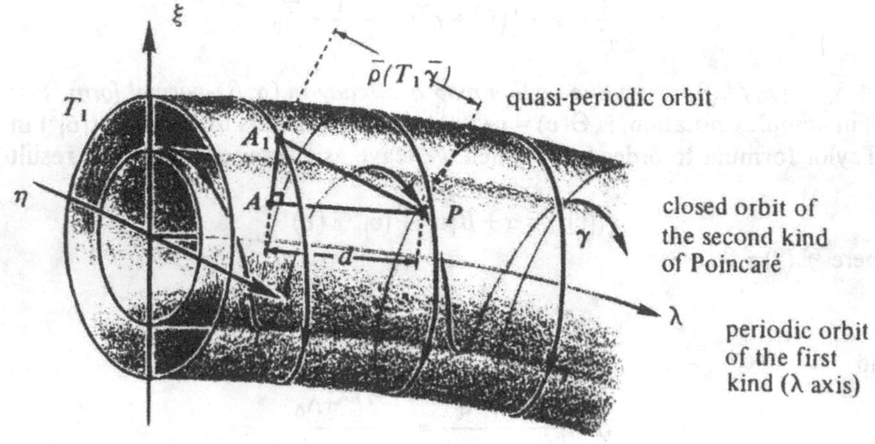

Figure 10.3-3

$\eta=0$, $\lambda=0$, $\bar{\Lambda}_0^{-3}=m/k$, $\xi_0\neq 0$ is $o^{\pm}$-unstable. As we saw in the proof of 10.3.3, this orbit γ is given by

$$t\mapsto\left(\xi_0 \sin t,\left(\bar{\Lambda}_0^{-3}+1\right)t,\xi_0 \cos t,\bar{\Lambda}_0\right)$$

and it lies on the cylinder (torus, if we identify in the λ-component 0 with 2π) in $\boldsymbol{R}^4$ given by the equations

$$\xi^2+\eta^2=\xi_0^2,\quad \bar{\Lambda}=\bar{\Lambda}_0$$

We prefer to look at this surface as a torus T_0 by regarding λ an angular variable. Let now $\bar{\Lambda}_1$ be arbitrarily close to $\bar{\Lambda}_0$ and such that $\bar{\Lambda}_1^{-3}/\pi$ is irrational. Then the orbit γ' with initial conditions $(0,0,\xi_0,\bar{\Lambda}_1)$ will lie on the torus T_1

$$\xi^2+\eta^2=\xi_0^2,\quad \bar{\Lambda}=\bar{\Lambda}_1$$

and it will be dense in T_1. If ρ denotes the Euclidean metric in $\boldsymbol{R}^4$ and $\bar{\rho}$ the corresponding Hausdorff metric, we have $\bar{\rho}(\gamma,\gamma')=\bar{\rho}(\gamma,T_1)$ by density of γ' in T_1. Let A be an arbitrary point on T_0, $A=(\eta_A,\lambda_A,\xi_A,\bar{\Lambda}_0)$ and let $d=\rho(A,\gamma)$. Denote by A_1 the point on T_1 closest to A, that is, $A_1=(\eta_A,\lambda_A,\xi_A,\bar{\Lambda}_1)$. Let $P\in\gamma$ be the point for which $\rho(A_1,\gamma)=\rho(A_1,P)$; $P=(\eta_P,\lambda_P,\xi_P,\bar{\Lambda}_0)$. It is clear then that $\rho(A_1,P)\geqslant\rho(A,P)$. Thus we can write

$$\begin{aligned}\rho(\gamma',\gamma)&=\bar{\rho}(T_1,\gamma)\geqslant\rho(A_1,\gamma)\\&=\rho(A_1,P)\geqslant\rho(A,P)\geqslant\rho(A,\gamma)=d\end{aligned}$$

The conclusion is that even though $\rho(\gamma(0),\gamma'(0))=|\bar{\Lambda}_0-\bar{\Lambda}_1|$ is arbitrarily small, $\bar{\rho}(\gamma,\gamma')\geqslant d$, that is, γ is $o^{\pm}$-unstable. ■

ISBN 0-8053-0102-X

EXERCISES

10.3A. Write down the equations of motion in model 3BIIH and show that they are unchanged when q^2, p_1, t are replaced by $-q^2, -p_1, -t$.

10.3B. Prove 10.3.2. [*Hint:* Use the formulas that give $(\eta, \lambda, \xi, \overline{\Lambda})$ as functions of $(g, l, \overline{G}, \overline{L})$ and the fact that $g=\beta$, $l=\alpha_0-\beta$, $\alpha_0=n\pi$.]

10.3C. Prove the last assertion of 10.3.3.

10.3D. Prove 10.3.4. [*Hint:* (i) Show first the openness assertions; you can use Lang [1972, p. 86]; (ii) Use the fact that $\mathfrak{D}^{\#}$ is open and γ is compact and inspire yourself by the proof of the existence and uniqueness theorem for Poincaré maps in Chapter 7.]

10.3E. In 10.3.5, show that in fact the perturbed orbit γ_μ is in a one-parameter family of closed orbits by considering the mapping $G\colon (\mu, t, \Lambda) \mapsto \lambda$-component of $F^\mu(0,0,0,\overline{\Lambda};\, t)$ as in the proof of 10.3.5.

10.3F. Prove 10.3.6. [*Hint:* Consider the mapping $G\colon (\mu, t, \overline{\Lambda}) \mapsto (\eta, \lambda)$, the first two components of $F^\mu(0,0,\xi_0,\overline{\Lambda};\, t)$.]

10.3G. Relate the Poincaré and Arenstorf orbits (10.3.5 and 10.3.6) to the inertial frame, that is, model 3BIH. Show that for $\mu=0$ the period is $2n\pi$, where $n=m+k$ is the order of γ_0 in the Poincaré model. (See Fig. 10.3-1.)

10.3H. Complete the proof of 10.3.7 by computing the Taylor formula to order four of $\nu(\xi^2+\eta^2)$. (*Hint:* See Barrar [1965b, p. 368]; watch out for the difference in sign conventions.)

10.3I. Prove the second to last statement of the section, that is, the statement in italics before 10.3.8. See Exercise 10.3B.

10.3J. Fill in the details about the orbital instability of Arenstorf's closed orbits as indicated at the beginning of the proof of 10.3.8.

10.4 TOPOLOGY OF THE PLANAR N-BODY PROBLEM*

Consider n particles of masses $m_1, \ldots, m_n$ moving in the plane $\boldsymbol{R}^2$ subject to Newton's gravitational law. If we remove collisions from the model (as we did earlier), we obtain an incomplete but smooth Hamiltonian vector field.†

The configuration space is the $2n$-dimensional manifold

$$M' = \left\{ x \in (\boldsymbol{R}^2)^n \,|\, x \notin \Delta \right\}$$

where

$$\Delta = \cup \left\{ \Delta_{ij} \,|\, 1 < i < j < n \right\}$$

and

$$\Delta_{ij} = \left\{ x \in (\boldsymbol{R}^2)^n \,|\, x_i = x_j \right\}$$

*Most of the results in this section are due to Smale [1970a, 1971b] with improvements due to Iacob [1973].

†As mentioned in Sect. 9.8, two-body collisions can be regularized and are relatively harmless. Using invariant manifolds and blowing up arguments, McGehee [1974, 1975] analyzes the behavior of the flow near triple collisions, extending the classic work of Sundman [1913] and Siegel [1941]; such collisions cannot be regularized and for the three-body problem incompleteness occurs only via collisions. For four bodies, this is not true, as shown by Mather and McGehee [1975]. The result uses the instability of near triple collisions to impart infinite velocity to a "shuttle craft" in finite time. Some complementary results are given in Saari [1972].

ISBN 0-8053-0102-X

Clearly $M'\subset(\boldsymbol{R}^2)^n$ is open. The velocity space is $TM'=M'\times(\boldsymbol{R}^2)^n$ with the structure of a trivial vector bundle. Define a Riemannian metric on M' by

$$g_x(u,v)=\langle u,v\rangle_x=\sum_{i=1}^{n} m_i u_i\cdot v_i$$

where $x\in M'$, $u,v\in T_xM'$, $u=(u_1,\ldots,u_n)$, $v=(v_1,\ldots,v_n)$, and $u_i\cdot v_i$ is the usual dot product in $\boldsymbol{R}^2$. The *kinetic energy* of the metric is

$$K(v)=\tfrac{1}{2}\sum_{i=1}^{n} m_i\|v_i\|^2$$

where $\|\cdot\|$ denotes the norm in $\boldsymbol{R}^2$. Define the *potential energy* of the problem by

$$V(x)=-\sum_{1\leqslant i<j\leqslant n}\frac{m_i m_j}{\|x_i-x_j\|}$$

and notice that $V\in\mathfrak{F}(M')$, Δ being the set of singularities of V.

Thus we have a Lagrangian system with Lagrangian L and energy function $E\in\mathfrak{F}(TM')$ given by $L=K-V\circ\tau_{M'}$, $E=K+V\circ\tau_{M'}$. Lagrange's equations for this system simply become

$$m_i\ddot{x}_i=-grad_i V(x),\qquad i=1,\ldots,n$$

where $grad_i$ is taken with respect to the usual Euclidean metric on $\boldsymbol{R}^2$ in the ith factor of $(\boldsymbol{R}^2)^n$. These represent the well-known Newton equations for the planar n-body problem.

We can pass to the Hamiltonian formulation. The Lagrangian $L=K-V\circ\tau_{M'}$ is hyperregular and the Legendre transformation is

$$FL: TM'\to T^*M',\quad FL|T_xM'=g_x^\flat$$

$$FL(v_1,\ldots,v_n)=(m_1v_1,\ldots,m_nv_n)\in T_x^*M'$$

where $x\in M'$, $(v_1,\ldots,v_n)=v\in T_xM'$. Thus the *kinetic energy in cotangent formulation*, which will be denoted for simplicity also by K, is

$$K(x,\alpha)=\tfrac{1}{2}\sum_{i=1}^{n}\frac{1}{m_i}\|\alpha_i\|^2$$

where $(x,\alpha)\in M'\times(\boldsymbol{R}^2)^n\simeq T^*M'$ and $\|\cdot\|$ denotes the usual Euclidean norm on $\boldsymbol{R}^2$. (Actually, the kinetic energy in cotangent formulation is $K\circ g^\sharp$.) The Hamiltonian of this mechanical system is $H\in\mathfrak{F}(T^*M')$

$$H=K+V\circ\tau_{M'}^*$$

$$H(x,\alpha)=\tfrac{1}{2}\sum_{i=1}^{n}\frac{1}{m_i}\|\alpha_i\|^2-\sum_{1\leqslant i<j\leqslant n}\frac{m_i m_j}{\|x_i-x_j\|}$$

ISBN 0-8053-0102-X

10.4.1 Definition. *The map* $C: (\mathbf{R}^2)^n \to \mathbf{R}^2$ *given by*

$$(x_1,\ldots,x_n) \mapsto \frac{\sum_{i=1}^{n} m_i x_i}{\sum_{i=1}^{n} m_i}$$

is called the ***center of mass*** *of the system.*

In order to simplify the problem, we fix the center of mass of the system at the origin. In other words, we want to consider the linear manifold

$$M = \left\{ x \in M' \,\middle|\, \sum_{i=1}^{n} m_i x_i = 0 \right\}$$

Since C has maximal rank, M has dimension $2n-2$. The tangent and cotangent bundle of M can be identified with

$$TM = \left\{ (x,v) \in TM' \,\middle|\, \sum_{i=1}^{n} m_i v_i = 0,\ \sum_{i=1}^{n} m_i x_i = 0 \right\}$$

$$T^*M = \left\{ (x,\alpha) \in T^*M' \,\middle|\, \sum_{i=1}^{n} \alpha_i = 0,\ \sum_{i=1}^{n} m_i x_i = 0 \right\}$$

10.4.2 Proposition. *T^*M is an invariant submanifold of T^*M' for the flow of the Hamiltonian vector field $X_H \in \mathfrak{X}(T^*M')$. The induced flow on T^*M is Hamiltonian with Hamiltonian $H|T^*M$.*

Proof. (The solution to Exercise 4.3C; Robinson [1975c]). We shall prove that T^*M is obtained from T^*M' using reduction by a Lie group leaving H invariant, and that $H|T^*M$ is the reduced Hamiltonian. This will then prove our claim according to 4.3.5.

We let the additive group $\mathbf{R}^2$ act on M' by translation, $\Phi: \mathbf{R}^2 \times M' \to M'$:

$$\Phi(g,x) = (g+x_1,\ldots,g+x_n)$$

The induced action on T^*M' is

$$\Phi^{T^*}: \mathbf{R}^2 \times T^*M' \to T^*M'$$

$$\Phi^{T^*}(g,(x,\alpha)) = ((g+x_1,\ldots,g+x_n),\alpha)$$

and clearly H is invariant under this action. The momentum mapping $J: T^*M' \to (\mathbf{R}^2)^*$ is given by (see 4.2.11) $J(x,\alpha)\cdot\xi = \alpha(\xi_{M'}(x))$. Since $exp: \mathbf{R}^2 \to$

ISBN 0-8053-0102-X

$\boldsymbol{R}^2$ is the identity,

$$\xi_{M'}(x)=\frac{d}{dt}\Phi_{\exp t\xi}(x)\bigg|_{t=0}=(\xi,\ldots,\xi)\in(\boldsymbol{R}^2)^n$$

so that

$$J(x,\alpha)\cdot\xi=\sum_{i=1}^{n}\alpha_i\cdot\xi_i$$

that is,

$$J(x,\alpha)=\sum_{i=1}^{n}\alpha_i.$$

J is a constant of the motion for X_H and clearly J is a submersion; $\mathbf{0}$ being a regular value for J,

$$J^{-1}(0)=\left\{(x,\alpha)\in T^*M'\,\middle|\,\sum_{i=1}^{n}\alpha_i=0\right\}$$

is a submanifold of T^*M', invariant under the flow of X_H.

The isotropy group of $\boldsymbol{R}^2$ at $\mathbf{0}\in\boldsymbol{R}^2$ under the coadjoint action is $\boldsymbol{R}^2$ and $\boldsymbol{R}^2$ acts properly and freely on $J^{-1}(0)$ so that $J^{-1}(0)/\boldsymbol{R}^2$ is a manifold with a unique symplectic form ω^0 satisfying $\pi_0^*\omega^0=i_0^*\omega_0$, where ω_0 is the canonical symplectic form on T^*M', $\pi_0\colon J^{-1}(0)\to J^{-1}(0)/\boldsymbol{R}^2$ is the canonical projection, and $i_0\colon J^{-1}(0)\hookrightarrow T'M'$ the inclusion. We shall define a diffeomorphism $\hat{f}\colon J^{-1}(0)/\boldsymbol{R}^2\to T^*M$ in the following way. Note that for each $(x,\alpha)\in J^{-1}(0)$, there is a unique $g\in\boldsymbol{R}^2$, namely,

$$g=-\left(\sum_{i=1}^{n}m_ix_i\right)\Big/\left(\sum_{i=1}^{n}m_i\right)\in\boldsymbol{R}^2$$

such that

$$\Phi^{T^*}(g;(x,\alpha))=((x_1+g,\ldots,x_n+g),\alpha)\in T^*M$$

Thus each orbit $\boldsymbol{R}^2\cdot(x,\alpha)$ contains a unique point

$$(x_0,\alpha)=\left(\left(\sum_{i=1}^{n}m_i(x_1-x_i)\Big/\sum_{i=1}^{n}m_i,\ldots,\sum_{i=1}^{n}m_i(x_n-x_i)\Big/\sum_{i=1}^{n}m_i\right),\alpha\right)\in T^*M$$

Define $f\colon J^{-1}(0)\to T^*M$ by $f(x,\alpha)=(x_0,\alpha)$ and notice that f so defined is smooth and invariant under the action Φ^{T^*}, that is, $f\circ\Phi_g^{T^*}=f$ for all $g\in\boldsymbol{R}^2$. It

ISBN 0-8053-0102-X

is easy to see that f is an open mapping and that f is a surjective submersion. Hence f defines a smooth map $\hat{f}: J^{-1}(0)/R^2 \to T^*M$ by $\hat{f} \circ \pi_0 = f$. It is easily verified that $\hat{f}$ is a diffeomorphism. In fact, $\hat{f}$ is symplectic, that is, $\hat{f}^* i^* \omega_0 = \omega^0$, where $i: T^*M \to T^*M'$ is the inclusion. Since π_0 is a surjective submersion, this is satisfied iff $\pi_0^* \hat{f}^* i^* \omega_0 = \pi_0^* \omega^0 = i_0^* \omega_0$, that is, iff $f^* i^* \omega_0 = i_0^* \omega_0$. This relation is proved with a short calculation, which we leave to the reader, using the formula

$$f(x,\alpha) = \left(\left(\sum_{i=1}^{n} m_i (x_1 - x_i) / \sum_{i=1}^{n} m_i, \ldots, \sum_{i=1}^{n} m_i (x_n - x_i) / \sum_{i=1}^{n} m_i \right), \alpha \right)$$

The conclusion is that $i^* \omega_0$ is the correct symplectic form on

$$T^*M = \left\{ (x,\alpha) \in T^*M' \,\middle|\, \sum_{i=1}^{n} m_i x_i = 0,\ \sum_{i=1}^{n} \alpha_i = 0 \right\}$$

The reduced Hamiltonian on $J^{-1}(0)/R^2$ satisfies $H_0 \circ \pi_0 = H \circ i_0$ (see 4.3.5). Let $[x,\alpha] = \pi_0(x,\alpha)$. We have

$$(H|T^*M \circ \hat{f})[x,\alpha] = (H|T^*M \circ \hat{f} \circ \pi_0)(x,\alpha)$$

$$= (H|T^*M \circ f)(x,\alpha) = H(x_0,\alpha)$$

$$(H \circ i_0)(x_0,\alpha) = (H_0 \circ \pi_0)(x_0,\alpha)$$

$$= H[x_0,\alpha] = H[x,\alpha]$$

since, as we saw before, $[x,\alpha] = [x_0,\alpha]$. Thus the image of H_0 by $\hat{f}^{-1}$ is exactly $H|T^*M$.

Thus we showed that T^*M with the induced symplectic form $i^*\omega_0$ from T^*M' is the reduced symplectic manifold and $H|T^*M$ the reduced Hamiltonian, so that by 4.3.5 $X_H|T^*M = X_{H|T^*M}$ is a Hamiltonian vector field. ■

From this argument we see that no topological information is lost in the reduction, so we shall define our mechanical system with symmetry starting on the reduced manifold. The mechanical system with symmetry in question will be (M,K,V,G), where:

$$M = \left\{ x \in M' \,\middle|\, \sum_{i=1}^{n} m_i x_i = 0 \right\}$$

with the Riemannian metric

$$g_x(u,v) = \sum_{i=1}^{n} m_i u_i \cdot v_i$$

ISBN 0-8053-0102-X

$u=(u_1,\ldots,u_n)$, $v=(v_1,\ldots,v_n)\in T_xM'$; $K\in\mathfrak{F}(T^*M)$ is the kinetic energy in cotangent formulation given by the metric g, that is,

$$K(x,\alpha)=\tfrac{1}{2}\sum_{i=1}^{n}\frac{1}{m_i}\|\alpha_i\|^2$$

where $\alpha=(\alpha_1,\ldots,\alpha_n)\in T_x^*M$; T^*M is identified with

$$\left\{(x,\alpha)\in T^*M' \;\middle|\; \sum_{i=1}^{n} m_i x_i=0,\ \sum_{i=1}^{n}\alpha_i=0\right\}$$

$\|\ \|$ denotes the Euclidean norm in R^2; V is the potential energy given by

$$V(x)=-\sum_{1<i<j<n}\frac{m_i m_j}{\|x_i-x_j\|}$$

$G=S^1=SO(2)$ acts on M' in the following way:

$$\Phi\colon S^1\times M'\to M'$$
$$\Phi(R_\theta,x)=(R_\theta x_1,\ldots,R_\theta x_n)$$

where $R_\theta x_k$ is the rotation by the angle θ of the vector $x_k\in R^2$, that is, if $(x_k^1,x_k^2)=x_k$,

$$R_\theta x_k=\left(x_k^1\cos\theta-x_k^2\sin\theta,\ x_k^1\sin\theta+x_k^2\cos\theta\right)$$

It is clear that M is invariant by Φ. Also $V\circ\Phi_{R_\theta}=V$ for any θ, as is easily seen. The induced action on the cotangent bundle is given by

$$\Phi^{T^*}\colon S^1\times T^*M'\to T^*M'$$
$$\Phi^{T^*}(R_\theta,(x,\alpha))=((R_\theta x_1,\ldots,R_\theta x_n),(\alpha_1\circ R_{-\theta},\ldots,\alpha_n\circ R_{-\theta}))$$

S^1 clearly acts on M', hence on M, by isometries.

The Hamiltonian $H\in\mathfrak{F}(T^*M)$ is given by

$$H=K+V\circ\tau_M^*$$

it is invariant under the action Φ of S^1 on M.

The momentum mapping $J\colon T^*M\to R$ is the usual angular momentum given by (see 4.2.11)

$$J(x,\alpha)\cdot\xi=\alpha(\xi_M(x))=(\alpha_1\cdot Ax_1+\cdots+\alpha_n\cdot Ax_n)\xi$$

ISBN 0-8053-0102-X

where

$$x=(x_1,\ldots,x_n)\in M\subset(R^2)^n$$

$$\alpha=(\alpha_1,\ldots,\alpha_n)\in(R^2)^n$$

$$A=\begin{pmatrix}0 & -1\\ 1 & 0\end{pmatrix},\quad \xi\in R$$

and "·" denoting the dot product in R^2. (Here and in what follows we identify R and R^*.)

Now we discuss the topology of the invariant and reduced invariant manifolds $I_{h,\mu}$ and $\hat{I}_{h,\mu}$ of the energy-momentum mapping $H\times J\colon T^*M\to R\times R$. We shall attempt here to fulfill the first two points of the topological program described in Sect. 4.5.

First, we want to determine the set $\Lambda\subset M$, where J_x fails to be surjective, that is, where the isotropy group S_x^1 of S^1 under the action Φ has nonzero dimension. But it is clear from the formula defining Φ that the action is free, so that for any $x\in M$, $S_x^1=\{1\}$, that is, $\Lambda=\varnothing$.

Next we compute the effective potential (in order to be able to determine the set $\Sigma'_{H\times J}$). Recall from Proposition 4.5.5 the formula for $\alpha_\mu\in\Omega^1(M)$:

$$\alpha_\mu(x)=g_x^b\circ\Xi_x\circ(J_x\circ g_x^b\circ\Xi_x)^{-1}(\mu)$$

where $J_x=J|T_x^*M$ and $\Xi_x\colon R\to T_xM$ is given by $\Xi_x(\xi)=\xi_M(x)$. We have $\xi_M(x)=(Ax_1,\ldots,Ax_n)\xi$, so

$$\begin{aligned}(J_x\circ g_x^b\circ\Xi_x)(\xi)&=\xi(J_x\circ g_x^b)(Ax_1,\ldots,Ax_n)\\&=\xi J_x(m_1Ax_1,\ldots,m_nAx_n)\\&=(m_1Ax_1\cdot Ax_1+\cdots+m_nAx_n\cdot Ax_n)\xi\\&=\left(\sum_{i=1}^n m_i\|x_i\|^2\right)\xi\end{aligned}$$

Thus $J_x\circ g_x^b\circ\Xi_x\colon R\to R$ is given by

$$J_x\circ g_x^b\circ\Xi_x=\sum_{i=1}^n m_i\|x_i\|^2$$

so that

$$(J_x\circ g_x^b\circ\Xi_x)^{-1}(\mu)=\frac{\mu}{\sum_{i=1}^n m_i\|x_i\|^2}$$

ISBN 0-8053-0102-X

and hence

$$\alpha_\mu(x)=\left(g_x^\flat\circ\Xi_x\right)\left(\frac{\mu}{\sum_{i=1}^n m_i\|x_i\|^2}\right)$$

$$=\frac{\mu}{\sum_{i=1}^n m_i\|x_i\|^2}\,g_x^\flat\,(Ax_1,\ldots,Ax_n)$$

$$=\frac{\mu}{\sum_{i=1}^n m_i\|x_i\|^2}(m_1Ax_1,\ldots,m_nAx_n)$$

Thus, the effective potential is

$$V_\mu(x)=(H\circ\alpha_\mu)(x)=K\left(\alpha_\mu(x)\right)+V(x)$$

$$=\tfrac{1}{2}\sum_{i=1}^n\frac{1}{m_i}\left(\frac{\mu}{\sum_{j=1}^n m_j\|x_j\|^2}\right)^2 m_i^2\|Ax_i\|^2+V(x)$$

$$=V(x)+\frac{\mu^2}{2\sum_{i=1}^n m_i\|x_i\|^2}$$

Since $\Lambda=\varnothing$, any $\mu\in\mathbf{R}$ is a regular value of J and hence by 4.3.8 the set of critical points $\sigma(H\times J)$ of $H\times J$ is exactly the set of relative equilibria of the mechanical system. Using the formula $\sigma(H|J^{-1}(\mu))=\alpha_\mu(\sigma(V_\mu))$ from Sect. 4.5, the set of relative equilibria is given by

$$\sigma(H\times J)=\bigcup_{\mu\in\mathbf{R}}\alpha_\mu\left(\sigma(V_\mu)\right)$$

Thus the specification of $\sigma(V_\mu)$ determines the set of relative equilibria.

In what follows, we give a somewhat more concrete physical interpretation of the relative equilibria.

10.4.3 Definition. *$x=(x_1,\ldots,x_n)\in M$ is called a **central configuration** if the force acting on x_i computed at x is proportional to m_ix_i for $1\leqslant i\leqslant n$, that is, if there exists $\lambda(x)\in\mathbf{R}$ such that*

$$grad_i\,V(x)=\lambda m_ix_i,\qquad 1\leqslant i\leqslant n$$

ISBN 0-8053-0102-X

where

$$grad_i V(x) = \left(\frac{\partial V(x)}{\partial x_i^1}, \frac{\partial V(x)}{\partial x_i^2} \right), \qquad x_i = (x_i^1, x_i^2)$$

The first thing to notice is that $\lambda(x)$ is uniquely determined by x. Indeed

$$\sum_{i=1}^{n} x_i \cdot grad_i V(x) = \lambda \sum_{i=1}^{n} m_i \|x_i\|^2$$

letting $Q(x) = \sum_{i=1}^n m_i \|x_i\|^2$, we obtain $\lambda(x) = -V(x)/Q(x)$ (since V is homogeneous of degree -1, $\sum_{i=1}^n x_i \cdot grad_i V(x) = -V(x)$). The following is taken from Iacob [1973].

10.4.4 Theorem. *(i) $x_0 \in M$ is a central configuration if and only if x_0 is a critical point of the map $x \mapsto Q(x) V^2(x)$, that is, $x_0 \in \sigma(QV^2)$.*

(ii) $x_0 \in M$ is a central configuration if and only if $\alpha_\mu(x_0)$ is a relative equilibrium for $\mu = \pm\sqrt{-V(x_0) Q(x_0)} \in R$.

(iii) $x_0 \in M$ is a central configuration if and only if $x_0 \in \sigma(V_\mu)$ for $\mu = \pm\sqrt{-V(x_0) Q(x_0)} \in R$.

Proof. (i) Since $V(x)$ and $Q(x)$ are never zero,

$$0 = d(QV^2)(x_0) = V^2(x_0)\, dQ(x_0) + 2V(x_0) Q(x_0)\, dV(x_0)$$

that is,

$$dV(x_0) = -\frac{V(x_0)}{2Q(x_0)} dQ(x_0)$$

or

$$grad_i V(x_0) = -\frac{V(x_0)}{2Q(x_0)} grad_i Q(x_0)$$

$$= -\frac{V(x_0)}{Q(x_0)} m_i x_i^0 = \lambda(x_0) m_i x_i^0$$

(ii) By (i), $x_0 \in M$ is a central configuration if and only if

$$dV(x_0) + \frac{V(x_0)}{2Q(x_0)} dQ(x_0) = 0$$

ISBN 0-8053-0102-X

We have

$$\begin{aligned}(K\circ g_x^b\circ\xi_M)(x)&=(K\circ g_x^b)(Ax_1,\ldots,Ax_n)\xi\\ &=K(\xi m_1Ax_1,\ldots,\xi m_nAx_n)\\ &=\xi^2\tfrac{1}{2}\sum_{i=1}^n\frac{1}{m_i}m_i^2\|Ax_i\|^2\\ &=\tfrac{1}{2}\xi^2Q(x)\end{aligned}$$

By 4.5.12(iv), $\alpha_\mu(x_0)$ is a relative equilibrium if and only if there exists $\xi\in\boldsymbol{R}$ satisfying $(J\circ g_{x_0}^b\circ\xi_M)(x_0)=\mu$ such that x_0 is a critical point of $V-K\circ g^b\circ\xi_M$, that is,

$$dV(x_0)-\tfrac{1}{2}\xi^2dQ(x_0)=0$$

The choice $\xi=\pm\sqrt{-V(x_0)/Q(x_0)}\in\boldsymbol{R}$ yields the equivalence of the two conditions, for

$$\begin{aligned}\mu&=\left(J\circ g_{x_0}^b\circ\xi_M\right)(x_0)\\ &=J\left(x_0,(\xi m_1Ax_1^0,\ldots,\xi m_nAx_n^0)\right)\\ &=\xi\sum_{i=1}^n\|Ax_i^0\|^2m_i\\ &=\xi Q(x_0)=\pm\sqrt{-V(x_0)Q(x_0)}\in\boldsymbol{R}\end{aligned}$$

(iii) By (ii), if x_0 is a central configuration, then for $\mu=\pm\sqrt{-Q(x_0)V(x_0)}$

$$dV(x_0)-\frac{\mu^2}{2Q^2(x_0)}dQ(x_0)=0$$

But $V_\mu(x)=V(x)+\mu^2/2Q(x)$ so that this condition is equivalent to $dV_\mu(x_0)=0$.

Conversely, if $x_0\in\sigma(V_\mu)$ with $\mu=\pm\sqrt{-Q(x_0)V(x_0)}$, then, since $\sigma(H\times J)=\cup_{\mu\in\boldsymbol{R}}\alpha_\mu(\sigma(V_\mu))$, $\alpha_\mu(x_0)$ is a relative equilibrium and by (ii), x_0 is a central configuration. ■

If we denote by C_n the set of central configurations in the planar n-body problem, we have

$$C_n=\sigma(V^2Q)=\sigma(V_\mu)\quad\text{for }\mu=\pm\sqrt{-V(x_0)Q(x_0)}$$

ISBN 0-8053-0102-X

10.4.5 Corollary. *$\Phi_{R_\theta}(C_n)=C_n$; if $a\in\mathbf{R}\backslash\{0\}$, $aC_n=\{ax|x\in C_n\}=C_n$ (R_θ is the rotation through an angle θ).*

Proof. Since $V\circ\Phi_{R_\theta}=V$, $Q\circ\Phi_{R_\theta}=Q$, we conclude $QV^2\circ\Phi_{R_\theta}=QV^2$ and taking the differential of this equality at x_0, since $T_{x_0}\Phi_{R_\theta}$ is an isomorphism, we conclude $x_0\in\sigma(QV^2)$ if and only if $\Phi_{R_\theta}(x_0)\in C_n$.

For the second equality, note that $V^2(ax)=(1/a^2)V(x)$, $Q(ax)=a^2Q(x)$, so that $(QV^2)(ax)=QV^2(x)$. Taking the differential of this relation at x_0, $ad(QV^2)(ax_0)=d(QV^2)(x_0)$, that is, $x_0\in\sigma(QV^2)$ if and only if $ax_0\in\sigma(QV^2)$. ■

The action of S^1 and the action of the multiplicative group $\mathbf{R}\backslash\{0\}$ commute, so $S^1\times(\mathbf{R}\backslash\{0\})$ acts on C_n. The orbit space of this action $\hat{C}_n=C_n/S^1\times(\mathbf{R}\backslash\{0\})$ is the set of equivalence classes of central configurations. Two central configurations are equivalent if and only if they are in the same $S^1\times(\mathbf{R}\backslash\{0\})$-orbit, that is, if and only if one is obtained from the other by a rotation and a homothety. We shall come back to the set $\hat{C}_n$ later when discussing Moulton's theorem.

We turn now to the determination of the sets $\sigma(V_\mu)$ and $\Sigma'_{H\times J}$. Let $S_Q^{2n-3}=\{x\in M|Q(x)=1\}$, a $(2n-3)$-dimensional manifold and define the map ϕ by

$$x\mapsto\left(\frac{x}{\sqrt{Q(x)}},\sqrt{Q(x)}\right)$$

which establishes a diffeomorphism $M\approx S_Q^{2n-3}\times(0,\infty)$. Let $V_S\colon S_Q^{2n-3}\to\mathbf{R}$ be the restriction of the potential V to $S_Q^{2n-3}\subset M$. We have

$$(V_\mu\circ\phi^{-1})(z,t)=V_\mu(zt)=\frac{1}{t}V_S(z)+\frac{\mu^2}{2Q(zt)}$$

$$=\frac{1}{t}V_S(z)+\frac{\mu^2}{2t^2}$$

since $Q(z)=1$.

10.4.6 Theorem. *(i) $\sigma(V_\mu)=\{x=\phi^{-1}(z,t)\in M\,|\,z\in\sigma(V_S)$ and $t=-(\mu^2/V_S(z))\}$;*

(ii) $\Sigma'_{V_\mu}=V_\mu(\sigma(V_\mu))=\{-(V_S^2(z)/2\mu^2)|z\in\sigma(V_S)\}$;

(iii) $\Sigma'_{H\times J}=\bigcup_{z\in\sigma(V_S)}\{(h,\mu)\in\mathbf{R}^2|2h\mu^2=-V_S^2(z)\}$.

ISBN 0-8053-0102-X

Proof. (i) $x\in\sigma(V_\mu)$ if and only if $dV_\mu(x)=0$ if and only if $d(V_\mu\circ\phi^{-1})(z,t)=0$ [where $x=\phi^{-1}(z,t)$] if and only if

$$\frac{1}{t}dV_S(z)+\left(-\frac{1}{t^2}V_S(z)-\frac{\mu^2}{t^3}\right)dt=0$$

if and only if $dV_S(z)=0$ and $t=-(\mu^2/V_S(z))$.

(ii)

$$\Sigma'_{V_\mu}=V_\mu(\sigma(V_\mu))$$

$$=\{(V_\mu\circ\phi^{-1})(z,t)\in R\,|\,z\in\sigma(V_S)\text{ and }t=-(\mu^2/V_S(z))\}$$

$$=\left\{\frac{1}{t}V_S(z)+\frac{\mu^2}{2t^2}\,\middle|\,z\in\sigma(V_S)\text{ and }t=-(\mu^2/V_S(z))\right\}$$

$$=\{-(V_S^2(z)/2\mu^2)\,|\,z\in\sigma(V_S)\}$$

(iii) From Sect. 4.5,

$$\Sigma'_{H\times J}=\{(h,\mu)\in R^2\,|\,h\in\Sigma'_{V_\mu}\}$$

$$=\{(h,\mu)\in R^2\,|\,2h\mu^2=-V_S^2(z),\,z\in\sigma(V_S)\}\quad\blacksquare$$

10.4.7 Corollary. *The set of central configurations C_n coincides with $\phi^{-1}(\sigma(V_S)\times(0,\infty))$, where $\phi\colon M\to S_Q^{2n-3}\times(0,\infty)$ is the diffeomorphism $\phi(x)=(x/\sqrt{Q(x)},\sqrt{Q(x)})$. Rephrased, $x=\phi(z,t)$ is a central configuration if and only if $z\in\sigma(V_S)$.*

Proof. By 10.4.4, $x\in M$ is a central configuration if and only if $x\in\sigma(V_\mu)$ for $\mu=\pm\sqrt{-V(x)Q(x)}$, so that by 10.4.6(i), if $(z,t)=\phi(x)$, $x\in C_n$ if and only if $z\in\sigma(V_S)$ and $t=V(x)Q(x)/V_S(z)$. But

$$V(x)Q(x)=V(tz)Q(tz)$$

$$=\frac{1}{t}V_S(z)t^2Q(z)=tV_S(z)$$

so that $x\in C_n$ if and only if $z\in\sigma(V_S)$ and $t>0$ is arbitrary. $\blacksquare$

Thus the set $\sigma(V_S)$ of critical points of $V_S\colon S_Q^{2n-3}\to R$ determines the relative equilibria, central configurations, and the critical part $\Sigma'_{H\times J}$ of the bifurcation set $\Sigma_{H\times J}$. We shall now investigate the map V_S further. Since $Q\circ\Phi_{R_\theta}=Q$ for any $R_\theta\in SO(2)$, the action leaves S_Q^{2n-3} invariant. Now

$$\Delta=\bigcup_{1\leqslant i<j\leqslant n}\Delta_{ij}$$

ISBN 0-8053-0102-X

where

$$\Delta_{ij}=\left\{x=(x_1,\ldots,x_n)\in(R^2)^n \mid x_i=x_j\right\}$$

Clearly Δ_{ij}, and hence Δ, is invariant under the action of S^1 on $(R^2)^n$ by rotations. Thus, we can conclude that S_Q^{2n-3} is diffeomorphic to the $(2n-3)$-dimensional sphere S^{2n-3} (it is actually an ellipsoid E^{2n-3}) in the $(2n-2)$-dimensional subspace $\{x\in(R^2)^n \mid \Sigma_{i=1}^n m_i x_i=0\}$ of $(R^2)^n$ with all the points of Δ removed, that is,

$$S_Q^{2n-3}=E^{2n-3}\backslash(E^{2n-3}\cap\Delta)\approx S^{2n-3}\backslash(S^{2n-3}\cap\Delta)$$

V_S is invariant as well, that is, $V_S\circ\Phi_{R_\theta}=V_S$, so that it defines a map

$$\hat{V}_S\colon S_Q^{2n-3}/S^1\to R$$

If we let $\pi_n\colon S_Q^{2n-3}\to S_Q^{2n-3}/S^1$ denote the canonical projection $\hat{\Delta}=\pi_n(E^{2n-3}\cap\Delta)$, and recalling that $E^{2n-3}/S^1\approx S^{2n-3}/S^1\approx CP^{n-2}$, complex $(n-2)$-dimensional projective space [see 4.3.4(iv)], we are led to the investigation of the set of critical points $\sigma(\hat{V}_S)$ of $\hat{V}_S\colon CP^{n-2}\backslash\hat{\Delta}\to R$. Since π_n is a surjective submersion and $\hat{V}_S\circ\pi_n=V_S$,

$$\sigma(V_S)=\pi_n^{-1}(\sigma(\hat{V}_S))$$

and thus $\sigma(V_S)$ is completely determined by $\sigma(\hat{V}_S)$. Thus by 10.4.6(iii)

$$\Sigma'_{H\times J}=\bigcup_{y\in\sigma(\hat{V}_S)}\{(h,\mu)\in R^2 \mid 2h\mu^2=-\hat{V}_S^2(y)\}$$

By 10.4.7, $\phi(C_n)=\sigma(V_S)\times(0,\infty)$. Using the diffeomorphism

$$\phi\colon M\to S_Q^{2n-3}\times(0,\infty),\qquad \phi(x)=\left(\frac{x}{\sqrt{Q(x)}},\sqrt{Q(x)}\right)$$

the action of $S^1\times(R\backslash\{0\})$ by rotation and homothety on M becomes:

$$\phi(\Phi(R_\theta,x))=\left[\frac{\Phi(R_\theta,x)}{\sqrt{Q(x)}},\sqrt{Q(x)}\right]$$

$$\phi(a\cdot x)=\left(\frac{a\cdot x}{\sqrt{Q(a\cdot x)}},\sqrt{Q(ax)}\right)$$

$$=\left(\frac{x}{\sqrt{Q(x)}},a\sqrt{Q(x)}\right)$$

ISBN 0-8053-0102-X

that is, S^1 acts on $\sigma(V_S)\times(0,\infty)$ by

$$(R_\theta,(z,t))\mapsto(\Phi(R_\theta,z),t)$$

and $R\setminus\{0\}$ acts on $\sigma(V_S)\times(0,\infty)$ by

$$(a,(z,t))\mapsto(z,at)$$

and clearly these two actions commute. $\phi\colon C_n\to\sigma(V_S)\times(0,\infty)$ becomes in this way an equivariant diffeomorphism. Thus, letting $\hat{C}_n=C_n/S^1\times(R\setminus\{0\})$, the set of equivalence classes of central configurations,

$$\hat{C}_n\overset{\hat{\phi}}{\approx}\sigma(V_S)\times(0,\infty)/S^1\times(R\setminus\{0\})=\sigma(V_S)/S^1=\sigma(\hat{V}_S)$$

since $(0,\infty)/(R\setminus\{0\})=\{\text{one point}\}$. We have proved the following result of Smale.

10.4.8 Corollary. (*i*) *For any $n\geq 2$ and any choice of the masses in the planar n-body problem, the set of equivalence classes of central configurations is diffeomorphic to the set of critical points of the map $\hat{V}_S\colon CP^{n-2}\setminus\hat{\Delta}\to R$, that is, $\hat{C}_n\approx\sigma(\hat{V}_S)$.*

(*ii*) $\Sigma'_{H\times J}=\bigcup_{y\in\hat{C}_n}\{(h,\mu)\in R^2|2h\mu^2=-\hat{V}_S^2(y)\}.$

The set $\hat{C}_n$ of equivalence classes of central configurations thus determines $\Sigma'_{H\times J}$ as well as the set $\sigma(H\times J)$ of relative equilibria.

Central configurations can be collinear and noncollinear. Collinearity means that the points $x_1,\ldots,x_n$ giving the positions of the n-bodies of masses $m_1,\ldots,m_n$ lie on the same line in the plane R^2, that is, $x_i^2=ax_i^1+b$, $i=1,\ldots,n$, with $a,b\in R$. But $\Sigma_{i=1}^n m_ix_i=0$, so that $b=0$, that is, the line on which the collinear central configurations lie passes through the origin. Since we will be ultimately interested in classes of central configurations, rotations of the plane do not matter and thus by making a rotation in R^2 we can assume that the collinear central configurations $x_1,\ldots,x_n\in R\times\{0\}$. We thus get an $(n-1)$-dimensional submanifold

$$M_{col}=\{x=(x_1,\ldots,x_n)\in M|x_i\in R\times\{0\},1\leq i\leq n\}$$

of M. Let $S_{Q\,col}^{n-2}=S_Q^{2n-3}\cap M_{col}$ and notice that $S_{Q\,col}^{n-2}$ is the part of the ellipsoid $\Sigma_{i=1}^n m_ix_i^2=1$ [of dimension $(n-2)$] lying in the $(n-1)$-dimensional subspace $\Sigma_{i=1}^n m_ix_i=0$ of R^n out of which we excluded all points $(x_1,\ldots,x_n)\in R^n$ for which $x_i=x_j$ for $i\neq j$. S^1 acts on S_Q^{2n-3} by rotations and the only rotation leaving $S_{Q\,col}^{n-2}$ invariant is R_π, the rotation through an angle π, and thus the group with two elements $Z/2Z$ acts on $S_{Q\,col}^{n-2}$. The orbit space is therefore $\hat{S}_{Q\,col}^{n-2}=RP^{n-2}$, real $(n-2)$-dimensional projective space. Regard $RP^{n-2}\subset CP^{n-2}$ and write $V_{col}=V_S|S_{Q\,col}^{n-2}$ and $\hat{V}_{col}=\hat{V}_S|RP^{n-2}\setminus(\hat{\Delta}\cap RP^{n-2})$.

ISBN 0-8053-0102-X

10.4.9 Proposition. $\{\hat{x}\in\hat{C}_n \mid x$ *collinear central configuration*$\}\approx\sigma(\hat{V}_S)\cap(\boldsymbol{RP}^{n-2}\backslash\hat{\Delta})=\sigma(\hat{V}_{col})$.

Proof. The first diffeomorphism is an immediate consequence of 10.4.8(i) and our previous considerations.

Now let $\hat{x}\in\sigma(\hat{V}_S)\cap(\boldsymbol{RP}^{n-2}\backslash\hat{\Delta})$, that is, $\hat{x}\in\boldsymbol{RP}^{n-2}\backslash\hat{\Delta}$ and $(d\hat{V}_S)(\hat{x})=0$. If $i:\boldsymbol{RP}^{n-2}\backslash\hat{\Delta}\to\boldsymbol{CP}^{n-2}\backslash\hat{\Delta}$ denotes the canonical embedding, then $\hat{V}_{col}=\hat{V}_S\circ i$ and hence $(d\hat{V}_{col})(\hat{x})=(d\hat{V}_S)(\hat{x})\circ T_{\hat{x}}i=0$ and so

$$\sigma(\hat{V}_S)\cap(\boldsymbol{RP}^{n-2}\backslash\hat{\Delta})\subset\sigma(\hat{V}_{col})$$

What remains to be shown is that a critical point of $\hat{V}_{col}$ is also a critical point of $\hat{V}_S$. Since the canonical projections $S_Q^{2n-3}\to\boldsymbol{CP}^{n-2}\backslash\hat{\Delta}, S_{Q\,col}^{n-2}\to\boldsymbol{RP}^{n-2}\backslash(\hat{\Delta}\cap\boldsymbol{RP}^{n-2})$ are surjective submersions it suffices to show that if $\boldsymbol{x}\in S_{Q\,col}^{n-2}$ is a critical point of V_{col}, then it is a critical point for V_S, too. If we denote by d_k the differential on the kth factor $\boldsymbol{R}^2$ in $(\boldsymbol{R}^2)^n$, we have for $\boldsymbol{v}=(\boldsymbol{v}_1,\ldots,\boldsymbol{v}_k)\in(\boldsymbol{R}^2)^n$,

$$dV(\boldsymbol{x})\cdot\boldsymbol{v}=\sum_{k=1}^{n}d_kV(\boldsymbol{x})\cdot\boldsymbol{v}_k$$

$$=-\sum_{1<i<j<n}m_im_j\sum_{k=1}^{n}d_k\left(\frac{1}{\|\boldsymbol{x}_i-\boldsymbol{x}_j\|}\right)\cdot\boldsymbol{v}_k$$

But

$$-\sum_{k=1}^{n}d_k\left(\frac{1}{\|\boldsymbol{x}_i-\boldsymbol{x}_j\|}\right)\cdot\boldsymbol{v}_k=\frac{1}{\|\boldsymbol{x}_i-\boldsymbol{x}_j\|^2}\sum_{k=1}^{n}d_k(\|\boldsymbol{x}_i-\boldsymbol{x}_j\|)\cdot\boldsymbol{v}_k$$

$$=\frac{1}{2\|\boldsymbol{x}_i-\boldsymbol{x}_j\|^3}\sum_{k=1}^{n}d_k(\|\boldsymbol{x}_i-\boldsymbol{x}_j\|^2)\cdot\boldsymbol{v}_k$$

$$=\frac{1}{2\|\boldsymbol{x}_i-\boldsymbol{x}_j\|^3}\sum_{k=1}^{n}2(\boldsymbol{x}_i-\boldsymbol{x}_j)\cdot d_k(\boldsymbol{x}_i-\boldsymbol{x}_j)\cdot\boldsymbol{v}_k$$

$$=\frac{(\boldsymbol{x}_i-\boldsymbol{x}_j,\boldsymbol{v}_i-\boldsymbol{v}_j)}{\|\boldsymbol{x}_i-\boldsymbol{x}_j\|^3}$$

so that

$$dV(\boldsymbol{x})\cdot\boldsymbol{v}=\sum_{1<i<j<n}\frac{m_im_j}{\|\boldsymbol{x}_i-\boldsymbol{x}_j\|^3}(\boldsymbol{x}_i-\boldsymbol{x}_j,\boldsymbol{v}_i-\boldsymbol{v}_j)$$

where (,) denotes the dot product in $\boldsymbol{R}^2$. The same formula will be true for V_S and $\boldsymbol{v}\in T_{\boldsymbol{x}}S_Q^{2n-3}$.

ISBN 0-8053-0102-X

Let

$$x=(x_1,\dots,x_n)\in S^{n-2}_{Q\,col}$$

that is,

$$x_i=(x_i^1,0)\in \mathbf{R}^2, \quad 1\leqslant i\leqslant n$$

$$\sum_{i=1}^{n} m_i(x_i^1)^2=1$$

and all the points with $x_i=x_j$ for $i\neq j$ are excluded. Let

$$w=(w_1,\dots,w_n)\in T_x\left(S^{n-2}_{Q\,col}\right)$$

that is, $\Sigma_{i=1}^n m_i w_i x_i^1=0$. Let

$$v=(v_1,\dots,v_n)\in T_x\left(S^{2n-3}_{Q}\right)$$

be a tangent vector to S_Q^{2n-3} at the *same* point $x\in S^{n-2}_{Q\,col}$; put $v_i=(v_i^1,v_i^2)$ and consider the vector

$$v^1=(v_i^1,\dots,v_i^1), \quad v\in T_x\left(S_Q^{2n-3}\right)$$

which must satisfy

$$0=\sum_{i=1}^{n} m_i x_i\cdot v_i=\sum_{i=1}^{n} m_i v_i^1 x_i^1$$

that is, $v^1\in T_x(S^{n-2}_{Q\,col})$. We have proved hence that if

$$v=(v_1,\dots,v_n)\in T_x\left(S_Q^{2n-3}\right), \quad v_i=\left(v_i^1,v_i^2\right), \quad x\in S^{n-2}_{Q\,col}$$

then

$$v^1=\left(v_i^1,\dots,v_i^1\right)\in T_x\left(S^{n-2}_{Q\,col}\right)$$

If x is a critical point of V_{col}, then necessarily $x\in S^{n-2}_{Q\,col}$, and hence by the expression for $dV_S(x)$ we see that $dV_S(x)\cdot v=dV_S(x)\cdot v^1$. Hence, if $dV_S(x)\cdot v^1=0$ for all $v^1\in T_x(S^{n-2}_{Q\,col})$, then $dV_S(x)\cdot v=0$ for all $v\in T_x S_Q^{2n-3}$ and $x\in \sigma(V_S)$. ■

The number of equivalence classes of collinear central configurations is given by the following.

ISBN 0-8053-0102-X

10.4.10 Theorem (F. R. Moulton). *For any choice of the masses in the planar n-body problem there are exactly $n!/2$ equivalence classes of collinear central configurations.*

Proof. (Smale). By 10.4.9 the number of equivalence classes of collinear central configurations is given by $\sigma(\hat{V}_{col})$. To count the points of $\sigma(\hat{V}_{col})$ in $RP^{n-2}\backslash(\hat{\Delta}\cap RP^{n-2})$ we proceed in three steps:

Step I: if $y\in\sigma(\hat{V}_{col})$, then y is necessarily a nondegenerate maximum;
Step II: Δ partitions S^{n-2}_{Qcol} into $n!$ diametrically opposed, connected components, so that $RP^{n-2}\backslash(\hat{\Delta}\cap RP^{n-2})$ has $n!/2$ connected components;
Step III: Since $\lim_{y\to\hat{\Delta}}\hat{V}_{col}(y)=-\infty$ for $y\in RP^{n-2}\backslash(\hat{\Delta}\cap RP^{n-2})$, $\hat{V}_{col}$ must have a maximum in each connected component. By Step I this is a nondegenerate maximum and hence is unique. Thus $\sigma(\hat{V}_{col})$ has as many critical points as connected components $RP^{n-2}\backslash(\hat{\Delta}\cap RP^{n-2})$ has, that is, $n!/2$.

Proof of Step I. The following computation is due to Th. Hangan (see D. Burghelea, Th. Hangan, H. Moscovici, and A. Verona [1973]). Let $u^1,u^2,\ldots,u^{n-1}$ be local coordinates on the ellipsoid S^{n-2}_{Qcol} and let $x^0=(x_1^0,\ldots,x_n^0)\in R^n$ be a critical point of V_{col}. The equation of the ellipsoid is

$$\sum_{i=1}^{n} m_i x_i^2=1$$

so that

$$\sum_{i=1}^{n} m_i x_i\frac{\partial x_i}{\partial u^\alpha}=0,\quad -\sum_{i=1}^{n} m_i x_i\frac{\partial^2 x_i}{\partial u^\alpha\,\partial u^\beta}=\sum_{i=1}^{n} m_i\frac{\partial x_i}{\partial u^\alpha}\frac{\partial x_i}{\partial u^\beta} \tag{1}$$

relations that will be useful later. As we saw in 10.4.9

$$\frac{\partial V_{col}}{\partial u^\alpha}=\sum_{1\le i<j\le n}\frac{m_i m_j}{|x_i-x_j|^3}(x_i-x_j)\left(\frac{\partial x_i}{\partial u^\alpha}-\frac{\partial x_j}{\partial u^\alpha}\right)=\sum_{k=1}^{n}\frac{\partial V}{\partial x_k}\frac{\partial x_k}{\partial u^\alpha} \tag{2}$$

The second partial derivatives are

$$\begin{aligned}\frac{\partial^2 V_{col}}{\partial u^\alpha\,\partial u^\beta}&=-2\sum_{1\le i<j\le n}\frac{m_i m_j}{|x_i-x_j|^3}\left(\frac{\partial x_i}{\partial u^\alpha}-\frac{\partial x_j}{\partial u^\alpha}\right)\left(\frac{\partial x_i}{\partial u^\beta}-\frac{\partial x_j}{\partial u^\beta}\right)\\&\quad+\sum_{1\le i<j\le n}\frac{m_i m_j}{|x_i-x_j|^3}(x_i-x_j)\left(\frac{\partial^2 x_i}{\partial u^\alpha\,\partial u^\beta}-\frac{\partial^2 x_j}{\partial u^\alpha\,\partial u^\beta}\right)\\&=b_{\alpha\beta}+a_{\alpha\beta}\end{aligned} \tag{3}$$

ISBN 0-8053-0102-X

At x^0, which is, by assumption, a critical point of V_{col}, that is, a collinear central configuration, we have

$$\frac{\partial V_{col}}{\partial x_i}(x^0) = \lambda m_i x_i^0 \tag{4}$$

Since V_{col} is a homogeneous function of degree -1,

$$-V_{col}(x^0) = \sum_{i=1}^{n} x_i^0 \frac{\partial V_{col}}{\partial x_i^0} = \lambda \sum_{i=1}^{n} m_i (x_i^0)^2 = \lambda > 0 \tag{5}$$

Thus, the second term of $\partial^2 V/\partial u^\alpha \partial u^\beta$ evaluated at x^0 becomes

$$a_{\alpha\beta} = \sum_{k=1}^{n} \frac{\partial V_{col}(x^0)}{\partial x_k} \frac{\partial^2 x_k}{\partial u^\alpha \partial u^\beta}(u^0)$$

$$= -V_{col}(x^0) \sum_{k=1}^{n} m_k x_k^0 \frac{\partial^2 x_k}{\partial u^\alpha \partial u^\beta}(u^0) \quad [\text{by (4) and (5)}]$$

$$= V_{col}(x^0) \sum_{k=1}^{n} m_k \frac{\partial x_k}{\partial u^\alpha}(u^0) \frac{\partial x_k}{\partial u^\beta}(u^0) \quad [\text{by (1)}]$$

If

$$v = (v_1, \ldots, v_n) = \sum_{\alpha=1}^{n-1} v'_\alpha \frac{\partial}{\partial u^\alpha}$$

$$w = (w_1, \ldots, w_n) = \sum_{\alpha=1}^{n-1} w'_\alpha \frac{\partial}{\partial u^\alpha}$$

denote two tangent vectors to the ellipsoid $S^{n-2}_{Q\,col}$ at x^0, then

$$v_i = \frac{\partial x_i}{\partial u^\alpha} v'_\alpha, \quad w_i = \frac{\partial x_i}{\partial u^\alpha} w'_\alpha \quad \text{(summation convention)}$$

and hence

$$D^2 V_{col}(x^0)(v, w) = a_{\alpha\beta} v'_\alpha w'_\beta + b_{\alpha\beta} v'_\alpha w'_\beta$$

$$= V_{col}(x_0) \sum_{k=1}^{n} m_k v_k w_k - 2 \sum_{1 \le i < j \le n} \frac{m_i m_j}{|x_i - x_j|^3} (v_i - v_j)(w_i - w_j)$$

so that

$$D^2 V_{col}(x^0)(v, v) = V_{col}(x^0) \sum_{k=1}^{n} m_k v_k^2 - 2 \sum_{1 \le i < j \le n} \frac{m_i m_j}{|x_i - x_j|^3} (v_i - v_j)^2$$

ISBN 0-8053-0102-X

which is clearly negative-definite (in particular nondegenerate). Thus x^0 is a nondegenerate maximum for V_{col}. The canonical projection $\pi: S^{n-2}_{Qcol} \to RP^{n-2} \backslash (\hat{\Delta} \cap RP^{n-2})$ being a surjective submersion, $\pi(x^0)$ will be a nondegenerate maximum for $\hat{V}_{col}$.

Proof of Step II. Associate to $x=(x_1,\ldots,x_n) \in S^{n-2}_{Qcol}$ the system of numbers $(i_1,\ldots,i_n)$, $i_j \in \{1,\ldots,n\}$ for all $1 \leq j \leq n$ such that $i_j < i_k$ if and only if $x_{i_j} < x_{i_k}$. Thus each connected component of S^{n-2}_{Qcol} corresponds to a certain ordering $(i_1,\ldots,i_n)$ of $\{1,\ldots,n\}$, and conversely. We conclude that S^{n-2}_{Qcol} has $n!$ components. Each component has a diametrically opposed one given by the opposite ordering (on S^{n-2}_{Qcol}, take $\{-x \mid x \in \text{component}\}$ to obtain the opposite one). ■

Unfortunately, there is no theorem like Moulton's for noncollinear central configurations. However, regarding the finiteness of $\hat{C}_n$, the following result of Smale holds.

10.4.11 Theorem. *If for a given choice of the masses $m_1,\ldots,m_n$ in the planar n-body problem $\hat{V}_S: CP^{n-2} \backslash \hat{\Delta} \to R$ has all its critical points nondegenerate, then $\hat{C}_n$ is finite.*

Proof. (Shub [1971]). If we show that V_S has no critical points in an (open) neighborhood N of Δ, then $\hat{V}_S$ will not have critical points in a neighborhood $\hat{N}$ (the projection of N) of $\hat{\Delta}$. Now $CP^{n-2} \backslash \hat{N}$ is closed and so is compact in CP^{n-2}. Moreover, by hypothesis, $\hat{V}_S | CP^{n-2} \backslash \hat{N}$ has all its critical points nondegenerate and hence isolated, so their number is finite.

Recall that the derivative of V is given by

$$dV(x)\cdot v = \sum_{1 \leq i < j \leq n} \frac{m_i m_j}{\|x_i - x_j\|^3} (x_i - x_j, v_i - v_j)$$

Let

$$x = (x_1,\ldots,x_{k_1}, x_{k_1+1},\ldots,x_{k_2},\ldots,x_{k_l},\ldots,x_n) \in \Delta$$

where we make the convention

$$x_1 = \cdots = x_{k_1}, x_{k_1+1} = \cdots = x_{k_2},\ldots,x_{k_l} = \cdots = x_n$$

that is, we regroup the equal R^2-components of x in consecutive groups. We may assume, by definition of Δ, that $k_1 \geq 2$ and $x_1^1 \neq x_n^1$, where $x_i = (x_i^1, x_i^2)$ for all $i = 1,\ldots,n$. Any point in a neighborhood of x can be written as $x + \delta$, where $\delta = (\delta_1,\ldots,\delta_n) \in (R^2)^n$. We shall prove that there is a vector $v(\delta) = (v_1(\delta),\ldots,v_n(\delta))$ defined in this neighborhood so that:

ISBN 0-8053-0102-X

(a) $\|v(\delta)\|$ is bounded;

(b) $\sum_{1 \leq i<j \leq n} \frac{m_i m_j}{\|x_i+\delta_i-(x_j+\delta_j)\|^3}(x_i+\delta_i-(x_j+\delta_j), v_i(\delta)-v_j(\delta)) \to +\infty$ as $\delta_i - \delta_j \to 0$ for $i,j \leq k_1$ and $i \neq \delta$; and

(c) $v(\delta)$ is in the tangent space to the ellipsoid

$$E^{2n-3} = \left\{ x \in (R^2)^n \,\middle|\, \sum_{i=1}^{n} m_i \|x_i\|^2 = 1 \text{ and } \sum_{i=1}^{n} m_i x_i = 0 \right\}$$

at $x+\delta \in E^{2n-3}$.

These conditions clearly imply that for each $x \in \Delta \cap E^{2n-3}$ there exists δ such that $dV_S(x) \neq 0$ [since we can find a $v(\delta)$ on which $dV_S(x) \cdot v$ is big]. Since $\Delta \cap E^{2n-3}$ is compact, the usual finite cover argument yields the desired neighborhood N of Δ in E^{2n-3}.

We shall prove (a), (b), and (c) by solving the following system of linear equations for v continuously in δ, for δ bounded.

(1) $\sum_{i=1}^{n} (m_i(x_i+\delta_i), v_i) = 0$

(2) $\sum_{i=1}^{n} m_i v_i = 0$

(3) $v_1 - v_j = \delta_1 - \delta_j$ for $2 \leq j \leq k_1$

(4) $v_{k_1+1} = \cdots = v_{k_l-1} = 0$

(5) $v_{k_l} = \cdots = v_n$

If this is done, continuity of $v(\delta)$ in δ yields that $\|v(\delta)\|$ is bounded, that is, (a). Equations (1) and (2) prove (c). Regarding (b) we have:

$$\sum_{1 \leq i<j \leq n} \frac{m_i m_j}{\|x_i+\delta_i-(x_j+\delta_j)\|^3}(x_i+\delta_i-(x_j+\delta_j), v_i - v_j)$$

$$= \sum_{1 \leq i<j \leq k_1} \frac{m_i m_j}{\|\delta_i - \delta_j\|^3}(\delta_i - \delta_j, v_i - v_j) \quad [\text{by (4) and (5)}]$$

$$= \sum_{1 \leq i<j \leq k_1} \frac{m_i m_j}{\|\delta_i - \delta_j\|^3}(\delta_i - \delta_j, (v_1 - v_j) - (v_1 - v_i))$$

$$= \sum_{1 \leq i<j \leq k_1} \frac{m_i m_j}{\|\delta_i - \delta_j\|^3}(\delta_i - \delta_j, \delta_i - \delta_j) \quad [\text{by (3)}]$$

$$= \sum_{1 \leq i<j \leq k_1} \frac{m_i m_j}{\|\delta_i - \delta_j\|}$$

This expression tends to $+\infty$ as $\delta_i - \delta_j \to 0$ and so (b) follows.

ISBN 0-8053-0102-X

We shall solve now the system (1)–(5) for the unknowns v_i. Consider the family of linear mappings

$$\sum_{i=1}^{n} m_i(x_i+\delta_i)\cdot v_i : (R^2)^n \to R$$

$$\sum_{i=1}^{n} m_i v_i : (R^2)^n \to R$$

$$v_1 - v_i : R^2 \to R^2 \quad \text{for} \quad 2 \le i \le k_1$$

restricted to the linear subspace of $(R^2)^n$ given by $v_{k_1+1} = \cdots = v_{k_l-1} = 0$ and $v_{k_l} = \cdots = v_n$. This defines then a linear map from a $2(k_1+1)$-dimensional vector space to a $(2k_1+1)$-dimensional vector space. The dependence on δ is smooth. We shall prove that at $\delta=0$ this map is surjective. It is thus surjective in a whole neighborhood of 0 in the δ-space and we can solve the system continuously in δ. To prove the desired surjectivity at $\delta=0$, restrict this map to the subspace defined by $v_n^2=0$. This linear map has determinant (the unknowns are now $v_1,\dots,v_{k_1},v_n^1$):

$$\begin{vmatrix} m_1x_1 & m_2x_2 & & m_{k_1}x_{k_1} & \overline{m}x_n^1 \\ m_1I & m_2I & & m_{k_1}I & \begin{pmatrix}\overline{m}\\0\end{pmatrix} \\ I & -I & \cdots & 0 & 0 \\ \vdots & \vdots & -I & \vdots & \vdots \\ I & 0 & \cdots & -I & 0 \end{vmatrix}$$

where $\overline{m} = \Sigma_{j=k_l}^{n} m_j$, I is the 2×2 identity matrix and the last column is an ordinary column of $2k_1+1$ numbers; in writing this determinant we used also the relations $x_1 = \cdots = x_{k_1}, \dots, x_{k_l} = \cdots = x_n$. Letting $m = \Sigma_{i=1}^{k_1} m_i$ and operating on the columns and rows of this determinant, we see that it equals

$$\begin{vmatrix} mx_1 & m_2x_2 & \cdots & m_{k_1}x_{k_1} & \overline{m}x_n^1 \\ mI & m_2I & \cdots & m_{k_1}I & \begin{pmatrix}\overline{m}\\0\end{pmatrix} \\ 0 & -I & \cdots & 0 & 0 \\ \vdots & \vdots & & \vdots & \vdots \\ 0 & 0 & \cdots & -I & 0 \end{vmatrix} = \pm \begin{vmatrix} mx_1 & \overline{m}x_n^1 \\ mI & \begin{matrix}\overline{m}\\0\end{matrix} \end{vmatrix}$$

$$= \pm m \begin{vmatrix} mx_1^1 & \overline{m}x_n^1 \\ m & \overline{m} \end{vmatrix}$$

$$= \pm m^2\overline{m}(x_1^1 - x_n^1) \neq 0$$

because by hypothesis $x_1^1 \neq x_n^1$ ■

ISBN 0-8053-0102-X

Summarizing, if, for a given choice of $m_1,\ldots,m_n$, $\hat{V}_S$ has all its critical points nondegenerate, $\hat{C}_n$ is finite and

$$\Sigma'_{H\times J}=\bigcup_{y\in\hat{C}_n}\left\{(h,\mu)\in \boldsymbol{R}^2 \mid 2h\mu^2=-\hat{V}_S^2(y)\right\}$$

is a finite collection of cubics in $\boldsymbol{R}^2$.

A system $(m_1,\ldots,m_n)$ of masses is called *critical* if $\hat{V}_S$ has at least one degenerate critical point. Call $\Sigma_n\subset(0,\infty)\times\cdots\times(0,\infty)$ (n times) the set of all such critical masses in the planar n-body problem. Smale [1976] asks, "What is the structure of Σ_n?" We shall see below that $\Sigma_3=\varnothing$. Palmore [1976b] proves that $\Sigma_n\neq\varnothing$ for all $n\geqslant 4$ and announces the result that the Lebesgue measure of Σ_n is zero.† A central configuration that gives a degenerate relative equilibria class is obtained in the following way: place $(n-1)$ unit masses at the vertices of a regular polygon with $n-1$ sides centered at the origin, and put an arbitrary mass $m_n>0$ at the origin. Combining 10.4.11 with Palmore's result, we have:

10.4.12 Theorem. *The set of $(m_1,\ldots,m_n)\in((0,\infty))^n$ for which the set $\hat{C}_n$ of relative equilibria classes in the planar n-body problem is not finite has Lebesgue measure zero.*

We turn now to the proof that $\Sigma_3=\varnothing$. In fact, we shall determine $\hat{C}_3$ completely. A first result in this respect is the following.

10.4.13 Theorem (Lagrange). *For any choice of the masses m_1, m_2, m_3 in the planar three-body problem, $x\in C_3$ is a noncollinear central configuration if and only if*

$$\|x_1-x_2\|=\|x_1-x_3\|=\|x_2-x_3\|=\left(-m\frac{Q(x)}{V(x)}\right)^{1/3}$$

where $m=m_1+m_2+m_3$. Thus the three bodies move in circles forming a fixed equilateral triangle.

Proof. By 10.4.4(i), $x\in C_3$ is equivalent to $x\in\sigma(QV^2)$, that is,

$$V(x)\,dQ(x)+2Q(x)\,dV(x)=0$$

We know that

$$dV(x)\cdot v=\sum_{1\leqslant i<j\leqslant 3}\frac{m_im_j}{\|x_i-x_j\|^3}(x_i-x_j,v_i-v_j)$$

and since

$$Q(x)=\sum_{i=1}^{3}m_i\|x_i\|^2$$

†Smale showed us his proof after the book was completed

ISBN 0-8053-0102-X

we have

$$dQ(x)\cdot v = \sum_{i=1}^{3} 2m_i(x_i, v_i)$$

Since $v \in T_x M$ iff $\Sigma_{i=1}^3 m_i v_i = 0$, a simple calculation shows that

$$dQ(x)\cdot v = \frac{2}{m} \sum_{1<i<j<3} m_i m_j (x_i - x_j, v_i - v_j)$$

Thus

$$\sum_{1<i<j<3} m_i m_j \left(\frac{1}{m} V(x) + Q(x) \|x_i - x_j\|^{-3} \right) (x_i - x_j, v_i - v_j) = 0$$

for all $v \in T_x M$. But since x_1, x_2, x_3 are not collinear, this relation is satisfied iff

$$\frac{1}{m} V(x) + Q(x) \|x_i - x_j\|^{-3} = 0$$

This is obtained making suitable choices for the vector v and using $\Sigma_{i=1}^3 m_i v_i = 0$.

For the last statement, recall that x is a central configuration iff $\alpha_\mu(x)$ is a relative equilibrium. But this means that $\Phi^{T^*}_{exp t\xi}(\alpha_\mu(x))$ is the integral curve of X_H through $\alpha_\mu(x)$. Hence the base integral curve will be $\Phi_{exp t\xi}(x)$ which is just rotation in circles. Since V, Q are invariant under Φ, the number $(-m(Q(x)/V(x)))^{1/3}$ is constant along the orbit. Thus the three bodies move in circles maintaining a distance $(-mQ(x)/V(x))^{1/3}$ between any two of them. (See Fig. 10.4-1.) ■

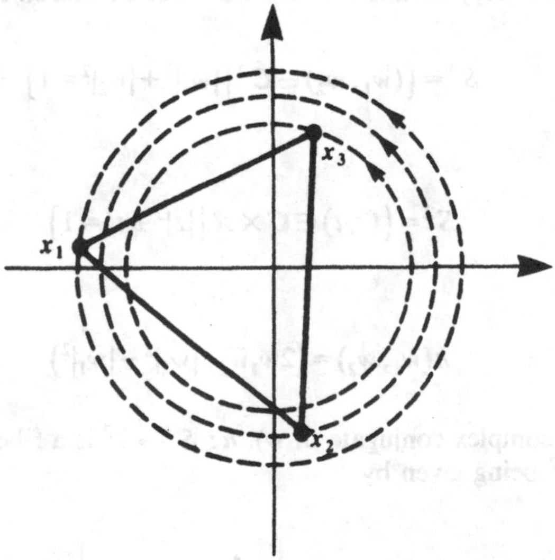

Figure 10.4-1

ISBN 0-8053-0102-X

If $\pi_3: S^3_Q \to CP^1 \backslash \hat{\Delta}$ denotes the canonical projection, then the set of equivalence classes of noncollinear central configurations is the set $\sigma(\hat{V}_S) \cap (CP^1 \backslash (\hat{\Delta} \cup RP^1))$, that is, the set of critical points of $\hat{V}_S$ that are *not* in RP^1 (see 10.4.9). Thus

$$\{y \in \hat{C}_3 \mid y \text{ noncollinear central configuration}\}$$
$$= \sigma(\hat{V}_S) \cap (CP^1 \backslash (\hat{\Delta} \cup RP^1))$$
$$= \pi_3\left(\left\{x \in S^3_Q \,\middle|\, \|x_1 - x_2\| = \|x_1 - x_3\| = \|x_2 - x_3\|\right\}\right)$$

by the above theorem.

We want now to compute $\hat{V}_S: CP^1 \backslash \hat{\Delta} \to R$. Recall that $CP^1 = S^2$ and clearly $(0,0,1) \in \hat{\Delta}$ using this identification. Denote by $p: S^2 \backslash \{(0,0,1)\} \to C$ the stereographic projection. If

$$S^2 = \{(z,t) \in C \times R \mid |z|^2 + t^2 = 1\}$$

p is given by

$$p(z,t) = z/(1-t)$$

Let $\bar{\Delta} = p(\hat{\Delta})$ and define $\bar{V}_S: C \backslash \bar{\Delta} \to R$ by $\bar{V}_S \circ p = \hat{V}_S$. Since p is a diffeomorphism $\sigma(\bar{V}_S) \approx \sigma(\hat{V}_S)$. In what follows we shall compute $\bar{V}_S: C \backslash \hat{\Delta} \to R$ explicitly and determine the set $\sigma(\bar{V}_S)$. (The following computations are taken from Iacob [1973].)

The *standard Hopf fibration* $h: S^3 \to S^2$ will be useful. If

$$S^3 = \{(w_1, w_2) \in C^2 \mid |w_1|^2 + |w_2|^2 = 1\}$$

and

$$S^2 = \{(z,t) \in C \times R \mid |z|^2 + t^2 = 1\}$$

h is given by

$$h(w_1, w_2) = (2w_1\bar{w}_2, |w_2|^2 - |w_1|^2)$$

(here $\bar{w}$ is the complex conjugate of w). $h: S^3 \to S^2$ is a fiber bundle, the fiber over $(z,t) \in S^2$ being given by

$$h^{-1}(z,t) = \left\{(w_1, w_2) \in C^2 \,\middle|\, |w_1|^2 = \frac{1-t}{2},\ |w_2|^2 = \frac{|z|^2}{2(1-t)},\ 2w_1\bar{w}_2 = z\right\} \approx S^1$$

ISBN 0-8053-0102-X

We have

$$(p \circ h)(w_1, w_2) = \frac{2w_1\bar{w}_2}{1-(|w_2|^2-|w_1|^2)} = \frac{w_1\bar{w}_2}{|w_1|^2}$$

Denote the variables in $C\backslash\bar{\Delta}$, the source of the map $\bar{V}_S$ that we want to compute, by ζ. Thus

$$\zeta = (p \circ h)(w_1, w_2) = \frac{w_1\bar{w}_2}{|w_1|^2}$$

and an easy computation yields the formulas

$$|w_1| = (1+|\zeta|^2)^{-1/2}, \quad |w_2| = |\zeta|(1+|\zeta|^2)^{-1/2} \tag{1}$$

$$w_1\bar{w}_2 + \bar{w}_1 w_2 = (\zeta+\bar{\zeta})(1+|\zeta|^2)^{-1}$$

which will be useful later.

Our next goal is to transform the quadratic form $Q(x)=\sum_{i=1}^3 m_i\|x_i\|^2$ into canonical form. This is accomplished using the *Jacobi coordinates* in

$$M = \left\{ x=(x_1, x_2, x_3)\in C^3 \,\middle|\, \sum_{i=1}^{3} m_i x_i = 0,\ x \not\in \Delta \right\}$$

These are given by

$$u_1 = x_1 - x_2, \quad u_2 = x_3 - \frac{m_1x_1+m_2x_2}{m_1+m_2}$$

Their physical significance is the following: u_i is the position vector of the body with mass m_{i+1} with respect to the center of mass of the bodies with masses $m_1,\ldots,m_i$, $i=1,2$. This interpretation will be of use later when we discuss briefly the relationship of the *full planar three-body problem* we are discussing now, to the *restricted three-body problem* dealt with in the previous sections. Since $\sum_{i=1}^3 m_i x_i = 0$, we get

$$\begin{aligned} x_1 &= n_1 u_1 - \frac{m_3}{m}u_2 \\ x_2 &= -n_2 u_1 - \frac{m_3}{m}u_2 \\ x_3 &= \left(1-\frac{m_3}{m}\right)u_2 \end{aligned} \tag{2}$$

ISBN 0-8053-0102-X

where

$$m=m_1+m_2+m_3, \quad n_1=\frac{m_2}{m_1+m_2}, \quad n_2=\frac{m_1}{m_1+m_2}$$

In Jacobi coordinates,

$$M=\{u=(u_1,u_2)\in C^2 \,|\, u_1\neq 0,\ u_2\neq -n_2u_1,\ u_2\neq n_1u_1\}$$

and the quadratic form Q becomes

$$Q(x)=\sum_{i=1}^{3} m_i|x_i|^2=\frac{m_1m_2}{m_1+m_2}|u_1|^2+\frac{m_3(m_1+m_2)}{m}|u_2|^2$$

Define then the diffeomorphism $\phi\colon C^2\to C^2$ by

$$\phi(u_1,u_2)=\left(\frac{m_1m_2}{m_1+m_2}u_1,\ \frac{m_3(m_1+m_2)}{m}u_2\right)$$

and notice that $(Q\circ\phi^{-1})(w_1,w_2)=|w_1|^2+|w_2|^2$ and hence $\phi\colon S_Q^3\to S^3\backslash\phi(\Delta)$ is a diffeomorphism. We also have from (2):

$$|x_1-x_2|=|u_1|=\sqrt{\frac{m_1+m_2}{m_1m_2}}\,w_1$$

$$|x_1-x_3|=|n_1u_1-u_2|=\left|n_1\sqrt{\frac{m_1+m_2}{m_1m_2}}\,w_1-\sqrt{\frac{m}{m_3(m_1+m_2)}}\,w_2\right|$$

$$|x_2-x_3|=|n_2u_1+u_2|=\left|n_2\sqrt{\frac{m_1+m_2}{m_1m_2}}\,w_1+\sqrt{\frac{m}{m_3(m_1+m_2)}}\,w_2\right|$$

ISBN 0-8053-0102-X

and hence using (1), we get

$$|x_1-x_2|^2=\frac{m_1+m_2}{m_1m_2}|w_1|^2$$

$$=\frac{m_1+m_2}{m_1m_2}(1+|\zeta|^2)^{-1}\neq 0 \text{ for all } \zeta\in C$$

$$|x_1-x_3|^2=(x_1-x_3)(\bar{x}_1-\bar{x}_3)$$

$$=n_1^2\frac{m_1+m_2}{m_1m_2}|w_1|^2+\frac{m}{m_3(m_1+m_2)}|w_2|^2$$

$$-n_1\left(\frac{m_1m_2m_3}{m}\right)^{-1/2}(w_1\bar{w}_2+\bar{w}_1w_2)$$

$$=\frac{m}{m_3(m_1+m_2)}\frac{1}{1+|\zeta|^2}$$

$$\left[n_1^2\frac{m_3(m_1+m_2)^2}{mm_1m_2}|\zeta|^2-n_1(m_1+m_2)\frac{m_3}{mm_1m_2}(\zeta+\bar{\zeta})\right]$$

$$=\frac{m}{m_3(m_1+m_2)}\frac{1}{1+|\zeta|^2}|\zeta-\zeta_1|^2$$

where

$$\zeta_1=n_1(m_1+m_2)\sqrt{\frac{m_3}{mm_1m_2}}=\sqrt{\frac{m_2m_3}{mm_1}}>0$$

Similarly,

$$|x_2-x_3|^2=\frac{m}{m_3(m_1+m_2)}\frac{1}{1+|\zeta|^2}|\zeta-\zeta_2|^2$$

where

$$\zeta_2=-n_2(m_1+m_2)\sqrt{\frac{m_3}{mm_1m_2}}=-\sqrt{\frac{m_1m_3}{mm_2}}<0$$

ISBN 0-8053-0102-X

Thus $\bar{\Delta}=\{\zeta_1,\zeta_2\}$ since in ζ-coordinates $x_1=x_3$ iff $\zeta=\zeta_1$ and $x_2=x_3$ iff $\zeta=\zeta_2$; $x_1\neq x_2$ always holds in ζ-coordinates as we saw above.

Remark. $CP^1\backslash\hat{\Delta}=S^2\backslash\hat{\Delta}=S^2\backslash\{\text{three points}\}$ since $\hat{\Delta}=\{(0,0,1),\ p^{-1}(\zeta_1),\ p^{-1}(\zeta_2)\}$.

The changes of coordinates that we defined are summarized in the following commutative diagram.

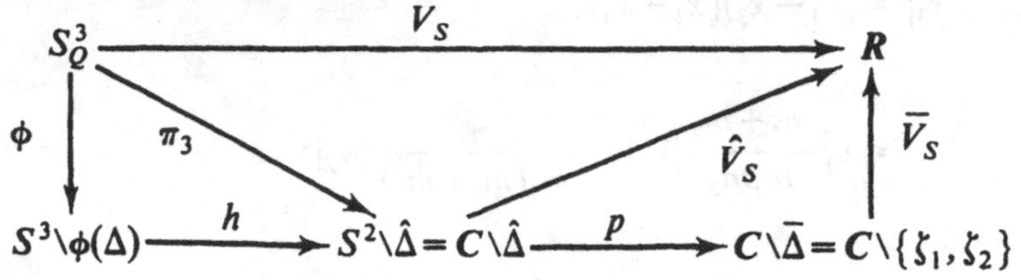

We have

$$\bar{V}_S(\zeta)=\left(\bar{V}_S\circ p\circ h\right)(w_1,w_2)=\left(\bar{V}_S\circ p\circ h\circ\phi\right)(u_1,u_2)$$

$$=(\hat{V}_S\circ\pi_3)(u_1,u_2)=V_S(u_1,u_2)=V_S(x_1,x_2,x_3)$$

the last equality being the Jacobi change of coordinates. Thus

$$\bar{V}_S(\zeta)=-\frac{m_1m_2}{|x_1-x_2|}-\frac{m_1m_3}{|x_1-x_3|}-\frac{m_2m_3}{|x_2-x_3|}$$

$$=-m_3\sqrt{m_1m_2(m_1+m_2)}\,\sqrt{1+|\zeta|^2}\left[\frac{1+\zeta_1\zeta_2}{(\zeta_1-\zeta_2)^2}-\frac{\zeta_2}{|\zeta-\zeta_1|}+\frac{\zeta_1}{|\zeta-\zeta_2|}\right]$$

We want to determine $\sigma(\bar{V}_S)$.

The critical points of $\sigma(\bar{V}_S)$ will correspond to equivalence classes of collinear and noncollinear central configurations. We shall consider them separately.

Noncollinear Central Configurations Classes. These are given (see comment following 10.4.13) in ζ-coordinates by

$$(p\circ\pi_3)\left(\left\{x\in S_Q^3\,\middle|\,\|x_1-x_2\|=\|x_1-x_3\|=\|x_2-x_3\|\right\}\right)$$

and hence having in view the above formulas that give $|x_1-x_2|$, $|x_1-x_3|$, and

ISBN 0-8053-0102-X

$|x_2 - x_3|$ as functions of ζ, we must have for these points

$$\sqrt{\frac{m_1+m_2}{m_1 m_2}}\ \frac{1}{\sqrt{1+|\zeta|^2}} = \sqrt{\frac{m}{m_3(m_1+m_2)}}\ \frac{1}{\sqrt{1+|\zeta|^2}}\,|\zeta-\zeta_1|$$

$$\sqrt{\frac{m_1+m_2}{m_1 m_2}}\ \frac{1}{\sqrt{1+|\zeta|^2}} = \sqrt{\frac{m}{m_3(m_1+m_2)}}\ \frac{1}{\sqrt{1+|\zeta|^2}}\,|\zeta-\zeta_2|$$

that is,

$$|\zeta-\zeta_1| = |\zeta-\zeta_2| = \sqrt{\frac{m_3}{m m_1 m_2}}\ (m_1+m_2)$$

In other words, the only noncollinear central configurations lie at the intersection of two circles of equal radii centered at ζ_1 and ζ_2, respectively. Denote these two critical points by λ_1 and λ_2. It is clear that $\lambda_2 = \bar{\lambda}_1$ by the above geometric interpretation and

$$\lambda_1 = \frac{1}{2}\sqrt{\frac{m_3}{m m_1 m_2}}\ \left[(m_2-m_1) + i\sqrt{3}\,(m_1+m_2)\right]$$

We also have

$$|\zeta_1-\zeta_2| = \sqrt{\frac{m_3}{m m_1 m_2}}\ (m_1+m_2)$$

$$|\lambda_1-\zeta_1| = \sqrt{\frac{m_3}{m m_1 m_2}}\ (m_1+m_2) = |\lambda_1-\zeta_2|$$

$$\bar{V}_S(\lambda_1) = \bar{V}_S(\lambda_2) = -(m_1 m_2 + m_1 m_3 + m_2 m_3)^{3/2} m^{-1/2} < 0$$

Hence the two critical points form two equilateral triangles $\zeta_1\lambda_1\zeta_2$ and $\zeta_1\lambda_2\zeta_2$ in C and they are in the same level set of $\bar{V}_S$. These two critical points are called, as in the case of the restricted three-body problem, *Lagrange critical points*.

We shall prove now that these critical points are *nondegenerate* maxima for $\bar{V}_S$.

ISBN 0-8053-0102-X

Letting $\zeta=\xi+i\eta$, a direct computation shows that

$$\frac{\partial \bar{V}_S}{\partial \xi}=-m_3\sqrt{m_1m_2(m_1+m_2)}\left\{\frac{\xi}{1+|\zeta|^2}\bar{V}_S(\zeta)\right.$$

$$\left.+\sqrt{1+|\zeta|^2}\left[\frac{\zeta_2(\xi-\zeta_1)}{|\zeta-\zeta_1|^3}-\frac{\zeta_1(\xi-\zeta_2)}{|\zeta-\zeta_2|^3}\right]\right\}$$

$$\frac{\partial \bar{V}_S}{\partial \eta}=-m_3\sqrt{m_1m_2(m_1+m_2)}\,\eta\left\{\frac{1}{1+|\zeta|^2}\bar{V}_S(\zeta)\right.$$

$$\left.+\sqrt{1+|\zeta|^2}\left[\frac{\zeta_2}{|\zeta-\zeta_1|^3}-\frac{\zeta_1}{|\zeta-\zeta_2|^3}\right]\right\}$$

$$\frac{\partial^2 \bar{V}_S}{\partial \xi^2}=-m_3\sqrt{m_1m_2(m_1+m_2)}\left\{\frac{1+\eta^2-\xi^2}{(1+|\zeta|^2)^2}\bar{V}_S(\zeta)\right.$$

$$+\frac{\xi}{1+|\zeta|^2}\frac{\partial \bar{V}_S(\zeta)}{\partial \xi}+\frac{\xi}{\sqrt{1+|\zeta|^2}}\left[\frac{\zeta_2(\xi-\zeta_1)}{|\zeta-\zeta_1|^3}\right.$$

$$\left.-\frac{\zeta_1(\xi-\zeta_2)}{|\zeta-\zeta_2|^3}\right]+\sqrt{1+|\zeta|^2}\left[\zeta_2\frac{|\zeta-\zeta_1|^2-3(\xi-\zeta_1)^2}{|\zeta-\zeta_1|^5}\right.$$

$$\left.\left.-\zeta_1\frac{|\zeta-\zeta_2|^2-3(\xi-\zeta_2)^2}{|\zeta-\zeta_2|^5}\right]\right\}$$

$$\frac{\partial^2 \bar{V}_S}{\partial \xi\,\partial \eta}=-m_3\sqrt{m_1m_2(m_1+m_2)}\,\eta\left\{\frac{-2\xi}{(1+|\zeta|^2)^2}\bar{V}_S(\zeta)\right.$$

$$+\frac{1}{1+|\zeta|^2}\frac{\partial \bar{V}_S(\zeta)}{\partial \xi}+\frac{\xi}{\sqrt{1+|\zeta|^2}}\left[\frac{\zeta_2}{|\zeta-\zeta_1|^3}-\frac{\zeta_1}{|\zeta-\zeta_2|^3}\right]$$

$$\left.+\sqrt{1+|\zeta|^2}\left[-3|\zeta-\zeta_1|^{-5}(\xi-\zeta_1)\zeta_2+3|\zeta-\zeta_2|^{-5}(\xi-\zeta_1)\zeta_1\right]\right\}$$

ISBN 0-8053-0102-X

and

$$\frac{\partial^2 \bar{V}_S}{\partial \eta^2} = -m_3\sqrt{m_1 m_2(m_1+m_2)}$$

$$\times\left\{\frac{1}{1+|\zeta|^2}\bar{V}_S(\zeta)+\sqrt{1+|\zeta|^2}\left[\frac{\zeta_2}{|\zeta-\zeta_1|^3}-\frac{\zeta_1}{|\zeta-\zeta_2|}\right]\right\}$$

$$-m_3\sqrt{m_1 m_2(m_1+m_2)}\ \eta\left\{\frac{-2\eta}{(1+|\zeta|^2)^2}\bar{V}_S(\zeta)\right.$$

$$+\frac{1}{1+|\zeta|^2}\frac{\partial \bar{V}_S}{\partial \eta}+\frac{\eta}{\sqrt{1+|\zeta|^2}}\left[\frac{\zeta_2}{|\zeta-\zeta_1|^3}-\frac{\zeta_1}{|\zeta-\zeta_2|^3}\right]$$

$$\left.+\sqrt{1+|\zeta|^2}\left[-3|\zeta-\zeta_1|^{-5}\eta\zeta_2+3|\zeta-\zeta_2|^{-5}\eta\zeta_1\right]\right\}$$

Using the fact that

$$\frac{\partial \bar{V}_S}{\partial \xi}(\lambda_1)=0, \quad \frac{\partial \bar{V}_S}{\partial \eta}(\lambda_1)=0$$

and the computed expressions for $|\lambda_1-\zeta_1|$, $|\lambda_1-\zeta_2|$, and $\bar{V}_S(\lambda_1)$ as well as the formulas:

$$1+|\lambda_1|^2=\frac{(m_1m_2+m_1m_3+m_2m_3)(m_1+m_2)}{mm_1m_2}$$

$$\xi_{\lambda_1}-\zeta_1=-\frac{1}{2}\sqrt{\frac{m_3}{mm_1m_2}}(m_1+m_2)$$

$$\xi_{\lambda_1}-\zeta_2=\frac{1}{2}\sqrt{\frac{m_3}{mm_1m_2}}(m_1+m_2)=\tfrac{1}{2}|\lambda_1-\zeta_1|=\tfrac{1}{2}|\lambda_1-\zeta_2|$$

ISBN 0-8053-0102-X

which can be checked by direct computation, we get

$$\begin{bmatrix} \dfrac{\partial^2 \overline{V}_S(\lambda_1)}{\partial \xi^2} & \dfrac{\partial^2 \overline{V}_S(\lambda_1)}{\partial \xi \partial \eta} \\ \dfrac{\partial^2 \overline{V}_S(\lambda_1)}{\partial \xi \partial \eta} & \dfrac{\partial^2 \overline{V}_S(\lambda_1)}{\partial \eta^2} \end{bmatrix}$$

$$= -3\left(\frac{m_1 m_2}{m_1+m_2}\right)^2 \sqrt{\frac{m}{m_1 m_2 + m_1 m_3 + m_2 m_3}} \begin{bmatrix} m_1+m_3 & -m_3 \\ -m_3 & m_2+m_3 \end{bmatrix}$$

which is a nonsingular matrix. Its eigenvalues are both negative and hence λ_1 is a nondegenerate maximum (i.e., λ_1 has index 2).

Collinear Central Configuration Classes. By Moulton's theorem, the number of these is $3!/2=3$. We shall call them *Euler critical points*. We also know that they lie in $\boldsymbol{RP}^1\backslash\hat{\Delta}$ and that they are *nondegenerate maxima* for $\hat{V}_{col}=\hat{V}_S|(\boldsymbol{RP}^1\backslash\hat{\Delta})$. Recall that $\boldsymbol{RP}^1$ was imbedded in $\boldsymbol{CP}^1$ by putting $Im\,x_1=Im\,x_2=Im\,x_3=0$. Using (2) this implies $Im\,u_1=Im\,u_2=0$, and by the definition of the diffeomorphism ϕ, $Im\,w_1=Im\,w_2=0$. Since $\zeta=w_1\overline{w}_2/|w_1|^2$, we have $Im\,\zeta=0$. Thus, the three Euler critical points of $\overline{V}_S$ lie on the line $Im\,\zeta=0$. Also, from the general theory in the proof of Moulton's theorem, we know that each critical point will be in exactly one component of $\boldsymbol{RP}^1\backslash\hat{\Delta}$, that is, in one component of $\{\zeta\in\boldsymbol{C}\,|\,Im\,\zeta=0\}\backslash\{\zeta_1,\zeta_2\}$. Denote the Euler critical points by $\varepsilon_1,\varepsilon_2,\varepsilon_3$, where the ordering is chosen in such a way that

$$\overline{V}_S(\varepsilon_1)\leqslant\overline{V}_S(\varepsilon_2)\leqslant\overline{V}_S(\varepsilon_3),\quad \varepsilon_i\in\boldsymbol{R}$$

Since $\varepsilon_1,\varepsilon_2,\varepsilon_3$ are nondegenerate maxima for $\overline{V}_S|Im\,\zeta=0$, we must have $\partial^2\overline{V}_S(\varepsilon_i)/\partial\xi^2<0$, $i=1,2,3$. A direct inspection of the formulas from above, using the fact that $\overline{V}_S(\zeta)\leqslant 0$ for any $\zeta\in\boldsymbol{C}$ and the fact that $\zeta_2<0$, $\zeta_1>0$, shows that

$$\frac{\partial^2\overline{V}_S(\varepsilon_i)}{\partial\eta^2}>0,\quad \frac{\partial^2\overline{V}_S(\varepsilon_i)}{\partial\xi\partial\eta}=0,\quad i=1,2,3$$

Hence the Euler critical points are nondegenerate; one eigenvalue of the Hessian is negative, the other one is positive. Thus the Euler critical points are nondegenerate saddle points of $\overline{V}_S$ (their index is 1).

We thus proved the following (compare with 10.2.5 and 10.2.6).

10.4.14 Theorem. *For any choice of the masses $m_1, m_2, m_3>0$ in the planar three-body problem, $\overline{V}_S$ has exactly five critical points which are all nondegenerate. Two of them are maxima (the Lagrange critical points) and three are saddle points (the Euler critical points).*

ISBN 0-8053-0102-X

Let $d_i = \overline{V}_S(\varepsilon_i)$, $i=1,2,3$ and $d_4 = \overline{V}_S(\lambda_1) = \overline{V}_S(\lambda_2)$. Thus, by 10.4.8(ii), we can write

$$\Sigma'_{H\times J} = \bigcup_{i=1}^{4} \{(h,\mu)\in R^2 \mid 2h\mu^2 = -d_i^2\}$$

It is clear now that the critical points in the full three-body problem have the same structure as the critical points in the restricted three-body problem (m_1, m_2 move in circles around the origin and $m_3=0$), discussed in Sect. 10.2. It should be noted here that any central configuration in the full three-body problem is a central configuration in a *certain* restricted three-body problem. To see this, it is enough to recall that if the three bodies with coordinates x_1, x_2, x_3 are in a central configuration in the full planar three-body problem, they necessarily move in circles around the origin, their center of mass. Neglecting the mass of the "lightest" among them amounts to the situation where the three bodies still move in circles, in a fixed configuration of equilateral triangle position, but are slightly displaced from their original positions when the mass of the light body was taken into account, too. But the situation just described represents a central configuration in the restricted three-body problem. It is tempting to try to prove stability results about the full three-body problem by regarding it as a perturbation of the restricted problem in the context just explained. However, this is probably difficult.

In what follows, we shall carry out points (i) and (ii) of the topological program discussed in Sect. 4.5 for the planar n-body problem. We shall then specialize to the planar three-body problem and write down explicitly all the invariant and reduced invariant manifolds.

Recall first some notation from Sect. 4.5. If (M,K,V,G) is a mechanical system with symmetry with $\Lambda=\varnothing$ (Λ is the set where J_x fails to be surjective) for $(h,\mu)\in R\times\mathfrak{g}^*$, $M_{h,\mu} = \{x\in M \mid V_\mu(x) \le h\}$, $E_{h,\mu} = \{\alpha\in T^*M \mid J(\alpha)=\mu, H(\alpha)\le h\}$, $\partial M_{h,\mu} = V_\mu^{-1}(h)$, $\partial E_{h,\mu} = I_{h,\mu}$, h being a regular value for V_μ and $J: T^*M\to\mathfrak{g}^*$ denoting the associated momentum mapping. If $J^{-1}(0)|M_{h,\mu}$ denotes the restriction of the vector subbundle $J^{-1}(0)\to M$ of T^*M to the manifold with boundary $M_{h,\mu}$, then the Invariant Manifold Theorem of Smale (4.5.9) states that

$$\alpha\left(J^{-1}(0)|M_{h,\mu}\right) \approx E_{h,\mu}$$

$$\beta\left(J^{-1}(0)|M_{h,\mu}\right) \approx I_{h,\mu}$$

Here α and β denote respectively the reduced unit disk and sphere bundle of the vector bundle $J^{-1}(0)|M_{h,\mu}$. Under the hypothesis that $J^{-1}(0)|M_{h,\mu} \cong M_{h,\mu} \times R^s$ is a trivial vector bundle, the Reduced Invariant Manifold Theorem of Smale assures that

$$I_{h,\mu} = \beta_s(M_{h,\mu}), \quad \hat{I}_{h,\mu} = \beta_s(\hat{M}_{h,\mu})$$

ISBN 0-8053-0102-X

where $\hat{M}_{h,\mu}=M_{h,\mu}/G_\mu$, G_μ being the isotropy subgroup of μ under the coadjoint action of G on $\mathfrak{g}^*$. Here β_s denotes $\beta(\cdot \times \boldsymbol{R}^s)$.

10.4.15 Theorem (Smale). *In the planar n-body problem for zero total momentum, the invariant and reduced invariant manifolds are given by:*

(*i*) $h>0$

$$I_{h,0}=(S^{2n-3}\backslash\Delta)\times\boldsymbol{R}\times S^{2n-4}$$

$$\hat{I}_{h,0}=(\boldsymbol{CP}^{n-2}\backslash\hat{\Delta})\times\boldsymbol{R}\times S^{2n-4}$$

(*ii*) $h<0$

$$I_{h,0}=(S^{2n-3}\backslash\Delta)\times\boldsymbol{R}^{2n-3}$$

$$\hat{I}_{h,0}=(\boldsymbol{CP}^{n-2}\backslash\hat{\Delta})\times\boldsymbol{R}^{2n-3}$$

Proof. (after Smale [1970a] and Iacob [1973]). (i) For $\mu=0$, the effective potential is $V_0(x)=V(x)$ [recall that $V_\mu(x)=V(x)+\mu^2/2Q(x)$]. Since $V(x)<0$ for $h>0$, $M_{h,0}=M\approx S_Q^{2n-3}\times(0,\infty)$ by $\phi\colon x\mapsto(x/\sqrt{Q(x)}\,,\sqrt{Q(x)}\,)$ as we saw in the discussion preceding 10.4.6. Now since $S_Q^{2n-3}\approx S^{2n-3}\backslash\Delta$, we have $M_{h,0}\approx(S^{2n-3}\backslash\Delta)\times\boldsymbol{R}$. Clearly $J^{-1}(0)\to M_{h,0}$ is a trivial vector bundle

$$J^{-1}(0)\cong M_{h,0}\times\boldsymbol{R}^{2n-3}$$

Thus, if $h>0$, $I_{h,0}=\beta(J^{-1}(0)|M)=\beta_{2n-3}(M)=M\times S^{2n-4}=(S^{2n-3}\backslash\Delta)\times\boldsymbol{R}\times S^{2n-4}$ (since $\partial M=\varnothing$, $\beta_k(M)=M\times S^{k-1}$). Clearly the S^1-action on M is transported (using the equivariant diffeomorphism ϕ) to an S^1-action on $S_Q^{2n-3}\times(0,\infty)$ that acts only on the first factor. Thus $M/S^1=(\boldsymbol{CP}^{n-2}\backslash\hat{\Delta})\times\boldsymbol{R}$ and for $h>0$ we conclude that $\hat{I}_{h,0}=\beta_{2n-3}(\hat{M}_{h,0})=\hat{M}_{h,0}\times S^{2n-4}=(\boldsymbol{CP}^{n-2}\backslash\hat{\Delta})\times\boldsymbol{R}\times S^{2n-4}$.

(ii) For $h<0$, we have

$$\begin{aligned}M_{h,0}&=\{x\in M\,|\,V_0(x)=V(x)\leqslant h\}\\&=\left\{(z,t)\in S_Q^{2n-3}\times(0,\infty)\,\middle|\,\frac{1}{t}V_S(z)\leqslant h\right\}\\&=\left\{(z,t)\in S_Q^{2n-3}\times(0,\infty)\,\middle|\,\frac{1}{h}V_S(z)\geqslant t\right\}\end{aligned}$$

Note that for $h<0$, $(1/h)V_S(z)>0$, so that the condition $(1/h)V_S(z)\geqslant t$ for $h<0$ fixed suggests the definition of the following map:

$$f_h\colon M_{h,0}\to S_Q^{2n-3}\times[0,\infty)$$

$$f_h(z,t)=\left(z,t-\frac{1}{h}V_S(z)\right)$$

ISBN 0-8053-0102-X

It is clear that f_h is a diffeomorphism so that for $h<0$ we can write

$$M_{h,0} \approx S_Q^{2n-3} \times [0, \infty) \approx (S^{2n-3} \backslash \Delta) \times [0, \infty)$$

$$\hat{M}_{h,0} \approx (CP^{n-2} \backslash \hat{\Delta}) \times [0, \infty)$$

Thus we have

$$\begin{aligned} I_{h,0} &= \beta_{2n-3}(M_{h,0}) \\ &= \beta_{2n-3}((S^{2n-3} \backslash \Delta) \times [0, \infty)) \\ &= (S^{2n-3} \backslash \Delta) \times \beta_{2n-3}([0, \infty)) = (S^{2n-3} \backslash \Delta) \times \boldsymbol{R}^{2n-3} \end{aligned}$$

since it is clear that $\beta_{2n-3}([0, \infty)) = \boldsymbol{R}^{2n-3}$. Similarly

$$\begin{aligned} \hat{I}_{h,0} &= \beta_{2n-3}(\hat{M}_{h,0}) \\ &= \beta_{2n-3}((CP^{n-2} \backslash \hat{\Delta}) \times [0, \infty)) \\ &= (CP^{n-2} \backslash \hat{\Delta}) \times \beta_{2n-3}([0, \infty)) = (CP^{n-2} \backslash \hat{\Delta}) \times \boldsymbol{R}^{2n-3} \end{aligned}$$ ■

10.4.16 Corollary. *In the planar three-body problem, the invariant and reduced invariant manifolds for zero momentum are given by:*

(*i*) $h>0$

$$I_{h,0} = S^1 \times (S^2 \backslash \{p_1, p_2, p_3\}) \times \boldsymbol{R} \times S^2$$

$$\hat{I}_{h,0} = (S^2 \backslash \{p_1, p_2, p_3\}) \times \boldsymbol{R} \times S^2$$

(*ii*) $h<0$

$$I_{h,0} = S^1 \times (S^2 \backslash \{p_1, p_2, p_3\}) \times \boldsymbol{R}^3$$

$$\hat{I}_{h,0} = (S^2 \backslash \{p_1, p_2, p_3\}) \times \boldsymbol{R}^3$$

where $p_1 = p^{-1}(\zeta_1)$, $p_2 = p^{-1}(\zeta_2)$, $p_3 = (0, 0, 1)$.

Proof. We established in the course of the discussion of $\hat{C}_3$ that $\hat{\Delta} = \{p_1, p_2, p_3\}$. Now look at the fiber bundle $\pi_3 | S_Q^3$: $S_Q^3 \to CP^1 \backslash \hat{\Delta}$ with fiber S^1 over each point. But the restriction of the fiber bundle π_3: $S^3 \to CP^1 = S^2$ to $\pi_3^{-1}(S^2\{(0,0,1)\})$ is trivial since the base space $S^2 \backslash \{(0,0,1)\} = \boldsymbol{C}$ is contractible. Thus $S_Q^3 = S^3 \backslash \Delta \approx S^1 \times (S^2 \backslash \{p_1, p_2, p_3\})$. The rest follows from 10.4.15. ■

ISBN 0-8053-0102-X

10.4.17 Theorem (Smale). *In the planar n-body problem for $\mu\neq 0$, $h>0$,*

$$I_{h,\mu}=(S^{2n-3}\backslash\Delta)\times \mathbf{R}^{2n-3}$$

$$\hat{I}_{h,\mu}=(CP^{n-2}\backslash\hat{\Delta})\times \mathbf{R}^{2n-3}$$

Proof.

$$M_{h,\mu}=\{x\in M\,|\,V_\mu(x)<h\}$$

$$=\left\{(z,t)\in S_Q^{2n-3}\times(0,\infty)\,\middle|\,\frac{1}{t}V_S(z)+\frac{\mu^2}{2t^2}<h\right\}$$

$$=\{(z,t)\in S_Q^{2n-3}\times(0,\infty)\,|\,t_0<t<+\infty\}$$

where t_0 is the unique positive solution of the quadratic equation $-2t^2h+2V_S(z)t+\mu^2=0$. Now as in 10.4.15 there is a diffeomorphism $f_{h,\mu}: M_{h,\mu}\to S_Q^{2n-3}\times[0,\infty)$ given by $f_{h,\mu}(z,t)=(z,t-t_0)$. Thus $M_{h,\mu}\approx(S^{2n-3}\backslash\Delta)\times[0,\infty)$ and the rest is exactly as in 10.4.15. ■

10.4.18 Corollary. *In the planar three-body problem, for $\mu\neq 0$, $h>0$,*

$$I_{h,\mu}=S^1\times(S^2\backslash\{p_1,p_2,p_3\})\times\mathbf{R}^3$$

$$\hat{I}_{h,\mu}=(S^2\backslash\{p_1,p_2,p_3\})\times\mathbf{R}^3$$

Finally we are left with the situation $\mu\neq 0$ and $h<0$, which is the most difficult one. As usual, the first task is to find $M_{h,\mu}$.

$$M_{h,\mu}=\{x\in M\,|\,V_\mu(x)<h\}$$

$$=\left\{(z,t)\in S_Q^{2n-3}\times(0,\infty)\,\middle|\,\frac{1}{t}V_S(z)+\frac{\mu^2}{2t^2}<h\right\}$$

Fix $z\in S_Q^{2n-3}$ and look at the map $t\mapsto(1/t)V_S(z)+\mu^2/2t^2$ whose graph is given in Fig. 10.4-2. Define the map

$$l_\mu: S_Q^{2n-3}\to\mathbf{R}$$

by

$$l_\mu(z)=\inf_{t\in\mathbf{R}}\left(\frac{1}{t}V_S(z)+\frac{\mu^2}{2t^2}\right)=-V_S^2(z)/2\mu^2$$

a smooth function for all $\mu\neq 0$. In the definition of $M_{h,\mu}$, the condition

$$\frac{1}{t}V_S(z)+\frac{\mu^2}{2t^2}<h$$

ISBN 0-8053-0102-X

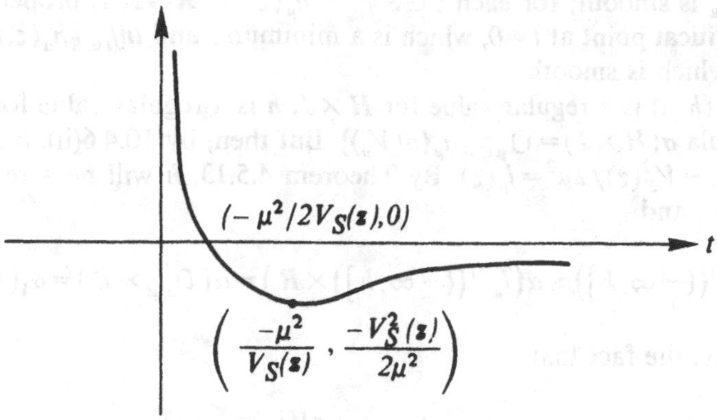

Figure 10.4-2

can hold for some $t>0$ if and only if $l_\mu(z)=-V_S^2(z)/2\mu^2<h$, that is, if and only if $z\in D_{h,\mu}=l_\mu^{-1}((-\infty,h])$. Hence

$$M_{h,\mu}=\left\{(z,t)\in D_{h,\mu}\times(0,\infty)\;\middle|\;\frac{1}{t}V_S(z)+\frac{\mu^2}{2t^2}<h\right\}$$

10.4.19 Lemma (Iacob). *If $(h,\mu)\in R^2\backslash\Sigma'_{H\times J}$, $h<0$, $\mu\neq 0$, then $M_{h,\mu}=\alpha_1(D_{h,\mu})$, where $D_{h,\mu}=l_\mu^{-1}((-\infty,h])$, and $l_\mu(z)=-V_S^2(z)/2\mu^2$.*

Proof. We shall use Theorem 4.5.13, which characterizes the manifolds α in terms of a function h satisfying certain properties. The proof will be done by exhibiting explicitly such a function.

Define the map

$$f: S_Q^{2n-3}\times(0,\infty)\rightarrow S_Q^{2n-3}\times R$$

$$f(z,t)=\left(z, \ln t-\ln(-\mu^2/V_S(z))\right)$$

It is easy to see that f is a diffeomorphism with inverse

$$f^{-1}: S_Q^{2n-3}\times R\rightarrow S_Q^{2n-3}\times(0,\infty)$$

$$f^{-1}(z,t)=\left(z, -\mu^2e^t/V_S(z)\right)$$

Define the map $h_\mu=V_\mu\circ f^{-1}: S_Q^{2n-3}\times R\rightarrow R$, where we regard $S_Q^{2n-3}\times R$ as a trivial bundle with one-dimensional fiber over S_Q^{2n-3}, the metric on the fiber R being given by the product. We have

$$h_\mu(z,t)=-\frac{V_S^2(z)}{2\mu^2}(2e^{-t}-e^{-2t})=l_\mu(z)(2e^{-t}-e^{-2t})$$

ISBN 0-8053-0102-X

Clearly h_μ is smooth; for each $z \in S_Q^{2n-3}$, $h_\mu(z,\cdot)$: $\boldsymbol{R} \to \boldsymbol{R}$ is proper and has a unique critical point at $t=0$, which is a minimum; and $\inf_{t\in \boldsymbol{R}} h_\mu(z,t) = h_\mu(z,0) = l_\mu(z)$, which is smooth.

Since (h,μ) is a regular value for $H \times J$, h is a regular value for V_μ [recall the formula $\sigma(H\times J) = \cup_{\mu\in \boldsymbol{R}} \alpha_\mu(\sigma(V_\mu))$]. But then, by 10.4.6(ii), h is a regular value for $-V_S^2(z)/2\mu^2 = l_\mu(z)$. By Theorem 4.5.13, h will be a regular value for h_μ, too, and

$$h_\mu^{-1}((-\infty,h]) \approx \alpha\left(l_\mu^{-1}((-\infty,h]) \times \boldsymbol{R}\right) = \alpha(D_{h,\mu} \times \boldsymbol{R}) = \alpha_1(D_{h,\mu})$$

Finally, the fact that

$$f|M_{h,\mu}: M_{h,\mu} \to h_\mu^{-1}((-\infty,h])$$

is a diffeomorphism, concludes the proof. ■

We can thus assert:

10.4.20 Theorem (Smale, Iacob). *In the planar n-body problem for* $(h,\mu) \in \boldsymbol{R}^2 \backslash \Sigma'_{H\times J}$, $h<0$, $\mu \neq 0$,

$$I_{h,\mu} = \beta_{2n-3}\left(\alpha_1(D_{h,\mu})\right), \quad \hat{I}_{h,\mu} = \beta_{2n-3}\left(\alpha_1(\hat{D}_{h,\mu})\right)$$

where $\hat{D}_{h,\mu} = D_{h,\mu}/S^1$.

In the case $n=3$ more can be said about the manifolds $D_{h,\mu}$ and $\hat{D}_{h,\mu}$. In general, if we denote by $\hat{l}_\mu$: $\boldsymbol{CP}^{n-2}\backslash\hat{\Delta} \to \boldsymbol{R}$ the map defined by $\hat{l}_\mu \circ \pi_n = l_\mu$, where π_n: $S_Q^{2n-3} \to \boldsymbol{CP}^{n-2}\backslash\hat{\Delta}$ is the canonical projection, that is, $\hat{l}_\mu(y) = -\hat{V}_S^2(y)/2\mu^2$, then it is easy to see that $\hat{D}_{h,\mu} = \hat{l}_\mu^{-1}((-\infty,h])$. Now, if $n=3$, $\boldsymbol{CP}^1\backslash\hat{\Delta} = S^2\backslash\hat{\Delta} \subset \boldsymbol{C}$, and hence the bundle over $S^2\backslash\{(0,0,1)\}$ is trivial and so

$$D_{h,\mu} = S^1 \times \hat{D}_{h,\mu}$$

and

$$\hat{D}_{h,\mu} = \left\{ y \in \boldsymbol{CP}^1\backslash\hat{\Delta} = S^2\backslash\hat{\Delta} \,|\, \hat{V}_S(y) \leqslant -\sqrt{-2h\mu^2} \right\}$$

$$= p^{-1}\left(\left\{ \zeta \in \boldsymbol{C}\backslash\{\zeta_1,\zeta_2\} \,|\, \overline{V}_S(\zeta) \leqslant -\sqrt{-2h\mu^2} \right\}\right)$$

where p: $S^2\backslash\{(0,0,1)\} \to \boldsymbol{C}$ is the stereographic projection.

10.4.21 Theorem (Smale, Iacob). *In the planar three-body problem for* $(h,\mu) \in \boldsymbol{R}^2\backslash\Sigma'_{H\times J}$, $h<0$, $\mu\neq 0$, *and the additional hypothesis* $d_1 < d_2 < d_3 < d_4$ (<0), *where* $d_i = \overline{V}_S(\varepsilon_i)$, $i=1,2,3$, $d_4 = \overline{V}_S(\lambda_1) = \overline{V}_S(\lambda_2)$, ε_i, $i=1,2,3$ *being the Euler critical points and* λ_j, $j=1,2$ *the Lagrange critical points, the following hold*

ISBN 0-8053-0102-X

(note that all unions are disjoint unions and the points subtracted are all interior to the manifolds i.e. they are not on the boundaries):

$a<2h\mu^2<b$		$\hat{D}_{h,\mu}$
$a=-\infty,$	$b=-d_1^2$	$(D^2\backslash\{0\})\cup(D^2\backslash\{0\})\cup(D^2\backslash\{0\})$
$a=-d_1^2,$	$b=-d_2^2$	$(D^2\backslash\{\textit{two points}\})\cup(D^2\backslash\{0\})$
$a=-d_2^2,$	$b=-d_3^2$	$D^2\backslash\{\textit{three points}\}$
$a=-d_3^2,$	$b=-d_4^4$	$(S^1\times I)\backslash\{\textit{three points}\}$
$a=-d_4^2,$	$b=0$	$S^2\backslash\hat{\Delta}=S^2\backslash\{\textit{three points}\}$

$a<2h\mu^2<b$		$I_{h,\mu};\hat{I}_{h,\mu}$
$a=-\infty,$	$b=-d_1^2$	$\begin{cases} S^1\times((S^5\backslash S^3)\cup(S^5\backslash S^3)\cup(S^5\backslash S^3)) \\ (S^5\backslash S^3)\cup(S^5\backslash S^3)\cup(S^5\backslash S^3) \end{cases}$
$a=-d_1^2,$	$b=-d_2^2$	$\begin{cases} S^1\times(((S^5\backslash(S^3\cup S^3))\cup(S^5\backslash S^3)) \\ (S^5\backslash(S^3\cup S^3))\cup(S^5\backslash S^3) \end{cases}$
$a=-d_2^2,$	$b=-d_3^2$	$\begin{cases} S^1\times(S^5\backslash(S^3\cup S^3\cup S^3)) \\ S^5\backslash(S^3\cup S^3\cup S^3) \end{cases}$
$a=-d_3^2,$	$b=-d_4^2$	$\begin{cases} S^1\times((S^1\times S^4)\backslash(S^3\cup S^3\cup S^3)) \\ (S^1\times S^4)\backslash(S^3\cup S^3\cup S^3) \end{cases}$
$a=-d_4^2,$	$b=0$	$\begin{cases} S^1\times(S^2\backslash\{\textit{three points}\})\times S^3 \\ (S^2\backslash\{\textit{three points}\})\times S^3 \end{cases}$

Figure 10.4-3 is helpful in understanding the first table. (Think of $\overline{V}_S$: $S^2\backslash\Delta\to\boldsymbol{R}$ as a "height function".)

Proof. The first table is obtained from Theorem 3.2.18; it is the standard Morse theory argument of attaching handles when passing a nondegenerate critical value of a given index (see Sect. 3.3).

For the second table, it is enough to notice that

$$\alpha_1(\{\text{point}\})=D^1;\quad \alpha_1(D^2)=D^3$$

$$\alpha_1(S^1\times I)=\alpha_1(S^1\times D^1)=S^1\times\alpha_1(D^1)=S^1\times D^2$$

$$\alpha_1(S^2\backslash\{\text{three points}\})=(S^2\backslash\{\text{three points}\})\times D^1$$

ISBN 0-8053-0102-X

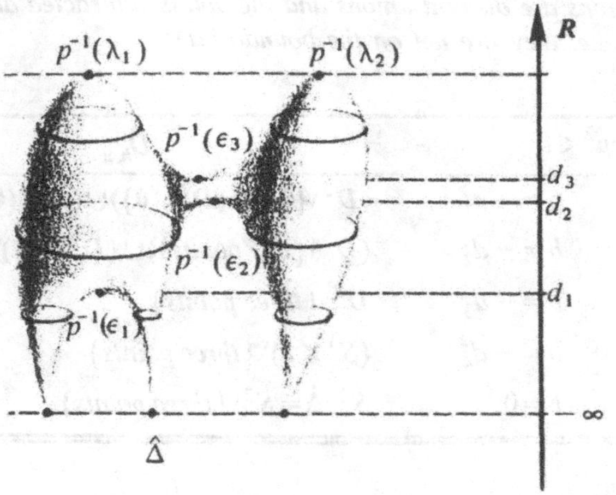

Figure 10.4-3

so that

$$\beta_3(\alpha_1(\{\text{point}\})) = \beta_3(D^1) = S^3$$

$$\beta_3(\alpha_1(D^2)) = \beta_3(D^3) = S^5$$

$$\beta_3(\alpha_1(S^1 \times I)) = \beta_3(S^1 \times D^2) = S^1 \times \beta_3(D^2) = S^1 \times S^4$$

$$\begin{aligned}\beta_3(\alpha_1(S^2\backslash\{\text{three points}\})) &= \beta_3((S^2\backslash\{\text{three points}\}) \times D^1)\\ &= S^2\backslash\{\text{three points}\} \times \beta_3(D^1)\\ &= (S^2\backslash\{\text{three points}\}) \times S^3 \quad \blacksquare\end{aligned}$$

A glance at 10.4.15, 10.4.17, and 10.4.20 shows the following.

10.4.22 Corollary. *The bifurcation set $\Sigma_{H \times J}$ contains the coordinate axes of* $\boldsymbol{R}^2$.

Remark. The topology of the two-body problem discussed in Sect. 9.8 can be recovered completely from 10.4.15, 10.4.17, and 10.4.20.

10.4.23 Theorem (Smale). (*i*) $H \times J|(H \times J)^{-1}(\{(h, \mu) \in \boldsymbol{R}^2 | h > 0, \mu \neq 0\})$ *is a trivial fiber bundle:*

$$\begin{aligned}&(H \times J)^{-1}(\{(h, \mu) \in \boldsymbol{R}^2 | h > 0, \mu \neq 0\})\\ &\simeq \{(h, \mu) \in \boldsymbol{R}^2 | h > 0, \mu \neq 0\} \times [(S^{2n-3}\backslash\Delta) \times \boldsymbol{R}^{2n-3}]\end{aligned}$$

ISBN 0-8053-0102-X

(ii) Let $(h_0, \mu_0) \in \mathbf{R}^2 \backslash \Sigma'_{H\times J}$, $h_0<0$, $\mu_0 \neq 0$. Then there exists a neighborhood U of (h_0, μ_0) in $\mathbf{R}^2$ and a diffeomorphism $\phi_{h,\mu}: I_{h,\mu} \to I_{h_0,\mu_0}$ smooth in (h,μ), that is, $((h,\mu),x) \mapsto \phi_{h,\mu}(x)$ is smooth.

Proof. **(i) In the proof of 10.4.17, we defined a diffeomorphism**

$$f_{h,\mu}: M_{h,\mu} \to S_Q^{2n-3} \times [0,\infty)$$

which is smooth in (h,μ). This induces a diffeomorphism [smooth in (h,μ)]

$$\beta_{2n-3}(f_{h,\mu}): \beta_{2n-3}(M_{h,\mu}) = I_{h,\mu} \to \beta_{2n-3}(S_Q^{2n-3} \times [0,\infty))$$

$$= (S^{2n-3} \backslash \Delta) \times \mathbf{R}^{2n-3}$$

and the map

$$F: \{(h,\mu) \in \mathbf{R}^2 | h>0, \mu \neq 0\} \times [(S^{2n-3} \backslash \Delta) \times \mathbf{R}^{2n-3}]$$

$$\to (H \times J)^{-1}(\{(h,\mu) \in \mathbf{R}^2 | h>0, \mu \neq 0\})$$

$$F((h,\mu),(z,t)) = \beta_{2n-3}(f_{h,\mu})^{-1}(z,t)$$

is the desired isomorphism of fiber bundles over the basis $\{(h,\mu) \in \mathbf{R}^2 | h>0, \mu \neq 0\}$.

(ii) For $(h_0, \mu_0) \in \mathbf{R}^2 \backslash \Sigma'_{H\times J}$ find an open neighborhood U such that $U \cap \Sigma'_{H\times J} = \varnothing$. Then for h close to h_0, and μ close to μ_0, $l_\mu((-\infty, h]) \approx l_{\mu_0}((-\infty, h_0])$, that is, using the family of diffeomorphisms $f|M_{h,\mu}$, $M_{h,\mu} \approx M_{h_0,\mu_0}$. Now take $\beta_{2n-3} \circ \alpha_1$ of this diffeomorphism to get the desired smooth family $\phi_{h,\mu}$. ■

10.4.24 Corollary. *In the planar n-body problem the bifurcation set is given by*

$$\Sigma_{H\times J} = \Sigma'_{H\times J} \cup (\mathbf{R} \times \{0\}) \cup (\{0\} \times \mathbf{R})$$

It should be mentioned that the methods by which we determined $I_{h,\mu}$ and $\hat{I}_{h,\mu}$ are not the only ones. R. Easton [1971] determines these manifolds using different techniques. We preferred Smale's method since it integrates naturally into the more general framework of the topological program discussed in Sect. 4.5.

It is legitimate to ask whether our discussion of the planar three-body problem is complete in view of the restrictions $d_1 < d_2 < d_3 < d_4$. The following conjecture of Smale would settle this doubt.

ISBN 0-8053-0102-X

10.4.25 Conjecture (Smale [1971b]). *For almost all choices of m_1, m_2, m_3 in the planar three-body problem, the numbers d_i, $i = 1, 2, 3, 4$ are distinct.*

We leave the discussion of $I_{h,\mu}$, $\hat{I}_{h,\mu}$, $D_{h,\mu}$, and $\hat{D}_{h,\mu}$ for other possible positions of d_4 among d_1, d_2, d_3 as well as cases in which some of them are equal as an exercise for the reader.

As we saw before, the manifolds $I_{h,\mu}$ are *not* compact and hence some solutions actually will "run off" the invariant manifolds in finite time. For the three-body problem, the reason for such bad behavior is due to collisions. For $\mu \neq 0$, triple collisions cannot occur.* There arises the natural question of how the integral manifolds $I_{h,\mu}$, $\mu \neq 0$ compactify and extend the flow to the compactification; that is, how do we regularize $I_{h,\mu}$? We refer the reader to the papers of Moser [1970], Easton [1972], and Lacomba [1975] for a discussion of the regularized submanifolds in the two, three, and restricted three-body problem, respectively. For $\mu = 0$, the results of McGehee (see the footnote at the beginning of this section) show that regularization is not possible, due to triple collisions.

Finally, it should be mentioned that an analysis of $I_{h,\mu}$ in the *spatial* three-body problem has been sketched by C. Simó [1975]. As was mentioned above, Palmore [1976b] has exhibited degenerate central configurations for the four-body problem. Relative equilibrium solutions in the four and restricted four-body problem are also discussed in C. Simó [1977].

*This result is due to Sundman and is proved in, for example, Pollard [1976, p. 66].

ISBN 0-8053-0102-X

APPENDIX*

The General Theory of Dynamical Systems and Classical Mechanics

A. N. Kolmogorov

INTRODUCTION

It came as a surprise to me that I would need to make an address at the final session of the Congress in this large hall, which up to now I had been familiar with more as a place for the performance of great musical masterpieces of the world under Mengelberg's conduction. The address that I have prepared, without taking into consideration the perspectives of such an esteemed position in the program of the present Congress, will be devoted to a rather specialized group of questions. My problem is to make clear the different paths that one may use to apply the basic ideas and results of present-day general measure theory and spectral theory of dynamical systems to the study of the conservative dynamical systems of classical mechanics. However, it seems to me that the theme that I have chosen can be of broader interest, since it is one of the examples of the birth of new, unexpected, and profound relationships between the different branches of classical and contemporary mathematics.

In his remarkable address at the Congress in 1900, Hilbert said that the unity of mathematics, the impossibility of dividing it into mutually independent branches, is a consequence of the very nature of our science. The most

---

***This appendix is an English translation of an address to the 1954 International Congress of Mathematicians by A. N. Kolmogorov [1957a], in which the first version of the stability theorem (8.3.6) was stated.**

ISBN 0-8053-0102-X

convincing confirmation of the validity of his view is the fact that, at every stage in the development of mathematics, there appear new joining points where, in the solution of quite specific problems, the concepts and methods of quite different mathematical disciplines become necessary and enter into a new interrelationship with each other. For the mathematics of the nineteenth century, one of these joining points was the complex question of integrating the systems of differential equations of classical mechanics, where the problems of mechanics and differential-equation theory were organically interwoven with the problems of the calculus of variations, many-dimensional differential geometry, the theory of analytic functions, and the theory of continuous groups.

After the appearance of Poincaré's works, the fundamental role of topology for this class of questions became clear. On the other hand, the Poincaré–Carathéodory recurrence theorem served as the starting point in the measure theory of dynamical systems, in the sense of the investigation of the properties of motions that take place at "almost all" initial states of a system. The "ergodic theory," which developed from this, has acquired various generalizations and has become an independent center of attraction and a junction in the web of methods and problems of various new divisions of mathematics (abstract measure theory, the theory of groups of linear operators in Hilbert and other infinite-dimensional spaces, the theory of random processes, etc.). At the preceding International Congress in 1950, the long address by Kakutani [23] was devoted to general questions in ergodic theory.

As we know, topological methods acquired significant applications in the theory of oscillations, in particular, in the solution of quite specific problems that arise in the study of automatic control systems, electrotechnology, etc. However, these real physical and technical applications deal primarily with nonconservative systems. Here, the problem usually reduces to finding individual asymptotically stable motions (in particular, stable rest points and stable limiting cycles) and to the study of pencils of integral curves that are attracted to these asymptotically stable motions.

In conservative systems, asymptotically stable motions are impossible. Therefore, the search for individual periodic motions, for example, has, for all its mathematical interest, only a restricted real physical interest in the case of conservative systems. Of special significance in the case of conservative systems is the measure-theoretic point of view, which enables us to study the properties of the basic set of motions. To this end, present-day general ergodic theory has produced a number of concepts that are extremely significant from a physical standpoint. However, our successes in an analytical sense from these contemporary points of view in handling the specific problems of classical mechanics have up to the present been more than restricted.

The question deals, in the first instance, with the following problem. Let us suppose that motion along an s-dimensional analytic manifold V^s is defined by a canonical system of differential equations with an analytic

ISBN 0-8053-0102-X

Hamiltonian function $H(q_1,\ldots,q_s,p_1,\ldots,p_s)$. Suppose also that there are k single-valued analytic first integrals $I_1,I_2,\ldots,I_k$ and that the conditions

$$I_1=C_1,\ldots,I_k=C_k$$

define an analytic manifold M^{2s-k} in the phase space Ω^{2s}. As we know, for almost all values of $C_1,\ldots,C_k$, there arises in a natural way an analytic invariant density on M^{2s-k}, which makes it possible to apply to the motions on M^{2s-k} the general principles of the measure theory of dynamical systems. It is natural to turn to more modern tools in cases in which, besides $I_1,\ldots,I_k$, there are no single-valued analytic first integrals independent of them, or in which the problem of finding them is too difficult and other classical analytic methods of carrying out the integration of the system are also inapplicable. In such cases, one must, by use of quantitative considerations, solve the question as to whether motion on M^{2s-k} is transitive or not (that is, whether almost all the manifold M^{2s-k} consists of a single unique ergodic set) and, in the case in which it is transitive, to determine the nature of the spectrum or, when it is not, study with accuracy up to a set of measure zero (or at least up to a set of small measure) the nature of the decomposition of M^{2s-k} into ergodic sets and the nature of the spectrum on these ergodic sets.

I know only two specific problems in classical mechanics in which this program has been completed to a greater or lesser degree:

1. For inertial motion along a closed surface V^2 with everywhere-negative curvature,* Hopf established in 1939 that motion on three-dimensional manifolds L_h^3 defined by the requirement that the energy $H=h$ be constant is transitive and that the spectrum is continuous (cf. [8]).

2. As will be shown later, in the case of inertial motion along an analytic surface that is sufficiently close to an ellipsoid in Euclidean three-space, the motion on L_h^3 is nontransitive and, up to a set of small measure, it can be decomposed into two-dimensional tori T^2 on each of which the motion is transitive and the spectrum discrete (cf. end of Sect. 2).

It seems to me, however, that the time has come when it should be possible to advance much more rapidly.

1 ANALYTIC DYNAMICAL SYSTEMS AND THEIR STABILITY PROPERTIES

The dynamical systems of classical mechanics constitute a special case of analytic dynamical systems with an integral invariant. The domain of such a dynamical system is an analytic n-dimensional manifold Ω^n (the phase space

*Perhaps it might be worthwhile to note that, in ordinary Euclidean space, one can define a closed surface V^2 of genus 1 and to place close to it a finite number of centers of attraction or repulsion that create on V^2 a potential of forces in such a way that the motion of a point mass on V^2 under the influence of these external forces will be mathematically equivalent to inertial motion in a metric possessing everywhere a negative curvature.

ISBN 0-8053-0102-X

of the system). Accordingly, admissible transformations of the coordinates $x_1,\ldots,x_n$ of a point $x \in \Omega^n$ will always be analytic.

The right-hand sides of the differential equations determining the motion

$$\frac{dx_\alpha}{dt} = F_\alpha(x_1,\ldots,x_n) \tag{1-1}$$

and the invariant plane generating the invariant measure

$$m(A) = \int_A M(x)\,dx_1 \cdots dx_n$$

will be assumed analytic functions of the coordinates.*

In line with what was said in the introduction, we shall concern ourselves primarily with canonical systems, systems in which $n=2s$, with a partition of the coordinates of the point $(q,p) \in \Omega^{2s}$ into two sets $q_1, q_2, \ldots, q_s$ and $p_1, \ldots, p_s$, with contact transformations as admissible transformations of coordinates, with equations of canonical form

$$\frac{dq_\alpha}{dt} = \frac{\partial H}{\partial p_\alpha}, \quad \frac{dp_\alpha}{dt} = -\frac{\partial H}{\partial q_\alpha} \tag{1-2}$$

and with invariant density

$$M(p,q) = 1$$

Particular attention will be given to the question as to what properties of dynamical systems, with "arbitrary" F_α and M (or an "arbitrary" function $H(q,p)$ in the case of canonical systems), are "typical" and which properties may occur only "exceptionally." However, this is quite a delicate question. An approach from the standpoint of the category of the corresponding sets in functional spaces of systems of functions $\{F_\alpha, M\}$ (or functions H) is, despite the known successes obtained in this direction in the general theory of abstract dynamical systems, interesting more as a means for proving existence than as a direct answer to arbitrarily stylized and idealized real inquiries by investigators in physics or mechanics. The approach from the standpoint of measure, on the other hand, is quite a sound and natural approach from the physical point of view (as was argued in detail, for example, by von Neumann [1]), but it runs into the problem of absence of a natural measure in functional spaces.

We shall follow two paths. In the first place, to obtain positive results stating that this or that type of dynamical system must be accepted as one of the essential, not "exceptional," systems, that cannot be "neglected" from any sensible point of view (similar to the way in which we neglect sets of measure

*Whenever we speak simply of "measure" without any other qualification, we mean the measure m.

ISBN 0-8053-0102-X

zero), we shall use the concept of stability in the sense of conservation of a given type of behavior of a dynamical system when there is a slight variation in the functions F_α and M or the function H. An arbitrary type of behavior of a dynamical system, for which there exists at least one example of its stable realization, must from this point of view be considered essential and may not be neglected. In accordance with the approach taken from the standpoint of analytic functions, "smallness" in the variation of the function $f_0(x)$ will be understood in the sense of change from a function $f_0(x)$ to a function of the form

$$f(x)=f_0(x)+\theta\varphi(x,\theta)$$

with a small value of the parameter θ, where the function φ is analytic with respect to the variables $x_1, x_2, \ldots, x_n, \theta$. Such an approach may be open to criticism, but by means of it one can obtain certain interesting results. When we may confine ourselves to closeness of the functions f_0 and f in the sense of closeness of their derivatives or arbitrary order, this will be pointed out.

To obtain negative results of the nonessential exceptional nature of a certain phenomenon, we shall apply only one somewhat artificial device: if on the class K of functions $f(x)$ it is possible to define a finite number of functionals

$$F_1(f), F_2(f), \ldots, F_r(f)$$

that in some sense or other may naturally be considered as assuming "generally speaking arbitrary" values

$$F_1(f)=C_1, \ldots, F_r(f)=C_r$$

in some region in the r-dimensional space of points $C=(C_1, \ldots, C_r)$, we shall consider an arbitrary phenomenon that takes place only when C is in a set of r-dimensional Lebesgue measure zero as exceptional and "negligible." I begin a survey of specific results with the application of this idea to the investigation of dynamical systems, the phase space of which is a two-dimensional torus.

2 DYNAMICAL SYSTEMS ON A TWO-DIMENSIONAL TORUS AND CERTAIN CANONICAL SYSTEMS WITH TWO DEGREES OF FREEDOM

In all that follows, by points on a torus T^2 we shall mean given circular coordinates x_1, x_2 (the point x does not change in the shift from x_α to $x_\alpha+2\pi$). The functions F_α in the right-hand members of the equations

$$\frac{dx_1}{dt}=F_1(x_1,x_2), \quad \frac{dx_2}{dt}=F_2(x_1,x_2)$$

ISBN 0-8053-0102-X

and the invariant density $M(x_1,x_2)$ will, in accordance with what was said above, be assumed analytic. We shall also assume that

$$F_1^2+F_2^2>0, \quad M>0 \tag{2-1}$$

For simplicity, we assume that the normalization condition $m(T^2)=1$ is satisfied. We introduce the mean frequencies of rotation

$$\lambda_1=\int_{T^2}F_1(x)\,dm, \quad \lambda_2=\int_{T^2}F_2(x)\,dm$$

A slight strengthening of the results of Poincaré, Denjoy, and Kneser lead in the present case to the conclusion that, by means of an analytic coordinate transformation, the equations of motion can be reduced to the form

$$\frac{dx_1}{dt}=\lambda_1 M(x_1,x_2), \quad \frac{dx_2}{dt}=\lambda_2 M(x_1,x_2)$$

It is well known that in the case of an irrational ratio

$$\gamma=\lambda_1/\lambda_2$$

all the trajectories are everywhere dense and the measure m is transitive. In addition, one can easily show, following Markov [2], that for irrational γ, a dynamical system is strongly ergodic; that is, it contains exactly one ergodic set E the points of which have with the appropriate measure, measure

$$\mu_e=cm$$

where c is a constant. The natural assertion that motions on a two-dimensional torus under conditions (2-1) possess "generally speaking" all the properties that we have just enumerated is already seen to apply to the principle, mentioned above, of neglecting cases in which some finite system of functionals (in the present case λ_1 and λ_2) assumes values in some set of measure 0 [in the present case, the set of points (λ_1,λ_2) with rational ratio γ].

In the article [3], I succeeded in proceeding somewhat further. Specifically, I showed that, under the assumption that there exist positive numbers c and h such that, for all integral r and s,

$$|r-s\gamma|\geqslant ch^s \tag{2-2}$$

the equations of motion can be reduced by an analytic transformation of coordinates to the form

$$\frac{dx_1}{dt}=\lambda_1, \quad \frac{dx_2}{dt}=\lambda_2 \tag{2-3}$$

As we know from the theory of Diophantine approximations, condition (2-2) is satisfied (for suitable c and h) for almost all irrational numbers γ.

ISBN 0-8053-0102-X

Thus, except for cases in which γ can be approximated "abnormally well" by fractions r/s, an analytic dynamical system with integral invariant on the torus T^2 under conditions (2-1) necessarily admits only almost-periodic and even more restrictively "conditionally periodic" motions with two independent frequencies λ_1 and λ_2.

As we know, many problems in classical mechanics with two degrees of freedom ($s=2$, $n=4$) in which the four-dimensional manifold Ω^4 is decomposed, with the exception of certain exceptional manifolds of no more than three dimensions, into the two-dimensional manifolds

$$L^2_{C_1C_2} = L^2(I_1 = C_1, I_2 = C_2)$$

because of the presence of two first integrals I_1 and I_2 that are single-valued on the entire manifold Ω^4. Since the four equations

$$\frac{\partial H}{\partial q_1} = \frac{\partial H}{\partial q_2} = \frac{\partial H}{\partial p_1} = \frac{\partial H}{\partial p_2} = 0$$

are satisfied at rest points, the set of these points in the case of an analytic function H is no more than countable. Therefore, they may fall into the manifold L^2 only as exceptions. From this we conclude that almost all compact manifolds L^2 are tori (since they are orientable, compact, two-dimensional manifolds admitting a vector field without zero vectors).

Problems of classical mechanics of the type that we have been considering are, as we know, always integrable. A qualitative investigation of the special problems of this type (motion under the influence of gravity along a surface of rotation, inertial motion along the surface of an ellipsoid in three-space, etc., the motion of a point along a plane under the influence of the Newtonian attraction of two immovable centers, etc.) also leads us to a large number of examples of the decomposition of the space Ω^4 basically into tori T^2 with windings that fill them everywhere densely from the trajectories of conditionally periodic motions with two independent frequencies λ_1 and λ_2. Among these tori there is, generally speaking, an everywhere dense set of tori that are, by virtue of the commensurability of the frequencies, decomposed into closed trajectories and a discrete set of singular manifolds of dimension <3 on which, in particular, rest points are placed and so-called asymptotic motions are set up. Consideration of these integrable problems yields a number of interesting examples of rather complicated partitions of the phase space Ω into ergodic sets with a remainder consisting of "nonregular points" that lie on the trajectories of asymptotic motions.*

In my article [3] referred to above, it is shown that, for exceptional irrational values of γ [that is, not satisfying condition (2-2)], there are indeed

*In connection with this, I mention that the extremely instructive qualitative analysis of the problem on the attraction by two immovable centers that was made in Charlier's well-known treatise has proven to be incomplete and partially erroneous. It has twice been corrected [4,5].

ISBN 0-8053-0102-X

a number of new possibilities, some of them rather unexpected for analytic systems (of this we shall speak later). However, in the problems of classical mechanics mentioned above, these exceptional cases fail to appear for an extremely simple reason: the transition to circular coordinates ξ_1, ξ_2 on the tori T^2 and to the parameters C_1 and C_2 of these tori in these problems is made by means of contact transformations. Therefore, the equations keep their canonical form

$$\frac{d\xi_\alpha}{dt} = \frac{\partial}{\partial C_\alpha} H, \quad \frac{dC_\alpha}{dt} = -\frac{\partial}{\partial \xi_\alpha} H$$

and since invariance of the tori T^2 is obtained only in the case

$$\frac{dC_1}{dt} = \frac{dC_2}{dt} = 0$$

then H depends only on C_1 and C_2, which leads, on each torus T^2, to equations (2-3) with constants λ_1 and λ_2 with no exceptions.

Therefore, the real significance for classical mechanics of the analysis that I have made of dynamical systems on T^2 depends on whether there are sufficiently important examples of canonical systems with two degrees of freedom that cannot be integrated by classical methods and in which invariant (with respect to the transformations S^t) two-dimensional tori play a significant role.

To show that such examples exist, we shall, following the study made by Birkhoff [6] of a neighborhood of an elliptic periodic motion, examine the system with circular coordinates q_1, q_2 and with momenta p_1, p_2 for which

$$H(q,p) = W(p)$$

The equations of motion take the form

$$\frac{dq_\alpha}{dt} = \frac{\partial W}{\partial p_\alpha}, \quad \frac{dp_\alpha}{dt} = 0$$

Obviously, the tori T_c^2 defined by the conditions

$$p_1 = c_1, \quad p_2 = c_2$$

are invariant and on each of them a periodic motion

$$\frac{dq_\alpha}{dt} = \lambda_\alpha(c) = \frac{\partial}{\partial c_\alpha} W(c_1, c_2)$$

arises, with two frequencies that are independent of C. Let us suppose that

ISBN 0-8053-0102-X

the Jacobian of the frequencies λ_α with respect to the momenta p_α is nonzero:

$$\left|\frac{\partial\lambda_\alpha}{\partial p_\beta}\right| = \left|\frac{\partial^2 W}{\partial p_\alpha \partial p_\beta}\right| \neq 0 \qquad (2\text{-}4)$$

It turns out that in this case, the partitioning of the region in question of the four-dimensional space Ω^4 into two-dimensional tori T^2 is basically stable with respect to small changes in H of the form

$$H(q,p,\theta) = W(p) + \theta S(q,p,\theta)$$

To obtain a precise formulation, let us consider a region $G \subset \Omega^4$ determined by the condition $p \in B$, where B is a bounded region in the plane of points p. Assuming that the functions W and S are analytic and that condition (2-4) is satisfied, we can prove that, for arbitrary $\varepsilon > 0$, there exist a $\delta > 0$ such that, for $|\theta| < \delta$, in the dynamical system

$$\frac{dq_\alpha}{dt} = \frac{\partial}{\partial p_\alpha} H(q,p,\theta), \qquad \frac{dp_\alpha}{dt} = -\frac{\partial}{\partial p_\alpha} H(q,p,\theta)$$

the entire region G except for a set of measure less than ε consists of invariant two-dimensional tori T^2 on each of which, in suitable (that is, depending analytically on (q,p)) circular coordinates ξ_1, ξ_2, the motion is determined by the equations

$$\frac{d\xi_1}{dt} = \lambda_1, \quad \frac{d\xi_2}{dt} = \lambda_2$$

where λ_1 and λ_2 are constant on each T^2, that is, they are conditionally periodic with two periods.

The proof consists in following the fate of the original tori T_c^2 with frequencies $\lambda_\alpha(c)$ satisfying condition (2-2) with varying θ and in showing that each such torus is not destroyed when θ is sufficiently small but is merely displaced in Ω keeping on itself the trajectories of conditionally periodic motions with constant frequencies λ_α.

Very likely, many of my listeners have already guessed that it is basically a matter of working out an idea already widely discussed in the literature on celestial mechanics, namely, the possibility of avoiding "abnormally small denominators" in calculating the perturbations of orbits. In contrast with the usual theory of perturbations, however, I obtain precise results instead of a conclusion as to the convergence of series of this or that approximation of finite order (with respect to θ). This is due to the fact that, instead of calculating the disturbed motion under fixed initial conditions, I modify the initial conditions themselves in such a way as to have motions with normal (in the sense of condition (2-2)) frequencies λ_α at all times when θ varies.

ISBN 0-8053-0102-X

I wish to make the following three remarks in connection with what has been said:

1. The theorem on the reducibility of motions on T^2 to the form (2-3) can be proven even under conditions of sufficiently high order of finite differentiability of the functions F_α and M (naturally with a corresponding weakening of the conclusion). The theorem on the conservation of tori in Ω^4, on the other hand, obviously has to require either that the functions $W(p)$ and $S(q,p,\theta)$ be analytic or that these functions have infinitely many derivatives satisfying certain restrictions on the order of their growth.

2. The exceptional set of measure less than ε foreseen in the second theorem may actually prove to be everywhere dense and, very likely, of positive measure for arbitrarily small θ. This is analogous to the "zones of instability" discovered by Birkhoff in his study of neighborhoods of elliptic periodic trajectories [6].

3. As one of the special cases to which all that has been said above applies, we may mention inertial motion along an analytic surface that is close to an ellipsoid in 3-space.

3 ARE DYNAMICAL SYSTEMS ON COMPACT MANIFOLDS "GENERALLY SPEAKING" TRANSITIVE, AND SHOULD WE CONSIDER A CONTINUOUS SPECTRUM AS THE "GENERAL" CASE AND A DISCRETE SPECTRUM AS THE "EXCEPTIONAL" CASE?

The hypothesis of the predominant occurrence of the transitive case and the case of a continuous spectrum (mixing) have been asserted more than once in connection with the "ergodic" hypotheses in physics. As applied to canonical systems, it is natural to consider both these hypotheses only for $(2s-1)$-dimensional invariant manifolds L_h^{2s-1}, which are defined by the requirement that the energy be constant:

$$H=h$$

and to apply them only to the case of compact manifolds L_h^{2s-1} since, on noncompact manifolds L_h^{2s-1}, in even the simplest problems there are "departing" trajectories (and they usually dominate from a standpoint of measure), of which we shall speak in Sect. 4. If the first hypothesis is relaxed, it is natural to apply the second not to the entire manifold Ω^n (or to L_h^{2s-1} in the case of canonical systems) but to those ergodic sets into which Ω^n is decomposed (neglecting, of course, ergodic sets the union of which is of measure zero).

In the application to analytic canonical systems, the answer to both questions is negative since the theorem on stability of the decomposition into tori that we asserted for systems with two degrees of freedom remains valid for an arbitrary number of degrees of freedom. If the equation

$$H(q,p,\theta)=W(p)+\theta S(q,p,\theta)$$

ISBN 0-8053-0102-X

holds in a $2s$-dimensional toroidal layer G of the phase space Ω^{2s}, then for $\theta=0$ this layer can be decomposed in an obvious manner into invariant s-dimensional tori T_p^s on each of which the motion is conditionally periodic with s periods. Also, if

$$\left|\frac{\partial^2 W}{\partial p_\alpha \partial p_\beta}\right| \neq 0$$

on almost all tori T_p^s, the periods are independent in the sense that

$$(n,\lambda)=\sum_\alpha n_\alpha \lambda_\alpha \neq 0$$

for arbitrary integers n_α. Therefore, the trajectories wind around the torus everywhere densely, the s-dimensional Lebesgue measure on T^s is transitive, and the entire torus constitutes a single ergodic set. Theorems 1 and 2 in my article [22] assert that, under the hypotheses described, this entire picture changes for small values of θ only in that certain tori corresponding to systems of frequencies for which the expressions (n,λ) decrease too rapidly with increasing

$$|n|=\sqrt{\sum n_\alpha^2}$$

may disappear. However, the majority of the tori T_p^s conserve the nature of the motions that arise on them and are only displaced in Ω^{2s}, continuing, for small values of θ, to fill G up to a set of small measure. Thus, for small changes in H, a dynamical system remains nontransitive and the region G remains, up to a remainder of small measure, partitionable into ergodic sets with discrete spectrum (with special nature mentioned).

In connection with this, it is interesting to note that certain physicists (see, for example, [7]) have made the hypothesis that the "general case" of a canonical dynamical system without departing trajectories is just the decomposition of Ω^{2s} into s-dimensional tori T^s on which there are conditionally periodic motions with s periods. Apparently, this idea is based only on the predominant attention that has been given to linear systems and to a restricted set of integrable classical problems. In any case, it should be noted that the methods of proving the theorem referred to above are connected in a very real way with just the problem of stratifying Ω^{2s} into tori T^s and are not applicable to stratifying it into tori of any other dimension $r>s$ or $r<s$.

The hypothesis stated above can hardly stand up in its general form since it is extremely likely that, for arbitrary s, there are examples of canonical systems with s degrees of freedom and with stable transitiveness and mixing on the manifolds L_h^{2s-1}. I have in mind motion along geodesics on compact manifolds V^s of constant negative curvature, that is, dynamical systems such

ISBN 0-8053-0102-X

that

$$H(q,p)=\sum_{\alpha\beta} g_{\alpha\beta}(q)p_\alpha p_\beta \tag{3-1}$$

where the q_α are coordinates on a compact manifold V^s of constant negative curvature and the $g_{\alpha\beta}$ are the components of a metric tensor on V^s.

The stability of negative curvature under small variations in the functions $g_{\alpha\beta}(q)$ requires no clarification. The difficulties consist only in the fact that variation of the functions $g_{\alpha\beta}(q)$ is not the only possible form of variation of the function $H(q,p)$, and the transitivity and mixing for $s>2$ remains proven only for the case of constant curvature whereas, with varying $g_{\alpha\beta}$, the curvature ceases to be constant. The second difficulty disappears in the case $s=2$, for which transitivity is proven even when the curvature is variable. However, the first of these is not significant if we confine ourselves to functions $H(q,p)$ of the form

$$H(q,p)=U(q)+\sum_{\alpha\beta} g_{\alpha\beta}(q)p_\alpha p_\beta \tag{3-2}$$

(with which classical mechanics is primarily concerned) since systems of the form (3-2) reduce to systems of the form (3-1) by a shift to a new metric.

If we remember what was said earlier regarding inertial motion along surfaces close to an ellipsoid in three-space, we conclude that, in even the simplest problems of classical mechanics, we need to consider as stable and hence worthy of equal and fundamental attention, at least the two cases that we have considered, one of which is connected with the transitivity on manifolds of constant energy and with continuous spectrum, the other with the absence of transitivity and with a primarily discrete spectrum.

I do not know of any analogous results regarding the stability of one general type of behavior or another of noncanonical dynamical systems with integral invariant and compact Ω''.

4 SOME REMARKS ON THE NONCOMPACT CASE

The distinctive feature of the noncompact case is the possibility of the existence of trajectories that depart, as $t\to+\infty$ or as $t\to-\infty$, from every compact subset of Ω. Here, I shall expound certain general facts from ergodic theory that are applicable for arbitrary continuous flows S^t in locally compact spaces Ω. Since a one-sided approach to infinity is possible only for trajectories constituting a set of measure zero, we first define a departing point x by the requirement that, for an arbitrary compact set K, there exists a T such that all points S^t_x, where $|t|>T$, lie outside K. We denote by Ω'' the set of all departing points. For purposes of detailed analysis of specific classical dynamical systems, it is expedient to construct "an individual ergodic theory," not in the purely metric variant expounded in the book of Hopf [9], but

ISBN 0-8053-0102-X

by following the earlier works by Hopf and Stepanov [10,11] and in certain places following directly the exposition in the memoir by Krylov and Bogolyubov [12], although this memoir deals also with the compact case.

In such an exposition, just as in the compact case, the concept of a regular point remains basic. A point x is said to be regular if there exists an invariant measure μ possessing the following properties:

1. $\mu(\Omega - I_x^c)=0$, where I_x^c is the closure of the trajectory passing through x.
2. $\mu(V_y)>0$ for an arbitrary neighborhood V_y of the point $y \in I_x$.
3. For arbitrary continuous functions $f(x)$ and $g(x)$ that are nonzero only on compact sets,

$$\lim \frac{\int_a^T f(S_x^t)\,dt}{\int_a^T g(S_x^t)\,dt} = \frac{\int_\Omega f\,d\mu}{\int_\Omega g\,d\mu}$$

provided

$$\int_\Omega g\,d\mu \neq 0$$

4. The measure μ is transitive.

Since there is no requirement of normalization, the measure μ is defined by a point only up to a constant factor. Nonetheless, we shall denote it by μ_x and shall call it the "individual measure" of the point x. Therefore, we make the following minor modification in the definition of ergodic sets: two points x and x' are said to belong to a single ergodic set if their individual measures coincide in the sense of coincidence up to a constant factor. Thus, the set Ω' of regular points can be represented as the sum of ergodic sets:

$$\Omega' = \sum \varepsilon$$

Of course, the measures μ_ε are defined by an ergodic set only up to a constant factor.

The individual ergodic theorem asserts that

$$\Omega = \Omega' + \Omega'' + N, \quad \text{where } \lambda(N)=0$$

with respect to an arbitrary invariant measure λ. Basically, however, the only thing that is essential for us is that $m(N)$ always be zero.

An arbitrary transitive invariant measure μ either is a measure μ_ε of some ergodic set ε or is of the form

$$\mu(A) = r_I(A \cap I)$$

where r_I is the "time" measure on the departing trajectory I. In contrast with

ISBN 0-8053-0102-X

the second trivial case, it is natural to call measures of the first type ergodic since corresponding to them is a set ε_μ, where

$$\mu_{\varepsilon_\mu} = \mu$$

Those considerations that, in the case of a compact space Ω, can be used to support the view that a compact dynamical system "of general type" is transitive, lead, when applied to noncompact dynamical systems, to the hypothesis that "in general" one or the other of two situations exists: Either the system is dissipative (that is, almost all its points depart), or the measure m is ergodic (obviously, in the second case, the departing points constitute only a set of measure zero).

Sometimes this hypothesis is also applied to individual classical problems in the following form. If a given problem has a certain number of first integrals and if there is no basis for expecting the discovery of new ones, then it seems likely that there is transitivity on the manifolds defined by giving the values of the known first integrals. In support of such a practice, it might be remarked that, from the investigations of Hedlund and Hopf, this alternative always holds for geodesic motions on spaces of constant negative curvature.

If it is known that a set of positive measure consisting of departing points exists, then, in accordance with what has been said, the hypothesis arises that the system is dissipative. Evidently, Birkhoff's assumption as to the dissipative nature of the three-body problem is based on considerations of this nature.

It seems probable, however, that it will prove possible to construct, by the methods indicated in Sect. 3 for canonical systems, examples of the stable simultaneous existence in Ω^{2s} of a dissipative subset of positive measure and a positive region G filled basically by s-dimensional invariant tori.

I mention the fact that, for the more elementary questions, specialists in the qualitative theory of differential equations have not occupied themselves to a great extent with specific problems dealing with departing trajectories of the different special types. A notable example of this is the fact that the refutation of Chazy's assertions regarding the impossibility of "exchange" and "capture" in the three-body problem [17, 18] was first done by the difficult (and logically unconvincing, without precise bounds for the errors!) method of numerical integration (cf. Becker [19] and Shmidt [20]) and only recently has an example of "capture" been constructed by Sitnikov quite simply and almost without numerical calculations [21].

5 TRANSITIVE MEASURES, SPECTRA, AND EIGENFUNCTIONS OF ANALYTIC SYSTEMS

We shall say that a measure μ in Ω^n is analytic if it can be written in the form

$$\mu(A) = \int_{V^k \cap A} f(\xi)\, d\xi, \ldots, d\xi_k$$

ISBN 0-8053-0102-X

where each V^k is an analytic manifold, locally closed in Ω^n, the dimension of which is $k \leqslant n$, and where f is an analytic function of the coordinates ξ_α on V^k (which depend analytically on the coordinates x_α in Ω^n).

The manifold V^k is uniquely determined by the measure μ (if it is not identically zero). Therefore, we may call the number k the dimension of the measure μ also.

We shall be especially interested in transitive measures. In this case, the manifold V^k must be invariant. Two invariant manifolds of the same dimension do not intersect, but two invariant manifolds of differing dimension can only be contained one in the other (specifically, the one of lower dimension in the one of higher dimension). Every invariant manifold carries on itself no more than one transitive measure. By virtue of what has been said, a system of analytic transitive measures has a relatively transparent structure.

Until a comparatively short time ago, only analytic transitive measures were known in analytic systems. Only recently, Grabar' [13], by constructing an analytic analog of an example of Markov (an analytic irreducible but not strictly ergodic dynamical system) gave an example of a nonanalytic transitive measure in an analytic system. However, it may prove that the union of all nonanalytic ergodic sets is always negligible in the sense of the basic measure m.

Ergodic sets are unambiguously defined by their measures μ_ε which, by their very definition, are transitive.

With regard to ergodic sets corresponding to analytical transitive measures (that do not reduce to the measure μ_ε of any trajectory), we know only that, in the case in which the measure μ_ε is analytic, an ergodic set is contained in the support V^r of the measure μ_ε since it is everywhere dense in it; however, even in certain simple classical examples, the difference $V^r - \varepsilon$ may also be everywhere dense in V^r.

The spectral properties of transitive measures on analytic systems have been only slightly studied.

Discrete spectra have as yet been obtained only with a finite basis of independent frequencies

$$\lambda_1, \lambda_2, \ldots, \lambda_x$$

Also, for analytic measures, the number of independent frequencies coincides in all known cases with the dimension.

A continuous spectrum has been completely determined only recently by Gel'fand and Fomin [14, 15] for certain cases of geodesic motions on surfaces of constant negative curvature. In these cases, it proved to be a Lebesgue spectrum of countable multiplicity.

The possibility is not excluded that only these cases (a discrete spectrum with a finite number of independent frequencies and a Lebesgue spectrum of countable multiplicity) are possible for analytic transitive measures or that they alone are the general typical cases in some sense or other.

ISBN 0-8053-0102-X

For nonanalytic transitive measures, it is more likely that their structure is completely arbitrary. This would be the case without doubt if someone were to establish an analytic analog of Kakutani's theorem [16] on isometric embedding of an arbitrary flow in the flow of a continuous dynamical system.

With regard to the eigenfunctions, we pause only for an example of an analytic dynamical system on a two-dimensional torus T^2 with discrete spectrum and everywhere-continuous eigenfunctions. Of course, this example, associated with a ratio $\gamma=\lambda_1/\lambda_2$ of average frequencies that can be approximated abnormally well by rational fractions r/s, indicates by its very origin that we are dealing not with a typical but with an exceptional phenomenon.

To clarify the question in greater detail, let us again look at the equations of motion on a two-dimensional torus, introducing into these equations a parameter θ that varies in some interval $[\theta_1,\theta_2]$:

$$\frac{dx_\alpha}{dt}=F_\alpha(x_1,x_2,\theta)$$

We shall assume that the functions $F_\alpha(x_1,x_2,\theta)$ are analytic. Obviously, the ratio of mean frequencies $\gamma(\theta)$ is also an analytic function of θ. If $\gamma(\theta)$ is not constant, then the set R of all θ for which it is possible to reduce the system analytically to the form

$$\frac{d\xi_\alpha}{dt}=\lambda_\alpha$$

will occupy almost all the interval $[\theta_1,\theta_2]$. The eigenfunctions

$$\varphi_{mn}=e^{i(m\lambda_1+n\lambda_2)}$$

when we return to the original coordinates x_1 and x_2 will, for $\theta\in R$, be analytic functions of x_1 and x_2. Generally speaking, however, even on R they will be everywhere discontinuous with respect to θ on that set. Also, this discontinuity cannot be removed by deleting from R a set of measure zero. These facts are considerably more significant than the fact that $\varphi_{mn}(x_1,x_2,\theta)$ can be defined even at certain points of the remainder set, $[\theta_1,\theta_2]\backslash R$ of measure zero, by virtue of the admissibility of their nonanalyticity and discontinuity with respect to x_1 and x_2.

It is possible that the dependence of $\varphi_{mn}(x_1,x_2,\theta)$ on the parameter θ on the set R is related to the class of functions of the type of monogenic Borel functions [24] and, despite its everywhere-discontinuous nature, will admit investigation by appropriate analytical tools.

CONCLUSION

I shall consider my purpose attained if I have succeeded in convincing my listeners that, despite the great difficulties and the restricted nature of the results obtained up to now, the problem posed of using general concepts of

ISBN 0-8053-0102-X

present-day ergodic theory for a qualitative analysis of motion in analytic and, in particular, canonical dynamical systems deserves considerable attention on the part of investigators who are capable of grasping the many-sided relationships with the most varied divisions of mathematics that are disclosed here. In conclusion, I wish to thank the organizing committee of the Congress for the opportunity presented to me of reading this paper and for the kind help in reproducing the abstract with formulas and bibliographic references, and all those present for the attention that they have shown me on this last day of our meetings, when everyone is already satiated with the enormous volume of addresses given on the preceding days.

BIBLIOGRAPHY

[1] J. von Neumann, *Mathematische Grundlagen der Quantenmechanik*, Berlin, 1932.
[2] A. A. Markov, *Trudy vtorogo vsesoyuznogo matematicheskogo s"yezda* [Proc. Second All-Union Math. Congr.], Vol. II, pp. 227–231 (1934).
[3] A. N. Kolmogorov, *Doklady Akad. nauk* **93**, No. 5, 763–766 (1953).
[4] H. J. Tallquist, *Acta Soc. Sci. Fennicae*, No. 3.A.T.1, No. 5 (1927).
[5] G. K. Badalyan, *Trudy vtorogo vsesoyuznogo matematicheskogo s"yezda* [Proc. Second All-Union Math. Congr.], Vol. II, pp. 239–241 (1934).
[6] G. D. Birkhoff, *Dynamical Systems*. Colloq. Publ. IX, Second Ed., Amer. Math. Soc., Providence, R.I., 1966.
[7] L. Landau and L. Pyatigorskiy, *Mekhanika*, 1940.
[8] E. Hopf, *Ber. Verh. Sächs. Akad. Wiss. Leipzig* **91**, No. 3, 261–304 (1939).
[9] E. Hopf, *Ergodentheorie*, Berlin, 1937.
[10] E. Hopf, *Math. Ann.* **103**, 710 (1930).
[11] V. V. Stepanov, *Compositio Math.* No. 3, 239 (1936).
[12] N. M. Krylov and N. N. Bogolyubov, *Ann. of Math.* **38** (1937).
[13] M. I. Grabar', *Doklady Akad. nauk* **95**, No. 1, 9–12 (1954).
[14] I. M. Gel'fand and S. V. Fomin, *Doklady Akad. nauk* **76**, No. 6, 771–774 (1951).
[15] I. M. Gel'fand and S. V. Fomin, *Uspekhi matem. nauk* **7**, No. 1, 118–137 (1952).
[16] S. Kakutani, *Proc. Nat. Acad. Sci. U.S.A.* **28**, No. 1, 16–21 (1942).
[17] I. Chazy, *J. de Math.* **8**, 353 (1929).
[18] I. Chazy, *Bull. Astr.* **8** (1952).
[19] L. Becker, *Monthly Notices* **80**, No. 6 (1920).
[20] O. Yu. Shmidt, *Doklady Akad. nauk* **58**, No. 2, 213–216 (1947).
[21] K. A. Sitnikov, *Matematich. sbornik* **32**, No. 3, 693–705 (1953).
[22] A. N. Kolmogorov, *Doklady Akad. nauk* **98**, No. 4 (1954).
[23] S. Kakutani, *Proc. Intern. Congr. Math.* **2**, 128–142 (1950).
[24] E. Borel, *Leçons sur les fonctions monogènes uniformes d'une variable complexe*, Paris, 1917.

ISBN 0-8053-0102-X

Bibliography

Ablowitz, M. J., Kaup, D. J., Newell, A. C., and Segur, H. 1974. The inverse scattering transformation, Fourier analysis for non-linear problems. *Studies in Appl. Math*. 53:249–315.

Abraham, R.

1963.a Transversality in manifolds of mappings. *Bull. Am. Math. Soc*. 69 (4):470–474.

1970. Bumpy metrics. *Proc. Symp. Pure Math*. 14:1–3.

1971. Predictions for the future of differential equations. In Chillingworth [1971]:163–166.

1972*a*. Introduction to morphology. In Ravatin [1972] 4:38–114.

1972*b*. Hamiltonian catastrophes. In Ravatin [1972] 4(1):1–37.

1978. Dynasim: Exploratory research in bifurcations using interactive computer graphics. *Annals. N.Y. Acad. Sci*. (to appear).

Abraham, R., and Robbin, J. 1967. *Transversal mappings and flows*. Benjamin-Cummings, Reading, Mass.

Abraham, R., and Smale, S. 1970. Nongenericity of Ω-stability. In Chern and Smale [1970] 14:5–8.

Adams, J. F. 1969. *Lectures on Lie groups*. Benjamin-Cummings, Reading, Mass.

Adler, M. 1977. Some finite dimensional integrable systems and their scattering behavior, *Comm. Pure Appl. Math*. 25:195–230. [1978–9] in preparation.

Adler, M., and Moser, J. 1977. On a class of polynomials connected with the Korteweg de Vries equation. MRC Preprint No. 1751.

Airault, H., McKean, H. P., and Moser, J. 1977. Rational and elliptic solutions of the Korteweg–De Vries equation and a related many-body problem. *Comm. Pure Appl. Math* 30(1):95–148.

Alekseev, V. 1970. Sur l'alluve finale du mouvement dans le problème des trois corps. *Actes du Congr. Intern. des Math*. 2:893–907.

Alexander, J. C., and Yorke, J. A. 1978. Global Hopf bifurcation. *Am. J. Math*. (to appear).

Almgren, F. J. 1966. *Plateau's problem: An invitation to varifold geometry*. Benjamin-Cummings, Reading, Mass.

Andronov, A. A., and Chaikin, C. E. 1949. *Theory of oscillations*. Princeton Univ. Press, Princeton, N.J.

Andronov, A. A., Chaikin, C. E., and Witt, A. 1966. *Theory of oscillations*. Princeton Univ. Press, Princeton, N.J.

ISBN 0-8053-0102-X

Andronov, A. A., and Leontovich, E. 1938. Sur la théorie de la variation de la structure qualitative de la division dú plan en trajectoires. *Dokl. Akad. Nauk. SSSR* 21:427–430.

Andronov, A. A., Leontovich, E., Gordon, I., and Maier, A. 1971. *Theory of bifurcations of dynamical systems in the plane.* Israel Program of Scientific Translations, Jerusalem (trans. Russian).

Andronov, A. A., and Pontriagin, L. 1937. Systèmes grossiers. *Dokl. Akad. Nauk. SSSR* 14:247–251.

·Andronov, A. A., and Witt, A. 1930. Sur la théorie mathematiques des autooscillations. *C.R. Acad. Sci. Paris* 190:256–258.

Anosov, D.

1962.*a* Roughness of geodesic flows on compact Riemannian manifolds of negative curvature. *Dokl. Akad. Nauk. SSSR* 145:707–709. *Sov. Math.* 3:1068–1069.

1967. Geodesic flows on compact Riemannian manifolds of negative curvature. Proc. Steklov Math. Inst. (90).

Apostol, T. 1957. *Mathematical analysis*. Addison-Wesley, Reading, Mass.

Appell, P. 1941, 1953. *Traité de mechanique rationelles*, Vols. I and II. Gauthier-Villars, Paris.

Arens, R.

1964*a*. Differential-geometric elements of analytic dynamics. *J. Math. Anal. Appl.* 9:165–202.

1964*b*. Normal form for a Pfaffian. *Pac. J. Math.* 14:1–8.

1966. Laws like Newtons. *J. Math. Phys.* 7:1341–1348.

1967. Quantum-mechanization of dynamical systems. *J. Math. Anal. Appl.* 19:337–385.

1970. A quantum dynamical relativistically invariant rigid body system. *Trans. Am. Math. Soc.* 147:153–201.

1971. Classical Lorentz invariant particles. *J. Math. Phys.* 12:2415–2422.

1972. An analysis of relativistic two-particle interactions. *Arch. Rat. Mech. Anal.* 47:255–271.

1973. Hamiltonian formalism for noninvariant dynamics. *Comm. Math. Phys.* 34:91–110.

1974. Motions of relativistic Hamiltonian interactions. *Nuovo Cimento* 21:395–409.

1975. Models of Hamiltonian several-particle interactions. *J. Math. Phys.* 16:1191–1198.

Arens, R. and Babbitt, D. G.

1965. Algebraic difficulties of preserving dynamical relations when forming quantum-mechanical operators. *J. Math. Phys.* 6:1071–1075.

1969. The geometry of relativistic *n*-particle interactions. *Pac. J. Math.* 28:243–274.

Arenstorf, R. F.

1963*a*. Periodic solutions of the restricted three-body problem representing analytic continuations of Keplerian elliptic motions. *Am. J. Math.* 85:27–35.

1963*b*. New regularization of the restricted problem of three-bodies. *Astronom. J.* 68:548–555.

1963*c*. Periodic trajectories passing near both masses of the restricted three-body problem. *Proc. Fourteenth Intern. Astronautical Congr., Paris* 4:85–97.

1966. A new method of perturbation theory and its application to the satellite problem of celestial mechanics. *J. Reine Angew. Math.* 221:113–145.

1967. New periodic solutions of the plane three-body problem. *Proc. Intern. Symp. on Differential Equations and Dynamical Systems (1965)*. Academic, New York.

Arms, J.

1977. Linearization stability of the coupled Einstein–Maxwell system. *J. Math. Phys.* 18:830–833.

1978. Linearization stability of gauge theories. Thesis, Univ. of Calif. Berkeley, Ca.

Arms, J., Fischer, A., and Marsden, J. 1975. Une approche symplectique pour des théorèmes de décomposition en géométrie ou relativité générale. *C.R. Acad. Sci. Paris*. 281:517–520.

Arnold, V. I.

1961. The stability of the equilibrium position of a Hamiltonian system of ordinary differential equations in the general elliptic case. *Sov. Math. Dokl.* 2:247–249.

1962. On the classical perturbation theory and the stability problem of planetary systems. *Dokl. Akad. Nauk. SSSR.* 145:481–490.

1963*a*. Proof of A. N. Kolmogorov's theorem on the preservation of quasi-periodic motions under small perturbations of the Hamiltonian. (Russian) *Usp. Mat. Nauk. SSSR*

ISBN 0-8053-0102-X

18:13–40; (English) *Russian Math. Surveys* 18:9–36.
1963*b*. Small divisor problems in classical and celestial mechanics. Russian *Usp. Mat. Nauk.* 18:91–192; (English) *Russian Math. Surveys* 18:85–192.
1964. Instability of dynamical systems with several degrees of freedom. *Dokl. Akad. Nauk. SSSR* 156:9–12.
1965. Sur une propiètè topologique des applications globelement canonique de la mécanique classique. *C.R. Acad. Sci. Paris* 26:3719–3722.
1966. Sur la geometrie differentielle des groupes de Lie de dimension infinie et ses applications a l'hydrodynamique des fluids parfaits. *Ann. Inst. Fourier Grenoble* 16(1):319–361.
1967. Characteristic class entering in quantization conditions. *Funct. Anal. Appl.* 1(1):1–13.
1968. Singularities of smooth mappings. *Russian Math Surveys* 23:1–43.
1969. One-dimensional cohomologies of Lie algebras of nondivergent vectorfields and rotation number of dynamic systems. *Funct. Anal. Appl.* 3:319–321.
1972. Lectures on bifurcations in versal families. *Russ. Math. Surveys* 27:54–123.
1976. Bifurcations of invariant manifolds of differential equations and normal forms in neighborhoods of elliptic curves. *Funct. Anal. Appl.* 10(4):1–12.
1978. *Mathematical methods of classical mechanics.* MIR (Moscow, 1975), Springer Graduate Texts in Math. No. 60 Springer-Verlag, New York.
Arnold, V. I., and Avez, A. 1967. *Théorie ergodique des systèmes dynamiques.* Gauthier-Villars, Paris (English edition: Benjamin-Cummings, Reading, Mass., 1968).
Arnowitt, R., Deser, S. and Misner, C. 1962. The dynamics of general relativity, in *Gravitation, an introduction to current research.* L. Witten (ed.) Wiley, New York.
Arraut, J. 1966. Note on structural stability. *Bull. Am. Math. Soc.* 72:542–544.
Artin, E. 1957. *Geometric algebra*, Wiley, New York.
Avez, A. 1974. Représentation de l'algèbre de Lie des symplectomorphismes par des opérateurs bornés, *C.R. Acad. Sci.* 279:785–787.
Avez, A., Lichnerowicz, A., and Diaz-Miranda, A. 1974. Sur l'algèbre des automorphismes infinitésimaux d'une variété symplectique. *J. Diff. Geom.* 9:1–40.
Bacry, H., and Levy-Leblond, J. M. 1968. Possible kinematics. *J. Math. Phys.* 9:1605–1614.
Bacry, H., Ruegg, H., and Souriau, J. M. 1966. Dynamical groups and spherical potentials in classical mechanics. *Commun. Math. Phys.* 3:323–333.
Ball, J. M.
1974. Continuity properties of nonlinear semigroups. *J. Funct. Anal.* 17:91–103.
1973*a*. Saddle point analysis for an ordinary differential equation in a Banach space and an application to dynamic buckling of a beam, in *Nonlinear elasticity*, pp. 93–160, J. Dickey (ed.). Academic, N.Y.
1973*b*. Stability theory for an extensible beam. *J. Diff. Equations.* 14:399–418.
1977. Convexity conditions and existence theorems in elasticity. *Arch. Rat. Mech. Anal.* 63:337–403.
Ball, J. M., and Carr, J. 1976. Decay to zero in critical cases of second-order ordinary differential equations of Duffing type. *Arch. Rat. Mech. Anal.* 63:47–57.
Banyaga, A.
1975*a*. Sur le groupe des difféomorphismes symplectiques, differential topology and geometry. *Lec. Notes Math.* 484:50–56.
1975*b*. Sur le groupe de difféomorphismes qui préservent une forme de contact reguliére, *C.R. Acad. Sci. Paris., Sér. A* 281:707–709.
Bargmann, V. 1954. On unitary ray representations of continuous groups, *Ann. of Math.* (2):1–46.
Barrar, R. B.
1965*a*. Existence of periodic orbits of the second kind in the restricted problem of three bodies. *Astron. J.* 70:3–5.
1965*b*. A new proof of a theorem of J. Moser concerning the restricted problem of three bodies. *Math. Ann.* 160:363–369.
1972. Periodic orbits of the second kind, *Indiana Univ. Math. J.* 22:33–41
de Barros, C. M. 1975. Systèmes mécaniques sur une variété Banachique. *C.R. Acad. Sci. Paris, Sér. A.* 280:1017–1020.
Bass, R. W. 1969. A characterization of central configuration solutions of the *n*-body problem,

ISBN 0-8053-0102-X

leading to a numerical technique for finding them. *Advan. Engin. Sci.* 1:323–330.

Bell, J. S. 1969. On the problem of hidden variables in quantum mechanics. *Rev. Mod. Phys.* 38:447–552.

Ben-Abraham, S. I., and Lonke, A. 1973. Quantization of a general dynamical system. *J. Math. Phys.* 14(12):1935–1937.

Benjamin, T. B.

1972. Stability of solitary waves. *Proc. Roy. Soc. London.* A328:153.

1974. Lectures on nonlinear wave motion: nonlinear wave motion. *Lec. App. Math.* 15:3–47.

Benjamin, T. B. and Ursell, F.

1954. The stability of a plane free surface of a liquid in vertical periodic motion. *Proc. Roy. Soc. (London) Ser. A.* 225:505–517

Benton, S. H. 1977. *The Hamilton–Jacobi equation.* Academic, New York.

Berezin, F. A. 1974. Quantization. *Izv. Akad. Nauk. SSSR* 38:1109–1165.

Berger, M.

1970. On periodic solutions of second-order Hamiltonian systems I. *J. Math. Anal. Appl.* 29(3):512–522.

1971. On periodic solutions of second-order Hamiltonian systems II. *Am. J. Math.* 93:1–10.

Bernard, P. See Ratiu and Bernard.

Bhatia, N. P., and Szegö, G. P. 1970. Stability theory of dynamical systems. *Grundlehren Math. Wiss.* Band 161. Springer-Verlag, Springer, New York.

Bibikov. J. N. 1973. A sharpening of a theorem of Moser. *Sov. Math. Dokl.* 14:1769–1773.

Bichteler, K. 1968. Global existence of spin structures for gravitational fields. *J. Math. Phys.* 9:813–815.

Birkhoff, G. D.

1913. Proof of Poincaré's geometric theorem. *Trans. Am. Math. Soc.* 14:14–22.

1915. The restricted problem of three bodies. *Rend. Circ. Mat. Palermo* 39:1–70.

1917. Dynamical systems with two degrees of freedom. *Trans. Am. Math. Soc.* 18:199–300.

1922. Surface transformations and their dynamical applications. *Acta Math.* 43:1–119.

1927. Stability and the equations of dynamics. *Am. J. Math.* 49:1–38.

1931. Proof of the ergodic theorem. *Proc. Nat. Acad. Sci.* 17:656–660.

1933. Quantum Mechanics and asymptotic series. *Bull. Am. Math. Soc.* 39:681–700.

1935. Nouvelles recherches sur les systèmes dynamiques. *Memoriae Pont. Acad. Sci. Novi Lyncaei* 1:85–216.

1950. *Collected mathematical papers.* Am. Math. Soc., Providence, R.I.

1966. *Dynamical systems.* Colloq. Publ. IX, 2nd ed. Am. Math. Soc., Providence, R.I.

Birkhoff, G. D., and Lewis, D. C. 1933. On the periodic motions near a given periodic motion of a dynamical system. *Annali di Matem.* 12:117–133.

Bishop, R., and Crittenden, R. 1964. *Geometry of manifolds.* Academic, New York.

Bishop, R., and Goldberg, S. 1968. *Tensor analysis on manifolds.* Macmillan, New York.

Blattner, R. J.

1973. Quantization and representation theory, in Harmonic Analysis on Homogeneous Spaces. *Proc. Symp. Pure Math.* (Am. Math. Soc.) 26:147–165.

1975*a*. Intertwining operators and the half density pairing. Coll. d'Analyse Harmonique Non-commutative. Marseille-Luminy. *Lec. Notes Math.* (Springer) 466:1–12.

1975*b*. Pairing of half-form spaces, in Proc. of "Colloque Symplectique," Aix-en Provence, CNRS Publ. no. 237, Paris.

1977. The metalinear geometry of non-real polarizations; Conf. on Diff. Geometric Methods in Math. Phys., Bonn. *Lec. Notes Math.* vol. 570. Springer, New York.

Bleuler, K., and Reetz, A. (eds.). 1977. *Differential geometrical methods in mathematical physics. Lec. Notes Math.* (Springer) 570.

Bochner, S., and Montgomery, D. 1945. Groups of differentiable and real or complex analytic transformations. *Ann. Math.* 46:685–694.

Bogoljubov, N. N., Mitropoliskii, J. A., and Samoilenko, A. M. 1976. *Methods of accelerated convergence in nonlinear mechanics.* Springer, New York.

Bohm, D., and Bub, J. 1966. A proposed solution of the measurement problem in quantum mechanics by a hidden variable theory. *Rev. Mod. Phys.* 38:453.

Bolza, O. 1973. *Lectures on the calculus of variations.* Chelsea, New York.

Bona, J. 1975. On the stability theory of solitary waves. *Proc. Roy. Soc. London* A344:363.

Bona, J., and Smith, R. 1975. The initial value problem for the Korteweg–de Vries equation. *Phil.*

ISBN 0-8053-0102-X

Trans. Roy. Soc. London A278:555–601.
Bonic, R., and Frampton, J. 1966. Smooth functions on Banach manifolds. *J. Math. Mech.* 16:877–898.
Bony, J. M. 1969. Principe du maximum, inegalité de Harnack et unicité du problème de Cauchy pour les operateurs elliptiques dégenerés. *Ann. Inst. Fourier, Grenoble* 19:277–304.
Born, M. 1949. Reciprocity theory of elementary particles. *Rev. Mod. Phys.* 21(3):463–473.
Bott, R.
1954*a*. Nondegenerate critical manifolds. *Ann. Math.* 60(2):248–261.
1954*b*. On manifolds all of whose geodesics are closed. *Ann. Math.* 60(3):375–382.
Bottkol, M. 1977. Bifurcation of periodic orbits on manifolds, and Hamiltonian systems. *Bull. Am. Math. Soc.* 83:1060–1062.
Bouligand, G. 1935. Sur la stabilité des propositions mathématiques. *Acad. Roy. Belg. Bull. Cl. Sci.* 21(5):277–282; 776–779.
Bourbaki, N. 1971. *Variétés differentielles et analytiques, Fascicule de résultats*, 33. Hermann, Paris.
Bowen, R.
1975. *Equilibrium states and the ergodic theory of Anosov diffeomorphisms. Lec. Notes Math.* 470. Springer, New York.
1977. *On Axiom A diffeomorphisms*, CBMS no. 35, AMS.
Brauer, R., and Weyl, H. 1935. Spinors in *n*-dimensions. *Am. J. Math.* 57:425–449.
Braun, M. 1973. On the applicability of the third integral of motion. *J. Diff. Equations* 13:300–318.
Bredon, G. E. 1972. *Introduction to compact transformation groups.* Academic, New York.
Brezis, H. 1970. On a characterization of flow invariant sets. *Commun. Pure Appl. Math.* 23:261–263.
Brjuno, A. D.
1970. Instability in Hamilton's system and the distribution of asteroids. *Math. USSSR Sb.* 12:271–312.
1971. The analytic form of differential equations. *Trans. Moscow Math. Soc.* 25:131–288.
1972. The analytic form of differential equations. *Trans. Moscow Math. Soc.* 26(1972): 199–239.
1974. Analytic integral manifolds. *Sov. Math. Dokl.* 15(3):768–772.
1975. Integral analytic sets. *Sov. Math. Dokl.* 16(1):224–228.
1976. Normal forms for nonlinear systems in *Fourteenth Intern. Congress of Theor. and Appl. Mech.* The Netherlands, August 1976.
Brouwer, D., and Clemence, G. M. 1961. *Methods of celestial mechanics*. Academic, New York.
Brown, E. W. 1911. On the new family of periodic orbits in the problem of three bodies, and on the oscillating orbits about triangular equilibrium points in the problem of three bodies. *Mon. Not. Roy. Astro. Soc.* 71:438–492.
Brunet, P. 1938. *Etude historique sur le Principe de la moindre Action.* Hermann, Paris.
Brunovsky, P.
1970. On one-parameter families of diffeomorphisms. *Comment. Math. Univ. Carolinae.* 11:559–582.
1971. One-parameter families of diffeomorphisms. In Chillingworth [1971]:29–33.
Bruns, H. 1887. Über die Integrale des Vielkörper-Problems. *Acta Math.* 11:25–96.
Bruslinskaya, N. N.
1961. Qualitative integration of a system of *n* differential equations in a region containing a singular point and a limit cycle. (Russian). *Dokl. Akad. Nauk., SSSR* 139:9–12.
1965*a*. The behavior of solutions of the equations of hydrodynamics when the Reynolds number passes through a critical value. (Russian) *Dokl. Akad. Nauk. SSSR* 6(4):724–728.
1965*b*. The origin of cells and rings at near-critical Reynolds numbers (Russian). *Uspekhi Mat. Nauk.* 20:259.
Buchner, M. 1970. On the generic nature of H1 for Hamiltonian vectorfields. In Chern and Smale [1970] 14:51–54.
Buchner, M., Marsden, J., and Schecter, S. 1978. A differential topology approach to bifurcation at multiple eigenvalues (to appear).
Burghelea, D., Albu, A., and Ratiu, T. 1975. *Compact Lie group actions* (in Romanian), Monografii Matematice. 5 (Universitatea Timisoara).
Burghelea, D., Hangan Th., Moscovici, H., and Verona, A. 1973. *Introduction to differential*

ISBN 0-8053-0102-X

topology(in Romanian) Editura Stiintifica, Bucuresti.
Burgoyne, N., and Cushman, R.
1974. Normal forms for real linear Hamiltonian systems with purely imaginary eigenvalues. *Cel. Mech.* 8:435–443.
1976. Normal forms for real linear Hamiltonian systems. Preprint No. 28, University of Utrecht.
Cabral, H. E. 1973. On the integral manifolds in the N-body problem. *Inv. Math.* 20:59–72.
Calabi, E. 1970. On the group of automorphisms of a symplectic manifold, in *Problems in Analysis*: A symposium in honor of Solomon Bochner, in Gunning, R. C., ed., Math. Series 31, Princeton Univ. Press, Princeton, N.J.
Carathéodory, C. 1965. *Calculus of variations and partial differential equations*. Holden-Day, San Francisco.
Caratù, G., Marmo, G., Saletan, E. J., Simoni, A., and Vitale, B. 1978 Invariance and symmetry in classical mechanics (preprint).
Cartan, E. 1922. *Leçons sur les Invariants intégraux.* Hermann, Paris.
Cartwright, M. L. 1964. From non-linear oscillations to topological dynamics. *J. London Math. Soc.* 39:193–201.
Casal, P. 1969. Cinématique des milieux continus. Marseille notes.
Chern, S. S. 1972. Geometry of characteristic classes, in *Proc. Thirteenth Biennial Sem. Can. Math Cong.* Halifax J. R. Vanstone, ed., 1–40.
Chern, S. S., and Smale, S. (eds.). 1970. *Proceedings of the Symposium in Pure Mathematics XIV, Global Analysis*, vols. 14, 15, 16. Am. Math. Soc., Providence, R.I.
Chernoff, P. 1973. Essential self-adjointness of powers of generators of hyperbolic equations, *J. Funct. Anal.* 12:401–414.
Chernoff, P., and Marsden, J.
1970. On continuity and smoothness of group actions. *Bull. Am. Math. Soc.* 76:1044.
1974. Properties of infinite dimensional Hamiltonian systems. *Lec. Notes Math. 425*. [Also Colloq. Intern. CNRS *237* (1975):313–330.]
Cherry, T. M. 1928. On the solution of Hamiltonian systems of differential equations in the neighbourhood of a singular point. *Proc. London Math. Soc.* 27(2):151–170.
Chevalley, C. 1946. *Theory of Lie groups.* Princeton Univ. Press, Princeton, N.J.
Chevallier, D. P.
1975*a*. Structures de variétés bimodelées et dynamique analytique des milieux continus. *Ann. Inst. H. Poincaré* 21A:43–76.
1975*b*. Variétés bimodelées et systèmes dynamiques en dimension infinie. Colloq. Intern. *CNRS*, no. 237, Paris.
Chichilnisky, G. 1972. Group actions on spin manifolds. *Trans. Am. Math. Soc.* 172:307–315.
Chillingworth, D. (ed.). 1971. Proceedings of the symposium on differential equations and dynamical systems. *Lec. Notes Math.* 206. Springer, New York.
Chillingworth, D. 1976. *Differential topology with a view to applications.* Pitman, London.
Chirikov, B. V. 1977. A universal instability of many-dimensional oscillator systems (preprint).
Choquet, G. 1969. *Lectures on Analysis* (3 vols.). Benjamin-Cummings, Reading, Mass.
Choquet-Bruhat, Y. 1968. *Géométrie differentielle et systèmes extérieurs.* Dunod, Paris.
Choquet-Bruhat, Y., Fischer, A., and Marsden, J. 1978. Maximal hypersurfaces and positivity of mass, in Proc. of 1976 Summer School of Italian Physical Society, J. Ehlers (ed.).
Chorin, A. J., Hughes, T. J. R., McCracken, M. F., and Marsden, J. E. 1978. Product formulas and numerical algorithms, *Comm. Pure Appl. Math.* 31:205–256.
Chow, S-N, and Mallet-Paret, J. 1977. The Fuller index and global Hopf bifurcation (preprint).
Chow, S-N, Mallet-Paret, J., and Yorke, J. A. 1977. Global Hopf bifurcation from a multiple eigenvalue (preprint).
Chu, B. Y. 1974. Symplectic homogeneous spaces. *Trans. Am. Math. Soc.* 197:145–159.
Chua, L. O., and McPherson, J. D. 1974. Explicit topological formulation of Lagrangian and Hamiltonian equations for nonlinear networks. *IEEE Trans. Circuit Systems.* 21(2):277–286.
Churchill, R. C., Pecelli, G., and Rod, D. L.
1975. Isolated unstable periodic orbits. *J. Diff. Equations* 17:329–348.
1977. Hyperbolic structures in Hamiltonian systems, *Rocky Mountain J. Math.* (to appear).
1978*a*. Poincaré maps, Hill's equation, and Hamiltonian systems. *J. Diff. Equations* (to appear).

ISBN 0-8053-0102-X

1978*b*. Survey of pathological Hamiltonian systems, in Stochasticity: Como Conference Proceedings on Stochastic Behavior in Classical and Quantum Hamiltonian Systems, J. Ford (ed.) (to appear).

Churchill, R. C., Pecelli, G., Sacolick, S., and Rod, D. L. 1977. Coexistence of stable and random motion. *Rocky Mountain J. Math.* (to appear).

Churchill, R. C., and Rod, D. L.

1976*a*. Pathology in dynamical systems I: general theory. *J. Diff. Equations* 21:39–65.

1976*b*. Pathology in dynamical systems II: applications. *J. Diff. Equations* 21:66–112.

1978. Pathology in dynamical systems III: analytic Hamiltonians. *J. Diff. Equations* (to appear).

Clauser, J. F., Horn, M. A., Shimony, A., and Holt, R. A. 1969. Proposed experiment to test local hidden variable theories. *Phys. Rev. Letters* 23:880–884.

Coddington, E., and Levinson, N. 1955. *Theory of ordinary differential equations*. McGraw-Hill, New York,

Calabi, E. 1970. On the group of automorphisms of a symplectic manifold, in *Problems in Analysis*: A symposium in honor of Solomon Bochner, in Gunning, R. C., ed., Math Series 31, Princeton Univ. Press, Princeton, N.J.

Conley, C. 1963. On some new long periodic solutions of the plane restricted three-body problem. *Comm. Pure Appl. Math.* 14:449–467.

Cook, A., and Roberts, P., 1970. The Rikitake two-disc dynamo system. *Proc. Cambridge Philos. Soc.* 68: 547–569.

Cook, J. M. 1966. Complex Hilbertian structures on stable linear dynamical systems. *J. Math. Mech.* 16:339–349.

Coppel, W. 1965. *Stability and asymptotic behavior of differential equations*. Heath, Boston.

Corben, H., and Stehle, P. 1960. *Classical mechanics*, 2nd ed., Wiley, New York.

Courant, R., and Hilbert, D. 1962. *Methods of mathematical physics* (I, II). Wiley, New York.

Crandall, M. G.

1967. Two families of periodic solutions in the plane four-body problem. *Am. J. Math* 89:275–318.

1972. A generalization of Pearo's existence theorem and flow invariance. *Proc. Am. Math. Soc.* 36:151–155.

Crumeyrolle, A.

1969. Structures spinorielles. *Ann. Inst. H. Poincare Sect. A* 11:19–55.

1971. Groupes de spinorialité. *Ann. Inst. H. Poincaré A* 14:309–323.

1972. Formes et opérateurs sur les champs de spineurs, *C. R. Acad. Sci. Paris* 274:90–93.

1976. Classes caractéristiques réelles spinorielles. *C. R. Acad. Sci. Paris A* 283(6):359–362.

Currie, D. G., and Jordan, T. F. 1967. Interaction in relativistic classical particle mechanics, in Boulder Lectures in Theor. Phys., 1967 Summer Inst. for Theoretical Phys., Univ. of Colorado, Boulder, Colorado.

Currie, D. G., Jordan, T. F., and Sudarshan, E. C. 1963. Relativistic invariance and Hamiltonian theories of interacting particles. *Rev. Mod. Phys.* 35:350–375.

Cushman, R.

1974. The momentum mapping of the harmonic oscillator, in *Inst. Nat. di Alta Mate., Symp. Math.* XIV: 323–342. Academic, London.

1975. The topology of the 2 : 1 resonance. *Notes*, Math. Inst. Utrecht.

1977. *Notes on topology and mechanics*. Math. Inst. Utrecht.

Cushman, R., and Duistermaat, J. J. 1977. The behavior of the index of a periodic linear Hamiltonian system under iteration. *Adv. Math*. 23:1–21.

Cushman, R., and Kelley, A. 1979. Strongly stable real infinitesimally symplectic mappings, J. Diff. Eqs. 31:200–223.

Cushman, R., and Zoldan, A. 1975. On the Morse index theorem (unpublished).

Daniel, J. W., and Moore, Ramon E. 1970. *Computation and theory in ordinary differential equations*. Freeman, San Francisco.

Dankel, T. G. 1970. Mechanics on manifolds and the incorporation of spin into Nelson's stochastic mechanics. *Arch. Rat. Mech. Anal* 37:192.

Darboux, G. 1882. Sur le problème de Pfaff. *Bull. Sci. Math.* (2)6:14–36; 49–68.

Delaunay, C. 1860. Théorie du mouvement de la lune. Mem. 28 (1860); 29 (1867). *Acad. Sci. France.* Paris.

Denjoy, A. 1932. Sur les courbes définies par les équations differentielles à la surface du tore. *J. Math. Pures Appl.* 11: 333–375.

ISBN 0-8053-0102-X

Deprit, A. and Deprit-Bartolomé, A. 1967. Stability of the triangular Lagrangian points. *Astron. J.* 72:173–179.

Deprit, A., and Henrard, J. 1968. A manifold of periodic orbits, in *Advances in Astronomy and Astrophysics*, 6:1–124.

Desoer, C. A., and Oster, G. F. 1973. Globally reciprocal stationary systems. *Int. J. Eng. Sci.* 11:141–155.

Destouches, J. 1935. Les espaces abstraits en logique et la stabilité des propositions. *Acad. Roy. Belg. Bull. Cl. Sci.* 21(5):780–786.

Devaney, R. L.

1976*a*. Reversible diffeomorphisms and flows. *Trans. Am. Math. Soc.* 218:89–113.

1976*b*. Homoclinic orbits in Hamiltonian systems. *J. Diff. Equations* 21:431–438.

1977. Collision orbits in the anisotropic Kepler problem (preprint).

1978*a*. Blue sky catastrophes in reversible and Hamiltonian systems (preprint).

1978*b*. Families of periodic solutions in Hamiltonian systems. *Ann. N. Y. Acad. Sci.* (to appear).

1978*c*. Homoclinic orbits to hyperbolic equilibria. *Ann. N.Y. Acad. Sci.* (to appear)

1978*d*. Morse-Smale singularities in simple mechanical systems. (preprint).

Dewitt, B. S. 1957. Dynamical theory in curved spaces, I. *Rev. Mod. Phys.* 26:377–397.

Dewitt, B. S., and Graham, N. 1973. *The many-worlds interpretation of quantum mechanics.* Princeton Univ. Press, Princeton, N.J.

Diaz-Miranda, A. 1971. Sur la structure symplectique et les lois de conservation associees à un lagrangien. *C. R. Acad. Sc. Paris* 272:1401–1404.

Dieudonné, J. 1960. *Foundations of modern analysis.* Academic, New York.

Diliberto, S. 1966. Formal stability of Hamiltonian systems with two degrees of freedom. Technical Report. Univ. of Calif., Berkeley, Calif.

Dirac, P. A. M.

1930. *The principles of quantum mechanics* (4th ed., 1958). Oxford Univ. Press, Oxford.

1950. Generalized Hamiltonian dynamics. *Can. J. Math.* 2:129–148.

1959. Fixation of coordinates in the Hamiltonian theory of gravitation. *Phys. Rev.* 114:924.

1964. *Lectures on quantum mechanics.* Belfer Graduate School of Sci., Monograph Series vol. 2. Yeshiva Univ., N.Y.

Djukic, D. D. 1974. Conservation laws in classical mechanics for quasi-coordinates. *Arch. Rat. Mech. Anal.* 56:79–98.

Dombrovski, P. 1962. On the geometry of the tangent bundle. *J. Reine Angew. Math.* 210:73–88.

deDonder, Th. 1927. *Théorie des invariants intégraux,* Gauthier-Villars Paris.

Droz-Vincent, P. 1975. Hamiltonian systems of relativistic particles. *Rep. Math. Phys.* 8:79–101.

Duff, G. 1956. *Partial differential equations.* University of Toronto Press, Toronto.

Dugas, R. 1955. *A history of mechanics.* Griffon, Neuchatel, Switzerland.

Duhem, P. 1954. *The aim and structure of physical theory.* Princeton Univ. Press, Princeton, N.J.

Duistermaat, J. J.

1970. Periodic solutions of periodic systems of ordinary differential equations containing a parameter. *Arch. Rat. Mech. Anal.* 38:59–80.

1972. On periodic solutions near equilibrium points of conservative systems. *Arch. Rat. Mech. Anal.* 45:143–160.

1973. *Fourier integral operators.* Courant Inst. Notes, New York.

1974. Oscillatory integrals, Lagrange immersions and unfolding of singularities. *Comm. Pure. Appl. Math.* 27:207–281.

1976*a*. On the Morse index in variational calculus. *Adv. Math.* 21:173–195.

1976*b*. Stable manifolds (preprint no. 40). Math. Inst. Utrecht.

Duistermaat, J. J., and Hörmander, L. 1972. Fourier integral operators II. *Acta Math.* 128:183–269.

Dumortier, F., and Takens, F. 1973. Characterization of compactness for symplectic manifolds. *Bol. Soc. Brasil. Mat.* 4(2):167–173.

Dyson, F. J. 1972. Missed opportunities. *Bull. Am. Math. Soc.* 78:635–652.

Easton, R.

1971. Some topology of the three-body problem. *J. Diff. Equations* 10:371–377.

1972. The topology of the regularized integral surfaces of the three-body problem. *J. Diff. Equations* 12:361–384.

ISBN 0-8053-0102-X

Ebin, D. G.
1970*a*. The manifold of Riemannian metrics, in *Proc. Symp. Pure Math. XV*, pp. 11–40. Am. Math Soc., Providence, R.I. (Also, *Bull. Am. Math. Soc.* 74(1968):1002–1004.
1970*b*. On completeness of Hamiltonian vector fields. *Proc. Am. Math. Soc.* 26:632–634.
1972. Espace des métriques riemanniennes et mouvement des fluides via les variétés d'applications. Lecture Notes VII. Ecole Polytechnique et Université de Paris.
1977. The motion of slightly compressible fluids viewed as a motion with strong constraining force. *Ann. Math.* 105:141–200.

Ebin, D. G., and Marsden, J. 1970. Groups of diffeomorphisms and the motion of an incompressible fluid. *Ann. Math.* 92:102–163. (Also, *Bull. Am. Math. Soc.* 75(1969):962–967.)

Edelen, D. G. B. 1960. The invariance group for Hamiltonian systems of partial differential equations. *Arch. Rat. Mech. Anal* 5:95–176.

Eells, J.
1955. Geometric aspects of currents and distributions. *Proc. Nat. Acad. Sci.* 41:493–496.
1958. On the geometry of function spaces, in *Symposium de Topologia Algebrica,* Mexico UNAM, Mexico City, pp. 303–307.
1966. A setting for global analysis. *Bull. Am. Math. Soc.* 72:751–807.

Egorov, Yu. V. 1969. On canonical transformations of pseudo-differential operators. *Uspehi. Mat. Nauk.* 25:235–236.

Ehresmann, Ch., and Lieberman, P. 1948. Sur les formes différentielles extérieures de degré 2. *C. R. Acad. Sci. Paris* 227:420–421.

Eilenberg, S., and Cartan, H. 1958. Foundations of fibre bundles, in *Intern. Sym. on Algebraic Topology,* UNAM, Mexico City, pp. 16–23.

Einstein, A., Podolsky, B., and Rosen, N. 1935. Can quantum mechanical description of physical reality be considered complete? *Phys. Rev.* 47:777–780.

Elhadad, J. 1974. Sur l'interpretation-de géométrie symplectique des états quantiques de l'atome d'hydrogéne, *Conv. di geom. simpl. e fisica math. I.N.D.A.M., Rome Symp. Math.* 14:259–291.

Elliasson, H. 1967. Geometry of manifolds of maps. *J. Diff. Geom.* 1:169–194.

Ellis, R. 1969. *Lectures on topological dynamics*, Benjamin-Cummings, Reading, Mass.

Estabrook, F. B., and Wahlquist, H. D. 1975. The geometric approach to sets of ordinary differential equations and Hamiltonian dynamics. *SIAM Review* 17(2):201–220.

Euler, L. 1767. De motu rectilineo trium corporum se mutuo attrahentium. *Novi Comm. Acad. Sci. Imp. Petrop.* 11:144–151.

Faddeev, L. D. 1970, Symplectic structure and quantization of the Einstein gravitation theory. *Actes du Congrès Intern. Math.* 3:35–40.

Faddeev, L. and Zakharov, V. E. 1971. Korteweg–de Vries as a completely integrable Hamiltonian system. *Funct. Anal. Appl.* 5:280.

Faure, R. 1956. Transformations isometriques en méchanique analytique et en méchanique ondulatoire. *C. R. Acad. Sci. Paris.* 242:2802–2803.

Fenichel, N.
1973. Exponential rate conditions for dynamical systems (in Peixoto [1973]).
1974*a*. Asymptotic stability with rate conditions for dynamical systems. *Bull. Am. Math. Soc.* 80:346–350.
1974*b*. Asymptotic stability with rate conditions. *Indiana Univ. Math. J.* 23(12).
1975. The orbit structure of the Hopf bifurcation problem. *J. Diff. Equations* 17(2):308–328.
1977. Asymptotic stability with rate conditions, II. *Indiana Univ. Math. J.* 26(1): 81–93.

Feynman, R. P., Leighton, R. B., and Sands, M. 1964. *The Feynman lectures on physics (II)*. Addison-Wesley, Reading, Mass.

Fischer, A. 1970. A theory of superspace, in *Relativity*, Camelli et al. (eds.). Plenum, New York.

Fischer, A., and Marsden, J.
1972. The Einstein equations of evolution—a geometric approach. *J. Math. Phys.* 13:546–568.
1973. Partial differential equations general relativity and dynamical systems. *Proc. Symp. Pure Math* 23:309–328.
1974. Global analysis and general relativity. *Gen. Rel.* and *Grav.* 5:89–93.
1975*a*. Deformations of the scalar curvature. *Duke Math. J.* 42:519–547.

ISBN 0-8053-0102-X

1975*b*. Linearization stability of non-linear partial differential equations. *Proc. Symp. Pure Math* 27:219–263.

1976*a*. A new Hamiltonian structure for the dynamics of general relativity. *J. Grav. and Gen. Rel.* 7:915–920.

1976*b*. Topics in the dynamics of general relativity, in Proc. Italian Summer School, Varenna 1976, J. Ehlers (ed.) (to appear).

1977. The manifold of conformally equivalent metrics. *Can. J. Math.* 29:193–209.

1978. Symmetry breaking in general relativity (to appear).

Flanders, H. 1963. *Differential forms*. Academic, New York.

Flaschka, H., and McLaughlin, D. W.

1976*a*. Some comments on Bäcklund transformations, canonical transformations and the inverse scattering method (in Miura [1976*b*]).

1976*b*. Canonically conjugate variables for the Korteweg–de Vries equation and the Toda lattice with periodic boundary conditions. *Prog. Theor. Phys.* 55:438–456.

Flaschka, H., and Newell, A. C. 1975. Integrable systems of nonlinear evolution equations (in Moser [1975*a*]).

Fleming, W. 1965. *Functions of several variables*. Addison-Wesley, Reading, Mass.

Fong, U., and Meyer, K. R. 1975. Algebras of integrals. *Rev. Colom. de Mathématicas* IX:75–90.

Franks, J.

1972. Differentiable Ω- stability. *Topology* 11:107–112.

1973. Absolutely structurally stable diffeomorphisms. *Proc. Am. Math. Soc.* 37:293–296.

Franks, J., and Robinson, R. C. 1976. A quasi-Anosov diffeomorphism that is not Anosov. *Trans. Am. Math. Soc.* 223:267–278.

Freedman, S. J., and Clauser, J. F. 1972. Experimental test of local hidden-variable theories. *Phys. Rev. Letters* 28:938–941.

Friefeld, C. 1968. One parameter subgroups do not fill a neighborhood of the identity in an infinite dimensional Lie (pseudo) group, in *Battelle Rencontes*, C. M. DeWitt and J. A. Wheeler (eds.). Benjamin-Cummings, Reading, Mass.

Froeschelé, C. 1968. Etude numerique de transformations ponctuelles planes conservant les aires. *C. R. Acad. Sci. Paris* 266:747–749. 846–848.

Fuller, F. B. 1967, An index of fixed point type for periodic orbits. *Am. J. Math.* 89:133–148.

Fulp, R. O., and Marlin, J. A.

Integrals of foliations on manifolds with a generalized symplectic structure (preprint).

Integrals and reduction of order (preprint).

Gaffney, M. P. 1954. A special Stokes theorem for complete Riemannian manifolds. *Ann. Math.* 60:140–145.

Gallissot, F.

1951. Application des formes extérieures du 2e ordre à la dynamique Newtonienne et relativiste. *Ann. Inst. Fourier (Grenoble)* 3:277–285.

1952. Formes extérieures en mécanique. Ibid. 4:145–297.

1958. Formes extérieures et la mécanique des milieux continus. Ibid. 8:291–335.

Garcia, P. L.

1974. The Poincaré–Cartan invariant in the calculus of variations, in *Symp. Math.*, vol. 14. Istituto Nazion. di Alta Mate. Roma: pp. 219–246.

1977. Gauge algebras, curvature and symplectic structure. *J. Diff. Geom.* 12:351–359.

Garcia, P. L., and Perez-Rendon, A. 1969. Symplectic approach to the theory of quantized fields., I. *Comm. Math Phys.* 13:22–44; II. *Arch. Rat. Mech. Anal.* 43:101–124.

Gardner, C. S., Greene, J. M., Kruskal, M. D., and Miura, R. M. 1967. Korteweg de Vries equation and generalizations. *Phys. Rev. Letters* 19:1095–7; *J. Math. Phys.* 9:1202–1204, 1204–1209, 10:536–1539, 11:952–960; *Comm. Pure Appl. Math* 27:97–133.

Garding, L., and Wightman, A. S. 1954. Representations of the anticommutation relations. *Proc. Natl. Acad. Sci. U.S.A.* 40:617–621; 622–630.

Gardner, C. S. 1971. Korteweg–de Vries equation and generalizations IV. The Korteweg-de Vries Equation as a Hamiltonian System. *J. Math. Phys.* 12:1548–1551.

Gawedzki, K.

1972. On the geometrization of the canonical formalism in the classical field theory. *Rep. Math. Phys.* 3:307–326.

1976. Fourier-like kernels in geometric quantization. *Dissertationes Mathematical* 128:1–83.

ISBN 0-8053-0102-X

Gelfand, I. M., and Fomin, S. 1963. *Calculus of variations*. Prentice-Hall, Englewood Cliffs., N.J.
Gelfand, I. M., and Lidskii, V. B. 1955. On the structure of regions of stability of linear canonical differential equations with periodic coefficients. *Uspeh. Mat. Nauk*. 10:3–40; (1958 transl. *Am. Math. Soc*. 8:143–182.
Genis, A. L. 1961. Metric properties of the endomorphisms of the *n*-dimensional torus (Russian). *Dokl. Akad. Nauk SSSR*. 138:991-993; (English) *Transl. Am. Math. Soc*. 2:750–752.
Germain, P., and Nayroles, B. (eds.). 1976. Applications of methods of functional analysis to problems in mechanics. *Lec. Notes Math*.503.
Giacaglia, G. E. O. 1972. Perturbation methods in nonlinear systems.*Appl. Math. Sci*. no. 8. Springer, New York.
Gillis, P. 1943. Sur les formes différentielles et la formula de Stokes. *Mém. Acad. Roy. Belgique* 20:3–95; 175–186.
Glasner, S. 1976. *Proximal flows. Lec. Notes. Math*. 517.
Godbillon, C.
1969. *Géometrie différentielle et mécanique analytique*. Hermann, Paris.
1971*a*. Formalisme Lagrangien (in Chillingworth [1971], pp. 9–11).
1971*b*. The principle of Maupertuis (in Chillingworth [1971], pp. 91–94).
Goffman, C. 1965. *Calculus of several variables*. Harper and Row, New York.
Goldschmidt, H. Z., and Sternberg, S. 1973. The Hamilton–Jacobi formalism in the calculus of variations. *Ann. Inst. Fourier* 23:203–267.
Goldstein, H. 1950. *Classical mechanics*. Addison-Wesley, Reading, Mass.
Golubitsky, M. 1975. Contact equivalence for Lagrangian submanifolds (in Manning [1975], pp. 71–73).
Golubitsky, M. and Guillemin, V. 1974. *Stable mappings and their singularities. Graduate Texts in Math*., vol. 14. Springer, New York.
Gordon, W. B.
1969. On the relation between period and energy in periodic dynamical systems. *J. Math. Mech*. 19:111–114.
1970*a*. The Riemannian structure of certain function space manifolds. *J. Diff. Geom*. 4:499–508.
1970*b*. On the completeness of Hamiltonian vector fields. *Proc. Am. Math. Soc*. 26:329–331.
1971. A theorem on the existence of periodic solutions to Hamiltonian systems with convex potential. *J. Diff. Equations* 10:324–335.
Gotay, M. J., Nester, J. E., and Hinds, G. 1978. Presymplectic manifolds and the Dirac-Bergmann theory of constraints (preprint).
Gottshalk, W. H. and Hedlund, G. A. 1955. Topological dynamics, in *Am. Math. Soc. Coll. Publ*., vol. 36. American Math. Society, Providence, R.I.
Graff, S. M. 1974. On the conservation of hyperbolic invariant tori for Hamiltonian systems, *J. Diff. Equations* 15:1–69.
Gray, J. 1959. Some global properties of contact structures. *Ann. Math*. 69(2):421–450.
Groenwold, H. J. 1946. On the principles of elementary quantum mechanics. *Physica* 12:405–460.
Gromoll, D., and Meyer, W. 1969. Periodic geodesics on compact Riemannian manifolds. *J. Diff. Geo*. 3:493–510.
Grosjean, P. V. 1974. The bisymplectic formalism and its applications to the relativistic and to the non-relativistic hydrodynamics of an adiabatic perfect fluid, in *Modern developments in thermodynamics*, pp. 181–193, B. Gal-or (ed.). Wiley, New York.
Gross, L. 1966. The Cauchy problem for the coupled Maxwell and Dirac equations. *Comm. Pure Appl. Math*. 19:1–15.
Guckenheimer, J.
1972. Absolutely Ω-stable diffeomorphisms. *Topology* 11:195–197.
1973*a*. Bifurcation and catastrophe (in Peixoto [1973], pp. 95–109).
1973*b*. One-parameter families of vector fields on two-manifolds: another nondensity theorem (in Peixoto [1973], pp. 111–127.
1973*c*. Catastrophes and partial differential equations. *Ann. Inst. Fourier. Grenoble* 23(2):31–59.
1974. Caustics and non-degenerate Hamiltonians. *Topology* 13:127–133.
1976*a*. A strange, strange attractor (in Marsden–McCracken [1976]).
1976*b*. An introduction to dynamical systems. (preprint).
Guillemin, V., and Pollack, A. 1974. *Differential topology*. Prentice-Hall, New Jersey.
Guillemin, V., and Sternberg, S. 1977. *Geometric asymptotics. Am. Math. Soc. Survey*, vol. 14.

ISBN 0-8053-0102-X

American Math. Society, Providence, R.I.
Gurtin, M. 1972. The linear theory of elasticity, in *Handbuch der Physik*, vol. II, S. Flügge (ed.).
Gustavson, F. G. 1966. On constructing formal integrals of a Hamiltonian system near an equilibrium point. *Astron. J.* 71:670–686.
Gutzwiller, M. C.
1967. Phase integral approximation in momentum space and the bound states of an atom I, *J. Math. Phys.* 8(10):1979–2000; II. *J. Math. Phys.* 10(6)(1969):1004–1021.
1973. The anisotropic Kepler problem in two dimensions, *J. Math. Phys.* 14: 139–152.
Haahti, H. 1977. Compact deformations of conformally flat structures on almost sympletic Banach-manifolds, (preprint).
Hagihara, Y. 1970. *Celestial mechanics* (5 vols.). M.I.T., Cambridge, Mass.
Hájek, O. 1968. *Dynamical systems in the plane*. Academic, New York.
Hale, J. K.
1961. Integral manifolds of perturbed differential systems. *Ann. Math.* 73:496–531.
1963. *Oscillations in nonlinear systems*. McGraw-Hill, New York.
1969. *Ordinary differential equations*. Wiley, New York.
1977. Lectures on generic bifurcation, in *Nonlinear analysis and mechanics*, R. Knops (ed.). Pitman, London.
Halmos, P. R. 1956. *Lectures on ergodic theory*. Chelsea, New York.
Halmos, P. R., and Von Neumann, J. 1942. Operator methods in classical mechanics. *Ann. of Math.* 43:332–350.
Hanson, A., Regge, T., and Teitelboim, C. 1976. *Constrained Hamiltonian systems*. Accad. Nazionale dei Lincei, Rome 22.
Harris, T. C. 1968. Periodic solutions of arbitrarily long periods in Hamiltonian systems, *J. Diff. Equations* 4:131–141.
Hartman, P.
1972. On invariant sets and on a theorem of Wazewski, in *Proc. Am. Math. Soc.* vol. 32, pp. 511–520.
1973. *Ordinary differential equations*. Wiley, New York.
Hasegawa, Y. 1970. C-flows on a Lie group for Euler equations. *Nagoya. Math. J.* 40:67–84.
Hausdorff, F. 1962. *Set theory*. Chelsea, New York.
Hayashi, C.
1953*a*. *Forced oscillations in nonlinear systems*. Nippon, Osaka.
1953*b*. Forced oscillations with nonlinear restoring force. *J. Appl. Phys.* 24:198–207.
1953*c*. Stability investigation of nonlinear periodic oscillations. *J. Appl. Phys.* 24:344–348.
1953*d*. Subharmonic oscillations in nonlinear systems. *J. Appl. Phys.* 24:521–529.
1964. *Nonlinear oscillations in physical systems*. McGraw-Hill, New York.
Helgason, S. 1962. *Differential geometry and symmetric spaces*. Academic, New York.
Hénon, M.
1976. A two-dimensional mapping with a strange attractor, *Comm. Math. Phys.* 50: 69–77.
1977. A relation in families of periodic solutions. *Celestial Mech.* 15:99–105.
Hénon, M., and Heiles, C. 1964. The applicability of the third integral of motion: some numerical experiments. *Astron. J.* 69:73.
Henrard, J. 1973. Liapunov's center theorem for a resonant equilibrium. *J. Diff. Equations* 14:431–441.
Hermann, R.
1962. Some differential geometric aspects of the Lagrange variational problem. *Ill. J. Math.* 6:634–673.
1965. E. Cartan's geometric theory of partial differential equations. *Adv. Math.* 1:265–317
1966. *Lie groups for physicists*. Benjamin-Cummings, Reading, Mass.
1968. *Differential geometry and the calculus of variations*. Academic, New York.
1969*a*. Quantum field theories with degenerate Lagrangians. *Phys. Rev.* 177:2453–2457.
1969*b*. Geometric Formula for current-algebra commutation relations. *Phys. Rev.* 177:2449–2452.
1970. *Lie algebras and quantum mechanics*. Benjamin-Cummings, Reading, Mass.
1971. *Lectures on mathematical physics* (I, II). Benjamin-Cummings, Reading, Mass.
1972*a*. Geodesics and classical mechanics on Lie groups. *J. Math. Phys.* 13:460–464.
1972*b*. Spectrum-generating algebras in classical mech. I, II. *J. Math. Phys.* 13:833, 878.

ISBN 0-8053-0102-X

1973. *Geometry, physics and systems*. Dekker, New York.
1976. *Geometric theory of non-linear differential equations, Bäcklund transformations and solitons*. Math. Sci. Press, Brookline, Mass.
Hicks, N. 1965. *Notes on differential geometry*. Van Nostrand, Princeton, N.J.
Hill, G. 1878. Researches in lunar theory. *Am. J. Math.* 1:5–26, 129–147, 245–260.
Hilton, P. (ed.). 1976. Structural stability, the theory of catastrophes, and applications in the sciences, in *Lec. Notes Math*, vol. 525. Springer, New York.
Hinds, G. 1965. Foliations and the Dirac theory of constraints, thesis. Dept. of Physics and Astro., Univ. of Maryland, Baltimore, Md.
Hirsch, M. W. 1976. *Differential topology. Grad. Texts. Math.*, vol. 33. Springer, New York.
Hirsch, M. W., Palis, J., Pugh, C. and Shub, M. 1970. Neighborhoods of hyperbolic sets. *Invent. Math.* 9:121–134.
Hirsch, M. W., and Pugh, C. 1970. Stable manifolds and hyperbolic sets. *Proc. Symp. Pure Math.* 14:133–163.
Hirsch, M. W., Pugh, C., and Shub, M. 1977. *Invariant manifolds, Lec. Notes Math.*, vol. 583. Springer, New York.
Hirsch, M. W., and Smale, S. 1974. *Differential equations, dynamical systems and linear algebra*. Academic, New York.
Hitchin, N. 1974. Harmonic spinors, *Advan. Math.* 14:1–55.
Hocking, J. and Young, G. 1960. *Topology*, Addison-Wesley, Reading, Mass.
Hodge, V. W. D. 1952. *Theory and applications of harmonic integrals*, 2nd ed. Cambridge Univ. Press., England.
Hoffman, K., and Kunze, R. 1961. *Linear algebra*. Prentice-Hall, Englewood Cliffs, N.J.
Holmes, P. J. 1977. Bifurcation to divergence and flutter in flow-induced oscillations...a finite-dimensional analysis, *J. Sound and Vib.* 53:471–503.
Holmes, P. J., and Marsden, J. E.
1978*a*. Bifurcation to divergence and flutter in flow-induced oscillations...an infinite-dimensional analysis. *Automatica* 14:367–384.
1978*b*. Bifurcations of dynamical systems and nonlinear oscillations in engineering systems., in Lec. Notes Math. 648, 163–206
Holmes, P. J., and Rand, D. A.
1976. The bifurcations of Duffing's equation: an application of catastrophe theory. *J. Sound and Vib.* 44:237–253.
1977. Bifurcations of the forced van der Pol oscillator. *Quart. Appl. Math.* 35(1978):495–509.
Hopf, E. 1942. Abzweigung einer periodischen Lösung von einer stationären Lösung eines Differential systems. *Ber. Math-Phys. Kl. Sächs. Acad. Wiss. Leipzig* 94:1–22; *Ber. Verh. Sächs. Acad. Wiss. Leipzig Math.-Nat. Kl.* 95(1):3–22.
Hörmander, L.
1963. *Linear differential operators*. Springer, Berlin.
1971. Fourier integral operators. *Acta. Math.* 127:79–183; 128 (1972):183.
Hsiang, W. 1966. On the compact homogeneous minimal submanifolds. *Proc. Nat. Acad. Sci.* 56(1):5–6.
Hurt, Norman E.
1968. Remarks on canonical quantization. *Il. Nuovo Cimento* 55:534–542; 58:361–362; *Lettre al Nuovo Cimento* 3(1970):137–138; 1(1971):473–4.
1970. Remarks on Morse theory in canonical quantization. *J. Math. Phys.* 11:539–551.
1971*a*. Deformation theory of quantizable dynamical systems. *Ann. di Mat. Pure and Appl.* (IV) LXXXIX:353–362.
1971*b*. A classification theory for quantizable dynamical systems. *Rep. on Math. Phys.* 2:211–220.
1971*c*. Differential geometry of canonical quantization. *Ann. Inst. H. Poincaré* 14:153–170.
1972*a*. Topology of quantizable dynamical systems and the algebra of observables. *Ann. Inst. H. Poincaré* 16(3):203–217.
1972*b*. Homogeneous fibered and quantizable dynamical systems. *Ann. Inst. H. Poincaré* 16(3):219–222.
Iacob, A.
1971. Invariant manifolds in the motion of a rigid body about a fixed point. *Rev. Roum. de*

ISBN 0-8053-0102-X

Math. Pures et Appl. Tome 16:1497–1521.
1973. *Topological methods in mechanics* (in Romanian). Editura Acad. Repub. Social. Romania, Bucharesti.
1975. Some geometry in the problem of the motion of a heavy rigid body with a fixed point, in *Géometrie Symplectique et Physique Mathématique*, J. M. Souriau (ed.). Colloq. Intern. C.N.R.S., vol. 237.
Ikebe, T., and Kato, T. 1962. Uniqueness of the self-adjoint extension of singular elliptic differential operators. *Arch. Rat. Mech. Anal.* 9(1):77–92.
Iooss, G.
1971*a*. Contribution à la theorie nonlinéaire de la stabilité des écoulements laminaires. Thèse, Faculté des Sciences, Paris VI.
1971*b*. Thèorie nonlinéaire de la stabilité des écoulements laminaires dans le cas de "l'échange des stabilités." *Arch. Rat. Mech. Anal.* 40:166–208.
1972*a*. Existence de stabilité de la solution periodique sécondaire intervenant dans les problèmes d'évolution du type Navier–Stokes. *Arch. Rat. Mech. Anal.* 49:301–329.
1972*b*. Bifurcation d'une solution T-periodique vers une solution nT-periodique, pour certains problèmes d'évolution du type Navier–Stokes, *C.R. Acad. Sci., Paris* 275. 935–938.
1973. Bifurcation et stabilité. Lecture notes, Université Paris XI.
1975. Bifurcation of a periodic solution of Navier–Stokes equations into an invariant torus. *Arch. Rat. Mech. Anal.* 58:35–56.
Iooss, G., and Chencinere, A. 1977. Bifurcation of T^2 to T^3 (preprint).
Iooss, G., and Joseph, D. D. 1977. Bifurcation and stability of nT-periodic solution branching from T-periodic solutions at points of resonance. *Arch. Rat. Mech. An.* 66:135–172.
Irwin, M. C. 1970. On the stable manifold theorem. *Bull. London Math. Soc.* 2:196–198.
Jacobi, C. G. J.
1837. Note sur l'intégration des équations differentielles de la dynamique, *C.R. Acad. Sci., Paris.* 5:61.
1884. *Vorlesungen über Dynamik*. Verlag G. Reimer, Berlin.
Jacobson, N. 1974. *Basic algebra I*. Freeman, San Francisco.
Jeffreys, W. 1965. Doubly symmetric periodic orbits in the three-dimensional restricted problem. *Astron. J.* 70:393–394.
John, F. 1975. *Partial differential equations*, 2nd ed., in *App. Math. Sci.*, vol. 1. Springer, New York.
Jorgens, K., and Weidmann, J. 1973. Spectral properties of Hamiltonian operators, in *Lec. Notes Math.*, vol. 313, Springer, New York.
Joseph, A. 1970. Derivations of Lie brackets and canonical quantization, *Comm. Math. Phys.* 17:210–232.
Joseph, D. D., and Sattinger, D. H. 1972. Bifurcating time periodic solutions and their stability. *Arch. Rat. Mech. Anal.* 45:79–109.
Jost, R. 1964. Poisson brackets (an unpedagogical lecture). *Rev. Mod. Phys.* 36:572–579.
Jost, R., and Zehnder, E. 1972. A generalization of the Hopf bifurcation theorem. *Helv. Phys. Acta.* 45:258–276.
Joubert, G. P., Moussa, R. P., and Roussaire, R. H. (eds.). 1975. *Differential topology and geometry, Lec. Notes Math.*, vol. 484. Springer, New York.
Kaiser, G. 1977. Phase-space approach to relativistic quantum mechanics I. Coherent-state representation for massive scalar particles. *J. Math. Phys.* 18:952–959.
Kaplan, J. and Yorke, J. 1978. Preturbulence: a régime observed in a fluid flow model of Lorenz (preprint).
Kaplan, W.
1940. Regular curve families filling the plane, I. *Duke Math. J.* 7:154–185.
1941. Regular curve families filling the plane, II. *Duke Math. J.* 8:11–46.
1942. Topology of the two-body problem. *Am. Math. Monthly* 49:316–323.
Katok, S. B. 1972. Bifurcation sets and integral manifolds in the problem of the motion of a heavy rigid body. *Usp. Math. Nauk.* 27(2). (Russian).
Katz, A. 1965. *Classical mechanics, quantum mechanics, field theory*. Academic, New York.
Kazhdan, D., Kostant, B., and Sternberg, S. 1978. Hamiltonian group actions and dynamical systems of Calogero type *Comm. Pure and Appl. Math.* 31:481–508.
Keller, J. B. 1958. Corrected Bohr–Sommerfeld quantum conditions for non-separable systems.

ISBN 0-8053-0102-X

Ann. Phys. 4:180–188.

Kelley, A.

1967*a*. The stable, center-stable, center, center unstable, and unstable manifolds. Appendix C in Abraham and Robbin [1967]; see also *J. Diff. Eqs.* 3(1967):546–570.

1967*b*. Stability of the center-stable manifold. *J. Math. Anal. Appl.* 18:336–344.

1967*c*. On the Liapounov sub-center manifold. *J. Math. Anal. Appl.* 18:472–478.

1969. Analytic two-dimensional subcenter manifolds for systems with an integral. *Pac. J. Math.* 29:335–350.

Kelley, J. 1975. *General topology*. Van Nostrand, Princeton, N.J.

Khilmi, G. F. 1961. *Qualitative methods in the many-body problem.* Gordon and Breach, New York.

Kiehn, R. M. 1974. An extension of Hamilton's principle to include dissipative systems. *J. Math. Phys.* 15:9–13.

Kijowski, J. 1974. Multiphase spaces and gauge in the calculus of variations. *Bull. Acad. Pol. des. Sci.* 22:1219–1225.

Kijowski, J., and Szczyrba, W. 1976. A canonical structure for classical field theories. *Comm. Math. Phys.* 46:183–206.

Klingenberg, W. 1977. *Lectures on closed geodesics*. Springer, Berlin.

Klingenberg, W., and Takens, F. 1972. Generic properties of geodesic flows. *Math. Ann.* 197:323–334.

Kneser, H. 1924. Reguläre Kurvenscharen auf den Ringflächen. *Math. Ann.* 91:135–154.

Kobayashi, S. 1973. *Transformations groups in differential geometry*. Springer, New York.

Kobayashi, S., and Nomizu, K. 1963. *Foundations of differential geometry*. Wiley, New York.

Kochen, S., and Specker, E. P. 1967. The problem of hidden variables in quantum mechanics. *J. Math. Mech.* 17:59–88.

Kolmogorov, A. N.

1954. On conservation of conditionally periodic motions under small perturbations of the Hamiltonian. *Dokl. Akad. Nauk. SSSR* 98:527–530.

1957*a*. General theory of dynamical systems and classical mechanics, in *Proc. of the 1954 Intern. Congress Math.*, pp. 315–333. North Holland, Amsterdam: (Russian; see Appendix for an English translation.)

1957*b*. Théorie générale des systèmes dynamiques de la mécanique classique. Sém. Janet, 1957–1958, no. 6. Fac. Sci., Paris, 1958. (Fr. transl. of preceding reference [1957*a*]).

Koopman, B.

1927. On rejection to infinity and exterior motion in the restricted problem of three bodies. *Trans. Am. Math. Soc.* 29:287–331.

1931. Hamiltonian systems and transformations in Hilbert space, *Proc. of the Nat. Acad. Sci.*, vol. 17. pp. 315–318.

Kopell, N., and Howard, L. 1974. Bifurcations under nongeneric conditions. *Adv. Math.* 13:274–283.

Korteweg, D. J., and de Vries, G. 1895. On the change of form of long waves advancing in a rectangular canal and on a new type of long stationary wave. *Phil. Mag.* 39:422–433.

Kostant, B.

1970*a*. Quantization and unitary representations, in *Lec. Notes Math.*, vol. 170. Springer-Verlag, New York.

1970*b*. Orbits and quantization theory. *Actes Cong. Intern. Mat.* 2:395–400.

1970*c*. On certain unitary representations which arise from a quantization theory, Group Representations in Math. and Phys. (Battelle, Seattle 1969 Rencontres), in *Lec. Notes. Phys.*, 6:237–253. Springer-Verlag, New York.

1974. Symplectic spinors, in *Conv. di geom. simp. e fisica math.. Symp. Math.* 14:139–152.

1975. On the definition of quantization, in *Colloq. Intern. C.N.R.S.*, vol. 237, pp. 187–210, Paris.

1978. Graded manifolds, graded Lie theory, and prequantization (preprint).

Kovalevskaya, S.

1889. Sur le problème de la rotation d'un corps solide autour d'un point fixe. *Acta Math.* 12:177–232.

Krein, M. G. 1950. A generalization of several investigations of A. M. Liapunov on linear differential equations with periodic coefficients. *Dokl. Akad. Nauk. SSSR* 73:445–448.

ISBN 0-8053-0102-X

Krupka, D., and Trautman, A. 1974. General invariance of Lagrangian structures. *Bull. Acad. Pol. Sci.* 22:207–211.

Kuchar, K.

1974. Geometrodynamics regained: A Lagrangian approach. *J. Math. Phys.* 15(6):708–715.

1976. The dynamics of tensor fields on hyperspace. *J. Math. Phys.* 17:777–791, 792–800, 801–820; 18:1589–1597.

Kuiper, N. H. (ed.). 1971. Manifolds—Amsterdam 1970, in *Lec. Notes Math.*, vol. 197. Springer.

Kunzle, H.

1969. Degenerate Lagrangian systems. *Ann. Inst. H. Poincaré (A)*. 11:393–414.

1972. Canonical dynamics of spinning particles in gravitational and electromagnetic fields. *J. Math. Phys.* 13:739–744.

Kupka, I.

1962. Stabilitéstructurelle, Sém. Janet, 1960–61, no. 7. Fac. Sci., Paris.

1963. Contribution à la théorie des champs génériques, *Contributions to Diff. Equations*. 2:457–484; also 3(1964):411–420.

Lacomba, E. A.

1973*a*. Mechanical systems with symmetry on homogeneous spaces. *Trans. Am. Math. Soc.* 185:477–491.

1973. Geodesics on homogeneous spaces with an invariant metric. *Ann. Acad. Brasil Cienc.* 45:77–82.

1975. Topology of regularized submanifolds in the restricted three-body problem (preprint).

Lagrange, J. L.

1808. Mémoire sur la théorie de la variation des éléments des planètes. *Mem. Cl. Sci. Math. Phys. Ins. France*: 1–72.

1809. Second mémoire sur la théorie de la variation des constantes arbitraires dans les problèmes de mécanique. *Mem. Cl. Sci. Math. Phys. Inst. France*: 343–352.

Lanczos, C. 1962. *The variational principles of mechanics*, 2nd ed. Univ. of Toronto Press.

Lanford, O. E. 1972. Bifurcation of periodic solutions into invariant tori: the work of Ruelle and Takens, in "Nonlinear Problems in Physical Sciences, Biology." *Lec. Notes Math.*, vol. 322.

Lang, S.

1962. *Introduction to differentiable manifolds*. Wiley, New York.

1964. *A second course in calculus*. Addison-Wesley, Reading, Mass.

1965. *Algebra*. Addison-Wesley, Reading, Mass.

1966. *Linear algebra*. Addison-Wesley, Reading, Mass.

1970. *Analysis (I, II)*. Addison-Wesley, Reading, Mass.

1972. *Differential manifolds*. Addison-Wesley, Reading, Mass.

LaSalle, J. P. 1976. The stability of dynamical systems, in Regional Conference Series in Applied Math., no. 25, SIAM.

LaSalle, J. P., and Lefschetz, S. (eds.).

1961. *Stability by Lyapunov's direct method with applications*. Academic Press, New York.

1963. *Nonlinear differential equations and nonlinear mechanics*. Academic Press, New York.

Laub, A. J., and Meyer, K. 1974. Canonical forms for symplectic and Hamiltonian matrices. *Celes. Mech.* 9:213–238.

Lawruk, B., and Tulczyjew, W. M. 1977. Criteria for partial differential equations to be Euler–Lagrange equations. *J. Diff. Equations*. 24:211–225.

Lawruk, B., Sniatychi, J., and Tulczyjew, W. M., 1975. Special symplectic spaces. *J. Diff. Equations*, 17:477–497.

Lawson, H. B. 1977. The qualitative theory of foliations, in *Am. Math. Soc., CBMS Series*, vol. 27. Am. Math. Soc., Providence, R.I.

Lax, P. D.

1957. Asymptotic solutions of oscillatory initial value problems. *Duke Math. J.* 24:627–646.

1968. Integrals of nonlinear equations of evolution and solitary waves. *Comm. Pure Appl. Math.* 21:467–490.

1973. *Hyperbolic systems of conservation laws and the mathematical theory of shock waves*. SIAM, Philadelphia.

1975. Periodic solutions of the *KdV* equation. *Comm. Pure Appl. Math.* 28:141–188.

Lefschetz, S.

1950. *Contributions to the theory of nonlinear oscillations*, Princeton Univ. Press, Princeton.

ISBN 0-8053-0102-X

1963. *Differential equations: geometric theory*, 2nd ed. Wiley, New York.

Leimanis, E. 1965. The general problem of the motion of coupled rigid bodies about a fixed point. Springer-Verlag, New York.

Lelong-Ferrand, J.

1958. Sur les groupes d'un parametre de transformations des variétés differentiables. *J. Math. Pures Appl.* 37(9):269–278.

1959. Condition pour qu'un groupe de transformations infinitésimales engendre un groupe global, vol. 249, pp. 1852–1854. C.R. Acad. Sci., Paris.

Leontovitch, A. M. 1962. On the stability of Lagrange's periodic solutions of the restricted three-body problem. *Dokl. Akad. Nauk. SSSR*. 143:525–528.

Leslie, J.

1967. On a differential structure for the group of diffeomorphisms. *Topology* 6:263–271.

1968. Some Frobenius theorems in global analysis. *J. Diff. Geom.* 2:279–297.

1971. On two classes of classieal subgroups of diff (M). *J. Diff. Geom.* 5:427–436.

Leutwyler, H. 1965. A no-interaction theorem in classical relativistic Hamiltonian particle mechanics. *Nuovo Cimento* 37:556–567.

Levi-Civita, T. 1920. Sur la régularisation du problème des trois corps. *Acta Math* 42:99–144.

Levine, H. I. 1959. *Singularities of differentiable mappings. I.* mathematisches Institut, Bonn.

Levy-Leblond, J. M. 1969. Group theoretical foundations of classical mechanics; the Lagrangian gauge problem. *Comm. Math. Phys.* 12:64–79.

Lewis, D. C. 1955. Families of periodic solutions of systems having relatively invariant line integrals, in *Proc. Am. Math. Soc.*, vol. 6, pp. 181–185. Am. Math. Soc., Providence, R.I.

Liapounov, A. M. 1947. Problème général de la stabilité du mouvement, *Ann. of Math. Studies*, N. 17. Princeton Univ. Press. Princeton, N.J.

Lichnerowicz, A.

1951. Sur les variétés symplectiques. *C.R. Acad. Sci. Paris.* 233:723–726.

1973. Algèbre de Lie des automorphismes infinitésimaux d'une structure de contact. *J. Math. Pures et Appl.* 52:473–508.

1975*a*. Variété symplectique et dynamic associée à une sous-variété. *C.R. Acad. Sci. Paris, Sér. A.* 280:523–527.

1975*b*. Structures de contact et formalisme Hamiltonian invariant. *C.R. Acad. Sci. Paris, Sér. A.* 281:171–175.

1976. Variétés symplectiques, variétés canoniques, et systèmes dynamiques, in *Topics in differential geometry*, H. Rund and W. Forbes (eds.). Academic Press, New York.

Lieberman, B. B. 1972. Quasi-periodic solutions of Hamiltonian systems. *J. Diff. Equations* 11:109–137.

Libermann, P.

1953. Forme canonique d'une forme différentielle extérieure quadratique fermée. *Acad. Roy. Belgique. Bull. Sci.* 39:846–850.

1959. Sur les automorphismes infinitésimaux des structures symplectiques et des structures de contact, in *Colloq. Géom. Diff. Globale*, (Brunelles, 1958) pp. 37–59. Centre Belge Rech. Math., Louvain.

Lindenstrauss, J., and Tzafriri, L. 1971. On the complemented subspace problem, *Israel J. Math.* 9:263–269.

Littlewood, J. E. 1952. On the problem of *n* bodies. *Meddel. Lunds Univ. Mat. Sem. Suppl. M. Riesz*, pp. 143–151.

Loomıs, L. H., and Sternberg, S. 1968. *Advanced calculus*. Addison-Wesley, Reading, Mass.

Lorenz, E.

1962. The statistical prediction of solutions of dynamic equations, in *Proc. Internat. Symp. Numerical Weather Prediction*, Tokyo, pp. 629–635.

1963. Deterministic nonperiodic flow. *J. Atmos. Sci.* 20:130–141.

Lovelock, D., and Rund, H. 1975. *Tensors, differential forms, and variational principles.* Wiley.

Mackey, G. W.

1962. Point realizations of transformation groups. *Illinois J. Math.* 6:327–335.

1963. *Mathematical foundations of quantum mechanics*. Benjamin-Cummings, Reading, Mass.

1968. *Induced representations of groups and quantum mechanics*. Benjamin-Cummings.

MacLane, S.

1968. Geometrical mechanics (2 parts). Dept. of Math., Univ. of Chicago, Chicago, Ill.

1970. Hamiltonian mechanics and geometry. *Am. Math. Monthly* 77:570–586.

ISBN 0-8053-0102-X

Magnus, W., and Winkler, S. 1975. *Hill's equation*, Interscience Tracts in Pure and Applied Math. No. 20, New York.

Malliavin, P. 1972. *Géometrie differentielle intrinsèque*. Hermann, Paris.

Manakov, S. V. 1976. Note on integration of Euler's equations of dynamics of *n*-dimensional rigid body, *Funct. An. Appl.* 1(4):93–94.

Mañé, R.

1975*a*. Absolute and infinitesimal stability (in Manning [1975], pp. 24–26).

1975*b*. On infinitesimal and absolute stability of diffeomorphisms (in Manning [1975], pp. 151–161).

Manning, A. (ed.). 1975. *Dynamical systems—Warwick* 1974, *Lec. Notes Math.*, vol. 468.

Markus, L.

1954. Global structure of ordinary differential equations on the plane. *Trans. Am. Math. Soc.* 76:127–148.

1960. Periodic solutions and invariant sets of structurally stable differential systems. *Bol. Soc. Mat. Mexicana* 5:190–194.

1961. Structurally stable differential systems. *Ann. Math.* 73:1–19.

1963. *Cosmological models in differential geometry*, Univ. of Minnesota Notes.

1971*a*. In *Lectures in differentiable dynamics, CBMS (3)*. Am. Math. Soc., Providence, R.I.

1971*b*. Ergodic Hamiltonian theory (in Chillingworth [1971], pp. 60–62).

1975. Dynamical systems-five years after (in Manning [1975], pp. 354–365).

Markus, L. and Meyer, K. R.

1969. Generic Hamiltonian systems are not ergodic, in Proc. 5th Internat. Conf.on Nonlinear Oscillation, Kiev, pp. 311–332.

1974. Generic Hamiltonian dynamical systems are neither integrable nor ergodic. *Mem. Am. Math. Soc.* 144.

Marle, G. M. 1976. Symplectic manifolds, dynamical groups and Hamiltonian mechanics, in *Differential geometry and relativity*. M. Cahen and M. Flato, (eds.). D. Reidel

Marmo, G., and Saletan, E. J., 1977. Ambiguities in Lagrange and Hamiltonian formalism: transformation properties. *Nuovo Cimento* 40:67–89.

Marmo, G., Saletan E. J., and Simon, A. 1978. Reduction of symplectic manifolds through constants of the motion (preprint).

Marsden, J.

1967. A correspondence principle for momentum operators. *Can. Math. Bull.* 10:247–250.

1968*a*. Hamiltonian one parameter groups. *Arch. Rat. Mech. Anal.* 28:362–396.

1968*b*. Generalized Hamiltonian mechanics. *Arch. Rat. Mech. Anal.* 28:326–362.

1969. Hamiltonian systems with spin. *Can. Math. Bull.* 12:203–208.

1972. Darboux's theorem fails for weak symplectic forms. *Proc. AMS.* 32:590–592.

1973*a*. On completeness of homogeneous pseudo–Riemannian manifolds. *Indiana Univ. Math. J.* 22:1065–1066.

1973*b*. On global solutions for nonlinear Hamiltonian evolution equations. *Comm. Math. Phys.* 30:79–81.

1974*a*. *Elementary classical analysis*. Freeman, San Francisco.

1974*b*. *Applications of global analysis in mathematical physics*. Publish or Perish, Boston.

1976. Well-posedness of equations of non-homogeneous perfect fluid. *Comm. P.D.E.* 1:215–230.

1978. Qualitative methods in bifurcation theory. Bull. Am. Math. Soc. 84:1125–1148.

Marsden, J. and Abraham, R. 1970. Hamiltonian mechanics on Lie groups and hydrodynamics, in *Proc. Pure Math.*, vol. 16, pp. 237–243. Am. Math. Soc., Providence, R.I.

Marsden, J., Ebin, D., and Fischer, A. 1972. Diffeomorphism groups, hydrodynamics and relativity, in *Proc. 13th Biennial Seminar of Can. Math. Congress*, J. R. Vanstone (ed.), Montreal, pp. 135–279.

Marsden, J., and Hughes, T.

1976. *A short course in fluid mechanics*. Publish or Perish, Boston, Mass.

1978. *Mathematical foundations of continuum mechanics* (lecture notes).

Marsden, J., and McCracken, M. 1976. The Hopf bifurcation and its applications, in *Notes in Applied Math. Sci.*, vol. 19. Springer, New York.

Marsden, J., and Weinstein, A. 1974. Reduction of symplectic manifolds with symmetry. *Rep. Math. Phys.* 5:121–130.

Marsden, J., and Wightman, A. S. 1967. *Lectures on statistical mechanics* (mimeographed notes, Princeton University).

ISBN 0-8053-0102-X

Martin, R. H. 1973. Differential equations on closed subsets of a Banach space. *Trans. Am. Math. Soc.* 179:399–414.

Maslov, V. P. 1965. *Théorie des perturbations et méthodes asymptotiques*. Dunod, Paris, 1972 (translated from Russian).

Massey, W. S. 1967. *Algebraic topology, an introduction*. Harcourt-Brace, New York.

Mather, J. 1967. Anosov diffeomorphisms (appendix in Smale [1967], pp. 792–795).

Mather, J. N., and McGehee, R. 1975. Solutions of the collinear four-body problem which become unbounded in a finite time (in Moser [1975*a*], pp. 573–597).

McGehee, R.

1973. A stable manifold theorem for degenerate fixed points with applications to celestial mechanics. *J. Diff. Equations* 14:70–88.

1974. Triple collision in the collinear three-body problem. *Inv. Math*. 27:191–227 (see also Moser [1975*a*]).

1975. Triple collision in Newtonian gravitational systems (in Moser [1975*a*], pp. 550–572).

McGehee, R., and Meyer, K. 1974. Homoclinic points of area preserving diffeomorphisms. *Am. J. Math*. 96:409–421.

McKean, H. P., and Trubowitz, E. 1976. Hill's operator and hyperelliptic function theory in the presence of infinitely many branch points. *Comm. Pure Appl. Math*. 29:143–226.

McKean, H. P., and van Moerbeke, P. 1975. The spectrum of Hill's equation. *Inv. Math.* 30:217–274.

McLaughlin, D. W. 1975. Four examples of the inverse method as a canonical transformation. *J. Math. Phys*. 16:96–99; 16:1704.

Menzio, M. R., and Tulczyjew, W. M. 1978. Infinitesimal symplectic relations and generalized Hamiltonian dynamics (to appear).

Mercier, A. 1959. *Analytical and canonical formalism in physics*. North Holland, Amsterdam; and Wiley, New York.

Meyer, K. R.

1970. Generic bifurcation of periodic points. *Trans. Am. Math. Soc.* 149:95–107.

1971*a*. Generic stability properties of periodic points. *Trans. Am. Math. Soc.* 154:273–277.

1971*b*. Bifurcations (in Chillingworth [1971], pp. 86–87).

1971*c*. Generic bifurcation (in Chillingworth [1971], pp. 128–129).

1973. Symmetries and integrals in mechanics (in Peixoto [1973], pp. 259–273).

1974. Normal forms for Hamiltonian systems. *Celest. Mech*. 9:517–522.

1975*a*. Homoclinic points of area-preserving maps (in Manning [1975], pp. 60–61).

1975*b*. Generic bifurcations in Hamiltonian systems (in Manning [1975], pp. 62–70).

Meyer, K. R., and Palmore, J.

1970. A generic phenomenon in conservative Hamiltonian systems (in Chern and Smale [1970], vol. 14, pp. 185–190).

1972. A new class of periodic solutions in the restricted three-body problem. *J. Diff. Equations* 8:264–276.

Meyer, K. R., and Schmidt, D. S. 1971. Periodic orbits near L_4 for mass ratios near the critical mass ratio of Routh. *Celest. Mech*. 4:99–109.

Michael, E. 1951. Topologies on spaces of subsets. *Trans. Am. Math. Soc.* 71:151–182.

Miller, J. G. 1976. Charge quantization and canonical quantization. *J. Math. Phys*. 17:643–646.

Milnor, J.

1956. On manifolds homeomorphic to the 7-sphere. *Ann. Math*. 64:399–405.

1963. *Morse theory*. Princeton Univ. Press, Princeton, N.J.

1965. *Topology from the differentiable viewpoint*. Univ. of Virginia Press, Charlottesville, Va.

1976. Curvatures of left invariant metrics on Lie groups, *Advan. Math* 21(3):293–329.

Minorsky, N.

1947. *Introduction to non-linear mechanics*. J. W. Edwards, Ann Arbor, Mich.

1974. *Nonlinear oscillations*. Krieger, Huntington, N.Y.

Mishchenko, A. S. 1970. Integral geodesics of a flow on Lie groups (Russian). *Funct. Anal. i Prilzen* 4:73–77; (English) *Funct. Anal. Appl*. 4:232–235.

Mishchenko, A. S. and Fomenko, A. T. 1978. Euler Equations on Finite Dimensional Lie Groups, Math. Izvestia, 12:371–389.

Misner, C., Thorne, K., and Wheeler, J. A. 1973. *Gravitation*. Freeman, San Francisco.

Misner, C. W., and Wheeler, J. A. 1957. Classical physics as geometry. *Ann. Phys*. 2:525–603.

Miura, R. M.

1968. Korteweg-de Vries equation and generalizations I. A remarkable explicit nonlinear

ISBN 0-8053-0102-X

transformation. *J. Math. Phys.* 9:1202–1204.

1976*a*. The Korteweg-de Vries equation: A survey of results. *SIAM Rev.* 18:412–459.

1976*b*. *Bäcklund transformations, Lec. Notes Math.*

Montgomery, D. 1936. On continuity in topological groups. *Bull. Am. Math. Soc.* 42:879.

Morlet, C. 1963. Le lemme de Thom et les théorèmes de plongement de Whitney, Sém. H. Cartan (1961–62) (4). Ecole Normale Supérieure, Paris.

Morosov, A. D. 1976. A complete qualitative investigation of Duffing's equations. *J. Diff. Equations* 12:164.

Morrey, C. B. 1966. *Multiple integrals in the calculus of variations*. Springer-Verlag, New York.

Moser, J.

1953. Periodische Lösungen des restringierten Dreikörperproblems, die sich erst nach vielen Umlaufen schliessen. *Math. Ann.* 126:325–335.

1955. Stabilitätsverhalten kanonischev Differentialgleichungssysteme. *Nachr. Akad. Wiss. Göttingen, Math. Phys.* K1.II:87–120.

1956. The analytic invariants of an area-preserving mapping near a hyperbolic fixed point. *Comm. Pure Appl. Math.* 9:673–692.

1958. New aspects in the theory of Hamiltonian systems. *Comm. Pure Appl. Math.* 9:81–114.

1962. On invariant curves of area-preserving mappings of an annulus. *Nachr. Akad. Wiss. Göttingen, Math. Phys.* K1.II:1–20.

1963*a*. Stability and nonlinear character of ordinary differential equations in nonlinear problems, in *Nonlinear problems*, pp. 139–149 (Langer, ed.). Univ. Wisconsin Press, Madison, Wis.

1963*b*. Perturbation theory for almost periodic solutions for undamped nonlinear differential equations, in *Intern. sym. on nonlinear differential equations and nonlinear mechanics*, pp. 71–79 (La Salle and Lefschetz, eds.). Academic, New York.

1964. On invariant manifolds of vector fields and symmetric partial differential equations, in *Differential analysis*, pp. 227–236. Oxford Univ. Press, Oxford.

1965. On the volume elements on a manifold. *Trans. Am. Math. Soc.* 120:286–294.

1966*a*. A rapidly convergent iteration method and nonlinear partial differential equations, I. *Ann. Scuola Norm. Sup. Pisa* (3)20:265–316; II. 20:499.

1966*b*. On the theory of quasi-periodic motions. *SIAM Rev.* 8:145–172.

1967. Convergent series expansions for quasi-periodic motions. *Math. Ann.* 169:136–176.

1968. Lectures on Hamiltonian systems, in *Memoirs Am. Math. Soc.*, vol. 81. Am. Math. Soc., Providence, R.I.

1969. On a theorem of Anosov. *J. Diff. Equations* 5:411–440.

1970. Regularization of Kepler's problem and the averaging method on a manifold. *Comm. Pure Appl. Math.* 23.

1973*a*. Stable and random motions in dynamical systems, with special emphasis on celestial mechanics, in *Ann. Math. Studies*, no. 77. Princeton Univ. Press, Princeton, N.J.

1973*b*. On a class of quasi-periodic solutions for Hamiltonian systems (in Peixoto [1973], pp. 281–288).

1975*a*. *Dynamical systems, theory and applications, Lec. Notes Phys.*, vol. 38. Springer, New York.

1975*b*. Three integrable Hamiltonian systems connected with isospectral deformations. *Adv. Math.* 16:197–220.

1976. Periodic orbits near an equilibrium and a theorem by Alan Weinstein. *Comm. Pure Appl. Math.* 29:727–747.

1977. The Birkhoff–Lewis fixed point theorem (Appendix 3.3 in Klingenberg [1977]).

1978. A fixed point theorem in symplectic geometry (preprint).

Moulton, F. 1902. *An introduction to celestial mechanics*. Macmillan, New York.

Myers, S. B., and Steenrod, N. 1939. On the group of isometries of a Riemannian manifold. *Ann. Math.* 40(2):406–416.

Nagumo, M. 1942. Über die Lage der Integralkurven gewöhnlicher Differentialgleichungen. *Proc. Phys.-Math. Soc. Jap.* 24:551–559.

Naimark, Y. I.

1959. On some cases of dependence of periodic motions on parameters. *Dokl. Akad. Nauk SSSR* 129:736–739.

1967. Motions closed to doubly asymptotic motion. *Sov. Math. Dokl.* 8:228–231.

Nash, J. 1956. The embedding problem for Riemannian manifolds. *Ann. Math.* 63:20–63.

Nehoroshev, N. 1972. Action-angle coordinates and their generalizations. *Trud. Mosk. Math.* 26.

ISBN 0-8053-0102-X

Nelson, E.
1964. Feynman integrals and the Schrödinger equation. *J. Math. Phys.* 5:332–343.
1967*a*. *Tensor analysis*. Princeton Univ. Press, Princeton, N.J.
1967*b*. *Dynamical theories of Brownian motion*. Princeton Univ. Press, Princeton, N.J.
1969. *Topics in dynamics I, flows*. Princeton Univ. Press, Princeton, N.J.

Nemytskii, V. V., and Stepanov, V. V. 1960. *Qualitative theory of differential equations*, Princeton Univ. Press.

Newhouse, S. E.
1974. Diffeomorphisms with infinitely many sinks, *Topology* 12:9–18.
1975*a*. Simple arcs and stable dynamical systems (in Manning [1975], pp. 53–55).
1975*b*. On simple arcs between structurally stable flows (in Manning [1975], pp. 209–233).
1976. Conservative systems and two problems of Smale (in Hilton [1976], pp. 104–110).
1977*a*. Quasi-elliptic periodic points in conservative dynamical systems (preprint).
1977*b*. The abundance of wild hyperbolic sets and non-smooth stable sets for diffeomorphisms (preprint).
1978. Topological entropy and Hausdorff dimension for area preserving diffeomorphisms of surfaces (preprint).

Newhouse, S. E., and Palis, J.
1973*a*. Hyperbolic nonwandering sets on two-dimensional manifolds (in Peixoto [1973], pp. 293–302).
1973*b*. Bifurcations of Morse–Smale dynamical systems (in Peixoto [1973], pp. 303–366).
1976. Cycles and bifurcations. *Astérisque* 31:43–141.

Newhouse, S. E., and Peixoto, M. 1977. There is a simple arc joining any two Morse-Smale flows (preprint).

Nickerson, H., Spencer, D., and Steenrod, N. 1959. *Advanced calculus*, Van Nostrand, Princeton.

Nirenberg, L., and Treves, F. 1970. On local solvability of linear partial differential equations. *Comm. Pure Appl. Math.* 22:1–38, 459–510.

Nitecki, Z. 1971. *Differentiable dynamics*. MIT Press, Cambridge, Mass.

Nomizu, K., and Kobayashi, S. 1963. *Foundations of differential geometry*, v. I, II. Wiley, New York.

Oden, J. T. (ed.). 1975. *Computational mechanics proceedings* 1974, *Lec. Notes Math.*, vol. 461. Springer, New York.

Oden, J. T., and Reddy, J. N. 1976. *Variational methods in theoretical mechanics*. Springer.

Ollongren, A. 1962. Three-dimensional galactic stellar orbits. *Bull. Astron. Inst. Neth.* 16:241.

Omori, H.
1970. On the group of diffeomorphisms on a compact manifold. *Proc. Symp. Pure Math.* 15:167–184.
1975. *Infinite dimensional Lie transformation groups*, *Lec. Notes Math.*, vol. 427.

Onofri, E., and Pauri, M.
1972. Dynamical quantization. *J. Math. Phys.* 13(4):533–543.
1973. Constants of the motion and degeneration in Hamiltonian systems. *J. Math. Phys.* 14:1106–1115.

Oster, G. F., and Perelson, A. S.
1973. Systems, circuits and thermodynamics. *Israel J. Chem., Katchalsky Memorial Issue* 11:445–478.
1974. Chemical reaction dynamics. *Arch. Rat. Mech. Anal.* 55:230–274; 57:31–98.

Ouzilou, R.
1970. Sur l'algèbre de Lie d'un groupe banachique. *C. R. Acad. Sci. Paris, Ser. A-B* 271:A1262–A1264.
1972. Expression symplectique des problèmes variationnels. *Symp. Math.* 14:85–98.

Oxtoby, J. C., and Ulam, S. 1941. Measure preserving homeomorphisms and metrical transitivity. *Ann. Math.* 42:874–920.

Palais, R.
1954. Definition of the exterior derivative in terms of the Lie derivative, in *Proc. Am. Math. Soc.* vol. 5, pp. 902–908. Am. Math. Soc., Providence, R.I.
1957. Global formulation of Lie theory of transformation groups, in *Mem. AMS.*, vol. 22.
1959. Natural operations on differential forms. *Trans. Am. Math. Soc.* 92:125–141.
1960. Extending diffeomorphisms, in *Proc. Am. Math. Soc.*, vol. 11, pp. 274–277.

ISBN 0-8053-0102-X

1963. Morse theory on Hilbert manifolds. *Topology* 2 (4):299–340.
1965. *Seminar on the Atiyah–Singer index theorem*, in *Ann. of Math. Studies*. Princeton Univ. Press, Princeton, N.J.
1968. *The Foundations of global non-linear analysis*. Benjamin-Cummings, Reading, Mass.

Ong Chong, P. 1975. Curvature and mechanics. *Adv. Math*. 15(3):269–311.

Palis, J.
1969. On Morse–Smale dynamical systems. *Topology* 8:385–405.
1971. Seminário de sistemos dinâmicos. I. M. P. A., Rio de Janeiro, 1971.
1974. Vectorfields generate few diffeomorphisms. *Bull. Am. Math. Soc.* 80:503–505.
1975. Arcs of dynamical systems: bifurcations and stability (in Manning [1975], pp. 48–53.

Palis, J., and deMelo, W. 1975. Introducão aos Sistemos dinâmicos. I. M. P. A., Rio de Janeiro, 1975.

Palis, J., and Pugh, C. (eds.), 1975. Fifty problems in dynamical systems (in Manning [1975], pp. 345–353).

Palis, J., Pugh, C., and Robinson, R. C. 1975. Nondifferentiability of invariant foliations (in Manning [1975], pp. 262–277).

Palis, J., Pugh, C., Shub, M., and Sullivan, D. 1975. Genericity theorems in topological dynamics (in Manning [1975], pp. 241–250).

Palis, J., and Smale, S. 1970. Structurally stability theorems (in Chern and Smale [1970], vol. 14, pp. 223–231).

Palmore, J.
1973. Classifying relative equilibria, I–III. *Bull. Am. Math. Soc.* 79(1973):904–908; 81(1975):489–491; *Lett. Math. Phys.* 1(1975):71–73.
1976*a*. New relative equilibria of the n-body problem. *Lett. Math. Phys.* 1:119–123.
1976*b*. Measure of degenerate relative equilibria, I. *Ann. Math* 104:421–429.

Pars, L. 1965. *A treatise on analytical dynamics*. Wiley, New York.

Pauli, W. 1953. On the Hamiltonian structure of non-local field theories. *Nuovo Cimento* 10:648–667.

Peixoto, M.
1959. On structural stability. *Ann. Math.* 69:199–222.
1962. Structural stability on two-dimensional manifolds. *Topology* 2:101–121.
1967*a*. Qualitative theory of differential equations and structural stability, in *Differential equations and dynamical systems* (Proc. Internat. Sympos., Mayaguez, P.R. 1965) pp. 469–480. Academic, New York.
1967*b*. On an approximation theorem of Kupka and Smale. *J. Diff. Equations* 3:214–227.
1973. *Dynamical systems*. (Ed.) Academic, New York.

Penot, J-P.
1967. Variété differentiables d'applications et de chemins. *C.R. Acad. Sci. Paris* 264:1066–1068.
1968. Une méthode pour construire des variétés d'applications au moyen d'un plangement, *C.R. Acad. Sci. Paris* 266:625–627.
1970*a*. *Geometrie des Variétés Fonctionnelles*. Thése, Paris.
1970*b*. Sur la théorem de Frobenius. *Bull. Math Soc. France* 98:47–80.
1972. Topologie faible sur des variétés de Banach. *C. R. Acad. Sci. Paris* 274:405–408.

Penrose, R. 1972. On the nature of quantum geometry, in *Magic without magic*, J. R. Klauder (ed.), vol. 333. Freeman, San Francisco.

Perron, O. 1937. Neue periodische Lösungen des ebenen Drei- und Mehrkörperproblems. *Math. Zeit.* 42:593–624.

Pliss, V. A. 1966. *Nonlocal problems of the theory of oscillations*. Academic, New York.

Plummer, H. C. 1901. Instability of the Lagrange collinear solutions. *Mon. Not. Roy. Astron. Soc.* 62:6–12.

Poenaru, V. 1975. The Maslov index for Lagrangian manifolds (in Manning [1975], pp. 70–71).

Poincare, H.
1880. Sur les courbes définies par les équations différentielles. *C.R. Acad. Sci. Paris* 90:673–675; see also *J. Math Pures Appl.* (3)7(1881):375–422; (3)8(1882):251–286; (4)1(1885):167–244; (4)2(1886):151–217; *Acta Math.* 13(1890):1–271.
1885. Sur l'équilibre d'une masse fluide animée d'un mouvement de rotation. *Acta Mathematica* 7:259–380.

ISBN 0-8053-0102-X

1892. *Les méthodes nouvelles de la mécanique céleste, 1, 2, 3.* Gauthier-Villars, Paris (1892); Dover, New York, 1957.

1912. Sur un théorème de géométriè. *Rend. del. Circ. Math. du Palermo* 33:375–407.

Pollard, H. 1976. *Introduction to celestial mechanics. Carus Mathematical Monographs.* No. 18. Math. Assoc. Am., Buffalo, N.Y.

Povzner, A. 1966. A global existence theorem for a nonlinear system and the defect index of a linear operator. *Transl. Am. Math. Soc.* 51:189–199.

Pugh, C. C.

1967. An improved closing lemma and a general density theorem. *Am. J. Math.* 89:1010–1021.

1968. The closing lemma for Hamiltonian systems, in Petrovsky, I. G., Intern. Cong. Mathematicians, Moscow 1966. "Mir", Moscow.

Pugh, C. C., and Robinson, R. C. 1977. The C^1 closing lemma, including Hamiltonians. *Pac. J. Math.* (to appear).

Pugh, C. C., and Shub, M.

1970*a*. The Ω-stability theorem for flows. *Invent. Math.* 11:150–158.

1970*b*. Linearization of normally hyperbolic diffeomorphisms and flows. *Invent. Math* 10:187–198.

Rabinowitz, P. H. 1978. Periodic solutions of Hamiltonian systems, *Comm. Pure Appl. Math.* 31:157–184.

Ratiu, T., 1979. Thesis, Berkeley.

Ratiu, T., and Bernard, P. (eds.). 1977. *Seminar on turbulence theory, Lec. Notes Math.*, vol. 615. Springer, New York.

Ravatin, J. (ed.), 1972. *Quatrième rencontre entre mathématicians et physicians.* Dept. de Math, Tome 9, Univ. de Lyon, Villeurbanne.

Rawnsley, J. H. 1974. De Sitter symplectic spaces and their quantization. *Proc. Comb. Phil. Soc.* 76:463–480.

Redheffer, R. M. 1972. The Theorems of Bony and Brezis on flow-invariant sets. *Am. Math. Monthly* 79:740–747.

Reeb, G.

1948. Sur les mouvements périodiques de certains systèmes mécaniques. *C.R. Acad. Sci. Paris* 227:1331–1332.

1949*a*. Sur les solutions périodiques de certains systèmes différentiels canoniques. *C.R. Acad. Sci. Paris* 228:1196–1198.

1949*b*. Quelques propriétés globales des trajectoires de la dynamique dues à l'existence de l'invariant intégral de M. Elie Cartan. *C.R. Acad. Sci. Paris* 229:969–971.

1951. Sur les solutions périodiques de certains systèmes différentiels perturbés. *Can. J. Math.* 3:339–362.

1952*a*. Sur certaines propriétés globales des trajectoires de la dynamique, dues à l'existence de l'invariant intégral de M. Elie Cartan, in Colloque de Topologie de Strasbourg, 1951, no. III. Univ. de Strasbourg, Strasbourg.

1952*b*. Sur la nature et la distribution des trajectoires périodiques de certains systèmes dynamiques, in C.R. Congr. Soc. Savantes Paris (Grenoble, 1952), Section des Sci., pp. 35–39. Gauthier-Villars, Paris.

1952*c*. Sur certaines propriétés topologiques des trajectoires des systèmes dynamiques. Acad. Roy. Belg. Cl. Sci. Mem. Coll. in 8° 27 (2), no. 9.

1952*d*. Remarques sur l'existence de mouvements périodiques de certains systèmes dynamiques excités. *Arch. Math.* 3:76–78.

1952*e*. Variétés symplectiques, variétés presque-complexes et systèmes dynamiques. *C.R. Acad. Sci. Paris* 235:776–778.

1954. Sur certaines problèmes relatifs aux variétés presque-symplectiques et systèmes dynamiques, in Convegno Intern. di Geometria Differenziale (Italia, 1953) Cromonese, Rome, pp. 104–106.

Reed, M., and Simon, B. 1974. *Methods of modern mathematical physics.* Vol. I; *functional analysis.* Vol. II: *selfadjointness and Fourier analysis.* Academic, New York.

Reeken, M. 1973. Stability of critical points under small perturbations. *Manuscripta Math.* 8:69–92.

Renz, P. 1971. Equivalent flows on smooth Banach manifolds. *Indiana Univ. Math. J.* 20:695–698.

ISBN 0-8053-0102-X

de Rham, G. 1955. *Variétés differentiables*. Hermann, Paris.
Riddell, R. C. 1970. A note on Palais' axioms for section functors. *Proc. Am. Math. Soc.* 25:808–810.
Riesz, F. 1944. Sur la théorie ergodique, *Comm. Math. Helv.* 17:221–239.
Robbin, J.
1968. On the existence theorem for differential equations. *Proc. Am. Math. Soc.* 19:1005–1006.
1971*a*. A structural stability theorem. *Ann. Math.* 94:447–493.
1971*b*. Stable manifolds of semi-hyperbolic fixed points. *Illinois J. Math.* 15:595–609.
1972*a*. Structural stability. *Bull. Am. Math. Soc.* 76:723–726.
1972*b*. Topological conjugacy and structural stability for discrete dynamical systems. *Bull. Am. Math. Soc.* 78:923–952.
1972*c*. La fonction géneratrice de Poincaré (d'après Weinstein) in Ravatin [1972], vol. 4(2):122–135).
1972*d*. Equilibre relatif dans les systèmes mécaniques (in Ravatin [1972], vol. 4(2):136–143).
1973. Relative equilibria in mechanical systems (see Peixoto [1973], pp. 425–441).
1974. *Symplectic mechanics, global analysis and its applications*, vol. 3, pp. 97–120. I. A. E. A., Vienna.
Robinson, R. C.
1970*a*. A global approximation theorem for Hamiltonian systems (in Chern and Smale, [1970], pp. 233–244).
1970*b*. Generic properties of conservative systems, I, II. *Am. J. Math.* 92:562–603; 897–906.
1971*a*. Lectures on Hamiltonian systems. *Monografias de Matematica*, no. 7. Inst. Mat. Pura Apl., Rio de Janeiro, Brazil.
1971*b*. Generic properties of conservative systems (in Chillingworth [1971], pp. 35–36).
1971*c*. Generic one-parameter families of symplectic matrices. *Am. J. Math.* 93:116–122.
1971*d*. Differentiable conjugacy near compact invariant manifolds. *Boletin de Sociedade Brasileira de Mat.* 2:33–44.
1973*a*. C^r Structural stability implies Kupka–Smale (in Peixoto [1973], pp. 443–449).
1973*b*. Closing stable and unstable manifolds on the two sphere. *Proc. Am. Math. Soc.* 41:299–303.
1974. Structural stability of vector fields. *Ann. Math.* 99:154–175.
1975*a*. Structural stability of C^1 flows (in Manning [1975], pp. 262–277).
1975*b*. Structural stability of diffeomorphisms (in Manning [1975], pp. 21–23).
1975*c*. Fixing the center of mass in the *n*-body problem by means of a group action. *Colloq. Intern.* C.N.R.S., vol. 237. Paris.
1976*a*. Structural stability of a C^1 diffeomorphism. *J. Diff. Equations* 22:28–73.
1976*b*. Structural stability theorems, in Cesari, L., Hale, J., and LaSalle, J., eds., *Dynamical systems, an international symposium*, vol. 2, pp. 33–36. Academic, New York.
1976*c*. A quasi-Anosov flow that is not Anosov. *Ind. Univ. Math. J.* 25:763–767.
1977*a*. The geometry of the structural stability proof using unstable disks. *Bol. Soc. Brasileira de Matematica* (to appear).
1977*b*. Global structural stability of a saddle mode bifurcation. *Trans. Am. Math. Soc.* (to appear).
1977*c*. Stability theorems and hyperbolicity in dynamical systems, in *Proc. of Nat. Sci. Found. Regional Conf. on Dynamical Systems, 1976* (to appear).
1977*d*. Stability, measure zero, and dimension two implies hyperbolicity (preprint).
1977*e*. Introduction to the closing lemma, in *Proc. of Nat. Sci. Found. Regional Conf. on the Structure of Attractors in Dynamical Systems,* 1977 (to appear).
1977*f*. Structural stability on manifolds with boundary (preprint).
Robinson, R. C., and Verjovsky, A. 1971. Stability of Anosov diffeomorphisms (in Palis [1971], IX-1-IX-22).
Robinson, R. C., and Williams, R. F.
1973. Finite stability is not generic (in Peixoto [1973], pp. 451–462).
1976. Classification of expanding attractors: an example. *Topology* 15:321–323.
1977. Wild insets (preprint).

ISBN 0-8053-0102-X

Rod, D. L. 1973. Pathology of invariant sets in the monkey saddle. *J. Diff. Equations* 14:129–170.

Rod, D. L., Pecelli, G., and Churchill, R. C. 1977. Hyperbolic periodic orbits. *J. Diff. Equations* 24:329–348.

Rodrigues, J. 1976. Sur les systèmes mécaniques généralisés. *C.R. Acad. Sci. Paris.* 282:1307–1309.

Roeleke, W. 1960. Über den Laplace-operator auf Riemannscher Manigfaltigkeiten mit diskontinuierlichen Gruppen. *Math. Nachr.* 21:132–149.

Roëls, J.

1971. An extension to resonant cases of Liapunov's theorem concerning the periodic solutions near a Hamiltonian equilibrium. *J. Diff. Equations* 9:300–324.

1973. Sur l'accouplement des sous-espaces invariants indécomposables des systèmes linéaires Hamiltonienes. *Acad. Roy. Belg. Bull. Cl. Sci.* 59:85–88.

1974. Sur la décomposition locale d'un champ de vecteurs d'une surface symplectique en un gradient et un champ Hamiltonien. *C.R. Acad. Sci. Paris* 278:29–31.

Roels, J., and Weinstein, A. 1971. On functions whose Poisson brackets are constant. *J. Math. Phys.* 12:1482–1486.

Roseau, M. 1966. *Vibrations non-linéaires et théorie de la stabilité, Springer Tracts in Nat. Phil.*, vol. 8. Springer, New York.

Rössler, O. E. 1976. An equation for continuous chaos. *Phys. Lett.* 57A:397–398.

Routh, E. J. 1955. *A treatise on the dynamics of a system of rigid bodies.* Dover, New York.

Royden, H. 1963. *Real analysis.* Macmillan, New York.

Rubin, H., and Ungar, P., 1957. Motion under a strong constraining force. *Comm. Pure Appl. Math* 10:65–87.

Ruelle, D.

1969. *Statistical mechanics: rigorous results.* Benjamin-Cummings, Reading, Mass.

1973. Bifurcations in the presence of a symmetry group. *Arch. Rat. Mech.* 51:136–152.

1976. *Dynamical systems and turbulence,* Duke Univ., Durham, N.C.

Ruelle, D., and Takens, F. 1971. On the nature of turbulence. *Comm. Math. Phys.* 20:167–192; 23:343–344.

Rund, H. 1966. *The Hamiltonian-Jacobi theory in the calculus of variations; its role in mathematics and physics.* Krieger, Huntington, N.J.

Saari, D.

1972. Singularities and collisions of Newtonian gravitational systems. *Arch. Rat. Mech. Anal.* 49:311–320.

1974. The angle of escape on the three-body problem. *Celest. Mech.* 9:175–181.

Sacker, R. 1965. A new approach to the perturbation theory of invariant surfaces. *Comm. Pure Appl. Math.* 18:717–732.

Sasaki, S. 1966. On almost contact manifolds, in *Proc. of U.S.–Japan Seminar in Differential Geometry*, Kyoto, pp. 128–136. Nippon Hyoronsha, Tokyo.

Satzer, W. J., Jr. 1977. Canonical reduction of mechanical systems invariant under abelian group actions with an application to celestial mechanics. *Indiana Univ. Math. J.* 26:951–976.

Schiefele, G., and Stiefel, E. 1971. *Linear and regular celestial mechanics*. Springer-Verlag, Berlin.

Schmidt, D. S. 1974. Periodic solutions near a resonant equilibrium of a Hamiltonian system. *Celestial Mech.* 9:81–103.

Schmidt, D. S. 1976. On the Liapunov sub-center theorem (in Marsden–McCracken [1976]).

Schmidt, D. S., and Sweet, D. 1973. A unifying theory in determining periodic families for Hamiltonian systems at resonance. *J. Diff. Equations* 14:597–609.

Schwartz, A. 1963. A generalization of a Poincaré–Bendixson theorem to closed two-dimensional manifolds. *Am. J. Math.* 85:453–458.

Schwartz, J. T. 1967. *Nonlinear functional analysis*. Gordon and Breach, New York.

Scott, A. C., Chu, F. Y. F., and McLaughlin, D. W. 1973. The soliton: a new concept in applied science. *Proc. IEEE* 61:1443–1483.

Segal, I. E.

1959. Foundations of the theory of dynamical systems of infinitely many degrees of freedom, I. *Mat-Fys. Medd. Danske. Vid. Selsk* 31(1959):12; II, *Can. J. Math* 13(1961):1–18; III, *Ill. J. Math.* 6(1962):500–523.

1960. Quantization of nonlinear systems. *J. Math. Phys.* 1:468–488.

ISBN 0-8053-0102-X

1963. *Mathematical problems of relativistic physics*. Am. Math. Soc., Providence, R.I.
1964. Quantum fields and analysis in the solution manifolds of differential equations, in *Analysis in function space*, Martin and Segal (eds.). MIT, Boston, Mass.
1965. Differential operators in the manifold of solutions of a nonlinear differential equation. *J. Math. Pures et Appl.* 44:71–105; 44:107–132.
1966. Conjugacy to unitary groups within the infinite dimensional symplectic group. Argonne Nat. Lab. Report No. 7216.
1974. Symplectic structures and the quantization problem for wave equations. *Symp. Math.* 14:9–117.

Seifert, H. 1947. Periodische Bewegungen mechanischer Systeme. *Math. Z.* 51:197–216.

Sell, G. R.
1971. *Topological dynamics and ordinary differential equations, Ann. Math. Studies*, no. 33. Von Nostrand–Reinhold, Princeton, N.J.
1977. Bifurcation of higher dimensional tori (preprint).

Sewell, M. J. 1977. On Legendre transformations and elementary catastrophes, *Math. Proc. Camb. Phil. Soc.* 82:147–163.

Shahshahani, S. 1972. Dissipative systems on manifolds *Inv. Math.* 16:177–190.

Shaw, R. 1978. Strange attractors, chaotic behavior, information flow, U. of Cal. Santa Cruz (preprint).

Shub, M.,
1971. Diagonals and relative equilibria (appendix to Smale [1971*b*]; in Kuiper [1971], pp. 199–1201).
1973. Stability and genericity for diffeomorphisms (in Peixoto [1973], pp. 493–514).
1975*a*.Homology theory and dynamical systems (in Manning [1975] pp. 36–39).
1975*b*.Topological entropy and stability (in Manning [1975], pp. 39–40).

Shub, M., and Smale, S. 1972. Beyond hyperbolicity. *Ann. Math.* 96:587–591.

Shub, M., and Sullivan, D. 1975. Homology theory and dynamical systems. *Topology* 14:109–132.

Shub, M., and Williams, R. F. 1975. Topological entropy and stability. *Topology* 14:329–338.

Siegel, C. L.
1941. Der Dreierstoss, *Ann. Math.* 42:127–168.
1956. *Vorlesungen über Himmelsmechanik*. Springer-Verlag, Berlin (see Siegel–Moser [1971]).
1964. *Symplectic geometry*. Academic Press, New York.

Siegel, C. L., and Moser, J. K. 1971. *Lectures on celestial mechanics*. Springer-Verlag, New York.

Simms, D. J. 1968. *Lie groups and quantum mechanics, Lec. Notes Math.*, vol. 52. Springer.

Simms, D. J., Woodhouse, N. M. J. 1976. Lectures on geometric quantization, *Lec. Notes Phys.*, vol. 53.

Simó, C.
1975. Topology of the invariant manifolds of the spatial three-body problem (preprint).
1977. Relative equilibrium solutions in the four-body problem (preprint).

Simon, C. P.
1972. A 3-dimensional Abraham-Smale example, *Proc. Am. Math. Soc.* 34:629–630.
1973. Bounded orbits in mechanical systems with two degrees of freedom and symmetry (in Peixoto [1973], pp. 515–525).
1974. A bound for the fixed point index of an area-preserving map with applications to mechanics. *Inv. Math.* 26:187–200; 32(1976):101.

Sinai, Y. 1970. Dynamical systems with elastic reflections, *Russian Math. Surveys*, 25:137–189.

Singer, I., and Thorpe, J. 1967. *Lecture notes on elementary topology and geometry*. Scott, Foresman, Glenview, Ill.

Slawianowski, J. J. 1971. Quantum relations remaining valid on the classical level. *Reports Math. Phys.* 2:11–34.

Slebodzinski, W.
1931. Sur les équations de Hamilton. *Bull. Acad. Roy. de Belg.* 17:864–870.
1970. *Exterior forms and their applications*. Polish Scientific Publishers, Warsaw.

Smale, S.
1959.*a* Diffeomorphisms of the two sphere. *Proc. Am. Math. Soc.* 10:621–626.
1963*a*. Stable manifolds for differential equations and diffeomorphisms, in *Topologia differenziale, CIME*. Cremonese, Rome; or *Ann. Scuola Norm. Sup. Pisa* 17(3):97–116.
1963*b*. Dynamical systems and the topological conjugacy problem for diffeomorphisms, in

ISBN 0-8053-0102-X

Proc. Intern. Congr. Math., Stockholm, (V. Stenstrom, Ed.) 1962 pp. 490–496. Institut Mittag-Leffler, Stockholm.

1964*a*. Diffeomorphisms with many periodic points, in *Diff. and Comb. Topo. Sym. in Honor of Marston Morse*. Princeton Univ. Press, Princeton, N.J.

1964*b*. Morse theory and a nonlinear generalization of the Dirichlet problem. *Ann. Math.* 80(2):382–396.

1966. Structurally stable systems are not dense. *Am. J. Math.* 88:491–496.

1967. Differentiable dynamical systems. *Bull. Am. Math. Soc.* 73:747–817.

1970*a*. Topology and mechanics (I, II). *Inv. Math.* 10:305–331; 11:45–64.

1970*b*. The Ω-stability theorem (in Chern and Smale [1970], pp. 289–298).

1971*a* Stability and genericity in dynamical systems, in *Séminaire Bourbaki* 1969–1970. *Lec. Notes Math.* vol. 180. Springer, New York.

1971*b*. Problems on the nature of relative equilibria in celestial mechanics (in Kuiper [1971], pp. 194–198).

1971*c*. Topology and mechanics (in Chillingworth [1971], pp. 33–34).

1972. On the mathematical foundation of electrical circuit theory. *J. Diff. Geom.* 7:193–210.

1973. Stability and isotopy in discrete dynamical systems (in Peixoto [1973], pp. 527–530).

Sniatycki, J.

1974. Dirac brackets in geometric dynamics. *Ann. Inst. H. Poincaré* 20(4):365–372.

1975*a*. Bohr–Sommerfeld conditions in geometric quantization. *Reports Math. Phys.* 7:303–311.

1975*b*. Wave functions relative to a real polarization. *Int. J. Theor. Phys.* 14:277–288.

1978. Geometric quantization and quantum mechanics (preprint).

Sniatycki, J., and Tulczyjew, W. M.

1971. Canonical dynamics of relativistic charged particles. *Ann. Inst. H. Poincaré* 15:177–187.

1972*a*. Generating forms of Langrangian submanifolds. *Indiana Univ. Math. J.* 22:267–275.

1972*b*. Canonical formulation of Newtonian dynamics. *Ann. Inst. H. Poincaré* 16(1): 23–27.

Sotomayor, J.

1964. Total stability of invariant surfaces (Spanish). *Univ. Nac. Inger. Inst. Mat. Puras Apl. Notas. Mat.* 2:163–169.

1968. Generic one-parameter families of vector fields. *Bull. Am. Math. Soc.* 74:722–726.

1973*a*. Structural stability and bifurcation theory (in Peixoto [1973], pp. 549–560).

1973*b*. Generic bifurcations of dynamical systems (in Peixoto [1973], pp. 561–582).

1974. Generic one-parameter families of vector fields in two-dimensional manifolds. *Publ. Math. I.H.E.S.* 43:5–46.

1977. Saddle connections of dynamical systems (to appear).

Souriau, J. M.

1964. *Géometrie et relativité*. Hermann, Paris.

1970*a*. *Structure des systèmes dynamiques*. Dunod, Paris.

1970*b*. Sur le mouvement des particules à spin en relativité générale. *C.R. Acad. Sci. Paris* 271:751–753.

1974. Sur la variété de Kepler. *Symp. Math.* 14:343–360.

1975*a*. Geometrié symplectique et physique mathématique, in Coll. Intern. du CNRS, no. 237, Paris.

1975*b*. Méchanique statistique, groupes de Lie et cosmologie (in Souriau [1975*a*], pp. 59–114).

1976. Construction explicite de l'indice de Maslov: applications, in 1975 *Fourth Intern. Coll. on Group Theoretical Methods in Physics. Lec. Notes Phys.*, 50:117–148. Springer, New York.

Spivak, M.

1965. *Calculus on manifolds*. Benjamin-Cummings, Reading, Mass.

1974. *A comprehensive treatise on differential geometry* (five vols.). Publish or Perish, Boston, Mass.

Steenrod, N. E. 1951. *The topology of fiber bundles*, Princeton Math. Series., vol. 14. Princeton Univ. Press, Princeton, N.J.

ISBN 0-8053-0102-X

Stein, P., and Ulam, S. 1964. Nonlinear transformation studies on electronic computers. *Rozprawy Mat.* 39:66.

Sternberg, S.

1961. Infinite Lie groups and the formal aspects of dynamical systems. *J. Math. Mech.* 10:451–474.

1964. *Lectures on differential geometry*. Prentice-Hall, Englewood Cliffs, N.J.

1969. *Celestial mechanics*, vols. I, II. Benjamin-Cummings, Reading, Mass.

1975. Symplectic homogeneous spaces. *Trans. Am. Math. Soc.* 212:113–130.

1977. On minimal coupling and the symplectic mechanics of a classical particle in the presence of a Yang–Mills field *Proc. Nat. Acad. Sci.* 74:5253–5254.

Sternberg, S., and Wolf, J. A. 1975. Charge conjugation and Segal's cosmology. *Nuovo Cimento*, 28:253–271.

Stoker, J. J. 1950. *Nonlinear vibrations*. Wiley, New York.

Stone, M. 1932. *Linear transformations in Hilbert space, Am. Math. Soc. Colloq. Publ.*, vol. 15.

Streater, R. F. 1966. Canonical quantization. *Comm. Math. Phys.* 2:354–374.

Streater, R. F., and Wightman, A. S. 1964. *PCT, spin, and statistics and all that*. Benjamin-Cummings, Reading, Mass.

Sudarshan, E. C. G., and Mukunda, N. 1974. *Classical dynamics; a modern perspective*. Wiley.

Sundman, K.

1913. Mémoire sur le problème des trois corps. *Acta. Math.* 36:105–179.

1915. Theorie der Planeten. *Enzycl. Math. Wiss.* (*Astronomie*) 6/2/1(15):729–807.

Swanson, R. C. 1976. *Lagrangian subspaces, intersection theory and the Morse index theorem in infinite dimension.* Thesis, Univ. of Calif., Santa Cruz.

Synge, J. L.

1926. On the geometry of dynamics. *Phil. Trans. A* 220:31–106.

1960. Classical dynamics. *Encycl. Phys.* 3:1–225.

Synge, J. L., and Griffith, B. A. 1959. *Principles of mechanics.* McGraw-Hill, New York.

Szczyryba, W. 1976. A symplectic structure on the set of Einstein metrics: A canonical formalism for general relativity. *Comm. Math. Phys.* 51:163–182.

Szebehely, V.

1967. *Theory of orbits*. Academic, New York.

1975. Regularization in celestial mechanics (in Oden [1975], pp. 257–263).

Szlenk, W. 1974. *Dynamical systems, Aahrus Univ. Lec. Ser., no.* 41, Aahrus, Denmark.

Takens, F.

1970*a*. Generic properties of closed orbits and local perturbations. *Math. Ann.* 188:304–312.

1971. A C^1-counterexample to Moser's twist theorem. *Nederl. Akad. Wetensch. Proc. Ser. A* 74 (or *Indag. Math* 33:378–386).

1972. Homoclinic points in conservative systems. *Inv. Math.* 18:267–292.

1973*a*. *Introduction to global analysis, Comm. no. 2 of Math. Inst.*, Utrecht.

1973*b*. Unfoldings of certain singularities of vectorfields: generalized Hopf bifurcations. *J. Diff. Equations* 14:476–493.

1974*a*. *Forced oscillations and bifurcations, Comm. no. 3 of Math. Inst.*, Utrecht.

1974*b*. Singularities of vectorfields. *Publ. Math. I.H.E.S.* 43:47–100.

1975*a*. Constrained differential equations (in Manning [1975], pp. 80–82).

1975*b*. Geometric aspects of non-linear R.L.C. networks (in Manning [1975], pp. 305–331).

1976. Constrained equations; a study of implicit differential equations and their discontinuous solutions (in Hilton [1976], pp. 143–234).

1977. Symmetries, conservation laws and variational principles, in Palis, J., and do Carmo, M., eds., *Geometry and Topology*, Lec. Notes Math. no. 597, Springer, New York.

Tatarinov, Ia. V.

1973. On the study of the phase space topology of compact configurations with symmetry. *Vestnik. Mosk. Univ. Math-Mech.* 5 (in Russian).

1974 Portraits of the classical integrals of the problem of rotation of a rigid body about a fixed point. *Vestnik Moscow Univ.* 6:99–105 (in Russian).

Taub, A. H. 1949. On Hamilton's Principle for perfect compressible fluids, in *Proc. Symp. Appl. Math. I.* pp. 148–157. Am. Math. Soc., Providence, R.I.

Taylor, M. E.

ISBN 0-8053-0102-X

1978. *Pseudo-differential operators and Fourier integral operators* (in preparation)
Thom, R. 1975. *Structural stability and morphogenesis: an outline of a general theory of models*. Benjamin-Cummings, Reading, Mass.
Thurston, W. 1976. Some simple examples of symplectic manifolds, *Proc. Am. Math. Soc.* 55:467–468.
Timoshenko, S. *et al.* 1974. *Vibration problems in engineering*, 4th ed. Wiley, New York.
Trautman, A. 1967. Noether equations and conservation laws. *Comm. Math. Phys.* 6:248–261.
Tripathy, K. C. 1976. Dynamical prequantization, spectrum generating algebras and the classical Kepler and harmonic oscillator problems, in 1975 Fourth Intern. Coll. on Group Theor. Methods in Phys., Nijmegen. *Lec. Notes in Phys.* vol. 50, pp. 199–209. Springer, New York.
Tromba, A. J.
1976. Almost-Riemannian structures on Banach manifolds, the Morse lemma and the Darboux theorem. *Can. J. Math.* 28:640–652.
1977. A general approach to Morse theory, *J. Diff. Geom.* 12:47–85.
Truesdell, C.
1968. *Essays on the history of mechanics*. Springer, New York.
1974. A simple example of an initial-value problem with any desired number of solutions. *Inst. Lombardo* (*Rend. Sci.*) *A* 108:301–304.
Truman, A. 1976. Feynman path integrals and quantum mechanics as $\hbar \to 0$. *J. Math. Phys.* 17:1852–1862.
Tulczyjew, W. M.
1974*a*. Hamiltonian systems, Lagrangian systems and the Legendre transformation. *Symp. Math.* 14:247–258.
1974*b*. Poisson brackets and canonical manifolds. *Bull. Acad. Pol. des Sci.* 22:931–934.
1975*a*. Sur la différentielle de Lagrange. *C.R. Acad. Sci.* 280:1295–1298.
1975*b*. Relations symplectiques et les équations d'Hamilton–Jacobi relativistes. *C.R. Acad. Sci.* Paris. 281:545–547.
1976*a*. Lagrangian submanifolds and Hamiltonian dynamics. *C.R. Acad. Sci. Paris* 283:15–18.
1976*b*. Lagrangian submanifolds and Lagrangian dynamics. *C.R. Acad. Sci. Paris* 283:675–678.
1977*a*. The Legendre transformation. *Ann. Inst. H. Poincaré* 27:101–114.
1977*b*. Symplectic formulation of relativistic particle dynamics. *Acta Phys. Polonica* 138:431–447.
1977*c*. A symplectic formulation of particle dynamics, in *Lec. Notes Math.*, vol. 570, pp. 457–463. Springer, New York.
1977*d*. A symplectic formulation of field dynamics, in *Lec. Notes Math.*, vol. 570, pp. 464–468. Springer, New York.
Urabe, M.
1971*a*. Subharmonic solutions to Duffing's equation (in Chillingworth [1971], pp. 62–63).
1971*b*. Numerical analysis of nonlinear oscillations (in Chillingworth [1971], pp. 158–160).
Vainberg, M. M. 1964. *Variational methods for the study of nonlinear operators*. Holden-Day, San Francisco.
Van De Kamp, P. 1964. *Elements of astromechanics*. Freeman, San Francisco.
Van Hove, L. 1951. Sur certaines représentations unitaires d'un groupe infini de transformations. *Mem. de l'Acad. Roy. de Belgique* (*Classe des Sci.*) *t. XXVI* 61–102.
Van Kampen, N. G. 1972. Transformation groups and the virial theorem. *Rep. Math. Phys.* 3(3):235–239.
Van Karman, J. 1940. The engineer grapples with nonlinear problems. *Bull. Amer. Math. Soc.* 46:615–675.
Varadarajan, V. S.
1968. *Geometry of quantum theory* (*I, II*). Van Nostrand, Princeton, N.J.
1974. *Lie groups, Lie algebras and their representations*. Prentice-Hall, Englewood, N.J.
Vinogradov, A. M., and Krasil'shchik, I. S. 1975. What is the Hamiltonian formalism? *Russian Math. Surveys* 30:1; 177–202.
Vinogradov, A. M. and Kupershmidt, B. A. (1977) The Structures of Hamiltonian Mechanics, Russ. Math. Surveys 32:177–243.
Von Neumann, J.
1932. Zur Operatorenmethode in der klassischen Mechanik. *Ann. Math.* 33:587–648, 789.

ISBN 0-8053-0102-X

1955. *Mathematical foundations of quantum mechanics.* Princeton Univ. Press, Princeton, N.J.
Wadati, M., Sanuki, H., and Konno, K. 1975. Relationship among inverse method, Bäcklund transformation and an infinite no. of conservation laws. *Prog. Theor. Phys.* 53:419–436.
Wall, G. E. 1963. On the conjugacy classes in the unitary, symplectic and orthogonal groups. *J. Austral. Math. Soc.* 3:1–62.
Wallach, N. R. 1977. *Symplectic geometry and Fourier analysis*, Math. Sci. Press, Brookline, Mass.
Warner, F. 1971. *Foundations of differentiable manifolds and Lie groups.* Scott, Foresman, Glenview, Ill.
Weinstein, A.
1971*a*. Perturbation of periodic manifolds of Hamiltonian systems. *Bull. Am. Math. Soc.* 77,5:814–818.
1971*b*. Symplectic manifolds and their Lagrangian submanifolds. *Adv. Math.* 6:329–346 (see also *Bull. Am. Math. Soc.* 75(1969):1040–1041).
1972. The invariance of Poincaré's generating function for canonical transformations. *Inv. Math.* 16:202–213.
1973*a*. Lagrangian submanifolds and Hamiltonian systems. *Ann. Math.* 98:377–410.
1973*b*. Normal modes for nonlinear Hamiltonian systems. *Inv. Math.* 20:47–57.
1974. Quasi-classical mechanics on spheres. *Symp. Math.* 14:25–32.
1977*a*. Symplectic *V*-manifolds, periodic orbits of Hamiltonian systems and the volume of certain Riemannian manifolds. *Comm. Pure Appl. Math* 30:265–271.
1977*b*. Lectures on symplectic manifolds, in C. B. M. S. Conf. Series., Am. Math. Soc., no. 29. Am. Math. Soc., Providence, R.I.
1978*a*. Bifurcations and Hamilton's principle *Math. Zeit.* 159:235–248.
1978*b*. Periodic orbits on convex energy surfaces (to appear).
1978*c*. A uninversal phase space for particles in Yang-Mills fields (preprint).
Weinstein, A., and Marsden, J. 1970. A comparison theorem for Hamiltonian vector fields. *Proc. Am. Math. Soc.* 26:629–631.
Weiss, B. 1967. Abstract vibrating systems. *J. Math. Mech.* 17:241–255.
Wheeler, J. A. 1964. Geometrodynamics and the issue of the final state, in *Relativity, groups, and topology*, C. DeWitt and B. DeWitt (eds.) Gordon and Breach, New York.
Whitney, H.
1933. Regular families of curves. *Ann. Math.* 34:244–270.
1944. On the extension of differentiable functions. *Bull. Am. Math. Soc.* 50:76–81.
Whittaker, E. T.
1896. On Lagrange's parentheses in the planetary theory. *Mess. Math.* 26:141–144.
1959. *A treatise on the analytical dynamics of particles and rigid bodies*, 4th ed. Cambridge Univ. Press, Cambridge.
Wigner, E.
1939. On unitary representations of the inhomogeneous Lorentz group. *Ann. Math.* 40(2):149–204.
1959. *Group theory and its application to the quantum mechanics of atomic spectra.* Academic, New York.
1964. *Unitary representations of the inhomogeneous Lorentz group including reflections, group theoretical concepts and methods in elementary particle physics*, pp. 37–80. Gordon and Breach, New York.
Williams, R.
1971. Expanding attractors (in Chillingworth [1971], pp. 125–127).
1974. Expanding attractors. *I.H.E.S. (Paris) Publ. Math.* 43:169–204.
1977. The structure of Lorenz attractors, in *Lec. Notes Math.*, vol. 615. Springer, New York.
Williamson, J.
1936. On the algebraic problem concerning the normal forms of linear dynamical systems. *Am. J. Math.* 58:141–163.
1937. On the normal forms of linear canonical transformations in dynamics. *Am. J. Math.* 59:599–617.
1939. The exponential representation of canonical matrices. *Am. J. Math* 61:897–911.
Willmore, T. J. 1960. The definition of Lie derivative. *Proc. Edin. Math. Soc.* 12(2):27–29.
Winkelnkemper, H. E. 1977. On twist maps of the annulus (preprint).

ISBN 0-8053-0102-X

Wintner, A.

1930. Über eine Revision der Sortentheorie des restringierten Dreikörper problems. *Sitzsber. Sächsischer Akad. Wiss. Leipzig* 82:3–56.

1934. On the linear conservative dynamical systems. *Ann. di Mat.* 13:105–112.

1936. On the periodic analytic continuations of the circular orbits in the restricted problem of three bodies. *Proc. Nat. Acad. Sci., U.S.A.* 22:435–439.

1941. *The analytical foundations of celestial mechanics.* Princeton Univ. Press, Princeton, N.J.

Whitham, G. B. 1974. *Linear and nonlinear waves.* Wiley, New York.

Wolf, J. A. 1964. Isotropic manifolds of indefinite metric. *Comment. Math. Helv.* 39:21–64.

Wu, F. W., and Desoer, C. A. 1972. Global inverse function theorem. *Trans. IEEE* 19:199–201.

Yano, K. 1957. *Theory of Lie derivatives.* North Holland, Amsterdam.

Yorke, J. A. 1967. Invariance for ordinary differential equations. *Math. Syst. Theory* 1:353–372.

Yosida, K. 1971. *Functional Analysis,* 3rd Ed. Springer, New York.

Zakharov, V. E., and Faddeev, L. D. 1972. Korteweg-de Vries equation: a completely integrable Hamiltonian system. *Funct. Anal. and Appl.* 5:280–287.

Zeeman, E. C. 1975. Morse inequalities for diffeomorphisms with shoes and flows with solenoids (in Manning [1975], pp. 44–47).

Zehnder, E.

1973. Homoclinic points near elliptic fixed points. *Comm. Pure Appl. Math.* 26:131–182.

1974. An implicit function theorem for small divisor problems. *Bull. Amer. Math. Soc.* 80:174–149.

1975. Generalized implicit functions theorems with applications to some small divisor problems, I. *Comm. Pure Appl. Math.* 28:91–140.

Ziegler, H. 1968. *Principles of structural stability.* Ginn–Blaisdell, Waltham, Mass.

ISBN 0-8053-0102-X

Index

A

F

I

J

T

GLOSSARY OF SYMBOLS

$E, F, \ldots$	finite-dimensional real vector spaces
$\|x\|$	norm of x
$L(E,F)$	continuous linear mapping of E to F
A' or $A^* \in L(F^*, E^*)$	transpose of $A \in L(E,F)$
$L^k(E,F)$	multilinear mappings
$L_a^k(E,F) \subset L^k(E,F)$	skew-symmetric mappings
$L_s^k(E,F) \subset L^k(E,F)$	symmetric mappings
$U \subset E$	open subset
$f\colon U \subset E \to F$	smooth (C^∞) mapping
$x \mapsto f(x)$	effect of f on x
$D^k f\colon U \subset E \to L_s^k(E,F)$	derivatives of f
$D_i f\colon U \subset E \to L_s^k(E_i,F)$	partial derivatives of f
$c'(t) = Dc(t) \cdot 1$	tangent to a curve
$M, N, \ldots$	C^∞ manifold
$\pi\colon E \to B$	vector bundle
$\Gamma^\infty(\pi)$	C^∞ sections of π
$T_m M$	tangent space at $m \in M$
$T_m f$ or $Tf(m)$	tangent of f at m
$\tau_M\colon TM \to M$	tangent bundle
$\tau_M^*\colon T^*M \to M$	cotangent bundle
$(\tau_M)_s^r\colon T_s^r(M) \to M$	tensor bundles
$\omega_M^k\colon \omega^k(M) \to M$	exterior form bundles
$f \in \mathcal{F}(M)$	C^∞ real-valued functions
$X \in \mathcal{X}(M) = \Gamma^\infty(\tau_M)$	vector fields
$\alpha \in \mathcal{X}^*(M) = \Gamma^\infty(\tau_M^*)$	one-forms
$t \in \mathcal{T}_s^r(M) = \Gamma^\infty((\tau_M)_s^r)$	tensor fields
$\otimes$	tensor product
$\omega \in \Omega^k(M) = \Gamma^\infty(\omega_M^k)$	k-forms
$\wedge$	exterior product
$f\colon M \to N$	mapping of manifolds
$f^*\colon \Omega^k(N) \to \Omega^k(M)$	pullback of forms
$\varphi\colon M \to N$	diffeomorphism of manifolds
$\varphi_s^r\colon T_s^r(M) \to T_s^r(N)$	induced tensor bundle isomorphism
$\varphi_*\colon \mathcal{T}_s^r(M) \to \mathcal{T}_s^r(M)$	induced tensor field isomorphism
$U \subset M$	open submanifold
$(U,\varphi), \varphi\colon U \to U' \subset E$	local chart
$e_1, \ldots, e_n$	basis of E
$\alpha^1, \ldots, \alpha^n$	dual basis of $E^* = L(E,R)$
$\underline{e}_1, \ldots, \underline{e}_n$	induced generators of $\mathcal{X}(U)$
$dx^1, \ldots, dx^n$	induced dual generators of $\mathcal{X}^*(U)$
$F_X\colon \mathcal{D}_X \subset M \times R \to M$	integral of vector field
L_X	Lie derivative
$[X, Y]$	Lie bracket

$\boldsymbol{d}$	exterior derivative
$\boldsymbol{i}_X$	inner product
Ω	volume n form
μ_Ω	measure of Ω
$div_\Omega X$	divergence of a vector field
$det_\Omega f$	determinant of a mapping
ω	symplectic form
$X^\flat = \omega_\flat(X)$	lowering action
$\alpha^\# = \omega_\#(\alpha)$	raising action
$X_H = (dH)^\#$	Hamiltonian vector field
$Sp(E,\omega)$	symplectic group
$-\theta_0$	canonical one-form on T^*M
$\omega_0 = -d\theta_0$	canonical two-form on T^*M
$\{f,g\}$	Poisson bracket of functions
$\{\alpha,\beta\}$	Poisson bracket of one-forms
$\mathfrak{X}_{\mathfrak{L}\mathfrak{H}}$	locally Hamiltonian vector fields
$\mathfrak{X}_{\mathfrak{H}}$	globally Hamiltonian vector fields
Σ_e	energy surface
$\boldsymbol{FL}$	Legendre transformation
ω_L	pullback of ω_0 by FL
$\Omega = -d\boldsymbol{\theta}$	symplectic form determined by a metric
(M,ω) or (P,ω)	symplectic manifold
$\Phi: G \times P \to P$	action of a Lie group G on P
$\mathfrak{g}$ or $\mathfrak{L}(G)$	Lie algebra of G
$J: P \to \mathfrak{g}^*$	momentum mapping
$\hat{J}(\xi)(p) = J(p)\cdot\xi$	dual momentum mapping
$P_\mu = J^{-1}(\mu)/G_\mu$	reduced phase space
$I_{h,\mu} = (H \times J)^{-1}(h,\mu)$	level surface of $H \times J$
V_μ	amended or effective potential
$\tilde{\omega}$	pullback of ω to $\boldsymbol{R} \times M$
$X: \boldsymbol{R} \times M \to TM$	time-dependent vector field
$\tilde{X}: \boldsymbol{R} \times M \to T(\boldsymbol{R} \times M)$	vector field associated to X
$\underline{t}$	unit time vector field on $\boldsymbol{R} \times M$
$\omega_H = \tilde{\omega} + dH \wedge dt$	Cartan form
$F: \boldsymbol{R} \times M \to \boldsymbol{R} \times M$	canonical transformation
$K_F: \boldsymbol{R} \times M \to \boldsymbol{R}$	generating function of F
$j^t: M \to \boldsymbol{R} \times M$	embedding at time t
S, W	Hamilton principal functions

Printed in the United States
by Baker & Taylor Publisher Services